Materials Science: Theory and Engineering

Nestor Perez

Materials Science: Theory and Engineering

 Springer

Nestor Perez
University of Puerto Rico
Mayaguez, Puerto Rico

Solution manual for the book is available as an electronic supplementary material at sn.pub/lecturer-material

ISBN 978-3-031-57154-1   ISBN 978-3-031-57152-7   (eBook)
https://doi.org/10.1007/978-3-031-57152-7

This Springer imprint is published by the registered company Springer Nature Switzerland AG
The registered company address is: Gewerbestrasse 11, 6330 Cham, Switzerland

If disposing of this product, please recycle the paper.

My late wife Neida
My daughters Jennifer and Roxie
My son Christopher

Foreword

Dr. Perez has four books to his credit, including this one, and they are all published by the most prestigious publishing house: Springer Verlag. All his four books deal with different but related domains of materials science and engineering. The first three books are in the areas of fracture mechanics, electrochemistry and corrosion, and phase transformation in solids. This current fourth book, *Materials Science: Theory and Engineering*, covers several ideas from his three previous volumes and focuses on some new ideas. This is an ideal textbook for both undergraduate and graduate students in the fields of mechanical engineering, materials science, and metallurgy. Besides, it is an excellent reference treatise for full-time practicing engineers as well as part-time and casual consultants.

There are several similar textbooks in the market. While most of them relate to classical mechanics and its application in materials science and engineering, Professor Perez has added in this book a full chapter on the fundamentals of quantum mechanics (Chap. 3). This is unique in a textbook for engineers!

Mechanical behaviors of solid materials are treated from their standpoints of both phenomenological and structural interpretations within the ranges of elasticity, plasticity, and fracture mechanics. In Chap. 5, the two-dimensional materials, such as graphene, are dealt with in detail. Both mechanical and electrochemical properties of such materials in the nano-scale range are of utmost importance in today's semiconductor industries. This is another very useful addition to this book. Furthermore, thermal phenomena and heat transfer during the deformation process of a solid material are treated in detail in two separate chapters (Chaps. 8 and 9). These additions make the book complete as a textbook in academia and as a strong reference material in industries. I feel honored for writing the forewords of all his four books!

University of Puerto Rico at Mayaguez (UPRM)   Jay (Jayanta) Banerjee
Department of Mechanical Engineering
Mayaguez, PR, USA
2023

Preface

Materials Science Theory and Engineering (MSTE) is a broad interdisciplinary field that integrates physical metallurgy, physics, thermodynamics, and corrosion along with engineering applications of materials, mostly in the form of crystalline solids. Objectively, MSTE focuses on the relation between atomic structure and mechanical or functional properties for designing, fabrication, and characterization of solid components with respect to performance, durability, and recyclability.

The purpose of this textbook is to present a compilation of research work on the aforementioned subjects into one comprehensive volume for readers who are engaged in materials science and engineering of phase transformation in metals. The first six chapters cover a variety of subjects, including atomic structure, crystallography, fundamentals of quantum mechanics, metallography and microscopy, two-dimensional materials, and crystal defects. The next five chapters focus on mass transport brought on by atomic diffusion, thermodynamics phase change, solidification heat transfer, solidification and phase diagrams, and solid-state phase transformation. Also, two chapters are devoted to mechanical behavior of solids and fracture mechanics. The last three chapters cover physical properties of solids, nondestructive methods, and electrochemical corrosion.

The textbook is primarily intended for engineering undergraduate and graduate students who are interested in learning about the principles of materials science and related engineering applications of solid materials, such as metals and their alloys, ceramics, and polymers. Students will benefit from learning the relevant theoretical background for each topic that is clearly explained with the help of illustrations, photos, figures, graphs, and schematic models, followed by the derivation of equations to quantify pertinent quantities.

Graduate-level coursework can cover advanced subjects including quantum mechanics, two-dimensional materials, fracture mechanics, nondestructive methods for evaluating structural integrity, and advanced analytical techniques in some appendices. A professor can support a student's learning process in the classroom by using example problems, illustrations, and problem sets. In particular, the textbook publisher provides professors access to the solution manual for the end-of-chapter problem sets, and professors instructing in the aforementioned fields are free to choose chapters and sections for undergraduate or graduate coursework.

The entire content of this textbook is unlikely to be covered in a single semester. Nonetheless, a professor is free to choose the portions of the textbook that best satisfy the requirements of a college course. Also, this textbook is a good resource for engineers and scientists working in the field of phase transformation in metals.

Mayaguez, PR, USA   Nestor Perez
2023

Contents

Contents   xxiii

Chapter 1
Atomic Structure

1.1 Introduction

This chapter considers the fundamentals of bonding force and binding energy, electron configuration and crystallography. This leads to a hard-sphere atomic model that assumes a periodic arrangement of atoms for determining crystal structures in metals, semiconductor, ceramics and polymers. These fundamentals are important in the engineering field since structural components with predetermined properties, performance and lifetime involve interdisciplinary fields, such as Materials Science and engineering, physics and so forth. The former deals with atomic structures and properties of matter, and the latter incorporates designing codes for fabricating structural components.

Regarding the arrangement of atomic particles (atoms, ions or molecules), crystallography has its prime place in materials science, whereas electrons are treated as wavelike particles being characterized by quantum crystallography using probabilistic wavefunctions.

The regularity of atom arrangements can be characterized in terms of the smallest repeating unit cells responsible for the symmetry of a crystal structure. Moreover, a particular crystallographic orientation-dependent property defines an anisotropic material (dissimilar properties in all directions). In such a case, certain solid properties, such as the critical resolved shear stress (τ_c) and electrical conductivity of a crystal, depend on the orientation of idealized unit cells in a crystal structure. This is an important concept in materials engineering, specifically in the semiconductor field, where computer chips made out of silicon-based wafers are designed having certain crystal orientations.

For comparison, quantum mechanics (QM) bisects the atom into smaller particles, such as electrons, protons, and neutrons. Actually, QM describes electrons as probabilistic matter waves and divides protons or neutrons into quarks. Materials science, on the other hand, divides a solid into organized crystalline and disorganized non-crystalline (amorphous) liquid-like crystal structures, and it bisects a

N. Perez, *Materials Science: Theory and Engineering*,
https://doi.org/10.1007/978-3-031-57152-7_1

crystalline crystal structure into small unit cells made up of closely packed atoms, ions, or molecules that represent an entire atomic crystal structure.

1.2   Atomic Entities of Crystals

Most of the theoretical background evolving in materials science is based on an atomic scale. The flowchart illustrated in Fig. 1.1 indicates the atomic entities to be covered in this chapter.

The diagram shown in Fig. 1.1 indicates how the this chapter is designed for introducing and explaining the relevant materials science features related to crystal structures and properties of solid materials.

1.3   The Atom

Specifically, Fig. 1.2a shows an aesthetic model for a hypothetical atom as per Clara Moskowitz Web site [1].

This atom model is also shown in Fig. 1.2b for emphasizing on the general arrangement of electrons orbiting the nucleus, which contains protons and neutrons treated as tiny spherical particles. A later section includes the electron lobes as the probable sites for finding electrons at random distances from the nucleus of an atom.

The atom may be defined as a small subatomic-size jiggling particle containing negatively charged electrons surrounding the nucleus composed of positively charged protons and neutrons. In particle physics, the protons and neutrons contain fundamental particles known as quarks. According to Richard Feynman, jiggling atoms have their own temperature-dependent speed. Thus, high intensity of the atom jiggling motion is perceived as heat.

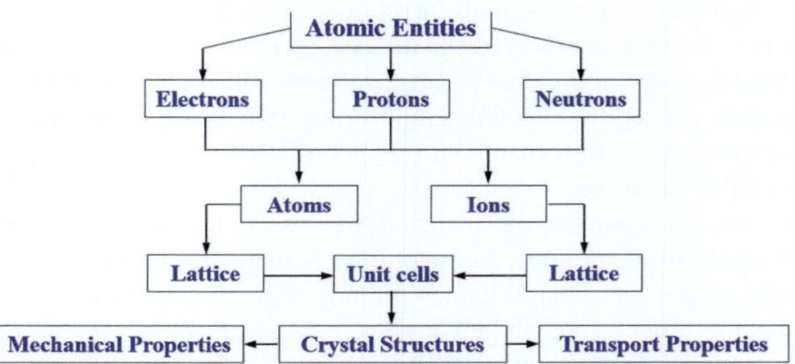

Fig. 1.1 Flowchart of the crystal structure and related properties of crystalline materials

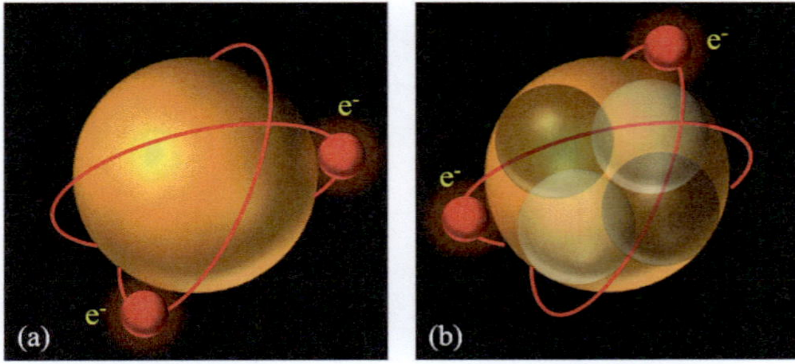

Fig. 1.2 (**a**) The atom showing the nucleus and electron orbitals and (**b**) the atom and arrangement of protons and neutrons in the nucleus

This approach makes the reader be aware of topics based on the atomic behavior of matter from a probabilistic quantum-mechanical concept that incorporates wave-like electrons in atoms and their positions (Hilbert space) in orbitals with discrete or quantized energy levels. It is expected that the reader reaches a good understanding of the bottom-up approach from quantum mechanics to engineering scales.

Particles like atoms or ions are made of electrons, protons, and neutrons. Despite that atoms contain empty spaces between electron orbitals, they are modeled as hard spheres.

The bonding of atoms in a crystal structure is attributed to the attractive forces between them. As a result, regular close-packed atoms form a crystalline solid from which idealized geometries called unit cells are extracted in order to define the type of crystal structure. Conversely, a non-regular packed atom forms an amorphous-structure without unit cells.

Atoms have a preferred arrangement in a three-dimensional (3D) space. This is called space lattice, which is the foundation of crystals.

1.4 Atomic Model

Historically, Niels Bohr in 1913 refined Rutherford's atomic model (published in 1911) by postulating that the condense nucleus of an atom is surrounded by negatively charged electrons located in discrete orbitals having specific quantized energy levels surrounding the nucleus of the atom (Fig. 1.3). This means that an electron (e^-) with negative charge $q_{e-} = 1.602 \times 10^{-19}$ C and mass $m_{e-} = 9.11 \times 10^{-31}$ Kg revolves about the nucleus, which contains protons (e^+) with a positive charge $q_{e+} = 1.602 \times 10^{-19}$ C and a mass $m_{e+} = 1.67 \times 10^{-27}$ Kg. On the other hand, the nucleus has neutrons with no charge.

Quantum mechanics treats electrons as wavelike particles having discrete or quantized energy levels. For instance, Fig. 1.3a shows a neutral hydrogen atom having one orbiting electron located in the ground quantum state defined by the

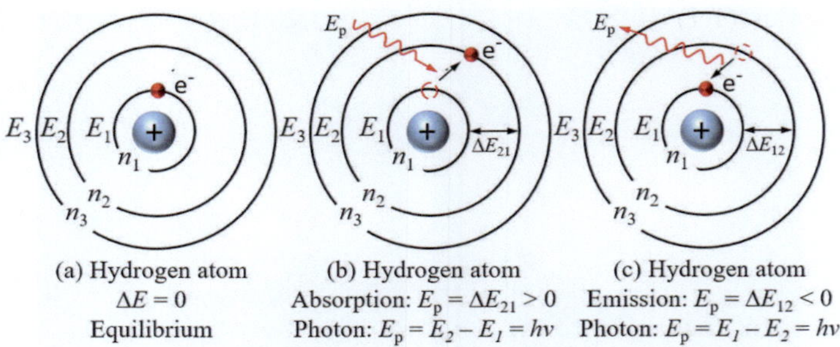

(a) Hydrogen atom   (b) Hydrogen atom   (c) Hydrogen atom

$\Delta E = 0$ Absorption: $E_p = \Delta E_{21} > 0$ Emission: $E_p = \Delta E_{12} < 0$

Equilibrium   Photon: $E_p = E_2 - E_1 = h\nu$ Photon: $E_p = E_1 - E_2 = h\nu$

Fig. 1.3 Bohr atomic model for the hydrogen atom. (**a**) Hydrogen at equilibrium. (**b**) An electron absorbing photon energy for jumping to a higher state. (**c**) An electron emitting energy for jumping to a lower state

quantum number n_1 and quantized energy level E_1. This model represents a hypothetical electron around an orbital.

According to Bohr, the electron can be excited by an incident beam of photons (Fig. 1.3b) with energy E_p, and as a result, the electron jumps to the next available higher-energy state n_2 with a quantized energy level E_2. This physical case is possible if Planck's energy is $E_p = \Delta E_{21} = (E_2 - E_1) > 0$, where ΔE_{21} is the energy difference between quantum states n_2 and n_1. Conversely, Fig. 1.3c shows that the electron at n_2 emits energy as a photon so that $E_p = \Delta E_{12} = (E_1 - E_2) < 0$. This implies that an electron can undergo a quantum state transition between orbital levels by emission or absorption of photon energy $E_p = |\Delta E|$. The energy difference between orbitals finds its place in optical spectra of excited hydrogen atoms (Levi [2, p. 82]).

1.4.1   Hydrogen Radius and Energy Levels

According to Broglie hypothesis, a wavelike particle has angular momentum (p) associated with the wavelength (λ) and Planck's constant (h), with its mass (m) and speed (υ). Thus,

$$p = \frac{h}{\lambda} = m\upsilon \tag{1.1}$$

For the Bohr circular orbital, the orbit circumference $2\pi r$ with radius r equals to the integer $n = 1, 2, 3, \ldots$ multiple of the Broglie wavelength (Le Bellac [3, p. 30]). Mathematically,

$$2\pi r = n\lambda \tag{1.2}$$

Combining Eqs. (1.1) and (1.2) along with the reduced Planck's constant $\hbar = h/2\pi$ yields

$$2\pi r = \frac{nh}{p} = \frac{nh}{mv} \tag{1.3a}$$

$$r = \frac{n\hbar}{p} = \frac{n\hbar}{mv} \tag{1.3b}$$

According to Newton's law, the force balance for a circular orbit dictates that the electrostatic force must be equal to the centripetal force (Levi [2, p. 81])

$$\frac{mv^2}{r} = \frac{1}{4\pi\epsilon_o}\frac{q_e^2}{r^2} = \left(\frac{q_e^2}{4\pi\epsilon_o}\right)\frac{1}{r^2} \tag{1.4}$$

which defines an energy balance between kinetic and coulomb energy gradients. Let the terms in parenthesis be

$$k_e = \frac{q_e^2}{4\pi\epsilon_o} = \frac{\left(1.602 \times 10^{-19}\,C\right)^2}{4\pi\left(8.8542 * 10^{-12}\,F/m\right)} = 2.31 \times 10^{-28}\,J.m \tag{1.5a}$$

$$k_e = \left(2.31 \times 10^{-28}\,J.m\right) / \left(1.6 \times 10^{-19}\,J/eV\right) \tag{1.5b}$$

$$k_e = 1.44 \times 10^{-9}\,eV.m \tag{1.5c}$$

Actually, k_e simplifies Eq. (1.4).

Inserting Eq. (1.5a) into (1.4) yields the electron velocity

$$v^2 = \frac{k_e}{mr} = \frac{q_e^2}{4\pi\epsilon_o mr} \tag{1.6a}$$

$$v = \left(\frac{k_e}{mr}\right)^{1/2} \tag{1.6b}$$

Combining Eqs. (1.3b) and (1.6) yields the Bohr radius with $m = m_e$

$$r = \left(\frac{\hbar^2}{mk_e}\right)n^2 = \left(\frac{4\pi\epsilon_o\hbar^2}{mq_e^2}\right)n^2 = \left(\frac{\hbar^2}{mk_e}\right)n^2 \tag{1.7a}$$

$$r = \left[\frac{\left(1.0546 * 10^{-34}\,J.s\right)^2}{\left(9.11 * 10^{-31}\,Kg\right)\left(2.31 \times 10^{-28}\,J^2s^2/Kg.m\right)}\right]n^2 \tag{1.7b}$$

$$r \simeq \left(5.30 \times 10^{-11}\,m\right)n^2 = (0.053\,nm)\,n^2 \tag{1.7c}$$

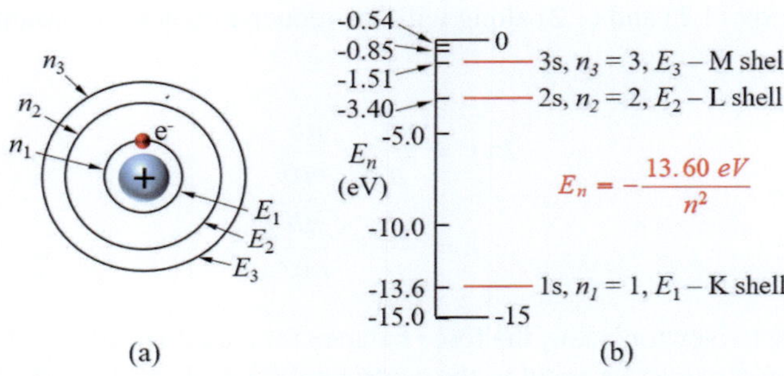

Fig. 1.4 (**a**) Bohr hydrogen model with three quantum levels and (**b**) quantized energy spectrum for the "s" sub-shells up to the quantum state $n = 3$

The general quantized energy level (E_n) for an electron is written in terms of kinetic (KE) and potential energies $U(r)$ (Levi [2, p. 81], Le Bellac [3, p. 30])

$$E_n = \frac{1}{2}mv^2 - U(r) = \frac{1}{2}mv^2 - \frac{q_e^2}{4\pi\epsilon_o}\frac{1}{r} \tag{1.8a}$$

$$E_n = \frac{1}{2}mv^2 - \frac{k_e}{r} = \frac{1}{2}\frac{k_e}{r} - \frac{k_e}{r} = -\frac{1}{2}\frac{k_e}{r} \tag{1.8b}$$

$$E_n = -\frac{mq_e^4}{32\pi^2\hbar^2\epsilon_o^2}\frac{1}{n^2} = -\frac{mq_e^4}{8h^2\epsilon_o^2}\frac{1}{n^2} \tag{1.8c}$$

Substituting Eq. (1.7a) into (1.8c) gives the quantized energy equation for the hydrogen atom

$$E_n = -\frac{13.60\,eV}{n^2} \tag{1.9}$$

Here, $R_e = 13.60\,\text{eV}$ is known as the Rydberg constant. The Bohr hydrogen model and related energy diagram based on Eq. (1.9) are shown in Fig. 1.4.

Now, the energy differences between orbitals related to Figs. 1.2b, c are written as

$$\Delta E_{21} = E_2 - E_1 = -(13.60\,eV)\left(\frac{1}{n_2^2} - \frac{1}{n_1^2}\right) > 0 \text{ (absorption)} \tag{1.10a}$$

$$\Delta E_{12} = E_1 - E_2 = -(13.60\,eV)\left(\frac{1}{n_1^2} - \frac{1}{n_2^2}\right) < 0 \text{ (emission)} \tag{1.10b}$$

The orbitals for hydrogen are shown in Fig. 1.4a, and the quantized energy levels, as per Eq. (1.9), are depicted in Fig. 1.4b for the "s" orbitals or sub-shells.

According to Eq. (1.9) and related quantized energy chart in Fig. 1.4b, an electron quantum jump from one orbit to another generates a change in an electron energy caused by either emission or absorption of a photon.

1.5 Quantum Mechanical Model of an Atom

The quantum mechanical model shown in Fig. 1.5a refines the classical shell model known as the Bohr model of an atom. It derives directly from quantum mechanics by using the Schrodinger wavefunction Ψ, which defines the probability amplitude of finding the likelihood of an electron at a certain position. Moreover, Fig. 1.5a also shows the wavefunctions Ψ_1 and Ψ_2 for hydrogen $1s$-orbital with a radius r for a quantum state $n = 4$. The geometrical shape of either wavefunction indicates that it folds back to itself since the electron behaves as a wavelike particle. Therefore, it can be treated as a quantum object conveniently represented as small sphere in three dimensions or as a circle in two dimensions (a tiny spot in Fig. 1.5a).

Since the Bohr model only works for hydrogen, consider an element with many electrons, such as iron (Fe). In this case, it is convenient to illustrate the quantized energy spectrum as depicted in Fig. 1.5b instead of the orbitals known as sub-shells in quantum mechanics. This element has 7 filled inner sub-shells with electron positions denoted with arrows $\uparrow\downarrow$ and one unfilled outer shell. These sub-shells belong to the principal shells related to their respective quantum numbers.

The quantum mechanical model for characterizing the behavior of electrons includes four quantum numbers being described below.

- n denotes the principal quantum number related to quantized energy level. It can be taken as an average distance from the nucleus.

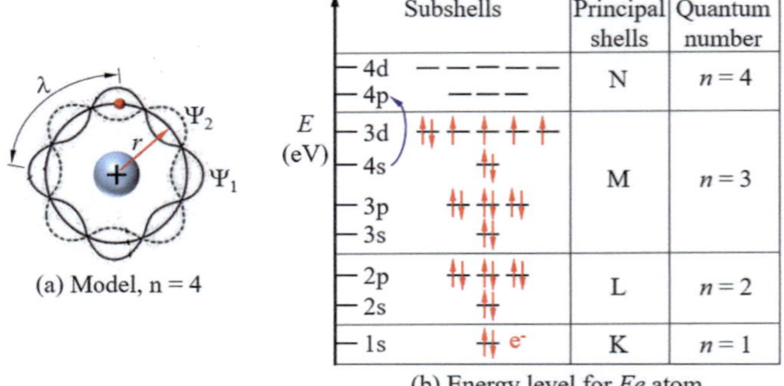

(a) Model, $n = 4$

(b) Energy level for Fe atom

Fig. 1.5 (a) Quantum mechanical model of an atom and (b) schematic quantized energy levels for shells and sub-shells. This figure shows the electron configuration for iron (Fe) with an atomic number of 26

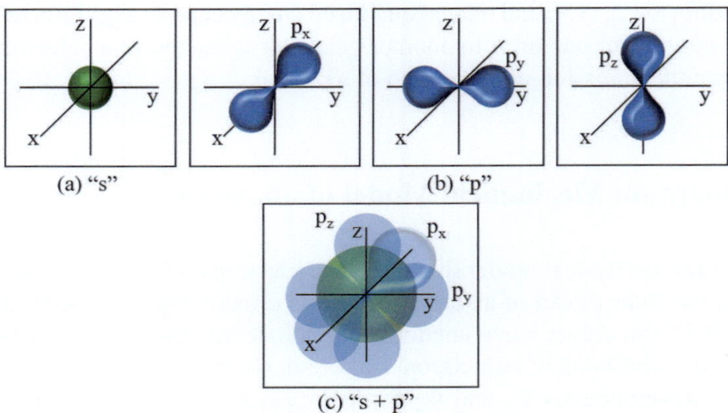

Fig. 1.6 Electron lobes for (**a**) the "s" orbital, (**b**) the "p" orbitals, and (**c**) the "s + p" orbitals together

- l denotes the angular momentum quantum number. It describes the shape of the orbital or shell, which can be divided into sub-shells with a similar l number. That is, the pair l-sub-shell with the maximum allowable number of electrons (superscript) can be $1\text{-}s^2$, $1\text{-}p^6$, $2\text{-}d^{10}$, $3\text{-}f^{14}$, and $4\text{-}g^{18}$.
- m_l denotes the magnetic quantum number.
- m_s denotes the spin quantum number. Electrons may have a clockwise spin ($\uparrow$) or a counterclockwise spin ($\downarrow$) analog to spin-up and spin-down.

The most common shapes of sub-shells or orbitals (quantum state of probabilities) are depicted in Fig. 1.6. The "s" orbital is circular as in the hydrogen atom (Fig. 1.6a), the "p" orbital is defined along the Cartesian coordinates in the form of pair lobes on opposite sides (Fig. 1.6b), and the "s + p" orbitals together as in neon (Fig. 1.6c). Moreover, a lobe represents a high probability density area of finding an electron.

The basic constituent of matter, in summary, is the atom that is composed of a nucleus surrounded by electrons. The former contains protons and neutrons, which, in turn, are constituted by small particles called up-quarks and down-quarks. Moreover, quarks are considered elementary particles.

Thus far, the orbitals in Fig. 1.6 represent models for the likelihood of electrons being found within these lobes. For hydrogen, the $1s^1$ lobe embodies the highest probability of finding the electron around the nucleus.

1.6   Potential Energy and Binding Force Models

Many atoms in single crystals have finite and continuous energy levels since atoms are arranged spatially to form crystalline structures, such as face-centered cubic

(FCC), hexagonal-close packed (HCP), and so forth. These crystalline structures are described in a later section.

An electron cloud is formed contributing significantly to the metallic bonding and related transport-property characteristics in semiconductors, crystalline metals, and their alloys. Moreover, a polycrystalline solid is made up of many grains, which, in turn, are relatively small regions of single crystals containing a long-range spatial order and preferred unit-cell orientations separated by grain boundaries.

Atomic Particles This is in reference to neutral atoms (equal numbers of e^- and e^+) forming a three-dimensional lattice due to the repetitive arrangement of atoms that form virtual geometries called unit cells. Thus, atoms are held together due to equal but opposite sign binding forces of attraction $F_A(r)$ and repulsion $F_R(r)$ at a interatomic spacing (distance) r. The former binding force is known as the Coulomb attraction $F_A(r)$ force, and the latter is due to the Pauli exclusion principle of repulsion $F_R(r)$. For an atom-pair at equilibrium, the total net binding force is $F(r_o) = 0$ at the equilibrium distance r_o since $F_A(r_o) = -F_R(r_o)$. The corresponding potential or binding energy is $U_0 = U_A(r_o) = U_R(r_o) < 0$, where $U_A(r_o)$, $U_R(r_o)$ are the attractive and repulsive potential energy components, respectively, at the equilibrium interatomic distance r_o. Here, $r_o = r_i + r_j$ and r_i, r_j are the radii of corresponding unlike particle-pair $(i\text{-}j)$ with opposite sign (such as $r_i, r_j = Na^+, Cl^-$). Other possibilities include like ion-pairs $(i\text{-}i)$ and $(i\text{-}j)$ so that $r_o = r_i + r_i$ and $r_o = r_j + r_j$.

Ionic Particles Basically, an atom with unequal numbers of electrons and protons defines an electrically charged ion. Positively charged ions are called cations $(H^+, Li^+, Cu^{2+}, \text{etc.})$, and negatively charged ions are called anions $(Cl^-, O^{-2}, \text{etc.})$. Specifically, ions are formed by the addition of electrons to neutral atoms, molecules, or other ions so that they become negatively charged particles called anions. Conversely, the removal of electrons from atoms forms positively charged cations. Exceptionally, when the hydrogen atom (H) loses the only electron in its ground state orbital, it becomes a positively charged hydrogen ion (H^+), which is the proton itself. In general, ions are formed when atoms lose or gain electrons, and as a consequence, ionic interactions arise from electrostatic attraction between an ionic pair of opposite charge (such as table salt Na^+Cl^-). This suggests that the strong coulombic attraction force between ions of opposite charges form crystal lattices of alternating cations and anions. This indicates that an ionic interaction is also associated with the Coulomb energy of the electrostatic attraction $U_A(r)$ and the Pauli exclusion principle energy of repulsion $U_R(r)$. For an unlike ion pair at equilibrium, the force and energy analysis are similar to those used for an atom pair. Thus, the energies are, again, $U_0 = U_A(r_o) = U_R(r_o) < 0$, and the total net binding force is $F(r_o) = 0$ since the atomic forces are $F_A(r_o) = -F_R(r_o)$.

1.6.1   General Potential Energy Function

This section is devoted to the analytical procedure for determining the binding force
and potential energy diagrams using an isolated hypothetical particle-pair model.
For clarity, a particle pair may be an atom pair, ion pair or a molecule pair in a solid.

For an atom pair, the repulsive force $F_R(r)$ arises from interactions between the
negatively charged electron cloud at an interatomic spacing r defined as the center-
to-center separation between two atoms or particles.

Mathematically, the net binding or electrostatic force equation for a particle pair
($P_1 P_2$-pair) with opposite charge is written as as

$$F(r) = F_A(r) + F_R(r) \qquad\qquad (1.11a)$$

$$F_A(r) = \frac{|(Z_1 q_e)(Z_2 q_e)|}{4\pi\epsilon_o}\frac{1}{r^2} \qquad\qquad (1.11b)$$

According to Coulomb's law of particle interactions, the interaction energy for a
particle pair is defined as the total or net potential energy defined by

$$U(r) = \int_\infty^r F(r)\,dr = \int_\infty^r [F_A(r) + F_R(r)]\,dr \qquad\qquad (1.12a)$$

$$U(r) = \int_\infty^r F_A(r)\,dr + \int_\infty^r F_R(r)\,dr \qquad\qquad (1.12b)$$

$$U(r) = U_A(r) + U_R(r) \qquad\qquad (1.12c)$$

where $U_A(r)$ and $U_R(r)$ are the potential energy terms (J or eV units) due to
attractive and repulsive forces, respectively, at an interatomic distance r. These
energy terms can be multiplied by Avogadro's number ($N_a = 6.022 \times 10^{23}$
$particle/mol$) to obtain, say, J/mol units.

Mathematically, Eq. (1.12c) can empirically be defined as

$$U(r) = -\frac{C_A}{r^m} + \frac{C_R}{r^n} \qquad\qquad (1.13a)$$

$$U_A(r) = -\frac{C_A}{r^m} \qquad\qquad (1.13b)$$

$$U_R(r) = \frac{C_R}{r^n} \qquad\qquad (1.13c)$$

Taking $dU(r)/dr$ from Eq. (1.13a) yields the binding force equation for a
particle pair (atom pair or ion pair)

$$F(r) = \frac{dU(r)}{dr} = \frac{mC_A}{r^{m+1}} - \frac{nC_R}{r^{n+1}} \qquad\qquad (1.14a)$$

$$F_A(r) = \frac{mC_A}{r^{m+1}} \tag{1.14b}$$

$$F_R(r) = -\frac{nC_R}{r^{n+1}} \tag{1.14c}$$

Here, C_R, C_A are constants, and n, m are exponents, $F_A(r)$ in Eq. (1.14b) is the Coulomb's electrostatic attractive force in an ion-pair system, Z_1, Z_2 are the oxidation states or valence of opposite sign, q_e is the electron charge, and ϵ_o is the vacuum permittivity.

For an $X_1^+ X_2^-$ ion-pair model, such as Na^+Cl^-, Eq. (1.14b) along with $m = 1$ becomes

$$F_A(r) = \frac{mC_A}{r^{m+1}} = \frac{|(Z_1 q_e)(Z_2 q_e)|}{4\pi\epsilon_o} \frac{1}{r^2} \tag{1.15a}$$

$$C_A = \frac{|(Z_1 q_e)(Z_2 q_e)|}{4\pi\epsilon_o} \tag{1.15b}$$

Plotting Eqs. (1.14a) and (1.13a) and their individual components for some hypothetical particle-pairs yield Fig. 1.7a and b, respectively.

Regarding crystal binding and related interatomic forces, solids or crystals are considered stable atomic structures due to atomic interactions holding atoms together. The resultant crystal structure has a specific arrangement of unit cells.

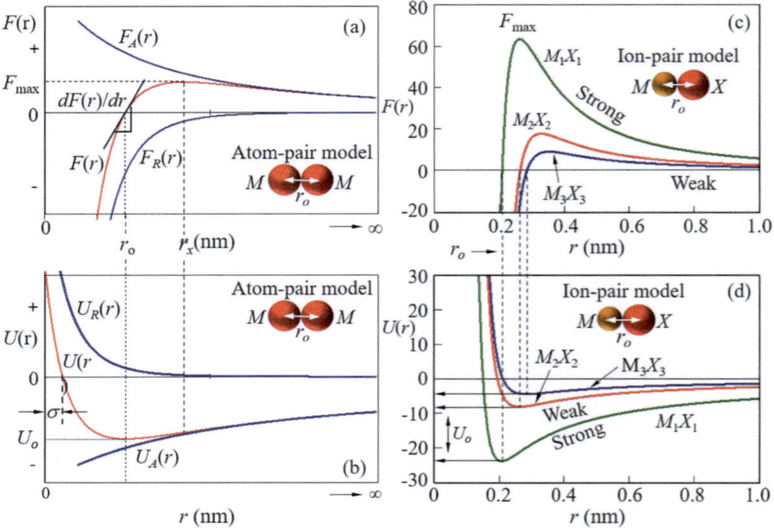

Fig. 1.7 Binding force and energy for hypothetical solids. (**a**) Binding force and (**b**) binding energy for an atom-pair model. (**c**) Comparison of binding force and (**d**) binding energy for three ion-pair cases in MX solid materials, where M is the metal and X is the non-metal

Moreover, taking into account an atom-pair or ion-pair model, the cohesive energy or crystal energy controls the modulus of elasticity and melting temperature.

At equilibrium, the net interatomic force and the minimum potential energy are $F(r_o) = 0$ and $U(r_o) = U_o$, respectively. Thus, the (r_o, U_o) point in Fig. 1.7b represents the equilibrium condition due to a force balance between the attractive and repulsive forces. Moreover, the tangent line through the equilibrium interatomic spacing r_o in Fig. 1.7a has a slope proportional to the modulus elasticity; $E \propto |dF(r)/dr|$.

The equilibrium distance r_o can be calculated by evaluating $F(r) = |dU(r)/dr|_{r-r_o} = 0$ so that

$$|F(r)|_{r=r_o} = \left| \frac{dU(r)}{dr} \right|_{r=r_o} = \frac{mC_A}{r^{m+1}} - \frac{nC_R}{r^{n+1}} = 0 \tag{1.16a}$$

$$r_o = \left(\frac{nC_R}{mC_A} \right)^{1/(n-m)} \tag{1.16b}$$

$$r_o^{n-m} = \left(\frac{nC_R}{mC_A} \right) \tag{1.16c}$$

$$r_o^n = r_o^m \left(\frac{nC_R}{mC_A} \right) \tag{1.16d}$$

Inserting Eq. (1.16d) into (1.13a) at $r = r_o$ and rearranging the result expression yield the equilibrium binding energy written as

$$U_o = U(r_o) = -\frac{C_A}{r_o^m} + \frac{C_R}{r_o^n} = -\frac{C_A}{r_o^m} + \frac{C_R}{r_o^m[nC_R/(mC_A)]} \tag{1.17a}$$

$$U_o = \frac{(m-n)C_A}{nr_o^m} < 0 \quad \text{for } n > m \tag{1.17b}$$

The equilibrium potential energy $U_o = U(r_o)$ in Eq. (1.17b) is known as the binding-energy, as shown in Fig. 1.7c for a general case and in Fig. 1.7d for specific ion pairs at the equilibrium interatomic spacing $r = r_o$. Regarding interatomic forces, the net binding force $F(r_o) = 0$ is directly associated with the minimum potential energy $U(r_o) = U_o < 0$ for a stable particle pair. This means that U_o is the energy well depth (minimum energy) that measures the bond strength, holding the two particles together. Moreover,

- if $r < r_o$, then the electron outer shells overlap, and the repulsive force tends to push the particle pair to its equilibrium physical state at $r = r_o$, which is the optimal separation of atoms (Fig. 1.7).
- For $r < r_o$, the repulsive force $F_R(r) > F_A(r)$ predominates the interatomic interaction.

- If $r > r_o$, then the atomic bond is elastically stretched or disassembled due to separation of the electron outer shells. Consequently, the attractive force tends to bring back the particle pair to its equilibrium condition.
- For $r > r_o$, the attractive force predominates; $F_A(r) > F_R(r)$.
- If $r = r_o$, equilibrium exists since $F_A(r_o) = -F_R(r_o)$, and the net interatomic force becomes zero; $F(r_o) = 0$. This suggests that the particle pair is most stable until an external force is exerted upon it.

Figure 1.7c depicts the non-linear binding force, and Fig. 1.7d illustrates the binding-energy behavior for three unlike ion-pair cases representing ceramic materials: M_1X_1, M_2X_2, and M_3X_3. Here, M is the metal, and X is the non-metal-like oxygen (O^{2-}) or chlorine (Cl^-).

The trendlines in Fig. 1.7c indicate that the M_1X_1, such as $Mg^{2+}O^{2-}$, ion pair is the strongest material and has enhanced properties because

$$F(M_1X_1)_{\max} > F(M_2X_2)_{\max} > F(M_3X_3)_{\max} \tag{1.18a}$$

$$U_o(M_1X_1)_{\min} < U_o(M_2X_2)_{\min} < U_o(M_3X_3)_{\min} \tag{1.18b}$$

$$T_m(M_1X_1) > T_m(M_2X_2) > T_m(M_3X_3) \tag{1.18c}$$

$$E(M_1X_1) > E(M_2X_2) > E(M_3X_3) \tag{1.18d}$$

where T_m and E are the melting temperature and the modulus of elasticity (mechanical property), respectively.

This theoretical treatment indicates that the equilibrium crystal energy or cohesive energy referred to as the equilibrium potential energy is $U_o < 0$, which is assumed to be lower than the energy of free atoms. According to this atom-pair or ion-pair theory, the nature of a stable crystal system is to reach a stable state, provided that the net binding force is $F(r) = 0$ at the equilibrium interatomic distance $r = r_o$. Nevertheless, in general, the particle-pair theory is the most common approach in materials science textbooks and university classrooms.

For an elastic deformation process at an atomic level, the maximum net force $F(r_x) = F_{\max}$ and its corresponding net potential energy $U(r_x)$ at $r_x > r_o$ (Fig. 1.7a or c) are derived next. Thus,

$$\left|\frac{dF(r)}{dr}\right|_{r=r_x} = -\frac{m(m+1)C_A}{r_x^{m+2}} + \frac{n(n+1)C_R}{r^{n+2}} = 0 \tag{1.19a}$$

$$r_x = \left[\frac{n(n+1)C_R}{m(m+1)C_A}\right]^{1/(n-m)} \tag{1.19b}$$

$$F(r_x) = \frac{mC_A}{r_x^{m+1}} - \frac{nC_R}{r_x^{n+1}} = F_{\max} \tag{1.19c}$$

$$U(r_x) = \int_{\infty}^{r_x} F(r_x)\,dr \tag{1.19d}$$

The terms $F(r_x)$ and $U(r_x)$ are assumed to cause bond deformation (stretching) or disassemble of a particle pair into its constituents (atoms or ions) since F_{max} overcomes the equilibrium electrostatic forces.

1.6.2   Lennard-Jones Potential Energy Equation

This is another potential energy model with known exponents but similar to Eq. (1.13a). Mathematically, the general Lennard-Jones potential energy functions for $U(r)$, $F(r)$, and k_u constant are written as

$$U(r) = k_u R_o \left[\left(\frac{\sigma}{r} \right)^n_R - \left(\frac{\sigma}{r} \right)^m_A \right] \tag{1.20a}$$

$$F(r) = k_u R_o \left[-n \left(\frac{\sigma^n}{r^{n+1}} \right)_R + m \left(\frac{\sigma^m}{r^{m+1}} \right)_A \right] \tag{1.20b}$$

$$k_u = \frac{n}{n-m} \left(\frac{n}{m} \right)^{m/(n-m)} \tag{1.20c}$$

The Lennard-Jones potential energy $U(r)$, and the net binding force $F(r)$, and relevant derivatives with $n = 12$ and $m = 6$ are

$$U(r) = 4R_o \left[\left(\frac{\sigma}{r} \right)^{12}_R - \left(\frac{\sigma}{r} \right)^6_A \right] = -\frac{C_R}{r^n} - \frac{C_A}{r^m} \tag{1.21a}$$

$$F(r) = \frac{dU(r)}{dr} = 4R_o \left[\frac{6\sigma^6}{r^7} - \frac{12\sigma^{12}}{r^{13}} \right] \tag{1.21b}$$

$$\frac{dF(r)}{dr} = \frac{d^2U(r)}{dr^2} = 4R_o \left[\frac{156\sigma^{12}}{r^{14}} - \frac{42\sigma^6}{r^8} \right] \tag{1.21c}$$

$$\frac{d^2F(r)}{dr^2} = 4R_o \left[\frac{336\sigma^6}{r^9} - \frac{2184\sigma^{12}}{r^{15}} \right] \tag{1.21d}$$

where $C_R = 4R_o\sigma^6$ and $C_A = 4R_o\sigma^{12}$ denote constants related to repulsive and attractive forces, respectively. Also, R_o denotes the Lennard-Jones constant, σ denotes the van der Waals radius (Fig. 1.7b) or simply an interatomic distance, and r denotes the interatomic distance between two adjacent particles. Nonetheless, Eqs. (1.13a) and (1.21a) define the potential energy that measures the strength of a stable particle pair at equilibrium.

Letting $|F(r)|_{r=r_o} = |U(r)/dr|_{r=r_o} = 0$ yields the equilibrium interatomic distance (spacing between nearest atoms and ions in a crystal)

$$r_o = 2^{1/6}\sigma = 1.1225\sigma = \left(\frac{2C_R}{C_A} \right)^{1/6} \tag{1.22a}$$

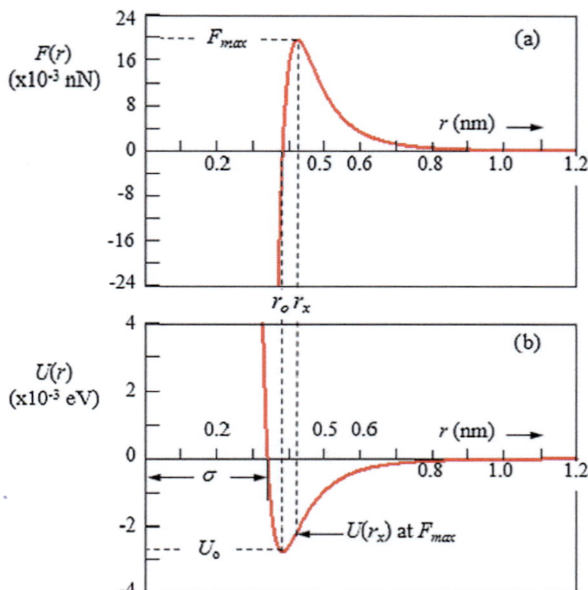

Fig. 1.8 (**a**) Binding force and (**b**) binding potential energy for graphite

$$\sigma = 2^{-1/6} r_o = 0.8909 r_o = 0.8909 \left(\frac{2C_R}{C_A} \right)^{1/6} \tag{1.22b}$$

Consider the Lebedeva et al. [4] and Popov et al. [5] published data for graphite, where $R_o = 0.00276\,\text{eV}$, $\sigma = 0.339\,\text{nm}$, and $r_o = 0.381\,\text{nm}$ are related to the carbon-carbon or C-C bond pair, which is analogous to one atom pair (i-i pair).

Plot $F(r)$ and $U(r)$ at $0 \le r \le 1.2\,\text{nm}$ yields Fig. 1.8a for the net binding force $F(r)$ with $F(r_x) = F_{\max}$ and Fig. 1.8b for the potential energy $U(r)$ with $U(\sigma) < U(r_o)$.

Figure 1.8a shows that $F(r) > 0$ is upwardly concave for $r \le r_x$, $F(r_o) = 0$ at equilibrium and $F(r_x) = F_{\max}$ since the second derivative yields a negative value; $d^2 F(r)/dr^2 < 0$. Conversely, Fig. 1.8b illustrates that the potential energy $U(r) < 0$ is downward concave (steep repulsive term) for $r \le r_o$, $U(r_o) = U_{\min} = U_o < 0$ at $r = r_o$, and then it becomes upwardly concave so that $U(r_x) > U_o$.

When the interatomic distance r approaches the equilibrium size ($r \to r_o$), the pair of atoms has the tendency to bind together due to an attraction energy. However, if $r < r_o$, then the repulsive energy dominates until the pair of atoms reach an equilibrium distance $r = r_o$.

Evaluating the potential energy and binding force expressions given by Eq. (1.21) reveals interesting results. For instance,

- Let $F(r)_{r=r_o} = dU(r)/dr = 0$ in Eq. (1.21b), and solve for $r = r_o$.
- Let $dF(r)/dr = 0$ in Eq. (1.21c), and solve for $r = r_x$ so that $F(r_x) = F_{\max}$.
- Compute $U(r_o)$ and $U(r_x)$ from (1.21a).

- Calculate $d^2 F(r_x)/dr^2 < 0$ from (1.21d) for a local $F(r_x) = F_{max}$. The results for graphite (Fig. 1.8) are

$$F(r_o) = 0 \text{ at } r_o = 0.381\,nm \tag{1.23a}$$

$$F(r) = F_{max} = 0.02\,N \text{ at } r = r_x = 0.42\,nm \tag{1.23b}$$

$$U(r_o) = U_o = -2.76\,meV \text{ at } r_o = 0.38\,nm \tag{1.23c}$$

$$U(r_x) = -2.21\,meV \text{ at } r_x = 0.42\,nm \tag{1.23d}$$

$$\frac{d^2 F(r)}{dr^2} = -11.04\,eV/nm^2 < 0 \text{ at } r_x = 0.42\,nm \tag{1.23e}$$

Theoretically, $F(r) \geq F_{max} = 0.02\,\text{nN}$ is the electrostatic force needed to overcome the weak binding force across the π-bonds in graphitehoneycomb planes. The change in potential energy $\Delta U(r) = |U(r_o) - U(r_x)| = 0.55\,\text{meV}$ is the driving force for stretching the C-C bond pair (π-bonds) at a binding force $F_{max} \approx 0.02\,\text{N}$. Fundamentally, the pair of carbon atoms is stable until an external force is exerted upon it.

For elastic bond deformation, $F(r_x) = F_{max}$ and $U(r_x) = [U_A(r_x) + U_R(r_x)] > U(r_o)$ at $r_x > r_o$ for an effective debonding or dissembling of the C-C pair. In general, during the elastic deformation process, bonds between adjacent atoms are deformed to a certain degree. This is also referred to as stretching the atomic bonding.

1.7   Characteristics of Crystals

Crystallography can be defined as the solid-state crystal science that deals with regular close-packed atoms forming a space lattice, which is the foundation of a crystal structure. In essence, a space lattice is defined as an infinite array of points (atoms) with identical surroundings in space and translation invariant characteristics (translation-symmetry operation).

The most common characteristics of crystals include crystal orientations, point coordinate x, y, z, crystallographic coordinates a, b, c, directions, and planes. Possibly, a habit plane can be defined as an invariant interface between two phases, such as austenite and martensite in some steels. The crystal orientation in terms of unit cells is related to slip systems on close-packed crystallographic (hkl) planes and $[uvw]$ directions.

A single crystal can be treated as a small or huge grain having dimensional boundaries. On the other hand, polycrystals have randomly oriented single crystals called grains, treated as isotropic materials (same properties in all directions) so that measured mechanical properties represent average values. Thus, physical and mechanical behavior of solids strongly depend on the crystal structure bonding mechanisms. This suggests that deformation of crystalline materials occurs by shear (sliding process) along the most close-packed (hkl) plane and $[uvw]$ direction.

1.8 Space Lattice Models

A lattice, in general, is a three-dimensional (3D) array of points describing the arrangement of particles that form crystals as illustrated in Fig. 1.9.

In the solid state, the array of particles are treated as atoms in metals, ions in ceramics, or molecules in polymers. Only metallic and ionic structures are considered hereafter.

Pure Metals The atomic packing of pure metals is based on the atom-pair (i-i pair) concept, which extends linearly along crystallographic planes and directions with preferred orientations. For comparison, Fig. 1.9 illustrates some common crystal lattices for metallic body-centered cubic (BCC) with its shaded unit cell (Fig. 1.9a), face-centered cubic (FCC) along with its shaded unit cell (Fig. 1.9b), and hexagonal close-packed with shaded basal plane (HCP) (Fig. 1.9c). In essence, these lattices are shown in their ideal packing conditions (Callister and Rethwisch [6, Chapter 3].

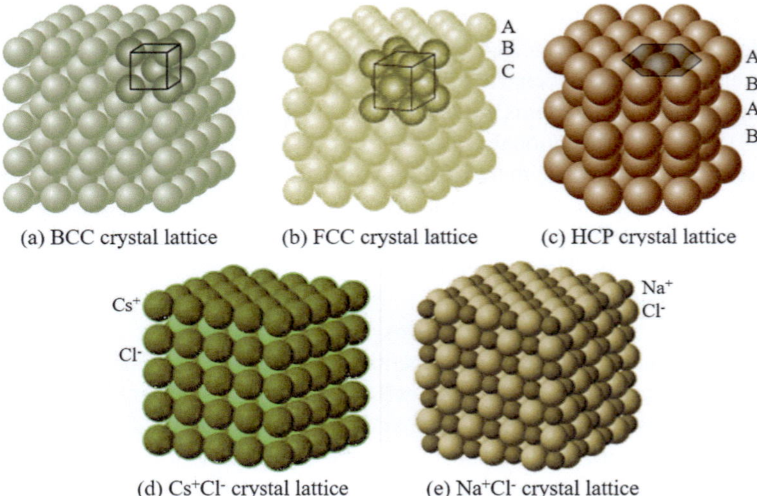

(a) BCC crystal lattice   (b) FCC crystal lattice   (c) HCP crystal lattice

(d) Cs$^+$Cl$^-$ crystal lattice   (e) Na$^+$Cl$^-$ crystal lattice

Fig. 1.9 Comparison of ideal atomic packing in metallic (**a**) body-centered cubic (BCC) shown the unit cell, (**b**) face-centered cubic (FCC) showing the unit cell, and (**c**) hexagonal closed-packed exhibiting the basal plane. After Callister and Rethwisch [6, Chapter 3]. Ceramics: ionic compounds. (**d**) $CsCl$ and (**e**) $NaCl$ lattices (Wikipedia.com)

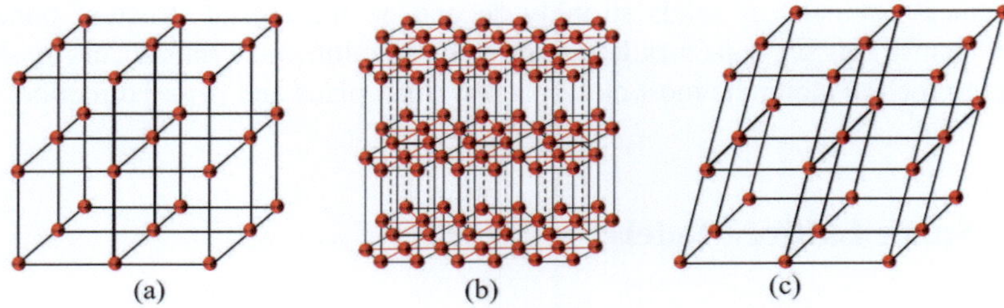

Fig. 1.10 Crystal lattices for (**a**) simple cubic, (**b**) hexagonal, and (**c**) triclinic crystals

Ceramics Similarly, ceramic materials are based on the ion-pair concept (i-j pair) with $i \neq j$ and radii $r_i \neq r_j$. Among many ionic compounds, the lattices or crystal structures of cesium chloride ($CsCl$) salt (Fig. 1.9d) and sodium chloride ($NaCl$) salt (Fig. 1.9e) exhibit close-packed atomic arrangements similar to metallic lattices, which can be found at Wikipedia.com.

In crystallography, the BCC structure (Fig. 1.9a) is considered to have a lack of closed-packed planes, and therefore, it does not have a stacking sequence. Only the FCC and HCP crystal structures are produced by the stacking of close-packed planes of atoms on top of one another, and as a result, the empty space among packed atoms is minimized. The close-packed stacking sequence for FCC and HCP are $ABCABC\ldots$ and $ABABAB\ldots$ as indicated in Fig. 1.9b and c, respectively. A small shift of this atomic arrangement is called a stacking fault, which may be considered as an intrinsic defect. Nonetheless, it can be assumed that stacking faults induce elastic strains in their vicinity.

For pure metals with atoms modeled as hard and perfect spheres, Fig. 1.10 depicts three (3) types of conventional crystal lattices, which constitute **crystal structures** or crystalline solids with symmetrical arrangements of atoms in a **space lattice** . By definition, a space lattice is an infinite three-dimensional array of points or atoms as depicted in Fig. 1.10a for cubic, Fig. 1.10b for hexagonal, and Fig. 1.10c for triclinic atomic arrangements. These three-dimensional (3D) array of points are based on repetitive conventional unit cells.

Notice that the above space lattices (Fig. 1.10) have their points connected with straight lines representing a repetitive pattern and the smallest rectangular or hexagonal shape defines the **unit cell**, which is described in the next section. Nonetheless, these atom arrangements, known as space lattices, dictate the behavior of engineering materials under specific service conditions.

1.9   Space Lattice and Unit Cell

In order to further clarify the concepts of crystallography and related unit cells, the space lattice shown in Fig. 1.11a exhibits one shaded rectangular shape representing

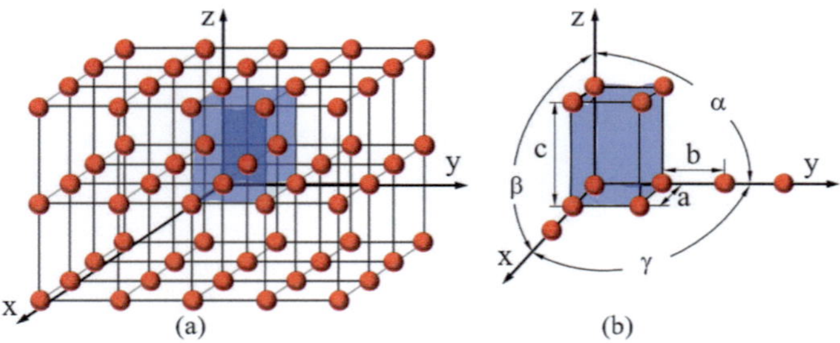

Fig. 1.11 Crystal system. (**a**) Bravais lattice and shaded unit cell and (**b**) the unit cell and crystallographic variables on Cartesian coordinates

the unit cell, which is extracted from the this space lattice and purposely isolated in Fig. 1.11b for showing the six constants related to the three-dimensional (3D) unit cell, which is characterized by a parallelepiped with cell edge lengths a, b, c edges and angles α, β, γ. Essentially, Fig. 1.11a is known as the Bravais lattice consisting of an infinite array of discrete points representing atoms or ions.

A 3D-lattice is described by three vectors $\mathbf{a}, \mathbf{b}, \mathbf{c}$ analogous to $\mathbf{i}, \mathbf{j}, \mathbf{k}$ in vector algebra. For clarity, the magnitude of $\mathbf{a}, \mathbf{b}, \mathbf{c}$ along with α, β, γ angles is depicted in Fig. 1.11b as the general representation of the morphology of a unit cell. Thus, a, b, c and α, β, γ are essential for determining the type of crystalline structures using the concept of unit cells.

Alternatively stated, in crystallography, the unit cell is the smallest volume element in a lattice. The unit cell is used to identify groups or family of crystal directions and planes denoted as $< uvw >$ and hkl, respectively. In essence, the unit cell symmetrically repeats throughout the crystal lattice.

There are seven (7) crystal systems and fourteen (14) 3D lattices (crystal lattices) represented as unit cells in Fig. 1.12. The 14 crystal lattices are grouped into seven lattice systems: three (3) cubic, two (2) tetragonal, four (4) orthorhombic and two (2) monoclinic systems.

1.9.1 Unit Cell Notations

The morphology (shape and size) of the unit cell is described by six variables, namely, a, b, c and α, β, γ. The former represent edges of the unit cell called lattice parameters, which, in turn, are the unit cell coordinate system. The latter are angles between the coordinate axes: $a \angle b = \alpha$, $b \angle c = \beta$, and $c \angle a = \gamma$. For comparison, the hexagonal unit cell is characterized by the Miller or Miller-Bravais edge coordinates, namely, a_1, a_2, a_3, c, where a_1, a_2, a_3 are located in the basal plane being introduced in a later section.

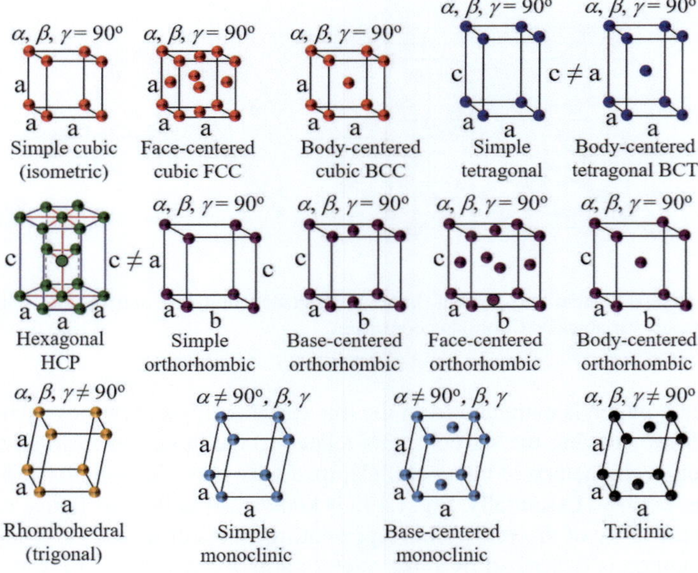

Fig. 1.12 Seven (7) types of crystal unit cells in color and fourteen (14) Bravais lattices

Indexing Lattice Planes and Directions

- a, b, c are intercepts so that their reciprocals define the Miller indices, namely, $h = 1/a, k/b$, and l/c. Each intercept may be unity or a fraction.
- h, k, l are integer numbers that represent intersects along the x, y, z Cartesian coordinates, respectively.
- h, k, l enclosed in parentheses, (hkl), is the Miller notation representing integer numbers for a single crystallographic plane. These integers enclosed in braces or curly brackets, $\{hkl\}$ represent the Miller notation for a group or family of crystallographic plane.
- $a/h, b/k, c/l$, represent crystallographic point coordinates along the x, y, z axes in Cartesian coordinates. Thus, (hkl) plane must satisfy the relationships $(x/a) + (y/b) + (z/c) = 1$ and $hx + ky + lz = 1$.
- u, v, w enclosed in square brackets, $[uvw]$, is the Miller notation representing integer numbers for a single crystallographic direction. The angle bracket notation, $< uvw >$, represents the Miller indices for a set or family of equivalent directions.
- $(hkl)[uvw]$ is the notation for a single slip system that represents the atomic mechanism of elastic deformation of a single crystal.
- $\mathbf{P} = [uvw]_{head} - [uvw]_{tail}$ is the notation for a vector difference, which describes the crystallographic direction from the tail to the head of a vector sketched as an arrow $\longrightarrow$.

1 H Hex																	2 He HCP
3 Li BCC	4 Be HCP											5 B Rho	6 C Hex	7 N Hex	8 O Gas	9 F SC	10 Ne FCC
11 Na BCC	12 Mg HCP											13 Al FCC	14 Si DC	15 P Orth	16 S Orth	17 Cl Orth	18 Ar FCC
19 K BCC	20 Ca FCC	21 Sc HCP	22 Ti HCP	23 V BCC	24 Cr BCC	25 Mn BCC	26 Fe BCC	27 Co HCP	28 Ni FCC	29 Cu FCC	30 Zn HCP	31 Ga Orth	32 Ge DC	33 As Rho	34 Se Hex	35 Br Orth	36 Kr FCC
37 Rb BCC	38 Sr FCC	39 Y HCP	40 Zr HCP	41 Nb BCC	42 Mo BCC	43 Tc HCP	44 Ru HCP	45 Rh FCC	46 Pd FCC	47 Ag FCC	48 Cd HCP	49 In Tetra	50 Sn Tetra	51 Sb Rho	52 Te Hex	53 I Orth	54 Xe FCC
55 Cs BCC	56 Ba BCC	57 La HCP	72 Zr HCP	73 Ta BCC	74 W BCC	75 Re HCP	76 Os HCP	77 Ir FCC	78 Pt FCC	79 Au FCC	80 Hg Rho	81 Tl HCP	82 Pb FCC	83 Bi Rho	84 Po SC	85 At FCC	86 Rn FCC
87 Fr BCC	88 Ra BCC	89 Ac FCC	104 Rf HCP	105 Db BCC	106 Sg BCC	107 Bh HCP	108 Hs HCP	109 Mt FCC	110 Ds BCC	111 Rg BCC	112 Cn BCC	113 Nh HCP	114 Fl FCC	115 Mc ?	116 Lv ?	117 Ts ?	118 Og FCC

Metals
Metalloids
Nonmetals

Fig. 1.13 Periodic table for crystal structures

- If $[uvw]$ direction is perpendicular to a plane normal (hkl), then the relation $[uvw] = (hkl)$ holds. This is denoted by the notation $[uvw] \perp (hkl)$.
- For $[uvw] \perp (hkl)$ condition, the relation $uh + vk + wl = 0$ holds.

1.9.2 Two-Dimensional Bravais Lattices

A Bravais lattice is a fundamental three-dimensional (3D) array of discrete points, such as atoms or ions, that describes the geometry of a periodic atomic arrangement of elements (Fig. 1.13).

2D Space In order to characterize the structure of crystalline solids and related point defects, one needs to consider, at least, one of the five two-dimensional (2D) Bravais lattices or symmetrical planes shown in Fig. 1.14.

Denote that $\mathbf{a}_1$ and $\mathbf{a}_2$ in these planes are discrete primitive vectors defining the shortest vectors of primitive lattices.

The symmetrical Bravais lattice $\{hkl\}$ planes are classified as

- *Oblique*: Monoclinic and triclinic $\{100\}$ planes
- *Square*: SC, FCC and BCC $\{100\}$ planes
- *Rectangular*: Tetragonal and orthorhombic $\{100\}$ planes
- *Centered rectangle*: BCC and body-centered orthorhombic (BCO) $\{110\}$ planes
- *Hexagon*: Hexagonal close-packed (HCP) with (0001) basal plane

The partial periodic table depicted in Fig. 1.13 includes the crystal structures for most elements.

Fig. 1.14 The five symmetrical Bravais lattices with (hkl) planes

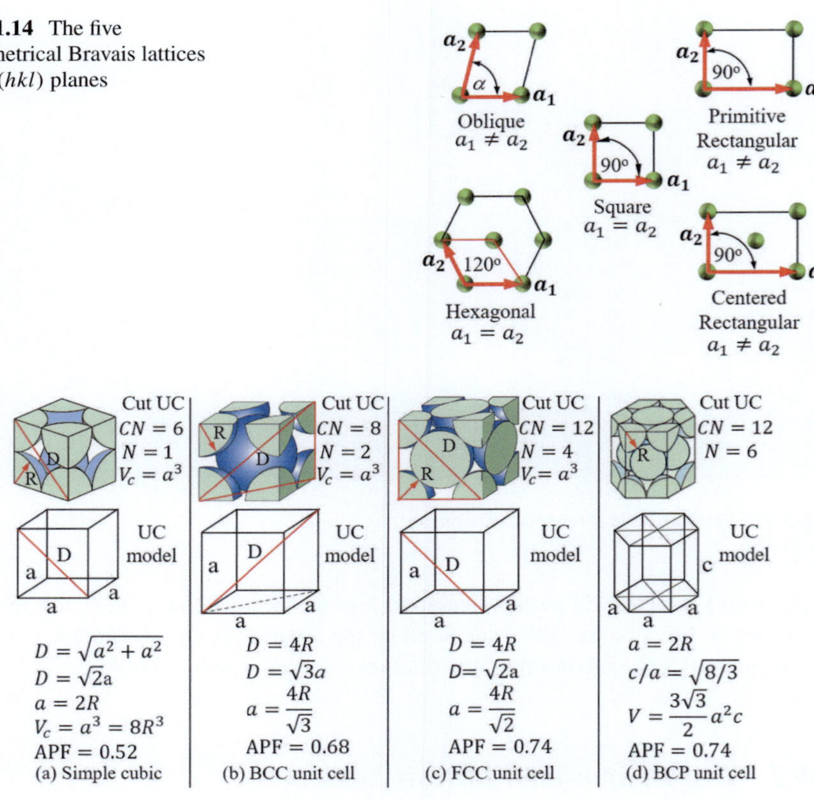

Fig. 1.15 Unit cell models and lattice parameter equations for common atomic structures. (**a**) Simple cubic (SC), (**b**) BCC, (**c**) FCC, and (**d**) HCP unit cells

Actually, Fig. 1.13 contains crystallographic data in an organized form, grouped in color for the reader's convenience.

1.9.3   Lattice Parameters

The physical dimensions of a unit cell (UC) in a crystal lattice is determined by the parameters a, b, c, which are experimentally determined using X-ray diffraction data. In crystallography, the lattice parameters are mathematically defined from the shape of a unit cell as shown in Fig. 1.15a for a simple cubic (SC), Fig. 1.15b for a BCC, Fig. 1.15c for a FCC, and Fig. 1.15d for an HCP crystal structures.

In materials science and chemistry, the coordination number (CN) is an integer based on the nearest neighbors bounds to an atom, and the lattice parameters a, b, c

are essential for determining the interplanar spacing (d-spacing or d_{hkl}) between parallel crystallographic planes.

The number of atoms (N) in a unit cell is the sum of atoms within the unit cell plus fractions of atoms on corners and faces. For the unit cells shown in Fig. 1.15,

$$N_{SC} = 8 \times \frac{1}{8} = 1\,atom \tag{1.24a}$$

$$N_{BCC} = 8 \times \frac{1}{8} + 1 = 2\,atoms \tag{1.24b}$$

$$N_{FCC} = 8 \times \frac{1}{8} + 6 \times \frac{1}{2} = 4\,atoms \tag{1.24c}$$

$$N_{HCP} = \left(12 \times \frac{1}{6}\right)_{corner} + \left(2 \times \frac{1}{2}\right)_{face} + (3)_{center} \tag{1.24d}$$

$$N_{HCP} = 6\,atoms \tag{1.24e}$$

Now, the volume of the unit cells (Fig. 1.15) are

$$V_{sc} = a^3 = (2R)^3 = 8R^3 \tag{1.25a}$$

$$V_{BCC} = a^3 = \left(\frac{4R}{\sqrt{3}}\right)^3 = \frac{64}{3\sqrt{3}}R^3 = 12.317R^3 \tag{1.25b}$$

$$V_{FCC} = a^3 = \left(\frac{4R}{\sqrt{2}}\right)^3 = 16\sqrt{2}R^3 = 22.627R^3 \tag{1.25c}$$

$$V_{hex} = \frac{3\sqrt{3}}{2}a^2c = 3\sqrt{2}a^3 = 24\sqrt{2}R^3 = 33.941R^3 \tag{1.25d}$$

where $c/a = \sqrt{8/3}$ and V_{hex} are important in crystallographic analysis.

Example 1.1 Derive the volume equation for an HCP unit cell starting with the basal plane area (A) shown below.

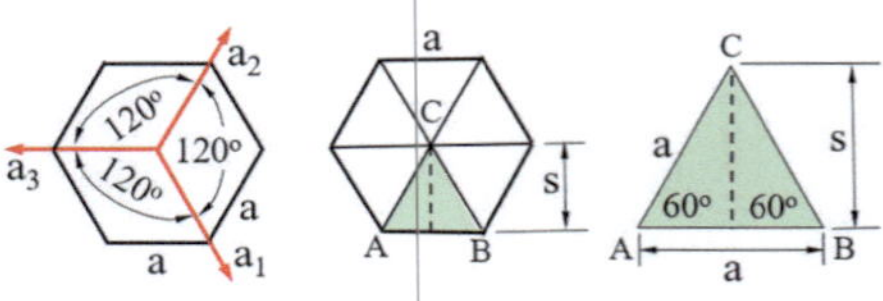

Solution The total basal plane area (ABC triangles) (A) and unit cell volume (V) are

$$A = 6\left[\frac{1}{2}(base)(height)\right] = 6\left[\frac{1}{2}(a)(s)\right] \tag{1.1E1a}$$

$$A = 6\left[\frac{1}{2}\,(a)\,(a\sin 60°)\right] = 6\left[\frac{1}{2}\,(a)\left(\frac{\sqrt{3}}{2}a\right)\right] = \frac{3\sqrt{3}}{2}a^2 \qquad (1.1\text{E}1\text{b})$$

$$V_{hex} = Ac = \frac{3\sqrt{3}}{2}a^2c \qquad (1.1\text{E}1\text{c})$$

where R is the atom radius and $a = 2R$ is the lattice parameter.

1.9.4   Atomic Packing Factor

John Dalton in 1803 suggested that atoms can be treated as hard or rigid solid spheres. Thus, the hard-sphere model allows one to determine the atomic packing factor (APF) as a volume ratio, which indicates how closely atoms are packed in a unit cell. Mathematically,

$$APF = \frac{NV_s}{V_c} = \left(\frac{4}{3}\pi R^3\right)\frac{N}{V_c} \qquad (1.26)$$

where N denotes the number of atoms in a unit cell and V_s and V_c denote the volume of a sphere and of a unit cell, respectively. According to the unit cells depicted in Fig. 1.15, the APF expressions for common structures are

$$APF_{SC} = \left(\frac{4}{3}\pi R^3\right)\left(\frac{N}{V_c}\right)_{sc} = \left(\frac{4}{3}\pi R^3\right)\frac{(1)}{8R^3} = \frac{\pi}{6} = 0.52 \qquad (1.27)$$

and

$$APF_{BCC} = \left(\frac{4}{3}\pi R^3\right)\left(\frac{N}{V_c}\right)_{BCC} = \left(\frac{4}{3}\pi R^3\right)\frac{(2)\,3\sqrt{3}}{64R^3} = 0.68 \qquad (1.28\text{a})$$

$$APF_{FCC} = \left(\frac{4}{3}\pi R^3\right)\left(\frac{N}{V_c}\right)_{FCC} = \left(\frac{4}{3}\pi R^3\right)\frac{4}{16\sqrt{2}R^3} = 0.74 \qquad (1.28\text{b})$$

$$APF_{hex} = \left(\frac{4}{3}\pi R^3\right)\left(\frac{N}{V_c}\right)_{hex} = \left(\frac{4}{3}\pi R^3\right)\frac{6}{24\sqrt{2}R^3} = 0.74 \qquad (1.28\text{c})$$

The term APF is also known as the packing efficiency of atoms to minimize the amount of voids. For instance, the FCC unit cell has 0.26 or 26% unoccupied space treated as interstitial voids within the packing arrangement of atoms.

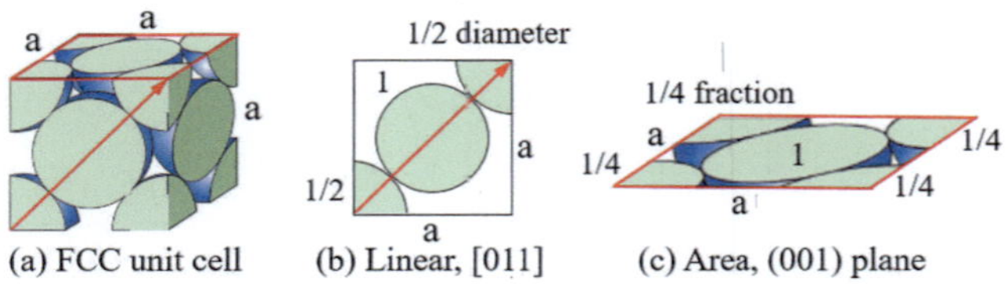

Fig. 1.16 Dimensional views for an FCC unit cell. (**a**) 3D view showing the [011] direction and the (001) plane (Callister and Rethwisch [6, p. 50]), (**b**) the [011] direction exhibiting a fractional atom diameters, and (**c**) the (001) plane with fractional atoms

1.9.5 Atomic Density

Consider Fig. 1.16 for determining the linear, planar, and volume densities as one-dimensional (1D), two-dimensional (2D), and three-dimensional (3D) analogs of the atomic packing factor in a unit cell.

Linear and planar densities are important variables relative to the plastic deformation process by a slip mechanism, which occurs on the most densely packed crystallographic $\{hkl\}$ planes and $< uvw >$ directions.

The relevant density (ρ) equations related to the hard-sphere atomic models are

$$\rho_L = \frac{N_{[uvw]}}{L_{[uvw]}} \quad \text{(linear density)} \tag{1.29a}$$

$$\rho_A = \frac{N_{(hkl)}}{A_{(hkl)}} \quad \text{(planar density)} \tag{1.29b}$$

$$\rho_V = \frac{N_c}{V_c} \quad \text{(mass density)} \tag{1.29c}$$

where $N_{[uvw]}, N_{(hkl)}, N_c$ denote the number of atoms touching each other in the $[uvw]$ direction, (hkl) plane, and unit cell, respectively. The denominators $L_{[uvw]}, A_{(hkl)}, V_c$ denote the length of $[uvw]$, planar area, and volume of the unit cell in terms of the lattice parameters or radius.

Figure 1.16a shows the cut FCC unit cell (Callister and Rethwisch [6, p. 50]) being selected for calculating the theoretical atomic densities. First of all, the number of equivalent atoms along the diagonal line ([011] direction) in Fig. 1.16b is based on the fractional amount of the intersected diameters, and that for a plane (Fig. 1.16c) is based on the fractional amount of intersected atoms by the (001) plane edges (1/4 at corners) and 1/2 at the center. Thus,

$$\rho_L = \frac{N_{[011]}}{L_{[011]}} = \frac{2 \times 1/2 + 1}{\sqrt{2}a} = \frac{2}{\sqrt{2}a} = \frac{2}{\sqrt{2}\left(4R/\sqrt{2}\right)} = \frac{1\,atom}{2R} \tag{1.30a}$$

$$\rho_A = \frac{N_{(100)}}{A_{(200)}} = \frac{1 + 4 \times 1/4}{a \times a} = \frac{2}{a^2} = \frac{2}{\left(4R/\sqrt{2}\right)^2} = \frac{1\,atom}{4R^2} \qquad (1.30b)$$

$$\rho_V = \frac{N_c}{V_c} = \frac{8 \times 1/8 + 6 \times 1/2}{a^3} = \frac{4}{a^3} = \frac{4}{\left(4R/\sqrt{2}\right)^3} = \frac{\sqrt{2}\,atom}{8R^3} \qquad (1.30c)$$

For iron (Fe) with $\rho = 7.64\,g/cm$ and $a = 0.365$ nm at $T = 925\,^{\circ}C$, Eq. (1.30) gives

$$\rho_L = 3.875\,atom/nm = 3.875 \times 10^7\,atom/cm \qquad (1.31a)$$

$$\rho_A = 15.012\,atom/nm^2 = 15.012 \times 10^{14}\,atom/cm^2 \qquad (1.31b)$$

$$\rho_v = 30.024\,atom/nm^3 = 8.226 \times 10^{22}\,atom/cm^3 \qquad (1.31c)$$

The theoretical mass (m) and atomic volume density (ρ_v) equations are of the form

$$m = \frac{NA_w}{N_a} \qquad (1.32a)$$

$$\rho_v = \frac{m}{V} = \frac{NA_w}{V_c N_a} \qquad (1.32b)$$

Here, A_w is the atomic weight (g/mol), $N_a = 6.022 \times 10^{23}\,atom/mol$ is the Avogadro's number, and $V = V_c$ is the volume of the unit cell. For iron (Fe) with $A_w = 55.845\,g/mol$,

$$m = \frac{NA_w}{N_a} = \frac{(4\,atom)\,(55.845\,g/mol)}{6.022 \times 10^{23}\,atom/mol} = 3.709 \times 10^{-22}\,g \qquad (1.33a)$$

$$\rho_v = \frac{m}{V} = \frac{3.709 \times 10^{-22}\,g}{\left(0.365 \times 10^{-7}\,cm\right)^3} = 7.63\,g/cm^3 \qquad (1.33b)$$

The slight discrepancy in the mass density may be attributed to crystal defects.

1.10   Crystallographic Directions

Crystallographic directions are important for determining, at least, one slip system, $(hkl)\,[uvw]$, as the failure mechanism of single crystals. Moreover, crystallographic directions are important because material properties, such as the modulus of elasticity (E), vary along different directions. This implies that crystals are anisotropic materials with respect to crystallographic directions.

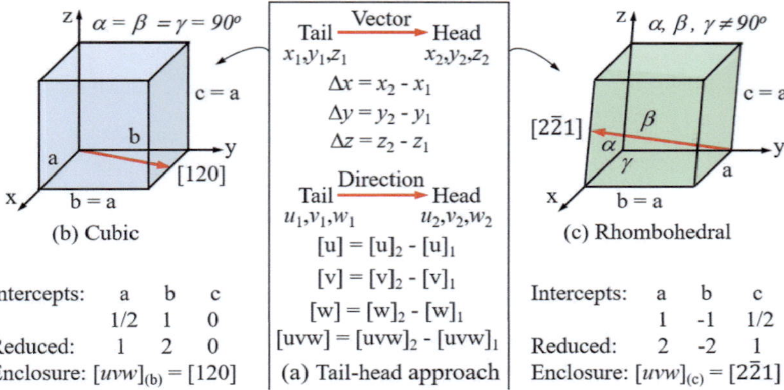

Intercepts: a   b   c
               1/2 1   0
Reduced: 1   2   0
Enclosure: $[uvw]_{(b)} = [120]$

Intercepts: a   b   c
               1   -1   1/2
Reduced: 2   -2   1
Enclosure: $[uvw]_{(c)} = [2\bar{2}1]$

Fig. 1.17 Procedure for determining a crystallographic direction using (**a**) the head-tail method, (**b**) intercept method for a cubic unit cell, and (**c**) intercept method for a rhombohedral unit cell

Three-Axis Coordinate System Notice that a, b, c lie on the x, y, z Cartesian coordinate system as shown in Fig. 1.17 for determining the Miller indices of a particular crystallographic $[uvw]$ direction. First of all, Fig. 1.17a illustrates the general head-tail method, and for comparison, Fig. 1.17b and c show the intercept method for cubic and rhombohedral unit cells.

Method A This is the vector tail-head approach related to point coordinates, which are x_1, y_1, z_1 for the vector head and (x_2, y_2, z_2) for the vector tail (Fig. 1.17a). The differences between these point coordinates are $\Delta x, \Delta y,$ and Δz. According to Callister and Rethwisch [6, p. 64], one can determine the Miller indices of crystallographic directions using normalized $\Delta x, \Delta y, \Delta z$ terms by the lattice parameters $a, b,$ and c, respectively. Thus,

$$u = \frac{n\Delta x}{a} = n\left(\frac{x_2 - x_1}{a}\right) \tag{1.34a}$$

$$v = \frac{n\Delta y}{b} = n\left(\frac{y_2 - y_1}{b}\right) \tag{1.34b}$$

$$w = \frac{n\Delta z}{c} = n\left(\frac{z_2 - z_1}{c}\right) \tag{1.34c}$$

where n is an integer number used to reduce $u, v,$ and w to integers.

Let's calculate the Miller indices for the crystallographic direction shown in Fig. 1.17b using Eq. (1.34). The partial results are

$$u = n\left(\frac{x_2 - x_1}{a}\right) = n\left(\frac{1a - 0a}{a}\right) = n \tag{1.35a}$$

$$v = n\left(\frac{y_2 - y_1}{b}\right) = n\left(\frac{-1b - 0b}{b}\right) = -n \tag{1.35b}$$

$$w = n\left(\frac{z_2 - z_1}{c}\right) = n\left[\frac{(1/2)\,c - 0c}{c}\right] = \frac{n}{2} \tag{1.35c}$$

Letting $n = 2$ reduces the results of Eq. (1.35) to the lowest integer numbers $u = 2$, $v = -2$, and $w = 1$. Subsequently, enclosure of these final results yields $[2\bar{2}1]$, which is exactly the same result shown in Fig. 1.17c.

Method B For clarity, the intercepts on the x, y, z coordinates are denoted as a, b, c

- Determine the value of each intercept, and fill the tables below unit cells in Fig. 1.17b, c.
- If the intercepts contain a fraction, multiply them by a small factor to convert them into integer numbers.
- Enclose the integer numbers in square brackets $[uvw]$. The result represents a single crystallographic direction. According to the conventional Miller notation, $[uvw]_{(b)} = [120]$ and $[uvw]_{(c)} = [2\bar{2}1]$ directions are illustrated in the tables in Fig. 1.17b, c. This is a direct method for determining the Miller indices of crystallographic directions.
- In crystallography, negative Miller indices or negative integer values are indicated with a bar over the integers. For example, $-1, -2, 1$ are written as $[\bar{1}\bar{2}1]$ direction.

Four-Axis Coordinate System For instance, graphite has a hexagonal structure with honeycomb-shaped layers of carbon atoms referred to as graphene sheets. This type of honeycomb structure does not have a central atom. Although graphite has an hexagonal structure, it is not considered an HCP material (Fig. 1.18a). Perhaps, this is one significant reason for the excellent transport properties of exfoliated graphene from graphite.

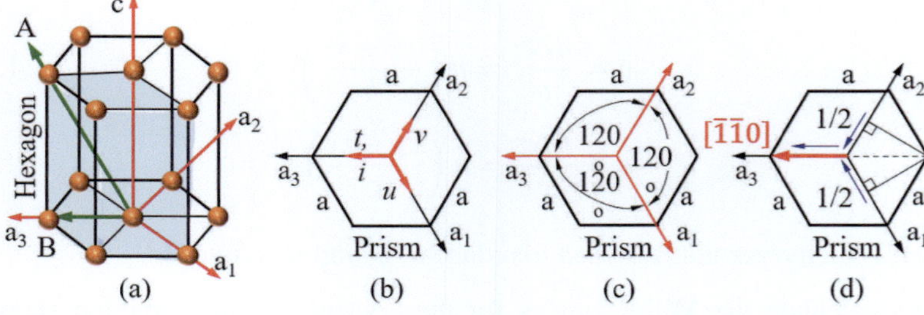

Fig. 1.18 Hexagonal unit cells. (**a**) A three-digit direction in the hexagon, (**b**) the basal plane with a_1, a_2, a_3 coordinates and unit vectors u, v, (t, i), w, (**c**) basal plane angles, and (**d**) intercept-paths as indicated by the short arrows to form a $[uvw]$ direction

For a conventional hexagonal close-packed (HCP) unit cell (Fig. 1.18a) with crystallographic coordinates a_1, a_2, a_3 on the basal plane and the c coordinate along the vertical z-axis, the origin of these crystallographic coordinates is located at the center of basal planes.

Crystallographically, the four-axis coordinate system is known as the Miller-Bravais index system, and it is designated the notation $[uvtw]$, which represents a particular crystallographic direction in a three-dimensional (3D) hexagonal crystal.

In order to effectively analyze the 4-index $[uvtw]$ system, one has to employ the hexagonal crystal a_1-a_2-c coordinates for determining the point-coordinates and consider some relevant features of the HCP unit cell. Thus:

- An hexagonal unit cell is constructed by three hexagons (parallelepiped or tetragonal-like shapes), one of which is delineated (shaded) in Fig. 1.18a containing a three-digit $[uvw]$ directions, namely, "A" and "B" in the shaded hexagon.
- The basal plane shown in Fig. 1.18b has the a_1, a_2, a_3 coordinates 120° degrees apart along with the vector $[uvtw]$ representing a direction with $t = -(u+v)$ and $(hkil)$ for a plane with $i = -(h+k)$.
- Use the conventional notation for the angles $a_1 \angle a_2 = 120°$, $a_2 \angle a_3 = 120°$, and $a_1 \angle a_3 = 120°$. Also, c is perpendicular to the basal plane.
- The basal plane in Fig. 1.18c shows the lattice parameter "a" along the a_1, a_2, a_3 basal coordinates.
- The vectors t, i in Fig. 1.18b arise due to the a_3 basal plane coordinate.
- The ideal $c/a = \sqrt{8/3}$ ratio is used for comparing real HCP materials (Chap. 2).
- For the direction in Fig. 1.18d, $-1a_1$, $-1a_2$, and $0c$ are the intercepts so that the 3-digit Miller notation becomes $\left[u'v'w'\right] = \left[\bar{1}10\right]$. This 3-digit Miller notation $\left[\bar{1}10\right]$ can be converted to a 4-digit Miller-Bravais notation $[uvtw] = \left[\bar{1}\bar{1}20\right]$ as illustrated below.

For hexagonal symmetry,

$$\left[u'v'w'\right] \rightarrow [uvtw] \tag{1.36a}$$

$$u = \frac{1}{3}\left(2u' - v\right) \tag{1.36b}$$

$$v = \frac{1}{3}\left(2v' - u\right) \tag{1.36c}$$

$$t = -(u+v) \tag{1.36d}$$

$$w = w' \tag{1.36e}$$

Conversely,

$$[uvtw] \rightarrow \left[u'v'w'\right] \tag{1.37a}$$

$$u' = \frac{1}{2}(3u + v) \tag{1.37b}$$

$$v' = \frac{1}{2}(3v + u) \tag{1.37c}$$

$$w' = w \tag{1.37d}$$

Sketched Direction For a certain $[uvw]$ direction, the Miller indices can be determined using equations based on the concept of head-to-tail point differences similar to Eq. (1.34). Thus,

$$u = 3n\left(\frac{a_1' - a_1''}{a}\right) \tag{1.38a}$$

$$v = 3n\left(\frac{a_2' - a_2''}{a}\right) \tag{1.38b}$$

$$t = 3n\left(\frac{a_3' - a_3''}{a}\right) \tag{1.38c}$$

$$w = 3n\left(\frac{z' - z''}{c}\right) \tag{1.38d}$$

For a given $[uvtw]$ direction, solve Eq. (1.38) for the vector head variables with n being a factor used to convert fractions to integer numbers. Thus,

$$a_1' = \frac{ua}{3n} + a_1'' \tag{1.39a}$$

$$a_2' = \frac{va}{3n} + a_2'' \tag{1.39b}$$

$$a_3' = \frac{ta}{3n} + a_3'' \tag{1.39c}$$

where the direction is sketched in a unit cell as an arrow that has a tail and head. The vector location is determined by the difference between the head a_1', a_2', a_3' and the tail a_1'', a_2'', a_3'' point coordinates.

Example 1.2 Determine **(a)** the crystallographic direction $[u'v'w']$ shown as vector **A** within the shaded hexagon in the given figure below, **(b)** convert $[u'v'w'] \rightarrow [uvtw]$. **(c)** What would you do if $[uvtw]$ is given and asked to sketch or plot this direction?

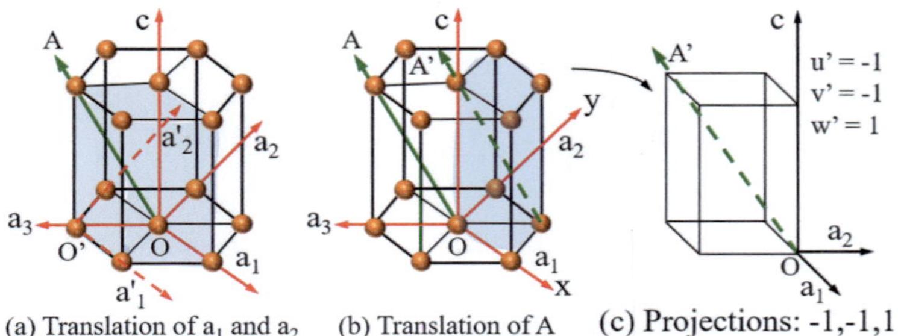

(a) Translation of a_1 and a_2 (b) Translation of A   (c) Projections: -1,-1,1

Solution

(a) Methods Point Coordinates According to the given figures given below, the principal point coordinates help determine the Miller indices of a particular crystallographic direction $\mathbf{A} = \left[u'v'w'\right]$ using the vector tail-head method given in Fig. 2.17a.

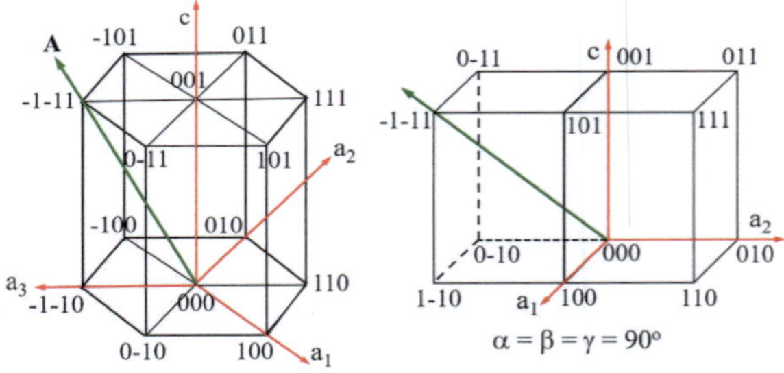

The vector tail-head method for vector A is

$$\left[u'v'w'\right] = \left[u'v'w'\right]_{head} - \left[u'v'w'\right]_{tail} \tag{1.2E1a}$$

$$\left[u'v'w'\right] = \left[u'v'w'\right]_{h2} - \left[u'v'w'\right]_{1} \tag{1.2E1b}$$

$$\left[u'v'w'\right] = \left[0\bar{1}1\right]_{head} - [000]_{tail} = \left[\bar{1}\bar{1}1\right] \tag{1.2E1c}$$

Hexagon Projections The $u'v'w'$ projections are interpreted as intercepts, which are converted to Miller indices using square bracket $\left[u'v'w'\right]$. The procedure for determining the three-digit Miller indices is illustrated below

$$
\begin{array}{lccc}
\text{Projections} = & a_1 & a_2 & c \\
= & -1 & -1 & 1 \\
\text{Reduction} = & -1 & -1 & 1 \\
\text{Enclosure} = & \left[u'v'w'\right] = \left[\bar{1}\bar{1}1\right] &
\end{array}
$$

Normalized Axes Using Eq. (1.34) yields the three-digit set describing the Miller indices in terms of the n factor. Thus,

$$u' = n\left(\frac{x_2 - x_1}{a}\right) = n\left(\frac{a_1' - a_1''}{a}\right) = n\left(\frac{-1a - 0a}{a}\right) = -n \tag{1.2E2a}$$

$$v' = n\left(\frac{y_2 - y_1}{b}\right) = n\left(\frac{a_2' - a_2''}{a}\right) = n\left(\frac{-1a - 0a}{a}\right) = -n \tag{1.2E2b}$$

$$w' = n\left(\frac{z_2 - z_1}{c}\right) = n\left(\frac{c' - c''}{c}\right) = n\left(\frac{1c - 0c}{c}\right) = n \tag{1.2E2c}$$

(b) Convert $\left[u'v'w'\right] = \left[\bar{1}\bar{1}1\right]$ to $[uvtw]$ so that

$$u = \frac{1}{3}\left(2u' - v'\right) = \frac{1}{3}\left[2\left(-1\right) - \left(-1\right)\right] = -\frac{1}{3} \tag{1.2E3a}$$

$$v = \frac{1}{3}\left(2v' - u'\right) = \frac{1}{3}\left[2\left(-1\right) - \left(-1\right)\right] = -\frac{1}{3} \tag{1.2E3b}$$

$$t = -\left(u + v\right) = -\left(-\frac{1}{3} - \frac{1}{3}\right) = \frac{2}{3} \tag{1.2E3c}$$

$$w = w' = 1 \tag{1.2E3d}$$

Multiply these results by 3, and enclosure the new result to get

$$\text{Fractions} = \left(-\frac{1}{3} \quad -\frac{1}{3} \quad \frac{2}{3} \quad 1\right) \times 3$$

$$\text{Integers} = \quad -1 \quad -1 \quad 2 \quad 3$$

$$\text{Enclosure} = [uvtw] = \left[\bar{1}\bar{1}23\right]$$

Therefore, the three-digit crystallographic direction $\left[\bar{1}\bar{1}1\right]$ is easy to draw or plot, and it is equivalent to the four-digit crystallographic direction $\left[\bar{1}\bar{1}23\right]$, which is not that easy to plot within the hexagonal unit cell.

(c) **Reverse Procedure** For $\left[\bar{1}\bar{1}23\right]$ direction, Eq. (1.39) gives the intercepts along with $n = 1$ (in this example) as shown below

$$a_1' = \frac{ua}{3n} + a_1'' = \frac{(-1)\,a}{3n} + 0a = -\frac{1}{3}a \tag{1.2E4a}$$

$$a_2' = \frac{va}{3n} + a_2'' = \frac{(-1)\,a}{3n} + 0a = -\frac{1}{3}a \tag{1.2E4b}$$

$$a_3' = \frac{ta}{3n} + a_3'' = \frac{(2)\,a}{3n} + 0 = \frac{2}{3}a \tag{1.2E4c}$$

$$w' = w = 1c \tag{1.2E4d}$$

Based on these results, divide the basal plane coordinates in intervals of 1/3, locate or mark the direction head point coordinates a_1', a_2', a_3', c, and proceed to sketch the direction (arrow) stating from the origin of the tail, $(0, 0, 0, 0)$, in this example.

If $[\bar{1}\bar{1}1]$ is used, then Eq. (1.39) along with the vector tail $a_2'', a_3'', c_2'' = 0$ gives the vector head

$$a_1' = \frac{ua}{n} + a_1'' = \frac{(-1)\,a}{1} + 0a = -a \tag{1.2E5a}$$

$$a_2' = \frac{va}{n} + a_3'' = \frac{(-1)\,a}{1} + 0a = -a \tag{1.2E5b}$$

$$c_2' = \frac{wc}{n} + c_2'' = \frac{(1)\,c}{1} + 0c = c \tag{1.2E5c}$$

Thus, the vector head is located at distances $a_1', a_2', c_2' = -1, -1, 1$. Vector "A" is shown in Fig. 1.18a within the delineate hexagon.

Example 1.3 Derive the ideal $c/a = \sqrt{8/3}$ lattice parameter axial ratio for an HCP unit cell using the drawings given below (Mongonon [7, p. 35]).

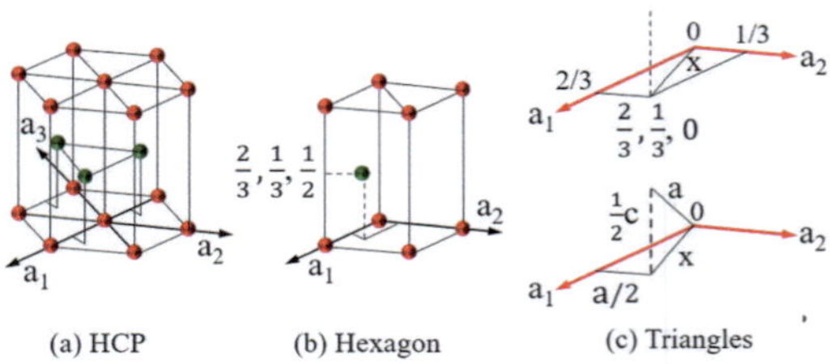

(a) HCP   (b) Hexagon   (c) Triangles

Solution In this example, Figs. Ex1.3a and Ex1.3b are reference sketches. From Fig. Ex1.3c,

$$\left(\frac{2}{3}a\right)^2 = x^2 + \left(\frac{1}{3}a\right)^2 \tag{1.3E1a}$$

$$x^2 = \left(\frac{2}{3}a\right)^2 - \left(\frac{1}{3}a\right)^2 = \frac{1}{3}a^2 \tag{1.3E1b}$$

From Fig. Ex1.3d,

$$a^2 = x^2 + \left(\frac{1}{2}c\right)^2 \tag{1.3E2}$$

Combining Eqs. (1.3E1b) and (1.3E2) yields

$$a^2 = x^2 + \left(\frac{1}{2}c\right)^2 \tag{1.3E3a}$$

$$a^2 = \frac{1}{3}a^2 + \left(\frac{1}{2}c\right)^2 \tag{1.3E3b}$$

$$\frac{c}{a} = \sqrt{\frac{8}{3}} = 1.633 \tag{1.3E3c}$$

which is an intrinsic property of an HCP structure. Normally, the c/a axial ratio for real HCP metals or alloys deviates from the ideal value. For hexagonal materials, $(c/a)_{alloy} > 1.633$ or $(c/a)_{alloy} < 1.633$ due to size of the elements being combined to form unit cells.

1.11 Crystallographic Planes

A crystallographic plane is identified by the Miller indices denoted as h, k and l having integer values. The procedure for determining the values of a crystallographic (hkl) plane is described next. Thus,

- Find the intercepts a, b, c using a given origin O with x, y, z coordinates. If a, b, c cannot be determined, define a new origin $O\prime$ with new coordinates x', y', z'.
- Find the intercept a from O to x, b from O to y, and c from O to z.
- If any of the a, b, c intercepts does not intersect its corresponding x, y or z coordinate, then an infinite ∞ value is assigned to that intercept.
- Take the reciprocal of intercepts, and clear any fraction so that the result is an integer number.
- Reduce the result to its minimum integer value.
- Enclose the result in parentheses: (hkl).
- Any negative integer is assigned a bar on top of it: $(\bar{h}kl)$, $(h\bar{k}l)$, $(\bar{h}\bar{k}l)$, and so forth.

Case 1: Type of Cubic Planes The (hkl) notation represents crystal planes that intersect the x, yz axes at distances of a/h,b/y,c/z as shown in Fig. 1.19. For instance, Fig. 1.19a shows the (001) face plane, Fig. 1.19b (111) shows a triangle plane, and Fig. 1.19c illustrates a diagonal (110) plane. The tabular procedure for finding (hkl) is given below each unit cell containing the drawn plane.

Case 2: Triangle Plane The tables below each unit cell in Fig. 1.20 illustrate three procedures used to identify the (121) plane. For Fig. 1.20a, subtract the shown coordinate points on the y-axis to determine the net intercept as $0, 1/2, 0$.

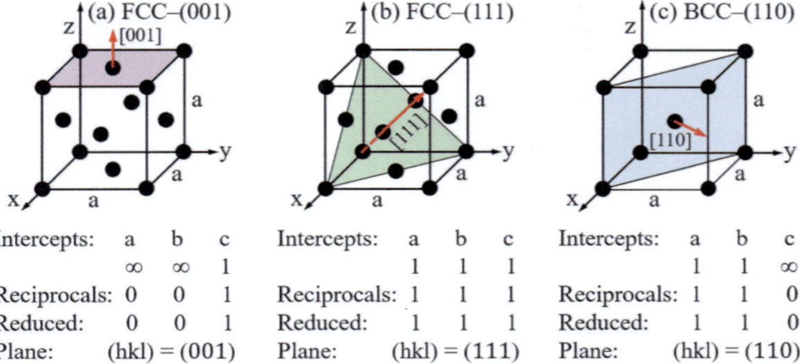

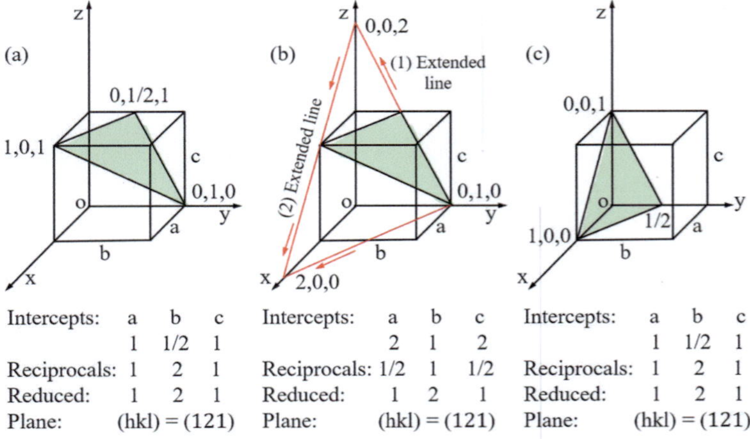

Fig. 1.19 Procedure for determining the Miller indices of selected planes in cubic unit cells. (**a**) FCC-(001), (**b**) FCC-(111), and (**c**) BCC-(110) planes. Notice that the corresponding perpendicular directions are also sketched

Fig. 1.20 Procedure for determining the Miller indices of selected planes in cubic unit cells. (**a**) FCC-(001), (**b**) FCC-(111), and (**c**) BCC-(110) planes. Notice that the corresponding perpendicular directions are also sketched

For Fig. 1.20b, extend the plane straight edges until they intersect the Cartesian coordinates as shown. This is a method shown in a video posted online by Dr. Shamberger [8]. The plane shown in Fig. 1.20c also gives the same Miller indices (121).

Case 3: Movable Origin of Coordinates When a (*hkl*) plane intercepts the origin $(0, 0, 0)$ of the x, y, x coordinates, either move the origin or move the plane to another location. Subsequently, determine the Miller indices as usual.

One particular case is shown in Fig. 1.21a for the family of cubic $\{\bar{1}21\}$ planes in which the coordinate origin is moved to two different locations in the unit cell. Thus, the procedure for determining the Miller indices is illustrated in the tables in

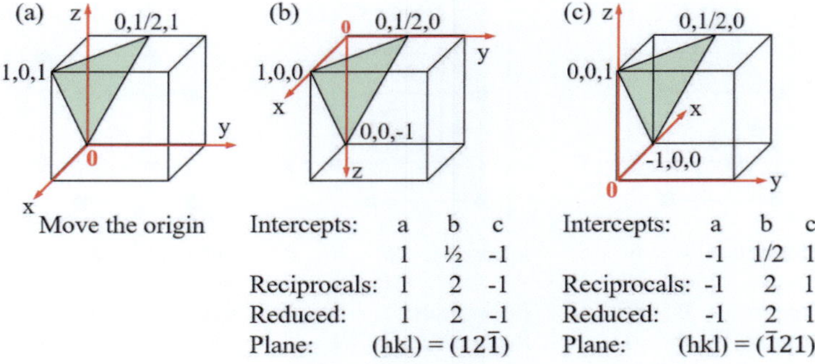

Fig. 1.21 An (hkl) plane intercepting the origin of the coordinates. (**a**) The (hkl) plane intercepts the origin, (**b**) the origin is moved upward along the z-axis leading to $(12\bar{1})$, and (**c**) the origin is moved forward along the x-axis leading to the $(\bar{1}21)$ plane

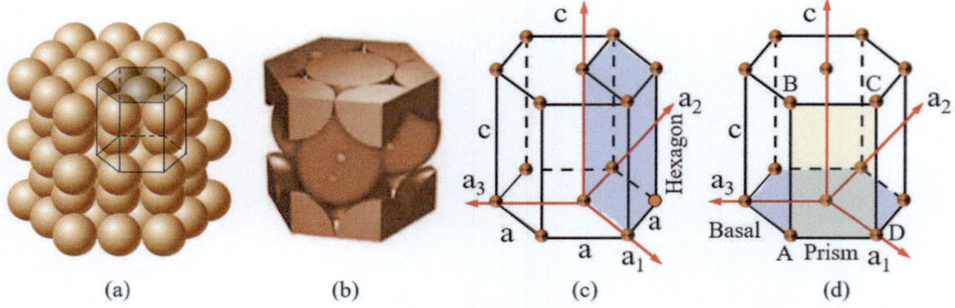

Fig. 1.22 Hexagonal close-packed (HCP) lattice. (**a**) HCP lattice showing a unit cell, (**b**) cut HCP unit cell, (**c**) HCP unit cell showing one hexagon, and (**d**) HCP unit cell showing the (0001) basal and the ABCD prism planes

Fig. 1.21b for $(12\bar{1})$ and Fig. 1.21c for $(\bar{1}21)$ planes. Notice that the plane does not change shape and remains stationary. Only the (hkl) indices change values due to the new origins being chosen.

Case 4: The (hkil) Hexagonal Planes Consider the hexagonal close-packed crystal lattice in Fig. 1.22 showing a unit cell. For convenience, the HCP unit cell is a term used hereafter. Notice the drawn unit cell (Fig. 1.22a), the ideal cut hexagonal close-packed (HCP) unit cell (Fig. 1.22b), the HCP unit-cell model and its Miller-Bravais 4-axis coordinate system (Fig. 1.22c) along with the drawn hexagon (shaded rectangular shape), and the shaded-basal plane (Fig. 1.22d) with 3-axis (a_1, a_2, a_3) coordinate system with $a_1 \angle a_2 = 120°$, $a_2 \angle a_3 = 120°$, and $a_3 \angle a_1 = 120°$ degrees.

This general hexagonal coordinate system is based on the 3-axis Miller index, and it is written as $hkil$ for planes or $[hktl]$ for directions. Hence, the Miller-Bravais system is characterized in a simplified form by using a 3-axis hexagon, which converted to a 4-axis system. Moreover, the $ABABAB \ldots$ stacking sequence represents alternating planes of HCP spheres. Regarding the packing of atoms as

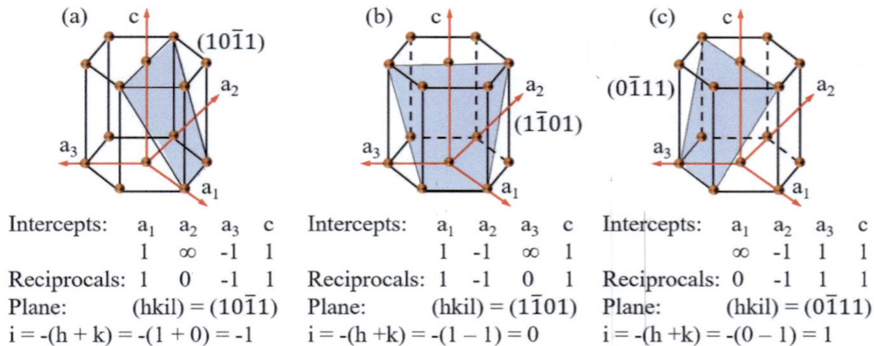

Intercepts: a_1 a_2 a_3 c   Intercepts: a_1 a_2 a_3 c   Intercepts: a_1 a_2 a_3 c
              1   ∞   -1   1   1   -1   ∞   1   ∞   -1   1   1
Reciprocals: 1   0   -1   1   Reciprocals: 1   -1   0   1   Reciprocals: 0   -1   1   1
Plane: $(hkil) = (10\bar{1}1)$ Plane: $(hkil) = (1\bar{1}01)$ Plane: $(hkil) = (0\bar{1}11)$
$i = -(h + k) = -(1 + 0) = -1$ $i = -(h + k) = -(1 - 1) = 0$ $i = -(h + k) = -(0 - 1) = 1$

Fig. 1.23 Procedure for determining some crystallographic planes in HCP unit cells. (**a**) $\left(10\bar{1}1\right)$, (**b**) $\left(1\bar{1}01\right)$, and (**c**) $\left(0\bar{1}11\right)$ planes

spheres, an individual atom away from the edges of A or B plane has $CN = 6$, and an individual atom in plane B touches three spheres in above and below A planes so that $CN = 3$.

Notice that the horizontal basal plane of the HCP unit cell is parallel to the a_1, a_2, a_3 axes and perpendicular to the c-axis with $c = \infty$, whereas the vertical ABCD prism plane has the opposite orientation with $a_1 = \infty$, $a_2 = \infty$, and $a_3 = \infty$. Moreover, the Miller-Bravais indices are the reciprocal of these axes, that is, $(hkil) = (0001)$ for the basal plane and $(hkil) = (1\bar{1}00)$ for the ABCD prism plane since $i = -(h + k)$.

Figure 1.23 illustrates the intercept procedure for determining the Miller indices of the selected HCP planes using four coordinates: a_1, a_2, a_3, and c. The shown planes are $\left(10\bar{1}1\right)$ in Fig. 1.23a, $\left(1\bar{1}01\right)$ in Fig. 1.23b, and $\left(0\bar{1}11\right)$ in Fig. 1.23c.

The reader is encouraged to analyze and study all the above crystallographic cases very carefully. It is understood that getting used to determine the Miller indices of crystallographic planes and directions in a hexagonal unit cell is a time-consuming process, but it is essential for understanding the basic concepts of crystallography with respect to crystals and their structure. So practice makes perfect.

1.12   Interstitial Sites in Crystal Structures

An interstice is the empty space called void or hole, where a FCC crystal structure can accommodate an interstitial (small size) atom (Fig. 1.24a). Hence, octahedral and tetrahedral (triangular pyramid) atomic geometries are the most common in textbooks and classrooms.

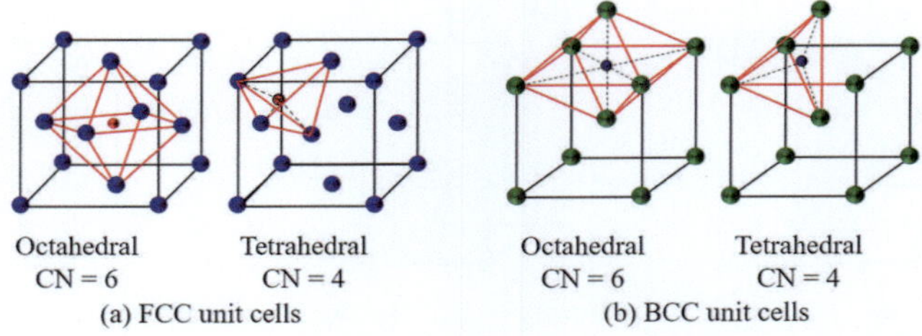

Octahedral   Tetrahedral   Octahedral   Tetrahedral

CN = 6   CN = 4   CN = 6   CN = 4

(a) FCC unit cells   (b) BCC unit cells

Fig. 1.24 FCC unit cells with (**a**) an octahedral geometry with coordination number $CN = 6$ and (**b**) a tetrahedral geometry with $CN = 4$

Table 1.1 Coordination numbers and radius ratio r/R (Callister and Rethwisch [6, p. 470])

CN	Radius ratio range	Void	Geometry
2	$0 < r/R < 0.155$	Linear	
3	$0.155 \leq r/R < 0.225$	Planar	
4	$0.225 \leq r/R < 0.414$	Tetrahedral	
6	$0.414 \leq r/R < 0.732$	Octahedral	
8	$0.732 \leq r/R < 1.000$	Cubic	

The coordination number (CN) is based on a central atom or ion bounded by other adjacent atomic spheres as elucidated in Table 1.1 containing the atomic radius ratio ranges (Callister and Rethwisch [6, p. 408]).

For instance, the FCC unit cell shown in Fig. 1.23a exhibits probable interstitial sites having a coordination number $CN = 6$ in the octahedral with a position at $(1/2, 1/2, 1/2)$ and $CN = 4$ in the tetrahedral interstices located at $(1/4, 1/4.1/4)$ or $(3/4, 3/4, 3/4)$ within the unit cells. This may be the case for iron-carbon (Fe-C) steels in which the carbon is an interstitial atom. Similarly, Fig. 1.24b depicts these interstices in BCC structures. Other crystals, such as an HCP, can have these interstices as well.

In steels, as an example, iron (Fe) has a greater solubility for carbon atoms in the FCC octahedral structure (called austenite) than in the BCC tetrahedral structure (known as ferrite) sites. This is essential or fundamental for hardening steels. In ceramic materials, there are also interstices in their ionic structures.

Example 1.4 Determine the limiting (minimum) radius ratio r/R for an FCC planar geometry with $CN = 3$. Deduce r/R for $CN = 2$. Recall that the radius ratio is useful in predicting the structure of ionic solids assuming that atoms or ions are hard spheres.

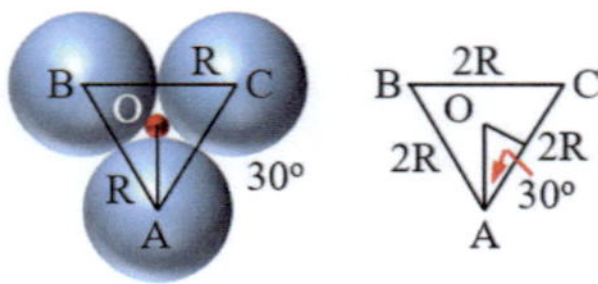

60° Triangle

Solution For $CN = 3$,

$$OA = r + R \tag{1.4E1a}$$

$$\cos(30°) = \frac{R}{OA} = \frac{R}{r + R} \tag{1.4E1b}$$

$$\frac{\sqrt{3}}{2} = \frac{R}{r + R} \tag{1.4E1c}$$

$$\frac{r}{R} = \frac{2}{\sqrt{3}} - 1 = 0.155 \tag{1.4E1d}$$

Hence,

$$0.155 \le \left(\frac{r}{R}\right)_{CN=3} < \left(\frac{r}{R}\right)_{CN=4} \tag{1.4E2}$$

For $CN = 2$,

$$0 < \left(\frac{r}{R}\right)_{CN=2} < \left(\frac{r}{R}\right)_{CN=3} = 0.155 \tag{1.4E3}$$

These ratio ranges match the values given in Table 1.1. Graphically, the coordination number CN of an atom is represented by a single atom or ion being bounded to immediate adjacent atoms. Mathematically, CN is clearly shown above that it is an integer number representing the surrounding atoms attached to a interstitial atom. Hence, the coordination number indirectly defines the atomic structure of crystalline materials and dictates the geometry formed by the atoms or ions.

Example 1.5 Use the figure given below to find the limiting (minimum) radius ratio r/R for a FCC tetrahedral site with $CN = 4$.

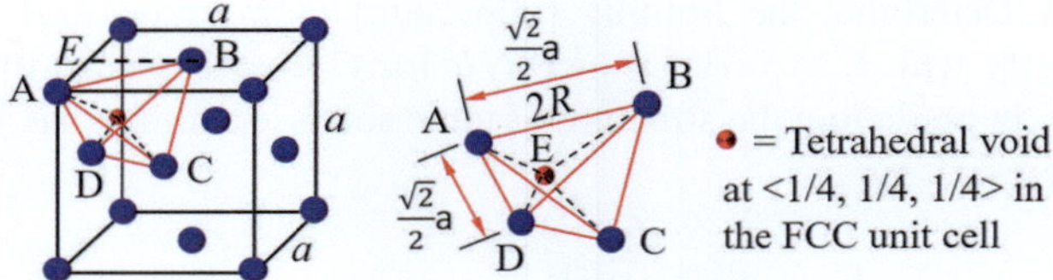

Solution Using the ABE triangle yields

$$\left(\frac{a}{2}\right)^2 + \left(\frac{a}{2}\right)^2 = (2R)^2 \tag{1.5E1a}$$

$$a = 2\sqrt{2}R \tag{1.5E1b}$$

From the ABD triangle on the tetrahedral geometry,

$$AB^2 + BD^2 = AD^2 \tag{1.5E2a}$$

$$\left(\sqrt{2}a\right)^2 + a^2 = AD^2 \tag{1.5E2b}$$

$$AD = \sqrt{3}a = \sqrt{3}\sqrt{2}R \tag{1.5E2c}$$

$$AD = \sqrt{6}R \tag{1.5E2d}$$

Then, the limiting (minimum) radius ratio for a tetrahedral geometry with $CN = 4$ is

$$AD = 2r + 2R \tag{1.5E3a}$$

$$\sqrt{6}R = 2(r + R) \tag{1.5E3b}$$

$$\frac{\sqrt{6}}{2} = \frac{r}{R} + 1 \tag{1.5E3c}$$

$$\frac{r}{R} = \frac{\sqrt{6}}{2} - 1 = 0.225 \tag{1.5E3d}$$

Hence,

$$0.225 \leq \left(\frac{r}{R}\right)_{CN=4} < \left(\frac{r}{R}\right)_{CN=6} \tag{1.5E4}$$

Example 1.6 During heat treatment of a carbon-steel specimen at $900\,^\circ C$, carbon atoms move (diffuse) within the FCC austenitic structure (γ-Fe phase). Assume that the specimen is slowly cooled in an inert environment within a furnace. **(a)** Use the relevant sketch given below to derive an equation for the theoretical radius r of an interstitial atom that can fit in an octahedral interstice with $CN = 6$. **(b)** Calculate r and compare its result with the experimental value $r_C = 0.071\,\text{nm}$ for carbon in the

γ-Fe ($r_{Fe} = 0.124$ nm) octahedral structure. Explain the main fundamental process for hardening steels.

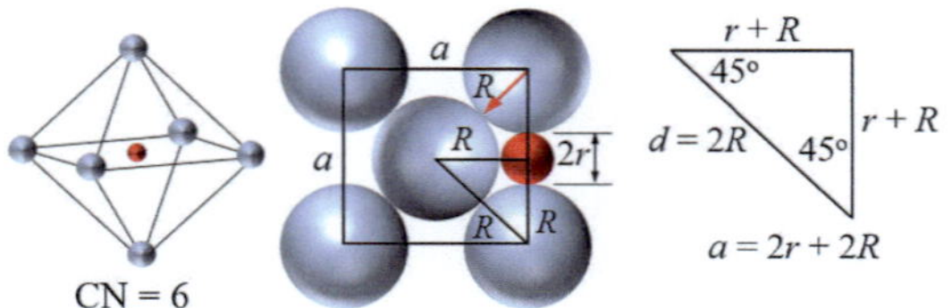

Solution

(a) Derivation: From the triangle,

$$d^2 = (r + R)^2 + (r + R)^2 = 2\,(r + R)^2 \tag{1.6E1a}$$

$$d^2 = 4R^2 \tag{1.6E1b}$$

$$4R^2 = 2\,(r + R)^2 \tag{1.6E1c}$$

$$2R^2 = r^2 + 2rR + R^2 \tag{1.6E1d}$$

$$2 = \frac{r^2}{R^2} + \frac{2r}{R} + 1 \tag{1.6E1e}$$

Then, the limiting (minimum) radius ratio for $CN = 6$ becomes

$$\frac{r}{R} = 0.414 \tag{1.6E2a}$$

$$r = 0.414R \tag{1.6E2b}$$

Also,

$$\cos\,(45°) = \frac{r + R}{2R} \tag{1.6E3a}$$

$$\frac{\sqrt{2}}{2} = \frac{r + R}{2R} \tag{1.6E3b}$$

$$\sqrt{2} = \frac{r}{R} + 1 \tag{1.6E3c}$$

$$\frac{r}{R} = \sqrt{2} - 1 = 0.414 \tag{1.6E3d}$$

$$r = 0.414R \tag{1.6E3e}$$

Hence,

$$0.414 \leq \left(\frac{r}{R}\right)_{CN=6} < \left(\frac{r}{R}\right)_{CN=8} \tag{1.6E4}$$

(b) Substitute $R = R_{Fe} = 0.124\,nm$ into Eq. (1.6E3e) so that

$$r = 0.414R = 0.414R_{Fe} \tag{1.6E5a}$$

$$r = 0.414\,(0.124\,nm) = 0.051\,nm \tag{1.6E5b}$$

$$r < r_C = 0.071\,nm \tag{1.6E5c}$$

The steel specimen expands upon heat treatment at $900\,°C$, and carbon atoms accommodate themselves at the octahedral sites by diffusion and get trapped there upon cooling.

1.13 Interplanar Spacing

The lattice interplanar spacing (d-spacing or d_{hkl}) is assumed to be a perpendicular distance between successive stack of parallel planes (Fig. 1.25a for a triangle plane and Fig. 1.25b for a rectangular plane) in a crystal structure under equilibrium.

The interplanar spacing d_{hkl} depends on the crystal system involved because it is a function of the plane indices (hkl) and the lattice constants $(a, b, c, \alpha, \beta, \gamma)$. Mathematically,

$$d_{hkl} = f\left[(hkl)\,,\,(a/h, b/k, c/l, \alpha, \beta, \gamma)\right] \tag{1.40}$$

which is a function of nine variables, but $h^2 + k^2 + l^2 = Integer$ is a factor related to the type of crystal system. Consequently, only certain (hkl) planes are relevant for determining the d_{hkl} analytical expressions. For orthogonal axes, $\alpha = \beta = \gamma = 90°$, and $d = d_{hkl}$ is defined at right angles to the (hkl) planes as shown in Figs. 1.25a, b. Mathematically,

$$d_{hkl} = f\left[(hkl)\,,\,(a/h, b/k, c/l)\right] \tag{1.41}$$

Fig. 1.25 Sets of parallel crystal planes. (**a**) triangle and (**b**) rectangular planes

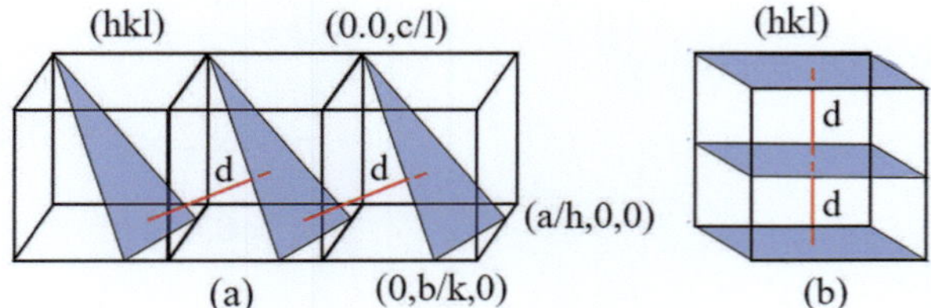

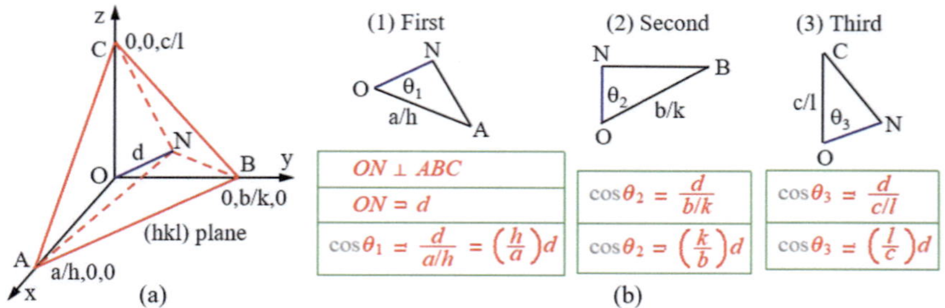

Fig. 1.26 Procedure for deriving the d-spacing denoted as d_{hkl} in simple unit cells. (**a**) Arbitrary (hkl) plane denoted ABC and (**b**) extracted triangles and related cosines

Assume an arbitrary (hkl) plane as illustrated in Fig. 1.26a with point coordinates $(a/h, 0, 0)$ on the x-axis, $(0, b/h, 0)$ on the y-axis, and $(0, 0, c/l)$ on the z-axis.

The objective now is to derive an expression for d_{hkl}, which is a perpendicular distance between successively adjacent crystallographic (hkl) planes having the same Miller indices.

Notice that the arbitrary (hkl) plane in Fig. 1.26a is divided into three triangles (Fig. 1.26b) in order to determine the cosine terms individually.

The d-spacing is derived from the reciprocal space-lattice analysis, and it is a relevant variable for indexing X-ray diffraction data through the Bragg's law (Chap. 2). As a result, the atomic structure or type of unit cell is determined, provided that diffraction occurs at an angle θ. Moreover, indexing X-ray diffraction data is out of the scope in this textbook, and the reader is encouraged to read a review paper by Liu et al. [9].

Regarding trigonometry as a computational component of geometry, the ON line in Fig. 1.26a is defined as the d-spacing ($d = ON = d_{hkl}$) in terms of the Miller indices h, k, l and the lattice parameters a, b, c for unit cells having orthogonal axes. Moreover, extract the OAN, OBN, and OCN triangles (Fig. 1.26b) from the plane in Fig. 1.26a with angles $\angle OAN = \theta_1$, $\angle OBN = \theta_2$, and $\angle OCN = \theta_3$. Thus, the d-spacing becomes the length of ON normal. Subsequently, find the cosine terms (Fig. 1.26b) for each triangle, and use the law of cosines. For orthogonal axes (Hammond [10, p. 137]),

$$\cos^2\theta_1 + \cos^2\theta_2 + \cos^2\theta_3 = 1 \tag{1.42a}$$

$$\left(\frac{h}{a}\right)^2 d^2 + \left(\frac{k}{b}\right)^2 d^2 + \left(\frac{l}{c}\right)^2 d^2 = 1 \tag{1.42b}$$

Consider the relevant conditions of the lattice parameters (a, b, c), and derive the d-spacing ($d = d_{hkl}$) for orthorhombic (Orthor), tetragonal (Tetra.), and cubic unit cells. From Eq. (1.42b),

$$\frac{1}{d^2} = \frac{h^2}{a^2} + \frac{k^2}{b^2} + \frac{l^2}{c^2} \quad \text{Orthor.-} \ a \neq b \neq c \tag{1.43a}$$

$$\frac{1}{d^2} = \frac{h^2 + k^2}{a^2} + \frac{l^2}{c^2} \quad \text{Tetra. -} \ a = b \neq c \tag{1.43b}$$

$$\frac{1}{d^2} = \frac{h^2 + k^2 + l^2}{a^2} \quad \text{Cubic -} \ a = b = c \tag{1.43c}$$

For comparison, the interplanar d-spacing for an hexagonal (Hex.) unit cell having non-orthogonal axes is written as

$$\frac{1}{d^2} = \frac{4}{3}\left(\frac{h^2 + hk + k^2}{a^2}\right) + \frac{l^2}{c^2} \quad \text{Hex.-} \ a \neq c \tag{1.44}$$

The remaining d-spacing expressions for the rhombohedral, monoclinic, and triclinic crystal systems can be found in Cullity's book [11, p. 501].

Thus far, only simple crystal structure cases have been included above using trigonometry as part of the geometry related to the mathematics of crystallography. For instance, Table 1.1 illustrates some geometry cases; however, there are many ways to mathematically relate different crystal structures.

It is a common practice in the field of crystallography to use geometry-based and symmetry-based concepts in order to compare crystal structures and characterize relevant similarities. In the end, the physical or mechanical behavior of crystals is strongly dependent on crystal structure having certain crystallographic geometry and symmetry.

Regardless of the crystal structure and related crystal geometry, the d-spacing defined by Eqs. (1.43) through (1.44) is a parameter that can be determined by X-ray diffraction data using an indexing procedure to reveal the crystal structure. Nonetheless, the d-spacing is a crystallographic variable that defines the interatomic distance between successive or parallel (hkl) planes as clearly illustrated in Fig. 1.25, but it strongly depends on the lattice parameters of the unit cells as depicted in Fig. 1.26 and as derived by Eqs. (1.43) and (1.44).

1.14   Ceramic Crystal Structures

Ceramic materials are as important as metals in certain engineering applications and are frequently produced as metal oxides, metal nitrides, and metal carbides. These materials can be classified as ceramic crystals being ionically or covalently bonded inorganic solids. For instance, Fig. 1.27 shows the unit cells for ionically bonded ceramics, such as sodium chloride $NaCl$ (Fig 1.27a), magnesium oxide MgO (Fig 1.27b), and aluminum oxide Al_2O_3 (Fig 1.27c). On the other hand, Fig 1.27d and e show the covalently bonded unit cells for diamond cubic (DC) and silicon

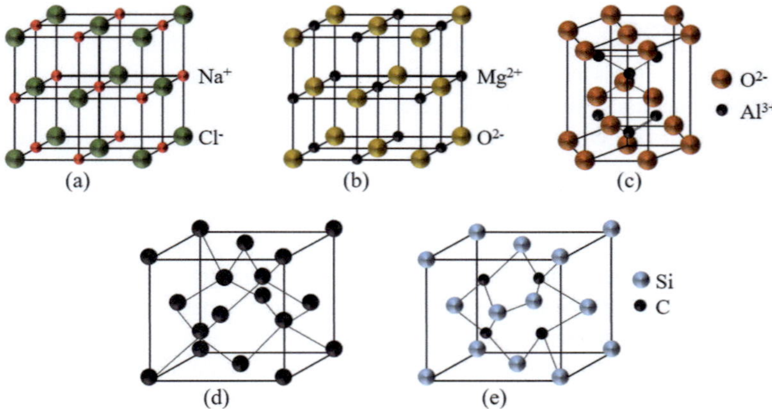

Fig. 1.27 Some ceramic structures. (**a**) $NaCl$, (**b**) MgO, (**c**) Al_2O_3, (**d**) diamond cubic, and (**e**) SiC unit cells

carbide (SiC), respectively. Actually, the SiC structure has many polycrystalline structures, such as FCC, HCP, and rhombohedral (rhom).

Recall that a ceramic structure is a stable crystalline material due to the electrostatic attraction between ions of opposite electric charge. However, some ceramic materials may have amorphous (non-crystalline) atomic structures. In essence, engineering ceramics are classified as the oxides, nitrides, and carbides.

Ceramic structures are also referred to as compounds having covalent and ionic bonds. Mechanically, covalent bonded ceramics are strong materials exhibiting high hardness and high compressive strength, and unfortunately, they have low ductility and low tensile strength. Moreover, the brittleness or absence of plasticity in crystalline ceramics is attributed to the nature of their binding characteristics.

Physically, ceramics are poor electrical and thermal conductors due to lack of free electrons in the conduction band. Hence, ceramics can be used as thermal and electrical insulators.

Ceramics have high-temperature melting points and high chemical stability, attributed to highly stable bonds. Regarding glasses, they are mostly amorphous silicates (silicon, oxygen, and metals) that can be considered as ceramics due to their common brittle characteristics. However, an amorphous glass can transform into a partially or fully crystalline structure during devitrification, which is a crystallization process used to produced devitrified ceramics.

1.15 Summary

Atomic bonding, electron configuration, and crystallography are topics that can be expanded in great details. However, their fundamentals being presented in this chapter guide the reader to reach an understanding of the atomic arrangement of atoms being modeled as hard spherical particles (atomic model). Thus, the atomic arrangement of atoms is a suitable geometrical condition for determining crystal structures in metals, semiconductor, ceramics, and polymers.

The atomic arrangement of atoms are essentially subsequently characterized with respect to the type of bonding. Essentially,

- Covalent bonding forms due to an electron-share condition at outer electron orbitals.
- Ionic bonding forms due to solid solution of atoms treated as positively and negatively charged atomic particles that form ceramic materials.
- Metallic bonding forms due to solid solution of atoms as neutral atomic particles for crystals with unit cells or amorphous solids having no unit cells.

Crystallography of the lattice structure provides means to classify crystalline materials with respect to the type of atomic structure, which influences material behavior and its properties related to crystallographic planes and directions.

Ceramics are crystalline, glassy, or crystalline-glassy brittle materials with complex crystal structures due to the nature of their bonds. Engineering applications of ceramics are mostly based on their thermal or electrical insulating characteristics.

Problems

1.1 Consider the electronic configuration for neutral iron (Fe) with an atomic number of 26 (26 protons and 26 electrons). **(a)** Write down the condensed electronic configuration for Fe and its valence electrons (electrons present in the outermost shell), and **(b)** write down the condensed electronic configuration for ferrous Fe^{2+} ion and ferric Fe^{3+} ion.

1.2 Consider the electronic configuration for neutral copper (Cu) with an atomic number of 29, which means that there are 29 protons and 29 electrons in the atom. **(a)** Write down the condensed electronic configuration for Fe and its valence electrons present in the outermost shell, and **(b)** write down the condensed electronic configuration for Cu^{1+} ion and Cu^{2+} ion.

1.3 An electron is confined to a small $L = 1.5$-mm-thin layer of silicon, where the effective mass of electrons is $0.25m_0$ with $m_0 = 9.11 \times 10^{-31}$ kg as the free-electron mass. Calculate **(a)** the kinetic energy and **(b)** the velocity of the electron at the ground state ($n = 1$). [Solution: (a) $E = 6.69 \times 10^{-7}$ eV and (b) $v = 968.81\, m/s$].

1.4 Consider a hypothetical crystal formed by HCP unit cells. Assume that the basal-plane lattice parameter is $a = 0.25$ nm and that the HCP volume $V_{hex} = 0.11$ nm^3. Use the sketches given in Example 1.2 to calculate the **(a)** area of the basal plane and **(b)** the c/a ratio of this hypothetical HCP unit cell. Is this hypothetical

HCP crystal larger than the HCP for carbon? [Solution: (a) $A = 0.16\,nm^2$ and (b) $c/a = 2.72$].

1.5 Calculate the theoretical mass (m) and atomic volume density (ρ_v) for FCC copper, which has a radius of 0.128 nm and atomic weight of $63.5\,g/mol$. Assume an experimental density of $\rho = 8.95\,g/cm^3$ at room temperature. Is there any discrepancy between ρ_v and ρ? Explain. [Solution: $m = 4.22 \times 10^{-22}\,g$ and $\rho_v = 8.90\,g/cm^3$].

1.6 Assume that a pure copper metal has a mass density of $8.90\,g/cm^3$ and atomic weight of $63.50\,g/mol$ and its cubic structure has a size of 0.362 nm. Calculate the number of equivalent atoms per unit cell (N_c), and determine the Bravais lattice. [Solution: $N_c = 4\,atoms$].

1.7 For an FCC-copper (Cu) crystal with a lattice parameter $a = 0.3628$ nm and d-spacing $d = d_{hkl} = 0.2565$ nm, determine the Miller indices for the corresponding (hkl) planes and the planar density (ρ_A). [Solution: $\{hkl\} = \{220\}$ and $\rho_A = 15.20\,atoms/nm^2$].

1.8 Assume an FCC-copper (Cu) crystal with a lattice parameter $a = 0.3628$ nm. If the d-spacing is $d = d_{hkl} = 0.3628$ nm, then determine the Miller indices for the corresponding (hkl) planes and the planar density (ρ_A). [Solution: $\{hkl\} = \{100\}$ and $\rho_A = 17.09\,atoms/nm^2$].

1.9 Assume an FCC-nickel (Ni) crystal with a lattice parameter $a = 0.3524$ nm and d-spacing $d = d_{hkl} = 0.3628$ nm. Determine the Miller indices for the corresponding (hkl) planes and the planar density (ρ_A). [Solution: $\{hkl\} = \{111\}$ and $\rho_A = 18.12\,atoms/nm^2$].

1.10 Consider the β-tin (β-Sn or white tin) metal with a body-centered tetragonal (BCT) unit cell with $a = 0.5832$ nm and $c = 0.3182$ nm at room temperature. If $p = 7.30\ g/cm^3$, $A_w = 118.69\ g/mol$, and $R = 0.151$ nm, then calculate the atomic packing factor (APF). [Solution: $APF = 0.53$].

1.11 Consider the given lattice parameters for a certain simple monoclinic unit cell taken from https://aip.scitation.org/doi/10.1063/1.5005813

$$a = 0.4950\,nm, \ \ b = 0.3358\,nm, \ \ c = 0.9079\,nm$$

$$\alpha = \gamma = 90^\circ \ \ \& \ \ \beta = 94^\circ$$

Calculate the atomic packing factor (APF) along with the unit cell equation $V_c = abc\sin(\beta)$. [Solution: $APF = 0.42$].

1.12 Determine **(a)** the crystallographic direction $B = \left[u'v'w'\right]$ shown within the shaded hexagon in the given figure in P1.12, and **(b)** convert $\left[u'v'w'\right] \rightarrow [uvtw]$. **(c)** What would you do if $[uvtw]$ is given and asked to sketch or plot this direction? [Solution: $B = [uvtw] = \left[2\bar{1}\bar{1}3\right]$].

1.13 Determine **(a)** the crystallographic direction $A = [u'v'w']$ shown within the shaded hexagon in the given figure, and **(b)** convert $[u'v'w'] \rightarrow [uvtw]$. **(c)** What would you do if $[uvtw]$ is given and asked to sketch or plot this direction? [Solution: $A = [uvtw] = [11\bar{2}3]$].

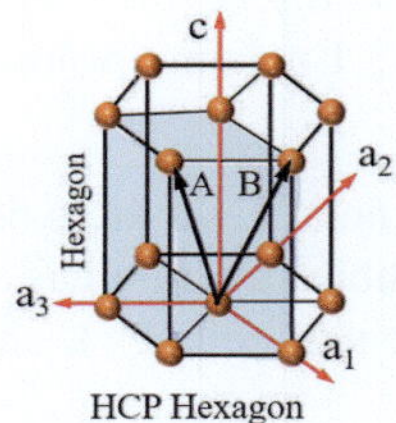

HCP Hexagon

1.14 Consider the (001) plane in an FCC unit cell with lattice parameter $a = 0.3531$ nm. Determine the **(a)** planar density ρ_A and **(b)** atomic packing factor (APF) of the FCC unit cell. [Solution: (a) $\rho_A = 1.60 \times 10^{15}\, atoms/cm^2$ and (b) $APF = 0.74$].

1.15 Determine the Miller indices for each of the (hkl) planes.

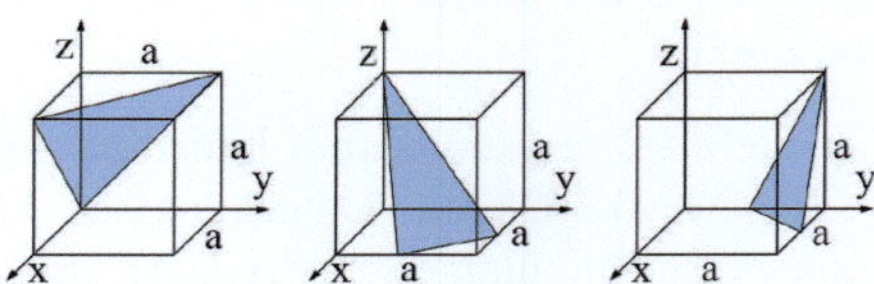

1.16 Consider an ion-pair of potassium K^+ and fluoride F^- with ionic radius of 0.13 nm and 0.14 nm, respectively. Based on this information, calculate **(a)** the binding forces $F_A(r_o)$ and $F_R(r_o)$ between this K^+F^- ion pair at equilibrium, where $m = 1$ and $n = 8$, **(b)** the attractive and repulsive constants C_A and C_R, **(c)** the equilibrium potential energy U_o at r_o, and **(d)** the maximum binding force F_{max} that causes debonding. [Solution: (a) $F_A(r_o) = 3.17\, nN$, (b) $C_A = 2.31 \times 10^{-28}\, N.m^2$ and $C_R = 3.02 \times 10^{-96}\, N.m^9$, (c) $U_o = -4.68$ eV, (d) $F_{max} = 1.60$ nN].

1.17 For a hypothetical X^+Y^- ion pair with ionic radii 0.125 nm and 0.132 nm, respectively, calculate the attractive force $F_A(r_o)$ between this ion pair at equilibrium. [Solution: $F_A(r_o) = 3.49$ nN].

1.18 Consider the X^+Y^- ion pair with ionic radii 0.123 nm and 0.135 nm, respectively. Calculate the energy to break up the X^+Y^- ion-pair bond. Let $m = 1$ and $n = 7$. [Solution: $U(r) \geq U_o = -4.87$ eV].

1.19 Assume the $Fe^{2+}O^{2-}$ ionic pair with assumed radii $r_{Fe^{2+}} = 0.07$ nm and $r_{O^{2-}} = 0.14$ nm. Calculate the potential energy U_o at equilibrium. Let $m = 1$ and $n = 7$. [Solution: $U_o = -24.38$ eV].

1.20 The Lennard-Jones potential energy between an atom-pair is

$$U\,(r) = \frac{2.51 \times 10^{-134}\ J.m^{12}}{r^{12}} - \frac{7.46 \times 10^{-77}\ J.m^6}{r^6}$$

Derive the net force equation, $F\,(r)$, for this hypothetical atom pair, and determine the equilibrium potential energy U_o. [Solution: $U\,(r_o) = -0.27\,eV$].

1.21 Determine the atomic packing factor for the diamond cubic structure shown below. Notice that each green carbon atom forms a tetrahedral geometry of four-coordinate carbon atoms. [Solution: APF $= 0.34$].

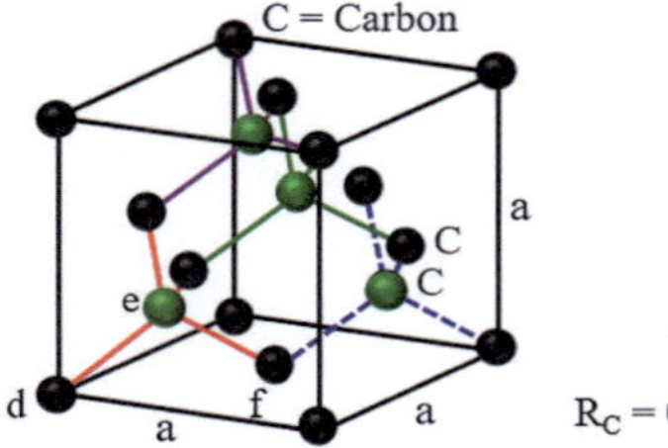

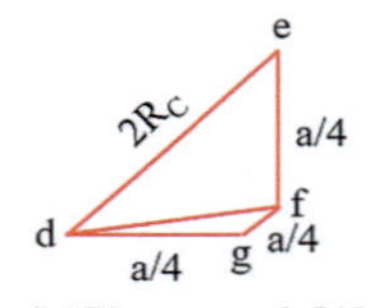

1.22 Determine the atomic packing factor for the diamond cubic gallium arsenide (GaAs) structure shown below. Notice that each green gallium (Ga) atom forms a tetrahedral geometry of four-coordinate arsenic (As) atoms. [Solution: APF $= 0.34$].

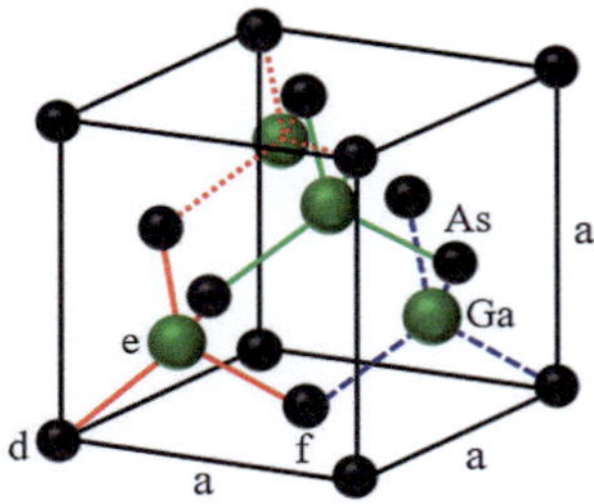

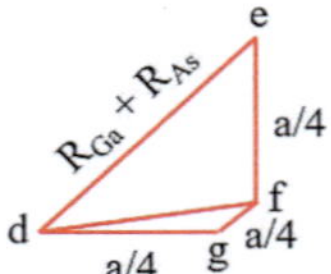

References

1. Clara Moskowitz. https://www.nbcnews.com/id/wbna43744729
2. A.F.J. Levi, *Applied Quantum Mechanics* (Cambridge University Press, New York, 2003)
3. M. Le Bellac, *Quantum Physics*, Translated by Patricia de Forcrand-Millard (Cambridge University Press, New York, 2006)
4. I.V. Lebedeva, A.A. Knizhnik, A.M. Popov, Y.E. Lozovik, B.V. Potapkin, Interlayer interaction and relative vibrations of bilayer graphene. Phys. Chem. Chem. Phys. **13**(13), 5687–5695 (2011)

5. A.M. Popov, I.V. Lebedeva, A.A. Knizhnik, Y.E. Lozovik, B.V. Potapkin. Barriers to motion and rotation of graphene layers based on measurements of shear mode frequencies. Chem. Phys. Lett. **536**, 82–86 (2012)
6. W.D. Callister, Jr., D.G. Rethwisch, *Materials Science and Engineering: An Introduction*, 9th edn. (Wiley, New York, 2014)
7. P.L. Mangonon, *The Principles of Materials Selection for Engineering Design* (Prentice Hall, New York, 1999)
8. Patric Shamberger. https://www.youtube.com/watch?v=RGCqLEU2IIM
9. Y. Liu, Y. Liu, M.G.B. Drew, Clarifications of concepts concerning interplanar spacing in crystals with reference to recent publications. SN Appl. Sci. **2**(4), 1–29 (2020)
10. C. Hammond, *The Basics of Crystallography and Diffraction*, 3rd edn. (Oxford University Press, New York, 2009)
11. B.D. Cullity, *Elements of X-Ray Diffraction*, 2nd edn. (Addison-Wesley Publishing, Reading, 1978)

Chapter 2
Algebraic Crystallography

2.1 Introduction

This chapter describes crystallography as an experimental science, characterizes crystal structures using X-ray diffraction, and includes vector algebra in orthogonal and non-orthogonal coordinate systems. In this context, crystallography is treated as the multidisciplinary science of crystals (crystalline solids) that deals with atomic structures responsible for properties of engineering materials.

Regarding crystallography and vector algebra, the phrase algebraic crystallography is introduced hereafter as a mathematical tool for determining crystallographic directions as translation vectors within unit cells.

Knowledge of X-ray diffraction and reciprocal lattice (three-dimensional parallelepiped) defined by a scattering vector g_{hkl} is essential to complement the analysis of crystallography and related atomic structure of a specimen surface. This includes the Ewald sphere methodology, Bragg's law, and reciprocal lattice as used in X-ray and electron diffraction.

From an engineering point of view, crystallography is a branch of science that provides means to determine the crystal structure of engineering materials subjected to mechanical deformation, solid-state phase transformation, and even a electrochemical surface process, where the crystallographic orientation of packed atoms can have a significant impact on the performance of electrochemistry-driven devices and relevant mechanisms (Alkire and Braatz [1]).

The main objectives of this chapter include basic knowledge for describing, understanding, and characterizing crystalline solids with respect to crystallography and vector algebra. Basically, this is an extension of Chap. 1 that leads to reciprocal lattice vector being determined by X-ray diffraction method. In essence, vector algebra in crystallography defines the vector coordinates related to the dot-product or cross product of crystallographic directions.

© The Author(s), under exclusive license to Springer Nature Switzerland AG 2024
N. Perez, *Materials Science: Theory and Engineering*,
https://doi.org/10.1007/978-3-031-57152-7_2

It should be mentioned that this chapter includes basic vector algebra and excludes advanced concepts of algebraic crystallography and the structure of space groups, which are made from 32 crystallographic point groups associated with 7 crystal lattice systems and 14 Bravais lattices.

2.2 Crystal Defects

Crystal structures and defects jointly controlled physical and mechanical properties of materials. Mainly, crystal defects can be categorized according to their dimensions, such as

- Zero-dimensional point defects, such as vacancies and impurity (extrinsic defects).
- One-dimensional linear defects, such as edge and screw dislocations.
- Two-dimensional defects, such as grain boundaries and twin boundaries.
- Three-dimensional volume defects, such as voids, pores, and inclusions.

2.3 X-ray Crystallography

X-ray crystallography (XRC) is essentially X-ray diffraction (XRD) of incident beams, and it is a tool for characterizing the size of unit cells and corresponding sets of crystallographic hkl planes and $\langle uvw \rangle$ directions.

The characterization of a crystal orientation is commonly carried out by using the Bragg diffraction or Laue diffraction technique that uses a beam of electrons known as X-rays being emitted from a cathodic source. Below are brief descriptions of these crystallographic techniques.

2.3.1 Bragg Diffraction Technique

Bragg law is based on a diffraction angle θ due to coherent and incoherent scattering from a crystal lattice. For diffraction to occur, the spacing in a crystal lattice must be comparable with the wavelength of an incident X-ray beam for coherent scattering.

This is schematically shown in Fig. 2.1, which illustrates a model with parallel planes of atoms in a crystal. If waves 1 and 2 in the incident beam strike atoms "H" and "Q," then the waves 1' and 2' are diffracted or scattered producing an X-ray pattern (Cullity [2, p. 83], Callister and Rethwisch [3, p. 76]).

The relationship between the interplanar spacing $d = d_{hkl}$ of successive (hkl) planes and diffraction angle θ is described by Bragg's law. For a fixed wavelength (λ), the triangle HAQ or HQT in Fig. 2.1 can be used to derive Bragg's law. Thus,

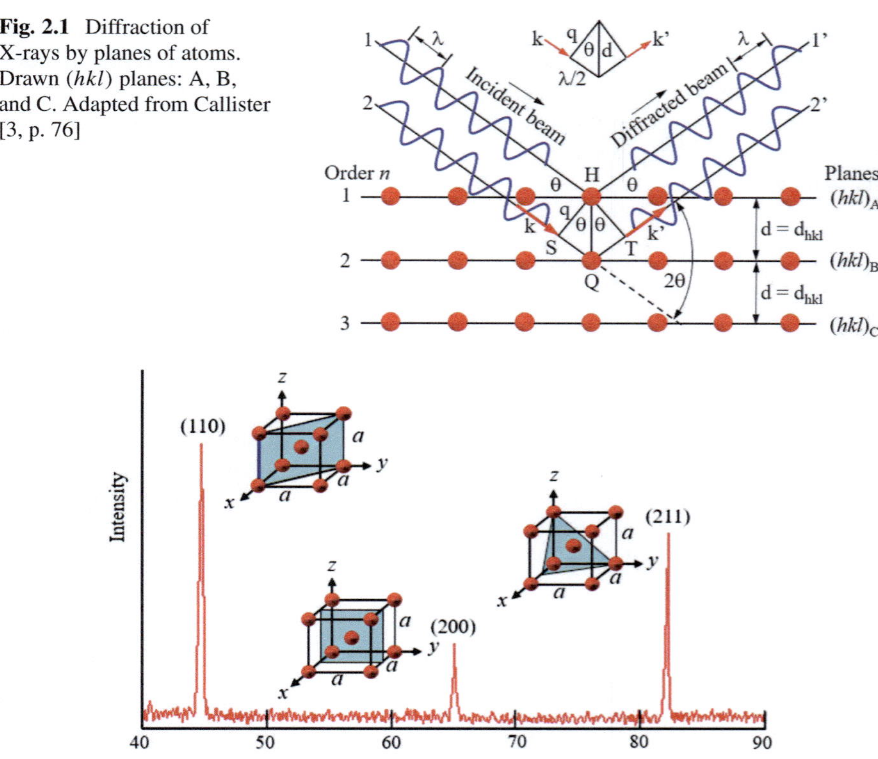

Fig. 2.1 Diffraction of X-rays by planes of atoms. Drawn (hkl) planes: A, B, and C. Adapted from Callister [3, p. 76]

Fig. 2.2 X-ray diffraction pattern from polycrystalline pure iron Fe (BCC α-phase or BCC ferrite). Adapted from Callister and Rethwisch [3, p. 78]

$$n\lambda = 2d \sin \theta \tag{2.1}$$

where $n = 1, 2, \ldots$ is the order of diffraction and λ is the wavelength. This expression implies that observable diffracted waves are possible at a Bragg angle θ with respect to a set of parallel (hkl) planes separated by a common d_{hkl} distance, which is related to a scattered vector $\overrightarrow{\mathbf{g}}_{hkl}$ of the diffracted beam. In fact, $\mathbf{g}_{hkl}$ is called the reciprocal lattice vector related to X-ray diffracted beams. See the Ewald sphere in Appendix 2A.

Many waves are commonly diffracted for a meaningful X-ray diffraction pattern, as partially shown in Fig. 2.2 by three peaks from pure iron (Fe) along with background noise (Callister and Rethwisch [3, p. 78]). This X-ray diffraction pattern identifies Fe as a body-centered cubic BCC, known as BCC α-phase or BCC ferrite.

This X-ray pattern is a plot of a function called intensity, $I = f(2\theta)$, due to the atom positions in the unit cell, where 2θ is twice the diffraction angle shown in Fig. 2.1. Moreover, the intensity of electromagnetic radiation is a measure of the rate of flow of energy per unit area ($joules/m^2.sec$) perpendicular to the direction of motion of the wave (Cullity [2, p. 5]).

The incident wavelength λ for diffraction to occur as an electromagnetic radiation phenomenon at a Bragg angle θ is related to the energy (E_e) of a moving wave-like particle. Hence,

$$E_e = hv = \frac{hc}{\lambda} = \frac{1.2398}{\lambda} \quad (\text{in } keV) \tag{2.2}$$

where h is the Planck's constant, v is the frequency, and c is the speed of light in the diffraction environment.

Example 2.1 Consider an electromagnetic radiation to be diffracted as a first-order incident wave (X-ray diffraction) of a hypothetical crystalline solid at room temperature. **(a)** Derive Bragg's equation using Fig. 2.1. In addition, determine **(b)** the d-spacing for the first peak of the X-ray pattern and **(c)** the energy of the incident wave. Assume a first-order X-ray diffraction with Cu-K_α monochromatic radiation and a single wavelength $\lambda = 0.1527 \ nm$. Assume a vacuum environment so that $c = 3 \times 10^8 \ m/s$.

Solution

(a) From Fig. 2.1, SQ and QT can be stated as

$$SQ = QT = d_{hkl} \sin \theta \tag{2.1E1a}$$

$$n\lambda = SQ + QT \tag{2.1E1b}$$

$$n\lambda = 2d_{hkl} \sin \theta \tag{2.1E1c}$$

(b) For the (110) plane, the peak is at $2\theta = 44.2°$, $\theta = 22.1°$, and $n = 1$ so that

$$d_{hkl} = \frac{\lambda}{2 \sin \theta} = \frac{0.1527 \ nm}{2 \sin (22.1°)} = 0.203 \ nm \tag{2.1E2a}$$

(c) From Eq. (2.2),

$$E_e = \frac{hc}{\lambda} = \frac{\left(6.63 \times 10^{-34} \ J.s\right)\left(3 \times 10^8 \ m/s\right)}{0.1527 \times 10^{-9} \ m} = 1.30 \times 10^{-15} \ J \tag{2.1E3a}$$

$$E_e = \left(1.30 \times 10^{-15} \ J\right) / \left(1.62 \times 10^{-19} \ eV/J\right) \tag{2.1E3b}$$

$$E_e = 8{,}025 \ eV = 8.025 \ keV \tag{2.1E3c}$$

Fig. 2.3 Transmission Laue
diffraction technique
producing a Laue X-ray
diffraction pattern with white
spots

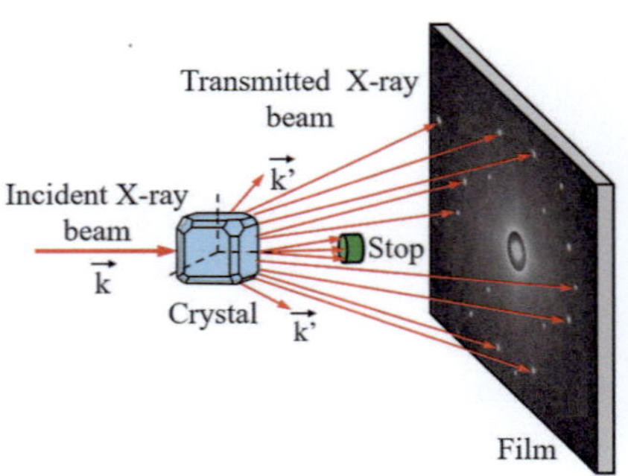

This result is within a suitable range, $100\ eV < E_e < 15\ keV$, for the incident
wave.

2.3.2 Laue Diffraction Technique

This technique is schematically illustrated in Fig. 2.3 for scattering of waves in the
process of diffraction by a crystal.

Actual measurements are made by using transmission or the back reflection
of X-ray beams. The Laue condition for diffraction is a constructive interference
related to the difference between incoming and outgoing waves (Fig. 2.1) that can
be represented by wavevectors $\vec{k}$ and $\vec{k'}$, respectively. The wavevector difference
defines the reciprocal lattice vector $\vec{g} = \vec{g}_{hkl}$

$$\vec{k'} - \vec{k} = \vec{g}_{hkl} \tag{2.3a}$$

$$\Delta \vec{k} = \vec{g} \tag{2.3b}$$

$$\vec{k} = \frac{2\pi}{\lambda}\,\vec{n} \tag{2.3c}$$

$$\vec{k'} = \frac{2\pi}{\lambda}\,\vec{n'} \tag{2.3d}$$

where $k' = \left|\vec{k'}\right| > k = \left|\vec{k}\right|$ and $\vec{n}$, $\vec{n'}$ are unit vectors. Moreover, the Laue
pattern shown in Fig. 2.3 is obtained as a regular array of white spots on a light-
sensitive film referred to as photographic emulsion (paper). Nonetheless, the Laue
pattern is used in a later section for describing the methodology related to crystal
orientation known as stereographic projection.

Example 2.2 Consider the wavevectors k and $k\prime$ shown below to derive the Bragg's
law.

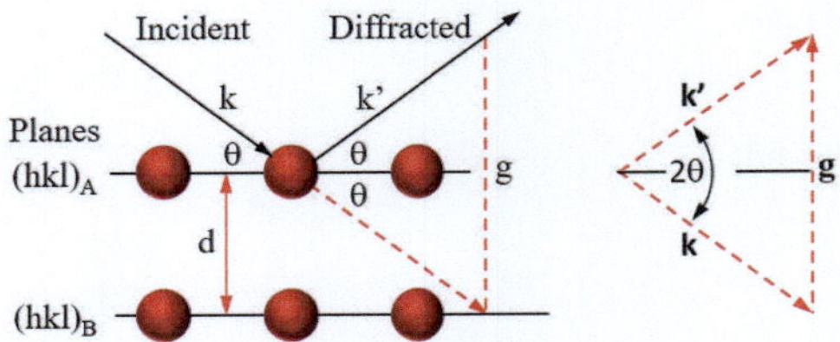

Solution From the triangle,

$$\sin\theta = \frac{\left|\vec{g}\right|/2}{\left|\vec{k'}\right|} = \frac{\left|\vec{g}\right|}{2\left|\vec{k'}\right|} \tag{2.2E1}$$

If the reciprocal lattice vector is $\left|\vec{g}\right| = 2\pi/d$, then Eq. (2.2E1) becomes

$$\frac{2\pi}{\left|\vec{k'}\right|} = 2d\sin\theta \tag{2.2E2a}$$

$$\lambda = 2d\sin\theta \tag{2.2E2b}$$

with

$$\lambda = 2\pi/\left|\vec{k'}\right| \tag{2.2E3}$$

Therefore, the Bragg's law, Eq. (2.2E2b), is satisfied only if Eq. (2.2E3) holds since the reciprocal lattice represents a set of vectors and a crystal is defined by discrete sets of points in a non-continuum manner.

2.4　X-Ray Attenuation

Attenuation, by definition, is the degradation or gradual loss of the signal strength of an incident energy beam or sound wave during the scattering and absorption process in a medium. In X-ray diffraction, X-ray attenuation is the reduction of a beam intensity (I) related to loss of energy at a depth or distance x in a solid material. This physical phenomenon arises due to interaction of X-ray photons with atoms of the specimen lattice.

The X-ray intensity is defined as an ordinary differential equation (ODE) of the form

$$dI(x) = -I(x)\,n\sigma\,dx \tag{2.4a}$$

$$\frac{dI(x)}{I(x)} = -n\sigma\, dx \tag{2.4b}$$

$$\int_{I_o}^{I} \frac{dI(x)}{I(x)} = -n\sigma \int_{0}^{x} dx \tag{2.4c}$$

$$\ln\left(\frac{I}{I_o}\right) = -n\sigma x \tag{2.4d}$$

Then, $I = f(x)$ (Lumbert's law) and $\alpha = f(I)$ functions are

$$I = I_o \exp\left(-n\sigma x\right) \tag{2.5a}$$

$$I = I_o \exp\left(-\alpha x\right) \tag{2.5b}$$

$$\alpha = n\sigma = -\frac{1}{x}\ln\left(\frac{I}{I_o}\right) \tag{2.5c}$$

$$\alpha_m = \alpha\rho \tag{2.5d}$$

where I_o = Initial X-ray intensity (W/m^2)
$\quad$ $I < I_o$ due to the decay behavior implied by Eq. (2.5b)
$\quad$ n = Atom density $(atoms/cm^3$ or $atoms/m^3)$
$\quad$ σ = Probability of a photon being scattered or absorbed within the solid
$\quad$ α = Linear attenuation coefficient or decay constant $(cm^{-1}$ or $m^{-1})$
$\quad$ α_m = Mass attenuation coefficient $(g/cm^4$ or $Kg/m^4)$
$\quad$ ρ = Mass density of the material $(g/cm^3$ or $Kg/m^3)$

The X-ray linear attenuation coefficient α describes the fraction of an X-ray beam being scattered or absorbed due to the intensity ratio $I/I_o < 1$.

Example 2.3 Assume that a hypothetical X-ray beam impacts a lead (Pb) plate perpendicularly. If the X-ray intensity changes from 2000 W/m^2 to 400 W/m^2 at a distance of 1 cm in the plate, then calculate the X-ray attenuation coefficient α.

Solution From Eq. (2.5c), the attenuation coefficient is

$$\alpha = -\frac{1}{x}\ln\left(\frac{I}{I_o}\right) = -\left(\frac{1}{1\,cm}\right)\ln\left(\frac{400}{2000}\right) = 1.61\,cm^{-1} \tag{2.3E1c}$$

This is a small attenuation coefficient representing a low probability per unit length that an X-ray beam goes through the plate.

2.5   Crystal Lattice

The crystal lattice (Fig. 2.4a) is a regular, ordered, and symmetrical three-dimensional array of points, such as atoms, ions, or molecules. In general, these

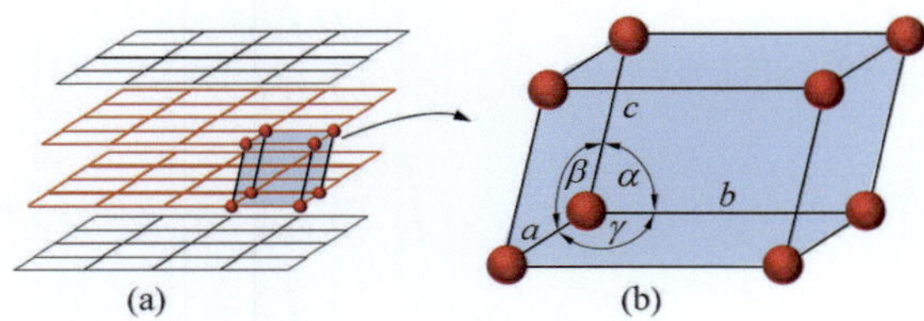

Fig. 2.4 (**a**) Crystal lattice and (**b**) conventional monoclinic or triclinic unit cell of a lattice structure

Table 2.1 Seven crystal structures and fourteen Bravais lattices (BL)

Unit cell	Cell edges	Cell angles	BL
Cubic	$a = b = c$	$\alpha = \beta = \gamma = 90°$	3
Tetragonal	$a = b \neq c$	$\alpha = \beta = \gamma = 90°$	2
Hexagonal	$a = b \neq c$	$\alpha = \beta = 90°, \gamma = 120°$	1
Orthorhombic	$a \neq b \neq c$	$\alpha = \beta = \gamma = 90°$	4
Trigonal	$a = b = c$	$\alpha = \beta = \gamma < 120°, \neq 90°$	1
Monoclinic	$a \neq b \neq c$	$\alpha \neq 90°, \beta = \gamma = 90°$	2
Triclinic	$a \neq b \neq c$	$\alpha \neq \beta \neq \gamma$	1

points can be treated as particles that have discrete translation symmetry described by a translation vector r_{hkl}, where (hkl) are integers related to crystallographic coordinates a, b, c of unit cells.

The unit cell (Fig. 2.4b) consists of atoms being modeled as perfect spheres with certain configuration based on the distance between two nearest-neighbor atoms in the a, b, and c axes, which are the edge lengths separated by the angles $\alpha = a \measuredangle c$, $\beta = b \measuredangle (c)$, and $\gamma = a \measuredangle b$.

Table 2.1 lists the seven types of conventional unit cells and their respective lattice parameters that define crystalline materials, some of which have additional atoms being centered or face-centered in the unit cells. Moreover, there are fourteen (14) Bravais lattices (BL) as indicated in Table 1.1 that are known as direct or real lattices, defining infinite crystalline arrangements of discrete atoms in three-dimensional space having periodic structures. Hence, they form unit cells of unique sizes and geometries described by the crystallographic vectors $(\mathbf{a}, \mathbf{b}, \mathbf{c})$.

The crystal systems in Table 2.1 are listed in order of decreasing rotational symmetry with respect to the crystal axes. In particular, a trigonal structure is known as rhombohedral, a crystal lattice has translation symmetry, and a crystal system has rotation symmetry. This implies that the cubic crystal structure is the most symmetrical system, whereas the triclinic structure is the least symmetrical crystal. This is illustrated in Fig. 1.12 and in Table 2.1 with respect to the cell edges (a, b, c) and axial angles (α, β, γ).

The cubic and hexagonal unit cells are considered henceforth for illustrating the importance of a mathematical description of crystallography using vector algebra only. The concept of tensors is excluded in this textbook.

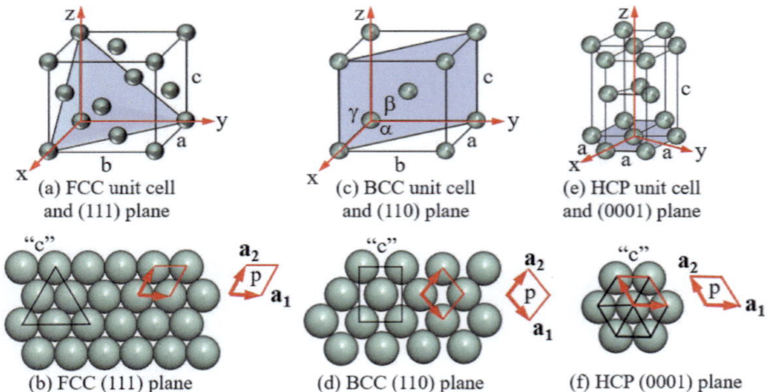

Fig. 2.5 Conventional unit cells and lattices for common crystal structures, primitive lattices "p" (smallest parallelogram with $\mathbf{a}_1, \mathbf{a}_2$ vectors) and conventional lattices "c." (**a**) FCC unit cell, (**b**) FCC plane, (**c**) BCC unit cell, (**d**) BCC plane, (**e**) HCP unit cell, and (**f**) HCP plane

2.5.1 Conventional Unit Cells and Lattice Planes

The unit cell is an arbitrary parallelepiped with edges a, b, c and related α, β, γ angles (Kelly et al. [4, p. 26]). These are the six crystallographic variables used to identify the type of atomic arrangement in the solid state, where a, b, c are treated as the crystallographic coordinates responsible the unit cell size.

Regarding the atomic arrangement of atoms, Fig. 2.5 depicts the FCC, BCC, and HCP units cells and related (hkl) planes of atoms having dense atomic arrangements: the triangle FCC (111) plane (Figs. 2.5a, b), rectangular BCC (110) plane (Figs. 2.5c, d), and an HCP (0001) basal plane (Figs. 2.5e, f).

An HCP unit cell has an ideal $c/a = \sqrt{8/3} = 1.633$ axial ratio, which is used to compared real or actual HCP materials. For instance,

- HCP Titanium: $(c/a)_{\text{real}} = (0.468\ nm) / (0.295\ nm) = 1.586 < (c/a)_{ideal}$
- HCP Zinc: $(c/a)_{\text{real}} = (0.495\ nm) / (0.266\ nm) = 1.861 > (c/a)_{ideal}$
- HCP Cadmium: $(c/a)_{\text{real}} = (0.562\ nm) / (0.298\ nm) = 1.886 > (c/a)_{ideal}$
- HCP Graphite: $(c/a)_{\text{real}} = (0.671\ nm) / (0.246\ nm) = 2.728 > (c/a)_{ideal}$

The particular (hkl) planes shown in Fig. 2.5b, d, f are small parallelograms containing oblique, square, and hexagonal lattice planes defined by an individual set of primitive (p) vectors ($\mathbf{a}_1, \mathbf{a}_2$). Hence, the concept of primitive "p" lattice plane arises with dimensions $\|\mathbf{a}_1\| = \|\mathbf{a}_2\| = \left(\sqrt{2}/2\right) a$ for FCC (Fig. 2.5b), $\|\mathbf{a}_1\| = \|\mathbf{a}_2\| = \sqrt{3}a/2$ for BCC (Fig. 2.5d), and $\|\mathbf{a}_1\| = \|\mathbf{a}_2\| = a$ for HCP (Fig. 2.5f) structures. Notice that Fig. 2.5 illustrates, in general, the set of vectors denoted as conventional ($\mathbf{a}, \mathbf{b}, \mathbf{c}$) and primitive ($\mathbf{a}_1, \mathbf{a}_2, \mathbf{a}_3$).

Example 2.4 Consider the ideal HCP unit cell for deriving **(a)** the ideal c/a axial ratio and **(b)** the volume equations.

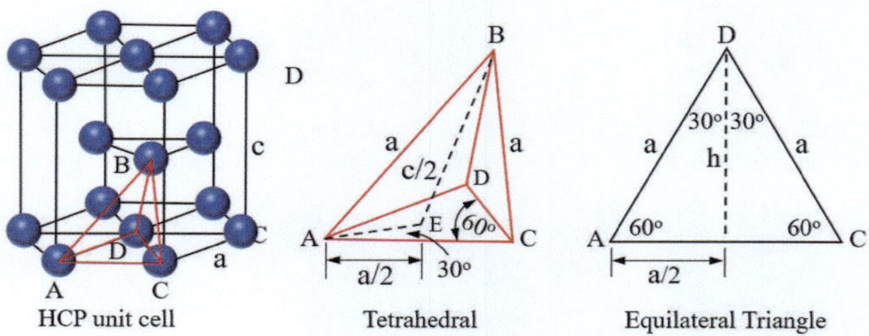

Solution

(a) Using the tetrahedral geometry and the Pythagorean theorem yields a simple relationship between the lattice parameters a and c

$$AB = AC = a = 2R \qquad (2.4\text{E1a})$$

$$BE = c/2 \qquad (2.4\text{E1b})$$

$$(AB)^2 = (AE)^2 + (BE)^2 \qquad (2.4\text{E1c})$$

$$a^2 = (AE)^2 + (c/2)^2 \qquad (2.4\text{E1d})$$

Then,

$$\cos(30°) = \frac{a/2}{AE} = \frac{\sqrt{3}}{2} \qquad (2.4\text{E2a})$$

$$AE = \frac{2}{\sqrt{3}}\frac{a}{2} = \frac{a}{\sqrt{3}} \qquad (2.4\text{E2b})$$

Combining Eqs. (2.4E1d) and (2.4E2b) yields the ideal axial c/a ratio for a hexagonal close-packed structure

$$a^2 = \left(\frac{a}{\sqrt{3}}\right)^2 + (c/2)^2 = \frac{a^2}{3} + \frac{c^2}{4} \qquad (2.4\text{E3a})$$

$$\frac{c}{a} = \sqrt{\frac{8}{3}} = 1.633 \qquad (2.4\text{E3b})$$

which is an intrinsic property used as a reference for characterizing actual hexagonal close-packed metallic materials.

(b) From the equilateral triangle,

$$\sin(60°) = \frac{h}{a} = \frac{\sqrt{3}}{2} \tag{2.4E4a}$$

$$h = \frac{\sqrt{3}}{2}a \tag{2.4E4b}$$

The volume of Hexagonal prism is

$$V = Area \times height = A_{hcp} \times c \tag{2.4E5a}$$

$$A_{hcp} = 3\,(apothem\ h)\,(side\ length\ a) \tag{2.4E5b}$$

$$V = 3hac \tag{2.4E5c}$$

Substituting Eq. (2.4E4b) and combining the resultant expression with Eq. (2.4E3b) along with $a = 2R$ give

$$V = \frac{3\sqrt{3}}{2}a^2 c = \frac{3\sqrt{3}}{2}\frac{\sqrt{8}}{\sqrt{3}}a^3 \tag{2.4E6a}$$

$$V = 3\sqrt{2}a^3 = 24\sqrt{2}R^3 \tag{2.4E6b}$$

Recall that R is the radius of an atom being modeled as hard sphere. The atomic packing factor for this HCP unit cell is

$$APF_{hcp} = \frac{6V_{sphere}}{V_{cell}} = \frac{6\,(4/3)\,\pi R^3}{24\sqrt{2}R^3} = 0.74 \text{ or } 74\% \tag{2.4E7}$$

This is an ideal atomic packing factor based on spherical atoms having the same radius R.

2.5.2 *Extraction of Unit Cells*

A selected example (borrowed from Callister and Rethwisch book [3, p. 46]) is depicted in Fig. 2.6a for an FCC crystal structure, which includes the surface plane shaded in yellow color.

Actually, this plane is shown in Fig. 2.6b in order to compare the conventional "c" and primitive "p" lattice planes.

Figure 2.6c is a practical sketch of the cut unit cell shown in Fig. 2.6a since it shows the fraction of corner atoms ($8 \times 1/8 = 1$) and face atoms ($6 \times 1/2 = 3$). Thus, there are four equivalent atoms contained in the FCC unit cell with diagonal length $D = \sqrt{2}a = 4R$ so that the lattice parameter becomes $a = 4R/\sqrt{2}$. This information is used to determine the atomic packing factor defined by

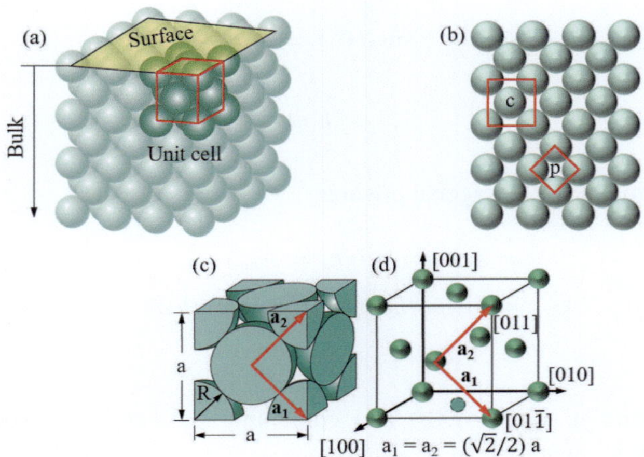

Fig. 2.6 Face-centered cubic (FCC). (**a**) Hard-sphere model of crystal structure, (**b**) dense lattice planes (conventional lattice "c" and primitive lattice "p"), (**c**) unit cell with lattice parameter "a" and radius "R," and (**d**) unit cell model (adapted from Callister and Rethwisch [3, p. 46])

$$APF_{FCC} = \frac{4V_{sphere}}{V_{cell}} = \frac{4\,(4/3)\,\pi\,R^3}{a^3} = \frac{4\,(4/3)\,\pi\,R^3}{\left(4R/\sqrt{2}\right)^3} = 0.74 \text{ or } 74\% \qquad (2.6)$$

The ideal FCC unit cell has 26% empty space in the form of holes, which are potential interstitial-atom sites during atomic diffusion or doping at a temperature T in the solid state. A similar result is obtained for an HCP unit cell. Nonetheless, this procedure can be used for other crystal structures.

Also shown in Fig. 2.6c, d is the chosen origin of the primitive lattice vectors $(\mathbf{a}_1, \mathbf{a}_2)$ on the (100) plane. The common magnitude of these vectors is determined using simple trigonometry. Hence,

$$\|\mathbf{a}_1\| = \|\mathbf{a}_2\| = \left(\frac{\sqrt{2}}{2}\right) a \qquad (2.7)$$

Figure 2.6d, on the other hand, is a model for the red unit cell shown in Fig. 2.6a where the selected lattice vectors $(\mathbf{a}_1, \mathbf{a}_2)$ are oriented along the crystallographic directions $\left[01\bar{1}\right]$ and $[011]$, respectively.

2.5.3 Vector Algebra in Crystallography

Consider the 3D Cartesian coordinates x, y, z with orthogonal characteristic as shown in Fig. 2.7a. Notice that two sets of vector coordinates are superimposed:

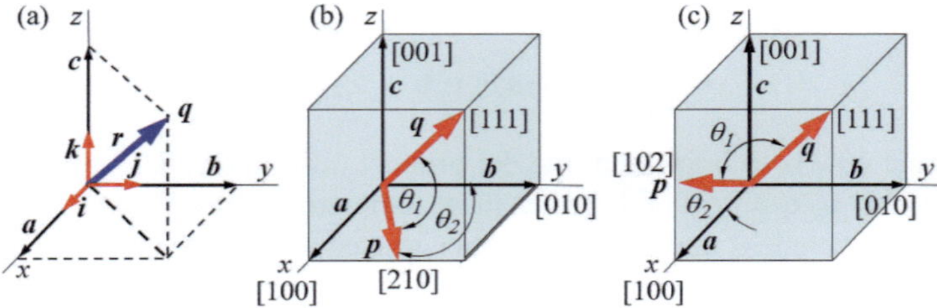

Fig. 2.7 Translation vectors in crystallography with a, b, c coordinates and $[uvw]$ directions in orthogonal unit cells. (**a**) Arbitrary vector $\mathbf{q} = [\mathbf{uvw}]$ along with $\mathbf{a}, \mathbf{b}, \mathbf{c}$ unit cell and conventional i, j, k unit vectors, (**b**) positive $\mathbf{p}$ and $\mathbf{q}$ directions (vectors) within the unit cell, and (**c**) positive $\mathbf{p}$ and $\mathbf{q}$ directions (vectors) at different angles

unit vectors ($\mathbf{i}, \mathbf{j}, \mathbf{k}$) and crystallographic translation ($\mathbf{a}, \mathbf{b}, \mathbf{c}$) vectors. Also included in Fig. 2.7a is an arbitrary conventional vector $\mathbf{r} = x\mathbf{i} + y\mathbf{j} + x\mathbf{k}$ ending at point q representing a particular direction in space.

The above algebraic aspects are used in crystallography for defining a $[uvw]$ direction related to the magnitudes of $\mathbf{a}, \mathbf{b}, \mathbf{c}$ vectors denoted as a, b, c lattice parameters, which in turn define the size of a 3D crystal element called unit cell. The edges of the unit cell are conventionally chosen to be right-handed and denoted by the cross product $\mathbf{a} \times \mathbf{b} = \mathbf{c}$. Here, $\mathbf{a} \times \mathbf{b}$ represents the direction of the vector $\mathbf{c}$.

Regarding the a, b, c lattice parameters, if $a = b = c$ and $\alpha = \beta = \gamma = 90^0 = \pi/2$, then, again, the cubic crystal is considered to have the highest symmetry and the triclinic crystal has the least symmetry among the crystal structures cited in Table 2.1.

For the purpose of clarity, a basis-vector can be designated as $\mathbf{r} = x\mathbf{i} + y\mathbf{j} + x\mathbf{k}$ and a lattice-vector as $\mathbf{r} = x\mathbf{a} + y\mathbf{b} + x\mathbf{c}$. These types of vectors are alternative approaches in vector space. However, lattice vectors are directly related to arrangement of atoms in a crystal. Moreover, a certain crystallographic direction can also be designated by $\mathbf{r} = u\mathbf{a} + v\mathbf{b} + w\mathbf{c}$.

For indexing a crystallographic direction as a spatial diagonal of any parallelepiped, Fig. 2.7b or c indicates that $[111]$, as an example, in 3D space is the sum of the coordinate directions; $[111] = [100] + [010] + [001]$. For a diagonal $[uvw]$ direction on a plane or crystal face (Fig. 2.7b), $[210] = [100] + [110]$.

Figure 2.7b, c illustrates a set of two position vectors, $\mathbf{p}$ and $\mathbf{q}$, representing crystallographic directions or lattice vectors denoted as

$$[u_1v_1w_1] = [210] \text{ and } [u_2v_2w_2] = [111] \quad \text{(Fig. 2.7b)} \tag{2.8a}$$

$$[u_1v_1w_1] = [102] \text{ and } [u_2v_2w_2] = [111] \quad \text{(Fig. 2.7c)} \tag{2.8b}$$

Recall that a vector has magnitude and direction, such as force. On the other hand, a scalar has magnitude.

Let $\mathbf{p}$ and $\mathbf{q}$ be the lattice vectors defined by

$$\mathbf{p} = u_1\mathbf{a} + v_1\mathbf{b} + w_1\mathbf{c} \tag{2.9a}$$

$$\mathbf{q} = u_2\mathbf{a} + v_2\mathbf{b} + w_2\mathbf{c} \tag{2.9b}$$

The goal now is to implement the dot product (also known as scalar product or inner product) $\mathbf{p} \cdot \mathbf{q} = \vec{p} \cdot \vec{q}$ between these non-zero vectors in order to determine the interplanar angle θ and some properties related to crystallography. Here, $\mathbf{p} \cdot \mathbf{q}$ means $\mathbf{p}$ dot $\mathbf{q}$.

Subsequently, the cross product $\mathbf{p} \times \mathbf{q} = \vec{p} \times \vec{q}$, on the other hand, produces a vector, and it defines the area of a parallelogram, namely, a crystallographic (hkl) plane. Both vector algebra tools are essential for determining the volume of a unit cell in real and reciprocal lattices. Other vector characteristics are implemented henceforth.

Dot Product The dot product (inner product), which produces a scalar quantity or number, is fundamentally defined by

$$\mathbf{p} \cdot \mathbf{q} = \|\mathbf{p}\| \, \|\mathbf{q}\| \cos\theta \tag{2.10}$$

According to Eqs. (2.9a), (2.9b), the dot product $\mathbf{p} \cdot \mathbf{q}$ gives

$$\mathbf{p} \cdot \mathbf{q} = (u_1\mathbf{a} + v_1\mathbf{b} + w_1\mathbf{c}) \cdot (u_2\mathbf{a} + v_2\mathbf{b} + w_2\mathbf{c}) \tag{2.11a}$$

$$\mathbf{p} \cdot \mathbf{q} = u_1 u_2 \, (\mathbf{a} \cdot \mathbf{a}) + u_1 v_2 \, (\mathbf{a} \cdot \mathbf{b}) + u_1 w_2 \, (\mathbf{a} \cdot \mathbf{c}) \tag{2.11b}$$

$$+ \, v_1 u_2 \, (\mathbf{b} \cdot \mathbf{a}) + v_1 v_2 \, (\mathbf{b} \cdot \mathbf{b}) + v_1 w_2 \, (\mathbf{b} \cdot \mathbf{c})$$

$$+ \, w_1 u_2 \, (\mathbf{c} \cdot \mathbf{a}) + w_1 v_2 \, (\mathbf{c} \cdot \mathbf{b}) + w_1 w_2 \, (\mathbf{c} \cdot \mathbf{c})$$

Regarding the dot product and orthogonality, the dot product measures an angle between vectors and computes the length of a vector, while orthogonality describes the generalization of perpendicularity to linear algebra.

The mathematical description of some vector properties are described below using general crystallographic coordinate systems. Thus:

- Two vectors, $\mathbf{p}$ and $\mathbf{q}$, are orthogonal at $90° = \pi/2$ if $\cos\theta = 1$ and $\mathbf{p} \cdot \mathbf{q} = 0$.
- Two vectors are parallel (codirectional: $\mathbf{p} \parallel \mathbf{q}$ at $0°$) if $\cos\theta = 0$ and $\mathbf{p} \cdot \mathbf{q} \neq 0$.
- Two vectors are non-orthogonal if $0 < \cos\theta < 1$ and $\mathbf{p} \cdot \mathbf{q} \neq 0$.
- Two unit vectors, $\mathbf{a}$ and $\mathbf{b}$, give $\mathbf{a} \cdot \mathbf{a} = \mathbf{b} \cdot \mathbf{b} = \mathbf{c} \cdot \mathbf{c} = 1$.
- Two unit vectors, $\mathbf{a}$ and $\mathbf{b}$, give $\mathbf{a} \cdot \mathbf{b} = \mathbf{b} \cdot \mathbf{a} = 0$, $\mathbf{a} \cdot \mathbf{c} = \mathbf{c} \cdot \mathbf{a} = 0$, and $\mathbf{b} \cdot \mathbf{c} = \mathbf{c} \cdot \mathbf{b} = 0$
- Absolute entities: $\|\mathbf{a}\| = a$, $\|\mathbf{b}\| = b$, and $\|\mathbf{c}\| = c$

Now, substituting these properties into Eq. (2.11b) yields

$$\mathbf{p} \cdot \mathbf{q} = (u_1 u_2)\, a^2 + (v_1 v_2)\, b^2 + (w_1 w_2)\, c^2 \tag{2.12}$$

with the magnitude terms defined by

$$\|\mathbf{p}\| = \sqrt{u_1^2\,\|\mathbf{a}\|^2 + v_1^2\,\|\mathbf{b}\|^2 + w_1^2\,\|\mathbf{c}\|^2} \tag{2.13a}$$

$$\|\mathbf{q}\| = \sqrt{u_2^2\,\|\mathbf{a}\|^2 + v_2^2\,\|\mathbf{b}\|^2 + w_2^2\,\|\mathbf{c}\|^2} \tag{2.13b}$$

where $\|\mathbf{a}\|^2 = a^2$ and $\|\mathbf{a}\|^2 = \|\mathbf{b}\|^2 = \|\mathbf{c}\|^2$. Then,

$$\|\mathbf{p}\| = \sqrt{u_1^2 a^2 + v_1^2 b^2 + w_1^2 c^2} \tag{2.14a}$$

$$\|\mathbf{q}\| = \sqrt{u_2^2 a^2 + v_2^2 b^2 + w_2^2 c^2} \tag{2.14b}$$

The interplanar angles are also defined by the dot products

$$\cos\alpha = \frac{\mathbf{b}\cdot\mathbf{c}}{\|\mathbf{b}\|\,\|\mathbf{c}\|} \tag{2.15a}$$

$$\cos\beta = \frac{\mathbf{c}\cdot\mathbf{a}}{\|\mathbf{c}\|\,\|\mathbf{a}\|} \tag{2.15b}$$

$$\cos\gamma = \frac{\mathbf{a}\cdot\mathbf{b}}{\|\mathbf{a}\|\,\|\mathbf{b}\|} \tag{2.15c}$$

The dot-product defined by Eq. (2.10) is solved for $\cos\theta$

$$\cos\theta = \frac{\mathbf{p}\cdot\mathbf{q}}{\|\mathbf{p}\|\,\|\mathbf{q}\|} \quad \text{for} \quad -\pi \le \theta \le \pi \tag{2.16a}$$

$$\cos\theta = \frac{(u_1\mathbf{a} + v_1\mathbf{b} + w_1\mathbf{c})\cdot(u_2\mathbf{a} + v_2\mathbf{b} + w_2\mathbf{c})}{\sqrt{u_1^2 a^2 + v_1^2 b^2 + w_1^2 c^2}\,\sqrt{u_2^2 a^2 + v_2^2 b^2 + w_2^2 c^2}} \tag{2.16b}$$

$$\cos\theta = \frac{(u_1 u_2)\,a^2 + (v_1 v_2)\,b^2 + (w_1 w_2)\,c^2}{\sqrt{u_1^2 a^2 + v_1^2 b^2 + w_1^2 c^2}\,\sqrt{u_2^2 a^2 + v_2^2 b^2 + w_2^2 c^2}} \tag{2.16c}$$

For non-orthogonal x, y, z axes (Fig. 2.4b), combine Eqs. (2.11b) and (2.15) to get

$$\mathbf{p}\cdot\mathbf{q} = (u_1 u_2)\,a^2 + (v_1 v_2)\,b^2 + (w_1 w_2)\,c^2 + (u_1 v_2 + v_1 u_2)\,ba\cos\gamma$$
$$+ (v_1 w_2 + v_1 w_2)\,bc\cos\alpha + (u_1 w_2 + w_1 u_2)\,ca\cos\beta \tag{2.17}$$

If $\alpha = \beta = \gamma = 90°$, Eq. (2.17) reduces to (2.12).

Cross Product The cross product $\mathbf{p}\times\mathbf{q}$, which produces a vector, is defined explicitly by

$$\mathbf{p}\times\mathbf{q} = \|\mathbf{p}\|\,\|\mathbf{q}\|\sin\theta \tag{2.18}$$

From Eq. (2.9),

$$\mathbf{p} \times \mathbf{q} = (u_1\mathbf{a} + v_1\mathbf{b} + w_1\mathbf{c}) \times (u_2\mathbf{a} + v_2\mathbf{b} + w_2\mathbf{c}) \tag{2.19a}$$

$$\mathbf{p} \times \mathbf{q} = u_1u_2\,(\mathbf{a} \times \mathbf{a}) + u_1v_2\,(\mathbf{a} \times \mathbf{b}) + u_1w_2\,(\mathbf{a} \times \mathbf{c}) \tag{2.19b}$$
$$+ v_1u_2\,(\mathbf{b} \times \mathbf{a}) + v_1v_2\,(\mathbf{b} \times \mathbf{b}) + v_1w_2\,(\mathbf{b} \times \mathbf{c})$$
$$+ w_1u_2\,(\mathbf{c} \times \mathbf{a}) + w_1v_2\,(\mathbf{c} \times \mathbf{b}) + w_1w_2\,(\mathbf{c} \times \mathbf{c})$$

Notice that the dot-product and the cross product are simple mathematical procedures covered in a traditional vector calculus course. Indeed, the former measures θ, and the latter defines a spanned area by two vectors.

Useful Vector Properties

- The cross product of two vectors, $\mathbf{p}$ and $\mathbf{q}$, is equal to the area of a parallelogram: $A = \|\mathbf{p} \times \mathbf{q}\|$
- Two vectors, $\mathbf{p}$ and $\mathbf{q}$, are orthogonal (perpendicular: $\mathbf{p} \perp \mathbf{q}$ at 90°) if $\sin\theta \neq 0$ and $\|\mathbf{p} \times \mathbf{q}\| > 0$
- Two vectors, $\mathbf{p}$ and $\mathbf{q}$, are non-orthogonal if $\sin\theta = 0$ and $\mathbf{p} \times \mathbf{q} = 0$ (also $\mathbf{p} \cdot \mathbf{q} \neq 0$)
- Entities:

$$\mathbf{a} \times \mathbf{b} = -\,(\mathbf{b} \times \mathbf{a}) = \mathbf{c} \tag{2.20a}$$

$$\mathbf{b} \times \mathbf{c} = -\,(\mathbf{c} \times \mathbf{b}) = \mathbf{a} \tag{2.20b}$$

$$\mathbf{c} \times \mathbf{a} = -\,(\mathbf{a} \times \mathbf{c}) = \mathbf{b} \tag{2.20c}$$

$$\mathbf{a} \times \mathbf{a} = \mathbf{b} \times \mathbf{b} = \mathbf{c} \times \mathbf{c} = 0 \tag{2.20d}$$

Applying Eqs. (2.20) to (2.19b) gives the cross product in a matrix form

$$\mathbf{p} \times \mathbf{q} = (v_1w_2 - w_1v_2)\,\mathbf{a} - (u_1w_2 - w_1u_2)\,\mathbf{b} + (u_1v_2 - v_1u_2)\,\mathbf{c} \tag{2.21a}$$

$$\mathbf{p} \times \mathbf{q} = \begin{bmatrix} \mathbf{a} & \mathbf{b} & \mathbf{c} \\ u_1 & v_1 & w_1 \\ u_2 & v_2 & w_2 \end{bmatrix} = \det \begin{bmatrix} \mathbf{a} & \mathbf{b} & \mathbf{c} \\ u_1 & v_1 & w_1 \\ u_2 & v_2 & w_2 \end{bmatrix} \tag{2.21b}$$

$$\mathbf{p} \times \mathbf{q} = \mathbf{a} \begin{bmatrix} v_1 & w_1 \\ v_2 & w_2 \end{bmatrix} - \mathbf{b} \begin{bmatrix} u_1 & w_1 \\ u_2 & w_2 \end{bmatrix} + \mathbf{c} \begin{bmatrix} u_1 & v_1 \\ u_2 & v_2 \end{bmatrix} \tag{2.21c}$$

with

$$\sin\theta = \frac{\mathbf{p} \times \mathbf{q}}{\|\mathbf{p}\|\,\|\mathbf{q}\|} \quad \text{for} \quad -\pi \le \theta \le \pi \tag{2.22}$$

Law of Cosines At times, this law is useful for finding the length of a vector $\mathbf{c}_1$, which connects $(\mathbf{a}_1, \mathbf{a}_2)$ vectors. Hence,

$$c_1^2 = a_1^2 + a_2^2 - 2a_1 a_2 \cos\theta \tag{2.23}$$

Below are some useful expressions related to dot-product and cross-product notations

$$(\mathbf{a}_1 + \mathbf{a}_2) \cdot (\mathbf{a}_3 - \mathbf{a}_4) = (\mathbf{a}_1 \cdot \mathbf{a}_3) - (\mathbf{a}_1 \cdot \mathbf{a}_4) + (\mathbf{a}_2 \cdot \mathbf{a}_3) - (\mathbf{a}_2 \cdot \mathbf{a}_4) \tag{2.24a}$$

$$\mathbf{a}_1 \cdot (\mathbf{a}_2 \times \mathbf{a}_3) = \mathbf{a}_2 \cdot (\mathbf{a}_3 \times \mathbf{a}_1) = \mathbf{a}_3 \cdot (\mathbf{a}_1 \times \mathbf{a}_2) \tag{2.24b}$$

$$\mathbf{a}_1 \times (\mathbf{a}_2 \times \mathbf{a}_3) = \mathbf{a}_2 (\mathbf{a}_1 \cdot \mathbf{a}_3) - \mathbf{a}_3 (\mathbf{a}_1 \cdot \mathbf{a}_2) \tag{2.24c}$$

$$(\mathbf{a}_1 \times \mathbf{a}_2) = -(\mathbf{a}_2 \times \mathbf{a}_1) \tag{2.24d}$$

$$\mathbf{a}_1 \times (\mathbf{a}_2 + \mathbf{a}_3) = (\mathbf{a}_1 \times \mathbf{a}_2) + (\mathbf{a}_1 \times \mathbf{a}_3) \tag{2.24e}$$

Additional cross-product terms using the Miller indices for crystallographic directions are

$$u \times v = -(v \times u) = w \tag{2.25a}$$

$$v \times w = -(w \times v) = u \tag{2.25b}$$

$$w \times u = -(u \times w) = v \tag{2.25c}$$

$$u \times u = v \times v = w \times w = 0 \tag{2.25d}$$

which are properties of the cross product related to crystallographic directions.

Example 2.5 Consider the cubic crystal and the crystallographic $[uvw]$ directions depicted in Fig. 2.7b, c for calculating **(a)** the angle θ_1 using the dot product and **(b)** angle θ_2 using both the dot product and the cross product. This example shows that these mathematical tools are very powerful not only in vector algebra but in crystallography. **(c)** Are these directions orthogonal? Recall that a cubic crystal has orthogonal x, y, z axes. **(d)** Assume that these two vectors represent an engineering scheme. In this case, discuss the application of the dot product.

Solution

(a) Dot product. For a cubic crystal, $a = b = c$, $[uvw]_1 = [210]_1$, and $[uvw]_2 = [111]_2$. Using Eq. (2.16c) gives the value of the angle θ_1 between vectors p and q. Hence,

$$\cos\theta_1 = \frac{a^2\,[210]_1 \cdot [111]_2}{a^2\sqrt{u_1^2 + v_1^2 + w_1^2}\,\sqrt{u_2^2 + v_2^2 + w_2^2}} \tag{2.5E1a}$$

$$\cos\theta_1 = \frac{(2)\,(1) + (1)\,(1) + (0)\,(1)}{\sqrt{2^2 + 1^2 + 0^2}\sqrt{1^2 + 1^2 + 1^2}} \tag{2.5E1b}$$

$$\cos\theta_1 = \frac{3}{\sqrt{5}\sqrt{3}} \tag{2.5E1c}$$

$$\theta_1 = 39.23° \tag{2.5E1d}$$

Thus, $\theta_1 < 90°$ indicates a condition that leads to $\mathbf{p} \cdot \mathbf{q} = [210] \cdot [111] > 0$ since these vectors are not orthogonal.

(b) **Cross product.** Let $[uvw]_1 = [210]_1$, $[uvw]_2 = [010]_2$, and $a = b = c$. In order to avoid confusion, let m be a vector along the y-axis. Thus,

$$\mathbf{p} = (u_1\mathbf{a} + v_1\mathbf{b} + w_1\mathbf{c}) = (u_1\mathbf{a} + v_1\mathbf{b}) \tag{2.5E2a}$$

$$\mathbf{m} = (u_2\mathbf{a} + v_2\mathbf{b} + w_2\mathbf{c}) = v_2\mathbf{b} \tag{2.5E2b}$$

$$\mathbf{p} \times \mathbf{m} = (u_1\mathbf{a} + v_1\mathbf{b}) \times (v_2\mathbf{b}) \tag{2.5E2c}$$

$$= u_1 v_2\,(\mathbf{a} \times \mathbf{b}) + v_1 v_2\,(\mathbf{b} \times \mathbf{b}) \tag{2.5E2d}$$

$$= u_1 v_2\mathbf{c} + v_1 v_2\,(0) = u_1 v_2\mathbf{c} \tag{2.5E2e}$$

$$\|\mathbf{p} \times \mathbf{m}\| = (2)\,(1)\,a^2 = 2a^2 \tag{2.5E2f}$$

and

$$\|\mathbf{p}\| = \sqrt{(2)^2\,a^2 + (1)^2\,a^2 + (0)^2\,a^2} = a\sqrt{5} \tag{2.5E3a}$$

$$\|\mathbf{m}\| = \sqrt{(0)^2\,a^2 + (1)^2\,a^2 + (0)^2\,a^2} = a \tag{2.5E3b}$$

From Eq. (2.22), the angle θ_2 between vectors $\mathbf{p}$ and $\mathbf{m}$ is

$$\sin\theta_2 = \frac{\|\mathbf{p} \times \mathbf{m}\|}{\|\mathbf{p}\|\,\|\mathbf{m}\|} = \frac{2}{\sqrt{5}\sqrt{1}} = \frac{2}{\sqrt{5}} \tag{2.5E4a}$$

$$\theta_2 = 63.44° \tag{2.5E4b}$$

Dot Product. From Eq. (2.16a),

$$\mathbf{p} \cdot \mathbf{m} = (u_1\mathbf{a} + v_1\mathbf{b} + w_1\mathbf{c}) \cdot (u_2\mathbf{a} + v_2\mathbf{b} + w_2\mathbf{c}) \tag{2.5E5a}$$

$$= (u_1\mathbf{a} + v_1\mathbf{b}) \cdot (v_2\mathbf{b}) \tag{2.5E5b}$$

$$\cos\theta_2 = \frac{\mathbf{p}\cdot\mathbf{m}}{\|\mathbf{p}\|\,\|\mathbf{m}\|} = \frac{(u_1\mathbf{a}+v\mathbf{b})\cdot(v_2\mathbf{b})}{a\sqrt{5}a\sqrt{1}} = \frac{v_1v_2\mathbf{b}^2}{a^2\sqrt{5}} \tag{2.5E5c}$$

$$= \frac{(1)(1)a^2}{a^2\sqrt{5}} = \frac{1}{\sqrt{5}} \tag{2.5E5d}$$

$$\theta_2 = 63.44^\circ \tag{2.5E5e}$$

Both Eqs. (2.22) and (2.16c) give the same result. Observe that $\theta_2 < 90^\circ$ gives $\mathbf{p}\cdot\mathbf{m} > 0$.

(c) These vectors, $\mathbf{p} = [uvw]_1 = [210]_1$ and $\mathbf{q} = [uvw]_2 = [010]_2$, are not orthogonal because $\mathbf{p}\cdot\mathbf{q} = 3 \neq 0$. For non-zero vectors, $\mathbf{p}\cdot\mathbf{q} = 0$ if and only if $\cos\theta = 0$ so that $\theta = 90^\circ$.

(d) In general, the dot product represents the projection of one vector onto another at θ as shown below in a general sense.

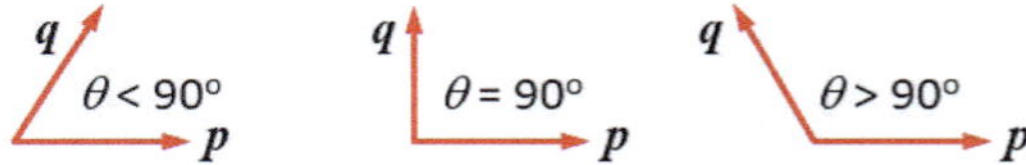

Practical applications of the dot product include finding the projection of a force in a given direction, determining the work done by a force, and, obviously, finding the angle between two vectors intersecting at an angle. The dot product can also be used in the production of solar energy by having the sun rays perpendicular to solar panels for maximum efficiency.

2.6 Rotation Matrix

Consider that a parallelogram is described by the general vectors $(\mathbf{a}_1, \mathbf{a}_2)$, and assume that $(\mathbf{b}_1, \mathbf{b}_2)$ are the resultant vectors after counterclockwise rotation at an angle θ about the origin of a vector system. In this case, the individual 2D matrices are

$$M_a = \begin{bmatrix} a_{11} & a_{12} \\ a_{21} & a_{22} \end{bmatrix} \tag{2.26a}$$

$$M_b = \begin{bmatrix} b_{11} & b_{12} \\ b_{21} & b_{22} \end{bmatrix} \tag{2.26b}$$

The general rotation matrix for $(\mathbf{a}_1, \mathbf{a}_2) \rightarrow (\mathbf{b}_1, \mathbf{b}_2)$ at a rotation angle θ can be described by

$$M_R = \begin{bmatrix} \cos\theta & \sin\theta \\ -\sin\theta & \cos\theta \end{bmatrix} \tag{2.27}$$

For (a_1, b_1) vertices, M_R, $\det M_R$, and $\cos \theta$ are, respectively (Hermann [5, p. 85])

$$M_R = \begin{bmatrix} b_1 & a_1 \\ -a_1 & (b_1 - a_1) \end{bmatrix} \tag{2.28a}$$

$$\det M_R = b_1 (b_1 - a_1) - (-a_1)(a_1) = a_1^2 - a_1 b_1 + b_1^2 \tag{2.28b}$$

$$\cos \theta = \frac{b_1 - a_1/2}{\sqrt{\det M_R}} \tag{2.28c}$$

where $\det M_R$ and $\sqrt{\det M_R}$ are the area and edge-length of a parallelogram, respectively. Here, $\det M_R$ stands for determinant of the square rotation matrix, and it is a scalar quantity.

In crystallography, the size of the parallelogram can be defined by using the general notation $\left(\sqrt{\det M_R} \times \sqrt{\det M_R}\right)$. This, then, identifies the crystallographic condition for a particular material X having a surface defined by the (hkl) plane.

Example 2.6 Consider the (001) plane at the bottom and top of the simple cubic (SC) unit cell shown below.

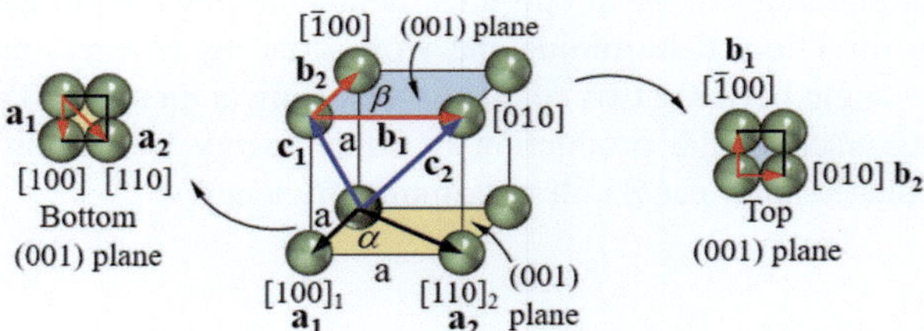

Calculate **(a)** the angles α and β between vectors $(\mathbf{a}_1, \mathbf{a}_2)$ and $(\mathbf{b}_1, \mathbf{b}_2)$, respectively, and **(b)** prove that $c_1 = a$, where a is the lattice parameter. **(c)** Determine which set of vectors defines a primitive lattice, which is a significant key in surface reconstruction of a substrate surface during adsorption or deposition, and **(d)** the matrix for each set of vectors and their vertices.

Solution By inspection, $\|\mathbf{a}_1\| = \|\mathbf{b}_1\| = \|\mathbf{b}_2\| = a$, $\|\mathbf{a}_2\| = \sqrt{2}a$. Recall that the lattice parameters a, b, c are equal for a cubic system. The angles are $\alpha = 45°$ and $\beta = 90°$. Let's prove that this is correct using the dot product and the law of cosines.

(a) For the bottom (001) plane, the dot product gives

$$\mathbf{a}_1 \cdot \mathbf{a}_2 = [uvw]_{a_1} a \cdot [uvw]_{a_2} a = [100]_{a_1} \cdot [110]_{a_2} a^2 \tag{2.6E1a}$$

$$\mathbf{a}_1 \cdot \mathbf{a}_2 = [1 \times 1 + 0 \times 1 + 0 \times 0] a^2 = a^2 \tag{2.6E1b}$$

The length of each vector is

$$\|\mathbf{a}_1\| = a_1 = \sqrt{u_1^2 + v_1^2 + w_1^2}\, a = \sqrt{1^2 + 0^2 + 0^2}\, a = a \qquad (2.6\text{E}2\text{a})$$

$$\|\mathbf{a}_2\| = a_2 = \sqrt{u_2^2 + v_2^2 + w_2^2}\, a = \sqrt{1^2 + 1^2 + 0^2}\, a = \sqrt{2}a \qquad (2.6\text{E}2\text{b})$$

From Eqs. (2.10) or (2.16a), the angle between these vectors is

$$\cos\alpha = \frac{\mathbf{a}_1 \cdot \mathbf{a}_2}{\|\mathbf{a}_1\|\,\|\mathbf{a}_2\|} = \frac{a^2}{(a)\left(\sqrt{2}a\right)} = \frac{1}{\sqrt{2}} \qquad (2.6\text{E}3\text{a})$$

$$\alpha = \arccos\left(\frac{1}{\sqrt{2}}\right) = 45° \qquad (2.6\text{E}3\text{b})$$

For the top (001) plane,

$$\mathbf{b}_1 \cdot \mathbf{b}_2 = [uvw]_{b_1}\, a \cdot [uvw]_{b_2}\, a = \left[\overline{1}00\right]_{b_1} \cdot [010]_{b_2}\, a^2 \qquad (2.6\text{E}4\text{a})$$

$$\mathbf{b}_1 \cdot \mathbf{b}_2 = [(-1) \times (0) + (0) \times (1) + (0) \times (0)]\, a^2 = 0 \qquad (2.6\text{E}4\text{b})$$

The length of each vector is

$$\|\mathbf{b}_1\| = b_1 = \sqrt{u_3^2 + v_3^2 + w_3^2}\, a = \sqrt{(-1)^2 + (0)^2 + (0)^2}\, a = a \qquad (2.6\text{E}5\text{a})$$

$$\|\mathbf{b}_2\| = b_2 = |\mathbf{a}_2| = \sqrt{u_4^2 + v_4^2 + w_4^2}\, a = \sqrt{(0)^2 + (1)^2 + (0)^2}\, a = a \qquad (2.6\text{E}5\text{b})$$

As expected, $\|\mathbf{b}_1\| = \|\mathbf{b}_2\| = a$. Now, the angle is

$$\cos\beta = \frac{\mathbf{b}_1 \cdot \mathbf{b}_2}{\|\mathbf{b}_1\|\,\|\mathbf{b}_2\|} = \frac{0}{(a)\,(a)} = 0 \qquad (2.6\text{E}6\text{a})$$

$$\beta = \arccos(0) = 90° \qquad (2.6\text{E}6\text{b})$$

(b) Using the law of cosines yields

$$c_1^2 = a_1^2 + a_2^2 - 2a_1 a_2 \cos\alpha \qquad (2.6\text{E}7\text{a})$$

$$c_1^2 = a^2 + \left(\sqrt{2}a\right)^2 - 2a\left(\sqrt{2}a\right)\left(\frac{1}{\sqrt{2}}\right) = a^2 + \left(\sqrt{2}a\right)^2 - 2a^2 \qquad (2.6\text{E}7\text{b})$$

$$c_1^2 = a^2 + 2a^2 - 2a^2 = a^2 \qquad (2.6\text{E}7\text{c})$$

$$c_1 = a \qquad (2.6\text{E}7\text{d})$$

Therefore, $c_1 = a$ is the vertical lattice parameter of the SC unit cell.

(c) Only the set of vectors $(\mathbf{b}_1, \mathbf{b}_2)$ defines a primitive lattice since $\|\mathbf{b}_1\| = \|\mathbf{b}_2\| = a$, which is the shortest distance on the (001) plane. On the other hand, $(\mathbf{a}_1, \mathbf{a}_2)$ give

$$\|\mathbf{a}_2\| = \sqrt{2}\,\|\mathbf{a}_1\| = \sqrt{2}a > \|\mathbf{b}_1\| = \|\mathbf{b}_2\| \tag{2.6E8}$$

Hence, $(a_1, a_2) = \left(1, \sqrt{2}\right) a$ and $(b_1, b_2) = (1, 1)\, a$.

(d) For the matrix notation, $(a_1 \times a_2) = \left(1 \times \sqrt{2}\right)$, $(b_1 \times b_2) = (1 \times 1)$ so that

$$M_a = \begin{pmatrix} 1 & 0 \\ 0 & \sqrt{2} \end{pmatrix} \tag{2.6E9a}$$

$$\det M_a = \sqrt{2} \tag{2.6E9b}$$

Also,

$$M_b = \begin{pmatrix} 1 & 0 \\ 0 & 1 \end{pmatrix} \tag{2.6E10a}$$

$$\det M_b = 1 \tag{2.6E10b}$$

Therefore, $(\mathbf{b}_1, \mathbf{b}_2)$ defines the SC primitive lattice since $\det M_b < \det M_a$. Mathematically, the determinant of a matrix is a scaling factor.

(e) **Extra** The dot product for $\mathbf{c}_1 = [101]$ and $\mathbf{a}_1 = [110]$ directions gives

$$\mathbf{c}_1 \cdot \mathbf{a}_2 = [uvw]_{c_1}\, a \cdot [uvw]_{a_2}\, a = [101]_{a_1} \cdot [110]_{a_2}\, a^2 \tag{2.6E11a}$$

$$\mathbf{c}_1 \cdot \mathbf{a}_2 = [1 \times 1 + 0 \times 1 + 1 \times 0]\, a^2 = a^2 \tag{2.6E11b}$$

The length of each vector is

$$\|\mathbf{c}_1\| = c_1 = \sqrt{u_1^2 + v_1^2 + w_1^2}\, a = \sqrt{1^2 + 0^2 + 1^2}a = \sqrt{2}a \tag{2.6E12a}$$

$$\|\mathbf{a}_2\| = a_2 = \sqrt{u_2^2 + v_2^2 + w_2^2}\, a = \sqrt{1^2 + 1^2 + 0^2}a = \sqrt{2}a \tag{2.6E12b}$$

From Eqs. (2.10) or (2.16a), the angle between these vectors is

$$\cos\theta = \frac{\mathbf{c}_1 \cdot \mathbf{a}_2}{\|\mathbf{c}_1\|\,\|\mathbf{a}_2\|} = \frac{a^2}{\left(\sqrt{2}a\right)\left(\sqrt{2}a\right)} = \frac{1}{2} \tag{2.6E13a}$$

$$\theta = \arccos\left(\frac{1}{2}\right) = 60° \tag{2.6E13b}$$

Similarly, the dot product for $\mathbf{c}_2 = [011]$ and $\mathbf{a}_1 = [110]$ directions yields $\theta = 60°$ since

$$\mathbf{c}_2 \cdot \mathbf{a}_2 = [uvw]_{c_2}\, a \cdot [uvw]_{a_2}\, a = [011]_{a_1} \cdot [110]_{a_2}\, a^2 \qquad (2.6\text{E}14\text{a})$$

$$\mathbf{c}_2 \cdot \mathbf{a}_2 = [0 \times 1 + 1 \times 1 + 1 \times 0]\, a^2 = a^2 \qquad (2.6\text{E}14\text{b})$$

$$\|\mathbf{c}_2\| = c_2 = \sqrt{u_1^2 + v_1^2 + w_1^2}\, a = \sqrt{0^2 + 1^2 + 1^2}\, a = \sqrt{2}a \qquad (2.6\text{E}14\text{c})$$

$$\|\mathbf{a}_2\| = a_2 = \sqrt{u_2^2 + v_2^2 + w_2^2}\, a = \sqrt{1^2 + 1^2 + 0^2}\, a = \sqrt{2}a \qquad (2.6\text{E}14\text{d})$$

2.7 Reciprocal Lattice in Crystallography

This section includes the theoretical background on reciprocal lattice of a perfect and symmetric arrangement of atoms called Bravais lattice in a three-dimensional space. The reciprocal lattice is based on Fourier transform of a primitive lattice, and it is a convenient mathematical technique to describe crystal structures having atomic periodicity.

Real and reciprocal lattices are connected via Fourier transformation. This is achieved by a vector algebra procedure for characterizing the geometry of unit cells based on crystallographic $\{hkl\}$ planes and $\langle uvw \rangle$ directions.

2.7.1 Primitive Lattice Vectors

Figure 2.8 depicts the most common conventional unit cells as large parallelepipeds with a set of translation lattice vectors $\mathbf{a}$, $\mathbf{b}$, $\mathbf{c}$ and a set of edge length a, b, c. If the crystallographic coordinates are orthogonal, then $(\mathbf{a}, \mathbf{b}, \mathbf{c}) = (\mathbf{x}, \mathbf{y}, \mathbf{z})$ in Cartesian coordinates.

These figures include common primitive unit cells as the small shared parallelepipeds, which represent the principal domain for translation symmetry of the lattice with primitive translation vectors $(\mathbf{a}_1, \mathbf{a}_2, \mathbf{a}_3)$ as the coordinate axes (Hermann [5, p. 19], Kelly and Knowles [6, p. 37], Waseda et al. [7, pp. 23-24]).

There is also another type of primitive unit cell known as Wigner-Seitz cell with a polyhedral shape in momentum space, and its reciprocal lattice is called the Brillouin zone. Detail features of the Wigner-Seitz cell can be found elsewhere (Waseda et al. [7, p. 187]).

The simple cubic SC (Fig. 2.8a) is its own primitive lattice, but the face-centered cubic (FCC) (Fig. 2.8b) and body-centered cubic (BCC) (Fig. 2.8c) have primitive cells as rotated parallelepiped cells. The hexagonal-closed packed (HCP) (Fig. 2.8d) has a hexagon as its primitive cell. Nonetheless, these unit cells represent three-

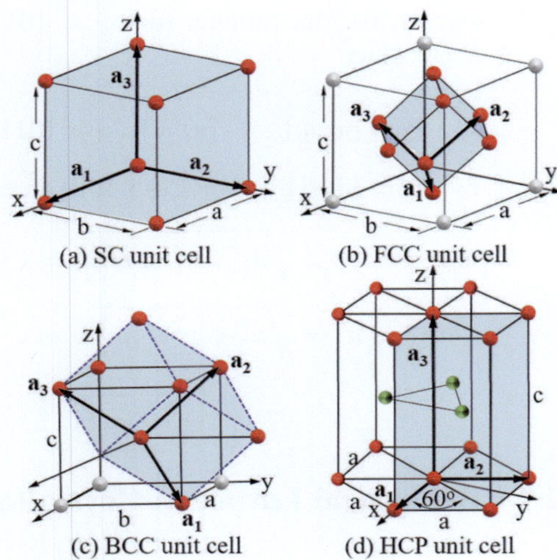

Fig. 2.8 Conventional and primitive unit cells (a) SC, (b) FCC, (c) BCC, and (d) HCP unit cells

dimensional symmetric arrangement of atoms described by their respective set of vectors.

A primitive unit cell contains only eight corner points and only one lattice point (one equivalent atom per unit cell) to choose as the origin of $(\mathbf{a}_1, \mathbf{a}_2, \mathbf{a}_3)$ vectors. In other words, the primitive cell has the lowest volume and one lattice point corresponding to one equivalent atom per unit cell.

For clarity, any of the primitive unit cells illustrated in Fig. 2.8 can be used to prove that this type of unit cell has exactly one lattice point. That is, multiply 8 corners by $1/8$ of an atom at each corner to get one equivalent atom, which translates to $8 \times 1/8 = 1$ lattice point. On the other hand, a non-primitive refers to a conventional unit cell with an additional body-centered or face-centered point. Therefore, conventional unit cells have more than one lattice point.

2.7.2 Reciprocal Lattice Vector

In order to describe a unit cell in the field of crystallography science or solid state physics, use of the reciprocal-lattice scattering vector $\mathbf{g}_{hkl}$ related to a crystallographic (hkl) plane can be described by a set of vectors $\mathbf{b}_i = (\mathbf{b}_1, \mathbf{b}_2, \mathbf{b}_3)$ (Kelly and Knowles [6, p. 440], Waseda et al. [7, Chapter 5]). Thus, any point in a lattice can be described by a translation vector (position vector) $\mathbf{r}_{uvw}$ depicted in Fig. 2.9a and defined by

$$\mathbf{r}_{uvw} = u\mathbf{a}_1 + v\mathbf{a}_2 + w\mathbf{a}_3 \tag{2.29}$$

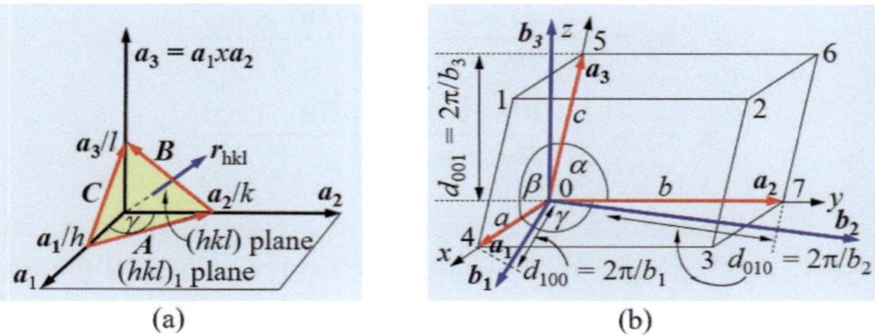

Fig. 2.9 Illustration of vector directions. (**a**) Cross product and (**b**) triclinic conventional and reciprocal lattice vectors

where u, v, w are integers called Miller indices of the crystal structure and $(\mathbf{a}_1, \mathbf{a}_2, \mathbf{a}_3)$ are vector directions.

These vectors span the crystal smallest parallelepiped along the conventional $\mathbf{a}_1, \mathbf{a}_2, \mathbf{a}_3$ vectors shown in Fig. 2.9a as orthogonal coordinates. Conversely, Fig. 2.9b illustrates the non-orthogonal coordinates for a triclinic unit cell.

The three-dimensional space defining the crystal structure is called the reciprocal lattice (direct space) with the corresponding reciprocal-lattice scattering vector $\mathbf{g}_{hkl}$, which in turn is related to the interplanar spacing d_{hkl} (Hermann [5, p. 68], Cullity [2, p. 482], Kelly and Knowles [6, p. 12]).

This vector algebra approach is useful in characterizing an X-ray diffraction pattern of a crystal (Fig. 2.2) since $d_{hkl} \propto 1/\|\mathbf{g}_{hkl}\|$, $\mathbf{g}_{(hkl)} \perp (hkl)$ and $d_{hkl} = f[\|\mathbf{a}_i\|, (hkl)]$. These are unique characteristic features of the reciprocal lattice approach related to vector algebra.

An X-ray diffraction indexing procedure would lead to the determination of an atomic arrangement of a crystal structure having a specific geometry and size described by the absolute values of $\|\mathbf{a}_i\|$ and $\|\mathbf{b}_i\|$ lattice vectors. In fact, an absolute value of a vector, say, $\mathbf{a}_1$, is called length, modulus, or magnitude.

The reciprocal lattice vector can be defined as $\mathbf{g}_{hkl} = k_1\mathbf{b}_1 + k_2\mathbf{b}_2 + k_3\mathbf{b}_3$, where k_1, k_2, k_3 are integers and $\mathbf{b}_1, \mathbf{b}_2, \mathbf{b}_3$ are the primitive lattice vectors. Actually, this expression resembles Eq. (2.29). Nonetheless, these primitive lattice vectors $\mathbf{b}_1, \mathbf{b}_2, \mathbf{b}_3$ depend on the conventional unit cell vectors $\mathbf{a}_1, \mathbf{a}_2, \mathbf{a}_3$ as analytically shown below. Moreover, the magnitude of $\mathbf{a}_1, \mathbf{a}_2, \mathbf{a}_3$ vectors have dimensions of length (nm), whereas $\mathbf{b}_1, \mathbf{b}_2, \mathbf{b}_3$ vectors have dimensions of reciprocal length (nm^{-1}). Similarly, the interplanar spacing d_{hkl} and the reciprocal lattice vector g_{hkl} are in nm and nm^{-1} units, respectively. For clarity, nm is a measure of length in the metric system that stands for nanometer.

Mathematically, the components of the reciprocal lattice vector denoted as $(\mathbf{b}_1, \mathbf{b}_2, \mathbf{b}_3)$ are defined in terms of the primitive lattice vectors $(\mathbf{a}_1, \mathbf{a}_2, \mathbf{a}_3)$ as per Eq. (2.30)

$$\mathbf{b}_1 = \frac{\omega(\mathbf{a}_2 \times \mathbf{a}_3)}{\mathbf{a}_1 \cdot (\mathbf{a}_2 \times \mathbf{a}_3)} = \frac{\omega(\mathbf{a}_2 \times \mathbf{a}_3)}{V} \tag{2.30a}$$

$$\mathbf{b}_2 = \frac{\omega\,(\mathbf{a}_3 \times \mathbf{a}_1)}{\mathbf{a}_1 \cdot (\mathbf{a}_2 \times \mathbf{a}_3)} = \frac{\omega\,(\mathbf{a}_3 \times \mathbf{a}_1)}{V} \tag{2.30b}$$

$$\mathbf{b}_3 = \frac{\omega\,(\mathbf{a}_1 \times \mathbf{a}_2)}{\mathbf{a}_1 \cdot (\mathbf{a}_2 \times \mathbf{a}_3)} = \frac{\omega\,(\mathbf{a}_1 \times \mathbf{a}_2)}{V} \tag{2.30c}$$

Here, $\omega = 2\pi$ is a factor (constant) related to angular wavevectors and Fourier transform of a $f(\kappa)$ function, $V = \mathbf{a}_1 \cdot (\mathbf{a}_2 \times \mathbf{a}_3)$ is the volume of the original unit cell, (i, j) are dummy indices, $\mathbf{b}_j \perp \mathbf{a}_i$ (perpendicular) for $i \neq j$, and $\mathbf{b}_j$ have dimensions of reciprocal length.

It can be stated that $(\mathbf{a}_1, \mathbf{a}_2, \mathbf{a}_3)$ and $(\mathbf{b}_1, \mathbf{b}_2, \mathbf{b}_3)$ are mutually reciprocal, provided that they satisfy the dot product $\mathbf{a}_i \cdot \mathbf{b}_j$ with the following Laue conditions using the Kronecker delta δ_{ij} notation

$$\mathbf{a}_i \cdot \mathbf{b}_j = \omega \delta_{ij} \ \text{ if } i = j \ \text{ and } \ \delta_{ij} = 1 \tag{2.31a}$$

$$\mathbf{a}_i \cdot \mathbf{b}_j = 0 \ \text{ if } \ i \neq j \tag{2.31b}$$

where the Kronecker unit matrix is defined by

$$\delta_{ij} = \begin{bmatrix} 1 & 0 & 0 \\ 0 & 1 & 0 \\ 0 & 0 & 1 \end{bmatrix} \tag{2.32}$$

Note that the Kronecker delta δ_{ij} is considered as an identity 3×3 matrix. Generally, Kronecker delta is a common function of variables i and j in the fields of mathematics, physics, and engineering. In essence, δ_{ij} is used to reduce the summation over the j variable, and it is treated as a symmetric tensor, which in turn describes the physical properties of scalars and vectors. Moreover, the Kronecker delta function δ_{ij} is also common in the field of tensor calculus or linear algebra. The reader can search the vast literature for additional insights on the generalized Kronecker delta δ_{ij} and related applications.

The angles $\alpha^* = \measuredangle\,(\mathbf{b}_2, \mathbf{b}_3)$, $\beta^* = \measuredangle\,(\mathbf{b}_3, \mathbf{b}_1)$, and $\gamma^* = \measuredangle\,(\mathbf{b}_1, \mathbf{b}_2)$ in the reciprocal lattice are defined by the given cosine functions of the form

$$\cos\alpha^* = \frac{\mathbf{b}_2 \cdot \mathbf{b}_3}{\|\mathbf{b}_2\|\,\|\mathbf{b}_3\|} \tag{2.33a}$$

$$\cos\beta^* = \frac{\mathbf{b}_3 \cdot \mathbf{b}_1}{\|\mathbf{b}_3\|\,\|\mathbf{b}_1\|} \tag{2.33b}$$

$$\cos\gamma^* = \frac{\mathbf{b}_1 \cdot \mathbf{b}_2}{\|\mathbf{b}_1\|\,\|\mathbf{b}_2\|} \tag{2.33c}$$

Accordingly, the Bravais primitive lattice vectors along with $V^* = \mathbf{b}_1 \cdot (\mathbf{b}_2 \times \mathbf{b}_3) = \omega^3/V$ can be determined by (Waseda et al. [7, p. 170])

$$\mathbf{a}_1 = \frac{\omega\,(\mathbf{b}_2 \times \mathbf{b}_3)}{\mathbf{b}_1 \cdot (\mathbf{b}_2 \times \mathbf{b}_3)} = \frac{\omega\,(\mathbf{b}_2 \times \mathbf{b}_3)}{V^*} \tag{2.34a}$$

$$\mathbf{a}_2 = \frac{\omega\,(\mathbf{b}_3 \times \mathbf{b}_1)}{\mathbf{a}_1 \cdot (\mathbf{b}_2 \times \mathbf{b}_3)} = \frac{\omega\,(\mathbf{b}_3 \times \mathbf{b}_1)}{V^*} \tag{2.34b}$$

$$\mathbf{a}_3 = \frac{\omega\,(\mathbf{b}_1 \times \mathbf{b}_2)}{\mathbf{a}_1 \cdot (\mathbf{b}_2 \times \mathbf{a}_3)} = \frac{\omega\,(\mathbf{b}_1 \times \mathbf{b}_2)}{V^*} \tag{2.34c}$$

By means of Fig. 2.9a, the cross product $\mathbf{a}_1 \times \mathbf{a}_2$ gives the direction of $\mathbf{a}_3$ vector, the perpendicular lattice translation $\mathbf{r}_{(hkl)}$ vector to the (hkl) plane is denoted as $\mathbf{r}_{(hkl)} \perp (hkl)$, and the point intercepts are $\mathbf{a}_1/h, \mathbf{a}_2/k$ and $\mathbf{a}_3/l$.

For the sake of clarity, the unit cell is simply an arbitrary parallelepiped with a, b, c edges and α, β, γ angles. By proper choice of these edges, a primitive cell can be constructed (Fig. 2.8) and characterized using Eq. (2.34). For a simple cubic, the magnitudes of the primitive vectors a_1, a_2, a_3 are equal to the lattice parameter a; that is, $|a_1| = |a_2| = |a_3| = a$.

In crystallography, the primitive unit cell (Fig. 2.8) consists of only one equivalent atom (one lattice point)), whereas a conventional unit cell (Fig. 2.4), such as an FCC, BCC, hexagonal, and so forth, have more than one equivalent atom. In this context, the repetitive pattern of the unit cell is the building block of crystals with translation symmetry. Thus, a pure crystalline solid is constituted by crystals known as grains, which are bounded by grain boundaries.

2.7.3 Interplanar Spacing

The interplanar spacing d_{hkl} between parallel (hkl) planes can be defined by the inverse scattering vector $\mathbf{g}_{hkl}$ (Kelly and Knowles [6, p. 441], Waseda et al. [7, p. 170])

$$d_{hkl} = \frac{\mathbf{a}_1}{h} \cdot \frac{\mathbf{g}_{hkl}}{\|\mathbf{g}_{hkl}\|} = \frac{\mathbf{a}_1 \cdot h\mathbf{b}_1}{h\,\|\mathbf{g}_{hkl}\|} = \frac{h\mathbf{a}_1 \cdot \mathbf{b}_1}{h\,\|\mathbf{g}_{hkl}\|} \tag{2.35a}$$

$$d_{hkl} = \frac{2\pi}{\|\mathbf{g}_{hkl}\|} = \frac{1}{\sqrt{h^2\,\|\mathbf{b}_1\|^2 + k^2\,\|\mathbf{b}_2\|^2 + l^2\,\|\mathbf{b}_3\|^2}} \tag{2.35b}$$

Using the dot product $\mathbf{g}_{hkl} \cdot \mathbf{g}_{hkl} = g_{hkl}^2 = \|\mathbf{g}_{hkl}\|^2$ yields (Kelly and Knowles [6, p. 469])

$$g_{hkl}^2 = (h\mathbf{b}_1 + k\mathbf{b}_2 + l\mathbf{b}_3) \cdot (h\mathbf{b}_1 + k\mathbf{b}_2 + l\mathbf{b}_3) \tag{2.36a}$$

$$g_{hkl}^2 = h^2\,(\mathbf{b}_1)^2 + k^2\,(\mathbf{b}_2)^2 + l^2\,(\mathbf{b}_3)^2 \tag{2.36b}$$

$$+\, 2kl\,(\mathbf{b}_2 \cdot \mathbf{b}_3) + 2lh\,(\mathbf{b}_3 \cdot \mathbf{b}_1) + 2hk\,(\mathbf{b}_1 \cdot \mathbf{b}_2)$$

Rearranging Eq. (2.35b) gives

$$g_{hkl}^2 / (2\pi)^2 = h^2 (b_1)^2 + k^2 (b_2)^2 + l^2 (b_3)^2 \tag{2.37}$$
$$+ 2klb_2b_3 \cos\alpha^* + 2lhb_3b_1 \cos\beta^* + 2hkb_1b_2 \cos\gamma^*$$

Combining Eqs. (2.32) and (2.33) yields the angles $\alpha^* = \measuredangle(\mathbf{b}_2, \mathbf{b}_3)$, $\beta^* = \measuredangle(\mathbf{b}_3, \mathbf{b}_1)$, and $\gamma^* = \measuredangle(\mathbf{b}_1, \mathbf{b}_2)$ in terms of α, β, γ

$$\cos\alpha^* = \frac{\cos\beta\cos\gamma - \cos\alpha}{\sin\beta\sin\gamma} \tag{2.38a}$$

$$\cos\beta^* = \frac{\cos\alpha\cos\gamma - \cos\beta}{\sin\alpha\sin\gamma} \tag{2.38b}$$

$$\cos\gamma^* = \frac{\cos\alpha\cos\beta - \cos\gamma}{\sin\alpha\sin\beta} \tag{2.38c}$$

These trigonometric functions are subsequently substituted into Eq. (2.37) for further analysis.

2.7.4 Practical Aspects of the Reciprocal Lattice

For practical purposes, the resultant mathematical definitions of the conventional or direct $\mathbf{a}_1, \mathbf{a}_2, \mathbf{a}_3$ and reciprocal primitive lattice $\mathbf{b}_1, \mathbf{b}_2, \mathbf{b}_3$ vectors for BCC, FCC, and HCP crystal structures, as shown in Fig. 2.8, are conveniently written as illustrated below

FCC Crystal

$$\mathbf{a}_1 = \frac{a}{2}(\mathbf{x} + \mathbf{y}) \qquad \mathbf{b}_1 = \frac{2\pi}{a}(-\mathbf{x} + \mathbf{y} + \mathbf{z}) \tag{2.39a}$$

$$\mathbf{a}_2 = \frac{a}{2}(\mathbf{y} + \mathbf{z}) \qquad \mathbf{b}_2 = \frac{2\pi}{a}(\mathbf{x} - \mathbf{y} + \mathbf{z}) \tag{2.39b}$$

$$\mathbf{a}_3 = \frac{a}{2}(\mathbf{x} + \mathbf{z}) \qquad \mathbf{b}_3 = \frac{2\pi}{a}(\mathbf{x} + \mathbf{y} - \mathbf{z}) \tag{2.39c}$$

Notice the coordinate dependency of these vectors.

BCC Crystal

$$\mathbf{a}_1 = \frac{a}{2}(\mathbf{x} + \mathbf{y} - \mathbf{z}) \qquad \mathbf{b}_1 = \frac{2\pi}{a}(\mathbf{x} + \mathbf{y}) \tag{2.40a}$$

$$\mathbf{a}_2 = \frac{a}{2}(-\mathbf{x} + \mathbf{y} + \mathbf{z}) \qquad \mathbf{b}_2 = \frac{2\pi}{a}(\mathbf{y} + \mathbf{z}) \tag{2.40b}$$

$$\mathbf{a}_3 = \frac{a}{2}\left(\mathbf{x} - \mathbf{y} + \mathbf{z}\right) \qquad \mathbf{b}_3 = \frac{2\pi}{a}\left(\mathbf{x} + \mathbf{z}\right) \tag{2.40c}$$

HCP Crystal

$$\mathbf{a}_1 = a\mathbf{x} \qquad\qquad \mathbf{b}_1 = \frac{2\pi}{a}\left(\mathbf{x} + \frac{1}{\sqrt{3}}\mathbf{y}\right) \tag{2.41a}$$

$$\mathbf{a}_2 = \frac{a}{2}\left(-\mathbf{x} + \sqrt{3}\mathbf{y}\right) \qquad \mathbf{b}_2 = \frac{2\pi}{a}\left(\frac{2}{\sqrt{3}}\mathbf{y}\right) \tag{2.41b}$$

$$\mathbf{a}_3 = c\mathbf{z} \qquad\qquad \mathbf{b}_3 = \frac{2\pi}{a}\left(\mathbf{z}\right) \tag{2.41c}$$

where

$$\mathbf{a}_2 = a\left[-\sin\left(30°\right)\mathbf{x} + \sin\left(30°\right)\mathbf{y}\right] = (a/2)\left(-\mathbf{x} + \sqrt{3}\mathbf{y}\right) \tag{2.42a}$$

$$\mathbf{a}_2 = (a/2)\left(-\mathbf{x} + \sqrt{3}\mathbf{y}\right) \tag{2.42b}$$

Using Eq. (2.35b) for three (hkl) face planes depicted in Fig. 2.9b gives the corresponding interplanar spacing d_{hkl} (d-spacing) as

$$d_{100} = \frac{2\pi}{\|\mathbf{g}_{100}\|} = \frac{2\pi}{b_1} \text{ for } (hkl)_1 = (100) \tag{2.43a}$$

$$d_{010} = \frac{2\pi}{\|\mathbf{g}_{010}\|} = \frac{2\pi}{b_2} \text{ for } (hkl)_2 = (010) \tag{2.43b}$$

$$d_{001} = \frac{2\pi}{\|\mathbf{g}_{001}\|} = \frac{2\pi}{b_3} \text{ for } (hkl)_3 = (001) \tag{2.43c}$$

with

$$(hkl)_1 = (100)_{1234} \,\|\, (100)_{5670} \ \& \ d_{100} = \frac{2\pi}{\|\mathbf{b}_1\|} \tag{2.44a}$$

$$(hkl)_2 = (010)_{4150} \,\|\, (010)_{3267} \ \& \ d_{010} = \frac{2\pi}{\|\mathbf{b}_2\|} \tag{2.44b}$$

$$(hkl)_3 = (001)_{4073} \,\|\, (001)_{1562} \ \& \ d_{001} = \frac{2\pi}{\|\mathbf{b}_3\|} \tag{2.44c}$$

The corresponding d_{hkl} expressions are based on Eq. (2.35b) as indicated next

$$d_{111} = \frac{2\pi}{\|\mathbf{g}_{(111)}\|} = \frac{2\pi}{\|\mathbf{b}_1 + \mathbf{b}_2 + \mathbf{b}_3\|} \tag{2.45a}$$

$$d_{110} = \frac{2\pi}{\|\mathbf{g}_{(110)}\|} = \frac{2\pi}{\|\mathbf{b}_1 + \mathbf{b}_2\|} \tag{2.45b}$$

Combine Eqs. (2.35a) and (2.36b) for the conventional crystal systems in Fig. 2.8a, b, c to get d_{hkl} for a cubic crystal

$$d_{hkl}^{cubic} = \sqrt{\frac{1}{\left(h^2 + k^2 + l^2\right) \|\mathbf{b}_1\|^2}} = \frac{a}{\sqrt{\left(h^2 + k^2 + l^2\right)}} \tag{2.46a}$$

$$\|\mathbf{b}_1\| = \|\mathbf{b}_2\| = \|\mathbf{b}_3\| = \frac{2\pi}{a} \quad \text{(cubic)} \tag{2.46b}$$

$$\alpha^* = \beta^* = \gamma^* = 90° \tag{2.46c}$$

For a hexagonal (hex) crystal (Fig. 2.8d), d_{hkl} becomes

$$d_{hkl}^{hex)} = \sqrt{\frac{1}{\left(h^2 + k^2 + hk\right) \|\mathbf{b}_1\|^2 + l^2 \|\mathbf{b}_3\|^2}} \tag{2.47a}$$

$$d_{hkl}^{hex)} = \sqrt{\frac{3 \|\mathbf{b}_1\|^2 \|\mathbf{b}_3\|^2}{\left(4h^2 + k^2 + hk\right) \|\mathbf{b}_3\|^2 + 3l^2 \|\mathbf{b}_1\|^2}} \tag{2.47b}$$

where

$$\|\mathbf{b}_1\| = \|\mathbf{b}_2\| = 2\pi \left[2/\left(a\sqrt{3}\right)\right] \quad \text{(hexagonal)} \tag{2.48a}$$

$$\|\mathbf{b}_3\| = 2\pi/c \tag{2.48b}$$

$$\alpha^* = \beta^* = 90° \text{ and } \gamma^* = 60° \tag{2.48c}$$

Additional details on crystallography or X-ray diffraction can be found elsewhere (Shmueli [8, pp. 25–98], Bricogne [9, pp. 25–98], Kittel [10, pp. 490–494], Wang and Lu [11, p. 58]).

Example 2.7 The computational sequence included hereafter is a simple procedure for indexing the diffraction peaks given in Fig. 2.2 as (hkl) planes for the BCC α-Fe unit cell. Determine **(a)** the Miller indices for the three (hkl) planes, **(b)** the lattice parameter "a" using Eqs. (2.1) and (2.46a) along with the diffraction angle 2θ for each diffraction peak and the wavelength $\lambda = 0.1527$ nm used to obtain the diffraction pattern, and **(c)** the volume of a unit cell.

Solution

(a) Let $K = \lambda/\left(2d_{hkl}\right)$ in Eq. (2.1) and tabulate the computations by (1) assigning a real number to K based on trial and error so that $K \sin^2 \theta$ is a small integer number, (2) assign this integer to $\left(h^2 + k^2 + l^2\right)$, and (3) find the Miller indices for the (hkl) planes. For a first order diffraction, $n = 1$ and

$$n\lambda = 2d_{hkl}\sin\theta \tag{2.7E1a}$$

$$d_{hkl} = \frac{a}{\sqrt{\left(h^2 + k^2 + l^2\right)}} \tag{2.7E1b}$$

$$\sin^2\theta = \frac{\lambda^2}{4a^2}\left(h^2 + k^2 + l^2\right) \tag{2.7E1c}$$

$$\left(h^2 + k^2 + l^2\right) = K\sin^2\theta \tag{2.7E1d}$$

This is an equation with unknown lattice parameter a and Miller index (hkl) plane. The computed partial results are

2θ	θ	$\sin^2\theta$	K	$K\sin^2\theta$	$\left(h^2 + k^2 + l^2\right)$	(hkl)
44.20	22.10	0.14154	14.130	2	2	110
64.40	32.20	0.28758	14.087	4	4	200
81.32	40.66	0.44097	14.133	6	6	211

(b) Now, calculate d_{hkl} from $K = \lambda/(2d_{hkl})$ and the lattice parameter a using Eq. (2.46)

θ	(hkl)	d_{hkl} (nm)	a (nm)
22.10	110	0.20294	0.28700
32.20	200	0.14328	0.28656
40.66	211	0.11718	0.28703

Take the average lattice parameter a so that

$$a = 0.28686\ nm \simeq 0.287\ nm \tag{2.7E2a}$$

$$a = b = c = 0.287\ nm \tag{2.7E2b}$$

$$\alpha = \beta = \gamma = 90° \tag{2.7E2c}$$

(c) The volume of the α-BCC unit cell is

$$V = abc = a^3 = 0.02364\ nm^3 \tag{2.7E3}$$

The reader should consult https://www.iucr.org/, the International Union of Crystallography (IUCr) Web page for indexing crystal structures. The IUCr Web site provides material of educational value.

Example 2.8 The primitive lattice vectors $(\mathbf{a}_1, \mathbf{a}_2, \mathbf{a}_3)$ for **(a)** SC, **(b)** BCC, and **(c)** FCC crystals are drawn below in orthogonal x, y, z axes. Determine the reciprocal lattices $(\mathbf{b}_1, \mathbf{b}_2, \mathbf{b}_3)$ for each crystal and the volume of the primitive unit cells. Explain.

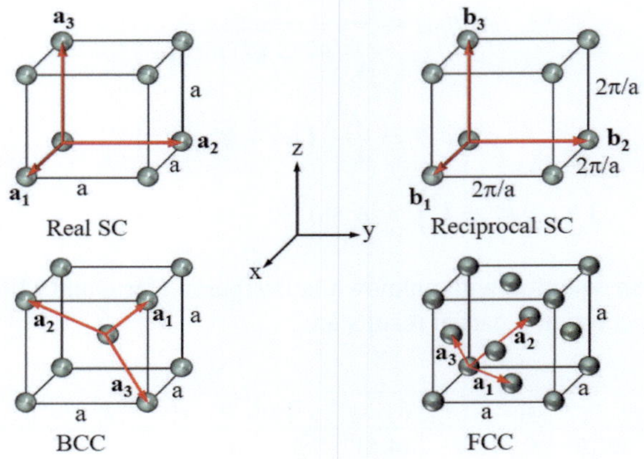

Solution

(a) The lattice vectors for the conventional SC unit cell with $a = b = c$ are

$$\mathbf{a}_1 = a\mathbf{x} \tag{2.8E1a}$$

$$\mathbf{a}_2 = a\mathbf{y} \tag{2.8E1b}$$

$$\mathbf{a}_3 = a\mathbf{z} \tag{2.8E1c}$$

The conventional SC unit cell geometry is itself the reciprocal unit cell. From Eq. (2.30) along with the volume of the primitive unit cell defined as $V_p = a_1 \cdot (\mathbf{a}_2 \times \mathbf{a}_3)$, the reciprocal lattice vectors along with $\omega = 2\pi$ and the volume of the conventional (V_c) unit cells are

$$\mathbf{b}_1 = \frac{\omega\,(\mathbf{a}_2 \times \mathbf{a}_3)}{\mathbf{a}_1 \cdot (\mathbf{a}_2 \times \mathbf{a}_3)} = \frac{\omega\,(a\mathbf{y}) \times (a\mathbf{z})}{a\mathbf{x} \cdot [(a\mathbf{y}) \times (a\mathbf{z})]} = \frac{\omega a^2\,(\mathbf{y} \times \mathbf{z})}{a^3 \mathbf{x} \cdot [\mathbf{y} \times \mathbf{z}]} = \frac{\omega}{a}\mathbf{x} \tag{2.8E2a}$$

$$\mathbf{b}_2 = \frac{\omega\,(\mathbf{a}_3 \times \mathbf{a}_1)}{\mathbf{a}_1 \cdot (\mathbf{a}_2 \times \mathbf{a}_3)} = \frac{\omega\,(a\mathbf{z}) \times (a\mathbf{x})}{a^3 \mathbf{x} \cdot [\mathbf{y} \times \mathbf{z}]} = \frac{\omega a^2\,(\mathbf{z} \times \mathbf{x})}{a^3 \mathbf{x} \cdot [\mathbf{y} \times \mathbf{z}]} = \frac{\omega}{a}\mathbf{y} \tag{2.8E2b}$$

$$\mathbf{b}_3 = \frac{\omega\,(\mathbf{a}_1 \times \mathbf{a}_2)}{\mathbf{a}_1 \cdot (\mathbf{a}_2 \times \mathbf{a}_3)} = \frac{\omega\,(a\mathbf{x}) \times (a\mathbf{y})}{a^3 \mathbf{x} \cdot [\mathbf{y} \times \mathbf{z}]} = \frac{\omega a^2\,(\mathbf{x} \times \mathbf{y})}{a^3 \mathbf{x} \cdot [\mathbf{y} \times \mathbf{z}]} = \frac{\omega}{a}\mathbf{z} \tag{2.8E2c}$$

$$V_c = abc = aaa = a^3 = V_p \tag{2.8E2d}$$

From Eqs. (2.30) and (2.35b),

$$\mathbf{g}_{hkl} = \frac{\omega}{a}(h\mathbf{x} + k\mathbf{y} + l\mathbf{z}) \tag{2.8E3a}$$

$$\mathbf{d}_{hkl} = \frac{\omega}{\|\mathbf{g}_{hkl}\|} = \frac{a}{\sqrt{h^2 + k^2 + l^2}} \tag{2.8E3b}$$

(b) For the BCC crystal with $V_c = a^3$ and $V_p = \mathbf{a}_1 \cdot (\mathbf{a}_2 \times \mathbf{a}_3) = a^3/2$,

$$\mathbf{a}_1 = \frac{a}{2}(\mathbf{x} + \mathbf{y} + \mathbf{z}) \tag{2.8E4a}$$

$$\mathbf{a}_2 = \frac{a}{2}(\mathbf{x} - \mathbf{y} + \mathbf{z}) \tag{2.8E4b}$$

$$\mathbf{a}_3 = \frac{a}{2}(\mathbf{x} + \mathbf{y} - \mathbf{z}) \tag{2.8E4c}$$

From Eq. (2.30), the reciprocal lattice vectors are

$$\mathbf{b}_1 = \frac{\omega(\mathbf{a}_2 \times \mathbf{a}_3)}{\mathbf{a}_1 \cdot (\mathbf{a}_2 \times \mathbf{a}_3)} = \frac{\omega}{a}(\mathbf{y} + \mathbf{z}) \tag{2.8E5a}$$

$$\mathbf{b}_2 = \frac{\omega(\mathbf{a}_3 \times \mathbf{a}_1)}{\mathbf{a}_1 \cdot (\mathbf{a}_2 \times \mathbf{a}_3)} = \frac{\omega}{a}(\mathbf{x} + \mathbf{z}) \tag{2.8E5b}$$

$$\mathbf{b}_3 = \frac{\omega(\mathbf{a}_1 \times \mathbf{a}_2)}{\mathbf{a}_1 \cdot (\mathbf{a}_2 \times \mathbf{a}_3)} = \frac{\omega}{a}(\mathbf{x} + \mathbf{y}) \tag{2.8E5c}$$

From Eqs. (2.30) and (2.35b),

$$\mathbf{g}_{hkl} = \frac{\omega}{a}[h(\mathbf{y} + \mathbf{z}) + k(\mathbf{x} + \mathbf{z}) + l(\mathbf{x} + \mathbf{y})] \tag{2.8E6a}$$

$$\mathbf{d}_{hkl} = \frac{\omega}{\|\mathbf{g}_{hkl}\|} = \frac{a}{\sqrt{h^2 + k^2 + l^2}} \tag{2.8E6b}$$

(c) For a conventional FCC crystal with $V_c = a^3$ and $V_p = \mathbf{a}_1 \cdot (\mathbf{a}_2 \times \mathbf{a}_3) = a^3/4$,

$$a_1 = \frac{a}{2}(\mathbf{x} + \mathbf{y}), \qquad a_2 = \frac{a}{2}(\mathbf{y} + \mathbf{z}), \qquad a_3 = \frac{a}{2}(\mathbf{x} + \mathbf{z}) \tag{2.8E7}$$

Then, the reciprocal lattice vectors become

$$\mathbf{b}_1 = \frac{\omega(\mathbf{a}_2 \times \mathbf{a}_3)}{\mathbf{a}_1 \cdot (\mathbf{a}_2 \times \mathbf{a}_3)} = \frac{\omega}{a}(-\mathbf{x} + \mathbf{y} + \mathbf{z}) \tag{2.8E8a}$$

$$\mathbf{b}_2 = \frac{\omega(\mathbf{a}_3 \times \mathbf{a}_1)}{\mathbf{a}_1 \cdot (\mathbf{a}_2 \times \mathbf{a}_3)} = \frac{\omega}{a}(\mathbf{x} - \mathbf{y} + \mathbf{z}) \tag{2.8E8b}$$

$$\mathbf{b}_3 = \frac{\omega(\mathbf{a}_1 \times \mathbf{a}_2)}{\mathbf{a}_1 \cdot (\mathbf{a}_2 \times \mathbf{a}_3)} = \frac{\omega}{a}(\mathbf{x} + \mathbf{y} - \mathbf{z}) \tag{2.8E8c}$$

From Eqs. (2.30) and (2.35b),

$$\mathbf{g}_{hkl} = \frac{\omega}{a} \left[h\left(-\mathbf{x} + \mathbf{y} + \mathbf{z}\right) + k\left(\mathbf{x} - \mathbf{y} + \mathbf{z}\right) + l\left(\mathbf{x} + \mathbf{y} - \mathbf{z}\right) \right] \tag{2.8E9a}$$

$$\mathbf{d}_{hkl} = \frac{\omega}{\|\mathbf{g}_{hkl}\|} = \frac{a}{\sqrt{h^2 + k^2 + l^2}} \tag{2.8E9b}$$

2.8 Stereographic Projection

The stereographic projection is a technique used in crystallography to determine angular projections between crystal flat faces, flat planes, or facets. For example, Fig. 2.10 illustrates three garnet crystals showing their natural facets having different orientations.

Figure 2.10a depicts the morphology of a red almandine, $Fe_3Al_2(SiO_4)$, single crystal (https://slideplayer.com/slide/3247982/); Fig. 2.10b depicts the andradite, $Ca3Fe_2Si_3O_{12}$, crystal exhibiting clear facets [12]; and Fig. 2.10c shows red spessartine, $Mn_3Al_2(SiO_4)_3$, facets (https://www.irocks.com/minerals/specimen/5862).

2.8.1 Derivation of the Stereographic Projection

The stereographic projection (stereogram) is a graphical technique used to plot crystal planes and directions in a three-dimensional (3D) crystal on a two-dimensional (2D) projection plane as shown in Fig. 2.11. Thus, the crystal $< uvw >$ directions being perpendicular to hkl plane normals are termed points or poles.

Notice that the Cartesian x, y, z coordinates in Fig. 2.11 are projected from the center of a crystal positioned at the center of a hollow sphere to intersect the sphere surface, and they become the principal poles. In essence, north N, south S, east E, and west W are the cardinal directions used in stereographic projections as reference poles.

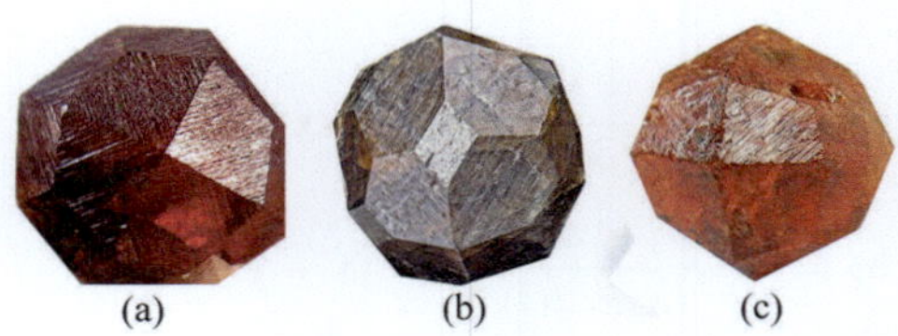

Fig. 2.10 Garnet single crystals showing facets. (**a**) Red Almandine $Fe_3Al_2(SiO_4)$ (https://slideplayer.com/slide/3247982/), (**b**) andradite $Ca3Fe_2Si_3O_{12}$ [12], and (**c**) red spessartine $Mn_3Al_2(SiO_4)_3$ (https://www.irocks.com/minerals/specimen/5862)

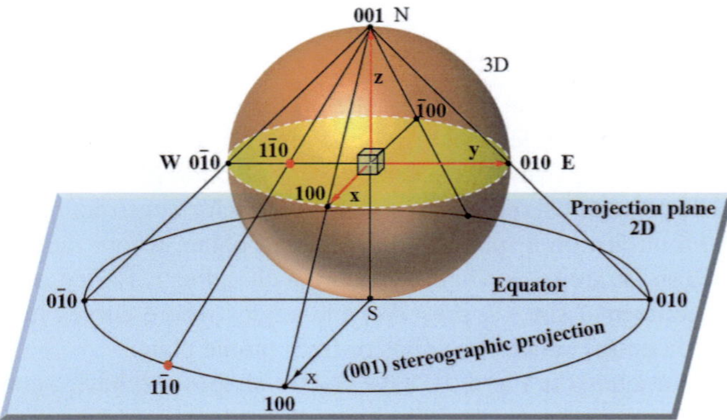

Fig. 2.11 3D illustration of the (001) stereographic projection from the north pole (N) using orthogonal points from the sphere onto a flat plane below the sphere

2.8.2 Drawing the Circle for a Stereographic Projection

Steps for drawing a circle representing the stereographic projection from a sphere of projection to a projection plane.

- Draw a sphere of projection and its corresponding equatorial plane with a common radius R, and draw a projection plane below the sphere of projection as shown in Fig. 2.11.
- Draw the x, y, z orthogonal Cartesian coordinates. Virtually place a rectangular-shaped crystal at the center of the sphere of projection at a convenient orientation. Let's work with the (001) stereographic projection as shown in Fig. 2.11.
- Take the top horizontal (001) plane of the crystal, and place the (001) pole on top of the sphere. Subsequently, project the (00$\bar{1}$) pole at the bottom surface of the sphere of projection along the $-z$-axis. This projected point becomes the southern hemisphere (south point) and the center point on the projection plane. Do the same for the (100), ($\bar{1}$00) and (010), (0$\bar{1}$0) poles along the x, y axes.
- Extend the $< uvw >$ directions perpendicular to the $\{hkl\}$ planes or faces along the Cartesian x, y, z orthogonal coordinates until intersecting the reference sphere surface. Thus, (100) $\perp$ x-axis and [100] $\parallel$ x-axis, (010) $\perp$ y-axis and [010] $\parallel$ y-axis, (001) $\perp$ z-axis and [001] $\parallel$ z-axis, where the symbols $\perp$ and $\parallel$ stand for perpendicular and parallel, respectively.
- Draw straight lines from the (001) pole through the (100), ($\bar{1}$00), (010), and (0$\bar{1}$0) poles until all straight lines intersect the projection plane. Draw a circle through the projected points on the projection plane. Now, the sphere and circle have a common radius R.
- If a straight line from the (001) through the (1$\bar{1}$0) pole on the equatorial plane projects on the circle, then the above procedure is correct.

- The stereographic projection of the unit sphere maps it to itself since its equator is the point at which it intersects the x-y plane.

All points on the projection sphere that lie in the northern hemisphere (north pole) will project within the circle known as a stereogram or stereographic projection, which in turn lies on the two-dimensional (2D) projection plane.

The preceding graphical procedure shows the step for constructing a stereogram, which maps all circles on a sphere to circles on a plane, where the project area is simply defined as the area of a circle; $A = \pi r^2$ with radius r. Hence, a stereographic projection is conformal since it preserves the angles where curves intersect and it translates circles on the sphere to circles or lines on the plane.

The crystal structures in Fig. 2.10 have similar shapes and have a common vitreous (glassy) luster appearance. Metallic crystals having a particular arrangement of atoms are treated as mineral crystals with respect to crystallographic crystal $\{hkl\}$ planes and $< uvw >$ directions.

A metallic crystal consists of packed atoms or ions in a cloud of mobile electrons. This arrangement is attributed to metallic bonding related to binding forces, leading to a zero net force, F_{net}, for a solid stable condition since the attractive (F_A) and repulsive (F_R) forces are equal and opposite to each other; $F_A = -F_R$.

Regarding the potential energy, a stable solid implies that there must exist a minimum energy $U(r_o) = U_o < 0$, where r_o is the equilibrium interatomic distance. This means that s solid reaches equilibrium when $F_{net} = F(r_o) = F_A + F_R = 0$ and $U(r_o) = U_o < 0$.

2.8.3   The Stereographic Projection of the Cubic System

Consider a cubic crystal located at the center of the sphere shown in Fig. 2.12a, where the crystallographic planes $(hkl)_x = (100)$, $(hkl)_y = (010)$, and $(hkl)_z = (001)$ are perpendicular (normal) to the orthogonal x, y, z axes, respectively.

The stereographic projection reveals symmetry relationships related to undistorted representation of angular projections in a x, y, z coordinate system. For instance, the z-axis can be taken as the normal to the $(hkl)_z$ plane of projection. This implies that the $([uvw]_z$ direction is the normal (perpendicular) vector to the $(hkl)_z$ plane having the same Miller indices. This is denoted as $(hkl)_z \perp (uvw)_z$.

Notice that the normal crystallographic directions to the crystal faces are denoted as $[uvw]_{y>0,N} = (010)$, $[uvw]_{y<0,N} = (0\bar{1}0)$, $[uvw]_{z>0,N} = (001)$, and $[uvw]_{z<0,N} = (00\bar{1})$, and they intersect the sphere at the equator. This information can also be denoted in a general form $(hkl) \perp [uvw]$ (perpendicular). Thus,

$$(100)_x \perp [100]_x \tag{2.49a}$$

$$(010)_y \perp [100]_y \tag{2.49b}$$

$$(001)_z \perp [001]_z \tag{2.49c}$$

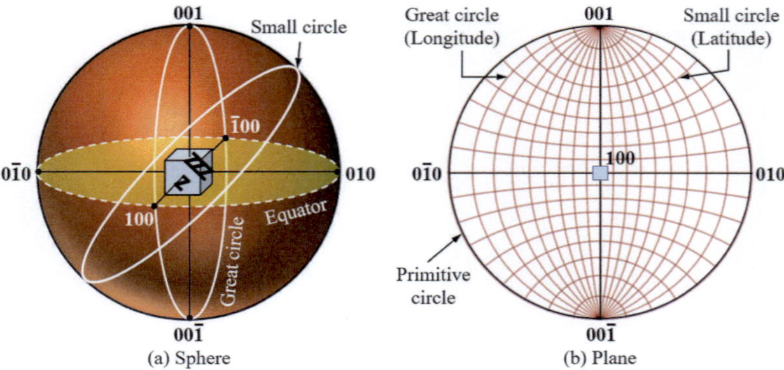

Fig. 2.12 Stereographic projection from a (**a**) sphere onto (**b**) a plane

In order to carry out a standard stereographic projection, one must assume the orientation of a crystal and take an (hkl) plane located at the center of a circle as shown in Fig. 2.12b for the $(hkl)_x = (100)$ plane. For a standard stereographic projection analysis of $(hkl)_y = (010)$ or $(hkl)_z = (001)$, one must rotate the crystal in order to see the taken plane at the center of the circle.

The principal faces (plane normals or facets) shown in Fig. 2.12 are geometrically at $90°$ from each other, and their coordinate normal vectors are orthogonal and are projected on the surface of a sphere as the initial step in stereographic projection analysis. For clarity, plane normals and their respective crystallographic directions have the same Miller indices, $(hkl) = [uvw]_N$, and are treated as poles or points (without parentheses) on a surface of the stereogram in Fig. 2.12b. In summary, the 3D-sphere is projected onto a 2D-circular plane.

The main angular projections of the selected (100) stereographic projection are $(100) \angle (010) = 90°$ and $(100) \angle (0\bar{1}0) = 90°$ on the equator, while $(100) \angle (001) = 90°$ and $(100) \angle (00\bar{1}) = 90°$ on the great circle (seen as a vertical straight line in 2D analysis called longitude or prime meridian). Moreover, other angular projections corresponding to a set of different crystal planes can be projected on the circular plane, provided that the longitudinal and latitude lines are separated by small-degree intervals. For the chosen stereogram in Fig. 2.12b, each line is separated (divided) by a $10°$ degree interval.

Stereogram poles can be determined by adding the Miller indices of crystal directions located on the circumference of a circle or a horizontal straight line (equatorial plane). For the upper right-hand quadrant in Fig. 2.12b, some new poles and angular projections are given below in bullet points. Thus,

- For $[010] + [001] = [011]$ on the edge of the circle, the dot product $[100] \cdot [011]$ yields $\cos\theta = 0$ and $\theta = 90°$.
- For $[100] + [010] = [110]$ on the equator, the dot product $[100] \cdot [110]$ gives the following results $\cos\theta = 1/\sqrt{2} = \sqrt{2}/2$ and $\theta = 45°$.

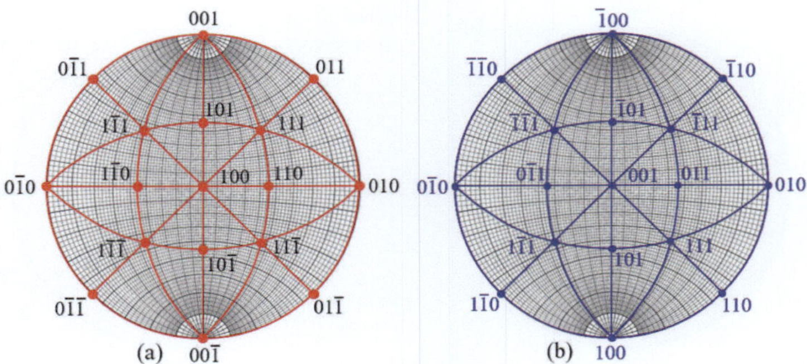

Fig. 2.13 Stereographic projections superimposed on the Wulff net for easy measurement of angles between poles. (**a**) the (100) and (**b**) (001) stereographic projections of a cubic crystal

- For $[100] + [001] = [101]$ on the equator, the dot product $[100] \cdot [101]$ gives the following results $\cos\theta = 1/\sqrt{2} = \sqrt{2}/2$ and $\theta = 45°$.
- For $[100] + [011] = [111]$ in the center of the quadrant, the dot product $[100] \cdot [111]$ gives $\cos\theta = 1/\sqrt{3} = \sqrt{3}/3$ and $\theta = 54.74°$.

The projections and angles are the key features for constructing stereographic projections. For instance, results are plotted on the stereogram or Wulff net shown in Fig. 2.13a along with results for other quadrants.

For comparison, the (001) stereographic projection is illustrated in Fig. 2.13b. Both (100) and (001) stereographic projections are superimposed on the Wulff net for easy measurement of angles between poles. The Wulff net has equally spaced 2°-thin lines and 10°-thick lines for precise plotting of the poles on a piece of paper.

2.8.4   Crystal Facets

The above procedure is clearly illustrated in Fig. 2.14a for the Andradite crystal, which is a mineral species of the Garnet group, with $(hkl)_x$, $(hkl)_y$, and $(hkl)_z$ points defining the principal poles at the sphere primitive great circle.

Figure 2.14b, on the other hand, shows a schematic metallic cubic crystal, say, a facet cube which has the principal axes identified as $(hkl)_x = (100)$, $(hkl)_y = (010)$, and $(hkl)_z = (001)$. These crystallographic planes in an actual metallic crystal structure can be imagined to arise from a 3D array of points (atoms or ions), which are geometrically and repetitively distributed in space. Thus,

- Draw a vector OP with the head ending on the sphere surface, and reflect it downward to the south pole of the sphere by drawing the PP'S line. This provides the OP' line on the equatorial plane, where point P' is the stereographic

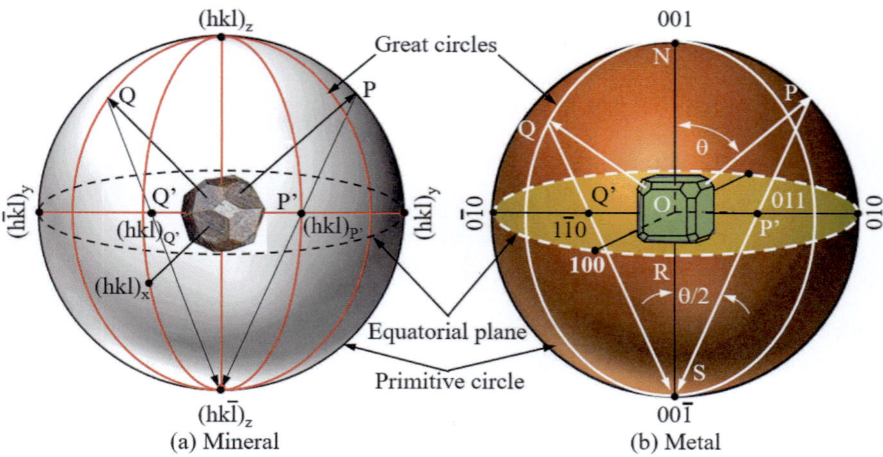

Fig. 2.14 Construction of stereographic projection for a (**a**) mineral, garnet single crystal, and (**b**) metal, cubic crystal

projection representing the direction $[uvw]_{P'}$. Similarly, point Q' is another stereographic projection with $[uvw]_{Q'}$.

- Repeat this procedure for as many stereographic projections needed for revealing the crystal symmetry on a 2D stereogram.

A stereographic projection leads to detailed angular projections of crystal planes. For instance, the distance OP', as an example, is written along with the sphere radius R as

$$OP' = R \tan(\theta/2) \tag{2.50}$$

The points of intersection with the equatorial plane (Fig. 2.14b) are defined as P' and Q', which correspond to the $1\bar{1}0$ and 011 projections, respectively.

2.9 Dot Product and Stereographic Projection

This section includes some additional mathematical aspects related to stereographic projections for cubic crystals. For instance, the angle θ between two crystallographic $[u_1v_1w_1]$ and $[u_2v_2w_2]$ directions can be determined using Eq. (2.16c) with lattice parameters $a = b = c$.

Let $\mathbf{p} = [u_1v_1w_1]$ and $\mathbf{q} = [u_2v_2w_2]$ in Eq. (2.10) to get the dot product defined by

$$[u_1v_1w_1] \cdot [u_2v_2w_2] = \|[u_1v_1w_1]\| \, \|[u_2v_2w_2]\| \cos\theta \tag{2.51a}$$

$$\|[u_1 v_1 w_1]\| = \sqrt{u_1^2 + v_1^2 + w_1^2} \tag{2.51b}$$

$$\|[u_2 v_2 w_2]\| = \sqrt{u_2^2 + v_2^2 + w_2^2} \tag{2.51c}$$

Solving Eq. (2.51a) for $\cos\theta$ gives

$$\cos\theta = \frac{u_1 u_2 + v_1 v_2 + w_1 w_2}{\sqrt{u_1^2 + v_1^2 + w_1^2}\sqrt{u_2^2 + v_2^2 + w_2^2}} \tag{2.52}$$

which is a suitable expression for crystallographic directions.

For cubic crystals, the angle θ between two $(h_1 k_1 l_1)$ and $(h_2 k_2 l_2)$ planes can also be determined using the dot product. Thus,

$$(h_1 k_1 l_1) \cdot (h_2 k_2 l_2) = \|(h_1 k_1 l_1)\| \, \|(h_2 k_2 l_2)\| \cos\theta \tag{2.53a}$$

$$\cos\theta = \frac{h_1 h_2 + k_1 k_2 + l_1 l_2}{\sqrt{h_1^2 + k_1^2 + l_1^2}\sqrt{h_2^2 + k_2^2 + l_2^2}} \tag{2.53b}$$

The dot product is a useful mathematical tool to verify the angle between two poles on a stereographic projection or stereogram.

Example 2.9 Find the angle θ for the pole "p" by inspection, dot product, and $[100]$-$[010]$-$[uvw]_p$ triangle for the (100) stereographic projection of a cubic crystal. By symmetry, $\theta = \theta_p = \theta_q$.

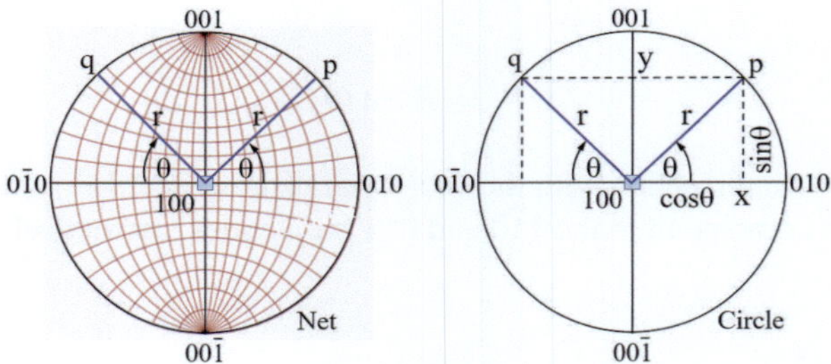

Solution From the 10°-interval stereogram, the angle θ for point "p" is

$$[uvw]_p = [001]_1 + [010]_2 = [011]_p \tag{2.9E1a}$$

$$\theta = \theta_p = 45° = \pi/4 \tag{2.9E1b}$$

This result reveals a particular characteristic of a stereogram. The (001) and (010) planes have their poles lying upon a great circle. Hence, these poles share a zone on the primitive great circle.

From the dot product, Eq. (2.52) with $[u_1 v_1 w_1] = [100]_2$ and $[u_2 v_2 w_2]_p = [011]_p$ directions yields

$$\cos\theta = \frac{[u_1 v_1 w_1]_2 \cdot [u_2 v_2 w_2]_p}{\|[u_1 v_1 w_1]_2\|\,\|[u_2 v_2 w_2]_p\|} \tag{2.9E2a}$$

$$\cos\theta = \frac{u_1 u_2 + v_1 v_2 + w_1 w_2}{\sqrt{u_1^2 + v_1^2 + w_1^2}\sqrt{u_2^2 + v_2^2 + w_2^2}} \tag{2.9E2b}$$

$$\cos\theta = \frac{[010]_2 \cdot [011]_p}{\|[010]_2\|\,\|[011]_p\|} \tag{2.9E2c}$$

$$\cos\theta = \frac{(1)\,(0) + (1)\,(1) + (0)\,(1)}{\sqrt{0^2 + 1^2 + 0^2}\sqrt{0^2 + 1^2 + 1^2}} = \frac{1}{\sqrt{2}} \tag{2.9E2d}$$

so that

$$\theta = \theta_p = \cos^{-1}\left(\frac{1}{\sqrt{2}}\right) = \frac{\pi}{4} = 45^\circ \tag{2.9E3}$$

Therefore, the angle between the (100) stereographic projection and the pole $[011]_p$ is clearly $\theta = 45^\circ$, which is confirmed by the dot product. Using the [100]-[010]-[011] triangle within the circle yields

$$\cos(45^\circ) = \frac{1}{\sqrt{2}} = \frac{\sqrt{2}}{2} = 0.70711 = \frac{x}{r} \tag{2.9E4a}$$

$$\sin(45^\circ) = \frac{1}{\sqrt{2}} = \frac{\sqrt{2}}{2} = 0.70711 = \frac{y}{r} \tag{2.9E4b}$$

from which $x = 0.70711r$ and $y = 0.70711r$ or

$$x = \frac{\sqrt{2}}{2}r \tag{2.9E5a}$$

$$y = \frac{\sqrt{2}}{2}r \tag{2.9E5b}$$

and

$$r = \sqrt{2}x \tag{2.9E6a}$$

$$r = \sqrt{2}y \tag{2.9E6b}$$

For cubic crystals, in general, the normal to the (hkl) set of planes is the pole representing the $[uvw]$ vector. Thus, the Wulff net is a projection of great circles representing lines longitude, whereas small circles represent lines of latitude on the sphere.

In crystallography, the Wulff net is an important tool for plotting crystallographic axes and planes since it is the three-dimensional (3D) equivalent of a protractor used to measure angles between poles. This means that a stereographic projection is a mapping method used to characterize angular relationships between (hkl) planes or faces and for projecting 3D information onto a 2D plane, where h, k, l for planes and u, v, w for directions are known as Miller indices.

2.10   Laue Pattern

The crystal orientation can be determined by using the Laue method to generate experimental X-ray diffraction data as a Laue X-ray diffraction pattern (Laue diagram or Laue reflections) consisting of spots as shown in Fig. 2.15a.

When a Laue transmission/back reflection of an X-ray beam strikes particular planes in a stationary crystal, a Laue X-ray diffraction pattern corresponding to the symmetry of the crystal is generated and observed as spots. This is clearly shown in Fig. 2.15 for a transmission Laue pattern recorded on a $120 \times 120 \, mm^2$ film (Weber [13]). The reader is encouraged to check Steffen Weber's Laue applet for generating Laue spots using Java Lauegram (Java simulation). Nonetheless, this Laue pattern resembles the cubic (100) stereographic projection.

Placing the circular plane and great circles on the Laue X-ray diffraction spots reveals the stereographic projection of the unknown crystal in Fig. 2.15b. Therefore, the Laue diagram is a stereographic projection of the crystal. Actually, the spots

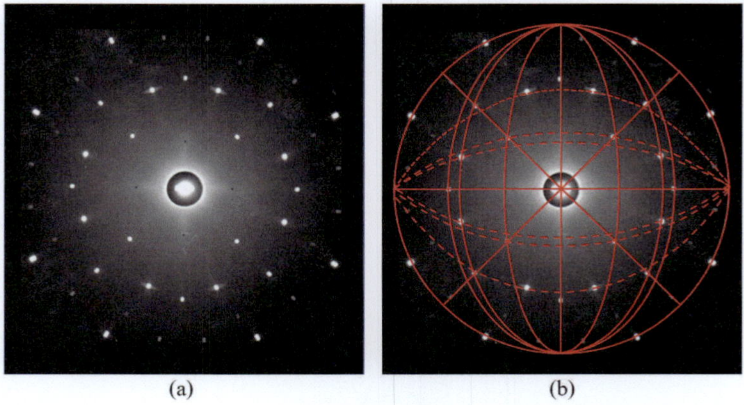

(a)   (b)

Fig. 2.15 (**a**) Laue image (in transmission mode) of an unknown crystal posted online by Weber [13] in July 1997 and (**b**) the same image with great circles which resembles the cubic (100) stereographic projection

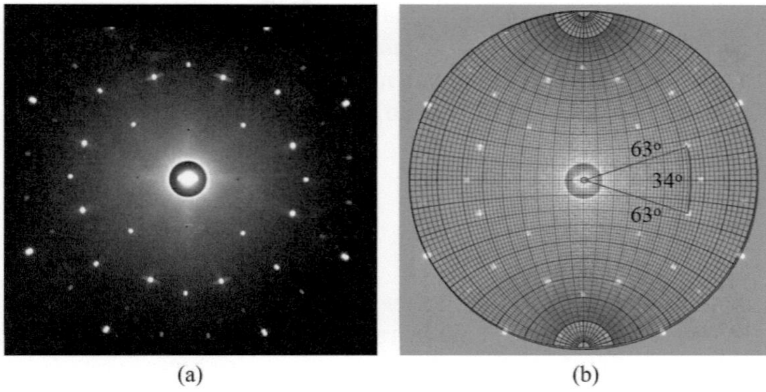

(a)   (b)

Fig. 2.16 Stereographic projection of an unknown crystal. (**a**) Laue X-ray diffraction pattern (Laue diagram). (**b**) The image is superimposed on the Wulff net for easy measurement of angles between poles

being connected by straight lines are called zones and are the common evidence of diffraction (reflection) from planes that share a common crystallographic direction $[uvw]$.

In a cubic system, if the $[uvw]$ direction lies in the (hkl) plane, then the Weiss zone relation must be satisfied

$$hu + kv + lw = 0 \tag{2.54}$$

For example, if $[01\bar{1}]$ lies on (011), then $(0)(0) + (1)(1) + (1)(-1) = 0$. For [011] and (011), Eq. (2.54) predicts that [011] does not lie on the (011) plane since $2 \neq 0$.

The Laue pattern in Fig. 2.15a is recast in Fig. 2.16a to measure the angles between poles. This can be accomplished by superimposing a transparent Wulff net on this Laue X-ray diffraction pattern. For clarity, one simple example is shown in Fig. 2.16b.

The Laue diffraction pattern is simply an experimental technique that provides information on a specimen crystallographic directions and crystal symmetry. Specifically, the Laue method can be used as a quality assurance for determining the orientation of single crystals in semiconductor wafers and single-crystal aircraft turbine blades.

The industrial importance of the single-crystal orientation in nickel-turbine blades is mainly related to the blades superior thermo-mechanical fatigue and creep properties.

Dynamic deformation processes related to high-cycle fatigue (HCF)-induced failures in aircraft gas-turbine engines are a serious concern for safety purposes. Further, it is evident in the literature that turbine blades are the components most likely to fail by this dynamic deformation process. This implies that there is a relationship between a stereographic projection and fatigue crystallographic orientation of tensile axis, which is the most dangerous direction that opens a crack. Therefore, knowing the crystal orientation in a single crystal and the direction of an applied load is important for avoiding fatigue failure.

2.11 Summary

The characterization of crystal structures using Laue diffraction, conventional X-ray, or electron diffraction is essentially related to the crystallography of an atomic arrangement of atoms. In essence, revealing the crystal structure of a material leads to a clear understanding of the deformation process of materials subjected to external forces. Thus, the mechanical behavior and properties of solid materials being subjected to external loads are dependent on the crystallography of crystalline materials and are affected by solid surface defects and the aggressiveness of the environment.

Crystallography is an experimental science, and it characterizes crystal structures using X-ray diffraction and vector algebra in orthogonal and non-orthogonal coordinate systems. A description of crystallography as a solid surface science implies that the crystal structure can be characterized using microscopy, X-ray diffraction, and stereographic projection.

X-ray diffraction is briefly described in order to elucidate how this technique provides means to reveal the atomic structure of crystalline solids. Hence, crystals subjected to static and dynamic forces, surface reactions, or a combination of these responds in a particular manner depends on the type and orientation of unit cells. Therefore, it is important to determine the type of crystal structure of a material prior to an application in a suitable environment. However, a crystalline material may be modeled as a defect-free structure, but this would be unrealistic because defects have a high degree of control on properties.

The general description of crystal structures plays a significant role on mechanical behavior, film formation, and corrosion. Thus, it is very important to characterize the crystallography of a crystalline materials prior to a suitable engineering application where the crystal orientation is a key factor for the structural integrity of a component during a design lifetime.

Appendix 2A Ewald Sphere Crystallography

The Ewald sphere is also known as the sphere of reflection with radius $1/\lambda$ passing through the origin "O" of the reciprocal lattice, where λ is the wavelength of the X-ray incident beam. For convenience, Fig. 2.17 shows the Ewald circle of radius $1/\lambda$ and the reciprocal lattice vector $\mathbf{CP} = \mathbf{g}_{hkl}$.

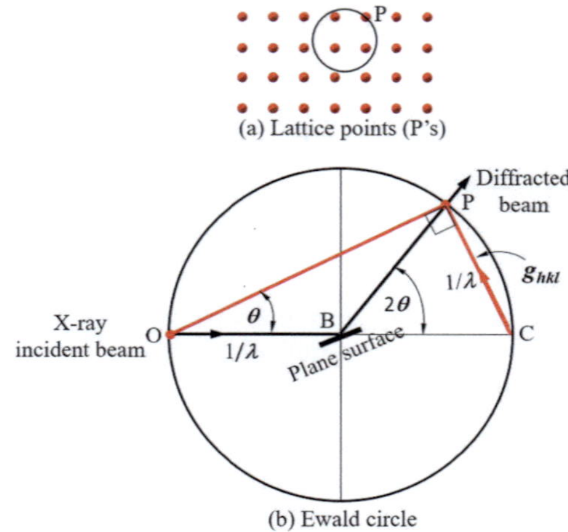

Fig. 2.17 Stereographic projection of an unknown crystal. (**a**) Laue X-ray diffraction pattern (Laue diagram). (**b**) image superimposed on the Wulff net for easy measurement of angles between poles

This circle is constructed such that the crystal plane is located at the center of the circle. Then,

- Draw a circle of radius $OB = 1/\lambda$
- The circumference must intercept a lattice point "P"
- Draw the BP line to define 2θ
- Draw the OP line to define θ
- Draw the CP line to define $\|g_{hkl}\| = 1/d_{hkl}$

Problems

2.1 (**a**) Derive the two-dimensional rotation matrix, Eq. (2.27), for the rotated coordinate system shown below, and (**b**) determine the reducible matrix for twofold $(2\pi/n = 2\pi/2 = \pi)$ and threefold $(2\pi/n = 2\pi/3)$ rotations and the eigenvalues λ. [Solution: (b) $\lambda = 0.36603$ for 3-fold rotation].

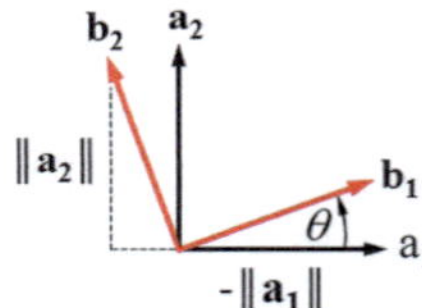

2.2 (**a**) Find the vector equations for the translations of the hexagonal space lattice in orthogonal Cartesian $(\mathbf{x}, \mathbf{y}, \mathbf{z})$ unit vectors. The primitive lattice is the small oblique parallelogram (red color) shown below. (**b**) Calculate the area (A) of the parallelogram with a lattice parameter of graphite $a = 25\ nm$. [Solution: (b) $A = 0.054\ nm^2$].

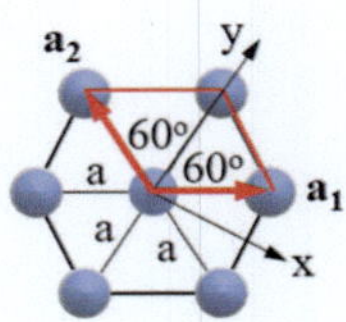

2.3 **(a)** Determine if the vectors $\mathbf{a_1} = 2i - j + 3k$ and $\mathbf{b_1} = 4i - j + 3k$ are perpendicular, parallel, or neither. Calculate the angle θ. **(b)** Derive the vector equation corresponding to the cross-product $\mathbf{a_1} \times \mathbf{b_1}$ using the given vectors. [Solution: $\theta = 19.36°$].

2.4 Using the dot product on the general unit cell shown in the sketch below, **(a)** derive a 3×3 matrix for the general tensor r_{ij} defined by

$$r_{ij} = \mathbf{a}_i \cdot \mathbf{a}_j = \|\mathbf{a}_i\|\,\|\mathbf{a}_j\| \cos\theta$$

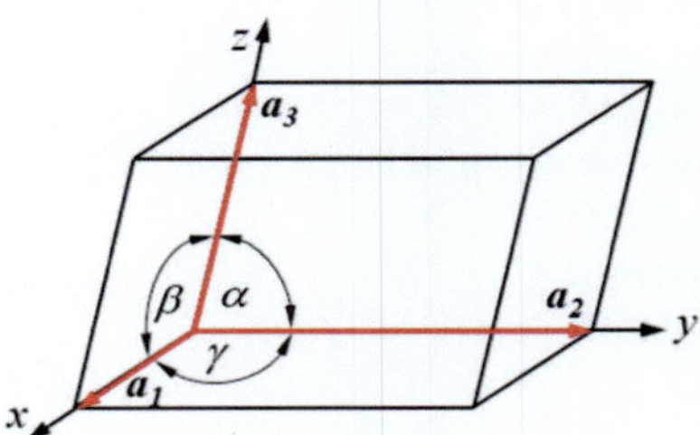

   (b) Find $\det r_{ij}$ for a BCC crystal if the lattice parameters are $a_1 = a_2 = a_3 = a = 1/3\ nm$. [Solution: $\det r_{ij} = 1.3717 \times 10^{-3}\ nm^2$].

2.5 Consider a hexagonal crystal structure described by the lattice vectors given below.

$$\mathbf{a_1} = a\mathbf{x} = a\,(1, 0, 0)$$

$$\mathbf{a_2} = \frac{a}{2}\mathbf{x} + \frac{\sqrt{3}a}{2}\mathbf{y} = a\left(\frac{1}{2}, \frac{\sqrt{3}}{2}, 0\right)$$

$$\mathbf{a_3} = c\mathbf{z} = c\,(0, 0, 1)$$

   Determine **(a)** the reciprocal lattice vectors $\mathbf{b_1}, \mathbf{b_2}, \mathbf{b_3}$ and **(b)** $\theta = \angle\,(\mathbf{a_1}, \mathbf{b_1})$. [Solution: $\theta = 30°$].

2.6 Calculate the angle θ between [110] and [111] directions (vectors) in a cubic crystal using the dot product. Recall that a cubic crystal has orthogonal x, y, z axes. [Solution: $\theta = 35.26°$].

2.7 Combine Eqs. (2.40) and (2.30) to derive the d_{hkl} expression for a simple cubic (SC).

2.8 Calculate length of $\|\mathbf{b}_i\|$, $\|\mathbf{g}_{hkl}\|$ and d_{hkl} associated with the $(hkl) = (200)$ plane in a cubic structure. Assume a lattice constant of $a = 0.287\ nm$. [Solution: $d_{111} = 0.166\ nm$].

2.9 It is well-known that diffraction of X-rays is described by the Bragg equation and that the reciprocal-lattice scattering vector $\mathbf{g}_{hkl}$ is related to the inverse of the interplanar spacing d_{hkl}. Using this information, calculate the diffraction angle θ for (110), (200), and (211) planes shown in Fig. 2.2. Determine the order of reflection taking into account if the Miller indices (hkl) are reducible or not. Let the lattice parameter and the electromagnetic radiation wavelength be $a = 0.287\ nm$ and $\lambda = 0.1527\ nm$, respectively. [solution: $\|\mathbf{g}_{110}\| \perp (110)\quad n = 1$, first order]

2.10 Consider an orthogonal coordinate system containing the (hkl) plane described by the vectors $\mathbf{a}_1$, $\mathbf{a}_2$ given below.

$$\mathbf{a}_1 = a\mathbf{x} = a\,(1, 0)\,; \qquad\qquad \mathbf{b}_1 = p_1\mathbf{x} + q_1\mathbf{y}$$

$$\mathbf{a}_2 = \frac{a}{2}\mathbf{x} + \frac{\sqrt{3}a}{2}\mathbf{y} = a\left(\frac{1}{2}, \frac{\sqrt{3}}{2}\right)\,; \quad \mathbf{b}_2 = p_2\mathbf{x} + q_2\mathbf{y}$$

(a) Find the coefficients p_i and q_i for the primitive reciprocal lattice vectors b_1, b_2, and **(b)** the angle $\theta = \sphericalangle\,(\mathbf{a}_1, \mathbf{b}_1)$. [Solution: (a) $p_1 = 2\pi/a$, $p_2 = 0$, $q_1 = -2\pi/\left(\sqrt{3}a\right)$, $q_2 = 4\pi/\left(\sqrt{3}a\right)$ and (b) $\theta = 30°$].

2.11 (a) Show that g_{hkl} is perpendicular, $g_{hkl} \perp (hkl)$, to nonparallel vectors $\mathbf{A}, \mathbf{B}, \mathbf{C}$ as shown in Fig. 2.9a. This implies that g_{hkl} and (hkl) are orthogonal so that the angle between them is $90° = \pi/2$ radians. This work requires that vectors $\mathbf{A}, \mathbf{B}, \mathbf{C}$ on the (hkl) plane edges be defined in terms of the intercept points. Subsequently, take the dot product $\mathbf{g}_{hkl} \cdot \mathbf{A}$, $\mathbf{g}_{hkl} \cdot \mathbf{B}$ and $\mathbf{g}_{hkl} \cdot \mathbf{C}$. **(b)** Derive a general expression for d_{hkl} for a cubic crystal, and **(c)** calculate values of $\|\mathbf{b}_i\|$, $\|\mathbf{g}_{hkl}\|$, and d_{hkl} for the three (hkl) planes shown in Fig. 2.2 if the lattice constant is $a = 0.287\ nm$. [Solution: (c) $\|\mathbf{b}_i\| = 21.893\ nm^{-1}$, $\|\mathbf{g}_{110}\| = 30.961\ nm^{-1}$, $d_{110} = 0.203\ nm$].

2.12 (a) Combine the Bragg equation and the reciprocal-lattice scattering vector $\mathbf{g}_{hkl}$ to calculate the diffraction angles for the (hkl) planes shown in Fig. 1.3. **(b)** Compare the results with those given in Example 1.4. Is $\mathbf{g}_{hkl}$ comparable with Bragg equation? **(c)** Deduce the order of diffraction n for these planes by reducing the Miller indices to their smallest integer values. [Solution: (a) $\theta_{110} = 22.10°$, (c) $n = 1$ for $(hkl) = (110)$].

2.13 For a simple cubic, the reciprocal lattice vectors are defined by

$$\mathbf{b}_1 = \frac{2\pi}{a}\,\vec{x}, \quad \mathbf{b}_2 = \frac{2\pi}{a}\,\vec{y}, \quad \mathbf{b}_3 = \frac{2\pi}{a}\,\vec{z}$$

(a) Derive an equation for the interplanar spacing d_{hkl} and **(b)** show that

$$\mathbf{b}_1 \cdot (\mathbf{b}_2 \times \mathbf{b}_3) = \frac{(2\pi)^3}{\mathbf{a}_1 \cdot (\mathbf{a}_2 \times \mathbf{a}_3)}$$

2.14 Prove that $\exp(i\mathbf{g}\cdot\mathbf{r}) = e^{i\mathbf{g}\cdot\mathbf{r}} = 1$ and n is an integer number.

2.15 Use the triangular (hkl) plane to show that $\mathbf{g}_{hkl} \perp (hkl)$.

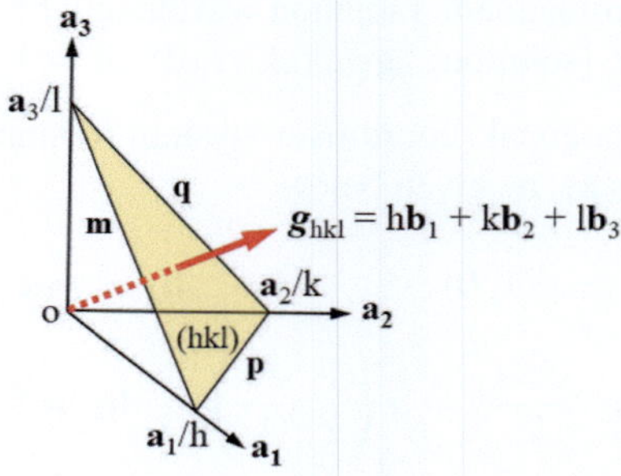

2.16 Use trigonometric identities as per the given figure to derive the cross product $\mathbf{p} \times \mathbf{q}$ equation, where

$$\mathbf{p} \times \mathbf{q} = \begin{bmatrix} p_x & p_y \\ q_x & q_y \end{bmatrix} \mathbf{k}$$

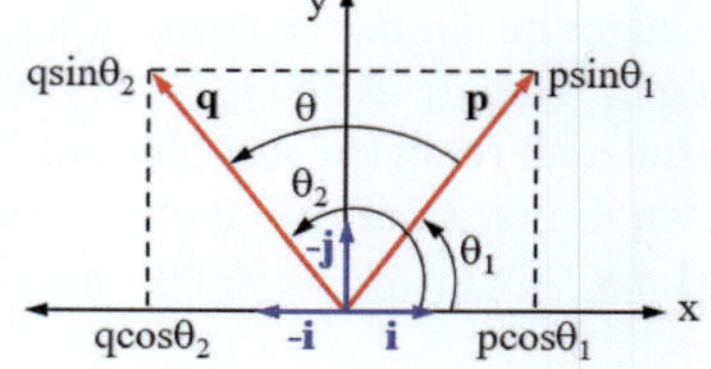

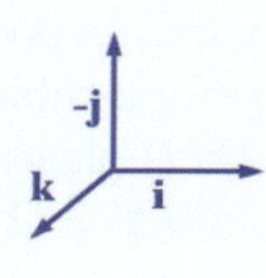

2.17 Let $\mathbf{p} = 2\mathbf{i} - 4\mathbf{j}$, $\mathbf{p} = n\mathbf{i} + 4\mathbf{j}$ and $\mathbf{p} \times \mathbf{q} = 40\mathbf{k}$. Find n. [Solution: $n = 8$].

2.18 Let $\mathbf{A} = \mathbf{i} - 3\mathbf{j}$, $\mathbf{B} = n\mathbf{i} + 4\mathbf{j}$ and $\mathbf{A} \times \mathbf{B} = 16\mathbf{k}$. Find n. [Solution: $n = 4$].

2.19 Consider a crystalline cubic material being analyzed by X-ray diffraction using K_α-radiation with a wavelength $\lambda = 0.15nm$. Determine the Miller indices of the (hkl) plane for an X-ray peak that obeys the Bragg's law at $\theta = 32.2°$. The lattice parameter is $a = 0.29\ nm$, and the order of reflection is $n = 1$. [Solution: $(hkl) = (200)$].

2.20 Find the angle $\theta = \theta_q$ for the pole "q" by inspection, dot product, and [100]-[010]-$[uvw]_p$ triangle for the (100) stereographic projection of a cubic crystal. The corresponding stereographic projection is given in Example 2.7. [Solution: $\theta = 45°$].

References

1. R.C. Alkire, R.D. Braatz, Electrochemical engineering in an age of discovery and innovation. AIChE J. **50**(9), 2000–2007 (2004)
2. B.D. Cullity, *Elements of X-Ray Diffraction*, 2nd edn. (Addison-Wesley Publishing Company, New York, 1978). ISBN 0-201-01174-3
3. W.D. Callister, Jr., D.G. Rethwisch, *Materials Science and Engineering: An Introduction*, 8th edn. (Wiley, New York, 2010)
4. A. Kelly, G.W. Groves, P. Kidd, *Crystallography and Crystal Defects*, Revised edition (Wiley, New York, 2000)
5. K. Hermann, *Crystallography and Surface Structure: An Introduction for Surface Scientists and Nanoscientists* (Wiley-VCH, Berlin, 2011)
6. A. Kelly, K.M. Knowles, *Crystallography and Crystal Defects*, 2nd edn. (Wiley, West Sussex, 2012)
7. Y. Waseda, E. Matsubara, K. Shinoda, *X-Ray Diffraction Crystallography: Introduction, Examples and Solved Problems* (Springer, Berlin, 2011)
8. U. Shmueli, Reciprocal space in crystallography, in *International Tables for Crystallography*, ed by U. Shmueli, Vol. B, 2nd edn., International Union of Crystallography 2001 (Published by Kluwer Academic Publishers, The Netherlands, 2006)
9. G. Bricogne, Fourier transforms in crystallography: Theory, algorithms and applications, in *International Tables for Crystallography*, ed. by U. Shmueli, Vol. B, 2nd edn. International Union of Crystallography 2001 (Published by Kluwer Academic Publishers, The Netherlands, 2006)
10. C. Kittel, *Introduction to Solid State Physics*, 8th edn. (Wiley, New York, 2005)
11. G.C. Wang, T.M. Lu, *RHEED Transmission Mode and Pole Figures: Thin Film and Nanostructure Texture Analysis* (Springer Science+Business Media, New York, 2014)
12. Mark Kucera, Website https://www.mindat.org/min-1651.html
13. S. Weber, (July 1997, update 10/27/2001) https://www.xtal.iqfr.csic.es/Cristalografia/parte_06-en.html

Chapter 3
Basic Quantum Mechanics

3.1 Introduction

This chapter introduces key concepts and basic aspects of quantum mechanics (QM) for describing the essential part of materials science related to atoms and their subatomic particles, such as electrons (e^-), protons (e^+), neutrons, and photons (packet of energy in the form of light). In reality, QM is a useful mathematical tool that helps understand the dynamic behavior of atomic particles.

The mathematical description of the particle motion incorporates quantization of energy, wave-particle duality, and the uncertainty principle. Regarding particle motion, the deterministic classical mechanics describes the particle at a macroscale, while the probabilistic quantum mechanics treats the particle at an atomic scale.

The main emphasis in the chapter is to describe the abstract behavior of a wavelike particle in a confined volume of a metal specimen using Schrodinger equation that describes, in particular, the three-dimensional (3D) time-dependent wavefunction $\Psi(r, t)$ with spatial (space $r = x, y, z$) and temporal (time t) variables. Subsequently, the 3D framework of $\Psi(r, t)$ is simplified to a one-dimensional (1D) time-independent wavefunction $\psi(x)$. This is important for understanding the concept of particle-in-a-box and wave-particle duality since electrons in solids behave as particles and waves.

The Heisenberg uncertainty principle (HUP) is also described as part of quantum mechanics related to the impossible simultaneous measurements of two variables, such as position and velocity, of a traveling particle in a quantum mechanical system.

This chapter includes basic concepts and mathematical tools of quantum mechanics needed for the reader to understand the behavior of materials at an atomic scale. Thus, the science of atomic and sub-atomic particles is presented hereafter in a simple form. In this context, the Schrodinger equation provides a framework for analyzing particle motion as a wavelike behavior.

© The Author(s), under exclusive license to Springer Nature Switzerland AG 2024
N. Perez, *Materials Science: Theory and Engineering*,
https://doi.org/10.1007/978-3-031-57152-7_3

3.2   Development of Quantum Mechanics

Historically, Max Planck in 1900 published his groundbreaking article on blackbody radiation, Niels Bohr in 1913 reported his atomic model related to atoms having electron orbitals (quantum state of probabilities), and Einstein in 1905 published his work on photoelectric effect related to light as quanta. These groundbreaking publications made them recipients of the Nobel Prize and the founders of quantum mechanics (quantum physics). In addition, Erwin Schrodinger in 1926 contributed a very significant aspect to probabilistic quantum theory; the quantum mechanical model that treats atomic particles (electrons) as wavelike matter having a oscillatory behavior described by a periodic quantum wavefunction $\Psi(r, t)$, $\Psi(x, t)$ or $\psi(x)$. Moreover, Paul Dirac in 1933 also contributed to the early development of quantum mechanics and shared the 1933 Nobel Prize for physics with Erwin Schrodinger for improving the atomic theory.

Werner Heisenberg in 1927 proposed the uncertainty principle related to the standard deviation of measurements in particle position and momentum. Experimentally, the century-old wave theory of light introduced by Thomas Young in 1801 is nowadays known as the double-slit experiment used to capture the interference effects caused by overlapping light waves at a distant screen.

3.3   The Wavefunction

For comparison purposes, Fig. 3.1a illustrates an example of Newton's deterministic classical mechanics for describing the motion of a baseball as a macroscopic particle.

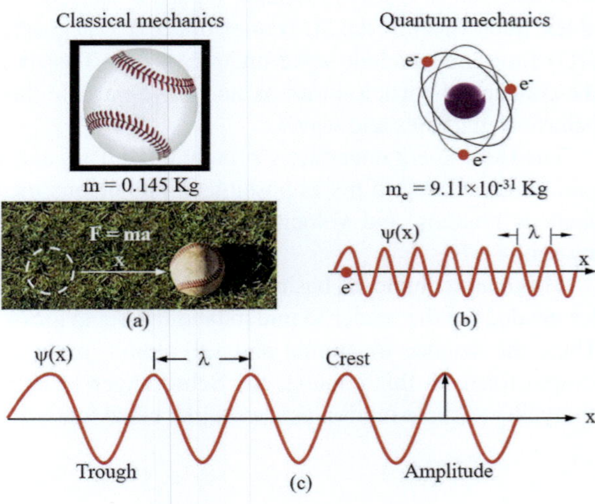

Fig. 3.1 Particle motion. (**a**) Classical mechanics, (**b**) quantum mechanics, and (**c**) nomenclature

A macroscopic object is a relativistic heavy particle (baseball). On the other hand, Fig. 3.1b schematically shows a probabilistic quantum mechanics for describing the motion of an electron as an atomic-size non-relativistic near massless particle. For example, the mass of the baseball is approximately $m_b = 0.145\ Kg$, and that of the electron is $m_e = 9.11 \times 10^{-31}\ Kg$. This implies that $m_b >> m_e$ or $m_b = \left(1.5917 \times 10^{29}\right) m_e$ due to the large number of electrons in the baseball. Moreover, the fundamental nomenclature of a wave is shown in Fig. 3.1c.

Despite that classical mechanics assures the existence of a macro particle at a specific place and time (Fig. 3.1a), it breaks down at an atomic scale. Instead, quantum mechanics assures a probabilistic domain over the atomic size particles, such as electrons (Fig. 3.1b).

In quantum mechanics, the three-dimensional (3D) time-dependent wavefunction $\Psi(r, t)$ is used to define the probability of finding a moving particle along the $r = (x, y, z)$ path at time t. Normally, a one-dimension (1D) time-dependent wavefunction $\Psi(x, t)$ or one-dimensional time-independent wavefunction $\psi(x)$ simplifies the analytical procedure for finding the solution of a Schrodinger equation.

3.4 Fundamental Definitions

According to Einstein photoelectric effect disseminated in 1905, light or electromagnetic radiation consists of a beam of particles called photons traveling at the speed of light $c = 3 \times 10^8\ m/s$ in a vacuum.

In quantum mechanics, a traveling particle is treated as wavelike particle with a wavelength λ and velocity $\upsilon << c$. Usually, the particle is an electron free to move within a confined volume $V = dxdydz$ or line L_x. This is the concept of a particle-in-a-box traveling in a region of constant potential energy U (electrical, gravitational, etc.).

Next, the particle quantized energy (E), momentum (p), de Broglie wavelength (λ), wave-particle frequency (v), and speed of light or photon (c) are given as a group or set of equations since they are basic expressions covered in undergraduate physics courses.

$$E = \frac{p^2}{2m} + U = K_E + P_E \tag{3.1a}$$

$$p = m\upsilon = \hbar k = 2\pi h k \tag{3.1b}$$

$$\lambda = \frac{h}{p} = \frac{h}{m\upsilon} \quad \text{(de Broglie's equation)} \tag{3.1c}$$

$$\lambda = \frac{hc}{\Delta E} = \frac{h}{p} \quad \text{(Photon)} \tag{3.1d}$$

$$p = \frac{E}{c} = \frac{hv}{c} = \frac{h}{\lambda} \quad \text{(photon)} \tag{3.1e}$$

$$E_n = \left(\frac{\pi^2 \hbar^2}{2mL^2} \right) n^2 = \left(\frac{h^2}{8mL^2} \right) n^2 \quad \text{(Quantized energy)} \tag{3.1f}$$

$$E = hv \quad \text{(Planck's equation)} \tag{3.1g}$$

$$E = mc^2 \quad \text{(Einstein's equation)} \tag{3.1h}$$

where $K_E = p^2/2m$ is the kinetic energy and $P_E = U$ is the potential energy of the particle in a medium. Actually, U is defined in Chap. 1, Eq. (1.12a), as the binding energy between an atom-pair, and here U is just the potential energy of an electron.

Combining Eqs. (3.1a), (3.1b), and (3.1d) yields the general wavenumber k (magnitude of the wavevector $\vec{k}$) and the total particle energy E equations written as

$$k = \sqrt{\frac{2m(E - U)}{\hbar^2}} \tag{3.2a}$$

$$E = \hbar\omega = \frac{\hbar^2 k^2}{2m} + U \tag{3.2b}$$

where $h, \hbar =$ Planck's constants

$h = 6.63 \times 10^{-34} \, m^2.kg/s = 6.63 \times 10^{-34} \, J.s$
$h = 4.1357 \times 10^{-15} \, eV.s$
$\hbar = h/2\pi = 1.0546 \times 10^{-34} \, J.s =$ Reduced Planck's constant
$m =$ Particle mass
$U =$ Space-dependent potential energy
$n =$ Quantum number (integer number)
$\omega = 2\pi v =$ Angular frequency
$v =$ Frequency
$\upsilon =$ Particle speed
$L =$ Length of a line in which a particle moves

Similarly, the spatial and temporal dependencies of the wavefunction $\Psi(r, t)$ are analytically shown in the subsequent sections, which can be treated as a practical quantum mechanics approach based on the Schrodinger equations and related wavefunction $\Psi(r, t)$ (Griffiths and Schroeter [1, p. 14]), Flugge [2]).

3.5 Wave-Particle Duality

For comparative purposes, quantum mechanics describes the behavior of the smallest objects or particles, such as electrons, photons, and the like, and engineering relates to the largest things, such as bridge or a power plant, in our universe. If one is to build a computer chip using n-type and p-type semiconductors, one must understand how electrons behave under specific conditions. Therefore, one has to deal with quantum mechanics.

In quantum mechanics, the wave-particle duality means that a quantum entity may be treated as a particle and as a wavelike particle. The behavior of a particle is described by a probability wavefunction $\Psi(r, t)$, $\Psi(x, t)$, or $\psi(x)$. Thus, the interpretation of a wavefunction is that its squared absolute value $|\Psi(r, t)|^2$, $|\Psi(x, t)|^2$, or $|\psi(x)|^2$ represents the probability density of finding the quantum mechanical particle at a position $r = (x, y, z)$ or $r = x$ (Griffiths and Schroeter [1, p. 14]).

3.6 Particle-in-a-Box Model

The particle-in-a-box is also (known as the infinite energy well) may be defined as the space between two-bounded atoms in a crystal plane. For example, in the sp^2 hybridization of carbon atoms forming a honeycomb plane, which is the graphene structure where the delocalized π-electron (free electron) is the most susceptible one to be boxed in a confined space. Moreover, small element from a piece solid metal can also be modeled as a box for characterizing an electron motion. Other applications of the particle-in-a-box model are found in electronic devices, such as quantum well infrared photodetector and in quantum well laser. For clarity, the word "laser" is an acronym for "light amplification by stimulated emission of radiation," and it emits a narrow beam of light.

Consider a three-dimensional (3D) boundary box in a rectangular metal component acting as an infinite potential well for an inner particle (Fig. 3.2a). The behavior of a wavefunction $\Psi(r, t)$ or $\psi(r)$ describing the particle motion is shown in Fig. 3.2b. Assume that one particle (electron) is trapped in a confined 3D box with volume $V = dxdydz$ (Fig. 3.2c) and finite dimensions (L_x, L_y, L_z), where $r = x, y, z$ is the arbitrary direction of the particle motion.

Using creative imagination, envision an electron having a particular wavelength (λ), defined by de Broglie theory, move freely back and forth in a particular direction r within the 3D box. This implies that there is no force acting on the electron and that the inner potential energy is $U(r) = U(x) = 0$. At any rate, the quantum mechanical problem at hand is to find the proper wavefunction $\Psi(r, t)$ or $\psi(r)$ based on the quantum mechanical boundary conditions.

The box walls, in general, are the potential energy barriers for the electron to penetrate or tunnel through them. This means that the electron energy, which is mostly kinetic energy, is less than the wall potential energy. Therefore, an electron moves freely in the box since $U(x) = 0$, but it bounces back at the walls. This means that the main constraints for $\psi(x, y, z)$ within the box walls are $\psi(r = 0) = \psi(r = L_r) = 0$. For clarity, the potential energy inside the box is $U(x) = 0$ and outside the box is $U(x) >> \infty$.

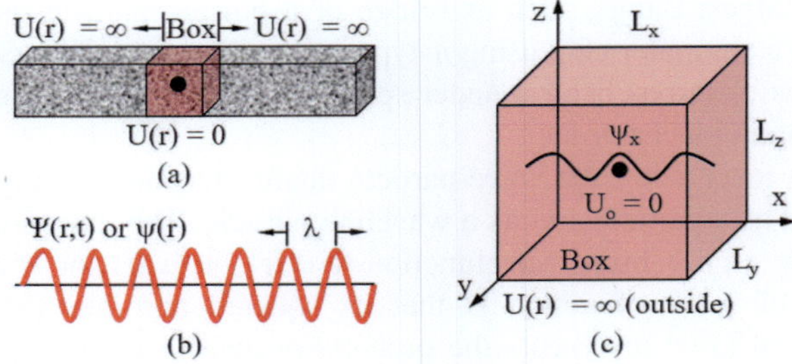

Fig. 3.2 Particle-in-a-box model. (**a**) A moving electron in a metal box with volume $V = dx\,dy\,dz$, (**b**) the electron wavefunction Ψ, and (**c**) trapped electron inside the box having finite dimensions $\left(L_x, L_y, L_z\right)$ and fixed potential energy; $U_o = 0$ inside the box and $U_o > 0$ out of the box. An electron travels in the x, y or z direction

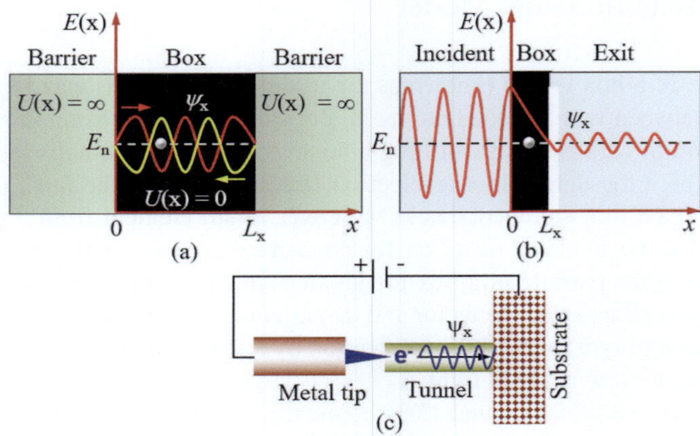

Fig. 3.3 A particle-in-a-box model for one-dimensional analysis (1D) using scanning tunneling microscopy (STM). (**a**) Trapped electron, (**b**) quantum tunneling through a metal oxide film, and (**c**) quantum tunneling in an electrochemical system

This model assumes that the electron can move freely (Fig. 3.2c) at a velocity $\upsilon < c$ since there is no force, $F(r) = -dU(r)/dr = 0$, acting upon it.

3.6.1 Infinite Potential Well

For one-dimensional analysis, the infinite potential-energy well shown in Fig. 3.3a is treated as a box containing one electron free to move within the boundaries denoted as $0 \leq x \leq L_x$. This is a one-dimensional confinement along the x-axis with the potential energy $U(x) = U(x)_{inside} = 0$ inside the box since there are no forces acting on the electron, which must have a finite kinetic energy (E_n) within the dimensional constraint $0 \leq x \leq L_x$. Physically, confinement is treated as a particle-in-a-box with thick enough walls that act as thick barriers, such as an atom, a piece of metal and an oxide film, for an electron to tunnel through it. Energetically, this implies that the electron cannot escape the box because its discrete energy level is $E_n << U(x) = U(x)_{outside} = \infty$ due to an electrical potential, gravitational potential, and the like.

Figure 3.3a schematically illustrates a sinusoidal behavior of the wavefunction $\psi(x) = \psi_a \sin(x)$ with a constant ψ_a for a wavelike particle, which cannot overcome the energy barrier $U(x) > 0$. The particle is free to move with $E_n > 0$ within the dimensional range $0 \leq x \leq L_x$, where $\psi(0) = \psi(L_x) = 0$ and at nodes.

If the barrier is thin enough, then a particle, such as an electron, may tunnel through it as illustrated in Fig. 3.3b. This event is known as a quantum tunneling phenomenon. Moreover, Fig. 3.3b suggests that an incident wavelike particle enters the box (tunnel) with an exponential decay behavior $\psi_x = \psi_b \exp(x)$ with constant ψ_b. After quantum tunneling, the wavelike particle shows a sinusoidal behavior at low energy. This is a quantum tunneling phenomenon associated with wave propagation through a potential barrier $E_n > U(x) >> 0$.

Quantum tunneling is a phenomenon that may also occur in scanning tunneling microscopy (STM) with a proper STM tip-substrate cell being electrically connected as shown in Fig. 3.3c. Electron tunneling is possible, provided that an applied bias voltage (ϕ_{bias}) or electric potential (voltage) and a specific tunneling current allow an electron with charge q_e to penetrate a nano tunnel (empty space) between electrodes (STM tip and STM substrate). As a result, electron tunneling occurs from the tip to a substrate unoccupied states (Kolasinski [3, p. 75]).

3.7 Time-Dependent Schrodinger Equation

Erwin Schrodinger in 1926 reported the mathematical model for describing the particle wavefunction $\Psi = \Psi(x, t)$, and Max Bohr in 1926 interpreted this wavefunction as the probability amplitude, which in turn defines the probability density function as $|\Psi|^2$ for determining the particle position. For the particle-in-a-box (Fig. 3.3a), the time-dependent boundary conditions are $\Psi(0, t) = \Psi(L_x, t) = 0$, whereas the probability is defined as $|\Psi|^2$.

The following analytical procedure is fundamentally based on a pure calculus approach, which can be found elsewhere (Wyatt and Dew-Hughes [4, pp. 75–82]). Consider the generalized wavefunction $\Psi(x, t)$ and corresponding derivatives

$$\Psi(x, t) = A \exp[i(kx - \omega t)] \tag{3.3a}$$

$$\frac{\partial \Psi(x, t)}{\partial t} = -i\omega A \exp[i(kx - \omega t)] = -i\omega \Psi(x, t) \tag{3.3b}$$

$$\frac{\partial \Psi(x, t)}{\partial x} = ikA \exp[i(kx - \omega t)] = ik\Psi(x, t) \tag{3.3c}$$

$$\frac{\partial^2 \Psi(x,t)}{\partial x^2} = (ik)^2 A \exp[i(kx - \omega t)] = -k^2 \Psi(x,t) \tag{3.3d}$$

where $i^{-1} = -i$ and

$$\omega = \frac{i}{\Psi(x,t)} \frac{\partial \Psi(x,t)}{\partial t} = 2\pi v \tag{3.4a}$$

$$k^2 = -\frac{1}{\Psi(x,t)} \frac{\partial^2 \Psi(x,t)}{\partial x^2} = \left(\frac{2\pi}{\lambda}\right)^2 \tag{3.4b}$$

Here, ω denotes the angular frequency, v denotes the frequency, k denotes the wavenumber, and λ denotes the wavelength.

The total energy E of a traveling particle is the sum of the kinetic energy $K_E = p^2/2m$ and the time-dependent potential energy $P_E = U(x,t)$. Mathematically, E, along with the particle momentum (p) and the wavenumber (k) equation, is

$$E = K_E + P_E = \frac{p^2}{2m} + U(x,t) \tag{3.5a}$$

$$E = hv = \frac{h\omega}{2\pi} \tag{3.5b}$$

$$p = \frac{h}{\lambda} = \frac{hk}{2\pi} \tag{3.5c}$$

Notice that E is governed by the magnitude of the $U(x)$ field in a quantum mechanical system. Now, Equating Eqs. (3.5a) and (3.5b) gives

$$\frac{h\omega}{2\pi} = \frac{p^2}{2m} + U(x,t) \tag{3.6a}$$

$$\frac{h\omega}{2\pi} = \frac{h^2 k^2}{8\pi^2 m} + U(x,t) \tag{3.6b}$$

Substituting Eqs. (3.4a), (3.4b) into (3.6b) along with the reduced Planck's constant $\hbar = h/2\pi$ yields the one-dimensional time-dependent Schrodinger equation,

$$\frac{ih}{2\pi} \frac{\partial \Psi(x,t)}{\partial t} = -\frac{h^2}{8\pi^2 m} \frac{\partial^2 \Psi(x,t)}{\partial x^2} + U(x,t)\,\Psi(x,t) \tag{3.7}$$

with boundary conditions

$$\Psi(0,t) = \Psi(L_x,t) = 0 \quad \text{(inside the box)} \tag{3.8a}$$

$$\Psi(x,t) = 0 \quad \text{(outside the box)} \tag{3.8b}$$

$$U(x,t) = \left\{ \begin{array}{ll} 0 & \text{at } 0 \le x \le L_x \quad \text{(inside the box)} \\ \infty & \qquad\qquad\qquad \text{(outside the box)} \end{array} \right\} \tag{3.8c}$$

Here, $r = (x, y, z)$, $\nabla (r) = \partial^2/\partial x^2 + \partial^2/\partial y^2 + \partial^2/\partial z^2$ is the differential operator (Laplacian operator) acting on $\Psi (r, t)$, $i = \sqrt{-1}$ is the imaginary number being used due to Euler's formulation, and in the end, only the real part of a complex formula makes physical sense. Moreover, Eq. (3.7) predicts stationary quantum states and serves as the foundation for deriving the time-independent Schrodinger equation for a single non-relativistic particle referred to as an electron (Pettifor [5, p. 29] and Baym [6, p. 52]).

Notice that the Schrodinger equation is a mathematical model that defines a linear second-order partial differential equation (PDE) related to energy and space-time. It governs the wavefunction $\Psi = \Psi (x, t)$ of a quantum mechanical system, such as a particle-in-a-box.

3.7.1 *Probability Density*

The particle-in-a-box can be modeled as a confined three-dimensional (3D) space with volume $V = dxdydz$ or as a confined one-dimensional line (1D) $0 \leq x \leq L_x$ at time t. Integrating the inner product $|\Psi|^2 = \Psi^*\Psi$ defines the probability density $P (r)$, where Ψ^* is the complex conjugate of Ψ. Thus, the probability density $P (r)$ for a 3D case and $P (x)$ for 1D analysis are defined as

$$P (r) = \int_{-\infty}^{+\infty} |\Psi (r, t)|^2 \, dxdydz = 1 \quad \text{(3D)} \tag{3.9a}$$

$$P (x) = \int_{-\infty}^{+\infty} |\Psi (x, t)|^2 \, dx = 1 \quad \text{(1D)} \tag{3.9b}$$

At present, $\Psi = \Psi (r, t)$ is considered to be normalized, provided that its inner product $|\Psi|^2$ with itself is equal to one (1) as shown by Eqs. (3.9a) and (3.9b).

For an isolated particle between 0 and ∞ points along the L_x line in the x-direction, $P (x)$ and the expectation value $\langle x \rangle$ of x are, respectively,

$$P (x) = \int_0^{\infty} |\psi (x)|^2 \, dx < 1 \tag{3.10a}$$

$$\langle x \rangle = \int_0^{\infty} x \psi (x) \, \psi^* (x) \, dx \tag{3.10b}$$

$$\langle x \rangle = \int_0^{\infty} x \, |\psi (x)|^2 \, dx \tag{3.10c}$$

Apparently, $P (x)$ in Eq. (3.10a) is a Hilbert-space condition, and $\langle x \rangle$ is the most probable value of the electron position since the particle is assumed to be in motion at all times. Moreover, a Hilbert space is a linear vector space with an infinite dimensional inner product space, where $\langle \psi, \psi \rangle = \int \psi^* \psi = \int |\psi|^2$. Conversely, Euclidean space refers to a finite dimensional linear space with an inner product.

The mathematical model describing the particle motion in the potential well is the Schrodinger wavefunction $\psi (x)$, which leads to the probability density function $p (x) = \psi^* (x) \psi (x)$ related to the likelihood of finding the particle at a position $0 \leq x \leq L_x$. In addition, the $\psi^* (x)$ function is the conjugate of $\psi (x)$.

The boundary conditions are $p (x) = 0$ and $\psi (x) = 0$ in the regions with $U (x) = \infty$. This means the $\psi (x) = 0$ at $x \leq 0$ and $x \geq L_x$. However, $\psi (x) \neq 0$ and $p (x) \neq 0$ at $0 < x < L_x$, which suggests that a particle is free to move in this interval and the probability density $P(x)$ is bounded at $0 \leq x \leq 1$.

3.8 Time-Independent Schrodinger Equation

Assume that the potential energy $U (x, t)$ becomes a time-independent variable $U (x)$. In this case, the separation of variables method defines the wavefunction as a product of two eigenfunctions; $\Psi (x, t) = \psi (x) \phi (t)$ (Flugge [2, p. 26]). For the purpose of clarity, an eigenfunction is the solution of a differential equation satisfying specified original and boundary conditions.

Substituting $\Psi (x, t) = \psi (x) \phi (t)$ into Eq. (3.7) changes the partial differential equation (PDE) into an ordinary differential equation (ODE). Rearranging the resultant expression yields the sought ODE

$$i\hbar \frac{d\psi (x) \phi (t)}{dt} = -\frac{\hbar^2}{2m} \frac{d^2\psi (x) \phi (t)}{dx^2} + U (x) \psi (x) \phi (t) \tag{3.11a}$$

$$i\hbar\psi (x) \frac{d\phi (t)}{dt} = -\frac{\hbar^2\phi (t)}{2m} \frac{d^2\psi (x)}{dx^2} + U (x) \psi (x) \phi (t) \tag{3.11b}$$

Divide Eq. (3.11b) by $\psi (x) \phi (t)$ and separate terms to get the time-independent Schrodinger equation

$$\frac{i\hbar}{\phi (t)} \frac{d\phi (t)}{dt} = -\frac{\hbar^2}{2m\psi (x)} \frac{d^2\psi (x)}{dx^2} + U (x) = E \tag{3.12a}$$

$$\frac{ih}{2\pi} \frac{1}{\phi (t)} \frac{d\phi (t)}{dt} = -\frac{h^2}{8\pi^2 m} \frac{1}{\psi (x)} \frac{d^2\psi (x)}{dx^2} + U (x) = E \tag{3.12b}$$

where E is the total particle energy corresponding to a particular particle quantum state. From, Eq. (3.11),

$$E = -\frac{\hbar^2}{2m\psi (x)} \frac{d^2\psi (x)}{dx^2} + U (x) \tag{3.13a}$$

$$E = \frac{i\hbar}{\phi(t)} \frac{d\phi(t)}{dt} \tag{3.13b}$$

The spatial dependency of the wavefunction $\Psi(x, t)$ can be derived by rearranging Eq. (3.13a) as homogeneous PDE

$$\frac{d^2\psi(x)}{dx^2} + \frac{2m}{\hbar^2}[E - U(x)]\psi(x) = 0 \tag{3.14}$$

from which

$$-\frac{\hbar^2}{2m}\frac{d^2\psi(x)}{dx^2} + U(x)\psi(x) = E\psi(x) \tag{3.15a}$$

$$K_E = -\frac{\hbar^2}{2m}\frac{d^2\psi(x)}{dx^2} \tag{3.15b}$$

$$P_E = U(x) \tag{3.15c}$$

$$\hat{H} = \left[-\frac{\hbar^2}{2m}\frac{d^2\psi(x)}{dx^2} + U(x)\right] \tag{3.15d}$$

The time-independent Schrodinger equation takes the general form

$$\hat{H}\psi(x) = E\psi(x) \tag{3.16}$$

The square bracket in Eq. (3.15d) is the Hamiltonian operator $\hat{H}$ defined as the sum of kinetic and potential energies, $\hat{H} = K_E + P_E$. Hence, $\hat{H}$ is an energy operator acting on $\psi(x)$. Here, Eq. (3.15a) is a quantum mechanics eigenvalue problem, where the wavefunction $\psi(x)$ is the **eigenfunction** of a Hamiltonian operator $\hat{H}$ and the total energy E of the particle is an **eigenvalue** of the eigenfunction $\psi(x)$. This means that the operator $\hat{H}$ acting on the wavefunction $\psi(x)$ produces an eigenvalue E times the wavefunction $\psi(x)$ as defined by Eq. (3.16).

The temporal dependency of the wavefunction $\Psi(x, t)$ is evaluated from Eq. (3.13b)

$$\frac{d\phi(t)}{\phi(t)} = -\frac{iE}{\hbar}dt \tag{3.17a}$$

$$\int \frac{d\phi(t)}{\phi(t)} = -\frac{iE}{\hbar}\int dt \tag{3.17b}$$

$$\ln[\phi(t)] = -\left(\frac{iE}{\hbar}\right)t + C \tag{3.17c}$$

$$\phi(t) = \exp\left[-\left(\frac{iE}{\hbar}\right)t + C\right] = \exp(C)\exp\left[-\left(\frac{iE}{\hbar}\right)t\right] \tag{3.17d}$$

Let $A = \exp(C)$ in Eq. (3.17d) so that

$$\phi(t) = A \exp\left[-\left(\frac{iE}{\hbar}\right)t\right] \tag{3.18}$$

which represents the time-dependent equation in a complex plane. Thus, the wavefunction takes the following mathematical form:

$$\Psi(x,t) = \psi(x)\phi(t) = \psi(x)A\exp\left(-\frac{iEt}{\hbar}\right) \tag{3.19}$$

with constant A, imaginary number $i = \sqrt{-1}$, and unknown eigenfunction $\psi(x)$. According to Euler's formula, Eq. (3.19) can be written as

$$\Psi(x,t) = \psi(x)A\left[\cos(Et/\hbar) - i\sin(Et/\hbar)\right] \tag{3.20}$$

from which the real part takes the form

$$\Psi(x,t) = \psi(x)A\cos(Et/\hbar) \tag{3.21}$$

Notice that $\psi(x)$ is still an unknown eigenfunction, but it is a partial solution of $\Psi(x,t)$. Moreover, the example given below illustrates the procedure for deriving an expression for the time-independent wavefunction $\psi(x)$.

Example 3.1 Use the one-dimensional time-independent Schrodinger equation given as a homogeneous second-order partial differential equation with $U(x) = 0$ inside an infinite potential well

$$-\frac{\hbar^2}{2m}\frac{\partial^2\psi(x)}{\partial x^2} = E\psi(x) \tag{3.1E1a}$$

$$\frac{\partial^2\psi(x)}{\partial x^2} + k^2\psi(x) = 0 \tag{3.1E1b}$$

with the wavenumber k defined as

$$k = \frac{\sqrt{2mE}}{\hbar} \tag{3.1E2a}$$

$$k^2 = \frac{2mE}{\hbar^2} \tag{3.1E2b}$$

The general solution of Eq. (3.1E1a) for $0 \leq x \leq L_x$ is written as

$$\psi(x) = A\sin(kx) + B\cos(kx) \tag{3.1E3}$$

Derive an expression for the constant A.

Solution Evaluating Eq. (3.1E3) using the boundary conditions, $\psi\left(0\right) = \psi\left(L_x\right) = 0$, yields

$$\psi\left(0\right) = A\sin\left(0\right) + B\cos\left(0\right) = 0$$
$$B = 0$$

and

$$\psi\left(L_x\right) = A\sin\left(kx\right) + B\cos\left(0\right) = 0 \tag{3.1E4a}$$
$$\psi\left(L_x\right) = A\sin\left(kx\right) + 0 = 0 \tag{3.1E4b}$$
$$\sin\left(kL_x\right) = \sin\left(n\pi\right) = 0 \tag{3.1E4c}$$
$$kL_x = n\pi \tag{3.1E4d}$$
$$k = \frac{n\pi}{L_x} \tag{3.1E4e}$$

Thus, the eigenfunction is

$$\psi\left(x\right) = A\sin\left(\frac{n\pi x}{L_x}\right) \tag{3.1E5a}$$
$$\psi\left(x\right) = A\sin\left(kx\right) \tag{3.1E5b}$$

Combining Eqs. (3.1E2a) and (3.1E4e) with $E = E_n$ gives the eigenenergy E_n for the particle

$$k = \frac{\sqrt{2mE}}{\hbar} = \frac{n\pi}{L_x} \tag{3.1E6a}$$
$$E_n = \frac{1}{2m}\left(\frac{\pi\hbar}{L_x}\right)^2 n^2 = \frac{\hbar^2 k^2}{2m} \tag{3.1E6b}$$

The quantization of a particle is achieved because of the quantum number n. For a real $\psi\left(x\right)$ function, normalization of $\psi\left(x\right)$ at $0 \leq x \leq L_x$ implies that $\psi^*\left(x\right) = \psi\left(x\right)$ and $\psi^*\left(x\right)\psi\left(x\right) = \psi^2\left(x\right)$. Thus, the probability density along with Eq. (3.1E5b) becomes

$$p\left(x\right) = \int_0^{L_x} \psi^*\left(x\right)\psi\left(x\right)dx = \int_0^{L_x} \psi^2\left(x\right)dx = 1 \tag{3.1E7a}$$
$$p\left(x\right) = \int_0^{L_x} \left[A\sin\left(kx\right)\right]^2 dx = A^2 \int_0^{L_x} \sin^2\left(kx\right)dx = 1 \tag{3.1E7b}$$

Let $\theta = kx$ and use the trigonometric identity $\sin^2(\theta) = (1/2)[1 - \cos(2\theta)]$ so that

$$\int_0^{L_x} \sin^2(kx)\,dx = \int_0^{L_x} \frac{1 - \cos(2kx)}{2}\,dx \tag{3.1E8a}$$

$$\int_0^{L_x} \sin^2(kx)\,dx = \frac{L_x}{2} - \frac{\sin(2kL_x)}{4k} \tag{3.1E8b}$$

Combining Eqs. (3.1E7b) and (3.1E8b) yields

$$p(x) = A^2\left[\frac{L_x}{2} - \frac{\sin(2kL_x)}{4k}\right] = 1 \tag{3.1E9a}$$

$$p(x) = A^2\left[\frac{L_x}{2} - \frac{\sin(2n\pi)}{4k}\right] = 1 \tag{3.1E9b}$$

$$p(x) = A^2\left[\frac{L_x}{2} - 0\right] = 1 \tag{3.1E9c}$$

$$A = \sqrt{\frac{2}{L_x}} \tag{3.1E9d}$$

Let $\psi(x) = \psi_n(x)$ and $k = k_n$ due to the principal quantum number n so that Eq. (3.1E5b) becomes

$$\psi_n(x) = \sqrt{\frac{2}{L_x}}\sin(k_n x) \tag{3.1E10a}$$

$$\psi_n(x) = \sqrt{\frac{2}{L_x}}\sin\left(\frac{n\pi x}{L_x}\right) \tag{3.1E10b}$$

This completes the solution of the one-dimensional time-independent Schrodinger equation.

Example 3.2 Shows the remaining algebraic steps for completing the method of separation of variables by treating $U(x) = 0$ inside the box or $U(x) > 0$ outside the box. Derive a simplified expression for $\psi(x)$ needed in Eq. (3.19).

Solution From Eq. (3.14), the ODE is defined as

$$\frac{d^2\psi(x)}{dx^2} = \frac{2m}{\hbar^2}[U(x) - E]\psi(x) \tag{3.2E1a}$$

$$\frac{d^2\psi(x)}{dx^2} = k^2\psi(x) \tag{3.2E1b}$$

where the wavenumber is

$$k^2 = \frac{2m}{\hbar^2}\left[U\left(x\right) - E\right] \tag{3.2E2}$$

The general solution of Eq. (3.2E1b) for a particle tunneling through a barrier is

$$\psi\left(x\right) = A\exp\left(kx\right) + B\exp\left(-kx\right) \tag{3.2E3}$$

which must satisfy the requirement $\psi\left(x\right) \to 0$ as $x \to \infty$. Here, A, B are unknown constants. If $x = \infty$, then $\exp\left(\infty\right) = \infty$ and $\exp\left(-\infty\right) = 0$. This implies that $\psi\left(x\right) \to \infty$ as $\exp\left(\infty\right) = \infty$, which is not realistic at all. Therefore, $A\exp\left(kx\right)$ must vanish from Eq. (3.2E3) so that

$$\psi\left(x\right) = B\exp\left(-kx\right) \tag{3.2E4}$$

Proof Take the derivatives of Eq. (3.2E4) so that

$$\frac{d\psi\left(x\right)}{dx} = -kB\exp\left(-kx\right) \tag{3.2E5a}$$

$$\frac{d^2\psi\left(x\right)}{dx^2} = k^2B\exp\left(-kx\right) \tag{3.2E5b}$$

$$\frac{d^2\psi\left(x\right)}{dx^2} = k^2\psi\left(x\right) \tag{3.2E5c}$$

which is exactly equal to Eq. (3.2E1b). Recall that the boundary conditions are $\psi\left(x\right) = \psi\left(L_x\right) = 0$ inside the box with $U\left(x\right) = 0$ and $\psi\left(x\right) = 0$ at $x > L_x$ outside the box with $U\left(x\right) = \infty$.

Example 3.3 Consider the 0.173-*nm*-thick copper oxide (CuO) film shown below and a finite energy barrier of $U\left(x\right) = 4\ eV$. Use the particle-in-a-box model to compute **(a)** the quantum number n for one $E_n = 5\text{-}eV$ electron, **(b)** the wavenumber k, and **(c)** the electron wavelength λ. The electron rest mass and the reduced Planck's constant are $m = 9.11 \times 10^{-31}\ Kg$ and $\hbar = 1.05 \times 10^{-34}\ Js$, respectively.

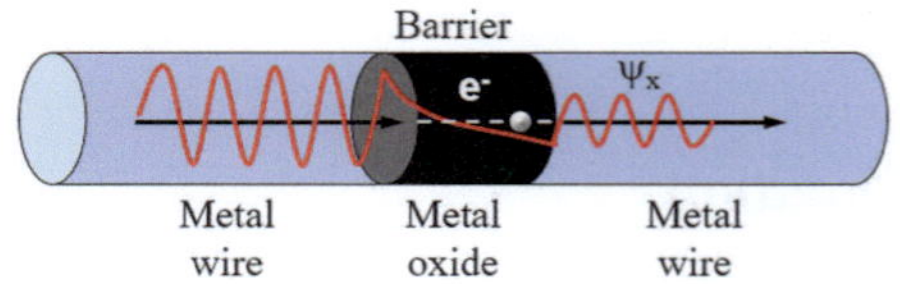

Solution

(a) Based on the given information, the finite potential energy well with $0 \le x \le L_x$ describes a one-dimensional particle-in-a-box model, where the length of the box is $L_x = 0.173\ nm$ and the electron quantized energy level is $E_n = 5\ eV$. From Eq. (3.1f), the quantum number n associated with E_n is

$$E_n = \left(\frac{\pi^2 \hbar^2}{2mL^2}\right) n^2 \tag{3.3E1a}$$

$$n = \left(\frac{L}{\hbar}\right) \sqrt{2m E_n} \tag{3.3E1b}$$

$$n = \left(\frac{0.173 \times 10^{-9}\ m}{1.05 \times 10^{-34}\ J.s}\right) \sqrt{2\left(9.11 \times 10^{-31}\ Kg\right)\left(5 \times 1.60 \times 10^{-19}\ J\right)} \tag{3.3E1c}$$

$$n = 2 \tag{3.3E1d}$$

The given quantized energy $E_2 = 5\ eV$ corresponds to the quantum number $n = 2$. This means that the electron moves along the L dimension within the box corresponding to the quantized energy level $E_2 = 5\ eV$ defined by the quantum number $n = 2$.

(b) From Eq. (3.2a) with $\hbar = 1.05 \times 10^{-34}\ J.s = 1.05 \times 10^{-34}\ \left(Kg.m^2/s^2\right) s$, the wavenumber $k = k_x$ is

$$k = \frac{1}{\hbar}\sqrt{2m\left(E - U\left(x\right)\right)} \tag{3.3E2a}$$

$$x = \sqrt{2\left(9.11 \times 10^{-31}\ Kg\right)\left(5 - 4\right)\left(1.60 \times 10^{-19}\ Kg.m^2/s^2\right)} \tag{3.3E2b}$$

$$k = \frac{x}{1.05 \times 10^{-34}\ \left(Kg.m^2/s^2\right) s} \tag{3.3E2c}$$

$$k = 5.15 \times 10^9\ m^{-1} = 5.15\ nm^{-1}\ \left(radians/nm\right) \tag{3.3E3d}$$

(c) The electron wavelength λ as per de Broglie expression, Eq. (3.3E1c), is

$$\lambda = \frac{2\pi}{k} = \frac{2\pi}{5.15\ nm^{-1}} = 1.22\ nm \tag{3.3E3}$$

which is close to a typical value of $1.23\ nm$ found in the literature.

3.9 The Wave Equation

For a wavelike particle traveling in the x-direction at a velocity υ, the one-dimensional time-dependent Schrodinger equation may be derived from the wave equation analogous to the heat equation (Pettifor [5, p. 29]). For clarity, the wave and Schrodinger equations have different dynamics, but the Schrodinger equation is considered as a fundamental equation analogous to Newton's second law in classical mechanics (Baym [6, pp. 50–53] and Phillips [7, p. 24]).

Recast Eq. (3.3a) and its second-order partial derivatives

$$\Psi\,(x, t) = A \exp\left[i\,(kx - \omega t)\right] \tag{3.22a}$$

$$\frac{\partial^2 \Psi\,(x, t)}{\partial t^2} = \omega^2 A \exp\left[i\,(kx - \omega t)\right] \tag{3.22b}$$

$$\frac{\partial^2 \Psi\,(x, t)}{\partial x^2} = k^2 A \exp\left[i\,(xk - \omega t)\right] \tag{3.22c}$$

Let

$$\frac{\partial^2 \Psi\,(x, t)}{\partial t^2} = C \frac{\partial^2 \Psi\,(x, t)}{\partial x^2} \tag{3.23}$$

where C is a constant. Substituting Eqs. (3.22b), (3.22c) into (3.23) gives

$$\omega^2 A \exp\left[i\,(kx - t\omega)\right] = C k^2 A \exp\left[i\,(xk - t\omega)\right] \tag{3.24a}$$

$$C = (\omega/k)^2 = v^2 \tag{3.24b}$$

Inserting Eq. (3.24b) into (3.23) yields the sought wave equation

$$\frac{\partial^2 \Psi\,(x, t)}{\partial t^2} = v^2 \frac{\partial^2 \Psi\,(x, t)}{\partial x^2} \tag{3.25}$$

This expression is used in Appendix 3A to derive the solution of a time-independent Schrodinger equation, Eq. (3.14), for a particle-in-a-box where $U\,(x) = 0$ at $0 \le x \le L_x$. Notice that this expression is a second-order partial differential equation (PDE) that describes the motion of a particle through a $\Psi(x, t)$ wave function.

Example 3.4   Verify that the partial derivatives $\partial \Psi\,(x, t)\,/\partial t$ and $\partial^2 \Psi\,(x, t)\,/\partial x^2$ of the complex wavefunction

$$\Psi\,(x, t) = A \exp\left[i\,(kx - \omega t)\right] \tag{3.4E1}$$

lead to Eq. (3.2b).

Solution   The derivatives of Eq. (3.4E1) are

$$\frac{\partial \Psi\,(x, t)}{\partial t} = -i\omega A \exp\left[i\,(kx - \omega t)\right] = -\omega t\,\Psi\,(x, t) \tag{3.4E2a}$$

$$\frac{\partial \Psi\,(x, t)\,(x, t)}{\partial x} = ik A \exp\left[i\,(kx - \omega t)\right] = ik\psi \tag{3.4E2b}$$

$$\frac{\partial^2 \psi}{\partial x^2} = i^2 k^2 A \exp\left[i\left(kx - \omega t\right)\right] = -k^2 A \exp\left[i\left(kx - \omega t\right)\right]$$

$$(3.4\text{E}2\text{c})$$

$$\frac{\partial^2 \psi}{\partial x^2} = -k^2 \Psi\left(x, t\right)$$

$$(3.4\text{E}2\text{d})$$

so that the Schrodinger equation becomes

$$i\hbar \frac{\partial \Psi\left(x, t\right)}{\partial t} = -\frac{\hbar^2}{2m} \frac{\partial^2 \Psi\left(x, t\right)}{\partial x^2} + U\left(x\right) \Psi\left(x, t\right) \qquad (3.4\text{E}3\text{a})$$

$$i\hbar\left[-i\omega \Psi\left(x, t\right)\right] = -\frac{\hbar^2}{2m}\left[-k^2 \Psi\left(x, t\right)\right] + U\left(x\right) \Psi\left(x, t\right) \qquad (3.4\text{E}3\text{b})$$

$$-i^2 \hbar \omega = \frac{\hbar^2 k^2}{2m} + U\left(x\right) \qquad (3.4\text{E}3\text{c})$$

But $i^2 = -1$ so that Eq. (3.4E3c) defining the energy of the particle becomes

$$E = \hbar \omega = \frac{\hbar^2 k^2}{2m} + U\left(x\right) \qquad (3.4\text{E}4)$$

which is exactly equal to Eq. (3.2b).

3.10 Derivation of the Particle Energy Equation

Regarding the particle quantized energy equation defined by Eq. (3.1f), it can be derived by recasting Eq. (3.1E10b) from Example 3.1 and taking its second derivative. Thus,

$$\psi_n\left(x\right) = \sqrt{\frac{2}{L_x}} \sin\left(\frac{n\pi x}{L_x}\right) \qquad (3.26\text{a})$$

$$\frac{d^2 \psi_n\left(x\right)}{dx^2} = -\sqrt{\frac{2}{L_x}} \left(\frac{n\pi}{L_x}\right)^2 \sin\left(\frac{n\pi x}{L_x}\right) \qquad (3.26\text{b})$$

which is related to Eq. (3.15a). In essence, the probability amplitude $\psi\left(x\right) = \psi_n\left(x\right)$ due to the quantum number n is just a differentiable sine function with an obvious continues sinusoidal oscillation.

From Eq. (3.15a) with $U\left(x\right) = 0$ and Eq. (3.26b), the corresponding quantized particle energy equation (E_n) becomes

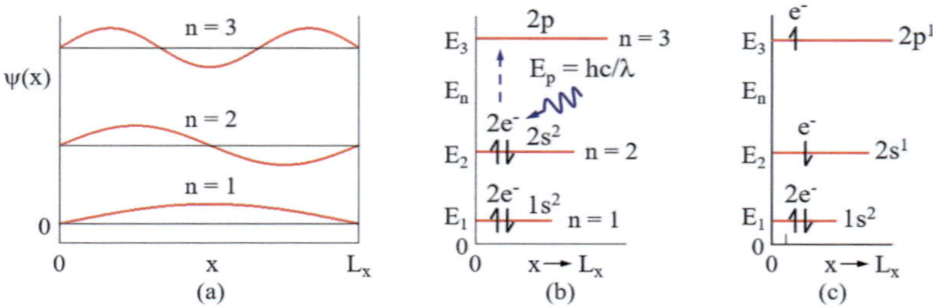

Fig. 3.4 A particle in a box with $0 \le x \le L_x$. (**a**) The standing wave function $\psi(x)$, (**b**) quantized energy levels. One photon with energy $E_p = hc/\lambda$ excites one $2s^2$ electron and (**c**) altered electron configuration

$$-\frac{\hbar^2}{2m}\frac{d^2\psi(x)}{dx^2} = E\psi(x) \tag{3.27a}$$

$$\frac{\hbar^2}{2m}\sqrt{\frac{2}{L_x}}\left(\frac{n\pi}{L_x}\right)^2 \sin\left(\frac{n\pi x}{L_x}\right) = E_n \sqrt{\frac{2}{L_x}} \sin\left(\frac{n\pi x}{L_x}\right) \tag{3.27b}$$

$$E_n = \left(\frac{\pi^2 h^2}{2mL_x^2}\right) n^2 \tag{3.27c}$$

$$n = n_x \qquad \text{(1D)} \tag{3.27d}$$

$$n = n_x + n_y + n_z \ \text{(3D)} \tag{3.27e}$$

This expression, Eq. (3.27c), gives the permitted, allowed, or quantized energy of the particle at $n = 1, 2, 3$ states. Particularly, $n > 1$ represents the excited states. It is clear now that the Schrodinger equation helps understand why electrons being confined to atoms have discrete energy levels described by E_n in Eq. (3.27c).

Plotting Eq. (3.26a) and (3.27c), as illustrated in Fig. 3.4, reveals the behavior of $\psi_n(x)$ and the quantized particle energy E_n states inside the box for $n = 1, 2, 3$. Moreover, the wavefunction $\psi_n(x)$ behaves like a standing wave due to its fixed ends along the x-axis at a stationary state n. Hence, $\psi_n(x)$ provides the path for defining the probability density distribution as $|\psi_n(x)|^2$.

The quantized energy changes between states are $\Delta E_{21} = E_2 - E_1$ and $\Delta E_{32} = E_3 - E_2$, where $\Delta E_{32} > \Delta E_{21}$. In general, Planck's quantum energy equation $E_p = \Delta E = h\upsilon$ is fundamentally a transition energy, where $\upsilon = c/\lambda$ is the frequency and c is the speed of light, can either be emitted or absorbed as a discrete (countable) packet called photon (massless particle). Therefore, $\Delta E = h\upsilon$ is the energy of a photon, and E_n is the particle quantized energy due to the constraint imposed by the quantum integer number n, which directly controls the energy level of the particle in motion.

The wavefunction $\psi(x)$ oscillatory behavior and the quantized particle energy E_n increase with increasing quantum number n as depicted in Fig. 3.4a, b, respectively. It is now appropriate to evaluate $\psi(x)$ at $n = 1, 2, 3$. Thus:

- For $n = 1$, the wavefunction is $\psi(0) = \psi(L_x) = 0$ along with the quantized energy $E_1 > 0$. This condition corresponds to lowest energy level (ground state), suggesting that $\psi(x) > 0$ without a node other than the boundary conditions; $x = 0$ and $x = L_x$.
- For $n = 2$, the wavefunction is $\psi(0) = \psi(L_x/2) = \psi(L_x) = 0$ along with the quantized energy level $E_2 > E_1$. As a result, $L_x/2$ becomes a node point.
- For $n = 3$, the wavefunction and the quantized energy are $\psi(0) = \psi(L_x/3) = \psi(4L_x/3) = \psi(L_x) = 0$ and $E_3 > E_2$. Thus, $L_x/3, 4L_x/3$ become node points. Therefore, the number of node points is $n - 1$ for a particle energy level E_n. Commonly, the particle is an always-moving delocalized electron e^- in a box.
- The particle energy levels E_n are quantized in quantum mechanics (quantum physics) and continuous in classical mechanics (classical physics). This means that in quantum mechanics, there is spacing between quantized energy levels as shown in Fig. 3.4b before and Fig. 3.4c after photon collisions.

Example 3.5 Consider the sketch given below for a hypothetical material constituted by conjugated chains of carbon-carbon (C-C and C=C) covalent bonds having delocalized π-bond electrons (e^-). Derive an expression for the maximum wavelength (λ) of a photon coming from a laser source to excite one π-electron from the highest occupied state (orbital) to the next lowest unoccupied state. Assume that e^- jumps to the nearest unoccupied energy state $n + 1$, the photon energy is defined as $E_p = hc/\lambda$ and that $\Delta E = 3.75\ eV$ is the energy difference between the corresponding adjacent states.

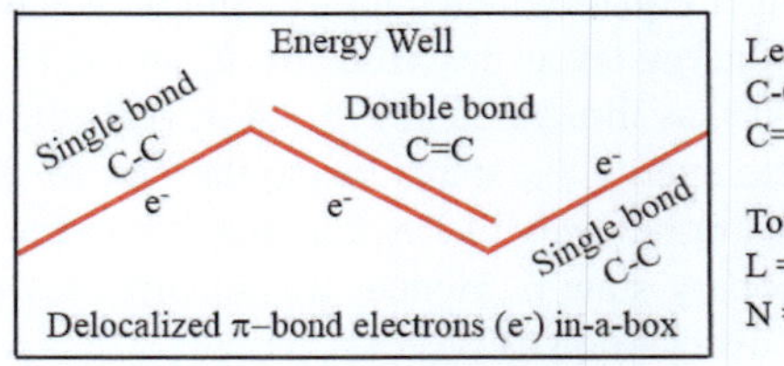

Solution From Eq. (3.27c), the electron quantized energy levels are

$$E_n = \left(\frac{\pi^2 h^2}{2mL_x^2}\right) n^2 \tag{3.5E1a}$$

$$E_1 = \frac{\pi^2 \left(1.05 \times 10^{-34}\ J.s\right)^2}{2\left(9.11 \times 10^{-31}\ Kg\right)\left(420 \times 10^{-12}\ m\right)^2} (1)^2 = 3.39 \times 10^{-19}\ J \tag{3.5E1b}$$

$$E_1 = 2.12 \; eV \tag{3:5E1c}$$

and

$$E_2 = E_1 \, (2)^2 = \left(3.39 \times 10^{-19} \; J\right) (2)^2 = 13.56 \times 10^{-19} \; J = 8.46 \; eV \tag{3.5E2a}$$

$$E_3 = E_1 \, (3)^2 = \left(3.39 \times 10^{-19} \; J\right) (3)^2 = 30.51 \times 10^{-19} \; J = 19.04 \; eV \tag{3.5E2b}$$

Thus, the difference in energy state is the photon energy

$$E_p = \Delta E = E_3 - E_2 = 16.95 \times 10^{-19} \; J = 10.58 \; eV \tag{3.5E3}$$

Then, the wavelength of the photon becomes

$$E_p = \frac{hc}{\lambda} = \Delta E \tag{3.5E4a}$$

$$\lambda = \frac{hc}{\Delta E} = \frac{\left(6.6261 \times 10^{-34} \; J.s\right) \left(3 \times 10^8 \; m/s\right)}{16.95 \times 10^{-19} \; J} \tag{3.5E4b}$$

$$\lambda = 1.1728 \times 10^{-7} \; m = 117.28 \; nm \; \text{(photon)} \tag{3.5E4c}$$

The above electron quantized energy levels are plotted in the figure below along with the electron configuration.

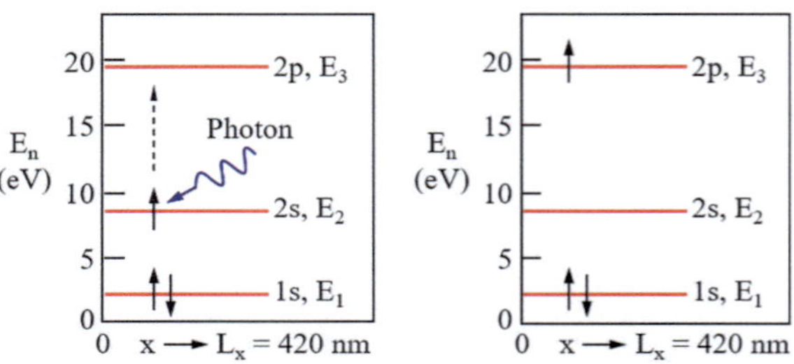

Notice that the photon energy is absorbed by one electron, and consequently, it moves from the $2s^1$ to the next highest energy state $2p^1$. In this hypothetical physical process, the electron absorbs $\Delta E = E_3 - E_2 = 10.58 \; eV$ from the photon, which is a particle-like quanta of electromagnetic radiation that travels at the speed of light. Eventually, the electron jumps to the $2p^1$ state, leaving behind an empty $2s$ state.

3.11 Heisenberg Uncertainty Principle

Werner Heisenberg [8] in 1927 published his work that became known as the Heisenberg uncertainty principle (HUP), which states that it is impossible to simultaneously measure the exact position and momentum of a particle or electron in a particular environment. In quantum mechanics, an electron being trapped in a confined volume $dx\,dy\,dz$, as shown in Fig. 3.2, is free to move within it, but its position and momentum are uncertain at time t. Nonetheless, Heisenberg's original uncertainty equation was based on the standard deviation of measurements σ_x for position and σ_p for momentum. Thus, the HUP is stated as an inequality of statistical value, and its derivation can be found in Kennard's book [9, pp. 105–107].

Mathematically, the original Heisenberg uncertainty principle is simply the product of standard deviations as indicated below

$$\sigma_x = \sqrt{\langle x^2 \rangle - \langle x \rangle^2} > 0 \tag{3.28a}$$

$$\sigma_p = \sqrt{\langle p^2 \rangle - \langle p \rangle^2} > 0 \tag{3.28b}$$

$$\sigma_x \sigma_p \geq \frac{\hbar}{2} = \frac{h}{4\pi} \tag{3.28c}$$

where $\langle x \rangle$ and $\langle p \rangle$ denote the average of measurements known as the expectation values of x and p, respectively. The uncertainty for quantum particles is now presented in a simple manner based on position x and momentum p_x.

Regarding measurements, the double-slit setup is used to determine the wavelike particle interference. This is illustrated in Fig. 3.5 for electrons passing through narrow slits. Here, Δx, and Δp represent the uncertainties of position and momentum, respectively.

Using trigonometry as per Fig. 3.5 yields expressions for the wavelength λ, Δx and $\Delta p = \Delta p_x$. Thus,

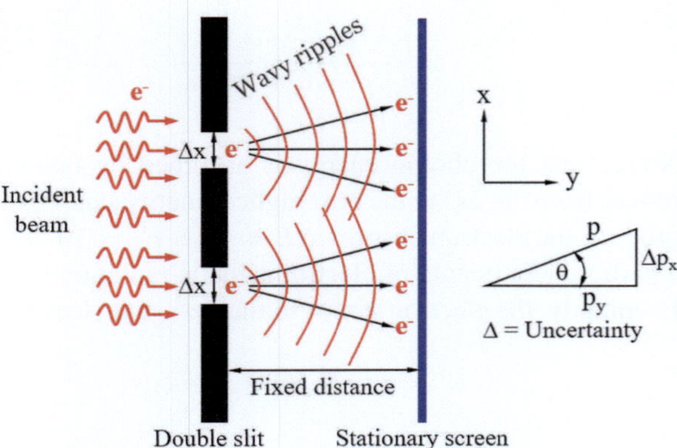

Fig. 3.5 Double-slit model and wave pattern due to particle wave split into separate waves. The particle can be an electron or photon with momentum p

$$\lambda = \Delta x \sin (\theta) \tag{3.29a}$$

$$\Delta p_x = p \sin (\theta) \tag{3.29b}$$

Solve Eq. (3.29b) for the momentum p of a particle

$$p = \frac{\Delta p_x}{\sin (\theta)} \tag{3.30}$$

which has a different mathematical form when compared to Eq. (3.1e) for light massless photons. Moreover, in quantum mechanics, light is quantized as energy quanta known as photon, and according to Fig. 3.5, the direction of a photon or electron with momentum p is along the x-axis. Nevertheless, a pair of photons is assumed to be in a continuous momentum entangled quantum state.

3.11.1   Uncertainty in Space Momentum

Combining Eqs. (3.29a), (3.29b), and (3.30) yields the **upper bound** of the Heisenberg uncertainty principle

$$\Delta x \Delta p = h \tag{3.31}$$

which suggests that any measurement of the expectation value Δx perturbs the expectation value of the particle momentum by an amount $\Delta p = h/\Delta x$. This uncertainty principle equation implies that the more certain one is about position, the less certain one is about momentum, or vice versa.

The **lower bound**, analytically derived by Gottfried [10, pp. 6, 26] using the root-mean-square approach of Δx and Δp, is of the form (Ozama [11])

$$\Delta x \Delta p \geq \frac{\hbar}{2} = \frac{h}{4\pi} \tag{3.32}$$

Combining Eqs. (3.30) and (3.31) yields a $\Delta x \Delta p$ range for two known limits

$$\frac{h}{4\pi} \leq \Delta x \Delta p \leq h \tag{3.33a}$$

$$5.2729 \times 10^{-35} \; J.s \leq \Delta x \Delta p \leq 6.6261 * 10^{-34} \; J.s \tag{3.33b}$$

$$3.2956 \times 10^{-16} \; eV.s \leq \Delta x \Delta p \leq 4.1413 \times 10^{-15} \; eV.s \tag{3.33c}$$

Recall that Δx is the uncertainty in position and Δp is the uncertainty in momentum of a particle.

3.11.2 Uncertainty in Space Speed

If the momentum is written as $p = mv$, then the uncertainty in momentum Δp and particle speed Δv are linearly related, and subsequently, the Heisenberg uncertainty principle, say Eq. (3.31), becomes a speed-dependent equation. Thus,

$$\Delta p = m\Delta v \tag{3.34a}$$

$$\Delta x \Delta v \geq \frac{h}{4\pi m} \tag{3.34b}$$

Notice that Eq. (3.34b) can be solved for the uncertainty in particle speed Δv based on the particle position measurements

$$\Delta v \geq \frac{h}{4\pi m \Delta x} \tag{3.35}$$

This expression depends on the particle mass m and position Δx. Next, some examples may clarify the use of Heisenberg position-momentum uncertainty principle, which is applied to position, momentum, energy, and time.

Example 3.6 According to the triangle depicted in Fig. 3.5 and Eq. (3.29a), $\sin\theta \approx \theta \approx \lambda/\Delta x$ describes the motion of an electron with a mass 9.11×10^{-31} Kg. Assume that the electron is traveling at 15×10^6 m/s speed. **(a)** Drive Eq. (3.31), and **(b)** find the uncertainty range for the electron position (Δx) using 0.8% uncertainty in momentum (Δp)

Solution

(a) The moment equation is

$$\sin\theta \approx \frac{\lambda}{\Delta x} = \frac{\Delta p_x}{p_y} \tag{3.6E1a}$$

$$p_y = \frac{h}{\lambda} \tag{3.6E1b}$$

Then, Eqs. (3.6E1a) and (3.6E1b) along with $\Delta p = \Delta p_x$ yield

$$\Delta x \Delta p = h \tag{3.6E2}$$

(b) The upper and lower uncertainties in positions are

$$p_y = mv = \left(9.11 \times 10^{-31}\ Kg\right)\left(15 \times 10^6\ m/s\right)$$

$$= 1.37 \times 10^{-23}\ Kg.m/s \tag{3.6E3a}$$

$$\Delta p_x = (0.8/100)\left(1.37 \times 10^{-23}\ Kg.m/s\right) = 1.10 \times 10^{-25}\ Kg.m/s$$

$$(3.6\text{E}3\text{b})$$

$$\Delta x_{upper} = \frac{h}{\Delta p} = \frac{6.63 \times 10^{-34}\ kg.m^2/s}{1.10 \times 10^{-25}\ Kg.m/s} = 6.03 \times 10^{-9}\ m = 6.03\ nm$$

$$(3.6\text{E}1\text{c})$$

$$\Delta x_{lower} = \frac{h}{4\pi\,\Delta p} = \frac{6.63 \times 10^{-34}\ kg.m^2/s}{4\pi\left(1.10 \times 10^{-25}\ Kg.m/s\right)}$$

$$= 4.8 \times 10^{-10}\ m = 0.48\ nm \qquad (3.6\text{E}3\text{d})$$

Hence,

$$0.48\ nm \le \Delta x \le 6.03\ nm \qquad (3.6\text{E}4)$$

This is a reasonable range of the uncertainty in position.

Example 3.7 **(a)** Drive an equation for the Heisenberg position-momentum uncertainty principle using the Gaussian wavefunction

$$\psi(x) = A \exp\left(-bx^2\right) = \left(\frac{2b}{\pi}\right)^{1/4} \exp\left(-bx^2\right) \qquad (3.7\text{E}1)$$

where A and b are constants. This function is taken from Griffiths and Schroeter book [1, p. 328]. **(b)** Calculate Δp if $\Delta x = 50\ pm$.

Solution

(a) Let $\psi = \psi(x)$ so that the probability density $|\psi|^2 = |\psi(x)|^2$ equation takes the form

$$\psi = \left(\frac{2b}{\pi}\right)^{1/4} \exp\left(-bx^2\right) \qquad (3.7\text{E}2\text{a})$$

$$\psi^* = \left(\frac{2b}{\pi}\right)^{1/4} \exp\left(+bx^2\right) \qquad (3.7\text{E}2\text{b})$$

$$\left|\psi^*\psi\right| = |\psi|^2 = \left(\frac{2b}{\pi}\right)^{1/2} \exp\left(-2bx^2\right) \qquad (3.7\text{E}2\text{c})$$

This expression describes probability as an exponential decay.

Defining and evaluating the expectation value $\langle x \rangle$ of an electron position x at $\pm\infty$ yields

$$\langle x \rangle = \int_{-\infty}^{+\infty} x \, |\psi|^2 \, dx = \int_{-\infty}^{+\infty} x \left(\frac{2b}{\pi} \right)^{1/2} \exp\left(-2bx^2 \right) dx = 0 \qquad (3.7\text{E}4)$$

This integral can be interpreted as a measurable average value of x or as the average value of position for a large number of particles described by a common wavefunction ψ. Next, the expectation value of position $\langle x^2 \rangle$ of the electron is written as

$$\langle x^2 \rangle = \int_{-\infty}^{+\infty} x^2 \, |\psi|^2 \, dx = \int_{-\infty}^{+\infty} x^2 \left(\frac{2b}{\pi} \right)^{1/2} \exp\left(-2bx^2 \right) dx \qquad (3.7\text{E}5\text{a})$$

$$\langle x^2 \rangle = \frac{1}{4b} \qquad (3.7\text{E}5\text{b})$$

The expectation value of momentum $\langle p \rangle$ for a moving electron in a particular direction is computed as

$$\langle p \rangle = \int_{-\infty}^{+\infty} p \, |\psi|^2 \, dx = \int_{-\infty}^{+\infty} p \left(\frac{2b}{\pi} \right)^{1/2} \exp\left(-2bx^2 \right) dx = 0$$
$$(3.7\text{E}6)$$

Similarly, the expectation value of momentum $\langle p^2 \rangle$ of the electron is defined by

$$\langle p^2 \rangle = \int_{-\infty}^{+\infty} p^2 \, |\psi|^2 \, dx = -\hbar^2 \left(\frac{2b}{\pi} \right)^{1/2} \int_{-\infty}^{+\infty} 2b \left(bx^2 - 1 \right) \exp\left(-2bx^2 \right) dx$$
$$(3.7\text{E}7\text{a})$$

$$\langle p^2 \rangle = b\hbar^2 \qquad (3.7\text{E}7\text{b})$$

Then, the equations for uncertainty in position and momentum are

$$\Delta x = \sqrt{\langle x^2 \rangle - \langle x \rangle^2} = \frac{1}{2\sqrt{b}} \qquad (3.7\text{E}8\text{a})$$

$$\Delta p = \sqrt{\langle p^2 \rangle - \langle p \rangle^2} = \sqrt{b}\,\hbar \qquad (3.7\text{E}8\text{b})$$

Multiplying Eqs. (3.7E8a) and (3.7E8b) defines the Heisenberg uncertainty principle in any measurement. The resultant expression containing position and momentum can be interpreted as a complementary or conjugate function of these variables. Thus,

$$\Delta x \, \Delta p \geq \frac{1}{2\sqrt{b}} \sqrt{b}\,\hbar = \frac{\hbar}{2} = \frac{h}{4\pi} \qquad (3.7\text{E}9)$$

(b) Substituting Planck's constant $h = 6.63 \times 10^{-34}\ kg.m^2/s$ into Eq. (3.7E9) yields expectation product as

$$\Delta x \Delta p \geq \frac{6.63 \times 10^{-34}\ kg.m^2/s}{4\pi} \tag{3.7E10a}$$

$$\Delta x \Delta p \geq 5.28 \times 10^{-35}\ kg.m^2/s \tag{3.7E10b}$$

For $\Delta x = 50\ pm = 50 \times 10^{-12}\ m$, the expectation value of momentum being treated as an average value is

$$\Delta p \geq \frac{5.28 \times 10^{-35}\ kg.m^2/s}{50 \times 10^{-12}m} = 1.06 \times 10^{-24}\ Kg.m/s \tag{3.10E11}$$

which has units of mass and velocity; $\Delta p = m\Delta v$. Regarding Eq. (3.7E9), the expectation product $\Delta x \Delta p$ is in the order of Planck's constant, and the given $\Delta x = 50\ pm = 50$ position represents the number of quantum systems with a common wavefunction ψ.

3.11.3 Uncertainty in Energy Time

The energy-time uncertainty principle derives from the position-momentum uncertainty principle. The former is a consequence of the latter, and it is written without proof as per Griffiths and Schroeter [1, p. 109].

$$\Delta E \Delta t \geq h \tag{3.36}$$

Consider an uncertainty in kinetic energy ΔK_E of a particle using the classical definition of kinetic energy change

$$\Delta K_E \geq \frac{1}{2}m\Delta v^2 = \frac{1}{2m}\Delta p^2 \tag{3.37}$$

Combining Eqs. (3.35) and (3.37) yields

$$\Delta K_E \geq \frac{1}{2m}\left(\frac{h}{4\pi \Delta x}\right)^2 \tag{3.38}$$

The uncertainty in potential energy of the particle (electron) is

$$\Delta U \geq -\frac{Zq_e^2}{4\pi \epsilon_o \Delta x} \tag{3.39}$$

where Z denotes the particle oxidation state, $q_e = 1.602 \times 10^{-19}$ C denotes the electron charge, and $\epsilon_o = 8.85 \times 10^{-12}$ F/m denotes the vacuum permittivity.

Now, the uncertainty in total energy ΔE of the particle becomes

$$\Delta E \geq \Delta K_E + \Delta U \tag{3.40a}$$

$$\Delta E \geq \frac{1}{2m} \left(\frac{h}{4\pi \Delta x} \right)^2 - \frac{Zq_e^2}{4\pi \epsilon_o \Delta x} \tag{3.40b}$$

Let $d\left(\Delta E\right)/d\left(\Delta x\right) = 0$ in Eq. (3.40b) so that

$$\Delta x = \frac{\epsilon_o h^2}{\pi m Z q_e^2} \tag{3.41}$$

which is treated as the Bohr radius, $r = \Delta x$, for hydrogen (Pillai and Pillai [12, pp. 72–73]).

Example 3.8 The application of the Heisenberg uncertainty principle is suitable at the atomic scale. Let the particle speed be 10^{-6} m/s, and calculate the uncertainty in position Δx for **(a)** an electron with $m = 9.11 \times 10^{-31}$ Kg and **(b)** a baseball with $m = 0.15$ Kg

Solution

(a) The electron at an atomic scale:

$$\Delta p \geq m \Delta \upsilon = \left(9.11 \times 10^{-31} \ Kg \right) \left(10^6 \ m/s \right) \tag{3.8E1a}$$

$$\Delta p \geq 9.11 \times 10^{-25} \ Kg.m/s \tag{3.8E1b}$$

and

$$\Delta x \Delta p \geq \frac{h}{4\pi} \tag{3.8E2a}$$

$$\Delta x \geq \frac{h}{4\pi \Delta p} = \frac{6.6261 \times 10^{-34} \ Js}{4\pi \left(9.11 \times 10^{-25} \ Kg.m/s \right)} \tag{3.8E2b}$$

$$\Delta x \geq \frac{h}{4\pi \Delta p} = \frac{6.6261 \times 10^{-34} \ \left(Kg.m^2/s^2 \right) s}{4\pi \left(9.11 \times 10^{-25} \ Kg.m/s \right)} \tag{3.8E2c}$$

$$\Delta x \geq 5.79 \times 10^{-11} \ m = 0.0579 \ nm \tag{3.8E2d}$$

(b) For the baseball at the macroscale,

$$\Delta p \geq m \Delta \upsilon = (0.15 \ Kg) \left(10^6 \ m/s \right) \tag{3.8E3a}$$

$$\Delta p \geq 1.5 \times 10^5 \; Kg.m/s \tag{3.8E3b}$$

and

$$\Delta x \Delta p \geq \frac{h}{4\pi} \tag{3.8E4a}$$

$$\Delta x \geq \frac{h}{4\pi \Delta p} = \frac{6.6261 \times 10^{-34} \; Js}{4\pi \left(1.5 \times 10^5 \; Kg.m/s\right)} \tag{3.8E4b}$$

$$\Delta x \geq \frac{h}{4\pi \Delta p} = \frac{6.6261 \times 10^{-34} \left(Kg.m^2/s^2\right) s}{4\pi \left(1.5 \times 10^5 \; Kg.m/s\right)} \tag{3.8E4c}$$

$$\Delta x \geq 3.52 \times 10^{-40} \; m \tag{3.8E4d}$$

Therefore, the Heisenberg uncertainty principle is not a suitable approach at a macroscale due to an immeasurably small position $\Delta x \geq 3.52 \times 10^{-40} \; m$.

3.12   Atoms in Crystal Structures

This is the case for a multielectron atom. It is known that the kinetic and potential energy of atoms derive from the electron motion along orbitals around nucleus of atoms. Particularly, the potential energy of an atom in a electrostatic field is

$$U(r) = \frac{1}{4\pi \epsilon_o} \frac{Zq_e^2}{r} \tag{3.42}$$

where Z is the valence of oxidation state of an atom in the solid state. Now, applying Eq. (3.15b) to a multielectron atom or solving the Schrodinger equation for an atom with $Z > 2$ is a complicated task.

Mathematically, recast the one-dimensional Schrodinger equations and the Hamiltonian operator $(\hat{H})$ for an atom with $Z \geq 1$ electrons to reveal the complexity of a multielectron quantum system (Griffiths and Schroeter [1, p. 209])

$$i\hbar \frac{\partial \Psi(r)}{\partial t} = \hat{H}\Psi(r) \quad \text{(time-dependent)} \tag{3.43a}$$

$$\hat{H}\psi(r) = E\psi(r) \quad \text{(time-independent)} \tag{3.43b}$$

$$\Psi(r) = \psi(r_1)\,\psi(r_2)\,\psi(r_3)\ldots\ldots\psi(r_N) \tag{3.43c}$$

$$\hat{H} = \sum_{j=1}^{Z}\left[-\frac{\hbar^2}{2m}\nabla_j^2 - \left(\frac{1}{4\pi\epsilon_o}\right)\frac{Zq_e^2}{r_j}\right] + \frac{1}{2}\left(\frac{1}{4\pi\epsilon_o}\right)\sum_{j\neq k}^{Z}\frac{q_e^2}{\left|r_j - r_k\right|} \tag{3.43d}$$

For hydrogen (H) with $Z = 1$ electron,

$$\hat{H} = -\frac{\hbar^2}{2m}\nabla_1^2 - \left(\frac{1}{4\pi\epsilon_o}\right)\frac{q_e^2}{r_1} + \frac{1}{2}\left(\frac{1}{4\pi\epsilon_o}\right)\frac{q_e^2}{r_1} \tag{3.44a}$$

$$\hat{H} = -\frac{\hbar^2}{2m}\nabla_1^2 \tag{3.44b}$$

and for helium (He) with $Z = 2$ electrons,

$$\hat{H} = \left[-\frac{\hbar^2}{2m}\nabla_1^2 - \left(\frac{1}{4\pi\epsilon_o}\right)\frac{2q_e^2}{r_1}\right] + \left[-\frac{\hbar^2}{2m}\nabla_2^2 - \left(\frac{1}{4\pi\epsilon_o}\right)\frac{2q_e^2}{r_2}\right] \tag{3.45}$$

$$+ \frac{1}{2}\left(\frac{1}{4\pi\epsilon_o}\right)\frac{q_e^2}{|r_1 - r_2|}$$

This demonstrates the complexity of a problem, which must be solved using a numerical or an approximation method for $Z > 2$. Imagine the number of terms in the series for an iron (Fe) atom with $Z = 26$ electrons. Therefore, the concept of one-electron Schrodinger equation prevails in this textbook.

3.13 Summary

Concepts and basic aspects of quantum mechanics (QM) have been described as an essential part of materials science related to atoms and their subatomic particles, such as electrons (e^-), protons (e^+), neutrons, or photons (packet of energy in the form of light). Additional quantum mechanical concepts are included in Chap. 14.

Despite the fundamentals of QM being introduced in this chapter, some advance mathematical concepts are included as an expansion for describing a wavelike particle traveling in the x-direction at a velocity v. This leads to the one-dimensional time-dependent Schrodinger equation, which is derived from the wave equation analogous to the heat equation. Moreover, the uncertainty principle is an important concept related to the measurement of a particle position.

Additionally, QM is an integral part of materials science sinceQM is everywhere because matter is made out of particles, which are the building block of particle physics. Apparently, particle size in quantum mechanics is in the order of Planck's constant and materials science is in the order of nanometers (nm).

It is amazing how quantum mechanics (QM) helps evolve nowadays modern technology. For instance, digital clocks in Global Positioning System (GPS) satellites, computer transistors made out of doped silicon, lasers that emit light with a well-defined wavelength, and so forth rely on QM. The reader is encouraged to visit www.youtube.com and watch pertinent videos on QM. Moreover, QM with applications to nanotechnology is an ongoing research field, where the fundamental properties of matter are studied at a nanoscale.

Appendix 3A Solution of the Wave Equation

The method of separation of variables is a powerful mathematical tool for converting a partial differential equation (PDE) to an ordinary differential equation (ODE). Thus, the wavefunction defined by Eq. (3.25) is rewritten as

$$\frac{\partial^2 \Psi(x,t)}{\partial t^2} = v^2 \frac{\partial^2 \Psi(x,t)}{\partial x^2} \tag{3.46}$$

with relevant initial condition (IC) and boundary condition (BC)

$$\psi = \Psi(x,t) \quad \text{for } 0 < x < L_x, \ t > 0 \tag{3.47a}$$

$$\Psi(x,t) = \Psi(L_x,t) = 0 \quad \text{for } t > 0 \tag{3.47b}$$

$$\Psi(x,t) = \Psi(x,0) \quad \text{for } 0 \le x \le L_x \tag{3.47c}$$

$$\Psi(x,0) = 0 \quad \text{for } 0 \le x \le L_x \tag{3.47d}$$

$$\Psi(x,t) = f(x,t) \quad \text{for } 0 < x < L_x \tag{3.47e}$$

The method of separation of variables for the wavefunction $\Psi(x,t)$ is

$$\Psi(x,t) = \psi(x)\,\phi(t) \tag{3.48a}$$

$$\frac{d\Psi(x,t)}{dx} = \phi(t)\frac{d\psi(x)}{dx} \quad \& \quad \frac{d\Psi(x,t)}{dt} = \psi(x)\frac{d\phi(t)}{dt} \tag{3.48b}$$

$$\frac{d^2\Psi(x,t)}{dx^2} = \phi(t)\frac{d^2\psi(x)}{dx^2} \quad \& \quad \frac{d^2\Psi(x,t)}{dt^2} = \psi(x)\frac{d^2\phi(t)}{dt^2} \tag{3.48c}$$

Substitute Eq. (3.48c) into (3.46) to get

$$\frac{\partial^2 \Psi(x,t)}{\partial t^2} = v^2 \frac{\partial^2 \Psi(x,t)}{\partial x^2} \tag{3.49a}$$

$$\psi(x)\frac{d^2\phi(t)}{dt^2} = v^2\phi(t)\frac{d^2\psi(x)}{dx^2} \tag{3.49b}$$

$$\frac{1}{v^2\phi(t)}\frac{d^2\phi(t)}{dt^2} = \frac{1}{\psi(x)}\frac{d^2\psi(x)}{dx^2} \tag{3.49c}$$

which implies that the time-dependent and space-dependent parts must be equal to a common constant C. Thus,

$$\frac{1}{v^2\phi(t)}\frac{d^2\phi(t)}{dt^2} = C \tag{3.50a}$$

$$\frac{1}{\psi(x)}\frac{d^2\psi(x)}{dx^2} = C \tag{3.50b}$$

Manipulating Eqs. (3.50a) and (3.50b) yields two homogeneous ordinary differential equations (ODEs)

$$\frac{d^2\phi(t)}{dt^2} - Cv^2\phi(t) = 0 \tag{3.51a}$$

$$\frac{d^2\psi(x)}{dx^2} - C\psi(x) = 0 \tag{3.51b}$$

The possible general solutions of Eq. (3.51b) depend on the constant C. Consider three general solutions written as

$$\psi(x) = A_1 + A_2 x \quad \text{for } C = 0 \tag{3.52a}$$

$$\psi(x) = A_3 \exp\left(\sqrt{C}.x\right) + A_4 \exp\left(-\sqrt{C}.x\right) \quad \text{for } C > 0 \tag{3.52b}$$

$$\psi(x) = A_5 \cos\left(\sqrt{-C}.x\right) + A_6 \sin\left(\sqrt{-C}.x\right) \quad \text{for } C < 0 \tag{3.52c}$$

The next step is to evaluate these equations with respect to the unknown constants.

Space-Dependent Part

For $C = 0$ and boundary conditions in a box, Eq. (3.52a) yields

$$\psi(0) = \psi(L_x) = 0 \tag{3.53a}$$

$$0 = A_1 + A_2(0) \tag{3.53b}$$

$$0 = A_1 + A_2(L_x) \tag{3.53c}$$

$$A_1 = 0 \quad \& \quad A_2 = 0 \tag{3.53d}$$

$$\psi(x) = 0 \tag{3.53e}$$

For $C > 0$ and boundary conditions in a box, Eq. (3.52b) gives

$$\psi(0) = \psi(L_x) = 0 \tag{3.54a}$$

$$0 = A_3 \exp\left(\sqrt{C}.0\right) + A_4 \exp\left(-\sqrt{C}.0\right) \tag{3.54b}$$

$$0 = A_3 \exp\left(\sqrt{C}.L_x\right) + A_4 \exp\left(-\sqrt{C}.L_x\right) \tag{3.54c}$$

From Eq. (3.54b),

$$0 = A_3 + A_4 \Longrightarrow A_4 = -A_3 \tag{3.55}$$

Combining Eqs. (3.54c) and (3.55) yields

$$0 = A_3 \exp\left(\sqrt{C}.L_x\right) - A_3 \exp\left(-\sqrt{C}.L_x\right) \tag{3.56a}$$

$$0 = A_3 \left[\exp\left(\sqrt{C}.L_x\right) - \exp\left(-\sqrt{C}.L_x\right)\right] \tag{3.56b}$$

$$A_3 = 0 \tag{3.56c}$$

and

$$0 = A_3 \exp\left(\sqrt{C}.L_x\right) + A_4 \exp\left(-\sqrt{C}.L_x\right) \tag{3.57a}$$

$$0 = (0) \exp\left(\sqrt{C}.L_x\right) + A_4 \exp\left(-\sqrt{C}.L_x\right) \tag{3.57b}$$

$$A_4 = 0 \tag{3.57c}$$

$$\psi(x) = 0 \tag{3.57d}$$

For $C < 0$ and boundary conditions in a box, Eq. (3.52c) yields

$$\psi(0) = \psi(L_x) = 0 \tag{3.58a}$$

$$0 = A_5 \cos\left(\sqrt{-C}.0\right) + A_6 \sin\left(\sqrt{-C}.0\right) \tag{3.58b}$$

$$0 = A_5 \cos\left(\sqrt{-C}.L_x\right) + A_6 \sin\left(\sqrt{-C}.L_x\right) \tag{3.58c}$$

From Eq. (3A13b) and (3.58c),

$$0 = A_5 \tag{3.59a}$$

$$0 = A_6 \sin\left(\sqrt{-C}.L_x\right) \tag{3.59b}$$

$$0 = \sin\left(\sqrt{-C}.L_x\right) \tag{3.59c}$$

It can be deduced from Eq. (3.59c) that $n = 1, 2, 3, \ldots$ and that

$$\sin\left(\sqrt{-C}.L_x\right) = \sin(n\pi) = 0 \tag{3.60a}$$

$$\sqrt{-C}.L_x = n\pi \tag{3.60b}$$

$$C = -\left(\frac{n\pi}{L_x}\right)^2 \tag{3.60c}$$

Then, Eq. (3.52c) along with (3.60c) becomes $\psi(x) = \psi_n(x)$ due to $n = 1, 2, 3, ..$

$$\psi_n(x) = A_6 \sin\left(\frac{n\pi x}{L_x}\right) \tag{3.61}$$

This expression is the same as Eq. (3.1E5a). According to Eq. (3.1E9d), $A_6 = B = \sqrt{2/L_x}$, and the one-dimensional time-independent Schrodinger equation becomes

$$\psi_n(x) = \sqrt{\frac{2}{L_x}} \sin\left(\frac{n\pi x}{L_x}\right) \tag{3.62}$$

which is exactly equal to Eq. (3.1E10b).

Time-Dependent Part

Let a complex exponential function be defined by

$$\phi(t) = A \exp\left[-\left(\frac{iEt}{\hbar}\right)\right] \tag{3.63}$$

which is the partial solution of Eq. (3.51a) with an unknown constant A. Inserting Eqs. (3.62) and (3.63) into (3.48a) with $A = A_n$ yields the solution of Eq. (3.46) in a complex variable form

$$\psi_n(x, t) = \psi(x)\phi(t) = A_n\sqrt{\frac{2}{L_x}} \sin\left(\frac{n\pi x}{L_x}\right) \exp\left[-\left(\frac{iEt}{\hbar}\right)\right] \tag{3.64}$$

It is convenient now to slightly modify Eq. (3.64) due to $n = 1, 2, 3, \ldots$ and E_n defined by Eq. (3.27c). Thus, general solution to Eq. (3.46) is

$$\psi_n(x, t) = \sum_{n=1}^{\infty} A_n\sqrt{\frac{2}{L_x}} \sin\left(\frac{n\pi x}{L_x}\right) \exp\left[-\left(\frac{iE_n t}{\hbar}\right)\right] \tag{3.65a}$$

$$\psi_n(x, t) = \sum_{n=1}^{\infty} A_n\sqrt{\frac{2}{L_x}} \sin\left[\left(\frac{n\pi}{L_x}\right)x\right] \exp\left[-\frac{i\hbar}{2m}\left(\frac{n\pi}{L_x}\right)^2 t\right] \tag{3.65b}$$

which implies that $\Psi(x, t)$ is a Fourier series and A_n are unknown coefficients.

Problems

3.1 Consider the particle-in-a-box model as an infinite potential energy well, and assume that the particle is an electron traveling along the x-axis with a dimensional constraint $0 \leq x \leq L_x$, where the inner and outer potential energy terms are $U(x) = 0$ and $U(x) = \infty$, respectively. Use Eq. (3.31a) to derive an expression for the expectation value $\langle x \rangle$ of the electron along the L_x line, regardless of the quantum number $n = 1, 2$ or 3 defining the quantized energy. [Solution: $\langle x \rangle = L_x/2$].

3.2 Use Eq. (3.5b) to compute the probability density $P(x)$ for an electron traveling between $a = 0.10L_x$ and $b = 0.50L_x$. **(a)** Derive $P(x)$ in terms of the quantum number n, and **(b)** mathematically plot $P(x) = f(n)$ as a continuous curve for $0 \leq n \leq 20$. **(c)** Quantum mechanically calculate and plot $P(x)$ when $n = 1, 2, 3, 4, 5$ as discrete values. **(d)** What can you conclude based on these calculations?

3.3 Assume that the particle-in-a-box model contains an electron traveling along the x-axis with a dimensional constraint $0 \leq x \leq L_x$, where the inner and outer potential energy terms are $U(x) = 0$ and $U(x) = \infty$, respectively. Use Eq. (3.31a) to derive an expression for the expectation value $\langle x \rangle$ of the electron along the L_x line corresponding to the quantized energy $E_n = E_1$ due to the quantum number $n = 1$. [Solution: $\langle x \rangle = L_x/2$].

3.4 Assume that an incident electron with $E = 1\ eV$ travels along the x-axis within a w-wide box (Fig. 2.2c), where the potential energy barrier is $U(x) = 2\ eV$. Use the transmission probability-gradient equation given below (without proof) for determining the tunnel width w of a given probability $T = 10^{-3}$. [Solution: $w \simeq 0.81\ nm$].

$$\frac{dT}{dw} = -a \exp(-2bw)$$

$$a = \frac{16E\,(U(x) - E)}{U(x)^2}\,\frac{\sqrt{2m\,[U(x) - E]}}{\hbar}$$

$$b = \frac{\sqrt{2m\,[U(x) - E]}}{\hbar}$$

3.5 Consider an electron being confined in a three-dimensional infinite potential well as shown in Fig. 2.1. Let the particle-in-a-box have equal sides $L = L_x = L_y = L_z$. Derive an expression for the number of states n and the electron energy $E \leq E_n$, where $i = 1, 2, 3, 4, .., \infty$, E is the electron energy, and $\Delta E = E_n - E$ is energy difference between states. Start with the time-dependent Schrodinger equation, which is easily solvable using the method of separation of variables. [Solution: $n \leq [L/(\pi \hbar)]\,\sqrt{2mE}$].

3.6 **(a)** Use Eq. (3.32c) to compute the ground-state energy of an electron for a cubic box with sides equal to 0.5 nm long. **(b)** Repeat the calculation for a box with 5 nm long each side. Compare results and make a conclusion. [Solution: (a) $E_1 = 1.49\,eV$ and (b) $E_1 = 0.0149\,eV$].

3.7 Assume that a 0.348-nm-thick CuO film on a copper substrate imposes a finite barrier of $U(x) = 4\,eV$. Use the particle-in-a-box model to compute **(a)** the quantum number n for one $E = 5$-eV electron, **(b)** the wavenumber k, and **(c)** the electron wavelength λ, and **(d)** compare this n value with the result given in Example 2.1. What can you conclude? The electron rest mass and the reduced Planck's constant are $m = 9.11 \times 10-31\,Kg$ and $\hbar = 1.05 \times 10-34\,Js$. [Solution: (a) $n = 4$, (b) $k = 5.15\,nm^{-1}$, (c) $\lambda = 1.22\,nm$].

3.8 Assume an electron in a one-dimensional box (Fig. 3.2a) is traveling along the $L_x = 4\,nm$ line. **(a)** Compute the probability $P_n(x)$, where $n = 1, 2, 3$ and 4, for finding the electron within the interval $\Delta x = 0.40\,nm$ from $x = L_x/2$. Recall that n is the quantum number related to the quantized electron energy E_n. Will $P_n(x)$ be the same for all four quantum states? **(b)** Calculate the wavelengths for first and second excited states when light (photon) strikes the electron. Data: $h = 6.6261 \times 10^{-34}\,J.s$, $m = 9.11 \times 10^{-31}\,Kg$, and $c = 3 \times 10^8\,m/s$. [Solution: (a) $P_1(x) = 0.20$ and (b) $\lambda_1 = 17,753\,nm$].

3.9 Assume that an electron moves within a box along the $0 \le L_x \le 1.20\,nm$ line and that laser beam provides the light with sufficient energy to excite the electron from its ground state to the next available state. Calculate **(a)** the absorbed energy (ΔE) by the electron and **(b)** the wavelength (λ) of the light. Data: $h = 6.6261 \times 10^{-34}\,J.s$, $m = 9.11 \times 10^{-31}\,Kg$, and $c = 3 \times 10^8\,m/s$. [Solution: (a) $\Delta E = 0.78\,eV$ and (b) $\lambda = 1,583.90\,nm$].

3.10 Apply the concept of a particle-in-a-box (infinite potential) model to a hypothetical double carbon-carbon bond 0.14 nm long, where the π electron from each carbon (C = C) atom is free to move. Let the potential energy inside the box be zero so that the π electrons will be trapped in the box. If the π electrons are excited by radiation from state 2 to state 3 by a photon beam, then calculate (a) the energy change ΔE and (b) the photon wavelength λ. [Solution: (a) $\Delta E = 6\,eV$ and (b) $\lambda = 206.96\,nm$].

3.11 If the position of an electron is measured to an accuracy of $0.012\,nm$, then determine **(a)** the momentum of the electron to an accuracy based on the Heisenberg uncertainty principle, **(b)** the uncertainty in speed of an electron Δv, and **(c)** the kinetic energy KE based on whether or not the calculated Δv is relativistic. [Solution: (a) $\Delta p = 4.394 \times 10^{-24}\,Kg.m/s$, (b) $\Delta v \ge 4.82 \times 10^6\,m/s$, and (c) $KE = 66.05\,eV$].

3.12 Assume that the lifetime of an excited atomic state is $8 \times 10^{-10}\,s$. Calculate the uncertainty in energy of the state. [Solution: $\Delta E \ge 4.11 \times 10^{-7}\,eV$].

3.13 Use the Heisenberg uncertainty principle to calculate **(a)** the minimum uncertainty in the electron speed (Δv) confined to a magnesium atom (Mg) within a radius of 173 pm and **(b)** the electron kinetic energy (KE). [Solution: (a) $\Delta \upsilon \geq 3.35 \times 10^5 \, m/s$ and (b) $KE = 0.32 \, eV$].

3.14 Use the Heisenberg uncertainty principle to calculate **(a)** the minimum uncertainty in the electron speed ($\Delta \upsilon$) confined to a nickel atom (Ni) within a radius of 163 pm and **(b)** the electron kinetic energy (KE). [Solution: (a) $\Delta \upsilon \geq 3.55 \times 10^5 \, m/s$ and (b) $KE = 0.36 \, eV$].

3.15 An incident beam of photons with wavelength 800 nm excite an electron in a deep box of width L_x (Fig. 2.2a). Calculate L_x for a possible change of energy state; $\Delta E = E_2 - E_1$. [Solution: $L_x = 0.85 \, nm$].

3.16 Consider a wavelike particle traveling in confined $0 \leq x \leq L_x$ line, and assume that the wavefunction $\psi(x)$ must satisfy the second-order homogeneous equation

$$\frac{d^2 \psi_n(x)}{dx^2} + n^2 \psi_n(x) = 0$$

$$\psi_n(x) = A \exp(i(kx - \omega t))$$

where n is the principal quantum number. Find an expression for A. [Solution: $A = \sqrt{1/2\pi}$].

3.17 Assume that an incident beam of photons with wavelength 850 nm excite an electron in a deep box of width L_x (Fig. 2.2a). Calculate L_x for a possible change of energy state; $\Delta E = E_2 - E_1$. [Solution: $L_x = 0.88 \, nm$].

3.18 Consider the Heisenberg uncertainty principle to calculate **(a)** the minimum uncertainty in the electron speed ($\Delta \upsilon$) confined to a nickel atom (Ni) within a radius of 160 pm and **(b)** the electron kinetic energy (KE). [Solution: (a) $\Delta \upsilon \geq 3.62 \times 10^5 \, m/s$ (b) $KE = 0.37 \, eV$].

3.19 **(a)** Calculate the ground-state energy of an electron for a cubic box with sides equal to 0.6 nm long. Use Eq. (3.32c). **(b)** Repeat the calculation for a box with 6 nm long each side. Compare results and make a conclusion. [Solution: (a) $E_1 = 1.04 \, eV$ and (b) $E_1 = 0.0104 \, eV$].

3.20 Use Eq. (3.32c) to **(a)** calculate the ground-state energy of an electron for a cubic box with sides equal to 0.4 nm long. Use Eq. (3.32c). **(b)** Repeat the calculation for a box with 4 nm long each side. Compare results and make a conclusion. [Solution: (a) $E_1 = 2.33 \, eV$ and (b) $E_1 = 0.0233 \, eV$].

References

1. D.J. Griffiths, D.F. Schroeter, *Introduction to Quantum Mechanics*, 3rd edn. (Cambridge University, New York, 2018)
2. S. Flugge, *Practical Quantum Mechanics* (Springer, Berlin, 1999)

3. K.W. Kolasinski, *Surface Science: Foundations of Catalysis and Nanoscience*, 2nd edn. (Wiley, West Sussex, 2008)
4. O.H. Wyatt, D. Dew-Hughes, *Metals, Ceramics and Polymers* (Cambridge University, Cambridge, 1974)
5. D.G. Pettifor, *Bonding and Structure of Molecules and Solids* (Oxford University Press Inc., New York, 1995)
6. G. Baym, *Lectures on Quantum Mechanics* (CRC Press, Taylor & Francis Group, New York, 2018)
7. A.C. Phillips, *Introduction to Quantum Mechanics* (Wiley, England, 2003)
8. W. Heisenberg, Über den anschaulichen Inhalt der quantentheoretischen Kinematik und Mechanik (in German). About the descriptive content of quantum theoretical kinematics and mechanics (Google translation in English). Z. Phys. **43**, 172–198 (1927)
9. E.H. Kennard, *Zur Quantenmechanik einfacher Bewegungstypen* (in German). *On quantum mechanics of simple types of motion* (Google translation in English). Z. Phys. **44**, 326–352 (1927)
10. K. Gottfried, Quantum mechanics, in *Fundamentals*, vol. I (CRC Press, Taylor & Francis Group, Boca Raton, 2018)
11. M. Ozawa. *Heisenberg's Original Derivation of the Uncertainty Principle and its Universally Valid Reformulations*. Curr. Sci. **109**(11), 2006–2016 (2015)
12. S.O. Pillai, S. Pillai, *Rudiments of Materials Science*, 3rd edn. (New Age International Publishers, New Delhi, 2012)

Chapter 4
Metallography and Microscopy

4.1 Introduction

Engineering applications of materials rely on relevant properties and environmental impact. The initial mechanical work for characterizing as-cast or heat-treated materials is based on metallographic techniques, followed by microscopy such as traditional optical microscope (OM) or light microscope (LM), scanning electron microscope (SEM), transmission electron microscope (TEM), and the like.

Metallography is essentially a metallurgical technique used to mechanically prepare a specimen with a flat and mirrorlike surface, which in turn is etched or chemically oxidized to reveal the surface microstructure. Subsequently, microscopy is used to reveal and characterize the microstructural features of an as-cast or metallurgically or mechanically processed metal. Hence, metallographic analysis and microscopy are used as tools to examine single and multiple phases within a metallurgical specimen material.

For comparison, metallography is used to reveal microstructural features and fractography for assessing fractured surfaces. In particular, black-and-white (B&W) or color metallography involves simple and sophisticated equipment for obtaining high-quality microstructures through the lens of science.

Some microscopic techniques may be complicated, and some may require training for obtaining and interpreting microstructural features. The main goal in the microscopy field in this chapter is to reveal microstructural features, such as grain size (d), inclusions, small secondary particles, and linear defects called dislocations. Regarding microscopic size particles, stereology provides means to perform quantitative metallography, and the American Society for Testing Materials (ASTM) recommends a comparative method for the grain size number.

The **microscope** is a conventional tool used to reveal images that can provide information on the mechanical, physical, and biological characteristics of specimens, and **microscopy** is the technical field of using microscopes to view real-space

© The Author(s), under exclusive license to Springer Nature Switzerland AG 2024   139
N. Perez, *Materials Science: Theory and Engineering*,
https://doi.org/10.1007/978-3-031-57152-7_4

magnified images of microstructural features, such as grain morphology and/or atomic arrangement on a specimen surface.

4.2 Microscopic Techniques

There is a wide range of microscopy techniques available for metals and life science. The most common microscopic techniques based on their field of view and resolution or magnification are described below in a general form.

Optical Microscopy (OM) It is used to view a specimen at relatively low magnifications through a series of lens with visible light. The optical microscope (OM) uses a small depth of field, and it is known as a light microscope (LM). An OM is commercially available as a simple, compound, digital, or stereo microscope. Essentially, an opaque specimen surface and the OM in light reflection mode are specific characteristics for observing microstructural features.

Digital Microscopy (DM) The DM uses optics, a large depth of field, and a digital camera for taking high-quality images of flat and slightly uneven surfaces at relatively low magnifications; mag. $\leq 1000x$.

Scanning Electron Microscopy (SEM) It is used to obtain three-dimensional images of a specimen surface topography by scanning the surface with a high-energy beam of electrons. SEM is a popular technique in forensic engineering.

Transmission Electron Microscopy (TEM) An electron beam is transmitted through a specimen to obtain an image of the surface features, such as dislocations and sub-sized precipitates. In essence, the TEM technique builds an image from the electrons that pass through a specimen.

Scanning Tunneling Microscopy (STM) It is a powerful technique based on quantum tunneling of electrons. It is a powerful tool for studying the topology of a solid surface at atomic scale. A nano distance between the STM tip and the solid substrate is essential for electron tunneling at a bias voltage.

Electrochemical Scanning Tunneling Microscopy (ESTM) It is a powerful tool for studying electrochemical problems.

Atomic Force Microscopy (AFM) It is used to characterize layered ion structures at a nanoscale distance between the AFM tip and specimen surface. It provides images of surface topography, even in aqueous solutions.

Dark-Field Scattering Microscopy (DFSM) It is used to detect an electrical double layer (EDL) between a electrolyte-solid interface at a nanoscale. An EDL consists of an ionic layer having a particular thickness classically known as the Helmholtz layer.

Field ion Microscopy (FIM) This is used to obtain an image of the atomic arrangement at a sharp metal tip.

Field Emission Microscopy (FEM) It is used to reveal the molecular structure on a sample surface.

Atom Probe Field ion Microscope (Atom Probe) It is used to identify individual atoms using needle-shaped specimens in an ultrahigh vacuum chamber. A high magnification is achieved due to a highly curved electric potential.

Apparently, OM, SEM, TEM, and AFM are the most common techniques in materials science.

4.3 Metallography

Metallography is the study of the microstructure of metals and alloys. The microstructure of a material is the arrangement of its grains, which are small crystals of the material. The size, shape, and orientation of the grains can have a significant impact on the material's properties, such as its strength, hardness, and ductility.

This section includes a simple technique used to prepare metallographic specimens and, subsequently, reveal the microstructure of crystalline materials under the microscope at a given magnification of the field of view. Thus, metallography is the science of examining (study of) the microscopic structures of solids, such as metals and alloys, using surface analysis tools like optical microscopy (OM) or metallurgical microscope (MM) and scanning electron microscopy (SEM).

Metallography is reliably used to characterize microstructural evolution in materials development, failure analysis, quantitative analysis for quality control, and so forth. Metallography is a useful technique for revealing the morphology or configuration of grains in multiple-phase analysis and the characteristics of a corroded metal surface, such as pitting morphology, metal oxide layers, and so forth. Apparently, the grain size is the most significant metallographic measurement since it is directly related to mechanical properties. Moreover, below are given some practical guidelines on what metallography has to offer experimentalists.

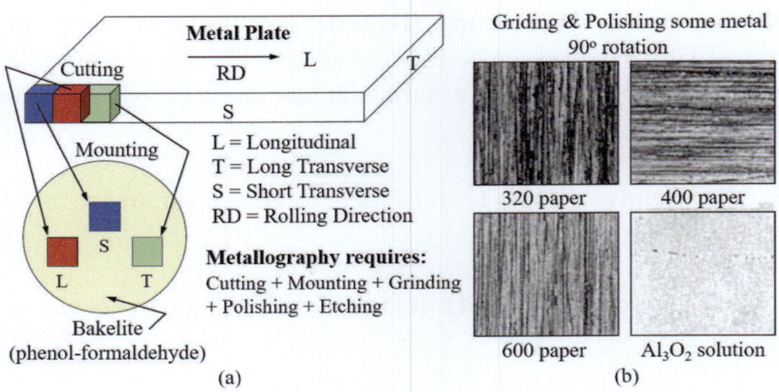

Fig. 4.1 Manual steps for mechanically preparing metallographic specimens. Subsequent steps require polishing and etching specimens for revealing microstructural features. The polished specimen surface (having black spots due to minor concentration of oxide impurities) is usually accomplished with a Al_2O_3-containing solution. Actual images of some metal alloy can be found elsewhere (Web site [1])

4.3.1   Specimen Preparation

Specimen preparation is an important part of metallography (Fig. 4.1). Proper specimen preparation can produce a specimen that is suitable for microscopic examination and that accurately represents the microstructure of the material. This information can then be used to understand the properties of the material and to determine why it may have failed.

Consider a small steel plate of unknown history or a cold rolled plate. Specimen preparation is the process of preparing a metallographic specimen for microscopic examination. The goal of specimen preparation is to produce a flat, smooth surface that is free of artifacts and that accurately represents the microstructure of a material. The specimen preparation process typically involves the following steps.

Cutting The first step is to cut a specimen from the material to be examined. The specimen should be cut to a size that is convenient for handling and that will fit on the microscope stage.

Mounting The next step is to mount the specimen in a mounting material. The mounting material should be hard, durable, and have a low coefficient of thermal expansion. The specimen is typically mounted in a hot or cold mounting resin. For instance, Fig. 4.1a illustrates some mounted specimens in hard Bakelite.

Grinding Once the specimen is mounted, it is ground to remove surface defects and to produce a flat, smooth surface. The grinding process is typically carried out using a series of progressively finer grades of abrasive paper, as shown in Fig. 4.1b. It is important to rotate the specimen 45° or 90° when changing abrasive paper for easy removal of previous scratches.

Polishing The final step in specimen preparation is polishing using $H_2O\text{-}Al_2O_3$ solution or commercial polishing paste. Polishing removes the scratches left by the grinding process and produces a surface that is free of defects. The polishing process is typically carried out using a series of progressively finer grades of polishing cloths and abrasives. At this stage, examination of a polished specimen may exhibit porosity, cracks, and nonmetallic inclusions. The resultant polished surface is ideally scratch-free and mirrorlike surface (bottom-right image in Fig. 4.1b). These images can be found elsewhere (Web site [1]).

Etching Once the specimen is polished, it is etched with a chemical solution. The etching solution selectively attacks different phases in the material, causing them to appear different colors under the microscope. This allows the microstructural features of the material to be seen more clearly.

Microscopic Examination After etching, the specimen is examined under a microscope. The microscope allows the microstructural features of the material to be seen in detail. The microstructural features can then be analyzed to determine the properties of the material. Moreover, metallographic etching by immersing or swabbing the polished metal surfaces reveals their microstructures under the microscope. In addition, electrolytic etching is used to prepare high-quality metallographic specimen or sample.

Example 4.1 How does the material hardness influence the sample removal technique for a microscopic study? Which synthetic polymer ()material) is suitable for mounting the metallographic specimen?

Solution To study a material's failure under a microscope, a small sample is removed from near the affected area. Soft materials are cut with a handsaw, while hard materials need a special cutting wheel, abrasive cutoff wheel. For mounting the specimen, use synthetic polymers known for their durability and resistance to heat and chemicals. For instance, thermosetting resin is a polymer that undergoes irreversible chemical cross-linking when heated. This process is suitable for a soft material. On the other hand, for a hard material, use Bakelite; it is a specific type of thermosetting resin formed by the reaction of phenol and formaldehyde.

4.4 Optical Microscopy

Optical microscopy is a light emission-reflection technique employed to closely view the surface of a sample at a small depth of field through the magnification of a lens with visible light using an optical microscope (instrument). The result of an optical work is a magnified image of the sample surface.

The optical microscope requires an opaque flat surface, where the emitted light beam is reflected on the surface at angle θ, which in turn depends on the roughness of a surface. Thus, the light reflection angle θ is the principal characteristic for studying microstructures at relatively low magnifications ($<1000x$).

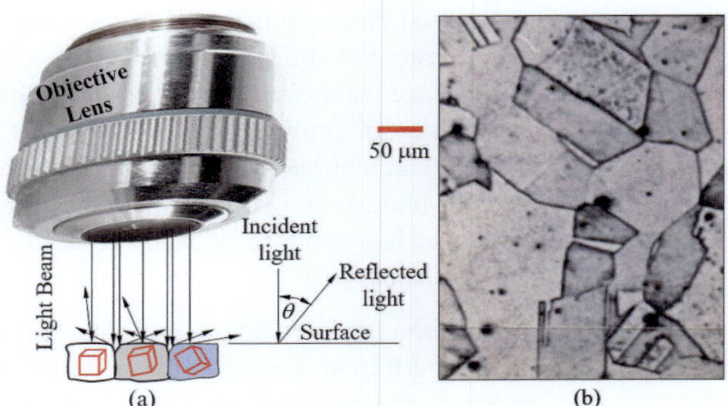

Fig. 4.2 Optical or light microscopy. (**a**) Light beam emitted from an objective lens and front view of grains with unit cell orientations and (**b**) actual grains with different darkness due to crystallographic orientations of the unit cells in a 304 stainless steel specimen

4.4.1   Metallographic Etching

Etching is a chemical or corrosion process used to enhance or reveal microstructural features, such as grain and particle morphology on the surface of a mirrorlike polished specimen. In order to reveal microstructural features of a single-phase or multiple-phase polycrystalline solid, metallographic specimens (Fig. 4.1a) are carefully ground, polished to a mirrorlike surface finish (Fig. 4.1b), and etched with an appropriate chemical solution (reagent).

Figure 4.2a illustrates the incident and reflected light beams for variations in darkness of grains due to differences in crystallographic orientations, which are symbolically represented by the units cells. Moreover, Fig. 4.2b depicts an actual microstructure of a polished and etched metallographic specimen.

It is clearly evident in Fig. 4.2b that the different dark tonalities of the grains is due to different light reflection angles. For $\theta \rightarrow 0°$ light reflection, the grains look white or bright due to a smooth grain surface. For $0° < \theta < 90°$ light reflections, the grains are gray due to slight rough grain surfaces attributed to the crystallographic orientations. On the other hand, for $\theta \rightarrow 90°$ light reflection, the grains look black due to rough specimen surface. Regarding Fig. 4.2b:

- The etched microstructure, as revealed under an optical microscope, exhibits grains with different surface texture due to distinct crystallographic orientations. This means that crystallographic orientations generate variant bright-reflective surfaces or different etching characteristics grains.
- The grain boundaries appear as irregular black or white lines since they represent grooves being formed during etching at a high rate. Basically, etching is a corrosion or metal dissolution process.

Fig. 4.3 (**a**) Zeiss Axio A1 optical microscope (Web site [2]) and (**b**) actual image of spangold alloy (76Au-19Cu-5Al) taken from George Vander Voort "Micrograph Gallery" available online [3]

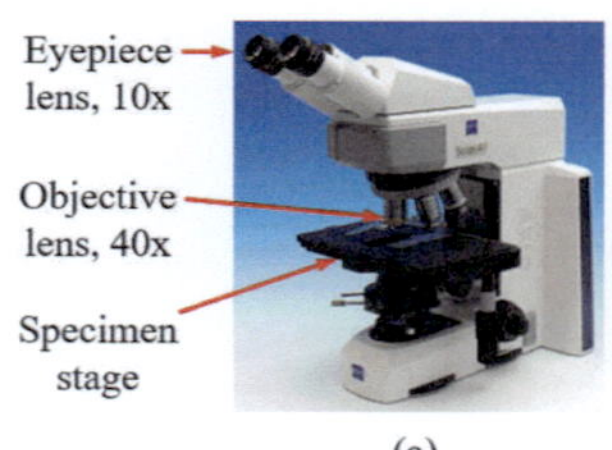

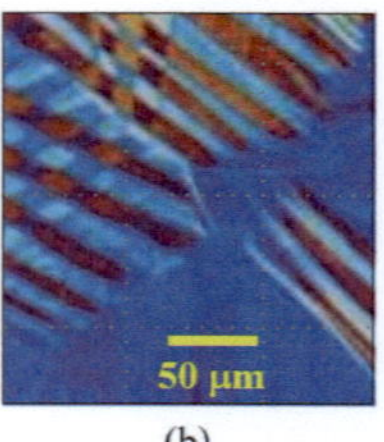

(a)　　　　　　　　　　　(b)

- If the optical microscope is equipped with a Nomarski effect (prism interference pattern), then one reveals a colorful microstructure. Also, tint-etching solutions are used as a microstructural coloration method.

Crystallographic orientations are treated as grain surface roughness that affect the light reflection angles and the darkness of grains. Obviously, grain boundaries contain defects, and consequently, most grain boundaries appear as irregular black lines in the form of grooves due to a severe chemical action of the etching solution.

4.4.2 Magnification and Scale Bar

The objective of this section is to illustrate a simple methodology for determining the scale bar on a photomicrograph or image if the magnification is known and vice versa. Thus, the measured scale-bar length is a reference for measuring the size of a feature under the microscope or on an image. Most modern microscopes provide a scale bar based on the magnification ($M = \#x$) of the field of view. This is important in determining the size of a particular feature on an image of a metallic particle embedded in a metal microstructure or image of a biological cell.

The magnification of a lens describes the ratio between image and object size as illustrated in Fig. 4.3a for a Zeiss Axio Scope A1 microscope (Web site [2]). The magnification for a spangold image ($76Au$-$19Cu$-$5Al$) in Fig. 4.3b taken from Vander Voort [3] "Micrograph Gallery" is shown as a L-line representing a 50-μm scale bar.

The results of black and white (B&W) and color metallography (tint etchants stain microstructural features) are suitably shown in Fig. 4.4 for two selected nonferrous alloys having unique microstructural features.

Tint etchants are sensitive to crystallographic orientation (Vander Voort [4]). Thus, color or tint etchants can reveal hidden microstructural features in the B&W microstructure, and it is, to an extent, aesthetic since it configures the perception of beauty in materials science and physical metallurgy (metal-making field).

For comparison purposes, Fig. 4.4a and b show the microstructure of typical brass in B&W and stained, respectively, having the same 50-μm line as the scale bar (Vander Voort [4]). Similarly, Fig. 4.4c (Levey et al. [5]) and Fig. 4.4d (Vander

Fig. 4.4 B&W and color micrographs. (**a**) B&W microstructure of wrought aluminum brass ($76Cu$-$22Zn$-$2Al$) annealed at $750°C$ (Vander Voort [4]), (**b**) color microstructure of wrought aluminum brass ($76\%Cu$-$22\%Zn$-$2\%Al$) annealed at $750°C$ (Vander Voort [4]), (**c**) B&W microstructure of Al-Au-Cu alloy (Levey et al. [5]), and (**d**) color microstructure of wrought aluminum brass Au-$19\%Cu$-$5\%Al$ (Vander Voort [4])

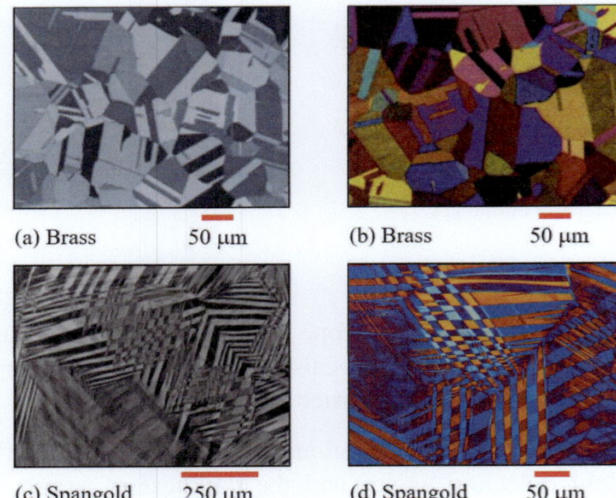

Voort [4]) compare the B&W and stained spangold microstructures, respectively, for their aesthetic characteristics at different scale bars.

Spangold microstructural features are created by heating the alloy at $100°C$ for 2 minutes and quenched in water to induce a martensitic phase transformation, revealed as small facets on surface of the alloy.

Prior to analyzing some real binary phase diagrams, it is convenient to illustrate microscopy calculations with respect to the magnification and scale bar on the image illustrated in Fig. 4.3b. Basically, knowing the scale bar (Fig. 4.3b) allows the calculation of an image magnification. In essence, to use a scale bar, you first need to measure the length of the line on the image. You can do this using a ruler or other measuring tool. Once you have measured the length of the line, you can compare it to the label on the scale bar to determine the actual size of the object in the image. Hence,

- Knowing the scale bar: Use a ruler to measure the length of the L-line below the 50 μm scale bar. Calculate the magnification M by dividing the image size L by the actual size $S = 50$ μm of the image; that is,

$$M = \frac{L}{S} \tag{4.1}$$

- Knowing the magnification: Assign a suitable value to S. Let $S = 100$ μm be the actual image size at M. Calculate the new length of the scale bar line

$$L = SM \tag{4.2}$$

It is convenient to use integer numbers for S and M. Thus, choose a suitable value for S so that the L-line is visible to the reader. It is common practice to label the scale bar line in multiples of millimeter (mm) units. Moreover, the scale bar is a measure of microstructural feature dimensions.

Magnification and scale bars are important tools for understanding the dimensions of objects. They allow objects to be seen in greater detail and to make measurements of objects that are too small to be seen with the naked eye.

It is clear now that magnification is the process of enlarging an object so that it can be seen in greater detail. In microscopy, magnification is the ratio of the size of the image to the size of the object. For example, if an object is magnified 10 times, then the image will be 10 times larger than the object.

Scale bar, on the other hand, is a graphic used to show the actual size of an object in a magnified image. It is typically a line with a label that indicates the length of the line in real-world units. For example, a scale bar might say "100 μm" (micrometers), which means that the line is 100 micrometers long in real life. Moreover, a scale bar is a graphic used to show the actual size of an object in a magnified image. For example, a scale bar might say "200 μm" (micrometers), which means that the line is 200 micrometers long in real life. In conclusion, a scale bar provides a reference point that we can be used to make accurate measurements.

4.5 Optical and Digital Microscopy

Microscopy allows revealing microstructural features; however, the type of microscope for such a purpose plays an important role in microstructural analysis. Actually, the digital microscope (DM) uses optics and a digital camera instead of an eyepiece. The acquisition of images can be accomplished at a large depth of field and at various angles, even on an uneven specimen surface.

Figure 4.5a shows the ferrite microstructure with a relatively equiaxed grain morphology as seen under an optical microscope. However, the same microstructure seen under a digital microscope reveals additional austenitic grains (white grains of

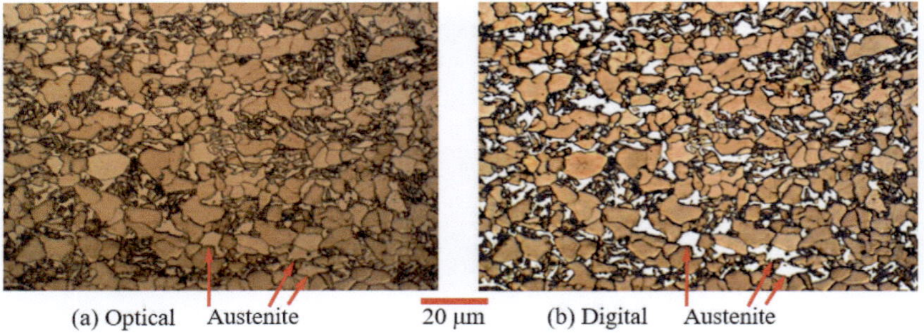

Fig. 4.5 Comparison of (**a**) optical (**b**) digital images of a common $0.24\%C$–$0.40\%Al$–$0.87\%Si$–$98.49\%Fe$ ferric steel microstructure (Grajcar et al. [6])

retained austenite) as illustrates in Fig. 4.5b (Grajcar et al. [6]). For comparison, notice the arrow in Figs. 4.5a, b.

The actual steel microstructure consists of ferritic and austenitic equiaxed grains (Grajcar et al. [6]). Thus, DM is an excellent tool for performing qualitative and quantitative analyses of the microstructural evolution in thermally and mechanically processes due to high-quality images.

4.6 Optical and Scanning Electron Microscopy

The main differences between an optical microscope (OM) and a scanning electron microscope (SEM) are the type of light beam applied to the specimen and the magnification capability. For example, Fig. 4.6a shows a commercial OM (Web site [7]) with a magnification range up to $1000x$, Fig. 4.6b exhibits an OM pearlite microstructure at relatively low magnification, Fig. 4.6c depicts a particular SEM unit with a magnification range that may extend to $100,000x$ (Web site [8]), and Fig. 4.6d shows an SEM image of a stress-corrosion cracking (SCC) specimen (Perez [9]). Moreover, OM requires a flat and polished specimen surface for obtaining a two-dimensional image, while SEM allows a smooth or rough specimen surface for revealing a three-dimensional image.

OM images of microstructures are used to correlate microstructural features, such as average grain size (d), and mechanical properties of polycrystalline solids. Particularly, the microstructure shown in Fig. 4.6b is called pearlite, which is a typical two-phase eutectoid steel consisting of orthorhombic cementite (Fe_3C) phase in the form of irregular black plate-like feature embedded in a white ferrite (α-Fe) matrix.

Fig. 4.6 (**a**) Optical microscope (OM) (Web site [7]), (**b**) pearlite microstructure in steel, (**c**) scanning electron microscope (SEM) (Web site [8]) and (**d**) fractured 304 stainless steel specimen under a stress-corrosion cracking (SCC) experiment (Perez [9])

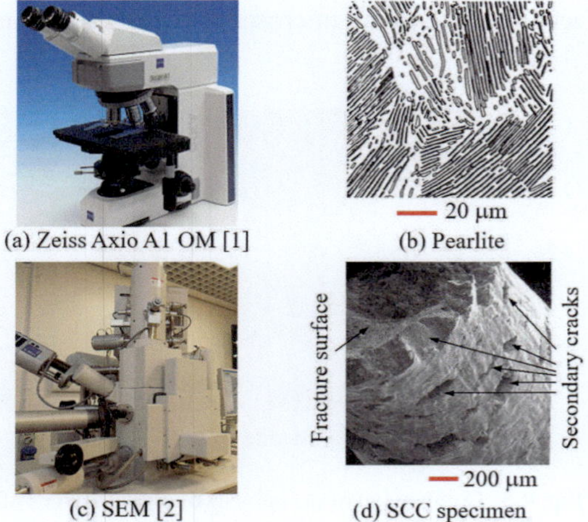

(a) Zeiss Axio A1 OM [1]						(b) Pearlite

(c) SEM [2]						(d) SCC specimen

SEM, on the other hand, produces three-dimensional images of objects, such as the one in Fig. 4.6d for a tensile stress-corrosion cracking (SCC) specimen, which is simultaneously exposed to a corrosive solution and to a tensile loading until fracture (failure). This SEM microphotograph shows secondary cracks as the typical evidence of an SCC experiment.

Microscopy and metallography are the essential techniques for analyzing the microstructural morphology of objects, images, and phases at certain magnifications. Additionally, polarized light microscopy (PLM) is another tool for microstructural analysis.

It is clear now that both optical microscopy (OM) and scanning electron microscopy (SEM) are two powerful tools used to study the microstructure of materials, but they have distinct differences.

4.7 Transmission Electron Microscope

Transmission electron microscopy (TEM) requires a small and thin specimen (less than $100-nm$ thick) so that an electron beam gets transmitted through a specimen to form an image of a particular specimen area at magnifications that may go up to $1,000,000x$. For example, Fig. 4.7a illustrates a Hitachi TEM HF-3300 model found at https://shorturl.at/jkwyS.

Figure 4.7b shows a TEM image of a dislocations network at $80,000x$ magnification in a 304 stainless steel specimen, which cannot be revealed using the conventional OM and SEM. These dislocations are linear defects formed during plastic deformation of specimens under mechanical loading for determining mechanical properties. In general, plastic deformation occurs due to active slip mechanisms.

For the sake of clarity, a dislocation is a small atomic plane having a Burgers vector $\vec{b}$, and it is seen as a line (linear defect) under the microscope. In reality, the dislocation motion is induced by an internal shear stress τ, which is related

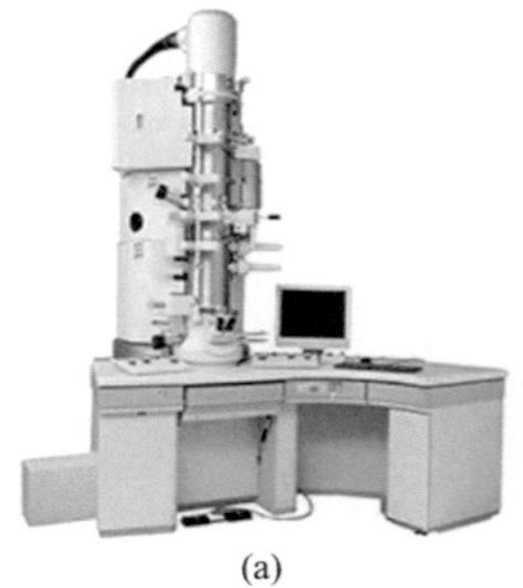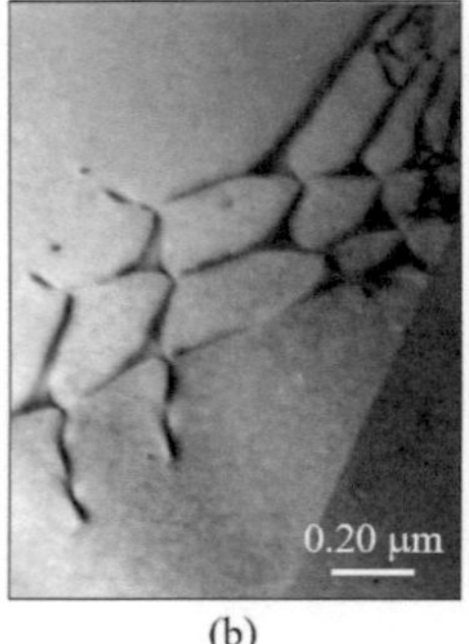

(a)   (b)

Fig. 4.7 (**a**) Hitachi TEM HF-3300 model and (**b**) TEM image showing a dislocation network in a 304 stainless steel specimen at 80,000x magnification

to a dislocation energy. If the dislocation moves, then a massive amount of atoms forming the atomic plane is subjected to an atomic displacement along the dislocation plane.

The theory of electron microscopy is complicated, but at this moment, the goal in this section is to show images of dislocations using microscopy. Nonetheless, dislocations emerge at a planar surface due to force distribution during plastic deformation and form their own morphology. Moreover, a theoretical background on dislocation theory and plasticity is included in a later chapter that deals with crystal defects.

Transmission electron microscope (TEM) is a powerful tool used to study the microstructure of materials at extremely high resolutions. Unlike optical microscopes, which use visible light, TEMs use a beam of electrons to illuminate the specimen. This allows them to achieve resolutions of up to 0.1 nm, making them ideal for observing individual microstructural features. The TEM ability to provide high-resolution images and detailed information about the specimen makes it a valuable instrument in a wide range of scientific and technological fields.

Applications of TEM in material science have a significant role for studying the microstructure of materials to improve their properties. In biology, TEM is used to study the structure of cells and organelles, while in medicine, TEM is suitable for diagnosing diseases by examining tissue samples.

4.7.1   Data Fitting

Regarding dislocation density, the dataset shown in Fig. 4.8 for a $CrMnFeCoNi$ high entropy alloy tested at ultralow temperature (15 K) exhibits interesting characteristics (Naeem et al. [10]).

The data in Fig. 4.8a shows a general curve fitting, and Fig. 4.8b illustrates the same dataset points, but the data curve fitting is carried out differently due to the

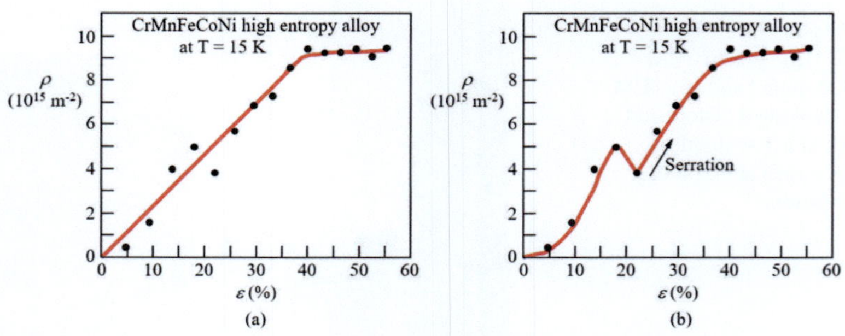

Fig. 4.8 Dislocation density as a function of strain at temperature of 15 K. (**a**) Fitting all data and (**b**) divided fitting at the initiation of serration (Naeem et al. [10])

reported serration behavior of the corresponding stress-strain diagram (Naeem et al. [10]).

In two-dimensional regression analysis of experimental data, the least squares method is the most common approach for finding the relationship between the independent (ε) and dependent (ρ) variables. This is mathematically represented by the generalized function $\rho = f(\varepsilon)$ for a given dataset (ε_i, ρ_i), where $i = 1, 2, 3, \ldots . N$. In essence, plotting $\rho = f(\varepsilon)$ reveals the nature of an experimental dataset: linear or nonlinear behavior or a combination of them. For instance, interpreting experimental data scatter in Fig. 4.8 is important in order to determine the dislocation density trend. Thus, regression analysis allows one to detect and quantify the relationship between two variables as exhibited in Fig. 4.8 for the $CrMnFeCoNi$ high entropy alloy (Naeem et al. [10]).

4.8 Scanning Tunneling Microscopy

Scanning tunneling microscopy (STM) is a powerful technique used to image surfaces at the atomic level. Developed in 1981 by Gerd Binnig and Heinrich Rohrer, it revolutionized our understanding of materials and opened up new avenues for research in physics, chemistry, and biology. As technology advances, one can expect to see STM become even more powerful and versatile. The STM has been used to study different materials. This will likely lead to even more discoveries and innovations in the years to come.

This section includes the powerful scanning tunneling microscopy (STM) technique for revealing and manipulating solid surfaces at an atomic scale. For instance, Fig. 4.9 shows the first STM invented in 1981 by Gerd Binnig and Heinrich Rohrer, who became 1986 Nobel Prize winners (Taken from the Web site [https://shorturl. at/eABM1]).

An electrical circuit is developed within the STM specimen chamber, where an electrical bias voltage is applied to a very sharp metal wire tip electrode or to a specimen (substrate) electrode. Subsequently, STM works by scanning the metal tip over the substrate surface.

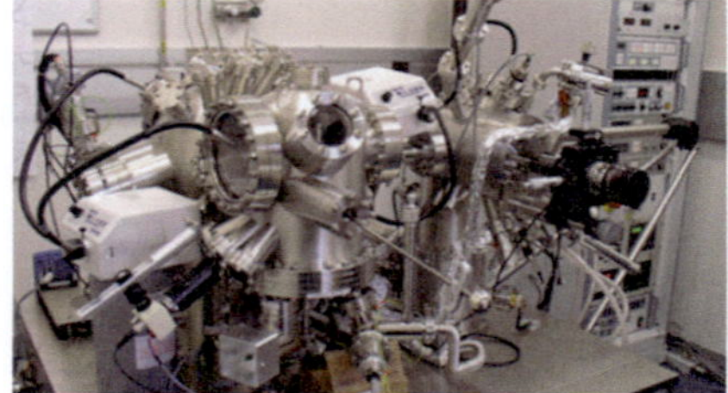

Fig. 4.9 The first STM invented in 1981 by Gerd Binnig and Heinrich Rohrer (1986 Nobel Prize winners) [https://shorturl.at/eABM1]

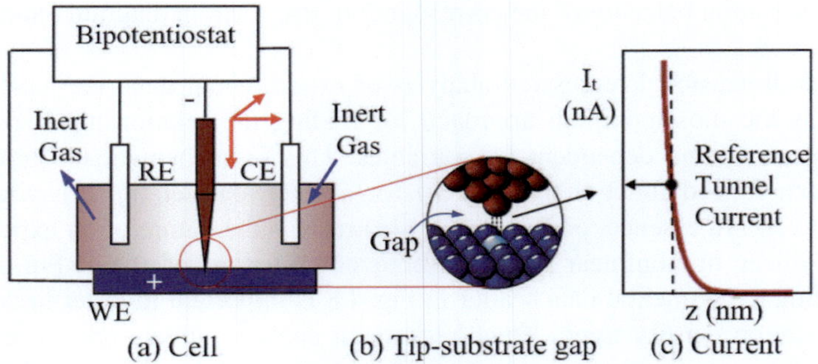

Fig. 4.10 (**a**) Schematic electrochemical cell for in situ EC-STM as per Zhang and Ulstrup [12]. Tip insert:"Figure, Michael Schmid, TU Wien," (**b**) tip-substrate gap and (**c**) the schematic plot based on data from Song et al. [13]

STM instrument is very useful for studying or characterizing the surface topology of substrates without electron or photon beams, and it allows the manipulation of atoms by scanning a sharp metal tip (wire) over the surface.

4.8.1 Electrochemical Cell

For an in situ electrochemical analysis, STM can be made into an electrochemical scanning tunneling microscope (EC-STM) as schematically illustrated in Fig. 4.10a, which depicts the conventional three-electrode electrochemical cell with a bipotentiostat that allows independent control of the electrochemical potential (voltage) of the tip and substrate relative to a reference electrode (RE). The working electrode (WE) is the specimen or substrate under investigation.

The basic principle of a STM unit is based on tunneling of electrons in a gap having a vertical distance in the range $0 < z < 5\ nm$. This tunneling distance is located between an electrically conductive sharp probe tip wire acting as an electrode ($Pt\text{-}Ir$ or tungsten W) and a working electrode (WE) or substrate (McBreen [11, p. 485]). When the EC-STM cell is excited in an ultrahigh vacuum (UHV) by an applied bias voltage (ϕ_{bias}), electrons flow across the gap (Fig. 4.10b) generating a measurable tunneling current I_t (Fig. 4.10c), which is a variable to be controlled during in situ STM or EC-STM work.

At a discrete vertical distance z, electrons can quantum mechanically tunnel through the gap between the STM tip and the substrate surface generating a tunneling current (I_t) described by an exponential decay function (Fig. 4.10c) (Song et al. [13])

$$I_t = I_o \exp\left(-\beta z\right) = \phi_{bias}/R_t \qquad (4.3)$$

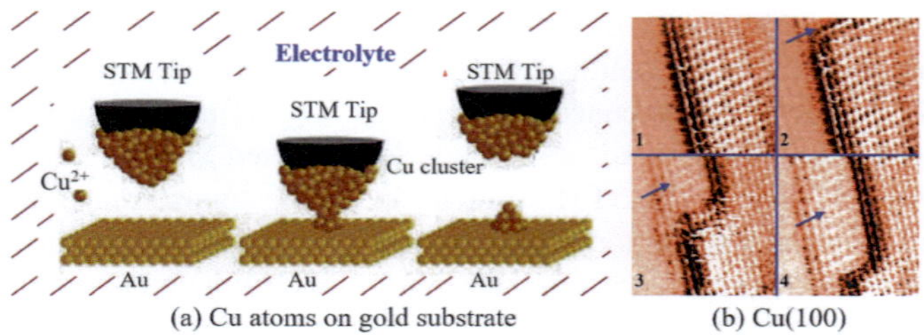

(a) Cu atoms on gold substrate   (b) Cu(100)

Fig. 4.11 (**a**) Tip-induced metal deposition of copper (Cu) on gold (Au) surface (Kolb and Schneeweiss [17]). (**b**) Atomic scale dynamic behavior of copper, Cu(100), dissolution in an 0.01 M HCl solution at $-0.23V$ vs. SCE (Magnussen et al. [18]). Images taken from Prof. Olaf Magnussen video-STM available online, http://www.atomic-movies.uni-kiel.de. These images (size: 5-*nm* × 5-*nm*) were captured using Windows Media

where R_t is the effective resistance of the tunneling gap, typically in the range $10^9 \ \Omega \leq R_t \leq 10^{11} \ \Omega$ (= Ohms) (Bard and Faulkner [14, p. 660]). Moreover, I_o and β are constants that must be determined through nonlinear curve fitting using experimental data; $I_t = f(z)$. The behavior of this function with respect to the reference tunneling current, however, implies that $T_t \rightarrow \infty$ as $z \rightarrow 0$. Therefore, I_t is very sensitive to the vertical distance (z) or gap between the STM tip and the WE.

Normally, the saturated calomel electrode (SCE) is used as the reference electrode (RE) with insignificant electrical current flow. The counter electrode (CE) is an auxiliary electrode (AE) that supplies the current required by the WE without limiting the measured response of the cell.

From a physics point of view, an applied positive potential known as a bias voltage ($\phi_{bias} > 0$), the STM tip causes electrons to tunnel from a tip to the WE. As a result of this electrochemical process, the potential in a gap decreases along with the energy level of the WE. These electrons have the tendency to flow toward empty (unfilled) states of the sample surface. Conversely, an applied negative potential ($\phi_{bias} < 0$) on the tip causes electrons to tunnel into the tip from the sample. Consequently, the energy gap and the energy level of the WE increase (von Bergmann [15], Chen [16, p. 30]).

Figure 4.11a schematically depicts a sequence of images for copper dissolution on a gold (Au) substrate surface (Kolb and M.A. Schneeweiss [17, pp. 26–30], Magnussen et al. [18]), and Fig. 4.11b shows a series of images illustrating copper (Cu) deposition in areas being marked with arrows. The images in Fig. 4.11b were captured from the atomic movie library or *video*-STM of Prof. Olaf Magnussen (https://www.atomic-movies.uni-kiel.de) using windows media player. These images are also found elsewhere (Engel and Reid [19, p. 284]).

Figure 4.11a schematically shows copper cations (Cu^{2+}) in an electrically conductive electrolyte solution being collected around the STM tip, and subsequently, the STM tip is brought down to contact the sample, leaving a cluster of Cu atoms on terraces. Then, Cu is reduced, $Cu^{2+} + 2e^- \rightarrow Cu$, at the STM tip, which moves when attached to a 3D piezoelectric scanner, which is subjected to be electrically charged for adjusting or controlling the position of a STM metal tip.

Most STM work to date concentrates on the crystalline surface structures of face-centered cubic (FCC) and hexagonal close-packed (HCP) materials (Duffe et al. [20, pp. 1–8]). If the STM work is for metal deposition, then an atomic reorganization initially forms a monolayer on the substrate. This monolayer can be defined as a pseudomorphic structure since the atoms being deposited have to adopt a lattice space of the substrate. For example, in the $Cu/Au(001)$ atomic coupling, Cu settles in the lattice spacing of the Au substrate in the (001) crystallographic plane.

The steady-state tunneling current, $I_{t,\infty}$ (referred to as Faradaic current), for an ultramicroelectrode (UME) or UME-disk with radius r is given by (Bard and Faulkner [14, p. 660], Bruckenstein and Janiszewska [21])

$$I_{t,disk} = 4zFCDr \qquad (4.4)$$

where r denotes the tip radius, $F = 96,500\ C$ denotes the Faraday constant, C denotes the Cu^{2+} ion concentration, and D denotes the diffusion coefficient of Cu^{2+} ions.

ESTM Feedback Mode The $ESTM$ technique can be in the feedback mode or collective mode represented as generalized electrochemical reaction of the form (Bard and Faulkner [14, p. 670])

$$O + ze^- \rightleftarrows R \qquad (4.5)$$

Here, O is the oxidized ion, such as Cu^{2+}; R is the reduced ion, such as Cu; and ze^- is the number of electrons being transferred between two electrodes. This expression represents an electrochemical reaction that is influenced by the passage of an electric current along with ion motion within an electrolyte that may be a multistep process.

4.9 Atomic Force Microscopy (AFM)

The atomic force microscopy (AFM) with a movable probe is a sophisticated technique widely used to obtain atomic-scale images of metal surfaces. For instance, Fig. 4.12a shows a commercial AFM from WITec Physik Instrument (WITec PI, www.pi.ws), and Fig. 4.12b illustrates the AFM model, including the specimen topology and components.

This instrument is a powerful tool for nanoscale measurements of deflection of the cantilever beam and the corresponding applied force by the probe tip. Notice that a laser beam and a position-sensitive photo diode (PSPD) are coupled so that any deflection of the cantilever beam toward or away from the substrate surface is detected or recorded by the PSPD.

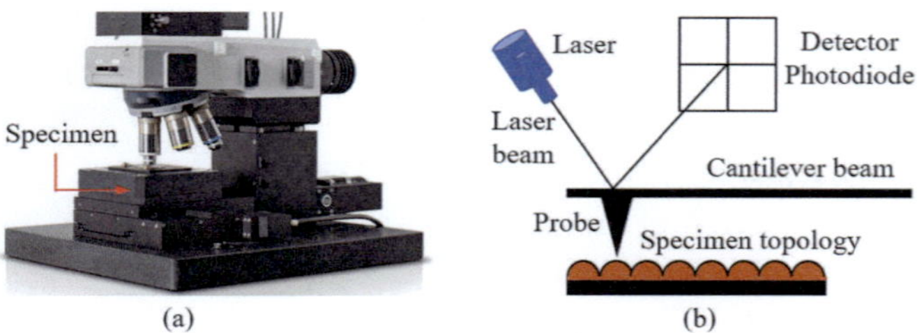

Fig. 4.12 (**a**) AFM from WITec Physik Instrument (WITec PI, www.pi.ws) and (**b**) AFM model

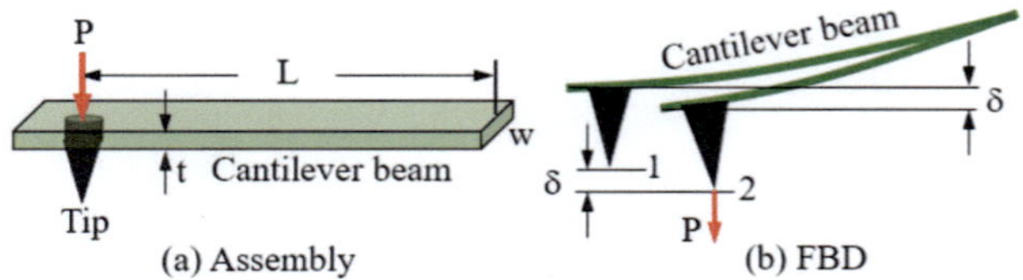

Fig. 4.13 (**a**) A straight plate-tip acting as an AFM assembly and (**b**) free-body diagram (FBD) of a cantilever beam

The AFM probe attached to a small cantilever beam (spring-like part) works as a surface sensor or as a raster scanning device across the specimen surface. It moves in the x, y, z directions along the specimen surface at a controlled distance (nanoscale) from the specimen surface to scan the specimen topology and obtain an AFM image as detected by a photodiode. The AFM can also be used as a source of mechanical bending force. Nonetheless, the probe tip acts a raster scanning device across the specimen surface.

Based on the information provided about the AFM cantilever assembly in Fig. 4.13a and its free-body diagram (FBD) in Fig. 4.13b, the AFM load or mechanical force (P) on a substrate surface in both tapping and continuous modes can be predicted using a first-order (Hooke's law) or a third-order equation.

According to Hooke's law (first order) and Lee et al. [22] (third order), equations for the AFM compressive point load P are written as

$$F = k\delta \qquad \text{Linear Hooke's law, macroscale} \qquad (4.6a)$$

$$k = \frac{Ewt^3}{4L^3} \qquad \text{Macroscale} \qquad (4.6b)$$

$$F = a_1\delta + a_2\delta^3 \qquad \text{Nonlinear, microscale} \qquad (4.6c)$$

where k is the cantilever spring constant (stiffness of the lever or spring-like stiffness), δ is the deflection, w, t, L are the cantilever dimensions, E is the modulus of elasticity, and a_1, a_2 are constants. According to Lee et al. [22], the P-δ behavior using the AFM technique is nonlinear for graphene; Eq. (4.6c). This is graphically depicted in Fig. 4.14 (Lee et al. [22]) for a monolayer and bilayer graphene samples (see Chap. 5).

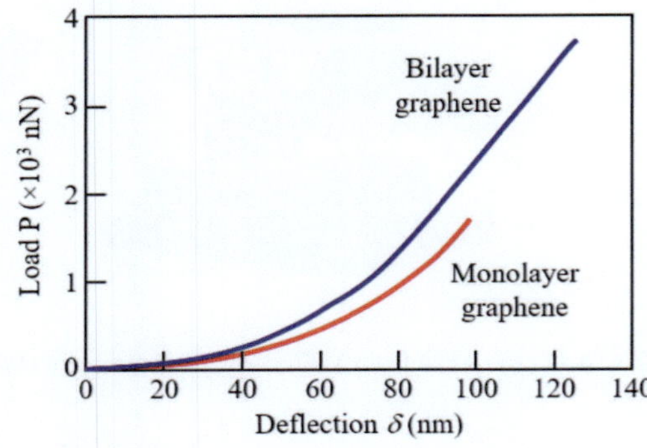

Fig. 4.14 Mechanical load deflection diagram for graphene (Lee et al. [22])

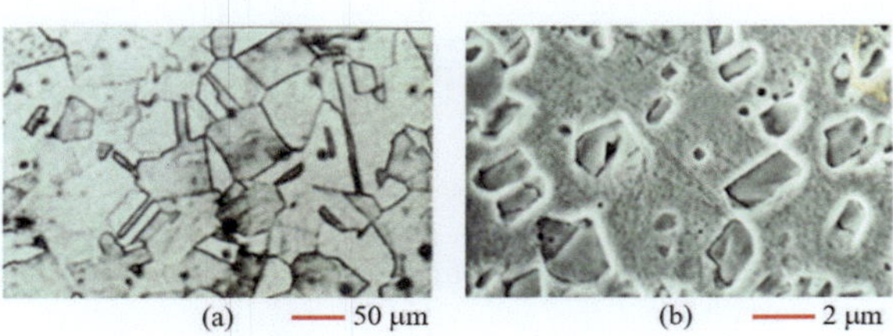

Fig. 4.15 (**a**) Crystalline microstructure of AISI 304 stainless steel and (**b**) partially amorphous microstructure $Ni_{53}Mo_{35}B_9Fe_2$ alloy containing embedded boride particles (Perez [9])

4.10 Optical and Transmission Electron Microscopy

By definition, grain boundaries and matrix-particle interfaces are two-dimensional defects that have the appearance of irregular lines under the microscope. For instance, Fig. 4.15a shows the microstructure of an AISI 304 stainless steel being annealed at $1000°C$ for 0.5 hour.

Notice the grain boundaries as dark lines because of the severe chemical attack using an Aqua Regia etching solution ($80\%HCl+20\%HNO_3$). Sometimes heating up (using safety precautions) an etching solution just a few degrees may resolve microstructural features very clear.

Figure 4.15b depicts a partially amorphous microstructure of $Ni_{53}Mo_{35}B_9Fe_{2\ 2}$ (Devitrium 7025) alloy being annealed for 24 hours at $1100°C$. This image shows crystalline particles etched with Marble's reagent. The AISI 304 material has an FCC microstructure, and the Devitrium 7025 alloy contains crystalline boride particles embedded in the Ni-Mo amorphous phase (no grain boundaries), as revealed by SEM and confirmed by X-ray diffraction (Perez [9]). The interfaces in Devitrium 7025 microstructure appear as bright areas due to optical effects.

The grain and interface boundaries in crystalline solids represent high-energy areas due to the atomic mismatch. These types of boundaries are microstructural defects that affect the electrical and thermal conductivity of the material and are the preferred sites for the onset of corrosion.

The dislocation network for the AISI 304 alloy is shown in Fig. 4.16a, and that for the $Ni_{53}Mo_{35}B_9Fe_2$ alloy is depicted in Fig. 4.16b. These dislocation networks represent linear defects generated during plastic deformation in tension mode (Perez [9]).

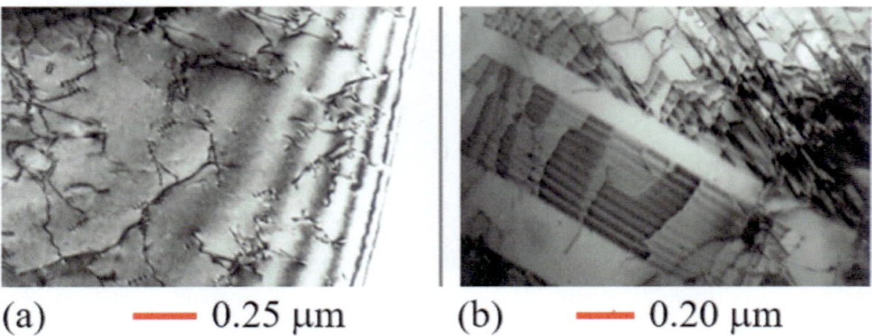

Fig. 4.16 Bright-field TEM photomicrographs showing dislocation networks for (**a**) AISI 304 stainless steel at 68,000x magnification and (**b**) $Ni_{53}Mo_{35}B_9Fe_2$ alloy at 80,000x magnification (Perez [9])

With respect to Fig. 4.16b, there is a clear sub-grain boundary shown as a dark horizontal line across the upper part of the TEM photomicrograph. By definition, the small white areas surrounded by dislocations are called sub-grains. Ultimately, the choice between OM and TEM hinges on the specific needs of the research since both techniques offer unique insights into the world of materials.

4.11 Stereology

Stereology in materials science includes a quantitative analysis, known as stereometric or quantitative metallography, for three-dimensional (3D) objects or phases, where the term "stereo" means solid. Moreover, stereology is based on systematic uniform random sampling.

The principal procedure in quantitative metallography involves the resolution of a microscopic image, conventionally known as a micrograph, the corresponding magnification for visualizing a relatively large number of grains in a single phase or multi-phase microstructure, and the microscopic technique used to obtain the microstructure. For instance, an optical micrograph (OM) provides a larger area of the microstructure than a SEM or TEM instrument. However, an OM micrograph may not show details seen under the SEM and TEM technique.

Nowadays, computer-assisted image analysis prevails among analysts. However, it is important for the reader to understand the statistical uncertainties related to systematic errors, methods for data analysis, and the methodology used to determine microstructural features, such as grain size, grain morphology, and spatial distribution of phases.

The importance of quantitative metallography is related, specifically, to mechanical behavior and properties of crystalline materials. The corrosion behavior of these materials should not be excluded from the common microstructural analysis. The main goal in this section is to introduce the methodology for grain analysis using the American Society for Testing Materials (ASTM) E112 and the International Organization for Standardization (ISO) 643 standard test procedures.

4.12 Metal Grain Geometry

By definition, a grain is a 3D particle with continuous stacking of unit cells forming a polyhedron of a specific geometry. For illustration purposes, Fig. 4.17 shows a sphere (Fig. 4.17a) and two polyhedron models of grains, such as a (1) tetrakaidekahedron with 14 faces, 24 corners, and 36 edges (Fig. 4.17b), and (2) a dodecahedron with 12 faces (Fig. 4.17c).

These polyhedron models in Fig. 4.17b, c represent the visual perception of spherical grains. This is the reason for using the term "average grain diameter" or "average grain size" in quantitative metallography.

In order to clarify the concept of grain geometry in a two-dimensional grain analysis using an optical microscopy (OM), Fig. 4.18a shows an equiaxed-grain

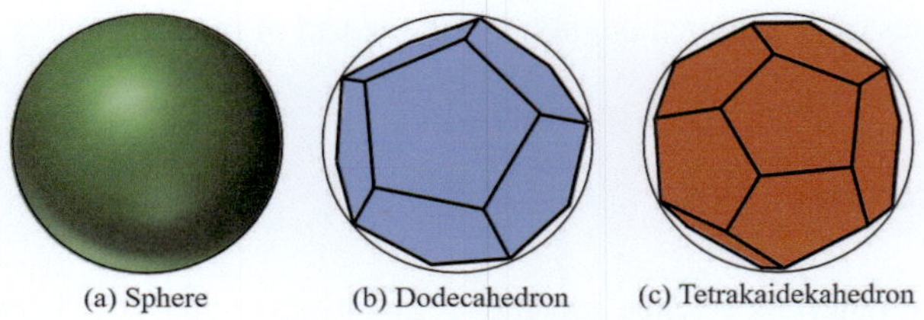

Fig. 4.17 Metal grain geometry model. (**a**) Sphere, (**b**) a dodecahedron within a sphere, and (**c**) a tetrakaidekahedron within a sphere

Fig. 4.18 Grain morphology. (**a**) Equiaxed grain microstructure of an annealed 70Cu-30Zn brass, (**b**) as-cast 80Cd-20Bi alloy showing a dendritic structure, and (**c**) cold worked 70Cu-30Zn brass exhibiting elongated grains along the rolling direction (RD). After Vander Voort [4], Vander Voort [23]

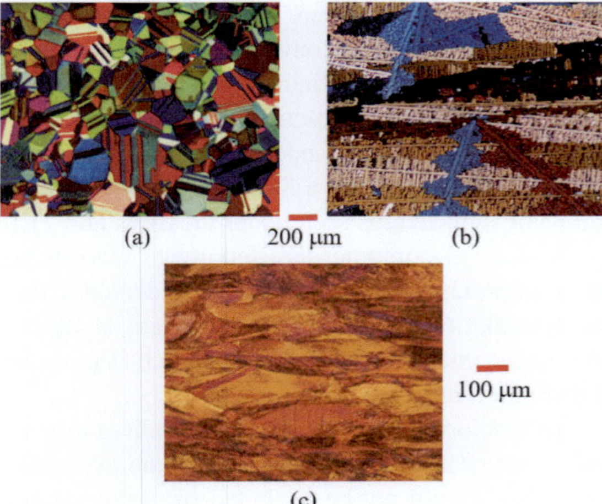

microstructure of an annealed $70Cu$-$30Zn$ brass sample, where the grains have similar coordinate lengths.

A non-equiaxed-grain geometry exhibits different coordinate lengths as shown in Fig. 4.18b for as-cast $80Cd$-$20Bi$ and Fig. 4.18c for a cold worked $70Cu$-$30Zn$ alloys. Actually, these microstructures are taken from Vander Voort [4] and Vander Voort [23] work on metallography posted online.

During mechanical deformation by cold working, the initial equiaxed grains are elongated as shown in Fig. 4.18c. This particular non-equiaxed grain geometry represents a relatively hard microstructure containing defects, such as dislocations, which in turn, are revealed under a TEM instrument.

4.13   ASTM Grain-Size Number

This section includes the procedure for determining the ASTM grain-size number (G) on the image of a microstructure projected at a magnification of $100\times$, followed by a comparison procedure with a series of graded standard grain-size charts as per ASTM E122 method. The comparative or qualitative method is used by metal foundries for a rapid quality control assurance during metal production. In essence, the ASTM grain-size number is a system used to quantify the average grain size of polycrystalline materials. It is a standardized method defined by the American Society for Testing and Materials (ASTM) in Standard Test Method E112. The reader is encouraged to consult the ASM Handbook Volume 9: Metallography and Microstructures.

Figure 4.19a shows an equiaxed-grain microstructure taken at $100\times$ magnification (Web site [24]), and Fig. 4.19b illustrates the ASTM chart (20-mm diameter reticle) for the grain-size number $1 \leq G \leq 10$ (Web site [25]). Also, Fig. 4.19c shows a table of ASTM grain size.

The image in Fig. 4.19a is the BCC-ferritic microstructure, which is the common phase in pure iron, and it is the dominant crystalline phase in low-carbon steels. For comparison, Fig. 4.19b shows the ASTM chart with a $20\ mm$ scale bar for determining the apparent ASTM grain-size number $G_o = 3$ (Fig. 4.19a).

Despite that the microstructure in Fig. 4.19a contains different grain sizes at a magnification M, the general perception of the grained structure is that the grains are similar; hence, this is an equiaxed grain structure.

A grain is essentially a small three-dimensional region (volume) that represents a single crystal having a certain continuous crystallographic orientation as a result of the liquid-solid phase transformation called solidification. In polycrystals, each grain has its own crystal orientation separated by grain boundaries, which are seen under the microscope as dark or black irregular lines. These lines are colored in Fig. 4.19a for decoration or aesthetic purposes.

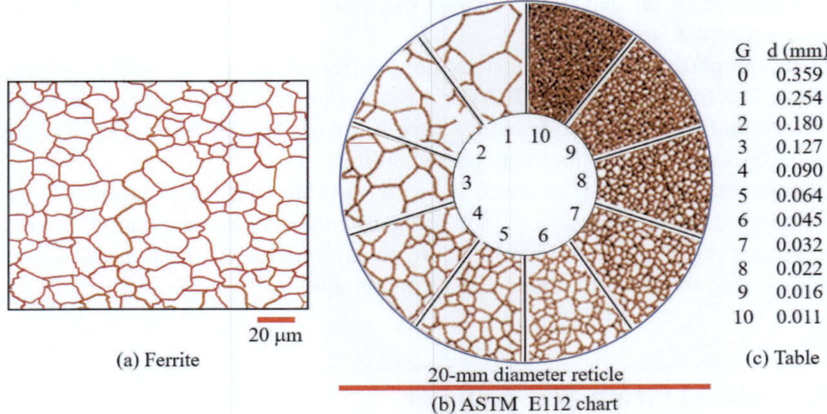

Fig. 4.19 Comparison method for the ASTM grain-size number E112 at 100×. (**a**) Ferrite in a low-carbon steel (Web site [25]), (**b**) ASTM chart (Web site [25]), and (**c**) table of ASTM grain size

Example 4.2 Determine (**a**) the magnifications of the ferrite image (M_i) shown in Fig. 4.19a and from the ASTM chart image (M_c) in Fig. 4.19b based on the printed scale bar lengths and (**b**) magnification adjustment factor Q.

Solution

(**a**) For the ferrite image in Fig. 4.19a,

$$M_i = \frac{L_{im}}{L_{ip}} = \frac{13.2\ mm}{20\ \mu m} = \frac{13.2 \times 10^3\ \mu m}{20\ \mu m} = 660\text{x} \tag{4.7a}$$

$$M_c = \frac{L_{cm}}{L_{cp}} \times (100\text{x}) = \frac{132\ mm}{20\ mm} \times (100\text{x}) = 660\text{x} \tag{4.7b}$$

Here, M_i is the ferrite image magnification (in Fig. 4.19a), M_c is the ASTM chart image magnification (in Fig. 4.19b), L_{im} is the image measured scale bar length with a ruler, L_{ip} is the printed scale bar length, L_{cm} is the measured scale bar length with a ruler, and L_{cp} is the printed scale bar length.

(**b**) The magnification adjustment factor Q is defined by the ASTM E112 standard as

$$Q = 2\log_2\left(\frac{M_i}{M_c}\right) = 2\left[\log_2\left(M_i\right) - \log_2\left(M_c\right)\right] \tag{4.8a}$$

$$Q = 2\left[\frac{\log\left(M_i\right)}{\log\left(2\right)} - \frac{\log\left(M_c\right)}{\log\left(2\right)}\right] = 2\left(\frac{\log\left(660\right)}{\log\left(2\right)} - \frac{\log\left(660\right)}{\log\left(2\right)}\right) = 0 \tag{4.8b}$$

$$Q = 6.6439\left[\log\left(M_i\right) - \log\left(M_c\right)\right] = 6.6439\left(\log\left(660\right) - \log\left(660\right)\right) = 0 \tag{4.8c}$$

Thus, the ASTM E112 grain-size number with $Q = 0$ is

$$G = G_o + Q = 3 + 0 = 3 \tag{4.9}$$

Accordingly, G is also calculated in a later section using the intercept method for determining the apparent grain density having units of $grains/in^2$ or $grains/mm^2$.

4.13.1   Logarithms

Logarithmic functions are used in physical and natural sciences and engineering. Specifically, the metallic grain and geological particle sizes are commonly characterized using this type of function. For convenience of the reader, below is a useful list of the logarithms

- $\log_a(y) = x$ is the general logarithm base "a" of y so that the inverse of the logarithmic operation yields a power function $y = a^x$, where $a = 2, 3, \ldots .10$.
- $\log_2(y) = x$ is the logarithm base "2" of y so that $y = 2^x$. For example, $\log_2(y) = 8$ implies that $y = 2^8 = 256$.
- $\log_3(y) = x$ is the logarithm base "3" of y so that $y = 3^x$. For example, $\log_3(y) = 9$ implies that $y = 3^9 = 19683$.
- $\log_{10}(y) = \log(y) = x$ is the common logarithm base "10" of y so that $y = 10^x$. For example, $\log(y) = 2$ implies that $y = 10^2 = 100$.
- $\log_e(y) = \ln(y) = x$ is the natural logarithm base "e" of y so that $y = e^x = \exp(x)$. For example, $\ln(y) = 1$ implies that $y = e^1 = e$.
- $\log_a(x) = \log_2(x) / \log_2(a)$ with a general base "a" is useful for logarithmic conversion.

Table 4.1 provides a list of some logarithmic properties. If the logarithm has a base or radix other than two, then one can use its property as a guide when working with logarithms.

Logarithms base 10, e, and 2 of y stand out in most applications. The binary logarithm $\log_2(x)$ is used in the metal industry for defining the ASTM grain size number G as part of the comparison method, and the common or base 10 logarithm $pH = \log\left[H^+\right]$ is used in chemistry for determining the acidity or alkalinity of a solution containing a concentration of hydrogen ions $\left[H^+\right]$, from which $pH = -log(H^+)$.

Table 4.1 Property of logarithms

$\log_2(N) = \log(N)/\log(2)$	$\log(2) = 0.30103$
$y = a^x, \log_a(y) = x$	$\log_a(a) = 1$
$\log_a(x^y) = y\log_a(x)$	$\log_2(2) = 1$
$\log_e(x) = \ln(x)$	$\log_{10}(10) = \log(10) = 1$
$e = 2.718281\ldots.$	$\log_2(10) = 3.3219$

4.13.2   Grain Density and ASTM Grain Size Number

Magnification M = 100x Let N be the grain density or grains per unit area $(grains/in^2)$ and G be the ASTM E112 grain-size number so that $N = f(G)$.

(1) For N in $grains/in^2$,

$$N = 2^{G-1} \quad (grains/in^2) \tag{4.10a}$$

$$G = 1 + \log_2(N) = 1 + \frac{\log(N)}{\log(2)} \tag{4.10b}$$

$$G = 1 + 3.3219 \log(N) \tag{4.10c}$$

(2) For N in $grains/mm^2$,

$$N = \left(\frac{1}{25.4}\right)^2 \left(2^{G-1}\right) = \left(1.55 \times 10^{-3}\right) 2^{G-1} \tag{4.11a}$$

$$G = -8.3335 + 3.3219 \log(N) \tag{4.11b}$$

Magnification M $\neq$ 100x The $N = f(G, M)$ and $G = f(N, M)$ relations are conveniently modified as shown below.

(3) For N in $grains/in^2$,

$$N_M = N \left(\frac{M}{100}\right)^2 = 2^{G-1} \tag{4.12a}$$

$$N = \left(\frac{100}{M}\right)^2 \left(2^{G-1}\right) \tag{4.12b}$$

$$G = 1 + \log_2\left[N\left(\frac{M}{100}\right)^2\right] = 1 + \log\left[N\left(\frac{M}{100}\right)^2\right] / \log(2) \tag{4.12c}$$

$$G = -12.2880 + 3.3219 \log(N) + 6.6438 \log(M) \tag{4.12d}$$

$$G \approx -35.2390 + 7.6503N + 15.3007M \tag{4.12e}$$

This is a convenient expression, Eq. (4.12d), for calculating the ASTM grain size number G at any magnification. This expression is linearized as defined by Eq. (4.12e) using the conversion $log(x) = 2.303 ln(x)$ and Taylor's series approximation $ln(x) \approx x - 1$.

(4) For N in $grains/mm^2$,

$$N\left(\frac{M}{100}\right)^2 = \left(\frac{1}{25.4}\right)^2 \left(2^{G-1}\right) \tag{4.13a}$$

$$N = \left(\frac{100}{M}\right)^2 \left(\frac{1}{25.4}\right)^2 \left(2^{G-1}\right) \tag{4.13b}$$

$$\log_2(N) = \log_2\left(\frac{15.5}{M^2}\right) + \log_2\left(2^{G-1}\right) \tag{4.13c}$$

$$G = 4.9542 + 3.3219\log(N) - 6.6439\log(M) \tag{4.13d}$$

$$G \approx 12.6048 + 7.6503N - 15.3009M \tag{4.13e}$$

This expression, Eq. 4.13e, has been linearized using the conversion $log(x) = 2.303ln(x)$ and Taylor's series approximation $ln(x) \approx x - 1$.

Magnification M = 1x Now, convert N to $grains/mm^2$ units by multiplying Eq. (4.10a) by 15.5 so that the ASTM E112 G becomes

$$N = \left(\frac{1}{25.4}\right)^2 \left(\frac{100}{1}\right)^2 \left(2^{G-1}\right) = 15.5 \times 2^{G-1} \tag{4.14a}$$

$$\log_2(N) = \log_2\left(15.5 \times 2^{G-1}\right) = \log_2(15.5) + \log_2\left(2^{G-1}\right) \tag{4.14b}$$

$$\log_2(10)\log(N) = \log_2(10)\log(15.5) + G - 1 \tag{4.14c}$$

$$G = -2.9542 + 3.3219\log(N) \quad \text{(ASTM E112)} \tag{4.14d}$$

$$G = -3 + 3.3219\log(N) \quad \text{(ISO 643)} \tag{4.14e}$$

For comparison purposes, the ASTM-E112 Eq. (4.14d) and ISO-643 Eq. (4.14e) at $M = 1x$ magnification are very similar since $\Delta G = |G_{ASTM} - G_{ISO}| = 0.0458$. The $N = f(G)$ power relationship is depicted in Fig. 4.20 at 100x magnification for a hypothetical material.

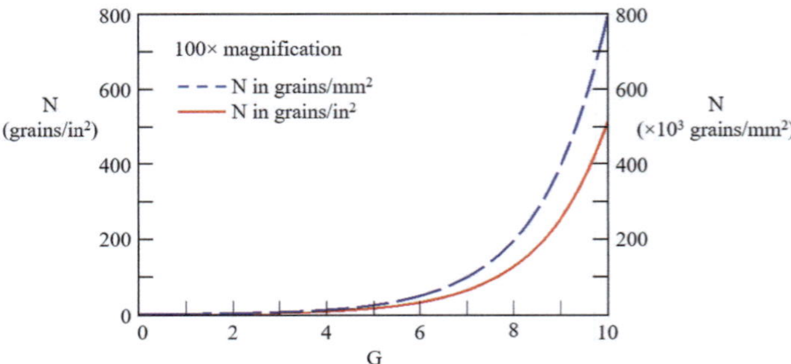

Fig. 4.20 Grain density function $N = f(G)$ at 100× magnification

It is clearly shown in Fig. 4.20 that the $N = f(G)$ function gives a nonlinear trend at $100x$ magnification and a significant comparison approach for characterizing the average grain size.

4.14   Average Grain Size Measurements

Measuring the average grain size of a material is crucial in various fields, including metallurgy, materials science, and engineering. It provides valuable insights into the material's properties and behavior, ultimately impacting its performance and applications.

4.14.1   ASTM Method

It is evident that the dimensional grain size (d) controls the mechanical properties of crystalline materials. In stereology, the quantitative metallography method provides means for predicting the average dimensional grain size of a microstructure.

Accordingly, $d = f(G)$ is written in terms of grain density (N) and ASTM grain-size number at a magnification M

$$d = \frac{1}{\sqrt{N}} = \frac{1}{\sqrt{2^{G-1}}} \quad (\text{in } inches, \text{ at } M = 100\text{x}) \tag{4.15a}$$

$$d = \frac{1}{M}\frac{1}{\sqrt{N_M}} = \frac{1}{100}\frac{1}{\sqrt{2^{G-1}}} \quad (\text{in } inches, \text{ at } 1\text{x}) \tag{4.15b}$$

$$d = \frac{1}{M}\frac{25.4}{\sqrt{N}} = \frac{0.254}{\sqrt{2^{G-1}}} \quad (\text{in } mm, \text{ at } 1\text{x}) \tag{4.15c}$$

Here, $M\sqrt{N_M} = M\sqrt{(100/M)^2 N} = 100\sqrt{N} = 100\sqrt{2^{G-1}}$ is a controlling factor in Eq. (4.15b). Moreover, relevant worked examples are given below for emphasizing the usefulness of quantitative metallography on equiaxed grains in single-phase structures.

4.14.2   Heyn Linear Intercept Method

Consider the image containing equiaxed grains shown in Fig. 4.21 with three straight lines (N_x) of equal length $L_x = 66 \, mm$ at a printed magnification M and the Heyn Linear Intercept Method.

Fig. 4.21 Ferritic microstructure at a magnification M and intercepted grains per line

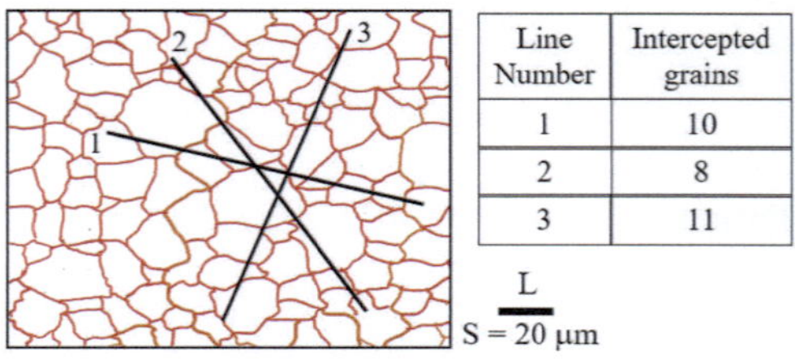

Line Number	Intercepted grains
1	10
2	8
3	11

This is the traditional mean linear-intercept approach for estimating a microstructure's average gain size (d) at a magnification of M. It was first reported in 1903. This technique, which counts intercepted grains by straight lines on an image, works well with single-phase microstructures comprising non-equiaxed (axes of varied length) grains. In essence, the Heyn Linear Intercept Method is a metallographic technique used to estimate the average grain size of a polycrystalline material.

Complications may arise when the grain boundaries are incomplete or not properly etched. Consequently, the linear intercept method would overestimate the grain size. Nonetheless, this method measures planar grain size since metallographic specimens are prepared to have a flat and smooth polished surface. In reality, grains are three-dimensional pack of atoms defined as the lattice with common dodecahedron and tetrakaidekahedron shapes (Fig. 4.17).

Measure the scale bar length (print bar code line), say, $L = 5$ mm and divide it by S. If the image in Fig. 4.21 is copied and placed in another file, then the image gets larger, and subsequently, the magnification changes. Nonetheless, the printed magnification equation is recast again

$$M = \frac{L}{S} = \frac{5 \times 10^3 \; \mu m}{20 \; \mu m} = 250\text{x} \qquad (4.16)$$

Measure the length of lines L_1, L_2, L_3 drawn on the printed microstructure (image), say, $L_x = L_1, L_2, L_3 = 32$ mm, and count the total number of grain-boundary intersections (P) per line. Thus, the total length (L_T) of the lines and total intersections (P) are

$$L_T = N_x L_x = (3)\,(32 \; mm) = 96 \; mm \qquad (4.17a)$$

$$P = 28 \text{ intersections} \qquad (4.17b)$$

(A) The mean intercept length is written as

$$\overline{L}_m = \frac{L_T}{MP} = \frac{96 \; mm}{(250)\,(28)} = 1.37 \times 10^{-2} \; mm \qquad (4.18)$$

Available empirical equations relating $G = f\left(\overline{L}_m\right)$ are convenient. Thus,

$$G = -6.6457 \log \left(\overline{L}_m \right) - 3.298 \quad (\overline{L}_m \text{ in } mm) \tag{4.19a}$$

$$G = -6.6353 \log \left(\overline{L}_m \right) - 12.60 \quad (\overline{L}_m \text{ in } inches) \tag{4.19b}$$

From Eq. (4.19a), the ASTM E112 for the grain-size number becomes

$$G = -6.6457 \log \left(1.37 \times 10^{-2} \right) - 3.298 = 9.08 \tag{4.20}$$

and from Eq. (4.15b), the average grain size (implicit average diameter of grains) is

$$d = \frac{0.254}{\sqrt{2^{G-1}}} = \frac{0.254}{\sqrt{2^{9.08-1}}} \tag{4.21a}$$

$$d = 1.54 \times 10^{-2} \, mm = 15.40 \, \mu m \tag{4.21b}$$

Mechanical properties depend on d, and specifically, the yield strength function $\sigma_{ys} = f(d)$ for most polycrystalline engineering materials is commonly defined as the Hall-Petch equation

$$\sigma_{ys} = \sigma_o + k_y d^{-1/2} \tag{4.22}$$

where σ_o and k_y are constants and σ_{ys} increases with decreasing d. Additional details on the Hall-Petch equation are included in a later chapter on mechanical properties.

(B) Counting grains is also a practical method for determining the average grain size d of a crystalline solid. The counting grains technique is a fundamental method used in materials science to determine the average grain size of polycrystalline materials. This information can be vital for understanding and predicting the material's properties and behavior. Let

$N_1 =$ Intercepted complete grains by a line L
$N_2 =$ Tangent line L touching a grain $= 0.5$
$N_3 =$ Line L ending within a grain $= 0.5$
$N_4 =$ Line L ending at a grain boundary $= 0.5$
$N_5 =$ Line L through a triple point $= 1.5$
$N_6 = \sum_{i=1}^{4} N_{i+1}$

Understanding the principles and limitations of this technique is crucial for researchers and engineers working with materials. By choosing the appropriate technique and considering the factors influencing grain size, we can gain valuable insights into the microstructure and properties of materials, leading to the development and optimization of materials for various applications.

The average grain size d can now be stated as

Table 4.2 Data for counting grain components per line as per Heyn linear intercept method

Line	N_1	N_2	N_3	N_4	N_5	N_6
L_1	9		1			1
L_2	6		1		1.5	2.5
L_3	8	0.5	1	0.5	4.5	6.5
L_4	5		1		4.5	5.5

$$d = \frac{1}{M} \sum_{i=1}^{k} \frac{L_i}{N_{1,i} + 0.5N_{6,i}} \quad (in\ inches) \tag{4.23a}$$

$$d = \frac{0.254}{M} \sum_{i=1}^{k} \frac{L_i}{N_{1,i} + 0.5N_{6,i}} \quad (in\ mm) \tag{4.23b}$$

For the fixed lines ($L_i = 32\ mm$) in Fig. 4.21, Table 4.2 lists the containing grain results

Constant:

$$X = \sum_{i=1}^{k=4} \frac{1}{N_{1,i} + 0.5N_{6,i}} = \frac{1}{N_{1,1} + 0.5N_{6,1}} + \frac{1}{N_{1,2} + 0.5N_{6,2}} \tag{4.24a}$$

$$+ \frac{1}{N_{1,3} + 0.5N_{6,3}} + \frac{1}{N_{1,4} + 0.5N_{6,4}}$$

$$X = \frac{1}{9 + 0.5\,(1)} + \frac{1}{6 + 0.5\,(2.5)} + \frac{1}{8 + 0.5\,(6.5)} + \frac{1}{5 + 0.5\,(5.5)} \tag{4.24b}$$

$$X = 0.46112 \tag{4.24c}$$

Substituting this result into Eq. (4.23b) yields the sought average grain size

$$d = \frac{0.254 L_i X}{M} \tag{4.25a}$$

$$d = \frac{(0.254)\,(32\ mm)}{250}\,(0.46112) = 1.50 \times 10^{-2}\ mm \tag{4.25b}$$

$$d = 15\ \mu m \tag{4.25c}$$

which is similar to the result given by Eq. (4.21b). Compare the results given by Eqs. (4.21b) and (4.25c) using a arithmetic average since there is no grain size reference. Thus, the percent error is computed as

$$error = \frac{|15.40 - 15|}{0.5\,(15.40 + 15)} \times 100\% = 2.63\% \tag{4.26}$$

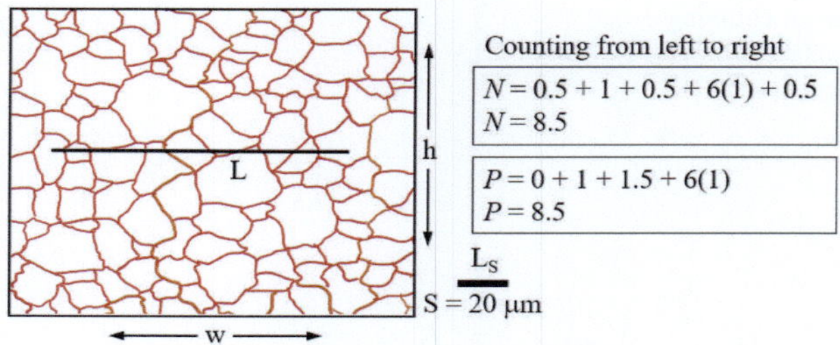

Fig. 4.22 The number of intercepts N and number of intersects P

This result is considered low and acceptable. Statistically, the higher the number of lines for counting grains, the lower the percent error. Moreover, Eq. (4.26) represents one statistical variable. For a complete statistical analysis, variables like the mean or average, range, variance, and standard deviation are the most commonly used quantitative data.

The aforementioned statistical variables and related statistical analysis are purposely excluded hereafter due to lack of pace.

(C) The ISO 643 Standard for determining an apparent grain size using a randomly drawn test line L of known length on an image.

Using intercepts Find the number of intercepts (N)

1. If L goes through a grain, then $N_1 = 1$
2. If L terminates within a grain, then $N_1 = 0.5$
3. If L is tangential to a grain boundary, then $N_3 = 0.5$

Using intersects Find the number of grain boundary intersects (P)

1. If L goes through a grain, then $P_1 = 1$
2. If L is tangential to a grain boundary, then $P_2 = 1$
3. If L goes through a triple point, then $P_3 = 1.5$
4. If L terminates within a grain, then $P_4 = 0$

For one line L on the ferritic microstructure in Fig. 4.22, the number of intercepts and intersects are $N = 8.5$ and $P = 8.5$, respectively,
From Fig. 4.22, the true line length L_t and the average grain size d are

$$L_t = \frac{SL_i}{L_s} = \frac{(20\ \mu m)\,(32\ mm)}{5\ mm} = 128\ \mu m \tag{4.27a}$$

$$d = \frac{L_t}{N} = \frac{L_t}{P} = \frac{128\ \mu m}{8.5} = 15.06\ \mu m \tag{4.27b}$$

Comparing the d-values as per ASTM E112-12 and ISO 643:2012 recommendations, the percent error between Eqs. (4.21b) and (4.27b) is

$$error = \frac{15.40 - 15.06}{0.5\,(15.40 + 15.06)} \times 100\% = 2.23\% \tag{4.28}$$

This result is considered a relatively small percent error, and therefore, ASTM E112 and ISO 643 standards give similar results for the average grain size d.

(D) The true area A_T, the print width w and print height h of the image (Fig. 4.22) is

$$A_T = \frac{w}{M}\frac{h}{M} = \frac{wh}{M^2} = \frac{(44)\,(35)}{250^2} = 0.02464\ mm^2 \tag{4.29}$$

and from Eq. (4.24b), the number of grain interceptions is

$$N_x = 9 + 0.5\,(1) + 6 + 0.5\,(2.5) + 8 + 0.5\,(6.5) \tag{4.30a}$$

$$N_x = 28\ grains/in^2 \tag{4.30b}$$

Use the unit conversions $1\ in = 0.254\ mm$ and $A_{ASTM} = 1\ in^2 = 0.0645\ mm^2$ at ASTM 100x so that the true number of interception becomes

$$N = N_x\left(\frac{A_{ASTM}}{A_T}\right) = (28)\left(\frac{0.0645\ mm^2}{0.02464\ mm^2}\right) \tag{4.31a}$$

$$N = 73.30\ grains/in^2 \tag{4.31b}$$

Notice that this result is the sought grain density needed to calculate the ASTM grain-size number G. Recasting Eq. (4.10a) yields the ASTM grain-size number G as

$$N = 2^{G-1} \tag{4.32a}$$

$$\log\,(N) = (G - 1)\log\,(2) \tag{4.32b}$$

$$G = 1 + \frac{\log\,(73.30)}{\log\,(2)} = 7.20 \tag{4.32c}$$

The true line length L_t and the average grain size d are

$$L_t = \frac{L_x}{M} = \frac{32\ mm}{250} = 0.13\ mm \tag{4.33a}$$

$$d = \frac{L_t}{P} = \frac{0.128\ mm}{8.5} = 1.53 \times 10^{-2}\ mm \tag{4.33b}$$

$$d = 15.30\ \mu m \tag{4.33c}$$

Notice that Eqs. (4.21b), (4.25c), (4.27b), and (4.33c) give similar results. It is convenient to calculate the average value. Thus, $d \approx 15.19)\ nm$ at $M = 250x$.

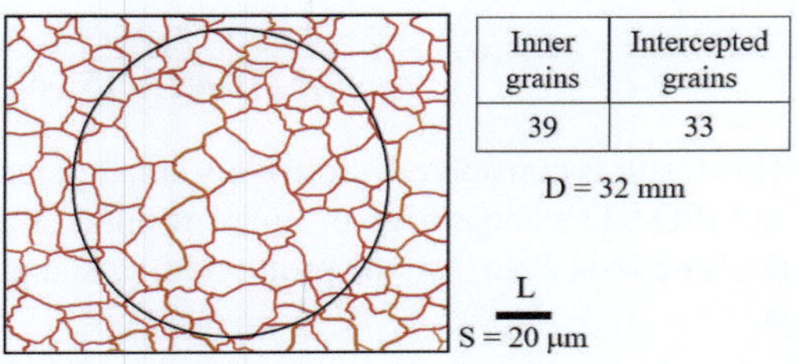

Fig. 4.23 Planimetric grain size method for determining the average grain size of ferrite (ferritic BCC phase) at the given magnification M

Factors affecting the accuracy of the Heyn Linear Intercept Method include the number of intercepts counted and grain structure. However, the accuracy of the Heyn Linear Intercept Method depends on the number of intercepts counted and the material's grain structure. More intercepts lead to a more accurate estimation of the average grain size, but the method is most accurate for materials with equiaxed grain structures. For materials with elongated or irregular grain structures, other methods might be more appropriate.

4.14.3 Planimetric Method

The planimetric method, known as the Jeffries method, is based on an inscribed circle or rectangle with a field of view with an ASTM recommended 5000-mm^2 area containing 50–100 grains at a magnification M. The main parameter is the grains/area relationship, from which the grain diameter d is determined as the average grain size of a microstructure. However, a different grains/area parameter can also be used for such a purpose as shown in Fig. 4.23.

The number of grains per unit area (N) and the Jeffries multiplier (F) are

$$N = F\left(N_{in} + 0.5 N_{inter}\right) \tag{4.34a}$$

$$F = \frac{M^2}{A} \tag{4.34b}$$

$$A = \frac{\pi}{4} D^2 \tag{4.34c}$$

where N_{in} denotes the number of grains inside the circle and N_{inter} denotes the number of intersected grains by the circumference of a circle or by the edges of a rectangle. The ferritic microstructure used above at a magnification M is analyzed hereafter. Notice that the inner grains (complete grains inside the circle) and intercepted grains by the circumference are included in Fig. 4.23 along with the circle diameter D and the scale bar S.

In order to be consistent with respect to the microstructure magnification $M = 550x$, the arbitrary printed diameter of the circle in Fig. 4.23 is, say, $D = 32\ mm$. Thus,

$$A = \frac{\pi}{4} D^2 = \frac{\pi}{4} (32 \; mm)^2 = 804.25 \; mm^2 \tag{4.35a}$$

$$F = \frac{M^2}{A} = \frac{(250)^2}{804.25 \; mm^2} = 77.71 \; mm^{-2} \tag{4.35b}$$

$$N = F \left(N_{in} + 0.5 N_{inter}\right) = \left(77.71 \; mm^{-2}\right)(39 + 0.5 \times 33) \tag{4.35c}$$

$$N = 4312.90 \; mm^{-2} \tag{4.35d}$$

Substituting Eq. (4.36b) into (4.15a) yields the average grain size as

$$d = \frac{1}{\sqrt{N}} = \frac{1}{\sqrt{4312.90 \; mm^{-2}}} \tag{4.36a}$$

$$d = 1.52 \times 10^{-2} \; mm = 15.20 \; \mu m \tag{4.36b}$$

This result is similar to previously calculated d-values.

4.15   Deep Etching Metallography

Deep etching is a process that uses a chemical solution or an electrical current to dissolve a material at a controlled rate for revealing the microstructure without significantly altering the bulk properties of the material.

One particular case is shown in Fig. 4.24 for a steel welding wire (Fe-$20Cr$-$2C$-$(1-4)Nb$-X), where X represents another element, using electron backscatter diffraction (EBSD) analysis or simply SEM analysis (Li et al. [26]).

Notice the conventional two-dimensional (2D) SEM microstructure in Fig. 4.24a consisting of austenite grains and embedded chromium carbide (M_7C_3) phase (white spots). After deep etching in Villela's reagent, the austenite matrix dissolved to a certain depth, and the lamellar M_7C_3 network clearly emerged (Fig. 4.24b) as a three-dimensional (3D) structure, which is partially shown in Fig. 4.24c as white structure at higher magnification.

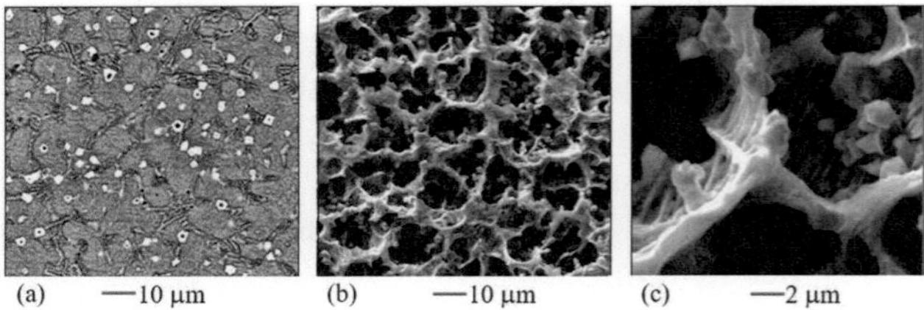

Fig. 4.24 SEM of a steel welding wire. (**a**) Without deep etching, (**b**) with deep etching, and (**c**) SEM higher magnification of the deep etched image. After Li et al. [26]

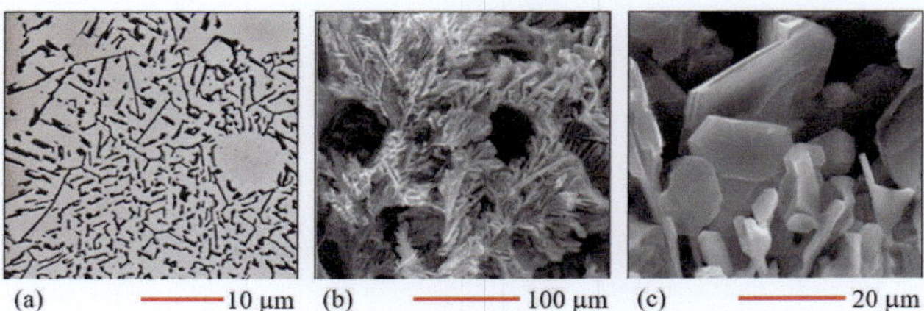

Fig. 4.25 Deep etching of an *Al-Zn-10Si-8Mg* cast alloy. (**a**) SEM eutectic α-matrix microstructure, (**b**) rosette-like silicon structure, and (**c**) SEM at higher magnification. After Tillova et al. [27]

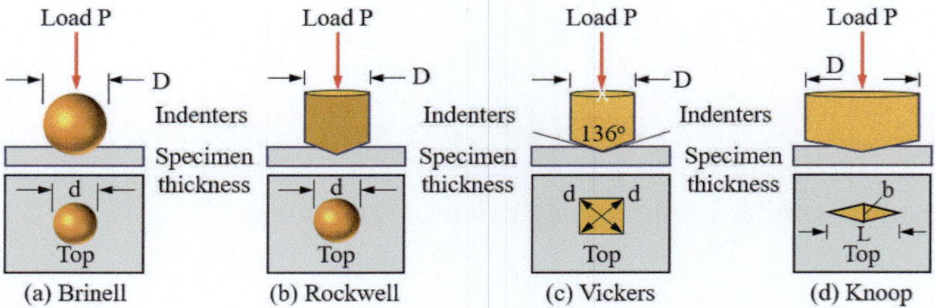

Fig. 4.26 Hardness testing techniques. (**a**) Brinell, (**b**) Rockwell, (**c**) Vickers, and (**d**) Knoop

Another deep etching example is shown in Fig. 4.25 for an *Al-Zn-10Si-8Mg* cast alloy (Tillova et al. [27]).

The typical microstructure of the as-cast alloy is shown in Fig. 4.25a. It consists of eutectic α-matrix with some embedded precipitates. Deep etching in diluted hydrochloric acid (HCl) solution reveals the 3D silicon phase as shown by the SEM image in Fig. 4.25b. Additional details can be revealed at a higher SEM magnification as depicted in Fig. 4.25c.

4.16 Hardness

Hardness is a measure of a material's resistance to deformation induced by an indenter at a specific compressive load P (Fig. 4.26). Hardness is an important property of metals, as it affects their wear resistance, strength, and toughness. There are many different ways to measure hardness, but the most common methods used in metallography are Brinell hardness testing, Rockwell hardness testing, Vickers hardness testing, and Knoop hardness testing.

Once the metal sample has been prepared, it can be examined under a microscope. The microstructure of the metal can be used to determine its properties and behavior. For example, the grain size of a metal can affect its strength and toughness. The presence of certain phases in a metal can affect its corrosion resistance.

Hardness testing is the process of measuring the hardness of a metal. There are many different methods of hardness testing, but the most common methods used in metallography are illustrated in Fig. 4.26 (after Heyden et al. [31, p. 12]).

Hardness is a vital material property that plays a major role in various applications across different industries. Several hardness testing techniques offer unique advantages and disadvantages depending on the specific material and desired outcome. The main advantages are classified as non-destructive and rapid testing. The disadvantages may be related to surface sensitivity and limited depth of penetration. Additional details are given below.

4.16.1 Brinell Hardness Testing

A hard steel or tungsten carbide ball is pressed into the metal sample with a known load. The diameter of the indentation is measured, and the hardness is calculated from the diameter and the load. Applications of Brinell hardness testing include measuring the hardness of relatively large and irregular metal, casting, and forging parts. Advantages include simple to use and wide range of materials. Disadvantages include destructive, large indentation, and sensitive to surface conditions.

The American Society for Testing and Materials, ASTM E10, defines the current standard Brinell hardness scale (Kg force per unit area). The Brinell hardness (HB) can be calculated using the following equation:

$$HB = \frac{2P}{\pi D(D - \sqrt{D^2 - d^2})} \tag{4.37}$$

where the Brinell number, HB, is in $Kg/mm^2 = 9.81\ MPa$, P is in kilogram (Kg), and D, d are in millimeter (mm).

4.16.2 Rockwell Hardness Testing

A diamond indenter is pressed into the metal sample with a known load P. The depth of the indentation is measured, and the hardness is calculated from the depth and the load. There are different Rockwell hardness scales, each with a different combination of indenter and load. The most common scales are Rockwell A (RA), which uses a diamond-tipped indenter and a major load of 60 kg for softer materials, such as annealed steels and aluminum alloys; Rockwell B (RB), which uses a steel ball indenter and a major load of 150 kg for harder materials, such as quenched and tempered steels and tool steels; and Rockwell C (RC), which uses a diamond-tipped indenter and a major load of 150 kg for the hardest materials, such as high-carbon steels and tungsten carbide. Advantages are fast, portable, and non-destructive for most materials. Disadvantages are destructive for very soft materials and limited depth of penetration.

The ASTM E18 defines the current standard Rockwell hardness scale, which depends on the magnitude of the compressive load P.

4.16.3 Vickers Hardness Testing

A diamond pyramid is pressed into the metal sample with a known load. This particular technique is known as Vickers microhardness test. The length of the diagonals of the indentation is measured, and the hardness is calculated from the diagonals and the load. Applications of Vickers hardness testing include measuring the hardness of metals, ceramics, plastics, case-hardened steels, small parts, and thin materials. Advantages are high accuracy, small indentation, and suitable for all materials. Disadvantages include requiring careful surface preparation, expensive equipment, and destructive nature. Moreover, the ASTM E92 or E384 defines the current standard Vickers hardness scale. The Vickers hardness can be calculated using the following equation (see Appendix 4A):

$$HV = \frac{P}{A_p} = \frac{2P sin(68°)}{d^2} = \frac{1.8544P}{d^2} \tag{4.38}$$

where HV is in Kg/mm^2, P is in Kg, A_p is the projected area of indentation, and d is in (mm). Conversion: $1\ Kg/mm^2 = 9.81\ MPa$.

4.16.4 Knoop Hardness Testing

The Knoop hardness test uses a rhombohedral-shaped diamond indenter instead of a square pyramidal indenter. From the ASTM E384, the Knoop hardness can be calculated using the following equation:

$$HK = \frac{P}{A_p} = \frac{P}{C_p L^2} \tag{4.39a}$$

$$C_p = \frac{tan(B/2)}{2tan(A/2)} = \frac{tan(130°/2)}{2tan(172.5°/2)} \tag{4.39b}$$

$$HK = \frac{14.2291P}{L^2} \tag{4.39c}$$

where HK is in Kg/mm^2, P is in Kg, A_p is the projected area of indentation, L is in (mm), $A = 172.5°$ is the included longitudinal edge angle, and $B = 130°$ is included transverse edge angle. Applications of Knoop hardness testing include measuring the hardness of thin materials, such as foils and case-hardened layers (coatings). Advantages are small indentation and good for thin materials. Disadvantages are requiring careful surface preparation, expensive equipment, and destructive.

4.17 Metallurgical Failure Analysis

The ultimate goal of failure analysis is to reveal the cause of problems and to provide solutions to prevent future failures. However, testing and evaluation of materials before and after failure are the key aspects of failure analysis under conditions simulating structural service life.

Metallographic etching techniques, chemical and electrolytic etching, play vital roles in metallurgical failure analysis by revealing microstructural features that aid in understanding the cause of material failure. The choice between these techniques depends on the specific material, desired information, available resources, and budget. Utilizing the appropriate technique can provide critical insights into the root cause of failure, leading to engineering solutions for preventing future occurrences.

In the metallurgical failure analysis field, metallography, visual and microscopic examinations, and hardness measurements are the most common and inexpensive techniques used to perform failure analysis. Regarding metallographic etching, **swabbing** is preferred for those metals and alloys (aluminum, nickel, stainless steels, etc.) that form a tenacious oxide surface layer. On the other hand, **submerging** is preferred for electrolytic etching using a suitable electrically conductive solution and applying electricity as direct current. Other techniques, such as selective etching, heat tinting, interference layer method, and electrolytic approach, can also be used for this purpose.

4.18 Summary

Despite the few microstructural features shown in this chapter, it is clear that the morphology of a microstructure depends on the processing condition of a material. However, revealing microstructural features, such as grain morphology and defects, depends on the microscopic technique. For instance, optical microscopy (OM) is a suitable technique for revealing the grain morphology at a relatively low magnification, determining the ASTM or ISO grain-size number G using the binary logarithm function as a comparison method and measuring the average grain size d of assumed spherical grains using either the intercept or planimetric method at relatively low magnifications.

Microscopic work, on the other hand, using the scanning electron microscopy (SEM) and transmission electron microscopy (TEM) techniques at high magnifications, allows the assessment of crystal or microstructural defects, which in turn are related to the material performance and properties. SEM provides three-dimensional images, while TEM produces two-dimensional images. Moreover, atomic force microscopy (AFM) technique is widely used to obtain the topology of solid surfaces at an atomic scale.

Hardness testing is an important tool for understanding the properties and behavior of metals. It is used in a wide variety of applications, including materials development, quality control, and failure analysis.

Appendix 4A Derivation of Vickers Harness Equation

The Vickers harness equation is derived using the sketches shown in Fig. 4.27. For instance, Fig. 4.27a shows the front view of the indenter, and Fig. 4.27b illustrates the top view of the indentation after removing the load P.

The Vickers hardness test uses a 136° pyramid diamond indenter. The resultant specimen surface deformation is ideally a squared indent with side a and diagonal length d. The Vickers harness equation, the lateral area (A) of the pyramidal impression, and the diagonal length are, respectively,

$$HV = \frac{P}{A} \tag{4.40a}$$

$$A = \frac{1}{2}(Perimeter)(height) = \frac{1}{2}(4a)(h) = 2ah \tag{4.40b}$$

$$d = \sqrt{2}a \tag{4.40c}$$

where

$$sin(\phi) = \frac{a/2}{h} \tag{4.41a}$$

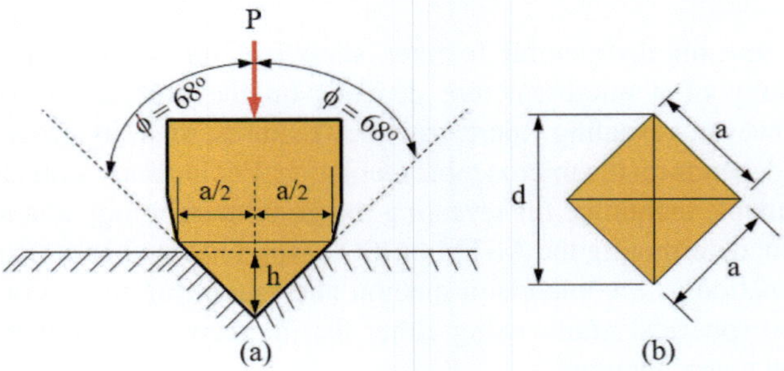

Fig. 4.27 Vickers hardness. (**a**) Indenter front view and (**b**) indentation top view

$$h = \frac{a/2}{sin(\phi)} \tag{4.41b}$$

$$A = \frac{2d}{\sqrt{2}} \frac{a/2}{sin(\phi)} = \frac{da}{\sqrt{2}sin(\phi)} \tag{4.41c}$$

$$A = \frac{d^2}{2sin(\phi)} \tag{4.41d}$$

The

$$HV = \frac{2Psin(\phi)}{d^2} = \frac{2Psin(68°)}{d^2} = \frac{2P(0.9272)}{d^2} \tag{4.42a}$$

$$HV = \frac{1.8544P}{d^2} \tag{4.42b}$$

Problems/Questions

4.1 According to Fig. 4.17, the grain size d can be defined quantitatively and unambiguously by its diameter. What is the base variable used as an alternative for defining d?

4.2 What type of microscope can be used to magnify an atom?

4.3 What is the primary use of metallography in the metal industry?

4.4 How important is the scale bar on an image?

4.5 Consider a $4 \times 5\text{-}in^2$ image of a microstructure. Determine **(a)** the ASTM grain-size number G if 30 $grains/in^2$ are observed at a magnification of $60\times$, **(b)** the number of grains N_x at 60x magnification, **(c)** N and N_x at 100x magnification using the calculated value of G in part (a), **(d)** N and N_x at $1\times$ magnification using the calculated value of G in part (a), and **(e)** d at 1x. [Solution: (a) $G = 4.43$, (b) $N_x = 600$ $grains$, (c) $N_x = 220$ $grains$, (d) $N_x \simeq 2 \times 10^6$ $grains$ and (e) $d = 77.50$ μm].

4.6 Determine **(a)** the ASTM grain-size number G if 35 $grains/in^2$ are observed at a magnification of 150x and **(b)** the average grain diameter d. [Solution: (a) $G = 7.30$ and (b) $d = 4.32$ mm].

4.7 Determine the average grain diameter d at a magnification of 200x if ASTM grain-size number is $G = 8$. [Solution: $d = 4.57$ mm]

4.8 For an ASTM grain-size number $G = 9$, determine **(a)** the average grain diameter d and **(b)** N at a magnification of 1x. [Solution: (a) $d = 15.80$ μm and (b) $N = 3968$ $grains/mm^2$].

4.9 For an ASTM grain density $N = 4000\ grains/mm^2$, calculate **(a)** the ASTM average grain size d and **(b)** the ASTM grain-size number at a magnification of 1x. [Solution: (a) $d = 15.80\ \mu m$ and (b) $G \simeq 9$].

4.10 For the given 50-μm image given below (taken from Majumdar et al. article [28]), calculate the average grain size d using **(a)** the comparison method where $d = f(G)$, **(b)** the mean intercept length $\overline{L}_m$ with $G = f\left(\overline{L}_m\right)$ and $d = f(G)$, and **(c)** the mean number of intersects (counting grains method) using three lines with a fixed length-value of your choice. [Solution: (a) $d = 7.87\ \mu m$, (b) $d = 7.46\ \mu m$ and (c) $d = 7.66\ \mu m$].

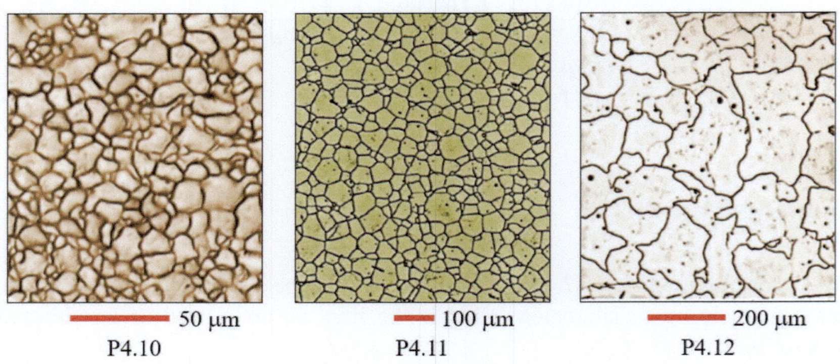

50 μm   100 μm   200 μm

P4.10   P4.11   P4.12

4.11 For the given 100-μm image in P4.11 above, calculate the average grain size d using the mean intercept length. [Solution: $d = 78.40\ \mu m$].

4.12 For the 200-μm image (taken from Callister and Rethwisch book [29, p. 288]), calculate the average grain size d using the mean intercept length. [Solution: $d = 117\ \mu m$].

4.13 For the 20000x image (taken from Brandon and Kaplan book [30, p. 465]), calculate the average grain size d using **(a)** the comparison method where $d = f(G)$, **(b)** the mean intercept length $\overline{L}_m$ with $G = f\left(\overline{L}_m\right)$ and $d = f(G)$, and **(c)** the mean number of intersects (counting grains method) using two lines with a fixed length-value of your choice. [Solution: (a) $d = 1.21\ \mu m$, (b) $d = 1.16\ \mu m$ and (c) $d = 1.07\ \mu m$].

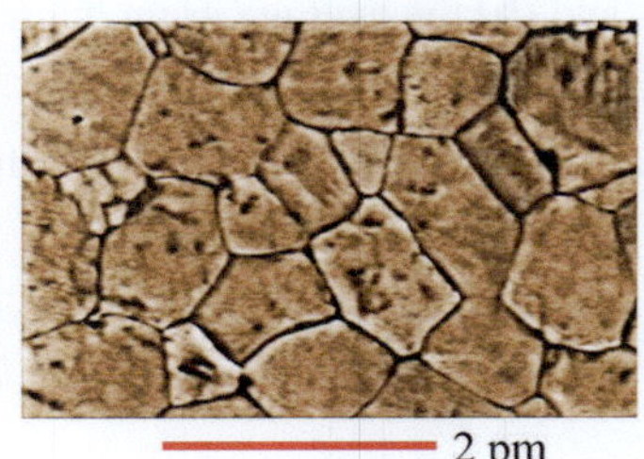

2 pm

4.14 For the given 200-μm image in P4.11 above, calculate the average grain size d using the mean intersections. [Solution: $d = 70.50\ \mu m$].

4.15 For the given 20000x image in P4.13 above, calculate the average grain size d using the grain-boundary intersections (P). [Solution: $d = 1.16\ \mu m$].

4.16 For the given 20000x image in P4.13 above, calculate the average grain size d using the mean intercept length. [Solution: $d = 1.16\ \mu m$].

4.17 For the given 20000x image in P4.13 above, calculate the average grain size d using the mean intersections. [Solution: $d = 1.07\ \mu m$].

4.18 For the 20000x image (taken from Brandon and Kaplan book [30, p. 465]), calculate the average grain size d using the planimetric method where $d = f\ (G)$. The inscribed circle is included for your convenience. [Solution: $d = 1.02\ \mu m$].

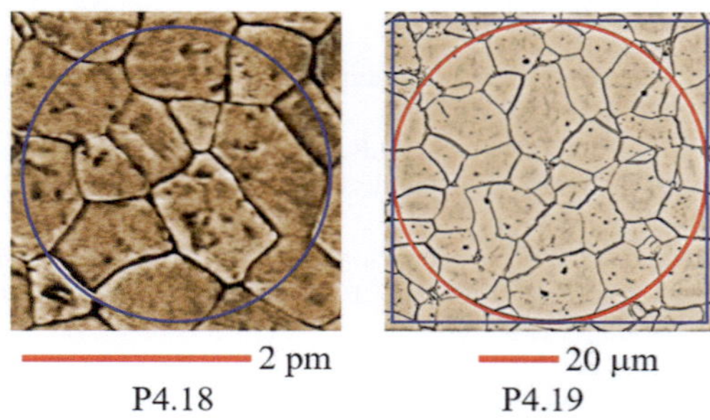

P4.18   P4.19

4.19 Consider the planimetric method to calculate the average grain size of ferrite using the inscribed circle onto the image given above (taken from Brandon and Kaplan book [30, p. 465]). [Solution: $d = 104\ \mu m$].

4.20 Consider the planimetric method to calculate the average grain size of ferrite using the inscribed rectangle onto the image given in Figure P4.19. [Solution: $d = 105\ \mu m$].

References

1. https://materialsection.files.wordpress.com/2011/09/2a2.pdf
2. https://www.labx.com/product/zeiss-axio-scopea1
3. G.F. Vander Voort, https://www.georgevandervoort.com/micrograph-gallery/
4. G.F. Vander Voort, *Color Metallography and Microstructures*, vol. 9 (ASM Handbook/ASM International, New York, 2004)
5. F.C. Levey, M.B. Cortie, L.A. Cornish, Determination of the 76 wt.% Au section of the Al-Au-Cu phase diagram. J. Alloys Compd. **354**(1–2), 171–180 (2003). Link: shorturl.at/dgBET
6. A. Grajcar, A. Kozłowska, K. Radwanski, A. Skowronek, Quantitative analysis of microstructure evolution in hot-rolled multiphase steel subjected to interrupted tensile test. Metals **9**(1304), 1–16 (2019)
7. shorturl.at/wyzAK

8. shorturl.at/cuLRX
9. N.L. Perez, *Evaluation of Thermally Degraded Rapidly Solidified Fe-Base and Ni-Base Alloys*, Ph.D. Dissertation (University of Idaho, Moscow, 1988)
10. M. Naeem, H. He, S. Harjo, T. Kawasaki, F. Zhang, B. Wang, S. Lan et al., Extremely high dislocation density and deformation pathway of CrMnFeCoNi high entropy alloy at ultralow temperature. Scr. Mater. **188**, 21–25 (2020)
11. J. McBreen, Physical methods for investigation of electrode surfaces, in *Fundamentals of Electrochemistry*, 2nd edn., ed. by V.S. Bagotsky (Wiley, New Jersey, 2006)
12. J. Zhang, J. Ulstrup, Oxygen-Free in situ scanning tunneling microscopy. J. Electroanal. Chem. **599**(2), 213–220 (2007)
13. M.B. Song, J.M. Jang, C.W. Lee, Electron tunneling and electrochemical currents through interfacial water inside an STM junction. Bull. Korean Chem. Soc. **23**(1), 71–74 (2002)
14. A.J. Bard, L.R. Faulkner, *Electrochemical Methods—Fundamentals and Applications*, 2nd edn. (Wiley, New York, 2001)
15. K. von Bergmann, *Iron Nanostructures Studied by Spin-Polarised Scanning Tunneling Microscopy*, Ph.D. thesis (Universität Hamburg, Hamburg, 2004)
16. C.J. Chen, *Introduction to Scanning Tunneling Microscopy*, 2nd edn. (Oxford University, New York, 2008)
17. D.M. Kolb, M.A. Schneeweiss, Scanning tunneling microscopy for metal deposition studies. Electrochem. Soc. Interface **8**(1), 26 (1999)
18. O.M. Magnussen, L. Zitzler, B. Gleich, M.R. Vogt, R.J. Behm, In-Situ atomic-scale studies of the mechanisms and dynamics of metal dissolution by high-speed STM. Electrochem. Acta **46**(24), 3725–3733 (2001)
19. T. Engel, P. Reid, *Physical Chemistry* (Pearson Education, Inc., Boston, 2013)
20. S. Duffe, T. Irawan, M. Bieletzki, T. Richter, B. Sieben, C. Yin, B. von Issendorff, M. Moseler, H. Hovel, Softlanding and STM imaging of Ag561 clusters on a C60 monolayer. Eur. Phys. J. **D 45**(3), 401–408 (2007)
21. S. Bruckenstein, J. Janiszewska, Diffusion currents to (Ultra) microelectrodes of various geometries: ellipsoids, spheroids and elliptical 'disks. J. Electroanal. Chem. **538–539**, 3–12 (2002)
22. C. Lee, X. Wei, Q. Li, R. Carpick, J.W. Kysar, J. Hone, Elastic and frictional properties of graphene. Phys. Status Solidi (b) **246**(11–12), 2562–2567 (2009)
23. G. Vander Voort, *Deformation and Annealing of Cartridge Brass* (Vac Aero International Inc., New York, 2015)
24. https://steel-guide.info/mikrostruktura-stali-ferrit?lang=en
25. https://pmpaspeakingofprecision.com/tag/mcquaid-ehn/
26. J. Li, R. Kannan, M. Shi, L. Li, Solidified microstructure of wear-resistant Fe-Cr-CB overlays. Metall. Mater. Trans. B **51**(4), 1291–1300 (2020)
27. E. Tillova, E. Durinikova, M. Chalupova, Characterization of phases in secondary AlZn10Si8Mg cast alloy. Mater. Eng. **18**, 1–7 (2011)
28. S. Majumdar, D. Bhattacharjee, K.K. Ray, On the micromechanism of fatigue damage in an interstitial-free steel sheet, in *The Minerals, Metals & Materials Society and ASM International* (2008), pp. 1676–1690
29. W. Callister, Jr., D.G. Rethwisch, *Materials Science and Engineering: An Introduction*, 10th edn. (Wiley, New York, 2018)
30. D. Brandon, W.D. Kaplan, *Microstructural Characterization of Materials*, 2nd edn. (Wiley, New York, 2008)
31. H.W. Heyden, W.G. Moffatt, J. Wulff, *The Structure and Properties of Materials*, vol. III. Mechanical Behavior (Wiley, New York, 1965)

Chapter 5
Two-Dimensional Materials

5.1 Introduction

This chapter contains a theoretical background on two-dimensional (2D) materials referred to as layered nanomaterials (thin sheets at a nanoscale), which can be grown on a substrate or exfoliated and placed on a substrate as a single-layer or N-layers.

Regarding semiconductor materials, metallic graphene, insulator boron nitride (hBN), and composite TMD (metal and chalcogen elements such as sulfur-S, selenium-Se, and tellurium-Te) are commonly treated as layered materials due to their relevant properties for the semiconductor industry. Specifically, 2D graphene and 2D transition-metal dichalcogenide (TMD) materials, such as MoS_2, WS_2, WSe_2, and the like, are the most studied nanomaterials. Consult the glossary online for diverse terminologies encountered throughout this chapter.

The general physical properties of TMDs are associated with their bandgaps, valence and conduction bands, and the possible heterojunctions due to the stacking of monolayers on semiconductor substrates. Consequently, characterizing 2D layered materials may be complicated because of the possible formation of van der Waals heterostructures caused by heterojunctions.

Most of 2D materials have a common hexagonal symmetry and suitable properties for designing futuristic engineering high-performance electronic devices, such as sensors, batteries, and supercapacitors. This implies that applications of 2D materials depend on their dimensional limitations in bulk, continuous atomic structure (commonly honeycomb planes), purity, and environment. Nevertheless, computer making, energy storage, and medicine, among others, appear to be the most relevant industries to use high-tech graphene-containing products.

The main emphasis hereafter is based on the research trends on 2D materials exposed to an electric field (E_f) in energy-producing devices (EPD). A theoretical analysis as per Lennard-Jones potential energy is included for predicting the maximum binding force needed to separate or exfoliate 2D atomic planes from bulk sources. Accordingly, the intent is to provide a combined theoretical

© The Author(s), under exclusive license to Springer Nature Switzerland AG 2024

N. Perez, *Materials Science: Theory and Engineering*,

https://doi.org/10.1007/978-3-031-57152-7_5

and experimental overview relevant to students or researchers interested in the fascinating world of 2D semiconductor materials.

5.2 Background on 2D Materials

This section is devoted to two-dimensional (2D) atomic structures, such graphene and transition-metal dichalcogenide (TMD) materials. Thus, 2D materials are N-layered nanomaterials due to their nano size dimensions and are mostly based on a hexagonal lattice being used as layers on substrates. From an engineering point of view, coupled 2D materials and substrates define composites materials with excellent properties for certain industrial applications.

In crystallography, a 2D arrangement of atoms form thin atomic planes referred to as layers or sheets. For instance, carbon atoms forming a honeycomb layer is the constituent of graphene planes (graphene sheets), whereas graphene layers are the constituents of the hexagonal graphite lattice. Thus, the graphene layer is a one-atom thick material (in the order of carbon diameter) with exceptional transport (electrical and thermal conductivity) and mechanical properties.

Historically, graphene sheets were mechanically exfoliated (mechanical exfoliation) from a graphite surface block by Novoselov et al. [1] in 2004 using a mechanical cleavage method (also known as the "Scotch-type method"). Thus, Andre Geim and Konstantin Novoselov became Nobel Prize recipients in 2010. Fundamentally, graphene is a highly conductive sheet of graphite since one-carbon atom has a delocalized electron (which is free to move along the hexagonal-shaped structure) and strong due to covalent σ-bonds.

The production of graphene sheets by mechanical exfoliation of hexagonal graphite is attributed to weak π-bonds between sheets due to weak van der Waals (vdW) forces. Thus, hexagonal graphite or rhombohedral graphite is crystallographically a stack of weakly coupled graphene planes. Nowadays, rhombohedral graphite is considered as a potential high-temperature material in the semiconductor industry (Kopnin et al. [2]).

Commercially, the general feeling is that (1) mechanical exfoliation of graphite to produce sp^2-bonded graphene may be irregular and time-consuming, (2) chemical reduction leads to toxic and irregular graphene sheets, and (3) chemical vapor deposition (CVD) is an apparent time-consuming and expensive process. However, the roll-to-roll graphene CVD reactor seems promising for scalable production of graphene (Kidambi et al. [3]) in 2018. Nonetheless, the goal in the graphene-sheet technology is to devise a cheap and scaling graphene production with a large-enough defect-free and continuous graphene sheet.

Two-dimensional (2D) transition-metal dichalcogenide (TMD) materials emerged after graphene as new research material sheets. Both graphene and some TMDs have crystallographic hexagonal structures with sp^2 hybridized bonding characteristic, where the strong covalent σ-bonding along the sheet plane and the weak π-bonding across basal planes make it possible for exfoliation (Novoselov et al. [1], Zhang [4], Kolobov and Tominaga [5]).

Again, 2D materials can also be produced using a high-power femtosecond laser (HPFL) (An et al. [6]) and characterized by Raman spectroscopy technique (Dong et al. [7], Cancado et al. [8]) and X-ray diffraction (XRD) (Seehra et al. [9]).

5.2.1 Types of Graphite

Graphene is exfoliated from treated hexagonal graphite (HG), which may contain the rhombohedral graphite (RG) phase. Subsequently, the latter is heat-treated at relatively high temperatures for a possible phase transition to the former as a high-temperature material in the semiconductor industry (Kopnin et al. [2]). Commercially, (1) mechanical exfoliation of graphite to produce sp^2-bonded graphene may be irregular and time-consuming, (2) chemical reduction leads to a toxic and irregular graphene sheets, and (3) chemical vapor deposition is a time-consuming and expensive process.

5.2.2 Methods for Producing Two-Dimensional Layers

Mechanical exfoliation of graphite is the pioneering technique for producing graphene sheets; however, chemical vapor deposition (CVD) seems to be the most preferable method for extracting carbon atoms by reduction from a carbon-rich environment at high temperatures. Essentially, peeling off or cleaving graphene sheets (hexagonal basal planes) from the bulk hexagonal graphite crystal using an "adhesive Scotch tape" is relatively a cost-effective method.

For comparison, An et al. [6] in 2018 have used a high-power femtosecond laser (HPFL) to exfoliate N-layer TMDs, and Kidambi et al. [3] in 2018 designed a scalable graphene sheet production at a high-rate (≥ 5 cm/s speed) using the roll-to-roll graphene CVD process.

5.2.3 Top-Down and Bottom-Up Concepts

The production of graphene, as described in the literature, is conceptually achieved using two different directions: top-down and bottom-up approaches. The former, a reduction of dimensions, uses graphite as the solid-state source for mechanical or liquid-phase exfoliation. The latter, a collection and consolidation of atoms or molecules into a structure, employs a carbon-based gas as the main source for epitaxial growth or chemical-vapor deposition (Kolobov and Tominaga [5], Mahmoudi et al. [10]).

5.3 Graphene Composite Surfaces

Properly deposited carbon atoms in honeycomb-shaped layers on a suitable sub-strate surface with a high-conductivity range is the main goal for fabricating graphene-MX_2/M surfaces, where MX_2 is an oxide or sulfate surface and M is any suitable metal or material. Subsequently, design guidelines to produce commercial electronic devices must be followed.

5.3.1 Hummers' Method

Apparently, Hummers et al. [11] developed a methodology in 1958 to chemically produce graphene oxide (GO) from an acid solution consisting of concentrated sulfuric acid (H_2SO_4), sodium nitrate ($NaNO_3$), potassium permanganate ($KMnO_4$), and graphite. Hence, the Hummers' method has been subjected to modifications (Bai et al. [12], Chen et al. [13]).

5.3.2 Chemical Vapor Deposition

This technique is common in the semiconductor field for producing high-quality thin films. In the case of graphene production, a graphene film in the form of single layer (monolayer) or N-layer to lay on a practical substrate as monocrystalline or polycrystalline material.

Chemical vapor deposition (CVD) has proven to be the most promising method for the synthesis of crystalline layers on different substrates, which in turn must have the appropriate crystallographic structure related to a limiting solubility level for, at least, monolayer growth at relatively high temperatures. Details on CVD methodology or technique are available in the literature (Pierson [14]).

5.4 Rhombohedral and Hexagonal Structures

Hexagonal graphite is not an hexagonal close-packed (HCP) structure like zinc (Zn), but it is a stable carbon-based structure since it can be subjected to sonication-assisted liquid-phase exfoliation, electrochemical exfoliation, mechanical exfoliation, and even oxidation exfoliation for producing graphene oxide (reduced graphene oxide). Nonetheless, hexagonal graphite is composed by graphene layers bonded by van der Waals forces (vdW), which are considered mechanically weak for exfoliation. On the other hand, rhombohedral graphite (3R) is also based on a

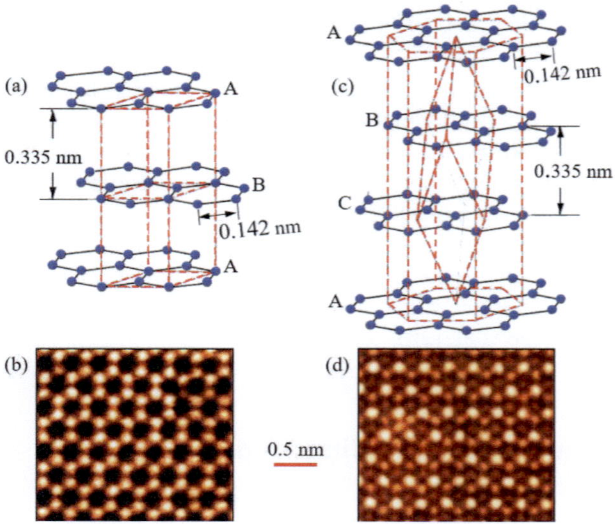

Fig. 5.1 Hexagonal symmetry and annular dark-field (ADF) image. (**a**) Hexagonal graphite (2H), (**b**) ADF image of CVD-grown hexagonal $MoSe_2$ monolayer, (**c**) rhombohedral graphite (3R), and (**d**) ADF image of $MoSe_2$ bilayer with 2H and 3R planes, respectively. Units cells: (**a**) and (**c**) after Krauss et al. [20]. ADF images: (**b**) and (**d**) after Lu et al. [21]

stacking sequence of graphene planes in graphite (Pierucci et al. [15], Torche et al. [16], Bernal [17], Latychevskaia et al. [18], Cong et al. [19]). Apparently, the principal material for producing graphene is hexagonal graphite with an $ABA\dots$ stacking (2H) since the rhombohedral graphite $ABCA$ stacking (3R) is more complicated.

Crystallographically, Fig. 5.1a shows the dimensions of the hexagonal graphite (2H) structure consisting of graphene layers having a specific stacking sequence (Krauss et al. [20]). Figure 5.1b depicts an annular dark-field (ADF) image of the morphology of CVD-grown hexagonal molybdenum diselenide ($MoSe_2$) film (Lu et al. [21]), which can be denoted as CVD-$MoSe_2$ two-dimensional TMD monolayer.

Notice that both graphite and $MoSe_2$ structures are not hexagonally close packed (HCP). Therefore, the honeycomb (hexagonal ring) is sufficiently large for impurity-induced disturbance in the conduction electron distribution and distortion of the crystal lattice.

Figure 5.1c illustrates the rhombohedral graphite (3R) structure and dimensions of the graphene layers having a specific stacking sequence similar to that of the hexagonal graphite (Krauss et al. [20]). Figure 5.1d exhibits an annular dark-field (ADF) image of bilayer $MoSe_2$ with 2H and 3R structures (Lu et al. [21]).

5.4.1   Carbon Phase Transition

The phase transformation from rhombohedral graphite (3R) to hexagonal graphite
(2H) is the preferred crystallographic condition for producing pure graphite. Crystallographically, the arrangement of carbon atoms can stack in a suitable manner to
form either an hexagonal graphite (2H) or a rhombohedral graphite (3R) structure
(Marsh and Rodriguez-Reinoso [22, p. 270], Qing et al. [23]). This implies that the
mechanism for graphite phase transition from the ABC to $ABAB$ stacking sequence
is a high-temperature phase transformation (Latychevskaia et al. in 2019 [18], Shi et
al. [24]). Moreover, graphite-to-diamond is also a high-temperature phase transition
(Fahy et al. [25], Isobe et al. [26], Xie et al. [27]).

5.5   Graphene Atomic-Bonding Model

Consider the hexagonal symmetry and the basal planes depicted in Fig. 5.2a. The
hexagonal graphite atomic structure has an ABA stacking sequence that forms the
unit cell (Callister and Rethwisch [28, p. 420], Yang et al. [30])).

Accordingly, the size of the hexagonal graphite unit cell is determined by the
lattice parameters $a = 0.25$ *nm* at the basal plane and $c = 0.67$ *nm* as the
perpendicular length with $c/a = 2.68$ ratio (Carter and Norton [29, p. 96]).
Moreover, chemical bonding is described by the sp^2 hybrid atomic bonding (Carter
and Norton [29, p. 63], Yang et al. [30]).

Figure 5.2b illustrates the simple hexagonal unit cell with the $ABAB$ stacking
sequence, Fig. 5.2c represents one-atom thick crystallographic plane known as the
(0001) basal plane, and Fig. 5.2d exhibits the fundamental or graphic visualization
of the formation of the sp^2 hybrid bonding in graphite and graphene. The π-bond
in sp^2 hybridization is perpendicular to the graphene plane, and it hybridizes to

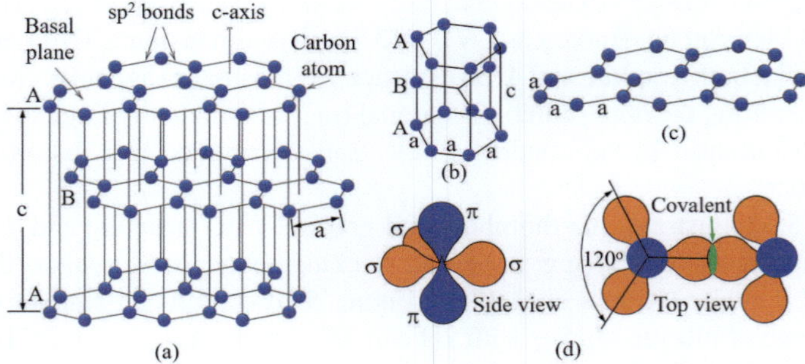

Fig. 5.2 Graphite and related features. (**a**) Structure of graphite (Callister and Rethwisch [28,
p. 420]), (**b**) hexagonal unit cell, (**c**) (0001) basal plane known as graphene honeycomb sheet, and
(**d**) sp^2 hybridization mode with an extended π-conjugation (Yang et al. [30])

form the π-conduction and σ-bond overlaps along the basal plane as illustrated in Fig. 5.2d (Pierson [31, p. 34]).

Carbon exists in various polymorphic forms such as rhombohedral graphite, hexagonal graphite, and amorphous diamond. For instance, graphite may exhibit both rhombohedral and hexagonal crystal stacking sequences. Nonetheless, the covalent σ-bond is stronger along the graphene plane than the π-bond, which is weak across graphene planes. According to Pierson [31, p. 45], the bonding energy and the binding forces in the sp^2 hybrid bonding are $U_\sigma = 524\ k/mol > U_\pi = 7\ kJ/mol$ and $F_\sigma > F_\pi$, respectively.

Crystallographically, stable hexagonal graphite (HG) has a staking sequence of $ABAB$, while the unstable rhombohedral graphite (HG) has the $ABCABC$ sequence, which is often intermixed with natural hexagonal graphite. Moreover, rhombohedral graphite can be produced by shear deformation (grinding) of hexagonal graphite (Marsh and Francisco Rodriguez-Reinoso [22, p. 270]). Subsequently, ABC can revert to $ABAB$ upon high-temperature heat treatment at approximately $1300°C$ or above (Gallego et al. [32]).

5.6   Hexagonal Symmetry

This section compares hexagonal symmetry between the hexagonal close-packed (HCP) structure as in Ti, Mg, and Co metals and hexagonal graphite (Fig. 5.3). It is assumed that the atomic arrangement in HCP is not a lattice of identical points as found in the face-centered cubic (FCC) (Carter and Norton [29, p. 79]).

Firstly, notice the different captions in Fig. 5.3a and b. Secondly, the atomic packing factor (APF) in HCP unit cell is 0.74, and the atomic planes from top to bottom accommodate in the hollows forming the ABA stacking sequence (2H)

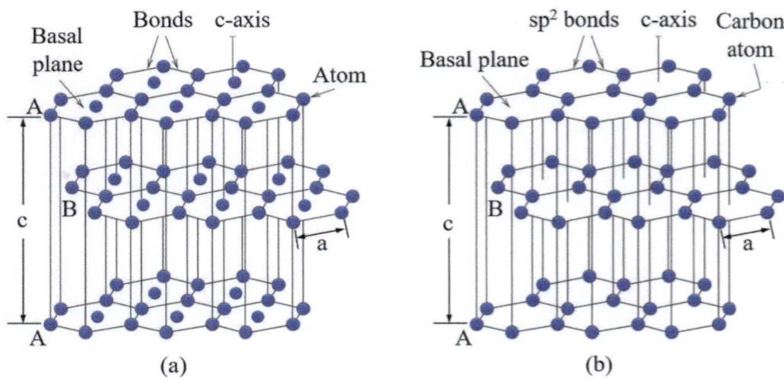

Fig. 5.3 Hexagonal symmetry. (**a**) Hexagonal close-packed (HCP) structure with $CN = 6$ in the basal plane, 17 atoms per unit cell, ABA staking, and (**b**) hexagonal graphite structure with $CN = 3$ in the basal plane, 15 atoms per unit cell, and ABA stacking

with strong metallic bond strength. The HCP structure has a center atom in the basal (0001) plane (hexagonal ring), making this structure very hard to exfoliate.

Hexagonal graphite, on the other hand, does not have a center atom in the basal (0001) plane, and as a result, this hexagonal plane (HP) is less dense than the common HCP plane, and its bonding strength between planes is considered weak. Thus, graphite exfoliation to separate the (0001) planes, known as graphene sheets, is possible due to an interplanar (interlayer) weak sp^2-bonding force, known as the van der Waals force (FvdW).

5.7 Sources of 2D Materials

Presently, two-dimensional layered materials known as transition-metal dichalcogenides (TMDs) on crystallographically comparable substrates have been studied as a search for finding alternative and highly efficient substrate composites. In Fact, TMDs emerged as post-graphene two-dimensional materials in the field of solid-state physics, where TMDs are known as topological insulators. For the sake of argument, graphene and MoS_2, WS_2, and $MoSe_2$ are two-dimensional TMDs materials that have a common hexagonal symmetry (Kolobov and Tominaga [5]).

For convenience of the reader, by definition, a chalcogen is an element appearing in Group 6A of the periodic table, whereas a dichalcogenide denotes an inorganic material containing two atoms of chalcogen per molecule or unit cell.

This section compares graphene and some transition-metal dichalcogenide (TMDs) having hexagonal symmetry depicted in Fig. 5.4 as bulk and honeycomb monolayer materials (Fujita et at. [33], Coleman et al. [34]).

Notice that graphene has the smallest honeycomb structure and, in theory, graphene and MoS_2, WS_2, BN can easily adhere to another atomic structure, forming monolayer-composites or nanocomposites. Apparently, silicon dioxide

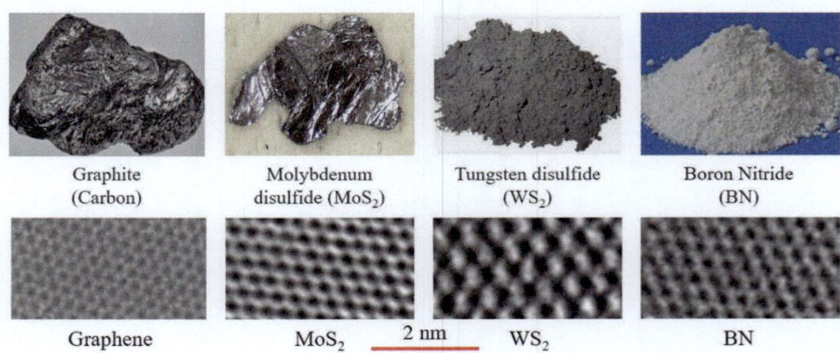

Fig. 5.4 Macroscopic images of bulk graphene and TMD sources, and microscopic TEM images of monolayers with hexagonal symmetry. References: graphene (Fujita et at. [33]) and MoS_2, WS_2, BN (Coleman et al. [34])

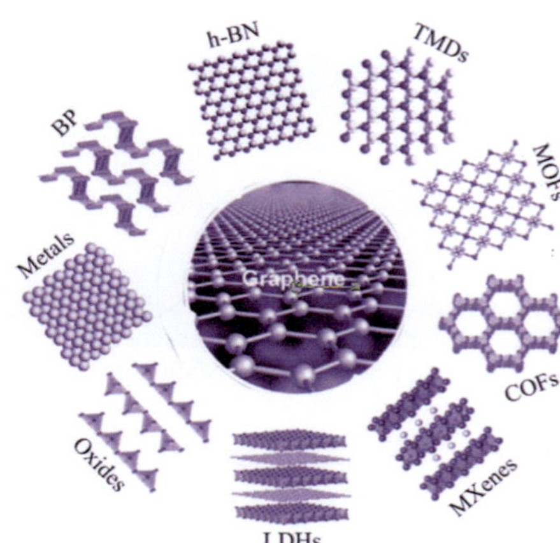

Fig. 5.5   Crystallography of ultrathin 2D nanomaterials (Zhang [4]

(SiO_2) grown on single silicon crystals (Si) and molybdenum disulfide (MoS_2) appear to be the most common substrates containing graphene.

5.7.1   Classification of 2D Nanomaterials

In order to comprehensively understand the structural features and outstanding properties of nanomaterials, the reader is encouraged to analyze Zhang's [4] crystallographic arrangements of atoms schematically shown in Fig. 5.5 as "Ultrathin 2D nanomaterials". In other words, this figure schematically shows the sub-fields and these nanomaterials.

Like graphene, TMDs can also be mechanically exfoliated, and as a result, the original 3D source undergoes a splitting or separation process. The resultant 2D samples usually contain structural defects, such as vacancies, dislocations, and the like. Actually, the assessment of these defects is important for determining the quality of a final product using, say, the Raman spectroscopy. This is detailed in a later section for a comprehensive understanding of 2D materials in suitable applications.

Raman spectroscopy has a vast range of applications across various disciplines, including non-destructive analysis of unknown compounds (chemistry), characterization of semiconductors (materials science), identification of minerals (geology), diagnosis of diseases (medicine), and analysis of drug quality (pharmaceutical industry).

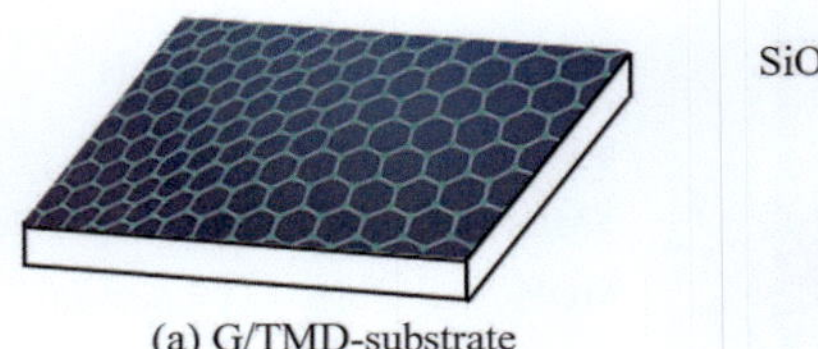
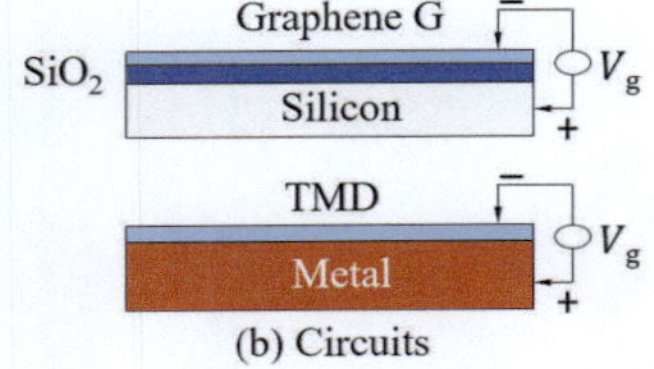

(a) G/TMD-substrate　　　　　　　　　　　(b) Circuits

Fig. 5.6 Schematic substrate-containing graphene electric system setup. (**a**) Schematic planar electrode or system containing a 2D layer of G/TMD-substrate composite (screenshot from https://www.youtube.com/watch?v=KhQrGtragXc) and (**b**) side view of a $G/SiO_2/Si$ circuit and a TMD layer/Metal circuit

5.7.2　Effects of Electric Field on Graphene-Substrate System

Let's generalize the concept of graphene-substrate system. For instance, Fig. 5.6a schematically shows a simple model for a substrate-containing graphene being exposed to an electric field strength E_x, which in turn has a significant influence on the graphene electron drift velocity v_x and mobility μ_x at an electron density n_e and temperature T.

Using M as the base-transition metal/material (Si, Mo, W, Cu, or conductive polymer), MX_2 becomes a transition metal dichalcogenide or simply the metal dioxide or sulfate (SiO_2, Si C, MoS_2, WS_2, WSe_2, etc.), and X is the chalcogenide material: sulfur (S), selenium (Se), or tellurium (Te).

This graphene-substrate system can also be subjected to a magnetic field for characterizing the Hall effect in graphene. Ideally, graphene-MX_2/M is the general notation for the substrate-containing graphene. Regarding Fig. 5.6b, V_g is the biased gate direct current (DC) voltage. Moreover, Raman spectroscopy (RS) and high-resolution transmission electron microscopy (TEM) are the most common techniques used for characterizing graphene-substrate systems. Details on these techniques are available in the literature.

Graphene-MX_2/M systems like graphene-SiO_2/Si and graphene-MoS_2 have been studied extensively (Liu et al. [35]). On the other hand, poly(methyl methacrylate) (PMMA), poly(dimethylsiloxane) (PDMS), and silicon carbide (Si C) substrates have been considered for characterizing the possibility of a polymeric substrate-graphene composite for energy-related devices (Layek et al. [36]), copper (Cu) substrate-graphene has also been characterized (Wood et al. [37]), graphene-WS_2 system (Georgiou et al. [38]), and so forth. There are selected references available in the vast literature on the subject matter.

Assume that the effect of E_x on v_x and μ_x can be described in a general power form

$$v_x = aE_x^{\beta} \tag{5.1a}$$

$$\mu_x = bE_x^{\gamma} \tag{5.1b}$$

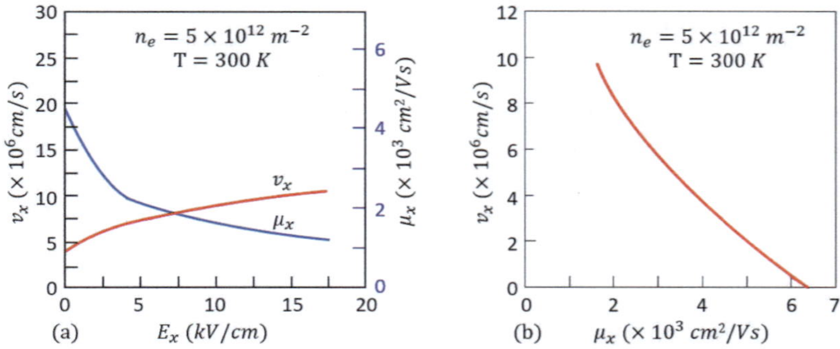

Fig. 5.7 (**a**) Effect of electric field strength (E_x) on the electron drift velocity (υ_x) and mobility (μ_x) for constant electron density (ρ_e) and temperature T. (**b**) Plot of $\upsilon_x = f(\mu_x)$ (Dong et al. [7])

$$\upsilon_x = c\mu_x^{\delta} \tag{5.1c}$$

where $\beta \neq \gamma \neq \delta = \beta/\gamma$ denote exponents and a, b, c denote constants with $c = ab^{-\delta}$. These equations can be linearized by taking the natural logarithm; $\ln(\upsilon_x) = \ln[f(E_x)]$, $\ln(\mu_x) = \ln[f(E_x)]$, and $\ln(\upsilon_x) = \ln[f(\mu_x)]$.

Recently, Dong et al. [7] observed through simulations that the electric field strength has a nonlinear non-ohmic effect on υ_x and μ_x in the SiO_2/Si substrate-containing graphene system. They concluded that graphene can be used to develop high-field high-speed electronic devices due to high υ_x and μ_x in relatively high electric fields (E_x) as shown in Fig. 5.7.

Despite that there may be significant differences among experimental and numerical results found in the literature, it is worthwhile to borrow Dong et al.'s [7] findings as a representative dataset for further analysis. Manipulating their plots for υ_x and μ_x as functions of the applied electric field strength (E_x) yields Fig. 5.7a (Dong et al. [7], Grann et al. [39], Walkowiak et al. [40], Jacoboni et al. [41], and Santos et al. [42]).

The functions $\upsilon_x = f(E_x)$ and $\mu_x = f(E_x)$ are deduced using nonlinear regression analysis. Subsequently, $\upsilon_x = f(\mu_x)$ trendline is derived in order to enhance the analysis of non-ohmic behavior of the SiO_2/Si substrate-containing graphene or graphene-SiO_2/Si system.

Accordingly, the trendlines in Fig. 5.7aq for υ_x and μ_x at $0 \leq E_x \leq 20\,kV/cm$ are given by

$$\upsilon_x = 3.80E_x^{1/3} \tag{5.2a}$$

$$\mu_x = 6400E_x^{-1/2} \tag{5.2b}$$

$$\upsilon_x = 1309.90\mu_x^{-2/3} \tag{5.2c}$$

Here, $\beta = 1/3$, $\gamma = -1/2$, $\delta = -2/3$, $a = 3.80$, $b = 6400$, and $c = 1309.90$. The reader should work out the units of a, b, c. Notice that the effective electron drift velocity in graphene (Fig. 5.7a) is in the order of $v_x = (300 \times 10^6 \ m/s)/300 = 10^6 \ m/s$, which is near the speed of light $c = 300 \times 10^6 \ m/s$.

Figure 5.7b exhibits the $v_x = f(\mu_x)$ function for the SiO_2/Si substrate-containing graphene. Obviously, this curve suggests that the electron drift velocity decreases as the electron mobility increases.

5.8 Mechanical Exfoliation

5.8.1 Potential Binding Energy and Binding Force

The empirical Lennard-Jones potential energy equation introduced in Chap. 1 is a relevant mathematical model for characterizing the theoretical binding energy $U(r)$ and related binding force $F(r)$ for mechanical exfoliation of graphite to produce graphene. For convenience, Eqs. (1.21a), (1.21b), (1.22a), and (1.22b) and Fig. 1.8 are recast in this section.

Mathematically,

$$U(r) = 4R_o \left[\left(\frac{\sigma}{r} \right)^{12} - \left(\frac{\sigma}{r} \right)^{6} \right] \tag{5.3a}$$

$$F(r) = \frac{dU(r)}{dr} = 4R_o \left[\frac{6\sigma^6}{r^7} - \frac{12\sigma^{12}}{r^{13}} \right] \tag{5.3b}$$

$$r_o = 2^{1/6}\sigma = 1.1225\sigma = \left(\frac{2\lambda_n}{\lambda_m} \right)^{1/6} \tag{5.3c}$$

$$\sigma = 2^{-1/6}r_o = 0.8909 r_o = 0.8909 \left(\frac{2\lambda_n}{\lambda_m} \right)^{1/6} \tag{5.3d}$$

In materials science, Eq. (5.3a) is also known as the Lennard-Jones function suitable for analyzing the concept of particle pair or interatomic pair potentials. It provides means to predict the potential energy and binding forces for an atom pair overlapping stability and separation (at center-center atom distance greater than the equilibrium length).

Graphically, Fig. 5.8 illustrates the theoretical driven force F_{max} (Fig. 5.8a) and potential energy $U(r)$ at F_{max} (Fig. 5.8b) induced by a hard metal wedge (Fig. 5.8c) acting destructively across the carbon atoms in the continuous honeycomb structure in graphite for separating a graphene plane by a shear force (Fig. 5.8d).

Potential energy is defined as the work required to move a system from its current configuration to a reference configuration. On the other hand, the binding force refers to the net force that holds a system together. Evaluating the potential

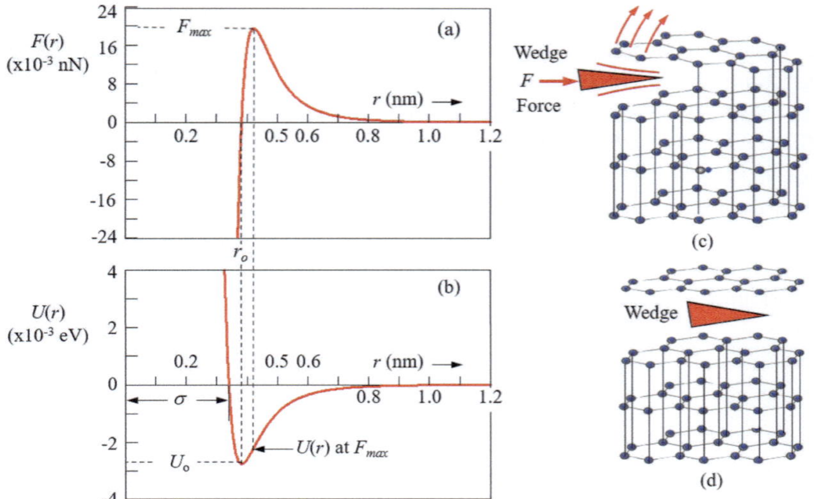

Fig. 5.8 Theoretical (**a**) binding force and (**b**) binding potential energy for a pair of carbon atoms. Model for (**c**) mechanical exfoliation of graphite structure by a wedge force and (**d**) complete separation of a graphene plane (exfoliated graphene sheet) induced by a mechanical shear force

energy and binding force expressions given by Eqs. (5.3b) and (5.3c), and related derivatives, (5.3d,e), reveals interesting results.

- Let $F(r)_{r=r_o} = dU(r)/dr = 0$ in Eq. (5.3b) and solve for the equilibrium interatomic distance $r = r_o$.
- Let $dF(r)/dr = 0$ and solve for $r = r_x$ so that $F(r_x) = F_{max}$.
- Compute $U(r_o)$ and $U(r_x)$ from Eq. (5.8a).
- Derive and calculate $d^2F(r_x)/dr^2$.
- Verify that $d^2F(r_x)/dr^2 < 0$ corresponds to $F(r_x) = F_{max}$.

An example can make these suggestions clear. For graphite with $R_o = 0.00276\ eV$ and $\sigma = 0.339\ nm$ (Lebedeva et al. [43] and Popov et al. [44]), the equilibrium interatomic distance becomes $r_o = 2^{1/6}\,(0.339\ nm) = 0.381\ nm$.

The forces, potential energies, and the force second derivative are

$$F(r_o) = 0 \text{ at } r_o = 0.381\ nm \tag{5.4a}$$

$$F(r) = F_{max} = 0.02\ N \text{ at } r = r_x = 0.42\ nm \tag{5.4b}$$

$$U(r_o) = U_o = -2.76\ meV \text{ at } r_o = 0.38\ nm \tag{5.4c}$$

$$U(r) = -2.21\ meV \text{ at } r_x = 0.42\ nm \tag{5.4d}$$

$$\frac{d^2F(r)}{dr^2} = -11.04\ eV/nm^2 < 0 \text{ at } r_x = 0.42\ nm$$

Theoretically, $F(r) \geq F_{max} = 19.51 \times 10^{-3}\ nN \approx 0.02\ nN$ is the mechanical force needed to overcome the weak binding force across the π-bonds in graphite honeycomb planes. Once this is achieved, graphite is exfoliated, and graphene sheets are produced. Moreover, the change in potential energy $\Delta U(r) = |U(r_o) - U(r)| = 0.55\ meV$ is also considered as the driving force for breaking or debonding the π-bonds at a binding force $F_{max} \approx 0.02\ N$.

The net force equation providing the trendline in Fig. 5.8a is $F(r) = F_A(r) + F_R(r)$, where $F_A(r)$ and $F_R(r)$ are the attractive and repulsive forces, respectively. At the equilibrium distance $r = r_o$, $F_A(r_o) = -F_R(r_o)$, and the pair of atoms is stable until an external force is exerted upon it as indicated in Fig. 5.8c, where $F(r) \geq F_{max}$ and $U(r) = ([U_A(r) + U_R(r)] > U(r_o)$ at $r > r_o$ for mechanical exfoliation to be effective in debonding at least one graphite plane (Fig. 5.8d).

5.9   Electrochemical Exfoliation

Electrochemical exfoliation is an attractive technique for producing graphene by ionic intercalation capable of forming a gaseous phase. Theoretically, this phase exerts a high pressure at the graphite interplanar location in an electrolyte under a specific applied DC bias potential (voltage).

The electrochemical graphite exfoliation to produce pristine functionalized graphene or graphite oxide is an ionic complex cathodic process since an external power supply charges the graphite electrode with a positive potential. Consequently, negative charged ions agglomerate at the opposite end of the graphite electrical contact and intercalate between the graphite planes.

This process may or may not allow functionalization (enhancement of the surface chemistry) to produce graphene sheets. It all depends on the electrolyte ionic constituents, cathodic potential, and, to an extent, temperature. However, oxygen or hydrogen evolution reaction may be detrimental to the exfoliation process, as it relates to atomic structural changes of the electrode surface. This is a general and simple explanation of the electrochemical exfoliation process.

Figure 5.9 schematically shows the basic electrochemical components (Fig. 5.9a) and the separation of one graphite plane process owing to high pressure exerted by ionic particles in the electrolyte (Fig. 5.9b). Ideally, a graphene sheet is produced using this basic electrochemical exfoliation mechanism (Fig. 5.9c). The idea for Fig. 5.9b is borrowed from Yu et al. [45].

During cathodic electrochemical exfoliation of graphite having a specific morphology (Fig. 5.9a), positively charged molecules or ionic particles, such as Li^+ in the electrolyte, are attracted to the negatively charged graphite cathodic surface. On the other hand, during electrochemical exfoliation, the negatively charged graphite anode attracts the positively charged ionic particles, such as SO_4^{2-} in the electrolyte containing dissolved sulfuric acid.

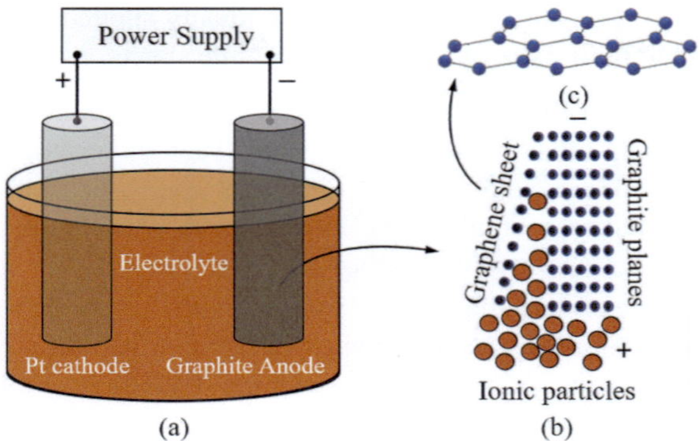

Fig. 5.9 Electrochemical exfoliation of graphite setup and mechanism of delamination of graphene sheets. (**a**) Basic electrochemical cell. (**b**) Mechanism during electrochemical exfoliation owing to intercalation of ionic particles, where the sp^2 hybridized bonding is due to strong planar covalent σ-bonds and weak interplanar π-bonds. (**c**) A graphene sheet

Ionic particles are assumed to intercalate the graphite planes and exert a high pressure on the π-bonds until broken in an assumed sequential manner. Eventually, a graphite plane gets detached from the graphite surface of the working electrode, and it becomes a graphene sheet; otherwise, the exfoliated graphite might be in the form of small graphene oxide (GO) particles due to the oxidative condition of the anodic electrode. Thus, electrochemical exfoliation is followed by filtration of the electrolyte containing suspended graphene sheets, conditioning, and application.

Experimentally, electrochemical exfoliation is based on a trial-and-error approach, followed by filtration of the electrolyte containing suspended graphene sheets, conditioning, and application.

Potassium Ferricyanide and Water Electrolyte The common electrolyte for electrochemical exfoliation (intercalation) is dissolved potassium ferricyanide $K_3[Fe(CN)_6]$, which prevents the oxidation of order graphite and slows electron transfer kinetics on the basal plane of graphite attributed to a relatively low reaction rate constant in the order of $K_o = 10^{-6} \ cm/s$ for the ferricyanide ions $Fe(CN)_6^{3-}$ or $Fe(CN)_6^{4-}$ (Cline et al. [46]). Conversely, stacking disorder in the hexagonal graphite increases electron-transfer rates and causes electronic perturbation. The disorder graphite behavior can be revealed through the D-band and D'-band in the Raman spectra (Pimenta et al. [47]). The Raman spectrum is described in a later section.

Potassium Hydroxide and Water Electrolyte According to Tripathi et al. [48], electrochemical exfoliation of hexagonal graphite with dissolved alkaline potassium hydroxide (KOH) electrolyte leads to order graphene, implying a high-quality few-layer graphene (FLG) graphene at a $10 \leq pH \leq 13$ range. The cell was preconditioned by applying a bias potential in the order of $+3$ volts for 100 seconds so that the hydroxyl OH^- ions begin to intercalate the graphite layers followed by ramping the bias potential to ± 10 volts, individually applied to the precondition stage of graphite to induce the exfoliation of graphite sheets.

Other Electrolytes Lastly, electrochemical exfoliation of graphite can successfully be achieved by using other types of electrolytes, such as aqueous sodium halide (such as NaF), sulfuric acid solution (H_2SO_4), inorganic salt ((NH_4)$_2SO_4$), and the like (Abdelkader et al. [49]). For instance, carbon in a sulfuric acid solution reacts and produces intercalating SO_4^{2-} anions between graphite planes. Accordingly, the size of the sulfate SO_4^{2-} anion is much bigger than the interatomic d-spacing ($d = 0.34\,nm$) between graphite planes, and therefore, electrochemical exfoliation occurs due to SO_4^{2-} anion intercalation (Tripathi et al. [48]).

5.10   Raman Spectroscopy and X-Ray Diffraction

5.10.1   Raman Spectroscopy

Raman spectroscopy is a powerful and useful non-destructive (noninvasive) vibrational technique used to characterize surface features of materials. For instance, selecting Cancado et al. [8] Raman spectra shown in Fig. 5.10a elucidates the most common vibrational bands denoted as D, D', 2D = G', 2D' = G" in graphene. Accordingly, Fig. 5.10b shows the effect of laser energy cases.

This technique produces valuable spectra using a monochromatic laser with a single wavelength λ_L to interact with atomic or molecular vibrational modes and phonons in a specimen. Raman spectroscopy is extremely sensitive to structural

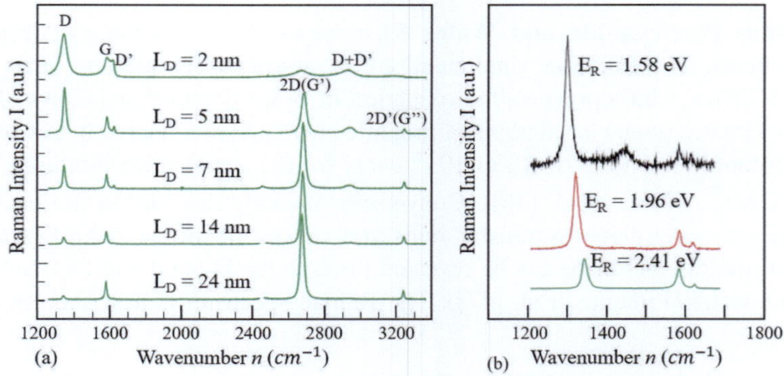

Fig. 5.10 Selected Raman spectra to show different types of vibrational bands due to defects in the graphene samples. (**a**) Ion bombarded single-layer graphene (SLG) measured at an excitation laser energy $EL = 2.41\,eV$ (with wavelength $\lambda_L = 514.5\,nm$) and (**b**) Raman spectra with three excitation laser energies for an ion-bombarded sample having an average defect distance $LD = 7\,nm$. After Cancado et al. [8]

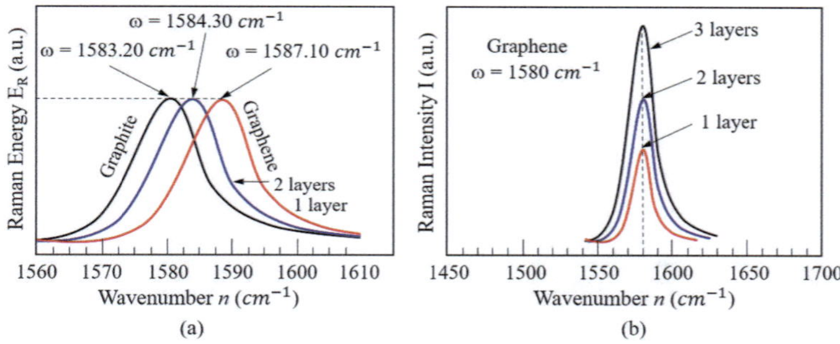

Fig. 5.11 Raman energy and intensity diagrams. (**a**) Raman energy E_R shift spectra and (**b**) Raman intensity I shift spectra. After Wang et al. [51]

morphology (geometric structure), type of bonding within atoms or molecules and defects (such as edge and grain boundaries), thickness of planes or films, thermal conductivity, and the like.

According to Dr. Mildred Dresselhaus in her "Raman Spectra of Graphene and Carbon Nanotubes" video conference posted in 2011 at https://www.youtube.com/watch?v=acvFIqtM_gU, there is also an M-band in a Raman spectrum not being discussed sufficiently (Boehm et al. [50]). The notation (letters) used for designating the type of vibrational bands for graphene are

- D-band due to defect concentration ($n = 1350\ cm^{-1}$)
- $D\prime$-band due to impurities ($n = 1620\ cm^{-1}$)
- G-band due to metallicity (sp^2) ($n = 1580\ cm^{-1}$)
- $G\prime$-band due to doping ($n = 2700\ cm^{-1}$)
- $D + D\prime$ due to impurities and defects ($n = 2940\ cm^{-1}$)
- $2D\prime(G")$ due to stacking order of graphene and impurities ($n = 3244\ cm^{-1}$)
- M-band for layer stacking ($n = 1875\ cm^{-1}$), not shown in the Raman spectra in Fig 5.10a.

Borrowing Wang et al.'s [51] idea, found in Wall's Application Note: 52252 [52], helps elucidate some unique Raman spectra (Raman spectroscopy) features depicted in Fig. 5.11, where the Raman energy may be defined as Planck's energy equation $E = hv$ with Plan's constant h and frequency v.

The Raman spectra depicted in Fig. 5.11a is designated as the 2D-band. Notice that the position of a 2D-band depends on the wavenumber k for graphite and graphene. For clarity, graphene designated as single layer or double layer has its corresponding peak wavenumber shifted to the right of a graphite band.

The wavenumbers for graphene depend on the specific vibrational mode of interest and the number of graphene layers. The position in a Raman 2D-band can be calculated using Wang et al.'s [51] empirical equation. For N-layer graphene, the wavenumbers are

$$k = 1581.60 + \frac{11}{1 + N^{1.6}} \tag{5.5a}$$

$$k_1 = 1581.6 + \frac{11}{1 + 1^{1.6}} = 1587.10 \; cm^{-1} \tag{5.5b}$$

$$k_2 = 1581.6 + \frac{11}{1 + 2^{1.6}} = 1584.30 \; cm^{-1} \tag{5.5c}$$

$$k_3 = 1581.6 + \frac{11}{1 + 3^{1.6}} = 1583.20 \; cm^{-1} \tag{5.5d}$$

Notice that the Raman intensity level at constant peak wavenumber shown in Fig. 5.11b is high for graphite and low for graphene. Therefore, the Raman intensity depends on the number of graphene layers.

Assessment of Defect Density A Raman spectrum is used for quantifying defects in 2D materials by using the Raman intensity ratio correlation for the D-band and G-band spectrum being investigated. Apparently, the defect density quantification has a foundation in the work of Tuinstra and Koenig [53].

Empirical defect density ρ_d equations for Raman-active modes in graphene is written in a general form (Cancado et al. [8], Bruna et al. [54])

$$\rho_d = \left(10^{14}\right) / \left(\pi L_d^2\right) \tag{5.6a}$$

$$L_d = \left(4.3 \times 10^3\right) \left(E_L^{-4}\right) (I_D/I_G)^{-1} \quad \text{in } nm^2 \tag{5.6b}$$

$$L_d = \left(1.8 \times 10^{-9}\right) \left(\lambda_L^4\right) (I_D/I_G)^{-1} \quad \text{in } nm^2 \tag{5.6c}$$

$$\rho_d = \alpha \, (I_D/I_G) \tag{5.6d}$$

where L_d denotes the crystallite size and the parameter α can be stated as

$$\alpha = \left(7.3 \times 10^9 \; cm^{-2}\right) E_L^- \tag{5.7a}$$

$$\alpha = \left(1.8 \times 10^{22} \; cm^{-2}\right) \lambda_L^{-4} \tag{5.7b}$$

Here, E_L denotes the laser excitation energy in eV units with a wavelength λ_L in nm units. For $\lambda_L = 532 \; nm$, Seehra et al. [9] reported $\alpha = 2.16 \times 10^{11} \; cm^{-2}$, but Eq. (5.7b) gives $\alpha = 2.25 \times 10^{11} cm^{-2}$ which is used in Eq. (5.6d) to get

$$\rho_d = \alpha \, (I_D/I_G) = \left(2.25 \times 10^{11} cm^{-2}\right) (I_D/I_G) \tag{5.8}$$

This empirical expression gives approximated values for the defect density ρ_d. For a Raman intensity ratio $I_D/I_G = 0.22$ (Seehra et al. [9]), Eq. (5.8) yields $\rho_d \approx 5 \times 10^{10} \; cm^{-2}$.

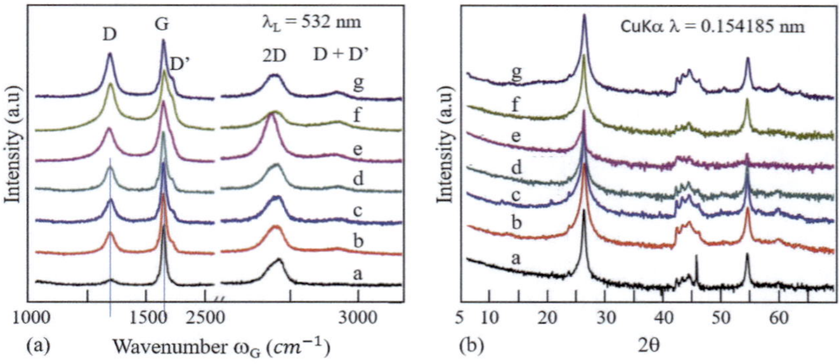

Fig. 5.12 Characteristic patterns of graphene and graphene-based commercial materials. (**a**) Raman spectra with designated bands. (**b**) X-ray diffraction diffractograms as reported by Seehra et al. [9]. Legend: [a]. 1-GR-NPL, [b] 8-GR-NP-ML-FL, [c] 16-GR-NPL-N2, [d] 15-GR-NPL-COOH, [e] 14-FL-GO-4-8L, [f] 6-GR-NP-12nm-FL, [g] 7-GR-NP-8nm-FL

5.10.2   X-Ray Diffraction Patterns

This section includes data found in Seehra et al.'s [9] review paper, which contains Raman spectroscopy (Fig. 5.12a) and X-ray diffraction (Fig. 5.12b) patterns for graphene.

The purpose of Fig. 5.12a is to elucidate further the importance of Raman spectroscopy for scrutinizing materials. Notice that the Raman spectra include additional D-band, $D + D\prime$-band, and even a M-band in graphene.

Figure 5.12b, on the other hand, exhibits X-ray diffraction characteristics in pattern behavior being dependent on the level intensities with arbitrary units (a.u.). The intensity peaks corresponding to the Bragg's angle (θ) remain nearly constant.

Using the sharpest peak at $2\theta = 26.1°$ and a CuK_α wavelength $\lambda = 0.154185$ nm (Seehra et al. [9]), one can calculate the d-spacing between honeycomb planes as per Bragg's equation with a diffraction angle $\theta = 13.05°$. Thus,

$$\lambda = 2d \sin (\theta) \tag{5.9a}$$

$$d = \frac{0.154185 \ nm}{2 \sin (26.1/2)} = 0.34 \ nm \tag{5.9b}$$

Here, $d = 0.34$ nm is the value reported in the literature for hexagonal graphite interplanar distance or simply the d-spacing between the hexagonal graphite basal planes. Recall that the hexagonal graphite interplanar distance refers to the spacing between individual layers of carbon atoms in the hexagonal crystal structure of graphite.

The distinct features of the XRD and Raman spectra (RS) patterns for the samples cited in Fig. 5.12 are relative evidence of the anisotropy in graphite.

5.11　Energy Bandgap

The electron energy is normally represented as an energy map of the conduction and valence bands. A three-dimensional (3D) plot of a general momentum-dependent energy function $E = f(k_x, k_y)$ provides a surface as shown in Fig. 5.13a for graphene (Beenakke [55]).

For 2D massless Dirac wavelike particles (electrons), the Dirac linear energy-wavevector function $E = f(k_x, k_y)$, at the Fermi energy, is attributed to the honeycomb structure of graphene or indirectly to the hexagonal symmetry of graphite. This plot exhibits the Dirac cones being connected at k-points on the k_x, k_y-plane.

The wavenumbers k_x, k_y in Fig. 5.13a are components of the wavevector $\vec{k}$. The upper and lower surfaces represent the conduction and valence bands, respectively. Notice that the (k_x, k_y)-plane has six k-points forming a hexagonal shape called the first Brillouin zone.

Figure 5.13b is a simplified 2D energy plot of the 3D surface shown in Fig. 5.13a for graphene. This particular case indicates that graphene has a zero-energy bandgap ($E_g = 0$) at this k-point corresponding to a position for the Fermi energy E_F, which is the highest energy state occupied by electrons in a material at absolute zero temperature. Therefore, graphene is treated as a semiconductor material. Conversely, Fig. 5.13c shows that $E_g > 0$ for exciting electrons in the valence energy band upon the addition of an impurity, such as Na and Na/Ir impurities

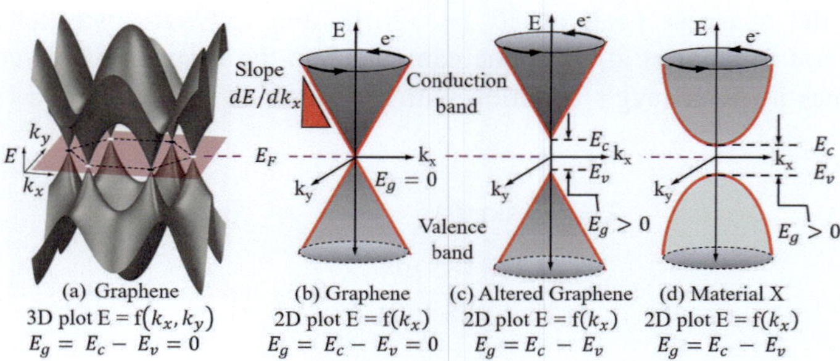

Fig. 5.13 Fundamental electron energy bandgap (E_g) and Dirac cones for graphene and parabola for a hypothetical material X. (**a**) 3D energy plot for pristine graphene (Beenakke [56]), (**b**) 2D energy plot for graphene with electron mass $m = 0$, (**c**) 2D energy plot from the k-point for altered graphene due to an addition of an impurity and (**d**) 2D parabola energy plot for a hypothetical material X with $m \neq 0$

(Papagno et al. [56]), and Fig. 5.13d is a parabola energy plot for a hypothetical material, namely, X.

The upper Dirac cone for graphene at $E_x > 0$ (positive direction) is assumed to partially fill up, and the electron kinetic energy (KE) increases. If $E_x < 0$ (negative or opposite direction), the lower Dirac cone is electron depleted, generating vacancies or "holes" as E_e decreases due to the addition of secondary electrons from another materials.

5.12   Graphene in the Semiconductor Field

The application of graphene in the semiconductor field is best illustrated by the n-type and p-type junction with a Fermi energy level E_F shown in Fig. 5.14 (Beenakker [55]).

The exploration and commercial use of graphene, as well as other forms of carbon allotropes, in the field of semiconductors is the key for producing electronic devices with high efficiencies and prolong energy-storage capabilities. Due to the extraordinary properties of honeycomb-shaped structure of graphene, the energy bandgap in graphene can be manipulated by a doping mechanism, which induces electrons and holes to move through the graphene layer with a velocity $\upsilon < c = 3 \times 10^8 \; m/s$.

Figure 5.14 schematically shows the basic Dirac cones and related conduction and valence bands having linear behavior indicated by the straight-edge paths. Thus, graphene can be classified as n-type or p-type semiconductor by inducing electron excitation in an electric field. As a result, a p-n junction can be engineered having a common Fermi level (E_F) at a temperature $T > 0 \; K$ and a non-zero energy bandgap $(E_g > 0)$. Notice that the position of the Fermi level determines the dominant carrier in a semiconductor. Explicitly, electrons dominate in n-type and holes in p-type semiconductor materials. Moreover, tunneling of an electron from the conduction band into a hole from the valence band is known as the Klein tunneling in a p-n junction (Beenakker [55]).

The energy level in the n-type is above the Dirac k-point into the conduction band, and the p-type energy level is just below the k-point into the valence band as

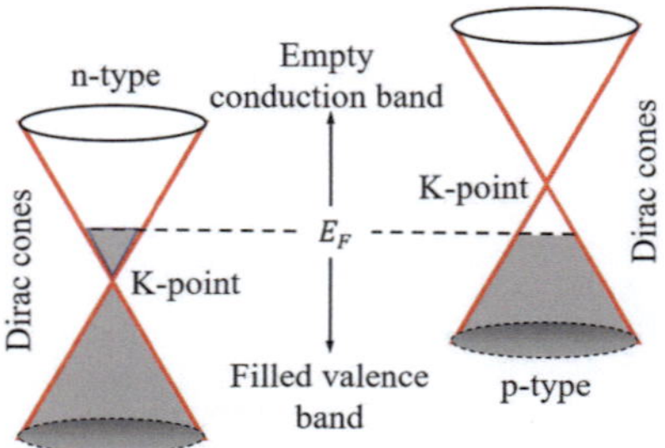

Fig. 5.14 Schematic to illustrate a p-n junction. After Beenakker [55]

in a graphene-substrate system. This implies that electrons in the valence band can be excited to move into the conduction band (n-type) or vice-verse to induce holes (p-type).

5.13 Summary

This chapter gives a condensed overview of the common trends in hexagonal two-dimensional (2D) atomic structures, such as graphene and TMD materials. Relevant methods in the field of 2D-material technology are purposely emphasized on mechanical and electrochemical exfoliation and chemical vapor deposition (CVD). In addition, the most used methods for characterizing these materials are Raman spectroscopy, X-ray diffraction, and transmission electron microscopy (TEM). Although details on how to generate data in a laboratory are not given, this overview chapter includes experimental results being gathered from the literature. Moreover, the intent in this chapter is to provide some unique datasets and how to interpret them from an experimentalist's prospective.

Mechanical exfoliation by the wedge model, electron drift velocity and mobility under an electric field strength range, electrochemical exfoliation by large ionic intercalation model, and the theoretical binding potential energy and binding force are briefly described and analyzed as part of the characterization of 2D materials. Moreover, the bandgap energy theory is also detailed for pristine and altered graphene surfaces using the Dirac cones and parabola models. Specifically, graphene in a semiconductor field application can acquire a n-type or p-type semiconductor behavior as illustrated in Fig. 5.14.

References

1. K.S. Novoselov, A.K. Geim, S.V. Morozov, D. Jiang, Y. Zhang, S.V. Dubonos, I.V. Grigorieva, A.A. Firsov, Electric field effect in atomically thin carbon films. Science **306**(5696), 666–669 (2004)
2. N.B. Kopnin, M. Ijas, A. Harju, T.T. Heikkila, High-temperature surface superconductivity in rhombohedral graphite. Phys. Rev. B **87**(14), 140503 (2013)
3. P.R. Kidambi, D.D. Mariappan, N.T. Dee, A. Vyatskikh, S. Zhang, R. Karnik, A.J. Hart, A scalable route to nanoporous large-area atomically thin graphene membranes by roll-to-roll chemical vapor deposition and polymer support casting. ACS Appl. Mater. Interfaces **10**(12), 10369–10378 (2018)
4. H. Zhang, Ultrathin two-dimensional nanomaterials. ACS Nano **9**(10), 9451–9469 (2015)
5. A.V. Kolobov, J. Tominaga, *Two-Dimensional Transition Metal Dichalcogenides* (Springer International Publishing, Switzerland, 2016)
6. S.J. An, Y.H. Kim, C. Lee, D.Y. Park, M.S. Jeong, Exfoliation of transition metal dichalcogenides by a high-power femtosecond laser. Sci. Rep. **8**(1), 1–6 (2018)
7. H.M. Dong, Y.F. Duan, F. Huang, J.L. Liu, Electron drift velocity and mobility in graphene. Front. Phys. **13**(2), 137203 (2018)

8. L.G. Cancado, A. Jorio, E.H.M. Ferreira, F. Stavale, C.A. Achete, R.B. Capaz, M.V. de Oliveira Moutinho, A. Lombardo, T.S. Kulmala, A.C. Ferrari, Quantifying defects in graphene via Raman spectroscopy at different excitation energies. Nano Lett. **11**(8), 3190–3196 (2011)

9. M.S. Seehra, V. Narang, U.K. Geddam, A.B. Stefaniak, Correlation between X-ray diffraction and Raman spectra of 16 commercial graphene-based materials and their resulting classification. Carbon **111**, 380–385 (2017)

10. T. Mahmoudi, Y. Wang, Y.B. Hahn, Graphene and its derivatives for solar cells application. Nano Energy **47**, 51–65 (2018)

11. W.S. Hummers Jr., S. William, R.E. Offeman, Preparation of graphitic oxide. J. Am. Chem. Soc. **80**(6), 1339 (1958)

12. H. Bai, C. Li, G. Shi, Functional composite materials based on chemically converted graphene. Adv. Mater. **23**(9), 1089–1115 (2011)

13. J. Chen, B. Yao, C. Li, G. Shi, An improved Hummers method for eco-friendly synthesis of graphene oxide. Carbon **64**, 225–229 (2013)

14. H.O. Pierson, *Handbook of Chemical Vapor Deposition* (Noyes Publications/William Andrew Publishing, LLC, New York, 1999)

15. D. Pierucci, H. Sediri, M. Hajlaoui, J.C. Girard, T. Brumme, M. Calandra, E. Velez-Fort, G. Patriarche, M.G. Silly, G. Ferro, V. Soulie're, M. Marangolo, F. Sirotti, F. Mauri, A. Ouerghi, Evidence for flat bands near the Fermi level in epitaxial rhombohedral multilayer graphene, ACS Nano **9**(5), 5432–5439 (2015)

16. A. Torche, F. Mauri, J.C. Charlier, M. Calandra, First-principles determination of the Raman fingerprint of rhombohedral graphite. Phys. Rev. Mater. **1**(4), 041001 (2017)

17. J.D. Bernal, The structure of graphite, in *Proceedings of the Royal Society of London. Series A, Containing Papers of a Mathematical and Physical Character*, vol. 106(740) (1924), pp. 749–773

18. T. Latychevskaia, S.K. Son, Y. Yang, D. Chancellor, M. Brown, S.t Ozdemir, I. Madan, G. Berruto, F. Carbone, A. Mishchenko, K. Novoselov, Stacking transition in rhombohedral graphite. Front. Phys. **14**(1), 13608 (2019)

19. C. Cong, T. Yu, K. Sato, J. Shang, R. Saito, G.F. Dresselhaus, M.S. Dresselhaus, Raman characterization of ABA-and ABC-stacked trilayer graphene. ACS Nano **5**(11), 8760–8768 (2011)

20. U.H. Krauss, H.W. Schmidt, H.A. Taylor Jr., D.M. Sutphin, *International Strategic Minerals Inventory Summary Report*. Natural Graphite, No. 930-H. US Department of the Interior, Geological Survey (1989)

21. X. Lu, M. Iqbal Bakti Utama, J. Lin, X. Gong, J. Zhang, Y. Zhao, S.T. Pantelides, et al., Large-area synthesis of monolayer and few-layer $MoSe_2$ films on SiO_2 substrates. Nano Lett. **14**(5), 2419–2425 (2014)

22. H. Marsh, F. Rodriguez Reinoso, *Activated Carbon* (Elsevier, New York, 2006). ISBN: 0080444636

23. J. Qing, V.L. Richards, D.C. Van Aken, Growth stages and hexagonal-rhombohedral structural arrangements in spheroidal graphite observed in ductile iron. Carbon **116**, 456–469 (2017)

24. Y. Shi, S. Xu, Y. Yang, S. Slizovskiy, S.V. Morozov, S.K. Son, S. Ozdemir, et al., *Electronic Phase Separation in Topological Surface States of Rhombohedral Graphite* (arXiv.org), arXiv preprint arXiv (2019), pp. 1911.04565

25. S. Fahy, S.G. Louie, M.L. Cohen, Pseudopotential total-energy study of the transition from rhombohedral graphite to diamond. Phys. Rev. B **34**(2), 1191 (1986)

26. F. Isobe, H. Ohfuji, H. Sumiya, T. Irifune, Nanolayered diamond sintered compact obtained by direct conversion from highly oriented graphite under high pressure and high temperature. J. Nanomater. **2013**, 15–15 (2013)

27. H. Xie, F. Yin, T. Yu, J.T. Wang, C. Liang, Mechanism for direct graphite-to-diamond phase transition. Sci. Rep. **4**(1), 1–5 (2014)

28. W.E. Callister, Jr., D.G. Rethwisch, *Materials Science and Engineering—An Introduction*, 10th edn. (Wiley, New York, 2018)

29. C.B. Carter, M.G. Norton, *Ceramic Materials: Science and Engineering*, vol. 716 (Springer, New York, 2007)
30. G. Yang, L. Li, W. Bun Lee, M. Cheung Ng, Structure of graphene and its disorders: a review. Sci. Technol. Adv. Mater. **19**(1), 613–648 (2018)
31. H.O. Pierson, *Handbook of Carbon. Graphite, Diamond and Fullerenes: Properties, Processing and Applications* (Noyes Publications, William Andrew, Park Ridge, 1993)
32. N.C. Gallego, C.I. Contescu, H.M. Meyer III, J.Y. Howe, R.A. Meisner, E.A. Payzant, M.J. Lance, S.Y. Yoon, M. Denlinger, D.L. Wood III, Advanced surface and microstructural characterization of natural graphite anodes for lithium ion batteries. Carbon **72**, 393–401 (2014)
33. J. Fujita, T. Hiyama, A. Hirukawa, T. Kondo, J. Nakamura, S. Ito, Near room temperature chemical vapor deposition of graphene with diluted methane and molten gallium catalyst. Sci. Rep. **7**(1), 1–10 (2017)
34. J.N. Coleman, M. Lotya, A. O'Neill, S.D. Bergin, P.J. King, U. Khan, K. Young, et al., Two-dimensional nanosheets produced by liquid exfoliation of layered materials. Science **331**(6017), 568–571 (2011)
35. N. Liu, P. Kim, J.H. Kim, J.H. Ye, S. Kim, C.J. Lee, Large-area atomically thin MoS_2 nanosheets prepared using electrochemical exfoliation. ACS Nano **8**(7), 6902–6910 (2014)
36. R.K. Layek, A.K. Nandi, A review on synthesis and properties of polymer functionalized graphene. Polymer **54**(19), 5087–5103 (2013)
37. J.D. Wood, S.W. Schmucker, A.S. Lyons, E. Pop, J.W. Lyding, Effects of polycrystalline Cu substrate on graphene growth by chemical vapor deposition. Nano Letters **11**(11), 4547–4554 (2011)
38. T. Georgiou, R. Jalil, B.D. Belle, L. Britnell, R.V. Gorbachev, S.V. Morozov, Y.J. Kim, et al., Vertical field-effect transistor based on graphene-WS_2 heterostructures for flexible and transparent electronics. Nat. Nanotechnol. **8**(2), 100–103 (2013)
39. E.D. Grann, S.J. Sheih, C. Chia, K.T. Tsen, O.F. Sankey, S.E. Guncer, D.K. Ferry, et al., Picosecond Raman studies of electric-field-induced nonequilibrium carrier distributions in GaAs-Based P-I-N nanostructure semiconductors. Appl. Phys. Lett. **64**(10), 1230–1232 (1994)
40. W. Walkowiak, Drift velocity of free electrons in liquid argon, in *Nuclear Instruments and Methods in Physics Research Section A: Accelerators, Spectrometers, Detectors and Associated Equipment*, vol. 449(1–2) (2000), pp. 288–294
41. C. Jacoboni, F. Nava, C. Canali, G. Ottaviani, Electron drift velocity and diffusivity in germanium. Phys. Rev. B **24**(2), 1014 (1981)
42. E.J.G. Santos, Electric field effects on graphene materials, in *Exotic Properties of Carbon Nanomatter* (Springer, Dordrecht, 2015), pp. 383–391
43. I.V. Lebedeva, A.A. Knizhnik, A.M. Popov, Y.E. Lozovik, B.V. Potapkin, Interlayer interaction and relative vibrations of bilayer graphene. Phys. Chem. Chem. Phys. **13**(13), 5687–5695 (2011)
44. A.M. Popov, I.V. Lebedeva, A.A. Knizhnik, Y.E. Lozovik, B.V. Potapkin, Barriers to motion and rotation of graphene layers based on measurements of shear mode frequencies. Chem. Phys. Lett. **536**, 82–86 (2012)
45. P. Yu, S.E. Lowe, G.P. Simon, Y.L. Zhong, Electrochemical exfoliation of graphite and production of functional graphene. Curr. Opin. Colloid Interface Sci. **20**(5–6), 329–338 (2015)
46. K.K. Cline, M.T. McDermott, R.L. McCreery, Anomalously slow electron transfer at ordered graphite electrodes: influence of electronic factors and reactive sites. J. Phys. Chem. **98**(20), 5314–5319 (1994)
47. M.A. Pimenta, G. Dresselhaus, M.S. Dresselhaus, L.G. Cancado, A. Jorio, R. Saito, Studying disorder in graphite-based systems by Raman spectroscopy. Phys. Chem. Chem. Phys. **9**(11), 1276–1290 (2007)
48. P. Tripathi, C. Patel, R. Prakash, M.A. Shaz, O.N. Srivastava, *Synthesis of High-Quality Graphene Through Electrochemical Exfoliation of Graphite in Alkaline Electrolyte*, arXiv preprint arXiv (2013), pp. 1310.7371

49. A.M. Abdelkader, A.J. Cooper, R.A.W. Dryfe, I.A. Kinloch, How to get between the sheets: a review of recent works on the electrochemical exfoliation of graphene materials from bulk graphite. Nanoscale **7**(16), 6944–6956 (2015)
50. H.P. Boehm, A. Clauss, G.O. Fischer, U. Hofmann, Dünnste kohlenstofffolien. Zeitschrift Für Naturforschung B **17**(3), 150–153 (1962)
51. H. Wang, Y. Wang, X. Cao, M. Feng, G. Lan, Vibrational properties of graphene and graphene layers. Journal of Raman Spectroscopy: An International Journal for Original Work in all Aspects of Raman Spectroscopy, Including Higher Order Processes, and also Brillouin and Rayleigh Scattering **40**(12), 1791–1796 (2009)
52. M. Wall, *The Raman Spectroscopy of Graphene and the Determination of Layer Thickness*, Application Note 52252 (Thermo Fisher Scientific, Madison, 2011). www.thermoscientific.com
53. F. Tuinstra, J.L. Koenig, Raman spectrum of graphite. J. Chem. Phys. **53**(3), 1126–1130 (1970)
54. M. Bruna, A.K. Ott, M. Ijäs, D. Yoon, U. Sassi, A.C. Ferrari, Doping dependence of the Raman spectrum of defected graphene. ACS Nano **8**(7), 7432–7441 (2014)
55. C.W.J. Beenakker, Colloquium: Andreev reflection and Klein tunneling in graphene. Rev. Mod. Phys. **80**(4), 1337 (2008)
56. M. Papagno, S. Rusponi, P.M. Sheverdyaeva, S. Vlaic, M. Etzkorn, D. Pacilé, P. Moras, C. Carbone, H. Brune, Large band gap opening between graphene dirac cones induced by Na adsorption onto an Ir superlattice. ACS Nano **6**(1), 199–204 (2012)

Chapter 6
Crystal Defects

6.1 Introduction

This chapter focuses on dislocation theory and inherently developed crystal defects that disrupt or dislocate the regular geometrical arrangement of atoms in crystalline solids. Essentially, crystal defects are imperfections or discontinuities that govern material properties.

The most common atomic discontinuities in engineering polycrystalline materials are the grain boundaries, shrinkage pores due the formation of gas bubbles during solidification, flaws and possibly inclusions in steels, such as brittle ceramic phases known as metal oxides (FeO, Al_2O_3), metal sulfates (MnS, Al_2S_3, Cr_2S_3), and chromite (MnO-Cr_2O_3). Thus, the most relevant inherent defects are

- Point defects due to missing atoms in the lattice, which is a periodic arrangement of atoms.
- Linear defects called dislocations, which consist of groups of atoms in irregular positions. Such dislocations are observed under the microscope as short irregular lines.
- Dislocations may be inherently present in a crystalline solid induced by rapid cooling (quenching) and may be induced during mechanical deformation.
- Planar defects, such as grain boundaries and stacking faults (SFs), are interfaces separating homogeneous and adjacent crystal grains or particles in solids.

The goal in this chapter is to characterize crystal defects in metallic and nonmetallic crystals at an atomic scale. By definition, crystals or crystalline solids have a long-range order (regular arrangement of atoms).

Amorphous solids, on the other hand, have a short-range order. Insights on theory and calculation of the material properties due to point defects can be found in the Proceedings of a Conference titled "Calculation of the Properties of Vacancies and Interstitial" published by National Bureau of Standards in 1966 [1].

N. Perez, *Materials Science: Theory and Engineering*,
https://doi.org/10.1007/978-3-031-57152-7_6

6.2   Point Defects

Considering the complex nature of solid structures, extensive studies are required for the proper understanding of the mechanisms related to crystal defects or surface reconstruction. Within this context, the phases "lattice points" and "lattice sites" are not the same. By definition, a lattice point refers to the atomic position of an atom within the periodic arrangement described by the Bravais lattice (see Fig. 1.1a), and the latter are available empty spaces between packed atoms forming the crystal structure.

Solid materials deviate from the ideal crystal structures due to point defects. Thus, any imperfection around a lattice point within a small region of disturbance is called a zero-dimensional (0D) point defect. For instance, Fig. 6.1 schematically elucidates the two-dimensional (2D) atomic structure in a metal (Fig. 6.1a) and in a ceramic material (Fig. 6.1b) containing 0D imperfections called crystal defects, such as Frenkel defects and Schottky defects. Moreover, crystalline solids are electronic conductors, and ceramics are ionic conductors.

Most crystals have, at least, inherent vacancies representing missing atoms or ions from their respective lattice points.

6.2.1   Stoichiometric Point Defects

In crystalline materials, stoichiometric point defects (Fig. 6.1a) are known as thermodynamic defects or **intrinsic point defects**, such as

- Vacancies being produced as empty spaces in the lattice sites during solidification or heat treatment at high temperatures. In ceramic materials, electrons may occupy a vacancy site to maintain electrical neutrality.
- Self-interstitial due to the random formation of atom clusters during solidification.

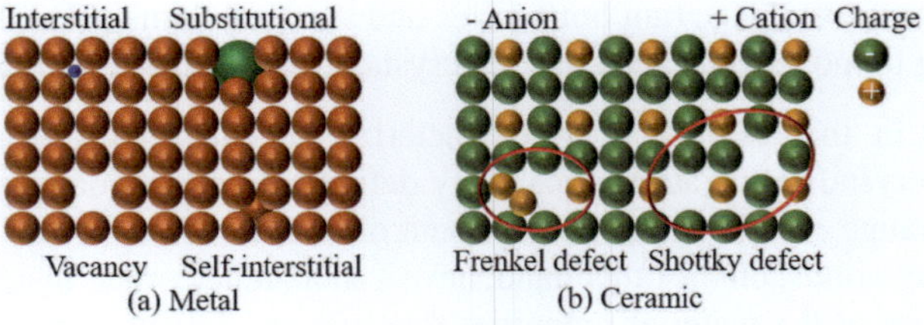

Fig. 6.1 Point defects in (**a**) metal and (**b**) ceramic materials

and **extrinsic point defects** such as

- Interstitial atoms being diffused into the base material structure by diffusion or being trapped as impurities during solidification.
- Dopant atoms being diffused into semiconductors.
- Substitutional atoms being introduced by solution hardening during the solidification.

Control of the solidification and heat treatment processes is important for producing metals and alloys, ceramics, and even polymeric materials with the least amount of intrinsic and extrinsic point defects. The main reasons for characterizing point defects in crystalline solids are related to material properties. Thus,

- Diffusion of interstitial carbon, hydrogen, and nitrogen enhances mechanical properties of steels.
- Doping boron and phosphorus by diffusion enhances electrical properties of semiconductors, such as silicon wafers (Si-wafers).

6.2.2   Nonstoichiometric Point Defects

This category defines metal excess defect and metal deficiency defect in ceramic materials (Fig. 6.1b). The metal excess defect in ceramic lattice sites is due to (1) missing anions (non-metal atoms like chlorine as in white sodium chloride $NaCl$). In this case, missing anions (negatively charged ion) represent vacancies or holes that are occupied by electrons for maintaining electrical neutrality. (2) The movement of extra cations (negatively charged ions like zinc in yellow zinc oxide— ZnO) and electrons at interstitial sites define a ceramic material containing metal excess defect. Thus, in general, the diversity of colors in ceramic materials is a consequence of the electron positions in the ceramic lattice sites.

Metal deficiency defect, as in red iron oxide (FeO), is due to the absence of metal ions from its lattice sites. In this case, the electrical charge is balanced by high positively charged cations (Fe^{2+}).

Vacancies and interstitial atoms as well as impurities are common metal defects, whereas Frenkel and Schottky defects are ionic point defects. Thus,

- Frenkel defects are based on misplacement of ions and vacancies
- Schottky defects are based on vacancies of cation ($+$) and anions ($-$)

6.2.3   Vacancy Concentration

The vacancy is an inherent point defect representing a missing atom from the periodicity of the perfect array of atoms. The number of vacancies per unit volume N_v (vac/m^3) depends on the temperature as defined by the Arrhenius type equation.

Thus, the non-linear form for N_v, vacancy concentration C_v and theoretical density for metallic or ceramic materials are

$$N_v = N_o \exp\left(-\frac{Q_v}{k_B T}\right) = N_o \exp\left(-\frac{Q_v}{RT}\right) \tag{6.1a}$$

$$C_v = \frac{N_v}{N_o} = \exp\left(-\frac{Q_v}{k_B T}\right) = \exp\left(-\frac{Q_v}{RT}\right) \tag{6.1b}$$

$$\rho = \frac{N A_w}{V N_A} \tag{6.1c}$$

$$N_o = \frac{N}{V} = \frac{\rho N_A}{A_w} \tag{6.1d}$$

where C_v denotes the concentration of vacancies or normalized vacancies, N_o denotes a constant, $R = 8.3145 \; J/mol.K$ denotes the universal gas constant $(mol/J.K)$, $k_B = 8.62 \times 10^{-5} \; eV/K = 1.38 \times 10^{-23} \; J/K$ denotes the Boltzmann constant, T denotes the absolute temperature (K), Q_v denotes the enthalpy change (ΔH_v) or activation energy for vacancy formation $(J/mol$ or $eV)$, ρ denotes the bulk mass density of the material (g/cm^3), V denotes volume (m^3) of material, $N_A = 6.022 \times 10^{23} \; atoms/moles$ denotes the Avogadro's number and A_w denotes the atomic weight of the material (g/mol).

The Arrhenius expression, Eq. (6.1a), can be based on number of moles or number of atoms. Hereafter, number of atoms is the preferred approach in the solid-state physics of crystalline materials.

The graphical representation of $N_v = f(T)$, Eq. (6.1a), is shown in Fig. 6.2a. The physical interpretation of the Eq. (6.1a) dictates that the formation of N_v is strongly dependent on a temperature range $0 < T < T_m$, where T_m is the melting temperature of the metal. Similarly, C_v is also a temperature-dependent factor.

Mathematically, (1) $N_v \rightarrow N_o$, $C_v \rightarrow N_v/N_o$ as $T \rightarrow T_m$ since $\exp(-0) = 1$ and (2) $N_v \rightarrow 0$, $C_v \rightarrow 0$ as $T \rightarrow 0$ since $\exp(-\infty) = 0$. Consequently, the formation of vacancies requires an incubation period. Also, if $T \rightarrow T_m \rightarrow \infty$, then $N_v \rightarrow N$ and $C_v \rightarrow 1$ due to $\exp(-0) = 1$.

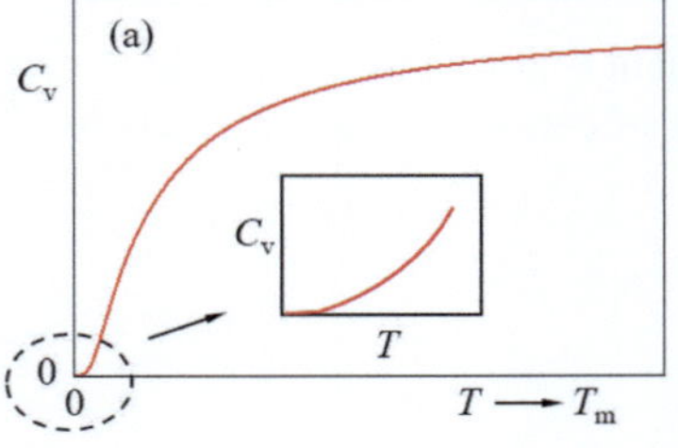

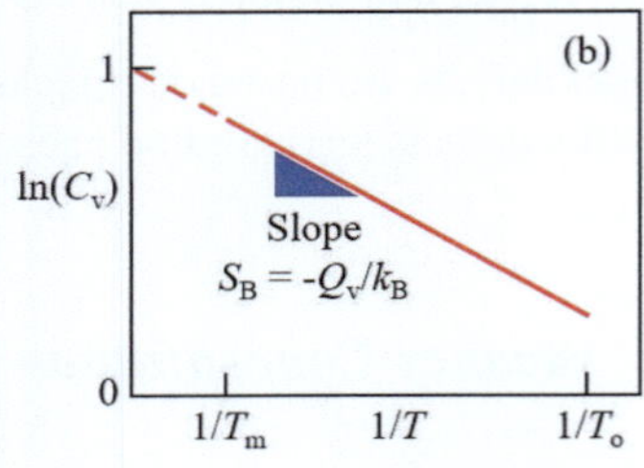

Fig. 6.2 Vacancy concentration based on the Arrhenius equation. (**a**) Non-linear $C_v = f(T)$ and (**b**) linear $C_v = f(1/T)$

Figure 6.2b exhibits a linear plot suggesting that it is a practical approach for analyzing the vacancy concentration data and determining the energy of formation Q_v through the slope S_B. This plot is obtained by taking the natural logarithm of Eqs. (6.1a), (6.1b). Thus,

$$\ln(N_v) = \ln(N_o) - \left(\frac{Q_v}{R}\right)\frac{1}{T} = I_R - S_R\left(\frac{1}{T}\right) \tag{6.2a}$$

$$\ln(C_v) = -\left(\frac{Q_v}{k_B}\right)\frac{1}{T} = S_B\left(\frac{1}{T}\right) \tag{6.2b}$$

where $S_R = Q_v/R$ is in units of $(J/mol)/(J/mol.K) = K$ and $S_B = Q_v/k_B$ has units of $(eV/atom)/(eV/atom.K) = K$.

Example 6.1   Calculate the equilibrium vacancy concentration (C_v) for copper at $T = 27\,°C$ and $900\,°C$. The given dataset includes the melting temperature $T_m = 1085\,°C$, activation energy $Q_v = 0.88\ eV/atoms$, mass density $\rho = 8.95\ g/cm^3$, atomic weight $A_w = 63.55\ g/mol$ and Avogadro's number $N_A = 6.022 \times 10^{23}$ $atoms/mol$.

Solution   From Eq. (6.1c),

$$N_o = \frac{\rho N_A}{A_w} = \frac{\left(8.95\ g/cm^3\right)\left(6.022 \times 10^{23}\ atoms/mol\right)}{63.55\ g/mol} \tag{6.1E1a}$$

$$N_o = 8.48 \times 10^{22}\ atoms/cm^3 \tag{6.1E1b}$$

and from Eq. (6.1a),

$$N_v = N_o \exp\left(-\frac{Q_v}{k_B T}\right) \tag{6.1E2a}$$

$$N_v = \left(8.48 \times 10^{22}\ atoms/cm^3\right)\exp\left[-\frac{0.88\ eV/atoms}{\left(8.62 \times 10^{-5}\ eV/atom.K\right)T}\right] \tag{6.1E2b}$$

Substituting the given temperatures in this equation yields an increasing number of vacancies

$$N_v = 1.41 \times 10^{15}\ atoms/cm^3\ at\ T = 27\,°C = 300\ K \tag{6.1E3a}$$

$$N_v = 1.41 \times 10^{19}\ atoms/cm^3\ at\ T = 900\,°C = 1173\ K \tag{6.1E3b}$$

Thus, Eq. (6.1b) gives an increasing vacancy concentration with increasing temperature

$$C_v = \frac{N_v}{N_o} = \frac{1.41 \times 10^{15}}{8.48 \times 10^{22}} = 1.66 \times 10^{-8} \text{ at } T = 300 \ K \tag{6.1E4a}$$

$$C_v = \frac{N_v}{N_o} = \frac{1.41 \times 10^{19}}{8.48 \times 10^{22}} = 1.65 \times 10^{-4} \text{ at } T = 1173 \ K \tag{6.1E4b}$$

Notice that these calculated C_v values are small with an increasing trendline as schematically depicted in Fig. 6.2a.

6.2.4 Overlayer Lattice Defects

Nowadays, it is common to conduct in-situ electrochemical experiments to reveal images of substrate crystallographic surfaces. Thus, the traditional current potential diagrams at a macroscale can be interpreted with a higher precision with the aid of

- Scanning transmission microscope (STM)
- Transmission electron microscope (TEM)
- Scanning electron microscope (SEM)
- Atomic force microscope (AFM)

which are imaging tools for structural characterization of solid surfaces. Therefore, knowledge of crystallography is essential to understanding adsorption by adherence and electrodeposition by bonding in a suitable environment at pressure P and temperature T. However, the electrochemical cell potential for adsorption, oxidation, and electrodeposition is a controlling factor during surface reconstruction or deterioration (corrosion).

The surface lattice, in general, has numerous defects, such as terraces, voids, adatoms, kinks, valleys, steps, and, to an extent, grain boundaries and particle grain interfaces. These defects control surface properties, such as surface energy (γ_s) and texture (roughness). Specifically, Fig. 6.3a illustrates some common metal surface defects and Fig. 6.3b shows some defects and a partial monolayer on an immersed substrate in a suitable electrolyte.

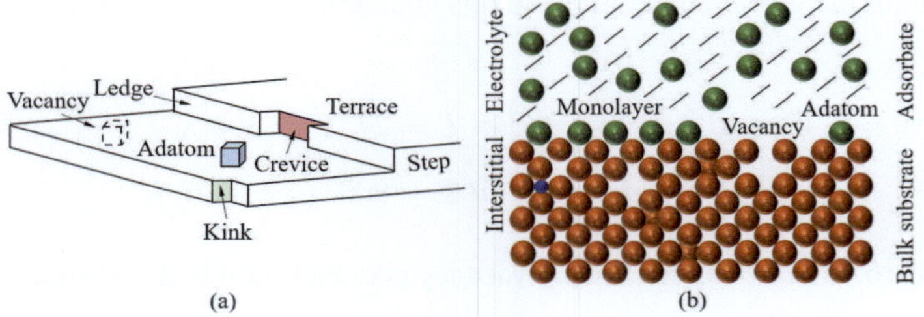

Fig. 6.3 Schematic surface defects on (**a**) irregular substrate and (**b**) electroplated surfaces

These surface defects are inevitably present in most cases and are potential sites for the initial adsorption or electrodeposition process. They can be eliminated or covered significantly by adsorption of different gas atoms or ions from the bulk lattice, leading to a surface reconstruction as an ion-induced monolayer having a specific (hkl) plane.

6.3   Linear Defects Dislocations

A linear defect is a one-dimensional (1D) edge of a distorted crystallographic (hkl) plane of atoms representing misaligned lattice points. This, then, leads to restricted rows of lattice points called dislocations, which are revealed under the microscope as straight and irregular lines with different orientations as indicated in Fig. 6.4a for RSA 304 and Fig. 6.4b for IM 304 alloys.

The historical concept of dislocations has a foundation on Frenkel's work [2] related to the theoretical strength of perfect crystals. According to Hirth and Lothe [3, p. 8], Masing and Polanyi [4], Prandtl [5], and Dehlinger [6] are the precursors of linear defects called dislocations. Coincidentally, Orowan [7], Polanyi [8], and Taylor [9] in 1934 published their work associated with the concept of edge dislocation.

The evolution of dislocations in a crystal structure or pattern depends on the material slip systems, $\{hkl\}\ \langle uvw \rangle$, and related loading modes of deformation, such as monotonic tension or compression, bending, fatigue, rotation, or a combination of them.

Dislocations are one-dimensional solid-state features that may be produced during solidification and rapid cooling due to significant solid contraction or shrinkage and mechanical deformation. On the other hand, non-crystalline or amorphous solids transmit stress, but they cannot flow as crystalline solids do due to their brittleness related to the frozen liquid-like state (Edwards and Mehta [10]). Moreover, dislocations may be sessile (known as Frank dislocations) because they are relatively immobile due to strong obstacles (barriers) within the solid material under deformation by external forces. The dislocation theory presented hereafter is based on glide or slip dislocations, which are mobile due to internal shear stresses.

Fig. 6.4 TEM images of dislocations. (**a**) RSA 304 and (**b**) IM 304 stainless steels

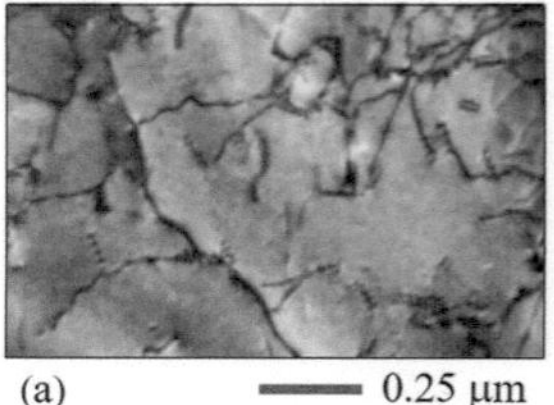
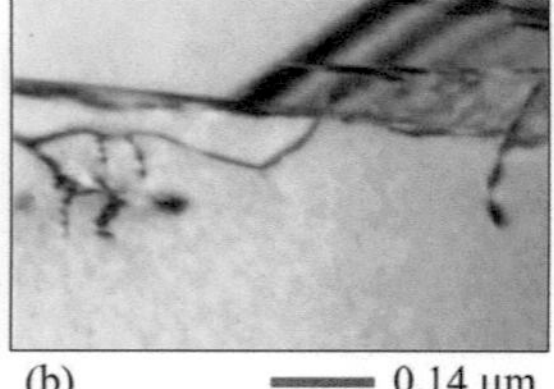

(a) ———— 0.25 μm   (b) ———— 0.14 μm

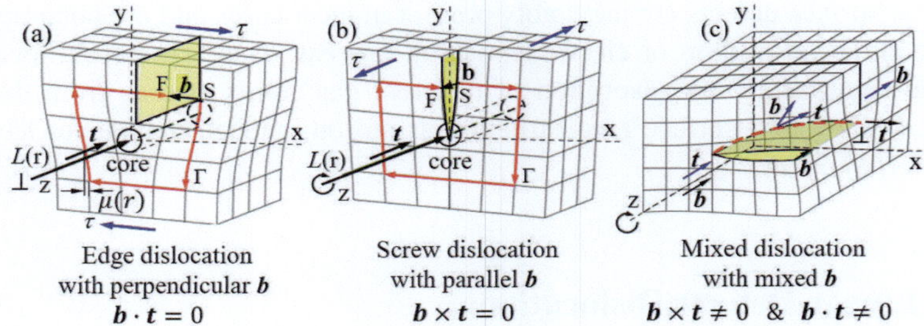

Fig. 6.5 Dislocation models for lattice distortion in a simple cubic lattice showing the Burgers circuit and line defect. (**a**) Edge dislocation and (**b**) screw dislocation showing the burgers circuit with a start "S" and finish "F". (**c**) Mixed dislocation. The schematic 3D lattice is borrowed from Blumenau's dissertation [11]

6.3.1 Dislocation Models

The dislocation-circuit models describing the concept of Burgers lattice distortion are shown in Fig. 6.5 (Hirth and Lothe [3, p. 23], Blumenau [11], Friedel [12, pp. 6–9], Kosevich [13, Chapter 1], Kovacs and Zsoldos [14, p. 12], and Read [15, chapter 3],). These models are denoted as the area of irregular rectangles for pure dislocations (Fig. 6.5a and b) and as a quarter circle-like for a mixed dislocation (Fig. 6.5c).

Dislocations are one-dimensional crystalline defects that divide the boundary between slipped and unslipped regions. The latter are elucidated in Fig. 6.5 as shaded areas.

Figure 6.5a schematically depicts the edge dislocation with a perpendicular Burgers vector, where the dot-product is $\mathbf{b} \cdot \mathbf{t} = \mathbf{0}$ and with a contour Γ enclosing the atomically deformed area, Fig. 6.5b shows the screw dislocation with a parallel Burgers vector, where the cross-product is $\mathbf{b} \times \mathbf{t} = \mathbf{0}$ and the distorted planes are enclosed by the contour Γ, and Fig. 6.5c illustrates a mixed dislocation with $\mathbf{b} \cdot \mathbf{t} \neq \mathbf{0}$ and $\mathbf{b} \times \mathbf{t} \neq \mathbf{0}$, except at its ends. Further, the dislocation Burgers vector $\mathbf{b}$ is a short distance that connects the ends of a deformed circuit, known as the Burgers circuit around the dislocation.

6.3.2 Edge Dislocation

Regarding Fig. 6.5a, it shows the most general model for a positive edge dislocation represented as an extra-half plane of atoms using the symbol ⊥, where the vertical stroke points to the inserted direction, and the horizontal stroke represents the slip plane.

In particular,

- According to Fig. 6.5a, the vector pointing from "S" to "F" (from the start or beginning to finish or end of the curve) is termed the Burgers vector denoted by **b**, and the line defect with a Burgers vector **b** $\neq$ 0 is known as a dislocation, which may be a straight or curved line as observed under the microscope.
- The Burgers circuit Γ is a representation of the localized crystallographic distortion due to a shear stress (τ) being induced by an external stress or force. The marked dislocation circuit has a missing step, which is replaced by the Burgers vector to complete the distortion circuit.
- Dislocations form either during solidification (growth process of a crystal), rapid cooling, or quenching from the solid state or plastic deformation induced by mechanical forces.
- Dislocation motion is related to displacement (μ), say, along the x-axis so that $\mu(x) << |\mathbf{b}| \neq 0$, and it occurs along the line $L(r)$ being located on an (hkl) slip plane. This motion occurs in the positive x-direction (to the right) until it reaches the crystal surface, leaving a step called slip line.
- The Burgers vector **b** for the edge dislocation is *perpendicular* to the dislocation unit tangent vector **t** denoted as **b** $\perp$ **t**, where the symbol $\perp$ is also used for such an algebraic representation and the angle between **b** and **t** is $\theta = \angle(\mathbf{b}, \mathbf{t}) = \pi/2$.
- If the distortion circuit contains only a dislocation, then the line defect is a perfect dislocation with a Burger vector $\mathbf{b} = (a/2)\langle uvw \rangle$ in cubic crystals.

It is fundamentally that a dislocation moves along its (hkl) glide plane or slip plane when the resolve shear stress (τ) reaches a critical value τ_c ($\tau \to \tau_c$) known as the phenomenological parameter or the Peierls stress (classical shear stress) $\tau_p = \tau$, written without prove as (Kovacs and Zsoldos [14, p. 116])

$$\tau_p = \frac{2G}{1-v} \exp\left[-\frac{2\pi w}{b(1-v)}\right] \quad \text{for } w \neq b \qquad (6.3a)$$

$$\tau_p = \frac{2G}{1-v} \exp\left[-\frac{2\pi}{(1-v)}\right] \quad \text{for } w = b \qquad (6.3b)$$

where G denotes the shear modulus, v denotes Poisson's ratio, b denotes the magnitude of a Burgers vector, and w denotes the dislocation width defined as the distance between atoms in the slip plane.

The dislocation width w, literately, measures the size of a dislocation core, and it can be approximated as $w = d/(1-v)$ for an edge dislocation) or $w = d$ for a screw dislocation, where d is the crystallographic d-spacing introduced in Chap. 1. Essentially, w in the above expressions represents small atomic distances that measure the degree of disruption of a perfect lattice (Kovacs and Zsoldos [14, pp. 119, 122]). Moreover, the dislocation width can also be approximated as $w = a/(1-v)$ for an edge dislocation, where "a" is the average lattice parameter of a cubic crystal structure (Friedel [12, p. 54]).

6.3.3   *Bubble Raft Model*

According to Friedel [12, p. 11], the bubble raft model was introduced by Bragg and Nye [16] in 1947. This model simply mimics the surface arrangement of atoms in a polycrystalline solid and related packing imperfections or defects, such as grain boundaries, voids, and dislocation. For instance, Fig. 6.6a shows a perfect crystalline raft, and Fig. 6.6b shows grain boundary raft of bubbles. Notice the three partial grains separated by grain boundaries.

Figure 6.6c shows a *bubble raft* model containing an array of dislocation pileup (darker points) within a grain (Hirth and Lothe [3]). This microscale linear defect can be defined as a dislocation pileup, which induces a pressure to the upper grain boundary. Here, θ denotes the tilt angle. If $\theta \leq 10°$, then the interface between these grains (Fig. 6.6c) is a low-angle grain boundary; otherwise, it is a high-angle boundary for $\theta > 10°$.

Figure 6.6d, on the other hand, exhibits the classic symmetrical low-angle grain boundary (θ) between two crystals. This is the edge dislocation model most used in common textbooks and classrooms (Read [15, p. 157]). Moreover, Fig. 6.6d schematically shows the 2D representation of the low-angle grain boundary, which is the classical model cited in many textbooks and university classrooms.

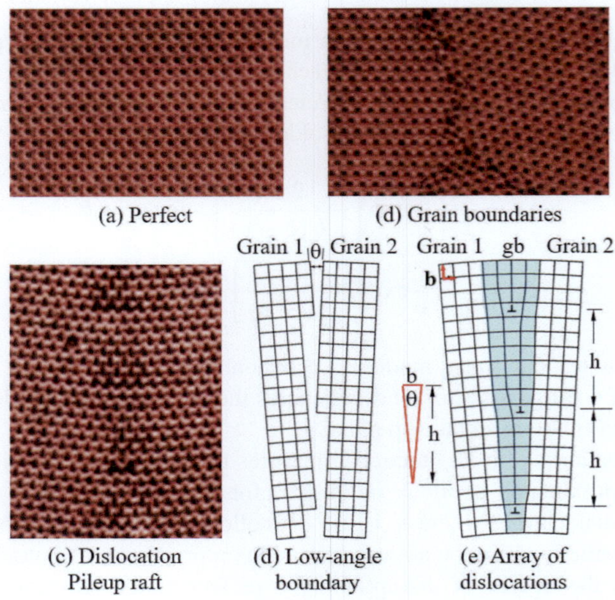

Fig. 6.6 Bubble raft model. (**a**) Perfect crystalline raft, (**b**) grain boundary raft (Bragg and Nye [16] in 1947. (**c**) Dislocation pileup raft (Hirth and Lothe [3]), (**d**) schematic low-angle (θ) grain boundary and (**e**) a schematic array of edge dislocations separated by a distance h in a hypothetical solid (Read [15, p. 157])

Figure 6.6e depicts the model for an array of edge dislocations with a common Burgers vector $\mathbf{b} = \vec{b}$ within a crystal (Read [15, p. 157], Masing and Polanyi [4, p. 157]). This array of dislocations represent the atomic misorientation of the lattice. For an array of edge dislocations (Fig. 6.6e), the distance between edge dislocations is approximated as

$$h = \frac{b}{\sin(\theta)} \simeq \frac{b}{\theta} \tag{6.4}$$

Increasing plastic deformation decreases the distance between dislocations, induces dissociation of primitive into partial dislocations with $b < (a/2)\langle uvw \rangle$ in cubic crystals, and increases dislocation density, which impedes easy dislocation motion because atoms are disabled to slide readily. This, then, enhances mechanical properties.

6.3.4 Screw Dislocation

Regarding Fig. 6.5b, it represents the pure screw dislocation with parallel Burgers vector ($\mathbf{b}$) and tangent vector ($\mathbf{t}$) denoted as $\mathbf{b} \parallel \mathbf{t}$, where the symbol $\parallel$ means *parallel* and the angle between $\mathbf{b}$ and $\mathbf{t}$ is $\theta = \angle(\mathbf{b}, \mathbf{t}) = 0$. Nonetheless,

- The dislocation circuit has the lattice plane shifted by one layer, inducing a step. Similarly, the interrupted distortion circuit has a start "S" and a finish "F" being connected by the Burgers vector $\mathbf{b}$.
- The deformation step represents the location of a Burgers vector, which closes the circuit.
- The localized distortion circuit is generated by a relative antiplane or off-plane displacement on the two sides of the dislocation line.
- The dislocation motion is vertically downward.
- The dislocation moves when $\tau \to \tau_p$.

Recall that the Burgers vector of a screw dislocation is parallel to the dislocation line (Fig. 6.5c), which is surrounded by planes of atoms in a helical form.

6.3.5 Mixed Dislocation

Figure 6.5c characterizes a mixed dislocation represented as an inner smooth curve within the crystal. Thus,

- This curved xy dislocation line has edge-dislocation and screw-dislocation characters at its ends.
- This type of dislocation has a resolved Burgers vector $\mathbf{b}$ defined on the xy-plane.
- The angle between $\mathbf{b}$ and $\mathbf{t}$ is $0 < \theta = \angle(\mathbf{b}, \mathbf{t}) < \pi/2$.

Within this context, dislocations cannot end within the crystal and must terminate at free surfaces, particles, grain boundaries, and even at other dislocations (Kosevich [13, p. 40]).

For a perfect dislocation, the Burgers vector **b** of the line defect around an arbitrary contour (Burgers circuit) "Γ" is written as the counterclockwise sum of the line integral enclosing the dislocation contour as shown in Fig. 6.5a (Friedel [12, p. 7], Kovacs and Zsoldos [14, p. 27]). Thus,

$$\mathbf{b} = \oint_{\Gamma} \frac{\partial \boldsymbol{\mu}(r)}{\partial s} ds \neq 0 \tag{6.5}$$

where $\boldsymbol{\mu}(r)$ is the elastic displacement vector, and ds is an arbitrary segment of the dislocation circuit core. This line integral quantitatively describes the Burgers vector **b** as proposed by Burgers in 1934 and the contour Γ defines the Burgers circuit. If **b** $\perp$ **t**, then the dot product is zero for the edge dislocation, and if **b** $\parallel$ **t**, then the cross product is zero for a screw dislocation. Mathematically,

$$\mathbf{b} \cdot \mathbf{t} = 0 \quad \text{(edge dislocation)} \tag{6.6a}$$

$$\mathbf{b} \times \mathbf{t} = 0 \quad \text{(screw dislocation)} \tag{6.6b}$$

$$\mathbf{b} \cdot \mathbf{t} \neq 0 \quad \text{(mixed dislocation)} \tag{6.6c}$$

Figure 6.5b, c are suitable for quantitative analysis of the dislocation core structure and related effects on the dislocation mobility, for deriving stress and strain entities using the theory of elasticity.

The most common slip planes for BCC and FCC crystal structures are $\{110\}\langle111\rangle$ and $\{111\}\langle110\rangle$, respectively, and their corresponding Burgers vectors are defined by

$$\mathbf{b} = \frac{a}{2}\langle111\rangle \quad (BCC) \tag{6.7a}$$

$$\mathbf{b} = \frac{a}{2}\langle110\rangle \quad (FCC) \tag{6.7b}$$

and their magnitudes become

$$b = \frac{a}{2}\sqrt{1^2 + 1^2 + 1^2} = \frac{\sqrt{3}a}{2} \quad (BCC) \tag{6.8a}$$

$$b = \frac{a}{2}\sqrt{(1)^2 + (1)^2 + 0} = \frac{a}{\sqrt{2}} = \frac{\sqrt{2}a}{2} \quad (FCC) \tag{6.8b}$$

Notice that $b_{BCC} > b_{FCC}$ because FCC is the most dense close-packed structure with $APF_{FCC} = 0.74$ (similar to $APF_{HCP} = 0.74$), whereas the BCC structure has an $APF_{BCC} = 0.68$ as defined by Eqs. (1.28b), (1.28c). Recall that a in these equations is the lattice parameter.

6.3.6 *Partial Dislocations*

During plastic deformation, the formation of a dislocation glide system depends on the direction of an applied external load, the orientation of a unit cells in a crystalline solid, and the internal resolved shear stress (τ_R) along a slip system $\{hkl\}\,\langle uvw\rangle$. As a result, dislocations are formed randomly; some may glide, some may split into partial dislocations, and some may become immobile called sessile dislocations (Read [15, Chapter 7]). This implies that dislocation motion causes plastic deformation along slip planes at random directions.

Dislocation motion can be modeled as shown in Fig. 6.7 for the formation of partial dislocations and the Burgers circuits (Friedel [12, p. 8]). For instance, Fig. 6.7a illustrates the case of a mixed dislocation, and Fig. 6.7b depicts a primitive or perfect dislocation that splits into two partial dislocations at a barrier called node A.

Notice that Fig. 6.7a shows a slightly curved dislocation line $L\,(r)$ with an arbitrary dislocation circuit and an arbitrary Burgers vector **b**. The direction of the traveling dislocation changes from $[uvw]_a$ to $[uvw]_b$. On the other hand, Fig. 6.7b depicts a primitive straight dislocation line $L_1\,(r)$ with a $[uvw]_1$ direction that splits at a note A into two straight partial dislocations with their own Burgers vectors b_2 and b_3. The corresponding Burgers circuit embraces the partial dislocations (Friedel [12, p. 8]). From vector algebra, the Burgers vectors form a vector triangle.

A dislocation with the primitive translation vector $\mathbf{b}_1$ decomposes into partial dislocations with $\mathbf{b}_2$ and $\mathbf{b}_3$ (Fig. 6.7b) if the sum of their energies is less than that of the original (Koseoich [13, p. 46]). This implies that each partial dislocation must have a Burgers vector smaller than a primitive translation vector; $|\mathbf{b}_2| < |\mathbf{b}_1|$ and $|\mathbf{b}_3| < |\mathbf{b}_1|$. Nonetheless, the area between $\mathbf{b}_2$ and $\mathbf{b}_3$ in Fig. 6.7b represents atom layers called stacking fault being induced by a local resolved shear stress (τ_R) and related to the stacking-fault energy (γ_{SF}) through the shear modulus (G) of the crystal being mechanically deformed.

Partial dislocations with Burgers vectors $\mathbf{b}_2$ and $\mathbf{b}_3$ are known as Shockley partials dislocations, and they derive from the dissociation (splitting) of a perfect

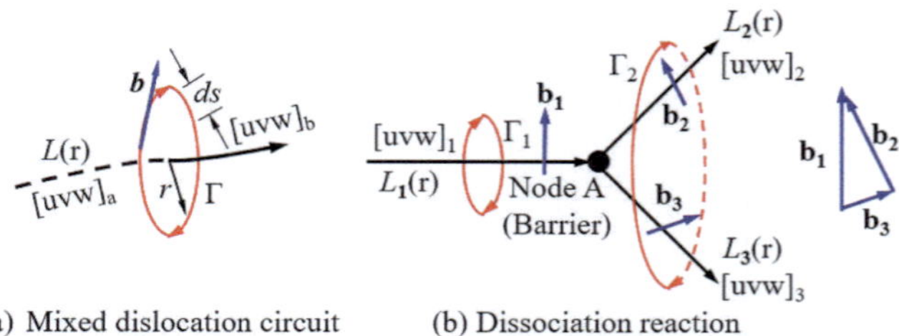

(a) Mixed dislocation circuit   (b) Dissociation reaction

Fig. 6.7 Dislocation circuits and partial dislocations. (**a**) Dislocation circuit with an arbitrary location of the Burgers vector and (**b**) decomposition of the primitive dislocation into partial dislocations. After Friedel [12, p. 8]

dislocation with $\mathbf{b}_1$. According to Fig. 6.7b, the dissociation of a perfect dislocation occurs at the node A represents a reaction. Thus,

$$|\mathbf{b}_1|^2 > |\mathbf{b}_2|^2 + |\mathbf{b}_3|^2 \quad \text{will dissociate} \tag{6.9a}$$

$$|\mathbf{b}_1|^2 = |\mathbf{b}_2|^2 + |\mathbf{b}_3|^2 \quad \text{remains stationary} \tag{6.9b}$$

$$|\mathbf{b}_1|^2 < |\mathbf{b}_2|^2 + |\mathbf{b}_3|^2 \quad \text{will not dissociate} \tag{6.9c}$$

It is fundamentally accepted that the dissociation of a mobile dislocation with $\mathbf{b}_1$ is attributed to an evolved internal resolved shear stress (τ_R) during plastic deformation. Accordingly, $\mathbf{b}_1$ represents the shortest vector within a unit cell.

Dislocations are diverse and can have different microscopic arrangements due to pinning points (nodes) imparted by dislocation pile-up, hard particles, and grain boundaries.

An example helps readers with a complete working knowledge of the properties of dislocations and with the proper vocabulary associated used in the field of imperfections in crystalline solids.

Example 6.2 Consider the $\{111\}\langle112\rangle$ slip system for an FCC crystal. Assume that a straight perfect dislocation with Burgers vector $\mathbf{b}_1$ dissociates (splits up) into two Shockley partial dislocations with Burgers vectors $\mathbf{b}_2$ and $\mathbf{b}_3$ in the (111) plane. The dislocation dissociation related to its corresponding stacking fault is written as a stationary reaction

$$\mathbf{b}_1 = \mathbf{b}_2 + \mathbf{b}_3 \tag{6.2E1a}$$

$$\frac{a}{2}[110] = \frac{a}{6}[211] + \frac{a}{6}[12\bar{1}] \tag{6.2E1b}$$

Determine **(a)** if the reaction is correct as written, **(b)** if it really can dissociate, **(c)** the splitting angle θ for $\mathbf{b}_2 \angle \mathbf{b}_3$ so that the perfect dislocation with Burgers vector $b_1 = (a/2)[110]$ is treated as θ-degree dislocation associated with the corresponding stacking fault and **(d)** write down the vector notation for the dislocation directions describing the stacking fault.

Solution

(a) Eliminate the lattice parameter a and expand the terms in the reaction so that

$$\frac{a}{2}[110] = \frac{a}{6}[211] + \frac{a}{6}[12\bar{1}] \tag{6.2E2a}$$

$$\frac{1}{2} + \frac{1}{2} + 0 = \left(\frac{2}{6} + \frac{1}{6} + \frac{1}{6}\right) + \left(\frac{1}{6} + \frac{2}{6} - \frac{1}{6}\right) \tag{6.2E2b}$$

$$1 = \left(\frac{4}{6}\right) + \left(\frac{2}{6}\right) = \frac{6}{6} = 1 \quad \text{(correct)} \tag{6.2E2c}$$

or

$$u\text{-axis: } \frac{a}{2} = \frac{2a}{6} + \frac{a}{6} = \frac{3a}{6} = \frac{a}{2} \quad \text{(correct)} \tag{6.2E3a}$$

$$v\text{-axis: } \frac{a}{2} = \frac{a}{6} + \frac{2a}{6} = \frac{3a}{6} = \frac{a}{2} \quad \text{(correct)} \tag{6.2E3b}$$

$$w\text{-axis: } 0 = \frac{a}{6} - \frac{a}{6} = 0 \quad \text{(correct)} \tag{6.2E3c}$$

Hence, the reaction is proven correct.

(b) Then,

$$|\mathbf{b}_1|^2 = \left[\frac{a}{2}\sqrt{(1)^2 + (1)^2 + 0}\right]^2 = \frac{a^2}{2} \tag{6.2E4a}$$

$$|\mathbf{b}_2|^2 = \left[\frac{a}{6}\sqrt{2^2 + 1^2 + 1^2}\right]^2 = \frac{a^2}{6} \tag{6.2E4b}$$

$$|\mathbf{b}_3|^2 = \left[\frac{a}{6}\sqrt{1^2 + 2^2 + (-1)^2}\right]^2 = \frac{a^2}{6} \tag{6.2E4c}$$

$$|\mathbf{b}_2|^2 + |\mathbf{b}_3|^2 = \frac{a^2}{6} + \frac{a^2}{6} = \frac{a^2}{3} \tag{6.2E4d}$$

Based on these calculations, Eq. (6.9a) yields

$$|\mathbf{b}_1|^2 > |\mathbf{b}_2|^2 + |\mathbf{b}_3|^2 \tag{6.2E5a}$$

$$\frac{a^2}{2} > \frac{a^2}{3} \tag{6.2E5b}$$

Therefore, dislocation splitting is favorable energetically because $a^2/2 > a^2/3$.

(c) The dislocation $\mathbf{b}_1$ splits into two partial dislocation $\mathbf{b}_2$ and $\mathbf{b}_3$. Now, the angle θ between dislocations with Burgers vectors $\mathbf{b}_2$ and $\mathbf{b}_3$ is determined by using the dot product

$$\cos\theta = \frac{\mathbf{b}_2 \cdot \mathbf{b}_3}{|\mathbf{b}_2|\,|\mathbf{b}_3|} = \frac{(a/6)\,[211] \cdot (a/6)\left[12\bar{1}\right]}{|(a/6)\,[211]|\,\left|(a/6)\left[12\bar{1}\right]\right|} \tag{6.2E6a}$$

$$\cos\theta = \frac{[211] \cdot \left[12\bar{1}\right]}{|[211]|\,\left|\left[12\bar{1}\right]\right|} = \frac{(2)\,(1) + (1)\,(2) + (1)\,(-1)}{\sqrt{(2)^2 + (1)^2 + (1)^2}\sqrt{(1)^2 + (2)^2 + (-1)^2}} \tag{6.2E6b}$$

$$\cos\theta = \frac{3}{\sqrt{6}\sqrt{6}} = \frac{3}{6} = \frac{1}{2} \tag{6.2E6c}$$

$$\theta = \cos^{-1}\left(\frac{1}{2}\right) = 60° = \pi/3 \tag{6.2E6d}$$

Hence, the perfect dislocation is a 60° dislocation since $\mathbf{b}_2 \angle \mathbf{b}_3 = 60°$. Here, the angles $\mathbf{b}_1 \angle \mathbf{b}_2 = 30°$ and $\mathbf{b}_1 \angle \mathbf{b}_3 = 30°$ make a total of 60° split. This means that the partial Burgers vectors make 30° angles with the perfect dislocation. Thus, the vector notations $\mathbf{b}_{30}$ and $\mathbf{b}_{60}$ are common in the literature.

Moreover, below is the graphical representation of the $(111)\langle 112\rangle$ slip system in an *FCC* unit cell (Fig. a) and in an FCC stacking fault (Fig. b).

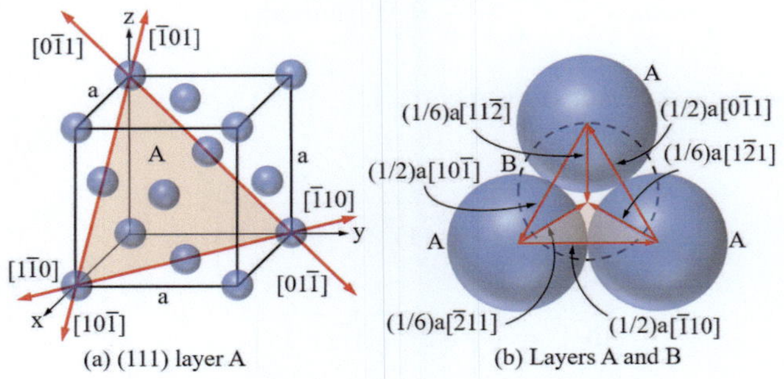

(a) (111) layer A   (b) Layers A and B

Only one partial dislocation system is shown in Fig. b, where "A" is the top (111) layer and "B" is the underneath (111) layer in a stacking fault. The layers move relative to each other inducing a dislocation splitting of the perfect dislocation with Burgers vector $\mathbf{b}_1 = (a/2)\langle 110\rangle$ into partial dislocations with $\mathbf{b}_2 = (a/6)\left[\bar{2}11\right]$ and $\mathbf{b}_3 = (a/6)\left[1\bar{2}1\right]$.

(d) The dislocation directions in vector notation are

$$[110] \rightarrow \mathbf{R}_{[110]} = \mathbf{x} + \mathbf{y} \tag{6.2E7a}$$

$$[211] \rightarrow \mathbf{R}_{[211]} = 2\mathbf{x} + \mathbf{y} + \mathbf{w} \tag{6.2E7b}$$

$$\left[12\bar{1}\right] \rightarrow \mathbf{R}_{\left[12\bar{1}\right]} = \mathbf{x} + 2\mathbf{y} - \mathbf{w} \tag{6.2E7c}$$

These vectors or line directions connect lattice points. Thus, each crystallographic direction is specified by the three Miller indices; (hkl) and the vector R_{hkl} along the corresponding Cartesian coordinates.

6.4 Stress Field Around Dislocations

For a straight dislocation, the average dislocation velocity (v) for a dislocation glide process during plastic deformation can empirically be defined by a simple mathematical equation, such as a power law or by an exponential equation (Kovacs and Zsoldos [14, pp. 215–216])

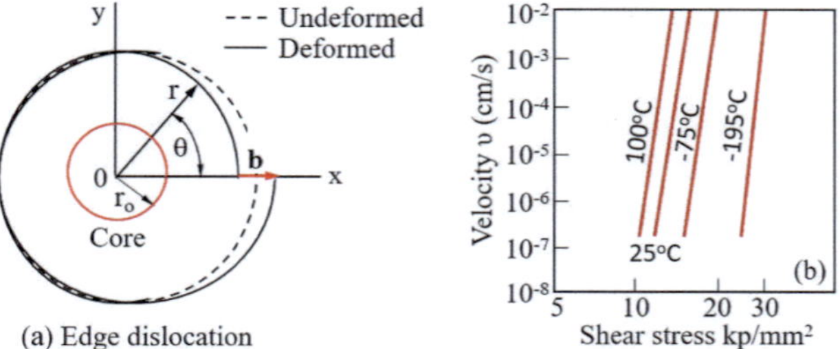

Fig. 6.8 The edge dislocation. (**a**) Model for deriving stresses around the dislocation and (**b**) edge dislocation velocity $\upsilon = f(\tau)$ in Fe-3%Si binary alloy at temperatures T. Log–log scales. Plot taken from Kovacs and Zsoldos [14, p. 215]

Fig. 6.9 Dislocation velocity terms in LiF crystals (FCC with $a = 0.40\ nm$ lattice parameter). After Johnston and Gilman [18]

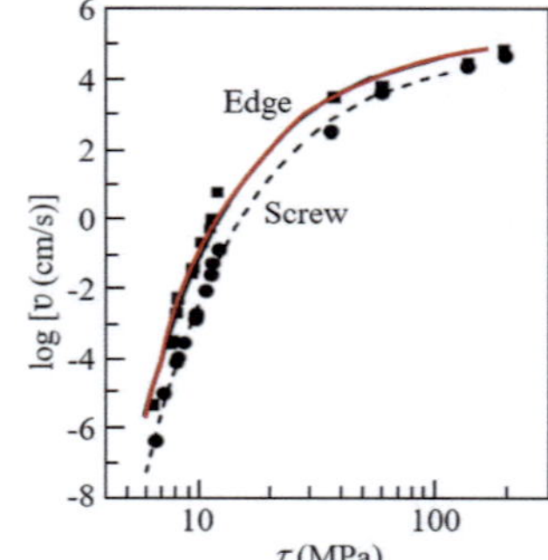

$$\upsilon = c_1 \tau^n \tag{6.10a}$$

$$\upsilon = c_2 \exp\left(-\frac{c_3 G}{\tau}\right) \tag{6.10b}$$

where c_1, c_2, c_3, n denote constants, τ denotes the shear stress that induces dislocation formation and motion, and G denotes the shear modulus.

Figure 6.8a shows the 2D edge dislocation model and Fig. 6.8b exhibits experimental data on a log-log scale as per Eq. (6.10a) at temperatures (Kovacs and Zsoldos book [14, p. 215]). Therefore, the dislocation velocity depends upon the applied stress Stein [17].

Johnston and Gilman [18] in 1959 showed that $\upsilon_{edge} > \upsilon_{screw}$ as depicted in Fig. 6.9 as per Eq. (6.10b) (Wang et al. [19, p. 239]).

This experimental datasets confirm that the evolved shear stress affects the dislocation velocity.

6.4.1　Stresses Around Edge Dislocations

The edge dislocation is also referred to as the glide dislocation. From Appendix 6A, the Airy stress function $\phi = f(x, y)$ along with the operator ∇ for a positive edge dislocation in Cartesian coordinates is defined by (Hirth and Lothe [3, p. 75], Lazar et al. [20])

$$\phi = -\lambda y \ln\left(x^2 + y^2\right) \tag{6.11a}$$

$$\lambda = \frac{Gb}{2\pi\,(1 - v)} \tag{6.11b}$$

$$\nabla^4 \phi = \left(\frac{\partial^2}{\partial x^2} + \frac{\partial^2}{\partial y^2}\right)^2 = 0 \tag{6.11c}$$

where $r = \sqrt{x^2 + y^2}$ denotes radius, G denotes the shear modulus, b denotes the magnitude of a Burgers vector, v denotes the Poisson's ratio, and ∇^4 denotes the biharmonic equation, which is the fourth-order partial differential equation operator.

From Eq. (6.11a), the elastic stress field around the edge dislocation in an isotropic solid is (Hirth and Lothe [3, p. 76], Weertman [21, p. 46])

$$\sigma_x = \frac{\partial^2 \phi}{\partial y^2} = -\frac{\lambda y\left(3x^2 + y^2\right)}{\left(x^2 + y^2\right)^2} = -\frac{\lambda \sin(\theta)\left[1 + 2\cos^2(\theta)\right]}{r} \tag{6.12a}$$

$$\sigma_y = \frac{\partial^2 \phi}{\partial x^2} = \frac{\lambda y\left(x^2 - y^2\right)}{\left(x^2 + y^2\right)^2} = \frac{\lambda \sin(\theta)\cos(2\theta)}{r} \tag{6.12b}$$

$$\tau_{xy} = \frac{\partial^2 \phi}{\partial x \partial y} = \frac{\lambda x\left(x^2 - y^2\right)}{\left(x^2 + y^2\right)^2} = \frac{\lambda \cos(\theta)\cos(2\theta)}{r} \tag{6.12c}$$

$$\sigma_z = v\left(\sigma_x + \sigma_y\right) = \frac{v\lambda y}{\left(x^2 + y^2\right)^2} \tag{6.12d}$$

$$\sigma_z = \frac{v\lambda \sin(\theta)\left[\cos(2\theta) - 1 - 2\cos^2(\theta)\right]}{r} = -\frac{2v\lambda \sin(\theta)}{r} \tag{6.12e}$$

where Eq. (6.12e) is simplified by using the trigonometric identity $\cos(2\theta) = 2\cos^2(\theta) - 1$. Similarly, the Airy stress function for a positive edge dislocation in polar coordinates, as defined in Appendix 6A, and the biharmonic equation are (Dundurs [22, p. 77],)

$$\phi = -\lambda r \ln(r) \sin(\theta) \tag{6.13a}$$

$$\nabla^4 \phi = \left(\frac{\partial^2}{\partial r^2} + \frac{1}{r}\frac{\partial}{\partial r} + \frac{1}{r^2}\frac{\partial^2}{\partial \theta^2}\right)^2 = 0 \tag{6.13b}$$

The stresses in polar coordinates are

$$\sigma_r = \frac{1}{r}\frac{\partial \phi}{\partial r} + \frac{1}{r^2}\frac{\partial^2 \phi}{\partial \theta^2} = -\frac{\lambda \sin (\theta)}{r} \tag{6.14a}$$

$$\sigma_\theta = \frac{\partial^2 \phi}{\partial r^2} = -\frac{\lambda \sin (\theta)}{r} \tag{6.14b}$$

$$\tau_{r\theta} = -\frac{\partial}{\partial r}\left(\frac{1}{r}\frac{\partial \phi}{\partial \theta}\right) = \frac{\lambda \cos (\theta)}{r} \tag{6.14c}$$

Notice that Eq. (6.14) predicts stress singularity denoted as $\sigma_{ij} \to \infty$ as $r \to 0$. Actually, r must be finite in order to avoid $\sigma_{ij} \to \infty$. The general notation σ_{ij} is conveniently simplified as $\sigma_{rr} = \sigma_r$, $\sigma_{\theta\theta} = \sigma_\theta$ and $\sigma_{r\theta} = \tau_{r\theta}$.

Let the strain energy of an edge dislocation be

$$U_e = \frac{1}{2}\int_{r_o}^r \tau_{r\theta} b\, dr = \frac{\lambda b \cos (\theta)}{2}\int_{r_o}^r \frac{dr}{r} \tag{6.15a}$$

$$U_e = \frac{\lambda b \cos (\theta)}{2}\ln\left(\frac{r}{r_o}\right) \tag{6.15b}$$

Letting $y = 0$, $\theta = 0$, $\cos (\theta) = 1$ and $\ln (r/r_o) \simeq 4\pi$ along the (hkl) slip plane yields

$$U_e = \frac{\lambda b}{2}\ln\left(\frac{r}{r_o}\right) = \frac{Gb^2}{4\pi (1-v)}\ln\left(\frac{r}{r_o}\right) \tag{6.16a}$$

$$U_e \simeq \frac{Gb^2}{1-v} \tag{6.16b}$$

Notice that $U_e \to \infty$ (diverges) as $r \to \infty$ or $r_o \to 0$ in Eq. (6.16a) and that it depends on the size of the crystal. This condition is not realistic, and therefore, one must introduce a finite dimension to r, such as the shortest distance from the free surface for one dislocation or an average distance between several dislocations (Hirth and Lothe [3, p. 63],). This implies that the theory of elasticity breaks down at an atomic level. Letting $\ln (r/r_o) \simeq 4\pi$ and $r_o \simeq b$ in Eq. (6.16a) yields Eq. (6.16b) as the expression used in most books and academic classrooms.

6.4.2   Stresses Around Screw Dislocations

The mathematical expressions for nonzero shear stress components (τ_{ij}) around a right-hand screw dislocation with positive Burgers vector **b** and related strain energy (U_s) can be found elsewhere (Weertman [21, p. 47]). The shear stresses are

$$\tau_{zx} = -\frac{Gb}{2\pi}\frac{x}{x^2 + y^2} = -\frac{Gb}{2\pi}\frac{\sin(\theta)}{r} \tag{6.17a}$$

$$\tau_{yz} = +\frac{Gb}{2\pi}\frac{y}{x^2 + y^2} = +\frac{Gb}{2\pi}\frac{\cos(\theta)}{r} \tag{6.17b}$$

According to Hooke's law, these stresses are $\tau_{zx} = G\gamma_{zx}$ and $\tau_{yz} = G\gamma_{yz}$. From elasticity theory, the simplified strain energy for the screw dislocation (U_s) is

$$U_s = \frac{1}{2G}\left(\tau_{zx}^2 + \tau_{yz}^2\right) = \frac{Gb^2}{8\pi}\left(\frac{1}{x^2 + y^2}\right) = \frac{Gb^2}{8\pi}\left(\frac{1}{r^2}\right) \tag{6.18a}$$

$$U_s = \int U dV = \int_{r_o}^{r}\frac{Gb^2}{8\pi}\left(\frac{1}{r^2}\right)(2\pi r dr) = \frac{Gb^2}{4\pi}\int_{r_o}^{r}\frac{dr}{r} \tag{6.18b}$$

$$U_s = \frac{Gb^2}{4\pi}\ln\left(\frac{r}{r_o}\right) \tag{6.18c}$$

$$U_s \simeq Gb^2 \tag{6.18d}$$

Again, the approximations $\ln(r/r_o) \simeq 4\pi$ and $r_o \simeq b$ are imposed on Eq. (6.18c). Moreover, it can be stated that the above dislocation strain energy equations have a mathematical foundation based on the assumption that dislocations are straight and stationary in an isotropic material. This implies that the strain energy, in general, arises from the crystal distortion caused by dislocations and that it is proportional to the Burgers vector.

Using polar coordinates with the nonzero stress $\sigma_{z\theta}$ yields the same result; that is,

$$\sigma_{z\theta} = \frac{Gb}{2\pi r} \tag{6.19a}$$

$$U_s = \sigma_{ij}d\varepsilon_{ij}dA = \int_0^{2\pi}d\theta\int_{r_o}^{r}\frac{\sigma_{z\theta}^2}{2G}rdr \tag{6.19b}$$

$$U_s = \int_0^{2\pi}d\theta\int_{r_o}^{r}\frac{Gb^2}{8\pi^2 r}dr = \frac{Gb^2}{8\pi^2}\ln\left(\frac{r}{r_o}\right)\int_0^{2\pi}d\theta \tag{6.19c}$$

$$U_s = \frac{Gb^2}{4\pi}\ln\left(\frac{r}{r_o}\right) \simeq Gb^2 \tag{6.19d}$$

where ε_{ij} is the notation for elastic strain and $U_s \to \infty$ (diverges) as $r \to \infty$ or $r_o \to 0$ (Hirth and Lothe [3, p. 63]).

6.4.3   *Mixed Dislocations: Stresses*

Superposition of edge and screw strain energies yields

$$U_{mix} = U_e + U_s = \frac{Gb^2}{4\pi\,(1-v)}\ln\left(\frac{r}{r_o}\right) + \frac{Gb^2}{4\pi}\ln\left(\frac{r}{r_o}\right) \tag{6.20a}$$

$$U_{mix} = \left(\frac{2-v}{1-v}\right)\frac{Gb^2}{4\pi}\ln\left(\frac{r}{r_o}\right) \simeq \left(\frac{2-v}{1-v}\right)Gb^2 \simeq \alpha Gb^2 \tag{6.20b}$$

$$\alpha = \left(\frac{2-v}{1-v}\right) \tag{6.20c}$$

For crystalline materials with $v = 1/3$, $\alpha = 5/2$, Eq. (6.20b) gives $U \simeq 2.5Gb^2$, which represents the energy equation of the two dislocations of edge and screw character.

Regarding dislocation motion, a defect-free single crystal called whisker may be strong as a polycrystalline material, but single crystals containing defects are less stronger than polycrystals because dislocations easily glide through the structure. Moreover, dislocation motion across grain boundaries in polycrystals is restricted due to changes in crystallographic direction of slip planes, and as a result, the polycrystal strength increases significantly. In this case, restricted dislocation motion is related to strain hardening beyond the yield point of a material undergoing elastic deformation.

External forces acting in the (hkl) glide plane and glide direction of a dislocation produce a shear stress component τ related to a force per unit length independent of the character of the dislocation (Mott and Nabarro [23]).

The force acting on the partial dislocations is written as

$$F = b\tau \tag{6.21}$$

where $\tau = \tau_p$ is the Peierls shear stress, Eq. (6.3), needed to move the dislocation along the (hkl) glide plane, where atomic bonds break and reform to an extent.

Read–Shockley Analytical Procedure According to this dislocation model (Prandtl [5]), the grain boundaries between, namely α and β phases must have a small orientation. This suggests that the model considers semi-coherent interface boundaries for deriving the grain boundary energy equation as a function of the angle of misfit (θ).

For convenience, only the cylindrical ring with core radius $r_o \geq b$, where $2b < r_o \leq 5b$, and length $L(r)$ for an edge dislocation is shown in Fig. 6.10 (Prandtl [5, p. 357]). Notice the dislocation energy regions; U_I denotes the core energy region (Fig. 6.10a), U_{II} denotes the cylinder energy region (Fig. 6.10b), dU_{II} denotes the infinitesimal dislocation elastic energy change and U_{III} denotes the outer elastic energy region.

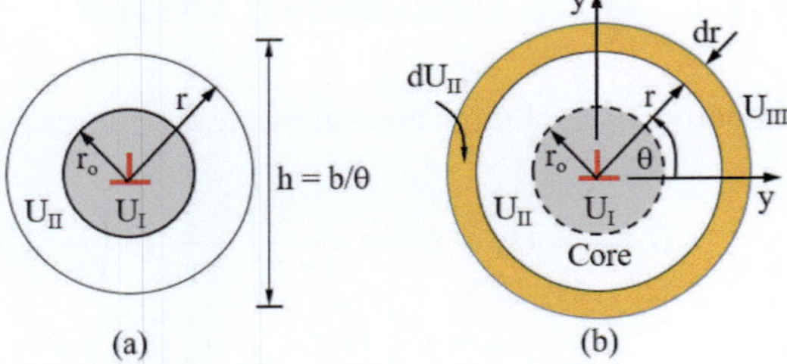

Fig. 6.10 Cylindrical ring model and dislocation core. (**a**) Dislocation core and related energy regions and (**b**) ring due to changes in dislocation energy dU_{II} in region U_{II} (Prandtl [5, p. 357])

Let the shear stress component on the slip plane along the slip direction be

$$\tau = \tau_o \left(\frac{b}{r} \right) \tag{6.22}$$

with

$$\tau_o = \frac{Gb^2}{2\pi} \quad \text{(screw dislocation)} \tag{6.23a}$$

$$\tau_o = \frac{Gb^2}{2\pi\,(1 - v)} \quad \text{(edge dislocation)} \tag{6.23b}$$

where τ_o denotes a constant or reference stress, G denotes the shear modulus, and v denotes the Poisson's ratio for monotonic loading. For clarity, monotonic loading refers to a quasi-static loading that increases continuously.

Let E_I or E_{II} be the energy per unit area of the boundary (region I or II in Fig. 6.10) and $U/b = U\theta/b$ be the dislocation energy per unit length outside the dislocation core. In this context, the energy balance in terms of the rate of change is written as

$$\frac{d}{d\theta} \left(\frac{U\theta}{b} \right) = \frac{dU}{d\theta} \tag{6.24}$$

Consider the cylindrical ring (shell) with radius r and infinite change in thickness dr depicted in Fig. 6.10b. The general elastic energy stored in the volume element dV is defined by

$$d\left(\frac{U}{\theta} \right) = \frac{1}{2}\tau b dr = \frac{1}{2}\tau_o b^2 \frac{dr}{r} = -\frac{1}{2}\tau_o b^2 \frac{d\theta}{\theta} \tag{6.25}$$

with

$$\frac{dr}{r} = \frac{dh}{h} = -\frac{d\theta}{\theta} \tag{6.26}$$

Integrating Eq. (6.25) yields the elastic energy U

$$\int d\left(\frac{U}{\theta}\right) = -\frac{1}{2}\tau_o b^2 \int \frac{d\theta}{\theta} \tag{6.27a}$$

$$\frac{U}{\theta} = \frac{\tau_o b^2}{2}\left[A - \ln\left(\theta\right)\right] = U_o\left[A - \ln\left(\theta\right)\right] \tag{6.27b}$$

$$U = U_o\theta\left[A - \ln\left(\theta\right)\right] \tag{6.27c}$$

$$U = U_o\theta\left[1 + \ln\left(\frac{b}{2\pi r_o}\right) - \ln\left(\theta\right)\right] \tag{6.27d}$$

where U_o is a constant defining the type of dislocation energy and A is the constant of integration. These constants are defined by (Dehlinger [6, p. 359])

$$U_o = \frac{\tau_o b^2}{2} \tag{6.28a}$$

$$A = 1 + \ln\left(\frac{b}{2\pi r_o}\right) + \frac{2}{\tau_o b^2}\left(E_I + E_{II}\right) \simeq 1 + \ln\left(\frac{b}{2\pi r_o}\right) \tag{6.28b}$$

Combining Eqs. (6.23a), (6.23b), and (6.27c) yields the elastic strain energy for edge and screw dislocations as functions of the angle of misfit (θ)

$$U_s = \frac{Gb^2\theta}{2\pi}\left[1 + \ln\left(\frac{b}{2\pi r_o}\right) - \ln\left(\theta\right)\right] \tag{6.29a}$$

$$U_e = \frac{Gb^2\theta}{2\pi\left(1 - v\right)}\left[1 + \ln\left(\frac{b}{2\pi r_o}\right) - \ln\left(\theta\right)\right] \tag{6.29b}$$

$$U_s = \left(1 - v\right)U_e \tag{6.29c}$$

Most metals have a Poisson's ratio $v = 1/3$. In such a case, the screw dislocation energy becomes $U_s = (2/3)\,U_e = 0.67U_e$. This simply implies that $U_s < U_e$.

If the dislocation core radius is $r_o > b$, which is usually the case, then $\ln\left(b/2\pi r_o\right) < 0$ and $\left[\ln\left(b/2\pi r_o\right) - \ln\left(\theta\right)\right] < 1$. This clearly indicates that the dislocation energy terms are always positive; $U_s > 0$ and $U_e > 0$.

Example 6.3 Consider the screw dislocation core in Fig. 6.10b to derive Eq. (6.18c) without any approximation using Hooke's law. The elastic shear strain (γ) for a screw dislocation is defined by

$$\gamma = \frac{b}{2\pi r} \tag{6.3E1}$$

Derive U_s using **(a)** Hooke's law and **(b)** the theory of dislocation. Calculate **(c)** the radius r if $\ln\left(r/r_o\right) \simeq 4\pi$ and $r_o = 5b$ and **(d)** the elastic energy U_s for copper in the

[110] direction of a screw dislocation. Assume that the lattice parameter of copper, the dislocation core radius, and the shear modulus are $a = 0.36\ nm$, $r_o = 5b$, and $G = 48\ GPa$, respectively.

Solution

(a) From Hooke's law and Eq. (6.3E1), the elastic shear stress and elastic energy per unit volume are, respectively

$$\tau = G\gamma \tag{6.3E2a}$$

$$U_s = \frac{1}{2}\tau\gamma = \frac{1}{2}G\gamma^2 \tag{6.3E2b}$$

$$U_s = \frac{Gb^2}{8\pi^2 r^2} \tag{6.3E2c}$$

(b) Multiplying Eq. (6.3E2c) by the volume of a cylindrical core, $V = 2\pi r dr$, and integrating yields Eq. (6.18c) without any approximation.

$$U_s = \frac{Gb^2}{4\pi}\int_{r_o}^{r}\frac{dr}{r} = \frac{Gb^2}{4\pi}\ln\left(\frac{r}{r_o}\right) \tag{6.3E3}$$

(c) For FCC copper, the burgers vector and its magnitude are

$$\mathbf{b} = \frac{a}{2}[110] \tag{6.3E4a}$$

$$b = \frac{a}{2}\sqrt{1^2 + 1^2 + 0} = \frac{\sqrt{2}}{2}a = \frac{\sqrt{2}}{2}(0.36\ nm) \tag{6.3E4b}$$

$$b = 0.25\ nm \tag{6.3E4c}$$

Then,

$$r_o = 5b = 5(0.25\ nm) = 1.25\ nm \tag{6.3E5}$$

The radius r is

$$r = r_o\exp(4\pi) = 5b\exp(4\pi) = \left(1.4338 \times 10^6\right)b \tag{6.3E6a}$$

$$r = (1.25)\exp(4\pi) \simeq 3.58 \times 10^5\ nm \tag{6.3E6b}$$

Check:

$$4\pi = 12.566 \tag{6.3E7a}$$

$$\ln\left(\frac{r}{r_o}\right) = \ln\left(\frac{3.58 \times 10^5}{1.25}\right) = 12.565 \qquad (6.3\text{E}7\text{b})$$

(d) For the $\mathbf{b}_1 = (a/2)\,[110]$ dislocation with $b = |\mathbf{b}_1|$, approximate Eq. (6.3E3) so that

$$U_s = \frac{Gb^2}{4\pi}\ln\left(\frac{r}{r_o}\right) \simeq Gb^2 \qquad (6.3\text{E}8\text{a})$$

$$U_s = G\left(\frac{a}{2}\right)^2 [110]^2 = G\left(\frac{a}{2}\right)^2 \left(\sqrt{1^2 + 1^2 + 0}\right)^2 \qquad (6.3\text{E}8\text{b})$$

$$U_s = \frac{Ga^2}{2} \qquad (6.3\text{E}8\text{c})$$

Thus, Eq. (6.3E8a) gives the magnitude of U_s as

$$U_s = \left(48 \times 10^9 \; Pa\right)\left(0.25 \times 10^{-9} \; m\right)^2 = 3 \times 10^{-9} \; Pa.m^2 \qquad (6.3\text{E}9\text{a})$$

$$U_s = 3 \times 10^{-9}\,\frac{N}{m^2}m^2 = 3 \times 10^{-9}\,\frac{N.m}{m} \qquad (6.3\text{E}9\text{b})$$

$$U_s = 3 \times 10^{-9} \; J/m \qquad (6.3\text{E}9\text{c})$$

Therefore, $U_s = 3 \times 10^{-9} \; J/m$ is needed to move a screw dislocation.

Hull–Bacon Analytical Procedure [24] Consider a closed clockwise curve known as the *Burgers circuit* around a line defect $L\,(r)$ depicted in Fig. 6.11a, and the dislocations modeled as cylindrical shells of radius r and thickness dr shown in Fig. 6.11b for a screw dislocation and Fig. 6.11c for an edge dislocation (Polanyi [8, pp. 67 & 70]).

The local Burgers vector of the linear defect around an arbitrary contour "C" is given by the line integral (Orowan [7]), and the direction of a Burgers vector is based on vector algebra. For characterizing dislocations,

$$\vec{b} = \oint \frac{\partial \mu\,(r)}{\partial s}\,ds \qquad (6.30\text{a})$$

$$\vec{b} \times \vec{L}\,(r) = 0 \quad (\text{screw} \circlearrowright, \text{parallel}) \qquad (6.30\text{b})$$

$$\vec{b} \cdot \vec{L}\,(r) = 0 \quad (\text{edge} \perp, \text{perpendicular}) \qquad (6.30\text{c})$$

The *Burgers vector* $\mathbf{b} = \vec{b}$ (Fig. 6.11b or c), defines a step required to close the Burgers circuit. Accordingly, the magnitude of a Burgers vector, $b = \lfloor \mathbf{b} \rfloor$, is the smallest interatomic spacing shown by the arrow between two atoms in a close-packed crystallographic direction.

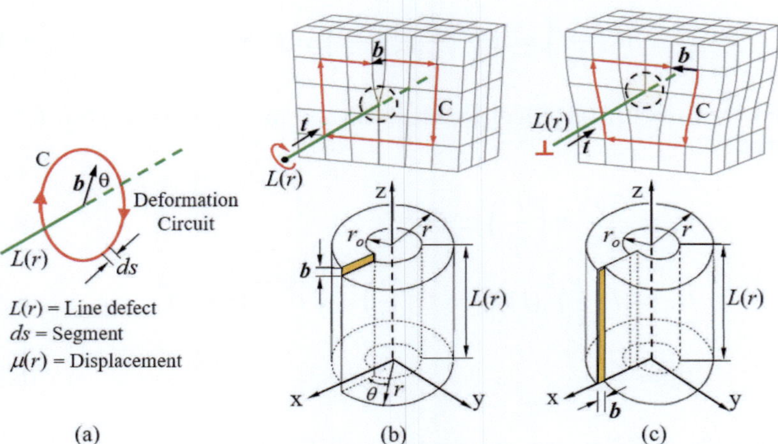

Fig. 6.11 Deformation circuits around dislocations. (**a**) The circuit around a dislocation line (Hirth and J. Lothe [3, p. 23]), (**b**) core of a screw dislocation and (**c**) core of an edge dislocation. After Hull and Bacon [24, p. 70]

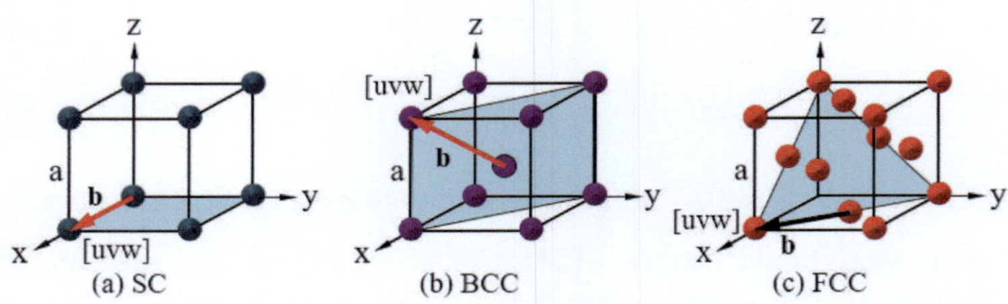

Fig. 6.12 The Burgers vector **b** in cubic crystals. (**a**) **b** along the x-axis on the (001) plane in a simple cubic (SC), (**b**) **b** on the (110) in a body-centered cubic (BCC) and (**c**) **b** on the (111) plane in a face-centered cubic

Notice that **b** has been assigned a selected location and direction in simple cubic SC (Fig. 6.12a), body-centered cubic BCC (Fig. 6.12b) and face-centered cubic FCC (Fig. 6.12c) unit cells. For clarity, **b** in Fig. 6.12 denotes the Burgers lattice vector of a perfect dislocation, which can dissociate or split into two partial dislocations with different Burgers vectors denoted by a dislocation reaction of the form $\mathbf{b} = \mathbf{b}_1 \rightarrow \mathbf{b}_2 + \mathbf{b}_3$, where $b_1 > b_2$, $b_1 > b_3$. Additional details on dislocation dissociation are given below.

Mathematically, the magnitude of a Burgers vector (b) in cubic crystals can be define by

$$\vec{b} = a\,[uvw] \quad (SC) \tag{6.31a}$$

$$b = a\sqrt{u^2 + v^3 + w^2} \quad (SC) \tag{6.31b}$$

$$\vec{b} = \frac{a}{2}[uvw] \quad (FCC \text{ or } BCC) \tag{6.31c}$$

$$b = \frac{a}{2}\sqrt{u^2 + v^3 + w^2} \quad (FCC \text{ or } BCC) \tag{6.31d}$$

Among excellent resources on dislocation theory available in the literature, a subsequent analytical procedure for deriving the energy of dislocations is found in the Hull and Bacon book [24, chapter 4].

Cartesian Coordinates The mathematical modeling of a *screw dislocation* is based on the elastic strain field within a dislocation cylinder (Fig. 6.11b), where the nonzero displacement (μ_z) and elastic strains (ϵ_{ij}) on a radial slit are

$$\mu_z = \frac{b\theta}{2\pi} = \arctan\left(\frac{y}{x}\right) \tag{6.32a}$$

$$\epsilon_{xz} = \epsilon_{zx} = \frac{1}{2}\left(\frac{\partial \mu_z}{\partial x} + \frac{\partial \mu_x}{\partial z}\right) = -\frac{b}{4\pi}\frac{y}{x^2 + y^2} = -\frac{b}{4\pi}\frac{\sin\theta}{r} \tag{6.32b}$$

$$\epsilon_{yz} = \epsilon_{zy} = \frac{1}{2}\left(\frac{\partial \mu_y}{\partial x} + \frac{\partial \mu_x}{\partial y}\right) = +\frac{b}{4\pi}\frac{x}{x^2 + y^2} = +\frac{b}{4\pi}\frac{\cos\theta}{r} \tag{6.32c}$$

and according to the theory of elasticity, the relevant nonzero elastic shear stresses are defined by

$$\tau_{xz} = \tau_{zx} = 2G\epsilon_{xz} = -\frac{Gb}{4\pi}\frac{\sin\theta}{r} \tag{6.33a}$$

$$\tau_{yz} = \tau_{zy} = 2G\epsilon_{xy} = +\frac{Gb}{4\pi}\frac{\cos\theta}{r} \tag{6.33b}$$

which are identical to Eqs. (6.17a), (6.17b). Here, the axial (E) and shear (G) moduli of elasticity are material's properties, and they are related as

$$E = 2G(1 + v) \tag{6.34}$$

The differential elastic energy for screw (dU_s) and edge (dU_e) dislocations can be defined as shown below

$$dU_s = \frac{1}{2}\tau_{yz}bdA = \frac{Gb^2}{4\pi}\frac{dr}{r} \tag{6.35a}$$

$$dU_e = \frac{1}{2}\tau_{xy}bdA = \frac{Gb^2}{4\pi(1 - v)}\frac{dr}{r} \tag{6.35b}$$

Integrating Eq. (6.35) yields the dislocation elastic energy terms

$$U_s = \frac{Gb^2}{4\pi} \int_{r_o}^{r} \frac{dr}{r} = \frac{Gb^2}{4\pi} \ln\left(\frac{r}{r_o}\right) \tag{6.36a}$$

$$U_e = \frac{Gb^2}{4\pi\,(1-v)} \int_{r_o}^{r} \frac{dr}{r} \tag{6.36b}$$

$$U_e = \frac{Gb^2}{4\pi\,(1-v)} \ln\left(\frac{r}{r_o}\right) \tag{6.36c}$$

$$U_s = (1-v)\,U_e \tag{6.36d}$$

Polar Coordinates   The nonzero shear stresses (τ_{ij}) and the shear strains (γ_{ij}) are

$$\tau_{rz} = +\tau_{xz} \cos\theta + \tau_{yz} \sin\theta \tag{6.37a}$$

$$\tau_{\theta z} = -\tau_{xz} \sin\theta + \tau_{yz} \cos\theta \tag{6.37b}$$

$$\gamma_{\theta z} = \gamma_{z\theta} = \frac{b}{4\pi r} \tag{6.37c}$$

For an element of volume dV, the elastic strain energy dU_ϵ in Cartesian and polar coordinates is defined by Hooke's law

$$dU_\epsilon = \frac{1}{2} dV \sum_{i=x,y,z} \sum_{j=x,y,z} \sigma_{ij}\epsilon_{ij} \tag{6.38a}$$

$$dU_\epsilon = \frac{1}{2} dV \sum_{i=r,\theta,z} \sum_{j=r,\theta,z} \sigma_{ij}\epsilon_{ij} \tag{6.38b}$$

where $\sigma_{ij}\epsilon_{ij} = \tau_{ij}\gamma_{ij}$ for $i \neq j$ and $\sigma_{ij}\epsilon_{ij} = \sigma_{ji}\epsilon_{ji}$ due to symmetry. Now, the screw dislocation energy change dU_s (Fig. 6.11b) is defined as the elastic energy (dU_ϵ) stored per unit dislocation length $L = L(r)$ (Fig. 6.11) within the volume element dV.

For a cylindrical ring, the infinitely differential volume dV, the general shear strain γ_{ij}, strain energy U, shear stress τ_{ij}, and strain energy per volume U_V are, respectively, written as

$$dV = 2\pi r L\,(r)\,dr \tag{6.39a}$$

$$\gamma_{ij} = \frac{b}{2\pi r} \tag{6.39b}$$

$$U = \frac{1}{2}\tau_{ij}\gamma_{ij} \tag{6.39c}$$

$$\tau_{ij} = G\gamma_{ij} \tag{6.39d}$$

$$U_V = \frac{1}{2}G\gamma_{ij}^2 = \frac{1}{2}\frac{Gb^2}{(2\pi r)^2} \tag{6.39e}$$

Thus,

$$dU = \left[\frac{1}{2}\frac{Gb^2}{(2\pi r)^2}\right]dV = \left[\frac{1}{2}\frac{Gb^2}{(2\pi r)^2}\right][2\pi r L\,(r)\,dr] \tag{6.40a}$$

$$dU = \frac{Gb^2 L\,(r)}{4\pi r}dr \tag{6.40b}$$

For a straight screw dislocation with $L = L\,(r)$, $dU_s = dU/L\,(r)$ so that

$$dU_s = \frac{1}{2}\,(2\pi r dr)\,(\tau_{\theta z}\gamma_{\theta z} + \tau_{z\theta}\gamma_{z\theta}) = (4\pi r dr)\,G\gamma_{\theta z}^2 \tag{6.41a}$$

$$dU_s = 4\pi r dr G\left(\frac{b}{4\pi r}\right)^2 = \frac{Gb^2}{4\pi}\frac{dr}{r} \tag{6.41b}$$

$$U_s = \frac{Gb^2}{4\pi}\int_{r_o}^{r}\frac{dr}{r} = \frac{Gb^2}{4\pi}\ln\left(\frac{r}{r_o}\right) \tag{6.41c}$$

and for a straight edge dislocation with $dU_e = dU/L\,(r) = dU/L$,

$$dU_e = \frac{1}{2}2\pi r dr\,(\tau_{\theta z}\gamma_{\theta z} + \tau_{z\theta}\gamma_{z\theta}) = 4\pi r dr G\gamma_{\theta z}^2 \tag{6.42a}$$

$$dU_e = \frac{4\pi r dr G}{1 - v}\left(\frac{b}{4\pi r}\right)^2 = \frac{Gb^2}{4\pi\,(1 - v)}\frac{dr}{r} \tag{6.42b}$$

$$U_e = \frac{Gb^2}{4\pi\,(1 - v)}\int_{r_o}^{r}\frac{dr}{r} = \frac{Gb^2}{4\pi\,(1 - v)}\ln\left(\frac{r}{r_o}\right) \tag{6.42c}$$

Dislocations in real crystalline materials are most commonly mixed in nature as schematically illustrated in Fig. 6.13 for pure screw, mix, and edge dislocations with a symmetrical lattice. Notice the curved line showing the lattice sections related to the dislocations Burgers vector orientations (Callister and Rethwisch [25, p. 104]).

The elastic strain energy of a mixed dislocation is $U_{mix} = U_s + U_e$ since the theory of elasticity does not consider the energy of a dislocation core. Thus,

$$U_{mix} = U_s + U_e = \left(\frac{2 - v}{1 - v}\right)\frac{Gb^2}{4\pi}\ln\left(\frac{r}{r_o}\right) \tag{6.43}$$

which compares with Eq. (6.20b).

Setting the limits $r = 10^5 b$ and $r_o = 0.349b$ (Hosford [26, p. 144]) yields $\ln\,(r/r_o) = 4\pi$ and Eqs. (6.41c), (6.42c) and (6.43) become

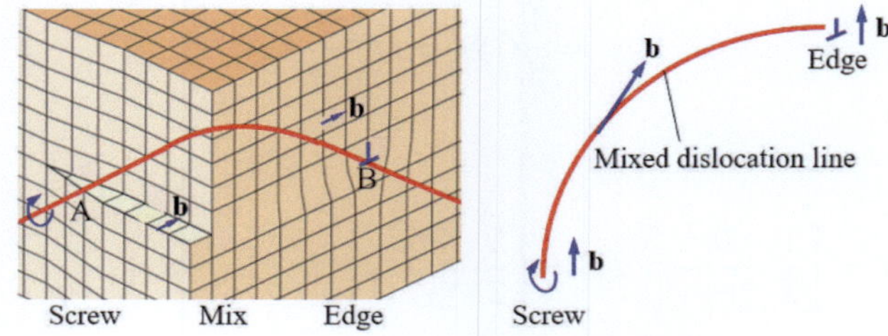

Fig. 6.13 Pure screw (A) → mixed (AB) → pure edge (B) dislocations with Burgers vector **b**. After Callister and Rethwisch [25, p. 104])

$$U_s \simeq Gb^2 \tag{6.44a}$$

$$U_e \simeq \frac{Gb^2}{1-v} \tag{6.44b}$$

$$U_s \simeq (1-v)\, U_e \tag{6.44c}$$

$$U_{mix} = \left(\frac{2-v}{1-v}\right) Gb^2 \tag{6.44d}$$

Despite that these expressions have been derived previously using a different analytical approach, the Hull–Bacon [24, chapter 4] analytical procedure confirms their mathematical definitions. Nonetheless, derivation of the dislocation energy terms assuming different procedures has been illustrated in detail for the reader convenience.

If the dislocation length $L = \left| \vec{L}\,(r) \right|$ is known, then the dislocation strain energy equations become

$$U_s \simeq Gb^2 L \tag{6.45a}$$

$$U_e \simeq \frac{Gb^2 L}{1-v} \tag{6.45b}$$

$$U_s \simeq (1-v)\, U_e \tag{6.45c}$$

$$U_{mix} = U_s + U_e = \left(\frac{2-v}{1-v}\right) Gb^2 L \tag{6.45d}$$

The mixed strain energy U_{mix} is significantly most realistic along with $L = \rho_d V$, where ρ_d denotes the dislocation density and V the volume of material containing the ρ_d.

Dislocations are linear defects in the crystalline lattice of a material. They are characterized by an extra half-plane of atoms that is either inserted or removed from the lattice. Dislocations play an important role in the mechanical properties of materials, as they can significantly affect the strength, ductility, and toughness of a material.

The strain energy of a dislocation is the energy stored in the lattice as a result of the distortion caused by the dislocation. The strain energy is a function of the Burgers vector of the dislocation, which is the vector that defines the extra half-plane of atoms. The strain energy also depends on the type of dislocation, such as edge dislocation or screw dislocation.

6.5   Dislocation Density

Dislocation density is a measure of the number of dislocations per unit volume in a material. In essence, dislocation density is a measure of the number of dislocations per unit volume in a material. Dislocations are linear defects in the crystalline lattice of a material that can significantly affect the mechanical properties of the material, such as its strength, ductility, and toughness.

The relationship between dislocation density and mechanical properties is complex and depends on a number of factors, including the type of material, the grain size, and the temperature. However, in general, an increase in dislocation density will lead to an increase in the strength and yield strength of a material. This is because dislocations act as obstacles to the movement of other dislocations, which makes it more difficult for the material to deform.

During plastic deformation of an elastic–plastic metal or alloy, the dislocation strain rate (Orowan [7]), and the dislocation shear stress (Taylor [27]) equations are, respectively,

$$\frac{d\gamma}{dt} = b\rho_d \upsilon \tag{6.46a}$$

$$\tau = \alpha G b \sqrt{\rho} \tag{6.46b}$$

where the dimensionless factor α is simply a multiplier; $\alpha = 0.2$ for FCC and $\alpha = 0.4$ for BCC metals (Nabarro et al. [28]).

Regarding crystal defects, Table 6.1 lists the vacancy concentration (Cv) and dislocation density (ρ_d) for selected FCC materials being deformed using the equal channel angular pressing (ECAP) technique as per Gubicza [29] overview article. The purpose of these datasets is to show the extent of these variables in crystalline materials. The actual units of ρ_d are dislocation lines per unit area ($lines/m^2$ or m^{-2}).

Table 6.1 Vacancy concentration and dislocation density (Gubicza [29])

Vacancy concentration	Dislocation density
$C_{v,316L} = 3 \times 10^{-5}$	$\rho_{d,316L} = 143 \times 10^{14}\ m^{-2}$
$C_{v,Ag} = 1.5 \times 10^{-4}$	$\rho_{d,Ag} = 46 \times 10^{14}\ m^{-2}$
$C_{v,Cu} = 20 \times 10^{-4}$	$\rho_{d,Cu} = 21 \times 10^{14}\ m^{-2}$
$C_{v,Ni} = 20 \times 10^{-4}$	$\rho_{d,Ni} = 18 \times 10^{14}\ cm^{-2}$

An increase in dislocation density can also lead to a decrease in the ductility and toughness of a material. This is because dislocations can act as nucleation sites for cracks, which can lead to brittle fracture.

The effect of dislocation density on the mechanical properties of a material can be controlled by a number of methods, such as grain refinement and heat treatment. Grain refinement is a process of reducing the grain size of a material. This can be done by a number of methods, such as cold working or recrystallization. Heat treatment is a process of heating a material to a certain temperature and then cooling it at a certain rate. This can be used to change the microstructure of a material, which can in turn affect the dislocation density.

The FCC crystallographic conditions for easy mechanical deformation by the slip mechanism are denoted as $\{111\} \langle 110 \rangle$. Some of these slip systems can be activated depending on the crystallographic direction of an applied external load (mechanical force). By definition, slip can be defined as the displacement of a crystallographic plane along a certain direction during plastic deformation. The concept of slip mechanism related to plastic deformation is characterized in a later chapter on mechanical behavior of solids.

Example 6.4 Consider edge, screw, and mixed dislocations in a copper alloy (Cu-alloy). Use the given data to calculate the dislocation strain energy, the general dislocation density, and the shear stress. Assume a constant dislocation length L, an average Burgers vector b, and a Poisson's ratio v.

$G = 45\,GPa$	$b = 0.29\,nm$	$v = 1/3$
$r = 10^5 b$	$r_o = 0.348b$	$L = 23\,nm$

Solution From Eq. (6.42c),

$$U_e = \frac{Gb^2}{4\pi\,(1-v)}\,\ln\left(\frac{r}{r_o}\right) \tag{6.4E1a}$$

$$U_e = \frac{\left(45 \times 10^9\,N/m^2\right)\left(0.29 \times 10^{-9}\,m\right)^2}{4\pi\,(1 - 1/3)}\,\ln\left(\frac{10^5}{0.348}\right) \tag{6.4E1b}$$

$$U_e = 5.68 \times 10^{-9}\,N = 5.68 \times 10^{-9}\,N.m/m = 5.68 \times 10^{-9}\,J/m \tag{6.4E1c}$$

$$U_e = \frac{5.68 \times 10^{-9}\,J/m}{1.60 \times 10^{-19}\,J/eV} = 3.55 \times 10^{10}\,eV/m \tag{6.4E1d}$$

From Eqs. (6.44c) and (6.45d), respectively

$$U_s = (1 - v)\,U_e = (1 - 1/3)\left(5.68 \times 10^{10}\,eV/m\right) \tag{6.4E2a}$$

$$U_s = 3.77 \times 10^{-9}\,J/m = 2.36 \times 10^{10}\,eV/m \tag{6.4E2b}$$

$$U_{mix} = U_e + U_s = 5.91 \times 10^{10} \ eV/m \tag{6.4E2c}$$

The general dislocation density is

$$\rho_d = 1/L^2 = 1/\left(23 \times 10^{-9} \ m\right)^2 = 18.90 \times 10^{14} \ m^{-2} \tag{6.4E3}$$

which reflects the magnitude in a heavily cold-rolled polycrystalline material, and it compares with the experimental value given in Table 6.1.

From Eq. (6.46b), the average internal shear stress needed to produce an average dislocation density $\rho_s = 18.90 \times 10^{14} \ m^{-2}$ in a mechanically deformed copper specimen is

$$\tau = \alpha G b \sqrt{\rho} \tag{6.4E5a}$$

$$\tau = (0.2)\,(45 \ GPa)\left(0.29 \times 10^{-9} \ m\right) \sqrt{18.90 \times 10^{14} \ m^{-2}} \tag{6.4E5b}$$

$$\tau = 0.11347 \ GPa = 113.47 \ MPa \tag{6.4E5c}$$

From an engineering point of view, this calculated shear stress represents a maximum value since dislocation density measurements are done after a specimen is fractured. In this example, the intrinsic number of vacancy concentration can only be predicted as per Table 6.1; that is, $C_{v,Cu} = 20 \times 10^{-4}$. Dislocations can act as obstacles to the movement of other dislocations, which makes it more difficult for the material to deform plastically. Therefore, an increase in dislocation density will typically lead to an increase in the strength, decrease the ductility and toughness of a material. By definition, toughness is a combination of strength and ductility of a material under a loading mode.

6.6 Planar Defects

Surface area, grain boundaries, stacking faults, and twin boundaries are known as planar defects. A grain boundary is the interface between two grains, and it is observed under the microscope as an irregular line with a certain width that represents an atomic mismatch, a twin boundary represents a displacement of a large quantity of atoms that move simultaneously during mechanical deformation. Regarding the surface area of a solid specimen, atoms lack of proper coordination number (CN) and structural geometry (unit cell). Therefore, the surface area is a 2D defect characterized as a high-energy or reactive plane containing mechanically induced defects introduced during machining.

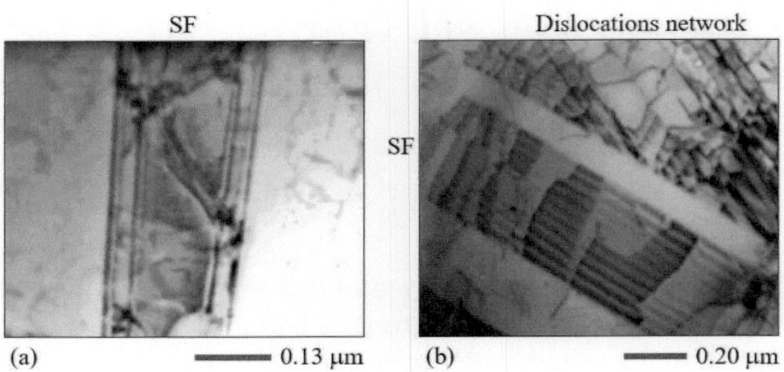

Fig. 6.14 Stacking faults in (**a**) AISI 304 stainless steel and (**b**) Ni-based alloy

6.6.1 Stacking Faults and Partial Dislocations

Actual Stacking Faults This section deals with stacking faults (SFs) as planar and dislocations as the linear crystal defects. For instance, the stacking fault (SF) in stainless steel type AISI 304 alloy is shown in Fig. 6.14a and that in a Ni-based alloy is depicted in Fig. 6.14b. Both alloys were mechanically deformed in tension loading.

Actual stacking faults are planar defects that have a characteristic fringe denoted by the //// pattern, where the orderly arrangement of atoms or the symmetrical stacking sequence of layers is disrupted locally during crystal growth or mechanical deformation. Thus, stacking faults are formed by a local deviation of the stacking sequence of layers in a crystalline solid.

Stacking Fault Model The stacking fault model illustrated in Fig. 6.15 represents a planar defect and denoted as A,B,C stacking sequence of atoms.

For instance, Fig. 6.15a shows the three-dimensional (3D) stacking sequence of atom layers (layer A being on top and layer C at the bottom of an atomic arrangement). Specifically, Fig. 6.15b schematically depicts the two-dimensional (2D) stacking fault due to the dissociation of a perfect dislocation with perpendicular Burgers vector $\mathbf{b}_1$ into two Shockley partial dislocations (or Shockley partials) with randomly oriented Burgers vectors $\mathbf{b}_2$ and $\mathbf{b}_3$. This type of planar defect is denoted by the vector notation $\mathbf{b}_1 \rightarrow \mathbf{b}_2 + \mathbf{b}_3$. Notice that the stacking fault in Fig. 6.15b is ideally surrounded by evenly space partial dislocations along the x-direction and the point where $\mathbf{b}_1$, $\mathbf{b}_2$, $\mathbf{b}_3$ coincide is called a nodal point.

Figure 6.15c, on the other hand, illustrates the initial stacking sequence, where for convenience, the layer A is chosen to be subjected to a small displacement in the xy space as indicated by the arrows and Fig. 6.15d illustrates the triangle vector notation defining the involved dislocations along their respective crystallographic directions, say, in a FCC crystal

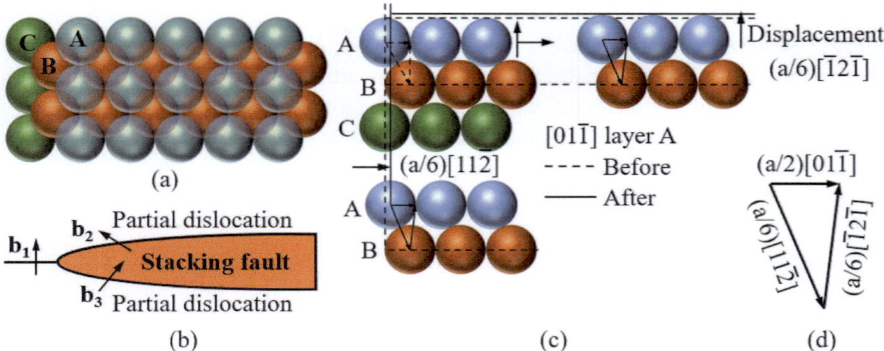

Fig. 6.15 Stacking fault model. (**a**) 3D top view of atom layers, (**b**) 2D stacking fault due to dissociation of a perfect dislocation $\mathbf{b}_1$, (**c**) 3D front view of the atom layers. Layer A is subjected to a small displacement in the directions indicated by arrows and (**d**) dislocation Burgers vectors forming a triangle

- Perfect dislocation with $\mathbf{b}_1 = (a/2)\left[01\bar{1}\right]$
- Partial dislocation with $\mathbf{b}_2 = (a/6)\left[11\bar{2}\right]$
- Partial dislocation with $\mathbf{b}_3 = (a/6)\left[\bar{1}2\bar{1}\right]$

The one-atom thick atomic layers are sequentially denoted as

$$ABCABCABCABC... \text{ perfect stacking sequence} \qquad (6.47a)$$

$$ABCABABCAB...... \quad \text{ stacking fault sequence} \qquad (6.47b)$$

$$\lfloor ABC\rfloor_{FCC} \lfloor ABAB\rfloor_{HCP} \lfloor CAB......\rfloor_{FCC} \qquad (6.47c)$$

where the boundary $ABAB$ is the disorderly stacking sequence representing the localized planar defect, which resembles the stacking sequence in HCP crystals. Thus, the $ABAB$ fault is denoted in Eq. (6.47c) as $\lfloor ABAB\rfloor_{HCP}$.

According to Ding et al. [30], a stacking of atoms in an HCP crystal may be complicated as shown below

$$ABABABAB....\text{perfect stacking sequence} \qquad (6.48a)$$

$$ABABBABA....\text{stacking growth fault - } BB \qquad (6.48b)$$

$$ABABCBCB....\text{stacking growth fault - } CB \qquad (6.48c)$$

$$ABABCACA....\text{intrinsic deformation fault - } CA \qquad (6.48d)$$

$$ABABCBAB....\text{extrinsic twin-like fault - } CB \qquad (6.48e)$$

These stacking faults, Eq. (6.48), represent possible atomic sequences that can be found in deformed HCP structures.

Zhao et al. [31] have indicated that the HCP-to-FCC transition in Ni-based alloys, such as $NiCoFeCrMn$ and $NiCoFeCrPd$ alloys, is a temperature-dependent crystallographic transition. This is attributable to the difference in vibrational (phonon) entropy ΔS between these structures. Moreover, the stacking sequence for the HCP-to-FCC transition is written as

$$|ABABAB...|_{HCp} \rightarrow |ABCABC...|_{FCC} \tag{6.49}$$

In particular, the temperature-dependent deformation mechanism and dislocation behavior in materials may be characterized by the stacking fault energy (γ_{SF} = SFE) due to the disruption of the stacking sequence of atoms during deformation. In the literature, γ_{SF} is also known as the generalized stacking fault energy.

6.6.2 Gamma Equation

The stacking fault surface energy (γ_{SF}) or energy density (in units of J/cm^2 or J/m^2) at a temperature T is related to partial dislocations with Burgers vectors $\mathbf{b}_2$ and $\mathbf{b}_3$. Quantitatively, γ_{SF} and the equilibrium spacing (d) between partial dislocations ($\mathbf{b}_2 = (a/n')\,[uvw]_2$ and $\mathbf{b}_3 = (a/n')\,[uvw]_3$) in a cubic crystal are inversely proportional. Thus,

$$\gamma_{SF} = \frac{E_{SF} - E_o}{A} \tag{6.50a}$$

$$\gamma_{SF} = \frac{G\,(\mathbf{b}_2 \cdot \mathbf{b}_3)}{2\pi d} = \frac{G}{2\pi d}\left[\left(\frac{a}{\sqrt{6}}\right)_{[11\bar{2}]}\right]\left[\left(\frac{a}{\sqrt{6}}\right)_{[12\bar{1}]}\right]\cos\theta \tag{6.50b}$$

$$d = \frac{G\,(\mathbf{b}_2 \cdot \mathbf{b}_3)}{2\pi\gamma_{SF}} \tag{6.50c}$$

Here, A is the localized area of the fault (cm^2 or m^2), E_{SF} is the stacking fault energy (J or eV), and E_o is the energy of a perfect crystal (J or eV) and $n' = 2, 4$ or 6. It is assumed that the dissociation of a perfect dislocation with a Burgers vector $\mathbf{b}_1$ occurs as per Eq. (6.9a). This is the reason for having the dot product $\mathbf{b}_2 \cdot \mathbf{b}_3$ in Eq. (6.50b).

For $\mathbf{b}_2 = (a/6)\left[11\bar{2}\right]_2$ and $\mathbf{b}_3 = (a/6)\left[\bar{1}2\bar{1}\right]_3$, Eq. (6.50b) yields

$$\gamma_{SF} = \frac{G\cos\theta}{2\pi d}\left[\frac{a}{n'}\sqrt{u^2 + v^2 + w^2}\right]_2\left[\frac{a}{n'}\sqrt{u^2 + v^2 + w^2}\right]_3 \tag{6.51a}$$

$$\gamma_{SF} = \frac{G\cos\theta}{2\pi d}\left[\left(\frac{a}{\sqrt{6}}\right)\right]_{[11\bar{2}]}\left[\left(\frac{a}{\sqrt{6}}\right)\right]_{[\bar{1}2\bar{1}]} \tag{6.51b}$$

Notice that if a perfect dislocation, as an example, with the Burgers vector $\mathbf{b}_1 = (a/2)\,[uvw]_1$ dissociates into two partial dislocation with $\mathbf{b}_2 = (a/n')\,[uvw]_2$ and $\mathbf{b}_3 = (a/n')\,[uvw]_3$, then γ_{SF} is minimized to an extent since $|\mathbf{b}_1| > |\mathbf{b}_2|,\,|\mathbf{b}_3|$.

The gamma surface is typically plotted as a three-dimensional surface, with one axis representing the shear displacement along a particular crystallographic plane and the other two axes representing the components of the Burgers vector. The energy associated with each stacking fault configuration is represented by the height of the surface at that point. The gamma surface is described below.

6.6.3 Gamma Surface

The result of a computer simulation of a stacking fault and related positions of partial dislocations is shown in Fig. 6.16 as a 3D gamma surface (γ-surface) for an FCC (111) plane of copper (after Y.N. Osetsky as cited in Hull–Bacon book [24, p. 88]).

The gamma surface is a three-dimensional view of the stacking fault energy (J/cm^2 or J/m^2), which is an energy concept used to characterize the deformation behavior of crystalline materials based on planar defects induced by dislocation interactions, which occur along certain crystallographic $[uvw]$ directions on specific crystallographic (hkl) planes.

Regarding the points "P" and "F" in Fig. 6.16, the point "P" stands for perfect dislocation with $\gamma_{SF,P} = 0$ representing a global minimum and "F" stands for intrinsic fault with $0 < \gamma_{SF,F} \ll \gamma_{SF,max}$, where $\gamma_{SF,max}$ corresponds to a peak on the stacking fault energy surface. Peaks represent maximum values (maxima)

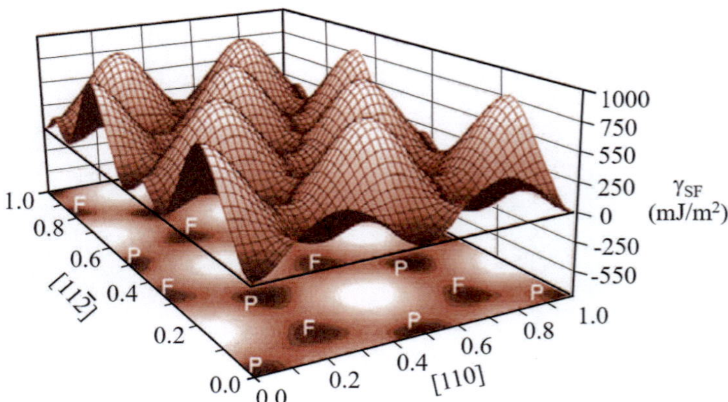

Fig. 6.16 Computer modeling of the stacking fault surface for an FCC (111) plane of copper after Y.N. Osetsky as cited in Hull-Bacon book [24, p. 88]. Here, μ is the displacement of atom layers in the stacking fault area along the selected dislocation directions; $[11\bar{2}]$ and $[110]$

of γ_{SF} being attributed to unstable dislocations. On the other hand, local minimum values (minima) of γ_{SF} surface being attributed to stable dislocations.

6.6.4 Gamma Surface Mathematical Model

For the purpose of clarity on how to plot the γ-surface as $\gamma = \gamma(x, y)$ function, Chen's dissertation [32] contains pertinent details on this matter. Mathematically,

$$\gamma(x, y) = \sum_{m,n} F_{mn} \cos\left(\frac{2\pi m x}{\sqrt{2}a}\right) \cos\left(\frac{2\pi n y}{a}\right) \tag{6.52a}$$

$$\gamma_1(x, y) \simeq A_1 \cos(B_1 x) \cos(C_1 y) \tag{6.52b}$$

$$\gamma_2(x, y) \simeq A_2 \sin(B_2 x) + C_2 \sin(D_2 y) \tag{6.52c}$$

where x, y denote the displacements in a lattice, F_{mn} denotes the Fourier coefficients with m, n being integer numbers in the series, and a denotes the usual lattice parameter of cubic crystals.

Plotting Eqs. (6.52b) and (6.52c) with unit constants yields schematic gamma surfaces denote by $\gamma_1(x, y)$-surface (Fig. 6.17a) and $\gamma_2(x, y)$-surface (Fig. 6.17b).

The $\gamma_1(x, y)$ function has a theoretical foundation since is similar to Eq. (6.52a), while $\gamma_2(x, y)$ has an empirical background. However, both functions depend on their respective constants and exhibit anisotropic characteristics.

The concept of a γ-surface can also be applied to study solid–melt interfaces (Spatschek et al. [33]). In addition, the gamma surface, also known as the generalized stacking fault energy surface (GSFE), is a graphical representation of the energy associated with different stacking fault configurations in a material. It is a powerful tool for understanding the mechanical properties of materials with complex crystal structures.

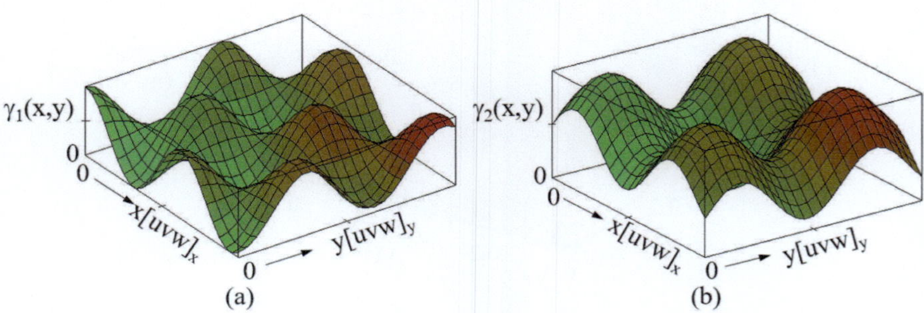

Fig. 6.17 Gamma surface as per (**a**) $\gamma_1(x, y)$ and (**b**) $\gamma_2(x, y)$

6.7   Volume Defects

Volume defects can be classified as (1) inclusions or precipitates that form by interdiffusion at heterogeneous interfaces, (2) shrinkage porosity that forms as an inherent defect in castings due to trapped gases during solidification, and (3) process porosity during consolidation and sintering of power metals (process known as PM or power metallurgy).

Conceptually, inclusions (such as precipitates and dispersed phase particles) are embedded secondary phases in the matrix or primary phase of a solid structure. Obviously, the combination of inclusions, clusters of voids and bubbles, porosity, dislocations, and point defects affect properties of crystalline solids.

6.7.1   Porosity

Porous metals are known as metal foam and are potential materials for certain industrial applications. For instance, Fig. 6.18 shows the morphology of selected porosity patterns in crystalline iron (Fe). Furthermore, porosity is a common defect in castings and consolidated metals, characterized by the presence of voids or cavities within the material. These voids are typically formed during the solidification

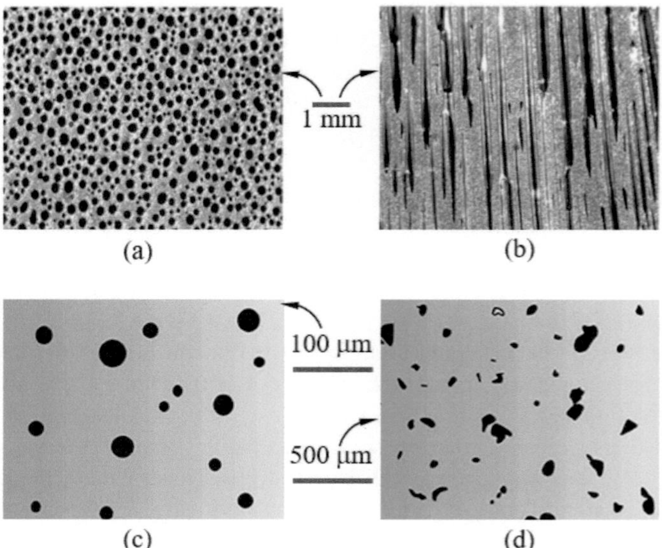

Fig. 6.18 For iron Fe, (**a**) cross-sectional of long pores and (**b**) longitudinal views (Hyun et al. [34]). For AlSi10Mg alloy, (**c**) entrapped gas porosity and (**d**) incomplete melting-induced porosity (Sola and a Nouri [35])

process when gases trapped in the molten metal fail to escape before solidification is complete.

For iron (Fe), Fig. 6.18a exhibits a cross-sectional and Fig. 6.18b a longitudinal views of porosity in consolidated iron Fe (Hyun et al. [34]). Moreover, Fig. 6.18c depicts entrapped gas porosity and Fig. 6.18d illustrates incomplete melting induced as characteristic pores in laser-based powder bed fusion (Sola and a Nouri [35]).

According to the open literature, the density ρ of a material is normally used to determined either the fraction or percentage of porosity p_s. Mathematically,

$$p_s = 1 - \frac{\rho_{measured}}{\rho_{theoretical}} \tag{6.53a}$$

$$p_s = 1 - \frac{\rho_{porous}}{\rho_{non-poroues}} \tag{6.53b}$$

Despite these expressions being defined in a simple form, they imply that increasing the content of porosity degrades mechanical properties.

Characterizing the effect of porosity on physical and mechanical properties of solid materials is a common practice in order to assure the safety and reliability of structural components.

Porosity is a critical concern in the production of castings and consolidated metals. By understanding the causes and effects of porosity, implementing effective mitigation strategies, and employing appropriate post-processing techniques, manufacturers can ensure the production of high-quality castings and consolidated metals with superior mechanical properties and performance.

In order to minimize porosity in castings and consolidated metals, a practicing engineer can employ melt degassing for removing gases from the molten metal before casting using methods such as vacuum degassing or inert gas purging, controlled solidification for controlling the solidification rate to allow gases to escape and prevent shrinkage voids, foundry architecture for designing the mold to promote efficient gas venting and minimize shrinkage and the like.

The relevant data shown in Fig. 6.19a exhibit the effect of shrinkage porosity (P_s) on the modulus of elasticity (E) of carbon steel castings (Ol'khovik [36]) and Fig. 6.19b shows the effect of sintering porosity on corrosion rate (C_R) of Ti-$10Mo$ alloy (Xu et al. [37]).

Observe that the experimental data in Fig. 6.19a can also be divided into two sets: set 1 for a linear relationship $E = f(p_s)$ at $0 \leq p_s \leq 4\%$ and set 2 for nonlinear relationship $E = f(p_s)$ at $4\% < p_s < 10\%$. Considering the anisotropy of steel castings, linear regression seems acceptable despite that the correlation coefficient (R^2) was not reported (Ol'khovik [36]). Nevertheless, the modulus of elasticity E is significantly affect by shrinkage porosity, which is a detrimental defect to mechanical behavior, specifically to fatigue since pores can act as stress concentration sources.

According to Xu et al. [37], curve fitted the C_R data as shown in Fig. 6.19b for some Ti-$10Mo$ alloys, curve fitting generates an exponential function of the form

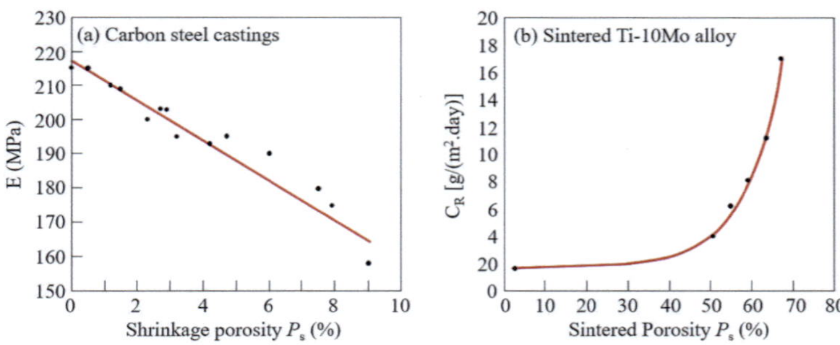

Fig. 6.19 Effect of (**a**) shrinkage porosity on the modulus of elasticity E of steel castings with carbon weight percent range $0.15\% \leq C \leq 0.28\%$ and yield strength range $292\,MPa \leq \sigma_{ys} \leq 247\,MPa$ (Ol'khovik [36]), and (**b**) sintered porosity on corrosion rate C_R of sintered Ti-$10Mo$ alloy (Xu et al. [37])

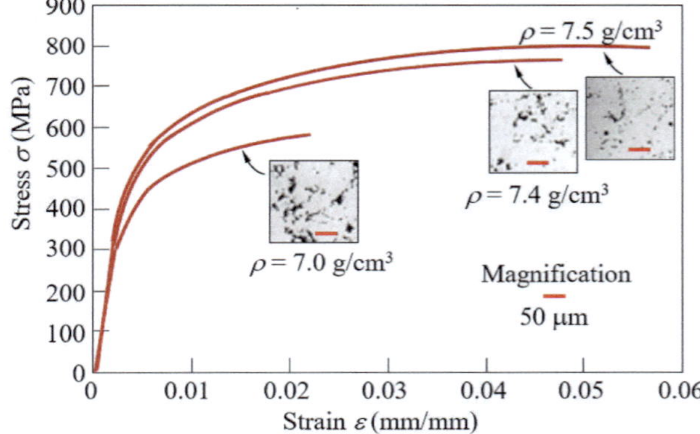

Fig. 6.20 Effect of porosity on the strength of Fe-Mo-Ni steel alloy. The stress–strain curves are identified with the corresponding alloy density. After Chawla and Deng [38]

$$C_R = 1.719211 + 0.01063 \exp\left(\frac{P_s}{0.09227}\right) \tag{6.54}$$

with a correlation coefficient of $R^2 = 0.98788 \simeq 1$. The effect of P_s on C_R is clearly accelerated at $P_s > 50\%$ by volume. The conformity of data points to an exponential function has a relatively high correlation coefficient for Ti-$10Mo$ alloys.

Powder metallurgy (PM) consolidation process generates metallic powder-packing microstructural characteristics and a degree of porosity that influence the mechanical, corrosion, and transport properties. For example, Fig. 6.20 depicts the effect of sintered porosity on Fe-Mo-Ni steel stress–strain curves (Chawla and Deng [38]). Notice that the stress–strain curves are displaced upward, and the fracture strains are increased as the sintering porosity is decreased, leading to an increase in the mass density (ρ) of the alloy.

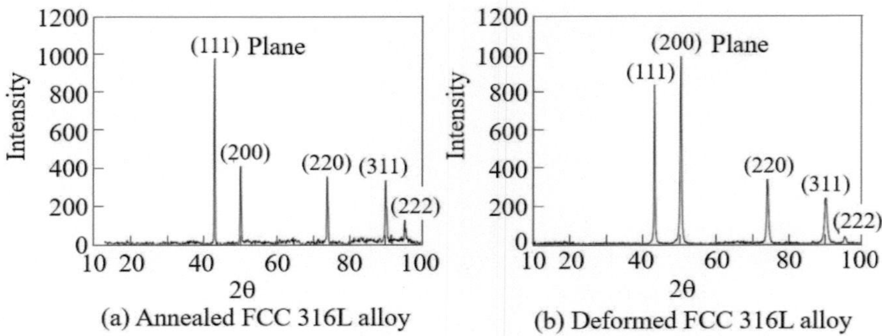

Fig. 6.21 Effect of dislocations on X-ray diffraction patterns of AISI 316L stainless steel samples. (**a**) Annealed and (**b**) mechanically deformed samples. After Pham et al. [39]

X-ray diffraction patterns can also be used to characterize the phases in a microstructure and the effect of mechanical deformation on position (2θ) and relative intensity (I) of peaks. For instance, Fig. 6.21a shows the X-ray diffraction pattern of an annealed 316L FCC austenitic stainless steel and compares with its mechanically deformed counterpart in Fig. 6.21b (Pham et al. [39]). Moreover, the X-ray diffraction patterns for the annealed and deformed samples have the same Bragg's angles corresponding to the identified (hkl) planes.

Clearly, the (111) plane in Fig. 6.21a prevails as the main diffraction peak in the annealed condition. After mechanically deforming the annealed sample, the most significant observation is the increase in intensity of the (200) plane, and a decrease in the (111) and other peaks (Fig. 6.21b). This is attributed to the formation of many dislocations in the deformed sample.

The apparent intensity of the X-ray beam is recorded as counts per second related to the number of electrons striking the target per second and to the spacing between rows of atoms in a specific (hkl) plane. The Bragg's equation relates the X-ray beam wavelength (λ), the d-spacing $d_{(hkl)}$ between parallel (hkl) planes and the diffraction angle (θ). From Eqs. (2.1) or (5.9a),

$$d_{(hkl)} = \frac{\lambda}{2\sin{(\theta)}} \qquad (6.55)$$

All indexed peaks in Fig. 6.21 obey this equation. Clearly, the X-ray patterns indicate that the AISI 316L designation is a crystalline material.

The intensity in both X-ray diffraction patterns is proportional to the number of scatters per unit area of the atomic planes, and the sharp peaks are an indication of the material crystallinity.

Calculation of the stored energy or lattice strain energy per unit volume $U_{\langle hkl \rangle}$ along crystallographic planes can mathematically be predicted using the Stibitz equation [40] for an isotropic material. Thus,

$$U_{\langle hkl \rangle} = \frac{3}{2} \left[\frac{E}{1 + 2v^2} \left(\frac{\Delta d}{d} \right)^2 \right]_{\langle hkl \rangle} = \frac{3}{2} \left[\frac{E \varepsilon^2}{1 + 2v^2} \right]_{\langle hkl \rangle} \qquad (6.56a)$$

$$\frac{\partial d}{\partial \theta} = \frac{\Delta d}{\Delta \theta} = -\frac{n\lambda \cos(\theta)}{2 \sin^2(\theta)} \qquad (6.56b)$$

where the change in d-spacing (Δd) can be determined from the broadening of X-ray diffraction peaks at half maximum. Moreover, $\varepsilon = \Delta d/d$ represents the local microstrain.

According to Zhao et al. [41], the average dislocation density (ρ_d) related to the average elastic-energy per unit length U_d can be calculated using Christien et al. equation [42]

$$\rho_d = \frac{U_{\langle hkl \rangle}}{U_d} = 3 \left[\frac{E}{(Gb^2)(1 + 2v^2)} \left(\frac{\Delta d}{d} \right)^2 \right]_{\langle hkl \rangle} \qquad (6.57a)$$

$$\rho_d = \frac{U_{\langle hkl \rangle}}{U_d} = 3 \left[\frac{E \varepsilon^2}{(Gb^2)(1 + 2v^2)} \right]_{\langle hkl \rangle} \qquad (6.57b)$$

$$U_d = \frac{1}{2} Gb^2 \qquad (6.57c)$$

where $U_{\langle hkl \rangle}$ is the elastic energy of a (hkl) plane.

6.7.2 Spherical Pore Growth Rate

Following Fredriksson and Akerlind idea [43, p. 263] analytical procedure and Liu et al. [44] copper porous images showing pore patterns (Fig. 6.22).

Copper was cast as vertical right-circular cylinders (50-mm diameter and 170-mm height) from the given pouring temperature. These images, published by Liu et al. [44], represent the porous microstructure of the cylinders being cut 20 mm from

Fig. 6.22 Porous copper castings obtained using different pouring temperatures, such as (**a**) $T_p = 1120\,°C$, (**b**) $T_p = 1150\,°C$ and (**c**) $T_p = 1180\,°C$. Images taken from Liu et al. [44]

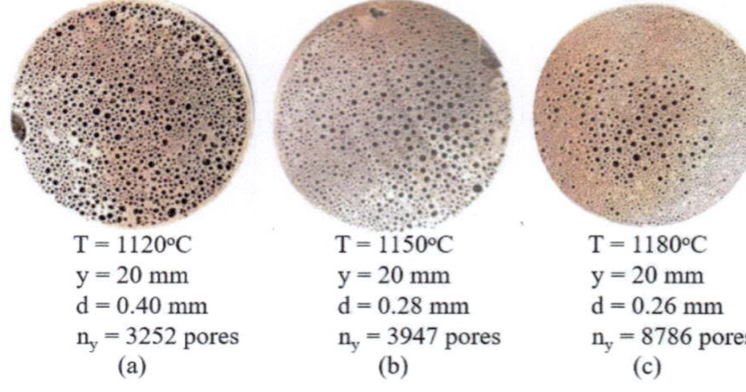

the bottom. The mean average and number of pores of these images are denoted as "d" and "n_y", respectively, as given below Fig. 6.22a for $T_p = 1120\,^\circ C$, Fig. 6.22b for $T_p = 1150\,^\circ C$ and Fig. 6.22c for $T_p = 1180\,^\circ C$.

Assume that dissolved monoatomic hydrogen lacks of equilibrium in the melt and consequently, it transforms to diatomic hydrogen as per the reversible chemical reaction $2H = H_2$, where H represents the dissolved monoatomic hydrogen in the melt. Initially, some molecules associate themselves to form spherical bubbles. Those dissolved hydrogen atoms having a non-equilibrium condition diffuse from the melt into the gas bubble and form hydrogen molecules. Hence, bubble growth at a pressure P_{gas}. This bubble formation and growth process stop near the melt surface as solidification ceases.

Undoubtedly, atomic hydrogen dissolved partially and molecular hydrogen forming bubbles are trapped during solidification of molten copper. These bubbles are revealed as pores on prepared metallographic samples. Moreover, a substantial amount of hydrogen bubbles is supposed to migrate to the melt surface and eventually to the local environment.

Example 6.5 Based on the information, use the non-equilibrium equation of state for gases and Fick's law of steady-state diffusion to derive an expression for the growth rate of spherical pores.

Solution The non-equilibrium equation of state for the number of molecular hydrogen related to a reference pressure P_o and surface energy γ takes the form

$$N_{H_2} = \frac{P_{gas} V_{sphere}}{R_{H_2} T} = \frac{P_{gas} V_s}{RT} \tag{6.5E1}$$

$$N_{H_2} = \frac{(P_o + 2\gamma/r)\left(4\pi r^3/3\right)}{RT} = \frac{4\pi\left(P_o r^3 + 2\gamma r^2\right)}{3RT} \tag{6.5E2}$$

$$\frac{dN_{H_2}}{dr} = \frac{4\pi\left(3P_o r^2 + 4\gamma r\right)}{3RT} \tag{6.5E3}$$

Notice that dN_{H_2}/dr resembles the hydrogen concentration gradient and depends on the bubble radius r and the temperature T.

For steady-state diffusion of monoatomic hydrogen, the diffusion flux is defined two mathematical forms. Thus,

$$J_H = \frac{dN_H}{A_{sphere} dt} = \frac{1}{A_s}\frac{dN_H}{dt} \tag{6.5E4}$$

$$J_H = -D_H \frac{\partial C_H}{\partial r} = -D_H \frac{C - \overline{C}_H}{r} = -D_H \frac{\left(X_H - \overline{X}_H\right)}{r V_m} \tag{6.5E5}$$

$$\frac{\partial C_H}{\partial r} = \frac{X_H - \overline{X}_H}{r V_m} \tag{6.5E6}$$

where V_m is the molar volume. Equating Eqs. (6.5E4) and (6.5E5) yields

$$\frac{dN_H}{dt} = -\frac{A_s D_H \left(X_H - \overline{X}_H\right)}{V_m} \frac{1}{r} = \frac{4\pi r^2 D_H \left(\overline{X}_H - X_H\right)}{V_m} \frac{1}{r} \tag{6.5E7}$$

$$\frac{dN_H}{dt} = \frac{4\pi r D_H \left(\overline{X}_H - X_H\right)}{V_m} \tag{6.5E8}$$

Accordingly, the hydrogen concentration is $C_H = X_H / V_m$, where X_H denotes the mole fraction in the melt, $\overline{X}_H$ denotes an average mole fraction close to the pore surface and V_m denotes the molar volume so that C_H acquires the *mol/volume* units.

Equating the number of hydrogen atoms is twice the number of hydrogen molecules, and using the Chain rule and Eq. (6.5E3) and the yields

$$\frac{dN_H}{dr} = 2\frac{dN_{H_2}}{dr} \tag{6.5E9}$$

$$\frac{dN_H}{dt} = \frac{dN_H}{dr}\frac{dr}{dt} = 2\left(\frac{dN_{H_2}}{dr}\right)\frac{dr}{dt} \tag{6.5E10}$$

Substituting Eq. (6.5E3) into (6.5E10) gives

$$\frac{dN_H}{dt} = \frac{8\pi \left(3P_o r^2 + 4\gamma r\right)}{3RT}\frac{dr}{dt} \tag{6.5E11}$$

Combining Eqs. (6.5E8) and (6.5E11) yields the growth rate of spherical pores

$$\frac{dr}{dt} = \frac{3RT D_H \left(\overline{X}_H - X_H\right)}{2V_m \left(3P_o r + 4\gamma\right)} \tag{6.5E12}$$

which is the theoretical or idealized growth rate of spherical pores. Moreover, the fraction of porosity is defined by (Liu et al. [44])

$$P_r = 1 - \frac{m}{\rho V} \tag{6.5E13}$$

The rate of steady-state nucleation of pores per unit volume (I_p) can be defined by

$$I_p = \frac{n_o k_B T}{h} \exp\left(-\frac{\Delta G_c}{k_B T}\right)\exp\left(-\frac{Q_d}{k_B T}\right) \tag{6.5E14}$$

In summary, dr/dt and I_p are relevant variables that describe some characteristics of the pore formation mechanism. For instance, the former is inversely proportional to the pore radius ($dr/dt \propto 1/r$) and both are strongly dependent on the temperature. Moreover, the reader can use the data provided below the images to perform nonlinear curve fitting.

6.8 Summary

It is clear now that inherent and developed crystal defects are imperfections or discontinuities that govern material properties. The most relevant crystal defects in engineering polycrystalline materials are the grain boundaries, shrinkage pores, inherent flaws, and inclusions in ferrous and non-ferrous. In essence, inclusions may be brittle ceramic phases known as metal oxides (FeO, Al_2O_3), metal sulfates (MnS, Al_2S_3, Cr_2S_3) and chromite (MnO-Cr_2O_3).

The most relevant inherent defects are point defects and linear defects (dislocations), which may be inherently present in a crystalline solid induced by rapid cooling and may be developed during fabrication or mechanical deformation. Moreover, pores are also defects induced by the formation of bubbles in the melt.

Appendix 6A Airy Stress Function

The goal in this Appendix is to derive the Airy stress function (ϕ) for an edge dislocation as shown in Fig. 6.11c. According to the theory of elasticity, the stress field around an edge dislocation is developed using the definition of Airy stresses in two dimensions.

$$\sigma_x = \frac{\partial^2 \phi}{\partial y^2} \quad \sigma_y = \frac{\partial^2 \phi}{\partial x^2} \quad \tau_{xy} = -\frac{\partial^2 \phi}{\partial x \partial y} \tag{6.58}$$

A selected Airy stress function must satisfy the governing fourth-order partial differential biharmonic equation or operator ∇^4 in rectangular (Cartesian) and polar coordinates given by

$$\nabla^4 = \frac{\partial^4 \phi}{\partial x^4} + \frac{\partial^4 \phi}{\partial x^2 \partial y^2} + \frac{\partial^4 \phi}{\partial y^4} = 0 \tag{6.59a}$$

$$\nabla^2 = \frac{\partial^2 \phi}{\partial r^2} + \frac{1}{r}\frac{\partial \phi}{\partial r} + \frac{1}{r^2}\frac{\partial^2 \phi}{\partial \theta^2} = 0 \tag{6.59b}$$

The general solution of the Airy stress function $\phi = \phi(x, y, \theta)$ or $\phi = \phi(r, \theta)$ can be found using the separation of variables method (SVM). Let $P = P(r)$ and $Q = Q(\theta)$ be arbitrary functions in polar coordinates. Thus, SVM dictates that

$$\phi(r,\theta) = P(r)\,Q(\theta) \tag{6.60a}$$

$$\phi = PQ \tag{6.60b}$$

where Eq. (6.60a) separates $\phi(r,\theta)$ into one-variable dependency functions; $P(r)$ and $Q(\theta)$. For clarity, Eq. (6.60b) is a convenient notation for developing the analytical procedure for finding the general solutions of ϕ and the corresponding stresses.

Take the second-order partial derivatives of Eq. (6.60) and substitute the resultant expressions into Eq. (6.59b) to get

$$Q\frac{\partial^2 P}{\partial r^2} + \frac{Q}{r}\frac{\partial P}{\partial r} + \frac{P}{r^2}\frac{\partial^2 Q}{\partial \theta^2} = 0 \tag{6.61}$$

Separating the terms gives

$$\frac{r^2}{P}\frac{\partial^2 P}{\partial r^2} + \frac{r}{P}\frac{\partial P}{\partial r} = -\frac{1}{Q}\frac{\partial^2 Q}{\partial \theta^2} = k^2 \tag{6.62}$$

where k^2 is an arbitrary constant so that

$$\frac{r^2}{P}\frac{\partial^2 P}{\partial r^2} + \frac{r}{P}\frac{\partial P}{\partial r} = k^2 \tag{6.63a}$$

$$-\frac{1}{Q}\frac{\partial^2 Q}{\partial \theta^2} = k^2 \tag{6.63b}$$

or

$$r^2\frac{\partial^2 P}{\partial r^2} + r\frac{\partial P}{\partial r} = k^2 P \tag{6.64a}$$

$$-\frac{\partial^2 Q}{\partial \theta^2} = k^2 Q \tag{6.64b}$$

Making these expressions as homogeneous equations yields

$$r^2\frac{\partial^2 P}{\partial r^2} + r\frac{\partial P}{\partial r} - k^2 P = 0 \tag{6.65a}$$

$$\frac{\partial^2 Q}{\partial \theta^2} + k^2 Q = 0 \tag{6.65b}$$

Method A The general solution of these homogeneous equations is

$$P(r) = \begin{cases} A_1 + A_2 \ln(r) & \text{for } k = 0 \\ B_1 r^k + B_2 r^{-k} & \text{for } k \neq 0 \end{cases} \tag{6.66a}$$

$$Q\left(\theta\right) = \begin{cases} C_1 + C_2\theta & \text{for } k = 0 \\ D_1 \cos\left(k\theta\right) + D_2 \sin\left(k\theta\right) & \text{for } k \neq 0 \end{cases} \tag{6.66b}$$

Hence, Eq. (6.60) becomes the general Airy stress function given by

$$\phi = \left[A_1 + A_2 \ln\left(r\right)\right]\left[C_1 + C_2\theta\right] \tag{6.67}$$

$$+ \left[B_1 r^k + B_2 r^{-k}\right]\left[D_1 \cos\left(k\theta\right) + D_2 \sin\left(k\theta\right)\right]$$

According to Fig. 6.11c, a qualitatively analysis indicates that $\phi = \sigma_x + \sigma_y$ has a maximum value at $\theta = -\pi/2$ and a minimum at $\theta = +\pi/2$, and $\phi = f\left(r\right)$ decreases with increasing r. This implies that Eq. (6.67) is significantly simplified by letting $k = 1$ so that $\sin\left(\pm\pi/2\right) = \pm 1$ and $\cos\left(\pm\pi/2\right) = 0$.

Taking the relevant partial derivatives of Eq. (6.67) gives

$$\frac{\partial\phi}{\partial\theta} = \left[A_1 + A_2 \ln\left(r\right)\right]\left[C_1 + C_2\right] \tag{6.68a}$$

$$+ \left[B_1 r^k + B_2 r^{-k}\right]\left[-kD_1 \sin\left(k\theta\right) + kD_2 \cos\left(k\theta\right)\right]$$

$$\frac{\partial\phi}{\partial r} = A_2 r^{-1}\left[C_1 + C_2\right] \tag{6.68b}$$

$$+ \left[B_1 k r^{k-1} - B_2 \frac{k}{r^{k+1}}\right]\left[D_1 \cos\left(k\theta\right) + D_2 \sin\left(k\theta\right)\right]$$

Suitable boundary conditions (BCs) based on the knowledge of a system are

$$\left[\frac{\partial\phi}{\partial\theta}\right]_{\theta=-\pi/2} = 0 \tag{6.69a}$$

$$\frac{\partial\phi}{\partial r} < 0 \tag{6.69b}$$

For $k = 1$ and $\theta = -\pi/2$,

$$\left[\frac{\partial\phi}{\partial\theta}\right]_{\theta=-\pi/2} = 0 = \left[A_1 + A_2 \ln\left(r\right)\right]\left[C_1 + C_2\right] + \left[B_1 r^k + B_2 r^{-k}\right]\left[kD_1\right]$$

$$\tag{6.70}$$

By inspection, Eq. (6.70) is satisfied if $C_1 = C_2 = D_1 = 0$. Thus, Eq. (6.67) becomes

$$\phi = \left[B_1 r^k + B_2 r^{-k}\right]\left[D_2 \sin\left(k\theta\right)\right] \tag{6.71}$$

which has three unknown components.

Differentiating Eq. (6.71) with respect to r yields for the limiting case

$$\frac{\partial \phi}{\partial r} = \left[B_1 k r^{k-1} - B_2 k r^{-k-1} \right] [D_2 \sin (k\theta)] = 0 \tag{6.72a}$$

$$0 = B_1 r^{k-1} - B_2 r^{-k-1} \tag{6.72b}$$

$$B_1 = \frac{B_2}{r^{k+1} r^{k-1}} = \frac{B_2}{r^{2k}} \tag{6.72c}$$

Substituting Eq. (6.72c) into (6.71) yields the Airy stress function as

$$\phi = 2 B_2 D_2 r^{-k} \sin (k\theta) = 2 D_3 r^{-1} \sin (k\theta) \tag{6.73}$$

Equating Laplace equation (harmonic equation) and Airy stress function

$$\nabla^2 \phi = \phi \tag{6.74a}$$

$$\frac{\partial^2 \phi}{\partial r^2} + \frac{1}{r} \frac{\partial \phi}{\partial r} + \frac{1}{r^2} \frac{\partial^2 \phi}{\partial \theta^2} = 2 D_3 r^{-1} \sin (k\theta) \tag{6.74b}$$

which is a non-homogeneous differential equation with a solution of the form

$$\phi = \left(\frac{D_3}{2} \right) r \sin (\theta) \ln (r) \tag{6.75}$$

In polar coordinates,

$$y = r \sin (\theta) \tag{6.76a}$$

$$r = \left(x^2 + y^2 \right)^{1/2} \tag{6.76b}$$

Inserting Eq. (6.76) into (6.75) gives the Airy function defined by

$$\phi = \left(\frac{D_3}{2} \right) y \ln \left[\left(x^2 + y^2 \right)^{1/2} \right] = \left(\frac{D_3}{2} \right) r \sin \theta \ln (r) \tag{6.77a}$$

$$\sigma_x = \frac{\partial^2 \phi}{\partial y^2} = \frac{D_3 y}{2} \left[\frac{3x^2 + y^2}{\left(x^2 + y^2 \right)^2} \right] \tag{6.77b}$$

$$\sigma_y = \frac{\partial^2 \phi}{\partial x^2} = \frac{D_3 y}{2} \left[\frac{y^2 - x^2}{\left(x^2 + y^2 \right)^2} \right] \tag{6.77c}$$

According to the elasticity theory, the plane-strain condition and the relevant nonzero elastic stress and strain are

$$\sigma_z = v\left(\sigma_x + \sigma_y\right) \tag{6.78a}$$

$$\varepsilon_x = v\left(\sigma_y + \sigma_z\right) \tag{6.78b}$$

$$\varepsilon_x = \frac{1}{E}\left[\sigma_x - v\left(\sigma_y + \sigma_z\right)\right] = \frac{1}{E}\left[\left\{\sigma_x - v\left[\sigma_y + v\left(\sigma_x + \sigma_y\right)\right]\right\}\right] \tag{6.78c}$$

$$\varepsilon_x = \frac{(1+v)}{E}\left[(1-v)\,\sigma_x - v\left(1+v\right)\sigma_y\right] \tag{6.78d}$$

Combining Eqs. (6.77b), (6.77c) and (6.78d) yields

$$\varepsilon_x = \frac{(1+v)}{E}\left[(1-v)\frac{\partial^2\phi}{\partial y^2} - v\frac{\partial^2\phi}{\partial x^2}\right] \tag{6.79}$$

and

$$\varepsilon_x = \frac{D_3\,(1+v)\,y}{2E}\left[\frac{2\,(1-v)}{\left(x^2+y^2\right)^2} + \frac{x^2 - y^2}{\left(x^2+y^2\right)^2}\right] \tag{6.80}$$

Defining the magnitude of a Burgers vector as the integral difference in elastic strains above and below the dislocation (hkl) slip plane gives

$$b = \int_{-\infty}^{\infty}\left(\frac{\partial\mu}{\partial s}\right)ds = \int_{-\infty}^{\infty}\left[\varepsilon_x\left(x,\,y\right) - \varepsilon\left(x,\,-y\right)\right]dx \tag{6.81a}$$

$$b = 2\int_{-\infty}^{\infty}\varepsilon_x\left(x,\,y\right)dx \tag{6.81b}$$

$$b = 2\int_{-\infty}^{\infty}\frac{D_3\,(1+v)\,y}{2E}\left[\frac{2\,(1-v)}{\left(x^2+y^2\right)^2} + \frac{x^2 - y^2}{\left(x^2+y^2\right)^2}\right]dx$$

Substituting Eq. (6.80) into (6.81b) with $E = 2G\,(1+v)$ and integrating yield

$$b = \frac{D_3}{2G}\int_{-\infty}^{\infty}y\left[\frac{2\,(1-v)}{\left(x^2+y^2\right)^2} + \frac{x^2 - y^2}{\left(x^2+y^2\right)^2}\right]dx \tag{6.82a}$$

$$b = -\frac{\pi\,(1-v)\,D_3}{G} \tag{6.82b}$$

$$D_3 = -\frac{Gb}{\pi\,(1-v)} \tag{6.82c}$$

Substitute Eq. (6.82c) into (6.75) to get (6.11a) as the simplified Airy stress function being used through the literature. Thus,

$$\phi = -\left[\frac{Gb}{2\pi\,(1-v)}\right] y \ln\left(x^2 + y^2\right) \tag{6.83a}$$

$$\phi = -\lambda y \ln\left(x^2 + y^2\right) \tag{6.83b}$$

$$\phi = -\lambda r \sin\left(\theta\right) \ln\left(r\right) \tag{6.83c}$$

Method B According to Hirth and Lothe [3, p. 74], the general Airy stress function can also be written as

$$\phi = \sum_k P_k\left(r\right) Q_k\left(\theta\right) \tag{6.84a}$$

$$\phi = \left[A_1 + A_2 \ln\left(r\right)\right] + \sum_{k=1}^{\infty}\left(B_k r^k + C_k r^{-k}\right) \sin\left(k\theta\right) \tag{6.84b}$$

$$+ \sum_{k=1}^{\infty}\left(D_k r^k + E_k r^{-k}\right) \cos\left(k\theta\right) \tag{6.84c}$$

Again, a qualitatively analysis of Fig. 6.11c suggests a maximum for $\phi = \sigma_x + \sigma_y$ at $\theta = -\pi/2$ and a minimum at $\theta = +\pi/2$, and $\phi = f\left(r\right)$ decreases with increasing r. This implies that Eq. (6.75b) can significantly be simplified by letting $k = 1$ so that $\sin\left(\pm\pi/2\right) = \pm 1$ and $\cos\left(\pm\pi/2\right) = 0$. Hence, the Airy stress function that obeys this behavior is of the form

$$\phi = C_1 r^{-1} \sin\left(\theta\right) \tag{6.85}$$

Hence,

$$\frac{\partial^2 \phi}{\partial r^2} + \frac{1}{r}\frac{\partial \phi}{\partial r} + \frac{1}{r^2}\frac{\partial^2 \phi}{\partial \theta^2} = C_1 r^{-1} \sin\left(\theta\right) \tag{6.86}$$

A particular solution of Eq. (6.86) is

$$\phi = \frac{C_1}{2} r \sin\left(\theta\right) \ln\left(r\right) \tag{6.87}$$

The corresponding Burgers vector is

$$b = -\int_{-\infty}^{+\infty} \left[\varepsilon_x\left(x, k\right) - \varepsilon_x\left(x, -k\right)\right] dx = -\frac{C_1}{2}\frac{2\pi\left(1-v\right)}{G} \tag{6.88a}$$

$$C_1 = -\frac{Gb}{\pi \, (1-v)} \tag{6.88b}$$

Inserting Eq. (6.88b) into (6.87) yields Eq. (6.11a) as

$$\phi = -\frac{Gb}{2\pi \, (1-v)} r \sin(\theta) \ln(r) = -\lambda r \sin(\theta) \ln(r) \tag{6.89a}$$

$$\phi = -\left[\frac{Gb}{2\pi \, (1-v)}\right] y \ln\left(x^2 + y^2\right) = -\lambda y \ln\left(x^2 + y^2\right) \tag{6.89b}$$

with the constant λ being defined by Eq. (6.11b) as

$$\lambda = \frac{Gb}{2\pi \, (1-v)} \tag{6.90}$$

This concludes the derivation of an Airy stress function for dislocations.

Problems

6.1 Assume that the vacancy formation at high temperature is mathematically determined by the Arrhenius equation, Eq. (6.1a). Show that $Q_v = \Delta H_v$.

6.2 Calculate **(a)** the number of vacancies per unit meter N_v and **(b)** the fractional vacancy concentration C_v in iron (Fe) at $860\,^\circ C$. Assume that the activation energy for vacancy formation is $1\ eV/atom$ and the mass density of iron at this temperature is $7.60\ g/cm^3$. The atomic weight for Fe is $55.85\ g/mol$. [Solution: (a) $N_v = 2.93 \times 10^{24}\ vac./m^3$ and (b) $C_v \simeq 4 \times 10^{-5}$].

6.3 Calculate the enthalpy of vacancy formation or the activation energy (in eV unit) for vacancy formation if the normalized vacancy fraction is $C_v = 4 \times 10^{-5}$ at $T_1 = 1300\,^\circ C$ and $C_v = 3 \times 10^{-4}$ at $T_2 = 2000\,^\circ C$. [Solution: $Q_v = 0.89\ eV$].

6.4 For $N_{v,1} = 2 \times 10^{22}\ vacancies/m^3$ at $T_1 = 750\,^\circ C = 1023\ K$ and $N_{v,1} = 3 \times 10^{24}\ vacancies/m^3$ at $T_1 = 950\,^\circ C = 1223\ K$, calculate **(a)** the number of vacancies $N_{v,3}$ and **(b)** the fractional vacancy concentration $C_{v,3}$ at $T_3 = 850\,^\circ C = 1123\ K$. [Solution: (a) $N_{v,3} = 3 \times 10^{23}\ vacancies/m^3$ and (b) $C_{v,3} = 7 \times 10^{-13}$].

6.5 Calculate **(a)** the equilibrium concentration of vacancies per cubic meter N_v and **(b)** the fractional vacancy concentration C_v in pure copper at $1000\,^\circ C$. Assume that the energy of formation of a vacancy in pure copper is $1\ eV$. **(c)** Calculate is the vacancy fraction at $1050\,^\circ C$ knowing that the melting point of copper $1084\,^\circ C$. Data: the Avogadro's number $N_A = 6.022 \times 10^{23}\ atoms/mol$, Boltzmann's constant $k_B = 8.62 \times 10-5\ eV/atom.K$, mass density of copper $\rho = 8.96\ g/cm^3$ and atomic weight $A_w = 63.54\ g/mol$. [Solution: (a) $N_v = 9.36 \times 10^{24}\ vac./m^3$, (b) $C_v \simeq 10^{-4}$ (c) $C_v \simeq 2 \times 10^{-4}$].

6.6 Assume that a hypothetical straight dislocation dissociates as indicated by the stationary reaction

$$\frac{a}{2}\,[110] = \frac{a}{4}\,[110] + \frac{a}{4}\,[1\bar{1}0] \qquad\qquad (6.2\text{E}1)$$

Determine **(a)** if the reaction is correct as written and **(b)** if it really can dissociate. [Solution: (a) correct and (b) yes].

6.7 Assume that the combination of the leading partial dislocation lying in the (111) plane with that which lies in the $(\bar{1}1\bar{1})$ plane forms another partial dislocation. Determine if the following dislocation reaction is correct as written [Solution: It is correct].

$$\frac{a}{6}\,[\bar{1}12] + \frac{a}{6}\,[2\bar{1}\bar{1}] \rightarrow \frac{a}{6}\,[101]$$

6.8 Determine if the following dislocation reaction is possible in BCC crystals. [Solution: It is correct].

$$\frac{a}{2}\,[1\bar{1}1] + \frac{a}{2}\,[11\bar{1}] \rightarrow a\,[100]$$

6.9 Consider the $[1\bar{1}0]$ direction of a dislocation in a copper specimen. Calculate **(a)** the magnitude of the Burgers vector, **(b) the radius** r if $\ln(r/r_o) \simeq 4\pi$ (if the lattice parameter of copper and the dislocation core radius are $a = 0.36\ nm$ and $r_o = 5b$ respectively), and **(c)** the elastic screw dislocation energy U_s when the shear modulus is $G = 48\ GPa$. [Solution: (a) $b = 0.25\ nm$, (b) $r = 3.58 \times 10^5\ nm$ and (c) $U_s = 3 \times 10^{-9}\ J/m$].

6.10 Consider edge, screw, and mixed dislocations in a diluted aluminum alloy. Use the given data to calculate **(a)** the radius r of a cylindrical dislocation circuit and **(b)** the energy U_{mix} for the mixed dislocation scheme. Here, U_e is the elastic edge-dislocation energy and L is the dislocation length. [Solution: (a) $r = 10^5 b = 30,000\ nm$ and (b) $U_{mix} = 6.67 \times 10^{-16}\ J$].

$G = 30\ GPa$	$b = 0.3\ nm$	$v = 1/3$
$U_e = 4 \times 10^{-9}\ J/m$	$r_o = 0.4b$	$L = 100\ nm$

6.11 Consider the splitting of $(a/2)\langle110\rangle$ in Al into Shockley partial dislocations. **(a)** Show mathematically that this is a 60°-dislocation and **(b)** calculate the width (d) of the stacking fault. Data: $a = 0.405\ nm$, $\gamma_{SF} = 2 \times 10^{-5}\ J/cm^2$ and $G = 25\ GPa$. [Solution: (b) $d = 0.544\ nm$].

6.12 Consider a hypothetical cubic crystal being subjected to an external load to produce plastic deformation. Subsequent determination of mechanical properties provided a shear modulus of 45 GPa. On the other hand, microscopic measurements revealed an average stacking fault layer having an equilibrium spacing of 2 nm and a lattice parameter of 0.37 nm. Assume that Shockley partial dislocations were produced on an intersecting slip plane according to the following reaction

$$\frac{a}{3}[111] \rightarrow \frac{a}{6}[101] + \frac{a}{6}[121]$$

Determine **(a)** whether or not this dislocation reaction occurs. If the reaction occurs (dislocation splitting), then determine **(b)** the angle θ between the partial dislocations and **(c)** the intrinsic stacking fault energy γ_{SF}. [Solution: (a) yes and (b) $\gamma_{SF} = 4.72 \times 10^{-2}\ J/m^2$].

6.13 Determine the intrinsic stacking fault energy γ_{SF} for two leading screw dislocation reactions in an FCC-aluminum (Al) specimen. Assume a constant stacking fault equilibrium spacing of 2.5 nm for the following reactions that produce Shockley partial screw dislocations

$$\frac{a}{2}\left[1\bar{1}0\right] = \frac{a}{6}\left[1\bar{2}1\right] + \frac{a}{6}\left[2\bar{1}1\right] \text{ on the (111) plane}$$

$$\frac{a}{2}\left[1\bar{1}0\right] = \frac{a}{6}\left[2\bar{1}1\right] + \frac{a}{6}\left[1\bar{2}\bar{1}\right] \text{ on the } \left(\bar{1}11\right) \text{ plane}$$

Also assume that the lattice parameter and the shear modulus are 0.405 nm and 24 GPa, respectively, and that the perfect screw dislocations in Al can split at a nodal point (hard particle) located on the (111) and $\left(\bar{1}11\right)$ planes. [Solution: $\gamma_{SF} = 3.48 \times 10^{-2}\ J/m^2$ on (111) plane and $\gamma_{SF} = 2.09 \times 10^{-2}\ J/m^2$ on $\left(\bar{1}11\right)$ plane].

6.14 Consider one screw dislocation. Calculate **(a)** the radius r of the distortion circuit by assuming that $\ln(r/r_o) \simeq 4\pi$ and $r_o = 6b$ and **(b)** the elastic energy U_s for copper in the [101] direction of the dislocation. Let the lattice parameter of copper, the dislocation core radius, and the shear modulus $a = 0.36\ nm$, $r_o = 4b$ and $G = 48\ GPa$, respectively. [Solution: (a) $r = 2.8675 \times 10^5\ nm$ and (b) $U_s = 3 \times 10^{-9}\ J/m$].

6.15 Consider the concept of a perfect dislocation with the shortest lattice vector **b** in cubic crystals. If the lattice parameter for Al and Fe are 0.405 nm and 0.287 nm, respectively, then determine **(a)** which metal has the largest $|\mathbf{b}|$ value. **(b)** If a mechanical force is applied along the z-axis, then draw the slip systems for each cubic structure using unit cells. [Solution: (a) $b_{FCC} = 0.286\ nm$].

6.16 Consider an FCC-Ni plate being plastically deformed at room temperature. It is well established that the mechanism of plastic deformation is by slip. Assume that three perfect screw dislocations with Burgers vectors $\mathbf{b}_{[100]}$, $\mathbf{b}_{[110]}$ and $\mathbf{b}_{[111]}$ may be form. **(a)** Draw these Burgers vectors in one FCC unit cell and subsequently, calculate the dislocation energy for each case. **(b)** Deduce which dislocation is the most likely to occur first. Recall that the Burgers vector points in the direction having the highest linear density. Once this is accomplished, find the dislocation energy expressions relative to this first dislocation for the other two dislocations. Data: $G = 72\ MPa$ and $a = 0.353\ nm$. [Solution: (b) $\mathbf{b} = a\,[110]$].

6.17 It is established that crystal plasticity is characterized by the onset of dislocation glide events. For cubic crystals, **(a)** draw side by side three unit cells representing for the SC, FCC, and BCC structures. Also draw the most likely lattice translation vector (Burgers vector $\mathbf{b}$) for the onset of a principal dislocation glide known as a perfect dislocation in each cubic crystal. **(b)** Write down the Burgers vectors and its corresponding magnitudes as mathematical expressions for each cubic crystal. **(c)** For perfect dislocations along the these Burgers vector pointing in the direction having the highest linear density, write down the mathematical expressions for the screw dislocation energy U_s and identify the highest screw dislocation energy.

6.18 Use the given data for a cold-worked copper alloy (Cu-alloy) to calculate **(a)** the mix dislocation strain energy and the shear stress. Assume constant average Burgers vector b and Poisson's ratio v. [Solution: (a) $U_{mix} = 6.33 \times 10^{10}\ eV/m$ (b) $\tau = 123.73\ MPa$].

$G = 45\ GPa$	$b = 0.30\ nm$	$v = 1/3$

6.19 Consider a plastically deformed 316L stainless steel specimen. If the dislocation density, the shear modulus, and the shear stress are $\rho_{316L} = 15 \times 10^{14}\ m^{-2}$, $78\ GPa$ and $125\ MPa$, respectively, calculate the screw dislocation energy U_s. [Solution: $U_s = 2.16 \times 10^9\ eV/m$].

6.20 Show that the screw dislocation energy is depends on the shear stress and the dislocation density. [Hint: $U_s \propto \tau/\sqrt{\rho}$].

6.21 Suppose that an FCC crystal is deformed to an extent so that an (hkl) has a dislocation along the $[\bar{1}01]$ and a Burgers vector $b = (a/2)\,[0\bar{1}1]$. **(a)** Determine the type of dislocation and **(b)** find the Miller indices of the (hkl) slip plane from the set of $\{111\}$ slip planes. [Solution: (a) mix dislocation and (b) $(hkl) = (111)$].

References

1. Calculation of the Properties of Vacancies and Interstitials, Proceedings of a Conference, National Bureau of Standards Miscellaneous Publication 287, Library of Congress Catalog Card Number: 67-60032. Shenandoah National Park, Va. (1966)
2. J.A. Frenkel, Zur theorie der elastizitätsgrenze und der festigkeit kristallinischer körper. Zeitschrift für Physik. Google translation from German: On the theory of the elastic limit and the strength of crystalline bodies. Mag. Phys. **37**(7–8), 572–609 (1926)
3. J.P. Hirth, J. Lothe, *Theory of Dislocations* (Wiley, New York, 1982)

4. G. Masing, M. Polanyi, Kaltreckung und Verfestigung, in *Ergebnisse der Exakten Naturwissenschaften*. Google translation from German: Cold stretching and consolidation, in *Results of Exact Science* (Springer, Berlin, 1923), pp. 177–245
5. L. Prandtl, A conceptual model to the Kinetic theory of solid bodies. Z. Angew. Math. Mech **8**, 85–106 (1928)
6. U. Dehlinger, Theory of recrystallization of pure metals. Ann. Phys. (Leipzig) **2**, 749 (1929)
7. E. Orowan, Plasticity of crystals. Z. Phys. **89**(9–10), 605–659 (1934)
8. M. Polanyi, On a kind of glide disturbance that could make a crystal plastic. Z. Phys. **89**, 660 (1934)
9. G.I. Taylor, Atomic mode of a dislocation - the mechanism of plastic deformation of crystals. Proc. R. Soc. A **145**, 362–387 (1934)
10. S.F. Edwards, A. Mehta, Dislocations in amorphous materials. J. de Physique **50**(18), 2489–2503 (1989)
11. A.T. Blumenau, The Modeling of Dislocations in Semiconductor Crystals. Dissertation, Department of Physics of the Faculty of Natural Sciences of the University of Paderborn, Paderborn, Germany, 2002
12. J. Friedel, *Dislocations* (Pergamon Press, Addison-Wesley Publishing Company, Reading, 1967)
13. A.M. Kosevich, Crystal dislocations and the theory of elasticity, in *Dislocations in Solids*, ed. by F.R.N. Nabarro (North-Holland Publishing Company, New York, 1979)
14. . I. Kovacs, L. Zsoldosi, *Dislocations and Plastic Deformation* (Pergamon Press Ltd, Newe York, 1973)
15. W.T. Read, *Dislocations in Crystals* (McGraw-Hill, New York, 1953)
16. L. Bragg, J.F. Nye, A dynamical model of a crystal structure, in *Proceedings of the Royal Society of London*. Series A, Mathematical and Physical Sciences, vol. 190, no. 1023 (1947) pp. 474–481
17. D.F. Stein, interview in *"This Week's Citation Classic"* (1980) http://garfield.library.upenn.edu/classics1980/A1980KB37200001.pdf
18. W.G. Johnston, J.J. Gilman, Dislocation velocities, dislocation densities, and plastic flow in lithium fluoride crystals. J. Appl. Phys. **30**(2), 129–144 (1959)
19. L. Wang, L. Yang, X. Dong, X. Jiang, *Dynamics of Materials: Experiments, Models and Applications* (Academic, New York, 2019)
20. M. Lazar, G.A. Maugin, E.C. Aifantis, Dislocations in second strain gradient elasticity. Int. J. Solids Struct. **43**(6), 1787–1817 (2006)
21. J. Weertman, *Dislocation Based Fracture Mechanics* (World Scientific Publishing Company, New York, 1996)
22. J. Dundurs, Elastic interaction of dislocations with inhomogeneities, in *Symposium on the Mathematical Theory of Dislocations* (1969), pp. 70–115
23. N.F. Mott, F.R.N. Nabarro, *Report on Strength of Solids* (The Physical Society, London, 1948) pp. 1–19
24. D. Hull, D.J. Bacon, *Introduction to Dislocations*, 5th edn. (Butterworth-Heinemann, Elsevier, New York, 2011)
25. W.D. Callister, Jr., D.G. Rethwisch, *Materials Science and Engineering: An Introduction*, 10th edn. (Wiley, New York, 2018)
26. W.F. Hosford, *Mechanical Behavior of Materials*, 2nd edn. (Cambridge University Press, New York, 2010)
27. G.I. Taylor, Plastic strain in metals. J. Inst. Metals **62**, 307–324 (1938)
28. F.R.N. Nabarro, Z.S. Basinski, D.B. Holt, The plasticity of pure single crystals. Adv Phys **13**, 193–323 (1964)
29. J. Gubicza, Lattice defects and their influence on the mechanical properties of bulk materials processed by severe plastic deformation. Mat. Trans. **60**(7), 1230–1242 (2019)
30. Z. Ding, S. Li, W. Liu, Y. Zhao, Modeling of stacking fault energy density in hexagonal-close-packed metals, in *Advances in Materials Science and Engineering* (2015), pp. 1–8

31. S. Zhao, G.M. Stocks, Y. Zhang, Stacking fault energies of face-centered cubic concentrated solid solution alloys. Acta Materialia **134**, 334–345 (2017)
32. Q. Chen, Evolution, Interaction, and Intrinsic Properties of Dislocations in Intermetallics: Anisotropic 3D Dislocation Dynamics Approach, Ph.D. Dissertation, Iowa State University, Ames, Iowa, 2008
33. R. Spatschek, A. Adland, A. Karma, Structural short-range forces between solid-melt interfaces. Phys. Rev. B **87**(2), 024109 (2013)
34. S.K. Hyun, T. Ikeda, H. Nakajima, Fabrication of lotus-type porous iron and its mechanical properties. Sci. Technol. Adv. Mat. **5**(1–2), 201–205 (2004)
35. A. Sola, A. Nouri, Microstructural porosity in additive manufacturing: the formation and detection of pores in metal parts fabricated by powder bed fusion. J. Adv. Manuf. Process. **1**(3), e10021 (2019)
36. E. Ol'khovik, Study of the effect of shrinkage porosity on strength low carbon cast steel, in *IOP Conference Series: Materials Science and Engineering*, vol. 91, no. 1 (2015), p. 012022
37. W. Xu, X. Lu, B. Zhang, C. Liu, S. Lv, S. Yang, X. Qu, Effects of porosity on mechanical properties and corrosion resistances of PM-fabricated porous Ti-10Mo alloy. Metals **8**(3), 188 (2018)
38. N. Chawla, X. Deng, Microstructure and mechanical behavior of porous sintered steels. Mat. Sci. Eng. A **390**, 98–112 (2005)
39. M.S. Pham, B. Dovgyy, P.A. Hooper, Twinning induced plasticity in austenitic stainless steel 316L made by additive manufacturing. Mat. Sci. Eng. A **704**, 102–111 (2017)
40. G.R. Stibitz, Energy of lattice distortion. Phys. Rev. **49**, 859 (1936)
41. G.H. Zhao, X.Z. Liang, B. Kim, P.E.J. Rivera-Díaz-del-Castillo, Modeling strengthening mechanisms in beta-type Ti alloys. Mater. Sci. Eng. A **756**, 156–160 (2019)
42. F. Christien, M.T.F. Telling, K.S. Knight, Neutron diffraction in-situ monitoring of the dislocation density during martensitic transformation in a stainless steel. Scripta Materialia **68**(7), 506–509 (2013)
43. H. Fredriksson, U. Akerlind, *Materials Processing During Casting* (Wiley, Chichester, 2006)
44. X. Liu, X. Li, Y. Jiang, J. Xie, Effect of casting temperature on porous structure of lotus-type porous copper. Procedia Eng. **27**, 490–501 (2012)

Chapter 7
Mass Transport

7.1 Introduction

The crystal structure of some crystalline materials, such as steel and silicon, is sensitive to heat treatment processes in specific gaseous environments since the atomic structure of thin surface layers can be altered by diffusion of interstitial atoms during heat treatment at relatively high temperatures and time $t > 0$.

Diffusion is a mass transport mechanism due to the motion of atoms, molecules, or ions into a medium. This is an atomic process related to the random walk theory, which is a statistical physics model used for understanding the kinetics of particles (atom, molecular, or ion) diffusion. The random walk model gives insight of the random collisions of species while traveling to a destination within a solid, liquid, and gas media, and a composite medium such as the human body. Nonetheless, the diffusion of a species j into a substrate is an interdiffusion process due to interstitial or substitutional atoms undergoing diffusion.

In physics and materials science, the mass transport by diffusion is a fundamentally random thermally activated motion of atoms in a solid crystal structure (lattice). The source of diffusing atoms can be from a solid in contact with a solid host or from a gas containing the diffusing solute. Thus, solid–solid or gas–solid diffusion is the most common terminology in the fields of science. Nonetheless, the mechanism for mass transport of impurity atoms into a solid is generally referred to as interdiffusion or impurity diffusion, which is classified as either interstitial diffusion if the impurity atoms are smaller than the host solid or substitutional diffusion if the impurity and host atoms have the same or similar size.

The main objective in this chapter is to introduce mathematical models for describing mass transport (1) by atomic diffusion under steady-state and transient conditions as defined by the Fick's laws of diffusion, (2) by ionic migration and electron flow within an electric field in electrochemical systems (electrical energy–related cells), and (3) by convective motion of a fluid in a certain environment. It

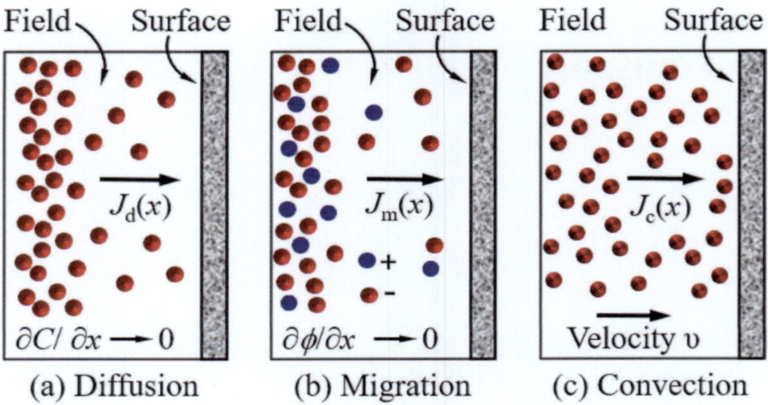

Fig. 7.1 Ideal modes of mass transport. One-dimensional (**a**) diffusion flux $J_d(x)$ due to a concentration gradient $\partial C/\partial x$, (**b**) migration due to a potential gradient $\partial \phi/\partial x$ and (**c**) convection due to a mass flow kinematic velocity υ. After J.T. Maloy [3]

should be mentioned that most of the theoretical background in this chapter can be found elsewhere (Perez [1, 2]).

7.2   Modes of Mass Transport

Consider the a one-dimensional mass transport of a species j under the steady-state assumption so that a one-dimensional analysis suffices to characterize the molar flux $J(x)$. This is schematically shown in Fig. 7.1a for diffusion, Fig. 7.1b for migration and Fig. 7.1c for convection, as the mass transport modes (after Maloy [3] as cited by Wang [4, p. 5]).

The total molar flux $J(x)$ or mass flux $J^*(x)$ can be defined as

$$J(x) = J_x = \sum J_k(x) \tag{7.1a}$$

$$J^*(x) = J_x^* = \sum J_k(x)\, A_{w,k} \tag{7.1b}$$

where $A_{w,k}$ is the atomic weight of the species, ion or particle k and $k = 1, 2, 3, 4, \ldots$ represents the number of species. For mass transport along the stationary x-direction under steady-state condition (Fig. 7.1), the concentration rate is $dC_j/dt = 0$ and the total molar flux J_x of a species j is defined as (Wang [4, p. 5], Bard and Faulkner [5, p. 29])

$$J_x = -D\frac{dC}{dx} - \frac{zFDC}{RT}\frac{d\phi}{dx} + C\upsilon \tag{7.2a}$$

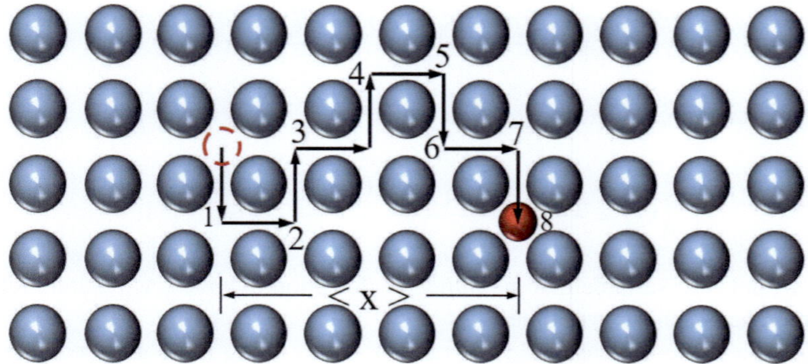

Fig. 7.2 Random walk model for atom jump

$$J_d = -D\frac{dC}{dx} \tag{7.2b}$$

$$J_m = -\frac{zFDC}{RT}\frac{d\phi}{dx} \tag{7.2c}$$

$$J_c = C\upsilon \tag{7.2d}$$

where J_d = Diffusion molar flux due to dC/dx $\left(mol/cm^2 \cdot s\right)$

J_m = Migration molar flux due to $d\phi/dx$ $\left(mol/cm^2 \cdot s\right)$
J_c = Convective molar flux due to fluid flow $\left(mol/cm^2 \cdot s\right)$
dC/dx = Concentration gradient $\left(mol/cm^4\right)$
$d\phi/dx$ = Potential gradient (V/cm)
D = Diffusivity or diffusion coefficient $\left(cm^2/s\right)$
$10^{-6}\ cm^2/s \leq D < 10^{-4}\ cm^2/s$ for most cations
C = Ionic concentration $\left(mol/cm^3\ or\ mol/liter\right)$
υ = Hydrodynamic velocity (cm/s)

7.3 Random Walks

Consider Fig. 7.2 for an interstitial particle with a random motion at a small time interval τ within an average distance $\langle x \rangle$.

Notice that the random walk of the red particle within the blue lattice points is based on random jumps. Thus, random walk of particles by succession of random steps is fundamentally based on probability and statistics to predict a system behavior.

The average distance for equal steps to the right and left can be deduced as

$$\langle x \rangle = \frac{\left[\sum (x_k)_{right} + \sum (-x_k)_{left} \right]}{k} = 0 \tag{7.3}$$

and the *mean-square distance* is defined by

$$\langle x^2 \rangle = \frac{\left[\sum (x_k)^2_{right} + \sum (-x_k)^2_{left} \right]}{k} = \lambda^2 \neq 0 \tag{7.4}$$

This is a nonzero variable that measures the deviation of a particle position relative to a reference position over time.

Consider a one-dimensional (1D) random walks along the x-axis. τ is the time step at $+x$ and $-x$, then the *mean-square distance* $\langle x^2 \rangle$ for a total random walk time t and the average statistical travel distance $\langle x \rangle$ are

$$\langle x^2 \rangle = \left(\frac{t}{\tau} \right) \lambda^2 = Dt \quad \text{(random)} \tag{7.5a}$$

$$\langle x \rangle = \sqrt{Dt} \quad \text{(random)} \tag{7.5b}$$

$$\langle x \rangle = \left(\frac{\lambda}{\tau} \right) t = \upsilon t \quad \text{(average)} \tag{7.5c}$$

where the variables D and υ are the diffusion coefficient and particle speed, respectively, in a certain medium subjected to a temperature T. Moreover, two-dimensional (2D) and three-dimensional (3D) random walks are excluded in this textbook.

From Eqs. (7.5a) and (7.5b), D, and the diffusion step length or depth L are

$$D = \frac{\lambda^2}{\tau} \tag{7.6a}$$

$$L = \sqrt{Dt} \tag{7.6b}$$

Now, the differences in particle speed between regular and random walk are compared below

$$\upsilon_{ran} = \frac{L}{t} = \frac{L}{L^2/D} = \frac{D}{L} \quad \text{(Random)} \tag{7.7a}$$

$$\upsilon_{ave} = \frac{\lambda}{\tau} \quad \text{(Average)} \tag{7.7b}$$

Combining Eqs. (7.6a) and (7.7b) yields

$$D = \lambda \upsilon_{ave} \tag{7.8}$$

This expression clearly illustrates that the diffusion coefficient or diffusivity D linearly depends on the particle speed υ during the random walk mechanism.

The particle displacement, in general, due to diffusion by random walk is $x = 2\langle x \rangle$ and from Eq. (7.5b), x is cast as the characteristic distance or length

$$x = 2\sqrt{Dt} \tag{7.9}$$

from which a common similarity dimensionless variable used in statistical physics, transport phenomena, and atomic diffusion is defined by

$$n = \frac{x}{2\sqrt{Dt}} \tag{7.10}$$

Mathematically, n is the solution characteristic or argument in the error function $\mathrm{erf}\,(n)$ and in the exponential function $\exp\left(-n^2\right)$ that occur in solutions to partial differential equations (PDEs). Nonetheless, the above mathematical approach is a general description of the relation between random walk and atomic diffusion.

Assume that the average translation kinetic energy and the internal energy per particle are equal. Thus,

$$\frac{m\upsilon_x^2}{2} = \frac{3k_B T}{2} \tag{7.11a}$$

$$\upsilon_x = \sqrt{\frac{3k_B T}{m}} \tag{7.11b}$$

where $m = $ Mass of the particle (atom, ion or molecule)

$\upsilon_x = $ Average velocity of a particle in the x-direction
$m = A_w/N_a = $ (Atomic weight)/(Avogadro's number)
$m = M_w/N_a = $ (Molecular weight)/(Avogadro's number)
$k_B = R/N_a = 1.38 \times 10^{-23}\ J/\,(K) = $ Boltzmann constant
$k_B = 1.38 \times 10^{-20}\ g \cdot m^2/\left(s^2 K\right) = 8.62 \times 10^{-5}\ eV/K$
$R = 8.3145\ J/\,(mol \cdot K) = $ Universal gas constant
$N_a = 6.022 \times 10^{23} = $ Avogadro's number
$k_B T = 4.14 \times 10^{-14}\ g \cdot cm^2/s^2 = 0.026\ eV$ at $300\ K$ $(25\,^\circ C)$

The product $k_B T$ can be taken as a scale factor for the amount of heat required to increase the thermodynamic entropy of an atomic system. Thermodynamically, the energy per particle is defined as $3k_B T/2$ (no proof is given), which may be referred to as thermal energy per particle. Actually, $k_B T$ represents an energy in transit during heat transfer by conduction, convection, or radiation.

Notice that the average kinetic energy is linearly proportional to the absolute temperature T. That is,

- The kinetic energy is defined as $KE = a_1 + b_1 T$ with slope $b = 3k_B/2$.

- The particle velocity is proportional the square root of T so that $\upsilon_x = a_2 + b_2\sqrt{T}$ with $b_2 = (3k_B/m)^{1/2}$.

The magnitude of b_1 or b_2 is in the order of the Boltzmann constant k_B.

Example 7.1 Calculate υ_x, D and τ at $T = 300\ K$ and $\lambda = 5 \times 10^{-10}\ cm$ for **(a)** carbon atoms in steel and **(b)** Albumin protein (a family of globular proteins). The atomic weight of carbon is $12\ g/mol$ and that of Albumin is $68,000\ g/mol$.

Solution

(a) For carbon, $A_w = 12\ g/mol$ at $T = 300\ K$, $\lambda = 5 \times 10^{-10}\ cm$ and

$$m = A_w/N_a = (12\ g/mol.atom) / \left(6.022 \times 10^{23}\ atom/mol\right) \quad (7.1\text{E1a})$$

$$m = 1.9927 \times 10^{-23}\ g \quad (7.1\text{E1b})$$

$$\upsilon_x = \sqrt{\frac{3k_B T}{m}} = \sqrt{\frac{3 \times 4.4 \times 10^{-14}\ g \cdot cm^2/s^2}{1.9927 \times 10^{-23}\ g}} = 81,389\ cm/s \quad (7.1\text{E1c})$$

and from Eq. (7.8) with $\upsilon_x = \upsilon_{ave}$

$$D = \lambda \upsilon_x = \left(5 \times 10^{-10}\ cm\right) (81,389\ cm/s) = 4.07 \times 10^{-5}\ cm^2/s \quad (7.1\text{E2})$$

From Eq. (7.6a),

$$\tau = \frac{\lambda^2}{D} = \frac{\left(5 \times 10^{-10}\ cm\right)^2}{2.35 \times 10^{-5}\ cm^2/s} = 6.14 \times 10^{-15}\ s \quad (7.1\text{E3})$$

(b) For an Albumin protein with $M_w = 68,000\ g/mol$ at $T = 300\ K$, $\lambda = 5 \times 10^{-10}\ cm$ and $\tau = 1\ ms$,

$$m = \frac{M_w}{N_a} = \frac{68,000\ g/mol.atom}{6.022 \times 10^{23}\ atom/mol} = 1.1292 \times 10^{-19}\ g \quad (7.1\text{E4a})$$

$$\upsilon_x = \sqrt{\frac{3k_B T}{m}} = \sqrt{\frac{3 \times 4.14 \times 10^{-14}\ g \cdot cm^2/s^2}{1.1292 \times 10^{-19}\ g}} = 1,048.80\ cm/s \quad (7.1\text{E4b})$$

and

$$D = \lambda \upsilon_x = \left(5 \times 10^{-10}\ cm\right) (1,048.80\ cm/s) = 5.24 \times 10^{-7}\ cm^2/s \quad (7.1\text{E5})$$

Subsequently, Eq. (7.6a) gives

$$\tau = \frac{\lambda^2}{D} = \frac{\left(5 \times 10^{-10} \ cm\right)^2}{5.24 \times 10^{-7} \ cm^2/s} = 4.77 \times 10^{-13} \ s \qquad (7.1E6)$$

The average percent error between the calculated velocities is

$$error = \frac{81,389 - 1,048.80}{(81,389 + 1,048.80)\,/2} \times 100 \simeq 195\% \qquad (7.1E7)$$

Therefore, these calculations are meant to show the reader that random diffusion varies from types of particles in diverse media. Denote that carbon atoms diffuse faster than albumin protein molecules.

7.4   Solid-State Diffusion

Vacancy Diffusion Consider the solid-state diffusion as a mass transport mechanism for introducing interstitial atoms in ferrous materials or dopant atoms into semiconductors. For instance, Fig. 7.3 schematically shows the main mechanisms known as vacancy diffusion or substitutional diffusion, which is the atomic motion of an atom from one lattice site in Fig. 7.3a to an adjacent vacancy in Fig. 7.3b.

Interstitial Diffusion, on the other hand, is the atomic motion of interstitial (small) atoms from one interstitial site (Fig. 7.3c) to another interstitial site (Fig. 7.3d). Common interstitial atoms are hydrogen, carbon, boron, and nitrogen. This type of diffusion is said to be faster than vacancy diffusion.

Steady-state diffusion and transient diffusion are the two conditions that must be considered in solving diffusion problems. Actually, the main objective is to harden a substrate or improve properties.

In the medical or clinical field, molecular transport by diffusion is commonly carried out by convective motion, and it is applied to various pathological conditions, including passive and active mass transport of drugs, and transport across membranes and even diffusion-weighted magnetic resonance imaging (DWI) based on the diffusion of water molecules in tissues.

7.4.1   Octahedral and Tetrahedral Sites

The preferred interstitial sites for small impurity atoms (solute) are the lattice empty spaces called voids or interstices among the substrate (solvent or host). As a result, the original chemical composition of the substrate is altered by interstitial diffusion of impurity atoms that may form a small diffusion layer from the substrate surface

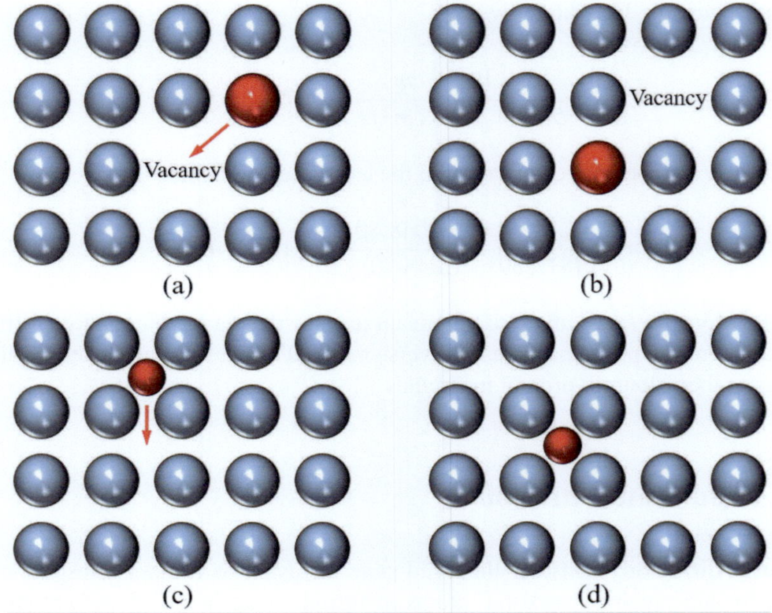

Fig. 7.3 Mass transport by diffusion. (**a**) Vacancy and substitutional atom, (**b**) substitutional diffusion, (**c**) interstitial atom and (**d**) interstitial diffusion

to a diffusion depth x, which in turn, sets the limit for interstitial solid solution with decreasing concentration.

For comparison, Fig. 7.4 shows two types of interstitial sites or voids (holes) known as octahedral and tetrahedral geometries in FCC and BCC crystal structures.

These voids are distinguished by the number of nearest-neighbor substrate (host) atoms called the coordination number (CN); $CN = 4$ for a tetrahedral and $CN = 6$ for octahedral sites. These interstitial sites are the obvious locations for impurity atoms during diffusion or solid-state solution.

The octahedral and tetrahedral geometries can also be drawn at different locations or be rotated as illustrated in Fig. 7.4a and b for FCC and Fig. 7.4c and d for BCC crystal structures. Moreover, the atom radius ratio $r/R = c_x$ or $r = c_x R$, where c_x is a constant cited in Fig. 7.4 for octahedral (oct) and tetrahedral (tet) holes or voids. For example, diffusion of carbon atoms into FCC iron (Fe) lattice at $T > 912\,°C$ is a practical industrial application for hardening steels, where the theoretical void radii in Fe with a radius $R = 0.124\,nm$ are

$$r_{Fe,oct} = (c_x R)_{oct} = 0.414\,(0.124\,nm) = 0.051\,nm \tag{7.12a}$$

$$r_{Fe,tet} = (c_x R)_{tet} = 0.255\,(0.124\,nm) = 0.032\,nm \tag{7.12b}$$

Fig. 7.4 Octahedral sites in
(**a**) FCC and (**b**) BCC unit
cells. Tetrahedral sites in (**c**)
FCC and (**d**) BCC crystal
structures. The interstitial
sites are represented by the
blue and red dots

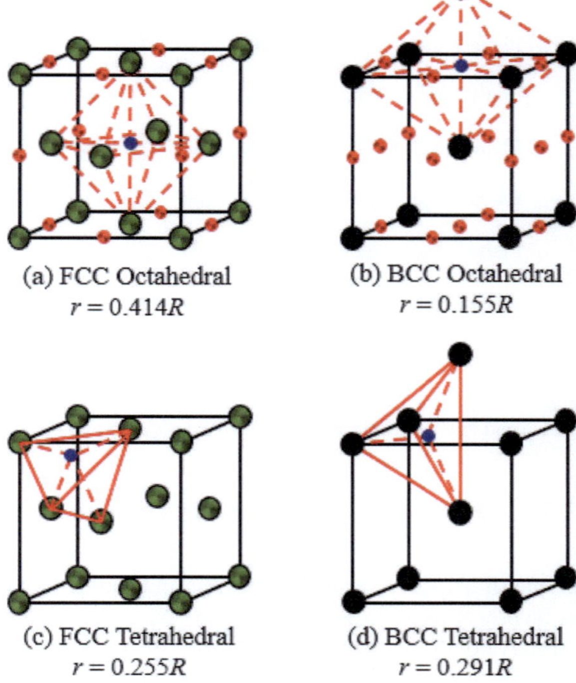

Since the carbon radius $r_C = 0.071\ nm > r_{Fe,oct} > r_{Fe,tet}$, then one can deduce that carbon atoms will occupy the octahedral sites inducing some lattice distortion.

7.5 Fick's Laws of Diffusion

Theoretically, atomic diffusion of a species j can occur in crystalline and amorphous substrates to a limited depth x or through a thin sheet, where the interstitial diffusion is faster than substitutional diffusion due to the impurity-substrate atom size effects. Regardless the type of diffusion, Fick's laws describe the time-independent and time-dependent mechanisms. Moreover, a particular model for interstitial diffusion is illustrated in Fig. 7.5 along with the theoretical concentration profile as a function of diffusion depth x.

This model indicates that interstitial atoms (small red dots) with radius $r > R$ agglomerates on the surface of a substrate (host) with atom radius R at a temperature T. Physically, diffusion of interstitial atoms into empty lattice spaces between (hkl) planes requires a small incubation time to proceed toward a certain depth x.

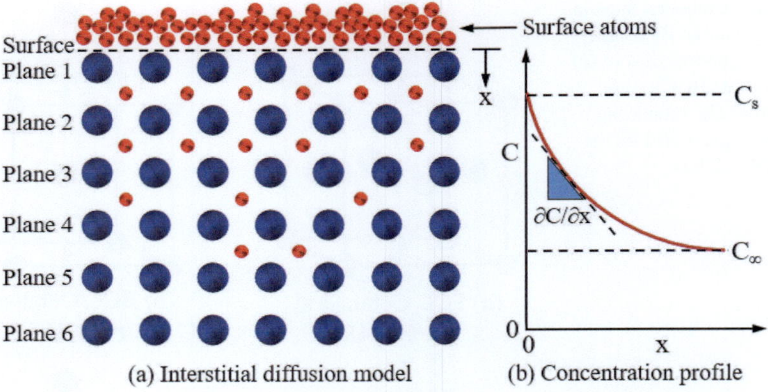

(a) Interstitial diffusion model   (b) Concentration profile

Fig. 7.5 (**a**) Model for interstitial diffusion within a cubic crystal plane and (**b**) concentration profile and its concentration gradient

7.5.1 *First Law: Steady-State Diffusion*

In one-dimensional analysis, the time-independent mass transport process is known as *steady-state diffusion*, where the concentration rate must be $\partial C/\partial t = 0$ as a primary condition at the substrate–environment interface. In essence, Fig. 7.5a illustrates two-dimensional model for interstitial atoms (red dots) having a concentration C and Fig. 7.5b shows the schematic profile as a function of diffusion depth x at time $t > 0$. Notice that the concentration gradient $\partial C/\partial x$ is the slope of a tangential line.

Mathematically, the mass transport by diffusion is quantified by the molar flux $J_x = J(x)$ of impurity atoms j, and it is defined by

$$J_x = C\upsilon \tag{7.13a}$$

$$J_x = \frac{m}{A_s t} \tag{7.13b}$$

$$C = \frac{N}{V} \tag{7.13c}$$

where $C = C(x, t)$ is the concentration of the impurity atom ($atoms/cm^3$), υ is the velocity of species j, m is the mass of a diffusing impurity into a substrate, A_s is the substrate exposed surface area, t is the diffusion time, N is the number of particles, and V is the system volume.

For diffusion of an impurity atom j in a specific homogeneous medium, Fick's first law of diffusion defines the molar flux as

$$J_x = -D\frac{\partial C}{\partial x} = \frac{1}{A_s}\frac{dm}{dt} \tag{7.14}$$

Here, $\partial C/\partial x$ is the concentration gradient that acts as the driving force for $J_x > 0$. Here, D is assumed to be constant within a small temperature range ΔT. The physical interpretation of the Eq. (7.14) dictates that the molar or mass flux J_x follows a similar behavior as D does because steady-state diffusion depends on the nature of the species j and the concentration gradient $\partial C/\partial x < 0$ in a time period. In particular, D is mathematically a multiplying factor for J_x.

Combining Eqs. (7.13a) and (7.14) yields the concentration gradient (Fig. 7.5b) related to the random walk speed

$$\frac{\partial C}{\partial x} = -\left(\frac{C}{D}\right) v_{ran} \tag{7.15}$$

Diffusion Coefficient It is now assumed that the diffusion coefficient D (also known as intrinsic diffusivity) obeys the Arrhenius-type rate equation so that

$$D = D_o \exp\left(-\frac{Q_d}{RT}\right) = D_o \exp\left(-\frac{Q_d}{k_B T}\right) \tag{7.16}$$

where D_o is a proportionality constant and Q_d is the activation energy (J/mol or eV) for diffusion. The product RT or $k_B T$ in Eq. (7.16) is a matter choice based on units. Anyway, $Q_d > k_B T$ so that an impurity atom j acquires a high mobility at high temperatures.

Convenient graphical representations of $D = f(T)$ and $D = f(1/T)$, as defined by Eq (7.16), are shown in Fig. 7.6a and b, respectively. Denote that D is strongly dependent on the temperature range $0 < T < T_m$, where T_m is the melting temperature of the substrate.

In solid-state diffusion, if $T \rightarrow 0$, then $D \rightarrow 0$ since $\exp(\infty) = 0$ and consequently, diffusion of a species j requires an incubation time in a particular environment. On the other hand, if $T \rightarrow T_m \rightarrow \infty$, then $D \rightarrow D_o$ due to

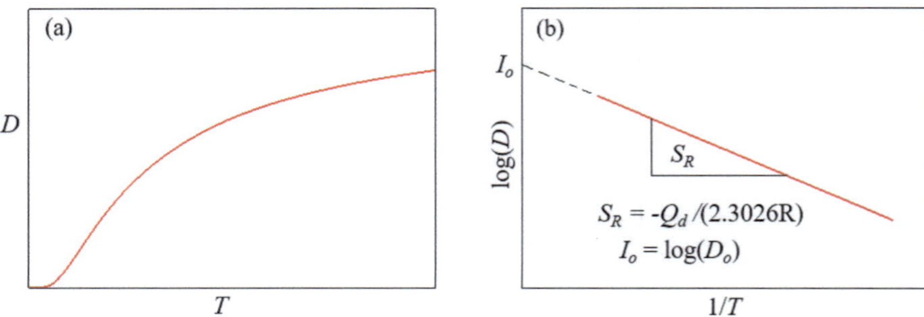

Fig. 7.6 (a) Nonlinear and (b) linear Arrhenius plots

$\exp(-0) = 1$. Therefore, mass transport by diffusion in the solid-state condition increases with increasing D and $T < T_m$.

Figure 7.6b exhibits a linear plot suggesting that it is a practical approach for analyzing diffusion data and for determining

- The thermal activation energy Q_d through the slope S_R
- The proportionality constant D_o.

This plot is obtained by taking the natural logarithm of Eq. (7.16) and then converted to base-10 logarithm. Hence,

$$\ln(D) = \ln(D_o) - \frac{Q_d}{R}\frac{1}{T} = I_R - S_R\left(\frac{1}{T}\right) \tag{7.17a}$$

$$\ln(D) = \ln(D_o) - \frac{Q_d}{k_B}\frac{1}{T} = I_R - S_R\left(\frac{1}{T}\right) \tag{7.17b}$$

Using base-10 logarithms yields

$$\log(D) = \log(D_o) - \left(\frac{Q_d}{2.3026R}\right)\frac{1}{T} = I_B - S_B\left(\frac{1}{T}\right) \tag{7.18a}$$

$$\log(D) = \log(D_o) - \left(\frac{Q_d}{2.3026k_B}\right)\frac{1}{T} = I_B - S_B\left(\frac{1}{T}\right) \tag{7.18b}$$

where

$$I_R = \ln(D_o) \tag{7.19a}$$

$$I_B = \log(D_o) \tag{7.19b}$$

$$S_R = \frac{Q_d}{R} = \frac{Q_d}{k_B} \tag{7.19c}$$

$$S_B = \frac{Q_d}{2.3026R} = \frac{Q_d}{2.3026k_B} \tag{7.19d}$$

The next example can illustrate the use of the diffusion and Arrhenius equations.

Example 7.2 Consider a thin sheet of metal 2-*mm* thick being exposed to atomic hydrogen on the sides as shown in the sketch below. First of all, adsorption and dissociation of gas A occur on the sheet surfaces at relatively high temperatures. Assume that the hydrogen concentration (in mol/m^3) and diffusion coefficient are hypothetically defined by

$$C_H = 73\sqrt{P}\exp\left(-\frac{12{,}000\ J/mol}{RT}\right) \tag{7.2E1a}$$

$$D_H = \left(5\ m^2/s\right) \exp\left(-\frac{40{,}000\ J/mol}{RT}\right) \tag{7.2E1b}$$

where pressure P is in MPa units. For pressures $P_1 = 4\ MPa$ and $P_2 = 0.1\ MPa$, $T = 400\,^{\circ}C$ and thickness $\Delta x = 1\ mm$, calculate (a) the hydrogen diffusion flux

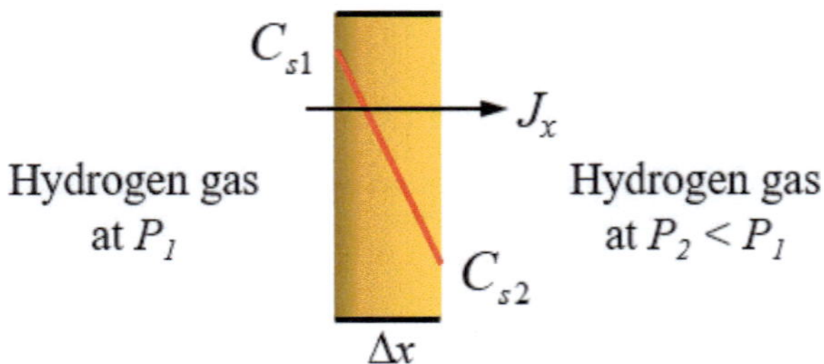

Solution

$$C_{H,P_1} = 73\sqrt{4}\exp\left(-\frac{12{,}000\ J/mol}{(8.3145\ J/mol.K)\,(400 + 273)\ K}\right) \tag{7.2E2a}$$

$$C_{H,P_1} = 17.10\ mol/m^3$$

$$C_{H,P_2} = 73\sqrt{0.1}\exp\left(-\frac{12{,}000\ J/mol}{(8.3145\ J/mol.K)\,(400 + 273)\ K}\right) \tag{7.2E2b}$$

$$C_{H,P_2} = 2.70\ mol/m^3$$

$$D_H = \left(5\ m^2/s\right)\exp\left(-\frac{40{,}000\ J/mol}{(8.3145\ J/mol.K)\,(400 + 273)\ K}\right) \tag{7.2E2c}$$

$$D_H = 3.93 \times 10^{-3}\ m^2/s \tag{7.2E2d}$$

Then,

$$J_x = -D_H\frac{\Delta C_H}{\Delta x} = -D_H\frac{C_{H,P_1} - C_{H,P_2}}{x_{P_1} - x_{P_2}} \tag{7.2E3a}$$

$$J_x = -\left(3.93 \times 10^{-3}\ m^2/s\right)\left[\frac{(17.10 - 2.70)\ mol/m^3}{0 - 1 \times 10^{-3}\ m}\right] \tag{7.2E3b}$$

$$J_x = 56.59\ mol/\left(m^2.s\right) \tag{7.2E3c}$$

which is relatively a large flux. Since the atomic or molecular weight of hydrogen is $A_w = 1g/mol$, the steady-state diffusion mass flux J_x is

$$J_m = A_w J_x = (1\ g/mol)\left(56.59\ mol/\left(m^2.s\right)\right) \tag{7.2E4a}$$

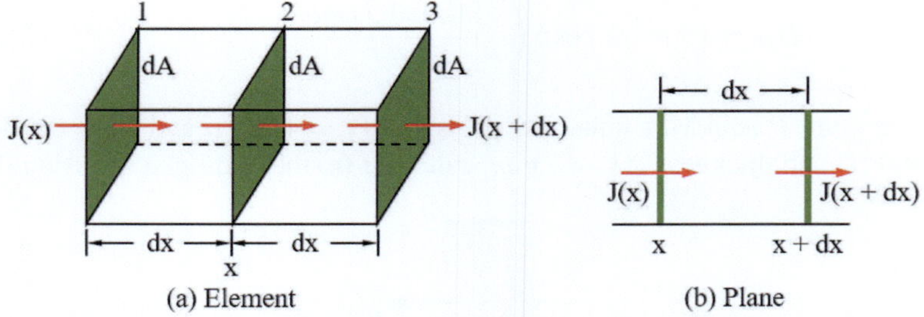

Fig. 7.7 Diffusion in a isotropic and homogeneous medium. (**a**) Rectangular element and (**b**) rectangular planes for one-dimensional diffusion flux J_x in the opposite direction of the concentration gradient dC/dx. After Crank [6, p. 3]

$$J_m = 56.59 \; g/\left(m^2.s\right) \tag{7.2E4b}$$

This is a simple example to follow, and it is essential for analyzing steady-state diffusion.

7.5.2   Second Law: Transient Diffusion

Rectangular Element Consider the central plane as the reference point in the rectangular volume element (Fig. 7.7a) and assume that the diffusing plane at position 2 moves along the x-direction at a distance x-dx from position 1 and $x+dx$ to position 3 (Crank [6, p. 3]).

The rate of diffusion (R_x) that enters the volume element (V) at position 1 and leaves at position 3 is defined by

$$R_x = dydz\left[J_x - \frac{\partial J_x}{\partial x}dx\right] - dydz\left[J_x + \frac{\partial J_x}{\partial x}dx\right] \tag{7.20a}$$

$$R_x = -2dxdydz\frac{\partial J_x}{\partial x} = -2dV\frac{\partial J_x}{\partial x} \tag{7.20b}$$

Similarly,

$$R_y = -2dxdydz\frac{\partial J_y}{\partial y} = -2dV\frac{\partial J_y}{\partial y} \tag{7.21a}$$

$$R_z = -2dxdydz\frac{\partial J_z}{\partial z} = -2dV\frac{\partial J_z}{\partial z} \tag{7.21b}$$

The concentration rate is defined as (Crank [6, pp. 4–5])

$$\frac{\partial C}{dt} = \frac{1}{2dV} \sum R_i \tag{7.22a}$$

$$\frac{\partial C}{dt} = -\frac{\partial J_x}{\partial x} - \frac{\partial J_y}{\partial y} - \frac{\partial J_z}{\partial z} \tag{7.22b}$$

Actually, Eq. (7.22b) is the equation of continuity for conservation of mass. For a three-dimensional analysis, the general flux equation is defined by

$$J = -D \left(\frac{\partial C}{\partial x} + \frac{\partial C}{\partial y} + \frac{\partial C}{\partial z} \right) \tag{7.23}$$

Combining Eqs. (7.22b) and (7.23) yields the *transient diffusion* equation known as Fick's second law in three dimensions with ∇^2 being the Laplace operator

$$\frac{\partial C}{dt} = D \left[\frac{\partial^2 C}{\partial x^2} + \frac{\partial^2 C}{\partial y^2} + \frac{\partial^2 C}{\partial z^2} \right] = D\nabla^2 C \tag{7.24a}$$

$$\frac{\partial C}{dt} = D \frac{\partial^2 C}{\partial x^2} \quad \text{(One- dimension)} \tag{7.24b}$$

Rectangular Plane Using Fig. 7.7b and Taylor series on $J(x + dx)$, the concentration rate can be defined by

$$\frac{\partial C}{dt} = \frac{J(x) - J(x + dx)}{dx} \tag{7.25}$$

The infinite-order Taylor-series expansion on $J(x + dx)$ is

$$J(x + dx) = \sum_{k=0}^{\infty} \frac{(dx)^k}{k!} f^k(x) \tag{7.26a}$$

$$J(x + dx) = J(x) + \frac{\partial J(x)}{\partial x} \frac{dx}{1!} + \frac{\partial^2 J(x)}{\partial x^2} \frac{dx^2}{2!} + \frac{\partial^3 J(x)}{\partial x^3} \frac{dx^3}{3!} + \cdots \tag{7.26b}$$

$$J(x + dx) \simeq J(x) + \frac{\partial J(x)}{\partial x} dx \tag{7.26c}$$

Substituting Eq. (7.26c) into (7.25) yields

$$\frac{\partial C}{dt} = \frac{J(x) - J(x) - [\partial J(x)/\partial x] dx}{dx} \tag{7.27a}$$

$$\frac{\partial C}{dt} = -\frac{[\partial J(x)/\partial x] dx}{dx} = -\frac{\partial J(x)}{\partial x} \tag{7.27b}$$

For practical purposes, Eqs. (7.24b) and (7.27b) are recast in order to elucidate the remarkable mathematical equations that describe Fick's second law of diffusion

$$\frac{\partial C}{\partial t} = -\frac{\partial J_x}{\partial x} \tag{7.28a}$$

$$\frac{\partial C}{\partial t} = D\frac{\partial^2 C}{\partial x^2} \tag{7.28b}$$

This expression, Eq. (7.28b), describes the transport of matter (atoms, ions or molecules) as a consequence of the concentration gradient $\partial C/\partial x$. Its solution depends on the initial condition (IC) before heat treatment (HT) and the boundary condition (BC) during HT.

Now, let N be the total amount or the quantity of diffusing particles (atoms, ions or molecules) in an isotropic medium of infinite length L, cross-sectional area A, and volume V. By definition, N and $C = C(x, t)$ take the form

$$N = \int_{-\infty}^{\infty} C(x, t)\, dx \tag{7.29a}$$

$$C = \frac{N}{V} \tag{7.29b}$$

The integral in Eq. (7.29a) can be solved once $C = C(x, t)$ is derived based on initial and boundary conditions.

7.6 Solution to Fick's Second law Equation

7.6.1 *Method of Separation of Variables*

The solution to one-dimensional Fick's second law of diffusion equation, Eq. (7.28b), is based on the transformation of a partial differential equation (PDE) to an ordinary differential equation (ODE). This can be achieved by the separation of variables method, which typically requires that the concentration $C(x, t) = f(x, t)$ of a species j is initially separated as $C(x, t) = f(x)\, g(t)$, where $f(x)$ and $g(t)$ are arbitrary functions that define $C(x, t)$.

Let $C = C(x, t) = f(n)$ with n being defined by Eq. (7.10) so that the gradient dn/dx and rate dn/dt are conveniently written as

$$n = \frac{x}{2\sqrt{Dt}} \tag{7.30a}$$

$$\frac{dn}{dx} = \frac{1}{2\sqrt{Dt}} \tag{7.30b}$$

$$\frac{dn}{dt} = -\frac{x}{4\sqrt{Dt^3}} = -\frac{x}{4t\sqrt{Dt}} \tag{7.30c}$$

Using the chain rule of differentiation yields

$$\frac{\partial C}{\partial t} = \frac{dC}{dn}\frac{\partial n}{\partial t} = -\frac{x}{4t\sqrt{Dt}}\frac{dC}{dn} \tag{7.31a}$$

$$\frac{\partial C}{\partial x} = \frac{dC}{dn}\frac{\partial n}{\partial x} = \frac{1}{2\sqrt{Dt}}\frac{dC}{dn} \tag{7.31b}$$

$$\frac{\partial^2 C}{\partial x^2} = \frac{d^2 C}{dn^2}\left(\frac{\partial n}{\partial x}\right)^2 + \frac{dC}{dn}\frac{\partial^2 n}{\partial x^2} = \left(\frac{1}{2\sqrt{Dt}}\right)^2 \frac{d^2 C}{dn^2} \tag{7.31c}$$

$$\frac{\partial^2 C}{\partial x^2} = \frac{1}{4Dt}\frac{d^2 C}{dn^2} \tag{7.31d}$$

where $\partial^2 n/\partial x^2 = 0$. Substitute Eqs. (7.31a) and (7.31d) into (7.28b) to get the sought ODE

$$-\frac{x}{4t\sqrt{Dt}}\frac{dC}{dn} = \frac{D}{4Dt}\frac{d^2 C}{dn^2} \tag{7.32a}$$

$$-2n\frac{dC}{dn} = \frac{d^2 C}{dn^2} \tag{7.32b}$$

Note that Eq. (7.32b) is an ODE containing one independent n only. Letting $f = dC/dn$ and $df = d^2 C/dn^2$ in Eq. (7.32b) yields

$$-2nf = df \tag{7.33a}$$

$$\frac{df}{f} = -2n \tag{7.33b}$$

$$\int \frac{df}{f} = -2\int n\,dn \tag{7.33c}$$

$$\ln(f) = -n^2 + K_x \tag{7.33d}$$

where K_x is a constant of integration.

Solving for f gives

$$f = \exp\left(-n^2 + K_x\right) = A\exp\left(-n^2\right) \tag{7.34a}$$

$$\frac{dC}{dn} = A\exp\left(-n^2\right) \tag{7.34b}$$

$$dC = A\exp\left(-n^2\right)dn \tag{7.34c}$$

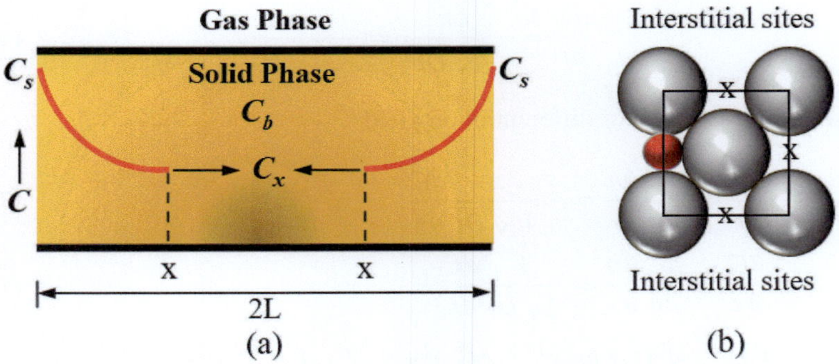

Fig. 7.8 Interstitial diffusion into a metallic plane. (**a**) Concentrations at different locations due to a diffusion process and (**b**) FCC plane showing positions of an impurity atom

In order to integrate Eq. (7.34c) and solve for the constant A, one needs to describe the diffusion problem and set the initial condition (IC) and boundary conditions (BCs).

Case 1: Bulk Concentration Model $C_b > 0$ Fig 7.8a schematically represents a specimen (steel, pure iron, or silicon plate), and assume that it is exposed to a diffusing interstitial atom (carbon, nitrogen, hydrogen, or boron) at a relatively high temperature for a prolong time t. Moreover, Fig. 7.8b elucidates the position of an interstitial atom (small red dot) and other possible interstitial sites marked with an "x" to accommodate the species j.

Carburizing It is a case-hardening process for hardening the surface of relatively low-carbon steels to a certain depth. Some examples are gears, firearm parts, engine camshafts, drilling screws, and the like. For the sake of clarity, assume that Fig. 7.5a represents the model for a low-carbon steel sample placed in a suitable furnace containing a carbon-rich gas (say, methane), which dissociates providing a constant concentration of carbon atoms on the steel specimen surface at constant temperature T and pressure P. Thus, carbon atoms diffuse into the FCC austenitic phase (γ-Fe) of the steel containing an initial amount of carbon termed bulk or background concentration $C(x, t) = C_b$. As a result, the carbon concentration is increased to a certain depth from the surface, and the steel hardens. In summary, carburizing increases surface hardness, improves wear resistance and fatigue strength.

Dopant Diffusion In the field of metallurgy, the Czochralski process is used to produce silicon (Si) single crystals in the form of ingots, which in turn, are sliced in thin substrates called wafers having a 25.4-mm or more in diameter. Firstly, these wafers are prepared to have smooth surfaces and subsequently, they are exposed to impurity-rich environments (AsH_3, PH_3, and B_2H_6) at relatively high temperatures for adjusting electrical properties by diffusion. Secondly, the wafers are subjected to a dopant diffusion process in two steps commonly known as predeposition or ion-implantation and drive-in diffusion. In essence, the role of As, P, and Sb (electron donors) in silicon wafers is to produce n-type semiconductors, while B, In, and Al (hole donors) in Si produce p-type extrinsic semiconductors.

Ion Implantation It is a technique used to introduce specific amounts of impurity ions (B^+, H^+, As^+, P^+) into a semiconductor substrate at an accelerated rate (high ion kinetic energy) at relatively low temperatures. Evidently, ions penetrate the substrate made out of silicon wafer by electronic collisions with electrons in the substrate lattice, creating dissipation of heat. This is a reason for having a controlled environment during ion implantation in order to produce either n-type or p-type semiconductors. Nowadays, semiconductor processors are commonly known as integrated circuits or electronic microchips, which can be exposed to an ion implantation process for doping impurity ions into silicon wafers. In effect, microchips are essential electronic devices in cell phones, car computers, tablets, desktop computers, and the like.

Mathematical Analysis It is important to analyze the diffusion process for hardening ferrous and for adjusting electrical properties in non-ferrous materials, such as silicon or germanium semiconductors. Figure 7.7a illustrates the general plane containing an initial concentration of an impurity species j (carbon, boron, phosphorous, and the like) that can be used to define meaningful conditions for solving certain diffusion problems based on a suitable initial condition denoted as IC and boundary condition denoted as BC for steady and transient states.

For an isotropic material treated as a semi-infinite ($0 < x < \infty$) medium,

$$\text{IC} \rightarrow C = C_b \text{ for } n = \infty \text{ at } x > 0, \ t = 0 \tag{7.35a}$$

$$\text{BC} \rightarrow C = C_s \text{ for } n = 0 \text{ at } x = 0, \ t > 0 \tag{7.35b}$$

$$\text{BC} \rightarrow C = C_b \text{ for } n = \infty \text{ at } x = \infty, t > 0 \tag{7.35c}$$

Integrating Eq. (7.34c) using the BC given by Eq. (7.35c) yields the constant A as

$$\int_{C_b}^{C_s} dC = A \int_0^\infty \exp\left(-n^2\right) dn = A \frac{\sqrt{\pi}}{2} \tag{7.36a}$$

$$A = \frac{2\left(C_s - C_b\right)}{\sqrt{\pi}} \tag{7.36b}$$

Integrate Eq. (7.34c) again with IC, Eq. (7.35a), to derive the solution of Fick's second law

$$\int_{C_b}^{C} dC = A \int_0^n \exp\left(-n^2\right) dn \tag{7.37a}$$

$$C - C_b = \left(C_s - C_b\right) \frac{2}{\sqrt{\pi}} \int_n^\infty \exp\left(-n^2\right) dn \tag{7.37b}$$

$$\mathrm{erfc}\,(n) = \frac{2}{\sqrt{\pi}} \int_{n}^{\infty} \exp\left(-n^2\right) dn \tag{7.37c}$$

where Eq. (7.37c) is known as the Gaussian complementary error function, which has no exact solution. Thus, Eq. (7.37b) becomes the general solution of Fick's second law of diffusion in terms of the dimensionless variable n

$$C - C_b = (C_s - C_b)\,\mathrm{erfc}\,(n) \tag{7.38a}$$

$$C - C_b = (C_s - C_b)\,[1 - \mathrm{erf}\,(n)] \tag{7.38b}$$

$$\frac{C - C_b}{C_s - C_b} = \mathrm{erfc}\,(n) = 1 - \mathrm{erf}\,(n) \tag{7.38c}$$

Here, $\mathrm{erf}\,(n) = 1 - \mathrm{erfc}\,(n)$ is the Gaussian error function of sigmoidal shape and $\mathrm{erfc}\,(n)$ is the complementary error function (Appendix 7A). Further, the Gaussian error function is defined by

$$\mathrm{erf}\,(n) = \frac{2}{\sqrt{\pi}} \int_{0}^{n} \exp\left(-n^2\right) dn = \frac{2}{\sqrt{\pi}} \int_{o}^{n} \sum_{k=0}^{\infty} \frac{(-1)^k\, n^{2k}}{k!}\, dn \tag{7.39a}$$

$$\mathrm{erf}\,(n) = \frac{2}{\sqrt{\pi}} \sum_{k=0}^{\infty} \frac{(-1)^k\, n^{2k+1}}{k!\,(2k+1)} \quad \text{(Maclaurin's series)} \tag{7.39b}$$

$$\mathrm{erf}\,(n) = \frac{2}{\sqrt{\pi}} \left(n - \frac{n^3}{3} + \frac{n^5}{10} - \frac{n7}{42} + \ldots \right) \tag{7.39c}$$

Substituting Eq. (7.30a) into Eq. (7.38c) yields (Appendix 7A)

$$\frac{C - C_b}{C_s - C_b} = \mathrm{erfc}\left(\frac{x}{2\sqrt{Dt}}\right) = 1 - \mathrm{erf}\left(\frac{x}{2\sqrt{Dt}}\right) \quad \text{(inwards)} \tag{7.40a}$$

$$C = C_b + (C_s - C_b)\,\mathrm{erfc}\left(\frac{x}{2\sqrt{Dt}}\right) \tag{7.40b}$$

From Eq. (7.40b), the concentration gradient is defined by

$$\frac{\partial C}{\partial x} = (C_s - C_b)\left(\frac{1}{\sqrt{\pi Dt}}\right) \exp\left(-\frac{x^2}{4Dt}\right) \tag{7.41}$$

It is important to note that if $\partial C / \partial x = 0$, then there is no diffusion flux in and out of the source. Assuming that $\partial C / \partial x \neq 0$ and inserting Eq. (7.41) into (7.28a) yields the diffusion flux, $J_x = f\,(x, t)$, written as

$$J_x = -D\,(C_s - C_b)\left(\frac{1}{\sqrt{\pi Dt}}\right) \exp\left(-\frac{x^2}{4Dt}\right) \tag{7.42}$$

From Eq. (7.41),

$$\frac{\partial^2 C}{\partial x^2} = -(C_s - C_b)\left(\frac{x}{2Dt\sqrt{\pi Dt}}\right)\exp\left(-\frac{x^2}{4Dt}\right) \tag{7.43}$$

Inserting Eq. (7.43) into (7.28b) gives the concentration rate $\partial C/\partial t = f(x,t)$

$$\frac{\partial C}{\partial t} = -\frac{(C_s - C_b)\,x}{2t\sqrt{\pi Dt}}\exp\left(-\frac{x^2}{4Dt}\right) \tag{7.44}$$

Case 2: Zero Bulk Concentration Model Assume that a pure metal or alloy specimen is subjected to diffusion of the species j, such as carbon into pure iron (Fe) or boron (B) into silicon. If the bulk concentration of a species j is initially $C_b = 0$, then Eqs. (7.40a) or (7.40b) simplifies to (see Appendix 7A)

$$\frac{C}{C_s} = 1 - \mathrm{erf}\left(\frac{x}{2\sqrt{Dt}}\right) = \mathrm{erfc}\left(\frac{x}{2\sqrt{Dt}}\right) \quad \text{(inwards)} \tag{7.45a}$$

$$C = C_s\left[1 - \mathrm{erf}\left(\frac{x}{2\sqrt{Dt}}\right)\right] = C_s\,\mathrm{erfc}\left(\frac{x}{2\sqrt{Dt}}\right) \tag{7.45b}$$

The first and second derivatives of Eq. (7.45b) are

$$\frac{\partial C}{\partial x} = -\frac{C_s}{\sqrt{4\pi Dt}}\exp\left(-\frac{x^2}{4Dt}\right) \tag{7.46a}$$

$$\frac{\partial^2 C}{\partial x^2} = \frac{x C_s}{4Dt\sqrt{\pi Dt}}\exp\left(-\frac{x^2}{4Dt}\right) \tag{7.46b}$$

then Eq. (7.28b) becomes

$$\frac{\partial C}{\partial t} = \frac{x C_s}{4t\sqrt{\pi Dt}}\exp\left(-\frac{x^2}{4Dt}\right) \tag{7.47}$$

Case 3: Outward Diffusion This is the case of interstitial atoms with a surface concentration $C_s = 0$ that simplifies Eqs. (7.40a) or (7.40b). Thus,

$$\frac{C}{C_b} = \mathrm{erf}(n) \quad \text{(outwards)} \tag{7.48}$$

Solving for C, using Eq. (7.30a) and comparing the resultant expression with Eq. (7.45b) yield the concentration of diffusing species j (substance)

$$C = C_b\,\mathrm{erf}\left(\frac{x}{2\sqrt{Dt}}\right) \quad \text{(outwards)} \tag{7.49a}$$

$$C = C_s \, \text{erf}\left(\frac{x}{2\sqrt{Dt}}\right) \quad \text{(inwards)} \tag{7.49b}$$

which differ on the initial concentration constants C_b and C_s.

7.6.2   Amount of Diffusing Solute

In dopant diffusion and ion implantation, the total amount N of the diffusing species j (dose) can be predicted by substituting Eq. (7.45b) into the integral in Eq. (7.29). Thus,

$$N = \int_0^\infty C\,dx = C_s \int_0^\infty \text{erfc}\left(\frac{x}{2\sqrt{Dt}}\right) dx \tag{7.50a}$$

$$N = \frac{C_s}{\sqrt{\pi}}\left(2\sqrt{Dt}\right) = 2C_s\sqrt{\frac{Dt}{\pi}} \tag{7.50b}$$

It is appropriate now to develop the analytical procedure for determining the effect of N on $C = C(x,t)$. According to Crank [6, p. 11], the solution of Eq. (7.28b) is of the form

$$C = \frac{B}{\sqrt{t}}\exp\left(-\frac{x^2}{4Dt}\right) = \frac{B}{\sqrt{t}}\exp\left(-n^2\right) \tag{7.51}$$

where B is an unknown constant. From Eq. (7.30a), the diffusion depth becomes

$$x = 2n\sqrt{Dt} \tag{7.52a}$$

$$dx = 2\sqrt{Dt}\,dn \tag{7.52b}$$

Substitute Eqs. (7.51) and (7.52b) into Eq. (7.29) and integrate to get

$$N = \int_0^\infty C\,dx = \int_0^\infty \frac{2B\sqrt{Dt}}{\sqrt{t}}\exp\left(-n^2\right) dn \tag{7.53a}$$

$$N = \left(2B\sqrt{D}\right)\int_0^\infty \exp\left(-n^2\right) dn = \left(2B\sqrt{D}\right)\frac{\sqrt{\pi}}{2} \tag{7.53b}$$

from which

$$B = \frac{N}{\sqrt{\pi D}} \tag{7.54}$$

Inserting Eqs. (7.54) and (7.30a) into (7.51) yields the concentration written as

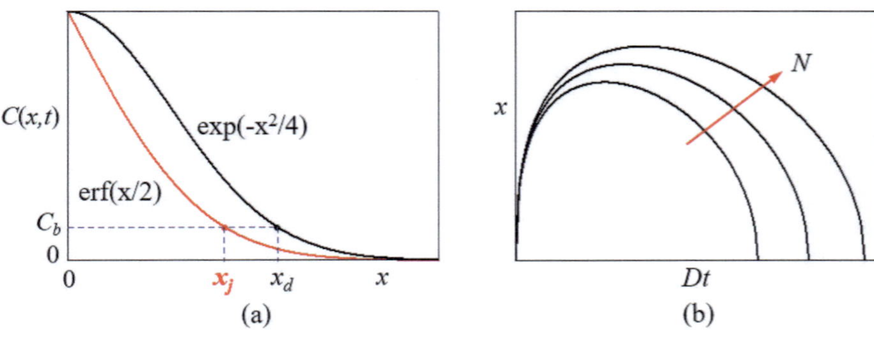

Fig. 7.9 Schematic plots for (**a**) for normalized concentration distribution using $\mathrm{erf}\, c(x/2)$ and $exp(-x^2/4)$ for interstitial diffusion in a semi-infinite medium, and (**b**) Interstitial diffusion and junction depth as function of $\sqrt{Dt}$ for increasing N

$$C = \frac{N}{\sqrt{\pi Dt}} \exp\left(-\frac{x^2}{4Dt}\right) \tag{7.55}$$

which represents a concentration decay with increasing diffusion depth, penetration, or junction depth x in the x-direction.

$$x = \sqrt{(4Dt)\ln\left(\frac{N}{C\sqrt{\pi Dt}}\right)} \tag{7.56}$$

Plotting Eqs. (7.49a) and (7.55) yields comparable $\mathrm{erf}\,(x/2)$ and $\exp\left(-x^2/4\right)$ nonlinear profiles. This is schematically depicted in Fig. 7.9a with $Dt = 1$.

Conversely, Fig. 7.9b illustrates the diffusion depth profile defined by $x = f(Dt)$ as per Eq. (7.55) with increasing N. The interpretation of these profiles suggests that diffusion ceases when x reaches a maximum. This can be a particular design condition in dopant diffusion technology.

Regarding the curves in Fig. 7.9a, $C = C(x, t)$ exhibits rapid and similar decay as x increases, indicating that diffusion in semi-infinite substrates is limited to a certain length or depth x denoted as $x = x_j$ in Eq. (7.49a) and $x = x_d$ in Eq. (7.55).

Example 7.3 Consider carburizing low-carbon steel (AISI 1025) at $950\,^{\circ}C$ and pure iron at $900\,^{\circ}C$ for 10 hours in a furnace containing a carbon-rich gas environment. Assume that semi-infinite ($0 < x < \infty$) specimens can be modeled as shown in Fig. 7.5 and that the surface carbon concentration in weight percent is kept constant at $C_s = 1\%$. Let the carbon diffusion coefficient be $D = 6 \times 10^{-6}\, mm^2/s$, $C_b = 0.25\%$ for steel and $C_b = 0\%$ for iron. Also assume that the controlled reactions for the mass transport by carburizing are

$$2CO \rightarrow C_{\gamma\text{-}Fe} + CO_2 \tag{7.3E1a}$$

$$CH_4 \rightarrow C_{\gamma\text{-}Fe} + 2H_2 \tag{7.3E1b}$$

$$CO + H_2 \rightarrow C_{\gamma\text{-}Fe} + H_2O \tag{7.3E1c}$$

where $C_{\gamma\text{-}Fe}$ is the carbon concentration in the specimens having a common FCC austenite phase ($\gamma\text{-}Fe$) denoted as a ferrous lattice. Once the heat treatment is complete, quench the specimens in water or oil from the carburizing temperature. **(a)** Plot $C = f(x)$ at $0 \leq x \leq 2\ mm$ for low-carbon steel (AISI 1025) and pure iron. **(b)** Calculate the concentration C at $x = 2\ mm$ for iron and steel. **(c)** If the density of iron or steel and the atomic weight of carbon are $7.87\ g/cm^3$ and 12 g/mol, respectively, convert weight percent to concentration of carbon in units of mol/cm^3. **(d)** Calculate the flux dN/dt in the specimens.

Solution

(a) For a steel specimen containing an initial bulk carbon concentration $C_b = 0.25\%$, Eq. (7.40a) along with the given data provides the concentration equation in weight percentage (%) written as

$$\frac{C - C_b}{C_s - C_b} = 1 - \mathrm{erf}\left(\frac{x}{2\sqrt{Dt}}\right) = \mathrm{erfc}\left(\frac{x}{2\sqrt{Dt}}\right) \tag{7.3E2a}$$

$$C = C_b + (C_s - C_b)\,\mathrm{erfc}\left(\frac{x}{2\sqrt{Dt}}\right) \tag{7.3E2b}$$

$$C = 0.25 + (1 - 0.25)\,\mathrm{erfc}\left(\frac{x}{2\sqrt{(6 \times 10^{-6}\ mm^2/s)(10 \times 60 \times 60\ s)}}\right) \tag{7.3E2c}$$

$$C = 0.25 + 0.75\,\mathrm{erfc}\,(1.0758x) \quad (\text{in } \%) \tag{7.3E2d}$$

which is a useful and simple nonlinear equation for determining the concentration of carbon atoms at a depth x. Similarly, for pure iron, Eq. (7.45a) yields the concentration equation in weight percentage (%) defined as

$$\frac{C}{C_s} = 1 - \mathrm{erf}\left(\frac{x}{2\sqrt{Dt}}\right) = \mathrm{erfc}\left(\frac{x}{2\sqrt{Dt}}\right) \tag{7.3E3a}$$

$$C = C_s\left[1 - \mathrm{erf}\left(\frac{x}{2\sqrt{Dt}}\right)\right] = C_s\,\mathrm{erfc}\left(\frac{x}{2\sqrt{Dt}}\right) \tag{7.3E3b}$$

$$C = (1)\,[1 - \mathrm{erf}\,(1.0758x)] = \mathrm{erfc}\,(1.0758x) \quad (\text{in } \%) \tag{7.3E3c}$$

The plot is given below. Notice that both curves become asymptotic at approximately $x = 1.8\ mm$. Obviously, $C_b = 0$ for iron and $C_b = 0.25\%$ for steel at $x = 1.8\ mm$.

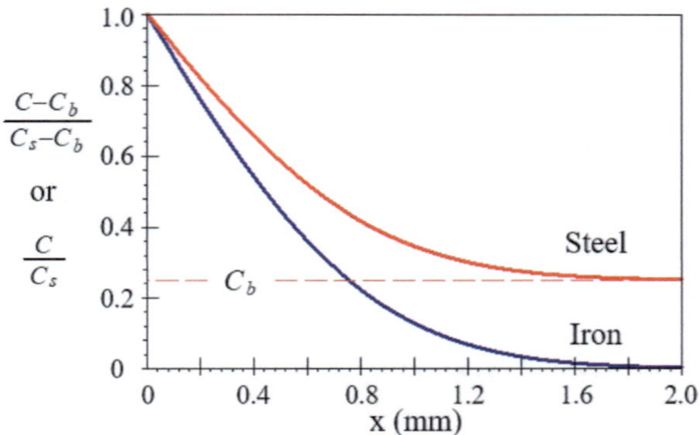

Notice that the concentration ratio exhibits a decay behavior and becomes asymptotic at $x > 1.6\ mm$ diffusion depth.

(b) The concentration values as per Eqs. (7.40a) and (7.45a) or from part (a) are

$$C_{steel} = 0.25 + 0.75\,\mathrm{erfc}\,[(1.0758)\,(2)] \simeq 0.25 \tag{7.3E4a}$$

$$C_{iron} = (1)\,\mathrm{erfc}\,[(1.0758)\,(2)] = 2.34 \times 10^{-3} \simeq 0 \tag{7.3E4b}$$

(c) The concentration of carbon in mol/cm^3 units on the surface of specimens along with Avogadro's number $N_a = 6.022 \times 10^{23}\ atoms/mol$ is

$$C_s = \left(\frac{7.87\ g/cm^3}{12\ g/mol}\right)\left(\frac{1\%}{100\%}\right) = 6.5583 \times 10^{-3}\ mol/cm^3 \tag{7.3E5a}$$

$$C_s = \left(6.5583 \times 10^{-3}\ mol/cm^3\right)\left(6.022 \times 10^{23}\ atoms/mol\right) \tag{7.3E5b}$$

$$C_s \simeq 4 \times 10^{21}\ atoms/cm^3 \tag{7.3E5c}$$

and that for the background concentration in the steel specimen becomes

$$C_b = \left(\frac{7.87\ g/cm^3}{12\ g/mol}\right)\left(\frac{0.25\%}{100\%}\right) = 1.6396 \times 10^{-3}\ mol/cm^3 \tag{7.3E6a}$$

$$C_b = \left(1.6396 \times 10^{-3}\ mol/cm^3\right)\left(6.022 \times 10^{23}\ atoms/mol\right) \tag{7.3E6b}$$

$$C_b \simeq 10^{21}\ atoms/cm^3 \tag{7.3E6c}$$

(d) From Eq. (7.50b), the amount of carbon on the surface of the specimens is

$$N = 2C_s\sqrt{\frac{Dt}{\pi}} \tag{7.3E7a}$$

$$N = (2)\left(6.5583 \times 10^{-3}\ mol/cm^3\right)$$

$$\times \sqrt{\left(6 \times 10^{-8}\ cm^2/s\right)\left(10 \times 60 \times 60\ s\right)/\pi} \tag{7.3E7b}$$

$$N = 3.4393 \times 10^{-4}\ mol/cm^2 \tag{7.3E7c}$$

$$N = \left(3.4393 \times 10^{-4}\ mol/cm^2\right)\left(6.022 \times 10^{23}\ atoms/mol\right) \tag{7.3E7d}$$

$$N \simeq 2.0711 \times 10^{20}\ atoms/cm^2 \tag{7.3E7e}$$

This is a large number of atoms per unit area.

Convert moles to grams so that

$$N = \left(3.4393 \times 10^{-4}\ mol/cm^2\right)(12\ g/mol) \tag{7.3E8a}$$

$$N = 4.1272 \times 10^{-3}\ g/cm^2 = 0.65\ g/mm^2 \tag{7.3E8b}$$

Additionally, assume a steady-state diffusion process so that the rate of carbon atoms per surface area per time can be approximated as

$$\frac{dN}{dt} = C_s\sqrt{\frac{D}{\pi t}} = \left(6.5583 \times 10^{-3}\ mol/cm^3\right)\sqrt{\frac{\left(6 \times 10^{-8}\ cm^2/s\right)}{\pi\left(10 \times 60 \times 60\ s\right)}} \tag{7.3E9a}$$

$$\frac{dN}{dt} = 4.78 \times 10^{-9}\ \frac{mol}{cm^2.s} = 5.74 \times 10^{-8}\ \frac{g}{cm^2.s} \tag{7.3E9b}$$

$$\frac{dN}{dt} \simeq 3 \times 10^{15}\ \frac{atoms}{cm^2.s} \tag{7.3E9c}$$

This is the flux dN/dt denoted as the rate of carbon atoms diffusing into the specimens.

Example 7.4 Consider a silicon (Si) n-type semiconductor wafer being doped with constant boron (B) at $1050\,^{\circ}C$ for one hour. If the amount of boron is 2.4×10^{20} $atoms/cm^2$, then calculate the diffusion depth x for a concentration of 6×10^{20} $atoms/cm^3$. Data: the activation energy and the proportionality constant for the Arrhenius equation are $3.45\ eV$ and $1.20\ cm^2/s$ are, respectively.

Solution The boron diffusion coefficient is

$$D = D_o \exp\left(-\frac{Q_d}{k_B T}\right) \tag{7.4E1a}$$

$$D = \left(1.20 \ cm^2/s\right) \exp\left[-\frac{3.45 \ eV}{\left(8.6173 \times 10^{-5} \ eV/K\right)\left(1050 + 273\right) \ K}\right] \tag{7.4E1b}$$

$$D = 8.65 \times 10^{-14} \ cm^2/s \tag{7.4E1c}$$

$$Dt = \left(8.65 \times 10^{-14} \ cm^2/s\right)(3{,}600 \ s) = 3.114 \times 10^{-10} \ cm^2 \tag{7.4E1d}$$

From Eq. (7.50b), the surface concentration of boron is

$$C_s = \frac{N}{2\sqrt{Dt/\pi}} = \frac{2.4 \times 10^{20} \ atoms/cm^2}{2\sqrt{\left(3.114 \times 10^{-10} \ cm^2\right)/\pi}} \tag{7.4E2a}$$

$$C_s = 1.2053 \times 10^{25} \ atoms/cm^3 \tag{7.4E2b}$$

Using Eq. (7.45a) yields the diffusion depth as

$$\frac{C}{C_s} = 1 - \operatorname{erf}\left(\frac{x}{2\sqrt{Dt}}\right) \tag{7.4E3a}$$

$$\operatorname{erf}\left(\frac{x}{2\sqrt{Dt}}\right) = \left(1 - \frac{C}{C_s}\right) = 1 - \frac{6 \times 10^{20}}{1.2053 \times 10^{25}} = 0.99995 \tag{7.4E3b}$$

From an error function table,

$$\frac{x}{2\sqrt{Dt}} = 2.86 \tag{7.4E4a}$$

$$x = 2\sqrt{3.114 \times 10^{-10} \ cm^2} \ (2.86) = 1.0094 \times 10^{-4} \ cm \tag{7.4E4b}$$

$$x \simeq 1.01 \ \mu m \tag{7.4E4c}$$

Using Eq. (7.56) and $Dt = 3.114 \times 10^{-10} \ cm^2$ yields

$$x = \left[(4Dt)\ln\left(\frac{N}{C\sqrt{\pi Dt}}\right)\right]^{1/2} \tag{7.4E5a}$$

$$A = \ln\left[\frac{2.4 \times 10^{20} \ atoms/cm^2}{\left(6 \times 10^{20} \ atoms/cm^3\right)\sqrt{\pi\left(3.114 \times 10^{-10} \ cm^2\right)}}\right] \tag{7.4E5b}$$

$$x = \sqrt{\left(4 \times 3.114 \times 10^{-10} \ cm^2\right)(A)} \tag{7.4E5c}$$

$$x = 1.0853 \times 10^{-4} \, cm \simeq 1.09 \, \mu m \tag{7.4E5d}$$

Therefore, there is approximately $(1.09 - 1.01) / ((1.09 + 1.01) /2) = 0.0762$ or 7.62% error in calculating the diffusion depth x from Eqs. (7.45a) and (7.56).

7.7   Kirkendall Effect

The Kirkendall effect is fundamentally based on diffusion by vacancy mechanism in binary alloy systems, where an A-B diffusion couple is in physical contact at a common interface known as the Kirkendall reference plane. In this case, the diffusion fluxes are $J_A = -J_B$, $J_A > J_B$ or $J_B > J_A$. It all depends of the diffusion coefficient D of the element with the lowest melting temperature (T_m) so that $D_A > D_B$ or $D_B > C_A$ and the corresponding concentration gradient $\partial C / \partial x$. Thus, the diffusion between A and B is referred to as interdiffusion since it is an exchange process, where Fick's first law is assumed to obey a vacancy diffusion mechanism. Thus,

$$J_A = -D_A \frac{\partial C_A}{\partial x} \tag{7.57a}$$

$$J_B = -D_B \frac{\partial C_B}{\partial x} \tag{7.57b}$$

This interdiffusion process occurs during a heat treatment called annealing at $T_A < T_{m,A}$ or $T_B < T_{m,B}$ of binary alloys, such as Ni-Ti, Cu-Ni, and Al-Si. Details on this type of diffusion can be found in the literature under Darken's analysis of binary systems [7] and Kirkendall effect (Smigelskas and Kirkendall [8]).

7.8   Effects of Defects on Diffusion

Diffusion is a structure-sensitive property where lattice irregularity or features, such as vacancies, grain and interface boundaries, and dislocations, are attributed to enhance diffusion by random walk. Therefore, diffusion in plastically deformed (cold-worked) metals is faster than in the annealed state due to these lattice defects and distorted crystallographic planes. If the diffusion coefficient D obeys the Arrhenius equation, Eq. (7.16), then the temperature-induced diffusivity increases with increasing number of lattice defects.

For self-diffusion and interdiffusion to occur, there must exist vacancies so that the direction of diffusing atoms within a substrate lattice is toward the adjacent vacancy sites. Conversely, interstitial diffusion needs empty lattice sites for this process to occur at a relative high temperature $T < T_m$. Moreover, interstitial diffusion is not always a beneficial process. For instance, hydrogen diffusion in steel pipes is a recognized phenomenon that causes embrittlement.

7.9 Measurements of Diffusion Coefficient

The measured diffusion coefficient D, in general, depends on the atomic structure of the sample, the environment, and the technique being used. For instance, radioactive tracers are the most common technique for such a purpose.

Chronologically, the work of Wasik and McCulloh in 1969 [9], Crank in 1975 [6], Murch and Nowick in 1984 [10], and Thrippleton et al. [11] in 2003 lead to experimental and theoretical details on how to measure the diffusion coefficient D.

The diffusion coefficient D measurements at relatively low temperatures, $T \ll T_m$, are important in phase precipitation during "aging" of hardenable non-ferrous alloys (Al-alloys and so forth), metal oxidation of metallic alloys (steels, brass, etc.), radiation damage, and the like.

Direct Method The experimental determination of diffusion coefficients can be accomplished by measuring the steady-state diffusion flux J_x and the concentration gradient dC/dx of a solute across a membrane. Subsequently, use Eq. (7.2b) to solve it for D as a direct method.

Indirect Method In this case, Fick's laws of diffusion are not used to determine D. Instead, a radioactivity approach is employed for such a purpose. Nonetheless, an indirect method depends on the concentration of an species. The reader is encouraged to search in the vast literature for published experimental techniques that measures D.

7.10 Mass Transport by Migration

The migration molar flux J_m develops due to an electric field, which causes a potential gradient in the solution. According to Faraday's law of electrolysis, the electric current density i is related to the migration flux. Thus, the generalized equation for the current density in an electrochemical cell is a product defined by the number of coulombs (zF) and the molar flux J_x

$$i = zFJ_x = zF \left(\frac{m}{A_w t A_s} \right) \tag{7.58}$$

where the rate of reaction process for metal reduction or oxidation is normally represented by the current density (i). Here, m is the mass loss or gain, and A_w is the atomic weight, t is time, and A_s is the electrode exposed area. The term $m/(A_w t A)$

represents the total ionic molar flux (J_m) of a particular species j in solution and has units of $mol/\left(cm^2.s\right)$.

Recall from physics that Faraday's law of electromagnetic induction differs from Faraday's law of electrolysis, despite that both are related to an electric field surrounding a conductor of length L. Actually, the conductor is an electrolyte in the former law and a solid in the latter law, which in turn is related to a magnetic flux density (magnetic induction) $\overrightarrow{\mathbf{B}}$ produced by currents in a magnetic field.

According to Faraday's law of electrolysis, an electrochemical process requires the passage of one faraday $(1\ F)$ unit of electricity to transfer one mole of electrons at the cathode and one mole at the anode. Thus, if one Faraday of electricity flows through an external circuit, then $N_A = 6.022x10^{23}$ electrons (Avogadro's number) pass through the circuit wire system, and subsequently, the well-known Faraday's constant can be calculated as follows:

$$F = q_e N_A = \left(1.6022x10^{-19}\ C/electrons\right)\left(6.022x10^{23}\ electrons/mol\right)$$

$$(7.59a)$$

$$F = 96,484\ C/mol \simeq 96,500\ C/mol \text{ of electricity} \tag{7.59b}$$

For one-dimensional analysis, the diffusion molar flux is

$$J_d = \frac{m}{A_{wt}A_S} \tag{7.60}$$

Combining Eqs. (7.58) and (7.60) along with $J_x = J_d$ yields the current density (i) due to species j in the electrolyte

$$i = +zFJ_x = -zFD\left(\frac{\partial C}{\partial x}\right)_{x=0} \quad \text{(Cathodic)} \tag{7.61a}$$

$$i = -zFJ_x = zFD\left(\frac{\partial C}{\partial x}\right)_{x=0} \quad \text{(Anodic)} \tag{7.61b}$$

The general expression for the current density can be include all three mass transport modes of matter in an electrolyte; that is,

$$i = zF\left[J_d + J_m + J_c\right] \tag{7.62}$$

Theoretically, this expression represents a general approach for kinetic studies, where the current density i in an electrolyte is influenced by diffusion, migration, and convection.

Substituting Eqs. (7.2b), (7.2c), (7.2d) into (7.62) yields

$$i = zF \left[-D\frac{dC}{dx} - \frac{zFDC}{RT}\frac{d\phi}{dx} + Cv \right] \tag{7.63}$$

The electrical potential gradient due to electrodiffusion or migration and due to an external electrical sources can be estimated as (Lorente [12], Begue and Lorente [13])

$$\left(\frac{d\phi}{dx}\right)_{migration} = \frac{RT}{xF} \tag{7.64a}$$

$$\left(\frac{d\phi}{dx}\right)_{external} = \frac{E}{L} \tag{7.64b}$$

Here, L is the distance from the electrode surface. Usually, J_d and J_m are coupled in the operation of electrochemical cells. This is a coupled mass transport process, which is referred to as **ambipolar diffusion**, where the migration molar flux J_m evolves due to a chemical potential gradient $d\mu/dx$, which in turn emerges as a result of the potential gradient $d\phi/dx$. However, other factors such as pressure gradient dP/dx and temperature gradient dT/dx are neglected as contributing factors to the total value of J_m.

7.11 Summary

The theory of mass transport is still an ongoing research topic due to either new discoveries in the vast field of materials science or solid-state physics and associated mathematical approaches to describe a particular atomic process. The mass transport by diffusion is a fundamentally random thermally activated motion of atoms in a solid crystal structure (lattice), and it is mathematically described by Fick's laws of diffusion.

Further, mass transport by diffusion and by migration is the most cited mechanism for motion of matter within a particular medium or phase. However, mass transport by convection is briefly mentioned, but it has an important place in the pharmaceutical and chemical industries. Moreover, mass transport by diffusion is described by Fick's laws. Mathematically, solutions of the second law of diffusion are important mathematical treatments in the fields of interstitial diffusion of carbon in steels and doping diffusion of boron in silicon wafers.

Most metallic substrates contain dislocations, grain boundaries, and lattice-inclusion interfaces. These defects are considered high-energy sites for hydrogen diffusion, which in turn can be the main cause for low steel toughness and yield strength. These properties are essential in designing codes for assuring structural integrity.

Appendix 7A Error Function Table

This appendix includes a convenient Table 7A.1 for selected values of the error function erf (n) and some mathematical details.

By definition, the error function is

$$\text{erf}(n) = \frac{2}{\sqrt{\pi}} \int_0^n \exp\left(-n^2\right) dn \tag{7.65a}$$

$$\text{erfc}(n) = \frac{2}{\sqrt{\pi}} \int_n^\infty \exp\left(-n^2\right) dn \tag{7.65b}$$

$$n = \frac{x}{2\sqrt{Dt}} \tag{7.65c}$$

$$\frac{\partial \, \text{erf}(n)}{\partial n} = \frac{2}{\sqrt{\pi}} \exp\left(-n^2\right) \tag{7.65d}$$

$$\frac{\partial^2 \, \text{erf}(n)}{\partial n^2} = \frac{4n}{\sqrt{\pi}} \exp\left(-n^2\right) \tag{7.65e}$$

Relations and selected values of error functions:

$$\text{erf}(n) + \text{erfc}(n) = 1 \tag{7.66a}$$

$$\text{erf}(-n) = -\,\text{erf}(n) \tag{7.66b}$$

$$\text{erfc}(-n) = 2 - \text{erf}(n) \tag{7.66c}$$

$$\text{erf}(0) = 0 \quad \& \quad \text{erfc}(0) = 1 \tag{7.66d}$$

$$\text{erf}(\infty) = 1 \quad \& \quad \text{erfc}(\infty) = 0 \tag{7.66e}$$

Table 7A.1 Values of the error function erf (n)

n	erf (n)	n	erf (n)	n	erf (n)
0	0	0.55	0.5633	1.30	0.9340
0.025	0.0282	0.60	0.6039	1.40	0.9523
0.05	0.0564	0.65	0.6420	1.50	0.9661
0.10	0.1125	0.70	0.6778	1.60	0.9763
0.15	0.1680	0.75	0.7112	1.70	0.9838
0.20	0.2227	0.80	0.7421	1.80	0.9891
0.25	0.2763	0.85	0.7707	1.90	0.9928
0.30	0.3286	0.90	0.7970	2.00	0.9953
0.35	0.3794	0.95	0.8209	2.10	0.9970
0.40	0.4284	1.00	0.8427	2.20	0.9981
0.45	0.4755	1.10	0.8802	2.30	0.9989
0.50	0.5205	1.20	0.9103	2.40	0.9993

Problems

7.1 Consider a particle traveling in the x-direction and calculate v_x, D_x and τ at $T = 300\ K$ and $\lambda = 5 \times 10^{-10}\ cm$ for **(a)** nitrogen and **(b)** hydrogen atoms in steel. Which element or particle diffuses faster? Why? [Solution: (a) $v_x = 42{,}200\ cm/s$, $D_x = 2.11 \times 10^{-5}\ cm^2/s$ and $\tau = 1.18 \times 10^{-14}\ s$].

7.2 Consider a single-crystal silicon wafer being exposed to a boron-rich environment in a furnace at relatively high temperatures for a two-step dopant diffusion process: predeposition and drive-in diffusion. The purpose of diffusing boron into silicon is to adjust the electrical properties of the wafer in an n-conductive integrated circuit (IC). Let the surface and background concentration be $7 \times 10^{25}\ atoms/m^3$ and $10^{20}\ atoms/m^3$, respectively. The diffusion coefficient is defined by

$$D = \left(2.55 \times 10^{-3}\ m^2/s\right) \exp\left[-\frac{4\ eV/atom}{\left(8.62 \times 10^{-5}\ eV/atom.K\right) T}\right]$$

According to the terminology used in the semiconductor industry, calculate **(a)** the amount of boron atoms on the surface of the silicon wafer during the predeposition process at $900\,^\circ C$ for 40 minutes and **(b)** the junction depth x_j in the drive-in diffusion process at $1100\,^\circ C$ for 2 hours. [Solution: (a) $N = 5 \times 10^{17}\ atoms/m^2$ and (b) $x_j = 1.21\ \mu m$].

7.3 (a) Derive Eq. (7.50b) using (7.45a) and Fick's first law of diffusion. **(b)** Calculate the diffusion depth if $Dt = 3 \times 10^{-10}\ cm^2$ and $C_s/C = 2 \times 10^4$. [Solution: (b) $x = 1.07\ \mu m$].

7.4 For interstitial diffusion, the concentration gradient is given by Eq. (7.45b). If $C_s \gg C_b$, then show that

$$\frac{C}{C_s} = \frac{1}{\sqrt{4\pi\,Dt}}\left(\frac{x^3}{12Dt} - x\right)$$

Hint: Use the Taylor's or the Maclaurin series to approximate the exponential function.

7.5 If interstitial diffusion of an impurity atom is to occur, then **(a)** derive an equation for the concentration rate $\partial C/dt$ in the x-direction at 1 mm. **(b)** Plot $\partial C/dt = f(t)$ at [0, 30 sec] interval. Is transient diffusion an instantaneous process? **(c)** Plot $\partial C/dt = f(t)$ at [0, 600 sec] interval. How long does it take $\partial C/dt$ to reach a maximum? Given data: $D = 10^{-5}\ cm^2/s$, $C_s = 10^{-4}\ mol/cm^3$.

7.6 Consider an AISI 1020 steel gear and a pure iron (Fe) gear exposed to a carburizing gas in a suitable furnace at $900\,^\circ C$. Assume that the diffusion coefficient and the carbon surface concentration are $6 \times 10^{-6}\ mm^2/s$ and 1%, respectively. For comparison purposes, calculate the heat treatment time **(a)** for steel (t_{steel}) and **(b)**

for pure iron (t_{Fe}) when the carbon concentration is 0.4% at a diffusion depth of 0.5 mm below the gear surface. Explain. [Solution: (a) $t = 4.37$ $hours$ and (b) $t = 8.15$ $hours$].

7.7 Consider an AISI 1025 steel gear and a pure iron (Fe) gear exposed to a carburizing gas in a suitable furnace at $900\,^\circ C$. Assume that the diffusion coefficient and the carbon surface concentration are 5.80×10^{-6} mm^2/s and 1%, respectively. For comparison purposes, calculate the heat treatment time **(a)** for steel (t_{steel}) and **(b)** for pure iron (t_{Fe}) when the carbon concentration is 0.4% at a diffusion depth of 0.5 mm below the gear surface. Explain. [Solution: (a) $t = 3.70$ $hours$ and (b) $t = 8.43$ $hours$].

7.8 Consider a 0.25%C-steel gear exposed to a carburizing gas in a suitable furnace at $900\,^\circ C$. Assume that the diffusion coefficient and the carbon surface concentration are 5.80×10^{-6} mm^2/s and 1%, respectively. Calculate the heat treatment time when the carbon concentration is 0.43% at a diffusion depth of 0.5 mm below the gear surface. [Solution: $t = 5.18$ $hours$].

7.9 Assume that the Arrhenius equation defines the diffusion coefficient D of carbon in steels (γ-$phase$) at an austenitic temperature range $1173\ K \le T \le 1373\ K$

$$D = \left(2 \times 10^{-5}\ m^2/s\right) \exp\left(-\frac{17{,}401\ K}{T}\right)$$

For a carburized steel sheet at $T = 1173\ K$ for 3.70 hours, determine the nominal carbon composition (bulk composition C_b) in a steel plate when the carbon content 0.40% (by weight) at a depth of 0.5 mm. Assume a carbon surface concentration 1%. [Solution: $C_b = 0.20\%$].

7.10 For a carburized steel sheet at $T = 1200\ K$ for 3 hours, determine the nominal carbon composition (bulk composition C_b) in a steel plate when the carbon content is 0.43% (by weight) at a depth of 0.5 mm. Use the diffusion coefficient equation given in problem 7.9 and assume a carbon surface concentration 1%. [Solution: $C_b = 0.20\%$].

7.11 Assume that a wafer made of silicon single crystal is exposed to a boron-rich environment in a furnace at relatively high temperatures for a two-step dopant diffusion process: predeposition and drive-in diffusion. Let the surface concentration and the background concentration be 7.5×10^{25} $atoms/m^3$ and 1.5×10^{20} $atoms/m^3$ boron (B) atoms, respectively. The diffusion coefficient is defined by

$$D = \left(2.55 \times 10^{-3}\ m^2/s\right) \exp\left[-\frac{46{,}404\ K}{T}\right]$$

Calculate **(a)** the amount of boron atoms on the surface of the silicon wafer during the predeposition process at $900\,°C$ for 40 minutes and **(b)** the junction depth x_j in the drive-in diffusion process at $1100\,°C$ for 2 hours. [Solution: (a) $N = 5 \times 10^{17}$ $atoms/m^2$ and (b) $x_j = 1.21\,\mu m$].

7.12 Assume that $D = 10^{-5}\,cm^2/s$ and $C_s = 10^{-4}\,mol/cm^3$ for interstitial diffusion of an impurity atom at 1-*hour* treatment. Based on this information and a diffusion depth of 1 *mm*, calculate **(a)** the concentration gradient $\partial C/dx$ and the diffusion flux J_x as per Fick's first law of diffusion, and **(b)** $\partial^2 C/dx^2$ and the concentration rate $\partial C/dt$ as per Fick's second law of diffusion. [Solution: (a) $J_x = 7.42 \times 10^{-10}\,mol/\left(cm^2.s\right)$ and (b) $\partial C/\partial t = 1.93 \times 10^{-9}\,mol/\left(cm^3.s\right)$].

7.13 Consider $D = 2 \times 10^{-5}\,cm^2/s$ and $C_s = 2 \times 10^{-4}\,mol/cm^3$ for interstitial diffusion of an impurity atom at 2-*hour* treatment. Based on this information and a diffusion depth of 2 *mm*, calculate **(a)** the concentration gradient $\partial C/dx$ and the diffusion flux J_x as per Fick's first law of diffusion, and **(b)** $\partial^2 C/dx^2$ and the concentration rate $\partial C/dt$ as per Fick's second law of diffusion. [Solution: (a) $J_x = 2.77 \times 10^{-9}\,mol/\left(cm^2.s\right)$ and (b) $\partial C/\partial t = 1.93 \times 10^{-9}\,mol/\left(cm^3.s\right)$].

7.14 Assume that a particle traveling in the x-direction. Calculate υ_x, D_x and τ at $T = 400\,K$ and $\lambda = 5 \times 10^{-10}\,cm$ for **(a)** nitrogen and **(b)** hydrogen atoms in steel. Which element or particle diffuses faster? Why? [Solution: (a) $\upsilon_x = 48{,}728\,cm/s$, $D_x = 9.12 \times 10^{-5}\,cm^2/s$ and $\tau = 2.74 \times 10^{-15}\,s$].

7.15 Calculate υ_x, D_x and τ at $T = 600\,K$ and $\lambda = 5 \times 10^{-10}\,cm$ for **(a)** nitrogen and **(b)** hydrogen atoms in steel. Which element or particle diffuses faster? Why? [Solution: (a) $\upsilon_x = 59{,}679\,cm/s$, $D_x = 2.98 \times 10^{-5}\,cm^2/s$ and $\tau = 8.39 \times 10^{-15}\,s$].

7.16 Assume that an AISI 1020 steel gear and a pure iron gear are exposed to a carburizing gas in a suitable furnace at $910\,°C$. Assume that the diffusion coefficient and the carbon surface concentration are $6.2 \times 10^{-6}\,mm^2/s$ and 1%, respectively. For comparison purposes, calculate the heat treatment time **(a)** for steel (t_{steel}) and **(b)** for pure iron (t_{Fe}) when the carbon concentration is 0.4% at a diffusion depth of 0.5 *mm* below the gear surface. Explain. [Solution: (a) $t = 4.23\,hours$ and (b) $t = 7.88\,hours$].

7.17 For a carburized steel sheet at $T = 1200\,K$ for 4 hours, determine the nominal carbon composition (bulk composition C_b) in a steel plate if $C_x = 0.43\%$ (carbon content by weight) at a depth of 0.5 *mm*. Use the diffusion coefficient equation given in problem P7.9 and assume a carbon surface concentration 1%. [Solution: $C_b = 0.20\%$].

7.18 Consider that $D = 2 \times 10^{-5}\,cm^2/s$ and $C_s = 2 \times 10^{-4}\,mol/cm^3$ for interstitial diffusion of an impurity atom at 1-*hour* treatment. Based on this information and a diffusion depth of 1 *mm*, calculate **(a)** the concentration gradient $\partial C/dx$ and the diffusion flux J_x as per Fick's first law of diffusion, and **(b)** $\partial^2 C/dx^2$ and the concentration rate $\partial C / \partial t$ as per Fick's second law of diffusion. [Solution: (a) $J_x = 2.97 \times 10^{-9} \, mol/\left(cm^2.s\right)$ and (b) $\partial C/\partial t = 1.41 \times 10^{-9} \, mol/\left(cm^3.s\right)$].

7.19 For a carburized steel sheet at $T = 1200 \, K$ for 2 hours, determine the nominal carbon composition (bulk composition C_b) in a steel plate if $C_x = 0.43\%$ (carbon content by weight) at a depth of 0.5 mm. Use the diffusion coefficient equation given in problem P7.9 and assume a carbon surface concentration 1%. [Solution: $C_b = 0.30\%$].

7.20 Consider a carburized steel sheet at $T = 1200 \, K$ for 2 hours. Determine the nominal carbon composition (bulk composition C_b) in a steel plate if $C_x = 0.43\%$ (carbon content by weight) at a depth of 0.5 mm. Use the diffusion coefficient equation given in problem P7.9 and assume a carbon surface concentration 1.10%. [Solution: $C_b = 0.27\%$].

7.21 Why convective mass transfer is independent of concentration gradient $\partial C/\partial x$ in Eq. (7.2d)?

7.22 Determine the concentration rate in the x-direction at 1 mm. Given data: $D = 10^{-5} \, cm^2/s$, $C_s = 0$, $C_b = 10^{-4} \, mol/cm^3$ and $t = 10 \, \sec$. [Solution: $\partial C/\partial t = 1.96x10^{-16} \, mol/cm^3.s$].

7.23 Show that $dC/dx = C_b/\sqrt{\pi Dt}$.

7.24 Prove that the diffusivity D is constant in the Fick's second law, Eq. (7.17).

7.25 If the migration flux is neglected in Eq. (7.2), approximate the total flux at low and high temperatures. Assume that $C_x >> C_o$ at a distance x from an electrode surface.

7.26 What does Fick's first law mean in terms of atoms or ions of a single phase?

7.27 What will Fick's first law mean if D does not vary with x or C?

7.28 Derive Fick's second law if the volume element in Fig. 6.2 has a unit cross-sectional area and the diffusing plane is located between x and $x + dx$, where J_x is the entering molar flux at x and J_{x+dx} is that leaving at $x + dx$.

7.29 Find an expression for x when

$$\frac{C_x - C_b}{C_o - C_b} = 0.5205 = \text{erf}\left(\frac{x}{\sqrt{4Dt}}\right)$$

References

1. N. Perez, *Phase Transformation in Metals: Mathematics, Theory and Practice* (Springer Nature Switzerland AG, Cham, 2020)
2. N. Perez, *Electrochemistry and Corrosion Science*, 2nd edn. (Springer International Publishing Switzerland, Cham, 2016)

3. J.T. Maloy, Factors affecting the shape of current-potential curves. J. Chem. Educ. **60**(4), 285–289 (1983)
4. J. Wang, *Analytical Electrochemistry* (Wiley-VCH, New York, 2001)
5. A.J. Bard, L.R. Faulkner, *Electrochemical Methods: Fundamentals and Applications*, 2nd edn. (Wiley, Hoboken, 2001)
6. J. Crank, *The Mathematics of Diffusion* (Oxford University Press, New York, 1975)
7. L.S. Darken, Diffusion, mobility and their interrelation through free energy in binary metallic systems. Trans. Met. Soc. AIME **175**, 184–201 (1948)
8. A.D. Smigelskas, E.O. Kirkendall, Zinc diffusion in alpha brass. Trans. AIME **171**, 130–142 (1947)
9. S.P. Wasik, K.E. McCulloh, Measurements of gaseous diffusion coefficients by a gas chromatographic technique. J. Res. Noti. Bureau Stand. A Phys. Chem. **73A**(2), 207–211 (1969)
10. G.E. Murch, A.S. Nowick (eds.), *Diffusion in Crystalline Solids* (Academic, New York, 1984)
11. M.J. Thrippleton, N.M. Loening, J. Keeler, A fast method for the measurement of diffusion coefficients: one-dimensional DOSY. Magn. Reson. Chem. **41**, 441–447 (2003)
12. S. Lorente, Constructal view of electrokinetic transfer through porous media. J. Phys. D Appl. Phys. **40**, 2941–2947 (2007)
13. P. Begue, S. Lorente, Migration versus diffusion through porous media: time-dependent scale analysis. J. Porous Media **9**, 637–650 (2006)

Chapter 8
Thermodynamics of Phase Change

8.1 Introduction

This chapter introduces classical thermodynamics of phase changes, which are referred to as phase transformations within the macroscopic science associated with energy change of matter during a process being influenced by temperature and pressure. In principles, thermodynamics describes the conservation of mass, conservation of energy, and the conversion of energy and defines the physical state of matter. Most phase changes henceforth are related to melting and solidification of metals and their alloys. Moreover, a phase change is also known as phase transition accompanied by energy transformation.

Thermodynamics is used to study the behavior of matter in the form of gas, liquid, solid, or a mixture undergoing a change from its equilibrium state and to determine the corresponding properties of states. For instance, a common phase change of matter is water, which can be treated as a substance that may exists as a solid phase (ice) undergoing melting to form a liquid phase at relatively high temperature $T > 0° C$.

This is a familiar heat transfer process that may occur at atmospheric pressure ($P = 1\ atm$). However, adding heat to the liquid water phase causes evaporation (gas) at $T > 100° C$. Simply stated, a pot of water placed on a burner boils and subsequently, evaporates under a normal household conditions. Eventually, metallic substances can also be subjected to phase changes at temperatures and pressures.

Commonly, a metal can undergo melting as a solid-to-liquid or solidification as a liquid-to-solid phase change carried out at atmospheric pressure.

In practice, a thermodynamic system can be designed and analyzed to produce energy as in a power plant or to study the thermodynamic process related to phase change as in solidification of pure metals and their alloys.

The goal in this chapter is to describe phase transformation (phase change) during solidification (freezing) using thermodynamic principles and formulations.

N. Perez, *Materials Science: Theory and Engineering*,
https://doi.org/10.1007/978-3-031-57152-7_8

This leads to the assessment of nucleation of particles and their atomic arrangement of atoms called lattice.

8.2 Thermodynamic Diagrams

A common thermodynamic diagram for a hypothetical material is shown in Fig. 8.1a as a map for the three-phase regions related to the general pressure function $P = f(T)$, where T is the temperature. These regions are identified as solid (S), liquid (L), and vapor (V) phases, which coexist at a fixed pressure–temperature triple point (TP).

Notice that Fig. 8.1a indicates the type of process for phase transition in the directions pointing by arrows, where the curves act as phase boundaries. For example, at low temperatures and pressures below the triple point (T-P), the phase transitions are

$$\text{Sublimation: } S \rightarrow V \tag{8.1a}$$

$$\text{Desublimation: } S \leftarrow V \tag{8.1b}$$

and those above the triple point are

$$\text{Melting: } S \rightarrow L \tag{8.2a}$$

$$\text{Solidification: } S \leftarrow L \tag{8.2b}$$

$$\text{Vaporization: } L \rightarrow V \tag{8.2c}$$

$$\text{Condensation: } L \leftarrow V \tag{8.2d}$$

Figure 8.1b, on the other hand, schematically illustrates the trendlines for temperature-dependent Gibbs energy (G) associated with the phases in Fig. 8.1a and the stable-phase regions at atmospheric pressure $P = 1\ atm$. Observe

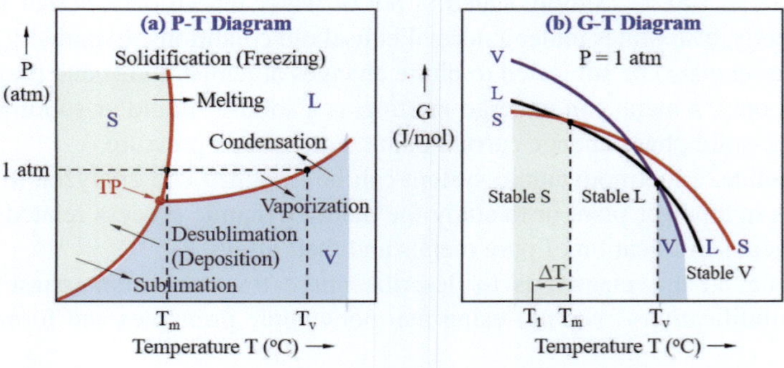

Fig. 8.1 Thermodynamic diagrams for a hypothetical material. (**a**) $P = f(T)$ and (**b**) $G = f(T)$ diagrams

the corresponding general position of the melting temperature T_m or freezing temperature $T_f = T_m$ and the vaporization temperature T_v, which divide the G-T diagram into stable-phase regions.

The phase change of matter, in general, is also a time-dependent process since some processes are slow and others are fast. In particular, a system may be in the liquidus, solidus, or gaseous state, and its thermodynamical conditions may be evaluated as functions of temperature, volume, pressure, and energy, leading to the conservation law of energy.

8.3   First and Second Laws of Thermodynamics

Mathematically, the first and second laws of thermodynamics for a system undergoing phase transformation can be written as

$$Q - W = \Delta H + \Delta KE + \Delta PE \quad \text{(First law)} \tag{8.3a}$$

$$Q = T \Delta S = T \, (S_l - S_s) \quad \text{(Second law)} \tag{8.3b}$$

$$W = \int P dV \tag{8.3c}$$

For the first law, $Q = \sum Q_i$ is the net heat transfer and Q_i may be the sum of heat transfer by conduction, convection, and radiation, $W = \sum W_i$ may be the sum of electrical, magnetic, mechanical work on the system or work done by the system, ΔH is the enthalpy change, ΔKE is the fluid velocity-dependent kinetic energy change and ΔPE is the distance-dependent potential energy change. Moreover, $Q - W$ in Eq. (8.3a) represents the net input energy. Actually, the first law of thermodynamics states that the quantity of energy is always preserved or conserved during an actual process.

For the second law, the heat transfer Q depends on the atomic disorder of the system since it is a form of disorganized energy and some disorganization (entropy) flow at a temperature T. This law states that the quality of energy is bound to decrease during an actual process and that entropy is a measure of the atomic disorder related to the uncertainty about an atomic position at a particular time and temperature, which is a measure of the average kinetic energy or motion of atomic particles (atoms or molecules).

The meaning of heat transfer and work done can be deduced as indicated below

- For $Q > 0$, the system absorbs heat from the environment, such as an induction or electric arc furnace where heat is converted to heat of fusion; $Q \rightarrow \Delta H_f$.
- For $Q < 0$, the system releases heat to the environment. This particular case is related to solidification of a melt, during which the latent heat of solidification ΔH_s being released at the liquid–solid interface is converted to heat; $\Delta H_s \rightarrow Q_s$. Subsequently, heat transfer occurs through the solidified melt to the local atmosphere; $Q_s \rightarrow Q_{atm}$.

- For $W < 0$, work is done on the system by the environment. This is the case of supplied electric energy to an induction furnace or to a pump.
- For $W > 0$, work is done by the system on the environment. This is the case of a turbine that produces mechanical work through a rotating shaft.

Gibbs Energy Change For steady-state solidification ($\Delta KE = 0$) with entropy change ΔS at constant temperature T and negligible height or depth effects ($\Delta PE = 0$), Eqs. (8.3a,b) give the combined thermodynamic laws and the Gibbs energy change ΔG

$$Q - W = \Delta H \tag{8.4a}$$

$$\Delta G = -W = \Delta H - Q = \Delta H - T\Delta S \tag{8.4b}$$

$$dG = dH - T dS \tag{8.4c}$$

$$G = H - TS \tag{8.4d}$$

where G is the path-independent Gibbs energy (known as Gibbs free energy or Gibbs function), ΔG is the change in Gibbs energy, dG is the infinitesimal change in Gibbs energy, W is the work done, Q is the amount of transient heat transfer, and ΔH is the enthalpy change.

During the liquid-to-solid (L-S) phase transformation, $\Delta H_f = \Delta H = H_l - H_s$ is the latent heat of fusion of a pure metal at the freezing temperature T_f and for comparison, ΔH_f^* is the latent heat of fusion of an alloy at the freezing temperature range $T_s < T_f^* < T_l$. Here, T_s is the solid temperature and T_l is the liquid temperature. Moreover, the latent heat of fusion ΔH_f and latent heat of solidification is interchangeable definition with $\Delta H_s = \Delta H_f$.

Further, the thermodynamic criteria for a macroscopic system are

- $\Delta G < 0$ for a spontaneous process
- $\Delta G = 0$ for a system in equilibrium
- $\Delta G > 0$ for a reverse spontaneous process

For a small undercooling $\Delta T \rightarrow 0$, $\Delta G \rightarrow 0$ so that $T \rightarrow T_f$, $\Delta H_f(T) = \Delta H_f(T_f)$ and $\Delta S_f(T) = \Delta S_f(T_f)$. Thus, Eq. (8.4b) yields the equations for entropy of fusion ΔS_f and the corresponding ΔG and ΔT equations

$$\Delta S_f = \frac{\Delta H_f}{T} \tag{8.5a}$$

$$\Delta G = \Delta H_f - T_f \frac{\Delta H_f}{T} = \Delta H_f \left(\frac{T - T_f}{T} \right) = -\Delta H_f \left(\frac{T_f - T}{T} \right) \tag{8.5b}$$

$$\Delta G = -\Delta H_f \left(\frac{\Delta T}{T} \right) \tag{8.5c}$$

$$\Delta T = \left(T_f - T \right) > 0 \tag{8.5d}$$

Here, the degree of undercooling ΔT is the driving force for liquid-to-solid ($L \rightarrow S$) transformation, the entropy of fusion ΔS_f is a measure of disorganized energy, the freezing temperature T_f is a measure of the atomic bond strength, and the $\Delta S_f T_f$ is a form of heat at $T = T_f$. This implies that the evolution of latent heat of fusion or solidification ΔH_f for a pure metal occurs at $T < T_f$ due to a moving L-S interface.

8.3.1 Thermodynamics Properties

It is important to understand that there are fundamental extensive and intensive thermodynamic properties that define characteristic features of a system at equilibrium. Specifically, fundamental properties are directly related to the fundamental laws of thermodynamics, extensive properties (mass, volume, etc.) are system size-dependent and intensive properties (density, temperature, pressure, etc.) are system size-independent. Moreover,

- An open system exchanges energy and matter with its surroundings.
- A closed system exchanges energy, not matter, with its surroundings.
- An Isolated system does not exchange energy and matter with its surroundings.

Briefly, a temperature difference induces heat transfer by conduction, convection, and radiation. Mainly the sun, ocean, and air are the most common natural sources of heat energy. On the other hand, fuel cells are man-made systems used to extract energy from fuels.

8.3.2 Specific Heats

The specific heat is the amount of heat required to raise the temperature of a substance by $1°C$ per unit mas unit (1 g). Similarly, heat capacity is defined as the quantity of heat necessary to raise the temperature of a general mass of a substance by $1°C$. Both are usually measured in kilo-joules per gram per degree Celsius ($kJ/g.°C$).

Consider dividing the above energy terms by the mass m of the working substance. Actually, the reason for doing so is to convert thermodynamic quantities into properties. Nevertheless, this leads to the specific internal energy $u = U/m = u(T, v)$ and to the specific enthalpy $h = H/m = h(s, P)$ with specific volume $v = V/m$ and specific entropy $s = S/m$.

For an isometric process ($dv = 0$) along with the first law in specific energy terms, the specific heat capacity or specific heat c_v is the energy required to raise the temperature of a unit mass of a substance by one degree. Letting $\Delta KE = \Delta PE = 0$ yields the first law as

$$\partial q - \partial w = du \tag{8.6a}$$

$$w = \int P dv = 0 \tag{8.6b}$$

$$u = u\,(T, v) \tag{8.6c}$$

$$du = \left(\frac{\partial u}{\partial T}\right)_v dT + \left(\frac{\partial u}{\partial v}\right)_T dv = \left(\frac{\partial u}{\partial T}\right)_v dT \tag{8.6d}$$

$$\frac{du}{dT} = \left(\frac{\partial u}{\partial T}\right)_v \tag{8.6e}$$

Dividing Eq. (8.6a) by dT and combining the resultant expression with Eq. (8.6e) yield the specific heat at a constant volume v

$$c_v = \left(\frac{\partial u}{\partial T}\right)_v = \left(\frac{\partial q}{\partial T}\right)_v \tag{8.7}$$

Similarly, for an isobaric process ($dP = 0$),

$$\partial q - \partial w = dh \tag{8.8a}$$

$$w = \int P dv = 0 \tag{8.8b}$$

$$h = h\,(T, P) \tag{8.8c}$$

$$dh = \left(\frac{\partial h}{\partial T}\right)_P dT + \left(\frac{\partial h}{\partial P}\right)_T dP = \left(\frac{\partial h}{\partial T}\right)_P dT \tag{8.8d}$$

$$\frac{dh}{dT} = \left(\frac{\partial h}{\partial T}\right)_P \tag{8.8e}$$

from which the specific heat capacity at constant pressure is defined by

$$c_p = \left(\frac{\partial h}{\partial T}\right)_P = \left(\frac{\partial q}{\partial T}\right)_P \tag{8.9}$$

It can be concluded that thermodynamics is a temperature-dependent energy science, and the specific heat defines the energy–temperature ratio as property. In other words, knowledge of the specific heat capacity allows the prediction of thermodynamic properties.

8.4   Solidification of Metals

Consider the solidification of a pure metal and assume a linear behavior of $H = H(T)$ and $G = G(T)$ as schematically shown in Fig. 8.2. The slopes of these plots are the specific heats c_l, c_s (Fig. 8.2a) and the entropy terms S_l, S_s (Fig. 8.2b) for the liquid and solid, respectively.

Figure 8.2b illustrates the phase stability regions, where $G_f = f\left(T_f\right)$ is the equilibrium Gibbs energy at the fusion or melting and ΔG is the driving force for $L \rightarrow S$ phase transformation. Specifically, solidification occurs when $\Delta G = \Delta G_m = (G_s - G_l) < 0$ at $T = T_s$, induced by an undercooling $\Delta T_u = \left(T_f - T_s\right) > 0$. Here, ΔG_m denotes the molar Gibbs energy of mixing. Conversely, melting occurs if $\Delta G = (G_l - G_s) < 0$ at $T = T_l$, where $\Delta T_s = \left(T_l - T_f\right) > 0$ denotes the degree of superheating the liquid.

8.4.1   Binary Solutions

Consider Fig. 8.3 showing a schematic plot of Gibbs energy function of the form $G = f(T, P, \sum X_i)$ for one A-B binary system with nominal composition X_o, where X is the mole fraction. This schematic plot is treated as a G-X diagram with $G = f(X)$ function at constant T and P.

According to the basic principles of solidification thermodynamics related to liquid-to-solid phase transformation, Gibbs energy G is a thermodynamics potential that is minimized at equilibrium with constant absolute temperature T and pressure P. Moreover, the driving force for solidification of binary or multicomponent systems is denoted as the Gibbs energy change ΔG.

Multicomponent System For a multicomponent system with $i = A, B, C, D, \ldots$ components, Gibbs energy G can have different mathematical forms, such as

- $G = f\left(T, P, \sum n_i\right)$ with n_i as the number of moles at temperature T and pressure P. In thermodynamics, T has an absolute unit, such as Kelvin degree.

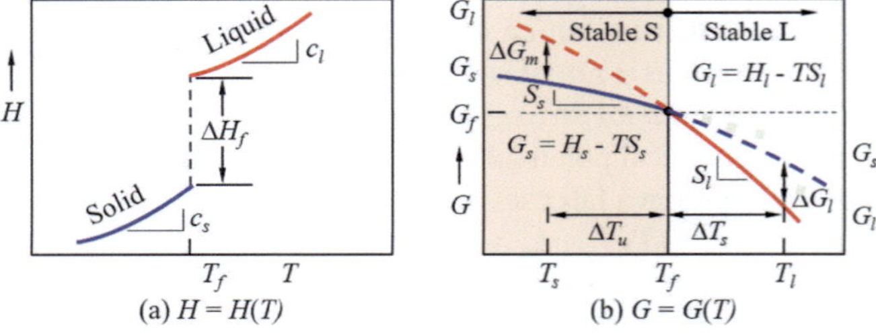

Fig. 8.2 Schematic energy trends at temperature T for liquid and solid phases. (**a**) Enthalpy and (**b**) Gibbs energy

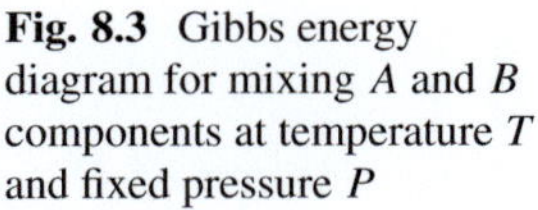

Fig. 8.3 Gibbs energy diagram for mixing A and B components at temperature T and fixed pressure P

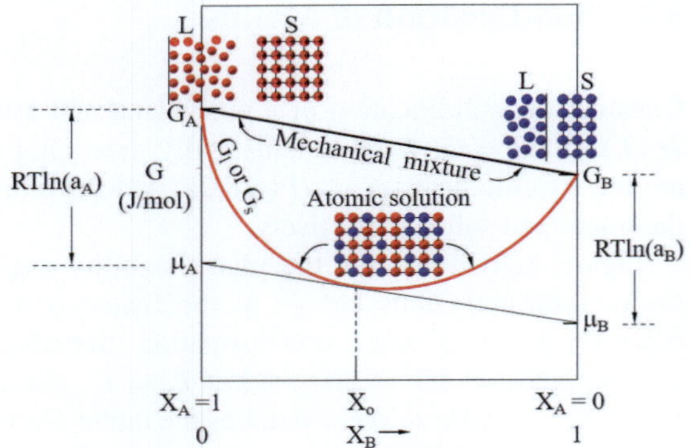

- $G = f\left(T, P, \sum X_i\right)$ with X_i as the mole fraction of $i = A, B, C, D, \dots$ components at T and P. Thus, a two-dimensional G-X diagram can be developed for ideal or regular solutions since $G = f(X)$ at T and P.

Binary System For an A-B binary system, the Gibbs energy functions are $G = f(T, P, X_A, X_B)$ or $G = f(T, P, n_A, n_B)$. In this case, a two-dimensional G-n or G-X diagram can be developed for ideal or regular solutions since $G = f(n)$ or $G = f(X)$ at constant T and P.

According to Fig. 8.3, $G = G_m$ is the Gibbs energy of mixing for A and B elements or components. The general Gibbs function is denoted as $G = f(X_B)$ since X_B is added to X_A for forming either an ideal or regular solution susceptible to undergo freezing under equilibrium conditions.

Regarding the G-X diagram in Fig. 8.3, connecting the solidification energies G_A and G_B represents a linear mechanical mixture with initial mole fractions denoted as $X_A = 1$ and $X_B = 1$. Let X_B be the alloying component to be mixed with X_A so that $X_A + X_B = 1$ prevails for an ideal solution of components.

After mixing A and B, the function $G = f(X)$ decreases and reaches a minimum value at $X_B = X_o$. Conversely, $G = f(X)$ increases at $X_B > X_o$. This is a nonlinear process denoted by a red curve in Fig. 8.3. Consequently, mixing A and B forms a single-phase solid solution having a certain lattice structure (Porter et al. [1, Chapter 1]).

An additional information can be extracted from the curve in Fig. 8.3. Use the mechanical mixture line as a reference. Draw a parallel line through the X_o point. As a result, $RT \ln(a_A)$ and $RT \ln(a_B)$ are extracted, where a_A, a_B are the activities and μ_A, μ_B are the chemical potential terms of A and B components. These a_A, a_B factors will be dealt with in a later section.

Regarding $G = f(T, P, n_A, n_B)$, differentiating this function yields the infinitesimal rate of change of the Gibbs energy (dG) related to chemical potentials (μ_i) with $i = A, B$

$$dG = \left(\frac{\partial G}{\partial T}\right)_{P,n_A,n_B} dT + \left(\frac{\partial G}{\partial P}\right)_{T,n_A,n_B} dP \qquad (8.10\text{a})$$

$$+ \left(\frac{\partial G}{\partial n_A}\right)_{T,P,n_B} dn_A + \left(\frac{\partial G}{\partial n_B}\right)_{T,P,n_A} dn_B$$

$$\mu_A = \left(\frac{\partial G}{\partial n_A}\right)_{P,T,n_B} \qquad (8.10\text{b})$$

$$\mu_B = \left(\frac{\partial G}{\partial n_B}\right)_{P,T,n_A} \qquad (8.10\text{c})$$

$$dG = \left(\frac{\partial G}{\partial T}\right)_{P,n_i} dT + \left(\frac{\partial G}{\partial P}\right)_{T,n_i} dP + \mu_A dn_A + \mu_B dn_B \qquad (8.10\text{d})$$

This expression, Eq. (8.10d), is very common in the literature.

8.4.2 Molar Gibbs Energy Change

Knowledge of chemical or potential energy is an important concept in the thermodynamics of materials science and engineering fields because it allows the prediction of thermodynamic properties, specifically the Gibbs energy of an entire system, at a given temperature and pressure.

The main objective of this section is to derive an expression for the molar Gibbs energy change in terms of potential energy and mole fraction of an A-B binary system. Moreover, the potential energy is an intensive property independent of system size.

For fixed T (isothermal) and P (isobaric), Eq. (8.10d) along with $dT = dP = 0$ and $dn = 0$ (for a fixed total number of moles n_i and mole fraction X_i in an A-B mixture) gives

$$dG = \mu_A dn_A + \mu_B dn_B \qquad (8.11\text{a})$$

$$\int dG = \mu_A \int dn_A + \mu_B \int dn_B \qquad (8.11\text{b})$$

$$X_A = \frac{n_A}{n_A + n_B} = \frac{n_A}{n} \quad \text{and} \quad X_B = \frac{n_B}{n_A + n_B} = \frac{n_B}{n} \qquad (8.11\text{c})$$

$$n_A = n X_A \quad \text{and} \quad n_B = n X_B \qquad (8.11\text{d})$$

$$dn_A = n d X_A + X_A dn = n d X_A = \frac{n_A}{X_A} d X_A \qquad (8.11\text{e})$$

$$dn_B = n d X_B + X_B dn = n d X_B = \frac{n_A}{X_B} d X_B \qquad (8.11\text{f})$$

$$\int_{G_1}^{G_2} dG = n_A \mu_A \int \frac{dX_A}{X_A} + n_B \mu_B \int \frac{dX_B}{X_B} \tag{8.11g}$$

$$\Delta G_m = G_2 - G_1 = n_A \mu_A \ln(X_A) + n_B \mu_B \ln(X_B) \tag{8.11h}$$

where G_1 and G_2 are the "before" and "after," respectively, Gibbs energy terms for the mixing process, Also, A, B denote the pure components in the A-B mixture, G denotes the Gibbs energy (J), $G_m = G/n$ denotes the molar Gibbs energy (J/mol). In the literature, $G_A, G_B = \mu_A^o, \mu_B^o$ implies that μ_A^o, μ_B^o are the standard or reference chemical potentials.

The total mole fraction in an A-B binary system is $X_A + X_B = 1$, which implies that $X_A < 1$ and $X_B < 1$. Conclusively, $\ln(X_A) < 0$ and $\ln(X_B) < 0$, and Eq. (8.11h) yields $\Delta G_m < 0$ as the criterion for an spontaneous mixture or reaction. Additionally, Eq. (8.11a) can be used to derive an expression for the changes in chemical potential for an A-B binary alloy system. The resultant expression is known as the Gibbs–Duhem equation for an A-B binary alloy system as illustrated in the example given below.

Example 8.1 For an A-B binary alloy with $n_A = 1.5\ mol$ and $n_B = 1\ mol$ at $298\ K$, calculate the enthalpy change ΔH of mixing for an ideal solid solution at atmospheric pressure $P = 1\ atm = 101.15\ kPa$.

Solution First of all,

$$n = n_A + n_B = 1.5\ mol + 1\ mol = 2.5\ mol \tag{8.1E1a}$$

$$X_A = \frac{n_A}{n_A + n_B} = \frac{1.5}{1.5 + 1} = 0.6 \tag{8.1E1b}$$

$$X_B = 1 - X_A = 0.4 \tag{8.1E1c}$$

This means that the nominal alloy composition is $0.60A$-$0.40B$. Then, the Gibbs energy of mixing is

$$\Delta G = nRT\left[X_A \ln(X_A) + X_B \ln(X_B)\right] \tag{8.1E2a}$$

$$= (2.5\ mol)(8.3145\ J/mol.K)(298\ K) \tag{8.1E2b}$$

$$\times\ [(0.60)\ln(0.60) + (0.40)\ln(0.40)]$$

$$= -4168.80\ J/mol \tag{8.1E2c}$$

which is a favorable quantity of energy for mixing because $\Delta G < 0$. This condition is essential for an spontaneous mixing of A and B atoms.

Further analysis yields the entropy of mixing is

$$\Delta S = -nR\left[X_A \ln(X_A) + X_B \ln(X_B)\right] \tag{8.1E3a}$$

$$= -(2.5\ mol)(8.3145\ J/mol.K) \tag{8.1E3b}$$

$$\times \left[(0.60) \ln (0.60) + (0.40) \ln (0.40)\right]$$

$$= 13.99 \ J/(mol.K) \tag{8.1E3c}$$

The enthalpy of mixing is

$$\Delta G = \Delta H - T \Delta S \tag{8.1E4a}$$

$$\Delta H = \Delta G + T \Delta S = -4168.80 \ J/mol \tag{8.1E4b}$$

$$+ (298 \ K)(13.99 \ J/mol.K)$$

$$\Delta H = 4169 \ J/mol \tag{8.1E4c}$$

Hence, it has been demonstrated that $\Delta H > 0$ for an ideal A-B binary alloy.

Example 8.2 Manipulate Eq. (8.12a) in order to derive a homogeneous ordinary differential equation (ODE) for the chemical potential of an A-B binary alloy system at constant T and P.

Solution From Eq. (8.11a),

$$dG = \mu_A dn_A + \mu_B dn_B \tag{8.2E1a}$$

$$\int_0^G dG = \mu_A \int_0^{n_A} dn_A + \mu_B \int_0^{n_B} dn_B \tag{8.2E1b}$$

$$G = \mu_A n_A + \mu_B n_B \tag{8.2E1c}$$

Differentiation Eq. (8.2E1c) yields

$$dG = \mu_A dn_A + n_A d\mu_A + \mu_B dn_B + n_B d\mu_B \tag{8.2E2}$$

Equating Eqs. (8.2E1a) and (8.2E2) at $dT = dP = 0$ gives the sought homogeneous ordinary differential equation (ODE)

$$\mu_A dn_A + n_A d\mu_A + \mu_B dn_B + n_B d\mu_B = \mu_A dn_A + \mu_B dn_B \tag{8.2E3a}$$

$$n_A d\mu_A + n_B d\mu_B = 0 \tag{8.2E3b}$$

Hence, Eq. (8.2E3b) is one of the mathematical forms known as the Gibbs–Duhem equation. Clearly, this mathematical model describes the relationship between chemical potential change $(d\mu_A, d\mu_B)$ and number of moles (n_A, n_B) for A and B components. It should be mentioned that the chemical potential is just the partial molar Gibbs energy at equilibrium conditions with constant T and P. This expression, Eq. (8.2E3b) is compared with other Gibbs–Duhem mathematical expressions in a later section.

8.4.3   Thermodynamic Criteria for Reactions

In theory, the Gibbs energy ΔG can be used as a tool or criterion for determining the direction and condition of a reaction as written. Consider the following generalized reactions

$$Cu \rightarrow Cu^{+2} + 2e^- \text{(oxidation reaction)} \tag{8.12a}$$

$$2NO + O_2 \rightarrow 2NO_2 \qquad \text{(reduction reaction)} \tag{8.12b}$$

$$L \rightarrow \alpha + \beta \qquad \text{(phase change reaction)} \tag{8.12c}$$

$$L \rightarrow \gamma + Fe_3C \quad \text{(phase change reaction)} \tag{8.12d}$$

$$\gamma \rightarrow \alpha + \beta \qquad \text{(phase change reaction)} \tag{8.12e}$$

where the arrow $\rightarrow$ indicates the direction of each reaction as written. For example, a copper atom Cu becomes an ion Cu^{+2} by giving up to two electrons $(2e^-)$.

Now, ΔG is used to predict the reaction spontaneity based on the following thermodynamic criteria

$$\Delta G < 0 \quad \text{for a spontaneous reaction at fixed } T \text{ and } P \tag{8.13a}$$

$$\Delta G > 0 \quad \text{for a non-spontaneous reaction at fixed } T \text{ and } P \tag{8.13b}$$

$$\Delta G = 0 \quad \text{for equilibrium at fixed } T \text{ and } P \tag{8.13c}$$

Notice that all three reactions above, Eq. (8.12), are considered to occur or proceed spontaneously as written, provided that $\Delta G < 0$. Otherwise, a reaction is non-spontaneous if it proceeds in the opposite direction $(\leftarrow)$ due to $\Delta G > 0$, or it will not proceed since the substance is in the equilibrium state with $\Delta G = 0$.

8.4.4   Atom-Pair Model

Consider the internal energy of atom pairs as the bonding energy that takes into account the interatomic distances in ideal solutions with $\Delta H = 0$ and regular (non-ideal) solutions with $\Delta H \neq 0$. Additional theoretical and analytical details on the subject matter can be found elsewhere (Dantzig and Rappaz [2, p. 43], DeHoff [3, p. 196], Gokcen [4, p. 81]).

The general two-dimensional (2-D) view of the concept of A-A, B-B, and A-B atom pairs (ij) in the solid state is illustrated in Fig. 8.4. These arrangements are initially in the unmixed state (Fig. 8.4a), and only A-A and B-B atom pairs are shown. After mixing (Fig. 8.4b) at a temperature T and pressure P, the A-B atom pairs appear randomly.

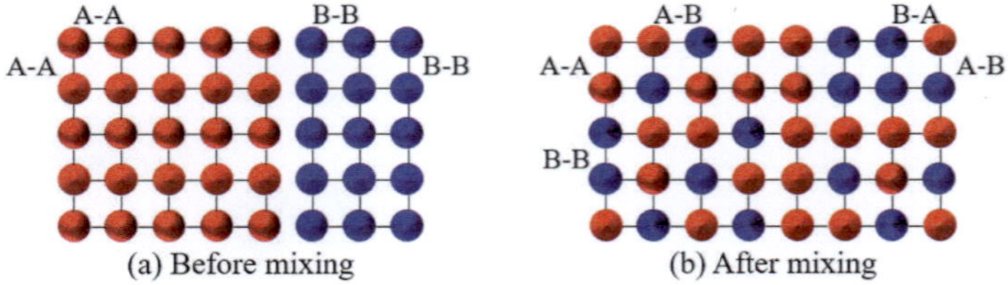

Fig. 8.4 Atomic arrangement (**a**) before mixing and (**b**) after mixing A and B atoms during phase transformation from the liquid phase at temperature T_f and pressure P to a solid phase at $T \ll T_f$ at P. The final product after solidification is a complete A-B binary alloy

This model serves as a guide to determine the coordination number (Z) of a particular atom pairs with binding energy (or cohesive energy) U_{AA}, U_{BB}, or U_{AB}.

The relevant bond fractions (f_{ij}) are f_{AA}, f_{BB}, and f_{AB}. Thus, the total internal energy (U_m) of mixing and the internal energy (U_{ij}) of the atom-pair model (sum of the attractive and repulsive atomic energies) are defined by

$$U_m = \sum f_{ij} U_{ij} = f_{AA} U_{AA} + f_{BB} U_{BB} + f_{AB} U_{AB} \tag{8.14a}$$

$$U_{ij} = -\left(\frac{A_1}{x^m}\right)_{\text{attraction}} + \left(\frac{A_2}{x^n}\right)_{\text{repulsive}} \tag{8.14b}$$

with

$$U_{ij}(x_o) = U_o^A < 0 \text{ at } x_o^a \quad \text{(solid state)} \tag{8.15a}$$

$$U_{ij}(x_o) = U_o^B < 0 \text{ at } x_o^b \quad \text{(solid state)} \tag{8.15b}$$

$$U_{ij}(x) \to 0 \text{ as } x \to \infty \quad \text{(vapor state)} \tag{8.15c}$$

$$U_{ij}(x) \to \infty \text{ as } x \to 0 \quad \text{(ionic state)} \tag{8.15d}$$

where A_1, A_2 denote constants for atom attraction and repulsion, respectively, x denotes the interatomic distance, m, n denote exponents.

Figure 8.5a shows the potential energies for strong and weak solids, and Fig. 8.5b shows the equilibrium interatomic distance $r = r_o$ for atom pairs (ij) in the solid state. For the purpose of clarity, $r_{o,AA} \neq r_{o,BB} \neq r_{o,AB}$.

In materials science or physical metallurgy, a solid phase is stable due to the chemical bonds between atoms being induced by an atomic force field. Hence, a crystal structure is a stable solid phase constituted by a packed atomic arrangement that has a potential energy, commonly known as internal energy for atomic bonding.

In order for the A-B alloys to achieve equilibrium in the solid state, the potential energy must be a minimum, which is designated as $U_{ij}(r_o) = U_o < 0$ in Fig. 8.5a for an A-B atom-pair in an ideal solid solution. In addition, recall from Chap. 1 that the net interatomic force and the minimum potential energy for a particle pair

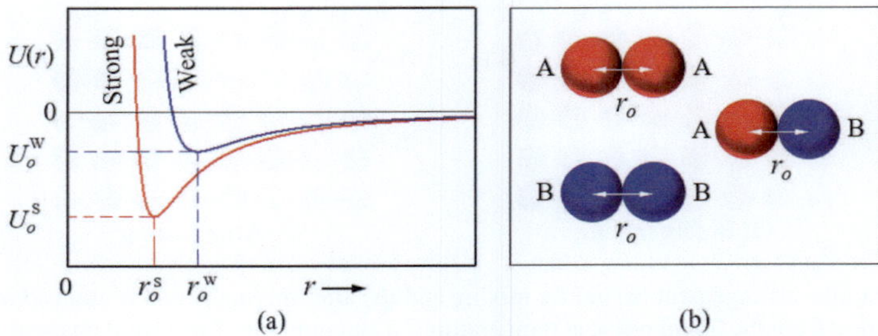

Fig. 8.5 (**a**) Internal energy $U(r) = U_{ij}$ as a function of interatomic distance r and (**b**) atom pairs for two hypothetical A and B atoms

at equilibrium are $F(r_o) = 0$ and $U(r_o) = U_o$, respectively. This means that the attractive and repulsive forces acting on the atom system are equal but of opposite signs, and the corresponding potential energy terms add up to reach a minimum value.

8.5   Enthalpy of Mixing

The enthalpy of mixing (ΔH_m) represents a thermodynamic quantity that measures the total heat content of the system. Actually, ΔH_m is used to classify solutions as indicated below

- $\Delta H_m > 0$ for an endothermic mixing due to absorption of heat during melting.
- $\Delta H_m < 0$ for an exothermic mixing due to release of heat during solidification.
- $\Delta H_m = 0$ for ideal solid solutions.

For one mole of disorder solution containing A and B atoms, the total mole fraction of the constituents (components) is $X_A + X_B = 1$ and the bonding-energy factors (bond fractions) f_{ij} per type of atomic bonding are defined by multiplying the number of bonds $(Z/2)\, X_i$ and the number of atoms $N_a X_i$. Thus,

$$f_{AA} = \frac{1}{2} Z X_A \left(N_a X_A\right) = \frac{1}{2} Z N_a X_A^2 = \frac{1}{2} Z N_a \left(1 - X_B\right)^2 \tag{8.16a}$$

$$f_{BB} = \frac{1}{2} Z X_B \left(N_a X_B\right) = \frac{1}{2} Z N_a X_B^2 \tag{8.16b}$$

$$f_{AB} = \frac{1}{2} \left[Z X_B \left(N_a X_A\right) + \left(Z X_A\right)\left(N_a X_B\right)\right] = \frac{1}{2} Z N_a \left(2 X_A X_B\right) \tag{8.16c}$$

where f_{AB} represents both $A\text{-}B$ an $B\text{-}A$ bonds, N_a denotes Avogadro's number, Z denotes the coordination number (CN) of an atom, and $1/2$ factor denotes the shared quantitative (it eliminates double counting nearest neighbors).

The potential energy of the A-B binary solution (U_m) of mixing is essentially the internal energy that depends on the A-B atomic bonding and the vibrational energy at finite temperature. Thus, the potential energy of the solution U_m is the sum of the product $f_{ij}U_{ij}$ along with $X_A + X_B = 1$, where $i, j = A, B$ (Dantzig and Rappaz [2, pp. 43–47])

$$U_m = f_{AA}U_{AA} + f_{BB}U_{BB} + f_{AB}U_{AB} \tag{8.17a}$$

$$U_m = \frac{1}{2}ZN_a\left(X_A^2 U_{AA} + X_B^2 U_{BB} + 2X_A X_B U_{AB}\right) \tag{8.17b}$$

$$U_m = \frac{1}{2}ZN_a\left(X_A U_{AA} + X_B U_{BB}\right) + \Omega_m X_A X_B \tag{8.17c}$$

where $U_{ij} = U_{AA}, U_{BB}, U_{AB}$. For a regular solution, $U_{AA} \neq U_{BB} \neq U_{AB}$ and for an ideal solution $U_{AA} = U_{BB} = U_{AB}$, $U_m = 0.5ZN_a U_{AA}$ and interaction parameter related to the mixing process becomes $\Omega_m = 0$.

For a regular solution,

- U_{AB} in Eq. (8.17b) measures the relative affinity of atoms A and B in solution. Actually, the concept of electron affinity is most appropriate for characterizing the A-B system in terms of potential energy. This implies that one component has a greater electron affinity than the other.
- If $U_{AB} > 0$ then demixing occurs due to a large interatomic distance $r >> r_o$ between A and B components.
- If $U_{AB} < 0$, then mixing occurs due to an equilibrium interatomic distance $r = r_o$ between A and B. Thus, the A-B binary system is assumed to form an ordered phase.

The exchange binding energy parameter Ω_m, also known as the interaction mixing parameter, due to the interaction between two atoms is defined as

$$\Omega_m = \frac{1}{2}ZN_a\left[2U_{AB} - (U_{AA} + U_{BB})\right] \tag{8.18a}$$

$$\Omega_m = ZN_a\left[U_{AB} - \left(\frac{U_{AA} + U_{BB}}{2}\right)\right] \tag{8.18b}$$

This parameter can have three energy characters like

- $\Omega_m < 0$ for atomic attraction in a regular solution, where A-B bonding is favorable causing the formation of a single intermetallic phase.
- $\Omega_m > 0$ for atomic repulsion in a regular solution, where A-B bonding is unfavorable causing the formation of clustering.
- $\Omega_m = 0$ for ideal solutions independent of atomic arrangements.

The interaction parameter Ω_m may be called the atomic binding parameter since it depends on binding energy terms like U_{AA}, U_{BB} and U_{AB} for a regular solution at a certain temperature T. Nonetheless, the enthalpy of mixing ΔH_m and related components are mathematically defined as (Dantzig and Rappaz [2, p. 43])

$$\Delta H_m = \Omega_m X_A X_B = \Omega_m \left(1 - X_B\right) X_B \tag{8.19a}$$

$$H_l = \Omega_l X_A X_B \tag{8.19b}$$

$$H_s = \Omega_s X_A X_B \tag{8.19c}$$

The enthalpy of mixing ΔH_m is essentially the heat being released or absorbed upon mixing components A and B.

Example 8.3   Derive Eq. (8.19a)

Solution   Expand the terms in Eq. (8.17b) in order to derive an expression for the exchange binding energy parameter Ω_m. Using $X_A + X_B = 1$ yields

$$X_A^2 U_{AA} = X_A \left(1 - X_B\right) U_{AA} = \left(X_A - X_A X_B\right) U_{AA} \tag{8.3E1a}$$

$$= X_A U_{AA} - X_A X_B U_{AA}$$

$$X_B^2 U_{BB} = X_B \left(1 - X_A\right) U_{BB} = \left(X_B - X_A X_B\right) U_{BB} \tag{8.3E1b}$$

$$= X_B U_{BB} - X_A X_B U_{BB}$$

Then,

$$X_A^2 U_{AA} + X_B^2 U_{BB} \tag{8.3E2}$$

$$= X_A U_{AA} + X_B U_{BB} - X_A X_B U_{AA} - X_A X_B U_{BB}$$

and

$$X_A^2 U_{AA} + X_B^2 U_{BB} + 2 X_A X_B U_{AB} \tag{8.3E3}$$

$$= X_A U_{AA} + X_B U_{BB} + X_A X_B \left(2 U_{AB} - U_{AA} - U_{BB}\right)$$

Inserting Eq. (8.3E3) into (8.22b) gives

$$U_m = \frac{1}{2} Z N_a \left[X_A U_{AA} + X_B U_{BB}\right] \tag{8.3E4}$$

$$+ \frac{1}{2} Z N_a \left[X_A X_B \left(2 U_{AB} - U_{AA} - U_{BB}\right)\right]$$

from which the interaction parameter takes the form

$$\Omega_m = \frac{1}{2} Z N_a \left[\left(2 U_{AB} - U_{AA} - U_{BB}\right)\right] \tag{8.3E5a}$$

$$\Omega_m = Z N_a \left[U_{AB} - \left(\frac{U_{AA} + U_{BB}}{2}\right)\right] \tag{8.3E5b}$$

Then, Eq. (8.3E4) becomes

$$U_m = \frac{1}{2}ZN_a\left(X_A U_{AA} + X_B U_{BB}\right) + \Omega_m X_A X_B \tag{80.3E6}$$

Example 8.4 Consider an A-B binary alloy in the liquid solution state and assume that the enthalpy of mixing can be defined in terms of partial enthalpy terms (shown below) for components A and B, provided that $X_A + X_B = 1$ and

$$\Delta H_m = X_A \Delta H_A + X_B \Delta H_B \tag{8.4E1}$$

with the linear functions $\Delta H_i = f\left(X_j\right)$ and $\Delta H_j = f\left(X_i\right)$ defined by

$$\Delta H_A = \Delta H_m + \frac{d\left(\Delta H_m\right)}{dX_A} X_B \tag{8.4E2a}$$

$$\Delta H_B = \Delta H_m + \frac{d\left(\Delta H_m\right)}{dX_B} X_A \tag{8.4E2b}$$

Based on this information, show that $\Delta H_i = \Omega_m X_j^2$ and $\Delta H_j = \Omega_m X_i^2$, where $i = A$ and $j = B$.

Solution Start with Eq. (8.19a) and take the derivatives with respect to X_A and X_B

$$\Delta H_m = \Omega_m X_A X_B = \Omega_m X_A\left(1 - X_A\right) \tag{8.4E3a}$$

$$\Delta H_m = \Omega_m\left(X_A - X_A^2\right) \tag{8.4E3b}$$

$$\frac{d\left(\Delta H_m\right)}{dX_A} = \Omega_m\left(1 - 2X_A\right) \tag{8.4E3c}$$

and

$$\Delta H_m = \Omega_m X_A X_B = \Omega_m\left(1 - X_B\right)X_B \tag{8.4E4a}$$

$$\Delta H_m = \Omega_m\left(X_B - X_B^2\right) \tag{8.4E4b}$$

$$\frac{d\left(\Delta H_m\right)}{dX_B} = \Omega_m\left(1 - 2X_B\right) \tag{8.4E4c}$$

Substituting Eqs. (8.4E3b) and (8.4E4b) into (8.4E2a,b), respectively, yields

$$\Delta H_A = \Delta H_m + X_B \frac{d\left(\Delta H_m\right)}{X_A} \tag{8.4E5a}$$

$$\Delta H_A = \Omega_m X_A X_B + \Omega_m X_B\left(1 - 2X_A\right) \tag{8.4E5b}$$

$$\Delta H_A = \Omega_m X_A\left(1 - X_A\right) + \Omega_m\left(1 - X_A\right)\left(1 - 2X_A\right) \tag{8.4E5c}$$

$$\Delta H_A = \Omega_m \left[X_A - X_A^2 + 1 - 2X_A - X_A + 2X_A^2 \right] \tag{8.4E5d}$$

$$\Delta H_A = \Omega_m \left(1 - X_A \right)^2 = \Omega_m X_B^2 \tag{8.4E5e}$$

Similarly, for component B

$$\Delta H_B = \Delta H_m + X_A \frac{d\left(\Delta H_m \right)}{X_A} \tag{8.4E6a}$$

$$\Delta H_B = \Omega_m X_A X_B + \Omega_m X_A \left(1 - 2X_B \right) \tag{8.4E6b}$$

$$\Delta H_B = \Omega_m X_A \left(1 - X_A \right) + \Omega_m X_A \left[1 - 2\left(1 - X_A \right) \right] \tag{8.4E6c}$$

$$\Delta H_B = \Omega_m X_A^2 = \Omega_m \left(1 - X_B \right)^2 \tag{8.4E6d}$$

8.6   Entropy of Mixing

The goal in this section is to show how the Gibbs energy function $G = f\left(X_B \right)$ at temperatures T is used to obtain data for constructing an A-B binary phase diagram. This implies that a T-X_B diagram can be determined from G-X_B diagram assuming an initially homogeneous A-B liquid solution, which in turn solidifies under controlled conditions to form a single phase as in isomorphous or two phases as in eutectic equilibrium phase diagrams. Aspects of equilibrium phase diagrams are included in Chap. 10.

First of all, the Gibbs energy change of mixing $\left(\Delta G_m \right)$ can be defined in terms of standard or reference free energies

$$\Delta G_m = G_2 - G_1 = X_A G_A + X_B G_B \tag{8.20}$$

and the entropy change of mixing $\left(\Delta S_m \right)$ for N_A and N_B number of atoms in the solidifying system can be defined using Boltzmann statistical definition of entropy.

The configurational entropy of mixing is defined by

$$\Delta S_m = k_B \ln\left(\omega \right) = k_B \ln \frac{\left(N_A + N_B \right)!}{N_A! N_B!} \tag{8.21}$$

where k_B denotes the Boltzmann's constant and ω denotes the probability parameter. Using the general Stirling approximation $\ln\left(N \right)! = N \ln\left(N \right) - N$ on Eq. (8.23) yields the configurational entropy of mixing as

$$\Delta S_m = k_B \left[\ln\left(N_A + N_B \right)! - \ln\left(N_A \right)! - \ln\left(N_B \right)! \right] \tag{8.22a}$$

$$\Delta S_m = -k_B \left[N_A \ln\left(\frac{N_A}{N_A + N_B} \right) + N_B \ln\left(\frac{N_B}{N_A + N_B} \right) \right] \tag{8.22b}$$

For one mole of A-B mixture, $N_a = N_A + N_B$, $k_B = R/N_a$, $N_A = X_A N_a$ and $N_B = X_B N_a$, where N_a is the Avogadro's number. In this context, Eq. (8.22b) reduces to a simplified form

$$\Delta S_m = -R\left[X_A \ln(X_A) + X_B \ln(X_B)\right] \tag{8.23}$$

It should be mentioned that ΔS_m is the entropy of mixing in a thermodynamic state of internal equilibrium, provided that the physical mixing process occurs without chemical reactions. In this case, there is no need to include individual reaction entropy components for the reactants and product. Hypothetically, if $A + B = C$ reaction occurs, then this reaction has the C component as the product, and consequently, an extra reaction entropy term would have to be incorporated in the foregoing analytical procedure.

Combining Eqs. (8.5b) and (8.23) yields ΔG_m for regular (non-ideal) and ideal solutions

$$\Delta G_m = \Delta H_m - T\Delta S_m \tag{8.24a}$$

$$\Delta G_m = \Delta H_m + RT\left[X_A \ln(X_A) + X_B \ln(X_B)\right] \quad \text{(Regular)} \tag{8.24b}$$

$$\Delta G_m = RT\left[X_A \ln(X_A) + X_B \ln(X_B)\right] \qquad \text{(ideal)} \tag{8.24c}$$

where $\Delta H_m = 0$ for an ideal and $\Delta H_m \neq 0$ for a non-ideal (regular) solutions. In general, a deviation from an ideal state defines a regular solution due to low temperatures and high pressure.

8.7 Gibbs Energy of Mixing

It is clear now that ΔG_m depends on the exchange binding energy term Ω_m and the disorder character of a regular solution through the entropy of mixing ΔS_m at constant pressure.

Inserting Eq. (8.19a) into (8.24b) yields ΔG_m for a non-ideal solution

$$\Delta G_m = \Omega_m X_A X_B + RT\left[X_A \ln(X_A) + X_B \ln(X_B)\right] \tag{8.25}$$

Here, $\Omega_m X_A X_B$ represents the deviation from ideality and $X_A \ln(X_A) + X_B \ln(X_B)$ represents the term for an ideal solution. For convenience,

$$\Delta G_m = \Delta H_m - T\Delta S_m \tag{8.26a}$$

$$\Delta G_m = \Omega_m X_A X_B - T\Delta S_m \tag{8.26b}$$

$$\Delta H_m = \Delta U_m + P\Delta V_m \tag{8.26c}$$

where ΔU_m is the change in bonding energy of mixing. Accordingly, the Gibbs energy, as defined by Eq. (8.26a), is directly related to the bonding energy, and this may be the reason why it is commonly treated as the driving force for the solidification process of stable crystal structures.

Mathematically, the simple definition of the Gibbs energy of mixing, Eq. (8.26a), seems appropriate for characterizing the physical behavior of mixing, where $\Omega_m \propto U_{ij}$ in Eq. (8.26b) and $\Delta H_m \simeq \Delta U_m$ when $\Delta V_m \to 0$ Eq. (8.26c).

The Gibbs energy of the liquid and solid phases with reference state $G_A = G_B = 0$ and $X_A = 1 - X_B$ can be defined as G_l, $G_s = f(X_B, T)$ for plotting a diagram. This can be accomplished by letting $\Omega_m = 0$ in the liquid phase and $G_s = \Delta G_v + G_l$ in the solid phase. Thus, the Gibbs energy for the liquid and solid phases is written as

$$G_l = (1 - X_B) G_A + X_B G_B + \Omega_l X_B (1 - X_B) \tag{8.27a}$$

$$+ RT [X_B \ln X_B + (1 - X_B) \ln (1 - X_B)]$$

$$G_s = (1 - X_B) \Delta S_{fA} (T - T_{fA}) + X_B \Delta S_{fB} (T - T_{fB}) \tag{8.27b}$$

$$+ RT [X_B \ln X_B + (1 - X_B) \ln (1 - X_B)] + \Omega_s X_B (1 - X_B)$$

Again, Ω_l, Ω_s denote the exchange binding energy terms for the regular solution. If Ω_l, $\Omega_s > 0$, then a repulsive interaction between atoms of components A and B occurs and if Ω_l, $\Omega_s < 0$, an attractive interaction prevails in the mixture.

8.8   Chemical Potential

Consider an A-B binary system undergoing solidification. Then, μ and G become directly related to the interactions of nearest-neighbor like A-A, B-B, and unlike A-B atom pairs at relatively high temperatures. Nonetheless, the state function of interest for $i = A, B$ components is the chemical potential of mixing defined as

$$\mu_i = \mu_i^o + RT \ln (\gamma_i) \tag{8.28}$$

where μ_i^o denotes the chemical potential for a pure component i and γ_i denotes the activity coefficient related to the interatomic action between A and B atoms to form a strong bond at equilibrium. Moreover, if $\gamma_i = 1$, the solution is ideal, and if $\gamma_i > 1$, the solution is regular, indicating a deviation from the ideal state.

It is essentially clear that Eq. (8.28) represents the partial Gibbs energy of mixing ($\mu_i = G_m$), which is related to the component chemical potentials (μ_A, μ_B). Thus,

$$G_m = X_A \mu_A + X_B \mu_B = (1 - X_B) \mu_A + X_B \mu_B \tag{8.29a}$$

$$\frac{\partial G_m}{\partial X_B} = \mu_B - \mu_A \tag{8.29b}$$

Combining Eqs. (8.29a,b) yields the chemical potentials representing the partial molar Gibbs energies

$$\mu_A = G_A = G_m - X_B \left(\frac{\partial G_m}{\partial X_B} \right) \tag{8.30a}$$

$$\mu_B = G_B = G_m + (1 - X_B) \left(\frac{\partial G_m}{\partial X_B} \right) \tag{8.30b}$$

The chemical potential function can also be defined as a function the chemical activity (a_i) of a component $i = A, B$. Thus,

$$\mu_i = G_i + RT \ln (a_i) \tag{8.31}$$

which is the classical thermodynamical activity (a_i) equation describing any deviation from the ideal state. It is written as

$$a_i = \gamma_i X_i \tag{8.32}$$

Note that the variables in Eq. (8.32) are dimensionless and that $0 < X_i < 1$ in regular solutions. If $\gamma_i = 1$, then $a_i = X_i$ for an ideal metallic liquid solution (melt). Actually, γ_i is the factor that determines the degree of deviation from an ideal solution.

For the binary regular solution case. Thus,

$$\Delta G_m = RT \left[X_A \ln (a_A) + X_B \ln (a_B) \right] \tag{8.33a}$$

$$\Delta H_m = RT \left[X_A \ln (\gamma_A) + X_B \ln (\gamma_B) \right] \tag{8.33b}$$

$$\Delta S_m = -RT \left[X_A \ln (X_A) + X_B \ln (X_B) \right] \tag{8.33c}$$

Combining Eqs. (8.19) and (8.33b) for pure components yields

$$RT \ln (\gamma_A) = \Omega_m X_B^2 \tag{8.34a}$$

$$RT \ln (\gamma_B) = \Omega_m X_A^2 \tag{8.34b}$$

from which

$$\Omega_m = \frac{RT \ln (\gamma_A)}{X_A^2} = \frac{RT \ln (\gamma_B)}{(1 - X_B)^2} = \text{Constant} \tag{8.35}$$

Note that Ω_m is a constant for all compositions X_B.).

Example 8.5 Consider the activity coefficient data from electrochemical measurements at $527° C$ $(800\,K)$ for $Zn\text{-}Cd$ alloys, where $A = Cd$ and $B = Zn$. The goal is to verify that Eq. (8.35) gives a constant value for a regular solution and to calculate

ΔG_m, ΔH_m and ΔS_m. Given data (Upadhyaya and Dube example [5, p. 150]):

X_{Cd}	0.2	0.3	0.4	0.5
γ_{Cd}	2.153	1.817	1.544	1.352

Solution From Eq. (8.34a) along with $A = Cd$ and $B = Zn$,

$$RT \ln(\gamma_{Cd}) = \Omega_m X_{Zn}^2 \tag{8.5E1a}$$

$$\frac{RT \ln(\gamma_{Cd})}{X_{Zn}^2} = \frac{RT \ln(\gamma_{Cd})}{(1 - X_{Cd})^2} = \Omega_m \tag{8.5E1b}$$

Calculations to the nearest integer with $R = 8.3145\ J/mol$ yield Ω_m in J/mol units

X_{Zn}	X_{Cd}	$(1 - X_{Cd})^2$	γ_{Cd}	Ω_m
0.8	0.2	0.64	2.153	8000
0.7	0.3	0.49	1.817	8106
0.6	0.4	0.36	1.544	8025
0.5	0.5	0.25	1.352	8024

Note that Ω_m is virtually a constant at the given alloy mole fractions of X_{Cd}. The average value is $\Omega_m = 8039\ J/mol$. Substituting this value into Eqs. (8.19a) and (8.26a) shows that ΔH_m, ΔG_m and ΔS_m are mathematically influenced by the interaction parameter Ω_m. Thus,

$$\Delta H_m = 8039 X_A X_B \tag{8.5E2a}$$

$$\Delta G_m = 8039 X_A X_B - T \Delta S_m \tag{8.5E2b}$$

$$T \Delta S_m = 8039 X_A X_B - \Delta G_m \tag{8.5E2c}$$

$$= 8039 X_A X_B - RT \left[X_A \ln(a_A) + X_B \ln(a_B) \right] \tag{8.5E2d}$$

Therefore, these functions of mixing depend on $\Omega_m = 8039\ J/mol$ and the solution is regular. Accordingly, Eq. (8.5E1b) can be used to determine γ_{Zn} and γ_{Cd}. Thus, the exponential functions for the activity coefficient terms are

$$\gamma_{Zn} = \exp\left[\frac{\Omega_m (1 - X_{Zn})^2}{RT} \right] = \exp\left[1.2087 (X_{Zn})^2 \right] \tag{8.5E3a}$$

$$\gamma_{Cd} = \exp\left[\frac{\Omega_m (1 - X_{Cd})^2}{RT} \right] = \exp\left[1.2087 (1 - X_{Cd})^2 \right] \tag{8.5E3b}$$

Plotting Eqs. (8.5E3a,b) along with the given data points yields the graph shown below.

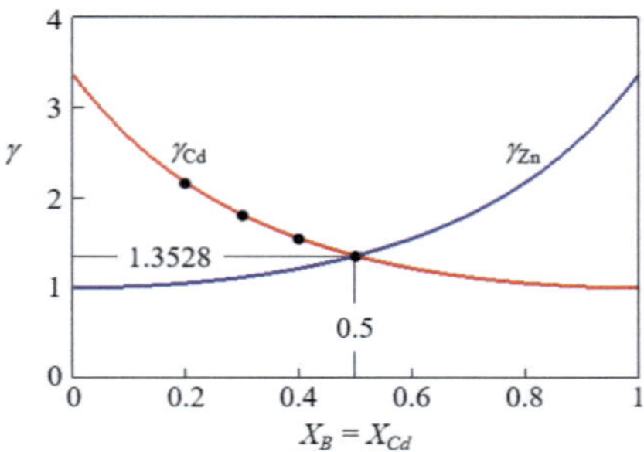

Polynomial fit on the given data yields a third-degree polynomial with a very high correlation coefficient R^2

$$\gamma_{Cd} = 2.942 - 4.155X_{Cd} + 3x^3 + 0.45X_{Cd}^2 \quad \text{with} \quad R^2 = 0.992 \tag{8.5E4}$$

From Eqs. (8.32) and (8.5E3), the activity expressions for Zn and Cd are

$$a_{Zn} = X_{Zn} \exp\left[1.2087 \left(X_{Zn}\right)^2\right] \tag{8.5E5a}$$

$$a_{Cd} = X_{Cd} \exp\left[1.2087 \left(1 - X_{Cd}\right)^2\right] \tag{8.5E5b}$$

Plotting Eq. (8.5E5) gives trendlines similar to $\gamma = f\left(X_B\right)$ above.

Example 8.6 Consider mixing a significant amount of Cu and Ni atoms to produce Cu-Ni alloys with $0 < X_B = X_{Ni} < 1$. Here, $A = Cu$ and $B = Ni$ so that A-B $= Cu$-Ni. **(a)** Derive expressions for U_{AA} and U_{BB} using the factor f_{ij} for A-A, B-B and A-B type of bonds and calculate **(b)** U_{AA}, U_{BB} for individual pairs of atoms before mixing, **(c)** Ω_{AA}, Ω_{BB} and Ω_{AB} using the available latent heat of fusion for copper $\Delta H_{f,Cu} = 1628 \ J/cm^3$ and for nickel $\Delta H_{f,Cu} = 2756 \ J/cm^3$, and **(d)** Ω_m and U_{AB} after mixing Cu and Ni atoms. The activity coefficient equations (curve fitting expressions) at $870 \ K \leq T \leq 1280 \ K$ interval, as reported by Oishi et al. [6] using electrochemical and comparative studies, are

$$\ln\left(\gamma_{Cu}\right) = 0.345 + \left(1.33 \times 10^3\right)\left(1/T\right) \tag{8.6E1a}$$

$$\ln\left(\gamma_{Ni}\right) = 0.691 + \left(604\right)\left(1/T\right) \tag{8.6E1b}$$

(e) Plot $\Omega_m = f\left(X_B\right)$ and $U_{AB} = g\left(X_B\right)$ in the interval $0 \leq X_B = X_{Ni} \leq 0.5$ at $T = 870$ and $1280 \ K$ temperatures.

Solution

(a) Before mixing Cu and Ni: For a pure metal, Eqs. (8.16), (8.17), and (8.18) with $i = A, B$ can be used to deduce the following:

$$U_{ii} \to \Delta H_{fi} \quad \text{for } X_i = 1 \tag{8.6E2a}$$

$$\Omega_{ii} \to U_{ii} < 0 \quad \text{for } X_i = 1 \tag{8.6E2b}$$

$$\Delta H_{fi} \to \Omega_{ii} < 0 \tag{8.6E2c}$$

The arrows in Eq. (8.6E2) stands for "approaching" or approximation. For instance, Eq. (8.6E2c) reads the enthalpy change ΔH_{fi} of formation approaches the value of interaction parameter $\Omega_{ii} < 0$, which acquires negative values during mixing Cu and Ni atoms.

Thus,

$$\Delta H_{f,A} = f_{AA} U_{AA} = \frac{1}{2} Z N_a U_{AA} \tag{8.6E3a}$$

$$\Omega_{AA} = -f_{AA} U_{AA} = -\frac{1}{2} Z N_a U_{AA} \tag{8.6E3b}$$

$$\Delta H_{f,A} = -\Omega_{AA} \tag{8.6E3c}$$

$$U_{AA} = \frac{2\Delta H_{f,A}}{Z N_a} \tag{8.6E3d}$$

and

$$\Delta H_{f,B} = f_{BB} U_{BB} = \frac{1}{2} Z N_a U_{BB} \tag{8.6E4a}$$

$$\Omega_{BB} = -f_{BB} U_{BB} = -\frac{1}{2} Z N_a U_{BB} \tag{8.6E4b}$$

$$\Delta H_{f,B} = -\Omega_{BB} \tag{8.6E4c}$$

$$U_{BB} = -\frac{2\Delta H_{f,B}}{Z N_a} \tag{8.6E4d}$$

(b) In the solid state, both Cu and Ni have an FCC crystal structure with a coordination number of $Z = 12$ and different atomic radii; that is, $R_{Cu} = 0.128$ nm and $R_{Ni} = 0.125$ nm. Recall that the Avogadro's number (physical property) is $N_a = 6.022 \times 10^{23}$ $atoms$. Thus, Eqs. (8.6E3d) and (8.6E4d) yield the sought internal energies

$$U_{AA} = U_{Cu} = -\frac{2\left(1628 \ J/cm^3\right)}{(12)\left(6.022 \times 10^{23}\right)} = -4.51 \times 10^{-22} \ J/cm^3 \tag{8.6E5a}$$

$$U_{BB} = U_{Ni} = -\frac{2\left(2756\ J/cm^3\right)}{(12)\left(6.022 \times 10^{23}\right)} = -7.63 \times 10^{-22}\ J/cm^3 \qquad (8.6\text{E}5\text{b})$$

(c) From Eqs. (8.6E3c) and (8.6E4c),

$$\Omega_{AA} = \Omega_{Cu} = -\Delta H_{f,A} = -\Delta H_{f,Cu} = -1628\ J/cm^3 \qquad (8.6\text{E}6\text{a})$$

$$\Omega_{BB} = \Omega_{Ni} = -\Delta H_{f,B} = -\Delta H_{f,Ni} = -2756\ J/cm^3 \qquad (8.6\text{E}6\text{b})$$

Unit conversion: Search for the atomic weight and mass density for Cu and Ni, and multiply Ω_{Cu}, Ω_{Ni} by

$$(A_w/\rho)_{Cu} = (63.55\ g/mol)\ /\ \left(8.93\ g/cm^3\right) \qquad (8.6\text{E}7\text{a})$$

$$= 7.1165\ cm^3/mol$$

$$(A_w/\rho)_{Ni} = (58.69\ g/mol)\ /\ \left(8.80\ g/cm^3\right) \qquad (8.6\text{E}7\text{b})$$

$$= 6.6693\ cm^3/mol$$

so that

$$\Omega_{AA} = -11,586\ J/mol \qquad (8.6\text{E}8\text{a})$$

$$\Omega_{BB} = -18,381\ J/mol \qquad (8.6\text{E}8\text{b})$$

$$\frac{\Omega_{AA} + \Omega_{BB}}{2} = -14,984\ J/mo \qquad (8.6\text{E}8\text{c})$$

(d) After mixing Cu and Ni: From Eq. (8.35),

$$\Omega_m = \frac{RT\ \ln\left(\gamma_B\right)}{\left(1 - X_B\right)^2} = \frac{RT\ \ln\left(\gamma_{Ni}\right)}{\left(1 - X_B\right)^2} \qquad (8.6\text{E}9\text{a})$$

$$\Omega_m = \frac{RT}{\left(1 - X_B\right)^2}\ [0.691 + (604)\ (1/T)] \qquad (8.6\text{E}9\text{b})$$

Equating Eqs. (8.6E9a) and (8.18a), and solving the resultant expression for U_{AB} yields

$$U_{AB} = \frac{1}{2}\left(U_{AA} + U_{BB}\right) + \frac{RT\ \ln\left(\gamma_B\right)}{Z N_a\left(1 - X_B\right)^2} \qquad (8.6\text{E}10)$$

Let $N_a\left(1 - X_B\right)^2 \gg 0$ and $RT\ \ln\left(\gamma_B\right)\ /\ \left[Z N_a\left(1 - X_B\right)^2\right] \to 0$ in the above expression, Eq. (8.6E10), and substitute Eq. (8.6E1b) into (8.6E10) to get

$$U_{AB} = U_{Cu\text{-}Ni} = \frac{U_{AA} + U_{BB}}{2} + \frac{RT}{ZN_a\,(1 - X_B)^2}\,[0.691 + (604)\,(1/T)]$$

(8.6E11)

Substitute the values of $R = 8.3145\ J/mol.K$, $Z = 12$ for FCC-Cu and FCC-Ni structures, $N_a = 6.022 \times 10^{23}\ atoms/mol$ and $(\Omega_{AA} + \Omega_{BB})\,/2$ into Eqs. (8.6E9b) and (8.6E11) to get Ω_m and U_{AB} in J/mol units. Thus,

$$\Omega_m = \frac{8.3145T}{(1 - X_B)^2}\,[0.691 + (604)\,(1/T)]$$

(8.6E12a)

$$U_{AB} = -14,984\ J/mol + \frac{1.1509 \times 10^{-24}}{(1 - X_B)^2}\,[0.691 + (604)\,(1/T)]$$

(8.6E12b)

Substituting the temperatures 870 K and 1280 K along with $X_B = X_{Ni}$ yields two two expressions for the interaction parameter

$$\Omega_m = \frac{10020\ J/mol}{(1 - X_{Ni})^2} \quad \text{at } T = 870\ K$$

(8.6E13a)

$$\Omega_m = \frac{12375\ J/mol}{(1 - X_{Ni})^2} \quad \text{at } T = 1280\ K$$

(8.6E13b)

and two expressions for the sought function $U_{AB} = U_{CuNi}$, Eq. (8.6E12b),

$$U_{Cu\text{-}Ni} = (-14,984\ J/mol) + \frac{1.5943 \times 10^{-24}\ J/mol}{(1 - X_{Ni})^2} \quad \text{at } T = 870\ K$$

(8.6E14a)

$$U_{Cu\text{-}Ni} = (-14,984\ J/mol) + \frac{1.3384 \times 10^{-24}\ J/mol}{(1 - X_{Ni})^2} \quad \text{at } T = 1280\ K$$

(8.6E14b)

(e) Plotting Eqs. (8.6E13) and (8.6E14) at the given temperatures yields the trendlines (behaviors) of $\Omega_m = f\,(X_{Ni})$ and $U_{Cu\text{-}Ni} = f\,(X_{Ni})$ functions.

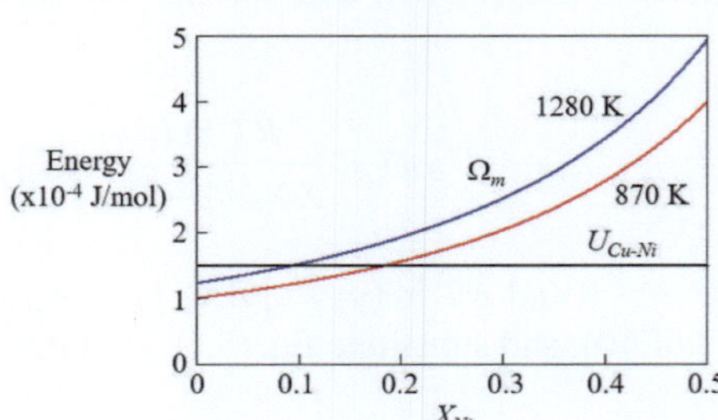

The reader is encouraged to produce the plots as an assignment and verify that $U_{AB} = U_{Cu\text{-}Ni} = -14,984 \; J/mol$ as $X_{Ni} \to 1$ and that Ω_m increases nonlinearly with increasing solute mole fraction X_B at both temperatures. In conclusion, Ω_m quantifies the nature of A-B atomic interactions in the melt at a temperature T. This plot clearly shows that the interaction parameter exhibits a common trendline shape at $870 \; K$ and $1280 \; K$. Moreover, $\Omega_m \to \infty$ and $U_{AB} = U_{Cu\text{-}Ni} = -14,984 \; J/mol$ as $X_{Ni} \to 1$. Note that the curves for Ω_m are intentionally limited to the equiatomic mole fraction; $X_{Cu}, X_{Ni} = 0.5$. In addition, the binary interaction parameter Ω_m is a measure of the interaction between A, B atoms. Notice that Ω_m increases nonlinearly with increasing solute mole fraction X_B at both temperatures. In conclusion, the interaction parameter $\Omega_m \to \infty$ quantifies the nature of A-B atomic interactions in the melt. Therefore, $\Omega_m \to \infty$ is an important fundamental interaction parameter in thermodynamic solidification.

8.9   Gibbs–Duhem Equation

Thermodynamics of solutions can be described by the conventional Gibbs–Duhem equation related to the energy of mixing, such as the Gibbs energy (G), enthalpy (H), entropy (S), and chemical potential energy (μ). In particular, the molar Gibbs energy, $G_m = G/n$ with n being the total amount of moles. The goal in this section is to derive the Gibbs–Duhem equation in different mathematical forms.

For an A-B solution, the Gibbs energy of mixing is related to partial molar energies $(\sum G_i)$ and mole fraction $(\sum X_i)$, where $i = A, B, C, D, \ldots$ are the components in a solution, For a multicomponent solution,

$$G = \sum_{1}^{N} X_i G_i = X_1 G_1 + X_2 G_2 + X_3 G_3 + X_4 G_4 + \ldots \qquad (8.36a)$$

$$dG = X_1 dG_1 + G_1 dX_1 + X_2 dG_2 + G_2 dX_2 + \ldots \qquad (8.36b)$$

For an A-B binary solution with $i = A, B$ components and fixed mole fractions X_A, X_B, Eqs. (8.36b) is simplified to

$$dG = X_A dG_A + X_B dG_B \qquad (8.37)$$

At equilibrium, $dG = 0$ in Eq. (8.37) gives the Gibbs–Duhem equation

$$X_A dG_A + X_B dG_B = 0 \qquad (8.38a)$$

$$dG_A = -\frac{X_B}{X_A} dG_B \qquad (8.38b)$$

Notice that $X_B/X_A < 1$ in Eq. (8.38b) is a multiplier. Therefore, the change of Gibbs energy dG_A for solvent component A is a negative quantity representing a fraction of the enthalpy change dG_B of the solute B since the amount of A decreases as more B is added to the in an A-B mix.

Mathematically, the chemical potential is defined as the slope of a partial molar Gibbs state function $G = f(X_i)$ at constant temperature T and pressure P with $i = 1, 2, 3, \ldots$ components. Thus,

$$\mu_i = \left(\frac{\partial G_i}{\partial X_i} \right)_{T,P} \tag{8.39}$$

For an A-B binary system along with $X_1 = X_A$, $X_2 = X_B$, $\mu_1 = \mu_A$, $\mu_2 = \mu_B$, the total Gibbs energy and its infinitesimally change are, respectively

$$G = X_A \mu_A + X_B \mu_B \tag{8.40a}$$

$$dG = X_A d\mu_A + \mu_A dX_A + X_B d\mu_B + \mu_B dX_B \tag{8.40b}$$

$$dG = X_A d\mu_A + X_B d\mu_B \quad \text{(at constant } X_A, X_B) \tag{8.40c}$$

If $dG = 0$, then Eq. (8.40c) becomes the Gibbs–Duhem equation in terms of chemical potential

$$X_A d\mu_A + X_B d\mu_B = 0 \tag{8.41a}$$

$$d\mu_A = -\frac{X_B}{X_A} d\mu_B \tag{8.41b}$$

$$\mu_A = -\int \frac{X_B}{X_A} d\mu_B \tag{8.41c}$$

For a multicomponent system,

$$\sum_{i=1}^{N} X_i d\mu_i = 0 \tag{8.42a}$$

$$X_1 d\mu_1 + X_2 d\mu_2 + \ldots = 0 \tag{8.42b}$$

which is applied to binary A-B solutions throughout this section.

8.9.1 *Activity and Activity Coefficient*

Consider a regular solution containing a certain ion j at a temperature T. Any deviation from the ideal solution is defined by an activity coefficient γ_i as indicated by Eq. (8.32).

The chemical potential and its first derivative can be defined in terms of the activity of component denoted as a_i. Thus,

$$\mu_i = RT \ln(a_i) \tag{8.43a}$$

$$d\mu_i = RT\, d\left[\ln(a_i)\right] \tag{8.43b}$$

Combine Eqs. (8.41b) and (8.43b) for an A-B binary solution to get the Gibbs–Duhem equation in terms of activity (a_i). The resultant expression is defined by

$$d \ln(a_A) = -\frac{X_B}{X_A} d \ln(a_B) \tag{8.44a}$$

$$\int_1^{0<X_A<1} \ln(a_A) = -\int_0^{0<X_B<1} \frac{X_B}{X_A} d \ln(a_B) \tag{8.44b}$$

$$\ln(a_A) - \ln(1) = -\int_0^{0<X_B<1} \frac{X_B}{X_A} d \ln(a_B) \tag{8.44c}$$

$$\ln(a_A) = -\int_0^{0<X_B<1} \frac{X_B}{X_A} d \ln(a_B) \tag{8.44d}$$

This integral, Eq. (8.44d), defines the area under a curve described by a nonlinear function $X_B/X_A = f(a_B)$. Actually, it can be solved graphically or analytically once an equation for $\ln(a_B) = f(X_A)$ is known.

Substituting the activity $a_i = \gamma_i X_i$, Eq. (8.32), into (8.44a) yields the Gibbs–Duhem equation in another mathematical form

$$d \ln(\gamma_A X_A) = -\frac{X_B}{X_A} d \ln(\gamma_B X_B) \tag{8.45a}$$

$$d \ln(\gamma_A) + d \ln(X_A) = -\frac{X_B}{X_A} \left[d \ln(\gamma_B) + d \ln(X_B)\right] \tag{8.45b}$$

$$d \ln(\gamma_A) + \frac{dX_A}{X_A} = -\frac{X_B}{X_A} \left[d \ln(\gamma_B) + \frac{dX_B}{X_B}\right] \tag{8.45c}$$

$$d \ln(\gamma_A) + \frac{dX_A}{X_A} = -\frac{X_B}{X_A} d \ln(\gamma_B) - \frac{X_B}{X_A} \frac{dX_B}{X_B} \tag{8.45d}$$

$$d \ln(\gamma_A) + \frac{dX_A}{X_A} = -\frac{X_B}{X_A} d \ln(\gamma_B) - \frac{dX_B}{X_A} \tag{8.45e}$$

But, $X_A + X_B = 1$ and $dX_A = -dX_B$, and consequently, Eq. (8.45e) becomes the Gibbs–Duhem equation in terms of activity coefficients (γ_i)

$$d \ln(\gamma_A) = -\frac{X_B}{X_A} d \ln(\gamma_B) \tag{8.46a}$$

$$\ln\left(\gamma_A\right) = -\int_0^{0<X_B<1} \frac{X_B}{X_A} d\ln\left(\gamma_B\right) \tag{8.46b}$$

Regarding the solution of Eq. (8.46b), it can be solved graphically or analytically once an appropriate expression for $\ln\left(\gamma_B\right) = f\left(X_A\right)$ is known.

8.9.2   The Alpha Function

Determining the activity or activity coefficient of a component A knowing the value of the component B in an A-B binary solution at relatively high temperature T and pressure P is a common practice considered by researchers using the Gibbs–Duhem equation, such as Eq. (8.44d) or (8.46b). It is recognized by the scientific and academic communities that these equations introduce uncertainty in calculating a_A or γ_A from experimentally determined a_B or γ_B due to the asymptotic behavior of these variables.

In order to improve the accuracy of a_A or γ_A, an α-function proposed by Darken and Gurry [7, p. 264] is used to modify the conventional Gibbs–Duhem equation. Mathematically, manipulating Eq. (8.35) yields

$$\ln\left(\gamma_A\right) = \frac{\Omega_m X_B^2}{RT} = \frac{\Omega_m X_B\left(1 - X_A\right)}{RT} \tag{8.47a}$$

$$\ln\left(\gamma_A\right) = -\frac{X_A X_B \Omega_m}{RT} + \frac{X_B \Omega_m}{RT} \tag{8.47b}$$

$$\ln\left(\gamma_A\right) = -\frac{X_A X_B \Omega_m}{RT} + \frac{\Omega_m}{RT}\left(1 - X_A\right) \tag{8.47c}$$

$$\ln\left(\gamma_A\right) = -\frac{X_A X_B \Omega_m}{RT} - \frac{\Omega_m}{RT}\left(X_A - 1\right) \tag{8.47d}$$

$$\ln\left(\gamma_A\right) = -\frac{X_A X_B \Omega_m}{RT} - \int_1^{X_A<1} \frac{\Omega_m}{RT} dX_A \tag{8.47e}$$

Again, from Eq. (8.35),

$$\ln\left(\gamma_B\right) = \frac{\Omega_m X_A^2}{RT} \tag{8.48a}$$

$$\Omega_m = \frac{RT\ln\left(\gamma_B\right)}{X_A^2} \tag{8.48b}$$

Insert Eq. (8.48b) into (8.47e) to get the Gibbs–Duhem equation in the form

$$\ln(\gamma_A) = -\frac{X_A X_B \Omega_m}{RT} - \int_1^{X_A<1} \frac{\Omega_m}{RT} dX_A \tag{8.49a}$$

$$\ln(\gamma_A) = -\frac{X_A X_B}{RT} \frac{RT \ln(\gamma_B)}{X_A^2} - \int_1^{X_A<1} \frac{1}{RT} \frac{RT \ln(\gamma_B)}{X_A^2} dX_A \tag{8.49b}$$

$$\ln(\gamma_A) = -\frac{X_A X_B \ln(\gamma_B)}{X_A^2} - \int_1^{0<X_A<1} \frac{\ln(\gamma_B)}{X_A^2} dX_A \tag{8.49c}$$

This integral can be solved graphically or analytically once an equation for $\ln(\gamma_B) = f(X_A)$ is known. Moreover, Eq. (8.49c) can be modified to improve its accuracy by introducing the alpha function (α-function) defined by Darken and Gurry [7, p. 512] as

$$\alpha_A = \frac{\ln(\gamma_A)}{X_B^2} \tag{8.50a}$$

$$\alpha_B = \frac{\ln(\gamma_B)}{X_A^2} \tag{8.50b}$$

Combining Eqs. (8.49c) and (8.50b) gives another Gibbs–Duhem equation in terms of the alpha function (α-function)

$$\ln(\gamma_A) = -\alpha_B X_A X_B - \int_1^{0<X_A<1} \alpha_B dX_A \tag{8.51}$$

Further, combining Eqs. (8.35) and (8.50) yields an expression relating the α-function for components A and B, and the interaction parameter Ω_m at a temperature T

$$\alpha_A = \alpha_B = \frac{\Omega_m}{RT} \tag{8.52}$$

An example can reveal some characteristics of the above Gibbs–Duhem equations and their particular accuracy for determining the activity (a_A) of the solvent A knowing the activity (a_B) of the solute B in an A-B binary solution at a relatively high temperature T and fixed pressure P.

Example 8.7 Consider the given dataset for Cu-Zn solutions at $T = 1060°\,C$ (Kubaschewski and Alcock book [8]). Based on this information, calculate **(a)** the activity of copper ($a_A = a_{Cu}$) at $X_{Cu} = 0.60$ and $X_{Zn} = 0.40$ using the available Gibbs–Duhem equations and **(b)** the a-function as per Eq. (8.52). The analytical solutions of the Gibbs–Duhem equations is via curve fitting expressions relating the activity a_{Cu} to mole fraction (composition) X_{Cu} and X_{Zn}.

X_{Zn}	1	0.45	0.30	0.20	0.15	0.1000	0.0500
P_{Zn} (atm)	4	1.28	0.60	0.24	0.12	0.0592	0.0289

Here, $P_{o,Zn} = 4$ atm (reference pressure) and $a_{Zn} = P_{Zn}/P_{o,Zn}$. The given plots [Fig. (a)–(d)] are conveniently constructed beforehand since they are the basis for calculating the activity of copper at $X_{Zn} = 0.40$ mole fraction.

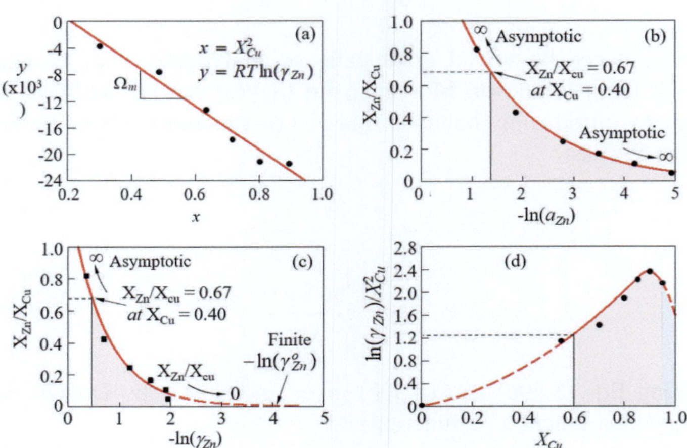

Assume that the correlation coefficient for each plot is close to absolute unity; $R^2 = |1|$.

Solution The given table below contains calculated data for plotting relevant relationships associated with the analysis of Gibbs–Duhem equations cited previously. The plots given are constructed using curve fitting on these tabulated results.

Notice that Fig. (a) depicts the interaction parameter Ω_m, and Figs. (b)–(d) show shaded areas representing the graphical solutions of the Gibbs–Duhem equations; Eqs. (8.44d), (8.46b), and (8.49c) or (8.51), respectively.

X_{Cu}	0.5500	0.7000	0.8000	0.8500	0.9000	0.9500
X_{Zn}	0.4500	0.3000	0.2000	0.1500	0.1000	0.0500
$a_{Zn} = P_{Zn}/P_{o,Zn}$	0.3200	0.1500	0.0600	0.0300	0.0148	0.0072
$\gamma_{Zn} = a_{Zn}/X_{Zn}$	0.7100	0.5000	0.3000	0.2000	0.1480	0.1440
$-\ln(a_{Zn})$	1.1394	1.8971	2.8134	3.5066	4.2131	4.9337
$-\ln(\gamma_{Zn})$	0.3425	0.6932	1.2040	1.6094	1.9105	1.9379

Method A: The Interaction Parameter [Fig. (a)] First of all, the pressure P_{Zn} and activity a_{Zn} trends for zinc are described by the following curve fitting equations (along with the correlation coefficients R^2) shown below.

From the activity equation, one obtains $\ln(a_{Zn}) = f(X_{Zn})$ and $d\ln(a_{Zn}) = g(X_{Zn})$ which are subsequently used to calculate the activity of copper (a_{Cu}) in the Cu-Zn alloy solution containing $X_{Zn} = 0.40$. This is done next.

Thus,

$$P_{Zn} = 4.2601\,(X_{Zn})^{1.7526} \text{ with } R^2 = 0.9862 \tag{8.7E1a}$$

$$a_{Zn} = 1.0659\,(X_{Zn})^{1.7534} \text{ with } R^2 = 0.9863 \tag{8.7E1b}$$

$$a_{Zn} = 1.0659\,(0.40)^{1.7534} = 0.21378 \tag{8.7E1c}$$

$$\ln(a_{Zn}) = \ln\left[1.0659\,(X_{Zn})^{1.7534}\right] \tag{8.7E1d}$$

$$d\ln(a_{Zn}) = \frac{1.7534}{X_{Zn}}dX_{Zn} \tag{8.7E1e}$$

From Eq. (8.35),

$$RT\ln(\gamma_{Cu}) = \Omega_m X_{Zn}^2 \tag{8.7E2a}$$

$$RT\ln(\gamma_{Zn}) = \Omega_m X_{Cu}^2 = \Omega_m\,(1 - X_{Zn})^2 \tag{8.7E2b}$$

Plotting $RT\ln(\gamma_{Zn}) = f\left(X_{Cu}^2\right)$ or $y = f(x)$ yields Fig. (a) along with data points and the interaction parameter Ω_m as the slope of the straight line

$$y = 6878.7 - 32,729x \text{ with } R^2 = 0.9691 \tag{8.7E3a}$$

$$\Omega_m = -32,729\ J/mol \quad \text{(slope)} \tag{8.7E3b}$$

From Eq. (8.7E2a) at $X_{Cu} = 0.60$, $X_{Zn} = 0.40$ and $T = 1,333\ K$, the activity coefficient γ_{Cu} and the activity a_{Cu} of copper follow, respectively

$$RT\ln(\gamma_{Cu}) = \Omega_m X_{Zn}^2 \tag{8.7E4a}$$

$$\ln(\gamma_{Cu}) = \frac{\Omega_m X_{Zn}^2}{RT} = \frac{(-32,729)\,(0.40)^2}{(8.3145)\,(1333)} = -0.47248 \tag{8.7E4b}$$

$$\gamma_{Cu} = \exp(-0.47248) = 0.62345 \simeq 0.62 \tag{8.7E4c}$$

$$a_{Cu} = \gamma_{Cu} X_{Cu} = (0.62)\,(0.60) = 0.37 \tag{8.7E4d}$$

which can be taken as the reference activity of copper (Cu), $a_{Cu,o} = a_{Cu} = 0.37$, in the alloy containing $X_{Zn} = 0.40$ mole fraction.

Method B: Interaction Parameter and Activity Combining Eqs. (8.7E2a,b) along with $a_{Cu} = \gamma_{Cu} X_{Cu}$ gives the activity for copper in the alloy containing $X_{Cu} = 0.60$ and $X_{Zn} = 0.40$ mole fractions

$$\ln\left(a_{Cu}\right) = \ln\left(X_{Cu}\right) + \frac{X_{Zn}^2}{X_{Cu}^2}\ln\left(a_{Zn}/X_{Zn}\right) \tag{8.7E5a}$$

$$\ln\left(a_{Cu}\right) = \ln\left(0.60\right) + \left(\frac{0.40}{0.60}\right)^2 \ln\left(0.21378/0.40\right) \tag{8.7E5b}$$

$$\ln\left(a_{Cu}\right) = -0.78928 \tag{8.7E5c}$$

$$a_{Cu} = \exp\left(-0.78928\right) = 0.45 \tag{8.7E5d}$$

The percentage error is calculated as

$$error = \left|\frac{a_{Cu,o} - a_{Cu}}{a_{Cu,o}}\right| = \left|\frac{0.37 - 0.45}{0.37}\right| \times 100 = 21.62\% \tag{8.7E6}$$

Notice that the copper activity $a_{Cu} = 0.45$, Eq. (8.7E5d), is slightly higher than $a_{Cu} = 0.37$ in Eq. (8.7E4d). At this moment, there is no experimental data for copper to compare the activity a_{Cu} values.

Method C: Gibbs–Duhem Equation and Activity [Fig. (b)] Combining Eqs. (8.44d) and (8.7E1e) with $a_A = a_{Cu}$ and $a_B = a_{Zn}$, the activity of copper is defined by an integral of the form

$$\ln\left(a_{Cu}\right) = -\int_0^{\ln(a_{Zn})\text{ at }X_{Zn}=0.40} \frac{X_{Zn}}{X_{Cu}} d\ln\left(a_{Zn}\right) \tag{8.7E7}$$

The limits of integration can be chosen as desired, but according to Eq. (8.7E7), the integral upper limit is $X_{Zn} = 0.40$ because $a_{Zn} = \gamma_{Zn}X_{Zn}$ with $\gamma_{Zn} > 0$, which represents the degree of deviation from the ideal condition. In essence, the lower limit of the integral is conveniently set to $X_{Zn} = 0$ so that $a_{Cu} = X_{Cu} = 1$ becomes a known quantity.

The graphical solution of the integral in Eq. (8.7E6) defines the area under the curve described by the function $X_{Zn}/X_{Cu} = f\left[-\ln\left(a_{Zn}\right)\right]$ or $y = f\left(x\right)$. Accordingly, the shaded area in Fig. (b) at $-5 < \ln\left(a_{Zn}\right) < -1$ and $\infty < X_{Zn}/X_{Cu} \leq 0.67$ represents the solution of an integral.

Notice that the $X_{Zn}/X_{Cu} = f\left[\ln\left(a_{Zn}\right)\right]$ function shows an asymptotic behavior to both x and y axes. Recall that an asymptotic behavior, in general, describes how a function behaves near a limit. This implies that

$$a_{Zn} = \gamma_{Zn}X_{Zn} = 0, \ \ln\left(a_{Zn}\right) \to \infty \text{ at } X_{Zn} = 0 \ \text{(x-axis)} \tag{8.7E8a}$$

$$\frac{X_{Zn}}{X_{Cu}} \to \infty \text{ as } X_{Cu} \to 0 \text{ and } X_{Zn} \to 1 \qquad \text{(y-axis)} \tag{8.7E8b}$$

This is an approximation method where $\ln\left(a_{Zn}\right)$ and X_{Zn}/X_{Cu} do not reach finite values, which are essential for evaluating the shaded area as the graphical solution of the integral. Consequently, the graphical method is not recommended in this part of the example because of the conditions defined by Eq. (8.7E8). Instead, an analytical solution is considered using the corresponding curve fitting equation. The analytical procedure is presented as a set of equations in order for the reader to have a logical sequence of expressions for arriving at the activity of copper (a_{Cu}), followed by a simple calculation to obtain the seeking result. Thus,

$$\ln(a_{Cu}) = -\int_0^{X_{Zn}=0.40} \frac{X_{Zn}}{X_{Cu}} d\ln(a_{Zn}) \tag{8.7E9a}$$

$$d\ln(a_{Zn}) = \frac{1.7534}{X_{Zn}} dX_{Zn} \quad [\text{from Eq. (8.7E1e)}] \tag{8.7E9b}$$

$$\ln(a_{Cu}) = -1.7534 \int_0^{X_{Zn}=0.40} \frac{dX_{Zn}}{X_{Cu}} \tag{8.7E9c}$$

$$= -1.7534 \int_0^{X_{Zn}=0.40} \frac{dX_{Zn}}{1 - X_{Zn}}$$

$$\ln(a_{Cu}) = -1.7534 \int_0^{X_{Zn}=0.40} \frac{dX_{Zn}}{1 - X_{Zn}} \tag{8.7E9d}$$

$$= -0.89568 \quad (\text{area})$$

$$a_{Cu} = \exp(-0.89568) = 0.41 \tag{8.7E9e}$$

The percentage error is calculated as

$$errror = \left| \frac{a_{Cu,o} - a_{Cu}}{a_{Cu,o}} \right| = \left| \frac{0.37 - 0.41}{0.37} \right| \times 100 = 10.81\% \tag{8.7E10}$$

which is lower than the result given by Eq. (8.7E6).

Method D: Gibbs–Duhem Equation and Activity Coefficient [Fig. (c)] The $X_{Zn}/X_{Cu} = f[-\ln(\gamma_{Zn})]$ function exhibits a finite value to the x-axis, but it shows an asymptotic behavior to the y-axis. Consequently, this method is recommended despite the fact that

$$\frac{X_{Zn}}{X_{Cu}} \to 0, \; \ln(\gamma_{Zn}) \to \ln\left(\gamma_{Zn}^o\right) \text{ as } X_{Cu} \to 1, \; X_{Zn} \to 0 \tag{8.7E11a}$$

$$\frac{X_{Zn}}{X_{Cu}} \to \infty \text{ as } X_{Cu} \to 0 \text{ and } X_{Zn} \to 1 \tag{8.7E11b}$$

From Eq. (8.46b) with $\gamma_A = \gamma_{Cu}$ and $\gamma_B = \gamma_{Zn}$, the activity coefficient and the corresponding curve fitting equation are

$$RT\ln(\gamma_{Cu}) = \Omega_m X_{Zn}^2 \tag{8.7E12a}$$

$$RT\ln(\gamma_{Zn}) = \Omega_m X_{Cu}^2 = \Omega_m (1 - X_{Zn})^2 \tag{8.7E12b}$$

$$\ln\left(\gamma_{Zn}\right) = \frac{\Omega_m \left(1 - X_{Zn}\right)^2}{RT} \tag{8.7E12c}$$

$$d\ln\left(\gamma_{Zn}\right) = \frac{2\Omega_m \left(X_{Zn} - 1\right)}{RT} dX_{Zn} \tag{8.7E12d}$$

which can be integrated and subsequently be combined with Eq. (8.7E7) in order to obtain a better result for the activity of copper in solution.

Fundamentally, the analytical procedure is very easy to follow as shown below in the form of a block or set of equations.

Thus,

$$\ln\left(\gamma_{Cu}\right) = -\int_0^{X_{Zn}=0.40} \frac{X_{Zn}}{X_{Cu}} d\ln\left(\gamma_{Zn}\right) \tag{8.7E13a}$$

$$= -\int_0^{0.40} \frac{X_{Zn}}{\left(1 - X_{Zn}\right)} d\ln\left(\gamma_{Zn}\right) \tag{8.7E13b}$$

$$\ln\left(\gamma_{Cu}\right) = -\int_0^{0.40} \frac{X_{Zn}}{\left(1 - X_{Zn}\right)} \frac{2\Omega_m \left(X_{Zn} - 1\right)}{RT} dX_{Zn} \tag{8.7E13c}$$

$$\ln\left(\gamma_{Cu}\right) = \int_0^{0.40} \frac{2\Omega_m X_{Zn}}{RT} dX_{Zn} \tag{8.7E13d}$$

$$= \int_0^{0.40} \frac{2\left(-32729\right) X_{Zn}}{\left(8.3145\right)\left(1333\right)} dX_{Zn} \tag{8.7E13e}$$

$$\ln\left(\gamma_{Cu}\right) = -0.47248 \tag{8.7E13f}$$

$$\gamma_{Cu} = \exp\left(-0.47248\right) = 0.62344 \simeq 0.62 \tag{8.7E13g}$$

$$a_{Cu} = \gamma_{Cu} X_{Cu} = \left(0.62\right)\left(0.60\right) = 0.37 \tag{8.7E13h}$$

Method E: Gibbs–Duhem Equation and α-function [Fig. (d)] Figure (d) shows finite values at $X_{Cu} = 0$ and $X_{Cu} = 1$, leading to a more accurate result for a_{Cu}. According to the alloy composition in this example, the integration limits are indicated in Fig. (d). From Eq. (8.51) with $\gamma_A = \gamma_{Cu}$ and $\gamma_B = \gamma_{Zn}$, the activity coefficient and related limits of integration are written as

$$\ln\left(\gamma_{Cu}\right) = -\alpha_{Zn} X_{Cu} X_{Zn} - \int_1^{X_{Cu}=0.60} \alpha_{Zn} dX_{Cu} \tag{8.7E14}$$

Once the value of the alpha-function for zinc is determined, the solution of the integral follows. From Eqs. (8.7E2b) and (8.52),

$$\ln\left(\gamma_{Zn}\right) = \frac{\Omega_m \left(1 - X_{Zn}\right)^2}{RT} \tag{8.7E15a}$$

$$\alpha_{Zn} = \frac{\ln(\gamma_{Zn})}{X_{Cu}^2} = \frac{1}{(1-X_{Zn})^2}\frac{\Omega_m(1-X_{Zn})^2}{RT} \tag{8.7E15b}$$

$$\alpha_{Zn} = \frac{\Omega_m}{RT} = -\frac{32729}{(8.3145)(1333)} = -2.9530 \tag{8.7E15c}$$

Then,

$$\int_1^{X_{Cu}=0.60} \alpha_{Zn}\, dX_{Cu} = \int_1^{X_{Cu}=0.60}(-2.9530)\, dX_{Cu} = 1.1812 \tag{8.7E16}$$

Inserting Eq. (8.7E16) into (8.7E14) yields

$$\ln(\gamma_{Cu}) = -\alpha_{Zn}X_{Cu}X_{Zn} - 1.1812 \tag{8.7E17a}$$

$$\ln(\gamma_{Cu}) = -(-2.9532)(0.60)(0.40) - 1.1812 \tag{8.7E17b}$$

$$= -0.47243$$

$$\gamma_{Cu} = \exp(-0.47243) = 0.62349 \simeq 0.62 \tag{8.7E17c}$$

$$a_{Cu} = \gamma_{Cu}X_{Cu} = (0.62)(0.60) = 0.37 \tag{8.7E17d}$$

The deviation from ideal solution is calculated by the activity coefficient $\gamma_{Cu} = 0.62$ or 62%.

Method F: Area Under the Curve [Fig. (b)] Curve fitting function $y = f(x)$, where y maps to X_{Zn}/X_{Cu} and x maps to $-\ln(a_{Zn})$, yields

$$y = 1.7236\exp(0.6797x) \tag{8.7E18a}$$

$$X_{Zn}/X_{Cu} = 1.7236\exp[0.6797\ln(a_{Zn})] \tag{8.7E18b}$$

$$X_{Zn}/X_{Cu} = 0.05761 \quad \text{at} \quad \ln(a_{Zn})_1 = -5 \tag{8.7E18c}$$

$$X_{Zn}/X_{Cu} = 0.67000 \quad \text{at} \quad \ln(a_{Zn})_2 = -1.3902 \tag{8.7E18d}$$

The integral area and activity at $X_{Zn} = 0.40$ or $X_{Zn}/X_{Cu} = 0.67$ are

$$\ln(a_{Cu}) = -\int \frac{X_{Zn}}{X_{Cu}}\, d\ln(a_{Zn}) \tag{8.7E19a}$$

$$\ln(a_{Cu}) = -\int_{-5}^{-1.39} 1.7236\exp[0.6797\ln(a_{Zn})]\, d\ln(a_{Zn}) \tag{8.7E19b}$$

$$\ln(a_{Cu}) = -0.90108 \tag{8.7E19c}$$

$$a_{Cu} = \exp(-0.90108) = 0.41 \tag{8.7E19d}$$

The percentage error is calculated as

$$error = \left|(a_{Cu,o} - a_{Cu})/a_{Cu,o}\right| = \left|\frac{0.37 - 0.41}{0.37}\right| \times 100 = 10.81\% \qquad (8.7E20)$$

For clarity, this percent error represents the difference between calculated and actual values. The goal is to minimize the error so that the computed error is zero or near-zero.

Method G: Area Under the Curve [Fig. (c)] Similarly, the curve fitting equation and limits of integration are

$$X_{Zn}/X_{Cu} = 1.3083 \exp\left[-1.4191 \ln(\gamma_{Zn})\right] \qquad (8.7E21a)$$

$$X_{Zn}/X_{Cu} \simeq 0.00 \quad at \quad -\ln(\gamma_{Zn})_1 = 4 \qquad (8.7E21b)$$

$$X_{Zn}/X_{Cu} = 0.67 \quad at \quad -\ln(\gamma_{Zn})_2 = 0.47 \qquad (8.7E21c)$$

The final calculations for the integral area and activity in a regular solution containing a mole fraction $X_{Zn} = 0.40$ or mole ratio $X_{Zn}/X_{Cu} = 0.67$ indicate that

$$\ln(\gamma_{Cu}) = \int \frac{X_{Zn}}{X_{Cu}} d\ln(\gamma_{Zn}) \qquad (8.7E22a)$$

$$\ln(\gamma_{Cu}) = \int_4^{0.47} 1.3083 \exp\left[-1.4191 \ln(\gamma_{Zn})\right] d\ln(\gamma_{Zn}) \qquad (8.7E22b)$$

$$\ln(\gamma_{Cu}) = -0.447003 \qquad (8.7E22c)$$

$$\gamma_{Cu} = \exp(-0.447003) = 0.62562 = 0.62 \qquad (8.7E22d)$$

$$a_{Cu} = \gamma_{Cu} X_{Cu} = (0.62)(0.60) = 0.37 \qquad (8.7E22e)$$

Hence, this example illustrates the use of different Gibbs–Duhem equations for calculating the activity (a_{Cu}) of copper in the $0.60Cu$-$0.40Zn$ or $60\%Cu$-$40\%Zn$ alloy. Moreover, the deviation from the ideal solution being calculated by the activity coefficient $\gamma_{Cu} = 0.62$ is equal to the result given in Method F.

8.10 Gibbs Energy and Binary Phase Diagrams

This section includes general features of equilibrium phase diagrams, also known as standard A-B binary phase diagrams that correlate with the Gibbs energy curves described by the general function $G = f(X_B)$, where the mole fraction of the alloying element or component is $X_B = 1 - X_A$. Moreover, a standard phase diagram is constructed at atmospheric pressure $P = 1\ atm$ and at low freezing rates (dT/dt) under controlled experimental conditions. With regard to the skewness (asymmetry) of the $G = f(X_B)$ curves, most are either uniform or altered Ushaped, as depicted in the plots given below, for individual liquid and solid phases. However, a mixture of phases requires the construction of a tangent line connecting the phase lines for defining the composition or mole fraction of the alloying element B within element A.

These standard phase diagrams are two-component maps that show the phase fields separated by phase field lines, and the general function that plots these phase field lines is also of the form $T = T(X_B)$. As a result, one can interpret these maps in order to predict the temperature or temperature range for stable phases. Only a few $G = f(X_B)$ curves at T are shown below: isomorphous and eutectic standard phase diagrams. The procedure for constructing standard phase diagrams is also shown.

8.10.1   Gibbs Energy and Isomorphous Phase Diagrams

Consider mixing components A and B to develop Gibbs energy diagrams using $G = f(X_B)$ as schematically shown in Figs. 8.6a,b,c and subsequently, construct a standard isomorphous phase diagram (Fig. 8.6d), which is the most simple A-B binary phase diagram.

Using the following dataset found in Gaskell and Laughlin book [9, p. 356], the Gibbs energy change $\Delta G = \Delta H - T\Delta S$, as defined by Eq. (8.5b), for components A and B at constant pressure is

$$\Delta G_A = (8000 - 10T)\left(10^{-3}\right) \quad in\ kJ/mol \tag{8.53a}$$

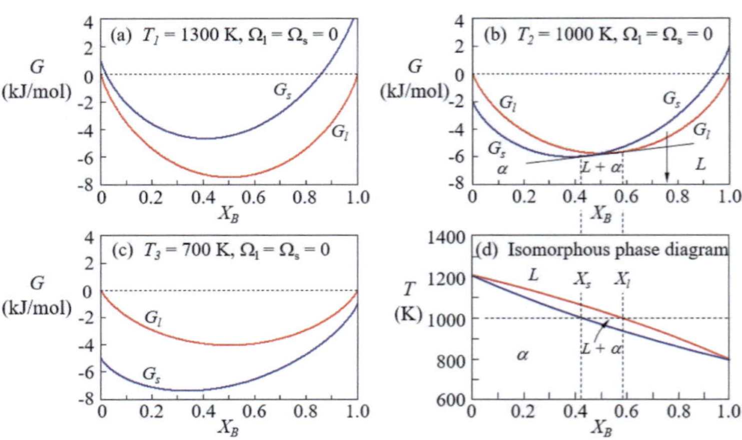

Fig. 8.6 Molar Gibbs energy diagrams at (**a**) 1300 K, (**b**) 1000 K, (**c**) 700 K and (**d**) isomorphous phase diagram

$$\Delta G_B = (12000 - 10T) \left(10^{-3}\right) \quad \text{in } kJ/mol \tag{8.53b}$$

$$T_{fA} = 1200 \ K \quad \& \quad T_{fB} = 1200 \ K \tag{8.53c}$$

from which

$$\Delta H_A = 8,000 \ J/mol \text{ and } \Delta H_B = 12,000 \ J/mol \tag{8.54a}$$

$$\Delta S_{fA} = \Delta S_{fB} = 10 \ J/mol \tag{8.54b}$$

Here, T_{fA}, T_{fB} denote the melting or freezing temperatures of components A and B, respectively. Apparently, this dataset is for a hypothetical binary alloy. The idea now is to use this dataset to plot $G = f(X_B)$ diagrams.

Based on the above information, Eqs. (8.27a,b) with $\Omega_l = \Omega_s = 0$ (for ideal solutions independent of atomic arrangements) can be used to plot $G_l, G_s = f(X_B)$ at selected temperatures T as illustrated in Figs. 8.6. Note that the plot in Fig. 8.6a is at $T > T_{fA}$, Fig. 8.6b is at $T_{fB} < T < T_{fA}$ and Fig. 8.6c is at $T < T_{fB}$.

The interpretation of Gibbs energy curves (Figs. 8.6a,b,c) and the isomorphous equilibrium phase diagram is given below.

- At temperature T_1 (Fig. 8.6a), the entire $G = f(X_B)$ curve for the liquid, $G_l(X_B)$, lies below that for the solid, $G_s(X_B)$, at any composition X_B. This implies that the liquid phase since $G_l < G_s$ for all X_B.
- At temperature T_2 (Fig. 8.6b), as the temperature decreases, the $G_l(X_B)$ curves intercept at $X_B = 0.5$, but the $G_l = f(X_B)$ and $G_s = f(X_B)$ have minimum values as depicted by the dashed vertical lines. Note that solid α-phase is stable at $X_B \leq X_s$ and at $X_s \leq X_B \leq X_l$. On the other hand, the liquid L phase is stable at $X_B \geq X_l$. Moreover, the minimum $G(X_B)$ values correspond to the liquidus and solidus intercepts at $T_2 = 1,000 \ K$.
- Temperature T_3 (Fig. 8.6c): The $G_s = f(X_B)$ curve for the solid α-phase is totally below the $G_l = f(X_B)$ curve for the liquid. Therefore, the single homogeneous solid α-phase is entirely stable since $G_s < G_l$ for all X_B. Eventually, the stable α-phase cools off to room temperature for subsequent use in a relevant solid shape.
- The isomorphous phase diagram (Fig. 8.6d): The $G_l = f(X_B)$ and $G_s = f(X_B)$ curves can be determine at the temperature range $800° C \leq T \leq 1200° C$ as shown in Fig. 8.6b. Notice that the curves in Figs. 8.6a,c are out of the this temperature range.

For clarity, $G = f(X_B)$ and $T = f(X_B)$ functions are associated with the $T = f(t)$ function, which provides relevant cooling curves at different A-B binary systems under a controlled solidification domain. This approach is included in Chap. 10 for constructing metallurgical equilibrium phase diagrams as graphical representations of liquid and crystalline phases. Moreover, thermodynamics in physical metallurgy is an important field because it provides means to predict the equilibrium. state of an alloy, such as Al-Si with a nominal composition C_o.

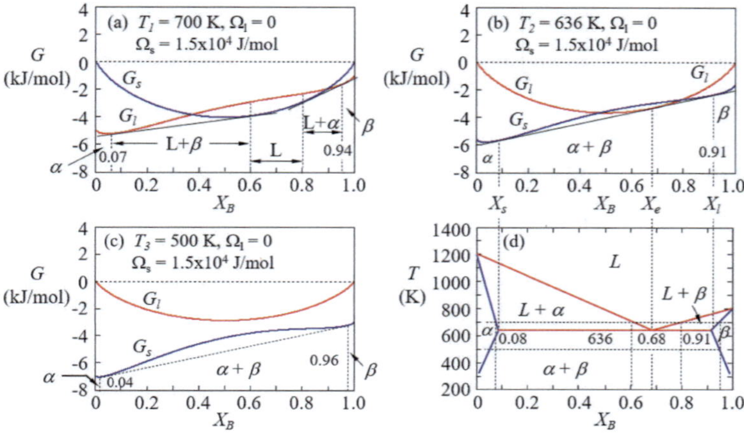

Fig. 8.7 Molar Gibbs energy diagrams at (**a**) 700 K, (**b**) 636 K, (**c**) 500 K and (**d**) schematic binary eutectic phase diagram

8.10.2 Gibbs Energy and Eutectic Phase Diagrams

Consider the function $G = f(X_B)$, as defined by Eqs. (8.27a,b), and assume that the components A and B form a binary eutectic phase diagram. Available thermodynamic data:

$\Delta H_A = 8,000\,J/mol$	$\Delta S_{fA} = 10\,J/mol$	$T_{fA} = 1200K$
$\Delta H_B = 12,000\,J/mol$	$\Delta S_{fB} = 10\,J/mol$	$T_{fB} = 1200K$
$\Omega_l = 0$	$\Omega_s = 1.5 \times 10^4\,J/mol$	

Use this dataset to plot $G_l = f(X_B)$ and $G_s = f(X_B)$ at temperature T and constant pressure $P = 1\,atm = 101.33\,kPa$. The resultant plots are thermodynamic diagrams; Fig. 8.7a at $T_1 = 700\,K$, Fig. 8.7b at $T_2 = 636\,K$ and Fig. 8.7c at $T_3 = 500\,K$. These diagrams are related to the schematic binary eutectic phase diagram shown in Fig. 8.7d.

Notice that Figs. 8.6d and 8.7d are represent different equilibrium phase diagrams that map out their respective phase regions. The former contains a single homogeneous solid α-phase, while the latter has small single α and β regions, and one large two-phase region ($\alpha + \beta$). Additionally, the first law of thermodynamics defines the conservation of energy approach, but it does not provide means to predict the direction of energy flow.

Thus far, only two simple equilibrium phase diagrams are constructed using the given data and the function $G = f(X_B)$ for liquid and solid phases. More complicated equilibrium phase diagrams require a substantial amount of data from the Gibbs energy approach.

The reader should be aware of sophisticated methods for mapping temperature versus composition known as equilibrium phase diagram. Thus, the Ab initio method is based on quantum mechanics and statistical analysis, and it may employ the Monte Carlo simulation technique in order to determine the Gibbs energy data and related parameters at a finite temperature T. Subsequently, an optimization method follows for plotting a given dataset as an equilibrium phase diagram.

8.11   Nucleation Theory

This section includes the classical analytical procedure used to quantitatively characterize the kinetics of nucleation of solid particles during liquid-to-solid ($L \rightarrow S$) phase transformation, commonly known as solidification. Thus, the classical nucleation theory is presented using two different theoretical approaches having the same physical meaning within the complex (intricate) solidification domain.

The formation of nuclei is accomplished when the liquid releases thermal energy known as the latent heat of fusion ΔH_f for a pure metal or ΔH_f^* for an alloy. This energy is transferred through the solid towards mold cavity walls and through the mold wall thickness to the mold surroundings.

For one-dimensional analysis, the solidification process ($L \rightarrow S$) is related to heat transfer denoted as the heat flux q_x. Subsequently, $\Delta H_f \rightarrow q_x$ for completion of the solidification process and cooling of the polycrystalline casting product.

8.11.1   Solidification Models

Consider a small undercooling ΔT for a randomly formation of round particles at a nucleation rate dN/dt. Assume that the particles consisting of unit cell clusters reach a critical radius r_c for subsequent growth at a rate dr/dt until solidification is complete. This is schematically shown in Fig. 8.8a, which elucidates the solidification model for the formation of critical size particles and equiaxed grains having preferred unit cell orientations (Rosenhain [10, p. 63]).

In theory, the solidified particle nuclei have critical radii $r - c$ and the grains have an average diameter d, which influence mechanical properties of polycrystalline solids. In materials science books and university classrooms, the Hall–Petch equation is the most cited correlation between d and the yield strength σ_{ys}; $\sigma_{ys} = f(d)$. This particular function is introduced in a later chapter.

The assumed round-particle growth is denoted by the radii $r_3 > r_2 > r_1 > r_c$ and relate rate of nucleation (dN/dt). The final stage of the solidification process is the formation of grain boundaries that separate grains. This is a simplified solidification model that describes the nature of grain formation.

Additional theoretical aspects of the microstructural evolution is modeled as particles (Fig. 8.8b) that must reach critical sizes for subsequent growth. Moreover,

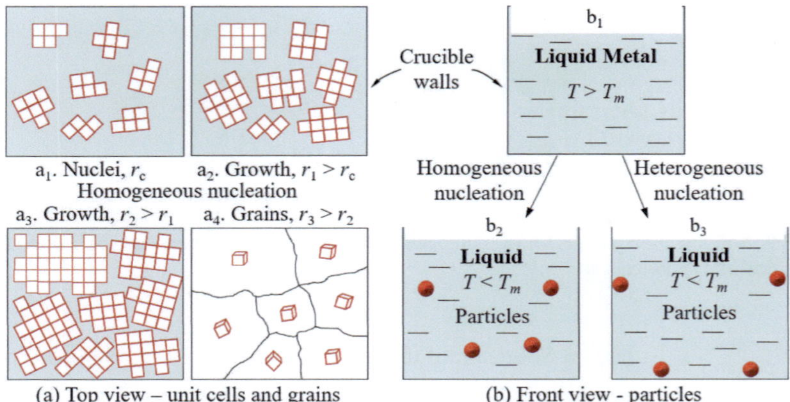

Fig. 8.8 Models for the solidification process. (**a**) Top view of the crucible for homogeneous nucleation of unit cells as nuclei (after Rosenhain [10, p. 63]), and (**b**) front view of the crucible for homogeneous and heterogeneous nucleation of spherical nuclei

the microstructural evolution as predicted by the classical theory of nucleation is characterized by homogeneous and heterogeneous processes.

Solidification starts when the liquid phase is cooled below the freezing temperature ($T_x < T_f$), and the rate of nuclei formation depends on the release rate of ΔH_f. Moreover, the solidification rate dT/dt, temperature gradient dT/dx and solidification velocity dx/dt are the main solidification parameters related to thermal stability and phase transformation that induce the microstructural evolution of crystalline alloys (Motaff et at. [11, Chapter 7] and Avner [12, Chapter 6]).

8.11.2 *Particle Formation Model*

Consider the model depicted in Fig. 8.9a for one-dimensional solidification of a pure metal. This model represents the onset of homogeneous nucleation in the form of a spherical cluster of atoms treated as a suspended sphere (α-phase) within the liquid metal at just below the equilibrium temperature T_f. Regarding particle size, an embryo has $r < r_c$ and a nucleus has $r \geq r_c$ for subsequent solidification.

Figure 8.9b illustrates the heterogeneous nucleation when a particle is in contact with a substrate β-phase, such as the metallic mold wall or an impurity particle, known as grain refiner. In this case, the cluster of atoms is modeled as a semi-elliptical cap treated as the nucleus subjected to grow as adjacent liquid atoms adjoin at the liquid–solid interface, where ΔH_f is released.

The notation related to the theoretical model shown in Fig. 8.9b is based on preexisting surfaces and the radius (r) of curvature of the droplet on a substrate β surface. The pertinent surface energy terms associated with nucleation and growth are

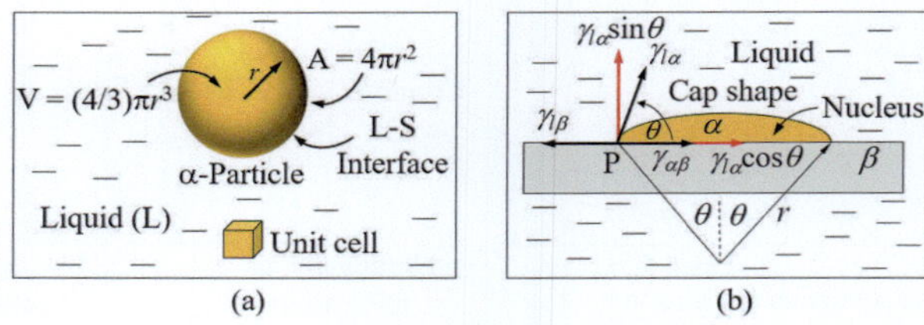

Fig. 8.9 Solidification models for (**a**) homogeneous nucleation of a sphere and (**b**) heterogeneous nucleation of a spherical cap

$$\gamma_{\alpha L} = \text{Solid-liquid interface surface energy}$$

$$\gamma_{\beta L} = \text{Agent-liquid interface surface energy}$$

$$\gamma_{\alpha \beta} = \text{Solid-solid interface surface energy}$$

$$\theta = \text{Wet or contact angle}$$

Phase transformation is a liquid-to-solid (L-S) process related to the energy release and to the kinetics of nucleation and growth of clusters of atoms, which are ideally modeled as perfect and hard spheres. Again, the smallest sphere is modeled as the unit cell (Fig. 8.9a).

8.11.3   *Homogeneous Nucleation of Metals*

Consider a cluster of atoms susceptible to form a spherical particle as shown in Fig. 8.9a and assume that the particle radius is $r < r_c$, where r_c is a particle critical radius. In this case, the particle can be treated as an unstable spherical embryo and from Eq. (8.5b) along with a small undercooling ΔT and the latent heat of fusion ΔH_f, the corresponding Gibbs energy density (ΔG_v) for a growing unstable embryo can be approximated as

$$\Delta G_v = -\Delta H_f \left(\frac{\Delta T}{T_f} \right) \tag{8.55}$$

which is linearly proportional to the solidification driving force ΔT due to release of latent heat of fusion ΔH_f, which is also known as latent heat of solidification.

For homogeneous nucleation of a spherical solid phase, the total Gibbs energy change is

$$\Delta G = V \Delta G_v + \Delta G_s = V \Delta G_v + A \gamma_{ls} \tag{8.56a}$$

$$\Delta G = \frac{4}{3} \pi r^3 \Delta G_v + 4 \pi r^2 \gamma_{ls} \tag{8.56b}$$

Table 8.1 Experimental dataset (Askeland et al. [13, p. 262])

Elements	Crystal	T_f ($^\circ C$)	ΔT ($^\circ C$)	ρ (g/cm^3)	ΔH_f (J/cm^3)	γ_{ls} (J/cm^2)
Cu	FCC	1085	236	8.96	1628	177×10^{-7}
Fe	FCC	1538	420	7.87	1737	204×10^{-7}
Ag	FCC	962	450	10.49	965	126×10^{-7}
Ni	FCC	1453	480	8.81	2756	255×10^{-7}

where ΔG is the net Gibbs energy change, V is the solid sphere volume, A is the solid sphere area, and ΔG is the change in Gibbs energy of the solid phase and γ_{ls} is the interfacial surface energy for the creation of a solid surface.

It is assumed that γ_{ls} is a temperature-independent and an isotropic energy term, which is listed in Table 8.1 for selected metals (Askeland et al. [13, p. 262]).

Substituting Eq. (8.55) into (8.56b) yields the homogeneous $\Delta G = \Delta G_{\mathrm{hom}}$

$$\Delta G = \Delta G_v + \Delta G_{ls} = -\left(\frac{4}{3}\pi r^3\right)\frac{\Delta H_f \Delta T}{T_f} + \left(4\pi r^2\right)\gamma_{ls} \tag{8.57a}$$

$$\frac{d\left(\Delta G\right)}{dr} = -\left(4\pi r^2\right)\frac{\Delta H_f \Delta T}{T_f} + (8\pi r)\,\gamma_{ls} \tag{8.57b}$$

$$\frac{d^2\left(\Delta G\right)}{dr^2} = -(8\pi r)\frac{\Delta H_f \Delta T}{T_f} + (8\pi)\,\gamma_{ls} \tag{8.57c}$$

Letting $d\left(\Delta G\right)/dr = 0$ at $r = r_c$ gives the critical radius of the nuclei for solidification to proceed

$$r_c = \frac{2\gamma_{ls} T_f}{\Delta H_f \Delta T} = \frac{2\gamma_{ls}}{\rho \Delta S_f \Delta T} = \frac{2\Gamma_{ls}}{\Delta T} \tag{8.58a}$$

$$\Gamma_{ls} = \frac{\gamma_{ls}}{\rho_s \Delta S_f} = \frac{\gamma_{ls} T_f}{\rho_s \Delta H_f} \tag{8.58b}$$

where Γ_{ls} denotes a material property known as the Gibbs–Thomson coefficient that describes the effect of L-S interface curvature on the local equilibrium temperature. Additional details on this matter can be elsewhere (Dantzig and Rappaz [2, p. 57]).

Substituting Eq. (8.58a) into (8.57a) yields the critical Gibbs energy change for nuclei with critical radius $r = r_c$

$$\Delta G_c = \frac{16\pi \gamma_{ls}^3}{3\left(\Delta H_f\right)^2}\left(\frac{T_f}{\Delta T}\right)^2 \tag{8.59}$$

Example 8.8 For copper (Cu), change $\Delta H_f = 1628\ J/cm^3$ to J/mol and then to eV. This is a convenient exercise for unit conversion.

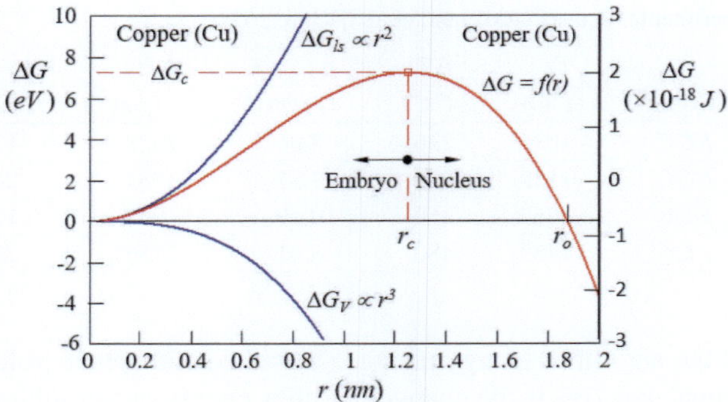

Fig. 8.10 Gibbs energy change profiles using based on the $\Delta G = f(r)$ function for theoretical solidification of pure copper. Conversion: $1\ J = 6.242 * 10^{18}\ eV$

$$\Delta H_f = \frac{\Delta H_f\ \left(J/cm^3\right) A_w\ (g/mol)}{\rho\ \left(g/cm^3\right)} \tag{8.8E1a}$$

$$= \frac{\left(1628\ J/cm^3\right)(63.55\ g/mol)}{8.92\ g/cm^3} = 11,599\ J/mol \tag{8.8E1b}$$

Then,

$$\Delta H_f = \frac{\Delta H_f\ (J/mol)\left(6.24181 \times 10^{18}\ eV/J\right)}{N_a} \tag{8.8E2a}$$

$$= \frac{(11599\ J/mol)\left(6.24181 \times 10^{18}\ eV/J\right)}{6.022 \times 10^{23}\ mol^{-1}} = 0.12\ eV \tag{8.8E2b}$$

For convenience, Eq. (8.57a) is plotted in Fig. 8.10 for copper (Cu).

From Table 8.1 and Eq. (8.59), $\Delta G_c = 7.4096 \times 10^{-19}\ J/cm^3 = 7.2442\ eV$ is a maximum energy at $r_c = 1.25\ nm$. Also shown in Fig. 8.10 are the individual energy changes due to effects of particle surface area and volume; $G_{ls} = \left(4\pi r^2\right)\gamma_{ls}$ and $G_v = \left(4\pi r^2\right) H_s \Delta T/T_f$ respectively. Eventually, these clusters reach critical and stable sizes as solidification proceeds.

Note that Eqs. (8.57a,c) give $\Delta G_c > 0$ and $d^2\left(\Delta G\right)/dr^2 < 0$ at $r = r_c$. This means that $\Delta G_c\ (r)_{r=r_c} = \Delta G_{max}\ (r_c)$ is a maximum, provided that $r \geq r_c$ for particle growth to occur.

Unstable Stage This occurs when $0 < \Delta G < \Delta G_{max}$ at $0 < r < r_c$ and therefore, nanosize particles (known as embryos, clusters and crystallites) may dissolve ($\Delta G \to 0$ and $r \to 0$)

Stable Stage For continues nucleation, $\Delta G_c = \Delta G_{max}$ at $r = r_c$ and ΔG_{max} is energy required for the formation of stable spherical nuclei. In fact, ΔG_{max} is the

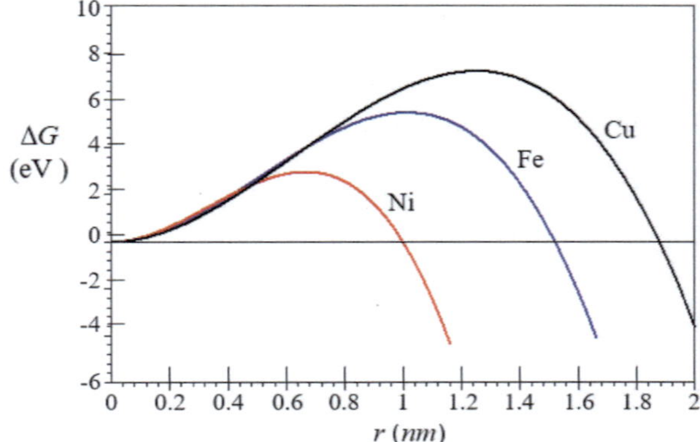

Fig. 8.11 Gibbs energy change profile using $\Delta G = f(r)$ for theoretical solidification of three pure elements as per data listed in Table 8.1

energy barrier or the thermodynamic driving force to the nucleation process. Thus, stable particle growth process when $\Delta G_c < \Delta G \leq -\infty$ at $r_c \leq r \leq \infty$. The final solid is a crystalline structure having a particular grain morphology.

Homogeneous Nucleation   It is characterized by a large undercooling ΔT to cause the formation of stable nuclei during directional or unidirectional solidification. Undoubtedly, the formation of nuclei requires heat transfer by conduction at the liquid–solid interface, which represents an evolving boundary being dependent on time and the type of thermal resistance within the solidification domain Γ.

The microstructural morphology, in general, controls the mechanical and physical properties of the material. However, solidified materials with mixture of different microstructural morphology have different responses to mechanical deformation, electrochemical behavior in aggressive media, and the like.

Figure 8.11 exhibits the total Gibbs energy change for the three pure elements listed in Table 4.1. Notice that $\Delta G_{c,Cu} > \Delta G_{c,Fe} > \Delta G_{c,Ni}$ and $r_{c,Cu} > r_{c,Fe} > r_{c,Ni}$. This is attributed to the magnitude of undercooling $\Delta T_{Cu} < \Delta T_{Fe} < \Delta T_{Ni}$. From Fig. 8.11 and Table 8.1, the lower ΔT the higher ΔG_c and the larger r_c.

In the Materials Science or Physical Metallurgy field, the effect of nucleation rate on the kinetics of phase transformation $L \rightarrow S$ depends on the rate of latent heat of fusion, which is part of the solidification heat transfer process (flow of heat or heat flow). In effect, heat transfer is of interest with respect to thermal resistance cases related to Fourier's law of conduction due to a temperature gradient, say, dT/dx. Whereas, Newton's law of cooling associated with a temperature difference between a specimen (body) and its surroundings. Cooling, in general, can be a natural or forced process. Nevertheless, these laws are well established in the field of heat transfer.

Example 8.9 Consider the homogeneous nucleation of pure iron (Fe) with a typical undercooling, $\Delta T = 420\ K$, latent heat of fusion $\Delta H_f = 1737\ J/cm^3$, surface energy $\gamma_{ls} = 204 \times 10^{-7}\ J/cm^2$, and freezing temperature $T_f = 1811\ K$. Use this information to calculate **(a)** the critical radius (r_c), **(b)** the critical Gibbs energy change (ΔG_c) at $r = r_c$, **(c)** the number of unit cells (N_{cell}) and **(d)** the number of atoms (N_{atom}) in a nucleus. Moreover, iron at T_f has an FCC crystal structure with 4 *atoms* in the unit cell and lattice parameter $a = 0.287\ nm$, which is normally determined using the X-ray technique.

Solution

(a) From Eq. (8.58a), the critical radius is

$$r_c = \frac{2\gamma_{ls}T_f}{\Delta H_f \Delta T} = \frac{2\left(204 \times 10^{-7}\ J/cm^2\right)(1811\ K)}{\left(1737\ J/cm^3\right)(420\ K)} = 1.01 \times 10^{-7}\ cm \tag{8.9E1a}$$

$$r_c = 1.10\ nm \tag{8.9E1b}$$

(b) From Eq. (8.59), the critical Gibbs energy change at $r = r_c$ is

$$\Delta G_c = \frac{16\pi \gamma_{ls}^3}{3\left(\Delta H_f\right)^2}\left(\frac{T_f}{\Delta T}\right)^2 = \frac{16\pi\left(204 \times 10^{-7}\ J/cm^2\right)^3}{3\left(1737\ J/cm^3\right)^2}\left(\frac{1811\ K}{420\ K}\right)^2 \tag{8.9E2a}$$

$$\Delta G_c = 8.77 \times 10^{-19}\ J \tag{8.9E2b}$$

(c) For the volume of the unit cell (V_{cell}), the volume of a sphere (V_s) and the number of unit cells (N_{cell}) in a nucleus are

$$V_{cell} = a^3 = (0.287\ nm)^3 = 2.36 \times 10^{-2}\ nm^3 \tag{8.9E3a}$$

$$V_s = \frac{4}{3}\pi r_c^3 = \frac{4}{3}\pi (1.01\ nm)^3 = 4.32\ nm^3 \tag{8.9E3b}$$

Thus,

$$N_{cell} = \frac{V_s}{V_{cell}} = \frac{4.32\ nm^3}{2.36 \times 10^{-2}\ nm^3/cells} = 183\ \text{cells} \tag{8.9E4}$$

(d) There are 4 atoms in a FCC unit cell and subsequently, the total number of atoms (N_{atom}) in a nucleus is

$$N_{atom} = 4N_{cell} = 4\,(183) = 732\ atoms \tag{8.9E5}$$

These results indicate that a copper cluster with $V_s = 4.32\ nm^3$ contains 183 unit cells and 732 atoms.

Example 8.10 Calculate **(a)** the number of unit cells, **(b)** the number of atoms in the critical nucleus (particle), **(c)** the entropy of fusion, and **(d)** the Gibbs energy for solidification of copper under homogeneous nucleation. Copper has an FCC crystal structure with 4 atoms per unit cell, a critical particle radius of 1.2512 nm, a density of 8.96 g/cm^3, and a lattice parameter of $a = 0.362\ nm$.

Solution

(a) The FCC unit cell volume is

$$V_{cell} = a^3 = (0.362\ nm)^3 = 4.74 \times 10^{-2}\ nm^3/unit\ cell \tag{8.12E1}$$

and the volume of a perfect spherical particle becomes

$$V_s = \frac{4}{3}\pi r_c^3 = \frac{4}{3}\pi\ (1.2512\ nm)^3 = 8.20\ nm^3 \tag{8.12E2}$$

Thus, the number of unit cells N_{cell} is

$$N_{cell} = \frac{V_s}{V_{cell}} = \frac{8.20\ nm^3}{4.7438 \times 10^{-2}\ nm^3/unit\ cell} = 173\ cells \tag{8.12E3}$$

(b) If there are 4 atoms in the FCC unit cell, then the number of atoms in the critical nucleus is

$$N_{atom} = 4N_{cell} = (4)\,(173) = 692\ atoms \tag{8.12E4}$$

Therefore, a copper cluster with $V_s = 8.20\ nm^3$ contains 173 unit cells and 692 atoms.

(c) From Eq. (8.5a) and Table 4.1, the entropy of fusion is

$$\Delta S_f = \frac{\Delta H_f}{T_f} = \frac{1628\ J/cm^3}{(1085 + 273)\ K} = 1.20\ J/\left(cm^3.K\right) \tag{8.12E5}$$

Divide this result with the density value to get

$$\Delta S_f = \left[1.20\ J/\left(cm^3.K\right)\right]/\left(8.96\ g/cm^3\right) \simeq 0.13\ J/g.K \tag{8.12E6}$$

Using conventional units for the entropy of fusion yields

$$V_s = \frac{4}{3}\pi r_c^3 = \frac{4}{3}\pi\ \left(1.2512 \times 10^{-7}\ cm\right)^3 \tag{8.12E7a}$$

$$= 8.20 \times 10^{-21}\ cm^3$$

$$\Delta S_f = \left(1.20\ J/cm^3.K\right)\left(8.20 \times 10^{-21}\ cm^3\right) \tag{8.12E7b}$$

$$= 9.84 \times 10^{-21}\ J/K$$

(d) From Eq. (8.5b), the critical Gibbs energy is

$$\Delta G_c = \Delta H_f - T_f \Delta S_f \tag{8.12E8a}$$

$$\Delta G_c = (1628)\left(8.20 \times 10^{-21}\right) - (1085 + 273)\left(9.84 \times 10^{-21}\right) \tag{8.12E8b}$$

$$\Delta G_c = -1.312 \times 10^{-20} \; J \tag{8.12E8c}$$

Note that a conversion of units is necessary in order to calculate the entropy of fusion and the Gibbs energy in proper units.

8.11.4 Heterogeneous Nucleation of Metals

Consider the classical heterogeneous nucleation model shown in Fig. 8.9b for solidification with the condition that α and β crystal structures are similar. Thus, the kinetics of heterogeneous nucleation is characterized by the energy balance

$$\Delta G_{het} = \left(-\frac{\pi r^3}{3}\Delta G_v + 4\pi r^2 \gamma_{l\alpha}\right) f(\theta) \tag{8.60a}$$

$$\Delta G_{het} = \Delta G_{\text{hom}} f(\theta) \tag{8.60b}$$

where $f(\theta)$ is the shape factor (geometric factor), $0 \leq f(\theta) \leq 1$, is derived in Appendix 8A as (Kurz and Fisher [14, p. 288])

$$f(\theta) = \frac{1}{4}\left(2 - 3\cos\theta + \cos^3\theta\right) = \frac{1}{4}(2 + \cos\theta)(1 - \cos\theta)^2 \tag{8.61}$$

Figure 8.12 shows the graphical representation of the nucleation process. For instance, Fig. 8.12a exhibits a sigmoidal behavior and Fig. 8.12b compares ΔG_{hom} and ΔG_{het} at a constant critical particle radius r_c.

The sigmoidal function $f(\theta)$ is evaluated at $0 \leq \theta \leq 180°$ ($0 \leq \theta \leq \pi$ radian) and its effect of $f(\theta)$ on ΔG_{het} is deduced as follows:

- If $f(\theta) = 1$, then $\Delta G_{het} = \Delta G_{\text{hom}}$ and homogeneous nucleation takes place due to lack of wetting effect between the α and β phases.
- It can be assumed that homogeneous nucleation arises when the heterogeneous nucleation model breaks down since the insoluble wetting agent (substrate β) has no effect on heterogeneous nucleation.
- During homogeneous nucleation, most nuclei formation occurs randomly within the molten metal.
- Ideally, $f(\theta) = 0$ yields $\Delta G_{het} = 0$ for a fully wetting effect between the α and β phases, provided that $r \geq r_c$.

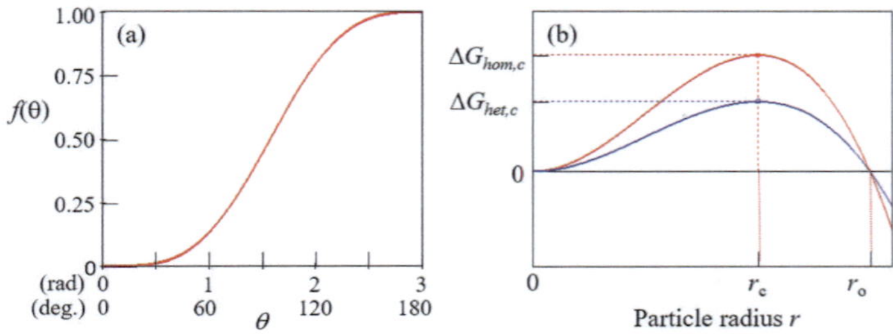

Fig. 8.12 (**a**) Profile of the shape factor $f(\theta)$ and (**b**) Gibbs energy change profiles for homogeneous and heterogeneous nucleation processes

- Actually, $0 < f(\theta) < 1$ induces a positive effect of wetting defined as the contact angle between the α and β phases. This implies that heterogeneous nucleation is enhanced by the driving force $\Delta G_{het} < \Delta G_{hom}$ as $f(\theta) \to 0$.

Heterogeneous nucleation is the most realistic mechanism since the mold internal surfaces and substrate particles act as insoluble agents so that $0 < f(\theta) < 1$ and $\Delta G_{het} < \Delta G_{hom}$. Moreover, heterogeneous nucleation constitutes the most practical aspect of foundry technology due to an enhanced solidification velocity for particles on the surface of high-temperature resistant preexisting seeds or agents.

Theoretical knowledge of the classical homogeneous and heterogeneous solidification models enhances the understanding of nucleation mechanisms cited in Figs. 8.8 and 8.9. Specifically, the temperature difference between the liquid metal and the impurity surface areas causes a liquid undercooling ΔT needed for phase transformation $L \to S$, making ΔT the driving force for solidification.

The characterization of heterogeneous nucleation provides a better fundamental understanding of particle nucleation under control conditions for an optimal solidification process of metals and their alloys. Moreover, heterogeneous nucleation represents a more realistic solidification process since the mold inner walls and/or suspended impurities in the melt (liquid) act as nucleating agents (catalyst or grain refiner).

8.11.5 Nucleation Sites

Nucleating agents in the melt represent preexisting surfaces as preferred nucleation sites. This implies that heterogeneous nucleation is a catalyst-induced process that depends on the nucleus contact angle θ and $\Delta T_{het} < \Delta T_{hom}$.

Adding grain refiners (such as TiB particles) to Al and Al-alloys is a common practice in the non-ferrous metal making industry for refining the microstructure and improving properties. This can be achieved by controlling the mechanical agitation of the melt and its cooling rate. The solidification response to adding grain refiners to the melt is to reduce the degree of undercooling ΔT, the liquidus temperature, and the critical Gibbs energy ΔG_v.

The interface surface energy (γ_{ls}), in principle, between two phases (liquid and solid) is the basic energy entity for the nucleation of embryos, which eventually grow to form stable nuclei. Methods for measuring interface energies can be found elsewhere (Fredriksson and Akerlind [15, p. 107], Ragone [16, p. 111]).

The nucleation sites are localized points within the melt where particles or atom clusters, in general, can randomly nucleate and grow as long as there is liquid metal in the solidification domain, where release of latent heat of fusion (ΔH_f or ΔH_f^*) must occur in a continuous and uniform manner.

8.12 Kinetics of Nucleation

Consider the homogeneous nucleation process where the number of stable nuclei n_s becomes a temperature-dependent function $n_s = f(T)$ with $r \geq r_c$ and the total enthalpy takes the form $n_s \Delta H$. From Eq. (8.5b), ΔG takes the form

$$\Delta G = n_s \Delta H - T \Delta S \tag{8.62}$$

The entropy change defined by Eq. (8.23) is now used to derive $n_s = f(T)$

$$\Delta S = k_B \ln(\omega) = k_B \ln \frac{(n_o + n_s)!}{n_o! n_s!} \tag{8.63a}$$

$$\Delta S = -k_B \left[n_s \ln \left(\frac{n_s}{n_o + n_s} \right) + n_o \ln \left(\frac{n_o}{n_o + n_s} \right) \right] \tag{8.63b}$$

where n_o denotes the number of atoms per volume in the liquid phase. Substituting Eq. (8.63b) into (8.62) yields

$$\Delta G = n_s \Delta H + k_B T \left[n_s \ln \left(\frac{n_s}{n_o + n_s} \right) + n_o \ln \left(\frac{n_o}{n_o + n_s} \right) \right] \tag{8.64}$$

For $d(\Delta G)/dn_s = 0$,

$$\ln \left(\frac{n_s}{n_o + n_s} \right) = -\frac{\Delta G}{k_B T} \tag{8.65}$$

where $T < T_f$ promotes a small undercooling $\Delta T > 0$. If $n_o >> n_s$, then Eq. (8.65) becomes an Arrhenius-type equation for stable nuclei formation

$$n_s = n_o \exp \left(-\frac{\Delta G}{k_B T} \right) \tag{8.66}$$

with ΔG as the apparent activation energy and n_o as the pre-exponential kinetic factor for heterogeneous.

The exponential function $\exp(-\Delta G/k_B T)$ can be treated as a phase-transformation factor or as a thermodynamic energy factor needed to solidify a group of atoms n_o from the liquid into a solid particle (cluster) n_s with a critical spherical radius r_c. This implies that $\Delta G = \Delta G_c$ acts as the effective activation energy that should remain constant during the course of a liquid-to-solid phase transformation process under the assumption of steady-state nucleation.

Linearizing Eq. (8.66) yields

$$\ln(n_s) = \ln(n_o) - \left(\frac{\Delta G}{k_B}\right)\frac{1}{T} \tag{8.67a}$$

$$n_s^* = n_o^* - k^* T^* \tag{8.67b}$$

where $n_s^* = \ln(n_s)$, $n_o^* = \ln(n_o)$, $k^* = \Delta G/k_B$, and $T^* = 1/T$. Thus, Fig. 8.13a shows the typical trend of an exponential function as defined by Eq. (8.66) and Fig. 8.13b plots Eq. (8.67) to find n_o and $\Delta G = \Delta G_c$ at a reciprocal temperature range $\Delta(1/T)$.

For $n_s = f(T) > 0$, groups of atoms in the liquid phase arrange themselves randomly as solidification progresses under steady-state conditions, and eventually, an atom attachment process transpires to what is known as nuclei of critical radius r_c predicted by Eq. (8.58a) when $\Delta G = \Delta G_c$ is met.

The number of atoms per molar volume (n_o) that can be attached to the nuclei is written as

$$n_o = \frac{N_a}{V_m} \tag{8.68a}$$

$$V_m = \frac{N_a V_{cell}}{N_{atom}} \simeq A_w/\rho \tag{8.68b}$$

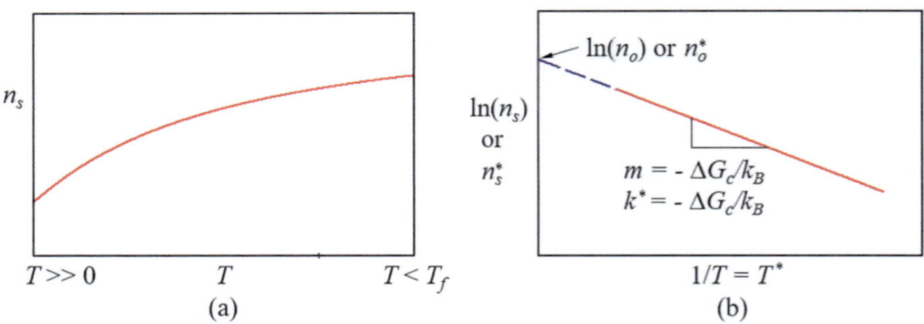

Fig. 8.13 Arrhenius plot for nucleation behavior. (**a**) Nonlinear behavior and (**b**) linearized Arrhenius equation

where $N_a = 6.022 \times 10^{23}\ atoms/mol$ denotes the Avogadro's number, V_m denotes the molar volume, A_w denotes the atomic weight, and ρ denotes the mass density of the pure metal.

Thus far, the Arrhenius empirical equation describes the nuclei behavior as illustrated in Fig. 8.13 using linear and nonlinear graphical approaches. It does not provide pertinent information for kinetic studies.

The kinetics of nucleation and growth processes are most realistic during phase transformation since the evolution of microstructural features dictates the mechanical behavior and related mechanical properties of the solidified metal.

8.12.1   Nucleation Rate

The kinetics of nucleation can the described by the nucleation rate (I_s) which strongly depends on the critical Gibbs energy ΔG_c for nuclei formation. Hence, the activation energy is $Q_d = \Delta G_d$ for diffusion. This implies that the evolution of nuclei per unit volume (I_s) with time at constant temperatures can analytically be modeled as an Arrhenius-type equation

$$I_s = \frac{dn_s}{dt} = I_o \exp\left(-\frac{\Delta G_c + Q_d}{k_B T}\right) \simeq I_o \exp\left(-\frac{\Delta G_c}{k_B T}\right) \tag{8.69a}$$

$$\ln\left(I_s\right) = \ln\left(I_o\right) - \left(\frac{\Delta G_c}{k_B}\right)\frac{1}{T} \tag{8.69b}$$

Here, I_o is a proportionality constant related to vibrational frequency of atoms $(1/s)$ in clusters per unit volume per unit time $[atom/\left(m^3.s\right)]$ as latent heat of fusion evolves during solidification, ΔG_c denotes the thermodynamic energy barrier for a cluster formation with a size r_c so that $\Delta G = \Delta G_c$ when $r = r_c$ at $\Delta T = \left(T_f - T\right) > 0$, Q_d is the activation energy for diffusion across the phase boundary.

Accordingly, the proportionality constant (pre-exponent factor) I_o in Eq. (8.69a) can be defined in terms of atoms per unit volume per time $(atoms/m^3.s)$ under the assumption that the mold wall and an nucleating agent are the potential sites for heterogeneous nuclei formation. Thus,

$$I_o = \frac{n_o k_B T}{h} \simeq v_o p_c n_o \tag{8.70}$$

where $v_o = D_l/\delta^2$ denotes the atomic vibrational frequency $(\simeq 10^{13}\ 1/s)$ at T_f, D_l denotes the diffusion coefficient in the melt (m^2/s), δ denoted the atomic jump distance (m), h denotes Planck's constant, p_c denotes the probability factor for attracting atoms to the nucleation sites and n_o denotes the particle density $(atoms/m^3$ or $particles/m^3)$ in the melt or the density of nucleation sites at the melt-solid interface $(sites/cm^2$ or $sites/m^2)$.

Substitute Eqs. (8.59) and (8.70) into (8.69a) to get the nucleation rate $I_s = f(\Delta T)$ in the most realistic form

$$I_s = \frac{n_o k_B T}{h} \exp\left[-\frac{16\pi \gamma_{ls}^3}{3k_B (\Delta H_f)^2} \frac{1}{T} \left(\frac{T_f}{\Delta T}\right)^2 - \frac{Q_d}{k_B T}\right] \qquad (8.71)$$

It is clearly shown in Eq. (8.71) that $I_s = f(T)$, but the driving force for nucleation is now the degree of undercooling $\Delta T = (T_f - T) > 0$. This means that phase transformation by the nucleation mechanism occurs at $T = T_f - \Delta T$.

Combining Eqs. (8.59) and (8.66) when $\Delta G = \Delta G_c$ yields the theoretical undercooled liquid temperature for nucleation

$$T^3 - 2T_f T^2 + T_f^2 T + \lambda = 0 \qquad (8.72)$$

with

$$\lambda = \frac{16\pi \gamma_{ls}^3 T_f^2}{3(\Delta H_f)^2 k_B \ln(n_s/n_o)} \qquad (8.73)$$

One of the three roots of Eq. (8.72) is the theoretical temperature of the undercooled liquid. An example can illustrate the relevant analytical procedure for finding such a temperature.

Example 8.11 Consider the nucleation of pure copper (Cu) at a slow cooling rate dT/dt under control conditions. Determine **(a)** the number of stable particles n_s per volume when the nuclei radius is 0.80 nm at the freezing temperature T_f, **(b)** the critical radius r_c and n_s at $\Delta T = 236$ K and **(c)** r_c and n_s at $\Delta T = 300$ K. Will homogeneous nucleation take place at the freezing temperature T_f and at undercooling ΔT? Use the given and relevant datasets from Table 4.1 to carry out all calculations. Recall that the stable nuclei n_s in the Cu-melt represent small solid particles having a crystalline structure surrounded by the solidifying liquid or melt.

$$T_f = 1085° \, C = 1358 \, K, \quad \Delta T = 236° \, C = 236 \, K$$

$$\Delta S_m = 10 \, J/mol.K, \quad \Delta H_f = 1628 \, J/cm^3$$

$$\gamma_{ls} = 177 \times 10^{-7} \, J/cm^2 = 0.177 \, J/m^2$$

$$V_m = 7 \times 10^{-6} m^3/mol$$

Solution

(a) For $\Delta T = 0$ K, there is no need to undercool the Cu-melt. This means that solidification is assumed to start at the freezing temperature $T = T_f = 1358$ K. In such a case, the contribution of the Gibbs energy per volume is $G_v = 0$, and subsequently, ΔG depends on the surface energy. From Eq. (8.56b) with $G_v = 0$,

$$\Delta G = -\frac{4}{3}\pi r^3 \Delta G_v + 4\pi r^2 \gamma_{ls} = 4\pi r^2 \gamma_{ls} \qquad (8.12\text{E1a})$$

$$\Delta G = 4\pi \left(0.80 \times 10^{-9}\ m\right)^2 \left(0.177\ J/m^2\right) \qquad (8.12\text{E1b})$$

$$\Delta G = 1.42 \times 10^{-18}\ J = 8.86\ eV \qquad (8.12\text{E1c})$$

which is a reasonable result. The constant n_o in Eq. (8.68a) is

$$n_o = \frac{N_a}{V_m} = \frac{6.022 \times 10^{23}\ clusters/mol}{7 \times 10^{-6}\ cm^3/mol} = 8.60 \times 10^{28}\ clusters/m^3 \qquad (8.12\text{E2})$$

From Eq. (8.66), the number of clusters having a radius of 0.8 nm is

$$n_s = n_o \exp\left(-\frac{\Delta G}{k_B T_f}\right) \qquad (8.12\text{E3a})$$

$$n_s = \left(8.60 \times 10^{28}\ clusters/m^3\right) \exp\left[-\frac{1.42 \times 10^{-18}\ J}{\left(1.38 \times 10^{-23}\ J/K\right)(1358\ K)}\right]$$
$$\qquad (8.12\text{E3b})$$
$$n_s = 1.06 \times 10^{-4}\ clusters/m^3 \qquad (8.12\text{E3c})$$

No homogeneous nucleation takes place since $n_s \simeq 0$.
(b) For $\Delta T = 236\ K$ and $T < T_f$, Eq. (8.58a) yields

$$r_c = \frac{2\gamma_{ls} T_f}{\Delta H_f \Delta T} = \frac{2\left(177 \times 10^{-7}\ J/cm^2\right)(1358\ K)}{\left(1628\ J/cm^3\right)(236\ K)} = 1.25\ nm \qquad (8.12\text{E4})$$

From Eq. (8.59), the critical Gibbs energy change at $r = r_c$ is

$$\Delta G_c = \frac{16\pi \gamma_{ls}^3}{3\left(\Delta H_f\right)^2}\left(\frac{T_f}{\Delta T}\right)^2 = \frac{16\pi \left(177 \times 10^{-7}\ J/cm^2\right)^3}{3\left(1628\ J/cm^3\right)^2}\left(\frac{1358\ K}{236\ K}\right)^2$$
$$\qquad (8.12\text{E5a})$$
$$\Delta G_c = 1.16 \times 10^{-18}\ J \qquad (8.12\text{E5b})$$

and at $T = T_f - \Delta T = 1358 - 236 = 1122\ K$

$$n_s = n_o \exp\left(-\frac{\Delta G_c}{k_B T}\right) \qquad (8.12\text{E6a})$$

$$n_s = \left(8.60 \times 10^{22} \; clusters/m^3\right) \exp\left[-\frac{1.16 \times 10^{-18} \; J}{\left(1.38 \times 10^{-23} \; J/K\right)\left(1122 \; K\right)}\right]$$

$$\tag{8.12E6b}$$

$$n_s = 2.50 \times 10^{-10} \; clusters/m^3 \tag{8.12E6c}$$

Therefore, homogeneous nucleation does not take place at an undercooling $\Delta T = 236 \; K$ since $n_s \simeq 0$.

(c) From Eq. (8.58a) with $\Delta T = 300 \; K$, the critical radius is

$$r_c = \frac{2\gamma_{ls}T_f}{\Delta H_f \Delta T} = \frac{2\left(177 \times 10^{-7} \; J/cm^2\right)\left(1358 \; K\right)}{\left(1628 \; J/cm^3\right)\left(300 \; K\right)} = 0.984 \; nm$$

$$\tag{8.12E7}$$

Again, from Eq. (8.59), the critical Gibbs energy change at $\Delta T = 300 \; K$ is

$$\Delta G_c = \frac{16\pi\gamma_{ls}^3}{3\left(\Delta H_f\right)^2}\left(\frac{T_f}{\Delta T}\right)^2 = \frac{16\pi\left(177 \times 10^{-7} \; J/cm^2\right)^3}{3\left(1628 \; J/cm^3\right)^2}\left(\frac{1358 \; K}{300 \; K}\right)^2$$

$$\tag{8.12E8a}$$

$$\Delta G_c = 7.18 \times 10^{-19} \; J \tag{8.12E8b}$$

and at $T = T_f - \Delta T = 1358 - 300 = 1058 \; K$,

$$n_s = n_o \exp\left(-\frac{\Delta G_c}{k_B T}\right) \tag{8.12E9a}$$

$$n_s = \left(8.60 \times 10^{22} \; clusters/m^3\right)\exp\left[-\frac{7.18 \times 10^{-19} \; J}{\left(1.38 \times 10^{-23} \; J/K\right)\left(1058 \; K\right)}\right]$$

$$\tag{8.12E9b}$$

$$n_s = 37.79 \; clusters/m^3 \simeq 38 \; clusters/m^3 \tag{8.12E9c}$$

Hence, nucleation takes place at $\Delta T = 300 \; K$ since 38 $clusters/cm^3$ of Cu are formed, each having a radius of 0.984 nm. Therefore, freezing the melt to cause L-S transformation is very sensitive to the degree of undercooling ΔT.

Example 8.12 Consider the solidification of pure aluminum (Al) with some properties taken from Dantzig and Rappaz [2, p. 275]).

T_f	$= 933 \; K$	ρ	$= 2{,}700 \; kg/m^3$
γ_{ls}^*	$= 0.093 \; J/m^2$	A_w	$= 26.98 \times 10^{-3} \; Kg/mol$
$\rho\Delta S_f^*$	$= 1.02 \times 10^6 \; J/\left(m^3.K\right)$	N_{atoms}	$= 4 \, atoms$
R	$= 1.43 \times 10^{-10} m$	a	$= 2\sqrt{2}R = 4.04 \times 10^{-10} \; m$

This example calls for the minimum degree of undercooling ΔT and the minimum undercooled liquid temperature $T < T_f$ for the onset of nucleation. For simplicity, assume a homogeneous nucleation process in order to determine ΔT and T. Hereinafter, use both ΔT and T for heterogeneous nucleation and compare results. Calculate **(a)** the proportionality constants n_o and I_o and the thermophysical properties ΔH_f, ΔS_f and Γ_{ls}, **(b)** ΔT and T for nucleating $1\ cluster/m^3$ homogeneously and **(c)** r_c, N_c, ΔG_c, n_s and I_s for homogeneous and heterogeneous nucleation using ΔT and T from part (b). Use the shape factor $\theta = \pi/2$ for heterogeneous nucleation and the probability factor $p_c = 1$.

Solution

(a) From Eq. (8.68b), the molar volume for FCC-*Al* structure along with the unit cell volume $V_{cell} = a^3$ is

$$V_m = \frac{N_a V_{cell}}{N_{atom}} \tag{8.12E1a}$$

$$V_m = \frac{\left(6.022 \times 10^{23}\ atoms/mol\right)\left(4.04 \times 10^{-10}\ m\right)^3}{4\ atoms} \tag{8.12E1b}$$

$$V_m = 9.93 \times 10^{-6}\ m^3/mol \tag{8.12E1c}$$

and from Eqs. (8.68a) and (8.70) with $T = T_f = 933\ K$, and Planck's constant $h = 6.63 \times 10^{-34}\ J.s$, the proportionality constants are

$$n_o = \frac{N_a}{V_m} = \frac{6.022 \times 10^{23}\ atoms/mol}{9.93 \times 10^{-6}\ m^3/mol} \tag{8.12E2a}$$

$$n_o = 6.06 \times 10^{28}\ atoms/m^3 \tag{8.12E2b}$$

$$I_o = \frac{n_o k_B T_f}{h} \tag{8.12E2c}$$

$$I_o = \frac{\left(6.06 \times 10^{28}\ atoms/m^3\right)\left(1.38 \times 10^{-23}\ J/K\right)(933\ K)}{6.63 \times 10^{-34}\ J.s} \tag{8.12E2d}$$

$$I_o = 1.18 \times 10^{42}\ atoms/\left(m^3.s\right) \quad \text{(typical value)} \tag{8.12E2e}$$

Letting $v_o = 10^{13}\ 1/s$ and $p_c = 1$ in Eq. (8.70) yields

$$I_o \simeq v_o p_c n_o = \left(10^{13}\ 1/s\right)(1)\left(5.80 \times 10^{28}\ atoms/m^3\right) \tag{8.12E3a}$$

$$I_o \simeq 6.06 \times 10^{41}\ atoms/\left(m^3.s\right) \quad \text{(approximate value)} \tag{8.12E3b}$$

Both calculated I_o values are within the typical range $10^{41} \le I_o \le 10^{43}$ cited in the literature. Therefore, use the average $\overline{I_o} = 8.93 \times 10^{41} \ atoms/\left(m^3.s\right)$ for calculating I_s.

Thermophysical Properties The entropy of fusion and latent heat of fusion at $T_f = 933 \ K$ are

$$\Delta S_f = \frac{1.02 \times 10^6 \ J/\left(m^3.K\right)}{2,700 \ kg/m^3} = 377.78 \ J/\left(Kg.K\right) \tag{8.12E4a}$$

$$\Delta H_f = T_f \Delta S_f = (933 \ K)\left(377.78 \ \frac{J}{Kg.K}\right) = 3.52 \times 10^5 \ J/Kg \tag{8.12E4b}$$

$$\Delta H_f = \left(2700 \ Kg/m^3\right)\left(3.52 \times 10^5 \ J/Kg\right) = 9.50 \times 10^8 \ J/m^3 \tag{8.12E4c}$$

and the Gibbs–Thomson coefficient, Eq. (8.58b), is

$$\Gamma_{ls} = \frac{\gamma_{ls}}{\rho \Delta S_f} = \frac{0.093 \ J/m^2}{1.02 \times 10^6 \ J/\left(m^3.K\right)} = 9.12 \times 10^{-8} \ K.m \tag{8.12E5}$$

The shape factor, Eq. (8.61), for heterogeneous nucleation with $\theta = \pi/2$ is

$$f\left(\theta = \pi/2\right) = \frac{1}{4}\left(2 - 3\cos\theta + \cos^3\theta\right) = 0.5 \tag{8.12E6}$$

(b) Finding the magnitudes of λ, ΔT and T: From Eq. (8.73), the constant λ with $n_s = 1 \ cluster/m^3$ (minimum) is

$$\lambda = \frac{16\pi \gamma_{ls}^3 T_f^2}{3\left(\Delta H_f\right)^2 k_B \ln\left(n_s/n_o\right)} \tag{8.12E7a}$$

$$\lambda = \frac{16\pi \left(0.093 \ J/m^2\right)^3 (933 \ K)^2}{3\left(9.50 \times 10^8 \ J/m^3\right)^2 \left(1.38 \times 10^{-23} \ J/K\right) \ln\left[1/\left(6.06 \times 10^{28}\right)\right]} \tag{8.12E7b}$$

$$\lambda = -1.4213 \times 10^7 \ K^3 \tag{8.12E7c}$$

Substitute λ and T_f into Eq. (8.72) to get

$$T^3 - 2T_f T^2 + T_f^2 T + \lambda = 0 \tag{8.12E8a}$$

$$T^3 - 1866T^2 + \left(8.7049 \times 10^5\right)T - 1.4213 \times 10^7 = 0 \tag{8.12E8b}$$

Solving this polynomial yields three roots

$$T_1 = 1049.40 \ K, \ T_2 = 16.937 \ K \text{ and } T_3 = 799.69 \ K \qquad (8.12E9)$$

- Root $T_1 \simeq 1049 \ K = 776° \ C$ is too high, and therefore, it is not feasible for nucleation because $T_1 > T_f = 933 \ K$.
- Root $T_2 \simeq 17 \ K = -256° \ C$ is much too low, and therefore, it is not feasible for nucleation because $T_1 << T_f = 933 \ K$.
- Root $T_3 \simeq 800 \ K = 527° \ C$ is a logical result and it is feasible for nucleation because $T_3 < T_f$ and ΔT is relatively small.

Nonetheless, using root T_3 yields a reasonable and a small degree of undercooling

$$\Delta T = T_f - T_3 = 933 \ K - 800 \ K = 133 \ K \qquad (8.12E10)$$

All relevant calculations carried out below are based on $T = T_3 = 800 \ K$ and $\Delta T = 133 \ K$. The results will yield interesting insights about the nucleation process and related theoretical background. Moreover, this theoretical approach suggests that $\Delta T = 133 \ K$ is the minimum degree of undercooling needed for the onset of homogeneous nucleation of 1 $cluster/volume$ contains N_c number of atoms, which in turn, under slow and steady-phase transformation, acquires a preferred crystal structure at a random position within the liquid phase.

Further, using these theoretical results for characterizing the heterogeneous nucleation process is an effective approximation for calculating the pertinent thermodynamic and kinetic entities required in this example problem.

(c) Using Eq. (8.58a) for both homogeneous and heterogeneous solidification at $\Delta T = 133 \ K$ yields the theoretical radius of a cluster

$$r_c = \frac{2\Gamma_{ls}}{\Delta T} = \frac{2\left(9.12 \times 10^{-8} \ K.m\right)}{133 \ K} = 1.37 \times 10^{-9} \ m \qquad (8.12E11a)$$

$$r_c = 1.37 \ nm \qquad (8.12E11b)$$

The number of atoms contained in a critical nucleus along with $n_o = 6.06 \times 10^{28} \ atoms/m^3$, Eq. (8.12E2b), is

$$N_c = \frac{N_a V_{sphere}}{V_m} = V_{sphere} n_o = \frac{4\pi}{3} r_c^3 n_o \qquad (8.12E12a)$$

$$N_c = \frac{4\pi}{3} \left(\frac{1.8240 \times 10^{-7} \ K.m}{\Delta T}\right)^3 \left(6.06 \times 10^{28} \ atoms/m^3\right) \qquad (8.12E12b)$$

$$N_c = \frac{1.5404 \times 10^9 \ atoms.K^3}{\Delta T^3} = \frac{1.54 \times 10^9 \ atoms.K^3}{(133 \ K)^3} \qquad (8.12E12c)$$

$$N_c = 655 \; atoms \tag{8.12E12d}$$

The homogeneous driving force, Eq. (8.59), is

$$\Delta G_{c,\mathrm{hom}} = \frac{16\pi \gamma_{ls}^3}{3\left(\Delta H_f\right)^2}\left(\frac{T_f}{\Delta T}\right)^2 = \frac{16\pi \left(0.093 \; J/m^2\right)^3}{3\left(9.50 \times 10^8 \; J/m^3\right)^2}\left(\frac{933 \; K}{\Delta T}\right)^2 \tag{8.12E13a}$$

$$\Delta G_{c,\mathrm{hom}} = \frac{1.2999 \times 10^{-14} \; J.K^2}{\Delta T^2} = \frac{1.2999 \times 10^{-14} \; J.K^2}{\left(133 \; K\right)^2} \tag{8.12E13b}$$

$$\Delta G_{c,\mathrm{hom}} = 7.35 \times 10^{-19} \; J \tag{8.12E13c}$$

and for heterogeneous nucleation, Eq. (8.60b) yields

$$\Delta G_{c,het} = \Delta G_{c,\mathrm{hom}}\, f\left(\theta\right) = \left(\frac{1.2999 \times 10^{-14} \; J.K^2}{\Delta T^2}\right)\left(0.5\right) \tag{8.12E14a}$$

$$\Delta G_{c,het} = \frac{6.4995 \times 10^{-15} \; J.K^2}{\Delta T^2} = \frac{6.4995 \times 10^{-15}}{\left(133\right)^2} \tag{8.12E14b}$$

$$\Delta G_{c,het} = 3.67 \times 10^{-19} \; J \tag{8.12E14c}$$

Subsequently, verify that $n_{s,\mathrm{hom}} = 1 \; clusters/m^3$ at $T = 800 \; K$

$$n_{s,\mathrm{hom}} = n_o \exp\left(-\frac{\Delta G_{c,\mathrm{hom}}}{k_B T}\right) = n_o \exp\left(-\frac{7.35 \times 10^{-19} \; J}{k_B T}\right) \tag{8.12E15a}$$

$$n_{s,\mathrm{hom}} = \left(6.06 \times 10^{28}\right)\exp\left[-\frac{7.35 \times 10^{-19} \; J}{\left(1.38 \times 10^{-23} \; J/K\right)\left(800 \; K\right)}\right] \tag{8.12E15b}$$

$$n_{s,\mathrm{hom}} = 1 \; clusters/m^3 \tag{8.12E15c}$$

and

$$n_{s,het} = n_o \exp\left(-\frac{\Delta G_{c,het}}{k_B T}\right) = n_o \exp\left(-\frac{3.67 \times 10^{-19} \; J}{k_B T}\right) \tag{8.12E16a}$$

$$n_{s,het} = \left(6.06 \times 10^{28}\right)\exp\left(-\frac{3.67 \times 10^{-19} \; J}{\left(1.38 \times 10^{-23} \; J/K\right)\left(8 \; K\right)}\right) \tag{8.12E16b}$$

$$n_{s,het} = 2 \times 10^{14} \; clusters/m^3 \tag{8.12E16c}$$

Similarly, using $\overline{I_o} = 8.93 \times 10^{41} \ atoms/\left(m^3.s\right)$ yields

$$I_{s,\text{hom}} = \overline{I_o} \exp\left(-\frac{\Delta G_{c,\text{hom}}}{k_B T}\right) = \overline{I_o} \exp\left(-\frac{7.35 \times 10^{-19} \ J}{k_B T}\right) \qquad (8.12\text{E}17\text{a})$$

$$I_{s,\text{hom}} = \left(8.93 \times 10^{41}\right) \exp\left[-\frac{7.35 \times 10^{-19} \ J}{\left(1.38 \times 10^{-23} \ J/K\right)(1106 \ K)}\right]$$
$$(8.12\text{E}17\text{b})$$

$$I_{s,\text{hom}} = 1.09 \times 10^{21} \ atoms/\left(m^3.s\right) \qquad (8.12\text{E}17\text{c})$$

and

$$I_{s,het} = \overline{I_o} \exp\left(-\frac{\Delta G_{c,het}}{k_B T}\right) = \overline{I_o} \exp\left(-\frac{3.67 \times 10^{-19} \ J}{k_B T}\right) \qquad (8.12\text{E}18\text{a})$$

$$I_{s,het} = \left(8.93 \times 10^{41}\right) \exp\left[-\frac{3.67 \times 10^{-19} \ J}{\left(1.38 \times 10^{-23} \ J/K\right)(1106 \ K)}\right]$$
$$(8.12\text{E}18\text{b})$$

$$I_{s,het} = 3.22 \times 10^{31} \ atoms/\left(m^3.s\right) \qquad (8.12\text{E}18\text{c})$$

Therefore, $\Delta G_{c,het} < \Delta G_{c,\text{hom}}$, and as expected, $n_{s,het} >> n_{s,\text{hom}}$ and $I_{s,het} >> I_{s,\text{hom}}$.

Plotting $r_c = f(\Delta T)$, $N_c = f(\Delta T)$, $\Delta G_c = f(\Delta T)$, $n_s = f(T)$ and $I_s = f(T)$ at the interval $0 \leq f(\theta) \leq 1$ should reveal interesting information and the procedure is purposely left out as an assignment for the reader. Nonetheless, the plotting equations are given below for convenience.

$$r_c = \left(1.824 \times 10^{-7}\right)/\Delta T \qquad (8.12\text{E}19\text{a})$$

$$N_c = \left(1.5404 \times 10^{9}\right)/\Delta T^3 \qquad (8.12\text{E}19\text{b})$$

$$\Delta G_{c,\text{hom}} = \left(1.2999 \times 10^{-14}\right)/\Delta T^2 \qquad (8.12\text{E}19\text{c})$$

$$\Delta G_{c,\text{hom}} = \left(6.4995 \times 10^{-15}\right)/\Delta T^2 \qquad (8.12\text{E}19\text{d})$$

and

$$n_{s,\text{hom}} = \left(6.06 \times 10^{28}\right) \exp\left(-53261/T\right) \qquad (8.12\text{E}20\text{a})$$

$$n_{s,het} = \left(6.06 \times 10^{28}\right) \exp\left(-26594/T\right) \tag{8.12E20b}$$

$$I_{s,\mathrm{hom}} = \left(8.93 \times 10^{41}\right) \exp\left(-53261/T\right) \tag{8.12E20c}$$

$$I_{s,het} = \left(8.93 \times 10^{41}\right) \exp\left(-26594/T\right) \tag{8.12E20d}$$

The reader is encouraged to plot these equations in order to reveal their trendlines. The result would be a phase map.

8.13   Critical Temperature for Nuclei Formation

It has been established that solidification starts by formation of nuclei with volume V and area A at random positions in the melt, and subsequently, crystal growth occurs from these nuclei, leading to the solidification reaction $L \rightarrow S$.

Now, assume a homogeneous nucleation process where the undercooling $\Delta T = T_f - T_c$ becomes the driving force for nuclei formation at T_c. Thus, the Gibbs excess energy $(-\Delta G_i)$ and the Gibbs molar energy $(-\Delta G_m)$ changes are mathematically approximated by

$$\Delta G_i = \frac{V}{V_m}(-\Delta G_m) + A\gamma_{ls} \tag{8.74a}$$

$$\Delta G_m = \frac{T_f - T_c}{T_c}(\Delta H_m) \tag{8.74b}$$

where $V_m = M/\rho$ denotes the molar volume $(m^3/kmol)$, M denotes the molar weight $(Kg/kmol)$, ρ denotes the particle density (Kg/m^3) and ΔH_m denotes the molar latent heat of fusion $(kJ/kmol)$.

If ΔG_c is defined as the activation energy for the formation of a nucleus with the critical size r_c, then (Fredriksson and Akerlind [15, p. 141])

$$\Delta G_c = \frac{16\pi}{3}\frac{V_m^2 \gamma_{ls}^3}{(\Delta G_m)^2} \tag{8.75}$$

For clarity, a particle with radius $r < r_c$ is defined as an embryo and if it grows by additions of adjacent atoms from the melt, then it becomes a nucleus with $r = r_c$ with ΔG_c at a critical temperature T_c.

The theoretical average activation energy for forming a nucleus at T_c is written as

$$\Delta G_c = 60 k_B T_c \quad \text{(Energy barrier)} \tag{8.76}$$

Combining Eqs. (8.74b), (8.75), and (8.76) yields the critical temperature T_c equation

$$60 k_B T_c = \frac{16\pi}{3} \left(\frac{T_f}{T_f - T_c} \right)^2 \frac{V_m^2 \gamma_{ls}^3}{(\Delta H_m)^2} \tag{8.77a}$$

which is a convenient expression for predicting T_c for homogeneous nucleation of a pure metal. Recall that the degree of undercooling has been defined as $\Delta T = T_f - T_c$.

For a heterogeneous nucleation, as in inoculation of carbon steels with $FeTi$ particles, the Gibbs molar energy ΔG_m depends on the mole fraction of the added elements. Additional details on this particular case can be found elsewhere (Fredriksson and Akerlind [15, pp. 142–143]).

Inoculation in casting metals and alloys is a practical method for controlling the solidification process and related microstructural features. Thus, the resultant microstructure controls mechanical behavior and, to an extent, corrosion.

8.14 Summary

Casting technology has evolved significantly for producing pure metals and their alloys using reliable sand and metal molds containing simple or intricate casting geometries. Undoubtedly, the mold thermophysical properties play a significant role for heat extraction during solidification.

In this chapter, the general definition of the Gibbs energy change, $\Delta G = \Delta H - T \Delta S$, as defined by Eq. (8.5b), is treated as the driving force for solidification since $\Delta G < 0$ at $T < T_f$ is a general condition that describes the physical behavior of the solidifying melt. Nevertheless, Gibbs energy function $G = f(X_B)$ for a binary A-B system when plotted at T provides significant information used to construct an equilibrium phase diagram (T-C diagram). Therefore, both G-X_B and T-C diagrams are essential for characterizing the solidification of binary alloys.

The classic theories of homogeneous and heterogeneous nucleation provide simple models for understanding solidification. Again, the Gibbs energy provides the basic analytical approach to determine the critical (maximum) energy change (ΔG_c) for obtaining the stable critical particle size (r_c) subjected to grow as solidification proceeds. From a macroscale point of view, the cooling curves (T-t) for pure metals or alloys define the onset and end of solidification and provide relevant data for constructing binary phase diagrams. Thus, both G-X_B and T-t curves are known methods for developing T-C diagrams, t is time.

Other theoretical aspects of the solidification process and relevant mathematical modeling of freezing (solidification) are a subject matter dealt with in the subsequent chapters.

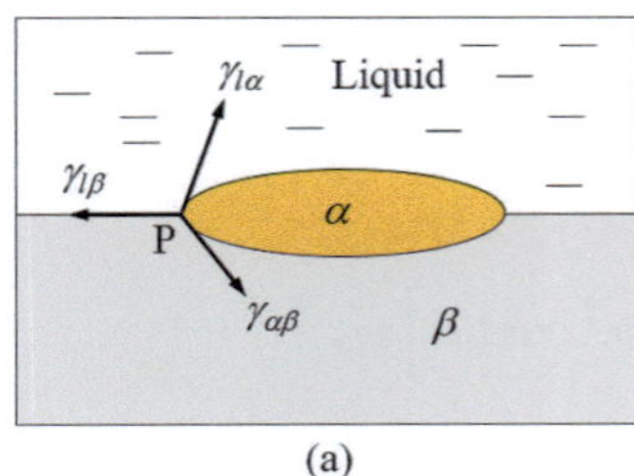
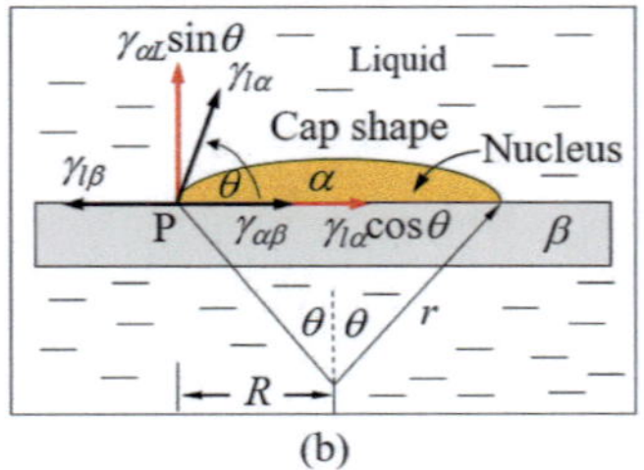

Fig. 8.14 Three-phase model for heterogeneous nucleation. (**a**) General model for one nucleus and (**b**) modified model showing the directions of surface tensions

Appendix 8A The Shape Factor

Consider the three-phase model illustrated in Fig. 8.14 for heterogeneous nucleation. For instance, Fig. 8.14a depicts a general elliptical nucleus α in contact with the liquid L and substrate β, and the directions of the surface energy γ terms at a triple junction "P." On the other hand, Fig. 8.14b illustrates a modified model along with trigonometric information. Nonetheless, the triple junction "P" imposes surface forces (surface tensions).

According to Newton's second law, the sum of forces is $\sum F_i = 0$ at mechanical equilibrium. The mathematical definitions of individual forces based on trigonometry are (Kurz and Fisher [14, Appendix A3])

$$\gamma_{\alpha\beta} = \gamma_{l\alpha} \cos(\theta) + \gamma_{l\beta} \cos(\theta) \tag{8.78a}$$

$$\gamma_{l\alpha} \sin(\theta) = \gamma_{l\beta} \sin(\theta) \tag{8.78b}$$

$$\gamma_{l\beta} = \gamma_{\alpha\beta} + \gamma_{l\alpha} \cos(\theta) \tag{8.78c}$$

The energy balance at the triple junction suggests that the Gibbs energy change is due to the creation of the interface energy for nucleation and the energy gained by the substrate β. Mathematically,

$$\Delta G_\gamma = \left(A_{l\alpha}\gamma_{l\alpha} + A_{\alpha\beta}\gamma_{\alpha\beta}\right) - A_{\alpha\beta}\gamma_{l\beta} = A_{l\alpha}\gamma_{l\alpha} + A_{\alpha\beta}\left(\gamma_{\alpha\beta} - \gamma_{l\beta}\right) \tag{8.79}$$

where $A_{l\alpha}, A_{\alpha\beta}, A_{\alpha\beta}$ denote surface areas and $\gamma_{l\alpha}, \gamma_{\alpha\beta}, \gamma_{l\beta}$ denote the surface energies. Combining Eqs. (8.78c) and (8.79) with $A_{\alpha\beta} = \pi r^2$ yields

$$\Delta G_\gamma = A_{l\alpha}\gamma_{l\alpha} - A_{\alpha\beta}\gamma_{l\alpha}\cos(\theta) = A_{l\alpha}\gamma_{l\alpha} - \pi r^2 \gamma_{l\alpha}\cos(\theta) \tag{8.80a}$$

$$\Delta G_\gamma = \left[A_{l\alpha} - \pi r^2 \cos(\theta)\right]\gamma_{l\alpha} \tag{8.80b}$$

The total Gibbs energy change (ΔG_{het}) for heterogeneous nucleation is the sum of Gibbs energy change (ΔG_f) per nucleus volume at the freezing temperature T_f

and the Gibbs energy change (ΔG_γ) due to the formation of nucleus surface area. Thus,

$$\Delta G_{het} = \Delta G_f + \Delta G_\gamma = V \Delta G_v + A \gamma_{l\alpha} \tag{8.81a}$$

$$\Delta G_{het} = V_\alpha \Delta G_v + \left[A_{l\alpha} - \pi r^2 \cos(\theta) \right] \gamma_{l\alpha} \tag{8.81b}$$

where $V = V_\alpha$ denotes the volume and $A = A_\alpha$ denotes the surface area of the nucleus α, and ΔG_f is written as

$$\Delta G_f = \Delta T \Delta S_f \tag{8.82}$$

The variables V_α and $A_{l\alpha}$ are derived in the next section.

Spherical Cap

Consider the general overview of a sphere on a substrate β surface and the thin α disk with radius a (Fig. 8.15a). On the other hand, Fig. 8.15b illustrates the proper nomenclature for deriving the surface energy $\gamma_{l\alpha}$ and the volume of the spherical cap V_a as per Eq. (8.81b).

The surface areas of interest along with the thin disk radius $a = r \sin\phi$ are defined by

$$A_{\alpha\beta} = \pi x^2 = \pi r^2 \left[1 - \cos^2(\theta) \right] \tag{8.83a}$$

$$A_{l\alpha} = \int_0^\theta 2\pi a r d\phi^2 \left[1 - \cos(\theta) \right] = 2\pi r^2 \left[1 - \cos(\theta) \right] \tag{8.83b}$$

The area A_a, the differential thickness dh, and the differential volume dV_α of the α disk are

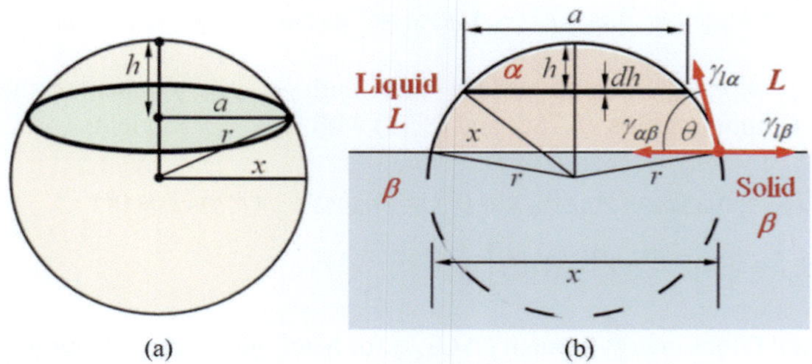

Fig. 8.15 (a) Sphere and (b) spherical cap with nomenclature

$$A_a = \pi r^2 \sin^2 \theta \tag{8.84a}$$

$$dh = (rd\theta)\sin(\theta) \tag{8.84b}$$

$$dV_\alpha = A_a dh \tag{8.84c}$$

Combining Eqs. (8.84a,b,c) and integrating the resultant expression yield the volume of a disk

$$V_\alpha = \int_0^\theta \left[\pi r^2 \sin^2(\phi)\right][rd\phi \sin(\phi)] = \pi r^3 \int_0^\theta \sin^3(\phi)\, d\phi \tag{8.85a}$$

$$V_\alpha = V_{het} = \pi r^3 \left[\frac{2}{3} - \frac{3}{4}\cos\theta + \frac{1}{12}\cos 3\theta\right] \tag{8.85b}$$

Use the trigonometric identity $\cos(3\theta) = 4\cos^3(\theta) - 3\cos(\theta)$ and V_{hom} to get

$$V_{het} = \pi r^3 \left[\frac{2}{3} - \cos(\theta) + \frac{1}{3}\cos^3(\theta)\right] \tag{8.86a}$$

$$V_{het} = \frac{\pi r^3}{3}\left[2 - 3\cos(\theta) + \cos^3(\theta)\right] \tag{8.86b}$$

The volume of a sphere during homogeneous nucleation is

$$V_{\text{hom}} = \frac{4\pi r^3}{3} \tag{8.87}$$

Thus, the shape factor is defined by

$$f(\theta) = \frac{V_{het}}{V_{\text{hom}}} = f_v = \text{Volume fraction} \tag{8.88a}$$

$$f(\theta) = \frac{2 - 3\cos(\theta) + \cos^3(\theta)}{4} = \frac{[2 + \cos(\theta)][1 - \cos(\theta)]^2}{4} \tag{8.88b}$$

Substituting Eqs. (8.83a) and (8.86b) into (8.81b) and rearranging the resultant expression yield

$$\Delta G_{het} = \left(-\frac{4\pi r^3}{3}\Delta G_v + 4\pi r^2 \gamma_{l\alpha}\right)\left[\frac{2 - 3\cos(\theta) + \cos^3(\theta)}{4}\right] \tag{8.89a}$$

$$\Delta G_{het} = \left(-\frac{\pi r^3}{3}\Delta G_v + 4\pi r^2 \gamma_{l\alpha}\right)f(\theta) \tag{8.89b}$$

$$\Delta G_{het} = \Delta G_{\text{hom}} f(\theta) \tag{8.89c}$$

$$f(\theta) = \frac{\Delta G_{het}}{\Delta G_{\text{hom}}} = \frac{V_{het}}{V_{\text{hom}}} \tag{8.89d}$$

This concludes the derivation of Eqs. (8.60a,b) and (8.61).

Problems

8.1 For the solidification of pure nickel (Ni) under homogeneous nucleation mechanism, calculate **(a)** the critical radius r_c and the critical Gibbs energy change ΔG_c (activation energy) for a stable nucleus having a perfect spherical shape. Use the data listed in Table 8.1. **(b)** If the atomic radius of nickel is $r = 0.125\ nm$, then compute the number of unit cells N_{cell} and the number of atoms N in the critical nucleus. Assume that the nucleus is stable and spherical. [Solution: (a) $r_c = 0.67$ nm and (b) $N_{atom} = 112\ atoms$].

8.2 For the solidification of pure iron (Fe), calculate **(a)** the critical radius r_c and the critical Gibbs energy change ΔG_c for a stable nucleus having a perfect spherical shape under homogeneous nucleation mechanism. Use the data listed in Table 8.1. **(b)** If the atomic radius of iron is $r = 0.124\ nm$, then compute the number of unit cells N_{cell} and the number of atoms N_{atm} in the critical nucleus. Assume that the nucleus is stable and spherical. [Solution: (a) $\Delta G_c = 1.27\ eV$ and (b) $N_{cell} = 101$ $cells$].

8.3 Show that $\Delta G = G_l - G_s = \Delta S_f \Delta T < 0$ for melting, provided that $\Delta H(T) = \Delta H(T_f)$ and $\Delta S(T) = \Delta S(T_f)$.

8.4 Show that $\Delta G = G_l - G_s = \Delta S_f \Delta T < 0$ for solidification, provided that $\Delta H(T) = \Delta H(T_f)$ and $\Delta S(T) = \Delta S(T_f)$.

8.5 Let $\Delta H_f = 8,000\ J/mol$ and $\Delta S_f = 10\ J/mol.K$. Determine **(a)** ΔG for $L \to S$ to occur spontaneously at $1,100\ K$ and **(b)** the wok done for such a phase transformation. [Solution: (a) $\Delta G = -3\ kJ/mol$ and (b) $W = 3\ kJ/mol$].

8.6 Let the fundamental Gibbs energy equation for mixing A and B elements be defined by

$$\Delta G = RT\left[X_A \ln X_A + X_B \ln X_B + \Omega X_A X_B\right] \tag{a}$$

where B is a constant. Here, $X_A = 1 - X_B$ and consequently, there is symmetry Based on this information, **(a)** find the critical values of $X_B = X_c$ and $\Omega = \Omega_c$ that leads to the symmetry inherent in this equation, **(b)** derive the critical ΔG_c and **(c)** plot $\Delta G/RT = f(X_B)$ to reveal that ΔG reaches a minimum at X_c. [Solution: (a) $\Omega_c = 2$ and (c) $\Delta G/RT = -0.19315 < 0$ at $X_c = 0.5$].

8.7 If pure nickel (Ni) is solidified at a slow cooling rate dT/dt, determine **(a)** the number of stable particles n_s per volume when the nuclei radius is 0.80 nm at the freezing temperature T_f, **(b)** the critical radius r_c and n_s at $\Delta T = 300\ K$, **(c)** r_c and n_s at $\Delta T = 480\ K$ and **(d)** the number of unit cells and atoms for $\Delta T = 480\ K$. Given data:

$$T_f = 1726 \ K, \quad \Delta S_m = 10 \ J/mol.K, \quad \Delta H_f = 2756 \ J/cm^3$$

$$\gamma_{ls} = 0.255 \ J/m^2, \quad V_m = A_w/\rho \simeq 7 \ cm^3/mol = 7 \times 10^{-6} \ m^3/mol$$

Will homogeneous nucleation take place at the freezing temperature T_f and undercooling ΔT? [Solution: (a) $n_s = 4 \times 10^{-15} \ clusters/cm^3$, (b) $r_c = 1.06$ nm and $n_s \simeq 0$, (c) $r_c = 0.665 \ nm$ and $n_s = 10 \times 10^{10} \ clusters/cm^3$, and (d) $N_{cell} = 28 \ cells$].

8.8 Consider the solidification of pure iron (Fe) at a slow cooling rate dT/dt. Determine **(a)** the number of stable particles n_s per volume when the nuclei radius is 0.80 nm at the freezing temperature T_f, **(b)** the critical radius r_c and n_s at $\Delta T = 400 \ K$ and **(c)** r_c and n_s at $\Delta T = 420 \ K$ and **(d)** the number of unit cells and atoms for $\Delta T = 420 \ K$. Will homogeneous nucleation take place at the freezing temperature T_f and undercooling ΔT? Use the molar volume $V_m = 7 \times 10^{-6}$ m^3/mol, the entropy of fusion $\Delta S_m = 10 \ J/mol.K$ and the relevant data from Table 8.1. [Solution: (a) $n_s = 3 \times 10^{-6} \ clusters/cm^3$, (b) $r_c = 1.06 \ nm$, (c) $r_c = 1.01 \ nm$ and (d) $N_{cell} = 185 \ cells$].

8.9 Assume that pure silver (Ag) solidifies at a slow cooling rate dT/dt. Determine **(a)** the number of stable particles n_s per volume when the nuclei radius is 0.80 nm at the freezing temperature T_f, **(b)** the critical radius r_c and n_s at $\Delta T = 200 \ K$ and **(c)** r_c and n_s at $\Delta T = 300 \ K$ and **(d)** the number of unit cells and atoms for $\Delta T = 300 \ K$. Will homogeneous nucleation take place at the freezing temperature T_f and undercooling ΔT? Use the entropy of fusion $\Delta S_m = 10 \ J/mol.K$ and the relevant data from Table 8.1. [Solution: (a) $n_s \simeq 0$ at $r = 0.80 \ nm$, (b) $r_c = 1.61 \ nm$ and $n_s \simeq 0$, and (c) $r_c = 1.08 \ nm$ and $n_s = 172 \ clusters/cm^3$, and (d) $N_{cell} = 78$ $cells$].

8.10 Consider the homogeneous nucleation of pure nickel (Ni) with a typical undercooling, $\Delta T = 480 \ K$, latent heat of fusion $\Delta H_f = 2756 \ J/cm^3$, surface energy $\gamma_{ls} = 255 \times 10^{-7} \ J/cm^2$, and melting or freezing temperature $T_f = 1726$ K. Use this information to calculate **(a)** the critical radius (r_c), **(b)** the critical Gibbs energy change (ΔG_c) at $r = r_c$, **(c)** the number of unit cells (N_{cell}) and **(d)** the number of atoms (N_{atom}) in a nucleus. Moreover, nickel at T_f has a FCC crystal structure with 4 *atoms* in the unit cell and an atomic radius $r = 0.125 \ nm$, which is normally determined using the X-ray technique. [Solution: (a) $r_c = 0.665 \ nm$, (b) $\Delta G_c = 2.96 \ eV$, (c) $N_{cell} = 28$ cells and (d) $N_{atom} = 4N_{cell} = 4 \ (28) = 112$ *atoms*].

8.11 This problem calls for the minimum degree of undercooling ΔT and the minimum undercooled liquid temperature $T < T_f$ for the onset of nucleation. For simplicity, assume a homogeneous nucleation process in order to determine ΔT and T. Hereinafter, use both ΔT and T for heterogeneous nucleation and compare results. Consider the solidification of pure copper (FCC-Cu) with some properties taken from Askeland et al. [13, *, p. 262]).

T_f	$= 1085° C$	ρ	$= 8940 Kg/m^3$
γ_{ls}^*	$= 0.177 J/m^2$	A_w	$= 63.55 \times 10^{-3} Kg/mol$
ΔH_f^*	$= 1628 J/cm^3$	N_{atoms}	$= 4\,atoms\,per\,unit\,cell$
R	$= 1.28 \times 10^{-10} m$	a	$= 2\sqrt{2}R = 3.62 \times 10^{-10} m$

Calculate **(a)** the proportionality constants n_o and I_o, the thermophysical properties ΔH_f, ΔS_f and Γ_{ls}, **(b)** ΔT and T for nucleating 1 $cluster/m^3$ homogeneously and **(c)** r_c, N_c, ΔG_c, n_s and I_s for homogeneous and heterogeneous nucleation using ΔT and T from part **(c)**. Use the shape factor $\theta = \pi/2$ for heterogeneous nucleation and the probability factor $p_c = 1$. [Solution: (a) $n_o = 8 \times 10^{28}\ atoms/m^3$, (b) $\Delta T = 252\ K$ and (c) $r_c = 1.17\ nm$].

8.12 Assume that melting pure lead (Pb) under controlled conditions and consider the processes schematically shown in the T-S diagram given below. If the latent heat of melting (ΔH_f) of lead (Pb) is 4810 J/mol at 600 K melting temperature and 1-atm pressure, then calculate the entropy, enthalpy, and Gibbs energy changes when 1 mole of liquid spontaneously freezes at 590 K. This means that the amount of undercooling is a small absolute temperature change set to $\Delta T = 10\ K$.

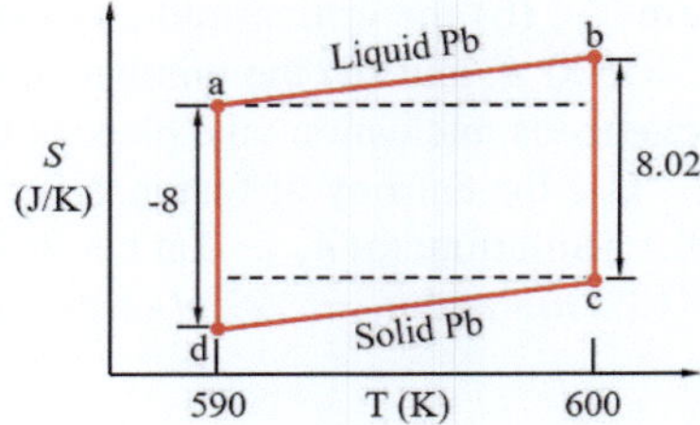

Here, the constant-pressure molar heat capacity of the liquid and the solid Pb phases and the entropy change cycle for solidifying Pb are adopted from Gaskell and Laughlin book [9, p. 87] given as

$$C_l = 32.40 - 3.10 \times 10^{-3} T \quad (in\ J/K)$$

$$C_s = 23.56 + 9.75 \times 10^{-3} T \quad (in\ J/K)$$

Calculate the relevant thermal energies, such as ΔS, ΔH and ΔG for each process; ab, bc, cd, da. Then, compute the entropy generation S_{gen} as the degree of irreversibility during solidification of lead under controlled conditions. Explain. [Solution: $\Delta S_{ab} = +0.51\ J/K$, $\Delta S_{bc} = -8.02\ J/K$, $\Delta S_{cd} = -0.49\ J/K$, $\Delta S_{da} = -8\ J/K$].

8.13 This problem calls for the minimum degree of undercooling ΔT and the minimum undercooled liquid temperature $T < T_f$ for the onset of nucleation. For simplicity, assume a homogeneous nucleation process in order to determine ΔT

and T. Hereinafter, use both ΔT and T for heterogeneous nucleation and compare results. Consider the solidification of pure copper (FCC-Cu) with some properties taken from Askeland et al. [13, p. 262]).

T_f	$= 1538^\circ C$	ρ	$= 7870 Kg/m^3$
γ_{ls}^*	$= 0.204 J/m^2$	A_w	$= 55.85 \times 10^{-3} Kg/mol$
ΔH_f^*	$= 1737 J/cm^3$	N_{atom}	$= 4\, atoms\, per\, unit\, cell$
R	$= 1.24 \times 10^{-10} m$	a	$= 2\sqrt{2}R = 3.51 \times 10^{-10} m$

Calculate (a) the proportionality constants n_o and I_o, the thermophysical properties ΔH_f, ΔS_f and Γ_{ls}, (b) ΔT and T for nucleating 1 $cluster/m^3$ homogeneously and (c) r_c, N_c, ΔG_c, n_s and I_s for homogeneous and heterogeneous nucleation using ΔT and T from part (b). Use the shape factor $\theta = \pi/2$ for heterogeneous nucleation and the probability factor $p_c = 1$. [Solution: (a) $n_o = 9 \times 10^{28}$ $atoms/m^3$, (b) $\Delta T = 338\ K$ and (c) $r_c = 1.25\ nm$].

8.14 Consider the solidification process for a hypothetical material and assume the pressure–temperature relation as defined by the Clapeyron equation given below along with the molar volume V_m.

$$\frac{dP}{dT} = \frac{\Delta H_f}{\Delta V_m T_f} \quad \& \quad m = \rho V_m$$

Calculate the melting temperature T_f at $P = 150\ kPa$ using the following data set [Solution: $T_f = 1006.40^\circ C$].

$\Delta H_f = 20 kJ/mol$	$P_1 = 100 kPa$	$T_1 = 1273 K$
$\rho_s = 8000 Kg/m^3$	$\rho_l = 900 Kg/m^3$	
$V_s = 0.020 L/mol$	$V_s = 0.018 L/mol$	

8.15 Consider a hypothetical pure metal undergoing solidification by releasing $\Delta H_f = 8{,}200\ J/mol$ at $1{,}100\ K$. Determine (a) ΔG for $L \rightarrow S$ to occur spontaneously when $\Delta S_f = 11\ J/mol.K$ and (b) the wok done for such a phase transformation. [Solution: (a) $\Delta G = -3.90\ kJ/mol < 0$ and (b) $W = 3.90\ kJ/mol$].

8.16 Consider a Zn-Cd electrochemical solution at a temperature of $800\ K$ with mole fractions $X_{Zn} = 0.8$ and $X_{Cd} = 0.2$, and activity coefficient

$$\gamma_{Cd} = 2.942 - 4.155 X_{Cd} + 0.45 X_{Cd}^2 + 3 X_{Cd}^3$$

Calculate the (a) the interaction mixing parameter Ω_m, (b) the activity coefficient γ_{Zn}, (c) the activities a_{Zn} and a_{Cd}, (d) the thermodynamic energies of mixing ΔH_m,

ΔG_m and ΔS_m. [Solution: (a) $\Omega_m = 7,970.10 \ J/mol$, (b) $\gamma_{Zn} = 1.05$, (c) $a_{Zn} = 0.84$ and (d) $\Delta S_m = 4.16 \ J/mol.K$].

8.17 Use the Gibbs–Duhem equation in the form of Eq. (8.46b) and assume that the following equation is valid above the melting temperature $T > T_f^*$

$$RT \ln (\gamma_2) = \Omega_m X_1^2 \tag{8.90}$$

(a) For a binary alloy, show that

$$RT \ln (\gamma_1) = \Omega_m X_2^2 \tag{8.91}$$

(b) For a suitable amount of liquid brass containing $X_1 = X_{Cu} = 0.70$, and $X_2 = X_{Zn} = 0.30$ mole fractions, calculate the activity coefficients $\gamma_2 = \gamma_{Zn}$ and $\gamma_1 = \gamma_{Cu}$) and the corresponding the activities ($a_2 = a_{Zn}$ and $a_1 = a_{Cu}$) at $T = 1,200$ K. Assume that $\Omega_m = -38,300 \ J/mol$ at $1,000 \ K \leq T \leq 1,500 \ K$. [Solution: (b) $\gamma_{Zn} = 0.15$ and $\gamma_{Cu} = 0.71$].

8.18 Consider the given dataset for Al-Cu alloys at $1,100° C$ ($1373 \ K$) taken from Upadhyaya and Dube [5, p. 166] example

X_{Al}	0.10	0.20	0.30	0.40	0.50	0.60	0.70	0.80
X_{Cu}	0.90	0.80	0.70	0.60	0.50	0.40	0.30	0.20
γ_{Cu}	0.96	0.76	0.49	0.30	0.16	0.13	0.07	0.05
a_{Cu}	0.86	0.61	0.34	0.18	0.08	0.05	0.02	0.01

(a) Perform nonlinear curve fitting using the activity coefficient function $\gamma_{Cu} = f(X_{Cu})$, plot the given data and the resultant equation. Determine the γ_{Al} and a_{Al} values in an Al-Cu alloy solution containing $X_{Al} = 0.60$ and $X_{Cu} = 0.40$ mole fractions. **(b)** Perform linear curve fitting using $y = RT \ln (\gamma_{Cu})$ and $x = X_{Al}^2$, and plot $y = f(x)$ with zero intercept, determine the interaction parameter Ω_m, γ_{Al}, a_{Al} for $X_{Al} = 0.60$ and $X_{Cu} = 0.40$ mole fractions. Compare the results by calculating the percentage error. **(c)** Examine the shape of the graph, split the dataset, and determine two linear curve fitting equations having a common intercepting point. Calculate a_{Al} at $X_{Al} = 0.05$ and $X_{Al} = 0.60$. [Solution: (a) $a_{Al} = 0.37$, (b) $\Omega_m = -60,917$ and (c) $a_{Al} = 0.0000535$ at $X_{Al} = 0.05$].

8.19 Assume that pure nickel (Ni) with a typical undercooling, $\Delta T = 450 \ K$ undergoes homogeneous nucleation. For latent heat of fusion $\Delta H_f = 2756 \ J/cm^3$, surface energy $\gamma_{ls} = 255 \times 10^{-7} \ J/cm^2$, and melting or freezing temperature $T_f = 1726 \ K$, calculate **(a)** the critical radius (r_c), **(b)** the critical Gibbs energy change (ΔG_c) at $r = r_c$, **(c)** the number of unit cells (N_{cell}) and **(d)** the number of atoms (N_{atom}) in a nucleus. Moreover, nickel at T_f has a FCC crystal structure with 4 *atoms* in the unit cell and an atomic radius $r = 0.125 \ nm$, which is normally determined using the X-ray technique. [Solution: (a) $r_c = 0.71$ nm, (b) $\Delta G_c = 5.38 \times 10^{-19}$ J, (c) $N_{cell} = 34$ cells and (d) $N_{atom} = 136$ $atoms$].

8.20 Show that the Gibbs–Duhem equation can be written as

$$X_1 \frac{d \ln (\gamma_1)}{d X_1} = X_2 \frac{d \ln (\gamma_2)}{d X_2} \tag{E1}$$

8.21 Use the given dataset **(a)** to determine the $\gamma_{Al} = f(X_{Al})$ and $a_{Al} = f(X_{Al})$ functions using least squares regression. The data is valid for liquid Al-Si alloys at $1,150\ K \le T \le 1,470\ K$. **(b)** Calculate Ω_m and a_{Al} at $X_{Si} = 0.18$ and $T = 1,300$ K. [Solution: (b) $\Omega_m = -18,544\ J/mol$].

X_{Si}	0	0.05	0.10	0.15	0.20	0.25	0.30	0.35	0.40
a_{Al}	1	0.92	0.83	0.77	0.69	0.62	0.55	0.49	0.43

8.22 The given schematic figures help visualize the aspects of thermodynamics for finding relationships between the activities a_i and the activity coefficients γ_i in A-B binary solutions at high temperatures. Manipulate Eqs. (8.25), (8.31), and (8.36) along with $X_A X_B = X_A^2 X_B + X_A X_B^2$ (Fredriksson and Akerlind [15, p. 20]) to derive Eq. (8.34) for the activity coefficients of components A and B.

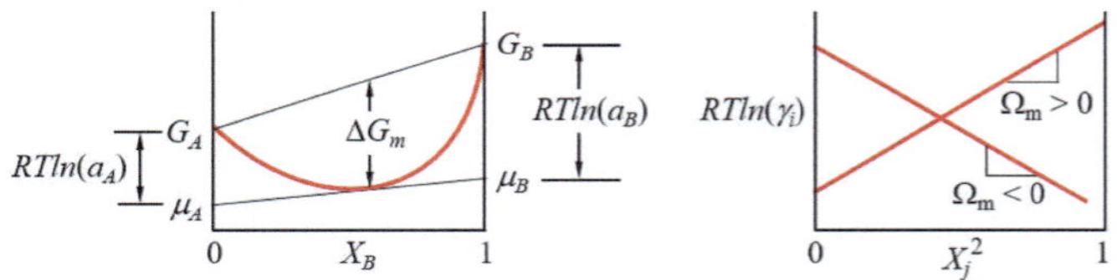

8.23 For an A-B binary solution, show that $\partial G_m / \partial X_B$ and $\partial G_m / \partial X_A$ are the linear slopes of the tangent lines (tangent rule) in Fig. 8.6 or P8.22, where ΔG_m denotes the Gibbs energy change of mixing and ΔG_A, ΔG_B denote the partial Gibbs energies.

8.24 For a binary Cu-Zn alloy solution containing $X_{Zn} = 0.40$ at $T = 1,060°\ C$, analytically calculate the activity of copper, a_{Cu}, using the data given in Example 8.5 and a curve fitting equation related to the area under the curve described by $X_{Zn}/X_{Cu} = f[-\ln(\gamma_{Zn})]$. [Solution: $a_{Cu} = 0.40833$].

8.25 Consider the thermodynamic dataset for regular Cr-Ti solutions at $T = 1,250°\ C$ ($1,523\ K$) given below to calculate **(a)** the area under the curve described by the power function $X_{Cr}/X_{Ti} = b[\ln(\gamma_{Cr})]^c$, where b, c are constants and **(b)** the activity a_{Ti} using the Gibbs-Duhem equation at $X_{Cr} = 0.53$ and $X_{Ti} = 0.47$. [Solution: (a) $Area = 0.51534$ and (b) $a_{Ti} = 0.78687$].

X_{Cr}	0.09	0.190	0.270	0.370	0.470	0.670	0.780	0.890
γ_{Cr}	3.356	2.800	2.444	2.103	1.745	1.248	1.106	1.018
a_{Cr}	0.302	0.532	0.660	0.778	0.820	0.836	0.863	0.906
$\ln(\gamma_{Cr})$	1.211	1.030	0.894	0.743	0.557	0.222	0.101	0.078
X_{Cr}/X_{Ti}	0.099	0.235	0.370	0.587	0.887	2.030	3.546	8.091

8.26 Consider the thermodynamic dataset for regular Cr-Ti solutions at $T = 1,250°\,C$ $(1,523\ K)$ given in P8.25 to calculate **(a)** the area under the curve described by the exponential function $X_{Cr}/X_{Ti} = b\exp\left[c\ln(\gamma_{Cr})\right]$, where b, c are constants and **(b)** the activity a_{Ti} using the Gibbs-Duhem equation at $X_{Ti} = 0.63$ and $X_{Cr} = 0.37$. [Solution: (a) $Area = 0.78159$ and (b) $a_{Ti} = 1.3765$].

8.27 Assume that hypothetical A-B binary alloy solutions at a relatively high temperature T are characterized by the equation

$$X_B/X_A = 0.40\left[\ln(\gamma_B)\right]^{-1.12}$$

Show that

$$\ln(\gamma_A) = 0.44126\,(X_B/X_A)^{0.10714}$$

References

1. D.A. Porter, K.E. Easterling, M.Y. Sherif, *Phase Transformations in Metals and Alloys*, 3rd edn. (Taylor & Francis Group, LLC, New York, 2009)
2. J.A. Dantzig, M. Rappaz, *Solidification*, 2nd edn. Revised & Expanded (EPFL Press, Switzerland, 2016)
3. R.T. DeHoff, *Thermodynamics in Materials Science* (McGraw-Hill, Inc., New York, 1993)
4. N.A. Gokcen, *Statistical Thermodynamics of Alloys* (Plenum Press, New York, 1986)
5. G.S. Upadhyaya, R.K. Dube, *Problems in Metallurgical Thermodynamics and Kinetics* (Pergamon Press Inc., New York, 1977)
6. T. Oishi, S. Tagawa, S. Tanegashima, Activity measurements of copper in solid copper-nickel alloys using copper-beta-alumina. Mater. Trans. **44**(6), 1120–1123 (2003)
7. L.S. Darken, R.W. Gurry, *Physical Chemistry of Metals* (McGraw Hill Book Company Inc., New York, 1953)
8. O. Kubaschewski, C.B. Alcock, *Metallurgical Thermochemistry*, 5th edn. (Pergamon Press, Oxford, 1979)
9. D.R. Gaskell, D.E. Laughlin, *Introduction to the Thermodynamics of Materials*, 6th edn. (Taylor & Francis Group, LLC, CRC Press, New York, 2018)
10. W. Rosenhain, *An Introduction to the Study of Physical Metallurgy* (D. Van Nostrand Company, New York, 1915)
11. W.G. Motaff, G.W. Pearsall, J. Wulff, Structure, in *The Structure and Properties of Materials*, vol. I (Wiley, New York, 1964)
12. S.H. Avner, *Introduction to Physical Metallurgy*, 2nd edn. (McGraw-Hill, Inc., New York, 1974)

13. D.R. Askeland, P.P. Fulay, D.K. Bhattacharya, *Essentials of Materials Science and Engineering*, 2nd edn. (Cengage Learning, Stanford, 2010)
14. W. Kurz, D.J. Fisher, *Fundamentals of Solidification*, 3rd edn. (Transactions on Technology Publications Ltd, Switzerland, 1992)
15. H. Fredriksson, U. Akerlind, *Solidification and Crystallization Processing in Metals and Alloys* (Wiley, West Sussex, 2012)
16. D.V. Ragone, *Thermodynamics of Materials*, vol. II (Wiley, New York, 1995)

Chapter 9
Solidification Heat Transfer

9.1 Introduction

The theoretical background on transient heat transfer of a phase-change is directly related to melting and solidification processes used in casting practices, food and pharmaceutical processing, latent heat energy storage, ice making, welding, and so forth. For a pure metal, the melting or solidification process occurs at a single temperature T_f, while an alloy undergoes either process at a temperature range $T_s < T_f^* < T_l$, where T_s and T_l are the solid and liquid temperatures, respectively. For clarity, the melting temperature (T_m) or fusion temperature (T_f) is used interchangeably so that $T_f = T_m$.

Regarding the rate of phase change, slow melting (SM) or slow solidification (SS) is treated as an equilibrium phase transformation related to equilibrium phase diagrams (see Chap. 10). Actually, phase transformation occurs when an amount of energy is either absorbed (melting) or released (solidification) at specific temperatures.

Melting is a heating process where solid-to-liquid transformation $(S \rightarrow L)$ is induced when the supplied thermal energy reaches the latent heat of fusion ΔH_f at the fusion temperature T_f. Conversely, solidification is a cooling or freezing process for converting a liquid into a solid $(L \rightarrow S)$ when the latent heat of solidification ΔH_s is released at the freezing temperature T_f or below. The use of ΔH_s compensates for the superheating effect $(\Delta T_s > 0)$ on solidification; otherwise, $\Delta H_s = \Delta H_f$ at $\Delta T_s = 0$. By definition, melting and solidification are energy absorption and release processes, respectively.

Rapid solidification (RS) of alloys is a deviation from equilibrium associated with a large amount of released ΔH_s at a short solidification time. One particular RS problem related to the time-temperature-transformation (TTT) diagram is included at the end of this chapter.

The goal in this chapter is to introduce analytical procedures and temperature diagrams for solving heat transfer problems related to phase transformation in

N. Perez, *Materials Science: Theory and Engineering*,
https://doi.org/10.1007/978-3-031-57152-7_9

"""

metals and alloys. This leads to a compilation of analytical solutions to the heat equation for one-dimensional melting and solidification problems in rectangular coordinates with moving liquid-solid (L-S) interfaces.

9.2 Statement of Heat Transfer Problems

Assume that heat transfer during melting or solidification occurs within a physical domain Γ containing the phase-change metal. Subsequently, the domain is divided into two semi-infinite regions by a sharp liquid-solid (L-S) interface with $\delta = 0$ thickness. However, if $\delta > 0$, then L-S interface is a mushy zone. For clarity, phase transformation is a transient heat transfer process, where fusion refers to melting the solid endothermically (addition of heat) and solidification refers to freezing the liquid or melt exothermically (release of heat). This means that the latent heat of fusion (ΔH_f) is the thermal energy for melting ($S \rightarrow L$) the solid, and latent heat of solidification (ΔH_s) is the thermal energy for freezing ($L \rightarrow S$) the liquid. Accordingly, transient heat transfer by conduction prevails as the main mechanism during phase transformation. However, heat transfer by radiation and convection may have significant contributions to the total heat transfer phenomenon.

9.2.1 Melting and Solidification

When a metal or alloy is subjected to a change of state, it is referred to as a phase transformation at a specific transformation point known as the melting temperature (T_m) upon melting (heating) a solid (S) phase or freezing temperature (T_f) upon solidification (cooling) a liquid (L) phase (Carslaw and Jaeger [1, pp. 282–296]. It is assumed hereafter that $T_m = T_f$ for a material undergoing phase change along with an evolved thermal energy. Recall that melting is represented by the reaction $S \rightarrow L$ and solidification by $L \rightarrow S$.

It is convenient to represent the involved thermal energy along with relevant thermal variables in the form of diagrams. Clearly, a mathematical treatment of the heat transfer theory related to Stefan problems and transcendental equations must be included for supporting the graphical form of datasets (Alexiades and Solomon [2, p. 48], Hahn and Ozisik [3, p. 452]). Moreover, the phase reaction $L \rightarrow S$ requires a transient conduction of heat, where an L-S interface (known as the solidification front) forms and propagates into the liquid phase at a velocity υ, which in turn depends on the rate of evolved latent heat of solidification ΔH_s (Gebhart [4, p. 228]).

9.3 Solidification Methods

The solidification methods schematically shown in Fig. 9.1 are common industrial practices, where the choice of sand or metallic mold material is an option that depends on the geometry and size of part to be produced. However, using metal molds implies that the liquid metal initially solidifies fast as a thin layer (metal skin) forming a gap at the interface due to mold expansion during absorption of heat and contraction (shrinkage) of the solidified layer during the release of latent heat of solidification. As a result, an interface resistance develops, and no physical bonding takes place at the solid-mold interface; otherwise, welding or soldering would occur naturally.

Firstly, Fig. 9.1a illustrates common symmetrical shapes in dimensionally complex castings. This suggests that solidification is faster in thin or small segments of castings, and consequently, control of the rate of solidification or freezing (solidification rate) is one of the key parameters for obtaining required solidified components or castings.

Secondly, Fig. 9.1b schematically shows a generalized view of a furnace for producing relatively large single crystals by forcing the start of a solidification process at a crystallographically prepared metal seed surface. This particular solidification technique requires pulling and rotational velocities of the growing single crystal.

Thirdly, Fig. 9.1c is also a generalized experimental setup for producing rapidly solidified metals and alloys. Actually, rapid solidification can be achieved by a fast extraction of heat from the melt. The resultant component is usually a ribbon or wire having amorphous, semi-amorphous, or crystalline microstructure.

This general introduction to solidification reflects the complexity involved during the microstructural evolution of metals and alloys. Further, a brief description of the above solidification methods enhances the understanding of a microstructural evolution using aspects of foundry technology and equilibrium phase diagrams.

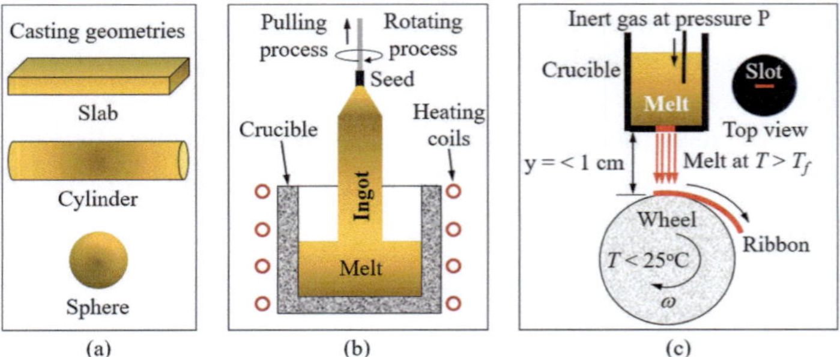

Fig. 9.1 Types of solidification methods. (**a**) conventional quasi-static solidification of common geometries, (**b**) quasi-static solidification of a single-crystal using the Czochralski process and (**c**) rapid solidification for producing ribbons. Here, ω denotes the rotating speed

9.3.1 Conventional Casting

Conventional casting is a common industrial practice for producing small and large amounts of alloys conventionally known as castings with simple (Fig. 9.1a) or intricate geometries.

In principle, the initial thermal contact at the start of solidification is influenced by the mold surface texture, gas interactions, casting geometric contours, and the like. Consequently, nucleation and growth stages in conjunction with the microstructural features, such as morphology and grain size, are strongly affected by the initial thermal solid-mold contact at the freezing temperature T_f and mainly by subsequent heat transfer conduction.

Solidification heat transfer is an energy transport phenomenon induced by the hot flowing molten metal into cold mold cavities. Then, the molten metal assumes the geometry within a mold, and it begins to cool, so that solidification initiates when it reaches the freezing temperature at $T < T_f$ related to a small undercooling ΔT.

Accordingly, a thermal contact between the molten metal and the mold inner surfaces causes heat transfer due to temperature differences. Sand molds, however, must have some thermal conductivity coupled with porosity for gases and air to escape. On the other hand, metallic molds with gas vents do not have porosity but high thermal conductivity for enhancing the heat removal during solidification.

9.3.2 Single Crystals

The single crystal method (Fig. 9.1b) is based on a directional solidification process (DSP) for producing continuous monocrystalline solids known as single crystals. The Czochralski process (Cz), discovered in 1916, is nowadays used to manufacture, in principle, impurity-free single crystals from a single nucleus at relatively slow upward speed (pulling rate), while the rod seed and the crucible rotate simultaneously in the opposite direction to minimize convection in the melt. Moreover, single crystals are monocrystalline solids with continuous lattice structures that lack grain boundaries, and typically, they are produced as cylinders having variable dimensions.

In the Czochralski process, a seed rod is simultaneously pulled upward and rotated in an inert environment in order to produce single crystals. This is an important directional solidification process for producing silicon (Si) single crystals used by the electronic industry.

9.3.3 Rapid Solidification

Rapid solidification (Fig. 10.1c) is achievable at a larger liquid metal undercooling ΔT compared to conventional solidification. This extended undercooling cause the melt to solidify as an amorphous, a semi-amorphous, or as a fine-grained microstructure. As a result, the metal casting is produced with enhanced properties.

Often amorphous alloys are produced in the form of thin ribbons, flakes, or wires by melt spinning, extraction, and the like. If a metallic powder is produced, then it is consolidated by hot extrusion or isostatic pressing (HIP) at relatively high temperatures. In any case, amorphous ribbons are subsequently chopped and consolidated.

Rapid solidification (RS) is a casting process carried out at high solidification or quenching rate dT/dt. For comparison, the range of solidification rate for rapid and slow conventional solidification techniques can be defined as indicated below

- For equilibrium or quasi-static solidification, $10^{-6}\,°C/s < dT/dt < 10^{+3}\,°C/s$ is the range in a slow solidification process.
- For nonequilibrium solidification, $10^{+3}\,°C/s < dT/dt < 10^{12}\,°C/s$ is the range in a rapid solidification process.

The rate of solidification dT/dt is related to the L-S interface velocity v_z, which in turn depends mainly on the rate of heat flow dq/dt by conduction from the liquid to the solid through an interface, where the latent heat of solidification ΔH_s is continuously released during the rapid steady-state solidification. Moreover, Appendix 9A contains valuable information for unit conversion. One particular example is shown for ΔH_s in J/cm^3 to J/mol units.

9.4 Thermal Energy Diagrams for Metals

For solidification, an amount of material is completely melted and conditioned at a pouring temperature T_p as shown at point "P" in Fig. 9.2a for a pure metal and Fig. 9.2b for an alloy, where $\Delta T_s = T_p - T_f$ is called superheat or superheating, where $T_p > T_f$. This is simply a change of state designated as solidification for which $L \to S$ occurs when the evolved thermal energy is the latent heat of fusion ΔH_f.

Notice that Fig. 9.2a schematically shows an isothermal freezing process at T_f for a pure metal, while Fig. 9.2b exhibits a non-isothermal process for an alloy that freezes at a temperature range $T_s \le T_f \le T_l$. Therefore, the latter is more complicated to analyze due to the formation of a planar or contoured L-S interface at an intermediate temperature; $T_s < T_i < T_l$. For modeling purposes, a mushy zone may be placed ahead of the interface.

The fundamental relation between heat transfer and temperature for phase changes can be represented as a temperature-energy $(T$-$Q)$ diagram as shown in

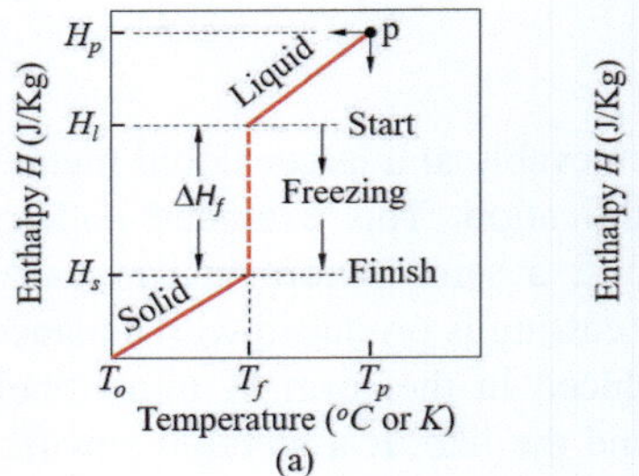
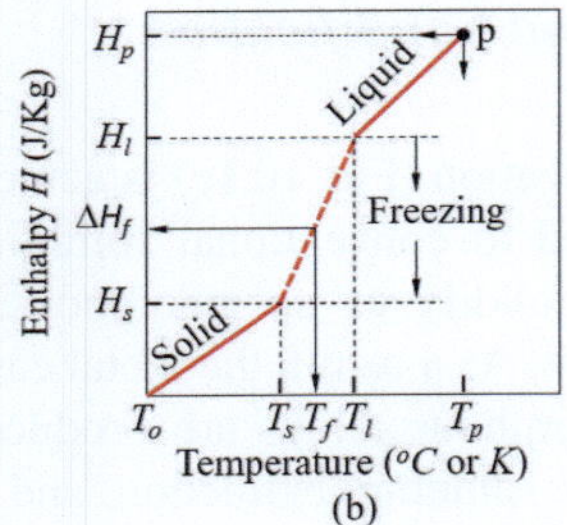

Fig. 9.2 Enthalpy diagram for freezing from $T_p \rightarrow T_f$. (**a**) Isothermal for pure metals and (**b**) non-isothermal for alloys and glassy substrates

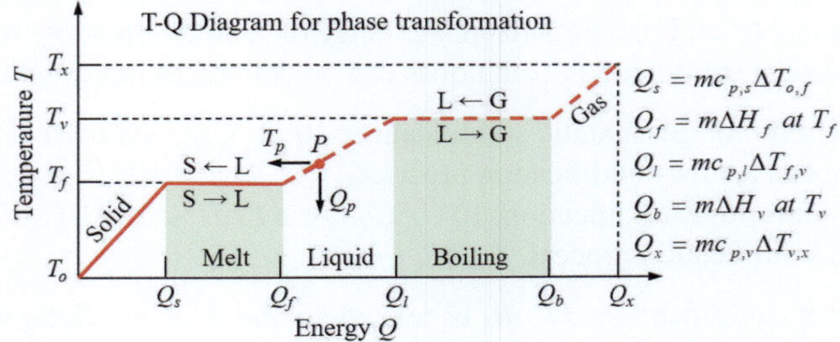

Fig. 9.3 Schematic temperature-energy (T-Q) diagram for phase transformation of a hypothetical metal or substance

Fig. 9.3 for easy interpretation of the amount of energy added to or extracted from a system. For convenience, this schematic diagram includes the thermal energy Q equations for the corresponding phase regions and the assumed linear temperature trends. Furthermore, a g the phase transitions shown in Fig. 8.1 and defined by Eqs. (8.1) and (8.2) is shown in Fig. 9.3.

Consider a pure solid being subjected to an endothermic process in a furnace. If the solid is initially at room temperature T_o, then slow heating causes absorption of thermal energy so that

- Melting occurs at the interchangeably fusion or melting temperature $T_m = T_f$ and the solid absorbs $Q_s \leq Q \leq Q_f$ thermal energy to become a liquidus phase.
- Boiling or vaporization occurs at $T_b = T_v$ for producing a gaseous phase at T_x.

Assume that a gas is subjected to an exothermic process in a furnace. If the gas is initially at temperature T_x, then slow cooling or freezing causes a release of thermal energy so that

- Condensation occurs at T_v and the gas releases $Q_l \leq Q \leq Q_b$ thermal energy to become a liquid phase.
- Solidification or freezing occurs at T_f for producing a solid phase.

Only melting and solidification processes are considered below. The relevant pouring temperature $T_p > T_f$ is located at point "P" in Fig. 9.3, and it is the point of departure for the solidification of either a pure metal or an alloy with composition C_o. The sequence of solidification events shown in Fig. 9.3 can be deduced with respect to the heat transfer Q in unit of joule (J), which is quantified as the change in thermal energy with measured heat capacity c_p at constant pressure P. Thus,

- For freezing the melt, $Q_p = mc_{p,l}\Delta T_s$ where $\Delta T_s = T_p - T_f$
- For local solidification, $Q_f = m\Delta H_s$ at $T = T_f$ and $t > 0$
- For total solidification, $Q_s = mc_{p,l}\Delta T_s + m\Delta H_s$ when $T_p \to T_f$ at $t > 0$
- For cooling the solid, $Q_c = mc_{p,s}\Delta T_c$ when $T_f \to T_o = 25\,^\circ C$ at $t >> 0$

where m denotes the mass of the sample (g), T denotes the sample temperature (K), t denotes time (sec), $c_{p,l}$ denotes the heat capacity ($J/g.K$) of the liquid phase at constant pressure P, $c_{p,s}$ denotes heat capacity of the solid phase at constant pressure, ΔT_s denotes the superheat, ΔT_c denotes the temperature range for cooling the solid phase, T_f denotes the material freezing temperature, and ΔH_s denotes the latent heat of solidification (J/mol or J/Kg).

9.5 Analytical Specific Heat Capacity

The general definition of the specific heat capacity c_p ($J/kg.K$) or heat capacity (J/K) at constant pressure (P) is the amount of heat energy required to raise the temperature of unit mass ($m = 1\,Kg$) of a material by one degree ($\Delta T = 1\,K$). Fundamentally, c_p is the slope or first derivative of the enthalpy function $H = H(T)$, which is related to the heat transfer induced by the evolved latent heat of solidification (ΔH_s) at T_f. Thus, c_p, H and Gibbs energy G are defined by

$$c_p = \left(\frac{dh}{dT}\right)_p = \left(\frac{\partial q}{\partial T}\right)_p \tag{9.1a}$$

$$ds = \frac{\partial q}{T} = \frac{c_p dT}{T} \tag{9.1b}$$

$$G = G(T) = H - TS \tag{9.1c}$$

$$S = -(dG/dT)_p \tag{9.1d}$$

Assume that $c_p = c_p(T)$ can be defined as a general polynomial

$$c_p = a_1 + a_2 T + a_3 T^2 + ... \tag{9.2}$$

where $i = 1, 2, 3..$ and a_i are coefficients. The physical behavior of the solidification process as per $H = H(T)$ and $G = G(T)$ functions for the liquid and solid phases is schematically shown in Fig. 9.4 for a pure metal.

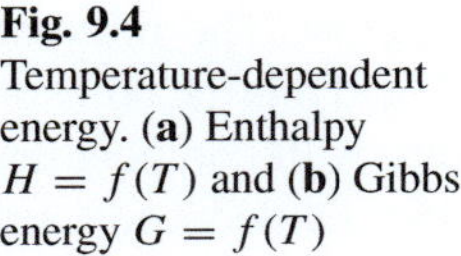

Fig. 9.4
Temperature-dependent
energy. (**a**) Enthalpy
$H = f(T)$ and (**b**) Gibbs
energy $G = f(T)$

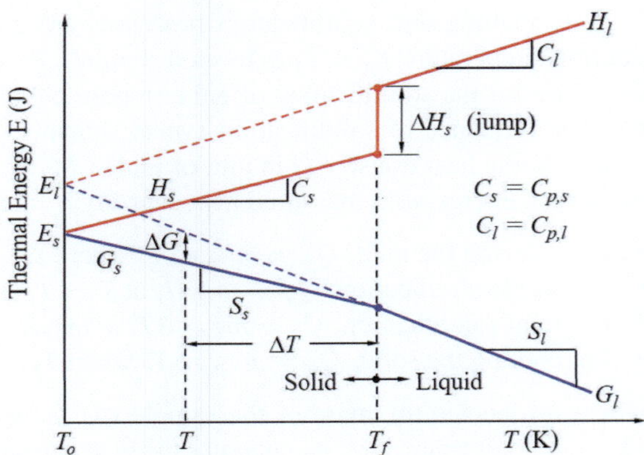

Note that Eqs. (9.1c), (9.1d) are linear expressions applied to individual phases, and as a result, the $H = H(T)$ plot reveals an enthalpy jump defining ΔH_s. Conversely, $G = G(T)$ shows a change in slope, which defines the entropy as illustrated in Fig. 9.4 and defined by Eq. (9.1d). The extended liquidus lines for H and G intercept at a temperature T_o, where the liquid thermal energy $E_l = H_l = G_l$. Similarly, for the solid phase, $E_s = H_s = G_s$ and $E_l > E_s$, since the liquid temperature is higher than that of the solid phase; otherwise, solidification would not occur.

9.6 Melting Problem in a Half-Space Region

Consider a two-phase solidification problem (Fig. 9.5) in a half-space containing a moving L-S interface, which divides the solid and liquid phases.

In rectangular coordinates, the half-space region with $x > 0$ is defined within a semi-infinite region where $0 < x < \infty$ with a L-S interface located at a characteristic length $x = s(t)$.

The plots in Fig. 9.5 represent idealized profiles of the physical temperature fields $T(x, t)$ during plain melting and solidification. For instance, Fig. 9.5a illustrates part of the crucible at constant temperature T_p, while heat flux q_x is supplied to the system to start the melting process at $x = 0$ and to finish it at location x_c in the positive x-direction.

Figure 9.5b shows the idealized temperature profiles related to solidification of pure materials being superheated at T_p, and Fig. 9.5c exhibits the temperature profile $T = f(t)$ in the mold and solid metal for the entire solidification processes.

Figure 9.5 directly illustrates that the temperature field is influenced by the release of latent heat ΔH_s at a rate defined by the solidification velocity υ_x in the opposite direction. Notice that Fig. 9.5a indicates that the melt is superheated

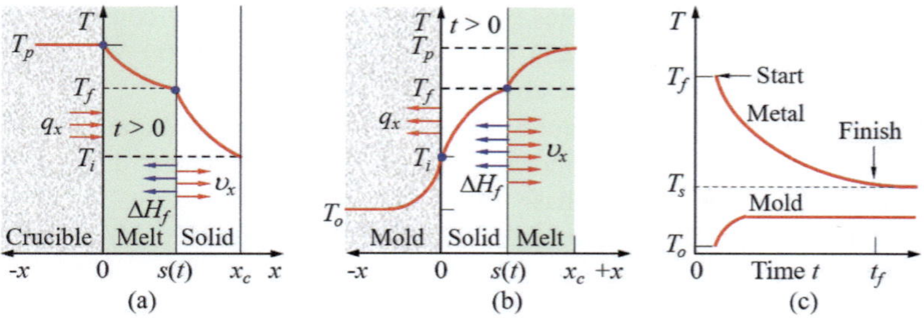

Fig. 9.5 Temperature profiles in a half-space $x > 0$ during (**a**) melting, (**b**) solidification and (**c**) for freezing the melt and heating the mold

at $T_p > T_f$ and Fig. 9.5b shows the effect of superheating the melt at pouring temperature $T_p > T_f$.

Consider a solid-phase metal initially at $T_i < T_f$ in a confined semi-infinite planar region $0 < x < \infty$, where a continuous supply of heat flux q_x suddenly raises the solid temperature to $T_p > T_f$ (Fig. 9.5a). As a result, the onset of melting starts at the crucible-liquid interface $x = 0$. At times $t > 0$, the liquid-solid (L-S) interface with velocity υ_x is located at $x = s(t)$, and the completion of the melting process is achieved when $s(t) = x_c$ at $t = t_f$. Moreover, $s(t)$ is the location of a L-S interface within the half-space region $0 < s(t) \leq x_c$, and it is called the characteristic length of the moving L-S interface in the positive x-direction.

9.6.1   Two-Phase Melting Problem

For one-dimensional analysis in the x-direction, the governing *heat equation* for the two-phase problem, continuity of temperature at the L-S interface, and the Fourier heat flux balance defined as $q_l - q_s = q_f$ for melting a slab in a half-space $x > 0$ in a semi-infinite region $0 < x < \infty$ are, respectively,

$$\frac{\partial T_l}{\partial t} = \alpha_l \frac{\partial^2 T_l}{\partial x^2} \quad \text{for } 0 < x < s(t), t > 0 \quad \text{(melt)} \tag{9.3a}$$

$$\frac{\partial T_s}{\partial t} = \alpha_s \frac{\partial^2 T_s}{\partial x^2} \quad \text{for } x < s(t), t > 0 \quad \text{(solid)} \tag{9.3b}$$

$$T_l = T_s = T_f \quad \text{at } x = s(t) \quad \text{(L-S interface)} \tag{9.3c}$$

$$k_s \frac{\partial T_s}{\partial x} - k_l \frac{\partial T_l}{\partial x} = \rho_l \Delta H_f \frac{ds(t)}{dt} \quad \text{at } x = s(t) \text{ (energy balance)} \tag{9.3d}$$

where q_f denotes the heat flux (W/m^2) due to release of latent heat of solidification (specific enthalpy at T_f), ρ_l denotes the density (Kg/m^3), k_l, k_s denotes the thermal conductivity $(W/m.K)$, c_p denotes the specific heat capacity at constant pressure $(J/Kg.K)$, ΔH_f denotes the latent heat of fusion (J/Kg), $ds(t)/dt$ denotes the rate of melting, and T denotes the absolute temperature.

The enthalpy change ΔH_f represents the amount of thermal energy that must be supplied to overcome the binding energy in the solid phase during melting or extracted from the liquid phase to allow atoms form a solid phase during solidification. Hence, the motion of a L-S interface depends on ΔH_f at the fusion temperature T_f. Nonetheless, Eq. (9.3d) is known as the Stefan condition, which follows the energy conservation law for a system of two different phases.

Finding the analytical solution of the heat partial differential equation (PDE), Eq. (9.3a) or (9.3b) requires that the PDE admits a similarity solution by reducing it to an ordinary differential equation (ODE).

Additional mathematical definitions are needed before solving the heat equation. These are thermal diffusivity α (m^2/s), the thermal inertia (thermal effusivity) γ $[J/(m^2.K\sqrt{s})]$, the heat flux q, the rate of removed energy (dQ/dt), and the amount of removed energy (Q) applicable to liquid and solid phases, and mold region are, respectively,

$$\alpha = \frac{k}{\rho c_p} \tag{9.4a}$$

$$\gamma = k/\sqrt{\alpha} = \sqrt{k\rho c_p} \tag{9.4b}$$

$$q = k\frac{\partial T}{\partial x} \tag{9.4c}$$

$$\frac{dQ}{dt} = -Ak\frac{\partial T}{\partial x} \tag{9.4d}$$

$$Q = -Ak\int \left(\frac{\partial T}{\partial x}\right) dt \tag{9.4e}$$

$$q_x = h(T_i - T_o) \quad \text{(Newton's law)} \tag{9.4f}$$

where Eqs. (9.4a), (9.4b) contain the thermophysical properties (Appendix 9A), q_x is the heat flux, h is the heat transfer coefficient, T_i is an intermediate temperature, and T_o is a reference temperature.

Using the separation of variables method found in Appendix 9B yields the general temperature equations for the liquid and solid phase

$$T_l = C_1 + C_2\,\mathrm{erf}\left(\frac{x}{2\sqrt{\alpha_l t}}\right) \tag{9.5a}$$

$$T_s = C_3 + C_4\,\mathrm{erfc}\left(\frac{x}{2\sqrt{\alpha_s t}}\right) \tag{9.5b}$$

which satisfy the heat equations, Eqs. (9.3a), (9.3b). Before finding the constants C_1, C_2, C_3, C_4, it is necessary to introduce some general mathematical definitions

when $x = s(t)$. In this case, the relevant system of equation is

$$n = \frac{x}{2\sqrt{\alpha t}} \quad \text{(similarity variable)} \tag{9.6a}$$

$$\lambda = \frac{s(t)}{2\sqrt{\alpha t}} \quad \text{for } \alpha = \alpha_l \text{ (liquid); } \alpha = \alpha_s \text{ (solid)} \tag{9.6b}$$

$$t = \frac{s(t)^2}{4\lambda^2\alpha} \quad \text{for } \alpha = \alpha_l \text{ (liquid); } \alpha = \alpha_s \text{ (solid)} \tag{9.6c}$$

$$s(t) = 2\lambda\sqrt{\alpha t} \quad \text{for } \alpha = \alpha_l \text{ (liquid); } \alpha = \alpha_s \text{ (solid)} \tag{9.6d}$$

$$\frac{ds(t)}{dt} = \lambda\sqrt{\frac{\alpha}{t}} \quad \text{for } \alpha = \alpha_l \text{ (liquid); } \alpha = \alpha_s \text{ (solid)} \tag{9.6e}$$

where n_l, n_s (Neumann similarity variables) and λ (solution characteristic) denote dimensionless parameters, t denotes the melting time, $s(t)$ denotes the characteristic length and the location of a L-S interface, and $ds(t)/dt$ denotes the rate of the L-S interface motion (interface velocity) or the rate of melting in the positive x-direction.

The similarity variable n defined by Eq. (9.6a) is a general expression, which can be applied to heat transfer problems related to liquid and solid phases at the solidification front. In this context, the similarity variable strongly depends on the thermal diffusivity α_s of the evolved solid phase during liquid-to-solid transformation.

The governing partial differential equations (PDEs) defined by Eqs. (9.3a), (9.3b) are transformed by a similarity transformation into a system of ordinary differential equations (ODEs), and the their general solutions are given by Eqs. (9.5a), (9.5b) and analytically derived in Appendix 9B.

9.6.2   Boundary Conditions

Finding the constants C_1, C_2 with boundary condition (BC):

$$T_l = T_l(x, t) = C_1 + C_2 \operatorname{erf}\left(\frac{x}{2\sqrt{\alpha_l t}}\right) \quad \text{(liquid)} \tag{9.7a}$$

$$\text{BC1 } T_l = T_l(0, t) = T_p \tag{9.7b}$$

$$T_p = C_1 + C_2 \operatorname{erf}(0) \rightarrow C_1 = T_p \tag{9.7c}$$

$$\text{BC2 } T_l = T_l[s(t), t] = T_f \tag{9.7d}$$

$$T_f = T_p + C_2 \operatorname{erf}\left(\frac{s(t)}{2\sqrt{\alpha_l t}}\right) = T_p + C_2 \operatorname{erf}(\lambda) \tag{9.7e}$$

$$C_2 = -\frac{\left(T_p - T_f\right)}{\mathrm{erf}\,(\lambda)} \tag{9.7f}$$

$$T_l = T_p - \frac{\left(T_p - T_f\right)}{\mathrm{erf}\,(\lambda)}\,\mathrm{erf}\left(\frac{x}{2\sqrt{\alpha_l t}}\right) \tag{9.7g}$$

Finding the constants C_3, C_4:

$$T_s = T_s\,(x, t) = C_3 + C_4\,\mathrm{erfc}\left(\frac{x}{2\sqrt{\alpha_s t}}\right) \quad \text{(solid)} \tag{9.8a}$$

$$\text{BC1 } T_s = T_l\,(0, t) = T_i \tag{9.8b}$$

$$T_i = C_3 + C_4\,\mathrm{erfc}\,(\infty) \rightarrow C_3 = T_i \tag{9.8c}$$

$$\text{BC2 } T_s = T_l\,[s\,(t)\,, t] = T_f \tag{9.8d}$$

$$T_f = T_i + C_4\,\mathrm{erfc}\left(\frac{s\,(t)}{2\sqrt{\alpha_s t}}\right) = T_i + C_4\left(\frac{s\,(t)}{2\sqrt{\alpha_s t}}\,\frac{\sqrt{\alpha_l}}{\sqrt{\alpha_l}}\right) \tag{9.8e}$$

$$C_4 = \frac{\left(T_f - T_i\right)}{\mathrm{erfc}\left(\lambda\sqrt{\alpha_l/\alpha_s}\right)} \tag{9.8f}$$

$$T_s = T_i + \frac{\left(T_f - T_i\right)}{\mathrm{erfc}\left(\lambda\sqrt{\alpha_l/\alpha_s}\right)}\,\mathrm{erfc}\left(\frac{x}{2\sqrt{\alpha_s t}}\right) \tag{9.8g}$$

From Eqs. (9.7g) and (9.8g),

$$\left(\frac{\partial T_l}{\partial x}\right)_{s(t)} = -\frac{\left(T_p - T_f\right)}{\mathrm{erf}\,(\lambda)\,\sqrt{\pi \alpha_l t}}\,\exp\left(-\lambda^2\right) \tag{9.9a}$$

$$\left(\frac{\partial T_s}{\partial x}\right)_{s(t)} = -\frac{\left(T_f - T_i\right)}{\mathrm{erfc}\left(\lambda\sqrt{\alpha_l/\alpha_s}\right)\,\sqrt{\pi \alpha_s t}}\,\exp\left(-\lambda^2\alpha_l/\alpha_s\right) \tag{9.9b}$$

Liquid at $T_f \leq T_l \leq T_p$ and Solid at $T_i \leq T_s \leq T_f$ Substituting Eqs. (9.9a) and (9.9b) along with (9.6e) into (9.3d) gives the transcendental equation with root λ for the melting process

$$\frac{\lambda\,\mathrm{erf}\,(\lambda)}{\exp\left(-\lambda^2\right)} + \frac{\gamma_s}{\gamma_l}\frac{c_l\left(T_f - T_i\right)}{\sqrt{\pi}\,\Delta H_f}\frac{\exp\left[\lambda^2\left(1 - \alpha_l/\alpha_s\right)\right]\mathrm{erf}\,(\lambda)}{\mathrm{erfc}\left(\lambda\sqrt{\alpha_l/\alpha_s}\right)} = \frac{c_l\left(T_p - T_f\right)}{\sqrt{\pi}\,\Delta H_f} \tag{9.10}$$

This complicated equation can be solved numerically by iteration.

Liquid at $T_f \leq T_l \leq T_p$ and Solid at $T_i = 0$ In this case, Eq. (9.10) becomes a simplified form of the transcendental equation with root λ for the melting process

$$\frac{\lambda \operatorname{erf}(\lambda)}{\exp\left(-\lambda^2\right)} + \frac{\gamma_s}{\gamma_l} \frac{c_l T_f}{\sqrt{\pi}\,\Delta H_f} \frac{\exp\left[\lambda^2\left(1 - \alpha_l/\alpha_s\right)\right]\operatorname{erf}(\lambda)}{\operatorname{erfc}\left(\lambda\sqrt{\alpha_l/\alpha_s}\right)} = \frac{c_l\left(T_p - T_f\right)}{\sqrt{\pi}\,\Delta H_f} \tag{9.11}$$

One-Phase Melting Problem of a Slab In this circumstance, Eq. (9.10) along with $T_i = T_f$ becomes a simplified transcendental equation (Hahn and Ozisik [3, p. 463])

$$\lambda \exp\left(\lambda^2\right) \operatorname{erf}(\lambda) = \frac{c_l\left(T_p - T_f\right)}{\sqrt{\pi}\,\Delta H_f} = \frac{S_{t,l}}{\sqrt{\pi}} \tag{9.12}$$

where $S_{t,l}$ denotes the Stefan number for the liquid phase. In effect, the preceding solutions are useful in melting problems, since $S_{t,l}$ is the principal parameters in Eq. (9.12) that depends on material properties (Appendix 9A) and temperature differences.

Example 9.1 Consider melting copper in a crucible using one-phase melting problem. If the slab thickness is $0.08\,m$, then calculate the melting time. Data: $\Delta H_f = 205{,}000\,J/Kg$, $k_l = 162\,W/m.K$, $\rho_l = 8020\,Kg/m^3$, $c_l = 570\,J/Kg.K$, $\alpha_l = 3.54 \times 10^{-5}\,m^2/s$, $T_f = 1083\,°C$ and $T_p = T_l = 1183\,°C$. Unit conversion

$$k \rightarrow 1\,W/m.K = 1\,W/m.°C \qquad dT/dx \rightarrow 1\,K/m = 1\,°C/m$$
$$c_s, c_l \rightarrow 1\,J/Kg.K = 1\,J/Kg.°C \quad dx/dt \rightarrow 1\,K/s = -272\,°C/s$$

Solution Using Eq. (9.12) with $s\,(t) = 0.08/2 = 0.04\,m$ (half-space region) yields

$$\lambda \exp\left(\lambda^2\right) \operatorname{erf}(\lambda) = \frac{c_l\left(T_p - T_f\right)}{\sqrt{\pi}\,\Delta H_f} = 0.15657 \tag{9.1E1a}$$

$$\lambda \exp\left(\lambda^2\right) \operatorname{erf}(\lambda) - 0.15657 = 0 \tag{9.1E1b}$$

$$\lambda = 0.35689 \tag{9.1E1c}$$

From Eq. (9.6c),

$$t = \frac{s\,(t)^2}{4\lambda^2\alpha_l} = \frac{(0.04\,m)^2}{4\,(0.35689)^2\left(3.54 \times 10^{-5}\,m^2/s\right).} \tag{9.1E2a}$$

$$t = 88.71\,s = 1.48\,\text{min} \tag{9.1E2b}$$

Therefore, melting a 8-cm thick slab at $T_l = 1183\,°C$ takes less than two minutes. For instance, if $1083\,°C < T_l = 1093\,°C < 1183\,°C$, then $t > 13\,min$. That is,

$$\lambda \exp\left(\lambda^2\right) \operatorname{erf}(\lambda) = \frac{c_l\left(T_p - T_f\right)}{\sqrt{\pi}\,\Delta H_f} = 0.015687 \tag{9.1E3a}$$

$$\lambda = 0.11737 \tag{9.1E3b}$$

$$t = \frac{s\,(t)^2}{4\lambda^2\alpha_l} = \frac{(0.04)^2}{4\,(0.11737)^2\,\left(3.54\times10^{-5}\right).} = 13.67\ \text{min} \tag{9.1E3c}$$

This is a reasonable result.

9.7   Solidification Problem in a Half-Space Region

9.7.1   Heat Transfer Through Planar Surfaces

In principle, heat transfer between plane (|) and contoured (convex $\smile$ and concave $\frown$) mold inner surfaces can be assumed similar (Geiger and Poirier [5, p.331], Poirier and Geiger [6, p. 331], Gaskell [7, p. 405], Stefanescu [8, pp. 48–49]). For instance, Fig. 9.6a illustrates a rectangular casting, and Fig. 9.6b shows its cross-sectional plane during unidirectional solidification, where the heat flux q_x crosses the liquid-mold (L-m) interface at $x > 0$ and flows in the x-direction at the freezing temperature T_f.

Figure 9.6c shows a hollow cylindrical casting, and Fig. 9.6d exhibits its annular cross-sectional area, illustrating the heat flux q_r flow in the radial direction at T_f. For spherical solidification, q_r also flows in the radial direction.

For one-dimensional solidification of a slab, cylinder, or sphere, heat fluxes q_x, q_r or the rate of thermal energy quantities $dQ_x/dt, dQ_r/dt = -A\,(q_x, q_{ls})$ give the amount of energy associated with a two-phase (liquid and solid) solidification problem (Alexiades and Solomon [2, p. 117]).

The thermal conductivity strongly depends on the heat interaction between two surfaces during solidification; mold-liquid metal and solid-liquid interface, where "solid" stands for solidified melt and "interaction" represents some degree of heat transfer resistance by the phase absorbing the thermal energy at the two-phase

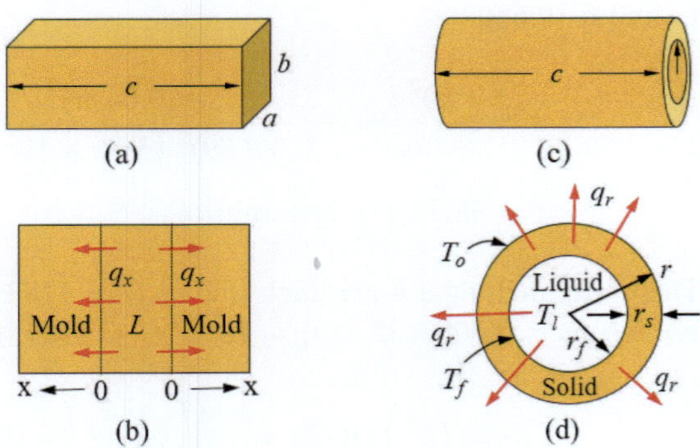

Fig. 9.6 Sketches of simple casting geometries and heat flux directions. (**a**) rectangular shape and (**b**) its cross-section. (**c**) Hollow cylindrical shape and (**d**) its cross-section as an annular ring

boundary, which evolves at $T < T_f$ and $t > 0$. This suggests that there must be temperature gradients (with or without L-S interface resistance) within the mold and the solidifying melt phases (Poirier and Geiger [6, pp. 338, 340]).

9.7.2 Solidification with Superheating

In principle, $\Delta T_s = (T_p - T_f) > 0$ is a measure of superheating the liquid metal. Actually, the entire mold cavity has to be completely filled in a relatively short time prior to the onset of solidification at $x > 0$, $t = 0$ and $T = T_f$; otherwise, premature solidification may take place in the gating or riser system and consequently, an incomplete casting geometry may be attained. Moreover, a high superheat saturates gases inducing the formation of metal oxides and melt penetration into the mold surface.

Superheat is related to pouring rate (volumetric rate) and heat dissipation. Fast pouring rates induce melt turbulence and erosion within the mold cavity. Conversely, too slow pouring rates cause premature solidification before filling the mold completely.

Once the melt completely fills the mold cavity in a very short time interval at the pouring temperature T_p, the temperature of a mold inner walls instantaneously increases to T_p at $t = 0$, and subsequently, solidification starts at T_f and the heat flux q_x flows in the x-direction (Fig. 9.6b) through the solid and mold walls.

9.7.3 Solidification with Supercooling

The magnitude of a supercooling or undercooling process is defined as ΔT_u, and it is the driving force required for nucleation of a solid phase in a mold cavity. Ideally, the mold is considered as a chemically nonreactive material at high temperatures, so that solidification is uniform and continuous along the evolved L-S interface positioned at $x = s(t) > 0$.

Two different supercooling conditions can be highlighted at a macroscale $\Delta H_s = \Delta H_f$ condition (Alexiades and Solomon [2, p. 85], Glicksman [9, p. 429])

$$T_f - \frac{\Delta H_f}{c_l} < T_i < T_f \tag{9.13a}$$

$$\Delta T = (T_f - T_i) < \frac{\Delta H_f}{c_l} \tag{9.13b}$$

where $T_i < T_f$ is the melt initial temperature prior to phase change and c_l is the specific heat of the liquid phase. Eventually, the L-S interface (solidification front) forms at a location $x = s\,(t)$ for planar solidification or $r = s\,(t)$ for radial solidification process.

9.8   Mold, Solid, and Liquid Thermal Resistances

Consider a pure metal undergoing solidification in a half-space region $0 \le x \le x_c$, and assume that the temperature profiles are as schematically shown in Fig. 9.7.

   The goal in this section is to derive a general transcendental equation that couples the mold, solid, and liquid as a solidification domain free of convective heat transfer effect on the solidification process and variable thermophysical properties (Appendix 9A).

9.8.1   *One-Dimensional Heat Equations*

For an ideal mold-solid contact surface area (chill zone), the governing heat equations, continuity of temperature at the L-S interface, and the Fourier heat flux balance for solidification are, respectively,

$$\frac{\partial T_m}{\partial t} = \alpha_m \frac{\partial^2 T_m}{\partial x^2} \quad \text{for} \ -\infty < x \le 0, t > 0 \tag{9.14a}$$

$$\frac{\partial T_s}{\partial t} = \alpha_s \frac{\partial^2 T_s}{\partial x^2} \quad \text{for} \ 0 < x \le s\,(t), t > 0 \tag{9.14b}$$

$$\frac{\partial T_l}{\partial t} = \alpha_l \frac{\partial^2 T_l}{\partial x^2} \quad \text{for} \ s\,(t) \le x < \infty, t > 0 \tag{9.14c}$$

$$T_s = T_l = T_f \quad \text{at} \ x = s\,(t) \tag{9.14d}$$

$$k_m \frac{\partial T_m}{\partial x} = k_s \frac{\partial T_s}{\partial x} = \rho_s \Delta H_f \frac{ds\,(t)}{dt} \quad \text{at} \ x = 0 \tag{9.14e}$$

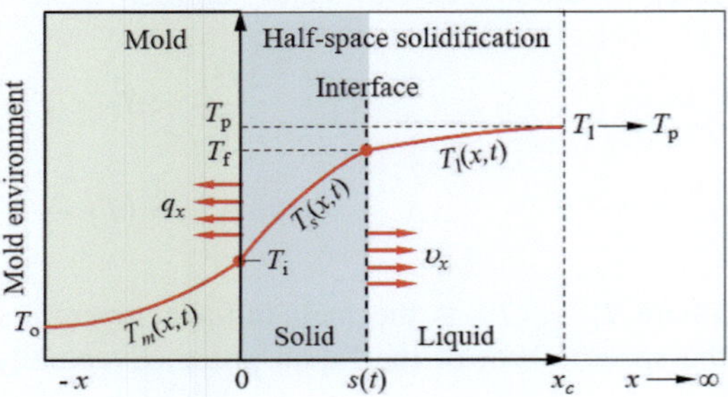

Fig. 9.7 Coupled mold-solid resistance problem with negligible liquid temperature gradient

$$k_s \frac{\partial T_s}{\partial x} - k_l \frac{\partial T_l}{\partial x} = \rho_s \Delta H_f \frac{ds\,(t)}{dt} \quad \text{at } x = s\,(t)\,, \; T_p = T_f \tag{9.14f}$$

$$k_s \frac{\partial T_s}{\partial x} - k_l \frac{\partial T_l}{\partial x} = \rho_s \Delta H_s \frac{ds\,(t)}{dt} \quad \text{at } x = s\,(t)\,, \; T_p > T_f \tag{9.14g}$$

This system of equations represents the mathematical domain for finding analytical solutions to simple second-order linear partial differential equation for the mold (subscript m), solid (subscript s), and liquid (subscript l) phases.

9.8.2   *Latent Heat of Solidification*

The latent heat of solidification (ΔH_s) is a special relationship that accounts for superheating and supercooling effects on the freezing melt (liquid phase). For a motionless superheated melt in a mold cavity, the effective latent heat of solidification (ΔH_s) is mathematically defined as

$$\Delta H_s = \Delta H_f + \int_{T_f}^{T_p} c_l dT \quad \text{(with } c_{pl} = c_l) \tag{9.15a}$$

$$\Delta H_s = \Delta H_f + c_l \left(T_p - T_f \right) = \Delta H_f + c_l \Delta T_s \tag{9.15b}$$

$$\Delta H_s = \Delta H_{f,B} + c_l \left| \Delta H_{f,B} - \Delta H_{f,A} \right| + c_l \Delta T_s \tag{9.15c}$$

where ΔH_f denotes, again, the latent heat of fusion that is released at T_f, $c_l \left(T_p - T_f \right)$ denotes the sensible specific heat due to melt superheating effect and c_l denotes the specific heat of the liquid phase at constant pressure P. For an A-B binary alloy, ΔH_s is predicted by Eq. (9.15c).

Firstly, the liquid-to-solid (L-S) phase transformation starts when ΔH_f is released on a smooth and flat mold-solid interface with the heat flux balance defined by Eq. (9.14f), which indicates that the melt is initially at $T_p = T_f$ and releases $\Delta H_s = \Delta H_f$ at $x = s\,(t)$. The mold-solid interface is known as the chill zone having a thin layer of the first solidified melt in the form of small equiaxed grains with similar axes lengths.

During solidification, ΔH_f is released, while the L-S interface evolves at $0 < s\,(t) < \infty$ in a time interval $0 < t \leq t_s$ for the local solidification time or $0 < t \leq t_f$ for the total solidification time due to the superheat effect. Hence, t_s is the real solidification time at $T = T_f$ and $t_f = t_{cool} + t_s$ is the apparent solidification time, since it takes into account the effect of superheat, since the melt has to freeze before it transforms to solid.

9.8.3 Analytical Solutions of the Heat Equations

For a one-dimensional analysis, the heat equations for solidification of pure metals in sand and metal molds admit similarity solutions. This is a mathematical approach for converting a partial differential equations (PDE) into an ordinary differential equation (ODE). In this context, the similarity solution is based on a similarity variable denoted as n. Moreover, this mathematical approach will be become clear to the reader as illustrated below for solidification heat transfer under ideal steady-state conditions.

The goal is to find a similarity solution for the temperature field. The resultant equation gives a suitable temperature profile in the solid or liquid phases.

Mold Region For a thick enough metal mold, modeled as a one-dimensional semi-infinite system ($-\infty < x \leq 0$) with interface thermal resistance, the general temperature field equation with boundary conditions is

$$T_m = C_1 + C_2 \operatorname{erfc}\left(\frac{-x}{2\sqrt{\alpha_m t}}\right) \quad \text{(mold)} \tag{9.16a}$$

$$\text{BC1 } T_m = T_m(-\infty, t) = T_o \tag{9.16b}$$

$$T_o = C_1 + C_2 \operatorname{erfc}(\infty) = C_1 \tag{9.16c}$$

$$\text{BC2 } T_m = T_m(0, t) = T_i \tag{9.16d}$$

$$T_i = T_o + C_2 \operatorname{erfc}(0) = T_o + C_2 \tag{9.16e}$$

$$C_2 = T_i - T_o \tag{9.16f}$$

where the error function yields $\operatorname{erfc}(\infty) = 0$ and $\operatorname{erfc}(0) = 1$. Substituting C_1, C_2 into Eq. (9.16a) yields the mold temperature field and the temperature gradient equations

$$T_m = T_o + (T_i - T_o)\operatorname{erfc}\left(\frac{-x}{2\sqrt{\alpha_m t}}\right) \quad \text{(mold)} \tag{9.17a}$$

$$\frac{\partial T_m}{\partial x} = \frac{T_i - T_o}{\sqrt{\pi \alpha_m t}} \exp\left(\frac{x^2}{4\alpha_m t}\right) \tag{9.17b}$$

$$\left(\frac{\partial T_m}{\partial x}\right)_{x=0} = \frac{T_i - T_o}{\sqrt{\pi \alpha_m t}} \tag{9.17c}$$

which is the equation describing an assumed uniform temperature gradient.

Solid Phase Similarly, the temperature equation and the boundary conditions for the solid phase are, respectively,

$$T_s = T(x,t)_s = D_1 + D_2 \operatorname{erf}\left(\frac{x}{2\sqrt{\alpha_s t}}\right) \quad \text{(solid)} \tag{9.18a}$$

$$\text{BC1} \; T_s = T_s(0,t) = T_i \tag{9.18b}$$

$$T_i = D_1 + D_2 \operatorname{erf}(0) \to D_1 = T_i \tag{9.18c}$$

$$\text{BC2} \; T_s = T_s[s(t),t] = T_f \tag{9.18d}$$

$$T_f = T_i + D_2 \operatorname{erf}\left(\frac{s(t)}{2\sqrt{\alpha_s t}}\right) = T_i + D_2 \operatorname{erf}(\lambda) \tag{9.18e}$$

$$D_2 = \left(T_f - T_i\right)/\operatorname{erf}(\lambda) \tag{9.18f}$$

Substituting D_1, D_2 into Eq. (9.18a) gives the temperature field equation, and subsequently, the rate of freezing and the temperature gradient equations

$$T_s = T_i + \frac{\left(T_f - T_i\right)}{\operatorname{erf}(\lambda)} \operatorname{erf}\left(\frac{x}{2\sqrt{\alpha_s t}}\right) \quad \text{(solid)} \tag{9.19a}$$

$$\frac{\partial T_s}{\partial t} = -\frac{\left(T_f - T_i\right)x}{2t\sqrt{\pi\alpha_s t}\,\operatorname{erf}(\lambda)} \exp\left(-\frac{x^2}{4\alpha_s t}\right) \tag{9.19b}$$

$$\frac{\partial T_s}{\partial x} = \frac{\left(T_f - T_i\right)}{\sqrt{\pi\alpha_s t}\,\operatorname{erf}(\lambda)} \exp\left(-\frac{x^2}{4\alpha_s t}\right) \tag{9.19c}$$

Evaluating $\partial T_s/\partial x$ at $x = 0$ and $x = s(t)$ and defining the dimensionless parameter λ and the rate of an evolved characteristic length $ds(t)/dt$ give a system of equations of the form

$$\left(\frac{\partial T_s}{\partial x}\right)_{x=0} = \frac{\left(T_f - T_i\right)}{\sqrt{\pi\alpha_s t}\,\operatorname{erf}(\lambda)} \tag{9.20a}$$

$$\left(\frac{\partial T_s}{\partial x}\right)_{s(t)} = \frac{\left(T_f - T_i\right)}{\sqrt{\pi\alpha_s t}\,\exp\left(\lambda^2\right)\operatorname{erf}(\lambda)} \quad \text{at } x = s(t) \tag{9.20b}$$

$$\lambda = \frac{s(t)}{2\sqrt{\alpha_s t}} \quad \& \quad s(t) = 2\lambda\sqrt{\alpha_s t} \tag{9.20c}$$

$$t = \frac{s(t)^2}{4\lambda^2 \alpha_s} \tag{9.20d}$$

$$\frac{ds(t)}{dt} = \lambda\sqrt{\frac{\alpha_s}{t}} \tag{9.20e}$$

$$\exp\left(\lambda^2\right)\operatorname{erf}(\lambda) = \frac{\left(T_f - T_i\right)}{\sqrt{\pi\alpha_s t}} \left[\left(\frac{\partial T_s}{\partial x}\right)_{s(t)}\right]^{-1} \tag{9.20f}$$

Notice that Eq. (9.20b) has been rearranged in the form of a transcendental equation defined by Eq. (9.20f) with constant temperature gradient $(\partial T_s / \partial x)_{s(t)}$.

Liquid Phase Similarly, the temperature equation and the boundary conditions for the liquid phase under superheating effects are, respectively,

$$T_l = T(x,t)_l = E_1 + E_2 \operatorname{erfc}\left(\frac{x}{2\sqrt{\alpha_l t}}\right) \quad \text{(liquid)} \tag{9.21a}$$

$$\text{BC1 } T_l = T_l(0,t) = T_p \tag{9.21b}$$

$$T_p = D_1 + C_2 \operatorname{erfc}(\infty) \rightarrow E_1 = T_p \tag{9.21c}$$

$$\text{BC2 } T_l = T_l[s(t),t] = T_f \tag{9.21d}$$

$$T_f = T_p + E_2 \operatorname{erfc}\left(\frac{s(t)}{2\sqrt{\alpha_l t}}\frac{\sqrt{\alpha_s}}{\sqrt{\alpha_s}}\right) = T_p + E_2 \operatorname{erfc}\left(\lambda\sqrt{\frac{\alpha_s}{\alpha_l}}\right) \tag{9.21e}$$

$$E_2 = \frac{(T_f - T_p)}{\operatorname{erfc}\left(\lambda\sqrt{\alpha_s/\alpha_l}\right)} \tag{9.21f}$$

Substituting the constants E_1, E_2 into Eq. (9.21a) yields the fully defined liquid-phase temperature field equation along with the temperature gradient equation evaluated at the location $s(t)$ of the L-S interface.

Thus,

$$T_l = T_p - \frac{(T_p - T_f)}{\operatorname{erfc}\left(\lambda\sqrt{\alpha_s/\alpha_l}\right)} \operatorname{erfc}\left(\frac{x}{2\sqrt{\alpha_l t}}\right) \quad \text{(liquid)} \tag{9.22a}$$

$$\frac{\partial T_l}{\partial t} = \frac{(T_p - T_f)\,x}{\sqrt{\pi \alpha_l t}\,\operatorname{erfc}\left(\lambda\sqrt{\alpha_s/\alpha_l}\right)} \exp\left(-\frac{x^2}{4\alpha_l t}\right) \tag{9.22b}$$

$$\left(\frac{\partial T_l}{\partial x}\right)_{s(t)} = -\frac{(T_p - T_f)}{\sqrt{\pi \alpha_l t}\,\operatorname{erfc}\left(\lambda\sqrt{\alpha_s/\alpha_l}\right)} \exp\left(-\frac{x^2}{4\alpha_s t}\frac{\alpha_s}{\alpha_l}\right) \tag{9.22c}$$

$$\left(\frac{\partial T_l}{\partial x}\right)_{s(t)} = -\frac{(T_p - T_f)\exp\left(-\lambda^2 \alpha_s/\alpha_l\right)}{\sqrt{\pi \alpha_l t}\,\operatorname{erfc}\left(\lambda\sqrt{\alpha_s/\alpha_l}\right)} \tag{9.22d}$$

Let the mold-solid interface temperature be T_i at $x = 0$, so that the heat flux balance or the conservation law of the heat fluxes gives T_i without any degree of superheating

$$q_m = q_s \tag{9.23a}$$

$$k_m\left(\frac{\partial T_m}{\partial x}\right)_{x=0} = k_s\left(\frac{\partial T_s}{\partial x}\right)_{x=0} \tag{9.23b}$$

$$\frac{k_m\,(T_i - T_o)}{\sqrt{\pi\,\alpha_m t}} = \frac{k_s\,(T_f - T_i)}{\sqrt{\pi\,\alpha_s t}\,\text{erf}\,(\lambda)} \tag{9.23c}$$

$$\gamma_m\,(T_i - T_o) = \frac{\gamma_s\,(T_f - T_i)}{\text{erf}\,(\lambda)} \tag{9.23d}$$

$$T_i = \frac{\gamma_s T_f + \gamma_m T_o\,\text{erf}\,(\lambda)}{\gamma_s + \gamma_m\,\text{erf}\,(\lambda)} \quad \text{(mold-solid)} \tag{9.23e}$$

$$T_i = \frac{\gamma_s T_p + \gamma_m T_o}{\gamma_s + \gamma_m} \quad \text{(mold-solid)} \tag{9.23f}$$

Inserting Eqs. (9.20b), (9.20e) and (9.22d) into (9.14f) and collecting dimensionless terms gives the general transcendental equation related to mold, solid, and liquid thermal resistances containing the thermal inertia (thermal effusivity) terms $\gamma_m, \gamma_s, \gamma_l$ (Dantzig and Rappaz [10, pp. 166–169])

$$\frac{S_t}{\sqrt{\pi}} = \left[\lambda \exp\left(\lambda^2\right) - \frac{\gamma_l}{\gamma_s}\frac{c_s\,(T_p - T_f)}{\sqrt{\pi}\,\Delta H_s} B\,(\lambda)\right]\left[\text{erf}\,(\lambda) + \frac{\gamma_s}{\gamma_m}\right] \tag{9.24a}$$

$$f\,(\lambda) = \left[\lambda \exp\left(\lambda^2\right) - \frac{\gamma_l}{\gamma_s}\frac{c_s\,(T_p - T_f)}{\sqrt{\pi}\,\Delta H_s} B\,(\lambda)\right]\left[\text{erf}\,(\lambda) + \frac{\gamma_s}{\gamma_m}\right] \tag{9.24b}$$

$$f\,(\lambda) = \frac{S_t}{\sqrt{\pi}} \tag{9.24c}$$

with

$$B\,(\lambda) = \frac{\exp\left[(1 - \alpha_s/\alpha_l)\,\lambda^2\right]}{\text{erfc}\,(\lambda\sqrt{\alpha_s/\alpha_l})} \tag{9.25a}$$

$$S_t = \frac{c_s\,(T_f - T_o)}{\Delta H_f} \tag{9.25b}$$

Appendix 9C illustrates the analytical procedure for deriving the transcendental expression Eq. (9.24a) as Eq. (9.125), which is confined to a two-region solidification domain as a Stefan condition for solidification with solid and liquid thermal resistances only.

Inserting Eqs. (9.20b) and (9.22d) into the heat flux (q_x) balance (9.14g) and integrating the resultant expression yield the solidification time t_q for a solidified thickness $s\,(t)$ under the assumption that solidification is controlled by heat conduction

$$t_q = \frac{\pi}{4}\left[\frac{\rho_s\,\Delta H_s\,\text{erf}\,(\lambda)\,\text{erfc}\,(\lambda\sqrt{\alpha_s/\alpha_l})}{B_1\,(\lambda) + B_2\,(\lambda)}\right]^2 s\,(t)^2 \tag{9.26a}$$

Table 9.1 Thermophysical data for pure aluminum (Dantzig and Rappaz [10, p. 170])

Material	T ($^\circ C$)	k ($W/m.K$)	ρ (Kg/m^3)	c_p ($J/Kg.K$)	ΔH_f (J/Kg)
Graphite mold	25	100	2200	1700	
Solid Al		211	2555	1190	3.98×10^5
Liquid Al	700	91	2368	1090	
$k \to 1\,W/m.K = 1\,W/m.{}^\circ C$			$dT/dx \to 1\,K/m = 1{}^\circ C/m$		
$c_p \to 1\,J/Kg.K = 1\,J/Kg.{}^\circ C$			$dx/dt \to 1\,K/s = -272\,{}^\circ C/s$		

$$B_1\left(\lambda\right) = \gamma_s \left(T_f - T_i\right) \exp\left(-\lambda^2\right) \operatorname{erfc}\left(\lambda\sqrt{\alpha_s/\alpha_l}\right) \tag{9.26b}$$

$$B_2\left(\lambda\right) = \gamma_l \left(T_p - T_f\right) \exp\left(-\lambda^2 \alpha_s/\alpha_l\right) \operatorname{erf}\left(\lambda\right) \tag{9.26c}$$

Continuing the analysis of solidification reveals the effect of superheating the liquid metal on the transcendental equation and other variables. For instance, the thermophysical data set for pure aluminum found in Dantzig and Rappaz book [10, p. 170] is intentionally adapted here for characterizing the effect of superheating on the solidification time as per Eqs. (9.26a) and (9.20d).

In this particular case, let the characteristic length and the pouring temperature range be $s\left(t\right) = 0.04\,m$ and $660\,{}^\circ C \leq T_p \leq 850\,{}^\circ C$. Now, use the selected data set listed in Table 9.1 to characterize the solidification process for pure aluminum at a macroscale.

Note that ΔH_f can be replaced by ΔH_s in Eq. (9.24a) in order to evaluate $f\left(\lambda\right)$ as a superheat-dependent dimensionless function as shown in Fig. 9.8a for $660\,{}^\circ C \leq T_p \leq 850\,{}^\circ C$. The $f\left(\lambda\right)$ curves are displaced downward as T_p increases (Fig. 9.8b). This implies that the magnitude of the pouring temperature influences the solidification process, since $f\left(\lambda\right)$ function significantly decreases with fixed λ. For instance, if $\lambda = 0.50$ in Fig. 9.8a, then $f\left(\lambda\right)$ approaches a small value. Moreover, increasing the melt T_p induces an increase in the latent heat of solidification ΔH_s during the solidification process.

The latent heat of solidification $\Delta H_s = f\left(T_p\right)$ shows a linear behavior (Fig. 9.8c) as per Eq. (9.15b). On the other hand, the solidification times, t_λ and t_q decrease with increasing transcendental parameter λ (Fig. 9.8d) and pouring temperature T_p. Notice that t_λ and t_q curves have negative slopes, but they deviate from each other very significantly at $T_p > T_f$. For convenience, Table 9.2 lists some results.

Solidification is rather a complex phase change process. For instance, Table 9.1 lists the pouring temperature T_p in ascending order, so that the reader captures the significance of this variable in the solidification field. As T_p increases, the solidification process becomes a more energetic source of energy production through the latent heat of solidification ΔH_s, which is released at the liquid-solid (L-S) interface. Thermodynamically, this means that the solidification process is a heat engine (HE) system.

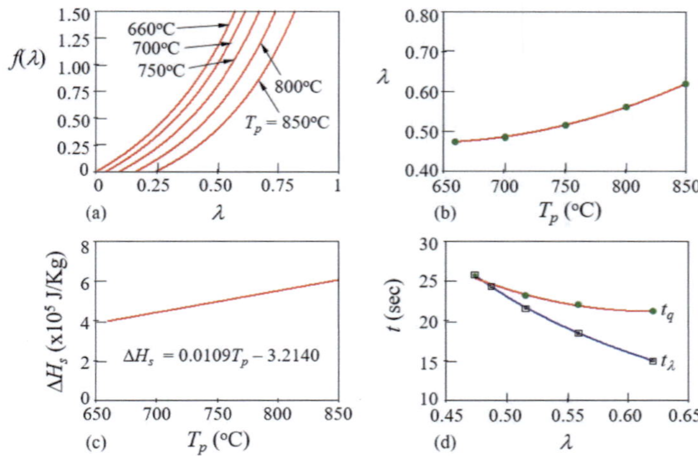

Fig. 9.8 Effect of melt superheating on solidification of pure aluminum. (**a**) The general transcendental equation, (**b**) the characteristic solution, (**c**) the latent heat of solidification (**d**) the solidification time. Pouring temperature range $660\,^\circ C \leq T_p \leq 850\,^\circ C$

Table 9.2 Results related to Fig. 9.8

T_p	ΔH_s		T_i	t_q	t_λ	
(°C)	$\left(\times 10^5 \, J/Kg\right)$	λ	(°C)	(sec)	(sec)	t_q/t_λ
660	3.98	0.47370	485.30	25.69	25.69	1.00
700	4.42	0.48715	482.25	24.29	24.28	1.00
750	4.96	0.51591	476.00	21.66	23.07	0.94
800	5.51	0.55926	467.21	18.43	22.19	0.83
850	6.05	0.62041	456.04	14.98	21.12	0.71

Notice that the solidification times t_q and t_λ decrease with increasing T_p because of the temperature difference between T_p and T_f; that is, $\Delta T = T_p - T_f$ and $\Delta T = T_p - T_i$. Actually, the L-S interface temperature T_i decreases with increasing T_p, and subsequently, the solidification heat transfer process is faster.

In this context, the final solidification product must be characterized using microscopy and mechanical testing procedures in order to confirm the heat transfer process reliability.

9.9   Mold and Solid Thermal Resistances

Consider a coupled mold-solid interface at $x = 0$ with an ideal planar L-S interface at T_i. Inserting Eq. (9.19c) at $x = s\,(t)$ into (9.14f) along with $\partial T_l/\partial x = 0$, Eqs. (9.4a) and (9.20c) yield the transcendental equation (Poirier and Geiger [6, p. 337])

$$\lambda \exp\left(\lambda^2\right) \text{erf}(\lambda) = \frac{c_s\left(T_f - T_i\right)}{\sqrt{\pi}\,\Delta H_f} \tag{9.27}$$

from which T_i takes a second definition

$$T_f - T_i = \sqrt{\pi}\,\lambda \exp\left(\lambda^2\right) \text{erf}(\lambda)\left(\frac{\Delta H_f}{c_s}\right) \tag{9.28a}$$

$$T_i = T_f - \sqrt{\pi}\,\lambda \exp\left(\lambda^2\right) \text{erf}(\lambda)\left(\frac{\Delta H_f}{c_s}\right) \tag{9.28b}$$

Combining Eqs. (9.26a) and (9.28a) T_i takes a third definition

$$T_i = T_o + \sqrt{\pi}\,\lambda \exp\left(\lambda^2\right) \frac{\Delta H_f}{c_s}\frac{\gamma_s}{\gamma_m} \tag{9.29a}$$

$$T_i = T_o + \sqrt{\pi}\,\lambda \exp\left(\lambda^2\right) \frac{\Delta H_f}{c_s}\sqrt{\frac{k_s\rho_s c_s}{k_m\rho_m c_m}} \tag{9.29b}$$

Substitute Eq. (9.29b) into Eq. (9.27) to get (Dantzig and Rappaz [10, p. 172])

$$\lambda \exp\left(\lambda^2\right)\left[\text{erf}(\lambda) + \sqrt{\frac{k_s\rho_s c_s}{k_m\rho_m c_m}}\right] = \frac{c_s\left(T_f - T_o\right)}{\sqrt{\pi}\,\Delta H_f} \tag{9.30a}$$

$$\lambda \exp\left(\lambda^2\right)\left[\text{erf}(\lambda) + \frac{\gamma_s}{\gamma_m}\right] = \frac{S_t}{\sqrt{\pi}} \tag{9.30b}$$

Once λ is numerically evaluated from Eq. (9.30a) or (9.30b), computations of variables follow.

Example 9.2 Assume that a 0.04-m thick cast iron slab is to be produced in a carbon steel mold and that there are temperature gradients in the mold and solid only. Use the thermophysical properties given below to calculate **(a)** the solidification time t, and the solidification velocity dx_s/dt, **(b)** the heat flux at $x = 0$, and **(c)** plot the temperature distribution at $-0.04 \leq x \leq 0\,m$ for the mold and $0 \leq x \leq 0.02\,m$ for the solid. Assume superheats 0, 100 and 200 °C. Cast iron data: $T_f = 1535\,°C$, $k_s = 73\,W/m.K$, $\rho_s = 7530\,Kg/m^3$, $c_s = 728\,J/Kg.K$ and $\Delta H_f = 247.20\,kJ/Kg$. Mold data: $T_o = 25\,°C$, $k_m = 35\,W/m.K$, $\rho_m = 7689\,Kg/m^3$ and $c_m = 669\,J/Kg.K$.

Solution

(a) Only one set of calculations is included below for $\Delta T_s = T_p - T_f = 100\,°C$. From Eq. (9.4a), the thermal diffusivity terms are

$$\alpha_s = \frac{k_s}{\rho_s c_s} = 1.3317 \times 10^{-5}\, m^2 s \tag{9.2E1a}$$

$$\alpha_m = \frac{k_m}{\rho_m c_m} = 6.8041 \times 10^{-6}\, m^2 s \tag{9.2E1b}$$

The latent heat of solidification due to a superheat $\Delta T_s = T_p - T_f$ with $c_l \simeq c_s$ is

$$T_p = 1635\,^\circ C = 1908\, K \tag{9.2E2a}$$

$$\Delta T_s = T_p - T_f = 100\,^\circ C = 100\, K \tag{9.2E2b}$$

$$\Delta H_s = \Delta H_f + c_l \left(T_p - T_f\right) = 247{,}200 + 728\,(100) \tag{9.2E2c}$$

$$\Delta H_s = 320{,}000\, J/Kg \tag{9.2E2d}$$

Replace ΔH_f by ΔH_s in Eq. (9.30a) to compensate for the superheat on solidification and solve for λ using iterations

$$\lambda \exp\left(\lambda^2\right)\left[\mathrm{erf}\,(\lambda) + \sqrt{\frac{k_s \rho_s c_s}{k_m \rho_m c_m}}\right] = \frac{c_s \left(T_f - T_o\right)}{\sqrt{\pi}\,\Delta H_s} \tag{9.2E3a}$$

$$\lambda \exp\left(\lambda^2\right)[\mathrm{erf}\,(\lambda) + 1.4909] = 1.9381 \tag{9.2E3b}$$

$$\lambda \exp\left(\lambda^2\right)(\mathrm{erf}\,(\lambda) + 1.4909) - 1.9381 = 0 \tag{9.2E3c}$$

$$\lambda = 0.62265 \tag{9.2E3d}$$

For comparison, the interface temperature is computed using Eqs. (9.28b) and (9.29b)

$$T_i = T_f - \sqrt{\pi}\lambda \left(\frac{\Delta H_s}{c_s}\right) \exp\left(\lambda^2\right) \mathrm{erf}\,(\lambda) = 1090.80\,^\circ C \tag{9.2E4a}$$

$$T_i = T_o + \sqrt{\pi}\lambda \frac{\Delta H_s}{c_s}\sqrt{\frac{k_s \rho_s c_s}{k_m \rho_m c_m}} \exp\left(\lambda^2\right) = 1090.80\,^\circ C \tag{9.2E4b}$$

From Eq. (9.20d) with $s\,(t) = 0.04/2 = 0.02\, m$, the solidification time t and the solidification velocity dx_s/dt are, respectively,

$$t = \frac{s\,(t)^2}{4\lambda^2 \alpha_s} = \frac{(0.02)^2}{4\,(0.62265)^2\,(1.3317 \times 10^{-5})} = 19.37\ \mathrm{sec} \tag{9.2E5a}$$

$$\frac{dx_s}{dt} = \lambda\sqrt{\frac{\alpha_s}{t}} = 5.16 \times 10^{-4}\, m/s = 1857.60\, mm/h \tag{9.2E5b}$$

(b) From Eq. (9.23c) at $x = 0$ and $\Delta T_s = 100\,^\circ C = 100\,K$, the heat flux flowing through the cast iron solid phase is calculated as

$$q_s = \frac{k_s\,(T_f - T_i)}{\mathrm{erf}\,(\lambda)\,\sqrt{\pi \alpha_s t}} = 1.83 \times 10^6\ W/m^2 \qquad (9.2\mathrm{E}6a)$$

$$q_s = 1.83\ MW/m^2 \qquad (9.2\mathrm{E}6b)$$

The heat flux crossing the mold-solid interface will eventually flow through the mold, and subsequently, it is dissipated into the local casting environment. Thus,

$$q_m = \frac{k_m\,(T_i - T_o)}{\sqrt{\pi \alpha_m t}} = 1.83 \times 10^6\ W/m^2 \qquad (9.2\mathrm{E}7a)$$

$$q_m = 1.83\ MW/m^2 \qquad (9.2\mathrm{E}7b)$$

According to the second law of thermodynamics, the entropy S of the local air in the casting environment increases due to this amount of heat transfer.

(c) From Eqs. (9.17a) and (9.19a), the temperature equations are, respectively,

$$T_m = T_o + (T_i - T_o)\,\mathrm{erfc}\left(\frac{-x}{2\sqrt{\alpha_m t}}\right) \quad \text{(mold)} \qquad (9.2\mathrm{E}8a)$$

$$T_m = 25\,^\circ C + (1065.80\,^\circ C)\,\mathrm{erfc}\,(-43.553x) \qquad (9.2\mathrm{E}8b)$$

$$T_s = T_i + \frac{(T_f - T_i)}{\mathrm{erf}\,(\lambda)}\,\mathrm{erf}\left(\frac{x}{2\sqrt{\alpha_s t}}\right) \quad \text{(solid)} \qquad (9.2\mathrm{E}8c)$$

$$T_s = 1090.80\,^\circ C + (714.79\,^\circ C)\,\mathrm{erf}\,(31.132x) \qquad (9.2\mathrm{E}8d)$$

Plotting these equations at $\Delta T_s = 0,\ 100$ and $200\,^\circ C$ reveals the effect of superheating the melt at a particular holding time, which is not included in the current solidification analysis. Note the opposite behavior of the temperature profile in the mold and solid.

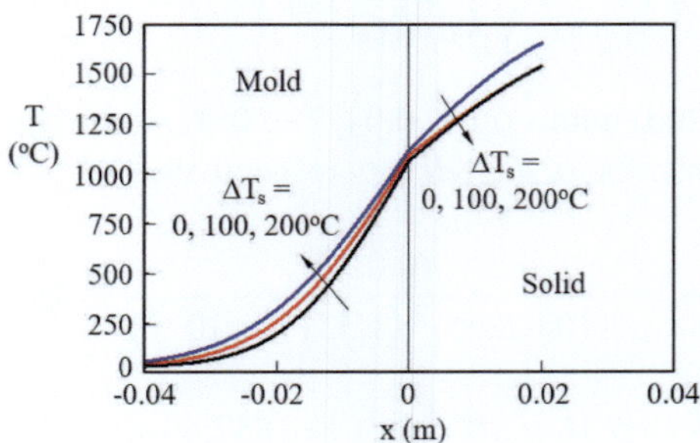

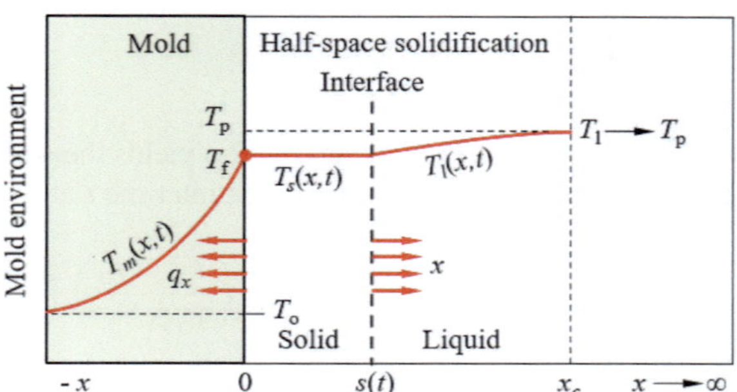

Fig. 9.9 Mold resistance and negligible solid and liquid thermal resistances

Therefore, the imposed superheating range $0 < \Delta T_s \leq 200\,^\circ C$ on the temperature fields in the steel mold and the solid (solidified melt) regions can be considered insignificant since $\Delta H_s \simeq \Delta H_f$. This may be the reason most authors use ΔH_f instead of ΔH_s in evaluating the solidification process of pure metals.

9.9.1   Sand Mold Thermal Resistance

Casting in sand molds is a conventional foundry practice, because it is cost-effective, and the molding sand can be recycled to an extent. Consider casting a metal slab into a low-thermal conductivity sand mold, and assume that mold resistance to heat transfer during solidification is controlled by the sand temperature gradient $\partial T_m/\partial x$. In this case, Fig. 9.9 schematically shows the corresponding temperature profiles for the mold, solid, and liquid phases. This implies that $(\partial T_m/\partial x) >> (\partial T_s/\partial x) \simeq (\partial T_l/\partial x) \simeq 0$. Neglect heat transfer by convection and radiation.

In this case, the governing heat equation, the continuity of temperature, and the heat flux balance are, respectively,

$$\frac{\partial T_m}{\partial t} = \alpha_m \frac{\partial^2 T_m}{\partial x^2} \quad \text{for} -\infty < x < 0, t > 0 \tag{9.31a}$$

$$T_s = T_l = T_f \quad \text{at } x = s\,(t) \tag{9.31b}$$

$$k_m \frac{\partial T_m}{\partial x} = \rho_s \Delta H_f \frac{ds\,(t)}{dt} \quad \text{at } x = s\,(t) \tag{9.31c}$$

The solution of Eq. (9.31a) is defined by (9.17a), but in this section, T_i is replaced by T_f so that

$$T_m = T_o + \left(T_f - T_o\right) \operatorname{erfc}\left(\frac{-x}{2\sqrt{\alpha_m t}}\right) \quad \text{(mold)} \tag{9.32a}$$

$$\frac{\partial T_m}{\partial x} = \frac{T_f - T_o}{\sqrt{\pi \alpha_m t}} \exp\left(\frac{-x}{2\sqrt{\alpha_m t}}\right) \tag{9.32b}$$

$$\left(\frac{\partial T_m}{\partial x}\right)_{x=0} = \frac{T_i - T_o}{\sqrt{\pi \alpha_m t}} \tag{9.32c}$$

Combining Eqs. (9.31c) and (9.32c) yields the solidification velocity, the characteristic length $s\,(t)$, and the solidification time t at $x = s\,(t)$ given by

$$\frac{ds\,(t)}{dt} = \frac{1}{\sqrt{\pi}} \left(\frac{T_f - T_o}{\rho_s \Delta H_f}\right) \left(\frac{k_m}{\sqrt{\alpha_m t}}\right) \tag{9.33a}$$

$$s\,(t) = \frac{2}{\sqrt{\pi}} \left(\frac{T_f - T_o}{\rho_s \Delta H_f}\right) \left(\frac{k_m}{\sqrt{\alpha_m}}\right) \sqrt{t} \tag{9.33b}$$

$$s\,(t) = \frac{2}{\sqrt{\pi}} \frac{\gamma_m \left(T_f - T_o\right)}{\rho_s \Delta H_f} \sqrt{t} \tag{9.33c}$$

$$t = \frac{\pi \alpha_m}{4} \left[\frac{\rho_s \Delta H_f}{k_m \left(T_f - T_o\right)}\right]^2 s\,(t)^2 \tag{9.33d}$$

$$t_2 = \frac{\pi}{4} \left[\frac{\rho_s \Delta H_s}{\gamma_m \left(T_i - T_o\right)}\right]^2 s\,(t)^2 \tag{9.33e}$$

The Amount of Thermal Energy According to Fourier's law, the heat flux q_x crossing the mold-solid interface and the rate of thermal energy dQ/dt through the mold are, respectively,

$$q_x = -k_m \left(\frac{\partial T}{\partial x}\right)_{x=0} \tag{9.34a}$$

$$\frac{dQ}{dt} = -A k_m \left(\frac{\partial T}{\partial x}\right)_{x=0} = \frac{A k_m \left(T_f - T_o\right)}{\sqrt{\pi \alpha_m t}} \tag{9.34b}$$

The heat transfer q_x denotes the amount of heat that crosses a unit area per unit time and A denotes the interface surface area. Integrating Eq. (9.34b) yields the amount of thermal energy through the mold in the negative x-direction

$$Q_t = \frac{A k_m \left(T_f - T_o\right)}{\sqrt{\pi \alpha_m}} \int_0^{t_s} \frac{dt}{\sqrt{t}} \tag{9.35a}$$

$$Q_t = \frac{2 A k_m \left(T_f - T_o\right)}{\sqrt{\pi \alpha_m}} \sqrt{t} \tag{9.35b}$$

For solidifying a liquid metal of mass m, the amount of thermal energy by conduction (Q_f) due to the release of latent heat of fusion ΔH_f is written as

$$Q_f = m \Delta H_f = \rho_s V \Delta H_f \tag{9.36}$$

where V is the casting volume. Equating Eqs. (9.35b) and (9.36) yields exactly (9.33b) in terms of volume-to-area ratio

$$s\left(t\right) = \frac{V}{A} = \frac{2}{\sqrt{\pi}} \left(\frac{T_f - T_o}{\rho_s \Delta H_f}\right) \left(\frac{k_m}{\sqrt{\alpha_m}}\right) \sqrt{t} \qquad (9.37)$$

which is a well-known equation in the foundry industry for convectional metal/sand molding and casting components.

9.9.2 Chvorinov's Rule

Again, $s\left(t\right)$ is the thickness of the solidified melt, which in turn represents the position of the L-S boundary (dual phase front) that has advanced into the liquid metal phase at a velocity defined as $ds\left(t\right)/dt$.

Solving Eq. (9.37) for t yields the Chvorinov's rule for a semi-infinite casting $(0 < x < \infty)$ (Poirier and Geiger [6, p. 331–332], Gaskell [7, p. 404])

$$t = \frac{\pi \alpha_m}{4} \left[\frac{\rho_s \Delta H_f}{k_m \left(T_f - T_o\right)}\right]^2 \left(\frac{V}{A}\right)^2 \qquad (9.38a)$$

$$t = B\left(M\right)^2 = B\left(\frac{V}{A}\right)^2 = Bs\left(t\right)^2 \qquad (9.38b)$$

$$B = \frac{\pi \alpha_m}{4} \left[\frac{\rho_s \Delta H_f}{k_m \left(T_f - T_o\right)}\right]^2 \qquad (9.38c)$$

where B is the Chvorinov's mold constant, the mold constant, or casting modulus (Stefanescu [8, p. 99]). Notice that the solidification time defined by Eq. (9.38b) depends on the *casting modulus M*, which is basically the characteristiclength $s\left(t\right)$.

According to Wlodawer [11, pp. 5–6], the casting modulus for rectangular, cylindrical, and spherical geometries can also be defined by

$$M = \frac{V}{A} = s\left(t\right) = \frac{A}{P} = \frac{\text{Cross-sectional area}}{\text{Perimeter of cross section}} \qquad (9.39a)$$

where V is the volume of a casting and A is the contact surface area of the metal-mold interface. According to Eq. (9.39a), the casting modulus is a ratio between the casting volume and its heat-emitting surface area. This ratio can be treated as the L-S interface location $s(t)$.

Rectangular Casting with Thickness x = a

$$s\left(t\right) = \frac{V}{A} = \frac{abc}{2bc} = \frac{a}{2} \tag{9.39b}$$

Cylindrical Casting with Radius r = a

$$s\left(t\right) = \frac{V}{A} = \frac{\pi r^2 c}{2\pi r c} = \frac{r}{2} \tag{9.39c}$$

Spherical Casting with Radius r = a

$$s\left(t\right) = \frac{V}{A} = \frac{\left(4/3\right)\pi r^3}{4\pi r^2} = \frac{r}{3} \tag{9.39d}$$

According to the Chvorinov's rule, solidification is considered complete when the casting modulus reaches the position of the characteristic length; $s\left(t\right) = M$, which denotes half the thickness of a rectangular casting geometry or the radius of a cylinder and sphere.

In fact, the Chvorinov's rule is subjected to theory-experiment discrepancy (Poirier and Geiger [6, p. 332], Flinn [12, p. 37]), but it relates the solidification phenomena to practical applications in the foundry industry for producing or casting pure metals and their alloys.

Example 9.3 Consider a 4-*cm* cubic sand-mold cavity being filled with a hypothetical alloy at T_p and assume that solidification is complete in 3 *minutes*. Based on this information, calculate the solidification time as per Chvorinov's rule for casting a 2-*cm* radius and 6-*cm* long cylinder and for a sphere with a 2-*cm* radius.

Solution From Eq. (9.38b) with $a = 4\,cm$,

$$t_{cube} = 3\,min \text{ and } s\left(t\right)_{cube} = V/A = a/6 = 0.33\,cm \tag{9.3E1a}$$

$$t_{cube} = B\left(\frac{V}{A}\right)^2 = B\left(\frac{a^3}{6a^2}\right)^2 = \frac{a^2 B}{36} = Bs\left(t\right)^2 \tag{9.3E1b}$$

$$B = \frac{36 t_{cube}}{a^2} = \frac{36\left(3\,\text{min}\right)}{\left(4\,cm\right)^2} = 6.75\ \text{min}\,/cm^2 \tag{9.3E1c}$$

For a cylinder,

$$s\left(t\right)_{cylinder} = \frac{V}{A} = \frac{r}{2} = 1\,cm \tag{9.3E2a}$$

$$t_{cylinder} = Bs\,(t)^2 = \left(6.75\,\frac{min}{cm^2}\right)(1\,cm)^2 = 6.75\,min \qquad (9.3E2b)$$

For a sphere,

$$s\,(t)_{sphere} = \frac{V}{A} = \frac{r}{3} = \frac{2}{3} \simeq 0.67\,cm \qquad (9.3E3a)$$

$$t_{sphere} = Bs\,(t)^2 = \left(6.75\,\frac{min}{cm^2}\right)\left(\frac{2}{3}\,cm\right)^2 = 3\,min \qquad (9.3E3b)$$

Therefore, the cylinder solidifies slower because $t_{cube} = t_{sphere} < t_{cylinder}$.

Example 9.4 Consider casting a pure copper (Cu) slab $(4 \times 6 \times 10\,cm^3)$ into a sand mold. Assume the effect of superheat $\Delta T_s = 10\,°C$ and use the data given (Appendix 9A) to calculate **(a)** the total solidification time t, **(b)** the solidification velocity $ds\,(t)\,/dt$, **(c)** the heating rate $\partial T_m/\partial t$ and mold thermal gradient $(\partial T_m/\partial x)_{x=0}$, **(d)** the amount of thermal energy Q and heat flux q_x through the mold, and **(e)** the Chvorinov's constant B. Assume that the mold external temperature remains at $25\,°C$. From Appendix 9A,

Copper (Cu)
$T_f = 1356\,K$
$\rho_s = 7525\,Kg/m^3$
$c_s = 429\,J/Kg.K$
$k_s = 400\,W/m.K$
$\Delta H_f = 205{,}000\,J/Kg$

Sand mold
$T_o = 298\,K$
$\rho_m = 1550\,Kg/m^3$
$c_m = 986\,J/Kg.K$
$k_m = 0.40\,W/m.K$

Unit conversion:

$$k_s, k_m \rightarrow 1\,W/m.K = 1\,W/m.°C \qquad dT/dx \rightarrow 1\,K/m = 1\,°C/m$$
$$c_s, c_m \rightarrow 1\,J/Kg.K = 1\,J/Kg.°C \qquad dx/dt \rightarrow 1\,K/s = -272\,°C/s$$

Solution

(a) Pouring temperature: $T_p = T_f + \Delta T_s = 1083 + 10 = 1093\,°C = 1366\,K$. The thermal diffusivity, Eq. (9.4a), is

$$\alpha_m = \frac{k_m}{\rho_m c_m} = \frac{0.40\,W/m.K}{\left(1550\,Kg/m^3\right)\left(986\,J/Kg.K\right)} = 2.6173 \times 10^{-7}\,m^2/s$$

$$(9.4E1)$$

For $x = 0.04\,m$ and $s\,(t) = x/2 = 0.02\,m$, use the effect of superheating so that Eqs. (9.15b) and (9.33d) give

$$\Delta H_s = \Delta H_f + c_s \Delta T_s = 205000 + (429)\,(10) = 209{,}290\,J/Kg$$

$$\tag{9.4E2a}$$

$$t = \frac{\pi \alpha_m}{4} \left[\frac{\rho_s \Delta H_s}{k_m \left(T_f - T_o\right)} \right]^2 s\,(t)^2 = 1138.70 \text{ sec} \tag{9.4E2b}$$

$$t \simeq 19 \text{ min} = 0.32\,h \tag{9.4E2c}$$

(b) From Eq. (9.33a), the solidification velocity is

$$\frac{ds\,(t)}{dt} = \frac{1}{\sqrt{\pi}} \left(\frac{T_f - T_o}{\rho_s \Delta H_s} \right) \left(\frac{k_m}{\sqrt{\alpha_m t}} \right) = 8.78 \times 10^{-6}\,m/s = 31.61\,mm/h$$

$$\tag{9.4E4}$$

This is a low solidification velocity.

(c) From Eq. (9.32a) and (9.32c),

$$\frac{\partial T_m}{\partial t} = \frac{\left(T_f - T_o\right) x}{2t \sqrt{\pi \alpha_m t}} \exp\left(-\frac{x^2}{4\alpha_m t} \right) = 0.16\,K/s \tag{9.4E5a}$$

$$\left(\frac{\partial T_m}{\partial x} \right)_{x=0} = -\frac{\left(T_f - T_o\right)}{\sqrt{\pi \alpha_m t}} = -34{,}576\,K/m \tag{9.4E5b}$$

(d) For $A = 2bc = 2\,(6 \times 10) = 120\,cm^2 = 0.012\,m^2$, Eqs. (9.35b) and (9.34a) give

$$Q = \frac{2Ak_m \left(T_f - T_o\right)}{\sqrt{\pi \alpha_m}} \sqrt{t} = 3.78 \times 10^5\,J = 378\,kJ \tag{9.4E6a}$$

$$q_m = -k_m \left(\frac{\partial T_m}{\partial x} \right)_{x=0} = 13{,}830\,W/m^2 \simeq 13.83\,kW/m^2 \tag{9.4E6b}$$

(e) From Eq. (9.38c), the Chvorinov's constant is

$$B = \frac{\pi \alpha_m}{4} \left[\frac{\rho_s \Delta H_s}{k_m \left(T_f - T_o\right)} \right]^2 = 2.8468 \times 10^6\,s/m^2 = 4.7447 \text{ min}/cm^2$$

$$\tag{9.4E7}$$

Using Eq. (9.38b) with $s\,(t) = 2\,cm$ gives

$$t_{slab} = Bs\,(t)^2 = \left(4.7447 \text{ min}/cm^2\right)(2\,cm)^2 \tag{9.4E8}$$

$$t_{slab} = 19 \text{ min}$$

Therefore, Eq. (9.38b) verifies that the solidification time is 19 min as per Chvorinov's rule.

9.9.3 Solid Thermal Resistance

For solidification of a pure metal in a **water-cooled (chill) metal mold**, a high thermal resistance develops in the solid (solidified melt) due to a temperature field profile $T_s(x, t)$ schematically shown in Fig. 9.7. In this particular case, the solid thermal resistance R_s predominates over other thermal resistances during solidification because $\partial T_m/\partial x \simeq \partial T_l/\partial x \to 0$ and $\partial T_s/\partial x >> 0$.

For a half-space region $x > 0$, the set of equations related to this thermal resistance case and to the conservation law of the heat fluxes are

$$\frac{\partial T_s}{\partial t} = \alpha_s \frac{\partial^2 T_s}{\partial x^2} \quad \text{for } -\infty < x < 0, t > 0 \tag{9.40a}$$

$$T_s = T_l = T_f \quad \text{at } x = s(t) \tag{9.40b}$$

$$-k_s \left(\frac{\partial T_s}{\partial x}\right)_{s(t)} = \rho_s \Delta H_f \frac{ds(t)}{dt} \quad \text{at } x = s(t) = 2\lambda\sqrt{\alpha_s t} \tag{9.40c}$$

The solution of Eq. (9.40a) is defined by (9.19a), which is recast here with $T_i = T_o$

$$T_s = T_o + \frac{(T_f - T_o)}{\operatorname{erf}(\lambda)} \operatorname{erf}\left(\frac{x}{2\sqrt{\alpha_s t}}\right) \tag{9.41}$$

Moving Boundary (MB) Theory The general moving $L\text{-}S$ boundary (MB) (known as free boundary, melting front or freezing front) theory is now used to model solidification. Accordingly, this is a two-phase subsystem within the solidification domain Γ that embraces the mold material and the casting geometry. Moreover, the $L\text{-}S$ boundary position $s(t) \to x_c$ is driven by a liquid temperature gradient (no supercooling below T_f). For convenience, the freezing temperature T_f is considered herein as the $L\text{-}S$ boundary temperature during the solidification process.

From Eq. (9.41) at $x = s(t)$,

$$\left(\frac{\partial T_s}{\partial x}\right)_{s(t)} = -\frac{(T_f - T_o)}{\sqrt{\pi \alpha_s t} \exp(\lambda^2) \operatorname{erf}(\lambda)} \tag{9.42}$$

Upon inserting Eq. (9.42) into (9.40c) and collecting terms yields the transcendental equation similar to Eq. (9.12)

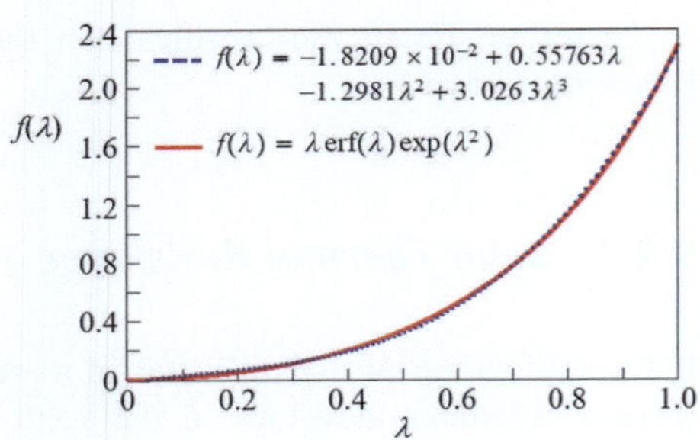

Fig. 9.10 Dimensionless transcendental nonlinear function $f(\lambda)$

$$\lambda \exp\left(\lambda^2\right) \operatorname{erf}(\lambda) = \frac{c_s\left(T_f - T_o\right)}{\sqrt{\pi}\,\Delta H_f} = \frac{S_t}{\sqrt{\pi}} \tag{9.43a}$$

$$f(\lambda) = \lambda \exp\left(\lambda^2\right) \operatorname{erf}(\lambda) \tag{9.43b}$$

Plotting Eq. (9.43b) yields a nonlinear trendline shown in Fig. 9.10. Once the right-hand side of Eq. (9.43a) is known, λ can be evaluated numerically or graphically. Subsequently, additional calculations follow in order to fully characterized the solidification process using conventional casting procedures.

Quasi-static Approximation (QSA) Theory Since one-dimensional solidification in conventional casting practices is a slow process, it seems appropriate to introduce quasi-static approximation (QSA) theory as a classical Stefan problem. First of all, the solution characteristic factor λ can be approximated using the exponential and error functions as power series

$$\exp\left(\lambda^2\right) = 1 + \frac{\lambda^2}{1!} + \frac{\lambda^4}{2!} + \frac{\lambda^6}{3!} + \ldots \simeq 1 \tag{9.44a}$$

$$\operatorname{erf}(\lambda) = \frac{2}{\sqrt{\pi}}\left(\lambda - \frac{\lambda^3}{3 \times 1!} + \frac{\lambda^5}{5 \times 2!} \ldots\right) \simeq \frac{2\lambda}{\sqrt{\pi}} \tag{9.44b}$$

$$\lambda \exp\left(\lambda^2\right) \operatorname{erf}(\lambda) \simeq \frac{2\lambda^2}{\sqrt{\pi}} \quad \text{for small } \lambda \tag{9.44c}$$

The significance of this approximation procedure is illustrated below for a one-dimensional solidification process of a binary alloy without the effects of an electric field and mechanical vibrations.

Combining Eq. (9.43a) and (9.44c) yields the solution characteristic λ as

$$\lambda = \sqrt{\frac{S_t}{2}} = \sqrt{\frac{c_s\left(T_f - T_o\right)}{2\Delta H_f}} \tag{9.45}$$

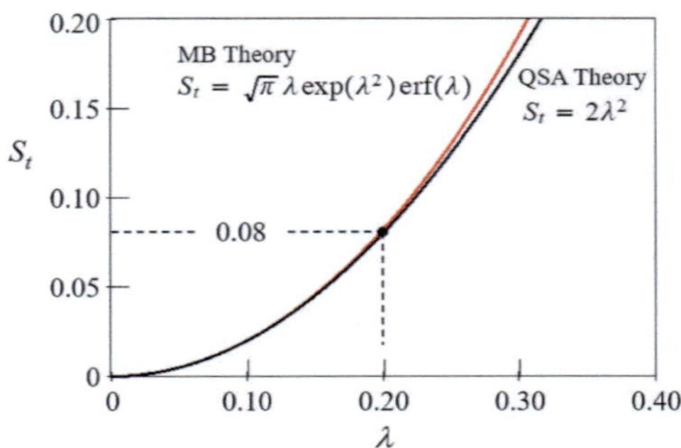

Fig. 9.11 Comparison of the moving boundary (MB) and the quasi-static approximation (QSA) theories for the Stefan number S_t

which strongly dependents on the release rate of the latent heat of solidification and the temperature difference $\Delta T = T_f - T_o$.

From Eq. (9.45), the Stefan number S_t and the solidification time t can be stated as

$$S_t = 2\lambda^2 \tag{9.46a}$$

$$t = \frac{s\,(t)^2}{2\alpha_s}\frac{1}{S_t} = \frac{s\,(t)^2}{2\alpha_s}\frac{\Delta H_f}{c_s\left(T_f - T_o\right)} \tag{9.46b}$$

Plotting Eqs. (9.43b) and (9.46a) reveals a nonlinear trendline as shown in Fig. 9.11. According to the scale used in Fig. 9.11, the Stefan number S_t, as defined by the MB and QSA theories, compare at $\lambda \leq 0.20$ since $S_t < 0.1$. For $S_t > 0.1$, the QSA theory is considered inadequate for describing the heat transfer process during unidirectional solidification (Glicksman [9, pp. 91–92]).

The objective now is to identify the significance of MB and QSA theories and rank them in order of importance for predicting the solidification time. It turns out that if $\lambda \leq 0.2$, these theories are in agreement, and the use of either theory is justified as long as $S_t \to 0$ and $\lambda \to 0$. Accordingly, the point $(S_t, \lambda) = (0.2, 0.08)$ in Fig. 9.11 serves as an upper limit for these theories to be in agreement.

An electric analogy with Ohm's law to heat flow by conduction used in solidification dictates that the temperature range ΔT is also a driving force for phase transformation $(L \to S)$ being restricted by an internal resistance R_x written as (Gaskell [7, p. 243])

$$R_x = \frac{\Delta T}{q_{s(t)}} = \frac{T_f - T_o}{q_{s(t)}} \tag{9.47}$$

where the heat flux crossing the L-S interface located at $s\,(t)$ is

$$q_{s(t)} = -k_s\left(\frac{dT_s}{dx}\right)_{s(t)} \tag{9.48a}$$

$$q_{s(t)} = \frac{k_s \left(T_f - T_o\right)}{\sqrt{\pi \alpha_s t} \, \exp\left(\lambda^2\right) \operatorname{erf}(\lambda)} \tag{9.48b}$$

Example 9.5 Use the moving boundary (MB) theory to evaluate the solidification of a $0.04\,m$-thick copper slab into a carbon steel mold. Initially, the copper melt is at a pouring temperature of $1093\,^{\circ}C$ and the mold is at $25\,^{\circ}C$. Calculate **(a)** the Stefan number S_t, **(b)** the solidification time t and the solidification velocity $ds\,(t)\,/dt$, **(c)** the temperature gradient $(\partial T/\partial x)_{x=0}$ and **(d)** the heat flux $q_{s(t)}$. Compare results using the QSA theory. Use the dataset for pure copper (Cu) from Appendix 9A.

Copper (Cu)
$T_f = 1083^o$
$\rho_s = 7525\,Kg/m^3$
$c_s = 429\,J/Kg.K$
$k_s = 400\,W/m.K$
$\Delta H_f = 205{,}000\,J/Kg$

Carbon steel mold
$T_o = 25\,^{\circ}C\,K$
$\rho_m = 7689\,Kg/m^3$
$c_m = 669\,J/Kg.K$
$k_m = 35\,W/m.K$

Solution
Moving Boundary (MB) Theory   The copper melt is superheated by pouring temperature is

$$\Delta T_s = T_p - T_f = 1093\,^{\circ}C - 1083\,^{\circ}C = 10\,^{\circ}C = 10\,K \tag{9.5E1}$$

From Eq. (9.4a), the thermal diffusivity terms for the mold and the solid phase are

$$\alpha_m = \frac{k_m}{\rho_m c_m} = 2.6173 \times 10^{-7}\,m^2/s \tag{9.5E2a}$$

$$\alpha_s = \frac{k_s}{\rho_s c_s} = 1.2391 \times 10^{-4}\,m^2/s \tag{9.5E2b}$$

(a)   For a half-space region, $x = 0.04\,m$ and $s\,(t) = x/2 = 0.02\,m$ and Eq. (9.15b) with compensated superheat effect gives

$$\Delta H_s = \Delta H_f + c_s \Delta T_s = 205000 + (429)\,(10) = 209{,}290\,J/Kg \tag{9.5E3}$$

From Eq. (9.30a), (9.30b),

$$S_t = \frac{c_s \left(T_f - T_o\right)}{\Delta H_s} = \sqrt{\pi}\,\lambda \exp\left(\lambda^2\right) \operatorname{erf}(\lambda) = 2.17 \tag{9.5E4}$$

(b) From Eq. (9.43a), the transcendental equation gives the characteristic length as

$$\lambda \exp\left(\lambda^2\right) \mathrm{erf}\left(\lambda\right) = \frac{c_s\left(T_f - T_o\right)}{\sqrt{\pi}\,\Delta H_s} = 1.2235 \tag{9.5E5a}$$

$$\lambda \exp\left(\lambda^2\right) \mathrm{erf}\left(\lambda\right) - 1.2235 = 0 \tag{9.5E5b}$$

$$\lambda = 0.82287$$

From Eqs. (9.20d) and (9.20e), respectively, the solidification time and the solidification velocity are, respectively,

$$t = \frac{s\left(t\right)^2}{4\lambda^2 \alpha_s} = \frac{\left(0.02\right)^2}{4\left(0.82287\right)^2\left(1.2391 \times 10^{-4}\right)} = 1.19 \ \mathrm{sec} \tag{9.5E6a}$$

$$\frac{ds\left(t\right)}{dt} = \lambda\sqrt{\frac{\alpha_s}{t}} = \left(0.82287\right)\sqrt{\frac{1.2391 \times 10^{-4}}{1.19}} = 8.40 \times 10^{-3}\,m/\,\mathrm{sec} \tag{9.5E6b}$$

Combining Eq. (9.40c) and (9.42) yields

$$\frac{ds\left(t\right)}{dt} = \left[\frac{1}{\sqrt{\pi \alpha_s t}\,\exp\left(\lambda^2\right) \mathrm{erf}\left(\lambda\right)}\right]\left[\frac{k_s\left(T_f - T_o\right)}{\rho_s \Delta H_s}\right] \tag{9.5E7a}$$

$$\frac{ds\left(t\right)}{dt} = 8.40 \times 10^{-3}\,m/\,\mathrm{sec} \tag{9.5E7b}$$

(c) From Eq. (9.42), the solid temperature gradient is

$$\left(\frac{\partial T_s}{\partial x}\right)_{s(t)} = -\frac{\left(T_f - T_o\right)}{\sqrt{\pi \alpha_s t}\,\mathrm{erf}\left(\lambda\right)}\exp\left(-\lambda^2\right) = -34{,}534\ K/m \tag{9.5E8a}$$

$$\left(\frac{\partial T_s}{\partial x}\right)_{s(t)} = -33{,}060\,^\circ C/m \tag{9.5E8b}$$

(d) The heat flux density at the *L-S* interface, $s\left(t\right) = 0.02\,m$, is calculated using Eq. (9.48b)

$$q_{s(t)} = \frac{k_s\left(T_f - T_o\right)}{\sqrt{\pi \alpha_s t}\,\exp\left(\lambda^2\right) \mathrm{erf}\left(\lambda\right)} = 1.322 \times 10^7\ W/m^2 \tag{9.5E9a}$$

$$q_{s(t)} = 13.22\ MW/m^2 \tag{9.5E9b}$$

or

$$q_{s(t)} = \rho_s \Delta H_s \frac{ds\,(t)}{dt} = (7525)\,(209290)\left(8.40 \times 10^{-3}\right) \qquad (9.5E10a)$$

$$q_{s(t)} = 13.22\,MW/m^2 \qquad (9.5E10b)$$

Quasi-static Approximation (QSA) Theory   Using Eq. (9.45) yields

$$\lambda = \sqrt{\frac{c_s\left(T_f - T_o\right)}{2\Delta H_s}} = 1.0413 \qquad (9.5E11)$$

and from Eq. (9.46a), $S_t = 2\lambda^2 = 2\,(1.0413)^2 = 2.17$. Then, Eqs. (9.20d), (9.20e) give

$$t = \frac{s\,(t)^2}{4\lambda^2 \alpha_s} = \frac{(0.02)^2}{4\,(1.5494)^2\left(1.1 \times 10^{-5}\right)} = 0.74\ \text{sec} \qquad (9.5E12a)$$

$$\frac{ds\,(t)}{dt} = \lambda\sqrt{\frac{\alpha_s}{t}} = (1.0413)\sqrt{\frac{1.2391 \times 10^{-4}}{0.74}} \qquad (9.5E12b)$$

$$\frac{ds\,(t)}{dt} = 1.35 \times 10^{-2}\,m/\text{sec} \qquad (9.5E12c)$$

Therefore, $t\,(MB) > t\,(QSA)$ and $[ds\,(t)/dt]_{MB} < [ds\,(t)/dt]_{QSA}$ are different, but the conventional Stefan number remains the same; $S_t = 2.17$.

During solidification, a phase transformation (L-S) imparts a two-phase boundary motion. For this reason, the evolved latent heat of solidification ΔH_s is justifiably a thermal energy in motion, and the theory of heat transfer sets the mathematical background for determining the solidification parameters defined above. Moreover, heat transfer during a casting process is an important parameter for quality casting products.

Example 9.6 Consider pouring an amount of copper melt at $1183\,^\circ C$ into a cooled metal mold having a slab geometry 0.08-m thick. Model the solidification process as a stable and planar L-S boundary that moves into the melt very slowly in the half-space region $x > 0$ until it reaches a location $x = s\,(t) = 0.04\,m$. **Case 1** Assume that a thermal conduction resistance to heat transfer develops in the solidified melt and that the melt cools to $T_f = 1083\,^\circ C$ and solidifies with an insignificant temperature gradient (no supercooling $\Delta T_s = \left(T_p - T_f\right) \to 0$). **Case 2** Now, assume that the melt undergoes a significant superheating $\Delta T_s = \left(T_p - T_f\right) >> 0$, and as result, solid and liquid temperature gradients develop during the solidification process. For both parts of the solidification problems, determine **(a)** the solidification time t and the rate of solidification $ds\,(t)/dt$, **(b)** the temperature gradients $(\partial T_s/\partial x)_{s(t)}$ and $(\partial T_l/\partial x)_{s(t)}$, **(c)** the heat fluxes and **(d)**

plot the temperature field equations $T_s(x, t)$ and $T_l(x, t)$ when $s(t) = 0.03\,m$. Use the data for pure copper from Appendix 9A.

Solution From Appendix 9A,

Mold	Pure copper	
$T_o = 25\,°C$	$T_p = 1183\,°C$	$T_f = 1083\,°C$
$\rho_m = 7689\,Kg/m^3$	$\rho_l = 8960\,Kg/m^3$ at T_f	$\rho_s = 8033\,Kg/m^3$
$c_m = 669\,J/Kg.K$	$c_l = 118\,J/Kg.K$	$c_s = 92\,J/Kg.K$
$k_m = 35\,W/m.K$	$k_l = 163\,W/m.K$	$k_s = 130\,W/m.K$
	$\Delta H_f = 209{,}000\,J\,Kg$	

From Eq. (9.4a), the thermal diffusivity terms are

$$\alpha_m = \frac{k_m}{\rho_m c_m} = 6.8041 \times 10^{-6}\,m^2 s \tag{9.6E1a}$$

$$\alpha_s = \frac{k_s}{\rho_s c_s} = 1.7590 \times 10^{-4}\,m^2 s \tag{9.6E1b}$$

$$\alpha_l = \frac{k_l}{\rho_l c_l} = 1.5417 \times 10^{-4}\,m^2 s \tag{9.6E1c}$$

The latent heat of solidification, Eq. (9.15b), due to a superheat $\Delta T_s = T_p - T_f$ is

$$T_p = 1183\,°C = 1456\,K \tag{9.6E2a}$$

$$\Delta T_s = T_p - T_f = 100\,°C = 100\,K \tag{9.6E2b}$$

$$\Delta H_s = \Delta H_f + c_l\left(T_p - T_f\right) = 209{,}000 + 728\,(1183 - 1083) \tag{9.6E2c}$$

$$\Delta H_s = 281{,}800\,J/Kg$$

The use of latent heat of solidification ΔH_s compensates for the superheating effect on the solidification process. First of all, evaluate the dimensionless parameter known as the solution characteristic λ using Eq. (9.30a). This is accomplished by using an iterative procedure (numerical analysis) based on an initial guess. For clarity, an iteration procedure is a repeated application of a function, where the output of each numerical step is used as the input for the next iteration until a controlling parameter is met. Moreover, Newton's method is a very common numerical procedure for iterating a function or finding a suitable solution of a function.

Thus,

$$\lambda \exp\left(\lambda^2\right)\left[\operatorname{erf}(\lambda) + \sqrt{\frac{k_s \rho_s c_s}{k_m \rho_m c_m}}\right] = \frac{c_s\left(T_f - T_o\right)}{\sqrt{\pi}\,\Delta H_s} \tag{9.6E3a}$$

$$\lambda \exp\left(\lambda^2\right) \left[\operatorname{erf}(\lambda) + 0.7305\right] = 0.19488 \tag{9.6E3b}$$

$$\lambda \exp\left(\lambda^2\right) \left(\operatorname{erf}(\lambda) + 0.7305\right) - 0.19488 = 0 \tag{9.6E3c}$$

$$\lambda = 0.19726 \tag{9.6E3d}$$

From Eq. (9.28b), the interface temperature T_i at $x = 0$ between the mold and solid interface is approximated as

$$T_i = T_f - \sqrt{\pi}\,\lambda \exp\left(\lambda^2\right) \operatorname{erf}(\lambda) \left(\frac{\Delta H_s}{c_s}\right) \tag{9.6E4a}$$

$$T_i = 838.34\,^{\circ}C \tag{9.6E4b}$$

Case 1 No Supercooling Effect of the Melt

(a) From Eq. (9.43a),

$$\lambda \exp\left(\lambda^2\right) \operatorname{erf}(\lambda) - \frac{c_s\left(T_f - T_o\right)}{\sqrt{\pi}\,\Delta H_s} = 0 \tag{9.6E5a}$$

$$\lambda \exp\left(\lambda^2\right) \operatorname{erf}(\lambda) - 0.19488 = 0 \tag{9.6E5b}$$

$$\lambda = 0.39438 \tag{9.6E5c}$$

From Eqs. (9.20d), (9.20e) with $s(t) = 0.03\,m$,

$$t = \frac{s(t)^2}{4\lambda^2\alpha_s} = \frac{(0.03)^2}{4\,(0.39438)^2\left(1.7590 \times 10^{-4}\right)} = 8.22 \text{ sec} \tag{9.6E6a}$$

$$\frac{ds(t)}{dt} = \lambda\sqrt{\frac{\alpha_s}{t}} = (0.39438)\sqrt{\frac{1.7590 \times 10^{-4}}{8.22}} \tag{9.6E6b}$$

$$\frac{ds(t)}{dt} = 1.82 \times 10^{-3}\,m/\sec \tag{9.6E6c}$$

Combining Eq. (9.40c) and (9.42) yields the rate of solidification at the planar L-S interface

$$\frac{ds(t)}{dt} = \left[\frac{1}{\sqrt{\pi\alpha_s t}\,\exp\left(\lambda^2\right)\operatorname{erf}(\lambda)}\right]\left[\frac{k_s\left(T_f - T_o\right)}{\rho_s\,\Delta H_s}\right] \tag{9.6E7a}$$

$$\frac{ds(t)}{dt} = 1.82 \times 10^{-3}\,m/\sec \tag{9.6E7b}$$

(b) From Eq. (9.42), the temperature gradients in the solid phase at $s\,(t) = 0.03\,m$ are

$$\left(\frac{\partial T_s}{\partial x}\right)_{s(t)} = -\frac{\left(T_f - T_o\right)}{\sqrt{\pi \alpha_s t}\,\exp\left(\lambda^2\right)\,\mathrm{erf}\,(\lambda)} = -31{,}767\,K/m = -31{,}767\,^\circ C/m$$

$$(9.6\mathrm{E}8\mathrm{a})$$

$$\left(\frac{\partial T_l}{\partial x}\right)_{s(t)} \to 0 \quad \text{(assumed)} \qquad\qquad (9.6\mathrm{E}8\mathrm{b})$$

(c) From Eqs. (9.48a), (9.48b), the heat flux balance $(q_s = q_f)$ at $s\,(t) = 0.03\,m$, is

$$q_s = -k_s\left(\frac{\partial T_s}{\partial x}\right)_{s(t)} = \frac{k_s\left(T_f - T_o\right)}{\sqrt{\pi \alpha_s t}\,\exp\left(\lambda^2\right)\,\mathrm{erf}\,(\lambda)} \simeq 4.12\,MW/m^2$$

$$(9.6\mathrm{E}9\mathrm{a})$$

$$q_f = \rho_s \Delta H_s \frac{ds\,(t)}{dt} = (8033)\,(281800)\left(1.82 \times 10^{-3}\right) = 4.12\,MW/m^2$$

$$(9.6\mathrm{E}9\mathrm{b})$$

(d) The solid and liquid temperature field equations up to $s\,(t) = 0.03\,m$ are

$$T_s = T_i + \frac{T_f - T_i}{\mathrm{erf}\,(\lambda)}\,\mathrm{erf}\left(\frac{x}{2\sqrt{\alpha_s t}}\right) = 838.34 + (578.42)\,\mathrm{erf}\,(13.149x)$$

$$(9.6\mathrm{E}10\mathrm{a})$$

$$T_l = T_f = 1083\,^\circ C \quad \text{(at interface)} \qquad\qquad (9.6\mathrm{E}10\mathrm{b})$$

Then, T_s at $0 \le x \le 0.03\,m$ and T_l at $0.03 \le x \le 0.04\,m$ intervals are shown in the figure at the end of this example.

Case 2: Supercooling Effect of the Melt

(a) In this case, the order of the calculation is altered for convenience. From Table 8.1, the undercooling temperature range is $\Delta T_u = 236\,^\circ C$, and the supercooled liquid temperature is

$$\Delta T_u = T_f - T_u \qquad\qquad (9.6\mathrm{E}11\mathrm{a})$$

$$T_u = T_f - \Delta T_u = 1083 - 236 = 847\,^\circ C \qquad\qquad (9.6\mathrm{E}11\mathrm{b})$$

$$T_u = 847\,^\circ C < T_f \qquad\qquad (9.6\mathrm{E}11\mathrm{c})$$

(b) For $s\,(t) = 0.03\,m$, Eqs. (9.17a), (9.19a) and (9.22a) at $t = 8.22$ sec and $ds\,(t)/dt = 1.82 \times 10^{-3}\,m/$ sec yield the temperature field equations as

$$T_m = T_o + (T_i - T_o)\,\mathrm{erfc}\left(\frac{-x}{2\sqrt{\alpha_m t}}\right) = 25 + (813.34)\,\mathrm{erfc}\,(-66.857x)$$

$$(9.6\text{E}12\text{a})$$

$$T_s = T_i + \frac{(T_f - T_i)}{\mathrm{erf}\,(\lambda)}\,\mathrm{erf}\left(\frac{x}{2\sqrt{\alpha_s t}}\right) = 838.34 + (578.42)\,\mathrm{erf}\,(13.149x)$$

$$(9.6\text{E}12\text{b})$$

$$T_l = T_u - \frac{(T_u - T_f)}{\mathrm{erfc}\,(\lambda\sqrt{\alpha_s/\alpha_l})}\,\mathrm{erfc}\left(\frac{x}{2\sqrt{\alpha_l t}}\right) = 847 + (428.05)\,\mathrm{erfc}\,(14.045x)$$

$$(9.6\text{E}12\text{c})$$

(c) from which

$$q_s = k_s\left(\frac{\partial T_s}{\partial x}\right)_{s(t)} = (130)\,(7345.20) = 0.95\,MW/m^2 \qquad (9.6\text{E}13\text{a})$$

$$q_l = k_l\left(\frac{\partial T_s}{\partial x}\right)_{s(t)} = (163)\,(-5680.30) = -0.93\,MW/m^2 \qquad (9.6\text{E}13\text{b})$$

$$q_f = q_s - q_l = 0.95 + 0.93 = 1.88\,MW/m^2 \qquad (9.6\text{E}13\text{c})$$

Then, $q_f = \rho_s \Delta H_s\,[ds\,(t)\,/dt]$ so that

$$\frac{ds\,(t)}{dt} = \frac{q_f}{\rho_s \Delta H_s} = \frac{1.88 \times 10^6\,W/m^2}{(8033)\,(281800)} = 8.31 \times 10^{-4}\,m/s \qquad (9.6\text{E}14\text{a})$$

$$t = 0.03/\left(8.31 \times 10^{-4}\right) = 36.10\,\text{sec} \qquad (9.6\text{E}14\text{b})$$

(d) The temperature distributions for T_m, T_s and T_l are given below

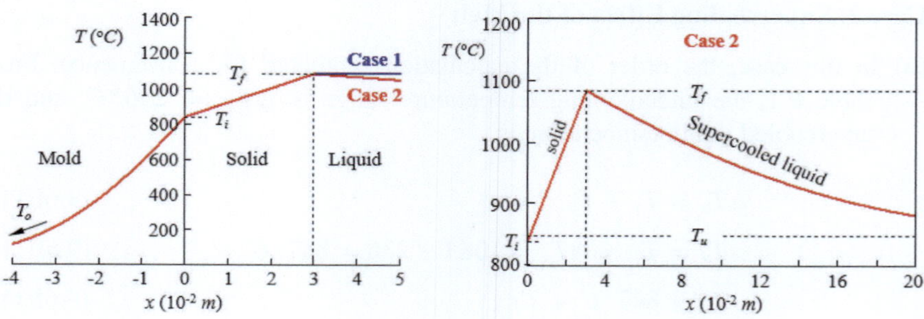

Therefore, expanding this region shows the nonlinearity of the supercooling as indicated by the bottom case 2 figure.

9.10 Phase-Change Numerical Methods

Mathematical modeling simple and intricate solidification problems can be feasible through the use of software packages that handle complicated algebra systems for solving ODEs and PDEs. This leads to an optimization procedure. Hence, computer algebra systems, such as MAPLE, MATLAB, and LINDO, are valuable tools for scientists and engineers. The literature is abundant in planar solidification numerical solutions of simple and intricate casting geometries.

The reader is encouraged to consult excellent review papers on cylindrical and/or spherical melting and solidification numerical methods, where cylindrical and spherical solidification processes include comparative analytical and numerical results (Alexiades and Solomon [2, p. 210–211], Hu and Argyropoulos [13], Caldwell and Ng [14, pp. 93–109].

9.11 Solidification Under Magnetic Fields

Applied magnetic fields, in general, on the solidification process may have positive or negative effects on the metal structure due to the presence of thermo-electromagnetic convection (TEMC). Characterizing the electromagnetic solidification processes includes the effects of weak or strong magnetic fields, electromagnetic vibration, and alternative electric current (AC) and so forth.

Magnetic fields may distort the L-S interface surface area and influence the freezing temperature of pure metals and, subsequently, affect to an extent the final microstructural features of metals and alloys. In essence, a dynamic magnetic field may induce flow motion of the melt due to alternating eddy currents. Consequently, melt motion gives rise to a magnetohydrodynamic problem that causes unstable solidification. On the other hand, a static magnetic field may reduce melt flow (local flow instability) during solidification and improve product quality. Thus, a magnetic field may be used to control the behavior of melts during solidification.

The reader is encouraged to search the literature for publications that characterize the effects of magnetic fields on solidification heat transfer (Poole [15], Boettinger et al. [16], Li et al. [17]).

9.12 Constitutional Supercooling

Consider the solidification of a binary alloy slab with nominal composition C_o. The mold cavity of rectangular shape is modeled as half-space regions in a finite one-dimensional domain accounting for constitutional supercooling (CS), which is an undesirable compositional change mechanism due to segregation of the solute B and impurities ahead of the planar L-S interface at time $t \geq 0$. Actually, CS is

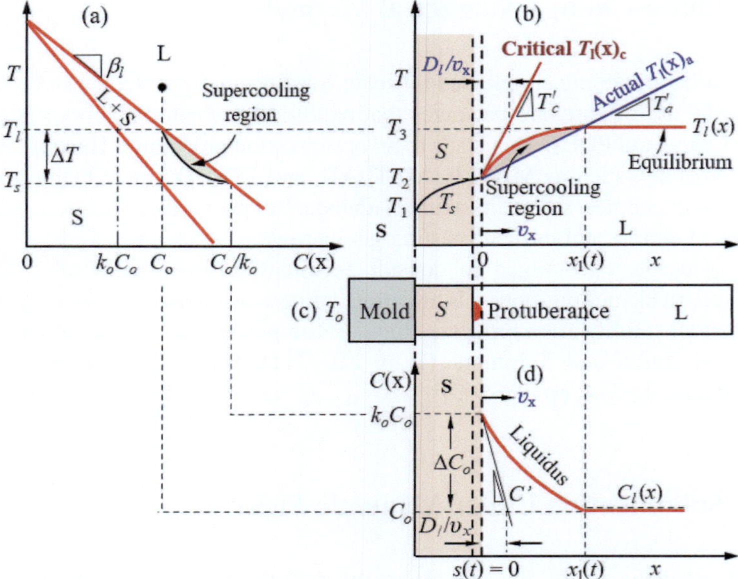

Fig. 9.12 Illustration of the constitutional supercooling phenomenon. (**a**) Partial equilibrium hypoeutectic phase diagram, (**b**) supercooling zone ahead of the *L-S* interface, (**c**) half-space model, and (**d**) concentration profile. After Dantzig and Rappaz [10, p. 339], Porter and Easterling [19, p. 215]

caused by the formation of a protuberance, where the solute is partitioned into the melt. Consequently, CS arises because of a heterogeneous distribution of solute *B* and impurities in the melt.

9.12.1   *Linear Stability Analysis*

The constitutional supercooling of the melt (Fig. 9.12) can be characterized by using Tiller's linear stability analysis (Tiller et al. [18]), which follows the conductive heat transfer and diffusion in the liquid phase. Basically, Tiller's linear stability analysis sets the rule for predicting the physical condition of the advancing planar interface along the *x*-axis.

Constitutional supercooling refers to composition changes, and it arises due to a positive temperature gradient $dT/dx > 0$. Thus, general criteria are

$$\left(\frac{dT_l(x)_a}{dx} \right)_{x=0} = \left(\frac{dT_l(x)_c}{dx} \right)_{x=0} \quad \text{(stability)} \qquad (9.49a)$$

$$\left(\frac{dT_l\,(x)_a}{dx}\right)_{x=0} < \left(\frac{dT_l\,(x)_c}{dx}\right)_{x=0} \qquad \text{(instability)} \qquad\qquad (9.49b)$$

The stability criterion, Eq. (9.49a), and the solidification domain can be described by the $T\,(C)$, $T_l\,(x)$, $T_l'\,(x) = dT_l\,(x)\,/dx$ and $C_l\,(x)$ fields or diagrams shown in Fig. 9.12 (Dantzig and Rappaz [10, p. 339], Porter and Easterling [19, p. 215]).

The information given in Fig. 9.12 is presented in bullet points for emphasizing important details on solidification and related constitutional supercooling. Thus,

- Initially, the onset of steady-state solidification starts without CS as a uniform thin-solid layer due to a high critical temperature gradient $T_l'\,(x)_c$ at the mold-liquid stable interface producing equiaxed grains at a relatively high solidification velocity v_x.
- A solidification heat transfer problem may arise due to a decrease of the solidification velocity v_x related to the heat transfer direction through the equiaxed-grain layer and due to perturbations (bumps-like) at the moving planar L-S interface. These perturbations may dissolve or grow as columnar or dendritic grain if the actual temperature gradient of the remaining melt is $T_l'\,(x)_a < T_l'\,(x)_c$.
- The $T\,(C)$ field (Fig. 9.12a) represents an equilibrium phase diagram for an A-B binary alloy, where B is the alloying solute. Constitutional supercooling is denoted as small deviation of the liquidus line for the binary C_o alloy and related partition coefficient k_o. Thus, $\Delta T = T_l\,(C) - T_s\,(C)$ is the degree of the driving force for the constitutional supercooling that causes a disturbance on the planar L-S interface.
- The $T_l\,(x)$ field (Fig. 9.12b) represents the liquid (melt) temperature field ahead of the initially advancing planar solidification front along the x-axis, where the formation of any protuberance ahead of the planar solidification front may be attributed to segregation of the solute B and impurities. Moreover, the temperature gradient $T_l'\,(x) = dT_l\,(x)\,/dx$ is divided into the actual $T_l'\,(x)_a$ and critical $T_l'\,(x)_c$ components, since constitutional supercooling occurs when $T_l'\,(x)_a < T_l'\,(x)_c$ and a nonlinear temperature distribution $T_l\,(x)$ at the supercooling region $0 < x \le x_l\,(t)$.
- A protuberance (Fig. 9.12c) may remelt or may grow causing instability of the planar solidification front due to constitutional supercooling of the liquid metal (L).
- The $C_l\,(x)$ field (Fig. 9.12d) at the supercooling region $s\,(t) < x \le x_l\,(t)$ represents the solute concentration behavior ahead of the advancing solidification front position $s\,(t) = 0$. The typical solute concentration distribution in the liquid phase ahead of the L-S interface is associated with the diffusion characteristic length or thickness of the boundary layer defined as $\delta_c = D_l/v_x$, where D_l is the diffusion coefficient of the solute B in the liquid phase and v_x is the velocity of the advancing interface.

Figure 9.13a illustrates the $T\,(x)$ diagram along with the constitutional supercooling region. Notice that Fig. 9.13b shows an equiaxed-grain microstructure due

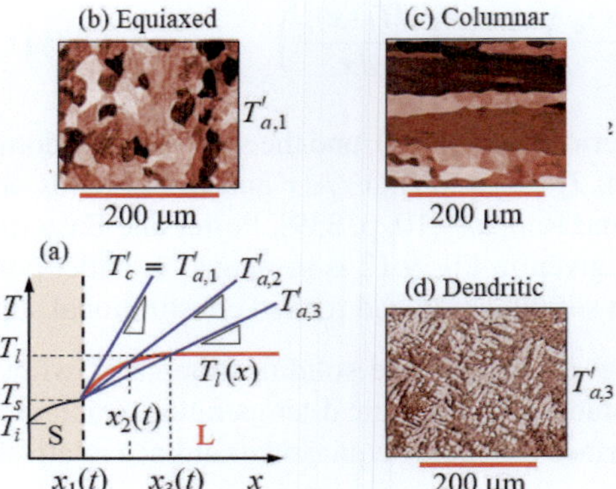

Fig. 9.13 Effect of actual temperature gradient T'_a on grain growth morphology of a binary alloy. (**a**) Constitutional supercooling diagram (Korner et al. [20]). (**b**) Stable planar front for equiaxed grain growth (Korner et al. [20]), (**c**) unstable planar front due to constitutional supercooling related to columnar grain growth (Ying et al. [21]) and (**d**) dendritic structure ahead of the interface due to unstable planar front (Ying et al. [21])

to the stability of a solidification front, where $T'_c = T'_{a,1}$ without constitutional supercooling (Korner et al. [20]). In this case, any protrusion or perturbation (bump-like at the solidification front) remelts and the planar solidification front continuous advancing at a steady-state velocity or rate.

A columnar grain structure (Fig. 9.13c after Korner et al. [20] in the form of needle-like grains evolves during solidification when $T'_{a,2} < T'_{a,1}$ at $x_1(t) < x(t) < x_2(t)$, since the solidification velocity is assumed to be $\upsilon_{x,2} < \upsilon_{x,1}$. Consequently, the binary A-B alloy exhibits liquid constitutional supercooling due to solute segregation in the melt adjacent to the solidification front causing a deviation from the stable planar L-S interface. Similarly, a dendritic grain structure (Fig. 9.13d after Ying et al. [21]) evolves for further increase of constitutional supercooling at $T'_{a,3} < T'_{a,2} < T'_{a,1}$ and $\upsilon_{x,3} < \upsilon_{x,2} < \upsilon_{x,1}$ at $x_1(t) < x(t) < x_3(t)$.

In summary,

- For a stable planar L-S interface, $T'_c = T'_a$ or $dT_l(x)_c = dT_l(x)_a$ which induces equiaxed grain formation (Fig. 9.13b). Recall that equiaxed grains are crystals that have similar axes lengths.
- For a mild unstable planar L-S interface, $T'_c \lesssim T'_a$ or $dT_l(x)_c \lesssim dT_l(x)_a$ which induces columnar grain formation (Fig. 9.13c). Columnar grains are long, thin, and coarse crystals.
- For a significant unstable planar L-S interface, $T'_c < T'_a$ or $dT_l(x)_c < dT_l(x)_a$ which induces dendritic grain formation (Fig. 9.13d). Dendritic grains form as tree-like crystals.
- Constitutional supercooling may be prevented by enhancing the melt mixing rate so that $dT_l(x)_a \to dT_l(x)_c$ or $dT_l(x)_a = dT_l(x)_c$.

9.13 Directional Solidification Process

Assume that one-dimensional quasi-static solidification is coupled to conduction-driven solidification and diffusion-driven liquid at the moving L-S interface. This implies that the mold stationary coordinate system (u, v) must be transformed to moving interface coordinate system (x, y) as schematically shown in Fig. 9.14 for left and right planar interfaces, smooth and flat planes.

Notice that the mold cavity is split into two equal half-space regions due to the inherent symmetry of the left and right L-S planar interfaces moving toward the center of a mold cavity at x_c for complete solidification.

The relationship between these coordinate systems is defined as the Galilean transformation $x = u + v_x t$ and $x = u - v_x t$, where x denotes the interface position at time $t > 0$, u denotes the casting dimension, and $dx/dt = \pm v_x$ denotes the normal interface velocity (Dantzig and Rappaz [10, pp. 181, 338], Asthana et al. [22, pp. 140–141]). Upon using the selected mold coordinate system, the one-dimensional diffusion is mathematically described by Fick's second law of diffusion as the concentration rate of the solute B in the liquid during solidification of a binary A-B alloy.

Thus,

$$\frac{\partial C_l}{\partial t} = D_l \frac{\partial^2 C_l}{\partial u^2} \tag{9.50}$$

which can be transformed to the moving coordinates using

$$dx/du = 1 \tag{9.51a}$$

$$\frac{\partial C_l}{\partial u} = \frac{\partial C_l}{\partial x} \frac{\partial x}{\partial u} = \frac{\partial C_l}{\partial x} \tag{9.51b}$$

$$\frac{\partial^2 C_l}{\partial u^2} = \frac{\partial^2 C_l}{\partial x^2} \tag{9.51c}$$

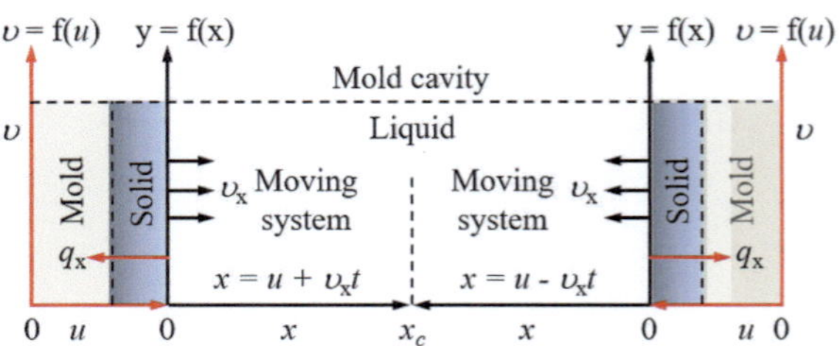

Fig. 9.14 Transformation of stationary coordinate (u, v) system to moving L-S interface coordinate system (x, y). The interface is ideally defined as a smooth and flat vertical plane moving in the x-direction with a velocity v_x

Hence,

$$\frac{\partial C_l}{\partial t} = D_l \frac{\partial^2 C_l}{\partial x^2} \qquad \text{(Fick's second law)} \qquad (9.52a)$$

$$\frac{\partial C_l}{\partial t} + \frac{\partial x}{\partial t}\frac{\partial C_l}{\partial x} = D_l \frac{\partial^2 C_l}{\partial x^2} \qquad \text{(chain rule)} \qquad (9.52b)$$

$$\frac{\partial C_l}{\partial t} - \upsilon_x \frac{\partial C_l}{\partial x} = D_l \frac{\partial^2 C_l}{\partial x^2} \qquad \text{(transformed)} \qquad (9.52c)$$

$$0 = \frac{\partial^2 C_l}{\partial x^2} + \frac{\upsilon_x}{D_l}\frac{\partial C_l}{\partial x} \qquad \text{(steady-state)} \qquad (9.52d)$$

Notice that the chain rule is a formula that differentiates a function of a function, and the result is a multiplication of two functions.

The solution to Eq. (9.52d) can be found elsewhere (Dantzig and Rappaz [10, p. 338], Asthana et al. [22, p. 141], Kurz and Fisher [23, p. 163]), but it is derived here for the sake of the reader.

Assume that the liquid concentration is described by the exponential function

$$C_l = C_l(x) = A_1 + A_2 \exp\left(-\frac{x}{D_l/\upsilon_x}\right) \qquad (9.53a)$$

$$\text{BC1} \quad C_l = C_l(0) = C_o/k_o \quad \& \quad \exp(0) = 1 \qquad (9.53b)$$

$$C_o/k_o = A_1 + A_2 \qquad (9.53c)$$

$$\text{BC2} \quad C_l = C_l(-\infty) = C_o \quad \& \quad \exp(-\infty) = 0 \qquad (9.53d)$$

$$C_o = A_1 \qquad (9.53e)$$

$$A_2 = C_o/k_o - A_1 = C_o/k_o - C_o \qquad (9.53f)$$

$$k_o = \frac{C_s}{C_l} \quad \text{at the } L\text{-}S \text{ interface} \qquad (9.53g)$$

Substituting A_1, A_2 into Eq. (9.53a) yields the solution to Eq. (9.52d)

$$C_l = C_o + \left(\frac{C_o}{k_o} - C_o\right)\exp\left(-\frac{x}{D_l/\upsilon_x}\right) \qquad (9.54a)$$

$$C_l = C_o + \Delta C_o \exp\left(-\frac{x}{D_l/\upsilon_x}\right) \qquad (9.54b)$$

$$C_l = C_o\left[1 + \frac{1-k_o}{k_o}\exp\left(-\frac{\upsilon_x x}{D_l}\right)\right] \qquad (9.54c)$$

from which the unknown concentration gradient dC_l/dx at $x = 0$ becomes inversely proportional to the diffusion characteristic length (D_l/υ_x).

The derivative of Eq. (9.54c) is

$$C' = \left(\frac{dC_l}{dx} \right)_{x=0} = -\left[\frac{(1-k_o)\,C_o}{k_o} \right] \frac{1}{D_l/v_x} \tag{9.55a}$$

$$C' = \left(\frac{dC_l}{dx} \right)_{x=0} = -\frac{(1-k_o)\,v_x C_o}{k_o D_l} \tag{9.55b}$$

The liquidus temperature (defined here as a critical entity) related to the solute concentration at any point x ahead of the interface is defined from slope of the liquidus line on the partial equilibrium phase diagram (Fig. 9.12a). Hence,

$$T_l\,(x)_c = T_f + \beta_l C_l\,(x) \quad \text{(liquidus)} \tag{9.56a}$$

$$T_l\,(x)_c = T_f + \beta_l C_o \left[1 + \frac{1-k_o}{k_o} \exp\left(-\frac{x}{D_l/v_x} \right) \right] \tag{9.56b}$$

$$\frac{dT_l\,(x)_c}{dx} = -\frac{\beta_l C_o\,(1-k_o)}{k_o D_l/v_x} \exp\left(-\frac{x}{D_l/v_x} \right) \tag{9.56c}$$

$$\frac{dT_l\,(x)_c}{dx} = -\frac{\beta_l C_o\,(1-k_o)\,v_x}{k_o D_l} \exp\left(-\frac{v_x}{D_l}x \right) \tag{9.56d}$$

Recall that $\beta_l < 0$ is the negative slope of the liquidus line from an equilibrium binary A-B phase diagram. Further analytical analysis of Eq. (9.56d) and Fig. 9.12a yields the constitutional supercooling parameter P_{cs} (Tarshis et al. [24]) and the growth restriction factor Q_{cs} written as (Maxwell and Hellawell [25]). Thus,

$$P_{cs} = \Delta T_o = -\beta_l \Delta C_o = T_l - T_s \tag{9.57a}$$

$$\Delta C_o = \frac{C_o\,(1-k_o)}{k_o} = \frac{C_o}{k_o} - C_o \tag{9.57b}$$

$$P_{cs} = \Delta T_o = -\frac{\beta_l\,(1-k_o)\,C_o}{k_o} \tag{9.57c}$$

$$Q_{cs} = k_o P_{cs} = -\beta_l\,(1-k_o)\,C_o \tag{9.57d}$$

$$Q_{cs} = k_o P_{cs} = \sum_1^i [-\beta_l\,(1-k_o)\,C_o]_i \tag{9.57e}$$

Notice that $dT_l\,(x)_c/dx$ in Eq. (9.56d) can be redefined in terms of P_{cs} and Q_{cs} parameters, which are inversely proportional to the average grain size $d \propto 1/Q_{cs}$ or $d \propto 1/P_{cs}$ and $d = f\,(Q_{cd})$ (Greer et al. [26]). Nonetheless, the solidification velocity depends on the solute enrichment or depletion at the L-S interface and the growth restriction factor Q_{cs}. Apparently, the role of Q_{cs} is to quantify the growth restriction.

For diluted multicomponent alloys, the growth restriction factor during solidification can be quantified as $Q_{cs} = \sum Q_{cs,i}$, Eq. (9.57e), where $i = 1, 2, 3 \ldots$ denotes the solute components in a certain alloy system.

The constitutional supercooling parameter P_{cs} and the growth restriction factor Q_{cs} are constants that may or may not correlated with the evolved average grain size d during solidification. However, $d = f(P_{cs})$ or $d = f(Q_{cs})$ can be quantified through curve fitting procedures which provide empirical expressions.

Possibly, $d = f(C_o)$ or $d = f(C_o, \beta_l)$ may explain some experimental observations for binary or multicomponent alloys.

Critical Temperature Gradient   Now, using the chain rule on the critical liquidus temperature gradient yields

$$\left(\frac{dT_l(x)_c}{dx} \right)_{x=0} = \frac{dT_l}{dC_l} \left(\frac{dC_l}{dx} \right)_{z=0} \quad \text{(critical)} \tag{9.58a}$$

$$T_c' = \left[\frac{dT_l(x)_c}{dx} \right]_{x=0} = -\frac{\beta_l C_o (1 - k_o)}{k_o D_l / \upsilon_x} \left(\frac{dC_l}{dx} \right)_{x=0} \tag{9.58b}$$

$$T_c' = \frac{\Delta T_o \upsilon_x}{D_l} \quad \text{(critical)} \tag{9.58c}$$

Note that T_c' is inversely proportional to the diffusion characteristic length D_l / υ_x. Physically, $T_c' \to 0$ as $\upsilon_x \to 0$ for a stationary or quasi-static L-S interface.

Actual Temperature Gradient   The actual liquid temperature ahead of the planar L-S interface is intentionally defined as a linear relationship in accord with Fig. 9.12b

$$T_l(x)_a = T_s + \left[\frac{dT_l(x)_a}{dx} \right] x \tag{9.59}$$

Note that the temperature gradient is simply the slope of Eq. (9.59), $dT_l(x)_a / dx > 0$, and it determines the steepness of the liquid temperature ahead of the L-S interface at a distance x.

9.13.1   Stability Criterion

Combining Eqs. (9.49a) and (9.56c) gives the stability criterion written as

$$\left[dT_l(x)_a / dx \right]_{x=0} \geq -\frac{(1 - k_o) \beta_l C_o \upsilon_x}{k_o D_l} = \frac{\Delta T_o \upsilon_x}{D_l} \quad \text{(stability)} \tag{9.60a}$$

$$\left[dT_l(x)_a / dx \right]_{x=0} < -\frac{(1 - k_o) \beta_l C_o}{k_o D_l} = \frac{\Delta T_o \upsilon_x}{D_l} \quad \text{(instability)} \tag{9.60b}$$

From Eq. (9.60a), the limit of a constitutional supercooling and the interface velocity are

$$\frac{\left[dT_l\,(x)_a\,/dx\right]_{x=0}}{\upsilon_x} = \frac{\Delta T_o}{D_l} \qquad \text{(limit)} \qquad (9.61a)$$

$$\upsilon_x = \frac{D_l}{\Delta T_o}\left[dT_l\,(x)_a\,/dx\right]_{x=0} \qquad \text{(critical)} \qquad (9.61b)$$

Accord with the nomenclature used in the literature, the actual liquid temperature gradient (G_a) in Eqs. (9.60a), (9.60b) and (9.61a), (9.61b) is redefined as (Tiller et al. [18])

$$\frac{G_a}{\upsilon_x} \geq \frac{\Delta T_o}{D_l} \qquad \text{(stability)} \qquad (9.62a)$$

$$\frac{G_a}{\upsilon_x} < \frac{\Delta T_o}{D_l} \qquad \text{(instability)} \qquad (9.62b)$$

$$\upsilon_x = \frac{D_l G_a}{\Delta T_o} \qquad \text{(critical)} \qquad (9.62c)$$

Theoretically, the stability criterion suppresses the formation of cellular and dendritic structures ahead of the L-S interface, and as a result, the liquid alloy solidifies as equiaxed grains with homogeneous solute distribution in the solid state. This microstructural feature is desirable for producing castings with uniform or isotropic mechanical properties.

The instability criterion, on the other hand, occurs by the formation of protrusions, and consequently, post-solidification heat treatment follows for homogenizing the microstructure morphology and solute distribution. Moreover, the classic stability criterion, Eq. (9.60a), for a planar L-S interface during the solidification of binary alloys strongly depends on the growth velocity (υ_x) of the interface, which, in turn, depends on the rate of release of the latent heat of solidification ΔH_s.

The position of a planar L-S interface is always at the coordinate origin $x = 0$, and the stability criterion is directly related to the conservation law of the heat fluxes at the interface with an assumed constant phase mass density $\rho = \rho_s = \rho_l$. Thus,

$$k_s\left(\frac{\partial T_s}{\partial x}\right)_{x_s(t)} - k_l\left(\frac{\partial T_l}{\partial x}\right)_{x_s(t)} = \rho\Delta H_s\frac{ds\,(t)}{dt} \qquad (9.63)$$

This partial differential equation (PDE) governs the evolution of solid morphology at a particular solidification velocity $\upsilon_x = ds\,(t)\,/dt$, which influences the solid properties. For clarity, Eq. (9.63) represents the energy conservation equation due to the evolution of latent heat of solidification ΔH_s during solidification, provided that $s\,(t) > 0$.

Example 9.7 Consider the solidification of a hypothetical A-B binary alloy with nominal solute composition $C_o = 0.10\,wt\%$, equilibrium partition coefficient $k_o = 0.02$, and liquidus line slope $\beta_l = -40\,K/wt\%$. If the solute liquid-diffusion coefficient and the critical liquid-temperature gradient ahead of the planar L-S interface are $D_l = 10^{-8}\,m^2/s$ and $T_c' = 10^5\,K/m$, respectively, calculate **(a)** the L-S interface velocity υ_x, **(b)** the characteristic length δ_c, **(c)** the time it takes the L-S interface to reach a length equals to its characteristic length, and **(d)** What does the stability criterion predict?

Solution

(a) From Eq. (9.56c), the L-S interface velocity is

$$T_c' = \left[\frac{T_l\,(x)_c}{dx}\right]_{x=0} = -\frac{\beta_l C_o\,(1-k_o)}{k_o D_l/\upsilon_x}$$

$$\upsilon_x = -\frac{k_o D_l T_c'}{\beta_l C_o\,(1-k_o)} = -\frac{(0.02)\,\left(10^{-8}\,m^2/s\right)\,\left(10^5\,K/m\right)}{(-40\,K/\%)\,(0.10\%)\,(1-0.02)}$$

$$\upsilon_x = 5.10 \times 10^{-6}\,m/s$$

(b) The characteristic length δ_c at L-S interface due to constitutional supercooling is

$$\delta_c = \frac{D_l}{\upsilon_x} = \frac{10^{-8}\,m^2/s}{5.10 \times 10^{-6}\,m/s} = 1.96 \times 10^{-3}\,m = 1.96\,mm$$

(c) The time for characteristic length δ_c is

$$t = \frac{\delta_c}{\upsilon_x} = \frac{1.96\,mm}{5.10 \times 10^{-3}\,mm/s} = 384.31\,\sec = 6.41\,min$$

(d) From Eqs. (9.57a), (9.57b), the concentration and temperature ranges are

$$\Delta C_o = \frac{C_o\,(1-k_o)}{k_o} = \frac{(0.1\%)\,(1-0.02)}{0.02} = 4.9\,\%$$

$$\Delta T_o = -\beta_l \Delta C_o = -(-40\,K/\%)\,(4.9\%) = 196\,K$$

From (9.58c), the critical temperature gradient is

$$T_c' = \frac{T_l\,(x)_c}{dx} = \frac{\Delta T_o \upsilon_x}{D_l} = \frac{(196)\,\left(5.10 \times 10^{-6}\right)}{10^{-8}}$$

$$T_c' = 10^5\,K/m = 10^5\,°C/m$$

and

$$T_a' = T_c' = 10^5 \, \frac{K}{m} = 10^5 \, \frac{{}^\circ C}{m} \quad stability!$$

Therefore, the stability criterion predicts a stable plane-front solidification, because the temperature gradients are equal.

Example 9.8 Consider the solidification of a Fe-0.08C steel with nominal solute composition $C_o = 0.10 \, wt\%$, equilibrium partition coefficient $k_o = 0.17$, and liquidus line slope $\beta_l = -84.91 \, K/wt\%$. If the solute liquid diffusion coefficient ahead of the planar L-S interface and the L-S interface velocity are $D_l = 2 \times 10^{-8} \, m^2/s$ and $v_x = 10^5 \, m/s$, respectively, calculate **(a)** the L-S interface characteristic length δ_c, **(b)** the time it takes the L-S interface to reach a length equals to its characteristic length and **(c)** What does the stability criterion predict? Use the given actual liquid temperature defined by

$$T_l \, (x)_a = 1540 - 142.96 \, \text{erfc} \, (4.4835x)$$

Solution

(a) From Eq. (9.56c), the characteristic length is

$$\delta_c = \frac{D_l}{v_x} = \frac{2 \times 10^{-8}}{10^{-5}} = 0.002 \, m$$

(b) The time for the advancing planar L-S interface to reach a length equals to its characteristic length is

$$t = \frac{\delta_c}{v_x} = \frac{0.002}{10^{-5}} = 200 \, \text{sec}$$

(c) The critical temperature is

$$T_c' = \left[\frac{T_l \, (x)_c}{dx} \right]_{x=0} = \frac{\beta_l C_o \, (1 - k_o)}{k_o D_l / v_x} = -\frac{(-84.91) \, (0.08) \, (1 - 0.17)}{(0.17) \, (2 \times 10^{-8}) / (8 \times 10^{-6})}$$

$$T_c' = 13{,}266 \, K/m$$

From the given actual liquid temperature equation, the actual temperature gradient is

$$T_l \, (x)_a = 1540 - 142.96 \, \text{erfc} \, (4.4835x)$$

$$\left[\frac{dT_l \, (x)_a}{dx} \right] = -\frac{2 \, (4.4835) \, (-142.96)}{\sqrt{\pi}} \exp\left[- (4.4835x)^2 \right]$$

$$\left[\frac{dT_l\,(x)_a}{dx}\right] = 723.25\exp\left(-20.102x^2\right)$$

$$T_a' = \left[\frac{dT_l\,(x)_a}{dx}\right]_{x=0} = 723.25\,K/m = 723.25\,°C/m$$

Therefore, the stability criterion predicts an unstable plane-front solidification because $T_a' < T_c'$. This means that the planar L-S interface breaks down and cellular or dendritic solidification takes place.

9.14   Round Single-Crystal Solidification

The floating zone (FZ process) is a suitable technique for purifying single crystals, Bridgman process (BP) for actuators and sensors, and the Czochralski process (CZ technique) for silicon wafers with or without the presence of a magnetic field. Nonetheless, these are metallurgical methods used to produce mainly semiconductor silicon (Si) or germanium (Ge) round single crystal ingots of different dimensions.

Only the Czochralski process is considered in this section for analyzing the mathematical formulation related to the manufacturing of semiconductors using the pulling rate technique. The sketch shown in Fig. 9.1b represents the crystal-melt-crucible arrangement for producing cylindrical single crystals under strict control at simultaneously constant pulling rate $\upsilon_p = dz/dt$ and rotational speed dr/dt.

The Czochralski process is nowadays used to manufacture high-quality cylindrical crystals, which are vertically pulled out of a crucible placed in an induction furnace. The resultant crystal is an intrinsic silicon ingot known as undoped silicon (without added impurities intentionally) or extrinsic silicon ingot referred to as doped silicon (with added impurities intentionally). Moreover, the solidifying melt takes on the crystal structure of the seed, and the rotating and pulling rates determine the diameter and length of a particular silicon ingot. A typical diameter range is in the order of 10 to 30 cm (about 4 to 12 inches).

In one-dimensional heat transfer analysis, the coupling of conductive, convective, and radioactive heat transfer modes in the crystal growth process is a boundary-value problem. The general governing heat flux balance, including the pulling rate term, is written as (Rea and Gleim [27])

$$\frac{\partial T}{\partial t} = \left(\alpha_s\frac{\partial^2 T}{\partial x^2}\right)_{cond} - \left(\frac{2h}{r}\,(T - T_o)\right)_{conv} - \left(\frac{2}{r}q_r\right)_{rad} - \left(\rho_s c_s \upsilon_p\frac{\partial T}{\partial x}\right)_{pull}$$

$$(9.64)$$

with particular initial and boundary conditions given by
where ε denotes the emissivity, $\sigma = 5.67 \times 10^{-8}$ $(W/m^2.K^4)$ denotes the Stafan-Boltzmann constant. This expression is subjected to approximations that follow certain assumptions and experimental observation, but the main purpose henceforth is to analytically define υ_p.

$$\text{IC:} \quad T = T(x, 0) = T_f$$
$$\text{BC:} \quad T = T(0, t) = T_f \quad \& \quad T = T(x, t) = T_o$$
$$dT/dz = \varepsilon\sigma\left(T^4 - T_o^4\right)/k_s - h\left(T - T_o\right)/k_s$$

For steady-state heat transfer, the transition heat transfer term is $\partial T/\partial t = 0$, and subsequently, Eq. (9.64) reduces to an ordinary differential equation for a planar L-S interface (Rea and Gleim [27])

$$\left(\alpha_s \frac{d^2 T}{dz^2}\right)_{cond} - \left[\frac{2h}{r}\left(T - T_o\right)\right]_{conv} - \left(\frac{2}{r}q_r\right)_{rad} - \left(\rho_s c_s \upsilon_p \frac{dT}{dz}\right)_{pull} = 0 \tag{9.65}$$

For cylindrical or spherical shapes, the radial temperature gradient during solidification can be defined by (Benmeddour and Meziani [28])

$$q_r = k_s \frac{dT}{dr} = -h\left(T - T_c\right) - \varepsilon_s\sigma\left(T^4 - T_o^4\right) \tag{9.66}$$

where T_c denotes the crucible wall temperature $T_w = T_c$ or the melt temperature $T_l = T_c$. Recall that the thermal conductivity k_s refers to the intrinsic ability of a material to transfer or conduct heat, whereas the product $k_s(dT/dr)$ measures the intrinsic thermal or heat flux q_r during phase transformation due to the effect of radiation. Therefore, dT/dr must be the driving force for heat transfer.

9.14.1   Conductive-Radioactive Heat Transfer

In the Czochralski process (Fig. 9.1b), the pulling rate ($\upsilon_p = dz/dt$) is directly related to the length of the crystal (with a desired radius) and speed of rotation (rpm) which are controlled during the growth process; otherwise, forced convection in the melt causes solidification problems.

The production of Czochralski crystals is subjected to variations in crystal pull rate or growth rate. However, a quantitative cost model is normally associated with Czochralski crystal dimensions. Multi-crystal production from a single crucible liner may reduce cost.

The theoretical maximum pulling rate in the Czochralski process can also be predicted by using the heat flux balance $q_s = q_l$

$$q_s = -k_s \left(\frac{dT}{dz}\right)_s \tag{9.67a}$$

$$q_l = -k_l \left(\frac{dT}{dz}\right)_l - \upsilon_p \Delta H_f \tag{9.67b}$$

Equate and rearrange Eqs. (9.67a), (9.67b) to get the Stefan condition for the evolving or moving L-S interface

$$k_s \left(\frac{dT}{dz}\right)_s - k_l \left(\frac{dT}{dz}\right)_l = \upsilon_p \rho_s \Delta H_f \tag{9.68}$$

Manipulating Eq. (9.68) yields the pulling rate or solidification velocity as

$$\upsilon_p = \frac{1}{\rho_s \Delta H_f}\left[k_s \left(\frac{dT}{dz}\right)_s - k_l \left(\frac{dT}{dz}\right)_l\right] \tag{9.69}$$

The Stefan-Boltzmann law describes the heat loss due to radiation as

$$dQ_{rad} = (2\pi r dz)\,\varepsilon\sigma T^4 \tag{9.70a}$$

$$\frac{dQ_{rad}}{dz} = (2\pi r)\,\varepsilon\sigma T^4 \tag{9.70b}$$

where dQ_{rad}/dz denotes the radiation energy gradient, $2\pi r dz$ denotes the radiating surface area of an increment length of the ingot, and ε denotes the emissivity for radioactive heat transfer from the crystal surface to the furnace surroundings.

From Fourier's law, the conductive heat energy and its corresponding derivative are

$$Q_{cond} = \left(\pi r^2\right) k_s \frac{dT}{dz} \tag{9.71a}$$

$$\frac{dQ_{cond}}{dz} = \left(\pi r^2\right) k_s \left[\frac{d^2 T}{dz^2} + \frac{dT}{dz}\frac{dk_s}{dz}\right] = \left(\pi r^2\right) k_s \frac{d^2 T}{dz^2} \tag{9.71b}$$

Here, dQ_{cond}/dz denotes the thermal energy gradient, πr^2 denotes the cross-sectional area of the ingot conducting the heat, k_s denotes the thermal conductivity of the solidified melt and dT/dz denotes the temperature gradient.

Now, equating (9.70b) and (9.71b) along with $k_s = k_l T_f / T$ at T_f yields

$$\frac{d^2 T}{dz^2} - \frac{2\varepsilon\sigma}{rk_s}T^4 = 0 \tag{9.72a}$$

$$\frac{d^2 T}{dz^2} - \frac{2\varepsilon\sigma}{rk_l T_f}T^5 = 0 \tag{9.72b}$$

which is a simplified equation representing the effect of radiation on radial solidification. Use a standard procedure to solve homogeneous ordinary differential equations (ODE) of nth order with constant coefficients.

The solution of Eq. (9.72b) is

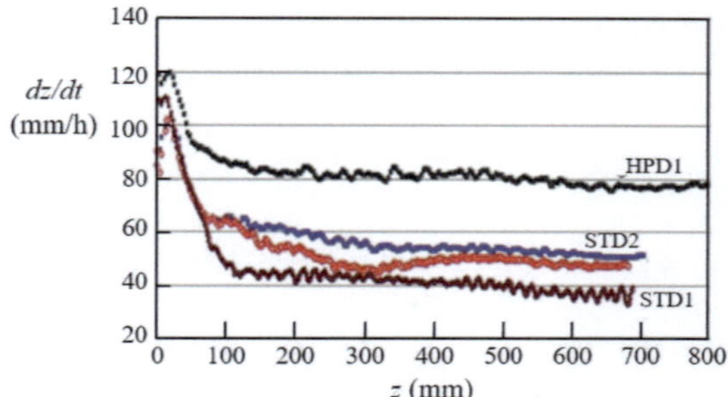

Fig. 9.15 Experimental pulling speed diagram for silicon (*Si*) single crystals. Legend: High-performance design (HPD1) and standard designs (STD1 & STD2) stand for different design conditions (Lan et al. [29])

$$\frac{dT}{dz} = -\frac{1}{2}\left(\frac{3rk_lT_f}{8\varepsilon\sigma}\right)^{1/4}\left[x + \left(\frac{3rk_l}{8\varepsilon\sigma T_f^3}\right)^{1/2}\right]^{-3/2} \tag{9.73a}$$

$$T = \left(\frac{3rk_lT_f}{8\varepsilon\sigma}\right)^{1/4}\left[x + \left(\frac{3rk_l}{8\varepsilon\sigma T_f^3}\right)^{1/2}\right]^{-1/2} \tag{9.73b}$$

$$\left(\frac{dT}{dz}\right)_{x=0} = \sqrt{\frac{2\varepsilon\sigma k_l T_f^5}{3r}} \tag{9.73c}$$

Substituting Eq. (9.73c) into (9.69) along with $dT_l/dz = 0$ gives the steady-state pulling rate or solidification velocity defined as

$$\upsilon_p = \frac{k_s}{\rho_s\Delta H_f}\left(\frac{2\varepsilon\sigma k_l T_f^5}{3r}\right)^{1/2} \tag{9.74}$$

For instance, Fig. 9.15 shows some $\upsilon_p = dz/dt$ experimental data obtained by hot-zone designs, coded as standard designs (STD1 and STD2) and as high-performance design (HPD1) (Lan et al. [29]). Here, z denotes the crystal length.

According to Lan et al. [29], the HPD1 curve is an improvement of the Czochralski process (Cz) process over the STD1 and STD2 conditions by controlling radiation to an extent. This leads to a lower power consumption and higher pulling rate. Notice that HPD1 curve gives $\upsilon_p = dz/dt \simeq 80\,mm/h$ for an ingot $z > 100\,mm$.

Example 9.9 For solidification of a silicon single crystal ingot using the Czochralski process, calculate **(a)** the pulling rate υ_p and time t to produce one 2-*m* long ingot with $r = 100\,mm$ radius using the coupled conductive-reactive heat transfer. **(b)** This part calls for conductive heat transfer in the radial direction. Thus, approximate the solidification velocity υ_r and the solidification time t in the radial direction by letting $\upsilon_p \rightarrow \upsilon_r$ and $dT/dz \rightarrow dT/dr$ in Eq. (9.69). Assume that the liquid silicon is at $T_l = 1850\,K = 1577\,^\circ C$ and the supplied inert gas is at $T_g = 700\,K = 427\,^\circ C$. Given data:

$T_f = 1685\,K$	$T_o = 600\,K$	$T_l = 1850\,K$
$\rho_s = 2400\,Kg/m^3$	$\rho_l = 2420\,Kg/m^3$	$k_s = 25\,W/m.K$
$c_s = 239\,J/Kg.K$	$c_l = 1000\,J/Kg.K$	$k_l = ?$
$\Delta H_f = 1{,}877{,}000\,J/kg$	$\sigma = 5.67 \times 10^{-8}\,W/\left(m^2.K^4\right)$	
$h = 7\,W/m^2.K$	$\varepsilon_s = 0.75$	$\varepsilon_l = 0.20$

Solution

(a) For $r = 0.1\,m$ and $L = 2\,m$, and

$$k_l = \frac{k_s T_f}{T_l} = \frac{(25)\,(1685)}{1850} = 22.77\,W/m.K \tag{9.9E1}$$

Equation (9.74) gives

$$\upsilon_p = \frac{1}{\rho_s \Delta H_f}\left(\frac{2\varepsilon_s \sigma k_l T_f^5}{3r}\right)^{1/2} = 6.57 \times 10^{-5}\,m/s = 236.52\,mm/h$$

$$\tag{9.9E2a}$$

$$t = \frac{L}{\upsilon_p} = \frac{2000\,mm}{236.52\,mm/h} = 8.46\,h \tag{9.9E2b}$$

(b) Combining Eq. (9.69) with $\upsilon_p \rightarrow \upsilon_r$ and $dT/dz \rightarrow dT/dr$ and (9.66) approximates the solidification velocity and the solidification time in the radial direction

$$\upsilon_r = -\frac{k_s}{\rho_s \Delta H_f}\frac{dT}{dr} = \frac{1}{\rho_s \Delta H_f}\left[h\left(T_f - T_g\right) + \varepsilon_s \sigma\left(T_f^4 - T_o^4\right)\right]$$

$$\tag{9.9E3a}$$

$$\upsilon_r = 7.64 \times 10^{-5}\,m/s = 275.04\,mm/h \tag{9.9E3b}$$

$$t = \frac{r}{\upsilon_r} = \frac{100\,mm}{275.04\,mm/h} = 0.36\,h = 21.60\,\text{min} \tag{9.9E3c}$$

Therefore, the above rough calculations suggest that the Czochralski process operating at an unknown rotational speed $\omega > 0$ (in this example) produces an ingot radius of 100-mm in 21.60 min at $\upsilon_r = 275.04\,mm/h$ and a length of 2-m in 8.46 hours at 236.52 mm/h.

9.14.2   Near-Field Temperature Distribution

Following Poirier and Geiger [6, pp. 356–358] analytical procedure, the conduction heat transfer in the hot crystal can be described by a partial differential equation in cylindrical coordinates. The equation has the form

$$\frac{\partial T}{\partial t} - \upsilon_p \frac{\partial T}{\partial z} = \alpha \left[\frac{\partial^2 T}{\partial z^2} + \frac{1}{r}\frac{\partial}{\partial r}\left(r\frac{\partial T}{\partial r} \right) \right] \tag{9.75}$$

Notice that the cooling or freezing rate strongly depends on the temperature gradients along the vertical z-direction $\partial T/\partial z$ and radial direction $\partial T/\partial r$.

Assuming a planar L-S interface at $z \to 0$ and at a constant crystal radius r, one-dimensional analysis under steady-state condition implies that $\partial T/\partial t = 0$ and $\partial T/\partial r = 0$. As a result, Eq. (9.75) is adopted to the solid (s) and liquid (l) phases along with the Stefan condition and boundary condition (BC) $T = T_f = T_l$ at $z = 0$

$$\alpha_s \frac{\partial^2 T_s}{\partial z^2} + \upsilon_p \frac{\partial T_s}{\partial z} = 0 \tag{9.76a}$$

$$\alpha_l \frac{\partial^2 T_l}{\partial z^2} + \upsilon_p \frac{\partial T_l}{\partial z} = 0 \tag{9.76b}$$

$$k_s \frac{dT_s}{dz} - k_l \frac{dT_l}{dz} = \upsilon_p \rho_s \Delta H_f \tag{9.76c}$$

where Eq. (9.68) has been renumbered as (9.76c) for convenience. The solutions of Eqs. (9.76a), (9.76b) for a near-field condition $(z \to 0)$ are (Poirier and Geiger [6, p. 357])

$$T_s = T_f + \frac{\alpha_s}{\upsilon_p}\frac{dT_s}{dz}\left[1 - \exp\left(-\frac{z\upsilon_p}{\alpha_s} \right) \right] \tag{9.77a}$$

$$T_l = T_f + \frac{\alpha_l}{\upsilon_p}\frac{dT_l}{dz}\left[1 - \exp\left(-\frac{z\upsilon_p}{\alpha_l} \right) \right] \tag{9.77b}$$

Mathematically, the far-field condition yields $z \to \infty$, $\exp(-\infty) = 0$ and by deduction, $dT_l/dz \to 0$. Thus, Eqs. (9.77a), (9.77b) become

$$T_s = T_f + \frac{\alpha_s}{\upsilon_p}\frac{dT_s}{dz} \tag{9.78a}$$

$$T_l = T_f + \frac{\alpha_l}{\upsilon_p}\frac{dT_l}{dz} = T_f \tag{9.78b}$$

From Eq. (9.69) or (9.76c), one can determine that

$$v_p = \frac{1}{\rho_s \Delta H_f} \left[k_s \left(\frac{dT}{dz} \right)_s - k_l \left(\frac{dT}{dz} \right)_l \right] > 0 \tag{9.79a}$$

$$k_s \frac{dT_s}{dz} > k_l \frac{dT_l}{dz} \tag{9.79b}$$

$$\frac{dT_l}{dz} < \frac{k_s}{k_l} \frac{dT_s}{dz} \tag{9.79c}$$

Using k_s, k_l values from Example 9.9 gives $v_p > 0$ if

$$dT_l/dz < (25/22.77)\, dT_s/dz = (1.0979)\, dT_s/dz \tag{9.80}$$

This condition, Eq. (9.79b) or (9.79c), is very important for industrial production of silicon (Si) single crystals using the Cz process.

9.15   TTT Diagram for Glass Formation

Conventional heterogeneous or slow solidification (SS) initiates on suspended particles and mold inner surfaces under the influence of small undercooling ΔT. On the other hand, rapid solidification (RS) of alloys initiates on a relatively cooled substrate surface at $\Delta T_{RS} >> \Delta T$, leading to glass (amorphous structure) formation. This type of casting is known as a rapidly solidified alloy (RSA).

For the sake of clarity, the TTT diagram schematically shown in Fig. 9.16a along with the schematic rapid solidification melt-spinning process setup in Fig. 9.16b is an important graphical representation of the kinetics for glass formation. Notice that "C-shape" curve in Fig. 9.16a denotes the L-S transition at temperature T.

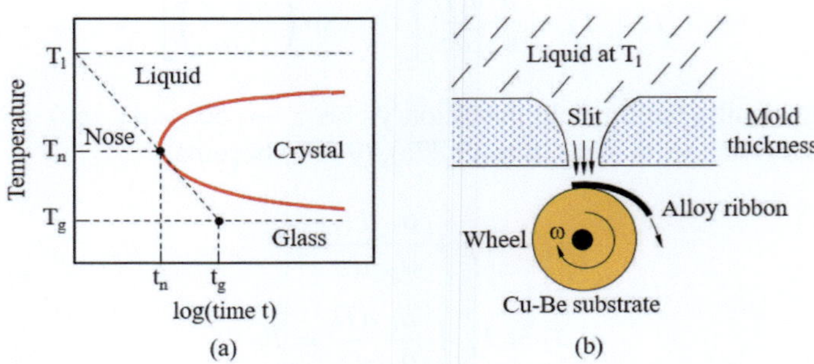

Fig. 9.16 Rapid solidification. (**a**) Schematic TTT diagram for glass formation and (**b**) melt spinning setup

Initially, the melt is a the pouring temperature $T_p = T_l$ and subsequent rapid freezing at deep undercooling $\Delta T_{RS} >> \Delta T$, the melt transforms into solid glass. Assuming that the freezing rate is linear, the T_l-T_g line intercepting the "nose" point at T_n can be used to define the respective critical or minimum cooling rate dT/dt for glass formation at T_g during rapid solidification.

Usually, thin ribbons can be produced using the RS technique by imposing a very large undercooling ΔT_{RS}, which in turn suppresses atomic diffusion and enhances heat extraction at an enormous rate from the alloy melt on a common $Cu\text{-}Be$ alloy wheel surface.

Example 9.10 Assume that a hypothetical $A_x B_y C_z$ alloy melt is initially at $T_l = 1540\,^\circ C$. Using the melt spinning technique (Fig. 9.16b) produces, say, amorphous $A_x B_y C_z$ strips with uniform thickness. Also assume that the heat transfer coefficient between the copper-beryllium $(Cu\text{-}Be)$ wheel substrate at $T_o = 20\,^\circ C$ and the melt at $1540\,^\circ C$ is $h = 18{,}000\,W/(m^2.K)$. Determine **(a)** the solidification rate and **(b)** the solidification time for a ribbon thickness of $\delta = 15\,\mu m$. Assume equal liquid and solid mass densities $\rho_l = \rho_s = 6800\,Kg/m^3$ and heat capacity $c_s = c_p = 660\,J/Kg.K$ at atmospheric pressure. Recall that an amorphous structure is treated as a metallic glass that lacks atomic symmetry and periodicity compared to crystalline structures.

Solution **(a)** The latent heat of solidification, Eq. (9.15a), is approximated as

$$\Delta H_s = c_s dT \simeq c_s \Delta T = c_s \left(T_p - T_o\right) \tag{9.10E1a}$$

$$\Delta H_s = (660\,J/Kg.K)\,(1540 - 20)\,K \tag{9.10E1b}$$

$$\Delta H_s = 10^6\,J/Kg \tag{9.10E1c}$$

The heat energy is released for solidification of the melt (liquid) per kilogram of the hypothetical $A_x B_y C_z$ alloy.

From Eq. (9.4d), the amount of thermal energy due to convective heat transfer at the melt-substrate interface is

$$q_{x=0} = h\,(T_i - T_o) \quad \text{(Newton's law)} \tag{9.10E2a}$$

$$\left[\frac{dQ}{dt}\right]_{x=0} = Ah\,(T_i - T_o) \tag{9.10E2b}$$

$$dQ_{x=0} = Ah\,(T_i - T_o)\,dt \tag{9.10E2c}$$

and the amount of thermal energy due to the release of latent heat of solidification ΔH_s is

$$Q_s = m\Delta H_s = V\rho_s \Delta H_s = As\,(t)\,\rho_s \Delta H_s \tag{9.10E3a}$$

$$dQ_s = As\,(t)\,\rho_s c_s dT \tag{9.10E3b}$$

For $dQ_{x=0} = dQ_s$, the thermal heat balance yields the solidification rate as

$$Ah\left(T_i - T_o\right)dt = \rho_s As\left(t\right)c_s dT \tag{9.10E4a}$$

$$\frac{dT}{dt} = \frac{h\left(T_p - T_o\right)}{s\left(t\right)\rho_s c_s} \tag{9.10E4b}$$

Using $h = 18{,}000\,W/\left(m^2.K\right) = 18{,}000\,J\left(m^2.K.s\right)$ and the remaining given data gives

$$\frac{dT}{dt} = \frac{\left[18{,}000\,J/\left(m^2.K.s\right)\right](1540 - 20)\,K}{\left(15 \times 10^{-6}\,m\right)\left(6800\,Kg/m^3\right)(660\,J/Kg.K)} = 0.41 \times 10^6\,K/s \tag{9.10E5}$$

(b) Integrating the above dT/dt equation yields

$$\frac{dT}{dt} = \frac{h\left(T_p - T_o\right)}{s\left(t\right)\rho_s c_s} \tag{9.10E6a}$$

$$T_p - T_o = \frac{h\left(T_p - T_o\right)}{s\left(t\right)\rho_s c_s}t \tag{9.10E6b}$$

from which the solidification time is approximated as

$$t = \frac{s\left(t\right)\rho_s c_s}{h} = \frac{\left(15 \times 10^{-6}\,m\right)\left(6800\,Kg/m^3\right)(660\,J/Kg.K)}{\left[18{,}000\,J/\left(m^2.K.s\right)\right]} \tag{9.10E7a}$$

$$t = 3.74 \times 10^{-3}\,s \tag{9.10E7b}$$

This result, $t = 3.74 \times 10^{-3}\,s$, implies that rapid solidification is a fast $L \rightarrow S$ phase transformation with a large evolved latent heat of solidification $\Delta H_s = 10^6\,J/Kg$.

9.16　Summary

Solidification heat transfer is a complicated process to model, but a quasi-static approximation (QSA) theory for the thermal field, known as a Stefan problem, is related to the moving L-S boundary (MB) theory. Thus, the mathematical solutions for predicting the solidification time t_f, solidification velocity $v = dx/dt$, temperature gradient dT/dx, and the amount of thermal energy Q removed from the casting through a mold wall were derived from transient heat transfer and evolved QSA and MB theories.

The simple analytical solutions and example problems on solidification of pure metals included in this chapter serve as a strong foundation for studying phase-change heat transfer. Different mathematical modeling on alloy solidification are also available, and fortunately, the reader interested on this topic has a vast literature field to search for specific publications that include solutions to one-phase and two-phase heat transfer problems.

The transient heat transfer provides a theoretical background for assessing the production of single crystals and rapidly solidified alloys. The given examples reflect the applications of these theories for solving complex solidification problems in a simplified manner.

Appendix 9A Thermophysical Properties

Consider the unit conversion in Table 9.3.

Average thermophysical properties for some metals at T_f and foundry molds are given in Table 9.4. The subscripts l, s, and m denote liquid, solid, and mold, respectively.

Example 9.11 For silver (Ag), change $\Delta H_f = 965 \, J/cm^3 \rightarrow \Delta H_f$ in J/mol

$$\Delta H_f = \frac{\Delta H_f \left(J/cm^3 \right)}{\rho \left(g/cm^3 \right)} A_w \left(g/mol \right) = \left(\frac{965 \, J/cm^3}{10.49 \, g/cm^3} \right)$$

$$\times \left(107.87 \, \frac{g}{mol} \right) = 9923.2 \, J/mol \qquad (9AE1b)$$

Table 9.3 Conversion table

Variable/property	Eng. units	Scientific units
Mass m	$1 \, lb_m$	$0.453592 \, Kg$
Thermal Energy Q	$1 \, Btu$	$1.05506 \, kJ$
Heat of fusion ΔH_f	$1 \, Btu/lb_m$	$2.326 \, kJ/Kg$
Density ρ	$1 \, lb_m/ft^3$	$16.0185 \, Kg/m^3$
Thermal conductivity k	$1 \, Btu/\left(hr.ft.°F \right)$	$1.7307 \, \left(J/m.s.K = W/mK \right)$
Thermal conductivity k	$1 \, cal/s.cm.°C$	$4.18 \, W/m.K$
Thermal Diffusivity α	$1 \, ft^2/hr$	$2.5807 \times 10^{-5} \, m^2/s$
Heat capacity c_p	$1 \, Btu/lbm.°F$	$4184 \, J/Kg.K$
Heat capacity c_p	$1 \, cal/g.°C$	$4188 \, J/kg.K$

Table 9.4 Thermophysical properties of some materials

Material	$T_f{}^a$	$\Delta H_f{}^a$	$\rho_s{}^b$	$\rho_l{}^b$	$c_s{}^b$	$c_l{}^b$	$k_s{}^b$	$k_l{}^b$
Ag	961	104	10,312	9329	240	310	362	175
Al	660	397	2700	2378	216	259	211	91
Cu	1083	209	8960	8033	92	118	130	163
Fe	1535	247	7870	7035	108	187	34	33
Mg	650	349	1740	1589	110	316	145	79
Ni	1450	292	8890	7890	131	157	70	60
Pb	330	23	11,300	10,700	0.13	0.13	37	31
Si	1412	1877	2329	2469	210	231	25	56
Sn	232	59	6700	7000	0.22	0.06	68	60
Steel mold			7689		669		35	
$\overline{C}$ mold^c			1800		720		65	
Sand mold			1550		986		0.40	

^{a} T_f in $^\circ C$ and ΔH_f in $\times 10^3\, kJ/Kg$ units
^{b} ρ_s, ρ_l in Kg/m^3, c_s, c_s in $J/Kg.K$ and k_s, k_l in $W/m.K$
^{c} $\overline{C}$ stands for graphite

Appendix 9B Solution of the Heat Equation

Method of Separation of Variables

For one-dimensional analysis, the heat equation can be solved by the separation of variables method in the form $T(x, t) = f(x) g(t)$. First of all, the heat equation is a partial differential equation written as

$$\frac{\partial T}{\partial t} = \alpha \frac{\partial^2 T}{\partial x^2} \tag{9.81}$$

The solution of the heat equation has a significant importance in physics and many engineering fields. The heat equation, Eq. (9.81), can be approximated using unidimensional analysis, and its solution can be derived by converting PDE to ODE by letting $T = T(x, t) = f(n)$ with the solution characteristic or similarity dimensionless variable n defined by

$$n = \frac{x}{2\sqrt{\alpha t}} \tag{9.82}$$

$$\frac{dn}{dx} = \frac{1}{2\sqrt{\alpha t}} \tag{9.83}$$

$$\frac{dn}{dt} = -\frac{x}{4\sqrt{\alpha t^3}} = -\frac{x}{4t\sqrt{\alpha t}} \tag{9.84}$$

Using the chain rule of differentiation yields

$$\frac{\partial T}{\partial t} = \frac{dT}{dn}\frac{\partial n}{\partial t} = -\frac{x}{4t\sqrt{\alpha t}}\frac{dT}{dn} \tag{9.85a}$$

$$\frac{\partial T}{\partial x} = \frac{dT}{dn}\frac{\partial n}{\partial x} = \frac{1}{2\sqrt{\alpha t}}\frac{dT}{dn} \tag{9.85b}$$

$$\frac{\partial^2 T}{\partial x^2} = \frac{d^2 T}{dn^2}\left(\frac{\partial n}{\partial x}\right)^2 + \frac{dT}{dn}\frac{\partial^2 n}{\partial x^2} = \left(\frac{1}{2\sqrt{\alpha t}}\right)^2\frac{d^2 T}{dn^2} \tag{9.85c}$$

$$\frac{\partial^2 T}{\partial x^2} = \frac{1}{4Dt}\frac{d^2 T}{dn^2} \tag{9.85d}$$

where $\partial^2 n/\partial x^2 = 0$. Substitute Eqs. (9.85a) and (9.85d) into (9.81) to get the sought ODE

$$-\frac{x}{4t\sqrt{\alpha t}}\frac{dT}{dn} = \frac{D}{4Dt}\frac{d^2 T}{dn^2} \tag{9.86a}$$

$$-2n\frac{dT}{dn} = \frac{d^2 T}{dn^2} \tag{9.86b}$$

For convenience, let $f = dT/dn$ so that $df = d^2 T/dn^2$ and Eq. (9.86b) becomes

$$-2nf = df \tag{9.87a}$$

$$\frac{df}{f} = -2n \tag{9.87b}$$

$$\int \frac{df}{f} = -2\int ndn \tag{9.87c}$$

$$\ln(f) = -n^2 + h \tag{9.87d}$$

where h is a constant of integration. Solving for f yields

$$f = \exp\left(-n^2 + h\right) = A\exp\left(-n^2\right) \tag{9.88a}$$

$$\frac{dT}{dn} = A\exp\left(-n^2\right) \tag{9.88b}$$

$$dT = A\exp\left(-n^2\right)dn \tag{9.88c}$$

In order to integrate Eq. (9.88c) and solve for A, one needs to describe the heat problem and set the initial condition (IC) and boundary conditions (BCs) as the limits of integration.

For an isotropic material as a semi-infinite $(0 < x < \infty)$ medium,

$$\text{IC} \rightarrow T = T_o \text{ for } n = \infty \text{ at } x > 0,\ t = 0 \tag{9.89a}$$

$$\text{BC} \rightarrow T = T_f \text{ for } n = 0 \text{ at } x = 0,\ t > 0 \tag{9.89b}$$

$$\text{BC} \rightarrow T = T_o \text{ for } n = \infty \text{ at } x = \infty,\ t > 0 \tag{9.89c}$$

Integrating Eq. (9.88c) using the BC given by Eq. (9.89c) yields the constant A as

$$\int_{T_o}^{T_f} dT = A \int_0^\infty \exp\left(-n^2\right) dn = A \frac{\sqrt{\pi}}{2} \tag{9.90a}$$

$$A = \frac{2\left(T_f - T_o\right)}{\sqrt{\pi}} \tag{9.90b}$$

Integrate Eq. (9.88c) again with IC defined by (9.89a) to get

$$\int_{T_o}^{T} dT = A \int_0^n \exp\left(-n^2\right) dn \tag{9.91a}$$

$$T - T_o = \left(T_f - T_o\right) \frac{2}{\sqrt{\pi}} \int_n^\infty \exp\left(-n^2\right) dn \tag{9.91b}$$

This integral defines the Gauss complementary error function, erfc (n). Thus,

$$T - T_o = \left(T_f - T_o\right) \text{erfc}\,(n) \tag{9.92a}$$

$$T - T_o = \left(T_f - T_o\right) \left[1 - \text{erf}\,(n)\right] \tag{9.92b}$$

Substituting Eq. (9.82) into (9.92) yields

$$T - T_o = \left(T_f - T_o\right) \text{erfc}\left(\frac{x}{2\sqrt{\alpha t}}\right) \tag{9.93a}$$

$$T - T_o = \left(T_f - T_o\right) \left[1 - \text{erf}\left(\frac{x}{2\sqrt{\alpha t}}\right)\right] \tag{9.93b}$$

$$T = T_o + \left(T_f - T_o\right) \text{erfc}\left(\frac{x}{2\sqrt{\alpha t}}\right) \tag{9.93c}$$

$$T = T_o + \left(T_f - T_o\right) \left[1 - \text{erf}\left(\frac{x}{2\sqrt{\alpha t}}\right)\right] \tag{9.93d}$$

The characteristics of the error function and its complementary part are erf (n) + erfc $(n) = 1$. Notice that the error function, as defined by the integral in Eq. (9.91b), does not have an exact solution. Thus,

$$\text{erf}(n) = \frac{2}{\sqrt{\pi}} \int_0^n \exp\left(-n^2\right) dn \tag{9.94a}$$

$$= \frac{2}{\sqrt{\pi}} \int_o^n \sum_{k=0}^{\infty} \frac{(-1)^k \, n^{2k}}{k!} dn \tag{9.94b}$$

$$\text{erf}(n) = \frac{2}{\sqrt{\pi}} \left(n - \frac{n^3}{3} + \frac{n^5}{10} - \frac{n^7}{42} + \ldots \right) \tag{9.94c}$$

The accuracy of erf (n) depends on the number of terms in the series, but it gives approximate values with negligible relative error for real values of n.

Similarity Solution

Below is the procedure for deriving the general similarity solution for the heat equation. This is just a similarity transformation of a partial differential equation (PDE) to an ordinary differential equation (ODE) by combining two independent variables (x and t) into a single independent variable n. In fact, x and t are known as independent invariant during transformation.

For one-dimensional heat equation, the Hellums-Churchill method is used hereafter with general initial condition (IC) and boundary conditions (BCs)

$$\frac{\partial T}{\partial t} = \alpha \frac{\partial^2 T}{\partial x^2} \quad \text{for } x > 0, t > 0 \tag{9.95a}$$

$$T(x,t) = T(x,0) = T_{IC} \quad \text{for } t = 0 \tag{9.95b}$$

$$T(x,t) = T(0,t) = T_{BC} \quad \text{for } t = 0 \tag{9.95c}$$

where T_{IC} denotes the initial condition temperature and T_{BC} denotes the boundary condition temperature. Now, making PDE, IC, and BC dimensionless, one can easily find the solution to Eq. (9.95a). Thus,

$$T^* = \frac{T - T_{IC}}{T_d}; \quad x^* = \frac{x}{x_o}; \quad t^* = \frac{t}{t_o} \tag{9.96a}$$

$$\frac{\partial T^*}{\partial t^*} = \left(\frac{\alpha t_o}{x_o^2} \right) \frac{\partial^2 T^*}{\partial x^{*2}} \tag{9.96b}$$

$$T^*\left(x^*, 0\right) = 0 \tag{9.96c}$$

$$T^*\left(0, t^*\right) = \frac{T_{BC} - T_{IC}}{T_d} = 1 \tag{9.96d}$$

$$T_d = T_{BC} - T_{IC} \tag{9.96e}$$

Now,

$$\frac{\alpha t_o}{x_o^2} = \frac{1}{4} \tag{9.97a}$$

$$n = \frac{x}{\sqrt{4\alpha t}} \tag{9.97b}$$

$$T(x, t) = T(n) \tag{9.97c}$$

where two independent variables, $x/\sqrt{t}$, are made into into a single independent variable n. Thus,

$$\frac{\partial n}{\partial x} = \frac{1}{\sqrt{4\alpha t}} \quad \& \quad \frac{\partial n}{\partial t} = -\frac{n}{2t} \tag{9.98}$$

and

$$\frac{\partial T}{\partial t} = \frac{dT}{dn}\frac{\partial n}{\partial t} = -\frac{n}{2t}\frac{dT}{dn} \quad \text{(Chain rule)} \tag{9.99a}$$

$$\frac{\partial T}{\partial x} = \frac{1}{\sqrt{4\alpha t}}\frac{dT}{dn} \tag{9.99b}$$

$$\frac{\partial^2 T}{\partial x^2} = \frac{1}{4\alpha t}\frac{d^2 T}{dn^2} \tag{9.99c}$$

This expression, Eq. (9.99c), reflects the usefulness of the chain rule for rearranging differential equations and for finding the derivatives of the composite functions.

Substituting Eqs. (9.99a), (9.99c) into (9.95a) and upon transformation of the PDE into an ODE yields

$$\frac{d^2 T}{dn^2} + 2n\frac{dT}{dn} = 0 \tag{9.100a}$$

$$\frac{\partial^2 T^*}{\partial n^2} + 2n\frac{\partial T^*}{\partial x} = 0 \tag{9.100b}$$

with boundary condition (BC) and initial condition (IC), respectively,

$$T^*(n) = 1 \text{ for } n = 0 \text{ at } x = 0, \ t > 0 \tag{9.101a}$$

$$T^*(n) = 0 \text{ for } n \to \infty \text{ at } x > 0, \ t = 0 \tag{9.101b}$$

Let $v = dT^*/dn$ so that

$$0 = \frac{dv}{dn} + 2nv \tag{9.102a}$$

$$\frac{dv}{v} = d\,(\ln v) = -2n\,dn \tag{9.102b}$$

$$v = C_1 \exp\left(-n^2\right) \tag{9.102c}$$

Then, the dimensionless temperature function along with BC and IC becomes

$$T^*(n) = C_1 \int_0^n \exp\left(-n^2\right) dn + C_2 \tag{9.103a}$$

$$T^*(n) = T^*(0) = 1 \quad \rightarrow C_2 = 1 \tag{9.103b}$$

$$T^*(n) = T^*(\infty) = 0 \tag{9.103c}$$

$$0 = C_1 \int_0^\infty \exp\left(-n^2\right) dn + 1 \tag{9.103d}$$

$$C_1 = -\frac{1}{\int_0^n \exp\left(-n^2\right) dn} \tag{9.103e}$$

Substituting C_1 and $C2$ results into (9.103a) gives an expression for the dimensionless temperature. Thus,

$$T^*(n) = -\frac{\int_0^n \exp\left(-n^2\right) dn}{\int_0^\infty \exp\left(-n^2\right) dn} + 1 = 1 - \mathrm{erf}(n) = \mathrm{erfc}(n) \tag{9.104a}$$

$$T^*(x, t) = \mathrm{erfc}(n) = \mathrm{erfc}\left(\frac{x}{\sqrt{4\alpha t}}\right) \tag{9.104b}$$

Combining Eqs. (9.96a), (9.96c) and (9.104b) yields the temperature similarity solution by transforming two independent variables (x and t) into one single variable n.

$$\mathrm{erfc}(n) = \frac{T - T_{IC}}{T_d} = \frac{T - T_{IC}}{T_{BC} - T_{IC}} \tag{9.105a}$$

$$T - T_{IC} = (T_{BC} - T_{IC})\,\mathrm{erfc}(n) \tag{9.105b}$$

Substitute Eq. (9.97b) into (9.105b) to get the dimensional temperature field equation as

$$T - T_{IC} = (T_{BC} - T_{IC})\,\mathrm{erfc}\left(\frac{x}{2\sqrt{\alpha t}}\right) \tag{9.106}$$

If $T_{IC} = T_o$ is the initial mold temperature and $T_{BC} = T_f$ is the melt freezing temperature, then Eq. (9.106) takes the form

$$T - T_o = \left(T_f - T_o\right)\operatorname{erfc}\left(\frac{x}{2\sqrt{\alpha t}}\right) \tag{9.107}$$

Solving Eq. (9.107) for T yields the field temperature equation written as the solution of Eq. (9.95a)

$$T = T(x, t) = T_o + \left(T_f - T_o\right)\operatorname{erfc}\left(\frac{x}{2\sqrt{\alpha t}}\right) \tag{9.108}$$

from which $\partial T/\partial x$ and $\partial T/\partial t$ are easily determined

$$\frac{\partial T}{\partial x} = -\frac{\left(T_f - T_o\right)}{\sqrt{\pi \alpha t}}\exp\left(-\frac{x^2}{4\alpha t}\right) \tag{9.109a}$$

$$\frac{\partial T}{\partial t} = \left(T_f - T_o\right)\left(\frac{x}{2t\sqrt{\pi \alpha t}}\right)\exp\left(-\frac{x^2}{4\alpha t}\right) \tag{9.109b}$$

$$\frac{\partial T}{\partial t} = \frac{\left(T_f - T_o\right)x}{2t\sqrt{\pi \alpha t}}\exp\left(-\frac{x^2}{4\alpha t}\right) \tag{9.109c}$$

where $\partial T/\partial x$ denotes the one-dimensional temperature gradient in the x-direction and $\partial T/\partial t$ denotes the rate of freezing, known as the rate of solidification.

Appendix 9C Transcendental Equations

The subsequent analytical procedure defines the transcendental equation related to solid and liquid thermal resistances as schematically shown by the temperature profile in Fig. 9.5b. This case is considered as the Neumann's solution to a Stefan problem when the liquid metal is initially at $T > T_f$ in the half-space region $x > 0$. Thus, $T = T_p$ in a practical casting work is defined as the pouring temperature.

Consider the solid temperature (T_s) equation as the general solution of the heat equation, the boundary, and initial conditions written as

$$\text{Solid}: T_s = C_1 + C_2 \operatorname{erf}\left(\frac{x}{2\sqrt{\alpha_s t}}\right) \tag{9.110a}$$

$$\frac{\partial T_s}{\partial t} = \alpha_s \frac{\partial^2 T_s}{\partial x^2} \quad \text{at } 0 < x < s(t), \; t > 0 \tag{9.110b}$$

$$T_s(x, t) = T_o \quad \text{at } x = 0, \; t > 0 \tag{9.110c}$$

$$T_s(x, t) = C_1 + C_2 \operatorname{erf}\left(\frac{x}{2\sqrt{\alpha_s t}}\right) = T_o + C_2 \operatorname{erf}\left(\frac{x}{2\sqrt{\alpha_s t}}\right) \tag{9.110d}$$

and

$$\text{Liquid:} T_l = C_3 + C_4 \operatorname{erf}\left(\frac{x}{2\sqrt{\alpha_s t}}\right) \tag{9.111a}$$

$$\frac{\partial T_l}{\partial t} = \alpha_l \frac{\partial^2 T_l}{\partial x^2} \quad \text{at } s(t) < x < \infty, \ t > 0 \tag{9.111b}$$

$$T_l(x,t) = T_p \quad \text{at } x > 0, \ t = 0 \tag{9.111c}$$

$$T_l(x,t) = C_3 + C_4 \operatorname{erfc}\left(\frac{x}{2\sqrt{\alpha_l t}}\right) = T_p + C_4 \operatorname{erfc}\left(\frac{x}{2\sqrt{\alpha_l t}}\right) \tag{9.111d}$$

The coupling conditions at $x = s(t)$ are

$$T_s(x,t) = T_l(x,t) = T_f \quad \text{at } x = s(t), \ t > 0 \tag{9.112a}$$

$$k_s \frac{\partial T_s}{\partial x} - k_l \frac{\partial T_l}{\partial x} = \rho \Delta H_f \frac{ds(t)}{dt} \quad \text{at } x = s(t), \ t > 0 \tag{9.112b}$$

Substituting Eqs. (9.110d) and (9.111d) at $x = s(t)$ into (9.112a) yields

$$T_o + C_2 \operatorname{erf}\left(\frac{s(t)}{2\sqrt{\alpha_s t}}\right) = T_p + C_4 \operatorname{erfc}\left(\frac{s(t)}{2\sqrt{\alpha_l t}}\right) = T_f \tag{9.113a}$$

$$T_o + C_2 \operatorname{erf}(\lambda) = T_p + C_4 \operatorname{erf}\left(\lambda\sqrt{\alpha_s/\alpha_l}\right) = T_f \tag{9.113b}$$

where

$$\lambda = \frac{s(t)}{2\sqrt{\alpha_s t}} \quad \& \quad s(t) = 2\lambda\sqrt{\alpha_s t} \tag{9.114a}$$

$$\frac{s(t)}{2\sqrt{\alpha_l t}} = \frac{s(t)}{2\sqrt{\alpha_l t}} \frac{\sqrt{\alpha_s}}{\sqrt{\alpha_s}} = \frac{s(t)}{2\sqrt{\alpha_s t}} \frac{\sqrt{\alpha_s}}{\sqrt{\alpha_l}} = \lambda\sqrt{\frac{\alpha_s}{\alpha_l}} \tag{9.114b}$$

From Eq. (9.113a),

$$C_2 = \frac{T_f - T_o}{\operatorname{erf}(\lambda)} \quad \& \quad C_4 = \frac{T_f - T_p}{\operatorname{erfc}(\lambda)} \tag{9.115}$$

Then Eqs. (9.110d) and (9.111d) become

$$T_s(x,t) = T_o + \frac{T_f - T_o}{\operatorname{erf}(\lambda)} \operatorname{erf}\left(\frac{x}{2\sqrt{\alpha_s t}}\right) \tag{9.116a}$$

$$T_l(x,t) = T_p + \frac{T_f - T_p}{\operatorname{erfc}\left(\lambda\sqrt{\alpha_s/\alpha_l}\right)} \operatorname{erfc}\left(\frac{x}{2\sqrt{\alpha_l t}}\right) \tag{9.116b}$$

Take $\partial T_s(x,t)/\partial x$ and $\partial T_l(x,t)/\partial x$, and substitute the resultant expressions into Eq. (9.112b) to get

$$\frac{\partial T_s\,(x,t)}{\partial x} = \frac{\left(T_f - T_o\right)}{\mathrm{erf}\,(\lambda)} \left(\frac{1}{\sqrt{\pi \alpha_s t}}\right) \exp\left(-\frac{x^2}{4\alpha_s t}\right) \tag{9.117a}$$

$$\frac{\partial T_l\,(x,t)}{\partial x} = \frac{T_f - T_p}{\mathrm{erfc}\left(\lambda\sqrt{\alpha_s/\alpha_l}\right)} \left(\frac{1}{\sqrt{\pi \alpha_l t}}\right) \exp\left(-\frac{x^2}{4\alpha_l t}\right) \tag{9.117b}$$

Subsequently, Eq. (9.112b) becomes

$$\rho\Delta H_f \frac{ds\,(t)}{dt} = k_s \frac{\partial T_s}{\partial x} - k_l \frac{\partial T_l}{\partial x} \quad \text{at } x = s\,(t),\ t > 0 \tag{9.118a}$$

$$\rho\Delta H_f \frac{ds\,(t)}{dt} = k_s \frac{\left(T_f - T_o\right)}{\mathrm{erf}\,(\lambda)} \left(\frac{1}{\sqrt{\pi \alpha_s t}}\right) \exp\left(-\frac{x^2}{4\alpha_s t}\right) \tag{9.118b}$$

$$- k_l \frac{T_f - T_p}{\mathrm{erfc}\left(\lambda\sqrt{\alpha_s/\alpha_l}\right)} \left(\frac{1}{\sqrt{\pi \alpha_l t}}\right) \exp\left(-\frac{x^2}{4\alpha_l t}\right)$$

But, $x = s\,(t)$ so that

$$\rho\Delta H_f \frac{ds\,(t)}{dt} = \frac{k_s\left(T_f - T_o\right)}{\sqrt{\pi \alpha_s t}} \frac{\exp\left[-s\,(t)^2/(4\alpha_s t)\right]}{\mathrm{erf}\,(\lambda)} \tag{9.119}$$

$$- \frac{k_l\left(T_f - T_p\right)}{\sqrt{\pi \alpha_l t}} \frac{\exp\left[-s\,(t)^2/(4\alpha_l t)\right]}{\mathrm{erfc}\left(\lambda\sqrt{\alpha_s/\alpha_l}\right)}$$

Substituting Eq. (9.114a) into (9.119) yields

$$\rho\Delta H_f \frac{ds\,(t)}{dt} = \frac{k_s\left(T_f - T_o\right)}{\sqrt{\pi \alpha_s t}} \frac{\exp\left(-\lambda^2\right)}{\mathrm{erf}\,(\lambda)} \tag{9.120}$$

$$- \frac{k_l\left(T_f - T_p\right)}{\sqrt{\pi \alpha_l t}} \frac{\exp\left(-\lambda^2\alpha_s/\alpha_l\right)}{\mathrm{erfc}\left(\lambda\sqrt{\alpha_s/\alpha_l}\right)}$$

Integrating yields

$$\int \frac{dt}{\sqrt{\pi \alpha_s t}} = \frac{2\sqrt{t}}{\sqrt{\pi \alpha_s}} \tag{9.121a}$$

$$\sqrt{\pi}\rho\Delta H_f s\,(t) = \frac{2k_s\left(T_f - T_o\right)\sqrt{t}}{\sqrt{\alpha_s}} \frac{\exp\left(-\lambda^2\right)}{\mathrm{erf}\,(\lambda)} \tag{9.121b}$$

$$- \frac{2k_l\left(T_f - T_p\right)\sqrt{t}}{\sqrt{\alpha_l}} \frac{\exp\left(-\lambda^2\sqrt{\alpha_s/\alpha_l}\right)}{\mathrm{erfc}\left(\lambda\sqrt{\alpha_s/\alpha_l}\right)}$$

and

$$\sqrt{\pi}\rho\Delta H_f = \frac{2\sqrt{t}}{s\left(t\right)\sqrt{\alpha_s}}\frac{k_s\left(T_f - T_o\right)}{\mathrm{erf}\left(\lambda\right)}\exp\left(-\lambda^2\right) \tag{9.122a}$$

$$-\frac{2\sqrt{t}}{s\left(t\right)\sqrt{\alpha_l}}\frac{k_l\left(T_f - T_p\right)}{\mathrm{erfc}\left(\lambda\sqrt{\alpha_s/\alpha_l}\right)}\exp\left(-\lambda^2\alpha_s/\alpha_l\right)$$

$$\sqrt{\pi}\rho\Delta H_f = \frac{2\sqrt{\alpha_s t}}{s\left(t\right)\alpha_s}\frac{k_s\left(T_f - T_o\right)}{\mathrm{erf}\left(\lambda\right)}\exp\left(-\lambda^2\right) \tag{9.122b}$$

$$-\frac{2\sqrt{\alpha_s t}}{s\left(t\right)\sqrt{\alpha_s}\sqrt{\alpha_l}}\frac{k_l\left(T_f - T_p\right)}{\mathrm{erfc}\left(\lambda\sqrt{\alpha_s/\alpha_l}\right)}\exp\left(-\lambda^2\alpha_s/\alpha_l\right)$$

Use the definition of thermal diffusivity for the solid and liquid phases written as $\alpha_s = k_s/\rho_s c_s$ and $\alpha_l = k_l/\rho_l c_l$ so that $c_s = k_s/\rho_s \alpha_s$ and let $\rho = \rho_s = \rho_l$ at the freezing temperature T_f. Manipulating Eq. (9.122b) using these definitions yields a convenient expression written as

$$\sqrt{\pi}\Delta H_f = \frac{k_s\left(T_f - T_o\right)}{\lambda\rho_s\alpha_s\,\mathrm{erf}\left(\lambda\right)}\exp\left(-\lambda^2\right) \tag{9.123}$$

$$-\frac{k_l\left(T_f - T_p\right)\exp\left(-\lambda^2\alpha_s/\alpha_l\right)}{\lambda\rho_s\sqrt{\alpha_s}\sqrt{\alpha_l}\,\mathrm{erfc}\left(\lambda\sqrt{\alpha_s/\alpha_l}\right)}$$

Again, manipulate Eq. (9.123) to get

$$\frac{\sqrt{\pi}\lambda\Delta H_f}{\left(T_f - T_o\right)} = \frac{c_s}{\mathrm{erf}\left(\lambda\right)}\exp\left(-\lambda^2\right) \tag{9.124a}$$

$$-\frac{\sqrt{\alpha_s}}{\sqrt{\alpha_s}}\frac{k_l}{\rho_s\sqrt{\alpha_s}\sqrt{\alpha_l}}\frac{\left(T_f - T_p\right)\exp\left(-\lambda^2\alpha_s/\alpha_l\right)}{\left(T_f - T_o\right)\mathrm{erfc}\left(\lambda\sqrt{\alpha_s/\alpha_l}\right)}$$

$$\frac{\sqrt{\pi}\lambda\Delta H_f}{\left(T_f - T_o\right)} = \frac{c_s\exp\left(-\lambda^2\right)}{\mathrm{erf}\left(\lambda\right)} - \frac{k_l\sqrt{\alpha_s}}{\rho_s\alpha_s\sqrt{\alpha_l}}\frac{\left(T_f - T_p\right)\exp\left(-\lambda^2\sqrt{\alpha_s/\alpha_l}\right)}{\left(T_f - T_o\right)\mathrm{erfc}\left(\lambda\sqrt{\alpha_s/\alpha_l}\right)} \tag{9.124b}$$

Using $\rho_s\alpha_s = k_s/c_s$ in Eq. (9.124b) yields

$$\frac{\sqrt{\pi}\lambda\Delta H_f}{\left(T_f - T_o\right)} = \frac{c_s\exp\left(-\lambda^2\right)}{\mathrm{erf}\left(\lambda\right)} - \frac{c_s k_l}{k_s}\sqrt{\frac{\alpha_s}{\alpha_l}}\frac{T_f - T_p}{T_f - T_o}\frac{\exp\left(-\lambda^2\alpha_s/\alpha_l\right)}{\mathrm{erfc}\left(\lambda\sqrt{\alpha_s/\alpha_l}\right)} \tag{9.125}$$

At last, the transcendental equation is

$$\frac{\exp\left(-\lambda^2\right)}{\mathrm{erf}\left(\lambda\right)} + \frac{k_l}{k_s}\sqrt{\frac{\alpha_s}{\alpha_l}}\frac{T_f - T_p}{T_f - T_o}\frac{\exp\left(-\lambda^2\alpha_s/\alpha_l\right)}{\mathrm{erfc}\left(\lambda\sqrt{\alpha_s/\alpha_l}\right)} = \frac{\sqrt{\pi}\lambda\Delta H_f}{c_s\left(T_f - T_o\right)} \tag{9.126a}$$

$$\frac{\exp\left(-\lambda^2\right)}{\mathrm{erf}\left(\lambda\right)} - \frac{\gamma_l}{\gamma_s}\frac{T_f - T_p}{T_f - T_o}\frac{\exp\left(-\lambda^2\alpha_s/\alpha_l\right)}{\mathrm{erfc}\left(\lambda\sqrt{\alpha_s/\alpha_l}\right)} = \frac{\sqrt{\pi}\lambda}{S_t} \tag{9.126b}$$

This is the transcendental equation for a two-region solidification domain.

Problems

9.1 Derive Eq. (9.3b) using the Fourier's law of heat conduction and the first law of thermodynamics.

9.2 Compare the solidification time using the Chvorinov's rule for casting a copper spheroid with $a = 2\,cm$ and $c = 4cm$, and a copper sphere with $r = 2\,cm$ in a sand mold with $B = 1.951\ \mathrm{min}/cm^2$. Which casting geometry solidifies first? Why? [Solution: $t_{spheroid} = 4.43$ min].

9.3 Consider a sand casting procedure for producing a flat $20 \times 10 \times 3\,cm^3$ AISI 1030 carbon-steel flat plate. Assume that the pouring and freezing temperatures of the steel are $1520\,^\circ C$ and $1510\,^\circ C$, respectively, and that the mold constant is $3\ \mathrm{min}/cm^2$. Based on this information, determine the solidification time for **(a)** the plate and **(b)** a 4-*cm* diameter sphere. [Solution: (a) $t = 3.21$ min and (b) $t = 1.33$ min].

9.4 Plot the given data on a log-log scale, and determine n and B as per Chvorinov's rule. Is $n = 2 \pm 0.05$? [Solution: $n = 2$ and $B = 2.5027$].

Geometry	$s\,(t) = M = V/A$ (cm)	t (min)
Cube	0.500	0.63
Sphere	0.67	1.11
Plate	1.03	2.68
Cylinder	1.33	4.44
Spheroid	1.51	5.67

9.5 Assume that the Czochralski method is used to produce single-crystal silicon ingots at a pulling velocity $v = 0.5\,mm/s$. In this solidification process, Q_{S-L} denotes the amount of heat released by the solid to a fluid phase, and Q_k denotes the amount of energy removed from the solid-liquid interface A_s by conduction through the ingot. Assume that $\rho_s = \rho_l = 2400\,Kg/m^3$, $T_f = 1687\,K$, $\Delta H_f = 1{,}877{,}000\,J/Kg$, $k_s = 25\,W/m.K$ and $L = 20\,cm$. Calculate **(a)** the L-S interface temperature T_i and **(b)** the amount of heat flux density q_k removed from it. [Solution: (a) $T_i = 1387\,K$ and (b) $q_k = 36{,}125\,W/m^2$].

9.6 Consider casting a 0.04-m thick copper slab into a mold with high thermal conductivity, and assume that the copper solid develops a high thermal resistance during solidification. The thermophysical properties are $\rho_s = 7525\,Kg/m^3$, $k_s = 400\,W/m.K$, $\alpha_s = 1.24 \times 10^{-4}\,m^2/s$ and $\Delta H_f = 205\,kJ/Kg$ at $T_f = 1083\,°C$. If initial mold temperature is $T_o = 25\,°C$, calculate **(a)** the solidification time t, **(b)** the solidification velocity $v = dx_s/dt$, and **(c)** the temperature gradient and the heat flux $(dT/dx)_{x_s}$ and the heat flux q_{x_s} at x_s. [Solution: **(a)** $t = 0.94$ sec, **(b)** $ds(t)/dt = 7 \times 10^{-3}\,m/sec\ v = 7 \times 10^{-3}\,m/s$ and **(c)** $q_{x_s} = 16.45\,MW/m^2$].

9.7 Assume that a 0.04-m thick iron slab casting is to be produced in carbon steel mold and that that there are temperature gradients in the mold and solid. Use the thermophysical properties given below to calculate **(a)** the solidification time t_s, **(b)** the solidification velocity dx_s/dt and **(c)** plot the temperature distribution for the mold and for the solid up to x_s. Assume a pouring temperature of 1555 $°C$, so that the superheat is 20 $°C$. Data for iron casting: $T_f = 1535\,°C$, $k_s = 73\,W/m.K$, $\rho_s = 7530\,Kg/m^3$, $c_s = 728\,J/Kg.K$ and $\Delta H_f = 247.20\,kJ/Kg$. Data for the steel mold: $T_o = 25\,°C$, $k_m = 35\,W/m.K$, $\rho_m = 7689\,Kg/m^3$ and $c_m = 669\,J/Kg.K$. [Solution: (a) $t_s = 15.97$ sec, (b) $dx_s/dt = 2253.60\,mm/h$].

9.8 Consider casting a 0.08-m thick iron slab in a 0.04-m thick copper mold walls. Using the given data, compute **(a)** the solidification time t, **(b)** the solidification velocity $ds(t)/dt$ if $(\partial T/\partial x)_{slab} > 0$, and $(\partial T/\partial x)_{mold} > 0$. Assume negligible solid-mold thermal resistance and superheat and constant air environmental temperature at 25 $°C$. **(c)** Plot $T = f(x, t_s)$ for the iron slab up to $x \leq s(t)$. Explain. Data for iron casting: $T_f = 1535\,°C$, $k_s = 73\,W/m.K$, $\rho_s = 7530\,Kg/m^3$, $c_s = 728\,J/Kg.K$ and $\Delta H_f = 247.20\,kJ/Kg$. Data for the steel mold: $T_o = 25\,°C$, $k_m = 35\,W/m.K$, $\rho_m = 7689\,Kg/m^3$, and $c_m = 669\,J/Kg.K$. [Solution: (a) $t_s = 60.62$ sec, (b) $dx_s/dt = 1188\,mm/h$].

9.9 Consider casting a constant metal volume $(0.02\,m^3)$ in the shape of a cube with a length c, a cylinder with radius r and $D = H$ (equal diameter and height), and a sphere with radius r. Use the Chvorinov's rule to determine which casting shape solidifies faster. Let the Chvorinov's constant be $B = 20,000$ min $/m^2$. [Solution: $t_{cube} = 41$ min $< t_{cylinder} < t_{sphere}$].

9.10 **(a)** Consider a cylindrical casting with height H and diameter D. Based on this information, minimize the casting modulus $M = V/A$, so that the solidification time as per Chvorinov's rule is also minimized. This implies that $V_{min} = f(r) < V_{actual} = f(r, H)$, so that $D = H$. **(b)** Let $r = 0.04\,m$ and $B = 20,000$ min $/m^2$, and calculate t_{actual} for $D = 1.5H$ and t_{min} for $D = H$. [Solution: (b) $t_{min} = 3.56$ min and $t_{actual} = 4.50$ min].

9.11 If it takes 4 $minutes$ to solidify a 5-cm cube in a sand mold, then calculate how long it will take to solidify a 2-cm radius and 6-cm long cylinder. Use the Chvorinov's rule. [Solution: $t_{cylinder} = 3.24$ min].

9.12 If it takes $3.5 \, minutes$ to solidify a $2\text{-}cm$ radius and $6\text{-}cm$ long cylinder, then calculate how long it will take to solidify a $5\text{-}cm$ cube in a sand mold. Use the Chvorinov's rule. [Solution: $t_{cube} = 4.32$ min].

9.13 Assume that a $0.04\text{-}m$ thick copper slab casting is to be produced in a steel mold and that there are temperature gradients in the mold and solid. Use the thermophysical properties given below to calculate **(a)** the solidification time t, **(b)** the solidification velocity dx_s/dt, **(c)** the heat flux q_s crossing the mold-solid interface at $x = 0$, and **(d)** plot the temperature distribution at $-0.04 \le x \le 0 \, m$ for the mold and $0 \le x \le 0.02 \, m$ for the solid. Assume superheats $0, 20$ and $100\,°C$. Data for copper casting: $T_f = 1083\,°C$, $k_s = 400 \, W/m.K$, $\rho_s = 7525 \, Kg/m^3$, $c_s = 429 \, J/Kg.K$ and $\Delta H_f = 205{,}000 \, kJ/Kg$. Data for metal mold: $T_o = 25\,°C$, $k_m = 7689 \, W/m.K$, $\rho_m = 35 \, Kg/m^3$ and $c_m = 669 \, J/Kg.K$. [Solution: (a) $t = 6.67$ sec, (b) $dx_s/dt = 5400 \, mm/h$, (c) $q_s = 2.72 \, MW/m^2$].

9.14 Assume that solidification occurs uniformly by a moving straight interface located at $x = s\,(t)$, and apply the boundary condition $T_s = T_s\,[s\,(t)\,,t] = T_f$ to Eq. (9.19a) to show that the similarity solution of the solid heat equation is of the form

$$s\,(t) = 2\lambda\sqrt{\alpha_s t}$$

9.15 Consider Neumann's solution for solidification of a pure metal using liquid and solid thermal resistances. Assume that the governing equations and boundary conditions are

$$\frac{\partial T_s}{\partial t} = \alpha_s \frac{\partial^2 T_s}{\partial x^2} \quad \text{for } 0 < x < s\,(t)\,, t > 0$$

$$\frac{\partial T_l}{\partial t} = \alpha_l \frac{\partial^2 T_l}{\partial x^2} \quad \text{for } x > s\,(t)\,, t > 0$$

$$T_s\,(x,t) = T_s\,(x,t) = T_f$$

$$k_s \frac{\partial T_s}{\partial x} - k_l \frac{\partial T_l}{\partial x} = \rho_s \Delta H_s \frac{ds\,(t)}{dt}$$

and

$$\text{BC} \quad T_s = T_s\,(0,t) = T_i \quad \& \quad T_l = T_l\,[s\,(t)\,,t] = T_f$$

$$T_s = T_s\,[s\,(t)\,,t] = T_f \quad \& \quad T_l = T_l\,(\infty,t) = T_p$$

$$\text{IC} \quad T_s = T_s\,(x,0) = T_i \quad \& \quad T_l = T_l\,(x,0) = T_p$$

Show that

$$T_s = T_i + \frac{T_f - T_i}{\text{erf}\,(\lambda)}\, \text{erf}\left(\frac{x}{2\sqrt{\alpha_s t}}\right)$$

$$T_l = T_p + \frac{T_f - T_p}{\operatorname{erfc}\left(\lambda\sqrt{\alpha_s/\alpha_l}\right)} \operatorname{erfc}\left(\frac{x}{2\sqrt{\alpha_l t}}\right)$$

and

$$\frac{\exp\left(-\lambda^2\right)}{\operatorname{erf}\left(\lambda\right)} - \frac{k_l}{k_s}\sqrt{\frac{\alpha_s}{\alpha_l}}\frac{T_p - T_f}{T_f - T_i}\frac{\exp\left(-\lambda^2\alpha_s/\alpha_l\right)}{\operatorname{erf}\left(\lambda\sqrt{\alpha_s/\alpha_l}\right)} = \frac{\sqrt{\pi}\lambda\Delta H_s}{c_s\left(T_f - T_i\right)}$$

9.16 Consider the solidification of a semi-infinite slab of aluminum (Al) into a graphite mold. Use the given data (Dantzig and Rappaz [10, p. 170]) below to carry out all required calculations. Initially, the mold and liquid Al are at temperatures indicated in the table, and the Al freezing temperature is $T_f = 660\,^\circ C$.

Material	T ($^\circ C$)	k ($W/m.K$)	ρ (Kg/m^3)	c_p ($J/Kg.K$)	ΔH_f (J/Kg)
mold	25	100	2200	1700	
Solid Al		211	2555	1190	3.98×10^5
Liquid Al	700	91	2368	1090	

Assume that there exists temperature gradients in the mold, solid Al and liquid Al regions and that the phase transformation process at hand is a two-phase solidification problem along with the mold thermal resistance. Set up the governing equations for each region, solve the corresponding heat equations (similarity solutions), and calculate **(a)** the mold-solid interface temperature T_i, **(b)** the solidification time t and **(c)** the solidification velocity $ds\left(t\right)/dt$. **(d)** Let the mold thickness be $x_m = 8s\left(t\right)$ and the half-space be $s\left(t\right) = 0.04\,m$. Plot the temperature profiles for $T_m(x, t)$, $T_s(x, t)$ and $T_l(x, t)$. [Solution: (a) $T_i = 482.80\,^\circ C$, (b) $t = 24.53$ sec, (c) $ds\left(t\right)/dt = 8.15 \times 10^{-4}\,m/\sec$].

9.17 Assume that the Chvorinov's constant is $4.50\,\min/cm^2$. How long will it take to solidify a $(2\,cm) \times (20\,cm) \times (30\,cm)$ flat steel plate in an enclosed sand mold? [Solution: $t = 3.31$ min].

9.18 Calculate the solidification time for casting pure iron (Fe) into a graphite mold using one 0.04-m thick plate. Use data from Appendix 9A. [Solution: $t = 1.57$ min].

9.19 Use the equilibrium Al-Si phase diagram to determine **(a)** the phase diagram parameters T_l, β_l and k_o for an Al-$7Si$ alloy, and **(b)** the constitutional supercooling parameter P_{cs} and the growth restriction factor Q_{cs}. [Solution: (a) $k_o = 0.1429$ and (b) $Q_{cs} = 38.60\,^\circ C$].

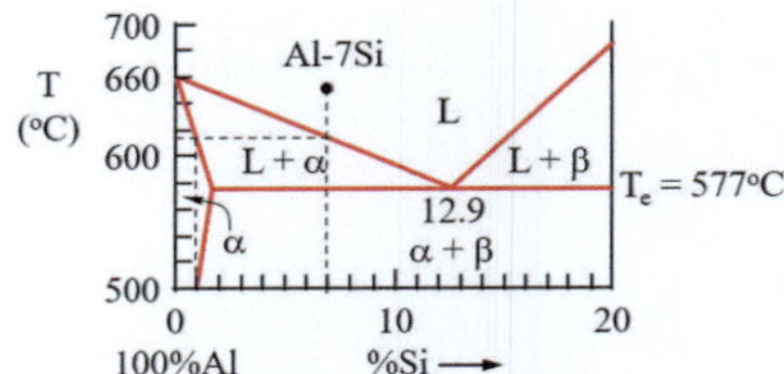

9.20 Use the equilibrium $Al\text{-}Cu$ phase diagram and a pouring temperature $T_p = 700\,°C$ to determine (**a**) the phase diagram parameters T_l, β_l, and k_o and (**b**) the constitutional supercooling parameter $P_{cs.}$ and the growth restriction factor $Q_{cs.}$ for the $Al\text{-}Cu$ alloys listed in the given table containing data as reported by Siqueira et al. [30]. [Solution: $P_{cs,2Cu} = 8.4\,°C$, $Q_{cs,2Cu} = 1.68\,°C$].

	v_x	G_l	dT/dt	d
Alloy	mm/s	K/mm	K/s	mm
$Al\text{-}2Cu$	463	0.526	0.243	3.15
$Al\text{-}5Cu$	0.312	0.715	0.223	3.96
$Al\text{-}8Cu$	0.419	0.542	0.227	3.75
$Al\text{-}10Cu$	0.282	0.754	0.213	3.73

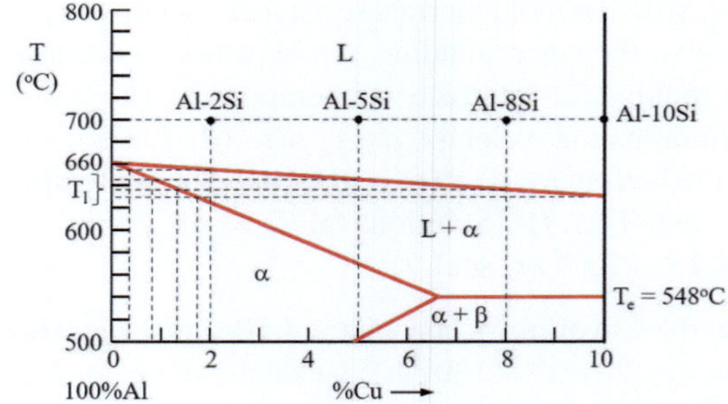

References

1. H.S. Carslaw, J.C. Jaeger, *Conduction of Heat in Solids*, 2nd edn. (Oxford University Press, Clarendon, London, 1959)
2. V. Alexiades, A.D. Solomon, *Mathematical Modeling of Melting and Freezing Processes* (Hemisphere Publishing Corporation, ISBN 1-56032-125-3, Washington DC, 1993)
3. D.W. Hahn, M.N. Ozisik, *Heat Conduction,* 3rd edn. (John Wiley & Sons, New York, 2012)
4. B. Gebhart, *Heat Conduction and Mass Diffusion* (McGraw-Hill, New York, 1993)
5. G.H. Geiger, D.R. Poirier, *Transport Phenomena in Metallurgy* (Addison Wesley Publishing Company, Reading, 1973)

6. D.R. Poirier, G.H. Geiger, *Transport Phenomena in Materials Processing* (Springer International Publishing Switzerland, 2016)
7. D.R. Gaskell, *An Introduction to Transport Phenomena in Materials Engineering*, 2nd edn. (Momentum Press, New York, 2013)
8. D.M. Stefanescu, *Science and Engineering of Casting Solidification*, 3rd edn. (Springer International Publishing Switzerland, 2015)
9. M.E. Glicksman, *Principles of Solidification: An Introduction to Modern Casting and Crystal Growth Concepts* (Springer, Berlin, 2011)
10. J.A. Dantzig, M. Rappaz, *Solidification*, 2nd edn., Revised & Expanded (EPFL Press, Switzerland, 2016)
11. R. Wlodawer, *Directional Solidification of Steel Castings* (Translated by L.D. Hewitt and English Translation Edited by R.V. Riley, Pergamon Press Ltd., New York, 1966)
12. R.A. Flinn, *Fundamentals of Metal Casting* (Addison-Wesley Publishing Company, Reading, 1963)
13. H. Hu, S.A. Argyropoulos, *Mathematical Modeling of Solidification and Melting: A Review.* Modelling and Simulation in Materials Science and Engineering, vol. 4 No. 4 (IOP Publishing, UK, 1996), pp. 371–396
14. J. Caldwell and Douglas K.S. Ng, *Mathematical Modeling: Case Studies and Projects* (Kluwer Academic Publishers, New York, 2004)
15. G.M. Poole, *Mathematical Modeling of Solidification Phenomena in Electromagnetically Stirred Melts*. Ph.D. Dissertation, The University of Alabama, Tuscaloosa, Alabama, 2014
16. W.J. Boettinger, F.S. Biancaniello, S.R. Coriell, Solutal convection induced macrosegregation and the dendrite to composite transition in off-eutectic alloys. Metall. Trans. A **12**(2), 321–327 (1981)
17. C.J. Li, H. Yang, Z.M. Ren, W.L. Ren, Y.Q. Wu, On nucleation temperature of pure aluminum in magnetic fields. Progr. Electromagn. Res. Lett. **15**, 45–52 (2010)
18. W.A. Tiller, K.A. Jackson, J.W. Rutter, B. Chalmers, The redistribution of solute atoms during the solidification of metals. Acta Metall. **1**(4), 428–437 (1953)
19. D.A. Porter, K.E. Easterling, *Phase Transformations in Metals and Alloys*, 2nd edn. (Chapman & Hall, New York, 1992)
20. C. Korner, H. Helmer, A. Bauereiß, R.F. Singer, Tailoring the grain structure of IN718 during selective electron beam melting, in *MATEC Web of Conferences*, vol. 14 (2014), p. 08001
21. L. Ying, X. Qingyan, L. Baicheng, A modified cellular automaton method for the modeling of the dendritic morphology of binary alloys. Tsinghua Sci. Technol. **11**(5), 495–500 (2006)
22. R. Asthana, A. Kumar, N.B. Dahotre, *Materials Processing and Manufacturing Science* (Elsevier, New York, 2005)
23. W. Kurz, D.J. Fisher, *Fundamentals of solidification*, 3rd edn. (Trans Tech Publications, Switzerland, 1992)
24. L.A. Tarshis, J.L. Walker, J.W. Rutter, Experiments on the solidification structure of alloy castings. Metall. Trans. **2**(9), 2589–2597 (1971)
25. I. Maxwell, A. Hellawell, A simple model for grain refinement during solidification. Acta Metall. **23**(2), 229–237 (1975)
26. A.L Greer, A.M. Bunn, A. Tronche, P.V. Evans, D.J. Bristow, Modeling of inoculation of metallic melts: application to grain refinement of aluminium by Al-Ti-B. Acta Mater. **48**(11), 2823–2835 (2000)
27. S.N. Rea, P.S. Gleim, Large area czochralski silicon. Texas Instrument Final Report No. 03-77-23, ERDA/JPL-954475-77/4, Texas Instruments, Dallas, 1977
28. A. Benmeddour, S. Meziani, Numerical study of thermal stress during different stages of silicon Czochralski crystal growth. Revue des Energies Renouvelables, Algeria **12**(4), 575–584 (2009)
29. C.W. Lan, C.K. Hsieh, W.C. Hsu, Czochralski silicon crystal growth for photovoltaic applications, in *Crystal Growth of Si for Solar Cells: Advances in Materials Research*, vol. 14, edited by K. Nakajima, N. Usami (Springer, Berlin, 2009), pp. 25–39
30. C.A. Siqueira, N. Cheung, A. Garcia, Solidification thermal parameters affecting the columnar-to-equiaxed transition. Metall. Mater. Trans. A **33**(7), 2107–2118 (2002)

Chapter 10
Solidification and Phase Diagrams

10.1 Introduction

Foundry technology and related casting process is introduced as the initial source for equilibrium or quasi-static solidification of pure metals and their alloys, and ceramic materials. As a result, solidification induces the evolution of microstructural features used to develop phase diagrams with the aid of cooling curves.

The most relevant aspect of foundry technology in this chapter is the pouring of a melt (liquid metal) into a sand or metal mold containing a hollow cavity and the subsequent liquid-to-solid (L-S) phase transformation, known as solidification due to an outward thermal energy flow and related thermal resistance within the solidification domain Γ. Therefore, optimal thermophysical properties of the mold and the solidifying melt are essential for making high-quality castings by nucleation and growth of the solid phase.

This chapter includes aspects of two-dimensional phase diagrams as graphical representations of the physical state of binary alloys using the temperature-composition relationship (T-C diagram) at a constant atmospheric pressure P_{atm}. Actually, phase diagrams are charts or phase maps that show regions for liquid, liquid-solid, and solid phases at equilibrium.

Understanding the microstructural evolution upon slow solidification of an alloy from its molten state is an essential task for assessing the atomic structure of phases and microstructure-property relationships. Thus, it is important to develop the procedure for constructing equilibrium phase diagram and to understand the microstructural evolution attributed to the nucleation and growth mechanism that form the phases cited in a phase diagram.

The as-cast microstructures are subjected to heat treatment in order to homogenize microstructures and enhance or adjust mechanical properties. This is a practical procedure because solidification time and heat treatment impart a large effect on the microstructural features and mechanical behavior of cast metals and their alloys. In addition, high pressure due to squeezing cast alloys is assumed to have a significant

© The Author(s), under exclusive license to Springer Nature Switzerland AG 2024

N. Perez, *Materials Science: Theory and Engineering*,

https://doi.org/10.1007/978-3-031-57152-7_10

effect on the microstructure and as-cast properties. This is out of the scope of this chapter.

10.2   Fundamentals of Foundry Technology

The common foundry sand is a mixture of sieved sand particles, clay, and water. Thus, its name green sand (foundry sand) because it is wet. The foundry sand can also be referred to as molding or casting sand, and it is usually silica or α-quartz sand (SiO_2), chromite sand ($FeCr_2O_4$), zircon sand ($ZrSiO_4$), or even a combination of these sands. The fundamental property of the sand molds is strength related to plasticity and permeability at relatively high temperatures.

Sand molds are poor thermal conductors with certain degree of porosity for high-temperature (T) gases to escape while quasi-static solidification takes place, mainly at the moving liquid-solid (L-S) interface. In practice, sand molds are commonly used as cost-effective foundry materials and high degree of recyclability. Metallic molds, on the other hand, are good thermal conductors that induce faster solidification than sand molds.

Figure 10.1a depicts a scanning electron microscopy (SEM) image of a natural subangular α-quartz sand grain morphology and Fig. 10.1b exhibits a typical SEM image of a clay-sand bridge between sand grain surfaces. This bridge is fundamentally the bonding between sand grains and the water-clay mixture needed for producing foundry sand molds with strength and plasticity suitable for sand castings. In fact, the grain shape affects porosity and permeability of the sand mold, and the bridging between grains (Perez [1], Perez [2]).

The schematic green-sand mold and related terminology are illustrated in Fig. 10.2. Denote the parting line that divides the mold in halves: cope and drag. Typically, green-sand molds are supported by flasks; otherwise, the molds would collapse during pouring liquid metals.

In regression analysis, three-dimensional curve fitting on sand properties provides a regression surface that may be difficult to assign as a particular mathematical model. However, a polynomial can give good results. For instance, Fig. 10.3

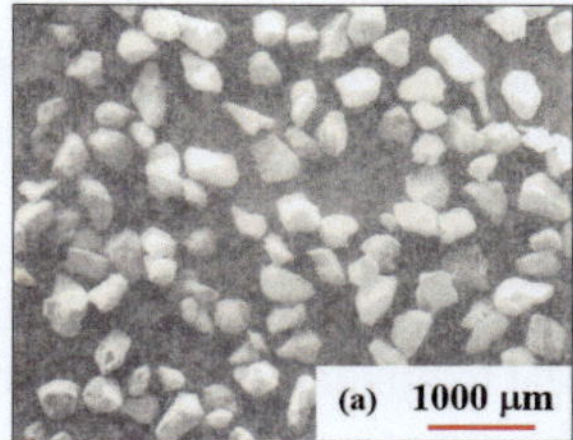

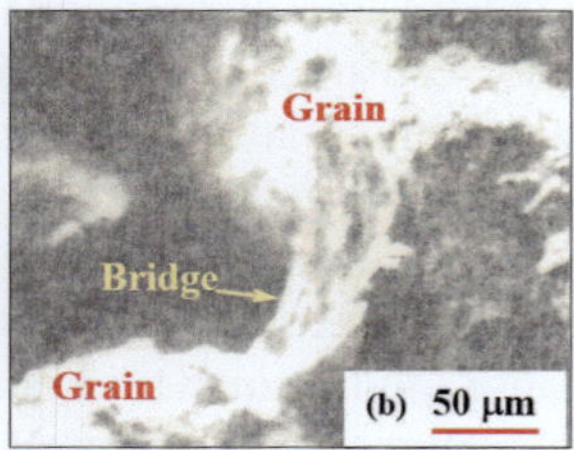

Fig. 10.1 SEM photomicrographs of (a) washed subangular α-quartz sand grains and (b) clay-grain bridge (Perez [2])

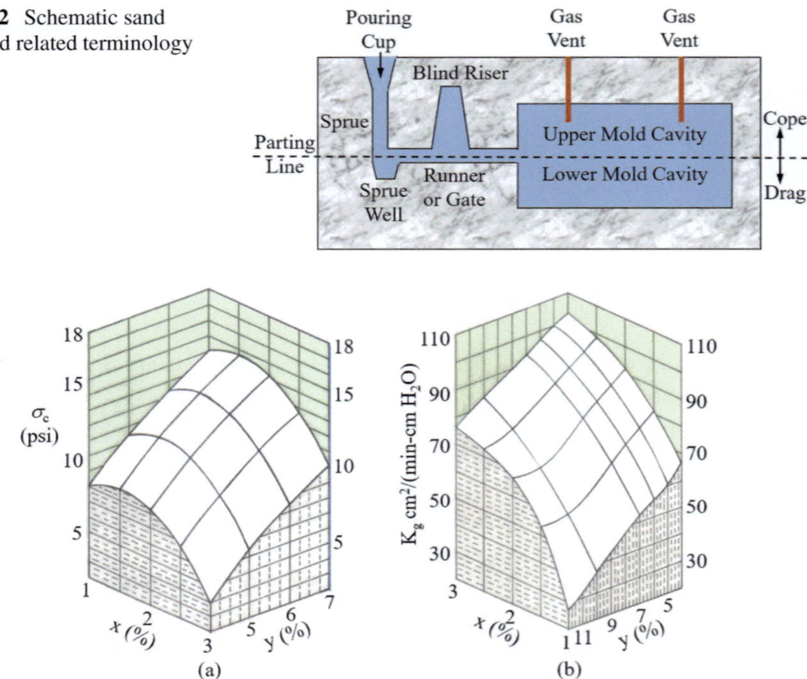

Fig. 10.2 Schematic sand mold and related terminology

Fig. 10.3 (**a**) Green compressive strength $\sigma_c = f(x, y)$ and (**b**) green permeability $K_g = g(x, y)$ of a foundry sand, where x and y are the percentage by weight of moisture (H_2O) and Na-bentonite, respectively. The regression surfaces are $\sigma_c = 4.4357 - 0.0038x + 1.4643y - 0.8134x^2 - 0.0681y^2 + 0.3347xy$ and $K_g = 50.0046 + 41.7005x - 5.2717y - 6.7403x^2 - 0.0289y^2 + 0.8124xy$ (Perez [1])

exhibits regression surfaces for green compressive strength σ_c (Fig. 10.3a) and green permeability K_g (Fig. 10.3b) as functions of percentage moisture x and added sodium (Na) and bentonite clay—denoted as Na-bentonite clay y. Moreover, dry strength is also a sand property used to prevent erosion by liquid metal during pouring since the inner mold surfaces dry out at $T \geq T_f$ (thermodynamic melting point). Nonetheless, the foundry sand should be plastic so that sand grains and clay particles are held together by electrostatic attraction forces. Thus, the basic mixture of these ingredients is a cost-effective approach in the foundry engineering field.

Green permeability K_g measures the amount of air and hot gases that flow through the mold walls. When $K_g \to \infty$, mold voids are accessibly large and the liquid metal easily penetrates the mold walls. Conversely, if $K_g \to 0$, then air and gases may not escape the mold, and consequently, gas entrapment may occur during metal solidification. Therefore, $0 < K_g < \infty$ is a large practical range, but it depends mainly on experience.

Fig. 10.4 Actual foundry practice for steelmaking castings. (**a**) Pouring liquid steel from an arc furnace into a ladle and (**b**) unfinished steel valve

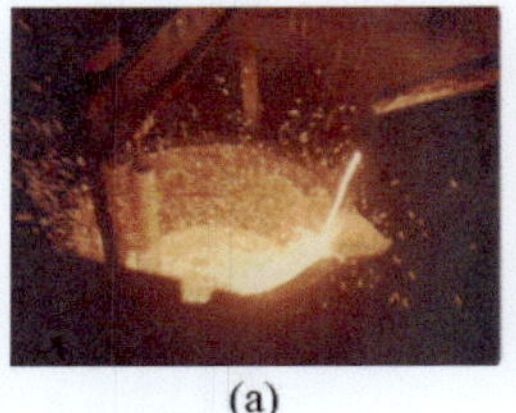

(a)   (b)

10.2.1   Foundry Practice

The purpose of this section is to give the reader insights on solidification-induced fluid flow problems that may arise in the making of high-quality castings. For the sake of simplicity, it is assumed that the solidification process is optimal and only the cooling curves related to solidification are considered in this chapter for constructing binary phase diagrams and for characterizing the microstructural evolution during solidification. Recall that Chap. 9 contains analytical procedures for assessing the solidification heat transfer and related solidification methods.

Figure 10.4a shows a ladle being filled with a molten steel to be poured into sand molds and Fig. 10.4b exhibits an actual semifinished steel valve that needs further processing to mechanically remove the excess solidified steel. Nonetheless, producing a casting requires the use of a thermally efficient ladle or crucible technology for energy savings, mold-material technology, fundamentals of fluid flow, mold-gating design, mathematics for conservation of mass and energy, and mathematics of heat transfer.

During the casting process, liquid metal is poured into the mold cavity, where it solidifies under a controlled fashion in order to obtain a desired as-solidification or as-cast microstructure, which, in turn dictates the properties of a casting product (Fig. 10.4b).

The making of castings is a complicated process because it involves sand or metal mold technology, melting metal scrap (recyclable steel) with relevant chemical composition, and quality control assurance prior to pouring the liquid metal into molds. This implies that understanding the fundamentals of fluid flow is a main requirement in order to produce sound castings since the quality of the casting is significantly governed by the pouring time of the melt into molds, molten metal flow in the molds, and heat transfer through the mold walls. In particular, the melt flow is influenced by the gating system and the local electromagnetic field that may cause electrohydrodynamic problems. Nowadays, assessing the solidification process of molten materials using external magnetic fields is widespread in the metal and semiconductor industries.

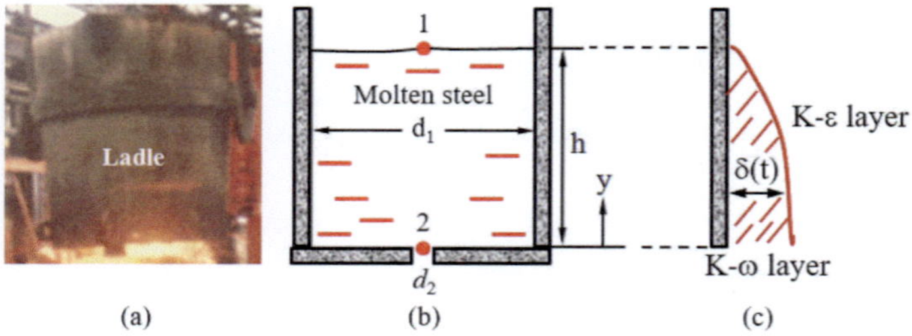

Fig. 10.5 Downward vertical fluid flow of molten steel from a ladle. (**a**) Ladle containing the molten steel, (**b**) cross section of the ladle, and (**c**) type of solidified layer at the inner ladle wall

10.2.2   Fluid Flow

The foundry case described below introduces a simplified mathematical form related to the fundamentals of fluid flow. The goal is to predict the time it takes to pour a certain amount of molten steel into molds. This is an important and yet simple procedure needed to prevent the molten material to freeze in the ladle. The main emphasis is on the use of fundamentals of thermodynamics for fluid flow problems related to the conservation of mass. One particular foundry practice is shown in Fig. 10.5a along the ladle sketch (Fig. 10.5b). Also, K-ω and K-ε layers (Fig. 10.5c) are included as references in fluid flow analysis for a steel casting.

From thermodynamics, the mass flow rate (dm/dt) and the flow velocity (v) as per Torricelli's theorem are

$$\frac{dm}{dt} = \frac{dm_1}{dt} - \frac{dm_2}{dt} = (\rho v A)_1 - (\rho v A)_2 \tag{10.1}$$

with

$$v = \frac{dy}{dt} = \sqrt{2gy} \tag{10.2}$$

where m denotes the mass of the molten material, t denotes the time, ρ denotes the steel density, v denotes the liquid steel velocity, A $(\pi/4)\, d_2^2$ denotes the cylindrical cross-sectional area, y denotes the height of the molten steel in the ladle, and $g = 9.81 \ m/s^2$ denotes the standard acceleration due to gravity.

Once the ladle is filled to a desired height h with liquid steel, the mass flow rate into the ladle is $dm_1/dt = dm_{in}/dt = 0$ and that out of the ladle is $dm_2/dt = dm_{out}/dt$.

For a cylindrical molten material within the ladle under a steady-state condition, the conservation of mass allows that $dm_2/dt = dm_{out}/dt$. Thus,

$$\frac{dm}{dt} = -(\rho v A)_2 = -\rho\sqrt{2gy}\left(\frac{\pi}{4}d_2^2\right) \tag{10.3a}$$

$$\frac{dm}{dt} = \rho \frac{dy}{dt} \left(\frac{\pi}{4} d_1^2 \right) \tag{10.3b}$$

Equating Eqs. (10.3a) and (10.3b) yields an expression for predicting the time to empty the ladle

$$-\frac{\pi}{4} d_2^2 \rho \sqrt{2gy} = \frac{\pi}{4} d_1^2 \rho \frac{dy}{dt} \tag{10.4a}$$

$$dt = -\left(\frac{d_1}{d_2} \right)^2 \frac{1}{\sqrt{2g}} \frac{dy}{\sqrt{y}} \tag{10.4b}$$

$$\int_0^t dt = -\left(\frac{d_1}{d_2} \right)^2 \frac{1}{\sqrt{2g}} \left[\int_{h_1}^{h_2} \frac{dy}{\sqrt{y}} \right] \tag{10.4c}$$

$$t = \left(\frac{d_1}{d_2} \right)^2 \frac{1}{\sqrt{2g}} \left[2 \left(\sqrt{h_1} - \sqrt{h_2} \right) \right] \tag{10.4d}$$

where h_1 is the initial height of the molten steel and $0 \le h_2 < h_1$ is the height of it after pouring. Any additional background on fluid mechanics and fluid dynamics is out of the scope of this textbook.

Example 10.1 Consider a 2-m-diameter ladle containing a 10-ton mass of molten steel at the pouring austenitic temperature $T_p > T_f$, where T_f is the upper freezing temperature of the molten steel. Calculate the time to empty the steel through an 8-cm-diameter hole at the bottom of the ladle. Neglect buoyancy forces induced by thermal gradients as the ladle cools and assume a ladle drainage operation of steel under a near-isothermal condition ($\Delta T \rightarrow 0$) with negligible vortex flow problems (flow revolving around an axis line). Also neglect the evolution of any thermal stratification (different temperatures coming into contact) of the molten steel contained by the ladle after standstill times. For clarity, the steel inner boundary layer can be characterized by the Reynolds K-ε model or by the shear stress transport K-ω model.

Solution For $d_1 = 2\,m$, $d_2 = 8\,cm$, $\rho = 7.80\,g/cm^3$, $m = 10\,ton = 10,000\,Kg$, the molten steel height in the ladle (depth of steel above the bottom hole) is determined from the mass density equation $\rho = m/V = m/A_1 h_1$

$$h_1 = \frac{m}{\rho A_1} = \frac{10 \times 10^6\,g}{\left(7.80\,g/cm^3 \right) \left(\pi/4 \right) \left(2 \times 10^2\,cm \right)^2} = 40.81\,cm \tag{10.1E1}$$

From Eq. (10.4d) with $h_2 = 0$ and $g = 9.81\,cm/s^2$,

$$t = \left(\frac{d_1}{d_2} \right)^2 \frac{1}{\sqrt{2g}} \left[2 \left(\sqrt{h_1} - \sqrt{h_2} \right) \right] \tag{10.1E2a}$$

$$t = \left(\frac{2 \times 10^2 \; cm}{8 \; cm}\right)^2 \left(\frac{2}{\sqrt{(2)\,(9.81 \times 10^2 \; cm/s^2)}}\right) \sqrt{40.81 \; cm} \qquad (10.1\text{E2b})$$

$$t = 180.28 \; s \simeq 3 \; \text{min} \qquad (10.1\text{E2c})$$

Therefore, it takes approximately 3 minutes to empty the ladle by pouring the molten steel into a mold having appropriate thermal properties, which must be suitable for achieving thermal stability and mechanical strength to hold the solidifying mass of molten steel in the mold cavity.

10.3 Thermodynamics of Phase Transformation

This section is an extension of Chap. 8 highlighting the influence of temperature (T) at constant pressure (P) on solidification or melting (phase transformation). Accordingly, fixing the pressure at $P = 1 \; atm = 101.33 \; kPa$, the Gibbs energy of mixing metallic liquid (G_l) and solid (G_s) phases are related to the enthalpy (H) and entropy (S) at T.

It is common to consider "close" and "open" thermodynamic systems. The former can exchange only energy with its surroundings because it does not allow transfer of matter in or out of the system, while the latter can exchange energy and matter with its surroundings. Thus, the thermodynamics of phase transformation is treated as an open system with homogeneous properties and composition.

The enthalpy function $H = H(T)$ is a measure of the heat content related to the internal energy, pressure, and volume. On the other hand, the Gibbs energy function $G = G(T)$ is the driving force for phase transformation since it is related to enthalpy and entropy $S = S(T)$, and it reaches the lowest value at equilibrium where $dG = 0$.

Thermodynamics of phase transformation (Chap. 8) and solidification heat transfer (Chap. 9) highlight important theoretical aspects of solidification. However, additional thermodynamical insights related to phase transformations can clarify the role of thermodynamics laws.

Mathematically, the first (conservation of energy) and second (concept of entropy) laws of thermodynamics and their combined version for a system undergoing phase transformation can conveniently be written (recast from Chap. 8) in a generalized form

$$Q - W = \Delta H + \Delta KE + \Delta PE \quad (\text{1st}) \qquad (10.5\text{a})$$

$$Q = T\Delta S \qquad (\text{2nd, rev.}) \qquad (10.5\text{b})$$

$$T\Delta S - W = \Delta H + \Delta KE + \Delta PE \quad (\text{Combined}) \qquad (10.5\text{c})$$

where Q denotes the heat transfer into or out of the system and it is a thermal energy in transit due to a temperature difference (freezing range) ΔT or a temperature gradient dT/dx, W denotes the work done on or by the system, ΔH denotes the enthalpy change of the substance in question, ΔKE denotes the kinetic energy change due to changes in internal energy, ΔPE denotes the potential energy change, ΔS denotes the entropy change due to atomic or molecular disorder, P denotes the pressure of the system, and V denotes the liquid or solid volume.

The meaning of heat transfer and work done can be deduced as indicated below:

- For $Q > 0$, the system absorbs heat from the environment during melting metals in an induction or electric arc furnace, where heat is converted to heat of fusion; $Q \rightarrow \Delta H_f$.
- For $Q < 0$, the system releases heat to the environment during solidification, where the latent heat of solidification (ΔH_s) being released at the liquid-solid (L-S) interface is converted to heat ($\Delta H_s \rightarrow Q_s$). Subsequently, heat transfer occurs through the solidified melt to the local atmosphere; $Q_s \rightarrow Q_{atm}$.

For steady-state solidification, $\Delta KE = \Delta PE = 0$ and Eq. (10.5a) is subsequently manipulated in order to define the Gibbs energy change (ΔG) of the system as

$$\Delta G = -W = \Delta H - Q \tag{10.6a}$$

$$\Delta G = \Delta H - T\Delta S \tag{10.6b}$$

from which the following thermodynamic criteria are relevant to a macroscopic system:

$$\Delta G < 0, \text{ spontaneous process} \tag{10.7a}$$

$$\Delta G = 0, \text{ equilibrium system} \tag{10.7b}$$

$$\Delta G > 0, \text{ reverse spontaneous process} \tag{10.7c}$$

For an isothermal process ($T = fixed, dT = 0$), the fundamental definition of the Gibbs energy can be used to derive its differential form for describing an infinitesimal (infinitely small) change induced by an infinitesimal change in entropy (dS). Thus,

$$G = H - TS \tag{10.8a}$$

$$dG = dH - TdS - SdT = dH - TdS \tag{10.8b}$$

Consider the solidification process as an open thermodynamic system, where $H = H(T)$ and $G = G(T)$ functions and their first derivative define the heat capacities $c_i = dH_i/dT$ (Fig. 10.6a) and entropy $S_i = dG/dT$ (Fig. 10.6b), where $i = l, s$ with "l" and "s" represent the liquid and solid phases, respectively. Notice

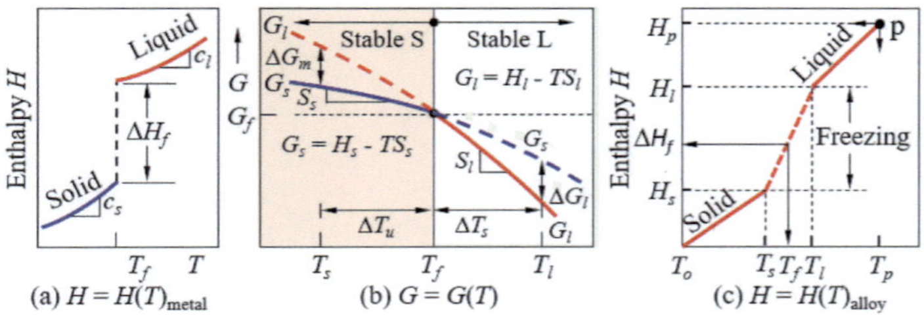

Fig. 10.6 Schematic energy trends at temperature T for liquid and solid phases. (a) Enthalpy diagram for a pure metal and (b) Gibbs energy diagram and (c) enthalpy diagram for an alloy

that the enthalpy jump is shown in Fig. 10.6a, which defines the latent heat of fusion $\Delta H_f = \Delta H = H_l - H$ at T_f.

For comparison, ΔH_f^* is the latent heat of fusion of an alloy at the freezing temperature range $T_s < T_f^* < T_l$ (Fig. 10.6c). Here, T_s is the solid temperature and T_l is the liquid temperature. Moreover, Fig. 10.6c exhibits a non-isothermal process for an alloy that freezes at a temperature range $T_s \leq T_f < T_l$.

Phase stability is graphically illustrated in Fig. 10.6b, where $G_f = f\left(T_f\right)$ is the equilibrium Gibbs energy at the fusion or melting temperature T_f. Notice that solidification occurs when $\Delta G = \Delta G_m = (G_s - G_l) < 0$ at $T = T_s$, induced by an undercooling $\Delta T_u = \left(T_f - T_s\right) > 0$. Here, ΔG_m denotes the Gibbs energy of mixing. For melting, $\Delta G = (G_l - G_s) < 0$ at $T = T_l$, where $\Delta T_s = \left(T_l - T_f\right) > 0$ denotes the degree of superheating the liquid.

For a small temperature change $\Delta T = \left(T - T_f\right) \rightarrow 0$ called undercooling, $\Delta G \rightarrow 0$ in Eq. (10.6b) at $T \rightarrow T_f$ so that $\Delta H_f(T) = \Delta H_f\left(T_f\right)$ and $\Delta S_f(T) = \Delta S_f\left(T_f\right)$. Thus, the entropy of fusion is defined as the normalized latent heat of fusion by the liquid temperature T undergoing solidification

$$\Delta S_f = \frac{\Delta H_f}{T} \tag{10.9}$$

Combining Eqs. (10.6b) and (10.9) gives

$$\Delta G = \Delta H_f - T_f \left(\frac{\Delta H_f}{T}\right) = \Delta H_f \left(\frac{T - T_f}{T}\right) \tag{10.10a}$$

$$\Delta G = -\left(\frac{\Delta H_f}{T}\right)\Delta T = -\Delta S_f \Delta T \tag{10.10b}$$

The degree of undercooling during L-S transformation is simply defined as

$$\Delta T = -\Delta G / \Delta S_f \tag{10.11a}$$

$$\Delta T = \left(T_f - T\right) > 0 \tag{10.11b}$$

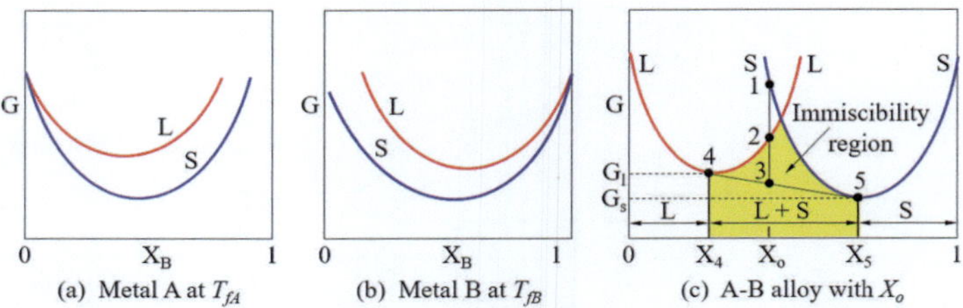

Fig. 10.7 Shape of $G = f(X_B)$ for two phases. **(a)** Stable liquid (L), **(b)** stable solid (S) and **(c)** alloy with nominal mole fraction X_o, solubility limits at X_4, X_5 and immiscibility region at $X_4 \leq X_B \leq X_5$

The entropy of fusion ΔS_f, in general, is a measure of disorganized thermal energy per temperature, while the freezing temperature T_f is a measure of the atomic bond strength. Moreover, the $\Delta S_f T_f$ product is related to solidification at $T = T_f$, while the $\Delta S_f (T - T_f)$ product indicates that solidification starts at a characteristic temperature $T < T_f$ due to an undercooling process. Mathematically, the entropy of fusion is $\Delta S_f = S_l - S_s$ during melting of an organized crystalline solid. Here, ΔS_f increases due to the disorganized atomic structure of a liquid phase.

10.3.1   Phase Separation

The Gibbs energy diagram (G-X diagram) has particular characteristics related to the concavity of the graph of $G_l, G_s = f(X_B)$ at T. For instance, the concave down shape of $G = f(X_B)$ function varies with temperature T as indicated in Fig. 10.7. The extreme conditions can be defined as $G_s(X_B) > G_l(X_B)$ at $T > T_f$ (Fig. 10.7a) and $G_s(X_B) < G_l(X_B)$ at temperature $T < T_f$ (Fig. 10.7b).

For solidifying a hypothetical *A-B* binary alloy with nominal mole fraction X_o (Fig. 10.7c), the Gibbs energies G_s, G_l have concave down shapes. The marked points indicate that:

- **Point 1** $G_l(X_o) = G_1(X_o)$ and only liquid phase exits at $T_l > T_f$.
- **Point 2** $G_s(X_o) = G_2(X_o)$ and only solid phase exits at $T_s < T_l$.
- **Point 3** $G_3(X_o) = G_4\left(X_B^s\right) + G_5\left(X_B^l\right)$ phase separation at $T_s < T > T_l$.
- **Points 4 and 5** These are connected by drawing a tangent line that acts as a tie-line (Lever rule which assumes infinite slow cooling), where $G_3(X_o)$ is the total Gibbs energy of mixing when phase separation occurs at $T_s < T > T_l$. Thus,

$$G_3(X_o) = G_4\left(X_B^s\right) + G_5\left(X_B^l\right) \quad \text{at } T_s < T > T_l \tag{10.12a}$$

$$G_3(X_o) = \left(\frac{X_B^l - X_o}{X_B^l - X_B^s}\right) G_s + \left(\frac{X_o - X_B^s}{X_B^l - X_B^s}\right) G_l \tag{10.12b}$$

$$G_3 (X_o) = f_s G_s + f_l G_l \tag{10.12c}$$

where the solid- and liquid-phase fractions, f_s and f_l, respectively, are defined by the Lever rule

$$f_s = \frac{X_B^l - X_o}{X_B^l - X_B^s} \tag{10.13a}$$

$$f_l = \frac{X_o - X_B^s}{X_B^l - X_B^s} \tag{10.13b}$$

with a generalized nominal composition X_o (mole fraction or weight percentage $wt\%$ or %) at a certain temperature T.

10.3.2   Characteristics of Energy Diagrams

There are different types of binary phase diagrams with specific phase fields representing phase regions, which have uniform chemical composition and physical properties. For the sake of clarity, below are some particular characteristics of the Gibbs energy diagram (Fig 10.7c) related to binary A-B phase diagrams.

Physical Behavior   The relevant physical behavior of the A-B solution is thermodynamically described by the Gibbs energy function, $G = f (X_B)$ at $0 \le X_B \le 1$ and temperature T, as hypothetically illustrated in Fig. 10.7, where the mole fraction relationship is $X_A + X_B = 1$.

Miscibility Gap   A particular A-B solution has a miscibility gap when it spontaneously decomposes into a mixture of two solutions with different concentrations, one of which is predominantly A-rich with some B in solution, and the other predominantly B-rich with some A in solution. This physical behavior occurs at point 3 in Fig. 10.7c. Subsequently, the $G = f (X_B)$ diagram is divided into three regions, which correspond to the mole fraction range $0 < X_B \le X_4$ for the liquid, $X_5 \le X_B \le 1$ for the stable solid, and $X_4 \le X_B \le X_5$ for mixture of L and S. Moreover, note that the tangent line intercepts the minimum Gibbs energy, $G_{\min} = G_l (X_4)$, for the liquid (L) and $G_{\min} = G_s (X_5)$ for the solid (S) phases. Hence, X_4, X_5 represent solubility limits of B atoms in the L and S phases composed of A, B atoms or components.

Immiscibility Region   If a single-phase solution decomposes into two different phases, then the A, B components become immiscible. Nonetheless, the immiscibility region is located at a mole fraction range $X_4 \le X_B \le X_5$ between $G_{\min} (X_4)$ and $G_{\min} (X_5)$ representing the immiscibility region, where there exists a mixture of L and S phases. Moreover, $G_l (X_4)$, $G_s (X_5)$ reach minimum values with energy gradient $dG/X_B = 0$ at X_4, X_5. Also, if $d^2G/X_B^2 = 0$, then $X_A = X_B$ at an inflection point.

Effect of Temperature When $T = T_{fA}$ (Fig. 10.7a), Eq. (10.6b) gives $\Delta G = \Delta H - T_{fA}\Delta S$ with $\Delta G_A = \Delta H_f$ and $\Delta S = 0$ since the entropy is $S_A = S_B$. Similarly, when $T = T_{fB}$ (Fig. 10.7b), the energy change becomes $\Delta G = \Delta H - T_{fA}\Delta S = \Delta G_B = \Delta H_f$ and $\Delta S = 0$ since the entropy terms are $S_A = S_B$. When $T_{fA} < T < T_{fB}$ (Fig. 10.7c), however, G_l and G_s curves intercept at some X_B value, and consequently, the minimum value of the $G_{\min,s} = f_s(X_B)$ curve is lower than that $G_{\min,l} = f_s(X_B)$ curve, and both L and S phases are stable in the immiscibility gap at $X_4 \leq X_B \leq X_5$. The opposite physical behavior may occur when $G_{\min,l} < G_{\min,s}$.

Effect of Composition When $X_4 \leq X_B \leq X_5$, the phase separation indicates that the liquid composition is given by $X_4 = X_{l,B}$ and the solid becomes B-richer since $X_5 = X_{s,B} < X_{l,B}$, and their mass fractions are given by the Lever rule, Eq. (10.13). Therefore, $G_{\min,l} < G_{\min,s}$ as before since $\Delta S_{\min,l} > \Delta S_{\min,s}$. Thus, if ΔS increases, then $\Delta G = \Delta H - T\Delta S$ decreases and vice versa.

Solubility Limit It is the maximum amount of a component that can be dissolved in a phase at a temperature T. For instance, carbon (C) has a limited solubility in iron (Fe) BCC-phase called ferrite, Cu in aluminum (Al) FCC-phase, and so forth.

Most A-B binary and A-B-C ternary phase diagrams are viewed as phase maps constructed using thermodynamic data obtained at atmospheric pressure. In effect, a phase diagram can be treated as a chart that provides conditions of an alloy. Thus, the goal in this chapter is to present a general procedure for constructing and interpreting an equilibrium phase diagram containing liquid- and solid-phase regions as function of temperature and chemical composition at constant pressure.

10.3.3   Microstructural Evolution and Temperature Gradient

For one-dimensional analysis, the directional solidification process (DSP) is explicitly related to the cooling rate dT_l/dt of the liquid phase (freezing rate), which can mathematically be described by the chain rule as

$$\frac{dT_l(x)}{dt} = \frac{dx}{dt}\frac{dT_l(x)}{dx} \tag{10.14}$$

where $v_x = dx/dt$ denotes the assumed steady-state solidification rate or solidification front velocity and dT/dx denotes the temperature gradient ahead of the assumed planar solidification front (L-S interface) along the x-direction.

This expression, Eq. (10.14), indicates that during very **slow solidification** (SS) or equilibrium solidification, the microstructural evolution depends on the molten metal undercooling (ΔT) and the solidification rate related to the evolution of a latent heat of solidification ΔH_s at the L-S interface.

The solidification heat transfer condition is generalized by the slow freezing rate characterized by the following mathematical statement (condition):

$$\frac{dT_l(x)}{dt} \to 0 \text{ as } \frac{dx}{dt} \to 0 \text{ at } \frac{dT_l(x)}{dx} > 0 \tag{10.15}$$

During ultra **rapid solidification** processing (RSP), on the other hand, the molten metal acquires a deep or large undercooling ΔT and the microstructural evolution depends on the fast freezing rate or cooling rate condition

$$\frac{dT_l(x)}{dt} \to \infty \text{ as } \frac{dx}{dt} \to \infty \text{ at } \frac{dT_l(x)}{dx} > 0 \tag{10.16}$$

and as a result the final casting product may consist of fine grains or amorphous structure. Moreover, any deviation from the equilibrium solidification stage is considered as a rapid solidification process. Most industrial solidification processes for producing alloys are carried out at intermediate solidification rates; $0 < dT/dt < \infty$.

The microstructural evolution during solidification, as described by Eq. (10.14), can be analyzed in a different manner. For instance,

- If $dT_l(x)/dt \to \infty$ is imposed on the solidification of a small casting dimension, such as a strip, then rapid solidification prevails as the phase transformation process (PTP), where the melt is thermodynamically unstable. Hence, the final microstructure can be entirely amorphous since there is not enough time for a phase partition (k_o) to develop. Hence, rapid solidification lacks microsegregation.
- If $0 < dT_l(x)/dt << \infty$, then the casting dimensions can be very large as in the case of ingot production and the solidification process can be thermodynamically stable. In this case, the final microstructure may be equiaxed, columnar, and dendritic or a combination of these microstructural morphologies. As a result, a phase partition (k_o) develops.

Manipulating or controlling the solidification rate dT_l/dt and the temperature gradient dT_l/dx during solidification leads to grain refinement associated with heterogeneous nucleation and minimum segregation effects. Theoretically, this leads to high-quality castings with optimal shape, proper microstructural morphology, optimum mechanical properties, and durability. In essence, the freezing or cooling rate dT_l/dt is directly related to the evolved latent heat of solidification ΔH_s and the heat transfer process through mold walls.

10.4   Cooling Curves and Phase Diagrams

For solidifying a pure metal A, the cooling curve shown in Fig. 10.8a is used to determine the temperature distribution related to the freezing process. Moreover, Fig. 10.8a elucidates a two-phase transformation (liquid and solid), where t and t_{total} denote the local and the total solidification times, respectively. The pouring and freezing temperatures denoted by T_p and T_f, respectively, are coupled with the phase diagram depicted in Fig. 10.8b.

The sequence of solidification or solidification paths for an A-B alloy with composition C_o at the pouring temperature $T_p > T_f$ (Fig. 10.8c) can be summarized in terms of heat transfer Q, specific heat capacity c_p at constant pressure P, and latent heat of fusion ΔH_f. Thus,

- For freezing the melt under controlled conditions, the heat transfer from the pouring temperature T_p is

$$Q_p = mc_{p,l}\Delta T_s \quad \text{where } \Delta T_s = T_p - T_f \tag{10.17a}$$

- For local solidification, the heat transfer is defined as

$$Q_f = m\Delta H_f \quad \text{at } T = T_f \text{ at } t > 0 \tag{10.17b}$$

- For total solidification, the total heat transfer is

$$Q_s = Q_p + Q_f \quad \text{when } T_p \to T_f \text{ at } t > 0 \tag{10.17c}$$

- For cooling the solid, the heat transfer is defined as

$$Q_c = mc_{p,s}\Delta T_c \quad \text{when } T_f \to T_o = 25\,^\circ C \text{ at } t \gg 0 \tag{10.17d}$$

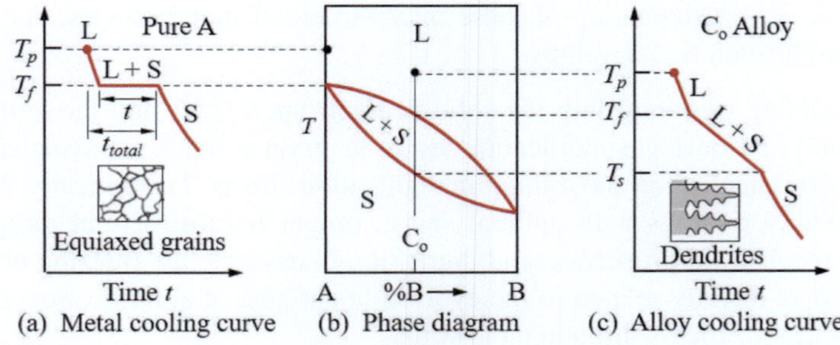

Fig. 10.8 Schematic cooling curves and related equilibrium phase diagram for a hypothetical A-B binary alloy. (**a**) Cooling curve for pure metal A showing the local solidification time t and total solidification time t_{total}, (**b**) isomorphous phase diagram, and (**c**) cooling curve for an A-B binary alloy with nominal composition C_o

Notice that the solidification of a C_o-alloy is a two-phase transformation, and as expected, the analysis of transient solidification is rather complicated because the latent heat of fusion ΔH_f^* is released at a temperature range $T_s < T_f < T_l$.

10.5   Melt Undercooling and Solidification Time

The goal is to obtain a solid material with properties related to specific microstructural features. In order to achieve this goal, the melt (metallic liquid phase) is initially at the pouring temperature $T_p > T_f$, where T_f is the freezing temperature, and subsequently, it begins to solidify at $T_x < T_f$. This means that the melt undergoes an undercooling or supercooling process, where the degree of undercooling is defined as a temperature range $\Delta T = (T_f - T_x) > 0$. This means that the melt cools to just below its freezing T_f for the formation of nuclei having a specific crystal structure. Thus, the Gibbs energy change, $\Delta G = (G_s - G_l.) < 0$, is taken as the solidification driving force for nuclei formation (solid phase) with a $G_s < G_l$. This means that ΔG must overcome the thermal resistance to form solid nuclei.

This is graphically shown in Fig. 10.9a. If the solidification process is slow, then the phase transformation produces a crystalline solid.

Figure 10.9b illustrates an ideal cooling curve (T-t diagram) starting from the pouring temperature T_p to room temperature T_o. The L curve from T_p to T_f represents the cooling liquid metal at a cooling rate (slope) dT/dt, the $L + S$ horizontal line represents the thermal arrest at T_f from the start to finish of the solidification process along with the undercooling $\Delta T = T_f - T_x$ with T_x being just below T_f, and the S curve represents the cooling solid down to T_o.

During the solidification process, latent heat of solidification ΔH_s or latent heat of fusion $\Delta H_f = \Delta H_s$ is released at the moving liquid-solid (L-S) interface with a velocity v_x and position $x = s\,(t)$. As a result, the molten metal solidifies at certain freezing rate dT/dt and local solidification time (t) as shown in Fig. 10.9b.

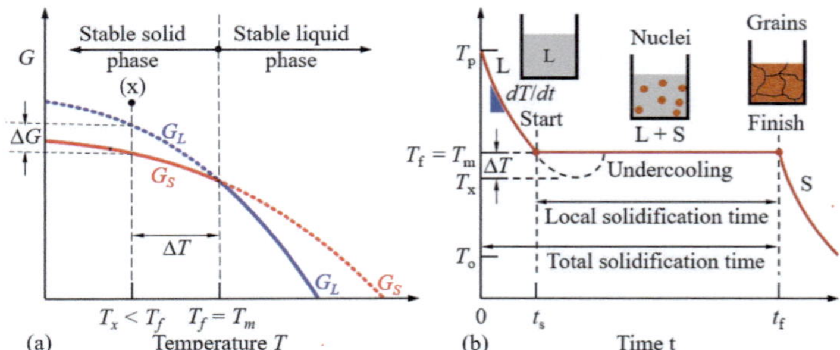

Fig. 10.9 (a) Stable phase regions and undercooling ΔT for solidification, and (b) cooling curve showing phase trend lines (L = liquid and S = solid)

This process has a start t_s and a finish t_f that defines the local solidification time as $t = t_f - t_s$. Also, one can define the total solidification time from the moment the liquid metal or molten metal is poured into a mold.

Interpretation of Fig. 10.9a reveals that:

- $\Delta G = (G_l - G_s) > 0$ is the driving force for solidification at $T < T_f$.
- $|\Delta T| > 0$ is the degree of undercooling for nucleation of particles.
- $G_s < G_l$ promotes a stable solid at $T < T_f$.

It is assumed that liquid undercooling always occurs during solidification.

10.5.1   Effect of Thermocouple Depth

The equilibrium solidification process is characterized by using the local solidification time and the degree of molten metal undercooling ΔT. This is accomplished by placing thermocouples in the mold cavities and measure the temperature as a function of time and fixed depth. In fact, thermocouples have become the industry-standard thermocouples and are cost-effective, highly accurate, and fundamentally safe.

Figure 10.10 shows two different and independent experimental cases in order to elucidate the concepts of solidification time and undercooling ΔT (X. Xu et al. [4]).

Firstly, Fig. 10.10a shows the effect of thermocouple position (depth) on the local solidification time of an $87Al$-$13Si$ alloy at a freezing cooling rate or solidification rate $dT = dt = 0.10 \ °C/s$ (Plotkowski [3]). Notice that the position of the finish solidification time t_f on the solidification curves is slightly apart. This is evident that the measured local solidification time strongly depends on the depth of a thermocouple tip.

Secondly, Fig. 10.10b illustrates a dendritic microstructure of a specimen made out of $80Co$-$20Pd$ alloy at $\Delta T = 20 \ K$ undercooling and Fig. 10.10c depicts an equiaxed microstructure for another specimen of the same alloy at $\Delta T = 340 \ K$ undercooling (x. Xu et al. [4, p. 43]). This suggests that a deeper undercooling level is associated with a relatively fast or intermediate solidification process, which represents a deviation from equilibrium solidification process (slow cooling or freezing).

Most industrial foundry practices are carried out at intermediate solidification velocities. However, the term "rapid solidification" is related to the production or either semi-amorphous or completely amorphous microstructures lacking grain boundaries. In this case, the solidified solid is called rapidly solidified metal (RSM) or rapidly solidified alloy (RSA).

Despite that dimensions and desired properties of as-cast solids for industrial or medical applications dictate the type of directional or nondirectional solidification process, the evolved microstructure is the main concern for a high-quality casting.

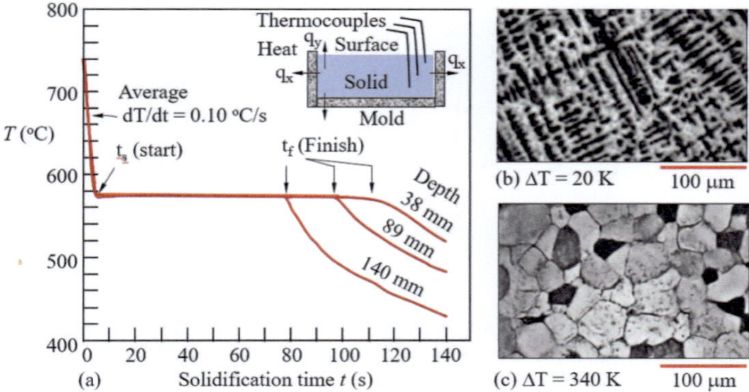

Fig. 10.10 (**a**) Effect of thermocouple depth on the solidification time of an $87Al$-$13Si$ alloy being cooled very slowly (Plotkowski [3]), (**b**) dendritic microstructure of a $80Co$-$20Pd$ alloy at $\Delta T = 20\,K$, and (**c**) equiaxed microstructure of the same $80Co$-$20Pd$ alloy (X. Xu et al. [4])

10.6   Construction of Equilibrium Phase Diagrams

The main focus in this section is to show the traditional methodology for constructing phase diagrams using simple cooling curves as functions of composition C_o at standard pressure ($1\,atm$). This is illustrated using two hypothetical A and B metals for developing A-B binary alloy systems, such as an isomorphous phase diagram having a single solid phase. Nonetheless, an isomorphous phase diagram refers to a homogeneous solid α-phase having the same crystal structure for all compositions.

A eutectic phase diagram having a eutectic point consists of three phases (L, α, β) at the eutectic temperature T_e. The Greek word "eutectic" means "easily melted" or "most fusible" at relatively low temperatures. A eutectic phase diagram having a eutectic point consisting of three phases (L, α, β) at the eutectic temperature T_e.

For clarity, an equilibrium phase diagram is a map of alloy phases, which are traditionally analyzed using metallography to reveal the microstructure morphology and X-ray diffraction for determining the crystal structure.

Nowadays a software called CALPHAD can be used to construct phase diagrams. For instance, CALPHAD stands for calculation of phase diagrams and it enables the development of binary and ternary phase diagrams using thermodynamic and property databases. Actually, (1) CALPHAD describes the phases as a function of composition, temperature, and pressure, and (2) it calculates atomic mobility, molar volume, thermal conductivity and diffusivity, viscosity and surface tension of liquids, electrical resistivity, and more (Saunders and Miodownik [5], Kattner [6]).

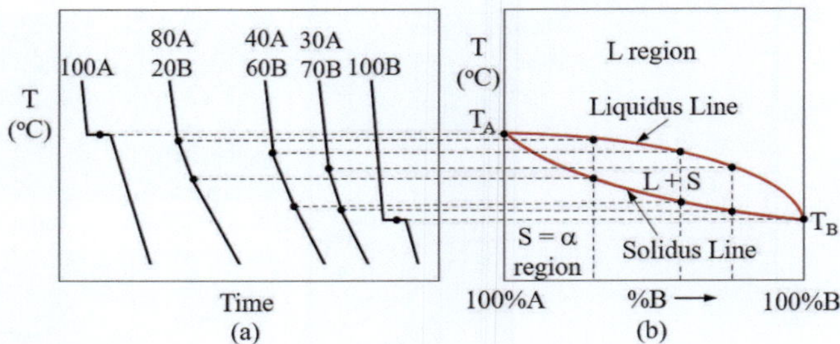

Fig. 10.11 Procedure for building a schematic isomorphous *A-B* phase diagram. (**a**) Cooling curves and (**b**) equilibrium isomorphous phase diagram

10.6.1   *Equilibrium Isomorphous System*

Firstly, pure metal *A* and pure metal *B* are solidified individually. Secondly, a series of *A-B* compositions forming binary alloys are melted and solidified under controlled environment. As a result, a series of cooling curves (Fig. 10.11a) are obtained and used to construct an isomorphous binary phase diagram as shown in Fig. 10.11b, which is a simple map of the liquid phase (*L*-phase) having completely soluble *A* and *B* metals, a substitutional solid solution α-phase having a homogeneous microstructure and a mushy zone composed of $L + S$ phases. Moreover, an isomorphous diagram is also called "lenticular diagram" because of its lens shape and it represents a complete solubility over the entire composition range.

Here, the freezing temperature (melting point) T_f for each element is denoted as T_A and T_B and that for each alloy is on the liquidus line.

10.7   **Equilibrium Eutectic Systems**

10.7.1   *Nomenclature of Binary Phase Diagram*

It is important to understand the terminology used to characterize phase diagrams. For instance, Fig. 10.12a represents the isothermal phase diagram as the simplest phase map with three phase regions: L, $L + \alpha$, and α.

Physically, mixing *A-B* atoms forms a liquid solution above the liquidus curve and a solid solution as a single α-phase below the solidus curve. This mixture between these curves consists of a slush solution (mud-like solution or mushy zone). Notice that the curves represent phase boundaries that separate phase regions since they denote phase transitions.

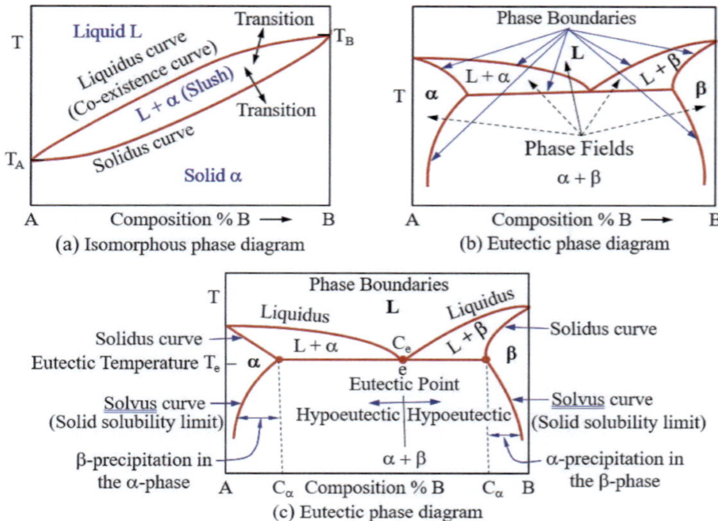

Fig. 10.12 Binary phase diagrams. **(a)** Isomorphous phase diagram, **(b)** eutectic phase diagram illustrating the phase boundaries and fields, and **(c)** eutectic phase diagram showing the triple points, the common nomenclature for phase boundaries, and types of alloys

Consider now a slightly more complicated eutectic diagram containing phase boundaries (Fig. 10.12b) that separate six (6) phase regions (Fig. 10.12c). Regarding Fig. 10.12b, it has two mushy zones: $L + \alpha$ and $L + \beta$.

This type of phase diagram bears its name due to the eutectic point "e," where liquid metal (known as molten metal) solidifies to a coupled α-β solid phase, $(\alpha + \beta)_e$, at just below the eutectic temperature T_e. This physical process represents the microstructural evolution through a single-phase reaction $L \rightarrow (\alpha + \beta)_e$ at $T = T_e - \Delta T$, where ΔT is in the order of $1\,^\circ C$ or less and the subscript "e" denotes eutectic.

Figure 10.12c illustrates the specific names assigned to the curve and straight lines, and the triple points (dots), where three phases coexist at a particular composition and eutectic temperature T_e denoted as (C_o, T_e). The final phase system of the solid product depends on the alloy composition C_o and the rate of solidification dT/dt. This implies that mixing the A and B atoms forms:

- A substitutional solid solution as a single α-phase, provided that A is the solvent component and B is the solute component, and they have similar atomic radii, which are used to define an atomic size factor in real A-B binary alloy systems. Notice that the single α-phase is within a region enclosed by the solidus and solvus boundary lines (left-hand side in Fig. 10.12c).
- A substitutional solid solution as a single β-phase, provided that B is the solvent component and A is the solute component. Both A and B have equal or similar

atomic radii. Similarly, the single β-phase is enclosed within the right-hand side solidus and solvus boundary lines (Fig. 10.12c).

- A eutectic structure composed of a coupled platelike α-β phases and composition $C_B = C_e$.
- A hypoeutectic alloy (dual-phase alloy) with composition $C_B < C_e$ and α-phase surrounded by eutectic grains.
- A hypereutectic alloy (dual-phase alloy) with $C_B > C_e$ and β-phase surrounded by grains.
- A dual-phase alloy with β-precipitate within α-grain structure when $C_{solvus} < C_B < C_\alpha$ or with α-precipitate within β-grain structure when $C_\beta < C_B < C_{solvus}$. Basically, a solvus curve is important for producing precipitation-hardening alloys due to the presence of microscopically small and hard β-particles embedded in soft (ductile) α-grains or vice versa.

Ferrous and non-ferrous eutectic A-B or A-B-X alloy systems with low melting points and high molten fluidity are suitable for casting and joining (welding and soldering) applications. These two characteristics make binary alloys very promising candidate materials for technological applications, such as mechanical processing (rolling, forging, etc.). Here, X represents additions of other metals in small quantities for adjusting the microstructure and related properties of castings. Further, low mass density and corrosion resistance are two additional characteristics that must be considered in certain applications (Robles-Hernandez et al. [7], Ruzbarsky [8]).

10.7.2   Binary Phase Diagram and Eutectic Alloys

Consider the construction of a binary phase diagram using many cooling curves. For clarity, Fig. 10.13a shows three schematic hypoeutectic cooling curves (one for a pure metal A, two for $C_{o,1} < C_e$ and $C_{o,2} > C_{o,1}$ alloy compositions); Fig. 10.13b illustrates the eutectic phase diagram with an invariant eutectic point "e," where the phase reaction $L \to \alpha + \beta$ occurs at just below the eutectic temperature T_e; and Fig. 10.13c shows three hypereutectic cooling curves (one for pure metal B, two for $C_{o,3} > C_e$ and $C_{o,4} > C_{o,3}$ alloy compositions). Here, hypoeutectic means less than and hypereutectic means greater than eutectic composition C_e at a relevant eutectic temperature T_e.

The A-B eutectic phase diagram in Fig. 10.13b is divided into hypoeutectic and hypereutectic regions. In principle, the melting/solidification temperature $T_m = T_f$ follows the liquidus curves, path 1 and path 3, under equilibrium conditions. Moreover, slow melting and solidification at atmospheric pressure should also follow these paths.

Temperature Range   Each type of binary alloy has a ΔT denoted as:

- For hypoeutectic alloys, $T_e < T_{C_{o,1}}, T_{C_{o,2}} < T_A$
- For hypereutectic alloys, $T_e < T_{C_{o,3}}, T_{C_{o,4}} < T_B$

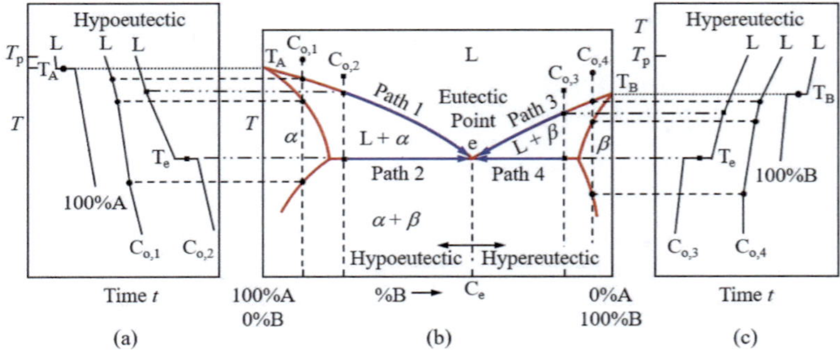

Fig. 10.13 Schematic A-B eutectic system. **(a)** Hypoeutectic compositions and their cooling curves, **(b)** equilibrium phase diagram, and **(c)** hypereutectic compositions and cooling curves

Phase Reactions All hypothetical alloys in Fig. 10.13b are initially at the liquid pouring temperature T_p (Fig. 10.13a and c). Upon slow freezing the molten alloys, the corresponding phase reactions representing the microstructural evolution are:

- For $C_{o,1}$ alloy, $L_1 \rightarrow \alpha + L \rightarrow \alpha_{grains} + \beta_{particles}$
- For $C_{o,2}$ alloy, $L_2 \rightarrow \alpha + L \rightarrow \alpha_{grains} + (\alpha + \beta)_e$
- For $C_{o,3}$ alloy, $L_3 \rightarrow \beta + L \rightarrow \beta_{grains} + \alpha_{particles}$
- For $C_{o,4}$ alloy, $L_4 \rightarrow \beta + L \rightarrow \beta + (\alpha + \beta)_e$

Figure 10.13b schematically illustrates the solidification paths related to the microstructural evolution in a hypoeutectic A-B binary alloy with composition $C_{o,2}$ and a hypereutectic alloy A-B with $C_{o,3}$. For comparison, the corresponding phase reactions for a pure metal and a hypothetical C_o-alloy are written in a general form

Hypoeutectic

$$L \rightleftarrows \alpha \qquad \text{at } T_f = T_A \tag{10.18a}$$

$$L \rightleftarrows L_2 + \alpha \quad \text{at } T_e \tag{10.18b}$$

Hypereutectic

$$L \rightleftarrows \beta \qquad \text{at } T_f = T_B \tag{10.18c}$$

$$L \rightleftarrows L_3 + \beta \quad \text{at } T_e \tag{10.18d}$$

where L denotes the initial liquid alloy at the pouring temperature T_p and L_2, L_3 denote the remaining liquid after a few solid α-phase nuclei (hypoeutectic $C_{o,2}$ alloy) and β-phase nuclei (hypereutectic $C_{o,3}$ alloy) are formed.

10.8   Microstructural Evolution of Eutectic Systems

10.8.1   Nucleation Mechanism

The liquid-to-solid phase transformation, Eq. (10.18), implies that there must exist a nucleation mechanism of solidification related to the preferred arrangement of atoms in the form of quasicrystals in the melt at T_f temperature and crystals below T_f. In essence, solidification is mediated by the amount of undercooling defined as the temperature range $\Delta T = T_f - T_l$, where $T_l < T_f$. Further details on the nucleation mechanism of solidification can be found elsewhere (Dantzig and Rappaz [9, p. 310]).

The solidification process continues until all the remaining liquid phase transforms to the eutectic coupled $\alpha + \beta$ solid phases as equiaxed or dendritic grains at slightly below the eutectic temperature T_e. Further, the solidification paths for the $C_{o,2}$ and $C_{o,3}$ alloys in Fig. 10.13b are marked with blue arrows and they converge to the eutectic point, where the reversible eutectic invariant reaction is written as

$$L \rightleftarrows \alpha + \beta \quad \text{at } T_e, C_e \tag{10.19}$$

where T_e is the eutectic temperature representing the eutectic thermal arrest and C_e is the eutectic composition. Thus, (C_e, T_e) defines the eutectic point "e" and Eq. (10.19) represents a particular eutectic $C_e = (A\text{-}B)_e$ mixture (composition) that freezes as $L \rightarrow \alpha + \beta$ or melts as $L \leftarrow \alpha + \beta$. In summary, Fig. 10.13 shows the most common three-phase reactions found in phase diagrams.

10.8.2   Equilibrium Solidification

Consider the schematic $A\text{-}B$ binary phase diagram shown in Fig. 10.14, which illustrates the corresponding schematic microstructural evolution for two $A\text{-}B$ alloys with compositions $C_{o,1}$ and $C_{o,2}$, where the element "A" is the solvent and "B" is the solute in this binary system (Moffatt et al. [10, Chapter 7], Avner [11, Chapter 6]).

This section includes simple analytical procedures related to solid and liquid mass fractions (f_s and f_l, respectively) using microsegregation models, such as (1) the linear "Lever rule" for determining the proportions (relative amounts) of liquid (L) and solid (S) phases present in a binary alloy at a solidification temperature range $T_s < T_f^* < T_l$. This model considers complete mixing in both liquid and solid phases and it is based on mass conservation. (2) The nonlinear Lever rule, known as the "Scheil equation" or Scheil model, considers complete or infinite mixing in the liquid phase and no diffusion in the solid phase. However, back-diffusion models based on microsegregation with diffusion in the solid phase during solidification are available in the literature (Dantzig and Rappaz [9, p. 457]). Hence,

- The equilibrium solidification induces diffusion in the solid and liquid phases.
- The Scheil model considers a deviation from the equilibrium solidification since it does not assume diffusion in the solid phase, but fast diffusion in the liquid phase. This solidification model is developed in a later section.

The microstructural evolution in pure metal A, eutectic $C_{o,2} = C_e$, and metal B occurs according to the phase reactions $L \rightarrow \alpha$ at T_A, $L \rightarrow \alpha + \beta$ at T_e, and $L \rightarrow \beta$ at T_B, respectively, as the constituent or alloying metal B is increased. On the other hand, the microstructural evolution in the $C_{o,1}$ and $C_{o,3}$ alloys occurs due to more complicated phase reactions

$$L_1 \rightarrow (L + \alpha)_{T_1} \rightarrow \alpha_{grains} \rightarrow \alpha_{grains} + \beta_{particles} \tag{10.20a}$$

$$L_3 \rightarrow (L + \beta)_{T_3} \rightarrow \beta_{grains} \rightarrow \beta_{grains} + \alpha_{particles} \tag{10.20b}$$

which suggests that the properties of these alloys are enhanced by the precipitation of (1) $\beta_{particles}$ within α_{grains} and (2) $\alpha_{particles}$ within β_{grains}.

Embedded particles in relatively ductile matrices harden alloys because they are hard solid barriers to dislocation motion during mechanical deformation and have their own crystal structures related to some type of interface coherency with the metal-base matrix.

Besides the microstructural evolution being driven by the release of latent heat of solidification (ΔH_s) at the liquid-solid (L-S) interface with a specific solidification velocity, the eutectic phase diagram in Fig. 10.14 contains important features such as:

- The eutectic composition C_e exhibits a lower melting point than the hypothetical elements A at T_A and B at T_B, and it is located at the (Ce, Te) point.
- The "solvus" sloping curve lines or simply the solubility limit lines indicate that the solubility of B in the A-rich single α-phase is limited at a composition

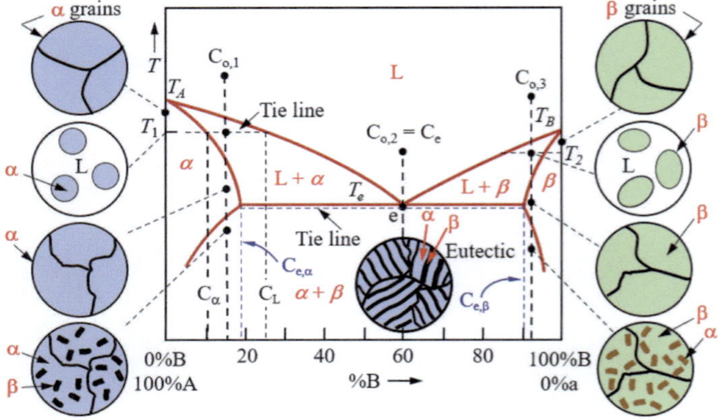

Fig. 10.14 Equilibrium A-B phase diagram and microstructural evolution in A-B binary alloys

$C_o < C_{e,\alpha}$. For instance, hypoeutectic alloy systems with $C_o \leq C_{e,\alpha}$ are heat treatable in order to induce precipitation hardening of the α-phase grains with the precipitated β-phase in the form of platelike particles. Conversely, the solubility of A in the B-rich single β-phase is limited at a composition $C_o > C_{e,\beta}$, and in theory, hypereutectic alloy systems with $C_o \geq C_{e,\alpha}$ are precipitation-hardenable materials.

- For hypoeutectic A-B alloy systems, the composition is $C_o < C_e$.
- For hypereutectic A-B alloy systems, the composition is $C_o > C_e$
- For a eutectic A-B alloy system, the composition is $C_o = C_e$.

The eutectic composition C_e either melts ($S \rightarrow L$) or freezes ($L \rightarrow S$) congruently at the eutectic temperature T_e.

10.8.3   Non-Equilibrium Solidification

A slight deviation from the equilibrium solidification rate, $(dT/dt)_e$, leads to a process called non-equilibrium solidification with a relatively faster solidification rate, $(dT/dt)_{ne} \gtrsim (dT/dt)_e$. This process induces a variation in chemical composition in the evolving solid phase.

Consider a hypoeutectic $C_{o,1}$ alloy solidifying from the molten state at some pouring temperature $T_p = T_l$ as shown in Fig. 10.15 (Moffatt et al. [10, Chapter 7] and Avner [11, Chapter 6]). The non-equilibrium solidification solidus path is shown as a dashed blue curve. The first evolved or nucleated solid is denoted as α_1 at T_1 on the equilibrium solidus curve. Subsequent solidification at T_2 is denoted as α_2 with composition B slightly below the equilibrium solidus composition.

It is assume now that solidification finishes at T_3 just below the equilibrium eutectic temperature T_e. This final solidification stage induces the last liquid

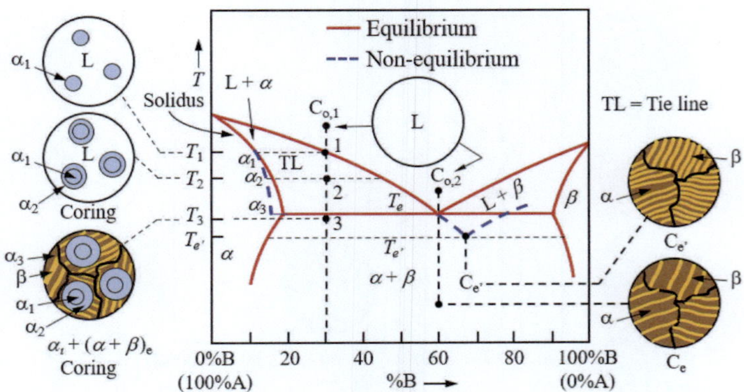

Fig. 10.15 Non-equilibrium solidification of two binary alloys of nominal (overall) compositions C_o, $C_{e'}$ and cored structures (Motaff et al. [10, pp. 181–182])

to solidify according to the phase reaction $L_3 \rightarrow \alpha_3 + \beta$, where α_3 has a composition at the equilibrium eutectic temperature T_e. Moreover, the equilibrium solid composition is $C_{\alpha_1} \simeq 11\%B$ and those for the non-equilibrium conditions are $C_{\alpha_2} \simeq 13\%B$ and $C_{\alpha_3} \simeq 16\%B$. Therefore, C_{α_2}, C_{α_3} have a lower content of B atoms in the non-equilibrium (dashed solidus line) than that corresponding to the equilibrium condition (solid solidus line).

This solidification phenomenon is known as "coring" or "dendritic segregation," and consequently, the $C_{o,1}$ alloy acquires a cored microstructure once solidification is complete. Usually, the cored microstructure is heat-treated in order to homogenize it for obtaining a relatively uniform or equiaxed microstructure.

For a eutectic $C_{o,2} = C_e$ alloy, a non-equilibrium solidification may occur, provided that the evolution of the β-phase is retarded. In this case, the eutectic point is displaced as shown in Fig. 10.15 with a non-equilibrium eutectic temperature $T_{e'} < T_e$ (dashed line) and eutectic composition $C_{\beta,e'} > C_{e,\beta}$. In essence, non-equilibrium solidification is likely to occur in real industrial solidification cases due to quick production of high-quality ingots or parts based on cost-effective and reliable processes.

10.9 The Gibbs Phase Rule

Consider a general A-B binary alloy system depicted in Fig. 10.16 for introducing the Gibbs phase rule (GPR), characterizing the thermodynamic state of the alloy and predicting the equilibrium condition of phases as a function of pressure (T), temperature (P), and composition (C_x).

The Gibbs phase rule implies that liquid and solid phases can coexist over a range of temperatures and compositions, while invariant reactions at triple points are associated with three phases in equilibrium.

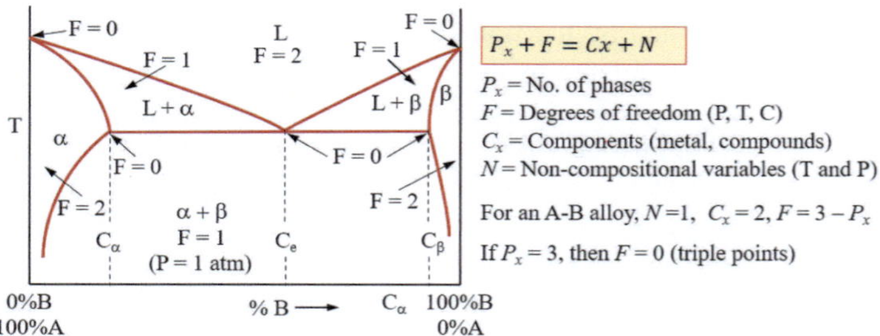

Fig. 10.16 The Gibbs phase rule and a eutectic phase diagram

10.9.1 Degrees of Freedom Formula

The number of degrees of freedom based on the number of independent intensive variables (independent on the size of a system) and it is defined by

$$P_x + F = C_x + N \tag{10.21}$$

where P_x denotes the number of phases, F denotes the degrees of freedom as the number of independent intensive variables (independent of each other) such as (P, T, C_x), C_x denotes the number of components such as metals and compounds, and N denotes the non-compositional variables such as T and P_x.

At constant pressure, the chemically independent components are $C_x = 2$ for metals A, B (binary) and $C_x = 3$ for A, B, C (ternary), and $N = 1$ (temperature since pressure is fixed at $P = 1\ atm$). Thus, Eq. (10.21) becomes

$$F = 3 - P_x \quad \text{(binary)} \tag{10.22a}$$

$$F = 4 - P_x \quad \text{(ternary)} \tag{10.22b}$$

In this case, the degrees of freedom F depends on the number of independent intensive variables such as (T, C_x) since the pressure is fixed. Hence, only T and C_x are the changing variables within a certain range.

Applying Eq. (10.22a) to an A-B binary alloy system with single- and two-phase regions and triple points yields the degrees of freedom F values shown in Fig. 10.16. Thus, the degrees of freedom for characterizing phase fields in a phase diagram are analyzed as follows:

- Single-phase field $\rightarrow L, \alpha, \beta$ with $P_x = 1$ phase and $F = 3 - P_x = 2$. Thus, two variables, T and C_x, can be specified to define the thermodynamic state of the system and they can be changed within the limits imposed on the eutectic phase diagram to maintain a single-phase field. For instance, $(L + \alpha)$ and $(L + \beta)$ phase fields are limited by the liquidus-solidus-eutectic triangle-like lines, and $(\alpha + \beta)$ phase field is limited by the eutectic-solvus lines.
- Two-phase field $\rightarrow (L + \alpha)$, $(L + \beta)$, $(\alpha + \beta)$ with $P_x = 2$ phases and $F = 3 - P_x = 1$. Thus, only one variable, T or C_x, can be specified to define the thermodynamic state of the system. This means that either T or C_x can be changed within a range in order to maintain the two-phase field in equilibrium.
- Three-phase (triple points) field $\rightarrow (L + \alpha + \beta)$ with $P_x = 3$ phases and $F = 3 - P_x = 0$. Thus, no variables, T and C_x, cannot be specified to define the thermodynamic state of the system since this is a fixed triple point. Actually, a triple point is invariant in nature since it is a fixed point in the phase field, where T and C_x are fixed variables. Moreover, an invariant point is related to an invariant reaction, such as the reversible eutectic reaction $L \rightleftarrows \alpha + \beta$ at the eutectic temperature T_e.

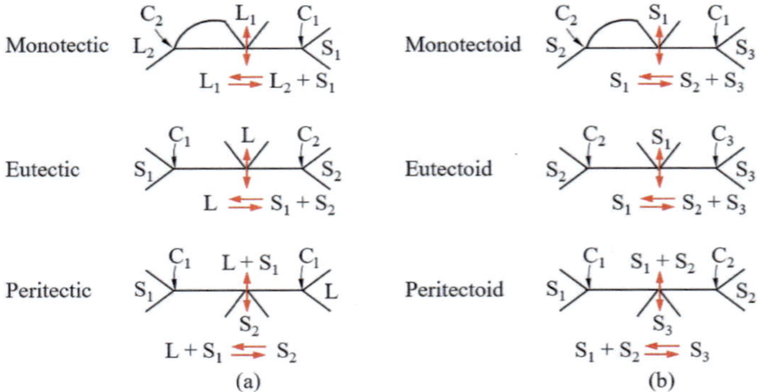

Fig. 10.17 The invariant phase reactions for **(a)** liquid and solid, and **(b)** all solids

The Gibbs phase rule is applicable to equilibrium phase diagrams with stable phases and is independent of magnetic, electric, and gravitational forces.

10.10   The Invariant Phase Reactions

It is important to understand the concept of invariant phase reactions that occur at specific triple points (three phases coexist) and temperatures. The most common invariant reactions are illustrated in Fig. 10.17a for liquid and solid reactions, and in Fig. 10.17b for all solid reactions.

The information given in Fig. 10.17 serves as guide for identifying the type of invariant reactions in phase diagrams.

10.11   The Lever Rule

10.11.1   Graphical Procedure

The Lever rule requires a tie-line drawn horizontally at T_f^* or at $T < T_f^*$ intercepting the liquidus and solidus lines as shown in Figs. 10.14 and 10.15. Nonetheless, this rule can be defined by the general temperature-dependent mean concentration $C_x(T)$

$$C_x(T) = f_l C_l(T) + (1 - f_l) C_s(T) \tag{10.23}$$

from which the liquid mass fraction (f_l) takes the form

$$f_l = \frac{C_s\,(T) - C_x\,(T)}{C_s\,(T) - C_l\,(T)} \tag{10.24a}$$

$$f_s = 1 - f_l \tag{10.24b}$$

where f_s denotes the frozen solid mass fraction during solidification and $C_l\,(T), C_s\,(T)$ denote the solute composition in the liquid and solid phases, respectively, at a temperature T. Hence, the use of Eqs. (10.23) and (10.24) is illustrated in the example below.

For the hypoeutectic $C_{o,1} = 85A\text{-}15B$ alloy at T_1 (Fig. 10.14), the characteristics of the melt and the Lever rule along with the alloying element "B" or solute "B" yield

Reaction at T_1: $L \rightarrow L_1 + \alpha$
Phases: L, α
Composition: $C_l = 75A\text{-}25B, C_s = C_\alpha = 90A\text{-}10B$
Partition coefficient: $k_o = C_s/C_l = 10/25 = 0.40 < 1$
Mass fraction:

$$f_\alpha = f_s = \frac{C_l - C_{o,1}}{C_l - C_\alpha} = \frac{25B - 15B}{25B - 10B} = 0.67$$

$$f_l = \frac{C_{o,1} - C_\alpha}{C_l - C_\alpha} = 1 - f_\alpha = 1 - 0.67 = 0.33$$

For the eutectic $C_{o,2} = C_e = 40A\text{-}60B$ alloy at just below T_e,
Reaction at T_e: $L \rightarrow (\alpha + \beta)_e$
Phases: α, β (plates)
Composition: $C_{s\alpha} = 82A\text{-}18B, C_{s\beta} = 10A\text{-}90B$
Partition coefficient: $k_o = C_s/C_l = 0$
Mass fraction:

$$f_\alpha = \frac{C_\beta - C_{o,2}}{C_\beta - C_\alpha} = \frac{90B - 60B}{90B - 18B} = 0.42 \text{ or } 42\%$$

$$f_\beta = \frac{C_{o,2} - C_\alpha}{C_\beta - C_\alpha} = 1 - f_\alpha = 1 - 0.42 = 0.58 \text{ or } 58\%$$

Hence, the eutectic microstructure consists of 42% α-phase and 58% β-phase at T_e, but it does not provide any information on the microstructural morphology, kinetics, and heat transfer phenomenon related to the solidification process of an alloy. Moreover, this quantitative approach for determining the proportions of phases in an $A\text{-}B$ binary alloy illustrates the usefulness of the Lever rule and equilibrium phase diagrams.

Example 10.2 Consider the isomorphous $Cu\text{-}Ni$ equilibrium phase diagram along with some schematic cooling curves and a $60\text{-}gram\ Cu$ and $40\text{-}gram\ Ni$ non-ferrous binary alloy being slowly solidified from the pouring temperature $T_p =$

$1300\,^\circ C$ to $T = 1250\,^\circ C$. Calculate the composition of each element in terms of the **(a)** number of moles X, **(b)** atomic percent N_x ($at\%$), and **(c)** weight percent W_x ($wt\%$) and **(d)** use the Lever rule to determine the compositions of the phases for an alloy with composition $C_o = 60Cu\text{-}40Ni$ at $1250\,^\circ C$. The atomic masses (amu) of copper (Cu) and nickel (Ni) are $A_{w,Cu} = 63.55\ g/mol$ and $A_{w,Ni} = 58.69\ g/mol$.

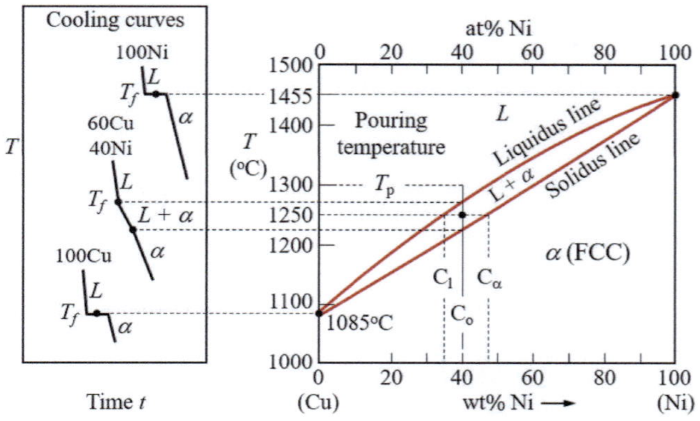

Solution

(a) The number of moles for each element is calculated as

$$X_{Cu} = \left(\frac{m}{A_w}\right)_{Cu} = \frac{60\ g}{63.55\ g/mol} = 0.94\ mol \tag{10.2E1a}$$

$$X_{Ni} = \left(\frac{m}{A_w}\right)_{Ni} = \frac{40\ g}{58.69\ g/mol} = 0.68\ mol \tag{10.2E1b}$$

(b) The atomic percent for each element is

$$N_{Cu} = \frac{X_{Cu}}{X_{Cu} + X_{Ni}} = \left(\frac{0.94}{0.94 + 0.68}\right)(100) \simeq 58\ at\% \tag{10.2E2a}$$

$$N_{Ni} = \frac{X_{Ni}}{X_{Cu} + X_{Ni}} = \left(\frac{0.68}{0.94 + 0.68}\right)(100) \simeq 42\ at\% \tag{10.2E2b}$$

(c) The weight percent of each element in the 100-$gram$ $60Cu\text{-}40Ni$ alloy is

$$W_{Cu} = \frac{m_{Cu}}{m_{Cu} + m_{Ni}} = \left(\frac{60\ g}{60\ g + 40\ g}\right)(100) = 60\ wt\% \tag{10.2E3a}$$

$$W_{Ni} = \frac{m_{Ni}}{m_{Cu} + m_{Ni}} = \left(\frac{40\ g}{60\ g + 40\ g}\right)(100) = 40\ wt\% \tag{10.2E3b}$$

(d) Using the Lever rule at $1250\,^{\circ}C$ yields the weight or mass fraction of the solid α-phase and the liquid L-phase

$$W_\alpha = f_\alpha = \frac{C_o - C_l}{C_l - C_\alpha} = \frac{40 - 35}{48 - 35} = 0.38 \qquad (10.2\mathrm{E}4\mathrm{a})$$

$$W_l = f_l = 1 - W_\alpha = 1 - f_\alpha = 1 - 0.38 = 0.62 \qquad (10.2\mathrm{E}4\mathrm{b})$$

The amounts of these phases based on a 100-$gram$ $60Cu$-$40Ni$ alloy at $1250\,^{\circ}C$ are

$$m_\alpha = W_\alpha m = (0.38)\,(100\,grams) = 38\,grams \qquad (10.2\mathrm{E}5\mathrm{a})$$

$$m_l = W_l m = (0.62)\,(100\,grams) = 62\,grams \qquad (10.2\mathrm{E}5\mathrm{b})$$

Equilibrium phase diagrams can be built or constructed using one of the relationships $T = f(X)$, $T = f(at\%)$, or $T = f(wt\%)$, provided that sufficient cooling curves are experimentally obtained under controlled slow solidification from suitable pouring temperatures.

10.11.2 Limited Solid Solubility and Eutectic Mechanism

Additional aspects of the microstructural evolution during equilibrium solidification is schematically shown in Fig. 10.18a for a slowly cooled hypothetical A-B binary alloy with C_o composition.

The mechanism for the formation of the eutectic microstructure consists of lamellar α-β solid phases as depicted in Fig. 10.18b, which indicates that the A-rich α-phase rejects excess B atoms and the B-rich β-phase rejects the excess A atoms at the L-S interface during the liquid-to-solid transformation with a certain interface velocity υ_z. This means that there is a limited solid solubility defined by the solvus curve line in Fig. 10.18a from $T = T_e$ to some $T \ll T_e$, making this condition a specific characteristic of the eutectic phase diagram.

Consider an A-B alloy with nominal composition C_o initially in the liquid state at T_1. Release of latent heat of fusion (ΔH_f) is an extraction process of heat which causes solidification. Assume that the liquid is continuously slow cooled from T_1 to T_5 in a suitably prepared solidification environment.

The equilibrium solidification steps shown below induce:

- Cooling (freezing) from T_1 to T_2 so that the A-B alloy liquid composition becomes $C_{l2} \lesssim C_o = 30\%B$ on the liquidus line and that for the first solid α-phase is $C_{\alpha2} \simeq 11\%B$ on the solidus line.
- Cooling the liquid and the first solid α_2 from T_2 to T_3 gives $C_{l3} \simeq 44\%B$ and $C_{\alpha2} \simeq 16\%B$. At this solidification stage, the solubility of B atoms in the α-phase increased approximately 5%, while the liquid phase increased 14% in

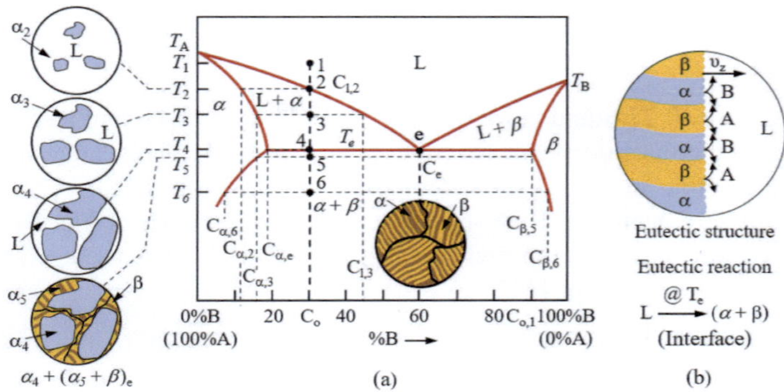

Fig. 10.18 **(a)** *A-B* binary phase diagram with microstructural evolution and **(b)** mechanism for the formation of a eutectic microstructure as alternating α and β layers (plates)

B atoms. This means that diffusion of B atoms is slower in the solid α-phase than in the liquid phase. Based on the information, the phase reaction, phases, composition, partition coefficient k_o, and phase fractions using the Lever rule are, respectively,

Reaction at T_3: $L_2 \rightarrow L_3 + \alpha_3$
Phases: L, α
Composition: $C_{l,} = 56A\text{-}44B,\ C_{\alpha,} = 84A\text{-}16B$
Partition coefficient: $k_o = C_s/C_l = 16/44 = 0.36 < 1$
Mass fraction:

$$f_{\alpha 3} = \frac{C_{l3} - C_o}{C_{l3} - C_{\alpha 3}} = \frac{44B - 30B}{44B - 16B} = 0.50$$

$$f_{l3} = \frac{C_o - C_{\alpha 3}}{C_{l3} - C_{\alpha 3}} = 1 - f_{\alpha 3} = 1 - 0.50 = 0.50$$

- Cooling the liquid-solid mixture (L-α_3) from T_3 to $T_4 = T_e$ yields the eutectic composition as $C_{l4} \simeq 60\%B$, $C_{\alpha 4} \simeq 19\%B$. At this solidification stage, the solubility of B atoms in the α-phase increased only 3% since diffusion in the solid state is getting slower than before. In this case, the remaining liquid must transform isothermally at $T_4 = T_e$. Thus,

Reaction at $T_4 = T_e$: $L_3 \rightarrow L_4 + (\alpha + \beta)_e$
Phases: L, α, β
Composition: $C_l = 40A\text{-}60B,\ C_\alpha = 81A\text{-}19B$
Partition coefficient: $k_o = C_s/C_l = 19/60 = 0.32 < 1$
Mass fraction:

$$f_{\alpha 4} = \frac{C_l - C_o}{C_l - C_{\alpha 4}} = \frac{60B - 30B}{60B - 19B} = 0.73$$

$$f_{l4} = \frac{C_o - C_{\alpha 4}}{C_l - C_{\alpha 4}} = 1 - f_{\alpha 4} = 1 - 0.73 = 0.27$$

Hence, $f_{l4} = 0.27$ liquid is available at $T_4 = T_e$ for subsequent transformation according to the eutectic phase reaction $L \rightarrow (\alpha + \beta)_e$. This suggests that the total solid fraction at $T < T_e$ is composed of 0.73 primary or proeutectic α-phase and 0.27 eutectic $(\alpha + \beta)_e$-phase, a mechanism shown in Fig. 10.18b. The eutectic structure may be composed of eutectic dendrites known as eutectic colonies (Dantzig and Rappaz [9, p. 414]).

- Now, cooling from $T_4 = T_e$ to T_5 (just below T_e) yields $C_{\alpha 5} \simeq 16\%B$ and $C_{\beta 5} \simeq 90\%B$ since the alloy composition is based on α and β solid phases known as the eutectic structure, which is denoted as the $\alpha + \beta$ field in Fig. 10.18a. Notice that the solubility of B atoms in the A-rich α-phase has decreased. The phase reaction, phases, and phase fractions using the Lever rule are, respectively,

$$
\begin{aligned}
&\text{Reaction at } T_5\text{:} \quad L_4 + \alpha_3 \rightarrow \alpha_3 + (\alpha_4 + \beta_4)_e \\
&\text{Phases:} \qquad\qquad \alpha, \beta \\
&\text{Composition:} \qquad C_{\alpha 5} = 16B, \, C_{\beta 5} = 90B \\
&\text{Mass fraction:}
\end{aligned}
$$

$$f_{\alpha 5} = \frac{C_{\beta 4} - C_o}{C_{\beta 4} - C_{\alpha 4}} = \frac{90B - 30B}{90B - 16B} = 0.81$$

$$f_{\beta 5} = \frac{C_o - C_{\alpha 5}}{C_{\beta 5} - C_{\alpha 5}} = 1 - f_{\alpha 5} = 1 - 0.81 = 0.19$$

- The α-phase in $(\alpha + \beta)_e$ is

$$f_{\alpha e} = f_{\alpha 5} - f_{\alpha 4} = 0.81 - 0.73 = 0.08$$

- Finally, cooling from T_5 to T_6 yields $C_{\alpha 6} \simeq 8\%B$ and $C_{\beta 6} \simeq 94\%B$, and therefore, there is a further limited solid solubility of B atoms in the α-phase and of A atoms in the β-phase. The phase reaction, phases, and phase fractions using the Lever rule are, respectively,

$$
\begin{aligned}
&\text{Phases:} \qquad\qquad \alpha, \beta \\
&\text{Composition:} \qquad C_{\alpha 6} = 92A\text{-}8B, \, C_{\beta 6} = 6A\text{-}94B \\
&\text{Mass fraction:}
\end{aligned}
$$

$$f_{\alpha 6} = \frac{C_{\beta 6} - C_o}{C_{\beta 6} - C_{\alpha 6}} = \frac{94B - 30B}{94B - 8B} = 0.74$$

$$f_{\beta 6} = \frac{C_o - C_{\alpha,6}}{C_{\beta 6} - C_{\alpha 6}} = 1 - f_{\alpha 6} = 1 - 0.74 = 0.26$$

The limited solid solubility problem is clearly evident in this analysis of the microstructural evolution predicted by a eutectic phase diagram.

All A-B eutectic alloy systems with composition $C_{e,\alpha} < C_{e,i} \leq C_e$ or $C_e < C_{e,i} \leq C_{e,\beta}$, where $i = 1, 2, 3, ...$, are technologically important due to the liquid-phase reaction $L \rightarrow \alpha_{grains} + (\alpha + \beta)_{eutectic}$ or $L \rightarrow \beta_{grains} + (\alpha + \beta)_{eutectic}$.

Consider now important concepts that are applicable to any phase diagram. Thus,

- The liquid-solid and solid-solid invariant phase reactions that occur at a specific temperature and solute composition.
- The variables for a thermodynamic state of an A-B binary alloy system using the Gibbs phase rule.
- The metallurgical conditions using the Hume-Rothery rules for predicting the type of A-B binary alloy based on the concepts of substitutional solid solution and interstitial solid solution.
- Again, the quantitative approach for determining the percentages of phases present in a hypothetical A-B binary alloy is a simple procedure and it is significantly essential for characterizing crystalline solids.

Additional Lever rule calculations based on mass fraction and phase partition may reveal certain solidification characteristics of an alloy. This is included in the next section using a modified version of the Lever rule.

10.11.3   Mathematical Procedure

First of all, Fig. 10.19a is a schematic A-B binary phase diagram illustrating the mushy zone, such as the $L + \alpha$ region for a hypoeutectic alloy and the $L + \beta$ region for a hypereutectic alloy. Here, α and β are the respective solid phases in the phase diagram.

Figure 10.19b depicts the hypoeutectic region defining the tie-line for the Lever rule at the C_o alloy freezing temperature T_f^* with a partition or segregation coefficient $k_o < 1$. This means that k_o is associated with solute compositions in the solid and liquid phases.

Similarly, Fig. 10.19c exhibits the tie-line for developing the Lever rule at T_f^* with $k_o > 1$, where k_o is mathematically defined below. Moreover, the term microsegregation stands for nonuniform distribution of rejected solute at the solidification front or by solute rejection throughout a mushy layer of solid dendrites during solidification.

10.11.4   Mass Fraction Method

The Lever rule is the most common method in any material science course that deals with weight or mass fraction of phases using an equilibrium phase diagram. For a

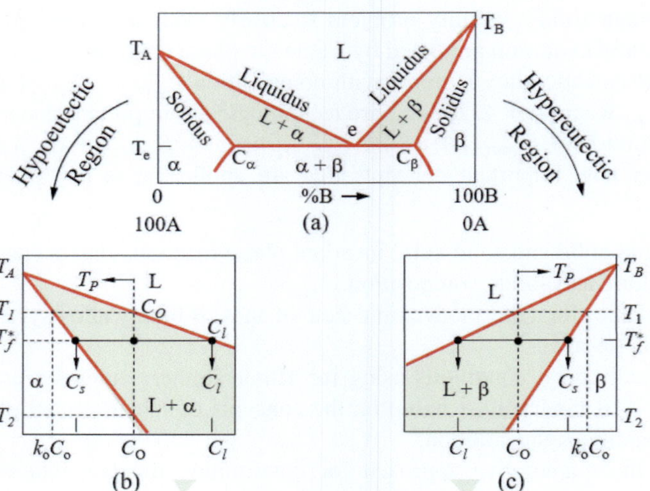

Fig. 10.19 Schematic *A-B* eutectic phase diagram. (**a**) Hypothetical equilibrium eutectic phase diagram, (**b**) partial hypoeutectic region, and (**c**) partial hypereutectic region

C_o alloy at T_f^* (Fig. 10.19b) having a total mass $m_o = m_s + m_l$ with $m_s = m_\alpha$ or $m_s = m_\beta$, the mass balance is written as

$$C_s + C_l = C_o \quad \text{at } T = T_f^* \tag{10.25a}$$

$$m_s C_s + m_l C_l = m_o C_o \tag{10.25b}$$

$$\frac{m_s}{m_o} C_s + \frac{m_l}{m_o} C_l = C_o \tag{10.25c}$$

where C_s, C_l denote the solid and liquid solute compositions and m_s, m_l denote the mass of solid and liquid phases. Let $f_s = m_s/m_o$ and $f_l = m_l/m_o$ be the solid and liquid mass fractions, respectively, with $f_s + f_l = 1$ so that Eq. (10.25c) becomes

$$f_s C_s + f_l C_l = C_o \tag{10.26a}$$

$$f_s C_s + (1 - f_s) C_l = C_o \tag{10.26b}$$

Combining Eqs. (10.26a), (10.26b) yields

$$f_s = \frac{C_l - C_o}{C_l - C_s} = 1 - f_l \tag{10.27a}$$

$$f_l = \frac{C_o - C_s}{C_l - C_s} = 1 - f_s \tag{10.27b}$$

which are exactly the same expressions deduced from the graphical-procedure section, Eq. (10.24). According to Fig. 10.19b, the first solid to form will have the composition $k_o C_o$, whereas the liquid concentration is $C_l \simeq C_o$ at $T = T_l$.

During solidification at $t > 0$, the solid-to-liquid concentration ratio C_s/C_l defines the partition coefficient k_o at L-S interface temperature (isothermal) T_f^* and pressure (isobaric) P. This ratio is metallurgically an important parameter that determines how solute B atoms or an impurity is distributed between molten and solidified alloy at a particular temperature. Hence,

$$C_s = k_o C_l \qquad \text{at } T_s < T_f^* < T_l \tag{10.28a}$$

$$k_o = \frac{C_s}{C_l} < 1 \quad \text{(hypoeutectic)} \tag{10.28b}$$

$$k_o = \frac{C_s}{C_l} > 1 \quad \text{(hypereutectic)} \tag{10.28c}$$

Combining Eqs. (10.26b) and (10.28a) yields the solid mass fraction with $k_o < 1$ for a hypoeutectic binary alloy written as

$$f_s = \frac{1}{1 - k_o} \left(1 - \frac{C_o}{C_l} \right) \quad \text{for } k_o < 1 \tag{10.29a}$$

$$\frac{C_o}{C_l} = 1 - (1 - k_o) f_s \qquad \text{for } k_o < 1 \tag{10.29b}$$

Manipulating Eq. (10.27a) yields

$$- f_s C_s = C_l - C_o - f_s C_l \tag{10.30}$$

Add $f_s C_o$ to Eq. (10.30) to get

$$f_s C_o - f_s C_s = C_l - C_o - f_s C_l + f_s C_o \tag{10.31a}$$

$$f_s (C_o - C_s) = (1 - f_s)(C_l - C_o) \tag{10.31b}$$

Multiplying to Eq. (10.31b) by dm yields the mass balance

$$f_s (C_o - C_s)\, dm = (1 - f_s)(C_l - C_o)\, dm \tag{10.32}$$

where dm denotes the increment solute mass in the solid phase during solidification.

For an ideal L-S interface (solidification front), the mass balance described by Eq. (10.32) physically means that $f_s (C_o - C_s)\, dm$ is the amount of solute rejected by the solid phase and $(1 - f_s)(C_l - C_o)\, dm$ is the fraction transferred to the liquid phase. Hence, solute segregation ahead of the L-S interface prevails as the mechanism for constitutional supercooling.

Use the idealized portion of the phase diagram in Fig. 10.19b to mathematically model the solidification paths of the liquidus and solidus as straight lines with the slopes β_l, β_s. Thus, the partition coefficient k_o is derived as

$$T_l = T_f + \beta_l C_l \tag{10.33a}$$

$$T_s = T_f + \beta_s C_s \tag{10.33b}$$

$$\beta_l = \frac{T_l - T_f}{0 - C_l} = -\frac{T_l - T_f}{C_l} < 0 \tag{10.33c}$$

$$\beta_s = \frac{T_s - T_f}{0 - C_s} = -\frac{T_s - T_f}{C_s} < 0 \tag{10.33d}$$

$$k_o = \frac{C_s}{C_l} = \frac{\beta_l}{\beta_s} \frac{T_s - T_f}{T_l - T_f} \tag{10.33e}$$

Now, Eq. (10.29a) can be defined in terms of temperature $f_s = f_s(T)$ along with its differential form. Thus,

$$f_l = 1 - f_s = 1 - \frac{1}{1 - k_o} \frac{T - T_l}{T - T_f} \tag{10.34a}$$

$$f_s = \frac{1}{1 - k_o} \frac{T - T_l}{T - T_f} \tag{10.34b}$$

$$\frac{df_s}{dT} = \frac{1}{1 - k_o} \left[\frac{1}{T - T_f} - \frac{T - T_l}{\left(T - T_f\right)^2} \right] \tag{10.34c}$$

where df_s/dT denotes the slope of the $f_s = g(T)$ function.

10.11.5　Volume Fraction Method

Stereology is the foundation of quantitative metallography method since it provides information on geometric quantities, such as volume fractions of two phases denoted as $f_s = f_\alpha$, $f_l = f_{melt}$, f_α and f_β. Thus, the quantitative metallography approach is significantly simplified.

For a solidifying material, the total volume (V) and the mass density (ρ) can be used to define the volume fraction of phases present at a certain temperature. For instance,

$$f_s + f_l = 1 \tag{10.35a}$$

$$V = \frac{m}{\rho} \tag{10.35b}$$

$$f_s = V_s/V \tag{10.35c}$$

$$f_l = V_l/V \tag{10.35d}$$

$$V = \frac{m_s}{\rho_s} + \frac{m_l}{\rho_l} \tag{10.35e}$$

This method is quite elementary, easy to understand, and widely used to determine the phase fractions of binary alloys.

10.12   Scheil Model

This model considers the mass balance between the amount of solute B in the solid phase and the amount of B that diffuses into the liquid. Thus, the L-S interface advances by an increment df_s and the liquid composition increases by dC_l. This solidification problem is implicitly illustrated in Fig. 10.20 by the hatched areas.

Solidification Problem   Consider the solidification of a finite slab ($0 \leq x \leq L$) of a binary alloy of composition C_o. Assume that the binary alloy melt is initially at the pouring temperature $T_p > T_f$ and that the mold is at room temperature T_o. The solidification conditions for the Scheil model are:

- Rapid freezing is due to heat conduction so that the solute diffusion occurs in the liquid only. This means that the diffusion coefficients are assumed to be $D_s = 0$ and $D_l > 0$.
- The moving solidification front is at $x = s\,(t)$, where the local equilibrium is reached.
- All thermophysical properties are independent of temperature and concentration. Specifically, the mass density $\rho_s = \rho_l$, specific heat $c_s = c_l$, and thermal conductivity $k_s = k_l$.

Fig. 10.20 Schematic distribution of the liquid and solid phases during solidification of a hypothetical binary alloy

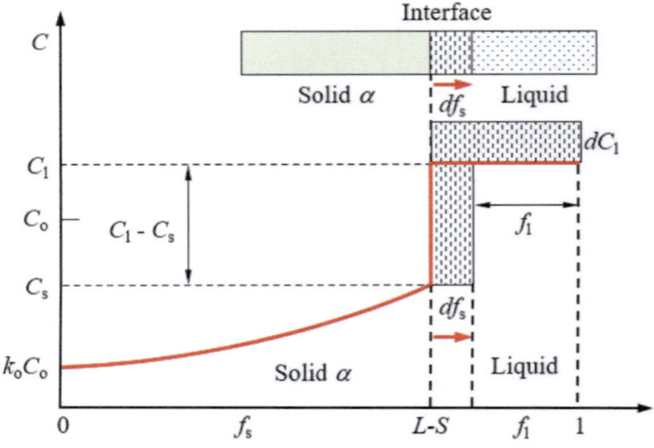

Mathematically, one-dimensional analysis of the solidification problem is directly related to the evolved solid at $x = s(t)$ and the latent heat of solidification ΔH_s. The analytical procedure for solving this two-phase solidification problem is found in Chap. 9, Eq. (9.3). For freezing a finite alloy slab with mixing of the melt (Alexiades and Solomon [12, p. 113]), the solidification mathematical problem is stated below.

Two-Phase Problem For one-dimensional analysis in the x-direction along with the freezing temperature range $T_l \leq T_f = T_f^* \leq T_s$ for a binary alloy with nominal composition C_o, the governing heat equations for the two-phase problem, continuity of temperature at the L-S interface, and the Fourier heat flux balance ($q_l - q_s = q_f$) for freezing a slab in a half-space $x > 0$ in a finite region $0 < x < L_x$ are, respectively,

$$\frac{\partial T}{\partial t} = \alpha_s \frac{\partial^2 T}{\partial x^2} \text{ for } 0 < x < s(t) \tag{10.36a}$$

$$T(x,t) = T_f \text{ at } s(t) \leq x \leq L_x, t > 0 \tag{10.36b}$$

$$T[(s(t),t)] = T_f \text{ at } t > 0 \tag{10.36c}$$

$$k_s \frac{\partial T_s}{\partial x} - k_l \frac{\partial T_l}{\partial x} = \rho_l \Delta H_s \frac{ds(t)}{dt} \text{ at } x = s(t), T_p = T_f \tag{10.36d}$$

$$(C_l - C_s) \frac{ds(t)}{dt} = [L_x - s(t)] \frac{dC(T_f)}{dx} \frac{dT(t)}{dx} \tag{10.36e}$$

$$s(0) = 0, \ T_f(0) = 0, \ T(x,0) = 0 \tag{10.36f}$$

$$T(0,t) = T_o \ \& \ k_s \frac{\partial T_s(L_x,t)}{\partial x} = 0 \tag{10.36g}$$

where L_x denotes a finite length, $\rho = \rho_s = \rho_l$ denotes the density (Kg/m^3), $k_l = k_s$ denotes the thermal conductivity ($W/m.K$), $c_p = c_s = c_l$ denotes the specific heat capacity at constant pressure ($J/Kg.K$), ΔH_f denotes the latent heat of fusion (J/Kg), $ds(t)/dt$ denotes the rate of freezing, and T denote the absolute temperature. Actually, the motion of the L-S interface depends on ΔH_s at the freezing temperature T_f. Moreover, Eq. (10.36d) is known as the Stefan condition.

Let's consider a dilute binary alloy, where the liquidus and solidus lines in the relevant equilibrium phase diagram can be approximated by straight lines. Thus, the mass balance with $f_s + f_l = 1, C_s = k_o C_l, C_s = k_o C_o$ (initial solidus concentration) and the f_s (fraction of the frozen or transformed solid) may be written as the solute conservation of mass equation (Moffatt et al. [10, p. 178])

$$(C_l - C_s) df_s = f_l dC_l \tag{10.37a}$$

$$(C_l - C_s) df_s = (1 - f_s) dC_l \tag{10.37b}$$

$$(C_l - k_o C_l)\, df_s = (1 - f_s)\, dC_l \tag{10.37c}$$

$$\frac{df_s}{1 - f_s} = \frac{dC_l}{(1 - k_o)\, C_l} \tag{10.37d}$$

Notice that the rate of change of composition dC_l depends on the rate of change of the solid fraction f_s induced by the evolved ΔH_f over a temperature range $T_1 < T_f^* < T_2$ for the A-B binary alloy with composition C_o.

Integrating Eq. (10.37c) yields

$$\int_0^{f_s} \frac{df_s}{1 - f_s} = \frac{1}{(1 - k_o)} \int_{C_o}^{C_l} \frac{dC_l}{C_l} \tag{10.38a}$$

$$-\ln (f_s - 1) = \frac{1}{(1 - k_o)} \ln \left(\frac{C_l}{C_o} \right) \tag{10.38b}$$

Reorganizing Eq. (10.38b) yields the non-equilibrium Lever rule known as the *Scheil equation* for the concentration of solute B in the liquid and solid phases with $C_s = k_o C_l$ (Fredriksson and Akerlind [13, p. 186])

$$C_l = C_o \, (1 - f_s)^{k_o - 1} \tag{10.39a}$$

$$C_s = k_o C_o \, (1 - f_s)^{k_o - 1} \tag{10.39b}$$

Solving Eq. (10.39b) for f_s yields the Scheil equation for the fraction of the solid phase

$$f_s = 1 - \left(\frac{C_l}{C_o} \right)^{1/(k_o - 1)} \tag{10.40a}$$

$$f_s = 1 - \left(\frac{C_s}{k_o C_o} \right)^{1/(k_o - 1)} \tag{10.40b}$$

Figure 10.21a illustrates the behavior of a solid concentration C_s defined by Eq. (10.40b) for an A-B binary alloy with composition C_o and constant k_o values. A similar graphical representation of solute concentration can be found elsewhere for a Ni-based superalloy (Seo et al. [14]).

Hypoeutectic Alloy For a solute partition coefficient $k_o < 1$, the liquidus and solidus lines have negative slopes and the alloy composition is in the range $0 < C \le C_e$. Notice that C_s increases with increasing f_s, while $C_s \to C_o$ very rapidly as $f_s \to 1$. Moreover, a negative solid slope and $k_o < 1$ condition suggests that the excess of solute B at the L-S mushy zone is rejected into the liquid phase. Consequently, solidification proceeds until the last liquid content reaches the eutectic composition at the eutectic temperature T_e.

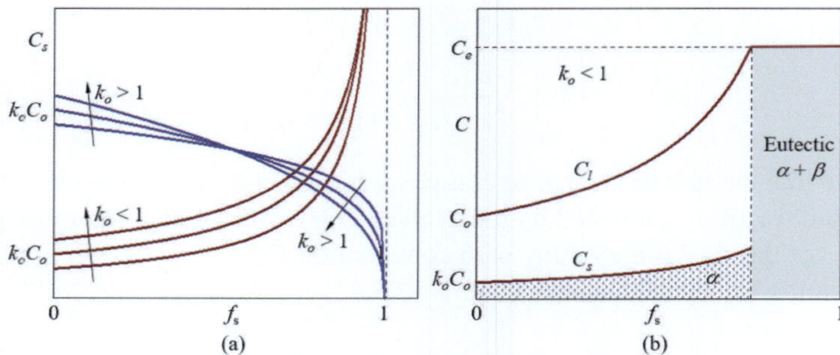

Fig. 10.21 Schematic concentration profiles. (**a**) Effect of partition coefficient k_o on C_s and (**b**) Scheil equations for the solid and liquid regions after solidification. The last liquid solidifies as a eutectic structure

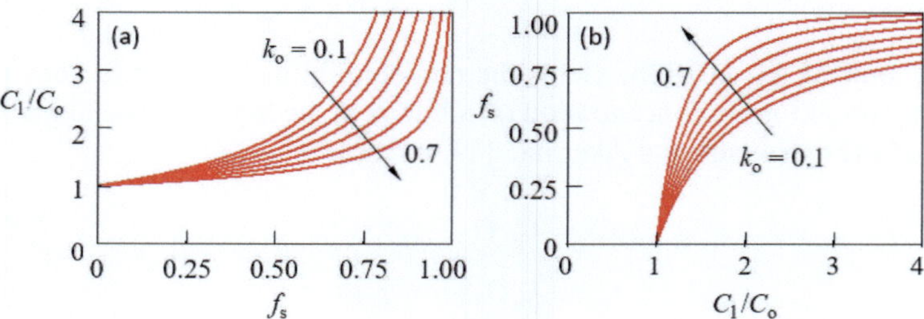

Fig. 10.22 Nonlinear Lever rule showing trends for (**a**) $C_l/C_o = g\,(f_s)$ and (**b**) $f_s = f\,(C_l/C_o)$

Hypereutectic Alloy For a solute partition coefficient $k_o > 1$, the liquidus and solidus lines have positive slopes and the alloy composition is in the range $C \geq C_e$. Obviously, $C_s \to 0$ as $f_s \to 1$ for complete solidification. Hence, plotting Eqs. (10.39a), (10.39b) reveals that C_l and C_s concentrations increase continuously with increasing f_s as shown in Fig. 10.21b, where the C_l curve becomes asymptotic by acquiring the eutectic composition C_e at T_e. During solidification of a hypereutectic alloy, the solid accommodates just the right amount of solute B, and consequently, the liquid solute content continues to decrease until $T = T_e$ and $L \to \alpha + \beta$.

Plotting Eq. (10.40a) as one particular example, the concentration ratio and solid fraction functions, $C_l/C_o = f\,(f_s)$ and $f_s = g\,(C_l/C_o)$, for some values of $k_o < 1$, reveals additional nonlinear trends as illustrated in Fig. 10.22a and b, respectively.

Eutectic Solid Fraction For an alloy with C_o and k_o (Fig. 10.22b), the first solid fraction is denoted as f_α. As solidification is near completion, the remaining liquid reaches the eutectic composition $C_l = C_e$, and subsequently, Eq. (10.38b) provides the eutectic-solid ($\alpha + \beta$) fraction f_e defined by

$$\ln\left(\frac{C_e}{C_o}\right) = (k_o - 1)\ln\,(f_e) \quad \text{for } C_o < C_e \tag{10.41a}$$

$$f_e = \exp\left[\frac{\ln(C_e/C_o)}{k_o - 1}\right] \tag{10.41b}$$

For convenience, let $T = T_f^*$ and $k_o < 1$ so that the Scheil equation and the linear Level rule expression are written as

$$f_s = 1 - \left(\frac{T - T_f}{T_l - T_f}\right)^{1/(k_o-1)} \qquad \text{(Scheil equation)} \tag{10.42a}$$

$$f_s = \frac{1}{1 - k_o}\frac{T - T_l}{T - T_f} \qquad \text{(Lever rule equation)} \tag{10.42b}$$

Differentiating Eqs. (10.42a), (10.42b) with respect to T and using the chain rule of differentiation yield the rate of formation of the solid fraction being dependent on the solidification rate (dT/dt)

$$\left(\frac{df_s}{dT}\right)_{Scheil} = -\frac{1}{(k_o - 1)(T_l - T_f)}\left(\frac{T - T_f}{T_l - T_f}\right)^{(2-k_o)/(k_o-1)} \tag{10.43a}$$

$$\left(\frac{df_s}{dT}\right)_{Lever} = \frac{2T - T_l - T_f}{(1 - k_o)(T - T_f)^2} \tag{10.43b}$$

$$\frac{df_s}{dt} = \frac{df_s}{dT}\frac{dT}{dt} \tag{10.43c}$$

A similar procedure can be used for finding df_s/dT and df_s/dt at $k_o > 1$. The chain rule defined by Eq. (10.43c) states that the rate of formation a the solid phase with respect to time (df_s/dt) is defined by the product relating a solid rate of change with respect to temperature (df_s/dT) times the temperature rate of change with respect to time (dT/dt).

The chain rule, as expressed in Eq. (10.43c), states that the rate of formation of the solid phase with respect to time (df_s/dt) is the product of the rate of change of the solid phase with respect to temperature (df_s/dT) and the rate of change of temperature with respect to time (dT/dt).

Example 10.3 Show that the concentration range can be defined as

$$\Delta C_o = C_o\left(\frac{1 - k_o}{k_o}\right) \tag{10.3E1}$$

Solution Letting $C_s = C_o$ and $\Delta C_o = C_l - C_s$ yields

$$\Delta C_o C_s = \Delta C_o C_o = C_o(C_l - C_s) \tag{10.3E2a}$$

$$\Delta C_o = \frac{C_o(C_l - C_s)}{C_s} = C_o\left(\frac{C_l}{C_s} - 1\right) \tag{10.3E2b}$$

$$\Delta C_o = C_o \left(\frac{1}{k_o} - 1 \right)$$
(10.3E2c)

$$\Delta C_o = C_o \left(\frac{1 - k_o}{k_o} \right)$$
(10.3E2d)

Actually, ΔC_o is used to solve some solidification problems.

Example 10.4 Consider the solidification of an $80Sn\text{-}20Bi$ binary alloy with nominal solute composition $C_o = 20\ wt\%Bi$, equilibrium partition coefficient k_o and liquidus line slope liquid β_l. If the solute liquid diffusion coefficient and the critical liquid temperature gradient ahead of the planar $L\text{-}S$ interface are $D_l = 10^{-8}$ m^2/s and $T_c' = 10^5\ K/m$, respectively, calculate **(a)** the $L\text{-}S$ interface velocity v_x, **(b)** the $L\text{-}S$ interface characteristic length δ_c, **(c)** the time it takes the $L\text{-}S$ interface to reach a length equals to its characteristic length, and **(d)** what the stability criterion predicts.

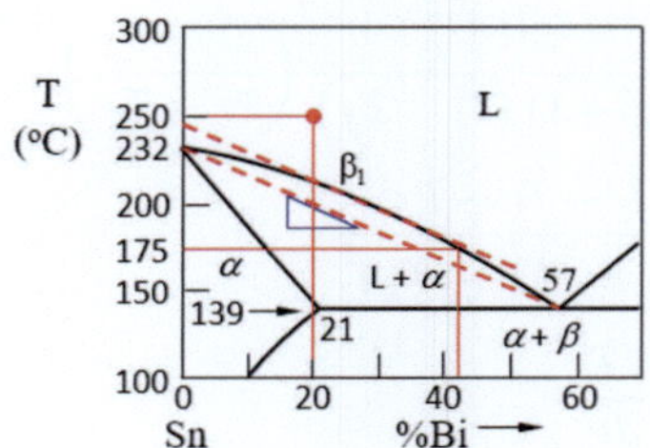

Solution From the given phase diagram, the Bi solute concentrations at $175\,^{\circ}C$ are

$$C_s \simeq 13\%Bi, \quad C_o = 20\%Bi, \quad C_l = 42\%Bi$$
(10.5E10b)

and the liquidus slope and partition coefficient are

$$\beta_l = \frac{232 - 139}{0 - 57} = -1.63\ ^{\circ}C/wt\% = -1.63\ K/wt\%$$
(10.5E2a)

$$k_o = \frac{C_s}{C_l} = \frac{13\%Bi}{42\%Bi} = 0.31$$
(10.5E2b)

(a) From Eq. (9.56c), the solidification velocity is

$$T_c' = \left[\frac{dT_l\,(x)_c}{dx} \right] = -\frac{\beta_l C_o\,(1 - k_o)}{k_o D_l / v_x} \exp\left(-\frac{x}{D_l / v_x} \right)$$
(10.5E3a)

$$T_c' = \left[\frac{dT_l\,(x)_c}{dx} \right]_{x=0} = -\frac{\beta_l C_o\,(1 - k_o)}{k_o D_l / v_x}$$
(10.5E3b)

$$v_x = -\frac{k_o D_l T_c'}{\beta_l C_o\,(1 - k_o)}$$
(10.5E3c)

$$v_x = -\frac{(0.31)\left(10^{-8}\ m^2/s\right)\left(10^5\ K/m\right)}{(-1.63\ K/\%)\,(20\%)\,(1-0.31)} \tag{10.5E3d}$$

$$v_x = 1.38 \times 10^{-5}\ m/s \tag{10.5E3e}$$

(b) The $L\text{-}S$ interface supercooling length is

$$\delta_c = \frac{D_l}{v_x} = \frac{10^{-8}\ m^2/s}{1.38 \times 10^{-5}\ m/s} = 7.25 \times 10^{-4}\ m = 0.725\ mm \tag{10.5E6}$$

(c) The time for a length δ_c is

$$t = \frac{\delta_c}{v_x} = \frac{7.25 \times 10^{-4}\ m}{1.38 \times 10^{-5}\ m/s} = 52.54\ \text{sec} \tag{10.5E37}$$

(d) From Eqs. (10.3E1) and (10.33a), the concentration and temperature ranges are

$$\Delta C_o = \frac{C_o\left(1-k_o\right)}{k_o} = \frac{(20\%)\,(1-0.31)}{0.31} = 44.52\ \% \tag{10.5E8a}$$

$$\Delta T_o = -\beta_l \Delta C_o = -\,(-1.63\ K/\%)\,(44.52\%) = 72.57\ K \tag{10.5E8b}$$

From (9.58c), the critical temperature gradient is

$$T_c' = \frac{dT_l\,(x)_c}{dx} = \frac{\Delta T_o v_x}{D_l} = \frac{(72.57\ K)\left(1.38 \times 10^{-5}\ m/s\right)}{10^{-8}\ m^2/s} \tag{10.5E9a}$$

$$T_c' = 1.00 \times 10^5\ K/m = 1.00 \times 10^5\ {}^\circ C/m \tag{10.5E9a}$$

Note that $T_a' = 10^5\ K/m$ (given) and T_c' compare

$$T_a' = T_c' = 10^5\ \frac{K}{m}\ \ \textit{stability!} \tag{10.5E10}$$

Therefore, the stability criterion predicts a stable planar solidification front because $T_a' = T_c'$.

10.13   The Hume-Rothery Rules

Apparently in the 1920s, the art of metallurgy became a science due to Hume-Rothery's work on solid solubility in $A\text{-}B$ alloys. The Mizutani's book [15] is a relevant source for insights in the Hume-Rothery rules (H-R rules) beyond the atomic size effect.

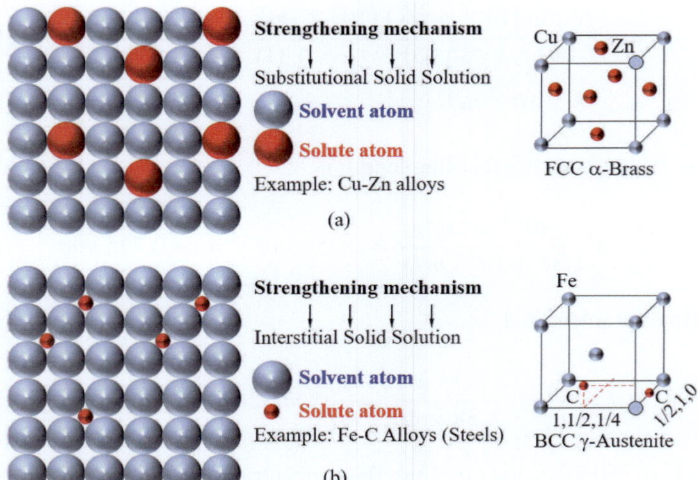

Fig. 10.23 Models for microstructural evolution due to (**a**) Cu-Zn substitutional solid solution and (**b**) Fe-C interstitial solid solution. Here, the unit cells and the radius of BCC iron (Fe) and hexagonal carbon (C) atoms are, respectively, $R_{Fe} = 0.124\ nm$ and $R_C = 0.071nm$. Also, carbon atoms C are at a tetrahedral site $1, 1/2, 1/4$ or at an octahedral site $1/2, 1, 0$

This section includes four classical H-R rules for substitutional and interstitial solid solutions with high solubility of B atoms in the A atomic structure for forming a single phase in A-B alloys; otherwise compound formation takes place during solidification. In general, the "A" component (metal) is the solvent (matrix) and "B" component is the solute, which is the alloying component in the A-B solid solutions (solid-state mixtures) with a single phase—otherwise, a single phase plus a secondary phase form. The rules are based on the concept of miscibility, which is basically the ability of two elements A and B to be partially or completely soluble in their solid state.

For a two-dimensional analysis, Fig. 10.23 schematically shows the respective crystallographic planes for a substitutional solid solution model (Fig. 10.23a) and for an interstitial solid solution (Fig. 10.23b). These types of solid solutions are strengthening mechanisms for the base metal or solvent. For example, an addition of solute zinc (Zn) atoms to a solvent plane of copper (Cu) atoms in the solid state forms a substitutional solid solution (Cu-Zn single-phase alloy), where zinc (Zn) atoms occupy Cu lattice sites. The FCC unit cell shows an order atomic arrangement since the Cu atoms are positioned at corners and Zn atoms at face-centered planes.

Figure 10.23b, on the other hand, illustrates the interstitial solid solution model for interstitial atoms (small atom) located at preferred interstitial sites in the lattice. This is the mechanism for enhancing the strength of an iron (Fe) matrix with carbon atoms. The FCC unit cell is for γ-Fe or γ-austenite at relatively high temperatures in the solid state being subjected to an appropriate cooling rate.

It is now pertinent to present the Hume-Rothery rules as a significant set of guidelines for predicting the degree of miscibility (mixing) or solid solution between A and B atoms in the solid state after a controlled solidification process.

The Hume-Rothery rules are described as follows:

(1) **The Atomic Size Factor** For substitutional and interstitial solid solutions, the basic characteristic size or factor S_f based on the perfect hard-sphere model for atoms is a percent radius difference between A and B elements or metals. It is defined as

$$S_f = \left| \frac{R_{solvent} - R_{solute}}{R_{solvent}} \right| \times 100\% \begin{cases} \leq 15\% \text{ for high solubility} \\ > 15\% \text{ for low solubility} \end{cases}$$

$$(10.44a)$$

$$S_f = \left| \frac{R_{solvent} - R_{instertitial}}{R_{solvent}} \right| \times 100\% \begin{cases} \leq 59\% \text{ for high solubility} \\ > 59\% \text{ for low solubility} \end{cases}$$

$$(10.44b)$$

(2) **The Electronegativity Factor** The electronegativity X_i values (see the periodic table of elements) of selected elements to form alloy solid solutions must be equal or similar. In this case, the X_i is considered as an invariant property of an atom, where $i = A, B$. Thus, the electronegativity effect e_i is defined by

$$e_{solute} \simeq e_{solvent} \text{ for high solubility} \qquad (10.45a)$$

$$e_{solute} < e_{solvent} \text{ for low solubility} \qquad (10.45b)$$

$$e_{solute} > e_{solvent} \text{ for low solubility} \qquad (10.45c)$$

(3) **The Atomic Structural Factor** The A and B elements must have the same or similar crystal structures (unit cells) in order to achieve complete solubility in the solid state.

(4) **The Relative Valence Factor** z The A and B elements must have the same oxidation state or valence "z" as indicated by the electrochemical reactions

$$A \rightarrow A^{+z} + ze^- \qquad (10.46a)$$

$$B \rightarrow B^{+z} + ze^- \qquad (10.46b)$$

where ze^- denotes the number of electrons in the above oxidation reactions. This rule suggests that the electronic configuration of the solute element B has an effect on the electron orbitals of the solvent A.

Compound Formation If the solid solubility of B is limited to form a complete substitutional A-B solid solution alloy, then some B atoms react with A atoms to form, hypothetically, a compound like $A_x B$ or $A_x B_y$.

In the case of steelmaking (Fe-C alloy system), the excess carbon (C) atoms form an Fe_3C compound called cementite. In carbon steels, Fe_3C forms from the austenite solid phase, and in cast irons, it forms from the liquid transformation. Relevant details are given in the section that deals with ferrous alloys related to the Fe-C phase diagram and its phase domains. In particular, Fe_3C has an orthorhombic unit cell with lattice parameters $a = 0.51\,nm$, $b = 0.67\,nm$, and $c = 0.45\,nm$.

10.14 Aluminum and Aluminum Alloy Systems

Aluminum-based alloys have practical or engineering applications across a wide spectrum of fields, such as the automotive, aeronautical, and aerospace industries.

Apparently, Al-Si, Al-Cu, Al-Mg, Al-Zn, and Al-Sn alloys are the most common systems since they have practical technological applications due to aluminum being a corrosion-resistant metal with a low density and high formability.

Most of the Al-alloy systems are precipitation-induced hardening from a single solid-phase solution, say α-phase, to precipitate β-phase as hard particles, usually in the form of small plates. Moreover, Al-Si alloys are also referred to as "silumin" because of their lightweight and high strength. Apparently, as-cast wrought Al-Si alloys are important engineering materials. The as-cast solidification condition may imply a conventionally or rapidly produced alloys under strictly controlled foundry practices.

Some Al-alloys are identified or classified using a series of four-number designations, such as:

- 1xxx for Al with a 99% minimum purity, non-heat treatable. Commercially pure aluminum has a variety of applications due to its corrosion-resistant characteristic.
- 2xxx for Al-Cu alloys, heat treatable for precipitation hardening. The most common alloy in this series is 2024-T351 for aircraft and sport goods applications.
- 3xxx for Al-Mn alloys, non-heat treatable. Apparently, 3003 alloy is used to make "aluminum cans" for beverages. This alloy is also used by the car and aircraft industries.
- 4xxx Al-Si alloys, heat treatable for precipitation hardening. Alloy 4043 seems to be the most practical material in this series for welding purposes, pump casings, cookware, and so forth.
- 5xxx for Al-Mg alloys, non-heat treatable. Apparently, 5182 alloy is a common alloy. The 5182 alloy is used by the "aluminum cans" and foundry industries.
- 6xxx for Al-Mg-Si alloys, heat treatable for precipitation hardening. The most common alloy in this series is 6061 alloy for boat parts.
- 7xxx for Al-Zn alloys, heat treatable for precipitation hardening. The most common alloy in this series is 7075 alloy, which is used in the aircraft industry.
- 8xxx for Al-X alloys, where X represents other metal. The most common alloy in this series is 8580 alloy.

- *Al-Li* alloys, where *Li* represents other metal. This type of alloy is used as an engine part by the car industry due to its significantly low mass density and relatively high strength due to strong precipitates (Al_3Li). Industrial applications of the *Al–Li* 2195 alloy include aircraft parts for weight reduction. Actually, the 2195 alloy has an active electrochemical behavior in seawater at room temperature.

An important feature of *Al*-alloy phase diagrams is the "solvus" sloping curve line for indicating the solubility range of the constituent or alloying element in the aluminum matrix related to precipitation-hardening mechanism.

10.14.1   Aluminum-Silicon Phase Diagram

One particular case is shown in Fig. 10.24 for a non-ferrous binary *Al-Si* alloy system (Guy and Hren [16, p. 302]). Notice the microstructural evolution during the solidification of slowly cooled *Al-Si* alloys with different compositions. Thus, the photomicrographs with given magnification and scale bar represent the microstructures at room temperature.

A few microstructural images are included in order for the reader to understand and easily interpret the connection between the equilibrium phase diagram and the corresponding equilibrium solidification process. Noticeably, an *Al-Si* composition induces a particular characteristic black-and-white (B&W) microstructural morphology. However, large β particles in 80*Al*-20*Si* and 50*Al*-50*Si* hypereutectic alloys dictate practical industrial applications.

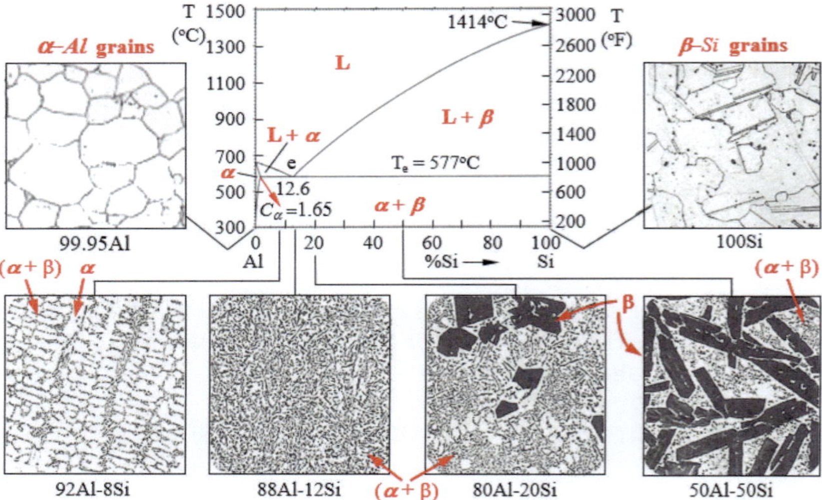

Fig. 10.24 Aluminum-silicon (*Al-Si*) equilibrium eutectic phase diagram and morphology of microstructure of pure *Al*, pure *Si*, and some *Al-Si* alloys (Guy and Hren [16, p. 302])

This alloy system has good castability (ease of forming a quality casting with minimum defects at low cost) and corrosion resistance. Hence, high castability can often be used as a principal condition for shape casting, provided that all solidification variables are controlled within a certain tolerance. Moreover, Al has an FCC and Si has a diamond-like crystal structures.

Example 10.5 Verify if the Al-Si alloy systems in Fig. 10.24 follows the Hume-Rothery rules. Microscopically, the Si-rich β-phase solidifies as platelike crystals with a common diamond-like cubic structure and the Al-rich α-phase solidifies as a face-centered cubic (FCC) phase. The radii of these elements are $R_{Al} = 143\,pm$ and $R_{Si} = 118\,pm$.

Solution

1. The atomic size effect is determined as the percent different (S_f) between the Al and Si radii

$$S_f = \left| \frac{R_{Al} - R_{Si}}{R_{Al}} \right| \times 100\% = \left| \frac{143 - 118}{143} \right| \times 100\% = 17.48\% \qquad (10.6\text{E1})$$

which is not within the range $0 < S_f \leq 15\%$ as defined by the first Hume-Rothery rule for forming complete substitutional solid solutions. Therefore, the Al-Si alloys form partial substitutional solid solubility because these elements are not comparable in size and $S_f = 17.48\%$ does not satisfy the Hume-Rothery size rule.

2. From any periodic table containing electronegativity values, $e_{Si} = 1.8$ and $e_{Al} = 1.5$, where silicon (Si) is treated as the solute metalloid with $e_{Si} < e_{Al}$. Therefore, Al-Si alloy systems form partial substitutional solid solutions and a compound formation prevails in the solid state.

3. Notice that Al and Si have similar crystal structures, and therefore, they form partial substitutional solid solutions, where Si has a low solubility in the Al matrix. Therefore, compound formation prevails in the solid state.

4. The oxidation reactions for Al and Si are obtained from any available electromotive force table of redox reactions

$$Al \rightarrow Al^{+3} + 3e^- \qquad (10.6\text{E2a})$$

$$Si \rightarrow Si^{+4} + 4e^- \qquad (10.6\text{E2b})$$

The valence difference leads to a partial substitutional solid solution. This is clearly shown in Fig. 10.24. Therefore, Al and Si are partially soluble in the solid state. Of particular interest is the eutectic reaction

$$L\,(12.6\%\,Si) \xrightarrow{\;577\,^\circ C\;} \alpha\,(1.65\%\,Si) + \beta\,(99.5\%\,Si) \qquad (10.6\text{E3})$$

Actually, the β-phase region and the β (99.5% Si)-limit are not visible in Fig. 10.24 due to the scale being used. Moreover, the α-phase region is also small, where the maximum solubility of Si in Al, $C_\alpha = 1.65\%$, occurs at the eutectic temperature $T - e = 577\,^\circ C$. This is a small amount of Si that can dissolve within the Al solvent without the formation of a second β-phase structure.

10.15   Copper and Copper Alloy Systems

Copper alloys are used as cast and wrought materials with specific designations found in the ASM Specialty Handbook (Davis [17]). Thus,

Cu pure copper with a natural reddish color
Cu-Al alloys have a wide range of applications.
Cu-Be alloys are used as non-sparking tools in the gas and petrochemical industries.
Cu-Ni alloys have high corrosion resistance.
Cu-Si alloys have high corrosion resistance.
Cu-Sn alloys are known as bronzes.
Cu-Sn-Zn-Pb alloys are used as gunmetals.
Cu-Zn alloys are known as brasses with a wide range of applications.

10.15.1   *Copper-Zinc Phase Diagram*

Consider the copper-zinc $(Cu$-$Zn)$ equilibrium phase diagram and some related microstructures shown in Fig. 10.25 (Cochrane [18, micrograph 422], Vander Voort [19]).

The most common Cu-Zn binary alloys in the non-ferrous industry are known as brass and bronze materials. For instance, brass is a Cu-Zn alloy with Zn being the alloying metal that induces variations in colors from red to yellow brasses. In particular,

- Figure 10.25a shows the Cu-Zn equilibrium phase diagramwith the α-phase region being the most useful for producing α-brass.
- Figure 10.25b exhibits a cast $70Cu$-$30Zn$ alloy with an α-brass dendritic microstructure (Cochrane [18, micrograph 422]).
- Figure 10.25c depicts an annealed $70Cu$-$30Zn$ alloy containing the α-brass or cartridge brass microstructure having equiaxed α-FCC grains (Vander Voort [19]).
- Figure 10.25d illustrates the microstructure of the as-cast $60Cu$-$40Zn$ alloy (Muntz metal) containing precipitated primary α-phase as dendrites with some β-phase in the interdendritic areas and grain boundaries (Cochrane [18, micrograph 441]). This cast microstructure is known as Widmanstatten microstructure due to its basket weave appearance.

Notice that the equilibrium Cu-Zn binary phase diagram has three binary invariant points at different temperatures. They are

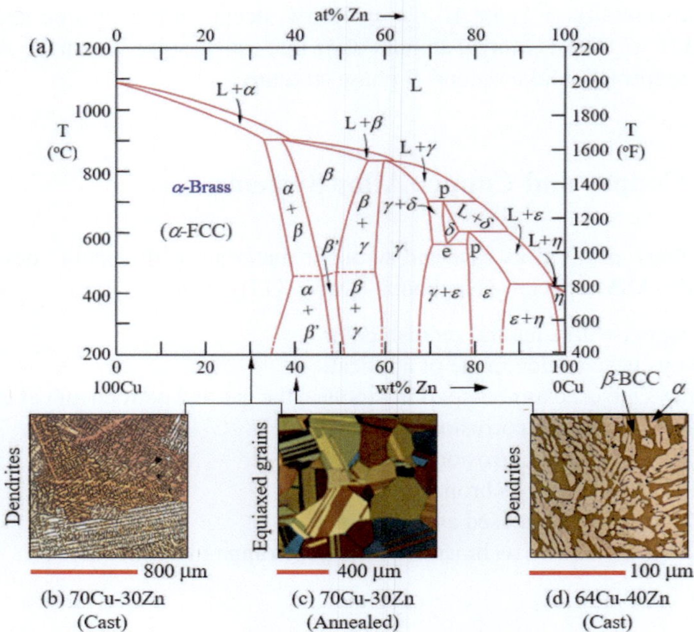

(b) 70Cu-30Zn
(Cast)

(c) 70Cu-30Zn
(Annealed)

(d) 64Cu-40Zn
(Cast)

Fig. 10.25 Copper-zinc phase diagram and microstructures. (**a**) Cu-Zn equilibrium phase diagram, (**b**) cast $70Cu$-$30Zn$ alloy known as cartridge brass (Cochrane [18, micrograph 422]), (**c**) annealed $70Cu$-$30Zn$ alloy (Vander Voort [19]), and (**d**) cast $60Cu$-$40Zn$ alloy (Muntz metal) with precipitated α-phase embedded in the β-phase (matrix) (Cochrane [18, micrograph 441])

$$\text{Peritectic point p: } L + \gamma \rightleftarrows \delta \text{ at } T_e = 700\,^{\circ}C \qquad (10.47a)$$

$$\text{Peritectic point p: } L + \delta \rightleftarrows \varepsilon \text{ at } T_e = 598\,^{\circ}C \qquad (10.47b)$$

$$\text{Eutectoid point e: } \delta \rightleftarrows \gamma + \varepsilon \text{ at } T_e = 560\,^{\circ}C \qquad (10.47c)$$

Further, polycrystalline metals or alloys may have dendritic grains in the as-cast (Fig. 10.25b) or elongated grains in the deformed condition. Therefore, either microstructural condition induces anisotropic properties and it is a common technological or engineering practice to heat-treat these alloys in order to homogenize the irregular microstructural morphology (size, shape, and orientation/texture) as depicted in Fig. 10.25b. As a result, the material becomes isotropic at a macroscale since the new regular grain structure is of equiaxial form as clearly shown in Fig. 10.25c through optical examination.

Actually, grains are three-dimensional polyhedral units separated by irregular grain boundaries, but optical examination provides a two-dimensional pattern of grains separated by irregular grain boundary lines. In addition, if the radii of FCC-Cu and HCP-Zn are $R_{Cu} = 0.128\ nm$ and $R_{Zn} = 0.133\ nm$, then one can determine the possibility of substitutional solid solution in Cu-Zn alloys. This can be done by calculating the percent radius difference as $\Delta R = [(0.133 - 0.128)/0.128](100\%) = 3.91\%$.

According to the Hume-Rothery atomic size effect rule, Cu and Zn form a substitutional solid solution in a single phase, say α-phase, since $\Delta R \leq 15\%$. However, the difference in crystal structures (FCC and HCP) implies the opposite, and consequently, Cu and Zn form a compound since Zn has limited solubility in the Cu-matrix.

Nevertheless, most Cu-Zn alloys are used as wrought brasses after being suitably deformed (rolling, forging, etc.) and heat-treated to obtain equiaxed grains (Fig. 10.25c). Thus, the common commercial alloy shapes are in the form of plates, tubes, wires, and the like.

Lastly, the composition of brasses can be modified by adding other metals. The resultant modified brasses are commonly known as Si-bronze, Sn-bronze, Al-bronze, and the like with useful properties, such as strength, ductility, or machinability. Moreover, brasses and bronzes are ductile and corrosion-resistant engineering materials. Regarding corrosion, these materials passivate through an opaque oxide film in most common industrial environments.

10.16   Lead and Lead Alloy Systems

Lead (Pb) properties include high toxicity, mass density, and ductility for ease of fabrication, absorption of high-energy radiation, and chemical stability in air and water. Technologically, Pb-alloys can be used in batteries as insoluble anodes, cable sheeting, and solders. Moreover, applications of lead alloys are based on their properties, such as high density, malleability, lubricity, flexibility, low melting point, corrosion resistance, and its effectiveness in shielding against X-rays and gamma radiation. For Pb-free soldering alloys see Li et al. [20] and for Pb-containing alloys see ASM Handbook [21, Vol. 2].

10.16.1   Lead-Tin Phase Diagram

Consider the lead-tin (Pb-Sn) equilibrium phase diagram and some related microstructures shown in Fig. 10.26.

The traditional Pb-Sn mixtures have been used as electrical solders (fusible metal alloys), but health concerns over the Pb content researchers have found alternatives based on lead-free (Pb-free) soldering alloys (Li et al. [20], Pang [22]) for applications in the electronic industry as printed circuit-board materials. Because of the abundance of published work, the Pb-Sn diagram and related microstructures are included in this section. The most common application of Pb-Sn alloys for soldering must have a composition $C_o \lesssim Ce$ for a relatively fast solidification at a joining point on an electric circuit board (ASM Handbook [21, Vol. 2]).

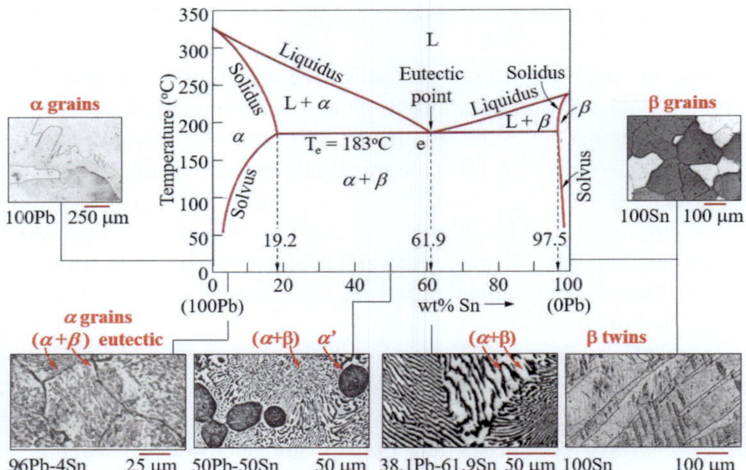

Fig. 10.26 Microstructures of lead-tin (*Pb-Sn*) alloys. (**a**) *Pb-Sn* equilibrium phase diagram, (**b**) 98*Pb*-2*Sn* hypoeutectic alloy, (**c**) 50*Pb*-50*Sn* hypoeutectic alloy, (**d**) 38.1*Pb*-61.9*Sn* eutectic alloy, and (**e**) 100*Sn* showing annealing twins due to a slight increase of the temperature during polishing (ASM Handbook [21, Vol. 2])

The *FCC-Pb* and tetragonal-*Sn* structures are crystallographically and dimensionally different since their atomic radii are $R_{Pb} = 0.175\ nm$ and $R_{Sn} = 0.151\ nm$. In essence, these metals can be mixed in the solid state to form a single solid solution, such as α and β. For *Pb-Sn* alloys with *Pb* as the solvent and *Sn* as the solute, the percent radius difference is $\%\Delta R = (100\%)(0.175 - 0.151)/0.175 = 13.71\%$, which means that *Pb-Sn* alloys form substitutional solid solutions. In this context, *Pb*-containing alloys including lead-tin (*Pb-Sn*), lead-arsenic (*Pb-As*), lead-antimony (*Pb-Sb*), and lead-calcium (*Pb-Ca*) are quite common alloy systems subjected to small additions of other metals (ASM Handbook [21, Vol. 2]).

Example 10.6 Assume a slow steady-state solidification process for a 100-gram 60*Pb*-40*Sn* alloy initially at $T_p = 300\,°C$ and consider a half-space region within a rectangular-shaped semi-infinite casting with the characteristic length defined as $s(t) = x_c = x/2 = 1\ cm$ for complete solidification. Use the given equilibrium phase diagram along with some schematic microstructures, thermophysical properties for the binary alloy, and sand mold to calculate (**a**) the mass fraction (f_s and f_l) and partition coefficient (k_o) of phases at temperature points "*A*" through "*e*" applying the Lever rule, (**b**) the heat transfer variables related to an ideal planar *L-S* interface, and (**c**) the rate of change of the solid fraction df_s/dt at just above the eutectic temperature T_e, say at $T = 184\,°C$.

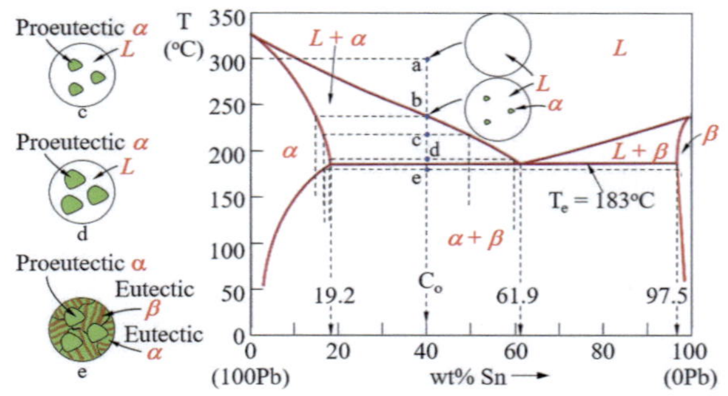

Material	$T_f{}^a$	$\Delta H_s{}^a$	$\rho_s{}^b$	$\rho_l{}^b$	$c_s{}^b$	$c_l{}^b$	$k_s{}^b$	$k_l{}^b$
60Pb-40Sn	460		9300	8700	100	100	20	20
Sand mold			1550		986		0.40	

Units of tabulated thermophysical data
[a] T_f in $°C$ and ΔH_s in kJ/Kg units
[b] ρ_s, ρ_l in Kg/m^3, c_s, c_s in $J/Kg.K$, and k_s, k_l in $W/m.K$

Solution

(a) Lever rule and its fundamental approach

(a_1) At $T_a = 300\,°C$, the alloy contains

Reaction	None, all liquid
Phases	L
Composition	$C_l = 40Sn,\ C_\alpha = 0$
Mass of phases	$f_l = 100\%\ \ \&\ \ f_\alpha = 0$
Partition coefficient	$k_o = C_\alpha/C_l = 0$

(a_2) At $T_b = 238\,°C$, the proeutectic or primary α-phase fraction becomes

Reaction	$L_a \rightarrow L + \alpha$
Phases	$L + \alpha$
Composition	$C_l \simeq 40.10Sn,\ C_\alpha = 15Sn,\ C_o = 40Sn$
Partition coefficient	$k_o = C_\alpha/C_l = 15/40.10 = 0.37$

The mass fraction of phases is

$$f_l = \frac{C_o - C_\alpha}{C_l - C_\alpha} \simeq \frac{40Sn - 15Sn}{40.10Sn - 15Sn} = 0.9960 \qquad (10.6\text{E1a})$$

$$f_\alpha = 1 - f_l = 1 - 0.9960 = 0.004 \qquad (10.6\text{E1b})$$

(a_3) At $T_c = 218\,^{\circ}C$, the proeutectic α-phase fraction becomes

Reaction	$L_b \to L + \alpha$
Phases	$L + \alpha$
Composition	$C_l \simeq 49.8Sn,\ C_\alpha = 16.50Sn,\ C_o = 40Sn$
Partition coefficient	$k_o = C_\alpha/C_l = 0.55$

The mass fraction of phases is

$$f_\alpha = \frac{C_l - C_o}{C_l - C_\alpha} = \frac{49.8Sn - 40Sn}{49.8Sn - 16.5Sn} = 0.29 \tag{10.6E2a}$$

$$f_\alpha = (100\ g)\,(0.29) = 29\ g \tag{10.6E2b}$$

$$f_l = \frac{C_o - C_\alpha}{C_l - C_\alpha} = \frac{40Sn - 16.5Sn}{49.8Sn - 16.5Sn} = 0.71 \tag{10.6E2c}$$

$$f_l = (100\ g)\,(0.71) = 71\ g \tag{10.6E2d}$$

(a_4) At $T_d = 184\,^{\circ}C$, the proeutectic or primary α-phase fraction becomes

Reaction	$L_c \to L + \alpha$
Phases	$L + \alpha$
Composition	$C_l = 59.6Sn,\ C_\alpha = 19Sn,\ C_o = 40Sn$
Partition coefficient	$k_o = C_\alpha/C_l = 0.32$

The mass fraction of phases is

$$f_\alpha = \frac{C_l - C_o}{C_l - C_\alpha} = \frac{59.6Sn - 40Sn}{59.6Sn - 19Sn} = 0.48 \tag{10.6E3a}$$

$$f_\alpha = (100\ g)\,(0.48) = 48\ g \tag{10.6E3b}$$

$$f_l = \frac{C_o - C_\alpha}{C_l - C_\alpha} = \frac{40Sn - 19Sn}{59.6Sn - 19Sn} = 0.52 \tag{10.6E3c}$$

$$f_l = (100\ g)\,(0.52) = 52\ g \tag{10.6E3d}$$

(a_5) At $T_{eutectic} = 183\,^{\circ}C$, the eutectic α-β-phase fraction becomes

Reaction	$L \to \beta' + (\alpha + \beta)_e$
Phases	$L,\ \beta,\ (\alpha + \beta)_e$
Composition	$C_e = C_l = 61.9Sn,\ C_{\alpha e} = 19.2Sn,\ C_o = 40Sn$
Partition coefficient	$k_o = C_{\alpha e}/C_l = 19.2/61.9 = 0.31$

The mass fraction of phases is

$$f_{\alpha'} = \frac{C_e - C_o}{C_e - C_\alpha} = \frac{61.9Sn - 40Sn}{61.9Sn - 19.2Sn} = 0.51 \tag{10.6E4a}$$

$$f_{\alpha'} = (100 \ g)(0.51) = 51 \ g \tag{10.6E4b}$$

$$f_e = 1 - f_{\alpha'} = 1 - 0.51 = 0.49 \tag{10.6E4c}$$

$$f_e = (100 \ g)(0.49) = 49 \ g \tag{10.6E4d}$$

$$f_e = f_l = 49 \ g \tag{10.6E4e}$$

At $T_{eutectic} = 183\,°C$, the eutectic mass fraction represents the remaining liquid phase, Eq. (10.6E4e), which transforms into eutectic phase $(\alpha + \beta)_e$ at $T < T_e = 183°$.

(a_6) At $T_d = 182\,°C$, the total α-phase fraction is

Reaction	$L_d \to \alpha + (\alpha + \beta)_e$
Phases	$\alpha + (\alpha + \beta)_e$
Composition	$C_\beta \simeq 97.6Sn, \ C_\alpha = 17Sn, \ C_o = 40Sn$
Partition coefficient	$k_o = C_s/C_l = 0$

Total mass fraction of phases is

$$f_\alpha = \frac{C_\beta - C_o}{C_\beta - C_\alpha} = \frac{97.6Sn - 40Sn}{97.6Sn - 17Sn} = 0.71 \tag{10.6E5a}$$

$$f_\alpha = (100 \ g)(0.71) = 71 \ g \quad (\text{total}) \tag{10.6E5b}$$

$$f_\beta = \frac{C_o - C_\alpha}{C_\beta - C_\alpha} = \frac{40Sn - 17Sn}{97.6Sn - 17Sn} = 0.29 \tag{10.6E5c}$$

$$f_\beta = (100 \ g)(0.29) = 29 \ g \tag{10.6E5d}$$

(a_7) The total mass fractions of α-phase and β-phase components in the eutectic structure are

$$f_{\alpha e} = f_\alpha - f_{\alpha'} = 0.71 - 0.51 = 0.20 \tag{10.6E6a}$$

$$f_{\alpha e} = (100 \ g)(0.20) = 20 \ g \tag{10.6E6b}$$

$$f_{\beta e} = f_e - f_{\alpha e} = 0.49 - 0.20 = 0.29 \tag{10.6E6c}$$

$$f_{\beta e} = 29 \ g \tag{10.6E6d}$$

At $T_d = 182\,^{\circ}C$, the $60Pb$-$40Sn$ alloy has $f_{\alpha'} = 51\%$ proeutectic (primary) α-phase and $f_e = 100\% - 51\% = 49\%$ $(\alpha + \beta)_e$ eutectic phase, from which $f_{\alpha e} = 20\%$ and $f_{\beta e} = 29\%$.

(a_8) At $T = 50\,^{\circ}C$, the total solid-phase fractions are

Reaction	$(\alpha + \beta)_e \rightarrow (\alpha + \beta)_{50\,^{\circ}C}$
Phases	$\alpha + \beta$
Composition	$C_\alpha = 3Sn,\ C_\beta = 98.5Sn,\ C_o = 40Sn$
Partition coefficient	$k_o = C_s/C_l = 0$

The mass fraction of phases is

$$f_\alpha = \frac{C_\beta - C_o}{C_\beta - C_\alpha} = \frac{98.5Sn - 40Sn}{98.5Sn - 3Sn} = 0.61 \tag{10.6E7a}$$

$$f_\alpha = (100\ g)\,(0.61) = 61\ g \tag{10.6E7b}$$

$$f_\beta = 1 - f_\alpha = 1 - 0.61 = 0.39 \tag{10.6E7c}$$

$$f_\beta = (100\ g)\,(0.39) = 39\ g \tag{10.6E7d}$$

Therefore, the $60Pb$-$40Sn$ alloy contains 61% of α-phase and 39% of β-phase. This means that $f_\alpha = 71\%\% - 61\% = 10\%$ α-phase reduction and $f_\beta = 39\% - 29\% = 10\%$ β-phase increment from $T_d = 182\,^{\circ}C$ to $T = 50\,^{\circ}C$. According to the Pb-Sn phase diagram, there is not any further phase transformation at $T \leq 50\,^{\circ}C$. Moreover, the reader is encouraged to plot the $k_o = f(T)$ relationship.

(b) Solidification heat transfer and its slightly advanced approach. This part of the example uses a similarity solution approach (see Chap. 9) for calculating the thermal diffusivity α and the thermal inertia (thermal effusivity) γ terms for the solid and mold materials. These terms are considered as thermophysical properties being defined by Eqs. (9.4a), (9.4b). Further, thermal inertia refers to as the property of a material related to the period of time the material takes to reach a local environmental temperature. Using the magnitudes of thermal conductivity k, density ρ, and specific heat capacity c_p yields the calculated thermal diffusivity terms

$$\alpha_s = \frac{k_s}{\rho_s c_s} = \frac{20\ W/m.K}{\left(9300\ Kg/m^3\right)(0.17\ J/Kg.K)} \tag{10.6E5a}$$

$$\alpha_s = 0.01265\ m^2/s \tag{10.6E5b}$$

$$\alpha_m = \frac{k_m}{\rho_m c_m} = \frac{0.40\ W/m.K}{\left(1550\ Kg/m^3\right)(0.986\ J/Kg.K)} \tag{10.6E5c}$$

$$\alpha_m = 2.6173 \times 10^{-7}\ m^2/s \tag{10.6E5d}$$

and the thermal inertia terms are

$$\gamma_s = \frac{k_s}{\sqrt{\alpha_s}} = \frac{20\ W/m.K}{\sqrt{0.01265\ m^2/s}} = 177.82\ \frac{W\sqrt{sec}}{m^2.K} \tag{10.6E6a}$$

$$\gamma_s = 177.82\ \frac{J}{m^2.K.s^{1/2}} \tag{10.6E6b}$$

$$\gamma_m = \frac{k_m}{\sqrt{\alpha_m}} = \frac{0.40\ W/m.K}{\sqrt{2.6173 \times 10^{-7}\ m^2/s}} = 781.87\ \frac{W\sqrt{sec}}{m^2.K} \tag{10.6E6c}$$

$$\gamma_m = 781.87\ \frac{J}{m^2.K.s^{1/2}} \tag{10.6E6d}$$

The latent heat of solidification is not given, but it can be calculated using Eq. (9.15b) with the fraction nominal composition $C_o = 0.20Sn$

$$\Delta H_s = H_{f,Pb} + C_o \left(H_{f,Sn} - H_{f,Pb} \right) + c_s \left(T_p - T_E \right) \tag{10.6E7a}$$

$$\Delta H_s = 23 + 40\,(56 - 23) + 100\,(300 - 183) \tag{10.6E7b}$$

$$\Delta H_s = 13{,}043\ J/Kg \tag{10.6E7c}$$

The suitable solution characteristic (λ) expression defined by Eq. (9.30b) becomes

$$\lambda \exp\left(\lambda^2\right) \left[\mathrm{erf}\,(\lambda) + \frac{\gamma_s}{\gamma_m} \right] = \frac{c_s \left(T_p - T_o \right)}{\sqrt{\pi}\,\Delta H_s} \tag{10.6E8a}$$

$$\lambda \exp\left(\lambda^2\right) \left[\mathrm{erf}\,(\lambda) + \frac{\gamma_s}{\gamma_m} \right] = \frac{(0.17)\,(300 - 25)}{\sqrt{\pi}\,(13{,}043)} \tag{10.6E8b}$$

$$\lambda \exp\left(\lambda^2\right) \left[\mathrm{erf}\,(\lambda) + \frac{177.82}{781.87} \right] = 2.0222 \times 10^{-3} \tag{10.6E8c}$$

$$\lambda \exp\left(\lambda^2\right) [\mathrm{erf}\,(\lambda) + 0.22743] = 2.0222 \times 10^{-3} \tag{10.6E8d}$$

Iteration yields the magnitude of the solution characteristic λ as

$$0 = \lambda \exp\left(\lambda^2\right) [\mathrm{erf}\,(\lambda) + 0.22743] - 2.0222 \times 10^{-3} \tag{10.6E9a}$$

$$\lambda = 8.5299 \times 10^{-3} \tag{10.6E9b}$$

An iterative method is an iteration procedure that requires repetition of a mathematical or computational procedure applied to a previous result. Actually, the result or outcome of an iteration is the starting point of the next iteration. An iterative procedure continues until an approximation to the solution of a problem is achieved using a control variable, say $\Delta\lambda = \lambda_i - \lambda_{i-1}$, in the range $10^{-4} < \Delta\lambda < 10^{-8}$. For example, if the last two results are $\lambda_i = 8.5299 \times 10^{-3}$ and $\lambda_i = 8.52989 \times 10^{-3}$, then $\Delta\lambda = 10^{-8}$ and the iteration stops.

From Eq. (9.23e), the mold-solid interface (intermediate) temperature along with the pouring temperature $T_p = 300 + 273 = 573\,K$ and $T_o = 25 + 273 = 298\,K$ is defined by

$$T_i = \frac{\gamma_s T_p + \gamma_m T_o \operatorname{erf}(\lambda)}{\gamma_s + \gamma_m \operatorname{erf}(\lambda)} \tag{10.6E10a}$$

$$T_i = \frac{177.82 \times 573 + 781.87 \times 298 \operatorname{erf}\left(8.5299 \times 10^{-3}\right)}{177.82 + (781.87) \operatorname{erf}\left(8.5299 \times 10^{-3}\right)} \tag{10.6E10b}$$

$$T_i = 561.83\,K = 288.83\,^\circ C \tag{10.6E10c}$$

The solidification time t is calculated applying two different approaches (methods) along with the characteristic length for complete solidification when $s(t) = x_c = x/2 = 1\,cm$. Using Eq. (9.6c) with the calculated solution characteristic $\lambda = 0.39769$ and Eq. (9.33e) with $T_i = 475.68\,^\circ C$ and $(T_i - T_o) = 288.83 - 25 = 263.83\,^\circ C = 263.83\,K$ yields, respectively,

$$t_1 = \frac{s(t)^2}{4\lambda^2 \alpha_s} = \frac{(0.01\,m)^2}{4\left(8.5299 \times 10^{-3}\right)^2 \left(0.01265\,m^2/s\right)} \tag{10.6E11a}$$

$$t_1 = 27.16\,\sec \tag{10.6E11b}$$

$$t_2 = \frac{\pi}{4}\left[\frac{\rho_s \Delta H_s}{\gamma_m (T_i - T_o)}\right]^2 s(t)^2 \tag{10.6E11c}$$

$$t_2 = \frac{\pi}{4}\left[\frac{(9300)(13043)}{(781.87)(263.83)}\right]^2 \left(10^{-2}\right)^2 = 27.16\,\sec \tag{10.6E11d}$$

These equations are used to verify the calculated solidification time $t = 27.16\,sec$, which is considered a reasonable result. Thus far, it has been assumed that solidification is continuous with a planar L-S interface having a velocity calculated next.

From Eq. (9.6e) along with $t = t_1 = t_2 = 27.16\,s$, the solidification velocity is

$$\left[\frac{ds(t)}{dt}\right] = \lambda\sqrt{\frac{\alpha_s}{t}} = \left(8.5299 \times 10^{-3}\right)\sqrt{\frac{0.01265\,m^2/s}{27.16\,s}} \tag{10.6E12a}$$

$$\upsilon_x = \left[\frac{ds(t)}{dt}\right] = 1.84 \times 10^{-4}\,m/s \tag{10.6E12b}$$

It is assumed that the solidification process does not experience constitutional supercooling because the solidification velocity is very small and that solidification is continuous with a planar liquid-solid (L-S) interface. Moreover, this solidification velocity is also assumed to represent an average velocity between atom jumps at the L-S and S-L interfaces.

The solid temperature gradient at the planar solidification front is approximated using Eq. (9.19c). Thus,

$$X = \frac{(T_p - T_i)}{\mathrm{erf}\,(\lambda)\,\sqrt{\pi \alpha_s t}} \tag{10.6E13a}$$

$$X = \frac{(300 - 288.83)}{\mathrm{erf}\,(8.5299 \times 10^{-3})\,\sqrt{\pi\,(0.01265)\,(27.16)}} = 1117.10\ ^\circ C/m \tag{10.6E13b}$$

$$Y = \exp\left(-\frac{s\,(t)^2}{4\alpha_s t}\right) \tag{10.6E13c}$$

$$Y = \exp\left(-\frac{(10^{-2})^2}{4\,(0.01265)\,(27.16)}\right) = 0.99993 \tag{10.6E13d}$$

Then,

$$\left(\frac{\partial T_s}{\partial x}\right)_{s(t)} = \frac{(T_p - T_i)}{\mathrm{erf}\,(\lambda)\,\sqrt{\pi \alpha_s t_1}}\exp\left(-\frac{s\,(t)^2}{4\alpha_s t}\right) = XY \tag{10.6E14a}$$

$$\left(\frac{\partial T_s}{\partial x}\right)_{s(t)} = (1117.10\ ^\circ C/m)\,(0.99993) \tag{10.6E14b}$$

$$\left(\frac{\partial T_s}{\partial x}\right)_{s(t)} = 1117\ ^\circ C/m \tag{10.6E14c}$$

Lastly, the magnitude of the heat flux q_s, Eq. (9.48a), with $s\,(t) = 1\ cm$ is

$$q_s = -k_s\left(\frac{\partial T_s}{\partial x}\right)_{s(t)} = -\,(9300\ W/m.K)\,(1117\ ^\circ C/m) \tag{10.6E15a}$$

$$q_s = -\,(9300\ W/m.K)\,(844\ K/m) \tag{10.6E15b}$$

$$q_s = -7.85 \times 10^6\ W/\left(m^2.s\right) = -7.85\ MJ/m^2 \tag{10.6E15c}$$

which represents the heat transfer through the solidified solid and through the mold walls, and eventually it dissipates to the local environment.

(c) For this part of the example, the temperature field equation $T_s = T_s\,(x)$ defined by Eq. (9.19a) and the rate of change of the solid fraction df_s/dt can be derived using the Scheil expression defined by Eq. (10.42a), where $T_l \leq T_f = T_f^* \leq T_e$ is the range of solidification. Thus,

$$T_s = T_i + \frac{\left(T_f - T_i\right)}{\operatorname{erf}(\lambda)} \operatorname{erf}\left(\frac{x}{2\sqrt{\alpha_s t}}\right) \quad \text{(solid)} \tag{10.6E16a}$$

$$f_s = 1 - \left(\frac{T - T_f}{T_l - T_f}\right)^{1/(k_o-1)} \quad \text{(Scheil equation)} \tag{10.6E16b}$$

For complete solidification, $T_f = T_l = T_e = 183\,°C$ and $t = 27.16$ sec

$$\frac{\partial T_s}{\partial t} = -\frac{\left(T_f - T_i\right)x}{2t\sqrt{\pi\alpha_s t}\,\operatorname{erf}(\lambda)} \exp\left(-\frac{x^2}{4\alpha_s t}\right) \tag{10.6E17a}$$

$$X_1 = -\frac{(183 - 288.83)\,(0.01)}{2\,(27.16)\,\sqrt{\pi\,(0.01265)\,(27.16)}\,\operatorname{erf}\left(8.5299 \times 10^{-3}\right)} \tag{10.6E17b}$$

$$X_1 = 1.9484 \tag{10.6E17c}$$

$$X_2 = \exp\left(-\frac{(0.01)^2}{4\,(0.01265)\,(27.16)}\right) = 0.99993 \tag{10.6E17d}$$

$$\frac{\partial T_s}{\partial t} = X_1 X_2 = (1.9484)\,(0.99993) \tag{10.6E17e}$$

$$\frac{\partial T_s}{\partial t} = 1.9483\ °C/s = 1.9483\ K/s \tag{10.6E17f}$$

and

$$\frac{\partial T_s}{\partial x} = \frac{\left(T_f - T_i\right)}{\sqrt{\pi\alpha_s t}\,\operatorname{erf}(\lambda)} \exp\left(-\frac{x^2}{4\alpha_s t}\right) \tag{10.6E18a}$$

$$X_3 = -\frac{(183 - 288.83)}{2\,(27.16)\,\sqrt{\pi\,(0.01265)\,(27.16)}\,\operatorname{erf}\left(8.5299 \times 10^{-3}\right)} \tag{10.6E18b}$$

$$X_3 = 194.84 \tag{10.6E18c}$$

$$\frac{\partial T_s}{\partial x} = X_3 X_2 = (194.84)\,(0.99993) \tag{10.6E18d}$$

$$\frac{\partial T_s}{\partial x} = 194.83\ °C/m = 194.83\ K/m \tag{10.6E18e}$$

From Eq. (10.6E16b) along with $k_o = 0.31$, $T_f = T_e = 183\,°C$, and $T_l = T_b = 238\,°C$, df_s/dT just above $T_e = 183\,°C$, say $T_l = 184\,°C$

$$\frac{df_s}{dT} = -\frac{1}{k_o - 1}\left(\frac{1}{T_l - T_f}\right)^{1/(k_o-1)}\left(T - T_f\right)^{(2-k_o)/(k_o-1)} = 482.41\ K^{-1}$$

$$\tag{10.6E19a}$$

From Eq. (10.43c), the chain rule gives the rate of change of the solid fraction df_s/dt

$$\frac{df_s}{dt} = \frac{df_s}{dT}\frac{dT}{dt} = \left(482.41\ K^{-1}\right)(1.9483\ K/s) \tag{10.6E20a}$$

$$\frac{df_s}{dt} = \left(482.41\ K^{-1}\right)(1.9483\ K/s) = 939.88\ s^{-1} \tag{10.6E20b}$$

This result is assumed to be slightly high due to the small amount of the alloy.

10.17   Classification of Eutectic Structures

The variety of geometrical arrangements of the α and β solid phases can be used to classify the eutectic microstructural morphology. Regardless of the type of alloy and magnification, Fig. 10.27 shows regular (Fig. 10.27a) and irregular (Fig. 10.27b) α-β lamellar (platelike) eutectic structures in Al-Si alloys (Elliott [23, pp. 16, 57]), Fig. 10.27c exhibits a cellular (hexagonal-like) α-β structure in a Pb-Sn alloy, and Fig. 10.27d depicts a dendritic structure in some alloy (Bailey [24, p. 9]).

The idea here is to show the reader the diversity in microstructural analysis one has to consider as a result of the complex solidification process. Moreover,

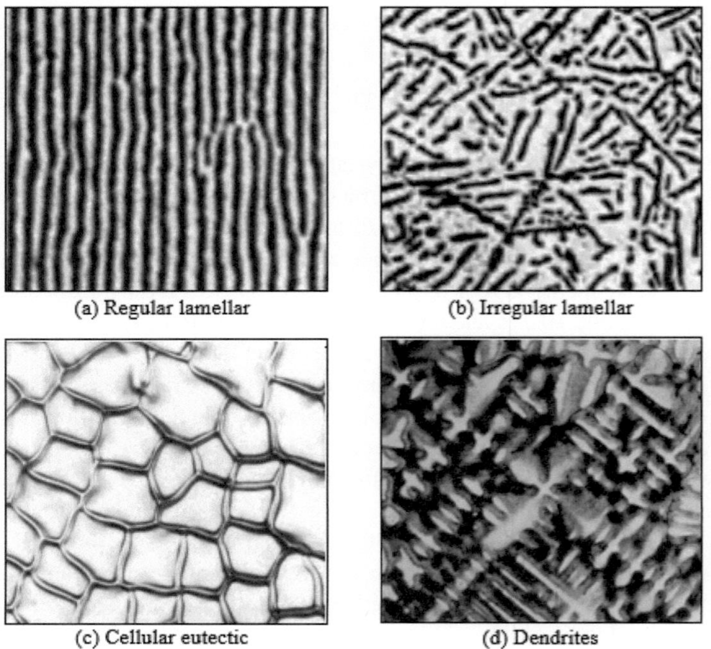

Fig. 10.27 Eutectic microstructures. **(a)** Regular Al-Si lamellar and **(b)** irregular eutectic structure in Al-Si (Elliott [23, p. 57]). **(c)** Cellular eutectic in dilute Sn-Pb alloy (Elliott [23, p. 16]), and **(d)** dendrites (Bailey [24, p. 9])

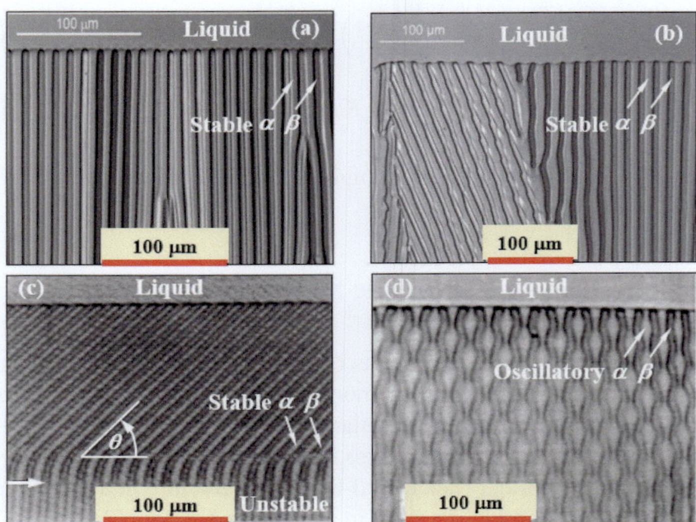

Fig. 10.28 Lamellar eutectic in directional solidification. For $N\ PG$-45.3DC samples (Witusiewicz et al. [25]), (**a**) and (**b**) regular lamellar showing different crystallographic orientations and for CBr_4-0.21%C_2Cl_6 samples (Karma and Sarkissian [26]), (**c**) regular transition (see arrow and angle θ) from $\upsilon_z = 0.85\,\mu m/s$ to stable tilted growth to $\upsilon_z = 3.30\,\mu m/s$ at $G_a = 80\,K/cm$, and (**d**) oscillatory lamellar structure at $\upsilon_z = 2\,\mu m/s$ and $G_a = 80\,K/cm$

both α and β solid phases in Fig. 10.27a seem to be fundamentally comparable dimensionally and geometrically.

Further, α and β solid phases are crystallographically different and their crystal structures are determined by X-ray diffraction. Moreover, Fig. 10.28a and b show snapshots of α-β lamellar structures during directional solidification in carbon tetrabromide and 0.21$wt\%$ hexachloroethane (CBr_4-0.21C_2Cl_6) (Witusiewicz et al. [25]), and neopentylglycol-(D)camphor (NPG-45.3DC) transparent organic alloy samples in Fig. 10.28c and d (Karma and Sarkissian [26]).

These images are ideal for research and academic purposes, and for clarity, the α (light gray) and β (dark gray) notations are assigned to the phases for identifying these eutectic structures. Some additional information on the conditions for eutectic solidification is given in the figure caption.

The well-aligned α-β solid phases shown in Fig. 10.28a are ideal for determining the lamellar λ-spacing as a function of growth velocity υ_z and liquid temperature gradient $G_l = dT_l/dz$. Specifically, the lamellar spacing $\lambda = f(\upsilon_z)$ and the undercooling $\Delta T = f(\lambda)$ functions appear to be the most studied analytically for characterizing the evolution of two-phase solidification microstructures in rod and lamellar morphology.

The amount and morphology of the eutectic structure, however, in a specific alloy dictate magnitude of properties. This is one of the reasons for characterizing the regular or irregular lamellar λ-spacing under the influence of a temperature gradient ($G_l = dT_l/dz$) and related alloy liquid (melt) undercooling $\Delta T = f(\lambda)$. Hence, $\lambda = f(v_z)$, $\Delta T = f(\lambda)$, and $G_l = dT_l/dz$ are important variables in the solidification field.

Additional details on these α-β eutectic structures can be found in the cited references (Tiller [27], Jackson and Hunt [28]).

10.17.1   Dendrite Growth

The onset of solidification starts at the mold-melt interface as a thin layer with inevitable defects, which are the potential sites for subsequent interface growth. For alloys, the planar interface morphology is commonly destabilized due to temperature gradients, solute segregation, and crystallographic anisotropy of the evolving microstructure. Consequently, dendrite growth dominates the evolution of the as-cast microstructure in most industrial cases.

Figure 10.29a shows an X-ray tomography dendritic image in a 1-mm-diameter sample cooled at $2\,^\circ C/min$ (Gibbs et al. [29]). Note that the shown vertical dendrite is composed of a predominantly solid cylindrical (stalk or stem) with secondary and tertiary dendrite arms with spherical caps.

For clarity, the blue- and red-colored areas are located at the shown distances from the dendrite tip as per the original authors. The main dendrite growth is assumed to occur in the z-direction at velocity v_z (primary stalk or stem) along with growing arms at v_x and v_{xz} into the liquid phase. For instance, the secondary arms grow at v_x along the $< 100 >$ crystallographic direction. Moreover, Fig. 10.29b depicts a particular as-cast dendritic microstructure in an unidentified Cu-based eutectic alloy taken from the 2005 Metals Conservation Summer Institute presentation (Napolita [30]).

Regarding NASA Isothermal Dendritic Growth Experiment (IDGE), it provides insights on the effect of microgravity on dendrite formation in order to understand

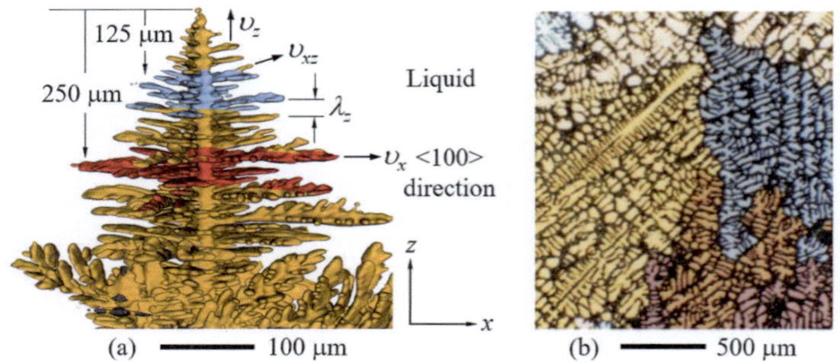

Fig. 10.29 Dendrite morphology. (a) Al-24Cu dendrites obtained by using X-ray tomographic technique on 1-mm-diameter sample cooled at $2\,^\circ C/min$ (Gibbs et al. [29]) and (b) as-cast dendritic structure in some metallic material (Napolita [30])

how heat and mass transfer affects materials processing. It is clearly evident in Fig. 10.29a that the morphology of dendrites consists of treelike branched crystals with solidification velocity $\upsilon_z > \upsilon_x, \upsilon_{xz}$ into the liquid phase. As solidification progresses, the solidification velocity υ_z and heat flow q_z occur in the opposite directions ($\upsilon_z \rightarrow$ and $q_z \leftarrow$).

In practice, as-cast components with dendritic structures are usually heat-treated to obtain equiaxed microstructures with uniform properties. Thus, the resultant microstructural evolution is strongly affected by the solidification velocity to produce either equiaxed or dendritic microstructure. In effect, dendrites form due to a deviation of the planar L-S interface instability accompanied with constitutional supercooling, where $dT_l(x)/dx < \beta_l \partial C_l / \partial x$. Here, β_l is the slope of a T-C diagram liquidus line.

10.18   The Iron-Carbon Phase Diagram

The iron-carbon (Fe-C) equilibrium phase diagramis also built using a series of cooling curves (not shown) of molten Fe-C_{Fe_3C} or Fe-$C_{graphite}$ alloys, where Fe_3C is a metastable phase called cementite (CM) or iron carbide (IC) with an orthorhombic unit cell. Graphite, on the other hand, is a natural hexagonal structure. Nonetheless, the Fe-C equilibrium phase diagram is a graphical representation of the physical states of several solid phases, which are widely used to characterize solid steel and cast iron products.

First of all, pure iron (Fe) solidifies exhibiting multiple solid-phase transformations until it reaches room temperature or below. This solid transformation process is crystallographically shown in Fig. 10.30a along with the schematic cooling curve (Fig. 10.30b).

Fig. 10.30 **(a)** Unit cells and **(b)** cooling curve related to the solidification process of pure iron (Fe)

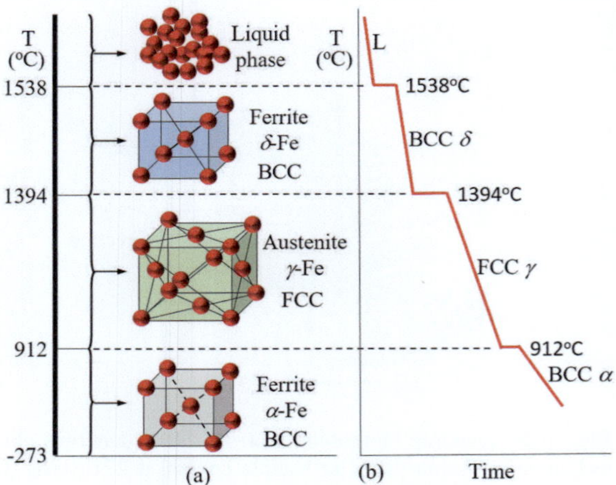

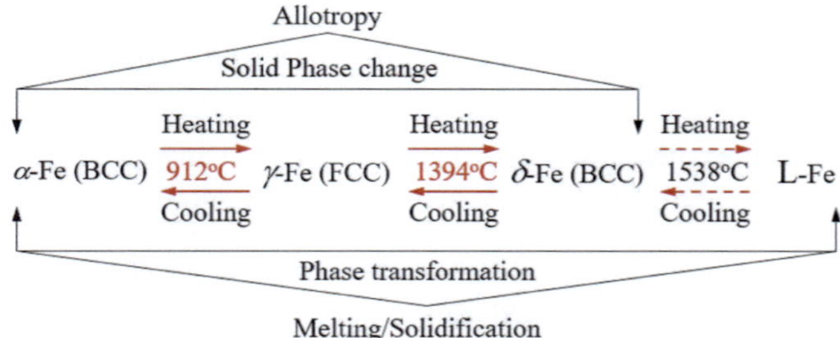

Fig. 10.31 Phase changes in solid-state allotropy and melting or solidification

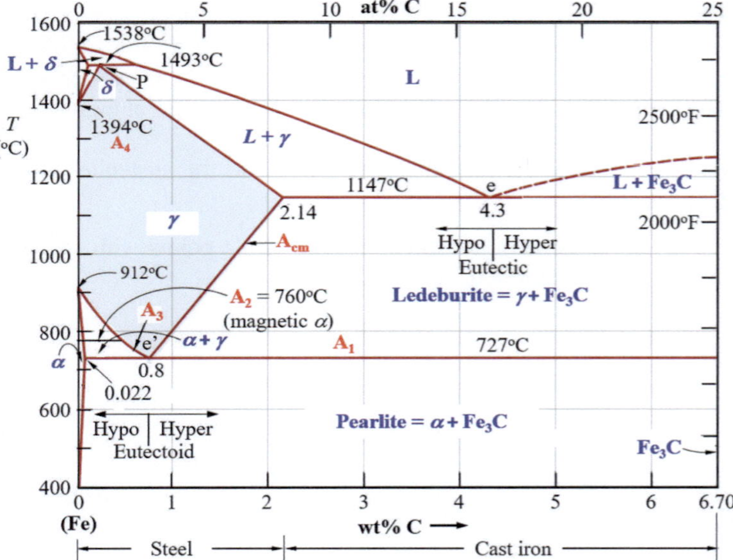

Fig. 10.32 The $Fe\text{-}C$ phase diagram showing the austenitic region. Adapted from Callister and Rethwisch [31], p. 287)

According to Fig. 10.30, pure Fe indicates that it is an allotropic metal and the corresponding process is illustrated in Fig. 10.31.

10.18.1   Iron-Carbon Phase Diagram

The $Fe\text{-}C$ equilibrium phase diagram is shown in Fig. 10.32, where the source of carbon in $wt\%\,C$ and $at\%\,C$ units is from stable cementite Fe_3C (Callister and Rethwisch [31, p. 287]).

One can find in the literature a superposition of $Fe\text{-}C$ diagrams using $Fe\text{-}C_{Fe_3C}$ and $Fe\text{-}C_{graphite}$ sources. Moreover, Fig. 10.32 has some specific multiple lines related to "thermal arrest" that act as phase boundaries, namely:

- A_o It is the temperature at which there is a magnetic change of cementite (Fe_3C) phase field from ferromagnetic to paramagnetic at approximately $210\,^\circ C$ (not shown).
- A_1: It is the eutectoid temperatureat which there is an upper limit of the ferrite-cementite ($\alpha\text{-}Fe_3C$) phase field called pearlite. For clarity, proeutectoid or primary α-ferrite forms at $A_1 < T < A_3$ and eutectoid α-ferrite incorporates in the pearlite phase at $T \le T_{e'}$.
- A_2: It is the Curie temperatureat which Fe in the α-phase loses its magnetism.
- A_3: The boundary between the γ-austenite and the austenite-ferrite ($\alpha\text{-}\gamma$) phase field. This boundary may shift upward or downward by the addition of other alloying metals, such as chromium (Cr), to plain $Fe\text{-}C$ carbon steels.
- A_4: The temperature at which γ changes to δ or vice versa at high temperatures; $T = 1394\,^\circ C$.
- A_{CM}: The boundary between the γ-austenite and the austenite-cementite ($\gamma\text{-}Fe_3C$) phase field.

The $Fe\text{-}C$ phase diagram has three invariant triple points with specific carbon compositions at temperatures. Thus,

$$\text{Peritectic point p:}\quad C_p = 0.22\%C \quad \text{at } T = 1493^\circ C \tag{10.48a}$$

$$\text{Eutectic point e:}\quad C_e = 4.30\%C \quad \text{at } T = 1147^\circ C \tag{10.48b}$$

$$\text{Eutectoid point e':}\quad C_{e'} = 0.80\%C \quad \text{at } T = 727^\circ C \tag{10.48c}$$

The corresponding invariant phase reactions and the mass fraction of phases (Lever rule) are

$$\text{Peritectic p:}\quad L + \delta \rightleftarrows \gamma \quad \text{at } T = 1493\,^\circ C \tag{10.49a}$$

$$f_l = \frac{0.16 - 0.10}{0.52 - 0.10} = 0.14 \tag{10.49b}$$

$$f_\delta = 1 - f_l = 1 - 0.14 = 0.86 \tag{10.49c}$$

$$\text{Eutectic e:}\quad L \rightleftarrows \gamma + Fe_3C \quad \text{at } T = 1147\,^\circ C \tag{10.49d}$$

$$f_\gamma = \frac{6.70 - 4.30}{6.70 - 2.14} = 0.53 \tag{10.49e}$$

$$f_{Fe_3C} = 1 - f_\gamma = 1 - 0.53 = 0.47 \tag{10.49f}$$

$$\text{Eutectoid e':}\quad \gamma \rightleftarrows p = \alpha + Fe_3C \quad \text{at } T = 727\,^\circ C \tag{10.49g}$$

$$f_\alpha = \frac{6.70 - 0.80}{6.70 - 0.022} = 0.88 \tag{10.49h}$$

$$f_{Fe_3C} = 1 - f_{Fe_3C} = 1 - 0.88 = 0.12 \qquad (10.49i)$$

In the metallurgical field, Fe-C alloys are classified in a number of ways. According to the Fe-C phase diagram in Fig. 10.32, the range of carbon content and related eutectoid and eutectic temperatures can be used to classify the ferrous Fe-C alloys as:

- Ferrous eutectoid Fe-C alloys: These are known as plain-carbon eutectoid steels if the carbon content is in the range $0\%C < C_o \leq 2.14\%C$. In practice, most commercially manufactured steel parts are in the range $0\%C < C_o < 2\%C$.
- Ferrous eutectic Fe-C alloys: These are known as cast irons if $2\%C < C_o \leq 6.70\%C$. Most commercially manufactured cast iron parts are in the range $2\%C < C_o \leq 4\%C$.

Employing the invariant points in Fig. 10.32, one can classify or divide eutectoid Fe-C alloys in groups, such as:

- Hypoeutectoid steel if $C_o < 0.80\%C$
- Eutectoid steel if $C_o = 0.80\%C$
- Hypereutectoid steel if $0.80\% < C_o \leq 2\%C$
- Hypoeutectic cast iron if $2\%C < C_o \leq 4.3\%C$
- Eutectic cast iron if $C_o = 4.3\%C$
- Hypereutectic cast iron if $4.3\%C < C_o < 6.7\%C$

Using foundry terminology, cast irons are ferrous Fe-C alloys known as:

- White cast iron containing 2–3.5%C, 0.5–3%Si, and 1–2%Mn
- Gray cast iron containing 2.5–4%C and 1–3%Si
- Ductile cast iron containing 3–3.6%C and 2–2.8%Si
- Malleable cast iron containing 2–5%C, 0.01–0.03%Cr, and some Cu and Ni
- Alloy cast iron containing 2–4%C, Si, and other elements

10.18.2   The Austenite Region

Figure 10.33 illustrates the nonmagnetic austenitic region γ along with the models for solid-state phase transformation upon slow cooling of a hypothetical hypoeutectoid carbon steel with composition C_{o1} and a hypereutectoid carbon steel with composition C_{o2}. Notice that the models assume that the solid-state phase transformation occurs along the austenite grain boundaries.

It is clearly shown in Fig. 10.33 that γ-austenite is an allotrope of iron capable of absorbing carbon from the iron-carbide phase (Fe_3C). Moreover, heating is a two-phase austenitization. Upon slow cooling, the γ-austenite starts to transform along its grain boundaries due to diffusion. However, quenching (rapid cooling) does not allow diffusion to take place and the phase transformation produces martensite.

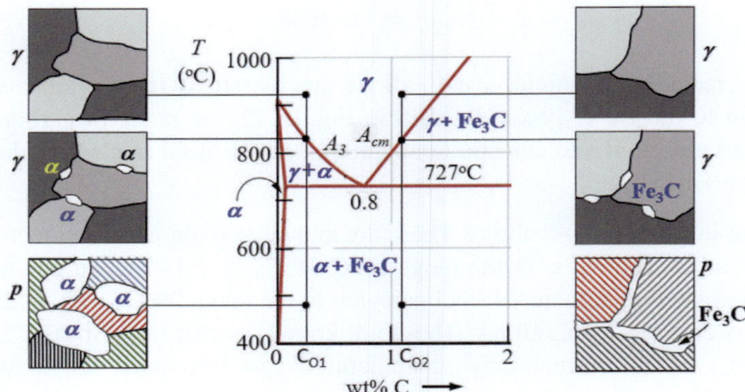

Fig. 10.33 Phase transformation of austenite in hypoeutectoid C_{o1} and hypereutectoid C_{o2} carbon steels. Here, $p = \alpha + Fe_3C$ is a dual phase called pearlite

The expected microstructures upon slow cooling of the FCC γ-austenite phase are described by the phase reactions cited below:

- $\gamma \;\rightarrow\; \gamma_o + \alpha_{grain} \;\rightarrow\; p + \alpha_{grain} \;=\; (\alpha + Fe_3C)_{grain} + \alpha_{grain}$ for the hypoeutectoid carbon steel with $C_{o1} \;<\; 0.80\%C$, where α-ferrite having a BCC crystal structure forms as the proeutectoid ductile solid phase at just $T_{e'} < T < A_3$. The remaining austenite γ_o-phase transforms into pearlite $p = (\alpha + Fe_3C)_{grain}$ according to the eutectoid reaction $\gamma_o \rightarrow (\alpha + Fe_3C)_{grain}$ at $T_{e'} = 727\,°C$. Moreover, cementite Fe_3C-phase is a hard orthorhombic crystal which alternates with elongated-like layers of α-ferrite.
- $\gamma \;\rightarrow\; \gamma_o + (Fe_3C)_{gb} \;\rightarrow\; p + (Fe_3C)_{gb} \;=\; (\alpha + Fe_3C)_{grain} + Fe_3C_{gb}$ for the hypereutectoid carbon steel with $C_{o2} > 0.80\%C$, where the cementite phase Fe_3C precipitates along the grain boundaries (gb) as the proeutectoid or primary phase at just $T > T_{e'} = 727\,°C$. Similarly, $\gamma_o \rightarrow (\alpha + Fe_3C)_{grain}$ at $T_{e'} < T < A_{cm}$.

Some cast steel and annealed steel (Fe-C steels) and pure iron (Fe) microstructures are shown in Fig. 10.34.

For the purpose of clarity, the microstructural features in Fig. 10.34 are clearly visible at different scale bars, which represent different magnifications. For instance,

- Figure 10.34a shows the as-cast SEM austenite microstructure (Website video [32]).
- Figure 10.34b clearly exhibits an α-ferrite phase (50 μm grain size) along with a few small black dots called inclusions (Pohl [33]).
- Figure 10.34c depicts the as-cast ferritic microstructure (α-ferrite grains) showing acicular ferrite and some nonmetallic inclusions in Fe-0.06%C steel (Ferreira de Oliveira et al. [34]).

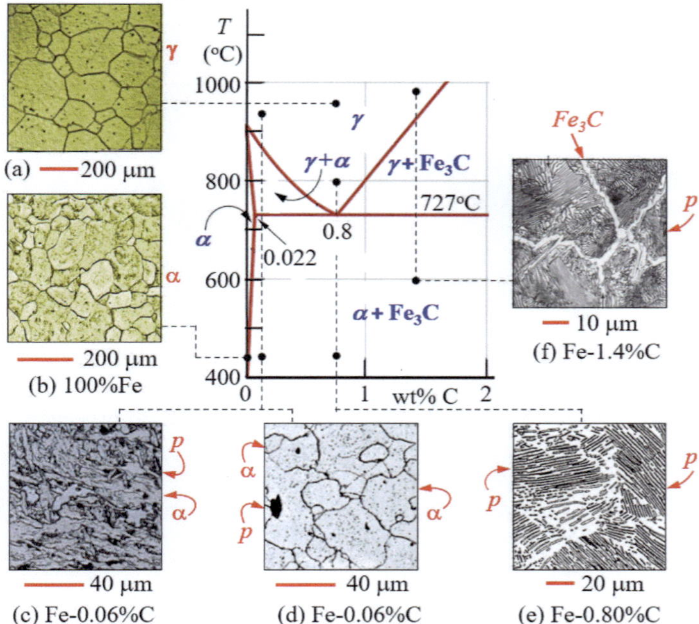

Fig. 10.34 Actual phase transformation of austenite in C_{o1} and C_{o2} cast carbon steels. Nomenclature: γ = austenite, α = ferrite, Fe_3C = cementite, and $p = (\alpha + Fe_3C)$ = pearlite. After Callister and Rethwisch [31, Chapter 9]

- Figure 10.34d illustrates the annealed steel microstructure composed of α-ferrite grains and some p-pearlite and nonmetallic inclusions in Fe-0.06%C steel (Ferreira de Oliveira et al. [34]).
- Figure 10.34e shows a typical p-pearlite phase in a 1080 eutectoid steel. This pearlite eutectoid structure is composed of alternate layers of light α-ferrite and dark Fe_3C-cementite phases (Metal Handbook [35, Fig. 37]).
- Figure 10.34f exhibits the microstructure of a Fe-1.4%C steel taken from Callister and Rethwisch book [31, p. 296]. This photomicrograph (Copyright 1971 by United States Steel Corporation) shows p-pearlitic grains separated by Fe_3C-cementite grain boundaries.

Regarding pearlite $p = (\alpha + Fe_3C)_{grain}$, it is a lamellar microstructure which contains a white hard lamellar cementite Fe_3C-phase constituent that alternates with the white ductile α-ferrite matrix. In essence, pearlite is a biphasic microstructure with alternating lamellae of ferrite and cementite phases.

Example 10.7 Use the small portions of the expanded microstructures for $C_o < 0.80\%C$ and $C_o > 0.80\%C$ Fe-C steels to identify the corresponding phases.

Solution

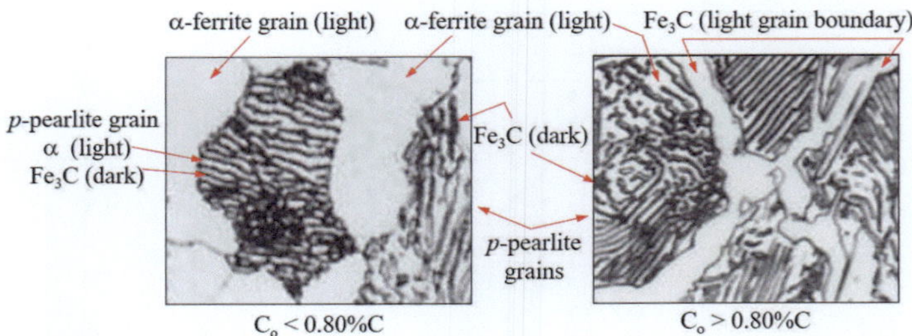

For the hypoeutectoid carbon steel with $C_o < 0.80\%C$ on cooling, the γ-austenite grains transform to pearlite $p = (\alpha + Fe_3C)$ and proeutectoid or primary ferrite BCC-α grains. The corresponding phase reaction is written to proceed from left to right

$$\gamma \rightleftarrows \gamma_o + \alpha_{grain} \rightleftarrows p + \alpha_{grain} = (\alpha + Fe_3C)_{grain} + \alpha_{grain} \qquad (10.7\text{E}1)$$

For the hypereutectoid carbon steel with $C_o > 0.80\%C$,

$$\gamma \rightleftarrows \gamma_o + (Fe_3C)_{gb} \rightarrow p + (Fe_3C)_{gb} = (\alpha + Fe_3C)_{grain} + Fe_3C_{gb} \qquad (10.7\text{E}2)$$

These phase reactions revert upon heating from right to left. Moreover, the mechanism for austenite to pearlite is described in details in Chap. 11.

Regarding cementite Fe_3C or iron carbide, it is a metastable compound being treated hereafter as a phase that may decompose into ferrite and graphite after a prolong period of time at $25\,°C < T < 727\,°C$. This implies that Fe_3C is stable at $T \leq 25\,°C$ and the T-C graphical relationship in Fig. 10.34 may be treated as the Fe-C metastable equilibrium phase diagram.

If a small amount of another metal X (such as Ni, Cr, Mn, V, etc.) is added to a plain-carbon Fe-C alloy system called "plain-carbon steel," then the nomenclature Fe-C-X can be used to designate an "alloy steel alloy"with $0\%C < C_o \leq 2\%C$ or as a "cast iron alloy" with $2\%C < C_o \leq 4\%C$.

10.18.3 Classification of Carbon Steels

Commonly, carbon steels are classified as cast steels and wrought steels, and according to their carbon, they are referred to as:

- Low-carbon steels: $C_o < 0.20\%C$
- Medium-carbon steels: $0.20 < C_o < 0.50\%C$
- High-carbon steels: $C_o > 0.50\%C$
- Eutectoid steel: $C_o > 0.80\%C$

Cast Steels These are the final products of steel foundry practices. Apparently, medium- and high-carbon-content cast steels are favorable in the agricultural and excavating fields as alternatives for intricate shapes with high strength (Guy and Hren [16, p. 336]).

Wrought Steels These are mechanically processed cast steels in order to improve mechanical properties. The most common and cost-effective mechanical deformation is the hot rolling, where the cast steel is deformed and heat-treated (normalized) simultaneously in the hot-rolling air surroundings at a suitable austenitic temperature.

Designations The following metal societies have their own designation system for classifying materials. Either numbers or a combination of letters and numbers represent materials designations. For example, the steel numbering systems for a eutectoid steel are:

- American Iron and Steel Institute (AISI), such as AISI 1080
- Society of Automotive Engineers (SAE), such as SAE 1080
- American Society for Testing and Materials (ASTM), such as ASTM A576 Grade 1080
- Unified Numbering System (UNS), such as UNS G10800

The reader is encouraged to consult the International Alloy Designations (IAD), chemical composition limits (CCL), and heat treatment procedures (HTP) for the types of recommended ferrous and non-ferrous alloys. These include some of the registered alloys subject of patent or patent applications.

10.19   Cast Irons

Most foundry practices optimize the solidification process to produce high-quality cast irons. Schematically, the microstructure morphology of cast irons is illustrated in Fig. 10.35 (Moffatt et al. [10, p. 195]) as found in Callister and Rethwisch book [31, p. 360].

These $Fe\text{-}C$ alloy systems are referred to as "cast irons" containing high concentrations of carbon atoms: $2.14\%C \leq C_o \leq 6.70\%C$ (theory) and $2.5\%C \leq C_o \leq 4\%C$ (practice). Cast irons can be classified as eutectic ferrous alloys due to the eutectic point shown in Fig. 10.35. Actually, commercial cast steels and cast irons have, at least, silicon (Si) and manganese (Mn), and traces of impurities such as sulfur (S) and phosphorus (P).

Cast irons have certain characteristics that make them different from other ferrous alloys. For example, (1) when cast iron is produced as a crude ingot, it is called pig, which can be remelted along with scrap and other alloying elements for adjusting the chemical composition and for producing required components, (2) cast iron is commonly called "gray cast iron" or "white cast iron" due to their appearance on

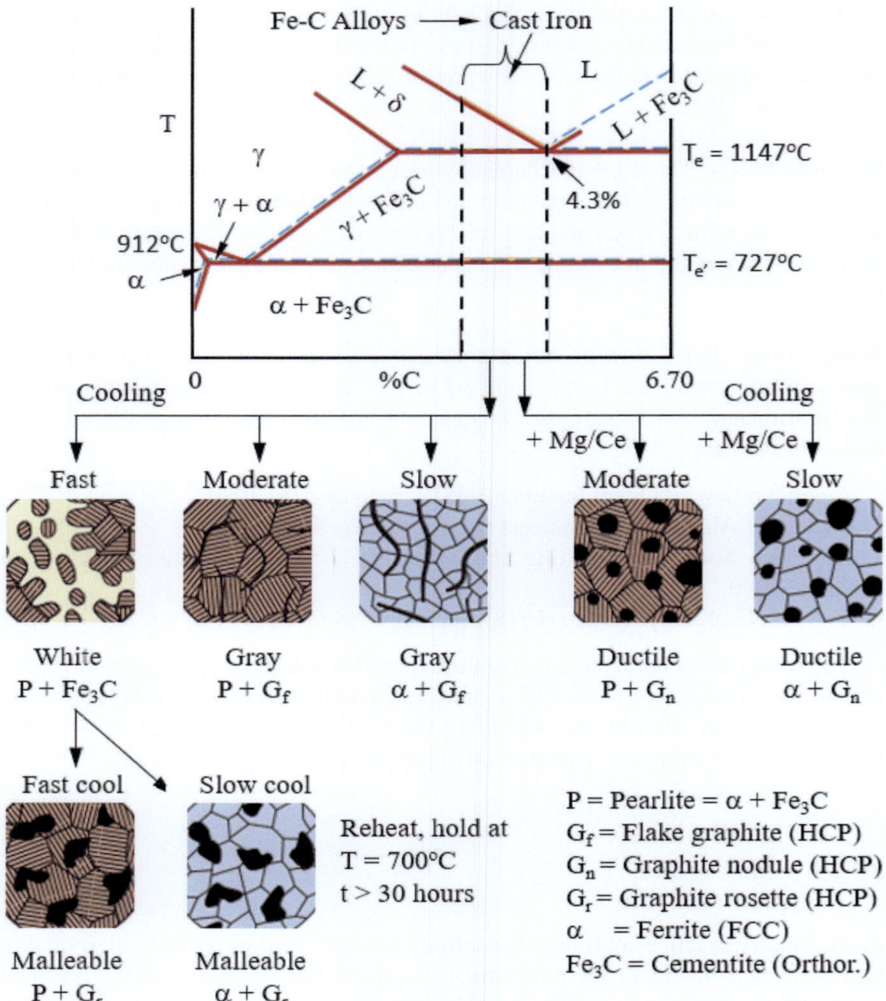

Fig. 10.35 Superimposed and partial *Fe-C* and *Fe-Fe₃C* schematic phase diagrams for a range of carbon composition, and schematic cast iron microstructures related to cooling rates. After Moffatt et al. [10, p. 195] and taken from Callister and Rethwisch [31, p. 360]

fracture surfaces, respectively, and (3) cast iron has a high mass density and it is naturally hard in the as-cast or heat-treated conditions.

Cast iron can be classified according to the graphite morphology into (1) flake, nodule, or rosette, compacted graphite or (2) white, gray, ductile, and malleable types (Moffatt et al. [10, p. 195], Elliott [36, pp. 5,12]). The American Society for Testing and Materials, ASTM A247, can be consulted for further information on cast iron.

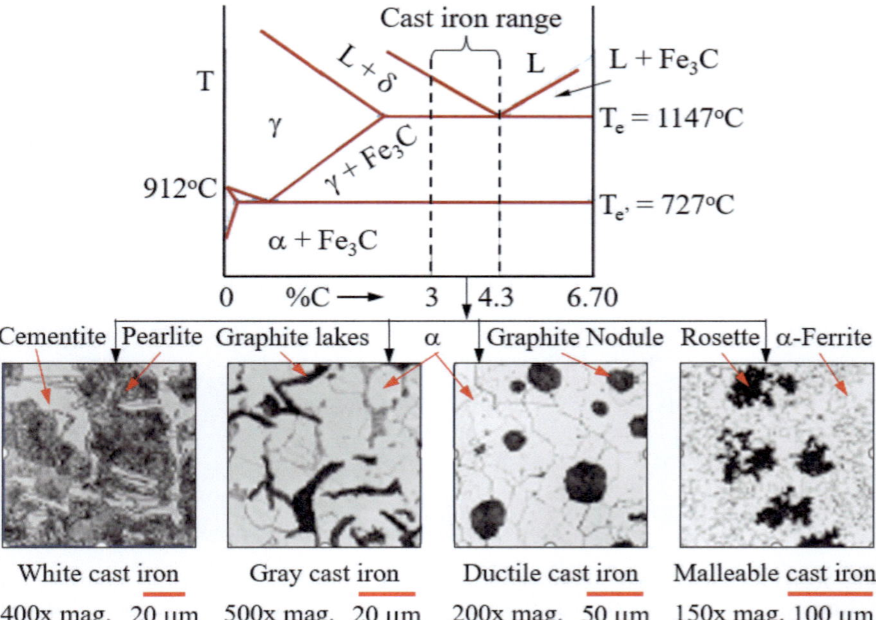

Fig. 10.36 Superimposed and partial $Fe\text{-}Fe_3C$ schematic phase diagram for a range of carbon composition and cast iron microstructures of commercial quality. Taken from Callister and Rethwisch [31, p. 357]

Figure 10.36 illustrates actual cast iron microstructures (Callister and Rethwisch [31, p. 357]). Nonetheless, revealing cast iron microstructures is an important metallographic work since their morphology dictates the mechanical properties and related to mechanical behavior of structural components.

One can deduce from the above microstructures that graphite morphology is sensitive to carbon content. For comparison, the ferrite α-phase is very sensitive to chemical composition and cooling rate.

Despite that cast irons are considered as high-corrosion-resistant materials attributed to their high carbon content, their industrial applications depend on cost, limitations on dimensions, and pitting and crevice corrosion. Engineering applications, such as automobile diesel pistons and connecting rods, prevail nowadays.

10.20   Metal Bonding

This section incorporates a common industrial application of diffusion principles in metal bonding and related phase diagrams for characterizing the microstructural evolution in engineering alloys susceptible to galvanizing, welding, and metal cladding.

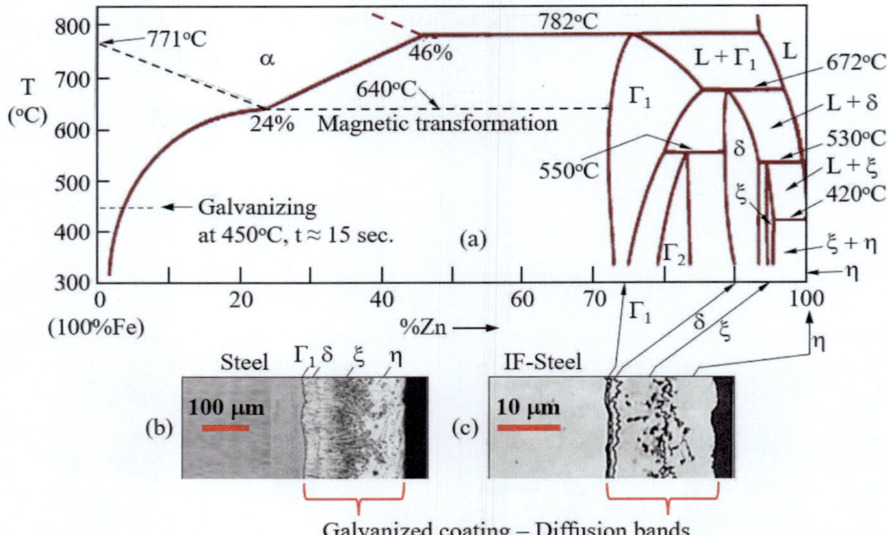

Fig. 10.37 (**a**) Equilibrium *Fe-Zn* phase diagram and related hot-dip galvanizing steel, (**b**) galvanized Zn coating after Guy and Hren [16, p. 393], and (**c**) galvanized Zn coating on IF-steel after Langill and Dugan [37, p. 93]

10.20.1   Galvanizing

This is a metallurgical process in which a steel specimen is immersed or hot-dip in a molten zinc bath at approximately $450\,°C$ (Fig. 10.37a). In this case, an ideal zinc coating deposits on the steel specimen surface and eventually it protects the steel from corrosion induced by harsh environments (Guy and Hren [16, p. 393], Langill and Dugan [37, p. 93]).

The microstructure of *Fe-Zn* intermetallic compounds evolves in hot-dip galvanized coating on a plain-carbon steel (Guy and Hren [16, p. 393]). For instance, IF-steel microstructures are shown in Fig. 10.37b, c in conjunction with the *Fe-Zn* phase diagram (Langill and Dugan [37, p. 93]). Moreover, notice the coating phases shown in Fig. 10.37b, c for the IF-steel hot-dip galvanized samples in a commercial grade zinc bath. Clearly, *Fe-Zn* intermetallic layers evolved as a function of dipping time by diffusion.

10.20.2   Fusion Welding

Welding, in general, involves fusion welding (arc welding, laser-beam welding, gas welding, etc.), pressure welding through plastic deformation, diffusion welding, explosive welding, flow welding, soldering, brazing, and so forth.

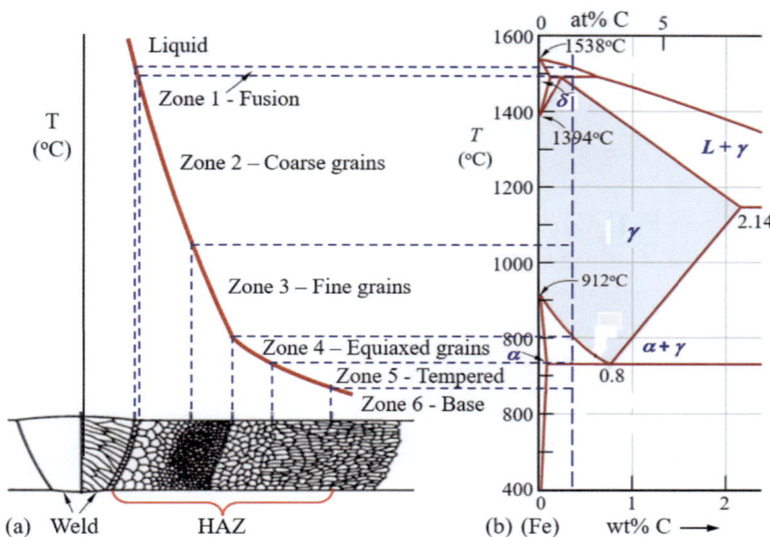

Fig. 10.38 **(a)** Schematic heat-affected zone (HAZ) morphology of a hypothetical welded low-carbon steel and **(b)** portion of the Fe-C phase diagram for the low-carbon steel (Easterling [38, p. 126])

The qualitative and quantitative analyses of welding characteristics are abundant in the literature and only aspects of welding solidification related to particular cases are included in this section. Actually, welding is a metal bonding process suitable for industrial application of localized melting/solidification related to atomic diffusion in the solid state, where grains grow in a variety of directions forming a heat-affected zone (HAZ) grain morphology as schematically illustrated in Fig. 10.38a for a hypothetical low-carbon steel (Fig. 10.38b) being welded (Easterling [38, p. 126]).

Fusion welding is a fast and pressure-independent localized melting and solidification process due to the coalescence of metals produced by heating to a suitable temperature. The joined and adjacent areas consist of heterogeneous microstructures as schematically illustrated in Fig. 10.38. The heterogeneity of the resultant microstructure adjacent to the welded area is referred to as the heat-affected zone (HAZ). Moreover, it is a common industrial practice to heat-treat the welded area for homogenizing the microstructure and indirectly adjust mechanical properties.

Among welding institutes, associations, and societies, the American Welding Society (AWS) and The Welding Institute (TWI) should be consulted for guidelines and requirements on the application of fusion welding. Additionally, common applications of fusion welding lead to the construction of steel tanks with different sizes, steel bridges, steel pipelines, and seagoing vessels (ships).

There are welding-related organizations such as the American Bureau of Shipping (ABS) and Lloyd's Register Group Services Limited that provide rules and guidelines for the design, construction, and maintenance of marine vessels and offshore structures.

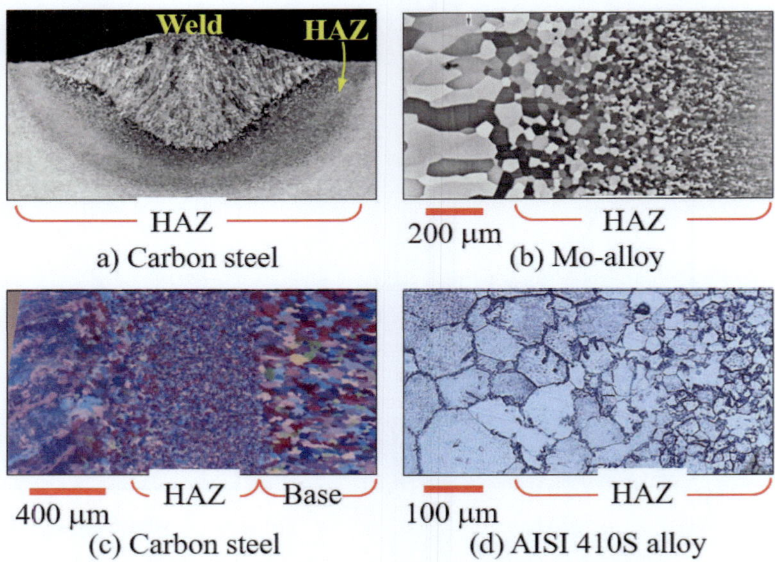

Fig. 10.39 Actual welded microstructures showing the welding and the heat-affected zone (HAZ). (**a**) Carbon steel (Thaulow et al. [39]), (**b**) *Mo*-alloy (Wadsworth et al. [40]), (**c**) carbon steel (Vander Voort [41]), and (**d**) AISI 410S stainless steel (Kose and Topal [42])

Consider the welding and heat-affected zone (HAZ) cases shown in Fig. 10.39a for carbon steel (Thaulow et al. [39]), Fig. 10.39b for molybdenum alloy (Wadsworth et al. [40]), Fig. 10.39c for carbon steel (Vander Voort [41]), and Fig. 10.39d for AISI 410S stainless steel (Kose and Topal [42]).

Notice the heterogeneous grain morphology at the welded regions or areas of joined materials. The black-and-white (B&W) and the color counterpart metallography clearly reveal details of the resulting microstructures.

Characteristics of the HAZ area depend on the properties of materials and the intensity of heat during fusion welding. These are critical aspects of fusion welding since the level of thermal diffusivity is related to the fast heating and cooling processes. Moreover, weld geometry and thermal diffusivity or conductivity of the base material are the general aspects of the resultant HAZ size.

Each HAZ area shown in Fig. 10.39 represents the non-melted parent metal that has experienced microstructural changes with respect to grain size. It is understood that mechanical properties are influenced by the grain morphology. This implies that mechanical properties vary with the grain size within the HAZ area, which is clearly divided into coarsened and grain-refined zones due to solid-state phase transformation during relatively fast cooling known as a localized quenching in air.

Another aspect of fusion welding is that the base material, commonly carbon steel, stainless steel, or aluminum alloy, is unaffected outside of HAZ areas. For instance, welded stainless steel shows visible colored bands or spots caused by surface oxidation known as heat tint.

10.21   Homogenization of Metals

Firstly, **inhomogeneities** in metals can arise from a variety of factors, such as the casting process, the solidification process, or the presence of secondary phases. Inhomogeneities in metals are non-uniformities in the distribution of elements or microstructural features within a metal or alloy. Actually, inhomogeneities can lead to a number of problems, such as:

- Embrittlement: This is a condition in which a material becomes brittle and susceptible to cracking.
- Corrosion: This is the deterioration of a material due to a reaction with its environment.
- Cracking: This is the formation of cracks in a material.

Secondly, **homogenization**, in the context of metallurgy, refers to the process of reducing or eliminating chemical and microstructural inhomogeneities in a metal or alloy. This is typically achieved through heat treatment, where the material is heated to a specific temperature for a certain period. The goal of homogenization is to create a more uniform microstructure (arrangement of crystals or grains), which can significantly improve the mechanical properties of the material. In essence, grains can vary in size, shape, and orientation. The microstructure of a metal can have a significant impact on its mechanical properties, such as strength, hardness, and ductility.

Homogenization can be used to reduce or eliminate these inhomogeneities. This is typically achieved through heat treatment, where the material is heated to a specific temperature for a certain period. The heat treatment causes the atoms in the material to diffuse, which can help to even out the distribution of elements and eliminate microstructural features such as grain boundaries.

The benefits of homogenization include:

- Improved strength: Homogenization can help to improve the strength of a metal by reducing the number of grain boundaries. Grain boundaries are weak points in a metal's microstructure, and they can act as sites for crack initiation.
- Improved ductility: Homogenization can also help to improve the ductility of a metal. Ductility is a material's ability to deform without cracking.
- Improved corrosion resistance: Homogenization can help to improve the corrosion resistance of a metal by eliminating microstructural features that can act as sites for corrosion initiation.

Homogenization is used in a variety of applications, including:

- The production of wrought alloys: Wrought alloys are alloys that have been deformed by rolling, forging, or extrusion. Homogenization is often used to improve the mechanical properties of wrought alloys.
- The production of castings: Castings are objects that have been formed by pouring molten metal into a mold. Homogenization is often used to eliminate casting defects and improve the mechanical properties of castings.

- The repair of welds: Welding can create inhomogeneities in a metal's microstructure. Homogenization can be used to repair these inhomogeneities and improve the mechanical properties of the weld.

Lastly, homogenization is an important process in the production and repair of metals. It can help to improve the mechanical properties of metals by reducing or eliminating inhomogeneities in their microstructure.

10.22   Ceramic Phase Diagrams and Structure

The main objective of this section is to enhance the preceding phase transformation by introducing ceramic phase diagrams (Fig. 10.40) and related atomic structures.

The configuration and interpretation of ceramic phase diagrams are similar to those for metal-metal systems. The general notion of ceramics is that they are refractory materials due to their physical and chemical stability at high temperatures.

Traditional ceramics include china, porcelain, clay and silica bricks, and tiles. On the other hand, engineering ceramics include:

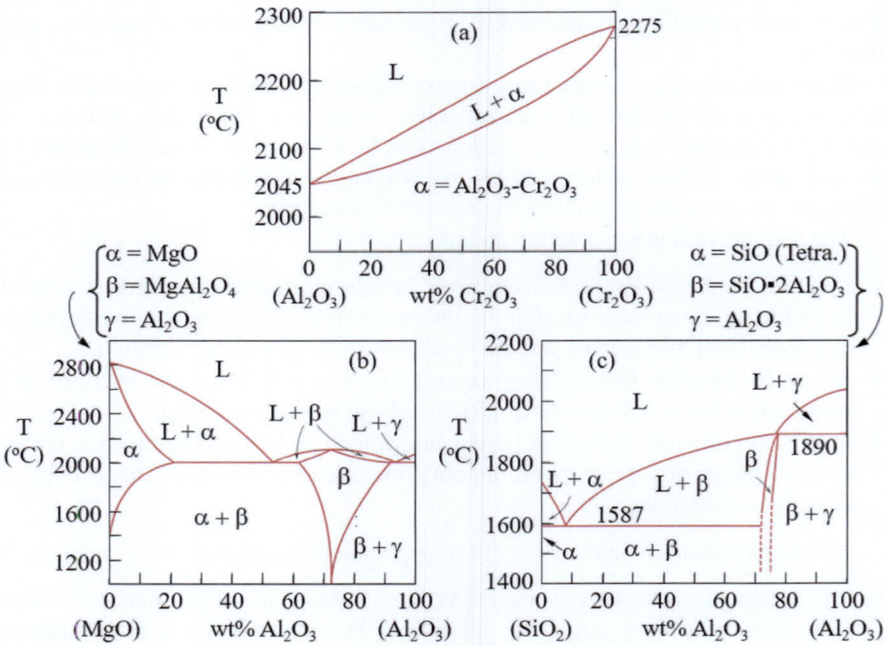

Fig. 10.40 Common ceramic phase diagrams and related phase fields. Taken from Callister and Rethwisch [31, pp. 487, 425–428]

- Metal oxides (MgO, Al_2O_3, and SiO_2). For instance, magnesium oxide (MgO) is known as magnesia, aluminum oxide (Al_2O_3) is alumina, sodium chloride ($NaCl$) is commonly known as table salt, and silicon dioxide (SiO_2) is silica.
- Metal carbides (SiC, TiC, and WC), which are interstitial compounds because carbon is a small atom that resides in octahedral or tetrahedral sites. In general, metal carbides have metallic-like properties and are refractory materials due to their high-temperature resistance. Moreover, silicon carbide (SiC) is a polytype compound: cubic β-SiC and hexagonal/rhombohedral α-SiC.
- Metal nitrides (Si_3N_4 and TiN). Despite that these ceramic materials have low electrical resistivity, they are thermally and chemically stable.

Figure 10.40 shows three common ceramic phase diagrams (Callister and Rethwisch [31, pp. 425–428]). For instance, the Al_2O_3-Cr_2O_3 binary system is an isomorphous phase diagram (Fig. 10.40a), while the MgO-Al_2O_3 (Fig. 10.40b) and SiO_2-Al_2O_3 (Fig. 10.40c) phase diagrams are eutectic systems, where the corresponding invariant phase reactions are $L \rightleftarrows \alpha+\beta$ and $L \rightleftarrows \alpha+\gamma$. Specifically, the SiO_2-Al_2O_3 alloy system has a congruent point, where $L \rightleftarrows \beta$ is the invariant reaction at relatively high temperature $T = 1890\,^\circ C$.

For clarity, each phase in Fig. 10.40 has its technological designation and atomic structure. That is,

- The α-Al_2O_3 structure (known as alumina) has a hexagonal unit cell with a rhombohedral primitive cell.
- In the Al_2O_3-Cr_2O_3 binary system, α-phase is a substitutional solid solution where Al^{3+} and Cr^{3+} exchange lattice positions. Both Al_2O_3 and Cr_2O_3 components have the same hexagonal crystal structure.
- In the MgO-Al_2O_3 binary system, the FCC α-phase is known as magnesia MgO, the cubic β-phase is called spinel with chemical formula $MgAl_2O_4$, and the γ-phase is a dual structure composed of alumina + spinel or $Al_2O_3 + MgAl_2O_4$.
- In the SiO_2-Al_2O_3 binary system, the FCC α-phase is called cristobalite SiO_2, the orthorhombic β-phase is known as mullite $3Al_2O_3$-$2SiO_2$, and the γ-phase is also a dual structure composed of alumina + mullite or $Al_2O_3 + SiO_2$. Moreover, SiO_2 can also exist as an amorphous structure known as silica glass, fused silica, or vitreous silica.

Figure 10.41 shows actual ceramic microstructures, which resemble metallic microstructures. This particular ceramic system shows the β-phase as an intermediate compound known as orthorhombic mullite (Guy [43, p. 87]).

Additional ceramic microstructures are shown in Fig. 10.42 for $99.5\%Al_2O_3$ alumina (Fig. 10.42a), $MgAl_2O_4$ spinel (Fig. 10.42b), and porcelain (Fig. 10.42c). Moreover, the alumina phase α-Al_2O_3 has wide engineering applications including powder-water solutions in metallography, electronic components, wear-resistant parts, corrosion-resistant parts, chemical processing scheme, and the like (taken from Callister and Rethwisch [31, p. 535] as a courtesy of H. G. Brinkies,

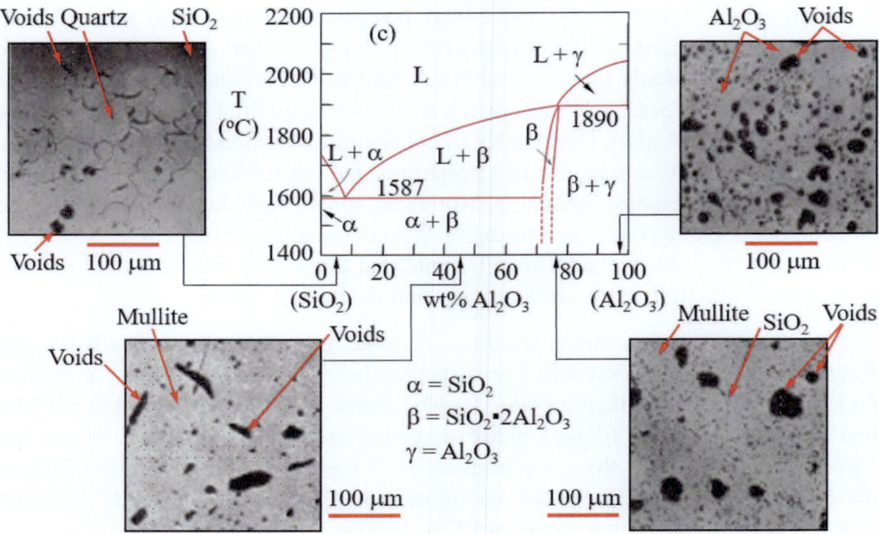

Fig. 10.41 The SiO_2-Al_2O_3 ceramic phase diagram illustrating phase fields and some alloy microstructures. After A.G.Guy [43, p. 87]

Swinburne University of Technology, Hawthorn Campus, Hawthorn, Victoria, Australia).

Ceramics are widely used in the medical and engineering fields. They, in general, can be classified as conventional or traditional ceramics, high-temperature refractories, cementitious materials (based on cements and mortars), and advanced ceramics (such as electroceramics and bioceramics for hip and knee implants in humans).

It is very important to know that cement is also classified as a ceramic material having a wide range of applications since the basic water-cement mixture forms a paste that sets and hardens. The most common ceramic cement is known as Portland cement in the civil engineering field for constructing reinforced bridges, buildings, and houses.

10.23 Ternary Phase Diagrams

This section includes a brief presentation and interpretation of the fundamentals of ternary phase diagrams and phase equilibria along with some examples.

It is a common practice to produce metallic, ceramic, and polymeric materials containing more than two components. For a ternary system at atmospheric pressure ($P = 1atm$), Fig. 10.43a shows the Cd-Bi-Pb ternary phase diagram based on binary eutectic systems (Clark and Varney [44, p. 61]). The development of this type

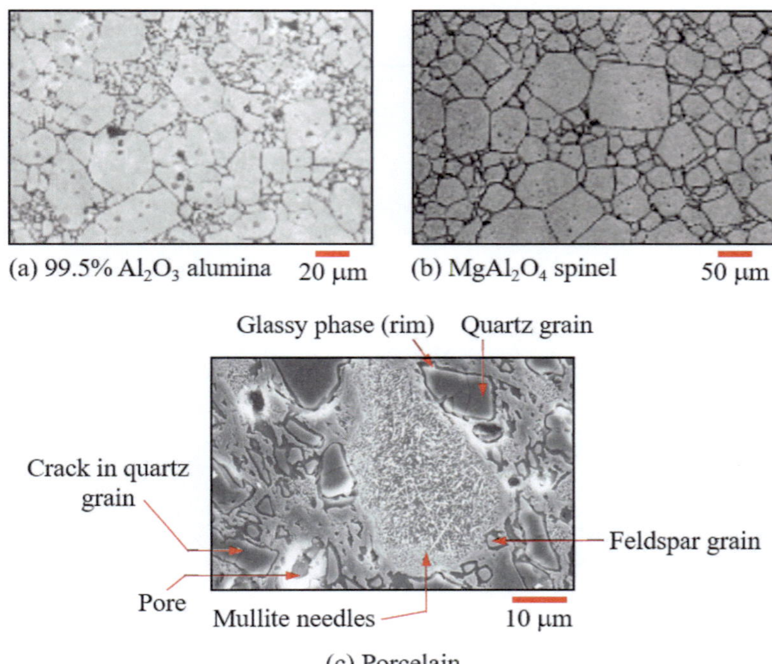

Fig. 10.42 Microstructure of ceramics. **(a)** Pure aluminum oxide (alumina) Al_2O_3, **(b)** magnesium aluminate spinel $MgAl_2O_4$, and **(c)** porcelain (taken from Callister and Rethwisch [31, p. 535] as a courtesy of H. G. Brinkies, Swinburne University of Technology, Hawthorn Campus, Hawthorn, Victoria, Australia)

of phase diagram yields an equilateral triangle configuration having, specifically, the eutectic isothermal curves (isotherms, boundary curves, or temperature contours) starting from the eutectic invariant point "e" in each eutectic phase diagram at T_e. Subsequently, the three binary eutectic curves become monovariant lines in the triangle and converge to the ternary eutectic point "E" (red point), where four phases coexist: $L + \alpha + \beta + \gamma$ at the ternary eutectic temperature T_E.

Figure 10.43b illustrates an isometric view (three-dimensional view or space surface) of the ternary system, where points 1,2,3 are hill-like tops and the isotherms (isothermal curves) from top to bottom are decreasing temperature contours down to the cliff-like toward the deepest valley-like ternary eutectic point E (red point).

Notice in Fig. 10.43b that points, corners, or vertices A, B, and C in the equilateral triangle (three-sided polygon with sides of equal length and three internal 60° vertex angles) represent 100%A, 100%B, and 100%C. Literally, Fig. 10.43b develops by folding and joining points 1-1, 2-2, and 3-3 from Fig. 10.43a, and subsequently, this yields a space surface with schematic isothermal curves toward the ternary eutectic point "E" (Nelson [45]).

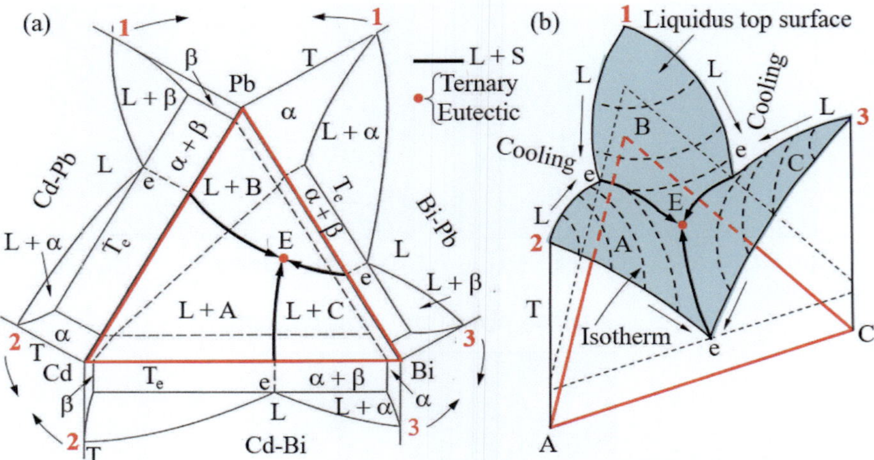

Fig. 10.43 The ternary phase diagram. **(a)** Two-dimensional view for the Cd-Bi-Pb system (Clark and Varney [44, p. 61]) and **(b)** three-dimensional view of a hypothetical *A-B-C* system showing liquid lines of descent called isotherms

During solidification, two-phase $L + A$, $L + B$, and $L + C$ ternary fields develop, where A, B, and C represent, in general, the solid phases related to the binary eutectic diagrams. Notice how the temperature boundary curves decrease toward the ternary eutectic point E (red dot), forming a space curve called binary recrystallization curve (Dantzig and Rappaz [9, p. 104]).

It seems most appropriate to exclude the drawings for the binary eutectic phase diagrams and illustrate the above idea as shown in Fig. 10.44a for an isometric view and Fig. 10.44b for a three-component plane having isothermal curves within the phase fields. Commonly, actual data is graphically displayed (presented) as an equilateral triangle known as the Gibbs triangle containing the eutectic space curve (red curves) merging at the ternary eutectic point E for complete solidification at just below T_E.

Recall that point "e" is the invariant binary eutectic point at T_e for each binary system and point "E" is the ternary eutectic point at T_E. The eutectic "e" curve evolving from each eutectic system converge to a ternary eutectic point E. In particular, the liquid line of descent "eE" is the univariant eutectic boundary for all three binary systems (Fig. 10.44).

Consider the procedure for determining the composition X of a hypothetical ternary alloy system in Fig. 10.45a using the equilateral triangle with a scale grid. Thus, it can be deduced from Fig. 10.45b that:

- All points on lines parallel to AB, BC, and CA have a fixed C, A, and B composition, respectively, thus, AB_i at C, BC_i at A, and CA_i at B, where i stands for any composition.

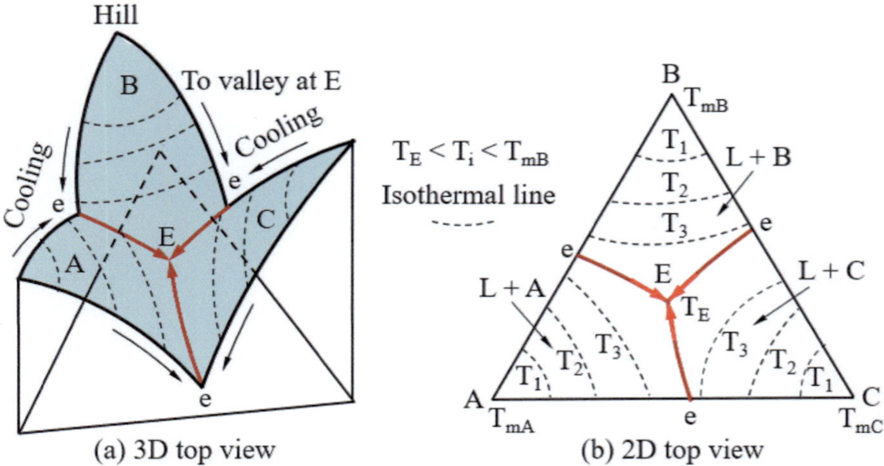

Fig. 10.44 Hypothetical A-B-C ternary phase diagram. (**a**) Top view of the three-dimensional surface and (**b**) two-dimensional plane showing isotherms: $T_E < T_i < T_m$, where $i = 1, 2, 3$ and $T_1 > T_2 > T_3$

- Total fraction of the three-component alloy system must sum to unity: $f_A + f_B + f_C = 1$.
- The composition of alloy X (Fig. 10.45a) in weight percent (%) is $30A$-$50B$-$20C$.
- From Fig. 10.45b, the phase fraction (f_i with $i = A, B, C$) for each component is

$$f_A = \frac{NX}{NA} = \frac{30}{100} = 0.30 \text{ or } 30\% A \tag{10.50a}$$

$$f_B = \frac{MX}{NB} = \frac{50}{100} = 0.50 \text{ or } 50\% B \tag{10.50b}$$

$$f_B = \frac{LX}{NC} = \frac{20}{100} = 0.20 \text{ or } 20\% C \tag{10.50c}$$

Figure 10.46 shows ternary phase fields at $650\,°C$ isotherm for the Fe-Cr-Ni alloy system (Raynor and Rivlin [46]). The corresponding phases are designated as $\alpha = (\alpha\text{-}Fe, Cr)$ and $\gamma = (\gamma\text{-}Fe, Ni)$.

The microstructural evolution during slow solidification for a selected Fe-Cr-Ni composition can be studied using this type of phase diagram with the aid of microscopy, X-ray diffraction, and so forth. Particularly, Fe-Cr-Ni are the principal elements in producing stainless steels for certain engineering applications.

Lastly, Fig. 10.47 illustrates a ceramic ternary phase diagram with descendant isotherms.

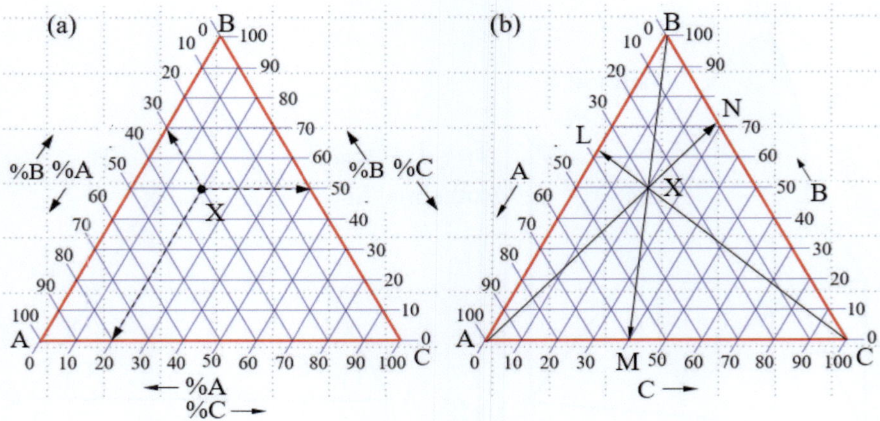

Fig. 10.45 Determining the (a) composition of alloy X and (b) fractions of components at X

Fig. 10.46 The *Fe-Cr-Ni* alloy system at $650\,^{\circ}C$ isotherm. The thick curves represent three-phase equilibria. After Raynor and Rivlin [46]

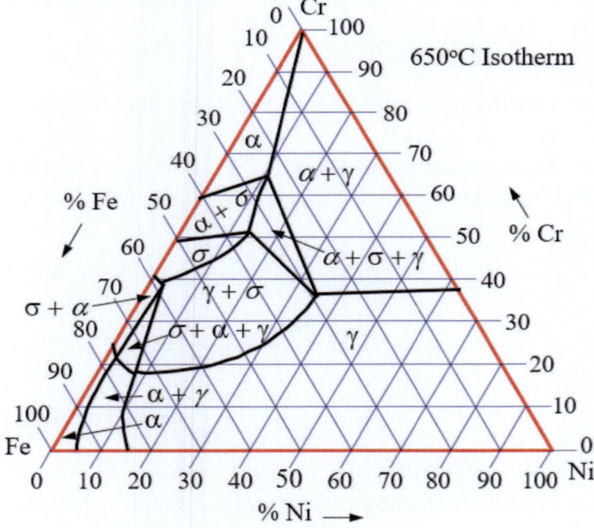

Techniques like differential thermal analysis, X-ray diffraction, and scanning electron microscopy are important tools for characterizing the phase transformation and microstructural features of ternary alloy systems. This implies that any alloy composition is directly related to the evolution of phases and subsequent phase precipitation at particular isotherms.

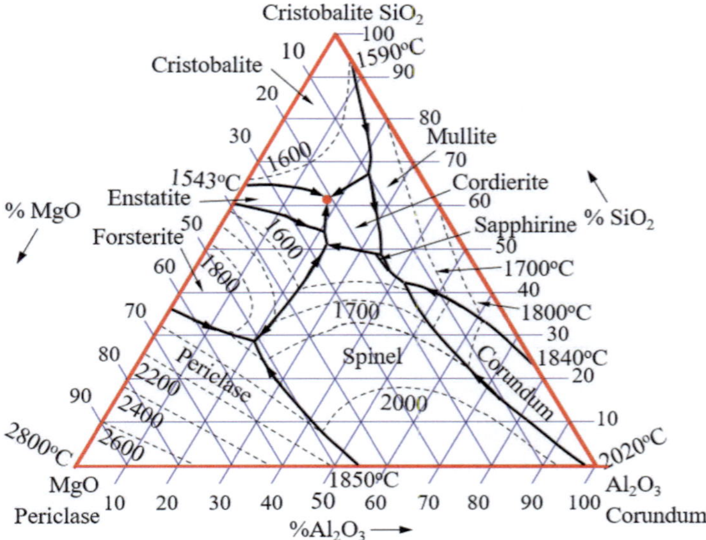

Fig. 10.47 The MgO-SiO_2-Al_2O_3 ternary system with liquid projections as isotherms. The thick curves represent three-phase equilibria with isotherms

10.24 Summary

Two-dimensional binary and ternary phase diagrams are graphical representations of the physical state of binary or ternary alloys. The binary phase diagram is the simplest representation of the temperature-composition relationship (T-C diagram) at constant atmospheric pressure P_{atm}. In essence, phase diagrams are charts or maps of discrete regions that represent liquid-solid and solid phases at equilibrium. These regions provide sufficient information for understanding the microstructural evolution (continuous or discontinuous grain growth) upon equilibrium solidification of an alloy.

Solidification is a phase transformation process from the liquid state (melt phase) to a solid state. This process is related to nucleation and growth mechanisms involving the rearrangement of atoms from short-range to long-range order to form the crystallographic lattice, while the melt releases latent heat of fusion ΔH_f at the moving liquid-solid (L-S) interface.

Control of the solidification process requires understanding the thermodynamics of the liquid (L) and solid (S) phases, and the kinetics of solidification, including the growth of nuclei and the velocity of the L-S interface. Hence, thermodynamics and heat transfer theories are the most relevant tools used in solidification and subsequent construction of phase diagrams.

An important aspect of this chapter is the introduction of thermodynamic phase diagrams denoted by the pressure $P = f(T)$ and Gibbs energy $G = f(T)$

functions, whereas the temperature-time $T = f(t)$ function represents a cooling curve of phase transitions or changes upon cooling a liquid phase.

Problems

Most of the problems being described below require the use of phase diagrams and the heat transfer equations developed in Chap. 9.

10.1 Defects and discontinuities in cast products are related to the cast shape and its dimensions. What are the most common casting defects?

10.2 What is the main purpose of the riser in casting design?

10.3 Calculate the theoretical volume change in percentage (%) for a polymorphic transformation of iron (Fe) from γ-Fe (austenite) to α-Fe (ferrite) or from γ-FCC to α-BCC structure. Assume the hard sphere atomic model. [Solution: $\Delta V / V_{FCC} = 8.82\%$]

10.4 Consider a 2-m-diameter ladle containing a mass of molten steel at the pouring austenitic temperature $T_p > T_f$, where T_f is the upper freezing temperature of the molten steel. If it takes 3.2 minutes to empty the ladle through an 8-cm-diameter hole at the bottom of the ladle, then calculate the mass of steel. Neglect buoyancy forces induced by thermal gradients as the ladle cools and assume a ladle drainage operation of steel under a near-isothermal condition ($\Delta T \rightarrow 0$) with negligible vortex flow problems (flow revolving around an axis line). Also neglect the evolution of the thermal stratification of the molten steel contained by the ladle after standstill times. [Solution: $m \simeq 6381\ Kg$]

10.5 Consider a quasi-static solidification of a Sn-$15Bi$ alloy with $C_o = 15\%$ (by weight) from the liquid state initially at $240\,^\circ C$ as shown in the equilibrium phase diagram given below.

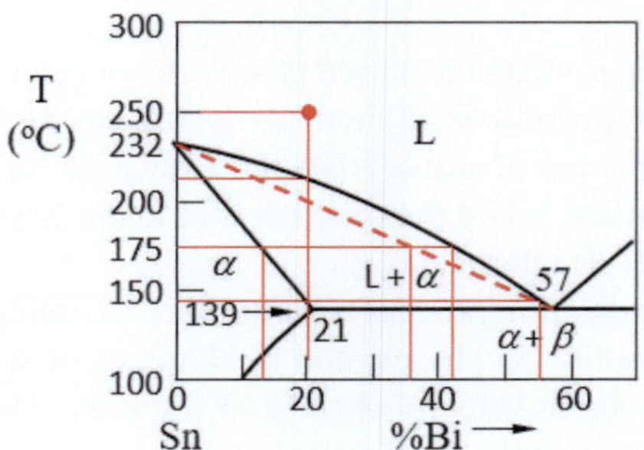

For a very slow solidification process, follow the equilibrium phase diagram and **(a)** determine the partition coefficient k_o at the onset of solidification ($\simeq 216\,^\circ C$), $200\,^\circ C$, and at the end of solidification ($\simeq 165\,^\circ C$). Is $k_o = f(T)$ a linear correlation? What can you conclude from these results? **(b)** Calculate the phase fractions at $200\,^\circ C$ and (c) show that

$$k_{o,c} = \frac{(f_s - 1)\,C_l + C_o}{f_s\,C_l}$$

and calculate $k_{o,c}$ at $200\,^\circ C$. [Solution: $k_{o,c} = 0.31$]

10.6 Consider the solidification of Pb-$10Sn$ alloy into a small slab shape. Assume that the diffusion flux J_x (Fick's first law) of solute Sn away from the moving L-S interface must equal the rate of solute Sn rejection $v_p \Delta C$ into the liquid phase, and that Fick's first law is applicable at the L-S interface at $x = 0$. For a constant L-S interface velocity of $10^{-4}\,m/s$, the actual liquid temperature gradient ahead of the interface is $dT_l/dx = 10^4\,K/m$ and Sn diffusion coefficient in the liquid is $D_l = 5 \times 10^{-9}\,m^2/s$. The liquidus slope β_l and the partition coefficient k_o must be determined from the equilibrium Pb-Sn diagram (not given). Based on this information, determine **(a)** the critical liquid temperature gradient and **(b)** the condition for an unstable interface under steady-state condition. What type of microstructure will be produced upon completion of solidification? [Solution: (a) $[\partial T_l\,(x)\,/\partial x]_{c,x=0} = 10^6\,K/m$]

10.7 Assume that the diffusion flux J_x (Fick's first law) of solute Sn away from the moving L-S interface must equal the rate of solute Sn rejection $v_x \Delta C$ into the liquid phase during solidification of a Pb-$10Sn$ alloy into a small slab shape and that Fick's first law is applicable at the L-S interface at $x = 0$. For a constant L-S interface velocity of $1.5 \times 10^{-4}\,m/s$, the actual liquid temperature gradient ahead of the interface is $dT_l/dx = 10^4\,K/m$ and the solute Sn diffusion coefficient in the liquid is $D_l = 5 \times 10^{-9}\,m^2/s$. The liquidus slope β_l and the partition coefficient k_o must be determined from the equilibrium Pb-Sn diagram (not given). Based on this information, determine **(a)** the critical liquid temperature gradient $\partial T_l/\partial x$ and **(b)** the condition for an unstable interface under steady-state condition. What type of microstructure will be produced upon completion of solidification? [Solution: (a) $\partial T_l/\partial x = 1.63 \times 10^6\,K/m$, (b) $[dT_l\,(x)\,/dx]_{a,x=0} = 10^4\,K/m$].

10.8 Assume that the diffusion flux J_x (Fick's first law) of solute Sn away from the moving L-S interface must equal the rate of solute Sn rejection $v_p \Delta C$ into the liquid phase during solidification of a Pb-$10Sn$ alloy into a small slab shape and that Fick's first law is applicable at the L-S interface at $x = 0$. For a constant L-S interface velocity of $2 \times 10^{-4}\,m/s$, the actual liquid temperature gradient ahead of the interface is $dT_l/dx = 10^4\,K/m$ and Sn diffusion coefficient in the liquid is $D_l = 5 \times 10^{-9}\,m^2/s$. The liquidus slope β_l and the partition coefficient k_o must be determined from the equilibrium Pb-Sn diagram (not given). Based on this information, determine **(a)** the critical liquid temperature gradient and **(b)** the condition for an unstable interface under steady-state condition. What type of microstructure will be produced upon completion of solidification? [Solution: (a) $\partial T_l/\partial x = 2.17 \times 10^6\,K/m$, (b) $[dT_l\,(x)\,/dx]_{a,x=0} = 10^4\,K/m$]

10.9 For the directional solidification of *Al-3Cu* and *Al-5Cu* alloys at $v_x = 0.10\,m/s$, use $k_o = 0.14$, $\beta_l = -2.6\,°C/wt\%$, and $D_l = 5 \times 10^{-5}\,m^2/s$ to determine the value of the critical temperature gradient for stability of the solidifying melt. Given *Al-Cu* phase diagram. [Solution: $[T_l(x)/dx]_{a,x=0} \geq 1.00 \times 10^6\,°C/m$ for Al-3Cu and $[T_l(x)/dx]_{a,x=0} \geq 1.74 \times 10^5\,°C/m$ for Al-5Cu]

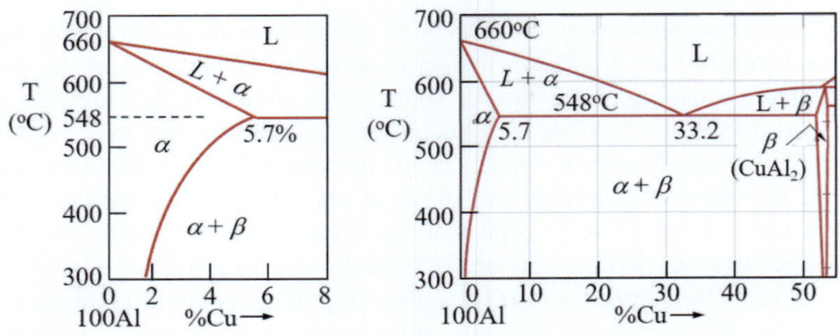

10.10 Consider the directional solidification of *Al-3Cu* and *Al-5Cu* at $v_x = 0.20\,m/s$. If $k_o = 0.14$, $\beta_l = -2.6\,°C/wt\%$, and $D_l = 5 \times 10^{-5}\,m^2/s$, then determine the value of the critical temperature gradient $[T_l(x)/dx]_{c,x=0}$ for stability of the solidifying melt. Which alloy requires a higher $[T_l(x)/dx]_{c,x=0}$? [Solution: $[T_l(x)/dx]_{a,x=0} \geq 2.00 \times 10^5\,°C/m$ for Al-3Cu and $[T_l(x)/dx]_{a,x=0} \geq 3.48 \times 10^5\,°C/m$ for Al-5Cu]

10.11 Assume that a 40-μm-thick *Al-20Si* ribbon is produced on a rotating copper roller (substrate). Calculate **(a)** the solidification time t, **(b)** the solidification velocity $ds(t)/dt$, **(c)** the solidification rate, and **(d)** the heat flux at the substrate-solid interface $x = 0$. The *Al-20Si* binary alloy is initially at its pouring temperature $T_p = 850\,°C$, while the copper mold is maintained at $T_o = 25\,°C$. The freezing temperatures of pure aluminum (*Al*) and pure silicon (*Si*) are $T_{Al} = 660\,°C$ and $T_{Si} = 1414\,°C$. The thermophysical properties and the *Al-Si* equilibrium phase diagram are

Material	k	ρ	$c_s = c_l$	ΔH_f
	$(W/m.K)$	(Kg/m^3)	$(J/Kg.K)$	(kJ/Kg)
Al-20Si alloy	161	2627	854	$\Delta H_{Al} = 396.67$
Copper substrate	398	8930	386	$\Delta H_{Si} = 1790.02$

Unit conversion:

$k \rightarrow 1\,W/m.K = 1\,W/m.°C$	$dT/dx \rightarrow 1\,K/m = 1\,°C/m$
$c_s \rightarrow 1\,J/Kg.K = 1\,J/Kg.°C$	$dx/dt \rightarrow 1\,K/s = -272\,°C/s$
$c_l \rightarrow 1\,J/Kg.K = 1\,J/Kg.°C$	$dx/dt \rightarrow 1\,°C/s = 274\,K/s$

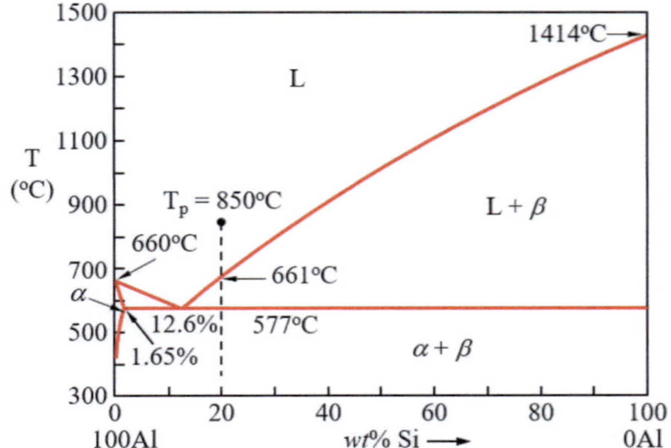

For comparison, consider the following temperature equations:

$$T_m = T_o + (T_i - T_o)\,\mathrm{erf}\left(-\frac{x}{2\sqrt{\alpha_m t}}\right) \quad \text{(mold)} \tag{a}$$

$$T_s = T_i + (T_p - T_i)\,\mathrm{erf}\left(\frac{x}{2\sqrt{\alpha_s t}}\right) \quad \text{(solid)} \tag{b}$$

These equations can be found elsewhere (C.L. Xu et al. [47]). [Solution: (a) $t_1 = 3.52 \times 10^{-5}$ sec, (b) $ds\,(t)/dt \simeq 0.57\,m/s$, (c) $\partial T_s/\partial t = -6.97 \times 10^6\,°C/$sec, (d) $q_s = -1.97\,GJ/(m^2.\mathrm{sec})$]

10.12 Assume an ideal solidification front having a planar L-S interface moving at a certain constant solidification velocity v_x. Suppose that you are asked to produce a 3-cm-thick Al-20Si flat plate in a stationary copper mold (enclosed substrate). The Al-20Si binary alloy is initially at its pouring temperature $T_p = 850\,°C$, while the copper mold is maintained at $T_o = 25\,°C$. The freezing temperatures of pure aluminum and pure silicon are $T_{Al} = 660\,°C$ and $T_{Si} = 1414\,°C$. The thermophysical properties and the Al-Si equilibrium phase diagram are given in P10.11. Calculate **(a)** the solidification time t, **(b)** the solidification velocity $ds\,(t)/dt$, **(c)** the solidification rate $\partial T_s/\partial t$, and **(d)** the heat flux q_s at the substrate-solid interface $x = 0$. [Solution: (a) $t = 4.96$ sec, (b) $ds\,(t)/dt = 1.51\,mm/$sec, (c) $\partial T_s/\partial t = -41.34\,°C/$sec, (d) $q_s = -5.25\,MJ/(m^2.\mathrm{sec})$]

10.13 Suppose that you are asked to produce a 2-cm-thick Al-20Si flat plate in a stationary copper mold (enclosed substrate). The Al-20Si binary alloy is initially at its pouring temperature $T_p = 850\,°C$, while the copper mold is maintained at

$T_o = 25\,^\circ C$. The freezing temperatures of pure aluminum (Al) and pure silicon (Si) are $T_{Al} = 660\,^\circ C$ and $T_{Si} = 1414\,^\circ C$. The thermophysical properties and the Al-Si equilibrium phase diagram are given in P10.11. Assume an ideal solidification front having a planar L-S interface moving at a certain constant solidification velocity υ_x. Calculate **(a)** the solidification time t, **(b)** the solidification velocity $ds\,(t)\,/dt$, **(c)** the solidification rate $\partial T_s/\partial t$, and **(d)** the heat flux q_s at the substrate-solid interface $x = 0$. [Solution: (a) $t = 2.20$ sec, (b) $ds\,(t)\,/dt = 2.27\,mm/$ sec, (c) $\partial T_s/\partial t = -111.44\,^\circ C/$ sec, (d) $q_s = -7.89\,MJ/\left(m^2.\,\text{sec}\right)$]

10.14 Consider casting medical devices made out of an equiatomic Ni-Ti shape memory alloy (SMA) or smart material with superelasticity and biocompatibility characteristics. Assume an ideal solidification front having a planar L-S interface moving at a certain constant solidification velocity υ_x. These medical devices are subsequently coated with a suitable material in order to prevent toxic Ni release. This problem considers the solidification of a 2-cm-thick flat plate made out of an equiatomic Ni-Ti alloy ($T_f = 1310\,^\circ C$) in a stationary copper mold (enclosed substrate at $T_o = 20\,^\circ C$) since the main goal is to characterize the solidification behavior of the Ni-Ti SMA specimen prior to any mass production of medical devices. Let the pouring temperature be $T_p = 1380\,^\circ C$ and use the data given below to calculate **(a)** the solidification time t, **(b)** the solidification velocity $ds\,(t)\,/dt$, **(c)** the solidification rate $\partial T_s/\partial t$, and **(d)** the heat flux q_s at the substrate-solid interface $x = 0$. [Solution: (a) $t = 2.57$ sec, (b) $ds\,(t)\,/dt = 1.95\,mm/$ sec, (c) $\partial T_s/\partial t = -47.45\,^\circ C/$ sec, (d) $q_s = -0.44\,MJ/\left(m^2.\,\text{sec}\right)$]

Material	k $(W/m.K)$	ρ (Kg/m^3)	c_s $(J/Kg.K)$	ΔH_f (kJ/Kg)
Ni-Ti alloy	18	6450	480	$\Delta H_{Ti} = 322.75$
Steel substrate	35	7845	500	$\Delta H_{Ni} = 297.67$

10.15 For a very slow quasi-static solidification of a Sn-$15Bi$ alloy (equilibrium phase diagram in P10.5) initially at the pouring temperature $T_p = 240\,^\circ C$, assume an ideal solidification front having a planar L-S interface moving at a certain constant solidification velocity υ_x. **(a)** Derive an expression for the partition coefficient as a function of the solid mass fraction, $k_o = F\,(f_s)$, and calculate its value at $200\,^\circ C$. **(b)** Determine the partition coefficient k_o based on solute composition only at the onset of solidification ($\simeq 216\,^\circ C$), $200\,^\circ C$, and at the end of solidification ($\simeq 165\,^\circ C$) in order to obtain an α-phase. [Solution: (b) $k_o = 0.31$ at $T = 200\,^\circ C$]

10.16 Consider casting a 1-cm-thick hypereutectic Al-$20Si$ alloy into a flat plate from the pouring temperature $T_p = 850\,^\circ C$ in a stationary copper mold (enclosed substrate) maintained at $T_o = 25\,^\circ C$ using a water-cooling system. The freezing temperatures of pure aluminum and pure silicon are $T_{Al} = 660\,^\circ C$ and $T_{Si} = 1414\,^\circ C$, and the thermophysical properties and the Al-Si equilibrium phase diagram are given in P10.11. Assume an ideal solidification front having a planar L-S interface moving at a certain constant solidification velocity v_x. Calculate **(a)** the partition coefficient k_o, **(b)** the mass fractions of the α-phase $(f_{\alpha,e})$ and β-solid phase $(f_{\beta,e})$ in the eutectic structure $(\alpha + \beta)_e$, **(c)** the latent heat of solidification ΔH_s and the solidification time t for initially transforming an amount of the liquid phase to primary β-phase and the remaining liquid to eutectic structure at the eutectic temperature $T_e = 577\,^\circ C$, **(d)** the solidification velocity $v_x = ds\,(t)\,/dt$ of the planar solidification front and the related solid temperature gradient $\partial T_s/\partial x$, and **(e)** the heat flux q_s at the substrate-solid interface $x = 0$. [Solution: (a) $k_o = 5$, (b) $f_{\alpha,e} = 0.89$ and $f_{\beta,e} = 0.11$, (c) $\Delta H_s \simeq 0.91\,MJ/Kg$, (d) $v_x = 2.27\,mm/\sec$, (e) $q_s = -7.89\,MJ/\left(m^2.\sec\right)$]

10.17 Consider casting $100\,grams$ of $30Pb$-$70Sn$ from $T_p = 250\,^\circ C$ into a sand mold having 1-cm-thick rectangular cavity and assume an ideal solidification front having a planar L-S interface moving at steady-state velocity v_x. The equilibrium phase diagram is given below. **(a)** Use the Lever rule to determine the balance of phase fractions at T_a, T_b, T_c, T_d shown in the given phase diagram. Also, include the analysis at the eutectic temperature T_e and at $T = 50\,^\circ C$.

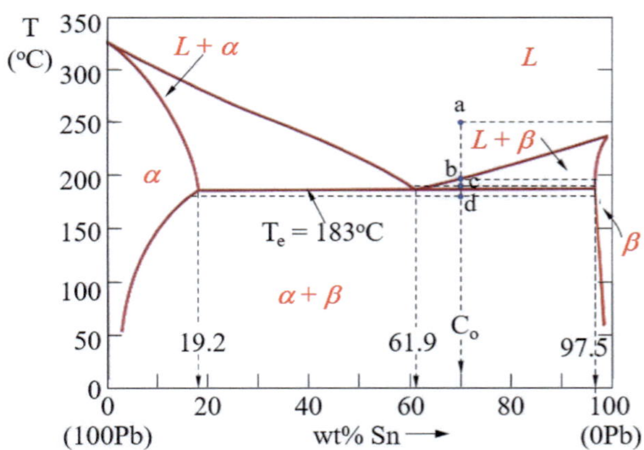

Use the given thermophysical properties for this alloy (Rocha et al. [48]) to calculate **(b)** the solidification time t, **(c)** the solidification velocity v_x, and **(d)** the heat flux q_s at $s\,(t)$. [Solution: (a) $f_e = 78\,g$ and $f_{\beta'} = 23\,g$ $183\,^\circ C$, (b) $t = 0.48$ sec, (c) $v_x = 1.05 \times 10^{-2}\,m/s$, (d) $q_s = -64.36\,MW/m^2$]

Material	$T_f{}^a$	$\Delta H_s{}^a$	$\rho_s{}^b$	$\rho_l{}^b$	$c_s{}^b$	$c_l{}^b$	$k_s{}^b$	$k_l{}^b$
$30Pb$-$70Sn$	180	50.50	8495	8085	134	223	57.4	32
Sand mold			1550		986		0.40	

Units of tabulated thermophysical data
[a] T_f in $^\circ C$ and ΔH_s in kJ/Kg units
[b] ρ_s, ρ_l in Kg/m^3, c_s, c_s in $J/Kg.K$, and k_s, k_l in $W/m.K$

10.18 Consider casting a $60Pb$-$40Sn$ alloy melt from the pouring temperature $T_p = 300\,°C$ into a $V = 1 \times 4 \times 10 = 40\,cm^3$ mold cavity and assume an ideal solidification front having a planar L-S interface moving at a steady-state velocity υ_x. Use the half-space approach under the ideal planar solidification front condition. Compare the mass fractions of the $60Pb$-$40Sn$ alloy. The pertinent equilibrium phase diagram is included in P10.17. Calculate **(a)** the mass fraction of phases at $T_e = 183\,°C$ and $T = 182\,°C$ using the Lever rule and the Scheil equation, **(b)** the solidification time t for transforming the melt to a solid composed of primary α-phase ($f_{\alpha'}$) and eutectic ($\alpha + \beta$)-phase (f_e), and **(c)** the solid mass (m_s) of the $60Pb$-$40Sn$ alloy. The alloy freezing range and other relevant temperatures are $183^{oC} \leq T_f^* \leq 238\,°C$, $T_s = 183\,°C$, $T_l = 238\,°C$, $T_p = 250\,°C$. The thermophysical properties of this alloy are assumed to be as given in the table below. Comment on the gain morphology that evolves during the assumed ideal solidification process. [Solution: (a) $f_{\alpha'} = 0.51$ (Lever rule) and $f_{\alpha'} = 0.44$ (Scheil model) at $T_e = 183\,°C$, (b) $t \simeq 374$ sec, (c) $m_s = 0.37\,Kg$]

Material	$\Delta H_s{}^{a}$	$\rho_s{}^{b}$	$\rho_l{}^{b}$	$c_s{}^{b}$	$c_l{}^{b}$	$k_s{}^{b}$	$k_l{}^{b}$
$60Pb$-$40Sn$	50	9270	8000	130	180	25	18
Sand mold		1550		986		0.40	

Units of tabulated thermophysical data
[a] T_f in $°C$ and ΔH_s in kJ/Kg units
[b] ρ_s, ρ_l in Kg/m^3, c_s, c_s in $J/Kg.K$, and k_s, k_l in $W/m.K$

10.19 Consider casting $1\,Kg$ of $90Al$-$10Cu$ melt from the pouring temperature $T_p = 700\,°C$ into a sand mold having 1-cm-thick rectangular cavity and assume an ideal solidification front having a planar L-S interface moving at a certain constant solidification velocity υ_x. Use the Lever rule to determine **(a)** the mass of the α'-phase as the primary solidification product and the mass of $(\alpha + \beta)_e$ eutectic structure at $T = 547\,°C$ and **(b)** the mass fractions of the α-phase ($f_{\alpha e}$) and β-phase ($f_{\beta e}$) in the eutectic structure, and **(c)** use the given thermophysical properties for this alloy (Sa et al. [49]) to calculate the solidification time t. [Solution: (a) $f_{\alpha'} = 0.76\,Kg$ and $f_e = 0.16\,Kg$, (b) $f_{\alpha e} = 0.07\,Kg$ and $f_{\beta e} = 0.09\,Kg$, (c) $t = 316.83$ sec]

Material	$T_f{}^{a}$	$\Delta H_s{}^{a}$	$\rho_s{}^{b}$	$\rho_l{}^{b}$	$c_s{}^{b}$	$c_l{}^{b}$	$k_s{}^{b}$	$k_l{}^{b}$
$90Al$-$10Cu$	626	377.83	2798	2683	1086.2	1027.8	185.7	85.6
Sand mold			1550		986		0.40	

Units of tabulated thermophysical data
[a] T_f in $°C$ and ΔH_s in kJ/Kg units
[b] ρ_s, ρ_l in Kg/m^3, c_s, c_s in $J/Kg.K$, and k_s, k_l in $W/m.K$

10.20 Consider casting $1\,Kg$ of $96Zn\text{-}4Al$ from $T_p = 450\,°C$ into a sand mold having $1\text{-}cm$-thick rectangular cavity and assume an ideal solidification front having a planar $L\text{-}S$ interface moving at a certain constant solidification velocity v_x. Determine **(a)** the mass for $f_{\eta e}$ and $f_{\beta e}$ when the alloy is at $T_e = 380\,°C$ (see the partial equilibrium phase diagram below) using the the Lever rule, and **(b)** the solidification time t using the given thermophysical properties for this alloy (Osorio and Garcia [50]). [Solution: (a) $f_{\eta e} = 2.80\,Kg$ and $f_{\beta e} = 1.50\,Kg$, (b) $t = 454.62\ sec$].

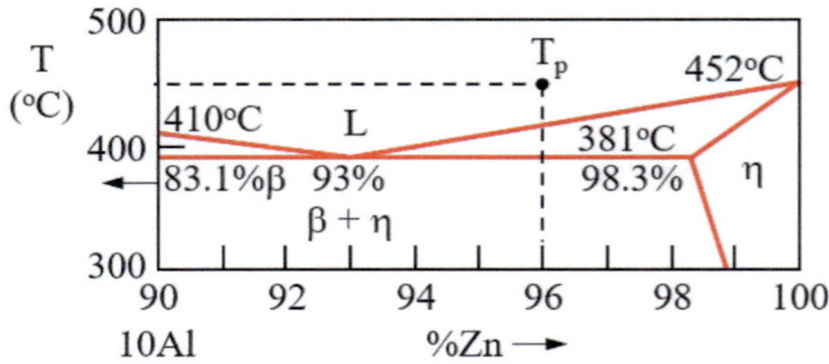

Material	$T_f{}^{\text{a}}$	$\Delta H_s{}^{\text{a}}$	$\rho_s{}^{\text{b}}$	$\rho_l{}^{\text{b}}$	$c_s{}^{\text{b}}$	$c_l{}^{\text{b}}$	$k_s{}^{\text{b}}$	$k_l{}^{\text{b}}$
$96Zn\text{-}4Al$	390	123.88	6431	6407	473.1	504.5	106	50.10
Sand mold			1550		986		0.40	

Units of tabulated thermophysical data
[a] T_f in $°C$ and ΔH_s in kJ/Kg units
[b] ρ_s, ρ_l in Kg/m^3, c_s, c_s in $J/Kg.K$, and k_s, k_l in $W/m.K$

10.21 Consider casting $1000\,grams$ of $85Al\text{-}15Cu$ from $T_p = 700\,°C$ into a sand mold having a $1\text{-}cm$-thick rectangular cavity and assume an ideal solidification front having a planar $L\text{-}S$ interface moving at a certain constant solidification velocity v_x. **(a)** Use the Lever rule to determine the mass of the α-phase ($f_{\alpha e}$) and the β-phase ($f_{\beta e}$) in the eutectic structure $(\alpha + \beta)_e$ at $T = 548\,°C$ and **(b)** use the given thermophysical properties for this alloy (Quaresma et al. [51]) to calculate the solidification time t. [Solution: (a) $f_{\alpha e} = 0.15\,Kg$, $f_{\beta e} = 0.20\,Kg$ and (b) $t \simeq 5.60\ min$]

Material	$T_f{}^{\text{a}}$	$\Delta H_s{}^{\text{a}}$	$\rho_s{}^{\text{b}}$	$\rho_l{}^{\text{b}}$	$c_s{}^{\text{b}}$	$c_l{}^{\text{b}}$	$k_s{}^{\text{b}}$	$k_l{}^{\text{b}}$
$85Al\text{-}15Cu$	615	374.27	2910	2760	1080	999	179	80
Sand mold			1550		986		0.40	

Units of tabulated thermophysical data:
[a] T_f in $°C$ and ΔH_s in kJ/Kg units
[b] ρ_s, ρ_l in Kg/m^3, c_s, c_s in $J/Kg.K$, and k_s, k_l in $W/m.K$

References

1. N. Perez, *Phase Transformation in Metals: Mathematics, Theory and Practice* (Springer Nature, Switzerland AG, 2020)
2. N. Perez, *Scientific Method in Analysis of Local Montana Semisynthetic Sands and their Suitability to Ferrous Foundry Usage*, M.S. Thesis, Montana Tech., Butte, Montana, USA (1983)
3. A.J. Plotkowski, *Refinement of the Cast Microstructure of Hypereutectic Aluminum-Silicon Alloys with an Applied Electric Potential*, M.S. Thesis, Grand Valley State University, Padnos College of Engineering and Computing (2012)
4. X. Xu, H. Hou, F. Liu, *Rapid Solidification of Undercooled Melts* (in Solidification, Chapter 2), InTech (2018). https://doi.org/10.5772/intechopen.70666
5. N. Saunders, A.P. Miodownik, *CALPHAD Calculation of Phase Diagrams - A Comprehensive Guide*. Pergamon Materials Series, vol. 1 (Elsevier, New York, 1998)
6. U.R. Kattner, The CALPHAD method and its role in material and process development. Tecnol Metal Mater Min. **13**(1), 3–15 (2016). https://www.ncbi.nlm.nih.gov/pmc/articles/PMC4912057/
7. F.C. Robles-Hernandez, J.M. Herrera Ramirez, R. Mackay, *Al-Si Alloys - Automotive, Aeronautical and Aerospace Applications* (Springer International Publishing, Switzerland AG, 2017)
8. J. Ruzbarsky, *Al-Si Alloys Casts by Die Casting: A Case Study* (Springer Nature, Switzerland AG (2019)
9. J.A. Dantzig, M. Rappaz, *Solidification*, 2nd edn., Revised & Expanded (EPFL Press, Switzerland, 2016)
10. W.G. Moffatt, G.W. Pearsall, G.W. Pearsall, J. Wulff, *Structure*. The Structure and Properties of Materials, Structure, vol. I (John Wiley & Sons, 1967)
11. S.H. Avner, *Introduction to Physical Metallurgy*, 2nd edn. (McGraw-Hill, New York, 1974)
12. V. Alexiades, A.D. Solomon, *Mathematical Modeling of Melting and Freezing Processes* (Hemisphere Publishing Corporation, Washington DC, 1993). ISBN 1-56032-125-3
13. H. Fredriksson, U. Akerlind, *Materials Processing During Casting* (John Wiley & Sons, The Atrium, Southern Gate, Chichester, West Sussex, 2006)
14. S.-M. Seo, J.-H. Lee, Y.-S. Yoo, C.-Y. Jo, H. Miyahara, K. Ogi, Solute redistribution during planar and dendritic growth of directionally solidified ni-base superalloy CMSX-10, *in 11th International Symposium on Superalloys 2008, TMS, The Minerals, Metals & Materials Society*, ed. by R.C. Reed, K.A. Green, P. Caron, T.P. Gabb, M.G. Fahrmann, E.S. Huron, S.A. Woodard. Pennsylvania TMS 2008 (John Wiley & Sons, Champion, 2008)
15. U. Mizutani, *Hume-Rothery Rules for Structurally Complex Alloy Phases* (CRC Press, Taylor and Francis Group, Boca Raton, 2011)
16. A.G. Guy, J.J. Hren, *Elements of Physical Metallurgy*, 3rd edn. (Addison-Wesley Publishing Company, Reading, 1974)
17. J.R. Davis (ed.), *Copper and Copper Alloys*. ASM Specialty Handbook (ASM International, Materials Park, 2001)
18. R.F. Cochrane (Contributor), *Dissemination of IT for the Promotion of Materials Science*. DoITPoMS, Micrographs (University of Cambridge, 2020). https://www.doitpoms.ac.uk/miclib/credits.php?id=22
19. G. Vander Voort, *Deformation and Annealing of Cartridge Brass*. Newslatter (Vac Aero International, Oakville, 2015)
20. C. Li, Y. Yan, T. Gao, G. Xu, The microstructure, thermal, and mechanical properties of Sn-$3.0Ag$-$0.5Cu$-xSb high-temperature lead-free solder. Materials **13**(19), 4443 (2020)
21. A.W. Worcester, J.T. O'Reilly (revised by), *Properties and Selection: Nonferrous Alloys and Special-Purpose Materials*. ASM Metals Handbook, vol. 2 (ASM International, Materials Park, 1992)
22. J.H.L. Pang, *Lead Free Solder* (Springer, Switzerland, 2012)

23. R. Elliott, *Eutectic Solidification Processing Crystalline and Glassy Alloys* (Butterworths, New York, 1983)
24. A.R. Bailey, *The Role of Microstructure in Metals*, 2nd edn. (Metallurgical Services Betchworth, Surrey, 1972)
25. V.T. Witusiewicz, L. Sturz, U. Hecht, S. Rex, Lamellar coupled growth in the neopentylglycol-(D) camphor eutectic. J. Crystal Growth **386**, 69–75 (2014)
26. A. Karma, A. Sarkissian, Morphological instabilities of lamellar eutectics. Metall. Mater. Trans. **27A**(3), 635–656 (1996)
27. W.A. Tiller, Polyphase solidification, in *Liquid Metals and Solidification: A Seminar Held During the Thirty-Ninth National Metal Congress and Exposition, Chicago 1957*, vol. 276 (ASM, Cleveland, 1958)
28. K.A. Jackson, J.D. Hunt, Lamellar and rod eutectic growth. Trans. Metal. Soc. AIME **236**(1129), 129–142 (1966)
29. J.W. Gibbs, K.A. Mohan, E.B. Gulsoy, A.J. Shahani, X. Xiao, C.A. Bouman, M. De Graef, P.W. Voorhees, The three-dimensional morphology of growing dendrites. Sci. Rep. **5**, 11824 (2015)
30. R.E. Napolita, *The Structure of Cast Metals*. PowerPoint presentation found online at shorturl.at/hjlAE (2005)
31. W. Callister, Jr., D.G. Rethwisch, *Materials Science and Engineering: An Introduction*, 10th edn. (Wiley, New York, 2018)
32. Video, *In-Situ Austenite-Pearlite Phase Transformation, recorded at the Chair of Ferrous Metallurgy* (Montanuniversity Leoben). https://www.youtube.com/watch?v=TMOsjtB73BQ
33. F. Pohl, Pop-in behavior and elastic-to-plastic transition of polycrystalline pure iron during sharp nanoindentation. Sci. Rep. **9**(1), 1–12 (2019)
34. B. Ferreira de Oliveira, M. Picanço Oliveira, L.A. Hernandez Terrones, M. Giardinieri de Azevedo, L. Barbosa Godefroid, Microstructure and mechanical properties of as-cast and annealed high strength low alloy steel. J. Mater. Sci. Res. **8**(4), 1 (2019)
35. ASM Handbook, *Alloy Phase Diagrams*, vol. 3 (ASM International, Materials Park, 1992)
36. R. Elliott, *Cast Iron Technology* (Butterworth, London, 1988)
37. T.J. Langill, Barry Dugan, Zinc materials for use in concrete, in *Galvanized Steel Reinforcement in Concrete*, Chapter 4, ed. by S.R. Yeomans (Elsevier, New York, 2004)
38. K. Easterling, *Introduction to the Physical Metallurgy of Welding*, 2nd edn. (Butterworth-Heinemann, London, 1992)
39. C. Thaulow, A.J. Paauw, K. Guttormsen, *The Heat-Affected Zone Toughness of Low-Carbon Microalloyed Steels*. Welding Research Supplement (International Institute of Welding, 1987), pp. 276–279
40. J. Wadsworth, G.R. Morse, P.M. Chewey, The microstructure and mechanical properties of a welded molybdenum alloy. Mater. Sci. Eng. **59**(2), 257–273 (1983)
41. G. Vander Voort, *Metallographic Examination of Welds Metallographic Examination of Welds*. https://www.georgevandervoort.com/wp-content/uploads/2012/06/Welding.pdf
42. C. Kose, C. Topal, Effect of post weld heat treatment and heat input on the microstructure and mechanical properties of plasma arc welded AISI 410S ferritic stainless. Mater. Res. Express **6**(6), 066517 (2019). https://iopscience.iop.org/article/10.1088/2053-1591/ab09b6
43. A.G. Guy, *Essentials of Materials Science* (McGraw-Hill, New York, 1976)
44. D.S. Clark, W.R. Varney, *Physical Metallurgy for Engineers*, 2nd edn. (D. Van Nostrand Company, New York, 1962)
45. S.A. Nelson, *Crystallization in Ternary Systems*. Lecture Notes for EENS 2120, Petrology (Tulane University, New Orleans, 2011). https://www.tulane.edu/\char126\relaxsanelson/eens212/ternaryphdiag.htm
46. G.V. Raynor, V.G. Rivlin, *Phase Equilibria in Iron Ternary Alloys*. The Institute of Metals, London, No. 4, 1988. Reference cited in Cr-Fe-Ni (Chromium-Iron-Nickel) Ternary Phase Diagrams, vol. 3 (ASM Handbook, 1992)
47. C.L. Xu, H.Y. Wang, F. Qiu, Y.F. Yang, Q.C. Jiang, Cooling rate and microstructure of rapidly solidified Al-20wt.%Si alloy. Mater. Sci. Eng. A **417**, 275–280 (2006)

48. O.L. Rocha, C.A. Siqueira, A. Garcia, Heat flow parameters affecting dendrite spacings during unsteady-state solidification of Sn-Pb and Al-Cu alloys. Metall. Mater. Trans. A **34**(4), 995–1006 (2003)
49. F. Sa, O.L. Rocha, C.A. Siqueira, A. Garcia, The effect of solidification variables on tertiary dendrite arm spacing in unsteady-state directional solidification of Sn–Pb and Al-Cu alloys. Mater. Sci. Eng. A **373**(1–2), 131–138 (2004)
50. W.R. Osorio, A. Garcia, Modeling dendritic structure and mechanical properties of Zn–Al alloys as a function of solidification conditions. Mater. Sci. Eng. A **325**(1–2), 103–111 (2002)
51. J. Quaresma, C.A. Santos, A. Garcia, Correlation between unsteady-state solidification conditions, dendrite spacings, and mechanical properties of Al-Cu alloys. Metall. Mater. Trans. A **31**(12), 3167–3178 (2000)

Chapter 11
Solid-State Phase Transformation

11.1 Introduction

This chapter initially introduces thermally induced or strain-induced phase transformation in the solid state. In essence, crystallography and phase transformation modeling are specifically coupled in order to explain the formation of new phases in the solid state. Recall that Chap. 1 introduces the principles of crystal lattice of a homogeneous phase as an essential part of crystallography.

This chapter includes a compilation of analytical solutions to one-dimensional phase transformation problems in the solid state with moving grain boundary migration. Moreover, the solid-state phase transformation is controlled by heating samples at temperatures T below the solidus temperatures T_s, followed by slow or rapid cooling in thermal-conductive media.

The phrase "solid-state phase transformation" refers to phase change from solid-to-solid denoted by invariant reactions representing allotropy, such as

Ferrite to austenite	$\delta\,(BCC) \to \gamma\,(FCC) = A$	for steels
Austenite to ferrite	$\gamma\,(FCC) \to \alpha\,(BCC) = F$	for steels
Austenite to pearlite	$\gamma\,(FCC) \to \alpha\,(BCC) + Fe_3C$	for steels
Austenite to martensite	$\gamma\,(FCC) \to \alpha'\,(BCT) = M$	for steels
Austenite to martensite	$\gamma\,(FCC) \to \alpha'\,(BCT) = M$	for Nitinol
Alpha to beta phase	$\alpha\,(HCP) \to \beta\,(BCC)$	for Ti-alloys
Beta to alpha	$\beta\,(BCT) \to \alpha\,(cubic)$	for Tin(Sn)

where A denotes austenite or austenitic phase, F denotes ferrite or ferritic phase, Fe_3C denotes cementite or iron carbide, and M denotes martensite or martensitic phase.

Regarding steels, they are iron (Fe)- and carbon (C)-based alloys being categorized as carbon steels, alloy steels, tools steels, and stainless steels. Thus, the general

N. Perez, *Materials Science: Theory and Engineering*,
https://doi.org/10.1007/978-3-031-57152-7_11

components of a steel are Fe-C-X, where X stands for additional elements, such as Ni, Mo, Cr, and so forth. In effect, the resultant Fe-based steel is a ferrous material with a nominal composition denoted as C_o.

11.2   Thermally Induced Phase Transformation

Solid-state diffusion during isothermal transformation induces nucleation and growth of a new atomic arrangement from heat-treated solid phases. This suggests that lattice defects, such as dislocations, point defects, and dimensional discontinuities, are potential sites for the onset of solid-state phase transformation.

Assume that a "parent" or "old" α-phase is the matrix to be transformed to β-phase at a particular critical temperature T_c upon slow cooling. Regardless of the physical conditions of the matrix, the phase change process occurs by diffusional transformation in the solid state associated with a growing α-β interface, while latent heat of transformation evolves at a critical temperature T_c.

11.2.1   Phase Transformation Crystallography

This section describes the effects of the crystallographic orientation relationship (OR) on the phase transformation of a metal or alloy lattice structure. Among the seven (7) types of crystal structures, the face-centered orthorhombic (FCO) or face-centered cubic (FCC) unit cell is a suitable atomic arrangement for illustrating the crystallography of solid-state transformation.

Consider a hypothetical thermally induced solid-state transformation upon cooling or heating, where an ideal α-phase with two FCO or FCC unit cells with x, y, z coordinates (Fig 11.1a) has a sought enclosed β-phase with a BCO, BCC or BCT unit cell with x', y', z' coordinates as a viable option.

For the transition of α-phase to an β-phase, the initial lattice parameters a_γ, b_γ, c_γ change to a_β, b_β, c_β. As a result, for example, the phase transformation may be represented by $(010)_\alpha \rightarrow (\overline{1}10)_\beta$ plane and $[101]_\alpha \rightarrow [111]_\beta$ direction as shown in Fig 11.1b.

For parallel or near parallel planes and directions,

$$(hkl)_\alpha \parallel (hkl)_\beta \quad \& \quad [uvw]_\alpha \parallel [uvw]_\beta \qquad (11.1a)$$

$$hu + kv + lw = 0 \quad \text{for } (hkl) \parallel [uvw] \qquad (11.1b)$$

This means that $(hkl)_\alpha$ plane is parallel ($\parallel$) to $(hkl)_\beta$ plane and $[uvw]_\alpha$ direction is parallel to $[uvw]_\beta$ direction. However, $[uvw]_\alpha$ must lie in the $(hkl)_\alpha$ plane, and,

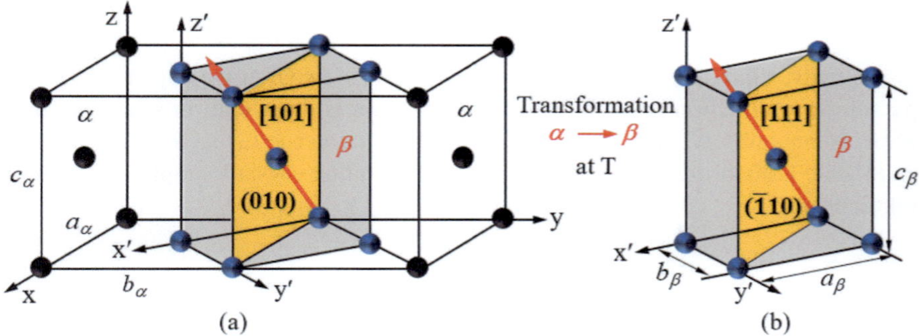

Fig. 11.1 Solid-state phase transformation of (**a**) a face-centered orthorhombic (*FCO*) or face-centered cubic (*FCC*) into (**b**) a *BCO*, *BCC*, or *BCT* unit cell

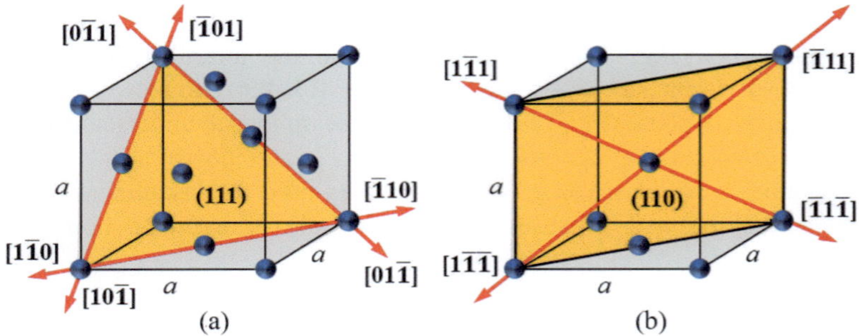

Fig. 11.2 Common crystallographic close-packed planes and directions in (**a**) *FCC* and (**b**) *BCC* unit cells

similarly, $[uvw]_\beta$ must lie in the $(hkl)_\beta$ plane. It can be deduced from Fig 11.1a that $(010)_\alpha \parallel (\bar{1}10)_\beta$ and $[101]_\alpha \parallel [111]_\beta$.

In particular, the Miller indices for the crystallographic planes shown in Fig 11.1 and the general notation given by Eq. (11.1a) indicate that $(hkl)_\alpha = (010)_\alpha$ and $[uvw]_\alpha = [101]_\alpha$. From Fig 11.1b, $(hkl)_\beta = (\bar{1}10)_\beta$ and $[uvw]_\beta = [111]_\beta$. Moreover, using the $(010)_\alpha$-$[101]_\alpha$ and $(\bar{1}10)_\beta$-$[111]_\beta$ systems in Eq. (11.1b) verifies that these directions are parallel to their respective planes.

Using the closest packed planes in the α and β structures, one can deduce that the α unit cell has *four* {111} close-packed planes and *three* ⟨110⟩ close-packed directions per each {111} plane (Fig. 11.2a). On the other hand, there are *six* {110} planes and *two* ⟨111⟩ close-packed directions (Fig. 11.2b) per each {110} plane in the β unit cell.

The above crystallographic approach for determining the orientation relationships (ORs) between planes and directions during solid-phase transformation is essentially ideal. However, slight misorientation in the parallelism of crystallographic directions is possible. Moreover, texture orientation and azimuthal orientation (angle between a reference direction) are also used for parallelism between planes and directions, respectively.

Actually, solid-state transformation is expected to occur at the expense of lattice parameter contraction or expansion. This situation is dealt with in a later section. Experimental observations using, say, X-ray diffraction can reveal the true nature of solid-state transformation, including the corresponding crystallographic orientation relationship.

11.2.2   Interface Coherency

This section includes phenomenological models that describe the logical interconnections or interface boundaries between two crystals having similar or different atomic arrangements. The general features of interface boundaries can be understood with aid of Fig. 11.3, which suggest that the interface boundaries (ib) between two different, say, α and β grains (Figs. 11.3a,b,c) and between β-particles embedded in the α-phase (Fig. 11.3d,e,f) are viable options to understand solid-state phase transformation.

These models are classical geometry sketches used for visualizing the atomic arrangement of α- and β-phases and related type of interface boundary (ib)

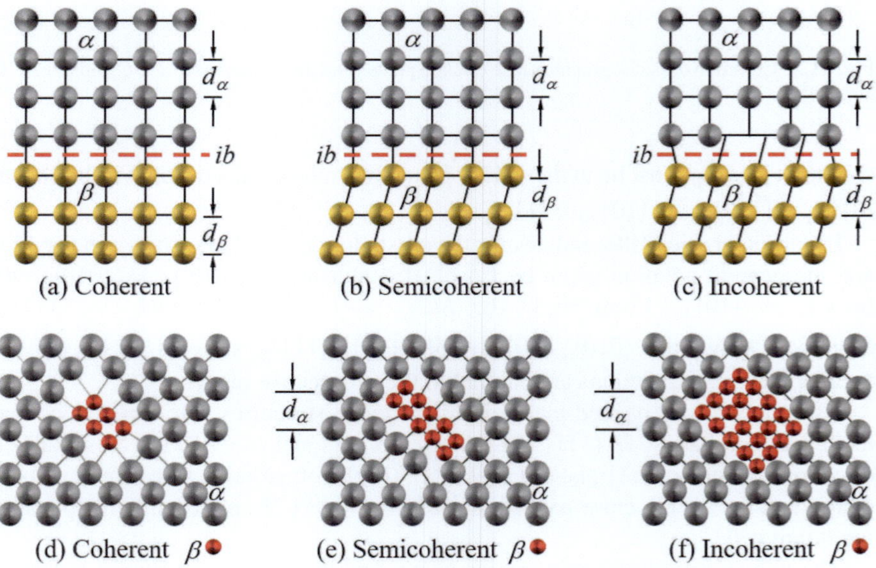

Fig. 11.3 Phase boundary models for (**a,b,c**) α-β-grains (Porter et al. [1, p. 147]) and (**d,e,f**) β-particles within α-grains (Guy and Hren [2, p. 422])

coherency due to diffusional transformation in solids (Porter et al. [1, pp. 146–149], Guy and Hren [2, p. 422]).

First of all, assume that the solid surfaces of two crystals are smooth and flat atomic planes having partial atomic bonding and that their surface planes repeat themselves underneath such surfaces.

During phase transformation in the solid state, the solid α-phase completely or partially changes to a new solid β-phase with a critical volume $V_{c,\beta}$ and a critical radius r_c, preferably, along grain boundaries (gb). Consequently, a α-β-interface with a total surface area A forms and moves (migrates), while the volume of the β-phase increases at the expense of the original volume of the α-phase.

Normally, the interplanar spacings (d_α and d_β) or the lattice parameters (a_α and a_β) and the surface energy $\gamma_{\alpha\beta}$ are the relevant macroscopic entities one can used to characterize the results of phase transformation. The interplanar spacing, d_α-spacing or d_β-spacing, defines the distance between atomic planes (Fig. 11.3).

Coherent Interface This interface (Fig. 11.3a,d) arises when there is a perfect atomic match (perfect registry of the lattices) between the two adjoining α and β metallic phases. However, a small lattice mismatch induces a small elastic strain in the adjacent crystals. This implies that the lattice parameters (a_α and a_β) are similar to an extent and the surface energy due to straining is $\gamma_{strain} \rightarrow 0$. Moreover, this type of interface boundary is found in some Cu-Si alloys as indicated by the crystallographic orientation relationship (OR) between α-FCC and β-HCP phases. The notation for an α-β orientation relationship is based on the parallelism between FCC and HCP structures (Porter et al. [1, p. 148]). Thus, $(111)_\alpha \parallel (0001)_\beta$ and $\left[\bar{1}10\right]_\alpha \parallel \left[11\bar{2}0\right]_\beta$, leading to $\epsilon_{\alpha\beta} \simeq 0$ since $a_\alpha \simeq a_\beta$. This indicates that the $(hkl)_\alpha$ plane of the α crystal lies parallel to the $(hkl)_\beta$ plane of the β crystal. This is a parallelism denoted by the symbol $\parallel$ or by the double-slant symbol $//$. Conclusively, a coherent interface forms when two different crystal lattices are continuous across the interface.

Semicoherent Interface The semicoherent interface (Fig. 11.3b,e) occurs when dislocations distort adjacent atomic planes, causing the rise of $\gamma_{strain} > 0$. This means that a slight lattice mismatch causes a periodic array of misfit dislocations, leading to a significant elastic strain.

Incoherent Interface The incoherent interface (Fig. 11.3c,f) represents the most antisymmetric phase boundary with the highest lattice distortion, leading to a high lattice misfit at the interface. Thus, the coherent surface energy $\gamma_{\alpha\beta}$ and related elastic strain $\epsilon_{\alpha\beta}$ due to elastic distortion caused by dimensional misfit of strain-free lattices at the α-β interface are given by

$$\gamma_{\alpha\beta} = \gamma_{chem} + \gamma_{strain} \tag{11.2a}$$

$$\epsilon_{\alpha\beta} = \frac{\lfloor d_\alpha - d_\beta \rfloor}{d_\alpha} \simeq \frac{\lfloor d_\alpha - d_\beta \rfloor}{d_\beta} \tag{11.2b}$$

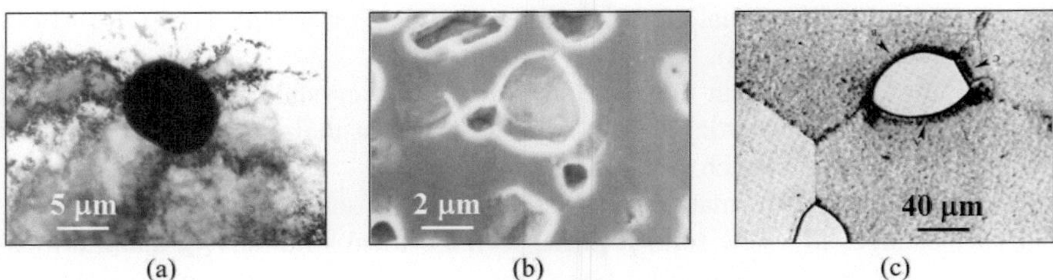

Fig. 11.4 Particle coherency in different metal matrices. (**a**) Heat-treated (HT) 304 stainless steel at $1000°C$ for $0.5\ h$ and air cooled, (**b**) HT Devitrium 7025 alloy at $1100°C$ for $24\ h$ and air cooled, and (**c**) HT Cu-In alloy (Porter et al. [1, p. 159])

$$\epsilon_{\alpha\beta} = \frac{\lfloor a_\alpha - a_\beta \rfloor}{a_\alpha} \simeq \frac{\lfloor a_\alpha - a_\beta \rfloor}{a_\beta} \tag{11.2c}$$

Here, γ_{chem} denotes the chemical interfacial energy, γ_{strain} denotes the interfacial strain surface energy due to dislocations at the interface, and a_α denotes the reference lattice parameter. Moreover, Eq. (11.2b,c) represents the lattice misfit at the interface. This suggests that the higher the $\epsilon_{\alpha\beta}$, the higher the array of dislocations and the higher the strain energy at the interface boundary.

Note that d_α and d_β depend on the lattice parameters determined by a diffraction technique, such as X-ray diffraction introduced in Chap. 2. Thus, knowledge of X-ray diffraction data is an essential complement to the analysis of crystallography and related atomic structure of solid phases.

Particle-Matrix Coherency Figure 11.4 shows distinguishable particles in three different alloys. At first glance, it is difficult to determine the atomic registry between the matrix and particles. However, one can deduce that dislocations may interact with particles during deformation.

Figure 11.4a depicts a near-circular $Cr_{23}C_6$ carbide particle located at a grain boundary triple point in a 304 stainless steel specimen, Fig. 11.4b clearly shows boride particles embedded in an amorphous Devitrium 7025 alloy matrix ($Ni_{53}Mo_{35}B_9Fe_2$ containing 1% impurities), and Fig. 11.4c illustrates a particle at a grain boundary triple point in a crystalline Cu-In alloy (Porter et al. [1, p. 159]).

11.3　Transformation of Austenite in Steels

This section is an extension of Chap. 10, and it is devoted to solid-state phase transformation in steels upon slow cooling the austenite γ-phase from an austenitic temperature T_γ. The γ-austenite region is the shaded field in the Fe-C equilibrium phase diagram (T-C diagram) shown in Fig. 11.5, adopted from Callister and Rethwisch book [3, p. 333] and found in Askeland et al. book [4, p. 495]. This

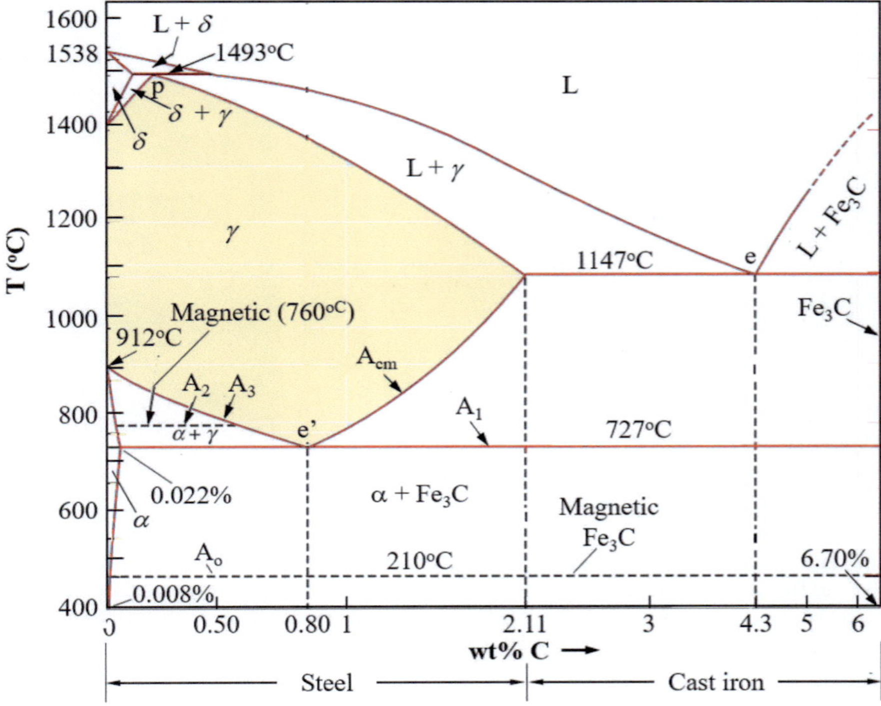

Fig. 11.5 The *Fe-C* phase diagram showing the austenitic region. Adapted from Callister and Rethwisch [3, p. 333]

phase diagram is useful in understanding phase transformation in steels being either annealed or normalized at a relatively slow transformation rate and low cooling rate dT_γ/dt.

The invariant reactions in Fig. 11.5 are defined by Eq. (10.48a,b,c):

$$\text{Peritectic p: } L + \delta \rightleftarrows \gamma \text{ at } 1493°C \tag{11.3a}$$

$$\text{Eutectic e: } L \rightleftarrows \gamma + Fe_3C \text{ at } 1147°C \tag{11.3b}$$

$$\text{Eutectoid e': } \gamma \rightleftarrows \alpha + Fe_3C \text{ at } 727°C \tag{11.3c}$$

For steels containing $0 < C < 2.11 \ \%wt$, the γ-austenitic phase is the *FCC* structure that must be homogenized (heat treated) prior to a slow cooling process in a suitable media, such as water, air, and so forth. The recommended heat treatment temperatures are

$$T_\gamma = A_3 + 50°C \quad \text{(for hypoeutectoid steels)} \tag{11.3d}$$

$$T_\gamma = A_{cm} + 50°C \quad \text{(for hypereutectoid steels)} \tag{11.3e}$$

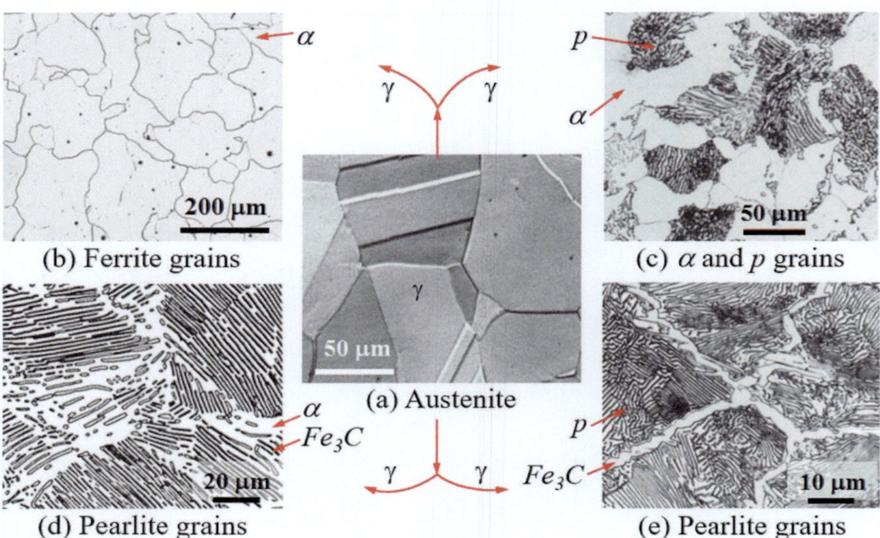

Fig. 11.6 Solid-state transformation of the austenite γ-phase upon slow cooling to room temperature. (**a**) High-temperature austenite, (**b**) pure Fe, (**c**) Fe-0.38C hypoeutectoid steel, (**d**) Fe-0.80C eutectoid steel at $T < T_e$ and (**e**) Fe-1.40C hypereutectoid steel at $T < T_e$. All microstructures are taken from Callister and Rethwisch book [3, chapter 10]

which can be altered, provided that T_γ is not too high in order to avoid high-temperature corrosion during heat treatment and other diffusion-related complications.

The phase transformation process adopted henceforth is based on cooling austenite at a particular cooling rate dT_γ/dt. For pure iron, the austenite-to-ferrite transformation, $\gamma_{FCC} \rightarrow \alpha_{BCC}$, occurs at $T < T_c = 1185\ K = 912°C$ (Fig. 11.5). A suitable microstructural representation of γ_{FCC} transformation is shown in Fig. 11.6. These microstructures can be found in Callister and Rethwisch book [3, Chapter 10].

Annealing Heat Treatment Consider pure iron (Fe), hypoeutectoid, eutectoid, and hypereutectoid plain-carbon wrought steel specimens being simultaneously heat treated at $T > T_{A_3}$ in a suitable furnace environment for a sufficient time period. Eventually, all specimens would have slight different austenitic microstructures, but consider the austenite γ-phase shown in Fig. 11.6a as the common high temperature microstructure subjected to undergo a prolong carbon diffusion-control phase transformation in the solid-state upon slow cooling in a furnace. This is a metallurgical process called annealing heat treatment. For instance, Fig. 11.6b-e is annealed microstructures as a result of the γ-phase austenitic transformation in different specimens.

For the pure Fe specimen, the microstructural evolution in the solid state is represented by γ-austenite $\rightarrow$ α-ferrite or γ-FCC $\rightarrow$ α-BCC. A suitably prepared metallographic specimen under optical microscopy reveals the ferritic microstructure shown in Fig. 11.6b, consisting of white or light ferrite grains. For clarity, the nomenclature used for this phase transformation upon slow cooling is illustrated in Fig. 11.6 as the (a)$\rightarrow$(b) process.

- For the hypoeutectoid steel with a composition $C_o = 0.38\%C < C_e = 0.80\%C$, (a)$\rightarrow$(c) in Fig. 11.6 means that austenite transforms to some proeutectoid ferrite (α) grains and to some pearlite $(\alpha + Fe_3C)_e$ grains (Fig. 11.6c). Firstly, a portion of γ-FCC austenite transforms to proeutectoid (primary) α-BCC ferrite (light grains) and secondly, the remaining γ-FCC austenite transforms to a two-phase eutectoid structure $(\alpha + Fe_3C)_e$ called pearlite (p), where $\alpha = \alpha_e$ is white and Fe_3C is black. The corresponding phase transformation is denoted by the reaction $\gamma \rightarrow \alpha + p = \alpha + (\alpha + Fe_3C)_e$, and the morphology of $(\alpha + Fe_3C)_e$ is based on alternating plates that form the pearlitic structure.
- For the eutectoid steel with $C_e = 0.80\%C$, the invariant reaction cited above is recast herein for convenience as $\gamma \rightarrow (\alpha + Fe_3C)_e$, which evolves just below the eutectoid temperature T_e. The final microstructure consists of 100% pearlite, where (a)$\rightarrow$(d) in Fig. 11.6d. Moreover, α is a soft white ferrite phase, and Fe_3C is a dark/black hard phase. Hence, pearlite is a metallic composite phase.
- For the hypereutectoid steel with $C_o = 1.40\%C > C_e$, the initial transformation occurs along the austenite grain boundaries as Fe_3C cementite (primary phase), followed by pearlite formation through the eutectoid invariant phase reaction $\gamma \rightarrow (\alpha + Fe_3C)_e$. The resultant microstructure, (a)$\rightarrow$(e) in Fig. 11.6e, is a pearlitic microstructure with hard and brittle cementite Fe_3C along the grain boundaries.

According to the Fe-C diagram, the steel microstructure depends on the $\%wt$ C or simply $\%C$ and the cooling rate dT_γ/dt for austenite transformation. The reversible ($\rightleftarrows$) invariant reactions associated with the solid-state phase transformation of the γ-phase in pure iron (Fe) at $T < 912°C$ and plain carbon steels (Fe-C) at $T < T_{e'} = 727°C$ are defined as

$$\gamma \rightleftarrows \alpha \quad \text{for } C = 0,\, T < 912°C \tag{11.4a}$$

$$\gamma \rightleftarrows \alpha + p \quad \text{for } 0 < C < 0.80\%,\, T < T_{e'} = 727°C \tag{11.4b}$$

$$\gamma \rightleftarrows p \quad \text{for } C = 0.80\%,\, T < T_{e'} = 727°C \tag{11.4c}$$

$$\gamma \rightleftarrows Fe_3C + p \quad \text{for } 0.80\% < C < 2.11\%,\, T < T_{e'} = 727°C \tag{11.4d}$$

where the arrows represent cooling ($\rightarrow$) and heating ($\leftarrow$). Note that pearlite (p) derives from the austenitic γ-phase as a dual-phase structure.

11.3.1   Solid-Phase Transformation Models

Modeling solid-phase transformation depends on the type of crystal symmetry subjected to a thermally induced or strain-induced phase transitions at a particular temperature in a suitable environment. For instance, Fig. 11.7 illustrates the models for transforming austenite upon slow cooling (SC) to obtain the microstructures depicted in Fig. 11.6b,c,d,e.

Theoretically, it can be assumed that diffusional transformation, indicated by the molar fluxes J_C and J_{Fe}, is the mechanism controlling the atomic arrangement in the new solid phase, where the excess carbon in the α-phase and excess Fe in the Fe_3C-phase are rejected as shown by the arrows in Figs. 11.7b,c,d.

The morphology of a transformed phase is not as uniform as depicted in Fig. 11.6, but the above models (Fig. 11.7) are simple and provide a starting point in understanding solid-phase transformation. Notice that only solid-solid phase transformations in inorganic materials, such as carbon steels, has been considered.

This physical phenomenon in steel alloys and other non-ferrous alloys are most difficult to model. In particular, annealing involves a diffusion process upon slow cooling, while quenching is a diffusionless mechanism upon rapid cooling.

In practice, the annealing process is a common heat treatment for homogenizing the microstructure through diffusion at an austenitic temperature $T_\gamma > A_3$, and subsequently, the homogenized microstructure is obtained at a slow cooling rate (dT/dt).

Ferrite Model The solid-state phase transformation from austenite $(\gamma\text{-}Fe)$ to ferrite $(\alpha\text{-}Fe)$ is represented by the pure iron (Fe) invariant reaction $\gamma\text{-}Fe \rightarrow \alpha\text{-}Fe$ at $T < T_c = 912°C$. The microstructures in Fig. 11.6a $\rightarrow$ Fig. 11.6b is modeled as shown in Fig. 11.7a. The resultant microstructure is composed of equiaxed ferrite grains in pure iron (Fe). The corresponding orientation relationship (OR) between the austenite and ferrite can be generalized as $\{111\}_\gamma \parallel \{101\}_\alpha$ planes and $\langle 110 \rangle_\gamma \parallel \langle 110 \rangle_\alpha$ directions with small misfit angles. Detail information on these ORs are known as Kurdjumov-Sachs (KS) or Nishiyama-Wasserman (NW) notations as per Nishiyama [5, pp. 22–24].

Ferrite-Pearlite Model This corresponds to the transformation of austenite in hypoeutectoid steels with $0\% < C < 0.8\%$ to ferrite/pearlite grains according to the phase reactions $\gamma \rightarrow \alpha + \gamma_1 \rightarrow \alpha + p$, where α is the equiaxed *proeutectoid ferrite* grains (light) and γ_1 is the remaining austenite that transforms to pearlite $p = \alpha + Fe_3C$ with a lamellar pearlitic structure (light α and dark Fe_3C). Thus, the corresponding microstructure due to diffusional transformation is denoted as Fig. 11.6a $\rightarrow$ Fig. 11.6c as per model in Fig. 11.7b. The resultant microstructure (Fig. 11.6c) is a combination of equiaxed *proeutectoid ferrite* α-grains and lamellar pearlite grains in a Fe-0.38C hypoeutectoid carbon steel.

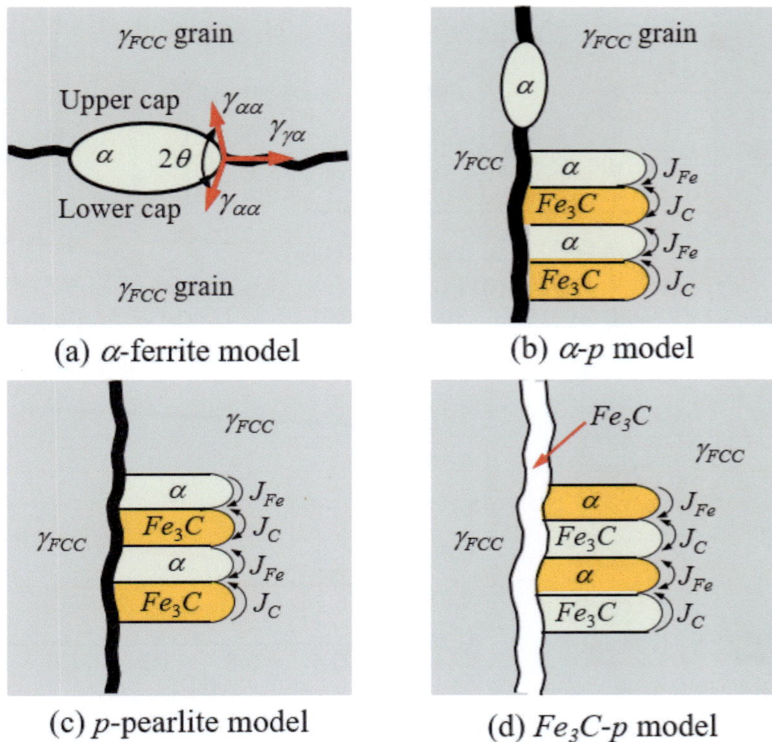

(a) α-ferrite model

(b) α-p model

(c) p-pearlite model

(d) Fe_3C-p model

Fig. 11.7 Phase transformation models for transforming austenite to (**a**) α-ferrite, (**b**) α-ferrite and p-pearlite, (**c**) p-pearlite, and (**d**) Fe_3C-cementite and p-pearlite

Pearlite Model This is related to the invariant reaction $\gamma \rightarrow p = \alpha + Fe_3C$ that occurs below the eutectoid temperature $T_{e'} = 727°C$. The microstructural evolution due to diffusional transformation is denoted as Fig. 11.6a $\rightarrow$ Fig. 11.6d as per model in Fig. 11.7c. The resultant microstructure (Fig. 11.6d) is composed of pearlite grains (light α and dark Fe_3C) in a eutectoid Fe-0.80C steel.

Pearlite and Cementite Model This corresponds to the phase reactions $\gamma \rightarrow Fe_3C + \gamma_1 \rightarrow Fe_3C + p$, where Fe_3C is the *proeutectoid cementite* phase as the light grain boundary surrounding the pearlite grains (pearlite colonies). Thus, Fig. 11.6a $\rightarrow$ Fig. 11.6e as per model in Fig. 11.7d by diffusional transformation. The resultant microstructure (Fig. 11.6e) is composed of pearlite grains (light α and dark Fe_3C) and light Fe_3C-cementite grain boundaries in a hypereutectoid Fe-1.40C carbon steel.

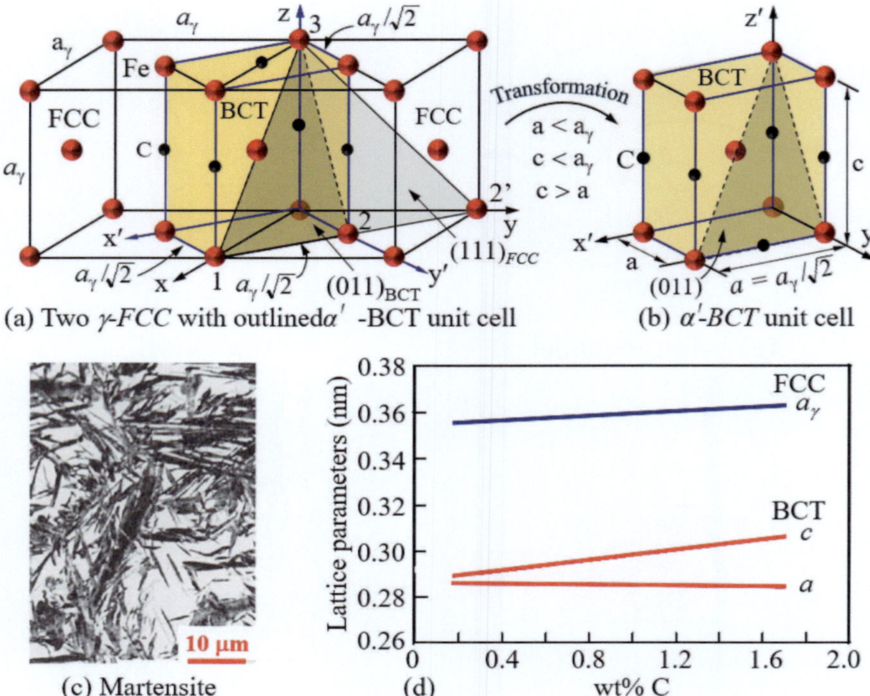

(a) Two γ-FCC with outlined α'-BCT unit cell   (b) α'-BCT unit cell

(c) Martensite   (d)

Fig. 11.8 Bain distortion model including (**a**) Austenite unit cells showing the unit cell to be transferred as a martensitic structure and (**b**) martensite unit cell showing lattice sites for carbon atoms. (**c**) Austenite and martensite lattice parameters as functions of weight percent carbon (wt% C). After Cullity and Stock [6, p. 350]. (**d**) Martensite with retained austenite (white) taken from Callister and Rethwisch [3, p. 324]

11.3.2 Martensite Formation

Following Cullity and Stock [6, p. 350], the martensitic transformation is based on the classical *Bain Distortion Model* for steels as schematically shown in Fig. 11.8a.

The martensitic phase, after Martens (1850–1914), derives from a soft face-centered cubic austenite (γ-FCC) as a hard metastable phase with a body-centered tetragonal (α'-BCT) lattice structure induced by a dimensionless mechanism at a high cooling rate (dT/dt).

The initial austenitic FCC unit cells are composed of iron (Fe) atoms (red) with embedded carbon atoms (black) in interstitial positions or sites. Notice that the drawn or enclosed (picked out) BCT unit cell is in yellow background between the two neighboring γ-FCC unit cells (Fig. 11.8a). For clarity, the coordinate transformation with fixed or rotation about the z-axis is denoted as $(x, y, z)_\gamma \rightarrow (x', y', z')_\gamma$ with $z = z'$ (Fig. 11.8b). The microstructure of martensite is shown in

Fig. 11.8c. In essence, Martensite is a hard, brittle, and metastable phase of steel that is formed when austenite is rapidly cooled.

Notice that the $(011)_{BCT}$ plane (1-2-3 triangle) in Fig. 11.8b is laying on the $(111)_{FCC}$ (1-2'-3 triangle). These are the crystallographic geometries used hereafter in order to illustrate the martensitic transformation with its corresponding unit cell lattice-parameters in Fig. 11.8d, from which curve fitting yields

$$a_\gamma = 0.35457 + \left(4.902 \times 10^{-3}\right)(\%C) \tag{11.5a}$$

$$c = 0.28675 + (0.01125)(\%C) \tag{11.5b}$$

$$a = 0.28637 - \left(1.3158 \times 10^{-3}\right)(\%C) \tag{11.5c}$$

$$c/a = 1.0012 + \left(4.4267 \times 10^{-2}\right)(\%C) \tag{11.5d}$$

where the lattice parameters are in *nm* units and these equations are valid for $0.20 \leq \%C \leq 1.70$. These equations may become useful for solving phase transformation problems.

The orientation relationship (OR) can be generalized using the $\{hkl\}$ closest-packed planes and related parallelism of $\langle uvw \rangle$ crystallographic directions. Thus,

$$\{hkl\}_\gamma \parallel \{hkl\}_{\alpha'} \tag{11.6a}$$

$$\langle uvw \rangle_\gamma \parallel \langle uvw \rangle_{\alpha'} \tag{11.6b}$$

where $\langle uvw \rangle_\gamma$ and $\langle uvw \rangle_{\alpha'}$ are the edge directions in the $\{hkl\}_\gamma$ and $\langle uvw \rangle_{\alpha'}$ planes, respectively. Crystallographically, note that the $(011)_{\alpha'}$ plane outlined within the two austenite γ-FCC unit cells (Fig. 11.8a) is shown in the martensite BCT unit cell as illustrated in Fig. 11.8b.

The martensitic transformation is characterized by the parallel crystallographic planes and directions written below as per most common cited work in the literature (Porter et al. [1, p. 393], Nishiyama [5, pp. 22–24])

$$(111)_\gamma \parallel (011)_{\alpha'} \quad \text{(Bain)} \tag{11.7a}$$

$$\left[\overline{1}01\right]_\gamma \parallel \left[\overline{1}\overline{1}1\right]_{\alpha'} \quad \text{(Kurdjumov-Sachs)} \tag{11.7b}$$

$$\left[\overline{1}\overline{1}2\right]_\gamma \parallel \left[0\overline{1}1\right]_{\alpha'} \quad \text{(Nishiyama-Wassermann)} \tag{11.7c}$$

$$\left[\overline{1}01\right]_\gamma \simeq\parallel \left[\overline{1}\overline{1}1\right]_{\alpha'} \quad \text{(Greninger-Troiano)} \tag{11.7d}$$

which are important crystallographic relationships related to the lattice parameters before and after solid-state transformation at a certain cooling rate (dT/dt) in a specific cooling medium.

These models provide significant insights into the mechanisms related to solid-state phase transformations of the γ-FCC austenitic structure during heat treatment procedures. The resultant phase is directly related to the mechanical properties and mechanical behavior of steels exposed to inert or aggressive environments.

Example 11.1 Consider the Bain distortion model illustrated in Fig. 11.8a for martensitic transformation in a 1040 carbon steel. Calculate (**a**) the μ_{1040} and μ_{1060} displacements, (**b**) the $\epsilon_{[100]_\gamma}$ and $\epsilon_{[001]_\gamma}$ strains experienced by iron (Fe) atoms during martensitic transformation denoted as $\gamma \rightarrow \alpha'$, and (**c**) deduce the orientation relationships corresponding to $[001]_\gamma$ and $[111]_\gamma$ crystallographic directions and (**d**) the cell volumes V_{FCC} and V_{BCT}. (e) Compare martensite and pearlite formation.

Solution

(**a**) From Eq. (11.5a,b,c,d) with $0.40\%C$ in the 1040 carbon steel,

$$a_\gamma = 0.35457 + \left(4.902 \times 10^{-3}\right)(0.40) = 0.357\ nm \qquad (11.1\text{E1a})$$

$$a = 0.28637 - \left(1.3158 \times 10^{-3}\right)(0.40) = 0.286\ nm \qquad (11.1\text{E1b})$$

$$c = 0.28675 + (0.01125)(0.4) = 0.291\ nm \qquad (11.1\text{E1c})$$

$$(c/a) = 0.291/0.286 = 1.018\ nm \qquad (11.1\text{E1d})$$

$$a = \left(\frac{0.286}{0.357}\right)a_\gamma = 0.801a_\gamma \simeq 0.80a_\gamma \qquad (11.1\text{E1e})$$

Then,

$$\Delta c = a_\gamma - c = 0.357 - 0.291 = 0.066\ nm \qquad (11.1\text{E2a})$$

$$\Delta a = a - \frac{a_\gamma}{\sqrt{2}} = 0.286 - \frac{0.357}{\sqrt{2}} = 0.034\ nm \qquad (11.1\text{E2b})$$

Now, the maximum displacement is approximated as

$$\mu = \sqrt{\Delta a^2 + \Delta c^2} = \sqrt{(0.034)^2 + (0.066)^2} = 0.074\ nm \qquad (11.1\text{E3})$$

(**b**) The lattice strains in tension (+) and compression (−) are

$$\epsilon_{[100]_\gamma} = \epsilon_{[010]_\gamma} = \frac{\sqrt{2}a - a_\gamma}{a_\gamma} = \frac{\sqrt{2}\,(0.286) - 0.357}{0.357} = 0.133 \qquad (11.1\text{E4a})$$

$$\epsilon_{[100]_\gamma} = \epsilon_{[010]_\gamma} \simeq 13.30\% \qquad (11.1\text{E4b})$$

$$\epsilon_{[001]_\gamma} = \frac{c - a_\gamma}{a_\gamma} = \frac{0.291 - 0.357}{0.357} = -0.185 \qquad (11.1\text{E4c})$$

$$\epsilon_{[001]_\gamma} = -18.50\% \quad \text{(compression)} \tag{11.1E4d}$$

and the Bain matrix representing the Bain strains or distortion is

$$B = \begin{pmatrix} \epsilon_{[100]_\gamma} & 0 & 0 \\ 0 & \epsilon_{[010]_\gamma} & 0 \\ 0 & 0 & \epsilon_{[001]_\gamma} \end{pmatrix} = \begin{pmatrix} 0.133 & 0 & 0 \\ 0 & 0.133 & 0 \\ 0 & 0 & -0.185 \end{pmatrix} \tag{11.1E5}$$

(**c**) The orientation relationships are

$$[001]_\gamma \parallel [001]_{\alpha'} \quad \& \quad [111]_\gamma \parallel [010]_{\alpha'} \tag{11.1E6}$$

These results indicate that the BCT unit cell contracts 18.50% along the crystallographic c-axis since $c < a_\gamma$ and $\epsilon_{[001]_\gamma} < 0$. On the other hand, BCT expands 13.30% along the a-axis since $a = a_\gamma/\sqrt{2} = 0.357/\sqrt{2} = 0.252\ nm$ (Fig. 11.8a or b) and $\epsilon_{([100]_\gamma} > 0$.

(**d**) For $a \simeq 0.80a_\gamma$, the unit cell volumes are

$$V_{FCC} = a_\gamma^3 = (0.357\ nm)^3 = 0.045\ nm^3 \tag{11.1E7a}$$

$$V_{BCT} = a^2c = (0.286)^2\,(0.291) = 0.024\ nm^3 \tag{11.1E7b}$$

Therefore, the martensite BCT unit cell is smaller than the austenite FCC unit cell. For illustration purposes, only the $(111)_\gamma \parallel (011)_{\alpha'}$ relationship is considered in this example.

(**e**) For comparison, the microstructural features of martensite and pearlite are different as shown in the figure below.

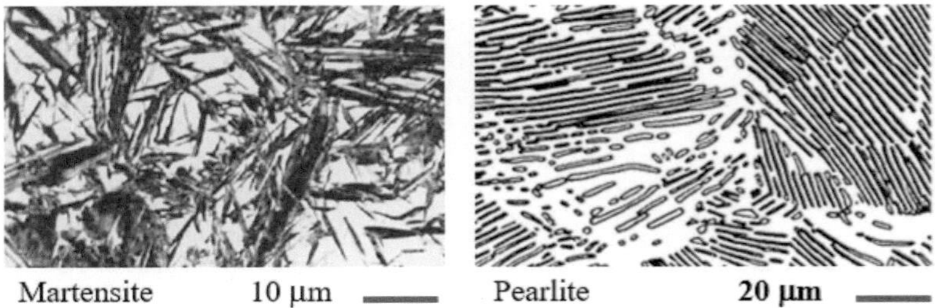

Martensite   10 µm ____   Pearlite   20 µm ____

Other differences indicate that the martensitic transformation is diffusionless at ultra high-transformation rate (time-independent), whereas the pearlitic transformation is due to atomic diffusion at relative low-transformation rate (time-dependent).

11.4   Thermodynamics of Transformation

Thermodynamics simplifies the characterization of solid-phase transformation by considering the microscopic model shown in Fig. 11.9a.

This model indicates that γ-grains transform into a α'-phase or α'-particle at γ grain boundaries. Notice that the surface energies $\gamma_{\gamma\gamma}, \gamma_{\gamma a'}$ arise as the γ-α' interface moves during the evolution of latent heat of transformation at a critical temperature T_c.

Figure 11.9a shows a typical model for the precipitation or formation of a coherent α'-phase along a grain boundary (gb) having two adjoined spherical caps. On the other hand, Fig. 11.9b, depicts the model for an arbitrary α'-phase growing at a triple junction with three adjoined γ-α' curved grain boundaries.

In the classical theory of nucleation, the invariant reaction $\gamma \rightarrow \alpha'$ for martensitic transformation is assumed to occur along a specific crystallographic plane known as habit (hkl) plane (Kaufman and Cohen [7, pp. 165–246], Olson and Cohen [8, p. 295]).

In this case, the Gibbs energy change (ΔG) for martensitic transformation is written as (Sinha [9, pp. 8.40–8.43])

$$\Delta G_{\text{hom}} = -V\Delta G_V + VE_s + A\gamma_{\gamma\alpha'} \tag{11.8a}$$

$$\Delta G_{het} = -V\Delta G_V + VE_s + A\gamma_{\gamma\alpha'} + \Delta G_d + \Delta G_i \tag{11.8b}$$

$$\Delta G_s = \Delta G_\varepsilon + \Delta G_p \tag{11.8c}$$

where V denotes the nucleus volume, A denotes the embryo surface area, E_s denotes the total strain energy due to the difference in volume of γ and α' phases, ΔG_ε denotes the elastic strain energy, ΔG_p denotes the elastic energy due to distortion of the habit plane, ΔG_v denotes the volumetric Gibbs energy change, ΔG_d denotes the release Gibbs energy term due the destruction or annihilation of defects (grain boundaries, vacancies, dislocations, etc.), and ΔG_i denotes the nucleus-defect interaction energy.

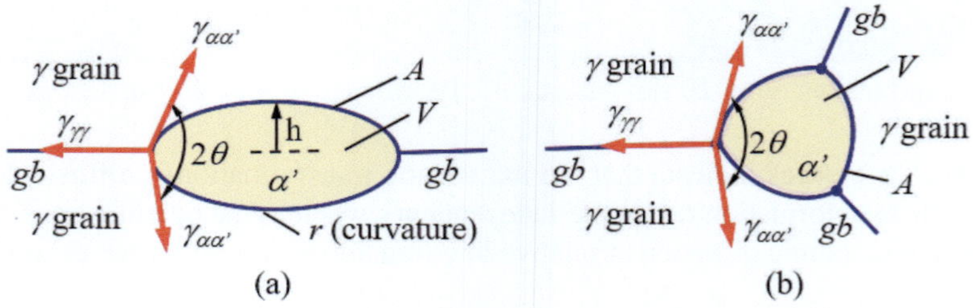

Fig. 11.9 Nucleation model for (a) an γ-nucleus growing along an γ-γ grain boundary (gb) with two adjoined spherical caps and (b) a α'-nucleus growing at an γ-grain triple junction with three adjoined γ-α' grain boundaries

Habit Plane This is fundamentally a flat or curved interface that defines a particular crystallographic plane between the martensitic and austenitic (parent or matrix) phases. In steels, the martensitic transformation is designated by a phase invariant reaction of the form $\gamma\text{-}FCC \longrightarrow \alpha'\text{-}BCT$, which is a representation of a highly strained martensite platelet (lath) related to an interfacial and strain energies.

From Olson and Cohen book [8, p. 303], E_s and the strain-energy proportionality factor K_ε can be defined as per Eshelby [10] elasticity approach for determining the elastic field of an ellipsoidal inclusion. The resultant mathematical equations are written as

$$E_s = \frac{K_\varepsilon h}{r} \tag{11.9a}$$

$$K_\varepsilon = \frac{\pi\,(2-v)}{8\,(1-v)}G_\varepsilon \gamma_\varepsilon^2 + \frac{\pi\,(1-v)}{4}G_\varepsilon \varepsilon_n^2 \tag{11.9b}$$

where h denotes the minor elliptical axis, $r >> h$, h/r denotes the aspect ratio, K_ε is in the order of $1\ GPa$, v denotes Poisson's ratio, G_ε denotes the isotropic shear modulus, γ_ε denotes the total transformation shear strain, and ε_n denotes the axial (normal) strain.

For homogeneous nucleation ($\Delta G_d = \Delta G_i = 0$) of an isolated martensitic particle (Fig. 11.9a) with radius r, semi-thickness h, volume $V = (4/3)\,\pi r^2 h$, and area $A = 4\pi r^2$, Eq. (11.8a) along with (11.8c) becomes the Kaufman and Cohen equation [7, pp. 165–246], which can be found in Olson and Cohen book [8, p. 356]:

$$\Delta G_{\text{hom}} = \frac{4}{3}\pi r^2 h \left(\Delta G_p - \Delta G_V\right) + \frac{4}{3}\pi r h^2 K_\varepsilon + 4\pi r^2 \gamma_{\gamma\alpha'} \tag{11.10}$$

Letting $|d\left(\Delta G_{\text{hom}}\right)/dr|_{r=r_c} = |d\left(\Delta G_{\text{hom}}\right)/dh|_{h=h_c} = 0$, solving the resultant expressions simultaneously for the critical radius $r = r_c$ of a stable α'-phase and the critical semi-thickness $h = h_c$, and evaluating Eq. (11.10) accordingly yield the critical Gibbs energy (nucleation energy barrier) ΔG_c.

Thus,

$$r_c = \frac{4K_\varepsilon \gamma_{\gamma\alpha'}}{\left(\Delta G_p - \Delta G_V\right)^2} \tag{11.11a}$$

$$h_c = -\frac{2\gamma_{\gamma\alpha'}}{\Delta G_p - \Delta G_V} \tag{11.11b}$$

$$\Delta G_{c,\text{hom}} = \frac{32\pi\, K_\varepsilon^2 \gamma_{\gamma\alpha'}^3}{3\left(\Delta G_p - \Delta G_V\right)^4} \tag{11.11c}$$

$$\Delta G_{c,het} < \Delta G_{c,\text{hom}} \quad \text{if} \quad \Delta G_{c,\text{hom}} > 60 k_B T \tag{11.11d}$$

Here, $k_B = 8.62 \times 10^{-5} eV/K = 1.38 \times 10^{-23} J/K$ is the Boltzmann constant, and the quantity $k_B T$ in Eq. (11.11d) is the thermal activation energy used as a transition energy for defining the limit between homogeneous and heterogeneous nucleation of martensite. Theoretically, this implies that the inequality $\Delta G_{c,het} \leq k_B T < \Delta G_{c,\mathrm{hom}}$ sets the limits for martensitic transformation, starting at the critical temperature $T = T_c$.

It is evident that the magnitude of $\Delta G_{c,\mathrm{hom}}$ is larger than the experimentally found barrier energy ($\Delta G_{c,\mathrm{hom}} > \Delta G_{c,\mathrm{exp}}$) for spontaneous martensitic transformation (Sinha [9, p. 8.42]). Therefore, martensite nucleation must be a heterogeneous process with nucleation at a lattice defect or inhomogeneity. In essence, $\Delta G_{c,het}$ defined by Eq. (11.11d) is the mathematical model that explicitly dictates that α'-phase growth occurs by lowering its the Gibbs energy. A comprehensive analysis of heterogeneous nucleation by defect dissociation can be found in Olson and Cohen work [8, p. 357].

Example 11.2 Consider quenching a small carbon steel plate in agitated water at $25°C$ and assume a homogeneous nucleation of martensite. Given data:

$$\Delta G_V = 180 MJ/m^3 \,\big|\, \gamma_{\gamma\alpha'} = 15 \times 10^{-3} J/m^2 \,\big|\, K_\varepsilon = 2.5 GPa$$

Based on this information, calculate the critical entities defined by Eqs. (11.9a,b,c) when (**a**) $\Delta G_p = 0$ and (**b**) $\Delta G_p = 50\ MJ/m^3$. Assume that the martensitic transformation starts at $T = 700\ K$ (427°C).

Solution

(**a**) If martensite homogeneous nucleation takes place without any distortion of the habit plane, then $\Delta G_p = 0$ and

$$r_c = \frac{4 K_\varepsilon \gamma_{\gamma\alpha'}}{\left(\Delta G_p - \Delta G_V\right)^2} = \frac{4 \left(2.5 \times 10^9\right)\left(15 \times 10^{-3}\right)}{\left(-180 \times 10^6\right)^2} \tag{11.2E1a}$$

$$r_c = 4.6296 \times 10^{-9}\ m = 4.63\ nm \tag{11.2E1b}$$

$$h_c = -\frac{2\gamma_{\gamma\alpha'}}{\Delta G_p - \Delta G_V} = -\frac{2\left(15 \times 10^{-3}\right)}{-180 \times 10^6} = 1.67 \times 10^{-10}\ m \simeq 0.17\ nm \tag{11.2E1c}$$

$$\Delta G_{c,\mathrm{hom}} = \frac{32\pi K_\varepsilon^2 \gamma_{\gamma\alpha'}^3}{3\left(\Delta G_p - \Delta G_V\right)^4} = \frac{32\pi \left(2.5 \times 10^9\right)^2 \left(15 \times 10^{-3}\right)^3}{3\left(-180 \times 10^6\right)^4} \tag{11.2E1d}$$

$$\Delta G_{c,\mathrm{hom}} = 6.73 \times 10^{-19} J = 4.21\ eV \tag{11.2E1e}$$

This result seems slightly high.

(b) For an elastic strain energy due to distortion of the habit plane during martensitic transformation, $\Delta G_p = 50\ MJ/m^3$ and

$$r_c = \frac{4 K_\varepsilon \gamma_{\gamma\alpha'}}{\left(\Delta G_p - \Delta G_V\right)^2} = \frac{4\left(2.5 \times 10^9\right)\left(15 \times 10^{-3}\right)}{\left(50 \times 10^6 - 180 \times 10^6\right)^2} = 8.88\ nm \tag{11.2E2a}$$

$$h_c = -\frac{2\gamma_{\gamma\alpha'}}{\Delta G_p - \Delta G_V} = -\frac{2\left(15 \times 10^{-3}\right)}{50 \times 10^6 - 180 \times 10^6} = 0.23\ nm \tag{11.2E2b}$$

$$\Delta G_{c,\text{hom}} = \frac{32\pi\, K_\varepsilon^2 \gamma_{\gamma\alpha'}^3}{3\left(\Delta G_p - \Delta G_V\right)^4} = \frac{32\pi\left(2.5 \times 10^9\right)^2\left(15 \times 10^{-3}\right)^3}{3\left(50 \times 10^6 - 180 \times 10^6\right)^4} \tag{11.2E2c}$$

$$\Delta G_{c,\text{hom}} = 2.47 \times 10^{-18}\ J = 15.47\ eV \tag{11.2E2d}$$

Additionally,

$$E_s = \frac{K_\varepsilon h_c}{r_c} = \frac{\left(2.5 \times 10^9\right)(8.88)}{0.23} = 96.52\ GJ/m^3 \tag{11.2E3a}$$

$$k_B T = \left(1.38 \times 10^{-23}\ J/K\right)(700\ K) = 9.66 \times 10^{-21}\ J \simeq 0.06\ eV \tag{11.2E3b}$$

These results indicate that martensitic transformation must take place under a heterogeneous nucleation process since $\Delta G_{c,\text{hom}} > k_B T = 0.06\ eV$ when $\Delta G_p = 0$ and $\Delta G_p = 50\ MJ/m^3$ at $T = 700\ K$.

11.5 Kinetics of Solid-State Transformation

Consider the homogeneous nucleation process with the number of stable nuclei being written as

$$n_s = n_o \exp\left(-\frac{\Delta G_{\text{hom}}}{k_B T}\right) \tag{11.12a}$$

$$n_s^* = n_o^* \exp\left(-\frac{\Delta G_{het}}{k_B T}\right) \tag{11.12b}$$

Similarly, the nucleation rate described by Eq. (8.69a) is adopted henceforth

$$I_s = I_o \exp\left(-\frac{\Delta G_c}{k_B T}\right) \exp\left(-\frac{Q_d}{k_B T}\right) \tag{11.13a}$$

$$I_s = I_o \exp\left[-\frac{16\pi \gamma_{ls}^3}{3k_B (\Delta H_t)^2} \frac{1}{T} \left(\frac{T_f}{\Delta T}\right)^2 - \frac{Q_d}{k_B T}\right] \tag{11.13b}$$

$$I_s = I_o \exp\left(-\frac{Q_x}{k_B T}\right) \tag{11.13c}$$

with

$$I_o = \frac{n_o k_B T}{h} \tag{11.14a}$$

$$Q_x = \frac{16\pi \gamma_{ls}^3}{3k_B (\Delta H_t)^2} \frac{1}{T} \left(\frac{T_f}{\Delta T}\right)^2 + Q_d \tag{11.14b}$$

where Q_x denotes the activation energy due to diffusion and ΔH_t denotes the evolved latent heat of transformation in the solid state.

The general expression for the critical Gibbs energy change is

$$\Delta G_c = \Delta H_t - T_c \Delta S_t = \int_o^{T_c} c_p (T)\, dT - T_c \int_o^{T_c} \frac{c_p (T)\, dT}{T} \tag{11.15}$$

Here, ΔS_t denotes the entropy change for $\gamma \rightarrow \alpha'$ transformation, and $c_p (T)$ denotes the temperature-dependent-specific heat capacity. Moreover, Eqs. (11.13a,b,c)–(11.15) are not restricted to martensitic transformation, and subsequently, they can be generalized as analytical models that can be applied to any solid-state phase transformation. Obviously, Gibbs energy and heat capacity are essential parameters for characterizing a thermal system.

11.5.1 Time-Temperature-Transformation Diagrams

In general, one can root the characterization of phase transformation into thermodynamics, physical metallurgy and crystallography, kinetics, and heat transfer theory. However, the analytical solutions to phase transformation problems are mostly based on heat transfer and thermodynamics. For instance, the $Fe\text{-}C$ phase diagram does not provide any information on the kinetics of phase transformation nor the cooling rate dT_γ /dt. In this case, the time-temperature-transformation (TTT) diagram is suited as the microstructural map for predicting the type and, to an extent, the morphology of the final microstructure upon solid-state transformation.

Additional details on the current subject matter can be found elsewhere (Simsir and Gur [11, p. 353] and Munirajulu et al. [12]).

Mathematically, the Johnson-Mehl-Avrami-Kolmogorov (JMAK) equation is usually cited in the literature as the Avrami equation [13], and it is a useful expression based on an isokinetic analysis (constant velocity) for determining the volume fraction of the solid phase (f_s) being transformed isothermally. The Avrami equation (known as Avrami kinetic equation and Avrami isothermal equation) is a transformation kinetics model that defines the solid fraction f_s being formed at a rate df_s/dt. Thus,

$$f_s = 1 - \exp\left(-k_x t^n\right) \tag{11.16a}$$

$$\frac{df_s}{dt} = nk_x t^{n-1} \exp\left(-k_x t^n\right) \tag{11.16b}$$

$$\ln\left[\ln\left(\frac{1}{1-f_s}\right)\right] = \ln\left(k_x\right) + n \ln\left(t\right) \tag{11.16c}$$

where k_x, n denote constants, t denotes the time at a constant temperature T, and i denotes the new solid phase. A compilation of equations for the current subject matter can be found elsewhere (Avramov and Sestak [14]).

Plotting Eq. (11.16a) with variables k_x, n yields the schematic sigmoidal transformation curves shown in Fig. 11.10a at fixed k_x and Fig. 11.10b at fixed n. If experimental datasets do not obey Eq. (11.16a), then the expressions given below can be used for curve fitting purposes

$$f_s = a_1 + a_2 \tanh\left(\frac{t - a_3}{a_4}\right) \tag{11.17a}$$

$$f_s = b_1 + b_2 \arctan\left(\frac{t - b_3}{b_4}\right) \tag{11.17b}$$

where a_i, b_i are coefficients to be determined numerically. Nonetheless, Fig. 11.11a depicts the use of Avrami expression, Eq. (11.16a), associated with a hypothetical TTT diagram (Fig. 11.11b).

Note that the rate of austenite-to-pearlite transformation ($\gamma \rightarrow p$) under isothermal conditions at T_1 and T_2 is $(df_i/dt)_1 > (df_i/dt)_2$. Only the start and finish of the isothermal transformation (IT) processes at T_1 and T_2 are shown in the corresponding TTT diagram (S-diagram). This is how a TTT diagram can be constructed using dilatometry data.

The enthalpy data published by Kulawik [15] provides a valuable graphical representation (Fig. 11.12a) of the enthalpy function $\Delta H_t = f(T)$ for characterizing the solid-state phase transformation of iron (Fe) and medium carbon steels (Fe-C) at a macroscale, where $\gamma \rightarrow \alpha, p, B$ or α'.

Notice that $\Delta H_t = \Delta H_{\gamma \rightarrow \alpha'}$ for martensite formation from austenite exhibits a linear trend at a small temperature range due to fast displacive atomic rearrangement

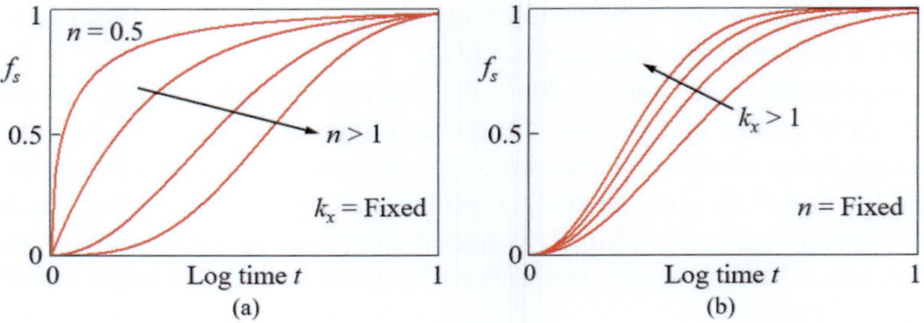

Fig. 11.10 Avrami equation. (**a**) Fixed k_x and (**b**) fixed n

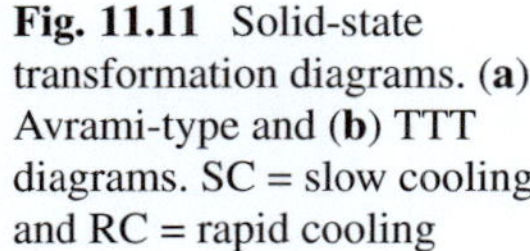

Fig. 11.11 Solid-state transformation diagrams. (**a**) Avrami-type and (**b**) TTT diagrams. SC = slow cooling and RC = rapid cooling

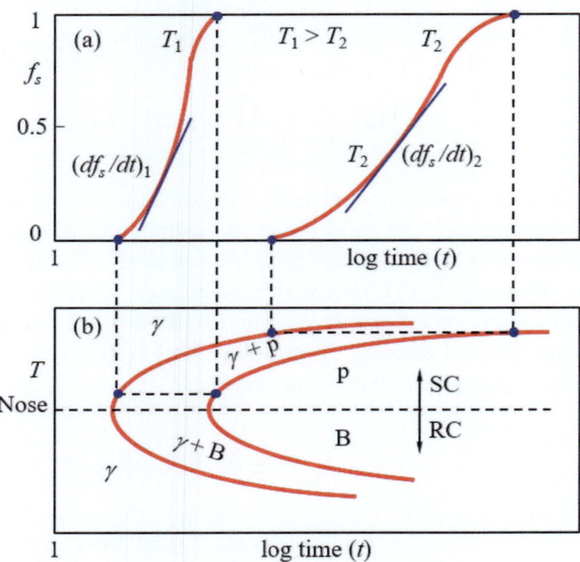

of the austenite crystal structure. Conversely, austenite transformation into α, p, B is nonlinear, attributed to the coupled heat-diffusion process, which in turn depends on the rate of evolved latent heat of transformation $d\left(\Delta H_t\right)/dt$.

Figure 11.12b, on the other hand, schematically shows the time-temperature-transformation (TTT) diagram for a eutectoid steel (Callister and Rethwisch [3, p. 378], Vander Voort [16, p. 16]). This is a simple and, yet, a valuable TTT diagram for characterizing or predicting solid-phase transformation.

Note that the temperature marked as "nose" divides the TTT diagram into a slow cooling (SC) region for pearlite formation, an intermediate cooling (IC) region for bainite transformation, and a rapid cooling (RC) region for martensite formation. Moreover, "s" and "f" stand for start and finish transformation lines.

Assume that three identical steel samples are austenitized at $900°C$ and, subsequently, cooled as indicated by the numbered lines designated a 1,2,3 in Fig. 11.12b. Thus,

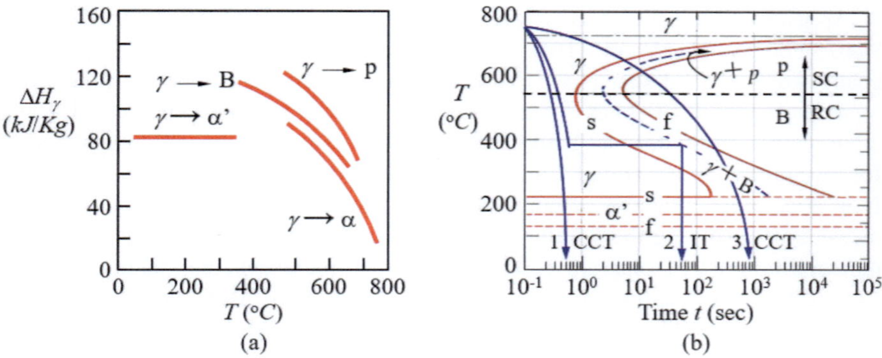

Fig. 11.12 Transformation diagrams. (**a**) Latent heat of transformation: $H\text{-}T$ diagram [Kulawik 18] and (**b**) TTT diagram (Callister and Rethwisch [3, p. 378] and Vander Voort [16, p. 16])

- *Line 1* represents the displacive transformation at a fast cooling rate or quenching rate $(dT/dt)_1$ in a suitable medium for a fast transformation of austenite to martensite. Thus, the TTT diagram indicates that γ transformation starts at $T > 200°C$ and finishes at $T < 200°C$. In fact, *Line 1* represents a continuous cooling transformation (CCT) process. Thus,

$$\gamma \rightarrow 100\% M \quad \text{at } t = 0.5 \text{ sec}$$

- *Line 2* denotes an isothermal transformation (IT) process, which indicates that a sample is quenched at $(dT/dt)_2$ and held at $T < T_{nose}$ for some time $t > 0$ until $\gamma \rightarrow B$ transformation is complete and, subsequently, quenched to room temperature. The transformed microstructure and transformation time are

$$\gamma \rightarrow 50\% B + 50\% M \quad \text{at } t = 50 \, sec$$

- *Line 3* is also a CCT process for $\gamma \rightarrow p = \alpha + Fe_3C$ at $(dT/dt)_3 << (dT/dt)_1$. Thus, the transformation process and the Lever rule introduced in Chap. 10 yield, respectively

$$\gamma \rightarrow 100\% p \quad \text{at } t = 20 \, sec$$

$$f_\alpha = \frac{6.67 - 0.8}{6.67 - 0.0218} = 0.883 \text{ or } 88.30\%$$

$$f_{Fe_3C} = 100 - f_\alpha = 100 - 88.30 = 0.117 \text{ or } 11.70\%$$

Appendix 11A includes some reference TTT diagrams for 1050, 1080, 4340, and 9440 steels (Vander Voort [16]). Particularly, TTT diagrams represent the kinetic changes during solid-state phase transformation upon heat treatment procedures.

A particular TTT phase transformation diagram can be constructed using dilatometry and molten salt bath techniques accompanied by metallographic work for revealing microstructural features, such as solid phases and grain morphology. It is a common practice to combine metallography, microscopy, and hardness measurements during the development of a graphical procedure for determining a nonequilibrium TTT diagram as a phase map that represents the kinetics of solid-state isothermal transformations at relevant temperatures. Moreover, the start and finish TTT curves (Figs. 11.11 and 11.12) represent phase boundaries used to predict the microstructure under nonequilibrium conditions.

11.6 Hardenability

Hardenability, in general, is a term used to describe the ability of steels to transform to martensite, and on the other hand, hardness defines the resistance to indentation or deformation in compression. Hence, hardenability of steel determines the depth and distribution of hardness induced by austenite transformation upon a rapid cooling (quenching) process.

Experimentally, hardenability is characterized by the Jominy end-quench test as per the American Society for Testing Materials (ASTM A 255). For instance, Fig. 11.13 shows the hardenability data and a portion of the TTT diagram for a eutectoid 1080 steel. In particular, Fig. 11.13a illustrates the Jominy bar and the Rockwell hardness scale-C (RHC) profile and some CCT curves superimposed on the AISI 1080 steel TTT diagram (Callister and Rethwisch [3, p. 387], Vander Voort [16, p. 11]). Moreover, Fig. 11.13a illustrates some correlations of the end-quench hardenability data along with some CCT curves. This is the usual graphical representation of the hardenability data for hardenable steels.

On the other hand, Fig. 11.13b depicts $HRC = f(x)$ profiles for some Jominy bars made out carbon and alloy steels containing $0.40\%C$. For instance, the AISI 1040 carbon ($Fe\text{-}0.4C$) steel and the AISI 4340 alloy ($Fe\text{-}0.40C\text{-}X's$) steel exhibit different hardness profiles due to the number of solid-solution elements in the latter steel.

In practice, the steel is austenitized and subsequently quenched at a specific rate to produce ferrite, pearlite, bainite, martensite, or a combination of these phases. If the final product is martensite, then a secondary heat treatment follows to obtain tempered martensite with a suitable toughness at the expense of some hardness.

It is evident nowadays that martensitic transformation also occurs in other materials, such as pure Ni, $Fe\text{-}Mn\text{-}Al$ alloys, $Ti\text{-}Mn$ alloys, and so forth. In general, the growth of martensite phase may occur at high velocity near or a fraction of the speed of sound due to diffusionless transformation at low temperatures.

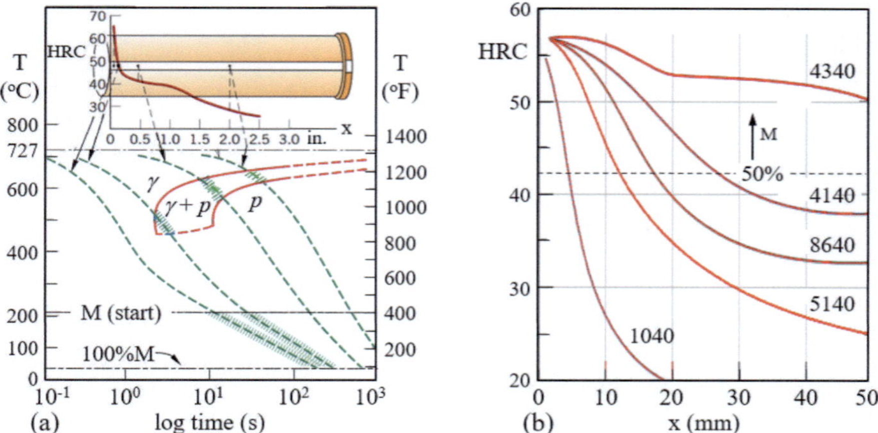

Fig. 11.13 Hardenability data of some steels. (**a**) Jominy test for hardness measurements and TTT diagram for 1080 steel (Callister and Rethwisch [3, p. 387], Vander Voort (editor) [16, p. 11]) and (**b**) hardness of some steels containing 0.40%C (Callister and Rethwisch book [3, p 388]

11.7   Solid Transformation Thermal Resistance

The goal in this section is to derive the dimensionless temperature (T_θ) equation (temperature ratio or normalized temperature) for one-dimensional analysis of solid-phase transformation in a plate, cylinder, or sphere initially at a temperature T_i and subsequently cooled in a fluid at a temperature T_o. This can be accomplished by transforming the dimensional heat equation to the dimensionless heat equation, which derives from dimensionless governing parameters. Subsequently, the effect of dimensionless temperature on solid-phase transformation is related to the Biot and Fourier numbers. One important aspect of the dimensionless temperature analysis is to derive analytical solutions independent of the specific variables.

For a plate, the governing one-dimensional heat equation, the continuity of temperature, initial condition (IC), and boundary condition (BC) are

$$\frac{\partial T}{dt} = \alpha \frac{\partial^2 T}{dx^2} \tag{11.18a}$$

$$T(x, 0) = T_i \quad \text{(IC)} \tag{11.18b}$$

$$\frac{\partial T(0, t)}{dx} = 0 \quad \text{(BC)} \tag{11.18c}$$

$$k\frac{\partial T(x_c, t)}{dx} + h\left[T(x_c, t) - T_o\right] = 0 \quad \text{(Energy balance)} \tag{11.18d}$$

where α denotes the thermal diffusivity, k denotes the thermal conductivity, h denotes the heat-transfer coefficient, and x_c denotes semithickness or half-thickness.

Following Geiger and Poirier [17, p. 294] and Poirier and Geiger [18, p. 290] with $\theta = T - T_o$, the heat equation becomes

$$\frac{\partial \theta}{dt} = \alpha \frac{\partial^2 \theta}{dx^2} \qquad (11.19a)$$

$$\theta(x, 0) = T_i - T_o = \theta_i \quad \text{(IC)} \qquad (11.19b)$$

$$\frac{\partial \theta(0, t)}{dx} = 0 \qquad \text{(BC)} \qquad (11.19c)$$

$$\frac{\partial \theta(x_c, t)}{dx} + \frac{h}{k}\theta(x_c, t) = 0 \quad \text{(Energy balance)} \qquad (11.19d)$$

The method of separation of variables denoted as a product function, $\theta(x, t) = F(x)\,G(t)$, yields the solution to Eq. (11.19a) (see Appendix 11B). Thus,

$$\frac{\theta}{\theta_i} = \frac{T - T_o}{T_i - T_o} = 2 \sum_{n=1}^{\infty} \frac{\sin(\lambda_n x_c)\cos(\lambda_n x)}{\lambda_n x_c + \sin(\lambda_n x_c)\cos(\lambda_n x_c)} \exp\left(-\lambda_n^2 \alpha t\right) \qquad (11.20a)$$

$$B_i = \lambda_n x_c \tan(\lambda_n x_c) \qquad (11.20b)$$

$$B_i = h x_c / k \quad \text{(Biot number)} \qquad (11.20c)$$

$$F_o = \alpha t / x_c^2 \quad \text{(Fourier number)} \qquad (11.20d)$$

Manipulating Eq. (11.20a) along with the Biot and Fourier numbers yields the normalized dimensionless temperature $T_\theta = f(\phi_n, B_i, F_o)$

$$T_\theta = \frac{T - T_o}{T_i - T_o} = 2 \sum_{n=1}^{\infty} \frac{\lambda_n x_c \sin(\lambda_n x_c)\cos[\lambda x_c (x/x_c)]}{(\lambda_n x_c)^2 + B_i \cos^2(\lambda_n x_c)} \exp\left(-(\lambda_n x_c)^2 F_o\right)$$

$$(11.21)$$

where $(n - 1) \le \phi_n = \lambda_n x_c \le n\pi$ are the roots of Eq. (11.20b) and x is the distance out from the center. If $n = 1$, $\phi = \lambda_1 x_c$ and $0 \le x_\theta = x/x_c \le 1$ in Eq. (11.21) yield the relative temperature equation for an isothermal solid-phase transformation as

$$T_\theta = \frac{T - T_o}{T_i - T_o} = \frac{2\phi \sin(\phi)\cos(\phi x_\theta)}{\phi^2 + B_i \cos^2(\phi)} \exp\left(-\phi^2 F_o\right) \qquad (11.22a)$$

$$\phi \tan(\phi) = (h/k) x_c = B_i \quad \text{(plate)} \qquad (11.22b)$$

Thus, plotting Eq. (11.22a) for a plate initially at T_i and subjected to heat transfer by convection in a fluid at T_o with $x_\theta = 0$ (center of a flat plate) yields Fig. 11.14. Note that $T_\theta = f(F_o)$ is an exponential-dependent function for fixed values of ϕ, B_i, where ϕ is an eigenvalue determined using Eq. (11.22b). Moreover, notice that Fig. 11.14a shows a series of curves for $0.01 \le B_i \le 0.1$ and Fig. 11.14b

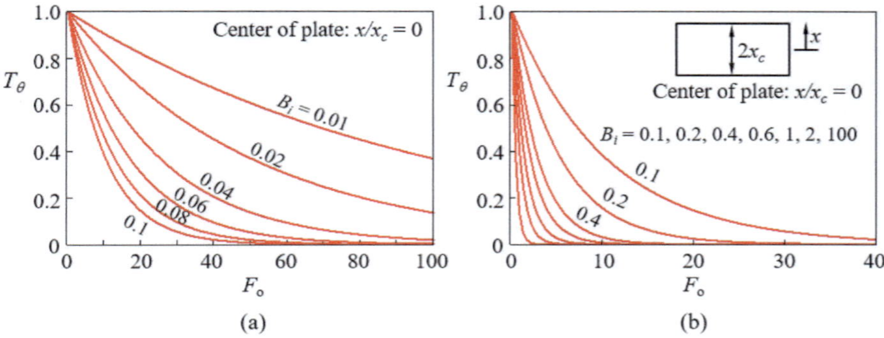

Fig. 11.14 Heisler charts for dimensionless temperature response of an infinite plate initially at T_i and subjected to cooling due to heat convection in a fluid at T_f. (**a**) $T_\theta = f(F_o)$ at $0.01 \le B_i \le 0.1$ and (**b**) $T_\theta = f(F_o)$ at $B_i \ge 0.1$

depicts curves for $0.1 \le B_i \le 100$. These plots indicate that T_θ decays very rapidly as $B_i \rightarrow \infty$.

For cylinders and spheres, Eqs. (11.20a,b) with $x = r \le r_c$ and $x_c = r_c$ yield

$$T_\theta = \frac{T - T_o}{T_i - T_o} = \frac{2 \sin(\lambda r_c) \cos[\lambda x_c (r/r_c)]}{\lambda r_c + (\lambda r_c/B_i) \sin(\lambda r_c)} \exp\left[-(\lambda r_c)^2 F_o\right] \tag{11.23a}$$

$$\lambda r_c \tan(\lambda r_c) = (h/k) r_c = B_i \quad \text{(cylinder or sphere)} \tag{11.23b}$$

Other plots can be generated for plates with $0 < x/x_c \le 1$ and contoured geometries with $0 \le r/r_c \le 1$. Moreover, these plots are referred to as Gurney and Lurie charts for small B_i and F_o or Heisler charts for $0.01 \le B_i \le \infty$ and $F_o \rightarrow \infty$.

The preceding Heisler charts are constructed using different dimensionless variables (Fig. 11.14) for a hypothetical flat plate with or without internal thermal resistance that leads to a high thermal conductivity. Evidently, the thermal engineering field uses Heisler charts as an important graphical analysis tool for evaluating one-dimensional transient heat transfer by conduction.

Lumped System Analysis (LSA) This is a transient heat transfer analysis with a particular criterion $Bi < 0.1$ and specific conditions, such as the temperature of a solid remains uniform within or throughout the body at all times and it changes with time only. In particular, Fig. 11.14a shows some $T_\theta = (F_o)_{B_i}$ curves for $Bi < 0.1$. Hence, $T = T(t)$ and $\partial T(x,t)/\partial x = 0$ when $Bi < 0.1$. This implies that heating or cooling a thick enough specimen to reach a uniform temperature takes times. In other words, the lumped system method is an analysis of transient heat transfer systems, where the temperature gradient is $\partial T(x,t)/\partial x = 0$ due to a uniform temperature.

Example 11.3 Consider conduction heat transfer in a solid with internal temperature variation as in quenching or cooling operations. Use the preceding analytical solution for a cooling problem related to an isothermal solid-phase transformations as indicated by the IT trajectories in the eutectoid TTT diagram given below.

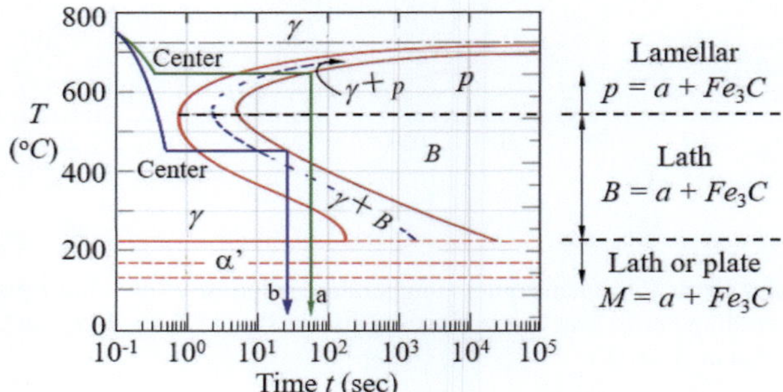

Use the given data for the heat transfer coefficient h, thermal diffusivity α, and the thermal conductivity k

$450°C \leq T \leq 650°C$
$h = 283.92 W/m^2.K$
$\alpha = 1.1871 \times 10^{-5} m^2/s$
$k = 34.614 W/m.K$

(a) For one 5-*cm*-thick 1080 steel plate initially at $750°C$ and immersed in a fluid at $T_o = 600°C$, calculate the time for complete pearlitic transformation ($\gamma \to p$) when the temperature at center of the plate is $T = 650°C$. **(b)** Repeat the calculation for bainitic transformation ($\gamma \to B$) in another 5-*cm*-thick AISI 1080 steel plate immersed in a similar fluid when $T_o = 400°C$ and $T = 450°C$. Approximate the plate surface temperatures with the aid of the TTT diagram.

Solution

(a) The semithickness is $x_c = (5/2 \, cm) = 2.50 \, cm = 0.025 \, m$, and the dimensionless temperature and the Biot number for austenite-to-pearlite transformation ($\gamma \to p$) are

$$T_\theta = \frac{T - T_o}{T_i - T_o} = \frac{650 - 600}{750 - 600} = 0.33 \qquad (11.3E1a)$$

$$B_i = hx_c/k = \frac{(283.92 \ W/m^2.K)(0.025 \ m)}{34.614 \ W/m.K} = 0.21 \qquad (11.3E1b)$$

Let $\phi = \lambda x_c$, $x_c = 0.025 \, m$, and $x/x_c = 0$ so that Eq. (11.23b) with $0 \leq \phi \leq \pi$ gives

$$\lambda x_c \tan (\lambda x_c) = (h/k) \, x_c = B_i \tag{11.3E2a}$$

$$\left[\begin{array}{c} \phi \tan (\phi) = 0.21 \\ \phi \in (0, \pi/2) \end{array} \right] \rightarrow \phi = \lambda x_c = 0.44282 \tag{11.3E2b}$$

The Fourier number, Eq. (11.23a) with $x/x_c = 0$ at center of the plate, is approximated by

$$F_o = \frac{1}{(\lambda x_c)^2} \ln \left\{ \frac{1}{T_\theta} \frac{2 \sin (\lambda x_c) \cos [\lambda x_c \, (x/x_c)]}{\lambda x_c + (\lambda x_c/B_i) \sin (\lambda x_c)} \right\} \tag{11.3E3a}$$

$$F_o = \frac{1}{(0.44282)^2} \ln \left(\frac{1}{0.33} \frac{2 \sin (0.44282) \cos [0.44282 \, (0)]}{0.44282 + (0.44282/0.21) \sin (0.44282)} \right) = 3.35 \tag{11.3E3b}$$

The time, Eq. (11.20d), for transforming austenite to pearlite, $\gamma \rightarrow p = \alpha + Fe_3C$, at the center of a $x_c = x/2 = 2.5$-*cm*-semithick plate is approximated as

$$t = \frac{x_c^2 F_o}{\alpha} = \frac{(0.025 \, m)^2 \, (3.35)}{1.1871 \times 10^{-5} \, m^2/\sec} = 176 \sec = 2.93 \min \tag{11.3E4}$$

From Eq. (11.20a) along with $\phi = \lambda x_c = 0.44282$ and $\lambda = \phi/x_c = (0.44282)/(0.025 \, m) = 17.713 \, 1/m$, and from the given TTT diagram, the time $t \simeq 53$ sec for complete surface transformation are now used to calculate the plate surface temperature. Using the first term of the series, Eq. (11.20a) along $x/x_c = 1$, the plate surface temperature is approximated by

$$\frac{T - T_o}{T_i - T_o} = \frac{2 \sin (\lambda_n x_c) \cos (\lambda_n x_c \, (x/x_c))}{\lambda_n x_c + \sin (\lambda_n x_c) \cos (\lambda_n x_c)} \exp \left(-\lambda^2 \alpha t \right) \tag{11.3E5a}$$

$$A = \frac{2 \sin (0.44282) \cos [0.44282 \, (1)]}{0.44282 + \sin (0.44282) \cos (0.44282)} = 0.93294 \tag{11.3E5b}$$

$$B = \exp \left[- (17.713 \, 1/m)^2 \left(1.1871 \times 10^{-5} \, m^2/s \right) (53 \, s) \right] = 0.82086 \tag{11.3E5c}$$

$$\frac{T - T_o}{T_i - T_o} = AB = (0.93294) \, (0.82086) = 0.77 \tag{11.3E5d}$$

$$T = T_o + 0.77 \, (T_i - T_o) = 600 + 0.77 \, (750 - 600) = 715.5°C \tag{11.3E5e}$$

(b) A similarly analysis for austenite-to-bainite transformation ($\gamma \rightarrow B$) at $T = 450°C$ requires that

$$T_\theta = \frac{T - T_o}{T_i - T_o} = \frac{450 - 400}{750 - 400} = 0.14 \tag{11.3E6a}$$

$$B_i = hx_c/k = \frac{\left(283.92 \ W/m^2.K\right)(0.025 \ m)}{34.614 \ W/m.K} = 0.21 \tag{11.3E6b}$$

Let $\phi = \lambda x_c$, $x_c = 0.025 \ m$, and $x/x_c = 1$ so that Eq. (11.23b) with $0 \le \phi \le \pi$ gives (by iteration)

$$\lambda x_c \tan (\lambda x_c) = (h/k) \, x_c = B_i \tag{11.3E7a}$$

$$\left[\begin{array}{c} \phi \tan (\phi) = 0.21 \\ \phi \in (0, \pi/2) \end{array} \right] \rightarrow \phi = \lambda x_c = 0.44282 \tag{11.3E7b}$$

The Fourier number along with $x/x_c = 0$ at the surface of the plate is

$$F_o = \frac{1}{(\lambda x_c)^2} \ln \left\{ \frac{1}{T_\theta} \frac{2 \sin (\lambda x_c) \cos [\lambda x_c \, (x/x_c)]}{\lambda x_c + (\lambda x_c/B_i) \sin (\lambda x_c)} \right\} \tag{11.3E8a}$$

$$F_o = \frac{1}{(0.3257)^2} \ln \left(\frac{1}{0.14} \frac{2 \sin (0.44282) \cos [0.44282 \, (0)]}{0.44282 + (0.44282/0.21) \sin (0.44282)} \right) \tag{11.3E8b}$$

$$F_o = 14.28 = \alpha t/x_c^2 \tag{11.3E8c}$$

For a $x_c = x/2 = 2.5$-cm-semithick plate to be transformed to bainite, $\gamma \rightarrow B$ is approximated by

$$t = \frac{x_c^2 F_o}{\alpha} = \frac{(0.025 \ m)^2 \, (14.28)}{1.1871 \times 10^{-5} \ m^2/s} = 752 \ sec = 12.53 \ min \tag{11.3E9}$$

From Eq. (11.20a) along with $\phi = \lambda x_c = 0.44282$ and $\lambda = \phi/x_c = (0.44282)/(0.025 \ m) = 17.713 \ 1/m$, and from the given TTT diagram, the time $t \simeq 22$ sec for complete surface transformation are now used to calculate the surface temperature. Using the first term of the series, Eq. (11.20a) along $x/x_c = 1$, the plate surface temperature is approximated as

$$\frac{T - T_o}{T_i - T_o} = \frac{2 \sin (\lambda_n x_c) \cos (\lambda_n x_c \, (x/x_c))}{\lambda_n x_c + \sin (\lambda_n x_c) \cos (\lambda_n x_c)} \exp \left(-\lambda^2 \alpha t \right) \tag{11.3E10a}$$

$$A = \frac{2 \sin (0.44282) \cos [0.44282 \, (1)]}{0.44282 + \sin (0.44282) \cos (0.44282)} = 0.93294 \tag{11.3E10b}$$

$$B = \exp \left[- (17.713 \ 1/m)^2 \left(1.1871 \times 10^{-5} \ m^2/s \right) (22 \ s) \right] = 0.92133 \tag{11.3E10c}$$

$$\frac{T - T_o}{T_i - T_o} = AB = (0.93294)\,(0.92133) = 0.92 \qquad (11.3\text{E}10\text{d})$$

$$T = T_o + 0.77\,(T_i - T_o) = 400 + 0.86\,(750 - 400) = 701°C$$
$$(11.3\text{E}10\text{e})$$

This result indicates that the time for phase transformation strongly depends on the specimen thickness and temperature of the cooling medium. Notice that the time for transforming austenite to pearlite, $\gamma \to p = \alpha + Fe_3C$, at the center of a $x_c = x/2 = 2.5\text{-}cm$-semithick plate is relatively short compared to austenite-to-bainite transformation. This is attributed to the high atom mobility at a high temperature.

11.8   Strain-Induced Phase Transformation

It is evident that certain crystalline materials undergo strain-induced phase transformation during plastic deformation as shown in Fig. 11.15. Martensitic transformation forms from the austenite habit plane via a displacive (dimensionless) atomic movement of atoms.

The martensitic phase forms as plates or needles being embedded in the austenitic parent phase (matrix) along certain crystallographic planes, inducing high residual stresses, lattice distortion, and even cracks. This phenomenon is denoted as phase reaction $\gamma \to \alpha'$, and it is a dynamic strain-hardening (work-hardening) process.

Figure 11.15a shows a typical stainless steel AISI 304 austenitic microstructure, and Fig. 11.15b depicts a partially transformed martensite during tension loading. Here, the austenite (light) and martensite (dark) grains are elongated along the

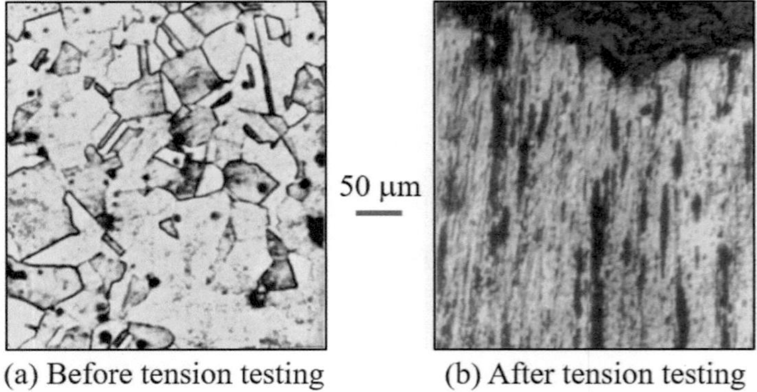

(a) Before tension testing   (b) After tension testing

Fig. 11.15 AISI 304 microstructures. (**a**) Austenite before tension testing and (**b**) austenite (light)-martensite (dark) after tension testing at room temperature. Images taken at a 200x magnification

tension direction. Moreover, the γ-α' microstructure in Fig. 11.15b, revealed out of a fractured specimen tested in tension at approximately 10^{-4} sec^{-1} slow strain rate, illustrates the effect of cold plastic deformation on the martensitic transformation.

Martensitic transformation is a solid-state phase transition induced by a rapid change of crystal structure referred to as displacive transformation, which is accompanied by an evolved elastic strain defined by Eq. (11.2c).

This type of crystal transformation may be irreversible as in certain quenched or mechanically deformed materials. Conversely, martensitic transformation may be reversible in shape memory alloys. However, martensite can be reverted to austenite, $\alpha' \rightarrow \gamma$, upon reheating at an austenitic temperature (phase diagram in Fig. 11.5). Moreover, alloying element content reduces the start temperature for martensitic transformation.

11.8.1   X-Ray Diffraction Patterns

Consider the X-ray patterns in Fig. 11.16a for thermally induced martensitic transformation in AISI 5160 steel and Fig. 11.16b for AISI 304 stainless steel.

This section introduces X-ray diffraction (XRD) patterns for comparing the orientation relationship (OR) theory on thermally induced and strain-induced martensitic transformations; γ-FCC $\rightarrow$ α'-BCT or γ-FCC $\rightarrow$ ε-hexagonal. For instance, Fig. 11.16a shows X-ray diffraction (XRD) patterns for AISI 5160 steel (0.56%-$0.64C$, 0.7-$0.9\%Cr$, 0.75-$1.00Mn$) being subjected to a high-speed quenching (HSQ) process in water at room temperature as per Martinez-Cazares et al. [19]. Notice that the strongest peaks clearly correspond to the $(111)_\gamma \parallel (110)_{\alpha'}$ orientation relationship (OR).

Figure 11.16b, on the other hand, depicts the same $(111)_\gamma \parallel (110)_{\alpha'}$ orientation relationship for a strain-induced martensitic transformation in an AISI 301 stainless steel ($0.15\%C$ max., 16%-$18\%Cr$, $2\%Mn$ max.) at variable strain ε values as per Celada-Casero et al. [20]. It is evident that there is a visible misfit angle 2θ between $(111)_\gamma \parallel (110)_{\alpha'}$ orientation relationship. Therefore, the above X-ray diffraction data confirm that there is an orientation relationship (OR) between γ and α' phases. Moreover, γ-FCC $\rightarrow$ ε-hexagonal peaks have a small peak intensity at $2\theta < 50°$.

11.8.2   Volume Fraction of Martensite

Among the relevant experimental data available in the literature, Fig. 11.17a shows the martensite volume fraction ($f_{\alpha'}$) of initially deformed AISI 301 stainless steel specimens in tension mode using light optical microscopy (LOM), magnetization measurements (SQUID = superconducting quantum interference device), and XRD (Celada-Casero [20]).

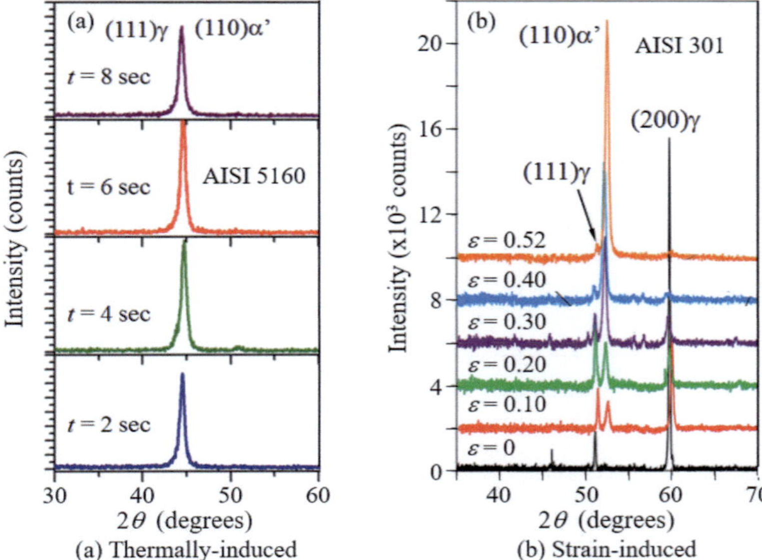

Fig. 11.16 Experimental X-ray patterns for thermally induced martensitic transformation in AISI 5160 steel. The data shows the orientation relationship (OR) of austenite and martensite peaks. After Martinez et al. [19]. (**b**) X-ray diffractograms for strain-induced martensitic transformation in AISI 301 stainless steel. After Celada-Casero et al. [20]

The kinetic curve fitting functions $f_{\alpha'} = f(\varepsilon)$ and $df_{\alpha'}/d\varepsilon$ for the strain ε data shown in Fig. 11.17a are defined as (Shin et al. [21])

$$f_{\alpha'} = a\left[1 - \exp\left(-k_x\varepsilon^n\right)\right] \tag{11.24a}$$

$$\frac{df_{\alpha'}}{d\varepsilon} = ak_x n\varepsilon^{n-1}\exp\left(-k_x\varepsilon^n\right) \tag{11.24b}$$

$$\frac{df_{\alpha'}}{d\varepsilon} = ak_x n\varepsilon^{n-1}\left(1 - f_{\alpha'}/a\right) \tag{11.24c}$$

where a, k_x, n are curve fitting constants. Here, $a = 0.92$, $k_x = 13.4$, and $n = 2.2$ (Celada-Casero [20]) are used in Eqs. (11.24a,b,c), and subsequently, curve fitting yields Fig. 11.17b, from which the functions $df_{\alpha'}/d\varepsilon = g(\varepsilon)$ and $df_{\alpha'}/d\varepsilon = g(f_{\alpha'})$ are evaluated as the rate of the α'-martensite phase transformation ($f_{\alpha'}$).

These rate functions coincide at point "p" where

$$df_{\alpha'}/d\varepsilon = g(\varepsilon = 0.22) \simeq 2.8$$
$$df_{\alpha'}/d\varepsilon = g(f_{\alpha'} = 0.22) \simeq 2.8$$

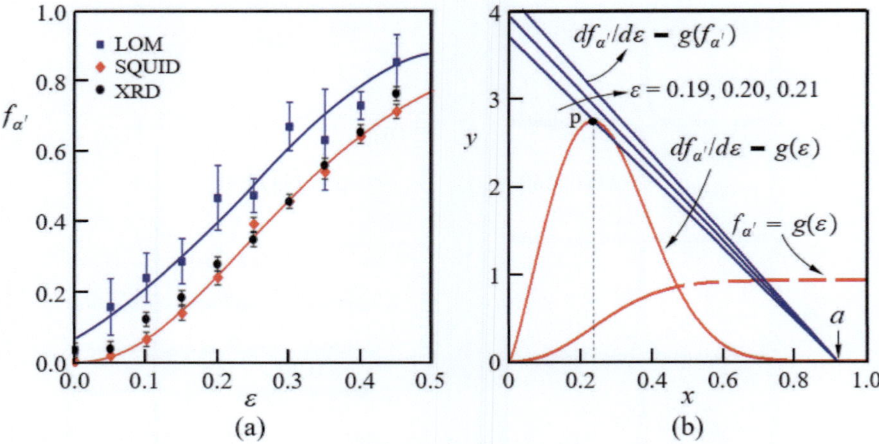

Fig. 11.17 (**a**) Room-temperature strain-induced martensite volume fraction in AISI 301 steel (Celada-Casero [20]) and (**b**) rate of martensite volume fraction

Notice that the $f_{\alpha'} = f(\sigma)$ and $df_{\alpha'}/d\sigma$ functions, where σ is the applied stress, can also be used to empirically correlate strain-induced phase transformation data.

11.8.3   Shape Memory Effect

This section describes the shape memory effect (SME) on crystalline materials due to a combination of the material atomic or molecular structure and a suitable external stimuli, such as heat or strain. As a result, materials that have the ability to recover a predetermined crystallographic shape induced by external stimuli are classified as shape memory materials (SMM) or shape memory alloys (SMA).

If the external stimulus is heat, then a SMA, such as the equiatomic Ti-Ni alloy referred to as nitinol, has a high-temperature phase called austenite (γ) and a low-temperature phase called martensite (α'). Thus, there must exist a critical or transformation temperature (T_c) so $\gamma \to \alpha'$ at $T > T_c$ and $\alpha' \to \gamma$ at $T < T_c$. This mechanism is illustrated in Fig. 11.18. According to the order of the above invariant reactions, the γ-austenite is the predetermined structure that represents the shape memory effect in some crystalline or polymeric materials. Obviously, the reverse order of invariant reactions makes α'-martensite the predetermined phase.

Figure 11.18a schematically shows thermally induced and strain-induced phase transformation cycles at T, while Fig. 11.18b schematically illustrates the counter-clockwise differential scanning calorimetry (DSC) thermograph for determining the heat flow, described by a general function $q = f(T)$, during the start "s" and finish "f" of austenite and martensite transformation. Note the downward arrows

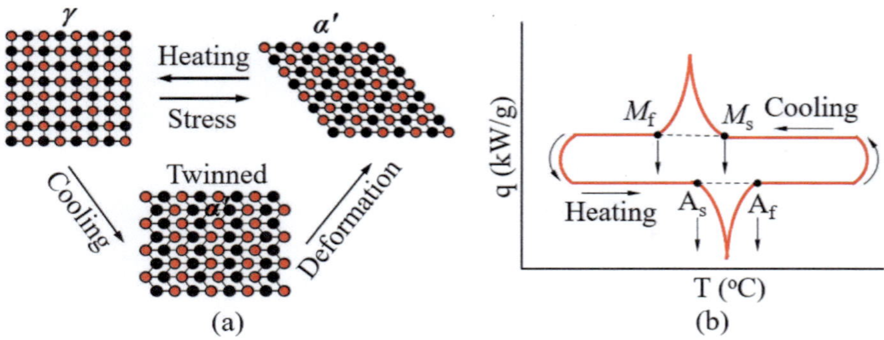

Fig. 11.18 Schematic nitinol phase transformation cycle (process). (**a**) Thermally-induced and (**b**) schematic DSC thermograph

are related to the temperature range for transformation and the thermally induced phase transformation is denoted by the phase reaction $\gamma \to \alpha'$ or $\alpha' \to \gamma$.

Note that $\gamma \to \alpha'$ and the reverse $\alpha' \to \gamma$ transformation do not take place at the same temperature. Therefore, T_c described above can be assigned to the peak temperature for either γ-phase or α'-phase. Conversely, the shape memory effect is evident during the process of restoring the original shape of a plastically deformed structure by heating it. Moreover, when comparing shape memory polymers (SMPs) and shape memory alloys (SMAs), one must consider the maximum mechanical deformation, cost-effective, thermal processing, and suitable applications.

11.8.4   Hysteresis Loop

A mechanical hysteresis promotes phase transformation (deformation) accompanied by energy dissipation due to internal friction, which, in turn, causes the generation of heat in a material during one loading-unloading cycle within the limits of an external stimuli. Moreover, a narrow hysteresis is beneficial for frequent phase transformation, whereas a wide hysteresis retards the phase transformation process at a temperature or stress range.

Figure 11.19 schematically shows the mechanical aspects of the strain-induced phase transformation in nitinol (Hartl and Lagoudas [22, p. 63]). For instance, Fig. 11.19a depicts the stress-strain (σ-ε) cycle during loading and unloading a specimen at a temperature T. As a result, nitinol undergoes strain-induced phase transformation.

Among the mechanical properties obtained from the macroscale σ-ε curve, the modulus of elasticity of austenite (E_γ) or martensite ($E_{\alpha'}$) is determined using Hooke's law $E = \sigma/\varepsilon$.

A solid-state phase deformation denoted by the phase reaction ($\gamma \to \alpha'$ or $\alpha' \to \gamma$) is induced by a mechanical plastic deformation. Moreover, Fig. 11.19b

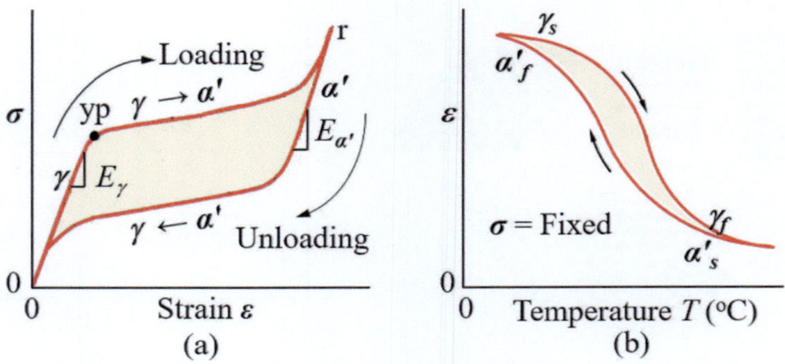

Fig. 11.19 SMA strain-induced transformation. (**a**) Quasi-static loading and (**b**) cyclic loading represents a hysteresis loop. After Hartl and Lagoudas [22, p. 63]

schematically shows a cyclic thermal loading referred to as a hysteresis loop for one SMA specimen undergoing solid-state phase change at a constant applied stress σ as the only external stimuli in a noncorrosive environment.

The area under a tension stress-strain (σ-ε) curve at a quasi-static strain rate ($d\sigma/d\varepsilon$) is known as the strain energy density (W) required to deform the material mechanically. Additionally, the total strain energy density is defined by

$$W = \int_{o}^{\varepsilon} \sigma d\varepsilon = \int_{o}^{\varepsilon_{yp}} \sigma d\varepsilon + \int_{\varepsilon_{yp}}^{\varepsilon} \sigma d\varepsilon \tag{11.25a}$$

$$W = \frac{\sigma_{ys}^2}{2E} + \int_{\varepsilon_{yp}}^{\varepsilon} \sigma d\varepsilon \tag{11.25b}$$

where Hooke's law ($\sigma = \varepsilon E$) is used to solve the first integral, σ in Eq, (11.25b) denotes a plastic stress, $\varepsilon \leq \varepsilon_{yp}$ denotes the elastic strain , ε_{yp} denotes the yield strain at the yield point (yp), σ_{ys} denotes the yield strength at ε_{yp}, E_γ denotes the modulus elasticity from zero to ε_{yp} for austenite, and $E_{\alpha'}$ denotes the modulus elasticity from point "r" for martensite. Nonetheless, once the plastic stress function $\sigma = f(\epsilon)$ is suitably defined, the integral in Eq. (11.25b) along with proper integral limits is easily solvable.

The engineering stress-strain (σ-ε) curve at $\varepsilon > \varepsilon_{yp}$ represents the $\gamma \rightarrow \alpha'$ transformation upon mechanical loading. However, the reverse invariant reaction $\alpha' \rightarrow \gamma$ occurs upon unloading from point "r," and eventually, the shape of nitinol is recovered at $\varepsilon = 0$.

According to Fig. 11.19a, the net strain energy density W_{net} for the phase transformation is a measure of the hysteresis loop mathematically defined by

$$W_{net} = W_{or} - W_{ro} = \left(\int_{o}^{\varepsilon_r} \sigma d\varepsilon\right)_{\gamma} - \left(\int_{o}^{\varepsilon_r} \sigma d\varepsilon\right)_{\alpha'} \tag{11.26}$$

where W_{or}, W_{ro} denote the areas under σ-ε curves. Further, Eq. (11.26) indicates that the martensitic transformation process can mathematically be characterized by considering elastic strain energy induced by the transformation strain and the plastic relaxation strain evolved during the relief of high internal stresses.

11.9   Quenching Heat Transfer

This section includes a mathematical approach for analyzing the transient heat flow when a hot specimen at T is suddenly immersed in a cold fluid at $T_o < T$. Thus, the heat transfer from the solid to the liquid resembles the common quenching procedure for either retaining a saturated phase at room temperature for subsequent precipitation hardening in non-ferrous alloys or for austenite-to-martensite phase transformation in steels. This implies that there is an intermediate temperature T_i at the solid-liquid interface, while the solid develops a temperature gradient during cooling.

This is a thermophysical process related to solid-state phase transformation and to the first law of thermodynamics due to the evolved transient heat transfer and thermal energy induced by a temperature variation.

Mathematically, the lumped capacity method is the simplest approach for analyzing the Newtonian cooling process since it considers heat transfer by conduction from within the solid to the solid surface and heat transfer by convection from the solid-fluid interface to the bulk of the fluid (melt).

Thermodynamics From Chap. 8, the thermodynamics internal energy (ΔU) in an adiabatic system related to quenching a solid specimen in a fluid is written as

$$\Delta U_{total} = 0 \tag{11.27a}$$

$$\Delta U_s + \Delta U_f = 0 \tag{11.27b}$$

$$m_s \Delta u_s + m_f \Delta u_f = 0 \tag{11.27c}$$

where m, Δu denote the mass and specific internal energy change, respectively. The subscript "s" stands for solid metal or alloy and "f" stands for fluid (liquid or gas). Moreover, the general definition of the specific internal energy is related to the specific heat capacity (c_v)

$$du = c_v\,(T)\,dT = \left(a_1 + a_2 T + a_3 T^2 + \ldots\right) dT \tag{11.28a}$$

$$\Delta u \simeq c_v \Delta T = c_v\,(T_i - T_o) \tag{11.28b}$$

Here, a_i with $i = 1, 2, 3, \ldots$ are coefficients, $c_v\,(T)$ denotes the specific heat capacity at constant volume being defined as a polynomial in Eq. (11.28a), $T_o < T_i < T$ denotes the intermediate temperature at the solid-liquid interface that develops during the initial stage of the quenching process, T_o denotes the cooling medium temperature (constant), and T denotes the specimen initial temperature. However, if $c_v(T)$ does not vary significantly at a temperature range ΔT, then Eq. (11.28b) is the fundamental choice for determining Δu.

Combining Eqs. (11.27c) and (11.28b), and solving for T_i, yields

$$0 = m_s c_s (T_i - T_o)_s + \left[m_f c_f (T - T_o) \right]_f \tag{11.29a}$$

$$T_i = \frac{m_f c_f T_{o,f} + m_s c_s T_{o,s}}{m_f c_f + m_s c_s} \tag{11.29b}$$

which is a useful expression in a heat treatment procedure: water quenching of a steel specimen.

Transient Heat Transfer Furthermore, the thermal energy related to transient heat flow is defined by the rate of enthalpy change (dH/dt) being coupled with the cooling rate (dT/dt). Thus, the heat transfer theory presented in Chap. 9 for solidification is the fundamental approach adapted hereafter with slight modifications for quenching (Gaskell [23, p. 366]). For instance, when a mass of solid $m_s = \rho_s V_s$ is suddenly immersed in a fluid, an amount of thermal energy of the solid is transferred to the fluid. This is represented below as an energy balance based on the rate of heat lost by the solid specimen and the rate of heat transfer to the fluid

$$-m_s \frac{dH}{dt} = \frac{h A_s (T - T_o)}{\rho_s V_s} \tag{11.30a}$$

$$\frac{dH}{dt} = -\frac{h A_s (T - T_o)}{m_s} = \frac{h A_s (T_o - T)}{\rho_s V_s} \tag{11.30b}$$

where c_s is the specific heat capacity at constant pressure, h is the heat transfer coefficient, A_s is the surface area, V is the volume of the solid specimen, and m is the mass of the solid phase.

By definition, the rate of enthalpy change can be written as

$$\frac{dH}{dt} = c_s \frac{dT}{dt} \tag{11.31}$$

Combining Eqs. (11.30b) and (11.31) yields

$$\frac{dT}{dt} = -\frac{h A_s (T - T_o)}{m_s c_s} = -\frac{h A_s (T - T_o)}{\rho_s V_s c_s} \tag{11.32a}$$

$$\frac{dT}{(T - T_o)} = \frac{d(T - T_o)}{(T - T_o)} = -\frac{h A_s}{\rho_s V_s c_s} dt \tag{11.32b}$$

Integrating Eq. (11.32b) gives an expression for the temperature field (profile)

$$\int_{T_i}^{T} \frac{dT}{T - T_o} = -\frac{hA_s}{\rho_s V_s c_s} \int_0^t dt \tag{11.33a}$$

$$\ln\left(\frac{T - T_o}{T_i - T_o}\right) = -\frac{hA_s}{\rho_s V_s c_s} t \tag{11.33b}$$

$$\frac{T - T_o}{T_i - T_o} = \exp\left(-\frac{hA_s}{\rho_s V_s c_s} t\right) \tag{11.33c}$$

$$T = T_o + (T_i - T_o) \exp\left(-\frac{hA_s}{\rho_s V_s c_s} t\right) \tag{11.33d}$$

Again, rearranging Eq. (11.33d) yields the time for cooling the solid specimen

$$t = -\frac{\rho_s V c_s}{hA_s} \ln\left(\frac{T - T_o}{T_i - T_o}\right) \tag{11.34a}$$

$$t = -\frac{m_s c_s}{hA_s} \ln\left(\frac{T - T_o}{T_i - T_o}\right) \tag{11.34b}$$

$$t = -\frac{\rho_s L_c c_s}{h} \ln\left(\frac{T - T_o}{T_i - T_o}\right) \tag{11.34c}$$

where $L_c = V_s/A_s$ is the characteristic length defined as (Gaskell [23, p. 369])

$$L_c = \frac{V_s}{2A_s} = \frac{A_s L}{2A_s} = L/2 \qquad \text{(slab)} \tag{11.35a}$$

$$L_c = \frac{V_s}{A_s} = \frac{\pi R^2 L}{2\pi RL} = \frac{R}{2} \qquad \text{(cylinder)} \tag{11.35b}$$

$$L_c = \frac{V_s}{A_s} = \frac{(4\pi/3) R^3}{4\pi R^2} = \frac{R}{3} \qquad \text{(sphere)} \tag{11.35c}$$

Multiplying Eq. (11.34b) by $(T_i - T_o)$ yields the amount of convective thermal energy Q_t being released during cooling of the solid in a fluid

$$(T_i - T_o) \ln\left(\frac{T - T_o}{T_i - T_o}\right) = -\frac{hA_s t}{\rho V_s c_s} (T_i - T_o) \tag{11.36a}$$

$$Q_t = hA_s (T_i - T_o) t \tag{11.36b}$$

$$Q_t = -\rho_s V_s c_s (T_i - T_o) \ln\left(\frac{T - T_o}{T_i - T_o}\right) \tag{11.36c}$$

Note that Eq. (11.36b) can be used to determine Q_t during solidification, where the argument of the natural logarithm is the dimensionless temperature (Perez [24, p. 262]).

Example 11.4 Consider quenching a 10-Kg AISI 1080 steel flat plate ($0.02 \times 0.08 \times 0.80 \ m^3$) being heat treated at $800°C$ in an inert tank containing $0.5 \ m^3$ of still water at $T_{o,f} = 25°C$. Assume that this volume of water has a constant density of $\rho = 10^3 \ Kg/m^3$, which is the inverse of the specific volume; $v = 1/\rho$. Calculate (**a**) the steel-water interface temperature T_i upon quenching, (**b**) the quenching rate or cooling rate using the AISI 1080 steel TTT diagram for austenite-to-martensite transformation, and (**c**) the evolved thermal energies during the cooling process. Data for the steel: $c_f = 4.18 \ kJ/Kg.°C$, $c_s = 0.45 \ kJ/Kg.°C$, $h = 20 \ W/m^2.K$, and $\rho = 7,800 \ Kg/m^3$. Unit conversion:

$k \rightarrow 1W/m.K = 1W/m.°C$	$dT/dx \rightarrow 1K/m = 1°C/m$
$c_s \rightarrow 1J/Kg.K = 1J/Kg.°C$	$dx/dt \rightarrow 1K/s = -272°C/s$
$c_l \rightarrow 1J/Kg.K = 1J/Kg.°C$	$dx/dt \rightarrow 1°C/s = 274K/s$

Solution

(**a**) The mass of the water is

$$m_f = (V\rho)_f = \left(0.5 \ m^3\right)\left(10^3 \ Kg/m^3\right) = 500 \ Kg \qquad (11.4E1)$$

and the solid-water interface temperature at the initial contact on two surface areas is

$$T_i = \frac{m_f c_f T_{o,f} + m_s c_s T_{o,s}}{m_f c_f + m_s c_s} \qquad (11.4E2a)$$

$$T_i = \frac{(500 \ Kg)\,(4.18 \ kJ/Kg.°C)\,(25°C) + (10 \ Kg)\,(0.45 \ kJ/Kg.°C)\,(800°C)}{(500 \ Kg)\,(4.18 \ kJ/Kg.°C) + (10 \ Kg)\,(0.45 \ kJ/Kg.°C)}$$
$$(11.4E2b)$$

$$T_i \simeq 27°C \qquad (11.4E2c)$$

Eventually, the steel mass reaches the initial water temperature in due time.

(**b**) From the AISI 1080 steel TTT diagram, the minimum quenching rates at $550°C$ and at $27°C$, respectively, for austenite-to-martensite transformation are

$$\left(\frac{dT}{dt}\right)_{TTT} > \frac{800°C - 550°C}{0 - 0.70 \ s} \simeq -357°C/s \quad \text{at } 550°C \qquad (11.4E3a)$$

$$\left(\frac{dT}{dt}\right)_{TTT} > \frac{800°C - 25°C}{0 - 20 \ s} = -38.75°C/s \quad \text{at } 25°C \qquad (11.4E3b)$$

From the TTT phase diagram, the martensitic plate reaches the initial water temperature $T_o = 25°C$ in $t >> 0.70 \ s$.

(**c**) For a 10-Kg steel plate, the volume, surface areas, and characteristic length L_c are

$$V_s = 0.02 \times 0.08 \times 0.80 \; m^3 = 0.00128 \; m^3 \qquad (11.4E4a)$$

$$2A_s = 2\left(0.08 \times 0.8 \; m^2\right) = 0.064 \; m^2 \qquad (11.4E4b)$$

$$L_c = \frac{V_s}{2A_s} = \frac{0.00128 \; m^3}{2\left(0.064 \; m^2\right)} = 0.01 \; m \qquad (11.4E4c)$$

Note that the characteristic length is in accord with Eq. (11.35a); that is, $L_c = L/2$ and it is simply the volume divided by the total surface area.

From Eq. (11.34b) along with $W \rightarrow J/s$ units, the time for the martensitic plate to reach $25°C$ under a slow cooling process is

$$t = \frac{m_s c_s}{h\left(2A_s\right)} \ln\left(\frac{T - T_o}{T_i - T_o}\right) \qquad (11.4E5a)$$

$$t = \frac{\left(10 \; Kg\right)\left(450 \; J/Kg.K\right)}{\left[20 \; \left(J/s\right)/\left(m^2.K\right)\right]\left(2 \times 0.064 \; m^2\right)} \ln\left(\frac{800 - 25}{27 - 25}\right) \qquad (11.4E5b)$$

$$t = 10,476 \; s \simeq 2.91 \; h \qquad (11.4E5c)$$

(d) The initial quenching or cooling rate upon solid-water contact is

$$\frac{dT}{dt} = -\frac{hA_s\left(T - T_o\right)}{m_s c_s} = \left(\frac{dT}{dt}\right)_{THT} \qquad (11.4E6a)$$

$$\left(\frac{dT}{dt}\right)_{THT} = -\frac{\left[20 \; W/\left(m^2.K\right)\right]\left(0.064 \; m^2\right)\left(800 - 25\right) \; K}{\left(10 \; Kg\right)\left(450 \; J/Kg.K\right)} \qquad (11.4E6b)$$

$$\left(\frac{dT}{dt}\right)_{THT} = -0.22 \; K/s = -0.22°C/s > \left(\frac{dT}{dt}\right)_{TTT} \quad \text{at } 25°C$$
$$(11.4E6c)$$

and the rate of enthalpy change representing the rate of the thermal energy release during cooling is

$$\frac{dH}{dt} = c_s \frac{dT}{dt} = \left[450 \; J/\left(Kg.°C\right)\right]\left(-273°C/s\right) \qquad (11.4E7a)$$

$$\frac{dH}{dt} = 99 \; J/s \qquad (11.4E7b)$$

The thermal energy being transferred is

$$Q_t = h\left(2A_s\right)\left(T_i - T_o\right)t \qquad (11.4E8a)$$

$$\Delta T = T_i - T_o = 27°C - 25°C = 2°C = 2 \; K \qquad (11.4E8b)$$

$$Q_t = \left(20\ W/m^2.K\right) \left(2 \times 0.064\ m^2\right) (2\ K) (10, 476\ s) \qquad \text{(11.4E8c)}$$

$$Q_t = 53637\ W.s = 53.637\ kJ \qquad \text{(11.4E8d)}$$

This example clearly shows the usefulness of the TTT diagram and the transient heat transfer (THT) theory in predicting the quenching rate for the metallurgical process related to austenite-to-martensite transformation. Both fields provide $(dT/dt)_{THT} = -0.22°C/s$ and $(dT/dt)_{TTT} > -38.75°C/s$ at $25°C$ for a relatively fast solid-state phase transformation.

11.10　Summary

In solid-phase transformation studies, thermally induced and strain-induced transformations are characterized using microscopy and X-ray crystallography. The latter is used as a tool to determine the atomic structure and reveal the orientation relationship (OR) between the initial and final phases. The corresponding phase transformation mechanism is either diffusion driven due to a concentration gradient $(\partial C/\partial x)$ or by a diffusionless (displacive) process in the solid-phase domain.

The solid-phase transformation is also characterized at a macroscale using TTT diagrams, which are mostly used for heat treating steels isothermally and continuously. Thus, isothermal transformation (IT) and continuous-cooling transformation (CCT) processes can be superimposed on a TTT diagram for designing the most suitable heat treatment process in order to obtain a desire microstructure. Moreover, thermally induced phase transformation is part of the hardenability of steels, shape memory polymers (SMPs), and shape memory alloys (SMAs), such as nitinol which is an equiatomic $Ni\text{-}Ti$ alloy.

Appendix 11A: TTT Diagrams

Below are some references for TTT phase diagrams adopted from Callister and Rethwisch [3, Chapter 10] and Vander Voort [16, p. 15]. For instance, Fig. 11.20 shows the TTT diagram for an AISI 1050 hypoeutectoid carbon steel (Vander Voort [16, p. 15]).

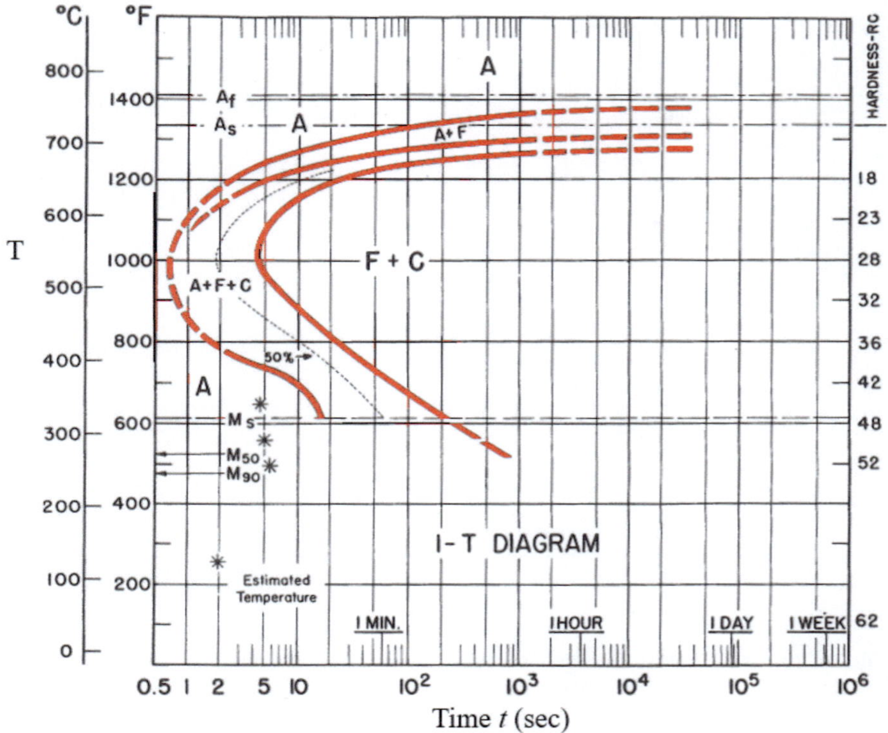

Time t (sec)

Fig. 11.20 The TTT phase diagram for a hypoeutectoid 1050 steel adapted from Callister-Rethwisch book [3, p. 378]

Nomenclature

Phase	Letter	Symbol
Austenite	A	γ
Bainite	B	B
Ferrite	F	α
Martensite	M	α'
Pearlite	P	p

Figure 11.21 depicts the eutectoid TTT phase diagram for an AISI 1080 eutectoid carbon steel (Callister and Rethwisch [3, p. 325], Vander Voort [16, p. 16]).

Figure 11.22 illustrates the TTT phase diagram for an AISI 4340 alloy steel (Callister and Rethwisch [3, p. 326], Vander Voort [16, p. 35]).

Figure 11.23 exhibits the TTT phase diagram for an AISI 9440 hypoeutectoid alloy steel (Vander Voort [16, p. 38]).

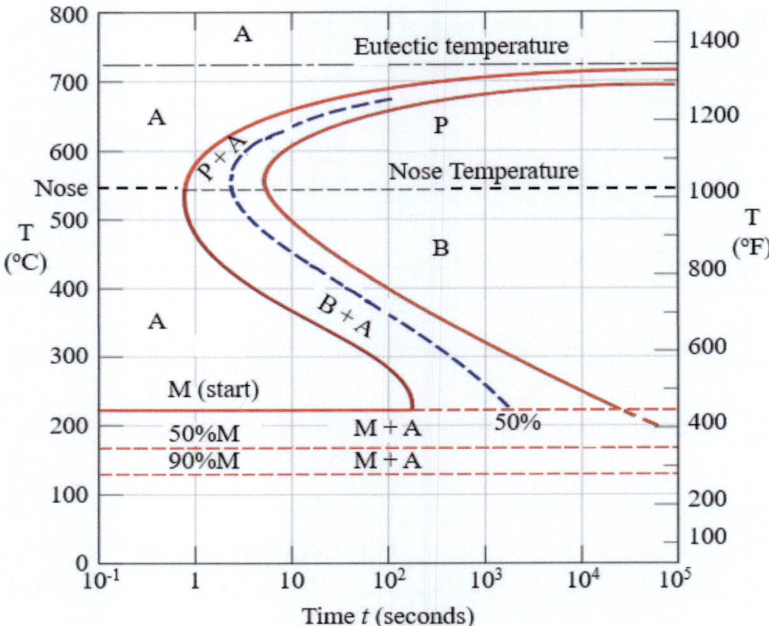

Fig. 11.21 The TTT phase diagram for a 1080 eutectoid steel adapted from references Callister and Rethwisch [3, p. 378] and Vander Voort [16, p. 16]

Appendix 11B: Separation of Variables

This Appendix includes the analytical procedure for deriving Eq. (11.20a). Following Poirier and Geiger [17, p. 294] and Poirier and Geiger [18, p. 290], let $\theta = T - T_o$ so that

$$\frac{\partial \theta}{dt} = \alpha \frac{\partial^2 \theta}{dx^2} \tag{11.37}$$

$$\theta(x, 0) = T_i - T_o = \theta_i \quad \text{(IC)} \tag{11.38}$$

$$\frac{\partial \theta(0, t)}{dx} = 0 \qquad \text{(BC)} \tag{11.39}$$

$$\frac{\partial \theta(x_c, t)}{dx} + \frac{h}{k}\theta(x_c, t) = 0 \quad \text{(Energy balance)} \tag{11.40}$$

In order to solve Eq. (11.37) by the method of separation of variables, the product $\theta(x, t) = F(x) \cdot G(t)$ is used for such a purpose. Thus, the derivatives of $\theta(x, t)$ are

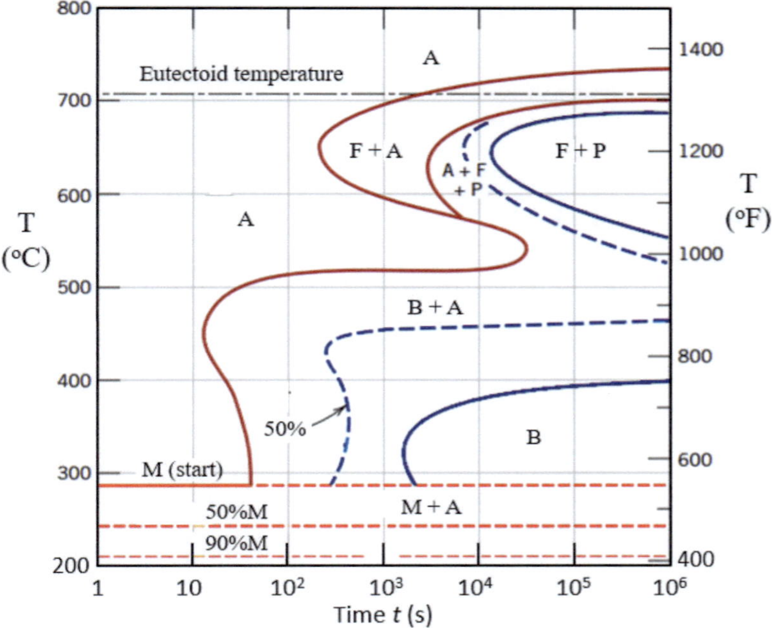

Fig. 11.22 The TTT phase diagram for a 4340 steel alloy adapted from Callister and Rethwisch [3, p. 326] and Vander Voort [16, p. 35]

$$\theta\,(x,t) = F\,(x) \cdot G\,(t) \tag{11.41}$$

$$\frac{\partial\theta\,(x,t)}{\partial x} = G\,(t)\,\frac{dF\,(x)}{dx} \tag{11.42}$$

$$\frac{\partial^2\theta\,(x,t)}{\partial x^2} = G\,(t)\,\frac{d^2F\,(x)}{dx^2} \tag{11.43}$$

and

$$\frac{\partial\theta\,(x,t)}{\partial t} = F\,(x)\,\frac{dG\,(t)}{dt} \tag{11.44}$$

Inserting Eqs. (11.43) and (11.44) into (11.37) with $\theta\,(x,t) = \theta$ yields the following expression:

$$\frac{1}{F\,(x)}\,\frac{d^2F\,(x)}{dx^2} = \frac{1}{\alpha G\,(t)}\,\frac{dG\,(t)}{dt} = -\lambda^2 \tag{11.45}$$

from which two homogeneous ordinary differential equations (ODEs) arise

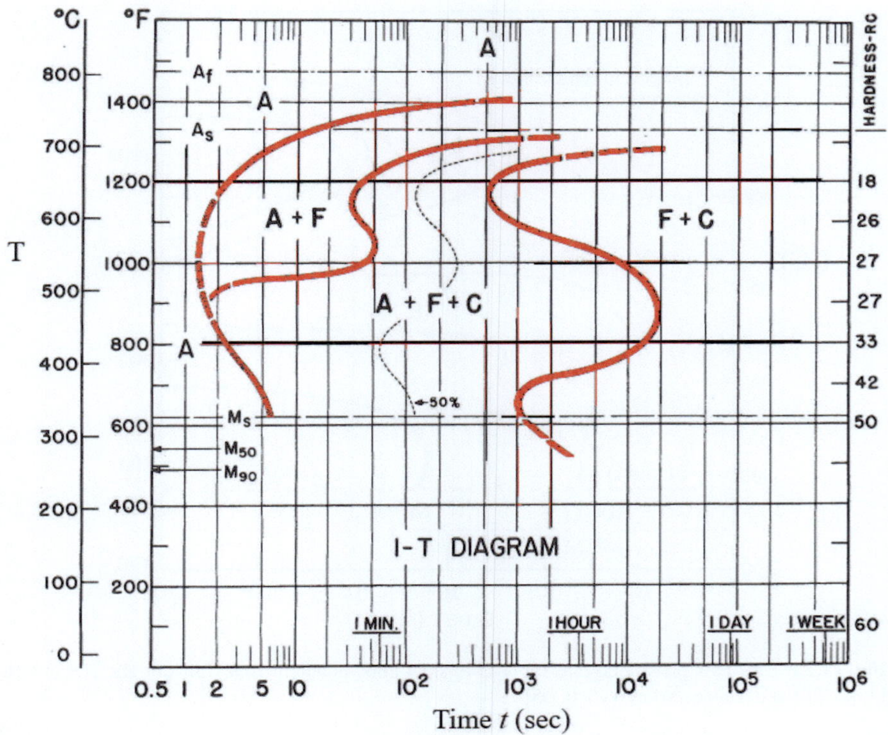

Fig. 11.23 The TTT phase diagram for an AISI 9440 alloy steel (Vander Voort [16, p. 38])

$$\frac{d^2 F(x)}{dx^2} + \lambda^2 F(x) = 0 \tag{11.46}$$

$$\frac{dG(t)}{dt} + \lambda^2 \alpha G(t) = 0 \tag{11.47}$$

The solutions of Eqs. (11.46,b) are

$$F(x) = a_1 \cos(\lambda x) + a_2 \sin(\lambda x) \tag{11.48}$$

$$G(t) = \exp\left(-\lambda^2 \alpha t\right) \tag{11.49}$$

where λ is the eigenvalue. Note that Eq. (11.48) satisfies (11.39) if $a_2 = 0$; that is,

$$\frac{\partial F(x)}{dx} = -\lambda a_1 \sin(\lambda x) + \lambda a_2 \cos(\lambda x) \tag{11.50}$$

$$\left[\frac{\partial F(x)}{dx}\right]_{x=0} = \lambda a_1 \sin(0) - \lambda a_2 \cos(0) = 0 \tag{11.51}$$

$$a_2 = 0 \tag{11.52}$$

Here, $a_2 = 0$ does not contribute to the solution of $F(x)$. Then, Eq. (11.48) becomes

$$F(x) = a_1 \cos(\lambda x) \tag{11.53}$$

$$\frac{dF(x)}{dx} = -\lambda a_1 \sin(\lambda x) \tag{11.54}$$

From Eq. (11.40),

$$\frac{dF(x_c, t)}{dx} + \frac{h}{k} F(x_c, t) = 0 \tag{11.55}$$

$$-a_1 \lambda \sin(\lambda x_c) + \frac{h}{k} a_1 \cos(\lambda x_c) = 0 \tag{11.56}$$

$$\frac{h}{k} \cos(\lambda x_c) = \lambda \sin(\lambda x_c) \tag{11.57}$$

$$\cot(\lambda x_c) = \frac{\lambda}{h/k} = \frac{\lambda x_c}{(h/k)\, x_c} = \frac{\lambda x_c}{B_i} \tag{11.58}$$

$$\lambda x_c \tan(\lambda x_c) = B_i \tag{11.59}$$

Substituting Eqs. (11.49) and (11.53) into (11.41) and generalizing the resultant expression for all values of λ_n satisfying (11.58) yield $\theta(x, t) = T - T_o$ and

$$\theta(x, t) = \sum_{n=1}^{\infty} F_n \cos(\lambda_n x) \exp\left(-\lambda_n^2 \alpha t\right) \tag{11.60}$$

Using the initial condition (IC) defined by Eq. (11.38) gives $\theta_i = T_i - T_o$ and

$$\theta_i = \sum_{n=1}^{\infty} F_n \cos(\lambda_n x) \tag{11.61}$$

Multiplying Eq. (11.61) by a weight function, say, cosine function $\cos(\lambda_m x)\, dx$, and integrating the resultant expression with appropriate limits of integration illustrate the application of the orthogonality integral approach. Thus,

$$\theta_i \int_0^{x_c} \cos(\lambda_m x)\, dx = \int_0^{x_c} \sum_{n=1}^{\infty} F_n \cos(\lambda_n x) \cos(\lambda_m x)\, dx \tag{11.62}$$

and for a nonzero integral, $m = n$ and $\lambda_n x_c = n\pi/x_c$. Then, Eq. (11.62) yields

$$\theta_i \left[\frac{1}{\lambda_n} \sin (\lambda_n x_c) \right] = F_n \left[\frac{x_c}{2} + \frac{1}{2\lambda_n} \sin (\lambda_n x_c) \cos (\lambda_n x_c) \right] \tag{11.63}$$

$$F_n = \frac{2\theta_i \sin (\lambda_n x_c)}{\lambda_n x_c + \sin (\lambda_n x_c) \cos (\lambda_n x_c)} \tag{11.64}$$

Substitute Eq. (11.64) into (11.60) to get the sought equation

$$\frac{\theta}{\theta_i} = \frac{T - T_o}{T_i - T_o} = 2 \sum_{n=1}^{\infty} \frac{\sin (\lambda_n x_c) \cos (\lambda_n x)}{\lambda_n x_c + \sin (\lambda_n x_c) \cos (\lambda_n x_c)} \exp \left(-\lambda_n^2 \alpha t \right) \tag{11.65}$$

which is defined as Eq. (11.20a) in the text.

Problems

11.1 Calculate the surface energy ratio $\gamma_{\alpha\alpha}/\gamma_{\alpha\beta}$ using the given the grain high angles below. [Solution: $\gamma_{\alpha\alpha}/\gamma_{\alpha\beta} = 1$].

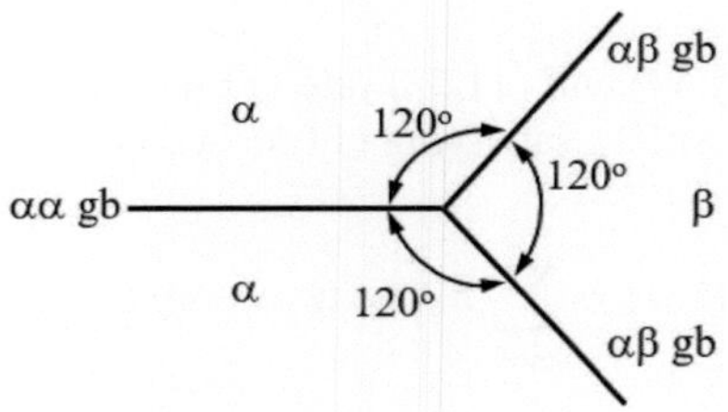

11.2 Calculate the displacement involved in martensitic transformation in a carbon steel plate containing 0.80% by weight. [Solution: $\mu = 0.0704 \ nm$].

11.3 If a 1080 steel bar undergoes martensitic transformation, then the lattice deformation occurs by contraction along the $[001]_\gamma$ crystallographic direction and the uniform expansion on the $(001)_\gamma$ plane. Determine (**a**) the respective Bain strains and (**b**) the maximum displacement; (**c**) write down the results in matrix format and (**d**) the orientation relationship between the austenite (parent) and martensite (product) lattices. [Solution: (a) $\epsilon_{[100]_\gamma} = 0.13$ and $\epsilon_{[001]_\gamma} = -0.20$, (b) $\mu = 0.07 \ nm$, (d) $[001]_\gamma \parallel [001]_{\alpha'}$]

11.4 Consider $(111)_\gamma \rightarrow (110)_{\alpha'}$ during martensitic transformation in a 1060 steel as shown in Fig. 11.8 and write down the Miller indices for the crystallographic directions corresponding to orientation relationships in the Bain model. [Solution: $[101]_\gamma \rightarrow [111]_{\alpha'}$, $[110]_\gamma \rightarrow [100]_{\alpha'}$ and $[112]_\gamma \rightarrow [011]_{\alpha'}$].

11.5 For a 1060 steel bar undergoing martensitic transformation, **(a)** calculate the respective Bain strains and **(b)** the maximum displacement, and write down the Bain matrix, and **(c)** determine the orientation relationship between the austenite (parent) and martensite (product) lattices. [Solution: (a) $\epsilon_{([100]_\gamma} = 0.13$ and $\epsilon_{[001]_\gamma} = -0.20$, (b) $\mu = 0.072\ nm$, (c) $[001]_\gamma \parallel [001]_{\alpha'}$].

11.6 For a 1030 steel bar undergoing martensitic transformation, **(a)** calculate the respective Bain strains and **(b)** the maximum displacement, and write down the Bain matrix, and **(c)** determine the orientation relationship between the austenite (parent) and martensite (product) lattices. [Solution: (a) $\epsilon_{([100]_\gamma} = 0.14$ and $\epsilon_{[001]_\gamma} = -0.20$, (b) $\mu = 0.074\ nm$, (c) $[001]_\gamma \parallel [001]_{\alpha'}$].

11.7 For a carbon steel plate being quenched in agitated water at $25°C$, calculate the critical entities defined by Eq. (11.11a,b,c,d) when **(a)** $\Delta G_p = 0$ and **(b)** $\Delta G_p = 50$ MJ/m^3, and assume a homogeneous nucleation of martensite without composition change. Assume that the martensitic transformation starts at $T = 700\ K\ (427°C)$. Explain. Data: [Solution: (a) $\Delta G_{c,\text{hom}} = 4.20\ eV$, (b) $E_s = 96.52\ GJ/m^3$].

$$\Delta G_V = 180 MJ/m^3 \quad\Big|\quad S_{\gamma\alpha'} = 15 \times 10^{-3} J/m^2 \quad\Big|\quad K_\varepsilon = 2.5 GPa$$

11.8 For a carbon steel plate being quenched in agitated water at $25°C$, calculate the critical entities defined by Eq. (11.11a,b,c,d) when **(a)** $\Delta G_p = 0$ and **(b)** $\Delta G_p = 60$ MJ/m^3. Assume that the martensitic transformation starts at $T = 700\ K\ (427°C)$ with no composition change for homogeneous nucleation of martensite. Explain. Use the following data [Solution: (a) $\Delta G_{c,\text{hom}} = 4.21\ eV$, (b) $E_s = 59.98\ GJ/m^3$].

$$\Delta G_V = 180 MJ/m^3 \quad\Big|\quad S_{\gamma\alpha'} = 15 \times 10^{-3} J/m^2 \quad\Big|\quad K_\varepsilon = 2.5 GPa$$

P11.9 Consider an isothermal solid-phase transformation in two 6-cm-thick 1080 steel plates as indicated by the IT trajectories in the eutectoid TTT diagram (Example 11.3). **(a)** If one plate is initially at $750°C$ and suddenly immersed in a fluid at $T_f = 600°C$, then calculate the time for complete pearlitic transformation $(\gamma \rightarrow p)$ when the temperature at center of the plate is $T = 650°C$, and determine the plate surface temperature using the time from the TTT diagram. **(b)** Repeat the calculations for bainitic transformation $(\gamma \rightarrow B)$ using another 6-cm-thick 1080 steel plate immersed in a similar fluid when $T_f = 400°C$ and $T = 450°C$. [Solution: (a) $t = 236\ sec$ and $T = 717°C$, (b) $t = 958\ sec$ and $T = 701°C$].

11.10 Plot **(a)** the Fourier number $F_o = f(\phi)$ at $0 \le \phi = \lambda x_c \le 0.8$, $B_i = 0.25$, and $x/x_c = 0, 0.3, 0.4, 0.5$, and **(b)** the Biot number $B_i = f(\phi)$ at $0 \le \phi \le 2$ and $F_o = f(B_i)$ at $0 \le B_i \le 2$. Let $T_\theta = 0.33$. You are on your own to characterize the nonlinearity of the curves.

11.11 For a quenched carbon steel plate in agitated water at $25°C$, calculate the critical entities defined by Eq. (11.11a,b,c,d) when **(a)** $\Delta G_p = 0$ and **(b)** $\Delta G_p =$

60 MJ/m^3. Use the following data $\Delta G_V = 182\ MJ/m^3$, $S_{\gamma\alpha'} = 16 \times 10^{-3}$ J/m^2, $K_\varepsilon = 2.6\ GPa$, and assume that the martensitic transformation starts at $T = 700\ K$ ($427°C$) with no composition change. Explain. [Solution: (a) $r_c = 5.02$ nm, $h_c \simeq 0.18\ nm$, $\Delta G_{c,\mathrm{hom}} = 5.28\ eV$, (b) $r_c = 11.56\ nm$, $h_c =, 0.27\ nm$, $\Delta G_{c,\mathrm{hom}} = 27.90\ eV$ and $E_s = 60.73\ MJ/m^3$].

P11.12 Assume that two 4-cm-thick 1080 steel plates are subjected to isothermal solid-phase transformation in a similar manner as illustrated in Example 11.3. **(a)** For one plate initially at $750°C$ and immersed in a fluid at $T_f = 600°C$, calculate the time for complete pearlitic transformation ($\gamma \to p$) when the temperature at center is $T = 650°C$ and the plate surface temperature along with the TTT diagram. **(b)** Repeat the calculations for bainitic transformation ($\gamma \to B$) when $T_f = 400°C$ and $T = 450°C$. Use the given dataset $h = 283.92\ W/m^2.K$, $\alpha = 1.1871 \times 10^{-5}$ m^2/s, and $k = 34.614\ W/m.K$. [Solution: (a) $t = 124\ sec$ and $T = 712.50°C$, (b) $t = 315\ sec$ and $T = 701°C$].

P11.13 Suppose that two 2-cmthick 1080 steel plates are initially at $750°C$ and are subjected to isothermal solid-phase transformation. Calculate the transformation time **(a)** for pearlitic transformation ($\gamma \to p$) when the fluid is at $T_f = 600°C$ and the temperature of one plate at center is $T = 650°C$ and the plate surface temperature along with the TTT diagram. **(b)** Repeat calculations for bainitic transformation ($\gamma \to B$) when the fluid is at $T_f = 400°C$ and the center of the other plate is at $T = 450°C$. Use the given dataset $h = 283.92\ W/m^2.K$, $\alpha = 1.1871 \times 10^{-5}\ m^2/s$ and $k = 34.614\ W/m.K$. [Solution: (a) $t = 32\ sec$ and $T = 690°C$, (b) $t = 125\ sec$ and $T = 680°C$].

11.14 Assume that hypothetical solid-phase transformations are 5% complete in 40 sec and 90% complete in 510 sec at a certain isothermal holding temperature. Calculate **(a)** the exponent n for time and rate constant k_x in the Avrami kinetic equation and **(b)** the time it takes for a 50% transformation. [Solution: (a) $n = 1.5$ and $k_x = 2 \times 10^{-4}\ \sec^{-n}$, (b) $t = 510$ sec].

11.15 Use the iron-iron carbide (Fe-Fe_3C) phase diagram for a 1%C (1 $wt\%$ carbon) hypereutectoid steel to determine **(a)** the %C in ferrite and cementite at $726°C$, **(b)** the mass fraction of ferrite and cementite at $726°C$ using the Lever rule, **(c)** the mass fraction of austenite and pro-eutectoid cementite at $T_e = 727°C$, and **(d)** the mass fraction of cementite in the eutectoid structure. [Solution: (a) $C_\alpha \simeq 0.02\%C$ and $C_{Fe_3C} \simeq 6.67\%C$, (b) $f_e = 0.85$ and $f_{(Fe_3C)'} = 0.15$, (c) $f_\gamma = 0.97$ and $f_{(Fe_3C)} = 0.03$, (d) $f_{Fe_3C,e} = 0.12$].

11.16 Assume that the kinetics of austenite-to-pearlite transformation obeys the Avrami equation. Calculate the time for 99% transformation if only (25%, 250 sec) and (90%, 1000 sec) dataset is available. [Solution: $t = 1,588$ sec].

11.17 Several 1080 steel specimens with identical shape and dimensions are heated at $800°C$ for one hour in furnace with controlled gaseous environment and then placed in another furnace at $650°C$ for isothermal solid-state phase transformation studies. Suppose that you are given (25%, 12 sec) and (90%, 40 sec) austenite-to-pearlite transformation points at $650°C$. Based on this information, determine (**a**) the time it takes for 95% transformation using the Avrami equation and the TTT diagram, and (**b**) write down chronologically the reactions along with the percentage of phases and related transformation time for the final microstructures after quenching the specimens in still water at $25°C$. Include the phase reaction and time for 100% complete austenite-to-pearlite transformation and the maximum time for 100% austenite-to-martensite transformation. [Solution: (a) $t = 47$ sec and $t_{TTT} = 46$ sec, (b) $100\%\gamma_o \rightarrow 100\%M$ at $t < 0.8\,s$].

11.18 Consider quenching a 1080 steel cylinder (radius $R = 0.01\,m$ and length $L = 0.10\,m$) being heat treated at $800°C$ in an inert tank containing still water at $T_{o,f} = 25°C$. Assume that this volume of water has a constant density of $\rho = 10^3\,Kg/m^3$, which is the inverse of the specific volume; $v = 1/\rho$. Calculate (**a**) the steel-water interface temperature T_i upon quenching, (**b**) the quenching rate or cooling rate using the 1080 steel TTT diagram for austenite-to-martensite transformation, (**c**) the time for the martensitic plate to reach $25°C$, and (**d**) the quenching rate and the evolved thermal energies during the cooling process. Data: $c_f = 4.18\,kJ/Kg.°C$, $c_s = 0.45\,kJ/Kg.°C$, $h = 20\,W/m^2.K$, and $\rho = 7800\,Kg/m^3$. [Solution: (a) $T_i = 27°C$, (b) $(dT/dt)_{TTT} = -84\,K/s$, (c) $t \simeq 5335\,s$, (d) $dT/dt = -357°C/s$, $dH/dt = -1.23 \times 10^5\,kJ$ and $Q_t = 1341\,J$].

11.19 Assume quenching a 1080 steel flat plate ($0.005 \times 0.30 \times 0.30\,m^3$) being heat treated at $800°C$ in an inert tank containing still water at $T_{o,f} = 25°C$. Assume that this volume of water has a constant density of $\rho = 10^3\,Kg/m^3$, which is the inverse of the specific volume; $v = 1/\rho$. Calculate (**a**) the Al–alloy/water interface temperature T_i upon quenching, and (**b**) plot the temperature and the rate of cooling profiles in order to reveal the trends for reaching equilibrium. Data: $c_f = 4.18\,kJ/Kg.°C$, $c_s = 0.45\,kJ/Kg.°C$, $h = 20\,W/m^2.K$, and $\rho = 7800\,Kg/m^3$. [Solution: (a) $T_i = 27°C$].

References

1. D.A. Porter, K. Easterling, M.Y. Sherif, *Phase Transformations in Metals and Alloys*, 3rd edn. (CRC Press, Taylor & Francis Group, LLC, Boca Raton, FL, USA, 2009)
2. A.G. Guy, J.J. Hren, *Elements of Physical Metallurgy*, 3rd edn. (Addison-Wesley Publishing Company, Reading, MA, 1974)
3. W.D. Callister, Jr., D.G. Rethwisch, *Materials Science and Engineering: An Introduction*, 10th edn. (Wiley, New York, USA, 2018)
4. D.R. Askeland, P.P. Fulay, W.J. Wright, *The Science and Engineering of Materials*, 6th edn. (Cengage Learning, New York, USA, 2010)
5. Z. Nishiyama, *Martensitic Transformation* (Academic Press, New York, 1978)
6. B.D. Cullity, S.R. Stock, *Elements of X-Ray Diffraction*, 3rd edn. (Pearson Education Limited, London, 2014)
7. L. Kaufman, M. Cohen, Thermodynamics and kinetics of martensitic transformations. Progr. Metal Phys. **7**, 165–246 (1958)

8. G.B. Olson, M. Cohen, *Dislocation Theory of Martensitic Transformations*, in *Dislocations in Solids*, vol. 7, ed. by F.R.N. Nabarro (NorthHolland Publishing, Amsterdam, 1986)
9. A.K. Sinha, *Physical Metallurgy Handbook* (The McGraw-Hill Companies, New York, 2003)
10. J.D. Eshelby, Elastic inclusions and inhomogeneities. Prog. Solid Mech. **2**, 89–140 (1961). Amsterdam
11. C. Simsir, C.H. Gur, *Simulation of Quenching*, in *Handbook of Thermal Process Modeling of Steels*, ed. by C. Hakan Gur, J. Pan (CRC Press, Taylor & Francis Group, LLC, Boca Raton, FL, USA, 2009)
12. M. Munirajulu, B.K. Dhindaw, A. Biswas, A. Roy, Modeling of eutectoid transformation in plain carbon steel. ISIJ Int. **34**(4), 355–358 (1994)
13. M. Avrami, Kinetics of phase change. I general theory. J. Chem. Phys. **7**(12), 1103–1112 (1939)
14. I. Avramov, J. Sestak, *Generalized Kinetics of Overall Phase Transition in Terms of Logistic Equation* [physics.chem-ph] (2015). https://arxiv.org/abs/1510.02250v1
15. A. Kulawik, Modeling of thermomechanical phenomena of welding process of steel pipe. Modelowanie Zjawisk Termomechanicznych Procesu Spawania Rury Stalowej. Arch. Metallur. Mater. **57**(4), 1229–1238 (2012)
16. G.F. Vander Voort (ed.), *Atlas of Time-Temperature Diagrams for Irons and Steels* (ASM International, Metals Park, OH, USA 1991)
17. G.H. Geiger, D.R. Poirier, *Transport Phenomena in Metallurgy* (Addison-Wesley Publishing Company, Reading, MA, USA, 1973)
18. D.R. Poirier, G.H. Geiger, *Transport Phenomena in Materials Processing*. The Minerals, Metals & Materials Society, A Publication of TMS (Springer, Switzerland, 2016)
19. G.M. Martinez-Cazares, D.E. Lozano, M.P. Guerrero-Mata, R. Colas, G.E. Totten, High-speed quenching of high carbon steel. Mater. Perf. Character. **3**(4), 256–267 (2014)
20. C. Celada-Casero, H. Kooiker, M. Groen, J. Post, D. San-Martin, In-situ investigation of strain-induced martensitic transformation kinetics in an austenitic stainless steel by inductive measurements. Metals **7**(7), 271 (2017)
21. H.C. Shin, T.K. Ha, Y.W. Chang, Kinetics of deformation induced martensitic transformation in a 304 stainless steel. Scripta Materialia **45**(7), 823–829 (2001)
22. D.J. Hartl, D.C. Lagoudas, *Thermomechanical Characterization of Shape Memory Alloy Materials*, in *Shape Memory Alloys: Modeling and Engineering Application*, ed. by D.C. Lagoudas (Springer Science, New York, USA, 2008)
23. D.R. Gaskell, *An Introduction to Transport Phenomena in Materials Engineering*, 2nd edn. (Momentum Press®, LLC, New York, 2013)
24. N. Perez, *Phase Transformation in Metals: Mathematics, Theory and Practice* (Springer Nature, Switzerland, 2020)

Chapter 12
Mechanical Behavior of Solids

12.1 Introduction

This chapter describes the mechanical behavior of initially crack-free solids under different loading conditions for determining their mechanical properties, which are important in designing structural components with suitable shapes and dimensional limitations. Thus, the determination of a load-bearing capacity of a material under tension, bending, compressive, torsion, or a combination of mechanical forces is critical in order to assure the integrity of a structural component.

Extensive details on mechanical strength and crystallography are coupled in order to characterize the atomic mechanisms of mechanical deformation, specifically, under quasi-static and fatigue loading conditions. In essence, the mechanical behavior of crack-free solids is mainly assessed through the conventional engineering stress-strain diagrams (curves) under quasi-static and fatigue loading. The interpretation of these curves is relatively easy, but they graphically reflect the relevant elastic-plastic deformation related to the slip mechanism and crystal defects.

Regarding mechanical elastic deformation, a solid containing an atom or molecule arrangement exhibits an elastic behavior under tension, compression, torsion, bending, flexural or a combination of these mechanical loading modes. This implies that the solid material obeys Hooke's law at a macroscale, and it recovers its original dimensions upon unloading. Conversely, plastic mechanical deformation leads to an irreversible atomic process since the solid material experiences a plastic behavior related to a permanent change of shape. Therefore, plastic deformation induces the formation of defects, such as dislocations, and alters the stress-strain relationship to an extent.

In practice, the mathematical characterization of the tensile plastic deformation, at a macroscale, can be achieved using some empirical stress-strain equations commonly known as power law relations. These equations are related to the concept

607

N. Perez, *Materials Science: Theory and Engineering*,
https://doi.org/10.1007/978-3-031-57152-7_12

of true stress-strain curves, which represent the plastic behavior between the yield strength and the ultimate tensile strength points.

12.2 Mechanical Properties and Testing Methods

Mechanical properties of solid materials can be determined according to a metal society or company design codes based on the shape of the specimen gage area, where mechanical deformation must take place monotonically. Commonly, a universal tension machine (Fig. 12.1a) is used to measure room-temperature mechanical properties of dimensionally limited solids by applying tensile forces through a cross head and strains using an extensometer (Fig. 12.1b). On the other hand, high-temperature mechanical properties are measured when the specimen is placed inside an induction furnace (Fig. 12.1c) using monotonic testing or static load for creep testing high-temperature deformation). Also, an open induction furnace (Fig. 12.1d) can be used for the same purpose. The images shown in Fig. 12.1a, c can be found at zwickroell.com [1], and those in Fig. 12.1b, d were published by MTS Systems Corp. [2].

Tensile Properties Tensile or tension tests are performed on universal testing machines (Fig. 12.1a) using brittle and ductile materials, such as metals and alloys, composites, and plastics. The specimens are loaded at low strain rates to obtain a tensile stress-strain curves up to failure. Subsequently, tensile properties are extracted or measured from these curves.

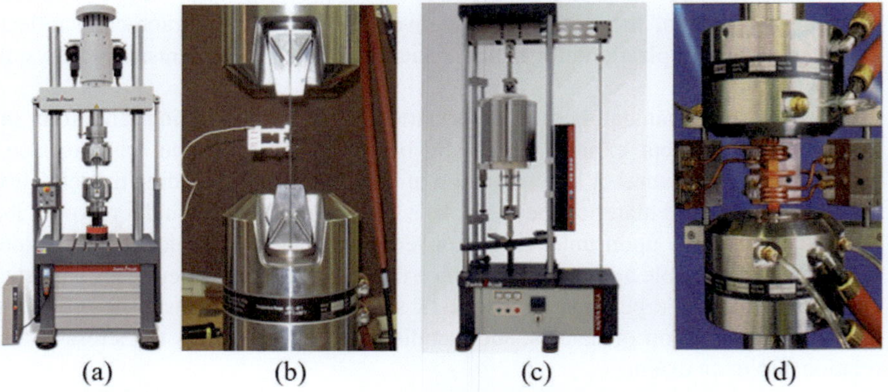

(a)   (b)   (c)   (d)

Fig. 12.1 (**a**) ZwickRoell servohydraulic testing machine for conventional mechanical properties (zwickroell.com [1]), (**b**) close-up hydraulic grips showing the extensometer on the specimen (MTS Systems Corp. [2]), (**c**) ZwickRoell testing machine with a closed induction furnace for creep properties (zwickroell.com [1]), and (**d**) close-up grips with an open induction furnace (MTS Systems Corp. [2])

Compression Properties A compression testing is performed on universal testing machines using brittle materials, such as concrete beams, metals and alloys, plastics, ceramics (including car windshield glass), composites, and corrugated materials like cardboard. This leads to the mechanical strength of these materials under compressive loading.

Torsional Properties Torsion tests are performed on specialized machines using any type of material. A specimen is twisted by applying torques or torsional forces until it breaks and torsion properties are measured or extracted from a torsion shear stress and shear strain curve. Common materials are wires, rods, screws, plastics, and the like.

12.3 Definitions

This section is concerned with some definitions the reader needs to assimilate before the fracture or deformation theories and mathematical definitions are introduced in a progressive manner. It is important to have a clear and precise definition of vital concepts in the field of applied mechanics so that the learning process for understanding plastic deformation becomes obviously easy. However, basic concepts such as stress, strain, safety factor, deformation, and the like are important in characterizing the mechanical behavior of solid materials subjected to forces or loads in service. Hence,

Deformation It is a measure of the movement of points in a solid body relative to each other.

Displacement It is a measure of the movement of a point along a vector quantity in a body subjected to a loading mode.

Strain It is a geometric quantity, which depends on the relative movement of two or three points in a body.

Stress It is the internal resistance of the body at a point due to an external force. Thus, the load (P) and the instantaneous, actual, or true cross-sectional area (A) after the material is elastically or plastically deformed are related as indicated by the equation of equilibrium of forces. Thus,

$$\sum F_y = P - \sigma A = 0 \tag{12.1a}$$

$$\sigma = \frac{P}{A_o} \quad \text{(engineering)} \tag{12.1b}$$

$$\sigma_t = \frac{P}{A} \quad \text{(true)} \tag{12.1c}$$

$$S_F = \frac{\sigma_{eng}}{\sigma_d} > 1 \tag{12.1d}$$

$$\sigma_d = \frac{\sigma}{S_F} = \frac{P}{S_F A_o} \tag{12.1e}$$

If A is replaced by the original cross-sectional area (A_o), then $\sigma = \sigma_{eng}$ is an engineering stress; otherwise, it is a local stress.

Safety Factor This is a parameter (S_F) to define a design stress for structural integrity. Simply stated, the safety is a design factor related to a design stress.

Design Stress It is an elastic stress σ_d defined as the maximum allowable stress acting on a machine part in order to prevent failure. Therefore, σ_d is the working stress.

Strength Represents a material's property, such as the yield strength σ_{ys}, and the stress σ is the variable being applied to a structure. The role of S_F in this simple relationship is to control the design stress so that $\sigma < \sigma_{ys}$. Usually, the safety factor is in the order of two, but its magnitude depends on the designer's experience or a design code.

Among all applied loads to structural components, the quasi-static tensile testing is the most common loading mode along with stress-strain curves representing the tensile mechanical behavior of engineering materials. The mechanical behavior represents itself the elastic or elastic-plastic deformation of a material at a given temperature and pressure.

12.4 Specimen Gage Shapes

The goal in this section is to introduce the specimen shapes shown in Fig. 12.2 as a practical guideline for fabricating specimen gage areas and for determining mechanical properties of materials according to a particular metal society.

Regarding specimen shapes, Fig. 12.2a is the most common rectangular or round specimen, and Fig. 12.2b is typically used for a welded rectangular specimen related to the architecture or structure of ongoing vessels (ships). Moreover, Fig. 12.2c–g schematically shows additional shapes of gage areas of interest along with the direction of the applied stress σ (Callister and Rethwisch [3, p. 145]).

According to Eq. (12.1c), the uniaxial true stress-strain relationship (σ_t-ε_t) is mathematically written as

$$\sigma_t = \frac{P}{A_o}\left(1 + \frac{\Delta L}{L_o}\right) = \sigma\,(1 + \epsilon) \tag{12.2a}$$

$$\varepsilon_t = \ln\,(1 + \epsilon) \tag{12.2b}$$

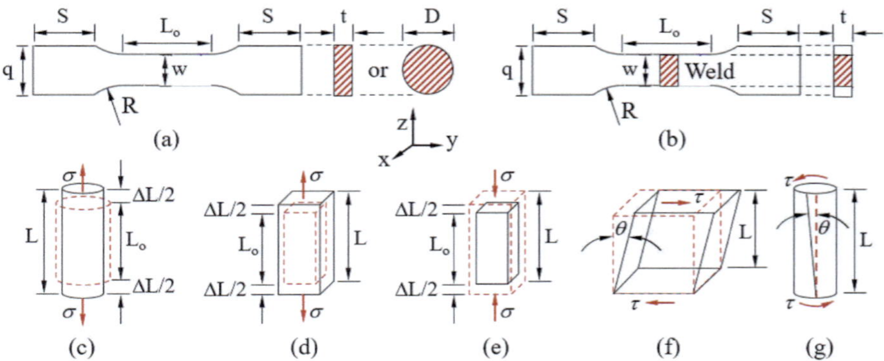

Fig. 12.2 Specimen models for mechanical testing of (**a**) rectangular or cylindrical, (**b**) welded rectangular specimens showing gage areas, where mechanical deformation occurs during (**c**) tension, (**d**) tension, (**e**) compression, (**f**) sliding, and (**g**) torsion testing. After Callister and Rethwisch [3, p. 145]

where P is the applied mechanical load (force), ϵ is the uniaxial strain, and ΔL, L_o are the change and original specimen gage lengths, respectively.

The suitable crack-free specimen nomenclature cited in Fig. 12.2 includes additional definitions. Thus:

- L_o is the original gage length, where the specimen experiences a local triaxial-state of stress (principally σ_x, σ_y, σ_z).
- $L > L_o$ is the instantaneous gage length.
- $\Delta L = L - L_o$ is the change in specimen gage length related to a change in gage cross-sectional area.
- $\sigma = P/A_o > 0$ is the uniaxial applied stress (tension in Fig. 12.2c,d) or $\sigma = P/A_o < 0$ in compression (Fig. 12.2e), where P is the applied load converted to stress.
- $\tau = P/A_o$ is the applied shear stress in sliding or torsion mode (Fig. 12.2f, g).
- θ is the angle representing deformation with respect to a reference point (Fig. 12.2f, g).
- $\varepsilon = \Delta L/L_o$ is the uniaxial strain, tensile engineering strain, or nominal strain.
- $\varepsilon_t = \ln(L/L_o)$ or $\varepsilon_t = \ln(1 + \varepsilon)$ is the tensile true or instantaneous strain related to the instantaneous loading action on the instantaneous cross-sectional gage area of a specimen. For a triaxial-state of strain analysis, the engineering strains are defined as $\varepsilon_x = \Delta L_x/L_o$, $\varepsilon_y = \Delta L_y/L_o$, and $\varepsilon_z = \Delta L_z/L_o$.
- $\gamma = \tan(\theta)$ is the shear strain or sliding shear strain in Fig. 12.2f or the torsional shear strain in Fig. 12.2g.
- $\sigma = P/A_o$ is the uniaxial stress, tensile engineering stress, or nominal stress with load P and original cross-sectional area $A_o = wL_o$ (rectangular, w = width) or $A_o = (\pi/4)\,d^2$ (round, d = diameter).
- $v = -\varepsilon_x/\varepsilon_z = -\varepsilon_y/\varepsilon_z$ is the Poisson's ratio (elastic ratio of the lateral and axial strains).

The above stress and strain definitions are directly related to conventional engineering stress-strain (σ-ε) diagrams, which, in turn, are used to extract mechanical properties, to classify solid materials, and to analyze elastic and plastic deformation. All these form an important domain in the analysis of mechanical behavior of materials.

12.5 Yielding Phenomenon in Crystalline Materials

Relevant engineering σ-ε curves for determining the mechanical properties of materials (treated as a defect-free continuum) in tension mode are schematically illustrated in Fig. 12.3.

A crystalline structure consists of many single crystals called grains. If the structure is ductile, then it eventually deforms plastically by slip, twinning or a coupled slip-twinning mechanism in many grains having different orientations. In contrast, if the structure is brittle, then it breaks at the onset of necking, which is attributed to a low load-bearing capacity of the structure to overcome the internal triaxial-state of stress.

Figure 12.3a It schematically shows different load $P = f(\mu)$ or $\sigma = f(\varepsilon)$ curves for brittle and elastic-plastic (ductile) materials, which have their respective modulus of elasticity E (Young's modulus) as a measure of their stiffness. Moreover, an elastic material exhibits a σ-ε straight line, which represents a linear behavior of a brittle solid (curve # 1 in Fig. 12.3a). Those materials showing linear and nonlinear behavior are referred to as elastic-plastic materials; however, some solids fracture at the maximum stress (curve # 2 in Fig. 12.3a), and some develop necking at the maximum stress, and consequently, the stress decreases until fracture occurs (curve # 3 in Fig. 12.3a).

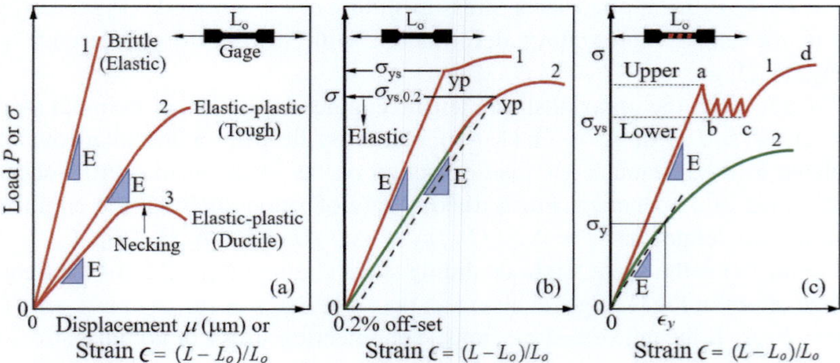

Fig. 12.3 Typical tensile load-strain curves for (**a**) brittle, tough, and ductile solids, (**b**) ductile solid and 0.2% offset yield point, and (**c**) ductile solid with upper and lower yield points and nonlinear ductile solids

Figure 12.3b The stress-strain transition from elastic to plastic behavior is defined as the elastic limit or the yield point, which is illustrated by curve #1 in Fig. 12.3b. In some materials, this transition is not easily determined with precision. In such a case, an approximation of a material's elastic limit called the 0.2% offset method is used by constructing a parallel line to the linear portion of the σ-ε curve as shown by curve # 2 in Fig. 12.3b. As a result, the 0.2% offset yield strength $\sigma_{ys,0.2}$ or σ_{ys} (0.2%) and the offset yield point (yp) are determined at the 0.2% offset line where the stress-strain curve is intercepted. Particularly, the offset yield strength $\sigma_{ys,0.2}$ is also called the proof stress and represents the stress at which the material is slight deformed plastically.

Figure 12.3c Some materials exhibit an irregular or non-continuous σ-ε curve with upper and lower yield points as shown by curve #1 in Fig. 12.3c. Notice that the stress drops from point "a" to point "b" leading to a fluctuating lower yield point or lower bound yield strength in the form of jogs from "b" to "c" representing the Luders bands on the σ-ε curve. The Luders bands cover the gage area prior to a nonlinear increase in stress from "c" to "d" points. Subsequently, the σ-ε curve resumes in a nonlinear fashion indicating plastic deformation until the material fractures. On the other hand, other materials show an entire nonlinear behavior as shown by curve #2 in Fig. 12.3c, which is used to determine a particular modulus of elasticity E at a selected strain ε_y.

12.5.1 Strain Softening and Strain Hardening

Figure 12.4a shows an engineering σ-ε curve for a hypothetical specimen tested, say, in tension. Notice that this material exhibits high strength ($E = 200\ GPa$, $\sigma_{ys} = 791\ MPa, \sigma_{ts} = 816\ MPa, \sigma_f = 451\ MPa$) and high ductility ($E_L = 12\%$ elongation). Here, E_L is a measure of the increase in specimen length (L) relative to its original gage length (L_o). Notice that the strength ratio $\sigma_{ts}/\sigma_{ys} \simeq 1.03$ is related to a narrow strain interval $\varepsilon_{ys} \leq \varepsilon \leq \varepsilon_{ts}$.

Most of the plastic deformation exhibited by this hypothetical material under tension loading is due to strain softening represented by the $\sigma_{ts} < \sigma < \sigma_f$ path, and consequently, the material stretches very significantly beyond the onset of necking. Based on this observation, strain softening is a degradation of the material strength with increasing steady strain at a constant strain rate. Therefore, strain softening is an indication of mechanical instability during plastic deformation.

Figure 12.4b, in contrast, illustrates an engineering σ-ε curve for a hypothetical specimen that exhibits strain hardening represented by the $\sigma_{ys} < \sigma < \sigma_{ts}$ path.

The overall shape of a tensile σ-ε curve depends on the history of a material relative to its microstructure. This means that a thorough analysis of the microstructure must be carried out prior to mechanical testing in a certain loading mode and environment.

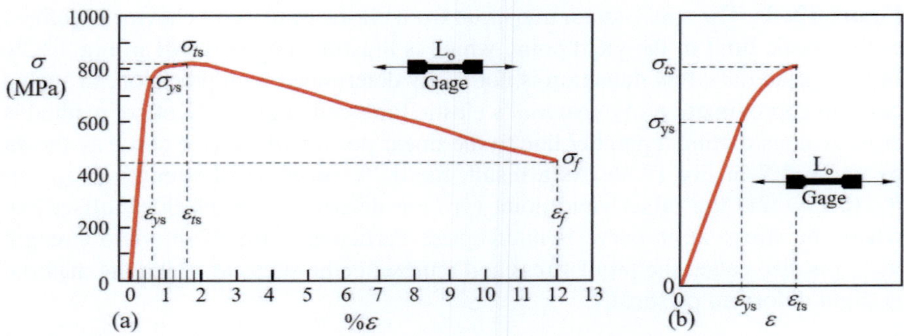

Fig. 12.4 Schematic engineering stress-strain curves for hypothetical specimens showing (**a**) strain softening and (**b**) strain hardening at a certain strain rate $d\varepsilon/dt$

12.5.2 Luders Bands

The Luders bands are shear or deformation bands related to the mechanical behavior of a crystalline solid undergoing plastic deformation, which may be referred to as the yield point phenomenon manifested as slip lines on the specimen gage length as depicted in Fig. 12.5a, which contains images taken as snap shots from a video posted online by Meier and Broumas [4].

Some metals, such as an annealed low-carbon steel (Fe-C) and aluminum-magnesium (Al-Mg) alloy sheets, exhibit upper yield point or upper bound yield stress as shown by curve # 1 in Figs. 12.3c and 12.5, where locked-in dislocations are released. Consequently, the stress drops from point "a" to point "b" leading to a fluctuating lower yield point or lower bound yield strength in the form of jogs from "b" to "c" representing the Luders bands on the σ-ε curve.

For the purpose of clarity, Fig. 12.5b schematically shows the σ-ε curve along with the assumed initial and final stages of the specimen gage length at the lower yield points.

The yielding phenomenon represents a localized heterogeneous transition from elastic to plastic deformation, which extends along the strain coordinate (ε) from point "a" to point "b" on curve # 1 in Fig. 12.3c. This phenomenon is called the yield-point elongation caused by atom movement (slip) to different locations in the lattice; therefore, slip is a lattice distortion.

Slip lines are known as Luders bands or stretcher strains that cover the gage length L_o. Once this is accomplished, the stress resumes in an increasing manner. Notice that the gage length L_o initially does not exhibit Luders bands, but they begin to appear as the specimen is steadily stretched in tension until the gage area is completely filled with these bands. Once the gage area is covered with Luders bands, the nonlinear stress increases as schematically indicated on curve # 1 in Fig. 12.3c from point "b" to point "c."

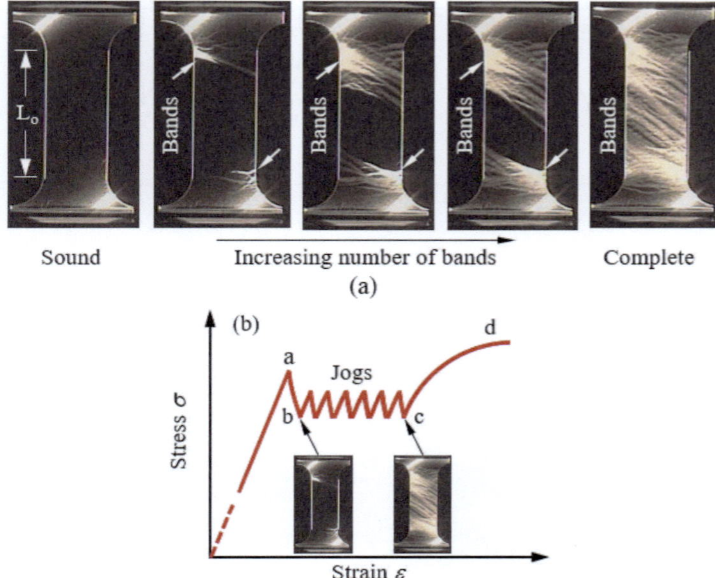

Fig. 12.5 (**a**) Luders bands (shear bands) in 1018 annealed steel. Images taken as snap shots from a video posted online by Mike Meier and Aaron Broumas (about 2009). Link: https://vimeo.com/4586024 [4] and (**b**) schematic stress-strain curves indicating the formation of jogs related to slip bands

The lower yield point is used to define the yield strength σ_{ys} as a material's property, and Luders bands are normally not acceptable aesthetically, but they can be removed by temper rolling that causes slight plastic deformation.

It is now appropriate to summarize the yielding phenomenon as per Fig. 12.3c. Thus,

- It is a localized heterogeneous transition from elastic to plastic deformation
- Some dislocations may torn at the upper stress level causing a drop in flow stress and a serrated (jogs) yielding process along the gage length of a specimen. This is the origin of lower yield stress.
- In polycrystalline materials, released dislocations at certain velocities tend to pile up at grain boundaries and hard particles (if any), producing localized stress concentrations.
- The jogs are referred to as a dynamic strain-aging process due to dislocation motion.
- The slip lines on the gage area are observed to be at approximately 45° with respect to the direction of the applied tensile or compressive stress.
- The upper yield point shown in Figs. 12.3c and 12.5b represents the proportional limit of the material.

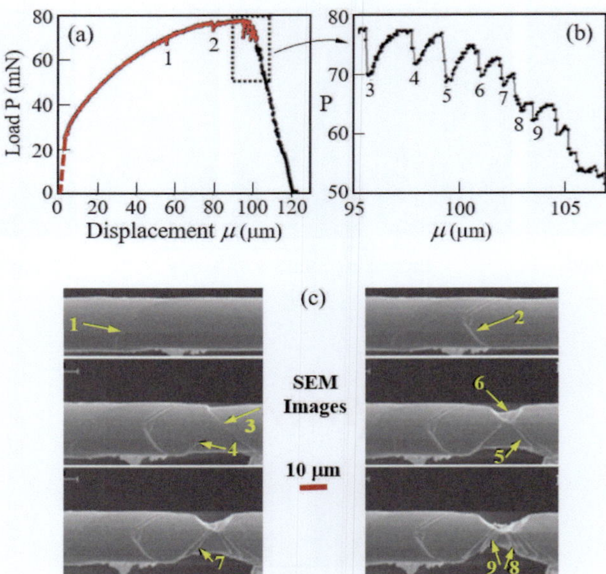

Fig. 12.6 Discontinuities known as pop-ins. (**a**) Load-displacement curve for a 25-μm-diameter annealed polycrystalline copper wire tested at room temperature. (**b**) Inset illustrating additional pop-ins after necking and (**c**) SEM images showing slip steps and necking. After Yang et al. [5]

A ductile crystalline material under tension loading may experience pop-ins shown as discontinuities on the load-displacement curve. One relevant example is shown in Fig. 12.6 for a 25-μm-diameter crystalline Cu wire tested at room temperature (Yang et al. [5]). The dashed straight line in Fig. 12.6a has been added to the original figure in order to show the linear portion of the load-displacement curve.

Notice the discontinuities of the P-μ curve in Fig. 12.6a and the inset in Fig. 12.6b. The shown discontinuities represent slip steps along the gage length (Fig. 12.6c). Eventually, the Cu wire (specimen) exhibits additional pop-ins after the necking process illustrated as a zigzag curve (Fig. 12.6b) and revealed as slip lines under a microscope: SEM images in Fig. 12.5c.

The experimental data depicted in Fig. 12.6 is a clear evidence of the plastic deformation mechanism in a crystalline material (Yang et al. [5]). This mechanism implies that an (hkl) slip plane is distorted by an external mechanical force, leading to plastic deformation along a dense $[uvw]$ crystallographic direction. Thus, a crystal deforms by the motion of a dislocation that is defined by the $(hkl)[uvw]$ slip system, which is further explained in a later section.

The combination of a slip plane and a slip direction constitute the slip mechanism denoted as $(hkl)[uvw]$, where the (hkl) slip plane contains a high atomic density and slips along the $[uvw]$ crystallographic direction. Moreover, experimental dataset shown in Fig. 12.6 is a relevant research finding for elucidating crystallographic slip mechanism as the most common deformation mechanism in crystalline materials.

12.6 Heterogeneity of Plastic Deformation

The heterogeneity of plastic deformation caused by slip and twinning mechanisms depends on the orientation of the unit cells with respect to the direction of the applied tensile or compressive stress that generates an internal shear stress. The most common crystallographic characteristics of plastic deformation are cited below.

- Slip is manifested by dislocation motion, and it is revealed under the microscope as a set of parallel thin lines due to slip steps corresponding to specific (hkl) slip planes.
- Mechanical twinning is seeing under the microscope as narrow bands or blocks of atoms with a needle-like morphology that move very rapidly at about the speed of sound (generation of sound waves) to acquire a different crystallographic orientation on specific twinning planes.
- Grain boundary sliding at high temperatures, where the grains are no longer strong barriers to dislocation motion, is the main cause of plastic deformation.
- Polycrystal solids containing many grains are mechanically stronger than single crystals at temperatures below the grain-boundary sliding temperature because grain boundaries are barriers to easy dislocation motion.

These four bullet points summarize the effect of individual mechanisms; however, a combination of them is possible.

12.7 Slip Systems

Close-packed crystallographic planes and directions are used to define particular slip systems representing the mechanism of lattice distortion (elastic or plastic deformation). One single slip system is denoted as $(hkl)\,[uvw]$. Many activated slip systems can be described by the general notation $\{hkl\}\,\langle uvw \rangle$, but it all depends on the loading mode (tension, torsion, etc.) and the orientation of the unit cells for deformation to occur. Thus, in general, plastic deformation of single crystals under tension initially occurs by slip-on parallel planes separated by the d-spacing (d_{hkl}). Essentially, slip is most susceptible to occur in close packed planes, where atoms move a certain distance producing a step. For instance, there are 48-slip systems in ferrite, which is an iron-based (α-Fe) phase having a BCC structure at room temperature, two-slip systems in an HCP structure, and 12-slip systems in an FCC structure. Among slip planes, the most probable slip systems in FCC and BCC crystals are denoted as $\{111\}\,\langle 110 \rangle$ and $\{110\}\,\langle 111 \rangle$, respectively.

12.8 Deformation of Single Crystals

Mechanically, slip occurs if an internal resolved shear stress ($\tau_r = RSS$) reaches a maximum or critical value ($\tau_r \rightarrow \tau_c = CRSS$), which depends entirely on the type of slip system and loading mode. Thus, slip is caused by a threshold value of stress called $CRSS$. During deformation of a single crystal, the slip planes reach a certain length, which are observed under the microscope as straight lines or usually as wavy lines. This suggests that slip occurs in close-packed planes by dislocation motion due to a resolved shear stress that overcomes, in general, the crystal resistance to plastic deformation. Further, relevant crystallographic information and mechanical behavior of single crystal are detailed in the next section.

12.8.1 Slip-Induced Plasticity

In materials science and physical metallurgy, slip is a lattice distortion representing plastic deformation due to one-dimensional dislocation motion along crystallographic shear planes and corresponding close-packed directions within the slip plane. Thus, the combination of relevant plane-direction, (hkl)-$[uvw]$, defines a slip system, and slip itself is the mechanism for plastic deformation.

Among many research articles on deformation of single crystals, Fig. 12.7a shows the slip lines in HCP zinc basal plane with a probable $(0001)\,[11\bar{2}0]$ slip system that evolves in tension mode (Smith [6, p. 279]).

Moreover, Fig. 12.7b illustrates an SEM image of a 5-μm-diameter FCC nickel (FCC-Ni) specimen (micropillar) exhibiting slip lines on the (111) plane under microcompression testing mode (Dimiduk et al. [7]). The set of slip lines represents the slipbands as indicative of plastic deformation induced by a shear stress (τ) along the close-packed crystallographic directions in the respective (hkl) slip plane.

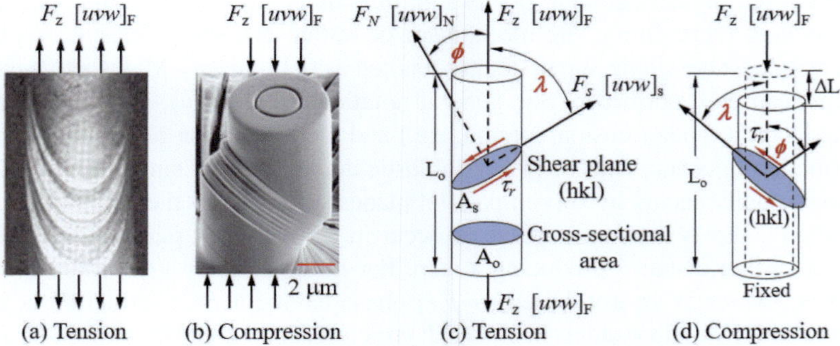

Fig. 12.7 Single crystals under force vector directions. (**a**) Slip lines under tension (Smith [6, p.279]), (**b**) parallel slip lines in the specimen loaded along the [269] direction in compression (Dimiduk et al. [7]), (**c**) tension loading model, and (**d**) compression loading model

Therefore, these images are evidence of the slip mechanism of plastic deformation of single crystals.

12.8.2   Schmid Factor

This law describes yielding of a single crystal when a critical resolved shear stress ($\tau_c = CRSS \neq 0$) activates at least one slip system under tension or compression loading. Recall that slip systems consist of crystallographic planes and directions denoted as $\{hkl\} \langle uvw \rangle$.

From the overwhelming literature on single crystals, consider modeling a single crystal bar loaded in tension (Fig. 12.7c) or in compression (Fig. 12.7d). These models show the loading $[uvw]_F$ direction along the z-axis with the axial force $F_z > 0$ in tension or $F_z < 0$ in compression (Mangonon [8, p. 275]), the shear force $[uvw]_s$ direction parallel to the shear (hkl) plane, the normal $[uvw]_N$ direction to the shear (hkl) plane, the original cross-sectional area A_o, the shear area A_s of the plane, and related ϕ, λ angles needed for deriving the resolved shear stress (τ_r) equation that initiates shear deformation. Subsequently, plastic deformation occurs when τ_r reaches a critical value $\tau_r = \tau_c$ (Mangonon [8, p. 276]).

For convenience, let's use the model in Fig. 12.7c so that the cosine equations are written as

$$\cos \phi = \frac{F_N}{F_z} = \frac{A_o}{A_s} \tag{12.3a}$$

$$A_s = A_o / \cos \phi \tag{12.3b}$$

$$\cos \lambda = \frac{F_r}{F_z} \tag{12.3c}$$

The resolved and normal forces are, respectively,

$$F_r = F_z \cos \lambda \tag{12.4a}$$

$$F_N = F_z \cos \phi \tag{12.4b}$$

and the theoretical shear stress along with Eqs. (12.3b) and (12.4a) is

$$\tau_r = \frac{F_r}{A_s} = \frac{F_z \cos \lambda}{A_o / \cos \phi} = \frac{F_z}{A_o} \cos \phi \cos \lambda \tag{12.5}$$

Introducing the applied stress $\sigma = F_z / A_o$ into Eq. (12.5) yields the Schmid factor defining the theoretical resolved shear stress (RSS) $\tau_r = \tau_{RSS}$ at σ and the critical resolved shear stress (CRSS) $\tau_c = \tau_{crss}$ for the onset of plastic deformation or yielding at $\sigma = \sigma_{ys}$

$$\tau_r = \sigma \cos \phi \cos \lambda = m\sigma \tag{12.6a}$$

$$\tau_r = \sigma \cos (\phi) \cos \left(\frac{\pi}{2} - \phi \right) = \frac{\sigma}{2} \sin (2\phi) \tag{12.6b}$$

$$\tau_c = \sigma_{ys} \cos \phi \cos \lambda = m\sigma_{ys} \tag{12.6c}$$

$$\tau_c = \frac{\sigma_{ys}}{2} \sin (2\phi) \tag{12.6d}$$

where $m = \cos(\lambda) \cos(\phi)$ is the Schmid factor with λ, ϕ angles between crystallographic directions and $\sigma = \sigma_{ys}$ is the yield strength of the single crystal at the onset of plastic deformation or fracture.

The normal stress to the slip plane is a stress component defined as

$$\sigma_N = \frac{F_N}{A_s} = \frac{F_z \cos \phi}{A_o / \cos \phi} \tag{12.7a}$$

$$\sigma_N = \sigma \cos^2 \phi \tag{12.7b}$$

There are some conditions to be considered about the deformation of a single crystal. Thus,

- Yielding occurs when $\tau_r = \tau_c$ at $\sigma = \sigma_{ys}$ along a $(hkl)\,[uvw]_s$ slip system. If $\tau_r \to \tau_c$, then slip occurs by dislocation motion.
- For $\phi = \lambda = 45°$, the applied stress is $\sigma = 2\tau_r$ or $\sigma_{ys} = 2\tau_c$ (Mangonon [8, p. 276]).
- For parallel stresses $\sigma_{ys} \parallel \tau_c$, the direction angles are $\phi = 90°$ and $\lambda = 0$. Consequently, $\tau_c = 0$ and the crystal fractures.
- For perpendicular stresses $\sigma_{ys} \perp \tau_c$, he direction angles are $\phi = 0°$ and $\lambda = 90°$. Therefore, $\tau_c = 0$ and the crystal fractures.

12.8.3   Vector Dot Product

The dot product of two vectors **A** and **B** is denoted as **A** · **B** (read A dot B), and it is defined as the product of the magnitudes of the two vectors ($|A|\,|B|$) and the cosine of the angle between the two vectors, say, $\cos(\theta)$. The magnitude or absolute value of a vector $|A|$ or $|B|$ is the square root of the sum of the square vector components. Moreover, if **A** · **B** = **0**, the vectors **A** and **B** are orthogonal (perpendicular) since $\cos(90°) = \cos(\pi/2) = 0$.

Mathematically, the general definition of the dot product is

$$\mathbf{A} \cdot \mathbf{B} = |A|\,|B| \cos (\theta) \tag{12.8a}$$

$$\cos(\theta) = \frac{\mathbf{A} \cdot \mathbf{B}}{|A|\,|B|} = \frac{[A_1 + A_2 + A_3] \cdot [B_1 + B_2 + B_3]}{\sqrt{A_1^2 + A_2^2 + A_3^2}\sqrt{B_1^2 + B_2^2 + B_3^2}} \tag{12.8b}$$

$$\cos(\theta) = \frac{A_1 B_1 + A_2 B_2 + A_3 B_3}{\sqrt{A_1^2 + A_2^2 + A_3^2}\sqrt{B_1^2 + B_2^2 + B_3^2}} \tag{12.8c}$$

In crystallography, these vectors become the vector directions so that

$$[\mathbf{uvw}]_1 \cdot [\mathbf{uvw}]_2 = |[\mathbf{uvw}]_1|\,|[\mathbf{uvw}]_2|\cos(\theta) \tag{12.9a}$$

$$\cos(\theta) = \frac{[\mathbf{uvw}]_1 \cdot [\mathbf{uvw}]_2}{|[\mathbf{uvw}]_1|\,|[\mathbf{uvw}]_2|} \tag{12.9b}$$

$$\cos(\theta) = \frac{[u_1 + v_1 + w_1] \cdot [u_2 + v_2 + w_2]}{\sqrt{u_1^2 + v_1^2 + w_1^2}\sqrt{u_2^2 + v_2^2 + w_2^2}} \tag{12.9c}$$

$$\cos(\theta) = \frac{u_1 u_2 + v_1 v_2 + w_1 w_2}{\sqrt{u_1^2 + v_1^2 + w_1^2}\sqrt{u_2^2 + v_2^2 + w_2^2}} \tag{12.9d}$$

Hence, Eq. (12.9c) or (12.9d) is a general expression that can be used to determine the angle between two crystallographic directions.

Example 12.1   Consider a small rectangular bar made out of an *FCC* single crystal with the unit cells oriented along with the (111) plane and probable shear directions as schematically shown below.

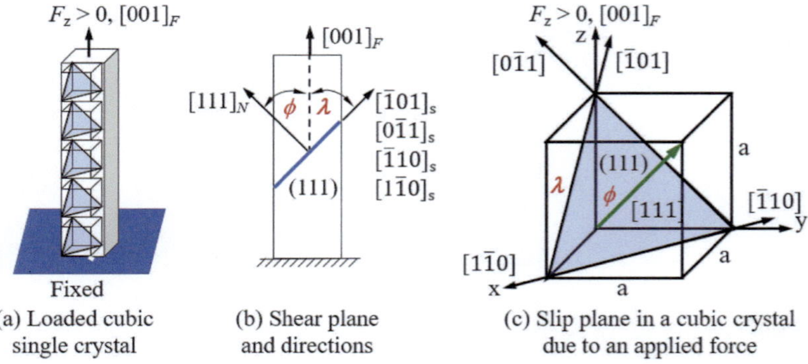

Will slip occur in all the shown directions? Evaluate all probable slip systems using the dot product.

Solution   For $[001]_F \angle [111]_N$ with $\theta = \phi$ in Eq. (12.9d),

$$\cos(\phi) = \frac{[\mathbf{uvw}]_F \cdot [\mathbf{uvw}]_N}{|[\mathbf{uvw}]_F|\,|[\mathbf{uvw}]_N|} \tag{12.1E1a}$$

$$\cos(\phi) = \frac{[001]_F \cdot [111]_N}{\sqrt{\left[(0)^2 + (0)^2 + (1)^2\right]_F}\sqrt{\left[(1)^2 + (1)^2 + (1)^2\right]_N}} \tag{12.1E1b}$$

$$\cos(\phi) = \frac{(0)(1) + (0)(1) + (1)(1)}{\sqrt{1}\sqrt{3}} = \frac{1}{\sqrt{3}} \tag{12.1E1c}$$

$$\phi = \cos^{-1}\left(\frac{1}{\sqrt{3}}\right) = 54.74^0 \tag{12.1E1d}$$

Similarly, for $[001]_F \measuredangle \left\langle \left[\bar{1}01\right]_s, \left[0\bar{1}1\right]_s, \left[\bar{1}10\right]_s, \left[1\bar{1}0\right]\right\rangle_s$ with $\theta = \lambda$,

$$\cos(\lambda) = \frac{[\mathbf{uvw}]_F \cdot [\mathbf{uvw}]_s}{|[\mathbf{uvw}]_F| \, |[\mathbf{uvw}]_s|} \tag{12.1E2a}$$

$$\cos(\lambda_1) = \frac{[001]_F \cdot \left[\bar{1}01\right]_s}{\sqrt{\left[(0)^2 + (0)^2 + (1)^2\right]_F}\sqrt{\left[(-1)^2 + (0)^2 + (1)^2\right]_s}} = \frac{1}{\sqrt{2}} \tag{12.1E2b}$$

$$\cos(\lambda_2) = \frac{[001]_F \cdot \left[0\bar{1}1\right]_s}{\sqrt{\left[(0)^2 + (0)^2 + (1)^2\right]_F}\sqrt{\left[(0)^2 + (-1)^2 + (1)^2\right]_s}} = \frac{1}{\sqrt{2}} \tag{12.1E2c}$$

$$\cos(\lambda_3) = \frac{[001]_F \cdot \left[\bar{1}10\right]_s}{\sqrt{\left[(0)^2 + (0)^2 + (1)^2\right]_F}\sqrt{\left[(0)^2 + (-1)^2 + (1)^2\right]_s}} = 0 \tag{12.1E2d}$$

$$\cos(\lambda_4) = \frac{[001]_F \cdot \left[1\bar{1}0\right]_s}{\sqrt{\left[(0)^2 + (0)^2 + (1)^2\right]_F}\sqrt{\left[(0)^2 + (-1)^2 + (1)^2\right]_s}} = 0 \tag{12.1E2e}$$

$$\lambda_{1,2} = \cos^{-1}\left(\frac{1}{\sqrt{2}}\right) = \frac{\pi}{4} = 45^\circ \tag{12.1E2f}$$

$$\lambda_{3,4} = \frac{\pi}{2} = 90^\circ \tag{12.1E2g}$$

Then, the Schmid factor becomes

$$\cos(\lambda_{1,2})\cos(\phi) = \left(1/\sqrt{2}\right)\left(1/\sqrt{3}\right) = 1/\sqrt{6} = 0.40825 \tag{12.1E3a}$$

$$\cos(\lambda_{3,4}), \cos(\phi) = (0)\left(1/\sqrt{3}\right) = 0 \tag{12.1E3b}$$

Therefore, the dot product yields only two activated slip systems: $[001]_F \left[\bar{1}01\right]_s$ and $[001]_F \left[0\bar{1}1\right]_s$ with $\tau_r > 0$.

Example 12.2 Consider the sketches below for a hypothetical *BCC* single crystal bar undergoing deformation along the slip systems $(110)_s\left[1\bar{1}1\right]_s$ and $(110)_s\left[\bar{1}11\right]_s$. Calculate the resolved shear stress τ_r for each case when the applied tensile stress is $9\ MPa$ along the z-axis. Determine if this material deforms plastically or fractures under the given conditions. Data: $\tau_c = 15\ MPa$.

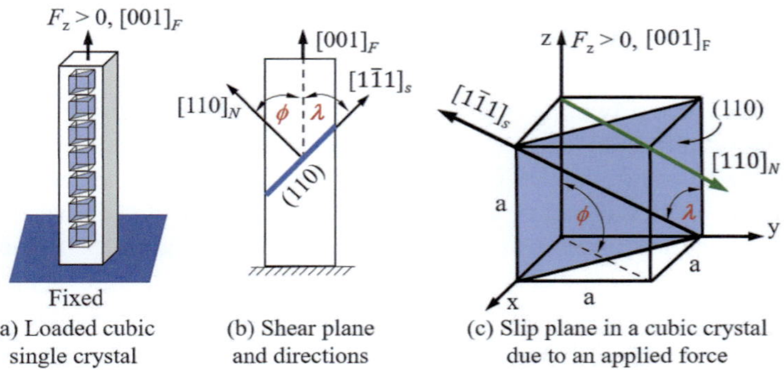

Solution The angle between the applied tensile stress σ along the $[001]_F$ direction and the $[110]_N$ normal direction to the (110) plane is denoted as $[001]_F\angle[110]_N$. From Eq. (12.9d) with $\theta = \phi$,

$$\cos(\phi) = \frac{[\mathbf{uvw}]_F\cdot[\mathbf{uvw}]_N}{|[\mathbf{uvw}]_F||[\mathbf{uvw}]_N|} = \frac{[001]_F\cdot[110]_N}{|[001]_F||[110]_N|} \tag{12.2E1a}$$

$$\cos(\phi) = \frac{(0)(1)+(0)(1)+(1)(0)}{\sqrt{(0)^2+(0)^2+(1)^2}\sqrt{(1)^2+(1)^2+(0)^2}} = 0 \tag{12.2E1b}$$

$$\phi = 90° \tag{12.2E1c}$$

From Eq. (12.9d), the angle $\theta = \lambda$ between the applied stress and the shear directions is denoted as $[001]_F\angle\left[1\bar{1}1\right]_s$, and its magnitude is

$$\cos(\lambda) = \frac{[\mathbf{uvw}]_F\cdot[\mathbf{uvw}]_s}{|[\mathbf{uvw}]_F||[\mathbf{uvw}]_s|} = \frac{[001]_F\cdot\left[1\bar{1}1\right]_s}{|[001]_F|\left|\left[1\bar{1}1\right]_s\right|} \tag{12.2E2a}$$

$$\cos(\lambda) = \frac{(0)(1)+(0)(-1)+(1)(1)}{\sqrt{(0)^2+(0)^2+(1)^2}\sqrt{(1)^2+(-1)^2+(1)^2}} \tag{12.2E2b}$$

$$\cos(\lambda) = \frac{1}{\sqrt{1}\sqrt{3}} = \frac{1}{\sqrt{3}} \tag{12.2E2c}$$

$$\lambda = \cos^{-1}\left(\frac{1}{\sqrt{3}}\right) = 54.74° \tag{12.2E2d}$$

Then, Eq. (12.6a) yields

$$\tau_r = \sigma \cos \phi \cos \lambda = (0)\left(2/\sqrt{6}\right) = 0$$

Therefore, the single crystal fractures because $\tau_r = 0$.

12.8.4   Twinning-Induced Plasticity

In materials science, twining is referred to as (1) **annealing twinning** which occurs during recrystallization and grain growth, mainly in FCC metals, such as α-brass, γ-Fe austenite, copper, gold, nickel, and the like. Fundamentally, annealing twinning involves nucleation and growth during heat treatment, where atomic diffusion is the basic mechanism in this process. (2) **Mechanical twinning** is a mechanical deformation mechanism that occurs in HCP and BCC crystals along definite crystallographic planes and specific directions. This mechanical process occurs due to a local critical resolved shear stress (τ_c) acting on a twinning plane with a different orientation with respect to the host lattice planes.

Comparing slip caused by dislocation motion in brass in Fig. 12.8a (Callister and Rethwisch [3, p. 201]) and by mechanical twinning in AISI 304 stainless steel (Fig. 12.8b), one can see that slip is defined by parallel thin lines and the mechanical twinning is defined by wide needle-like lines. Moreover, Fig. 12.8c shows a simple model for plastic deformation by slip mechanism, and Fig. 12.8d schematically illustrates the model for plastic deformation by mechanical twinning (Christian and Mahajan [9]).

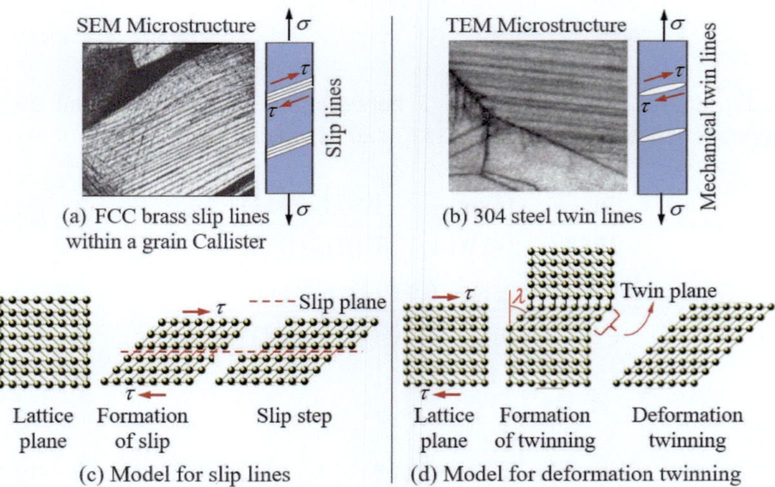

Fig. 12.8 Comparison of slip and mechanical twinning. (**a**) Slip lines in a brass grain (Callister and Rethwisch [3, p. 201], (**b**) deformation twins in AISI 304 stainless steel, (**c**) slip mechanism model, and (**d**) deformation twinning model (Christian and Mahajan [9])

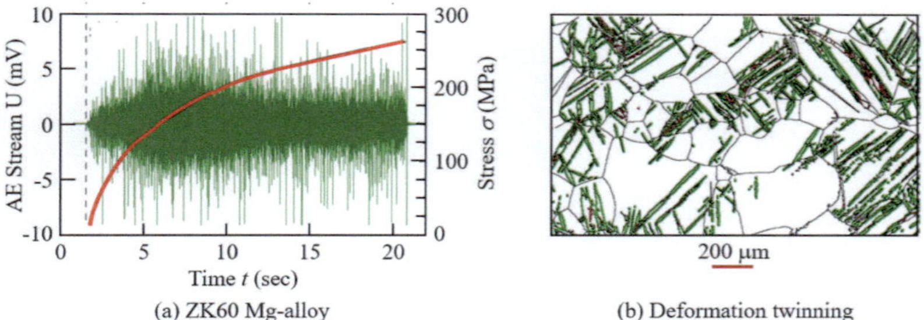

(a) ZK60 Mg-alloy   (b) Deformation twinning

Fig. 12.9 (**a**) Acoustic emission (AE) data synchronized with a tension loading curve of a ZK60 Mg-alloy specimen during deformation by twinning (Vinogradov et al. [10]) and (**b**) EBSD orientation map of titanium (Ti) deformation twinning along $\{11\bar{2}2\}$ planes (Ghaderi and Barnett [11])

(a)   (b)

Fig. 12.10 Comparison of (**a**) annealing twins in a Co-based alloy (Vander Voort [12]) and (**b**) mechanical twins in a zirconium specimen (Dimiduk et al. [7])

It is evident that deformation by mechanical twinning is an audible process (called Tin cry) when twin bands acquire a new crystallographic orientation very rapidly. This implies that a group of the atoms move simultaneously quite rapidly on particular twin planes, generating sound waves (vibrations) at a high strain rate. For instance, Fig. 12.9a shows experimental results for a synchronized tension loading and acoustic emission (AE) on a magnesium (Mg) alloy (Vinogradov et al. [10]), while Fig. 12.9b exhibits mechanical twins (green color) in a 204-μm grain size titanium (Ti) specimen being deformed at a strain rate $d\varepsilon/dt = 0.14\ s^{-1}$ (Ghaderi and Barnett [11]).

For comparison, Fig. 12.10a shows annealing twins with a rectangular-like shape within FCC brass grains (Vander Voort [12]) and mechanical twins (Fig. 12.10b) with a needle-like shape with diverse lengths in HCP zirconium Zr (Dimiduk et al. [7]).

Effect of Solidification Methods   Figure 12.11 shows mechanical twins in an AISI 304 stainless steel being produced by rapid solidification (RSA 304 in Fig. 12.11a)

Fig. 12.11 Mechanical twins in 304 stainless steel being deformed in tension. (**a**) RSA 304 and (**b**) IM 304 (Perez [13])

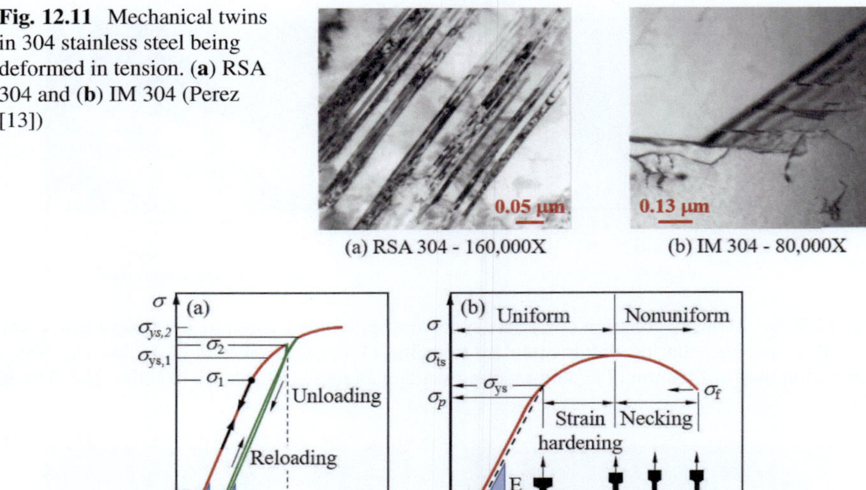

Fig. 12.12 Characteristics of stress-strain curves. (**a**) Loading-unloading stress trendline and (**b**) necking process

and by conventional solidification (ingot metallurgy, IM 304 in Fig. 12.11b) (Perez [13]). Their chemical composition is the same, but their grain morphology is slightly different. Nonetheless, plastic deformation by mechanical twinning prevails in both *FCC* austenitic materials.

Clearly, a few dislocations are revealed in Fig. 12.11b. Other microstructural features and mechanical behavior in diverse environments can be found elsewhere (Perez [13]).

12.8.5   Loading-Unloading Stress

This section is an extension of the stress-strain relationship described above at a macroscale. Firstly, Fig. 12.12a shows a hypothetical stress-strain curve with loading-unloading paths at $0 \leq \sigma_1 < \sigma_{ys,1}$. The elastic σ-ε path upon unloading from σ_1 returns to zero stress, and as a result, the total elastic strain is recovered. Theoretically, during elastic deformation, the atoms are displaced a small amount without acquiring new atomic positions, and upon unloading, the atoms return to their original positions.

Secondly, when this material is unloaded from $\sigma_2 > \sigma_{ys,1}$, the elastic strain ε_e is partially recovered, and the material retains a fraction of plastic strain ε_p. The loading-unloading trajectory has a slope or modulus of elasticity $E_2 \simeq E_1 = E$. In fact, E_2 is slight less than E_1. Reloading the material again, the σ-ε path is similar to the unloading path, and it connects to the original or virgin stress-strain curve, generating a second yield point for a new yield strength $\sigma_{ys,2}$ due to strain hardening caused by the formation of dislocations (line defects). Consequently, the dislocation density (ρ) increases from its initial value at σ_{ys} until it reaches a saturated magnitude at σ_{ts}. This process is induced by a shear strain rate $(d\gamma/dt)$ related to dislocation velocity (v).

Figure 12.12b shows a smooth elastic-plastic transition on the stress-strain curve, where the actual onset of plastic deformation at a macroscopic level is called proportional limit σ_p, but it is difficult to measure and reveal it graphically. Conventionally, the elastic-plastic transition is determined on a macroscopic level using the commonly known 0.2% offset method, previously introduced in Fig. 12.3b and again shown in Fig. 12.12b. It is determined by drawing a (dashed) straight line parallel to the elastic portion of the stress-strain curve until it intersects the σ-ε curve, establishing the yield point (yp) related to the yield strength σ_{ys} of the material. Other aspects of the engineering σ-ε curve is related to the strain hardening process represented by the σ-ε curve at $\sigma_{ys} \leq \sigma \leq \sigma_{ts}$, where the specimen permanently changes shape and dislocations are massively produced.

From Fig. 12.12b, the nomenclature of relevant significance about the stress-strain (σ-ε) curve are summarized below. Thus,

- The proportional limit σ_p is the stress limit defining the elastic-plastic transition on a stress-strain curve.
- The yield strength σ_{ys} is the conventional engineering elastic limit, which, in turn, is the maximum stress used by Hooke's law or Hooke's equation.
- The tensile strength σ_{ts} is the maximum stress known as the ultimate tensile strength or UTS. A specimen loaded at $(\varepsilon, \sigma)_{ts}$ is under a local triaxial-state of stress, and as a consequence, the specimen fractures or continues to deform plastically by necking.
- The fracture strength σ_f is the stress at fracture or failure.
- The modulus of elasticity $E = d\sigma/d\varepsilon \simeq \Delta\sigma/\Delta\varepsilon$ is the linear slope of the σ-ε curve that describes the stiffness of a material.

These are the most common mechanical properties used to characterize engineering materials. Specifically, σ_{ys} and σ_{ts} are the most relevant properties in engineering design schemes, where the design or working stress for a particular structural part is simply defined as $\sigma_d = \sigma_{ys}/S_f$ or $\sigma_d = \sigma_{ts}/S_f$ with a safety factor $S_f > 1$.

12.8.6 Strain Hardening Region

The gradual transition from elastic-plastic flow indicated in Fig. 12.12b is related to:

- The hardening process of the lattice structure treated as the crystallographic matrix.

- An induced triaxial-state of stress may cause rapid failure at onset of necking, where the stress-strain relationship is treated as an instability point at $\sigma = \sigma_{ts}$.
- Increasing dislocation density as $\sigma \to \sigma_{ts}$
- Increasing plasticity since ductility is exhausted in this region.

The plastic region on a σ-ε curve occurs at $\sigma_{ys} \leq \sigma \leq \sigma_{ts}$, and it represents a gradual increase in stress since the material hardens due to the interaction of dislocations from different crystallographic planes. Consequently, the dislocation density (lines/area) increases as the load-bearing capacity of the material decreases, and eventually, the stress reaches a maximum value called the tensile strength, which represents a point of instability for the material to break because it cannot withstand the internal triaxial state of stress or to deform further by necking at the gage length weakest point.

12.8.7   Necking at Instability Point

According to the Considere criterion, the onset of necking occurs at the ultimate or maximum tensile stress σ_{ts} on a stress-strain curve and continuous in the region where $\sigma_{ts} \leq \sigma \leq \sigma_f$ (Fig. 12.12b). Here, σ_{ts} is the transition stress level for developing the instability point, where necking of the gage length begins at a high-stress concentration point, where the strain is known as the strain-hardening exponent or work-hardening coefficient ($\varepsilon_{ts} = n$). This is attributed to the material's defects, and eventually, the material's load-bearing capacity is affected by the reduction of the cross-sectional area. Regarding the strain hardening exponent n, its value depends on the stress power laws described in the next section.

12.9   Analytical Stress-Strain Relationships

This section includes an analytical procedure for assessing the strain hardening region on the tensile stress-strain curve (Fig. 12.13a) based on the concepts of true (instantaneous) and engineering strains at $\varepsilon_{ys} \leq \varepsilon \leq \varepsilon_{ts}$. By definition, strain hardening is called work hardening, and it is related to plastic deformation due to an increasingly saturated gage length (L_o) by linear defects called dislocations.

The area under the engineering σ-ε curve (Fig. 12.13a) for uniform deformation at $0 < \varepsilon \leq \varepsilon_{ts}$ can be divided into two regions; the uniform elastic-deformation region described by Hooke's law (Hooke's equation) at $0 < \varepsilon \leq \varepsilon_{ys}$ and the uniform plastic-deformation region described by a power law equation (flow stress expression) at $\varepsilon_{ys} < \varepsilon \leq \varepsilon_{ts}$ and $\sigma_{ys} < \sigma \leq \sigma_{ts}$.

The most common power law relationships are defined by Hollomon's equation [14], Ludwik's equation [15], and Ramberg-Osgood equation [16]. For clarity, the Ludwik's equation is cited in Hollomon's paper [14]. Nevertheless, the

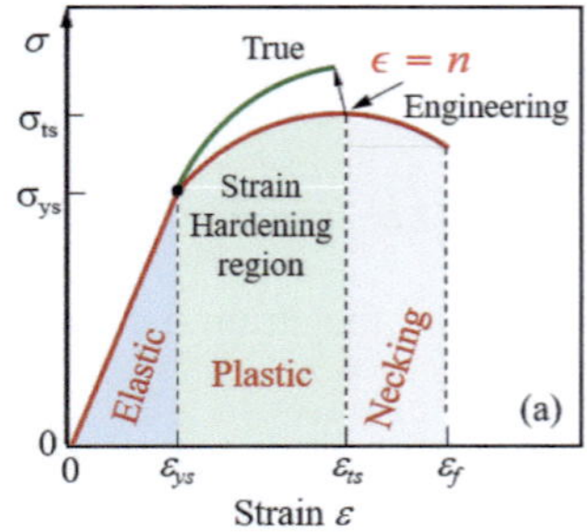

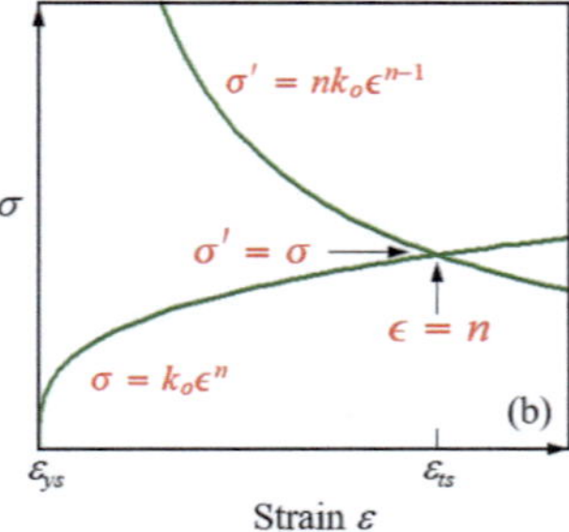

Fig. 12.13 Stress-strain curves for characterizing (**a**) the strain hardening region based on true and engineering stresses and (**b**) the point of instability

equations are

$$\sigma = \varepsilon E \quad \text{for } 0 < \varepsilon \leq \varepsilon_{ys} \qquad \text{(Hooke's equation)} \qquad (12.10a)$$

$$\sigma = k_o \varepsilon^n \quad \text{for } \varepsilon_{ys} < \varepsilon \leq \varepsilon_{ts} \qquad \text{(Hollomon's equation)} \qquad (12.10b)$$

$$\sigma = \sigma_{ys} + k_o \varepsilon^n \quad \text{for } \varepsilon_{ys} < \varepsilon \leq \varepsilon_{ts} \quad \text{(Ludwik's equation)} \qquad (12.10c)$$

$$\varepsilon = \frac{\sigma}{E} + k_o \left(\frac{\sigma}{E}\right)^{1/n} \quad \text{(Ramberg-Osgood equation)} \qquad (12.10d)$$

$$\sigma = C_o \left(\frac{d\varepsilon}{dt}\right)^m \qquad (12.10e)$$

Here, k_o is a constant called the strength coefficient, and $0 \leq n \leq 1$ is the strain hardening exponent range with $n = \varepsilon = \varepsilon_{ts}$ at the σ_{ts}-ϵ_{ts} instability point due to a triaxial-state of stress, where the material either undergoes necking or fractures, C_o is also a strength coefficient (constant), m is the strain-rate sensitivity exponent, and $d\varepsilon/dt$ is the true strain rate.

The exponent n is a measure of the capacity of a material in gaining strength during plastic deformation due to an increase in dislocation density. The magnitude of n can be obtained from the slope of the stress-strain curve in log-log scale.

The variables σ and ε are elastic entities as per Hooke's law and plastic entities as per Hollomon's and Ludwik's empirical models. The terms $\varepsilon_e = \sigma/E$ and $\varepsilon_p = k_o (\sigma/E)^{1/n}$ in Ramberg-Osgood equation stand for elastic and plastic strain components, respectively. These can be modified to find the best fit of experimental data. For instance, Hollomon's equation can be substituted into the plastic term of the Ramberg-Osgood equation.

The physical meaning of $n = 0$ implies that the material is perfectly plastic, $n = 1$ is for a perfectly elastic material, and $0 < n < 1$ is the proper range for most engineering materials. Moreover, the rate of strain hardening ($d\sigma/d\varepsilon$) at $\sigma_{ys} < \sigma \leq \sigma_{ts}$ and $\varepsilon_{ys} < \varepsilon \leq \varepsilon_{ts}$ is derived from Eq. (12.10b) or (12.10c). For simplicity, let's use Hollomon's equation so that

$$\frac{d\sigma}{d\varepsilon} = \sigma' = nk_o\varepsilon^{n-1} = nk_o\varepsilon^n\varepsilon^{-1} = \frac{nk_o\varepsilon^n}{\varepsilon^1} \tag{12.11a}$$

$$\frac{d\sigma}{d\varepsilon} = \frac{n\sigma}{\varepsilon} \tag{12.11b}$$

and at the instability point, the strain is $\varepsilon = \varepsilon_{ts}$ and the strain hardening exponent becomes

$$\frac{d\sigma}{d\varepsilon} = \sigma' = \sigma \tag{12.12a}$$

$$\frac{n\sigma}{\varepsilon} = \sigma \tag{12.12b}$$

$$n = \varepsilon = \varepsilon_{ts} \tag{12.12c}$$

The trendlines given by Eqs. (12.10b) and (12.11b) are clearly shown in Fig. 12.13b, and their intersection point defines $n = \varepsilon$ and $\sigma' = \sigma$ at instability induced by an internal triaxial state of tress represented by the principal σ_x, σ_y and σ_z entities.

The strain hardening exponent n is defined as the slope of a plastic stress-strain curve as shown in the example below.

Example 12.3 Use Hollomon's equation to graphically show how to determine the strain hardening exponent n for a strain hardenable material that exhibits a hypothetical engineering stress-strain curve (recast Fig. 12.13a).

Solution From Eq. (12.10b), Hollomon's equation gives

$$\sigma = k_o\varepsilon^n \quad \text{for } \varepsilon_{ys} < \varepsilon \leq \varepsilon_{ts} \tag{12.3E1a}$$

$$\ln(\sigma) = \ln(k_o) + n\ln(\varepsilon) \tag{12.3E1b}$$

$$n = \frac{\ln(\Delta\sigma)}{\ln(\Delta\varepsilon)} \tag{12.3E1c}$$

$$k_o = \frac{\sigma}{\varepsilon^n} \quad \text{for } \sigma_{ys} < \sigma \leq \sigma_{ts} \tag{12.3E1d}$$

and from Eq. (12.10c), Ludwik's equation yields

$$\sigma = \sigma_{ys} + k_o\varepsilon^n \quad \text{for } \varepsilon_{ys} < \varepsilon \leq \varepsilon_{ts} \tag{12.3E2a}$$

$$\ln(\sigma - \sigma_{ys}) = \ln(k_o) + n\ln(\varepsilon) \tag{12.3E2b}$$

$$n = \frac{\ln\left[\Delta(\sigma - \sigma_{ys})\right]}{\ln(\Delta\varepsilon)} \tag{12.3E2c}$$

$$k_o = \frac{\sigma - \sigma_{ys}}{\varepsilon^n} \quad \text{for } \sigma_{ys} < \sigma \leq \sigma_{ts} \tag{12.3E2d}$$

The sketches given below show the true and engineering σ-ε curves and the linearized approach for determining the n and k_o variables. This graphical approach is suitable for a smooth and continuous engineering σ-ε curve at $\varepsilon_{ys} < \varepsilon \le \varepsilon_{ts}$.

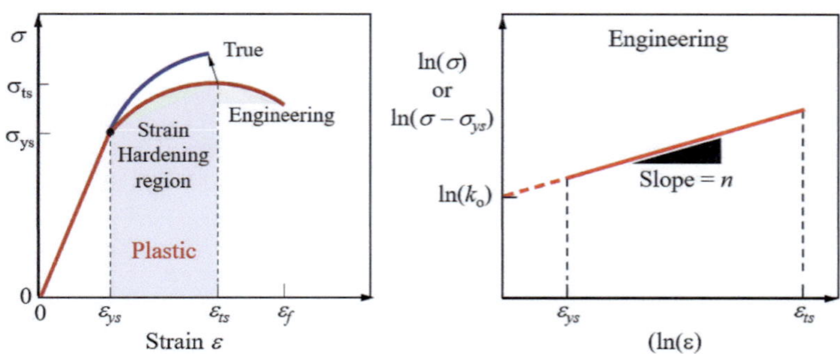

According to linear approach, two or more data points within the strain range $\varepsilon_{ys} < \varepsilon \le \varepsilon_{ts}$ are needed for determining the strain hardening exponent n. Notice that the true stress-strain curve is above the engineering curve because the former uses the instantaneous cross-sectional area $A < A_o$ at $\varepsilon_{ys} < \varepsilon \le \varepsilon_{ts}$ to determine the true stress $\sigma_t = P/A$, while the engineering stress, $\sigma_e = P/A_o$, is based on the original cross-sectional area. Therefore, the true stress is $\sigma_t > \sigma_e$ at $\varepsilon_t > \varepsilon_{ys}$ since is $A_o > A$.

True Strain, Engineering Strain Following Hollomon's theoretical work [14], the stress (σ) is defined as the mechanical load (P) divided by the instantaneous cross-sectional area (A) of a specimen. Thus, the load expression and its derivatives are

$$P = A\sigma \tag{12.13a}$$

$$dP = Ad\sigma + \sigma dA \tag{12.13b}$$

$$0 = Ad\sigma + \sigma dA \tag{12.13c}$$

$$\frac{d\sigma}{\sigma} = -\frac{dA}{A} \tag{12.13d}$$

Now, using the concept of volume constancy yields

$$AL = A_o L_o = C \tag{12.14a}$$

$$AdL + LdA = 0 \tag{12.14b}$$

$$\frac{dA}{A} = -\frac{dL}{L} = -d\varepsilon \tag{12.14c}$$

where C denotes a constant. Integrating Eq. (12.14c) yields the true or effective strain equation cited in Hollomon's paper [14] which was already suggested by

Ludwik [15] in 1909

$$\int_0^{\varepsilon_t} d\varepsilon = -\int_{A_o}^{A} \frac{dA}{A} = \int_{L_o}^{L} \frac{dL}{L} \tag{12.15a}$$

$$\varepsilon_t = \ln\left(\frac{A_o}{A}\right) = \ln\left(\frac{L}{L_o}\right) \tag{12.15b}$$

The engineering strain equation, on the other hand, is defined as

$$\varepsilon = \int_{L_o}^{L} \frac{dL}{L_o} = \frac{1}{L_o}\int_{L_o}^{L} dL = \frac{L - L_o}{L_o} \tag{12.16a}$$

$$\frac{L}{L_o} = 1 + \varepsilon \tag{12.16b}$$

Substituting Eq. (12.16b) into (12.15b) gives the coupled true and engineering strains

$$\varepsilon_t = \ln\left(1 + \varepsilon\right) \tag{12.17a}$$

$$\varepsilon = \exp\left(\varepsilon_t\right) - 1 \tag{12.17b}$$

True Stress, Engineering Stress   For the engineering stress, solve Eq. (12.14a) for A_o/A, and combine it with Eq. (12.13a) so that

$$\frac{A_o}{A} = \frac{L}{L_o} \tag{12.18a}$$

$$\sigma_t = \frac{P}{A} = \frac{P}{A}\frac{A_o}{A_o} = \frac{P}{A_o}\frac{A_o}{A} = \sigma\left(\frac{A_o}{A}\right) \tag{12.18b}$$

$$\sigma_t = \sigma\left(\frac{L}{L_o}\right) \tag{12.18c}$$

These equations are common in university classrooms and relevant textbooks on materials science and engineering materials.

Substituting Eq. (12.16b) into (12.18c) gives the coupled true-engineering stress relationship

$$\sigma_t = \sigma\left(1 + \varepsilon\right) \tag{12.19}$$

Combining Eqs. (12.17b) and (12.19) yields the true power law expression as

$$\sigma_t = \sigma\exp\left(\varepsilon_t\right) \tag{12.20}$$

which suggests that $\sigma_t > \sigma$ as $\varepsilon_t > \varepsilon$ at $\varepsilon_{ys} < \varepsilon \le \varepsilon_{ts}$.

12.10   Strain Energy Density

The physical event for mechanical deformation of a solid with a particular shape can be described by the action of an apply load P for elongating or stretching the solid along a specific coordinate direction. Thus, the work done W by the load P causes mechanical deformation. Initially, the uniform linear elastic is followed by nonlinear plastic deformation.

12.10.1   Load and Stress Curves

Consider a specimen with an original gage length L_o and an original uniform cross-sectional area A_o being loaded in tension as shown in Fig. 12.14a. Assume that the load-displacement diagram shown in Fig. 12.14b corresponds to the uniform deformation of the specimen. Thus, the work done (W) by the load P causes the specimen to elongate a small amount dx as P continuously increases at a steady-state condition. The shaded strip in Fig. 12.14b is the region bounded by the given function $P\,(x)$, and it is the area under the curve. A similar approach can be used for the stress function $\sigma\,(\varepsilon)$ (Fig. 12.14c).

For clarity, a load-displacement curve is not reproducible for different specimen dimensions of the same material. Changing the specimen dimensions for the same material requires another load-displacement curve. Instead, a stress-strain curve is ideally reproducible due to the normalized load by the original cross-sectional area $(\sigma = P/A_o)$ and the specimen extension/compression by the original gage length $(\Delta L/L_o)$ of the specimen. Thus, the use of stress-strain curves eliminates the dependency on specimen dimensions. The goal now is to derive the strain-energy density equation for mechanical deformation. Two methods are given below.

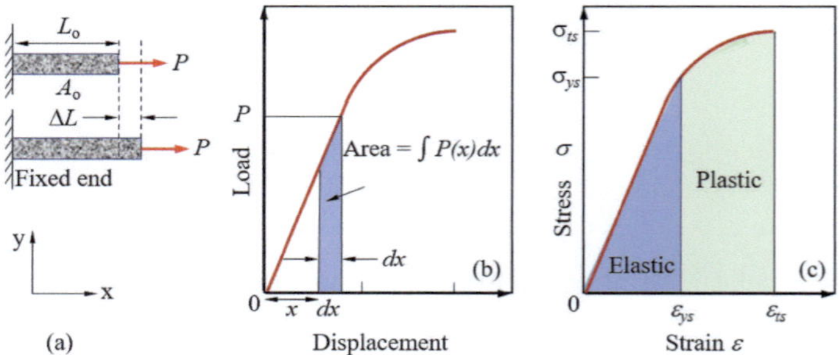

Fig. 12.14 Strain-energy density as the area under the curve. (**a**) Specimen under tension loading, (**b**) load-displacement diagram shown an element of area A, and (**c**) elastic and plastic regions for area assessment under the curve

Calculus Approach From elementary calculus, the area under the $P = f(x)$ curve (Fig. 12.14b) is denoted as $A_p = W$ and that for $\sigma = f(\varepsilon)$ curve (Fig. 12.14c) is $A_\sigma = U$. Dividing the areas A_p and A_σ into N rectangles yields

$$W = A_p = \sum_1^N P(x)\,\Delta x \tag{12.21a}$$

$$U = A_\sigma = \sum_1^N \sigma(\varepsilon)\,\Delta\varepsilon \tag{12.21b}$$

from which the areas under the P-x and σ-ε curves become definite integrals, respectively

$$W = \int P(x)\,dx \tag{12.22a}$$

$$U = \int \sigma(\varepsilon)\,d\varepsilon \tag{12.22b}$$

Applying Eq. (12.22b) to the total shaded areas in Fig. 12.14b gives

$$U_e = \int_{elastic} \sigma(\varepsilon)\,d\varepsilon \tag{12.23a}$$

$$U_p = \int_{plastic} \sigma(\varepsilon)\,d\varepsilon \tag{12.23b}$$

$$U = U_e + U_p = \int_0^{\varepsilon_{ys}} \sigma(\varepsilon)\,d\varepsilon + \int_{\varepsilon_{ys}}^{\varepsilon_{ts}} \sigma(\varepsilon)\,d\varepsilon \tag{12.23c}$$

Mechanical Work Approach The work done W by the load P for a small displacement x is written as the area under the curve as shown in Fig. 12.14b. This area represents the strain energy for uniform deformation. Thus,

$$dW = Pdx \tag{12.24a}$$

$$W = \int_0^x Pdx \tag{12.24b}$$

For a linear elastic deformation in Hooke's law regime, the load-displacement is described by a linear equation $P = kx$ with a constant slope (spring constant) k and displacement or change in length x. In this case, Eq. (12.24b) becomes

$$W = \int_0^x kxdx = \frac{1}{2}kx^2 \tag{12.25}$$

Mathematically it is worth noting that Eq. (12.25) is another common expression for the elastic-strain energy due to elastic deformation schemes.

Assume a constant specimen volume defined as $V = AL$ so that the work done W is converted to strain energy density U as shown below

$$U = \frac{W}{V} = \int_0^x \frac{Pdx}{V} = \int_0^x \frac{Pdx}{AL} \tag{12.26a}$$

$$U = \int_0^x \sigma d\varepsilon = \int_0^x \sigma(\varepsilon)\,d\varepsilon \tag{12.26b}$$

since $\sigma = \sigma(\varepsilon) = P/A$ and $d\varepsilon = dx/L$ act as a response to mechanical deformation by the applied load P.

12.10.2   Modulus of Resilience

In material science, Eq. (12.23a) or (12.26b) is the mathematical definition of resilience, which is the ability of a material to absorb strain energy during elastic deformation at a particular strain rate $d\varepsilon/dt$ and to release it upon unloading within Hooke's law regime. The maximum strain energy during elastic deformation is known as the modulus of resilience $U_e(\varepsilon)_{\max}$ up to the yield point. Similarly, Eq. (12.23b) defines the strain energy being absorbed by a material during plastic deformation, and only a portion of it is released upon unloading.

12.10.3   Modulus of Toughness

Regarding Eq. (12.23c), it defines the modulus of toughness as an indicative of an inelastic behavior since the unloaded material releases a fraction of the elastic strain (Fig. 12.12a). In essence, the modulus of toughness being defined by Eq. (12.23c) represents the total strain energy per unit volume at a strain ε. Hence, modulus of toughness or simply toughness is the ability of a crack-free material to absorb strain energy to deform or fracture under applied strains. For clarity, fracture toughness (see Chap. 13) is the ability of a crack-containing material to absorb strain energy, specifically, at the crack tip to cause crack growth or propagation under applied stresses.

Substituting Eqs. (12.10a), (12.10b) into (12.23c) yields the maximum strain-energy density known as modulus of toughness for linear elastic (Hooke's equation) and nonlinear plastic stress flow (Hollomon's equation)

$$U(\varepsilon) = U_e(\varepsilon) + U_p(\varepsilon) = \int_0^{\varepsilon_{ys}} E\varepsilon d\varepsilon + \int_{\varepsilon_{ys}}^{\varepsilon_{ts}} k_o\varepsilon^n d\varepsilon \tag{12.27a}$$

$$U\left(\varepsilon\right) = E \int_{0}^{\varepsilon_{ys}} \varepsilon d\varepsilon + k_o \int_{\varepsilon_{ys}}^{\varepsilon_{ts}} \varepsilon^n d\varepsilon \tag{12.27b}$$

$$U\left(\varepsilon\right) = \frac{1}{2}\sigma_{ys}\varepsilon_{ys}^2 + \frac{k_o\left(\varepsilon_{ts} - \varepsilon_{ys}\right)^{n+1}}{n+1} \tag{12.27c}$$

$$U\left(\varepsilon\right) = \frac{\sigma_{ys}^2}{2E} + \frac{k_o\left(\varepsilon_{ts} - \varepsilon_{ys}\right)^{n+1}}{n+1} \tag{12.27d}$$

where $U_e\left(\varepsilon\right)$ denotes the elastic modulus of toughness and $U_p\left(\varepsilon\right)$ the plastic modulus of toughness. This implies that $U\left(\varepsilon\right) \simeq U_e\left(\varepsilon\right)$ for brittle and $U\left(\varepsilon\right) \simeq U_p\left(\varepsilon\right)$ ductile materials since $U\left(\varepsilon\right)$ is the area under the σ-ε curve. During deformation in tension mode, most engineering materials exhibit a combination of elastic and plastic behavior.

The modulus of toughness is:

- A measure of the strain energy per unit volume or strain energy density that a specimen under, say, tension loading can absorb until it fractures and
- Determined as the area under the stress-strain curve at ε_f.

The net modulus of toughness denoted by the notation $U\left(\varepsilon\right)$ can theoretically be predicted up to a strain corresponding to the interval $0 < \varepsilon \leq \varepsilon_f$ since beyond the mechanical instability point (necking), the strain is nonuniform along the gage length, although the specimen continues to plastically deform until it fractures.

Further analysis of a fracture specimen is commonly achieved by conducting fractographic work on the fracture surfaces under a microscope at relatively low magnifications. In this case, microscopy is known as fractography since it is used to study the fracture surfaces of materials. Thus, fractography is a typical method for conducting failure analysis known as forensic engineering.

12.11   Gage Length and Strain Rate

Tensile testing a specimen (Fig. 12.15a) is a simple procedure for producing a σ-ε curve until fracture and for determining the mechanical properties of a material. However, quasi-static mechanical properties are dependent on the gage length L_o and strain rate $d\varepsilon/dt$ at temperatures T as schematically shown below.

Figure 12.15b schematically shows a nonlinear trendline for $\varepsilon = f\left(L_o\right)$ representing the effect of a gage length L_o on the strain. Ideally, the specimen starts to deform at the mid-section of the gage length, where L_o is small and the strain is relatively high. However, $\varepsilon = f\left(L_o\right)$ decreases as L_o increases. Therefore, a $2\text{-}in = 5.08\text{-}cm\ L_o$ is recommended by the ASTM E8 as a standard practice consistency in determining mechanical properties.

The effects of strain rate $d\varepsilon/dt$ (Fig. 12.15c) and temperature T (Fig. 12.15d) must be determined for simulating the effect of corrosive service conditions on the

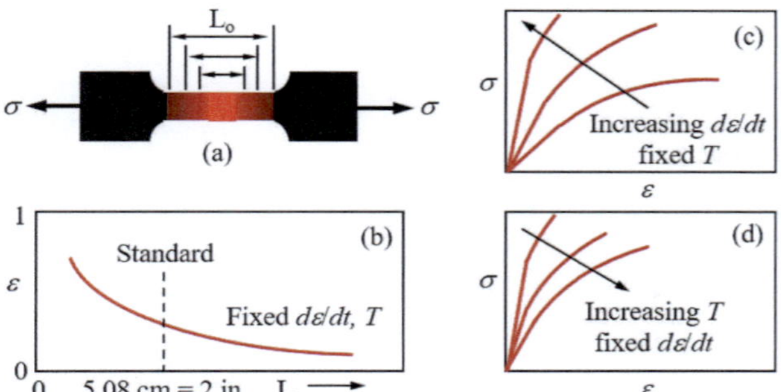

Fig. 12.15 Tensile testing. (**a**) Different gage length at fixed $d\varepsilon/dt$, (**b**) strain $\varepsilon = f(L_o)$, at fixed T and $d\varepsilon/dt$, (**c**) effect of strain rate at fixed T and L_o, and (**d**) effect of temperature at fixed $d\varepsilon/dt$ and L_o on the tensile curves

material. In order to optimize a tensile or any other mode of testing procedure to obtain reliable mechanical properties, one has to choose or select a sound (crack-free) material and the proper L_o and $d\varepsilon/dt$ at T.

Ductility It is the ability of a material to undergo a dimensional change or shape. It is calculated using (1) the total elongation E_L and (2) the reduction in cross-sectional area R_A at fracture. These ductility factors are defined by

$$E_L = \frac{\Delta L}{L_o} \times 100\% = \frac{L_f - L_o}{L_o} \times 100\% = (100\%)\,\varepsilon_f \qquad (12.28a)$$

$$R_A = \frac{\Delta A}{A_o} \times 100\% = \frac{A_o - A_f}{A_o} \times 100\% \qquad (12.28b)$$

where L_f, L_o denote the final and original gage lengths and A_o, A_f denote the original and final (at fracture) specimen cross-sectional areas, respectively. At a macroscale, brittle materials exhibit $E_L > 0$ and $R_A \to 0$ due to an elastic deformation domain, and ductile materials show $E_L >> 0$ and $R_A > 0$ due to a combined elastic-plastic deformation upon loading until fracture.

12.12 Creep Deformation

Common creep data is usually plotted as schematically illustrated in Fig. 12.16a for strain response and Fig. 12.16b for rupture stress response to rupture time intervals.

Creep is a time temperature-dependent mechanical deformation at a constant stress (isostress), and it can be considered as a quasi-static mechanical behavior of materials. Usually creep tests are carried out at a temperature in the range $0.3T_m \leq T \leq 0.6T_m$, where T_m is the melting temperature of a material. Obviously, different amorphous and crystalline solid materials have different creep behavior,

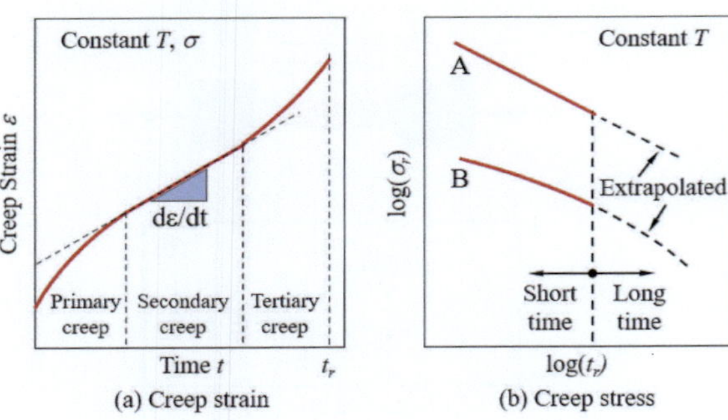

(a) Creep strain   (b) Creep stress

Fig. 12.16 Schematic creep data plots. (**a**) Creep strain regions and (**b**) creep rupture stress for hypothetical A and B materials, and ideal extrapolation

and commonly engineering creep data is extrapolated, to an extent, for predicting creep deformation at a prolong design rupture lifetime (t_r) in hours or years.

Materials used in high-temperature applications, such as turbine engines, reactors, stream generators, and chemical plants, exhibit creep behavior at relatively high temperatures. As a general rule, the effect of temperature on the mechanical properties of crystalline solids is detected at a homologous temperature (normalized temperature) $T/T_m > 0.3$.

The reader is encouraged to consult reliable sources, such as ASTM E139 and ASTM E292 standard or recommended creep test procedures for aircraft turbine blades, nuclear reactors and ASME boiler & pressure vessel design codes, and publications related to creep rupture of aluminum/steel alloys subjected to house and building fires.

The general creep analysis can be elaborated into three different states of deformation at relatively high temperatures. According to Fig. 12.16a,

Primary Creep It is a transient deformation that ideally starts with an instantaneous strain upon loading (stressing) a specimen and continuous gradually toward a steady-state deformation due to work hardening at relatively low temperatures.

Secondary Creep It is a steady-state deformation process at a constant strain rate $d\varepsilon/dt$ due to a balance between work hardening and in situ annealing. Moreover, $d\varepsilon/dt$ is an essential engineering design parameter for long-life applications at high temperatures.

Tertiary Creep It is an accelerated creep deformation related to instability that leads to rupture or fracture of a specimen. This mechanical acceleration is attributed to formation of voids, micro-cracks at grain boundaries, and possibly specimen necking.

Figure 12.16 elucidates the extrapolation method from short to long creep testing for A and B materials. Extrapolating the data to long terms at low stresses overestimate the actual performance. For instance, Fig. 12.17a shows creep rupture stress curves for 2024-T851 and Fig. 12.17b for 5454-O aluminum alloys at temperatures, where the dashed lines represent extrapolation paths (Kaufman [17, pp. 42,94]).

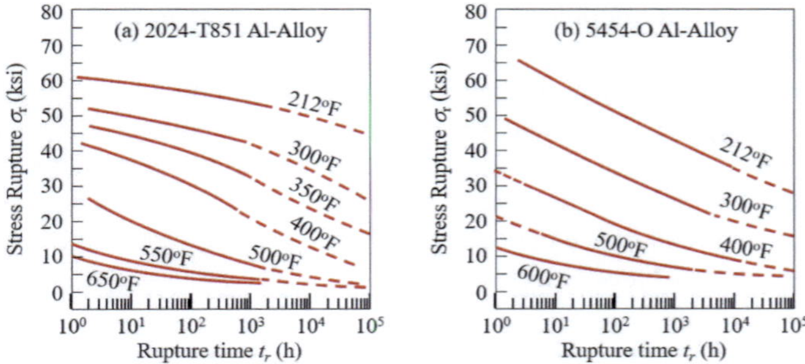

Fig. 12.17 Experimental creep data based on $\sigma_r = f(t_r)$ for (**a**) 2024-T851 and (**b**) 5454-O aluminum alloys (Kaufman [17, pp. 42,94]

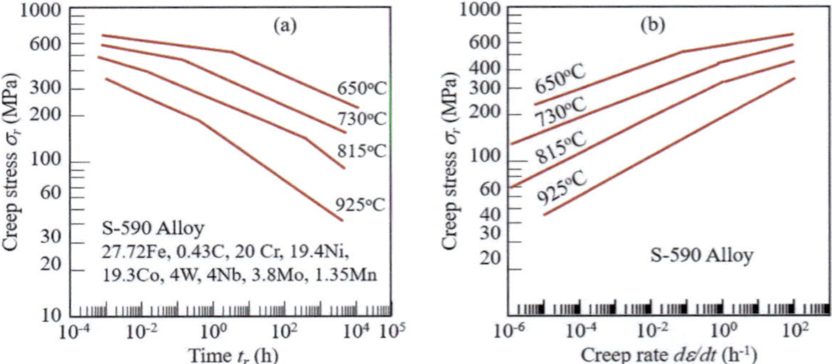

Fig. 12.18 Experimental creep data for S-590 alloy based (**a**) $\sigma_r = f(t_r)$ and (**b**) $\sigma_r = f(d\varepsilon/dt)$. After Goldhoff [18]

Figure 12.18 shows creep data for *S*-590 alloy (Goldhoff [18]). For instance, Fig. 12.18a shows the $\sigma_r = f(t_r)$ relationship along with the transition (t_r, σ_r) points at temperatures. This leads to stress rupture or creep stress curves with two slightly different negative slopes at the (t_r, σ_r) points. Moreover, Fig. 12.18b elucidates the $\sigma_r = f(d\varepsilon/dt)$ relationship, where $d\varepsilon/dt$ is the strain rate.

For crystalline materials, the creep mechanisms are diffusion creep due to diffusion of vacancies and dislocation creep due to dislocation motion at high temperatures. For amorphous materials, the typical creep mechanism is called viscous creep due to liquid-like behavior.

12.12.1 Strain Rate Creep

High-temperature creep is a thermally activated process controlled by diffusion of atoms. This suggests that plastic flow of polycrystalline solids is mainly due to slip caused by dislocation motion, grain boundary sliding, or a combination of these mechanisms.

A thorough mathematical analysis can be found in publications by Ruano and Sherby [19], Przystupa [20], and Viernstein and Kozeschnik [21]. But, the Norton power-law creep equation (Norton [22]) representing the stress dependency and the Arrhenius exponential function (Arrhenius [23]) for temperature dependency are common empirical mathematical models for the steady-state strain rate ($d\varepsilon/dt$), known as creep rate. Thus,

$$\frac{d\varepsilon}{dt} = K_1\sigma^n \qquad \text{at constant } T \qquad (12.29a)$$

$$\frac{d\varepsilon}{dt} = K_2 \exp\left(-\frac{Q_c}{RT}\right) \qquad \text{at constant } \sigma \qquad (12.29b)$$

$$\frac{d\varepsilon}{dt} = K_1\sigma^n K_2 \exp\left(-\frac{Q_c}{RT}\right) \qquad \text{(combined)} \qquad (12.29c)$$

$$\frac{d\varepsilon}{dt} = K_3\sigma^n \exp\left(-\frac{Q_c}{RT}\right) \qquad \text{(simplified)} \qquad (12.29d)$$

$$\ln(d\varepsilon/dt) = \ln(K_3) + n\ln(\sigma) - \frac{Q_c}{RT} \qquad \text{(linearized)} \qquad (12.29e)$$

where K_1, K_2, $K_3 = K_1 K_2$, n are fitting constants, Eq. (12.29a) is the power governing empirical expression and Eqs. (12.29b-e) temperature-dependent Arrhenius type creep expressions.

Plotting Eq. (12.29e) reveals the creep rate dependency on stress σ (Fig. 12.19a) and reciprocal temperature $1/T$ (Fig. 12.19b). Moreover, theoretical or experimental analysis of creep data obtained under increasing stress and temperature conditions provides linear correlations between the natural logarithm of the creep strain rate $d\varepsilon/dt$, applied stress σ, and reciprocal temperature $1/T$.

The slope n in Fig. 12.19a at a stress range $\sigma_{low} \leq \sigma \leq \sigma_{high}$ and Q_c/R in Fig. 12.19b at a reciprocal temperature range $1/T_{high} \leq 1/T \leq 1/T_{low}$ are assumed to be constant or nearly constant. Moreover, the creep data in Fig. 12.19 is described by an equation of a straight line.

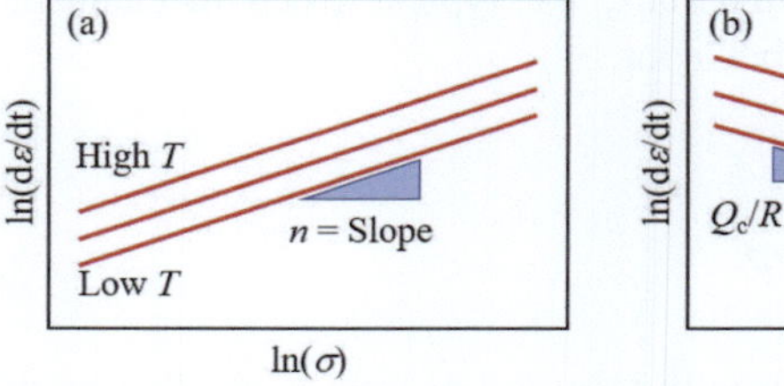

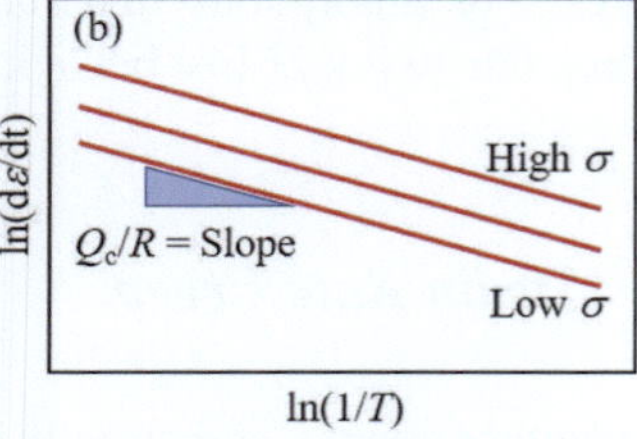

Fig. 12.19 Schematic strain rate $d\varepsilon/dt$ creep dependency (**a**) on tress σ at temperature T and (**b**) on reciprocal temperature $1/T$ at stress σ

12.12.2 *Parametric Methods*

For time-dependent mechanical deformation at high temperatures, creep is assumed to be a thermally activated process governed by an exponential function described as an Arrhenius-type equation. Thus, the creep rate $r = 1/time = 1/t_r$ and creep strain rate $d\varepsilon/dt$ expressions are written as

$$\frac{1}{t_r} = A \exp\left(-\frac{Q_c}{RT}\right) = A \exp\left(-\frac{P_{LM}}{T}\right) \tag{12.30a}$$

$$\frac{d\varepsilon}{dt} = B \exp\left(-\frac{Q_c}{RT}\right) = B \exp\left(-\frac{P_{LM}}{T}\right) \tag{12.30b}$$

$$\ln(t_r) = -\ln(A) + \frac{Q_c}{RT} \tag{12.30c}$$

$$\ln(d\varepsilon/dt) = \ln(B) - \frac{Q_c}{RT} \tag{12.30d}$$

where Q_c is the creep activation energy (J/mol), $P_{LM} = Q_c/R$ [in units of K log ($hours$)] is known as the Larson-Miller parameter (Larson and Miller [24]), $R = 8.3145\ J/mol.K$, and $C_A = \ln(A)$, $C_B = \ln(B)$, P_{LM} are slopes.

In the creep literature, the natural logarithm (base e) is converted to common logarithm (base 10) due to prolong rupture creep time; $\ln(x) = 2.3026 \log(x)$.

As indirectly stated in the introductory background on creep through Eq. (12.29e), creep data (Figs. 12.16 and 12.17) are directly related to three main variables, namely, time (t), temperature (T), and stress (σ). This leads to a P_x parameter, which combines the effects of these principal variables (Kaufman [17, p. 3)].

The general definition of the P_x parameter can mathematically be defined by

$$P_x = f(t, T, \sigma) = \frac{\sigma^{Q_c} \ln(t) - \ln(t_A)}{(T - T_A)^{R_x}} \tag{12.31}$$

where Q_c, t_A, T_A, R_x are material constants related to high-temperature processes, such as creep rupture performance and diffusion.

Finding a curve fitting equation for a particular dataset depends on the experimental data values being measured. Thus, Eq. (12.31) can be manipulated in order to accomplish such a purpose.

Larson-Miller Parameter (P_{LM}) Solving Eqs. (12.30c) and (12.30d) gives expressions for the Larson-Miller parameter P_{LM} (Larson and Miller [24]) used to extrapolate creep data for high temperature creep applications. The most common Larson-Miller parameter equation is deduced from Eq. (12.30c) with $C_A = \ln(A)$, and also from Eq. (12.31) with the assumptions that $R_x = -1$, $Q_c = 0$, $T_A = 0$, $C_A = -\ln(t_A)$, and $\ln(t_r) = \ln(t)$. Thus,

$$P_{LM} = T \left[\ln (t_r) + C_A \right] \tag{12.32a}$$

$$\ln (t_r) = -C_A + \frac{P_{LM}}{T} \tag{12.32b}$$

Manson-Haferd Parameter (P_{MH}) The general equation defining the Manson-Haferd parameter based on Eq. (12.31) with $R_x = 1$ and $Q_c = 0$ is of the form

$$P_{MH} = \frac{\ln (t_r) - \ln (t_A)}{T - T_A} \tag{12.33a}$$

$$\ln (t_r) = \ln (t_A) + P_{MH} (T - T_A) \tag{12.33b}$$

which can be treated as a linear relationship.

Sherby-Dorn Parameter (P_{SD}) For comparison, curve fitting creep data can also be accomplished using the Sherby-Dorn parameter P_{SD} (Sherby and Dorn [25])

$$P_{SD} = t_r \exp \left(-\frac{Q_c}{RT} \right) \tag{12.34a}$$

$$\ln (P_{SD}) = \ln (t_r) - \frac{Q_c}{RT} \tag{12.34b}$$

$$\ln (t_r) = \ln (P_{SD}) + \frac{Q_c}{RT} \tag{12.34c}$$

Mathematically, if $P_{LM} = Q_c/R$, then Eq. (12.34c) gives $\ln (t_r) = f (P_{SD}, P_{LM})$ as probable function for fitting creep data

$$\ln (t_r) = \ln (P_{SD}) + \frac{P_{LM}}{T} \tag{12.35}$$

The graphical representation of these parametric methods is schematically shown in Fig. 12.20 along with isostress lines. Creep data is often nonlinear, which can make it difficult to analyze and interpret. Linearizing the data can make it easier to see trends and patterns in the data. Using these parametric equations in the form $ln(t_r) = f(1/T)$ fits linear curves to creep datasets: Fig. 12.20a for Larson-Miller and Fig. 12.20c for Sherby-Dorn equations. Similarly, using $ln(t_r) = f(T)$ gives straight lines for the Manson-Haferd equation as shown in Fig. 12.20b.

Regarding Larson-Miller parameter (Larson and Miller [24]), curve fitting creep data using Eqs. (12.32a) and (12.32b) yields straight isostress lines as schematically illustrated in Fig. 12.21a for $\ln (t_r) = f (1/T)$ and Fig. 12.21b for $\ln (d\varepsilon/dt) = f (1/T)$ at σ and Fig. 12.21c for $\sigma_r = f (P_{LM})$ at temperature T.

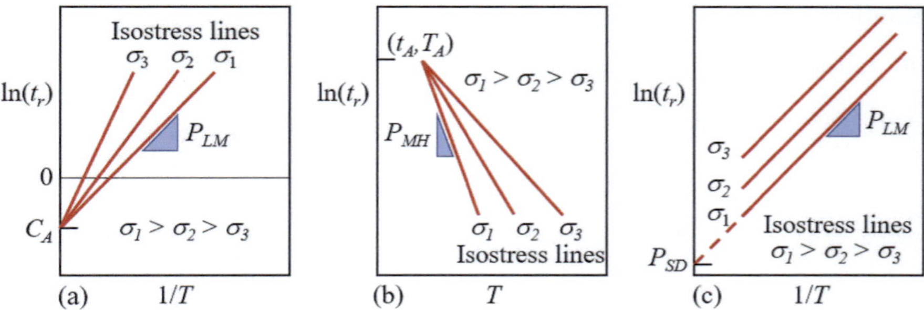

Fig. 12.20 Schematic isostress lines. (**a**) Larson-Miller (Larson and Miller [24]), (**b**) Manson-Haferd, and (**c**) Sherby-Dorn methods

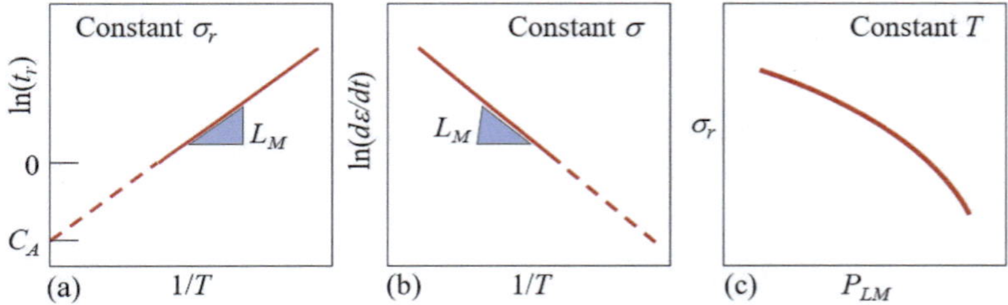

Fig. 12.21 Creep data plots. (**a**) $\ln(t_r) = f(1/T)$, (**b**) $d\varepsilon/dt = f(1/T)$, and (**c**) $\sigma_r = f(P_{LM})$

The Larson-Miller parametric method seems to be the most broadly used extrapolation technique for predicting creep life of metallic materials. Moreover, to compare creep behavior of different materials, linearized creep data can make it easier to compare the creep behavior of different materials.

Example 12.4 Use the given creep data for 15CrMo steel (Zhu et al. [26]) to perform curve fitting. (**a**) Determine the Larson-Miller parameter at $\sigma_r = 65\ MPa$. What is the magnitude of the activation energy Q_c? How long will it take to rupture a suitable specimen at $T = 680°C$? (**b**) Find the best curve fitting equation for the $\sigma_r = f(t_r)$ data taken at $T = 550°C$. The average tensile strength is approximately $\sigma_{ts} = 540\ MPa$.

No.	T (°C)	σ_r (MPa)	t_r (h)
1	550	200	46
2	550	180	205
3	550	150	960
4	550	140	1440
5	550	130	6048
6	550	120	11088

No.	T (°C)	σ_r (MPa)	t_r (h)
7	690	65	77
8	670	65	218
9	650	65	672
10	630	65	1673
11	610	65	5032

Solution

(a) Linear curve using fitting $\ln(t_r) = f(1/T)$ at $\sigma_r = 65\,MPa$ yields

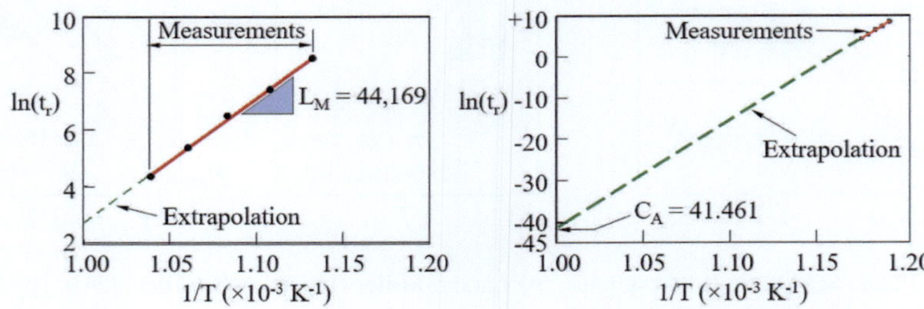

The resultant curve fitting equation with a high correlation coefficient $R^2 \simeq 0.99$ in the close interval $[610°C, 690°C]$ is

$$\ln(t_r) = 44169\left(10^{-3}\right)(1/T) - 41.461 \tag{12.4E1a}$$

$$C_A = 41.461 \text{ in units of } \log(h) \tag{12.4E1b}$$

$$P_{LM} = 44169 \text{ in units of } K.\log(h) \tag{12.4E1c}$$

$$Q_c = 1.82 \times 10^5\ J/mol \tag{12.4E1d}$$

Therefore, $C_A \neq 20$ as commonly cited in the literature. The corresponding rupture time at $T = 680°C = 953\,K$ and $1/T = 1/953 = 1.0493 \times 10^{-3}$ is

$$\ln(t_r) = 44169\left(10^{-3}\right)(1.0493) - 41.461 = 4.8855 \tag{12.4E2a}$$

$$t_r = \exp(4.8855) = 132.36\,h \tag{12.4E2b}$$

(b) Curve fitting the $\sigma_r = f(t_r)$ data at $T = 550°C$ yields a power law equation with a high correlation coefficient $R^2 \simeq 0.98$

$$t_r = \left(6 \times 10^{25}\right)(\sigma_r)^{-10.45} \tag{12.4E3}$$

Plotting the given data and the rupture time $t_r = f(\sigma_r)$ expression, Eq. (12.4E3), gives

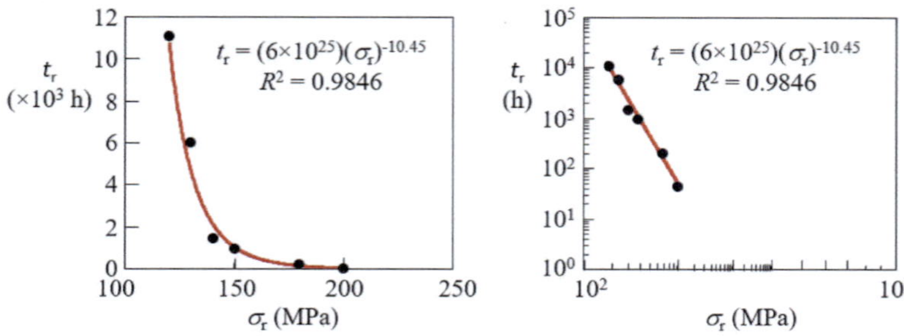

Notice that $\sigma_r = f(t_r)$ in log-log scale compacts the wide range of data into a nearly straight line with a curve fitting correlation coefficient denoted as $R^2 \simeq 0.98$, which suggests a good fit using the power law equation for the rupture stress-ratio range $0.22 \le \sigma_r/\sigma_{ts} \le 0.37$.

Example 12.5 This example compares creep data for two steels. **(a)** Use the Larson-Miller approach with $C_A = 20$ to calculate the creep rupture time for S-590 (Fe-$21Cr$-$20Ni$-$20Co$ steel) (Larson and Miller [24]) and $800H$ alloys (Fe-$21Cr$-$32Ni$-Al, Ti) components (Tamura et al. [27]) subjected to 100 MPa (20 ksi) at $800°C$ (1073 K), $600°C$ (873 K), and $27°C$ (300 K). Compare results since the S-590 does not contain Al and Ti. **(b)** Calculate the fracture load for two identical specimens having a $50\,mm \times 50\,mm$ cross-sectional area subjected to a 2.5-*year* service at $800°C$ (1073 K). **(c)** Use $T_m \simeq 1500°C$) for S-590 and $T_m \simeq 1350°C$ for $800H$ alloys and explain the meaning of results.

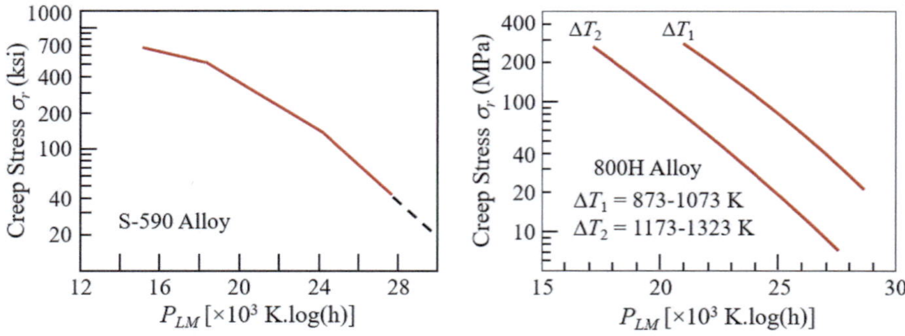

Solution

(a) From the plots at $\sigma_r = 100\ MPa$, the Larson-Miller parameter value is $P_{LM} = 25\times10^3\ K.\log(h)$ for S-590 alloy and $P_{LM} = 21.50\times10^3\ K.\log(h)$ for $800H$ alloy. From Eq. (12.32b), the rupture lifetime or creep time t_r is calculated at different testing temperatures.

At $T = 800°C = 1073\ K$ ($T = 0.53T_m$ for S-590 and $T = 0.59T_m$ for $800H$), t_r is

$$\ln(t_r) = -C_A + \frac{P_{LM}}{T} = 3.2992 \quad (S\text{-}590) \tag{12.5E1a}$$

$$t_r = \exp(3.2992) = 27.09\,h \quad (S\text{-}590) \tag{12.5E1b}$$

and

$$\ln(t_r) = -C_A + \frac{P_{LM}}{T} = 3.7279 \times 10^{-2} \quad (800H) \tag{12.5E2a}$$

$$t_r = \exp\left(3.7279 \times 10^{-2}\right) = 1.04\,h \quad (800H) \tag{12.5E2b}$$

At $T = 600°C = 873\,K$ ($T = 0.40T_m$ for S-590 and $T = 0.44T_m$ for $800H$), t_r is

$$\ln(t_r) = -C_A + \frac{P_{LM}}{T} = 8.6369 \quad (S\text{-}590) \tag{12.5E3a}$$

$$t_r = \exp(8.6369) = 5635.8\,h \quad (S\text{-}590) \tag{12.5E3b}$$

and

$$\ln(t_r) = -C_A + \frac{P_{LM}}{T} = 4.6277 \quad (800H) \tag{12.5E4a}$$

$$t_r = \exp(4.6277) = 102.28\,h \quad (800H) \tag{12.5E4b}$$

At $T = 27°C = 300\,K$ ($T = 0.20T_m$ for S-590 and $T = 0.22T_m$ for $800H$), t_r is

$$\ln(t_r) = -C_A + \frac{P_{LM}}{T} = 63.333 \quad (S\text{-}590) \tag{12.5E5a}$$

$$t_r = \exp(63.333) = 3.20 \times 10^{27}\,h \quad (S\text{-}590) \tag{12.5E5b}$$

$$t_r \to \infty \quad (S\text{-}590) \tag{12.5E5c}$$

and

$$\ln(t_r) = -C_A + \frac{P_{LM}}{T} = 51.667 \quad (800H) \tag{12.5E6a}$$

$$t_r = \exp(51.667) = 2.7460 \times 10^{22}\,h \quad (800H) \tag{12.5E6b}$$

$$t_r \to \infty \quad (800H) \tag{12.5E6c}$$

Among the two high-temperature resistant alloys, the S-590 alloy seems to be the most suitable for high temperature applications because it takes longer to reach failure. Literately, specimens with appropriate dimensions loaded at $\sigma = 100\,MPa$ and $T = 300\,K$ will never fail since $t_r \to \infty$.

(b) Using Eq. (12.32a) yields the Larson-Miller parameter

$$P_{LM} = T\,[\ln\,(t_r) + C_A] = (1073)\,[\ln\,(2.5) + 20] \qquad (12.5E7a)$$

$$P_{LM} = 22.44 \times 10^3 \qquad (12.5E7b)$$

From the given figures at $P_{LM} = 22.44 \times 10^3$ (creep parameter), the creep stress for the alloys is

$$\sigma_r = 200\ MPa \quad (S\text{-}590\ \text{alloy}) \qquad (12.5E8a)$$

$$\sigma_r \simeq 190\ MPa \quad (800H\ \text{alloy}) \qquad (12.5E8b)$$

Despite that these alloys have different melting temperatures, these results suggests that the alloys have similar creep resistance (creep strength or creep limit) or creep rate under the given conditions.

The applied loads at $\sigma_r = 200\ MN/m^2 = 200\ MPa$ are

$$A_o = \left(50 \times 10^{-3}\ m\right)\left(50 \times 10^{-3}\ m\right) = 0.0025\ m^2 \qquad (12.5E9a)$$

$$P = A_o\sigma_r = 500\ kN \quad (S\text{-}590) \qquad (12.5E9b)$$

$$P = A_o\sigma_r = 475\ kN \quad (800H) \qquad (12.5E9c)$$

Hence, both equations yield similar results for the fracture load P under the given quasi-static conditions.

(c) The S-590 alloy seems to be the most suitable material for high temperature applications because it withstand a higher creep load. Moreover, a thorough analysis of creep data is of technological importance for high-temperature alloys in aerospace, metallurgy, military, nuclear reactors, and electronics applications.

It is worth noting that the S-590 and $800H$ alloys have relatively high melting temperatures, such as $T_m \simeq 1500°C$ and $T_m \simeq 1350°C$, respectively. A practical aspect of creep testing or creep studies is the range of testing temperature. It seems appropriate to set $0.5T_m \leqslant T < T_m$ as a suitable range for creep testing temperature. However, in this example, this particular criterion is not met at low temperature data $(T = 27°C)$, but it is significantly important to show the effect of sustained or fixed applied stress on the mechanical deformation time.

12.13 Fatigue in Metals

Fatigue in metals is a type of damage that occurs when a metal is subjected to repeated cyclic loading. This can cause the metal to crack and eventually fail, even if the stresses are well below the ultimate tensile strength of the material. Furthermore, fatigue is a major problem in engineering, as it can lead to the failure of critical components in machines, structures, and vehicles. It is estimated that fatigue is responsible for up to 90

Fatigue in solid materials subjected to repeated cyclic loading can be defined as a progressive failure due to crack initiation (stage I), crack growth (stage II), and crack propagation (stage III or instability stage). In materials science, fatigue crack initiation of initially crack-free solids is a mechanical surface phenomenon caused by dislocation formation and dislocation motion at matrix-defect interfaces.

12.13.1 *Fluctuating and Cyclic Loads*

Fatigue is a form of failure caused by fluctuating or cyclic loads over a short or prolong period of time. Therefore, fatigue is a time-dependent failure mechanism related to microstructural features. The fluctuating loading condition is not a continuous failure process as opposed to cyclic loading. The former is manifested in bridges, aircraft, and machine components, while the latter requires a continuous constant or variable stress amplitude until fracture occurs.

It is also important for the reader to know that fatigue failure or fracture can occur at a maximum stress below the quasi-static yield strength (σ_{ys}) of a particular material. Obviously, temperature and environmental effects must be considered in fatigue failure characterization. From an engineering point of view, predicting fatigue lifetime is major a requirement.

12.13.2 *Cyclic Stress History*

Fatigue is a complex phenomenon that can be difficult to predict and prevent. However, by understanding the factors that affect fatigue and taking steps to mitigate them, engineers can design and maintain structures that are safe and reliable.

The schematic cyclic stress history diagram in Fig. 12.22a shows a constant stress amplitude with positive maximum and negative minimum stress levels, while the minimum stress in Fig. 12.22b is zero. This implies that the mean stress is a key stress term for fatigue analysis. Both Fig. 12.21a, b represent a symmetric loading with constant stress amplitude, whereas Fig. 12.22c elucidates a random loading as a function of time.

These schematic curves represent fatigue stress histories from which the number of cycles are accounted for fatigue lifetime in terms of number of cycles (N) consumed by a specimen prior to failure or fracture.

Any general fatigue model must be able to predict fatigue damage, which is related to a stress spectrum or history. Thus, fatigue damage can be axial due to

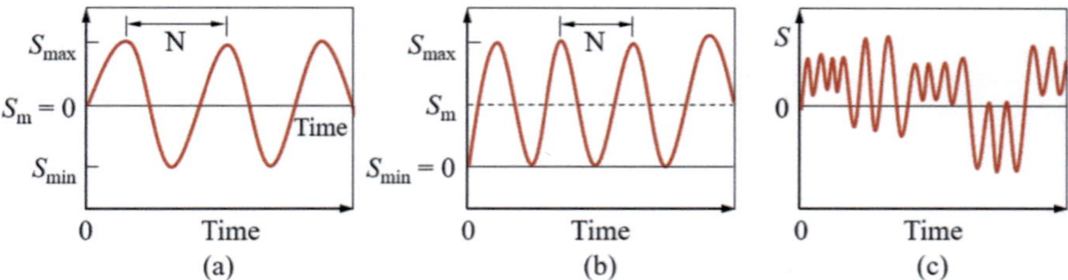

Fig. 12.22 Schematic cyclic stress histories. (**a**) Symmetrical with $R_x = -1$, (**b**) symmetrical with $R_x = 0$, and (**c**) asymmetrical (random)

tension-compression, flexural due to bending, and torsional due to twisting applied stress.

From Fig. 12.22a, the stress range ($\Delta\sigma$) is the algebraic difference between the maximum and minimum stresses in a cycle expressed as $\Delta\sigma = \sigma_{max} - \sigma_{min}$. Other common stress parameters extracted from a stress spectrum are the mean stress (σ_m), alternating stress (σ_a), the stress ratio (R_x), and the stress amplitude (A_s). Thus,

$$\sigma_m = \frac{\sigma_{max} + \sigma_{min}}{2} \tag{12.36a}$$

$$\sigma_a = \frac{\Delta\sigma}{2} = \frac{\sigma_{max} - \sigma_{min}}{2} \tag{12.36b}$$

$$R_x = \frac{\sigma_{min}}{\sigma_{max}} = \frac{K_{min}}{K_{max}} \tag{12.36c}$$

$$A_s = \frac{\sigma_a}{\sigma_m} = \frac{\sigma_{max} - \sigma_{min}}{2\sigma_m} = \frac{1 - R_x}{1 + R_x} \tag{12.36d}$$

These stress parameters, Eqs. (12.36a), (12.36b), (12.36c), and (12.36d), can be varied while conducting fatigue tests for characterizing materials having specific geometries, weldment, or microstructural features. In fact, the stress ratio is the most common parameter in determining the fatigue behavior of crack-free and cracked specimens. Moreover, preventing fatigue includes reducing the applied stress range, surface treatment, and inspection using ultrasonic testing, radiography, or dye penetrant testing.

12.13.3   Structural Integrity

The mechanics of fatigue and fracture is concerned with the reliability and effectiveness of structural components in engineering applications. Thus, a precise determination of fatigue crack growth rate (da/dN), remaining fatigue life (N), and a detailed understanding of how materials respond to fluctuating or cyclic

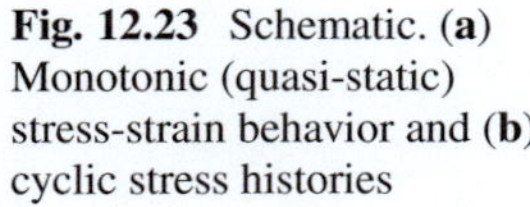

Fig. 12.23 Schematic. (**a**) Monotonic (quasi-static) stress-strain behavior and (**b**) cyclic stress histories

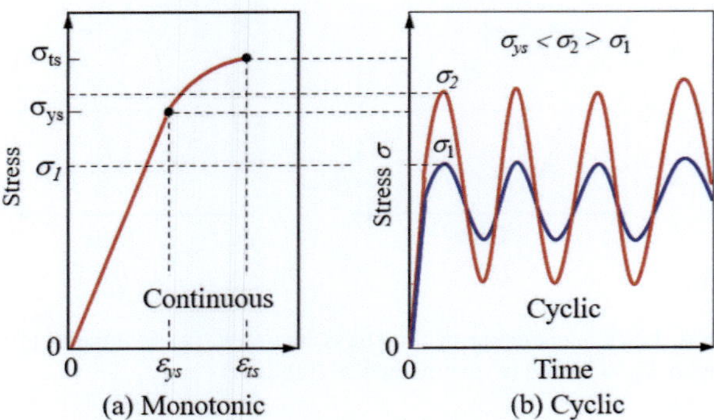

Fig. 12.24 Schematic fatigue S-N curves

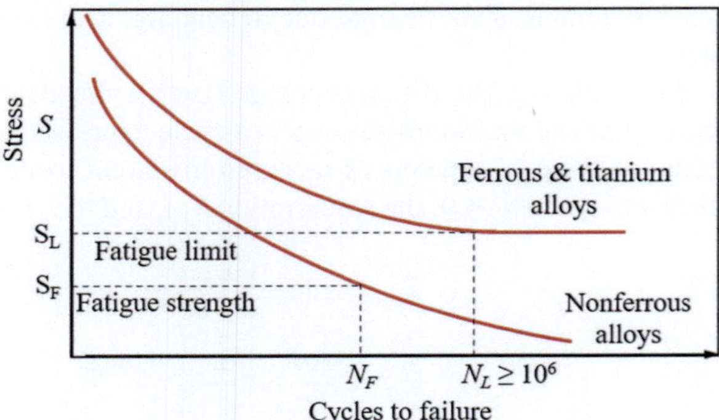

loads are important conditions for assuring structural integrity. In reality, structural components are generally subjected to a wide spectrum of stresses over their lifetime that strongly affects fatigue lifetime. The wide spectrum on a structural component may consist of constant stress amplitude (CSA) and variable stress amplitude (VSA) blocks as schematically shown in Fig. 12.22c.

The stress level from the monotonic quasi-static stress-strain curve in Fig. 12.23a is connected to the dynamic stress-time cyclic history in Fig. 12.23b. For instance, it is expected that the number of cycles to failure is $N_1 > N_2$ because the maximum fatigue stresses are $\sigma_1 < \sigma_2$, where $\sigma_1 < \sigma_{ys}$ and at $\sigma_{ys} < \sigma_2 < \sigma_{ts}$. At $\sigma_{ys} < \sigma_2 < \sigma_{ts}$, the material is in the plastic regime, and consequently, cracking is accelerated, and the material consumes less cycles before failure.

For crack or notch-free specimens, the usual characterization of fatigue behavior is through a stress-cycle curve, commonly known as a *S-N* diagram. For example, Fig. 12.24 shows schematic *S-N* curves for two hypothetical materials.

Fatigue in Metals!crack-free Specimen Finite fatigue lifetime can be defined by the total number of consumed cycles, $N = N_i + N_p$, where (N_i) is the fatigue-crack initiation life and $N = N_p$ is the remaining fatigue life cycles till the crack reaches a critical length for crack propagation to occur without any warning.

Preexisting Crack Specimens   In this case, the finite fatigue life is reduced because $N_i = 0$ and $N = N_p$.

It should be pointed out that the fatigue limit S_L (for infinite life) is the stress level below which fatigue failure does not occur. Generally, ferrous alloys such as steels exhibit a stress limit (Fig. 12.24), while nonferrous do not show a fatigue limit with increasing cycles to failure. The latter materials are characterized by determining the fatigue strength at a specific life (N). This stress is meaningless if the specific life is not identified.

Typically, fatigue life exhibits data scatter, and therefore, the difference in failure response of test specimens is due to microstructural defects and machining defects. For instance, a particular material having a fine-grained microstructure exhibits superior fatigue properties over coarse-grained microstructure.

Fatigue failure may be due to discontinuities such as inadequate design, improper maintenance, and so forth. Nonetheless, fatigue failure can be prevented by:

- Avoiding sharp surfaces caused by punching, stamping, and the like.
- Preventing the development of surface discontinuities during processing.
- Avoiding misuse, abuse, assembling errors, and improper maintenance.
- Using proper material and heat treatment procedures.
- Using inert environments whenever possible.

Normally the nominal stresses in most structures are elastic or below the static yield strength of the base material. In pertinent cases, the strain-life $(\epsilon\text{-}N)$ can be determined instead of the stress-life $(S\text{-}N)$ in high-/low-cycle fatigue schemes.

The $S\text{-}N$ curve, also known as the Woehler curve, is a graphical representation of the relationship between the stress amplitude and the number of cycles to failure for a given material. It is a fundamental tool in fatigue analysis and is used to predict the fatigue life of components subjected to cyclic loading. In particular, fatigue failures typically occur at stress concentrations, such as sharp corners or cracks. The fracture surface will exhibit a characteristic pattern of striations, which are microscopic ridges that are formed by the growth of the crack over time. The striations are oriented perpendicular to the direction of crack growth, and they can be used to determine the rate of crack growth.

12.13.4   Fractography and Fatigue Mechanism

Understanding the surface features of fatigue fractures or failures is important in characterizing the dynamic mechanical behavior of solids. This can be accomplished by using fractography, which is the study of fracture surfaces to determine the cause of failure. It is a valuable tool for investigating fatigue failures, which are caused by repeated cyclic loading.

For instance, fractography is commonly used to reveal surface fracture features, such as beach marks, ratchet marks, striations, and smooth surfaces that may develop gradually during a cyclic scenario.

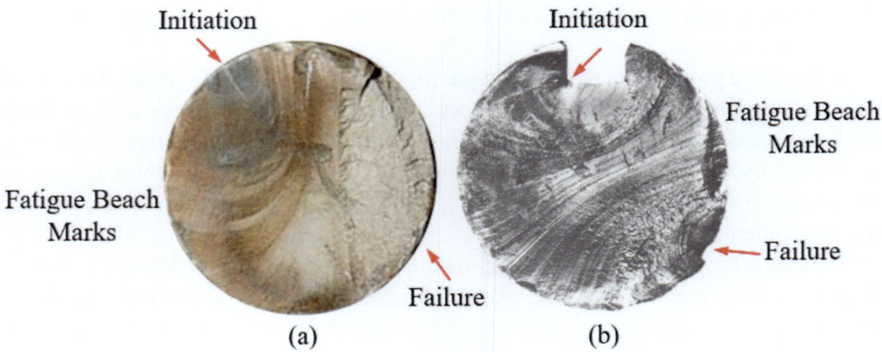

Fig. 12.25 Torsional fatigue and beachmarks on shaft fractured surfaces. (**a**) Solid shaft (Phillips [28]) and (**b**) shaft with a keyway (ASM Vol. 12, Atlas of Fractographs, Figure 194 [29])

Crystallographically, fatigue crack initiation may occur along the slip direction due to a local maximum shear stress. After the consumption of many cycles, the crack may change direction when the maximum principal normal stress (in the vicinity of a crack tip) governs crack growth.

Regarding fatigue stage II, some materials develop striations and beachmarks, as common fatigue surface features revealed using fractography, which is the study of fracture surfaces for revealing the microstructural features related to fatigue crack initiation and growth.

Consider a common torsional fatigue beachmark morphology on shaft fractured surfaces clearly shown in Fig. 12.25. The main difference between these shafts is the crack initiation sites: surface initiation site in Fig. 12.25a (Phillips [28]) and stress-concentration site in Fig. 12.25b (ASM Vol. 12, Atlas of Fractographs, Figure 194 [29]) marked with arrows.

Fatigue crack initiation is mostly favorable on the surface of a specimen due to fewer restrictions with respect to stress concentration sites or surface conditions; however, inclusion-matrix interfaces are potential cracking sites. On the other hand, fatigue crack growth is influenced by the material bulk properties and environment.

Crack growth or extension is a decohesion-cohesion mechanism related to the formation of beachmarks (interruptions in crack growth) and striations (the ridge of microplastic deformation between beachmarks). Nonetheless, a fatigue crack is macroscopically smooth and flat; microscopically, it is not. But, in general, a fatigue crack extends along slip bandscrystallographically defined by the general $\{hkl\} \langle uvw \rangle$ slip systems. By definition, one striation represents the microscopic crack advance in one load cycle.

The microscopic evidence of fracture mechanism can be obtained through fractography, which is a classical method in failure analysis. Initially, the microscopic features of fatigue damage are lip bands acting as the source of small nucleating crack sites on specimen surfaces.

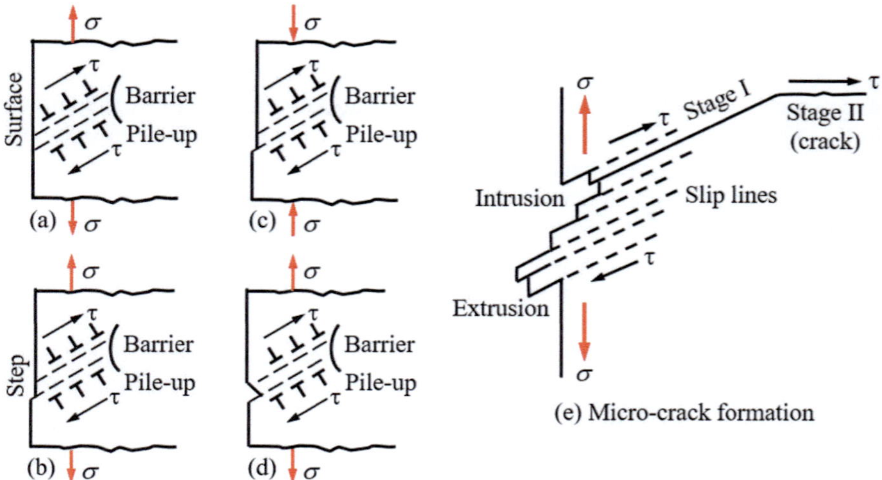

Fig. 12.26 Cottrell-Hull fatigue mechanism in ductile materials presented as a sequence of events from (**a**) through (**e**). After Cottrell and Hull [30]

The lip bands produce slip steps called intrusions and extrusions which are the microscopic features of stages I and II fatigue damage. Specifically, intrusions are caused by reversed slip due to load reversal at the source for crack initiation, which may consume most of the solid fatigue lifetime before crack growth and subsequent crack propagation. In essence, crack propagation velocity may be used as a variable for characterizing materials.

12.13.5   Cottrell-Hull Model

Consider a polycrystalline solid with a smooth surface being subjected to an elastic-cyclic stress range, in which $\sigma_{max} < \sigma_{ys}$, but σ_{max} is high enough to activate a slip mechanism forming extrusion-intrusion deformation known as the Cottrell-Hull modified mechanism shown in Fig. 12.26 after the original work of Cottrell and Hull [30] in 1957, followed by Tanaka and Mura [31] in 1981. Additional fatigue mechanisms are compiled by J. Man et al. [32] in their 2009 article.

Assume a stress ratio be $R = -1$ for a fully reversed cyclic load system causing irreversible damage after many cycles. For convenience, take the slip planes A and B so that dislocation pile-ups occur on both sides of the planes, but having opposite signs as indicated in Fig. 12.26a. When the slip system $\{hkl\} < uvw >$ is activated due to a local maximum shear stress τ, a surface step is created at $\sigma > 0$ (Fig. 12.26b). Subsequently, dislocation motion is reversed at $\sigma_{min} > 0$ (Fig. 12.26c). In this case, the upper part moves toward its original position, leaving an inward step (Fig. 12.26d) called intrusion in the order of the Burgers vector $b = 3$

mm (Felbeck and Atkins [33]). Eventually, the specimen is subjected to intrusions and extrusions as shown in Fig. 12.26e.

This mechanism is repeated many times until a deeper intrusion acts as a microcrack. During this stage I, many life cycles are consumed before crack growth in the direction perpendicular to the local principal normal tension stress, which governs the crack growth behavior, and consequently, the Cottrell mechanism no longer applies in a simple manner. This mechanism can take place in a few grains before the crack changes direction. Once this occurs other dislocation mechanisms may take place, possibly as described by the Frank-Read source. When multiple cross-slip occurs, the Frank-Read source may not complete a loop cycle (Dieter [34, p. 178]).

Once the microcrack becomes long enough it causes an increase in the stress concentration at the crack tip, and subsequently, the local stress is truncated to the yield strength of the solid.

12.13.6   Characteristics of Crack Growth

The characteristics of crack growth may be **transcrystalline** either by progressive plastic straining, which causes typical fatigue striations, or by cleavage at low temperatures due to the presence of brittle inclusions. On the other hand, crack growth may be **intercrystalline** due to bonding deficiency, aggressive environment or due to initiation and coalescence of voids within or between grains. Then, the final stage III of fatigue fracture is caused by a dynamic crack propagation mechanism when the slowly growing crack size a reaches a critical length $a > a_c$.

Example 12.6 Use the given room-temperature maximum stress-cycle (S-N) data for a 15-*mm*-diameter hypothetical steel rod. **(a)** Compute the mechanical tension load (P_d) to ensure infinite fatigue life and **(b)** the bar diameter d when $N = 4 \times 10^4$ *cycles* to avoid fatigue failure at the maximum load from part (a). Use a safety factor of 2 for both parts.

N (Cycles)	σ(MPa)
14800	468
19650	400
28000	351
40000	321
60000	305
100000	300
200000	300
360000	302
680000	300

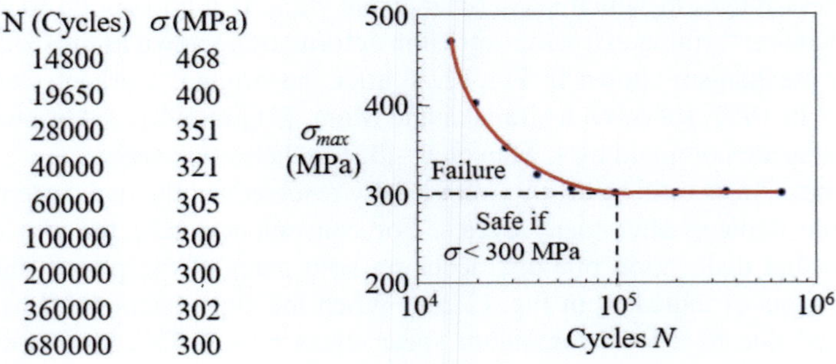

Solution

(a) From the plot, the fatigue limit is $\sigma_L = S_L = 300\ MPa$ at $N \geq 10^5\ cycles$, from which the design stress is $\sigma_d = \sigma_L/S_F = (300\ MPa)/2 = 150\ MPa$. Thus, the design force is

$$\sigma_d = \frac{P_d}{A_o} = \frac{\sigma_L}{S_F} \tag{12.6E1a}$$

$$P_d = \frac{A_o \sigma_L}{S_F} = \frac{(\pi/4)\,d^2 \sigma_L}{S_F} \tag{12.6E1b}$$

$$P_d = \frac{(\pi/4)\left(15 \times 10^{-3}\ m\right)^2 \left(300 \times 10^3\ kN/m^2\right)}{2} \tag{12.6E1c}$$

$$P_d = 26.51\ kN \tag{12.6E1d}$$

(b) From the given plot, $\sigma_{\max} = 330\ MPa$ at $N = 4 \times 10^4$ cycles. Then,

$$\sigma = \frac{\sigma_{\max}}{S_F} = \frac{330\ MPa}{2} = 165\ MPa \tag{12.6E2a}$$

$$\sigma = \frac{P_d}{A_o} = \frac{P_d}{(\pi/4)\,D^2} \tag{12.6E2b}$$

$$d \geq \sqrt{\frac{P_d}{(\pi/4)\,\sigma}} = \sqrt{\frac{26.51 \times 10^{-3}\ MN}{(\pi/4)\left(165\ MN/m^2\right)}} \tag{12.6E2c}$$

$$d \geq 1.43 \times 10^{-2}\ m = 14.30\ mm \tag{12.6E2d}$$

12.14 Mechanical Property Correlations

This section includes a brief reference compilation related to correlations between mechanical properties and particle/grain size. The precursors of these correlations are Hall [35] and Petch [36], who independently noticed in the early 1050s that the yield strength (σ_{ys}) and the grain size (d) have a power law relationship of the form $\sigma_{ys} = f\left(d^{-1/2}\right)$ for polycrystalline solids with grain size ranging from 1 μm to $10^3\ \mu m$. Basically, the Hall-Petch equation correlates a macroscale mechanical property, such as the yield strength σ_{ys}, and a microscale variable known as the average grain size d.

One can find in the vast literature many supporting studies of the applicability of the Hall-Petch equation to pure metals, single-phase alloys, multiphase alloys, and intermetallic-phase containing alloys. In fact, the Hall-Petch equation is an empirical fit to experimental data. Hence, the Hall-Petch equation and its general power law form are mathematically defined by

$$\sigma_{ys} = \sigma_o + k_y d^{-1/2} \quad \text{(Hall-Petch)} \tag{12.37a}$$

$$\sigma_{ys} = \sigma_o + k_y d^n \quad \text{(general)} \tag{12.37b}$$

where σ_o denotes the frictional stress that opposes dislocation motion, k_y denotes the dislocation locking-term related to grain boundary resistance in spreading yielding from one grain to an adjacent grain, n denotes an exponent, and d denotes either the average grain size or particle size (equivalent diameter).

The reason for introducing a general power law equation is that some polycrystalline solids do not follow $\sigma_{ys} = f\left(d^{-1/2}\right)$, but $\sigma_{ys} = f\left(d^n\right)$ with $n \neq -1/2$ instead. However, one can mathematically determine that Eq. (12.37a) or (12.37b) predicts $\sigma_{ys} \to \sigma_o$ as $d^{-1/2} \to 0$ or $d^n \to 0$ which represent a decay behavior.

Despite that σ_o, k_y, n are curve fitting parameters, they have meaningful definitions. For instance, the frictional stress σ_o is mathematically defined as

$$\sigma_o = f\left(\sigma_{ss}, \sigma_i, \sigma_{gb}, \sigma_p, \sigma_{sg}, \ldots.\right) \tag{12.38}$$

where σ_{ss} denotes a stress due to solid solution, σ_i denotes the incoherent friction stress that may be defined as the Peierls-Nabarro stress required to move a dislocation from its equilibrium state, σ_{gb} denotes a stress due to grain boundary strength, σ_p denotes the stress due to particles (precipitates) embedded in the matrix, σ_{sg} denotes a stress due to secondary boundary resistance, and the like. Moreover, it is difficult to isolate these secondary stresses, and consequently, σ_o becomes a curve fitting parameter representing the plateau term in Eq. (12.37).

12.14.1 Particle-Size and Grain-Size Strengthening

In theory, solid solution, grain refinement, precipitation hardening, and work hardening are strengthening mechanisms of engineering materials. Regarding grain refinement and precipitation hardening mechanisms, the smaller grain and particle sizes, the higher the yield strength of the material since grain boundaries and particle-matrix interfaces act as barriers (impediments) to dislocation motion. Conversely, grain growth and particle coarsening mechanisms have an opposite effect of material properties.

In engineering, yield strength and grain/particle size correlations can be evaluated using the Hall-Petch empirical equation, which is based on dislocation pile-up mechanism at grain boundaries during plastic deformation. Moreover, tangled dislocations and other complicated dislocation network are portions of the total dislocation interactions.

It is evident in Fig. 12.27 that the yield strength (σ_{ys}) exhibits a decay behavior with increasing particle and grain sizes (Perez [13]). The $Ni_{53}Mo_{35}Fe_9B_2$ alloy

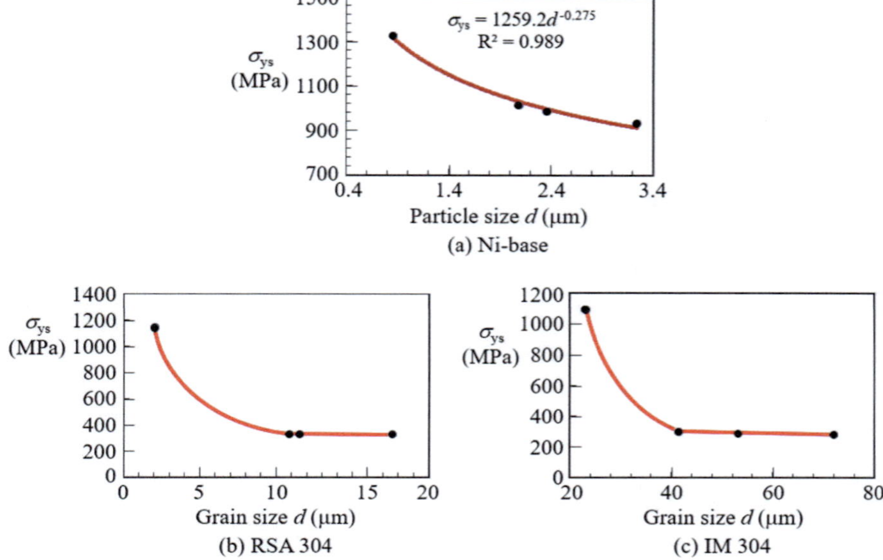

Fig. 12.27 Yield strength $\sigma_{ys} = f(d)$ for (**a**) precipitate-base $Ni_{53}Mo_{35}Fe_9B_2$ alloy, (**b**) grain-base RSA 304 alloy, (**c**) grain-base IM 304 alloy. Experimental data taken from Perez [13]

shows a continuous decrease with increasing particle size (Fig. 12.27a), whereas RSA 304 (Fig. 12.27b) and IM 304 (Fig. 12.27c) alloys show a drastic σ_{ys} decrease accompanied with a plateau.

Notice that the nearly horizontal trendlines in Fig. 12.27b,c indicate that grain growth is not significant in RSA 304 and IM 304 austenitic stainless steels due to their relatively high ductility, which, in turn, is favorable for easy dislocation motion in these low strength materials.

Example 12.7 The dataset in Fig. 12.27c exhibits a decay and plateau trends. Suggest a possible curve fitting equation and calculate the σ_{ys} for $d = 30 \; \mu m$.

Solution A better fitting equation includes an additional plateau term. Thus,

$$\sigma_{ys} = ax^{-b} + c \tag{12.7E1a}$$

$$\sigma_{ys} = 6.9406 \times 10^{11} d^{-6.559600} + 284.11 \quad \left(R^2 = 1 \right) \tag{12.7E1b}$$

$$\sigma_{ys} = 6.9406 \times 10^{11} (30)^{-6.559600} + 284.1100 = 426.04 \; MPa \tag{12.7E1c}$$

where a is the span, b is the decay factor, and c is the plateau. The reader is encouraged to plot Eq. (12.7E1b) and compare it with Fig. 12.27c. In addition, measured grain size dataset for the $Ni_{53}Mo_{35}Fe_9B_2$ alloy obeys the Hall-Petch equation, Eq. (12.37a), with a relatively high correlation coefficient R^2 as shown in

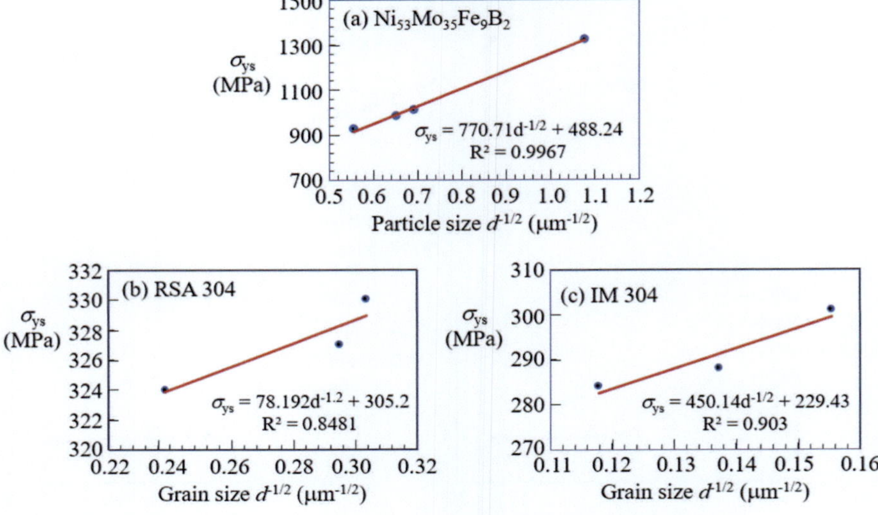

Fig. 12.28   Hall-Petch plots. (**a**) Ni-base, (**b**) RSA 304, and (**c**) IM 304 alloys

Fig. 12.28a. Alloys RSA 304 (Fig. 12.28b) and IM 304 (Fig. 12.28c) exhibit lower correlation coefficients.

Notice that RSA 304 and IM 304 alloys have different grain sizes, but similar trendlines with low correlation coefficients.

Example 12.8 Use the given data to determine the constants for the Hall-Petch relationship defined by Eq. (12.37a) and σ_{ys} for $d = 14\ \mu m$.

$d\ (\mu m)$	$d^{-1/2}\ (\mu m^{-1/2})$	$\sigma_{ys}\ (MPa)$
10	0.31623	117.53
15	0.25820	117.07
20	0.22361	116.79
25	0.20000	116.60

Solution The constants are

$$\sigma_{ys} = \sigma_o + k_y d^{-1/2} \tag{12.8E1a}$$

$$117.53 = \sigma_o + k_y (10)^{-1/2} \tag{12.8E1b}$$

$$117.07 = \sigma_o + k_y (15)^{-1/2} \tag{12.8E1c}$$

Subtracting Eqs. (12.8E1b,c) yields $k_y = 7.9271\ MPa.\mu m^{-1/2}$ and then $\sigma_o = 115.02\ MPa$. Thus, $\sigma_{ys} = 117.14\ MPa$ for $d = 14\ \mu m$

12.15   Deformation in Compressive Mode

Most commonly used industrial metalworking processes, such as hot-working and cold-working in a large-scale production, induce plastic deformation along with an enhancement of strength and a reduction of ductility.

Among the metalworking processes, plastic deformation of brass by cold rolling (CR) and hot rolling (HR) are schematically compared in Fig. 12.29 (Guy and Hren [37, p. 458]). Cold rolling is a work hardening process induced by a mechanism called strain hardening due to the formation of line defects known as dislocations. Hot rolling, on the other hand, deforms and simultaneously heat treats the material, leading to an annealed microstructure.

Only cold rolling (CR) is described, to an extent, in order to emphasize on plastic deformation by compressive forces, which, in turn, induce stress flow of the working specimen, and changes in specimen shape and microstructural features.

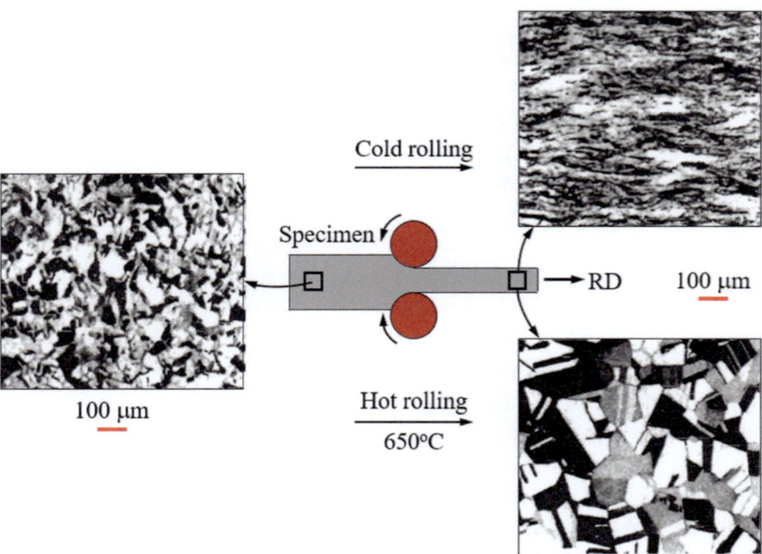

Fig. 12.29 Mechanical deformation in compression by cold rolling (CR) and hot rolling (HR) along the horizontal rolling direction (RD). Adapted from Guy and Hren [37, p. 458]

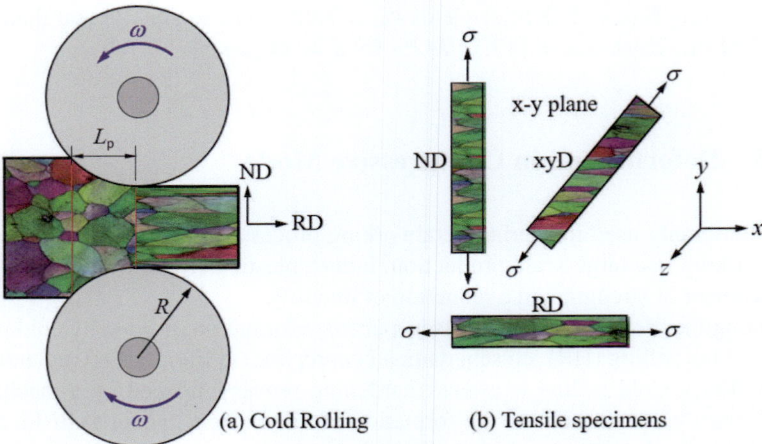

Fig. 12.30 Schematic cold rolling (CR) showing (**a**) an equiaxed grain structure being elongated along the rolling direction (RD) and (**b**) recommended tensile specimen orientations for mechanical property determination with reference to the RD, normal direction (ND), and x-y plane at angle θ

12.15.1 Flat Cold Rolling Process

The sketch in Fig. 12.30a schematically illustrates the flat cold rolling process and related specimen orientations for determining mechanical properties of cold rolled materials (Dieter [34, p. 608]). For aesthetic purposes, the IF-steel plate between rollers is part of a color-map microstructure taken from Wauthier-Monnin et al. paper [38]. Notice the symmetrical rollers and the plate being deformed due to the upper and lower contact lengths relate to the contact arc. Thus, there must exist a neutral point N, while the local upper and lower compressive forces induce reduction of the plate thickness.

Regarding the entrance velocity v_o and the exit velocity v_x of the plate in the deformation zone, cold working is possible if $v_x > v_o$. Moreover, Fig. 12.30b depicts recommended specimen orientations for determining the mechanical properties of a cold worked material.

First of all, if the width (b) of the plate remains fixed, then the fraction of cold working (C_w) per pass or cycle is achieved when the new plate thickness is $h > h_o$. The outcome of cold working using cold rolling is a strain hardening process commonly used in the metal forming industry. Mathematically,

$$C_w = \frac{A_o - A}{A_o} = 1 - \frac{A}{A_o} = \frac{bh_o - bh}{bh_o} \tag{12.39a}$$

$$C_w = \frac{h - h_o}{h_o} = \frac{\Delta h}{h_o} \tag{12.39b}$$

which imply that cold working is a strain hardening process and it is a function of strain mathematically defined as $C_w = f(\varepsilon)$.

Using the volume consistency, Eq. (12.40a), and the definition of engineering strain ε, Eq. (12.40b), and manipulating the algebra along with Eq. (12.39a) yield an equation for determining the fraction of cold working (C_w)

$$A_o L_0 = AL \tag{12.40a}$$

$$\varepsilon = \frac{L - L_o}{L_o} = \frac{L}{L_o} - 1 \tag{12.40b}$$

$$C_w = \frac{\varepsilon}{\varepsilon + 1} \tag{12.40c}$$

This expression, Eq. (12.40c), is given by Callister and Rethwisch [3, p. P-29] as a problem 7.27 assuming a tensile specimen with constant volume during mechanical deformation.

The corresponding roller radius (R) and the contact length (L_p) during cold working are correlated by

$$R^2 = L_p^2 + \left(R - \frac{\Delta h}{2} \right)^2 \tag{12.41a}$$

$$L_p^2 = R^2 - \left(R - \frac{\Delta h}{2} \right)^2 = R^2 - \left[R^2 - 2R\Delta h + \left(\frac{\Delta h}{2} \right)^2 \right] \tag{12.41b}$$

$$L_p \simeq \sqrt{R\Delta h} = \sqrt{R(h_o - h_f)} = \sqrt{Rh_o C_w} \tag{12.41c}$$

For convenience, the Hollomon equation, Eq. (12.10b), can be used to determine the flow (σ_{flow}). Thus,

$$\varepsilon = \ln \left(\frac{h_o}{h_f} \right) \tag{12.42a}$$

$$\sigma = k_o \varepsilon^n \tag{12.42b}$$

$$\sigma_{flow} = \sigma_{avg} = \frac{1}{\varepsilon} \int \sigma d\varepsilon = \frac{1}{\varepsilon} \int k_o \varepsilon^n d\varepsilon \tag{12.42c}$$

$$\sigma_{flow} = \frac{k_o \varepsilon^n}{n + 1} \tag{12.42d}$$

$$\sigma_{flow} = \frac{\sigma_{ys} + \sigma_x}{2} = \frac{\sigma_1 + \sigma_2}{2} \tag{12.42e}$$

The corresponding single roller force (F_{cr}) and its transmitted power (W_{cr}) are

$$F_{cr} = bL_p \sigma_{flow} \tag{12.43a}$$

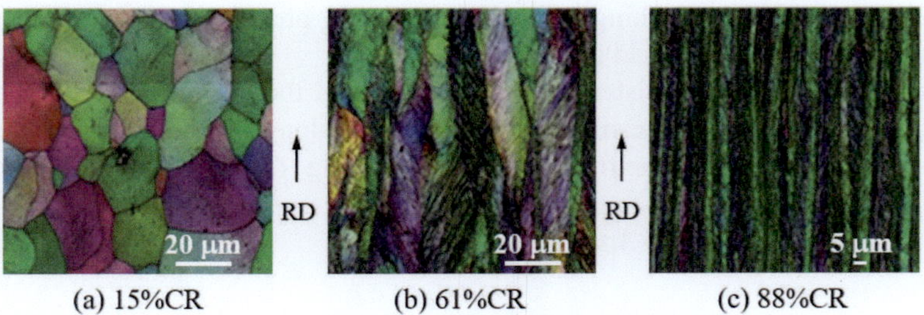

Fig. 12.31 Color map of microstructures of mechanically deformed IF-steel by cold rolling (CR). (**a**) 15%CR, (**b**) 61%CR, and (**c**) 88%CR. The arrow points the direction of cold rolling CR. After Wauthier-Monnin et al. [38]

$$W_{cr} = L_p F_{cr} N \qquad (12.43b)$$

For plastic deformation, ε is calculated using Eq. (12.42a), and $\sigma_x = \sigma_2$ is determined from a stress-strain (σ-ε) curve at ε. Additional details and derivations are included in Appendix 12A.

Figure 12.31 depicts three color maps of cold-rolled (CR) IF-steel microstructures (Wauthier-Monnin et al. [38]). In essence, the IF-steel has a low carbon content ($\simeq 0.02\%C$), high ductility, and BCC ferrite-phase microstructure.

The microstructures and the corresponding percentage cold rolling are: described as

- 15%CR equiaxed-microstructure (Fig. 12.31a) containing a low dislocation density has a slightly elongated α-BCC structure with a hardness, say, H_{15CR}. Moreover, the microstructure has a pancake-shaped grain morphology.
- 61%CR elongated-microstructure (Fig. 12.31b) showing elongated grains and a higher dislocation density (hardly visible thin lines per area) with a hardness, say, $H_{61CR} > H_{15CR}$.
- 88%CR elongated-microstructure (Fig. 12.31c) exhibiting severely deformed grains and a high dislocation density with a hardness, say, $H_{88CR} > H_{61CR}$.

The sequence of the grain morphology in Fig. 12.31 is a good example of the outcome of the cold rolling process related to a practical enhancement of mechanical properties and surface finish with increasing mechanical deformation, which indirectly is associated with increasing dislocation density. Moreover, cold rolling occurs below the metal or alloy recrystallization temperature $T << T_m$ (usually at room temperature), where T_m is the material melting temperature.

The analysis of a cold rolling (CR) process requires use of tensile properties and data for Eq. (12.42b). In this case, Kalpakjian and Schmid [39, p. 63] compilation of true stress-strain curves (Fig. 12.32) can be used as an open research data for, at least, solving examples and problems related to CR at room temperature.

Accordingly, the true stress-strain curves in Fig. 12.32 are correspond to plastic stresses and strains as per Eq. (12.42b). This means that the validity of this equation

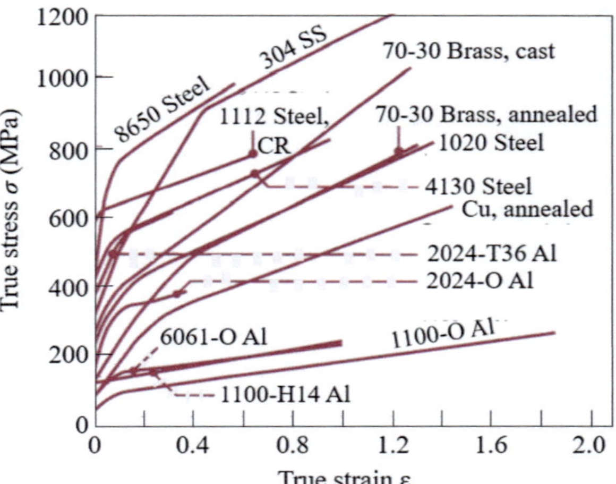

Fig. 12.32 Tensile true stress-strain curves at room temperature representing plastic behavior in a graphical form for selected engineering materials. Taken from Kalpakjian and Schmid [39, p. 63]

is set as $\sigma_{ys} \leq \sigma \leq \sigma_{ts}$, where σ_{ts} is the tensile strength known as the ultimate tensile strength (UTS). Moreover, this stress inequality starts at the onset of the yielding phenomenon and ends at the onset of necking, where the tested material experiences a triaxial-state of stress.

For clarity, a triaxial-state of stress refers to normal stresses acting on the specimen under tension. If the material resistance to plastic deformation overcomes the triaxial-state of stress, then necking begins, and the material contains a ductile matrix (ductile atomic lattice). Otherwise, the material fails, or fractures at the onset of necking where the fracture stress is $\sigma_f = \sigma_{ts}$.

For convenience, Table 12.1 contains data based for Eq. (12.42b) as an open source from the materials world (Kalpakjian and Schmid [39, p. 62]).

Example 12.9 Consider the schematic flat rolling process shown below for a strengthened an annealed copper strip.

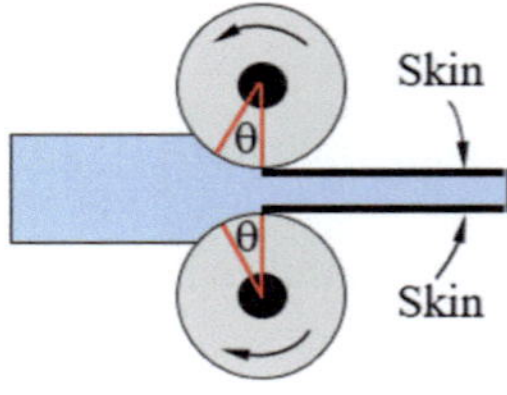

The Cu strip dimensions are $b = 228\ mm$, $h_o = 25\ mm$, and $h_f = 20\ mm$, and roller radius is $R = 305\ mm$. If the rollers rotate at $N = 100\ rpm$, then use the "Cu, annealed" curve in Fig. 12.32 to extract the required stress σ at a strain ε to calculate C_w, ε, μ, θ, F_{total}, and W_{total}. Further, rpm stands for revolutions per minute.

Table 12.1 Power constants for $\sigma = k_o \varepsilon^n$ (Kalpakjian and Schmid [39, p. 62])

Material	k_o (MPa)	n
1100-O *Al*-alloy	180	0.20
2024-T4 *Al*-alloy	690	0.16
5052-O *Al*-alloy	210	0.13
6061-O *Al*-alloy	205	0.20
6061-T6 *Al*-alloy	410	0.05
7075-O *Al*-alloy	400	0.17
70-30 Brass, annealed	895	0.49
Phosphor bronze, annealed	720	0.46
Molybdenum, annealed	725	0.13
1045 steel, hot rolled	965	0.14
1112 steel, annealed	760	0.19
4135 steel, annealed	1015	0.17
4340 steel, annealed	640	0.15
17-4 P-H, annealed	1200	0.05
304 stainless steel (SS), annealed	1275	0.45
410 stainless steel (SS), annealed	960	0.10

Solution Cold rolling with $\Delta h = h_o - h_f = 25\ mm - 20\ mm = 5\ mm$. Thus,

$$C_w = \frac{h_o - h_f}{h_o} = \frac{25 - 20}{25} = 0.20 \text{ or } 20\% \tag{12.9E1a}$$

$$\varepsilon = \ln{(25/20)} = 0.22 \tag{12.9E1b}$$

From Fig. 12.32, $\sigma_{ys} = 83\ MPa$ at $\varepsilon = 0$ and $\sigma_x = 280\ MPa$ at $\varepsilon = 0.22$ and

$$\sigma_{flow} = 0.5 \left(\sigma_{ys} + \sigma_x \right) = 0.5\,(83\ MPa + 280\ MPa) \tag{12.9E2a}$$

$$\sigma_{flow} = 181.50\ MPa \tag{12.9E2b}$$

The contact length is

$$L_p = \sqrt{R h_o C_w} = \sqrt{(305\ mm)\,(25\ mm)\,(0.20)} = 39.05\ mm \tag{12.9E3}$$

The angle of contact (θ) up to the neutral point is

$$L_p = R \tan{(\theta)} \tag{12.9E4a}$$

$$\theta = \tan^{-1}\left(\frac{L_p}{R} \right) = \left(\frac{180}{\pi} \right) \tan^{-1}\left(\frac{39.05}{305} \right) = 7.41° \tag{12.9E4b}$$

For a single roller,

$$F_{cr} = bL_p\sigma_{flow} = \left(228 \times 10^{-3} \; m\right)\left(39.05 \times 10^{-3} \; m\right)\left(181.50 \; \frac{MN}{m^2}\right)$$
$$\text{(12.9E5a)}$$

$$F_{cr} = 1.62 \; MN \tag{12.9E5b}$$

$$F_{total} = 2F_{cr} = 2\,(1.62 \; MN) = 3.24 \; MN \tag{12.9E5c}$$

It is now convenient to convert revolutions per minute (rpm) to revolutions per second (rps). Thus, $N = 100 \; rpm = 100/60 \; rps$ so that

$$W_{cr} = \tag{12.9E6a}$$

$$W_{cr} = \left(39.05 \times 10^{-3} \; m\right)(1.62 \; MN)\,(100/60) \tag{12.9E6b}$$

$$W_{CR} = 0.11 \; \frac{MN.m}{s} = 0.11 \; MJ/s = 0.11 \; MW \tag{12.9E6c}$$

$$W_{total} = 2W_{cr} = 2\,(0.11 \; MW) = 0.22 \; MW \tag{12.9E6d}$$

$$W_{total} = \left(1341.02 \; \frac{hp}{MW}\right)(0.22 \; MW) = 295 \; hp \tag{12.9E6e}$$

Therefore, $W_{total} = 295 \; hp$ drives the two rollers to mechanically deform the copper strip.

12.16 Summary

Regarding mechanical deformation and related properties, the fundamentals of mechanical behavior of single crystals and polycrystals are well-known by the engineering community. In this chapter, tensile and compressive quasi-static conditions prevail as the loading modes for characterizing the mechanical behavior of crystalline solids. Conversely, metal fatigue is a dynamic behavior due to either fluctuating or cyclic loading.

The load-bearing capacity of a material under tension, bending, compressive, torsion, or a combination of forces is critical in order to assure the integrity of a structural component. Thus, the assessment of the mechanical behavior of a material is through a series of stress-strains curves, and the respective interpretation of the mechanical deformation can be accompanied with relevant mathematical models for crack-free. The mechanical behavior of initially cracked specimens is detailed in Chap. 13.

From a materials science point of view, the crystallography of slip and mechanical twinning mechanisms is of great importance in understanding plastic deformation of metals and alloys at relatively low temperatures; otherwise, grain boundary sliding at high temperatures $T \rightarrow T_m$ becomes the dominant mechanism in the creep field, which is important in high-temperature applications. Here, T_m is the melting temperature of the materials under mechanical deformation. Hence, mechanical properties of solid materials are important in designing structural components subjected to either low, intermediate or relatively high temperatures.

Plastic deformation by compressive forces induced by cold rolling is most commonly used in the manufacturing metal industry for a large-scale production using hot-working and cold-working methods.

Appendix 12A Cold Rolling

This appendix includes a two-roller setup for cold rolling operation and related nomenclature as shown in Fig. 12.33 (Dieter [34, p. 608]).

Equilibrium of Forces Using the nomenclature in Fig. 12.33a and b gives the force balance for a half-element as

$$\sum F_x = (\sigma_x + d\sigma_x)(h + dh) - \sigma_x h \pm (Rd\theta)\mu P_x \cos\theta \qquad (12.44)$$

$$-2(Rd\theta)P_x \sin\theta = 0$$

$$\frac{d(\sigma_x h)}{d\theta} = 2PR(\sin\theta \mp \mu\cos\theta) \qquad (12.45)$$

and for a small θ, $\sin\theta \simeq \theta$, $cos\theta \simeq 1$, and Eq. (12.45) becomes

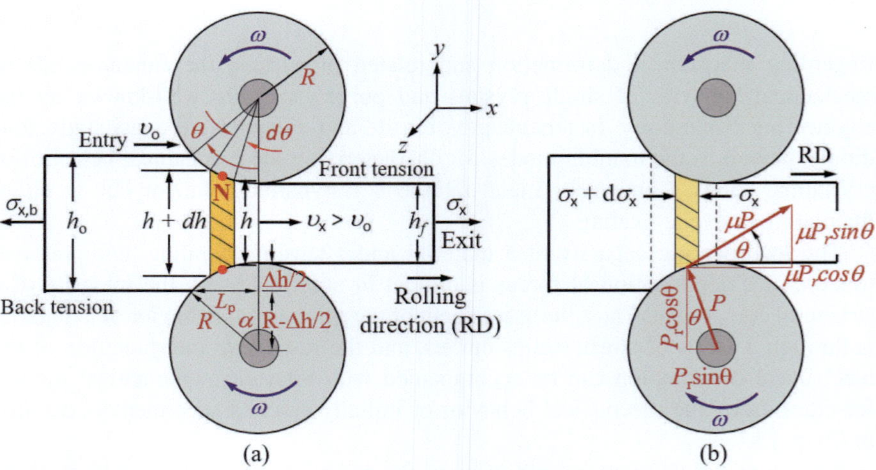

Fig. 12.33 Schematic two-roller setup for cold rolling (CR). (**a**) Specimen and roller nomenclature and (**b**) roller pressure components. After Dieter [34, p. 608]

$$\frac{d\,(\sigma_x h)}{d\theta} = 2PR\,(\theta \mp \mu) \tag{12.46}$$

where $\pm$ occurs since the direction of the friction force changes at the neutral point (red dot), σ_x denotes the horizontal compressive stress (MPa), P_r denotes the radial pressure along the contact arc, $\mu = \tan\theta = L_p/R$ denotes the coefficient of friction, L_p denotes the contact length, R denotes the roller radius, $\tau = \mu P_r$ denotes the tangential shear stress, μP denotes the friction force, $\mu P_r\, Rd\theta$ denotes the tangential friction force, $\mu P_r\, Rd\theta cos\theta$ denotes the horizontal component of the tangential friction force, $\sigma_N = P_r\, Rd\theta$ denotes the normal stress, and $P_r\, Rd\theta sin\theta$ denotes the horizontal component of the normal stress.

For convenience, the Von Mises plasticity model for plane-strain condition is adopted hereafter. The mathematical model for an isotropic material undergoing strain hardening during cold rolling is

$$P - \sigma_x = \frac{2}{\sqrt{3}}\sigma_{ys} = \frac{2}{\sqrt{3}}\sigma_{flow} \tag{12.47}$$

Combining Eqs. (12.47) and (12A.4) yields

$$\frac{d\left[(P - \sigma_{flow})h\right]}{d\theta} = 2PR\,(\theta \mp \mu) \tag{12.48}$$

$$\frac{d\left[P/\sigma_{flow} - 1\right]}{d\theta} = \frac{2R\,(\theta \mp \mu)}{h}\left(P/\sigma_{flow}\right) \tag{12.49}$$

Let $f = P/\sigma_{flow}$ and $h = h_f + R\theta^2$ so that

$$\frac{df}{f} = \frac{2R\theta}{h}d\theta \mp \frac{2R\mu}{h}d\theta \tag{12.50}$$

$$\frac{df}{f} = \frac{2R\theta}{h_f + R\theta^2}d\theta \mp \frac{2R\mu}{h_f + R\theta^2}d\theta \tag{12.51}$$

Integrating Eq. (12.51) yields

$$\int\frac{df}{f} = 2R\int\frac{\theta d\theta}{h_f + R\theta^2} \mp 2R\mu\int\frac{d\theta}{h_f + R\theta^2} \tag{12.52}$$

$$\ln\,(f) = \ln\left(\frac{h}{R}\right) \mp 2\mu\sqrt{\frac{R}{h_f}}\,\tan^{-1}\left(\theta\sqrt{\frac{R}{h_f}}\right) + \ln\,(C_1) \tag{12.53}$$

$$\ln\left(P/\sigma_{flow}\right) = \ln\left(\frac{h}{R}\right) \mp 2\mu\sqrt{\frac{R}{h_f}}\,\tan^{-1}\left(\theta\sqrt{\frac{R}{h_f}}\right) + \ln\,(C_1) \tag{12.54}$$

where C_1 is the constant of the integration. Let

$$H = 2\sqrt{\frac{R}{h_f}}\,\tan^{-1}\left(\theta\sqrt{\frac{R}{h_f}}\right) \tag{12.55}$$

so that

$$P/\sigma_{flow} = \frac{C_1 h}{R}\exp\left(\mp\mu H\right) \tag{12.56}$$

At the **entry zone**, $\theta = \alpha$, $H = H_o$, $h = h_o$, and $P/\sigma_{flow} = 1$. Hence,

$$C_1 = \frac{R}{h_o}\exp\left(-\mu H_o\right) \tag{12.57}$$

$$H_o = 2\sqrt{\frac{R}{h_f}}\,\tan^{-1}\left(\alpha\sqrt{\frac{R}{h_f}}\right) \tag{12.58}$$

then,

$$P/\sigma_{flow} = \left(\frac{h}{h_o}\right)\exp\left[\mu\left(H_o - H\right)\right] \tag{12.59}$$

$$P = \sigma_{flow}\left(\frac{h}{h_o}\right)\exp\left[\mu\left(H_o - H\right)\right] \tag{12.60}$$

Let the flow stress and friction displacement be

$$\sigma_{flow} = \frac{2}{\sqrt{3}}\sigma_{ys} \tag{12.61}$$

$$\mu = \tan\left(\frac{L_p}{R}\right) \tag{12.62}$$

Thus, Eq. (12.60) becomes

$$P = \left(\frac{2}{\sqrt{3}}\sigma_{ys}\right)\left(\frac{h}{h_o}\right)\exp\left[\mu\left(H_o - H\right)\right] \tag{12.63}$$

Exit Zone For $\theta = 0$, $H_f = 0$, $H_o = 0$, $h_o = h_f$, and $P/\sigma_{flow} = 1$,

$$C_1 = \frac{R}{h_o}\exp\left(-\mu H\right) \tag{12.64}$$

$$H = 2\sqrt{\frac{R}{h_f}}\,\tan^{-1}\left(\theta\sqrt{\frac{R}{h_f}}\right) \tag{12.65}$$

Then, the roller pressure (P), the mean pressure (P_m), and the mean thickness (h_m) expressions are

$$P = \left(\frac{2}{\sqrt{3}}\sigma_{ys}\right)\left(\frac{h}{h_f}\right)\exp\left(\mu H\right) \tag{12.66}$$

$$P_m = -\frac{\sigma_{flow}h_m}{\mu L_p}\left[1 - \exp\left(\frac{\mu L_p}{h_m}\right)\right] \tag{12.67}$$

$$h_m = \frac{1}{2}\left(h_o + h_f\right) \tag{12.68}$$

The force F, work done U, and power W_{CR} required for plastic deformation using one roller are

$$F_{cr} = bL_p\sigma_{flow} \tag{12.69}$$

$$U_{cr} = 2\pi L_p F_{cr} \tag{12.70}$$

$$W_{cr} = L_p F_{cr} N \tag{12.71}$$

where $\omega = 2\pi N$ is the roller angular velocity, N is in revolutions per minute (rpm) or revolutions per second (rps), and L_p is treated as the moment arm.

Problems

12.1 Suppose that a hypothetical alloy bar is tested in tension at a low strain rate ($d\varepsilon/dt$) and exhibits an engineering stress-strain behavior as shown below. If the original gage length is 350 mm long and the modulus of elasticity is 100 GPa, calculate **(a)** the elastic strain ε at $\sigma = 400\ MPa < \sigma_{ys}$, **(b)** the instantaneous gage length L, and **(c)** the gage length growth ΔL. [Solution: (a) $\varepsilon_y = 0.40\%$, (b) $L = 351.40\ mm$, (c) $\Delta L = 1.40\ mm$].

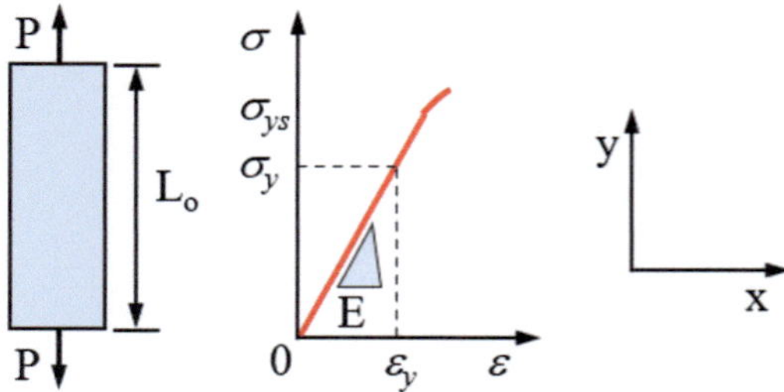

12.2 A cylindrical brass rod with an original diameter $d_o = 10\ mm$ under elastic deformation experiences a reduction in diameter $\Delta d = -2 \times 10^{-3}\ mm$. Calculate the **(a)** lateral and axial strains (ε_x and ε_y), **(b)** the elastic stress σ_x, and the corresponding mechanical load P_x. Data: $E = 97\,GPa$, $\sigma_{ys} = 145\,MPa$, and $v = 1/3$. [Solution: (a) $\varepsilon_x = 0.0006$ and $\varepsilon_y = -0.0002$, (b) $P_x = 4,571\,N$].

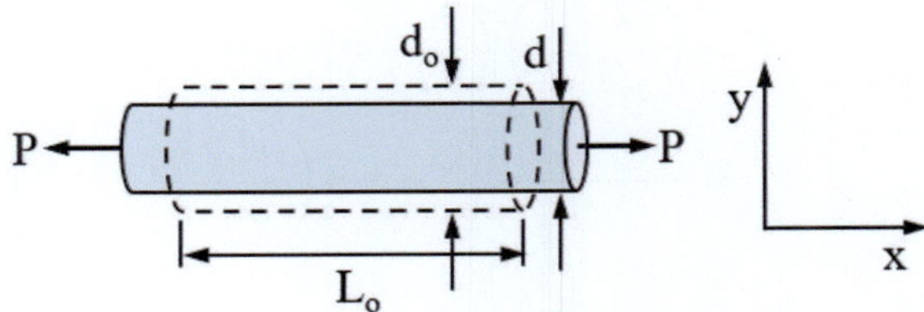

12.3 One 1045 steel bar was tested in tensile mode at a strain rate of $10^{-4}\,s^{-1}$ in a room temperature environment. Use the given data to determine **(a)** the modulus of elasticity E; **(b)** the parameters of the Hollomon's equation, $\sigma = k_o\varepsilon^n$, at $\varepsilon_{ys} < \varepsilon \le \varepsilon_{ts}$; and **(c)** the duration of the test. Data: $\sigma_{ys} = 412\,MPa$ and $\sigma_{ts} = 585$. [Solution: (a) $E \simeq 200\,GPa$, (b) $k_o = 966.03\,MPa$ and $n = 0.14029$, (c) $t \simeq 280\,s$].

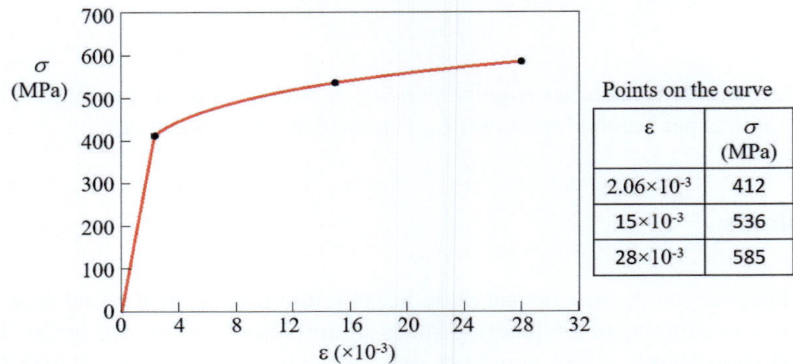

Points on the curve

ε	σ (MPa)
2.06×10^{-3}	412
15×10^{-3}	536
28×10^{-3}	585

12.4 Consider one 1045 steel bar being tested in tensile mode at a strain rate of $10^{-4}\,s^{-1}$ and at room temperature. Use the given data to determine **(a)** the modulus of elasticity E, **(b)** the parameters of the Ludwik's equation, $\sigma = \sigma_{ys} + k_o\varepsilon^n$, at $\varepsilon_{ys} < \varepsilon \le \varepsilon_{ts}$ and **(c)** the duration of the test. Data: $\sigma_{ys} = 412\,MPa$ and $\sigma_{ts} = 585$. [Solution: (a) $E \simeq 200\,GPa$, (b) $k_o = 1895\,MPa$ and $n = 0.660$, (c) $t = 4.67$ min].

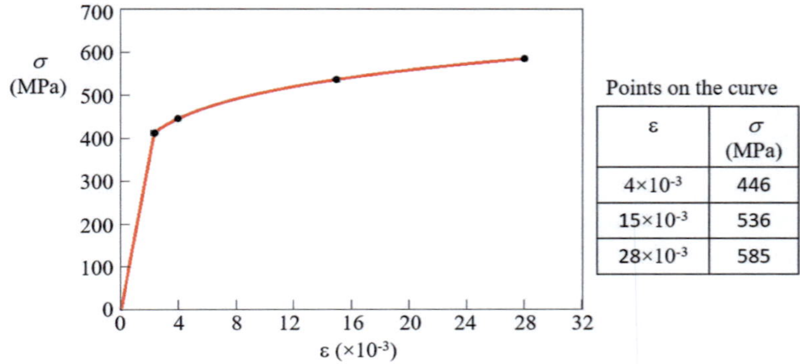

ε	σ (MPa)
4×10^{-3}	446
15×10^{-3}	536
28×10^{-3}	585

12.5 Derive the modulus of toughness $U(\varepsilon)$ using Ludwik's equation. Compare the resultant expression with Eq. (12.27d).

12.6 Use the load-deformation tensile curve shown in Fig. 12.4 for a hypothetical steel round bar tested at a temperature T to determine **(a)** the specimen final gage length (L_f) and the change in gage length $(\Delta L_{0.2\%})$ at 0.2% strain due to mechanical deformation; **(b)** the mechanical forces at yielding, ultimate and fracture conditions; and **(c)** the total modulus of toughness. Explain and use the given data. [Solution: (a) $L_f = 39.2\ mm$ and $\Delta L_{0.2\%} = 0.07\ mm$ (b) $P_f = 15.85\ kN$, (c) $U(\varepsilon) = 19.01\ MJ/m^3$].

$$d_o = 6.35\ mm = 0.25\ inch,\ L_o = 35\ mm = 0.1375\ inch$$

$$d\varepsilon/dt = 2 \times 10^{-4}\ s^{-1}$$

12.7 Determine (a) the mechanical properties using the given stress-strain diagram for a hypothetical 2022-X alloy (b) the total modulus of toughness $U(\varepsilon)$ up to the UTS point. Given data: $L_o = 10\ cm,\ d_o = 1\ cm,\ d\varepsilon/dt = 2 \times 10^{-4}\ s^{-1}$, and $T = 23°C$. [Solution: (a) $\sigma_{ys} = 260\ MPa,\ \sigma_{ts} = 460\ MPa,\ E = 100\ GPa$, (b) $U(\varepsilon) = 114.39\ MJ/m^3$].

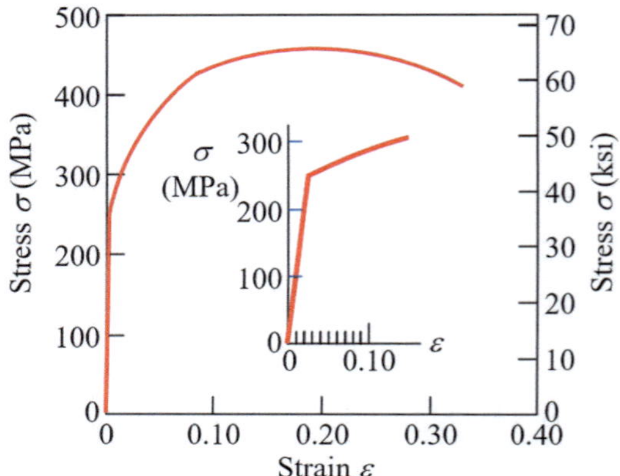

12.8 Calculate **(a)** the total modulus of toughness $U(\varepsilon)$ using Hooke's and Ludwik's equations and assuming the 0.2% offset method for determining the yield strength σ_{ys} and **(b)** the tensile strength. Given data: $\sigma_{ys} = 200 \ MPa$, $E = 100 \ GPa$, $n = 0.5$, and $k_o = 500 \ MPa$. [Solution: (a) $U(\varepsilon) = 216.94 \ MJ/m^3$, (b) $\sigma_{ts} = 553.55 \ MPa$].

12.9 Calculate **(a)** the total modulus of toughness $U(\varepsilon)$ using Hooke's and Ludwik's equations and assuming the 0.2% offset method for determining the yield strength σ_{ys} and **(b)** the tensile strength. Given data: $\sigma_{ys} = 392 \ MPa$, $E = 200 \ GPa$, $n = 0.15$, and $k_o = 995 \ MPa$. [Solution: (a) $U(\varepsilon) = 154.54 \ MJ/m^3$, (b) $\sigma_{ts} = 1140.60 \ MPa$].

12.10 Assume that the σ-ε curve for a hypothetical alloy is described by the Hollomon equation given below:

$$\sigma = k_o \varepsilon^n$$

$$n = 0.25$$

$$k_o = 500 \ MPa$$

Calculate **(a)** the mechanical properties (σ_{ys} as per the 0.2% method, σ_{ts} and E), and **(b)** plot the equation within a valid strain range, **(c)** the change in gage length ΔL and the gage length L at the ultimate tensile strain ε_{ts} if the original gage length is $L_o = 50.80 \ mm$, and **(d)** the maximum load for a round specimen with a $d_o = 6.35\text{-}mm$ diameter. [Solution: (a) $\sigma_{ys} = 400 \ MPa$, (c) $\Delta L = 4.064 \ mm$, (d) $P_{\max} = 17,038 \ N$].

12.11 Assume that 0.82-MPa tension stress is applied to a round single crystal of pure copper ($Cu\text{-}FCC$) along the [112] slip direction. Calculate **(a)** the resolved shear stress τ_r to activate the (111) [110] slip system by dislocation motion and **(b)** the yield strength if the critical resolved shear stress is $\tau_c = 0.6 \ MPa$. [Solution: (a) $\tau_r = 0.45 \ MPa$, (b) $\sigma_{ys} = 1.91 \ MPa$,].

12.12 Consider the plastic deformation of a molybdenum ($BCC\text{-}Mo$) single crystal at room temperature by activating the (110) $[\bar{1}11]$ slip system with a [110] normal direction to the (110) plane. Assume that the specimen is loaded in tension at $\sigma = 50 \ MPA$ level along the [011] direction. Calculate the resolved shear stress τ_r. Data: $E = 330 \ GPa$, $\sigma_{ys} = 70 \ MPa$, $\sigma_{ts} = 320 \ MPa$. [Solution: $\tau_r = 20.41 \ MPa$].

12.13 Consider the plastic deformation of a molybdenum ($BCC\text{-}Mo$) single crystal at room temperature by activating the (110) $[1\bar{1}1]$ slip system with a [110] normal direction to the (110) plane. Assume that the specimen is loaded in tension at $\sigma = 50 \ MPA$ level along the [011] direction. Calculate the resolved shear stress τ_r. Data: $E = 330 \ GPa$, $\sigma_{ys} = 70 \ MPa$, $\sigma_{ts} = 320 \ MPa$. [Solution: $\tau_r = 0$].

12.14 If one hypothetical specimen, loaded in tension, underwent rupture in 15 *hours* at $1000°C$ and $100 \ MPa$, then use the Larson-Miller and Sherby-Dorn parameter equations to calculate the time to rupture a similar specimen subjected to $900°C$ and $100\ MPa$. Compare results. Data: $Q_c = 238\ kJ/mol$ and $C_A = 20$. [Solution: $t_r\ (LM) \simeq 104\ h$ and $t_r\ (SD) = 102\ h$].

12.15 Assume that one S-590 alloy specimen is exposed to a $650°C$ environment. Use the creep data in Fig. 12.18 to determine **(a)** the t_r, $d\varepsilon/dt$, P_{LM} with $C_A = 20$ and Q_c at $\sigma_r = 300\ MPa$ and $\sigma_r = 300\ MPa$, and **(b)** n and K_1 for the governing expression defined by Eq. (12.29a). [Solution: (a) $P_{LM} = 24630\ K$, (b) $n = 12.82$ and $K_1 = 1.75 \times 10^{-36}\ h^{-1}$].

12.16 Consider the flat rolling process shown in Example 12.9 to strengthen an annealed copper strip. Use the "Cu, annealed" curve in Fig. 12.32 to extract the required stress σ value at a strain ε. The Cu strip dimensions are $b = 228\ mm$, $h_o = 26\ mm$, and $h_f = 20\ mm$, and roller radius is $R = 300\ mm$. If the rollers rotate at $N = 100\ rpm$, then calculate C_w, ε, μ, θ, F_{total}, and W_{total} for two rollers. [Solution: $C_w = 0.23$, $\varepsilon = 0.26$, $W_{total} = 1475\ hp$].

12.17 Use the Hollomon equation and the available data $(0.16,\ 645\ MPa)$ and $(0.20,\ 674\ MPa)$ to calculate the true stress corresponding to the true strain of 0.18 for a steel specimen. [Solution: $\sigma \simeq 660\ MPa$].

12.18 Consider the given room-temperature maximum stress-cycle (S-N) data for a hypothetical steel rod of $16\ mm$ in diameter. Calculate **(a)** the mechanical tension load (P_d) to ensure infinite fatigue life and **(b)** the bar diameter d when $N = 4 \times 10^4$ *cycles* to avoid fatigue failure at the maximum load from part (a). Use a safety factor of 2 for both parts. Here, N is in cycles to failure and σ is in MPa units. [Solution: (a) $P_d = 26.16\ kN$, (b) $d = 14.40\ mm$].

N	14820	19647	40003	60019	10×10^4	20×10^4	30×10^4
σ	470	402	323	304	300	300	300

12.19 Plot the S-N data given in P12.18 and determine the fatigue limit as a material property. According to previous fatigue failure schemes, a 50-mm-diameter rod may be subjected to a cyclic load with a 400-MPa maximum stress and a 0-MPa minimum stress. Determine **(a)** the fatigue limit σ_L, **(b)** the rod fatigue life N at $400\ MPa$, **(c)** the design stress σ_d for a prolong fatigue life using a safety factor of 1.6, and **(d)** the design fatigue life N at σ_d. [Solution: (a) $\sigma_L = 300\ MPa$, (b) $N = 20,000\ cycles$, (c) $\sigma_d = 250\ MPa$, (d) $N = \infty$].

12.20 Use the given steady-state creep rate $(d\varepsilon/dt)$ data for a hypothetical alloy at $1000°C$ ($1273\ K$) to calculate the $d\varepsilon/dt$ at $900°C$ and $25\ MPa$. The activation energy is $Q_c = 300\ kJ/mol$ and $R = 8.3145\ J/mol.K$. [Solution: $d\varepsilon/dt = 9.56 \times 10^{-5}\ s^{-1}$].

$$d\varepsilon/dt\,(s^{-1}) \qquad \sigma\,(MPa)$$
$$1.98 \times 10^{-4} \qquad 30$$
$$3.92 \times 10^{-5} \qquad 20$$

12.21 **(a)** Use the data in Fig. 12.17a to fit $\sigma_r = f(T)$ at $t = 10\,hours$ and $t = 100\,hours$. Is there a linear relationship? **(b)** Calculate the rupture stress σ_r and **(c)** the Larson-Miller parameter P_{LM} with $C_A = 20$ at $450°F$ for both cases. [Solution: (b) $\sigma_r = 26.90\,ksi$ at $t = 10\,h$, (c) $P_{LM} = 10,036$ at $t = 10\,h$].

References

1. zwickroell.com catalog advertised online
2. MTS Systems Corp., *Advanced Materials & Processes*, vol. 161(8) (2003)
3. W.D. Callister, Jr., D.G. Rethwisch, *Materials Science and Engineering: An Introduction*, 10th edn. (Wiley, New York, 2018)
4. M. Meier, A. Broumas, Luders bands in steel, in *Video Showing Luders Band Formation and Propagation in a Low Carbon Steel* (Department of Chemical Engineering and Materials Science, University of California, California, 2001). https://vimeo.com/4586024
5. B. Yang, C. Motz, W. Grosinger, W. Kammrath, G. Dehm, Tensile behaviour of micro-sized copper wires studied using a novel fibre tensile module. Int. J. Mater. Res. **99**(7), 716–724 (2008)
6. W.F. Smith, *Principles of Materials Science and Engineering*, 3rd edn. (McGraw-Hill Inc., New York, 1996)
7. D.M. Dimiduk, M.D. Uchic, T.A. Parthasarathy, Size-affected single-slip behavior of pure nickel microcrystals. Acta Mater. **53**(15), 4065–4077 (2005)
8. P.L. Mangonon, *The Principles of Materials Selection for Engineering Design* (Prentice Hall, New York, 1999)
9. J.W. Christian, S. Mahajan, Deformation twinning. Prog. Mater. Sci. **39**(1–2), 1–157 (1995)
10. A. Vinogradov, E. Agletdinov, D. Merson, Mechanical twinning is a correlated dynamic process. Sci. Rep. **9**(1), 1–13 (2019)
11. A. Ghaderi, M.R. Barnett, Sensitivity of deformation twinning to grain size in titanium and magnesium. Acta Mater. **59**(20), 7824–7839 (2011)
12. G.F. Vander Voort, online subject: *Microstructure, Its Evolution, Microstructure—As-Cast, Cold Worked and Annealed, Hot Worked and Annealed and Heat Treated Part 1* (Research & Technology, Buehler Ltd., Lake Bluff, 2000). https://www.georgevandervoort.com/images/
13. N.L. Perez, *Evaluation of Thermally Degraded Rapidly Solidified Fe-Base and Ni-Base Alloys*, Ph.D. Dissertation (University of Idaho, Moscow, 1988)
14. J.H. Hollomon, Tensile deformation. Aime Trans. **12**(4), 1–22 (1945)
15. P. Ludwik, *Elemente der Technologischen Mechanik*. Google translator: *Elements of Technological Mechanics* (Springer, Berlin, 1909)
16. W. Ramberg, W.R. Osgood, *Description of Stress-Strain Curves by Three Parameters*, Technical Note No. 902 (NASA National Advisory Committee for Aeronautics, Washington DC, 1943)
17. J.G. Kaufman, *Parametric Analyses of High-Temperature Data for Aluminum Alloys* (ASM International, Materials Park, 2008). www.asminternational.org
18. R.M. Goldhoff, *Which method for extrapolating stress-rupture data?*. Mater. Des. Eng. **49**(4), 93–97 (1959)
19. O.A. Ruano, O.D. Sherby, On constitutive equations for various diffusion controlled creep mechanisms. Revue Phys. Appl. **23**(4), 625–637 (1988)

20. M.A. Przystupa, *Microstructural characterization of dislocation networks during harper-dorn creep of FCC, BCC and HCP metals and alloys*. Final Report (Grant # DE-FG02-03ER46077) (Department of Materials Science and Engineering, University of California, Los Angeles, 2007)
21. B. Viernstein, E. Kozeschnik, Integrated physical-constitutive computational framework for plastic deformation modeling. Metals **10**(7), 869 (2020)
22. F.H. Norton, *The Creep of Steels at High Temperatures* (McGraw-Hill, New York, 1929)
23. S.A. Arrhenius, *Über die Dissociationswärme und den Einfluss der Temperatur auf den Dissociationsgrad der Elektrolyte*, Google translation: *About the Heat of Dissociation and the Influence of Temperature on the Degree of Dissociation of the Electrolytes*. Physical Chemistry Journal, Zeitschrift für physikalische Chemie **4**(1), 96–116 (1889)
24. F.R. Larson, J. Miller, A time-temperature relationship for rupture and creep stresses. Trans. ASME **74**, 765–775 (1952)
25. O.D. Sherby, J.E. Dorn, Creep correlations in alpha solid solutions of aluminum. J. Miner. Met. Mater. Soc. JOM **4**(4), 959–964 (1952)
26. X. Zhu, H. Cheng, M. Shen, J. Pan, Determination of C parameter of Larson-Miller equation for 15CrMo steel. Adv. Mater. Res. **791–793**, 374–377 (2013)
27. M. Tamura, F. Abe, K. Shiba, H. Sakasegawa, H. Tanigawa, Larson-miller constant of heat-resistant steel. Metall. Mater. Trans. A **44**(6), 2645–2661 (2013)
28. J. Phillips, *Improper Design Leads to Fatigue Failure In Blower Shaft, Warren Forensics Blog, Case Studies* (2017). https://www.warrenforensics.com/2017/04/27/
29. Introduction: atlas of fractographs, in *Fractography, ASM Handbook*, vol. 12 (ASM International, New York, 1992)
30. A.H. Cottrell, D. Hull, Extrusion and intrusion by cyclic slip in copper, in *Proceedings of the Royal Society of London. Series A. Mathematical and Physical Sciences*, vol. 242(1229) (1957), pp. 211–213
31. K. Tanaka, T. Mura, A dislocation model for fatigue crack initiation. J. Appl. Mech. **48**(1), 97–103 (1981)
32. J. Man, K. Obrtlık, J. Polak, Extrusions and intrusions in fatigued metals: Part 1, state of the art and history. Philos. Mag. **89**(16), 1295–1336 (2009)
33. D.K. Felbeck, A.G. Atkins, *Strength and Fracture of Engineering Solids*, 2nd edn. (Prentice Hall, Inc., Upper Saddle River, 1996)
34. G.E. Dieter, *Mechanical Metallurgy* (McGraw-Hill, Inc., New York, 1986)
35. E.O. Hall, The deformation and ageing of mild steel: III Discussion of results. Proc. Phys. Soc. Sect. B **64**(9), 747–755 (1951)
36. N.J. Petch, The cleavage strength of polycrystals. J. Iron Steel Inst. **174**(1), 25–28 (1953)
37. A.G. Guy, J.J. Hren, *Elements of Physical Metallurgy*, 3rd edn. (Addison-Wesley Publishing Company, Inc., Reading, 1974)
38. A. Wauthier-Monnin, T. Chauveau, O. Castelnau, H. Regie, B. Bacroix, The evolution with strain of the stored energy in different texture components of cold-rolled if steel revealed by high resolution X-ray diffraction. Mater. Charact. **104**, 31–41 (2015)
39. S. Kalpakjian, S.R. Schmid, *Manufacturing Engineering and Technology*, 6th edn. (Prentice Hall, New York, 2009)

Chapter 13
Fracture Mechanics

13.1 Introduction

This chapter describes fracture toughness as the mechanical resistance or the ability of a crack-containing material to absorb strain energy at the crack tip for subsequent crack growth and propagation under an applied stress. For comparison, the modulus of toughness or simply toughness (see Chap. 12) is the ability of an initially crack-free material to absorb strain energy to deform or fracture under applied strains.

In materials science and metallurgy, the modulus of toughness or toughness is a short term for the strain-energy density (U) that defines the ability of a material to absorb energy and plastically deform without fracturing. It is graphically illustrated as the area under a stress-strain curve and mathematically defined as an integral defined by Eq. (12.23c) under quasi-static conditions. On the other hand, Charpy impact strain energy (absorbed energy by the specimen during fracture) is also referred to as the impact fracture toughness of a notched specimen under a dynamic condition evaluated at a temperature range.

In the fracture mechanics field or continuum mechanics, aspects of quasi-static, fatigue, and impact fracture toughness of polycrystals or engineering materials are thoroughly defined for comparison purposes, since they are exposed to varying loads during service. Moreover, quasi-static condition prevails at low loading rates describing linear-elastic fracture mechanics (LEFM) of brittle materials, which allow a very small plastic zone ahead of crack tips. The main emphasis is to define the quasi-static plane-strain fracture toughness as a dimensional-independent property for several crack configurations, the crack growth rate, and the impact toughness of cracked specimens.

Fracture mechanics is presented as a mathematical and graphical field for illustrating the fundamental aspects of crack growth under quasi-static and dynamic loading conditions for a specimen having a perpendicular or inclined crack at an angle relative to an applied mechanical load. This requires an analysis of an internal stress field ahead of a crack tip.

N. Perez, *Materials Science: Theory and Engineering*,
https://doi.org/10.1007/978-3-031-57152-7_13

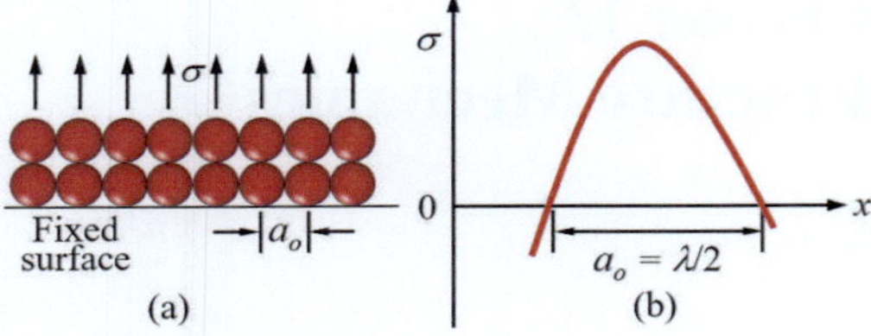

Fig. 13.1 Theoretical strength model (**a**) Atomic planes under tension and (**b**) sinusoidal stress-strain curve

13.2 Theoretical Strength of Solids

Consider the atomic model shown in Fig. 13.1a, and assume that an external stress induces elastic deformation of the atomic bonds, where $a_o = \lambda/2$ in Fig. 13.1b is the atomic separation and λ is the wavelength. In effect, the atomic bonds on the plane have the tendency to break at a critical strain induced by an external mechanical load.

Assume a sinusoidal stress σ behavior in the tensile elastic domain so that

$$\sigma = \sigma_{\max} \sin\left(\frac{\pi x}{\lambda/2}\right) \simeq \left(\frac{\pi x}{a_o}\right) \sigma_{\max} \tag{13.1a}$$

$$\sigma = \frac{dW}{dx} \tag{13.1b}$$

where W is the potential or strain energy density, x is the displacement, and σ is the theoretical strength of a perfect or defect-free solid that defines the maximum possible stress a specimen can withstand prior to fracture.

In the engineering field, the goal is to design a solidification process for producing materials exhibiting strength close to the theoretical fracture stress.

From Hooke's law, the tensile modulus of elasticity E, known as the Young's modulus, measures the resistance to elastic deformation. Thus,

$$E = \frac{\sigma}{\varepsilon} = \frac{\sigma}{x/a_o} = \frac{1}{x/a_o}\left(\frac{\pi x}{a_o}\right)\sigma_{\max} \tag{13.2}$$

Combining Eqs. (13.1) and (13.2) yields the maximum theoretical strength

$$\sigma_{\max} = \frac{E}{\pi} = 0.32E \tag{13.3}$$

The total strain energy density is defined as the area under the stress-displacement curve (see Chap. 12). Mathematically,

$$W = \int_0^{\lambda/2} \sigma \, dx = \int_0^{\lambda/2} \sigma_{\max} \sin\left(\frac{\pi x}{\lambda/2}\right) dx \tag{13.4a}$$

$$W = \frac{\lambda}{\pi} \sigma_{max} \tag{13.4b}$$

which defines the total strain energy density at σ_{max} due to inherent crystal defects, such as flaws or cracks.

13.3   Griffith Crack Theory

The development of linear-elastic fracture mechanics (LEFM) started with Griffith [1] work on glass in 1921 using an energy balance approach to predict the fracture stress of glass specimens containing surface defects. Griffith defined the crack resistance (R_c) as twice the specific surface energy (γ_s), $R_c = 2\gamma_s$, due to the existence of two crack surfaces in pure brittle glass. However, most engineering brittle materials undergo, to an extent, plastic deformation at the crack tip so that $R_c = 2\left(\gamma_s + \gamma_p\right)$ as suggested by Orowan [2, p. 139]. Here, γ_p is the specific surface energy due to plasticity ahead of the crack tip.

Consider the specimen shown in Fig. 13.2a containing a through-thickness central crack configuration with two crack tips being remotely loaded in tension mode. Plasticity ahead of the tips is plastic zones modeled as circles representing idealized energy release areas. The small circles represent plastic zones, where plastic deformation dictates the behavior of crack growth.

Figure 13.2b, on the other hand, illustrates a specimen having the single-edge crack and the plastic zone being modeled as an ellipse with a stress element (red square) on the elastic-plastic boundary ahead of the crack tip at r and θ. This stress element is also shown in Fig. 13.2c with additional details.

A crack tip acts as a stress concentration site adjacent to the process zone (plastic zone) during deformation of the atomic structure. Subsequently, atomic bonds fracture locally, and crack growth occurs under the effect of an applied remote stress σ.

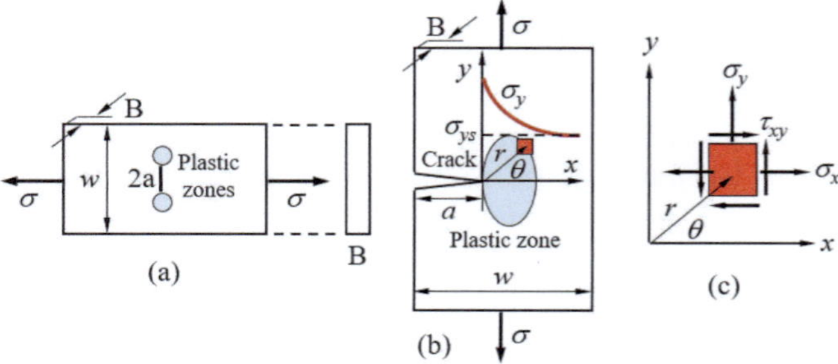

Fig. 13.2 (**a**) Large brittle plate containing one through-the-thickness central crack, with two idealized energy release areas ahead of the crack tips, (**b**) a large brittle plate containing a single-edge crack with an idealized plastic zone, and (**c**) stress element at the crack tip

Local crack tip stress domain induces plastic deformation, leading to negligible energy consumption at the crack tip, and as a result, crack propagation occurs at relatively high velocity in brittle materials. On the other hand, crack tip blunting occurs along with a large energy consumption of strain energy in ductile materials.

According to the theoretical background for crack growth in brittle materials, the Griffith theory states that the magnitude of the total surface energy increases due to the creation of two new crack surfaces during crack growth. This implies that the material potential energy decreases and the stable crack size increases until it reaches a critical size a_c at a critical stress σc. Therefore, the cracked material reaches a critical state, and the crack propagates at a relatively high velocity leading to a final mechanical fracture.

13.3.1   Elastic-Strain Energy

When the elastic or brittle solid body (specimen) is remotely loaded from the crack faces, the product of the released elastic-strain energy density ($\int \sigma d\epsilon$) and the cylindrical volume element ($2\pi a^2 B$) at the crack tips (Fig. 13.2a) yields the elastic-strain energy as

$$W_e = -2\left(\pi a^2 B\right) \int \sigma d\epsilon = -2\left(\pi a^2 B\right) \int E' \epsilon d\epsilon \qquad (13.5a)$$

$$W_e = -2\left(\pi a^2 B\right) \left(\frac{E' \epsilon^2}{2}\right) = -\left(\pi a^2 B\right) \left(\frac{\sigma^2}{E'}\right) \qquad (13.5b)$$

where $\sigma = E'\epsilon$ = Hooke's law

$\quad E' = E$ for plane-stress condition
$\quad E' = E/\left(1 - v^2\right)$ for plane-strain condition
$\quad E$ = Modulus of elasticity (MPa)
$\quad \epsilon$ = Elastic strain
$\quad \sigma$ = Applied remote stress (MPa)
$\quad a$ = One-half crack length (mm)
$\quad v$ = Poisson's ratio
$\quad 4aB = 2(2aB)$ = Total surface crack area (mm^2)
$\quad B$ = Specimen thickness (mm)

According to Griffith [1], the elastic-surface energy for creating new crack surfaces during crack growth is defined by

$$W_s = 2\left(2aB\gamma_s\right) \qquad (13.6)$$

where γ_s is the specific surface energy for atomic bond breakage. The factor of 2 in Eq. (13.6) is due to two crack tips.

For an elastically stressed solid body, the total potential energy is

$$W = W_s + W_e = 2\,(2aB\gamma_s) - \left(\pi a^2 B\right)\left(\frac{\sigma^2}{E'}\right) \tag{13.7}$$

Recall that E' is a control modulus of elasticity for plane-stress or plane-strain condition.

13.3.2   Potential Energy

Letting $U = W/B$ gives the total energy gradient or total potential energy per unit thickness as

$$U = U_s + U_e \tag{13.8a}$$

$$U = 2\,(2a\gamma_s) - \frac{\pi a^2 \sigma^2}{E'} \tag{13.8b}$$

where U_s denotes the elastic surface energy and U_e denotes the released elastic energy per unit thickness. Taking the second derivative of Eq. (13.8b) with respect to the crack length a yields

$$\frac{d^2\,(U)}{da^2} = -\frac{2\pi\sigma^2}{E'} \tag{13.9}$$

which implies that $a \rightarrow a_{\max} = a_c$ when $d^2U/da^2 < 0$ and the cracked specimen under load is an unstable system. Consequently, the crack will always grow and reach a critical length $a \rightarrow a_c$.

13.3.3   The Strain-Energy Release Rate

If $dU/da = 0$ in Eq. (13.8b), then $2\gamma_s E' = \pi a\sigma^2$ from which the applied stress (σ), the crack length (a) or the strain-energy release rate (G_I) for brittle solid materials are derived as

$$\sigma = \sqrt{\frac{(2\gamma_s)E'}{\pi a}} \tag{13.10a}$$

$$a = \frac{(2\gamma_s)E'}{\pi\sigma^2} \tag{13.10b}$$

$$G_I = 2\gamma_s = \frac{\pi a\sigma^2}{E'} \quad \text{(pure elastic)} \tag{13.10c}$$

$$G_I = 2\left(\gamma_s + \gamma_p\right) = \frac{\pi a\sigma^2}{E'} \quad \text{(elastic-plastic)} \tag{13.10d}$$

Notice that Eq. (13.10d) contains the plastic specific surface energy (γ_p) to compensate for plasticity ahead of the crack tip (Orowan [2, p. 139] and Irwin [3, pp. 551–590]).

13.4 Linear-Elastic Fracture Mechanics

13.4.1 Modes of Loading

According to the modes of remote loading shown in Fig. 13.3a, the mechanical behavior of solids containing cracks of specific geometry and size can be predicted by evaluating the stress-intensity factor K_I, K_{II} or K_{III} near a crack tip.

Although a combined loading can be encountered in structural components, but K_I is the most studied and evaluated experimentally for determining its critical value called plane-strain fracture toughness K_{IC} as a material's property.

Figure 13.3a shows mode I, mode II, and mode III crack-coordinate systems and the principal-stress components near the crack tip at a distance r and angle θ. Accordingly, Fig. 13.3b depicts the crack tip, the plastic zone, and a stress element ahead of the crack tip at distance r and at an angle θ with respect to the crack plane along the x-axis.

In fracture mechanics, a is the crack length or size, and $r << a$ is the plastic zone size, which forms due to a localized plastic deformation related to dislocation mechanism, and it can have an arbitrary shape.

13.4.2 Stress-Intensity Factors and Stress Fields

The stress field ahead of a crack tip plays a central role in characterizing the fracture parameters governing crack growth. In order to quantify the intensity of the stress field ahead of the crack tip as per Irwin's work [3, pp. 551–590], the concept of stress-intensity factor can be used to characterize crack growth; K_I for mode I

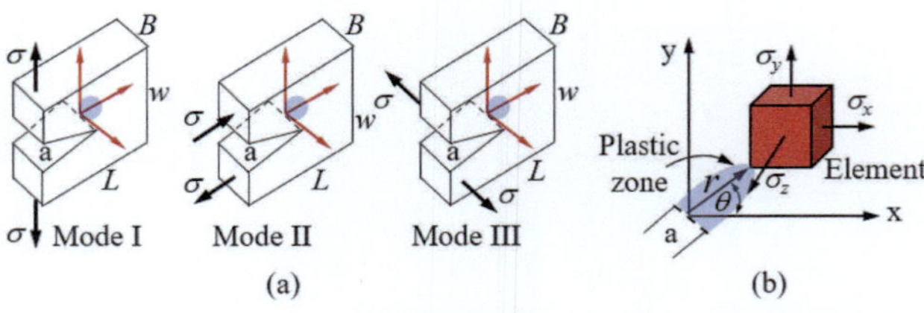

Fig. 13.3 (a) Stress loading modes and (b) crack tip coordinate system with a principal stress element ahead of the crack tip

(opening), K_{II} for mode II (sliding), K_{III} for mode III (tearing), or a combination of these modes of loading (mixed-mode fracture mechanics). Moreover,

- The stress σ_y in Fig. 13.2b or 13.3a for mode I is a local component of the stress field ahead of the crack tip responsible for crack opening since $\sigma_y \rightarrow \infty$ as $r \rightarrow 0$. Actually, $\sigma_y \rightarrow \infty$ is known as a singularity, but it is truncated by the yielding process at the elastic-plastic boundary ahead of the crack tip.
- The plastic zone is a small area referred to as the process zone with an arbitrary shape and size r. The atomic arrangement in this zone is locally destroyed by the effect of the opening stress σ_y.
- During crack growth, the specimen potential energy per unit thickness (ΔU) decreases, and the crack surface energy per unit thickness (U_s) increases, since new crack surfaces are developed during crack growth.

The stress field ahead of a stationary crack tip can be quantified at suitable temperatures, provided that the material exhibits some degree of brittleness; otherwise, the linear-elastic fracture mechanics theory is not applicable.

The analysis of stress, strain, and displacement fields near the crack tip represents an important part of fracture mechanics for brittle and ductile materials. However, linear-elastic fracture mechanics prevails in this textbook. An engineering approach for characterizing plastic fracture mechanics is out of the scope of this textbook.

The stress distribution, in general, ahead of the crack tip is a three-dimensional scheme for each mode of loading. Only the most relevant stress field components associated with individual stress-intensity factors K_i and trigonometric functions $f_i(\theta)$, where $i = I, II, III$, are written as

$$\sigma_{ij} = \frac{K_i}{\sqrt{2\pi r}} f_{ij}(\theta) \quad \text{(general)} \tag{13.11a}$$

$$\sigma_y = \frac{K_I}{\sqrt{2\pi r}} \cos\frac{\theta}{2}\left(1 + \sin\frac{\theta}{2}\sin\frac{3\theta}{2}\right) \tag{13.11b}$$

$$\tau_{xy} = \frac{K_{II}}{\sqrt{2\pi r}} \cos\frac{\theta}{2}\left[1 - \sin\frac{\theta}{2}\cos\frac{3\theta}{2}\right] \tag{13.11c}$$

$$\tau_{yz} = \frac{K_{III}}{\sqrt{2\pi r}} \cos\frac{\theta}{2} \tag{13.11d}$$

$$\sigma_z = v(\sigma_x + \sigma_x) = \frac{2vK_I}{\sqrt{2\pi r}} \cos\frac{\theta}{2} \quad \text{(plane strain)} \tag{13.11e}$$

$$\sigma_z \simeq 0 \quad \text{(plane stress)} \tag{13.11f}$$

Again, $i = I, II, III$ are loading modes, $K_i = K_I, K_{II}, K_{III}$ are the stress-intensity factors (stress-intensity parameters), and $f_i(\theta) = f_I(\theta), f_{II}(\theta), f_{III}(\theta)$ are the trigonometric functions related to K_i.

13.4.3   *Plastic Zone Size*

For a steady-state crack growth along the crack plane, the plane-stress plastic zone size (r) in Eq. (13.11b) with $\theta = 0$ and $\sigma_y = \sigma_{ys}$ is compared with the plane-strain and with the general $r = f(\sigma/\sigma_{ys})$ counterparts. Thus, the plastic zone size for mode I is defined by

$$r = \frac{1}{2\pi}\left(\frac{K_I}{\sigma_{ys}}\right)^2 \quad \text{(plane stress, } \sigma_z \simeq 0\text{)} \tag{13.12a}$$

$$r = \frac{1}{6\pi}\left(\frac{K_I}{\sigma_{ys}}\right)^2 \quad \text{(plane strain, } \sigma_z > 0\text{)} \tag{13.12b}$$

$$r = \frac{a}{2}\left(\frac{\sigma}{\sigma_{ys}}\right)^2 \quad \text{(general)} \tag{13.12c}$$

Normally, r has a very small magnitude in brittle materials, and theoretically, it can be used to define an effective crack length $a_e = a + r$ and an effective stress-intensity factor.

According to the theory of linear-elastic fracture mechanics, $r << a$ is the plasticity condition for brittle materials. Regarding to Eq. (13.12c), the applied stress is $\sigma < \sigma_{ys}$ so that $r < a$, which a main requirement for linear-elastic fracture mechanics.

13.4.4   *Fracture Mechanics Criteria*

Rearranging Eq. (13.10c) yields the stress-intensity factor K_I (pronounced "kay one") for mode I loading (Irwin [3, pp. 551–590]) and the corresponding strain-energy release rate G_I for dimensionally unconstrained specimens

$$\sigma\sqrt{\pi a} = \sqrt{(2\gamma_s + \gamma_p)E'} = \sqrt{G_I E'} \tag{13.13a}$$

$$K_I = \sigma\sqrt{\pi a} \quad \text{(applied)} \tag{13.13b}$$

$$K_I = \sigma\sqrt{\pi a_e} \quad \text{(applied)} \tag{13.13c}$$

$$G_I = \frac{K_I^2}{E'} \quad \text{(released)} \tag{13.13d}$$

$$G_{IC} = \frac{K_{IC}^2}{E'} \quad \text{(critical state)} \tag{13.13e}$$

In linear-elastic fracture mechanics (LEFM), Eqs. (13.13d), (13.13e) are important relations, since G_I represents the material energy dissipation during crack growth, while (R_c) is the resistance to crack extension. Actually, G_I is known as the crack driving force and K_I is the intensity of the stress field at the crack tip.

At crack propagation, $a = a_c$ is the critical crack length, $K_I = K_{IC}$ (pronounced "kay one c") is the critical stress-intensity factor known as the plane-strain fracture toughness for thick specimens, and $K_I = K_C$ is the plane-stress fracture toughness for thin specimens. Moreover, the plane-strain fracture toughness K_{IC} of a brittle material is analogous to the yield strength σ_{ys}, since both are material properties and both are used to define the design constraints for structural components, $K_{Id} = K_{IC}/S_F$ and $\sigma_d = \sigma_{ys}/S_F$, where the magnitude of the safety factor S_F is dictated by design codes or experience. Nonetheless, the critical stress-intensity factor, K_C or K_{IC}, defines fracture toughness as a material property, and it is independent of specimen dimensions. However, the stress-intensity factor, say K_I, depends on specimen dimensions and crack configuration, since it indicates the intensity of the stress that induces crack growth.

The most important aspects of K_{IC} and G_{IC} are related to the following fracture mechanics criteria for failure. Thus,

- Fracture occurs by crack propagation when $K_I = K_{IC}$, $K_{II} = K_{IIC}$ or $K_{III} = K_{IIIC}$, and $a = a_c$ at a stress σ.
- Failure occurs by the onset of crack growth when $G_I = C_{IC}$ and $a < a_c$ at a stress σ.

Accordingly, Eq. (13.13d) couples the applied stress-intensity factor and related strain energy release rate at the crack tip, while Eq. (13.13e) defines the fracture-toughness condition as $G_I = G_{IC}$ for $K_I = K_{IC}$ under plane-strain state or $G_I = G_C$ for $K_I = K_C$ under plane-stress state.

The critical relationship defined by Eq. (13.13e) is graphically shown in Fig. 13.4a for plane-strain and plane-stress conditions with respect to the effect of specimen thickness on the stress-intensity factor (Stephens et al. [4, p. 137]). For plane-strain condition in elastic-plastic materials, $K_I = K_{IC}$ for a thick specimen thickness.

Figure 13.4b, on the other hand, illustrates the crack length at two critical points, where G_{IC} is the critical strain-energy release rate for the onset of stable crack growth and K_{IC} is the plane-strain fracture toughness for the onset of crack propagation (fast crack growth leading to fracture). Moreover, Fig. 13.4a) depicts $K_I = f(B)$ from which the critical stress-intensity factor is defined as a thickness-independent material property called the plane-strain fracture toughness (K_{IC}) for mode I.

The American Society for Testing Materials (ASTM E399) imposes specimen size requirements for a valid and reliable fracture mechanics test under plane-strain conditions. The specimen size requirements are

$$a,\, B,\, (w - a) \geq 2.5 \left(K_{IC}/\sigma_{ys}\right)^2 \tag{13.14}$$

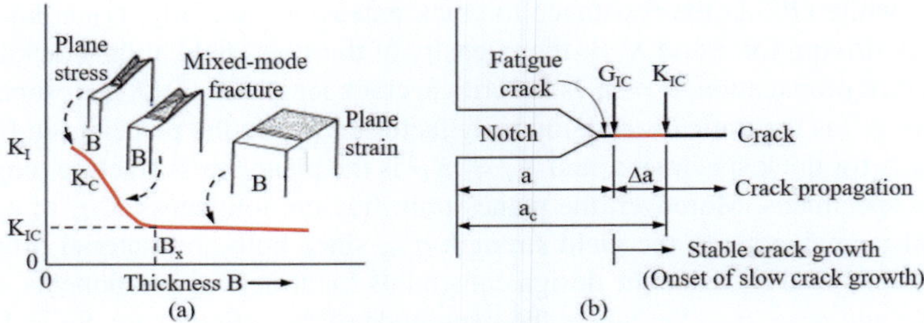

Fig. 13.4 (a) The fracture surface appearance and the effect of specimen thickness on the stress-intensity factor, $K_I = f(\sigma)$, which becomes the plane-strain fracture toughness at a fixed specimen thickness (Stephens et al. [4, p. 137]) . (b) The growth of a crack length and (locations identifying the fracture toughness in terms of energy (G_{IC}) and stress-intensity (K_{IC}) for mode I loading

where a is the initial crack length, $B_x \geq B$ is the specimen thickness, and $b = w - a$ is the ligament. This implies that K_{IC} is independent of specimen thickness at some value of $B_x \geq B$ (Fig. 13.4a). This can be set to $B_x \geq B_{ASTM}$.

Example 13.1 A large and wide brittle plate containing a single-edge crack fractures at a tensile stress of 4 MPa. The strain-energy release rate (G_C) and modulus of elasticity (E) are 4 J/m^2 and 65,000 MPa, respectively. Assume plane stress condition, and include the thickness $B = 3$ mm in all calculations. **(a)** Plot the theoretical total surface energy (U_s), the released strain energy (U_e), and the total potential energy change (W). Interpret the energy profiles. Determine **(b)** the critical crack length, **(c)** the maximum potential energy change ($W_{\max}$), **(d)** Will the crack grow unstably? **(e)** What is the critical stress-intensity factor for this brittle plate?

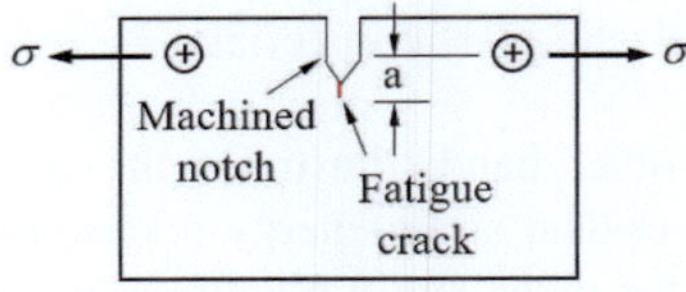

Solution
Given data

$$\sigma = 4\ MPa; \quad E' = E = 65{,}000\ MPa \quad \text{(plane stress)}$$

$$G_c = 4\ J/m^2 = 4 \times 10^{-6}\ J/mm^2$$

$$\gamma_s = G_c/2 = 2\ J/m^2 = 2 \times 10^{-6}\ J/mm^2$$

(a) For a single-edge crack with one crack tip, Eq. (13.6) is divided by 2. Thus, the surface energy becomes

$$W_s = 2aB\gamma_s = (2)\left(3 \times 10^{-3}\ m\right)\left(2\ J/m^2\right)a \tag{13.1E1a}$$

$$W_s = \left(12 \times 10^{-3}\ J/m\right)a = \left(12 \times 10^{-6}\ J/mm\right)a \tag{13.1E1b}$$

On the other hand, the released strain energy is also divided by 2 so that

$$W_e = -\left(\pi a^2 B\right)\left(\frac{\sigma^2}{2E}\right) \tag{13.1E2a}$$

$$W_e = -\frac{(\pi)\left(3 \times 10^{-3}\ m\right)(4\ MPa)^2\,a^2}{(2)\,(65000\ MPa)} \tag{13.1E2b}$$

$$W_e = -\left(1.16 \times 10^{-6}\ MPa.m\right)a^2 = -\,(1.16\ Pa.m)\,a^2 \tag{13.1E2c}$$

$$W_e = \left(1.16\ J/m^2\right)a^2 = -\left(1.16 \times 10^{-6}\ J/mm^2\right)a^2 \tag{13.1E2d}$$

Thus, the total potential energy becomes

$$W = W_s + W_e \tag{13.1E3a}$$

$$W = \left(12 \times 10^{-6}\ J/mm\right)a - \left(1.16 \times 10^{-6}\ J/mm^2\right)a^2 \tag{13.1E3b}$$

The energy profiles are given below. In essence, stable crack growth occurs when $W < W_{max}$, but $W > 0$. At the maximum potential energy, the crack length reaches a critical value $a = a_c$, and the slope becomes $dW/da = 0$. The total potential energy associated with crack growth is simply the sum of the surface energy and the released strain energy. The former energy is needed for creating new crack surfaces by allowing the breakage of atomic bonds. The latter energy is negative, because it is released during crack growth, and it is needed for unloading the regions near crack flanks.

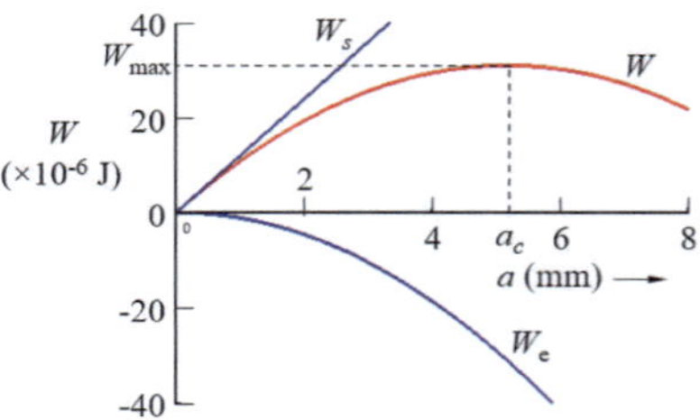

(b) The critical crack length is determined by

$$W = \left(12 \times 10^{-6}\ J/mm\right)a - \left(1.16 \times 10^{-6}\ J/mm^2\right)a^2 \tag{13.1E4a}$$

$$\left. \frac{dW}{da} \right|_{a=a_c} = \left(12 \times 10^{-6} \ J/mm \right) - \left(2.32 \times 10^{-6} \ J/mm^2 \right) a_c = 0$$

$$\text{(13.1E4b)}$$

$$a_c = \frac{12 \times 10^{-6}}{2.32 \times 10^{-6}} = 5.17 \ mm = 0.20 \ inches \qquad \text{(13.1E4c)}$$

Therefore, $a_c = 5.17 \ mm$ is the maximum allowable crack length in the system.

(c) The maximum potential energy is

$$W_{\max} = \left(12 \times 10^{-6} \ J/mm \right) a_c - \left(1.16 \times 10^{-6} \ J/mm^2 \right) a_c^2 \qquad \text{(13.1E5a)}$$

$$W_{\max} = \left(12 \times 10^{-6} \ J/mm \right) (5.17 \ mm) - \left(1.16 \times 10^{-6} \ J/mm^2 \right) (5.17 \ mm)^2$$

$$\text{(13.1E5b)}$$

$$W_{\max} = 31.03 \times 10^{-6} \ J \qquad \text{(13.1E5c)}$$

(d) From part (a),

$$W = \left(12 \times 10^{-6} \ J/mm \right) a - \left(1.16 \times 10^{-6} \ J/mm^2 \right) a^2 \qquad \text{(13.1E6a)}$$

$$\frac{dW}{da} = \left(12 \times 10^{-6} \ J/mm \right) - \left(2.32 \times 10^{-6} \ J/mm^2 \right) a \qquad \text{(13.1E6b)}$$

$$\frac{d^2W}{da^2} = -2.32 \times 10^{-6} \ J/mm^2 \qquad \text{(13.1E6c)}$$

Therefore, the crack will grow unstable because $d^2W/da^2 < 0$.

(e) The critical stress-intensity factor or the plane-strain fracture toughness for the brittle plate is determined by using Eq. (13.13b) at fracture, where $K_I = K_{IC}$. Thus,

$$K_{IC} = \sigma \sqrt{\pi a_c} \qquad \text{(13.1E7a)}$$

$$K_{IC} = (4 \ MPa) \sqrt{\pi \left(5.17 \times 10^{-3} \ m \right)} \qquad \text{(13.1E7b)}$$

$$K_{IC} = 0.51 \ MPa\sqrt{m} \qquad \text{(13.1E7c)}$$

From Eq. (13.13a) along with $K_I = K_{IC}$ and $\gamma_p \simeq 0$ for a brittle material,

$$K_{IC} = \sqrt{2\gamma_s E} = \sqrt{2 \left(2 \ J/m^2 \right) \left(65000 \times 10^6 \ Pa \right)} \qquad \text{(13.1E8a)}$$

$$K_{IC} = \sqrt{2 \left(2 \ Pa.m \right) \left(65000 \times 10^6 \ Pa \right)} \qquad \text{(13.1E8b)}$$

$$K_{IC} = 5.1 \times 10^5 \; Pa\sqrt{m} \tag{13.1E8c}$$

$$K_{IC} = 0.51 \; MPa\sqrt{m} \tag{13.1E8d}$$

Also, combining Eqs. (13.10b) and (13.13b) yields

$$K_{IC} = \sqrt{\frac{2E}{a_c B}} \, (2a_c B \gamma_s - W_{max}) = 0.51 \; MPa\sqrt{m} \tag{13.1E9}$$

Therefore, $K_{IC} = 0.51 \; MPa\sqrt{m}$ and $G_{IC} = 4 \; J/m^2$ (given) are very small values, which are typical for a brittle material like glass.

Example 13.2 Consider a hypothetical $(1\text{-}m) \times (100\text{-}mm) \times (3\text{-}mm)$ metal sheet (with welded upper-end) containing one $2a = 0.4 \; mm$ long central crack (slit) loaded in tension by hanging a mass of $800 \; Kg$ ($220.46 \; lb_m$). Analyze **(a)** the initial and **(b)** the critical conditions of the metal sheet. **(c)** Calculate the reduction in strength if the crack-free fracture stress is $165 \; MPa$. The plane-strain fracture toughness and the modulus of elasticity of the metal are $K_{IC} = 0.80 \; MPa\sqrt{m}$ and $E = 60 GPa$, respectively. Assume a Poisson's ratio of 0.3 and plane strain conditions.

Solution

(a) Initial conditions of the metal sheet: The applied mass has to be converted to mechanical load, which in turn is divided by the cross-sectional area of the sheet as if the crack is not present. The applied stress σ is

$$\sigma = \frac{P}{A_o} = \frac{mg}{A_o} = \frac{(800 \; Kg)\left(9.81 \; m/s^2\right)}{\left(100 \times 10^{-3} \; m\right)\left(3 \times 10^{-3} \; m\right)} \tag{13.2E1a}$$

$$\sigma = 26.16 \times 10^6 \; \left(Kg/m.s^2\right)/m^2 = 26.16 \times 10^6 \; Pa \tag{13.2E1b}$$

$$\sigma = 26.16 \; MPa \tag{13.2E1c}$$

Now, the modulus of elasticity E', stress-intensity factor K_I and strain-energy release rate G_I are

$$E' = E/\left(1 - v^2\right) = 60/\left(1 - 0.3^2\right) = 65.934 \; MPa \tag{13.2E2a}$$

$$K_I = \sigma\sqrt{\pi a} = (26.16 \; MPa)\sqrt{\pi\left(0.2 \times 10^{-3} \; m\right)} = 0.66 \; MPa\sqrt{m} \tag{13.2E2b}$$

$$G_I = \frac{K_I^2}{E'} = \frac{\left(0.66 \; MPa\sqrt{m}\right)^2}{65.934 \times 10^3 \; MPa} = 6.61 \times 10^{-6} \; MPa \tag{13.2E2c}$$

$$G_I = 6.61 \; Pa = 6.61 \; J/m^2 \tag{13.2E2d}$$

Therefore, the metal sheet will not fracture because $K_I < K_{IC}$ when $a = 0.2$ mm and $\sigma = 26.16 \, MPa$. Notice that the stress-intensity factor K_I is used as a fracture criterion, since K_{IC} is a material property. Actually, K_{IC} is an intrinsic thickness-independent property of a material.

(b) Critical conditions of the metal sheet: The metal sheet is loaded at $\sigma = 26.16 \, MPa$. The critical condition for crack propagation requires that $K_I = K_{IC}$, $G_I = G_{IC}$ and $a = a_c$. Thus,

$$K_{IC} = \sigma \sqrt{\pi a_c} \tag{13.2E3a}$$

$$a_c = \frac{1}{\pi} \left(\frac{K_{IC}}{\sigma} \right)^2 = \frac{1}{\pi} \left(\frac{0.80 \, MPa\sqrt{m}}{26.16 \, MPa} \right)^2 \tag{13.2E3b}$$

$$a_c = 0.30 \, mm > a = 0.20 \, mm \tag{13.2E3c}$$

This means that $2a_c = 0.60 \, mm$ is the maximum total crack length the metal sheet will allow prior to crack propagation. From Eq. (13.13d), the plane-strain condition also yields the critical driving force as

$$G_{IC} = \frac{K_{IC}^2}{E'} = \frac{\left(0.80 \, MPa\sqrt{m} \right)^2}{65.934 \times 10^3 \, MPa} \tag{13.2E4a}$$

$$G_{IC} = 9.71 \times 10^{-6} \, MPa \cdot m = 9.71 \, J/m^2 \tag{13.2E4b}$$

Therefore, the metal sheet will not fracture because $G_I < G_{IC}$. From Eq. (13.13b) with $a = 0.2 \, mm = 0.2 \times 10^{-3} \, m$ at fracture, the critical stress or fracture stress is

$$\sigma_c = \frac{K_{IC}}{\sqrt{\pi a}} = \frac{0.80 \, MPa\sqrt{m}}{\sqrt{\pi \left(0.2 \times 10^{-3} \, m \right)}} \tag{13.2E5a}$$

$$\sigma_c = 31.92 \, MPa \tag{13.2E5b}$$

Once again, the metal sheet will not fracture because $\sigma < \sigma_c$. Now, the maximum allowable or critical mass is

$$\sigma_c = \frac{m_c g}{A_o} \tag{13.2E6a}$$

$$m_c = \frac{A_o \sigma_c}{g} \tag{13.2E6b}$$

$$m_c = \frac{A_o \sigma_c}{g} = \frac{\left(100 \times 10^{-3} \, m \right) \left(3 \times 10^{-3} \, m \right) \left(31.92 \times 10^6 \, N/m^2 \right)}{\left(9.81 \, m/s^2 \right)} \tag{13.2E6c}$$

$$m_c = 976.15 \, Kg \tag{13.2E6d}$$

(c) The reduction in strength due to a small crack is

$$\frac{165 - 31.92}{165} x100\% \simeq 81\% \tag{13.2E7}$$

In conclusion, the metal sheet will not fracture because $K_I < K_{IC}$, $G_I < G_{IC}$ and $m < m_c$, but the presence of a small crack (defect) reduces the strength of the plate by 81%. This example establishes the importance of fracture mechanics in designing parts subjected to develop cracks during service. Moreover, for crack propagation to take place at high speed, the notation $a \rightarrow a_{\max} = a_c$ implies that the original crack length a approaches its maximum allowable crack-length value, $a \rightarrow a_{\max}$, which in turn represents a stable crack growth, while the cracked specimen is an unstable system. On the other hand, the notation a_c simply suggests that crack propagation (fast crack growth) leads to failure or fracture at an applied stress σ.

13.5 Geometry-Correction Factors

The geometry-correction factor (GCF = β) is an important parameter in fracture mechanics that accounts for the influence of specimen geometry on the stress intensity factor (SIF = K_I). The SIF is a measure of the stress concentration at the tip of a crack and is a key factor in determining the fatigue life and fracture toughness of a component.

There are several finite geometry-correction factors $\beta = f(a/w)$ listed in Table 13.1 for plates containing through-thickness cracks.

The geometry-correction factors (GCFs) are used to modify the SIF calculated for ideal geometries to account for the actual geometry of the component. They are dimensionless multipliers that are applied to the ideal SIF to give the true SIF for the specific geometry.

The stress-intensity factor for infinite rectangular specimen dimensions (uncorrected) and finite specimen dimensions (corrected) under mode I, respectively, are written as

$$K_I = \sigma\sqrt{\pi a} \qquad \text{(Uncorrected)} \tag{13.15a}$$

$$K_I = \beta\sigma\sqrt{\pi a} \qquad \text{(Corrected)} \tag{13.15b}$$

where β = Specimen finite geometry-correction factor

$\beta = f(a/w)$ for plates

$\beta = f(d_i/d_o)$ for a cylindrical component with a surface crack

a = Crack length (also called crack size)

w = Specimen width

d_i, d_o = Small and large diameters, respectively

Table 13.1 Through-thickness crack configurations for solid specimens (Tada et al. [5])

Center Cracked Specimen Let $x = a/w$ and $K_I = \alpha\sigma\sqrt{\pi a}$
$\sigma = P/(wB)$ $\alpha = \sqrt{sec(\pi x)}$ for $x \le 0.7$
$\alpha = (1 - 0.50x + 0.37x^2 - 0.04x^3)/\sqrt{1-x}$ for $x < 1$
Single-edge Cracked Specimen Let $x = a/w < 0.6$ and $K_I = \alpha\sigma\sqrt{\pi a}$
$\sigma = P/(wB)$
$\alpha = 1.12 - 0.23x + 10.55x^2 - 21.71x^3 + 30.38\,x^4$
Double-edge Cracked Specimen Let $x = a/w < 0.7$ and $K_I = \alpha\sigma\sqrt{\pi a}$
$\sigma = P/(wB)$
$\alpha = (1.12 - 0.561x - 0.205x^2 + 0.471x^3 - 0.19x^4)/\sqrt{1-x}$
Three-Point Bend Specimen (ASTM E399)
Let $x = a/w < 0.7$ and $K_I = \alpha\sigma\sqrt{\pi a}$
$y = 1.12 - 0.561x - 0.205x^2 + 0.471x^3 - 0.19x^4$
$\alpha = y/\sqrt{1-x}$ $\sigma = P/(wB)$
Compact Tension Specimen
Let $x = a/w < 0.7$ and $K_I = \alpha\sigma\sqrt{\pi a}$
$\sigma = P/(wB)$ $0.45 \le x \le 0.55$
$\alpha = \dfrac{2+x}{(1-x)^{1/2}}(0.886 + 4.64x - 13.32x^2 + 14.72x^3)$
Disk-Shaped Compact Specimen
Let $x = a/w < 0.7$ and $K_I = \alpha\sigma\sqrt{\pi a}$
$\sigma = P/(wB)$ $\alpha = \dfrac{(2+x)y}{(\pi x)^{1/2}(1-x)^{3/2}}$
$y = (0.76 + 4.80x - 11.58x^2 + 11.431x^3 - 4.08x^4)$

(continued)

Table 13.1 (continued)

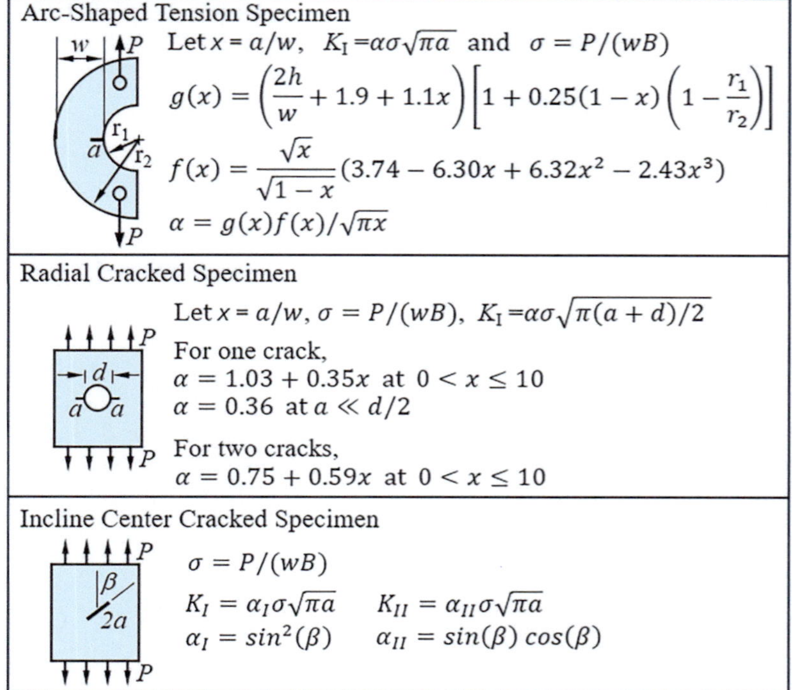

Arc-Shaped Tension Specimen

Let $x = a/w$, $K_I = \alpha\sigma\sqrt{\pi a}$ and $\sigma = P/(wB)$

$$g(x) = \left(\frac{2h}{w} + 1.9 + 1.1x\right)\left[1 + 0.25(1-x)\left(1 - \frac{r_1}{r_2}\right)\right]$$

$$f(x) = \frac{\sqrt{x}}{\sqrt{1-x}}(3.74 - 6.30x + 6.32x^2 - 2.43x^3)$$

$$\alpha = g(x)f(x)/\sqrt{\pi x}$$

Radial Cracked Specimen

Let $x = a/w$, $\sigma = P/(wB)$, $K_I = \alpha\sigma\sqrt{\pi(a+d)/2}$

For one crack,

$\alpha = 1.03 + 0.35x$ at $0 < x \leq 10$

$\alpha = 0.36$ at $a \ll d/2$

For two cracks,

$\alpha = 0.75 + 0.59x$ at $0 < x \leq 10$

Incline Center Cracked Specimen

$\sigma = P/(wB)$

$K_I = \alpha_I\sigma\sqrt{\pi a}$ $K_{II} = \alpha_{II}\sigma\sqrt{\pi a}$

$\alpha_I = \sin^2(\beta)$ $\alpha_{II} = \sin(\beta)\cos(\beta)$

In fracture mechanics, the K_I is typically calculated for ideal geometries, such as infinite plates or bars with simple crack configurations. However, real-world components often have finite dimensions and complex geometries, which can significantly affect the stress distribution around the crack tip and, consequently, the stress-intensity factor (SIF).

For a rectangular-shaped specimen, the geometry-correction factor $\beta = f(a/w)$ is also known as the boundary-correction factor that accounts for the effect of crack configuration and boundary conditions. It is used as a multiplier or correction factor in Eq. (13.15b) account for geometry effects on the stress field and stress-intensity factor. Notice that if $a/w \rightarrow 0$, then $\beta = f(0) \rightarrow 1$, and Eq. (13.15b) reduces to (13.15a). Actually, $\beta = f(a/w)$ is constraint which dictates large specimen dimensions, a deep crack, and very small plastic zone ($r << a$) for brittle materials.

The K_I expression, Eq. (13.15a), for an infinite (large) plate is a linear function of the applied stress (σ), and it increases with initial crack size (a). On the other hand, Eq. (13.15b) is for a finite plate, and it is corrected due to specimen geometry effects. The onset of crack propagation is a **critical condition**, implying sudden crack extension by tearing in a shear-rupture mode or by cleavage at high velocity. This means that the crack is unstable when a critical condition exists at a certain applied load since $a \rightarrow a_c$ and $K_I \rightarrow K_{IC}$.

Example 13.3 Consider a AISI 4340 steel plate with $K_{IC} = 41\ MPa\sqrt{m}$ and $\sigma_{ys} = 1900\ MPa$. Use some of the specimen configurations given in Table 13.1 to determine the least allowable applied tension stress (σ) for common specimen dimensions; $w = 10\ mm$, $B = w/2$, $S = 2.4w$, $a = 5\ mm$ and $x = a/w = 0.5$.

Solution For a center-cracked specimen,

$$\beta = \left(1 - 0.5x + 0.37x^2 - 0.044x^3\right)/\sqrt{1-x} = 1.1837 \tag{13.3E1a}$$

$$\sigma = K_{IC}/\left(\beta\sqrt{\pi a}\right) \tag{13.3E1b}$$

$$\sigma = \left(41\ MPa\sqrt{m}\right)/\left((1.1837)\sqrt{\pi\left(5*10^{-3}\ m\right)}\right) \tag{13.3E1c}$$

$$\sigma = 276.36\ MPa \tag{13.3E1d}$$

In summary, the geometric correction factor is significant because $\beta > 1$. It is a crucial tool in fracture mechanics for accurately assessing the behavior of a crack and designing components for safe and reliable operation. It provides a bridge between ideal geometry and real-world component.

For a single-edge tension specimen,

$$\beta = 1.12 - 0.23x + 10.55x^2 - 21.70x^3 + 30.38x^4 = 2.8288 \tag{13.3E2a}$$

$$\sigma = K_{IC}/\left(\beta\sqrt{\pi a}\right) \tag{13.3E2b}$$

$$\sigma = \left(41\ MPa\sqrt{m}\right)/\left[2.8288\sqrt{\pi\left(5*10^{-3}\ m\right)}\right] \tag{13.3E2c}$$

$$\sigma = 115.64\ MPa \tag{13.3E2d}$$

For a double-edge tension specimen,

$$\beta = 1.12 - 0.56x - 0.21x^2 + 0.47x^3 - 0.19x^4/\sqrt{1-x} = 0.82946$$
$$\tag{13.3E3a}$$

$$\sigma = K_{IC}/\left(\beta\sqrt{\pi a}\right) \tag{13.3E3b}$$

$$\sigma = \left(41\ MPa\sqrt{m}\right)/\left[0.82946\sqrt{\pi\left(5*10^{-3}\ m\right)}\right] \tag{13.3E3c}$$

$$\sigma = 394.39\ MPa \tag{13.3E3d}$$

For a single-edge bend specimen,

$$\beta = \left(\frac{3S}{w\sqrt{\pi}}\right)\frac{1.99 - x\left(1-x\right)\left(2.15 - 3.93x + 2.70x^2\right)}{2\left(1+2x\right)\left(1-x\right)^{3/2}} \tag{13.3E4a}$$

$$= 5.0985 \tag{13.3E4b}$$

$$\sigma = K_{IC}/\left(\beta\sqrt{\pi a}\right) \tag{13.3E4c}$$

$$\sigma = \left(41\ MPa\sqrt{m}\right)/\left[5.0985\sqrt{\pi\left(5*10^{-3}\ m\right)}\right] \tag{13.3E4d}$$

$$\sigma = 64.16\ MPa \tag{13.3E4e}$$

For a disk-shaped compact specimen,

$$\beta = \frac{(2+x)\left(0.76+4.8x-11.58x^2+11.43x^3-4.08x^4\right)}{(\pi x)^{1/2}(1-x)^{3/2}} \tag{13.3E5a}$$

$$\beta = 8.1173 \tag{13.3E5b}$$

$$\sigma = K_{IC}/\left(\beta\sqrt{\pi a}\right) \tag{13.3E5c}$$

$$\sigma = \left(41\ MPa\sqrt{m}\right)/\left[8.1173\sqrt{\pi\left(5*10^{-3}\ m\right)}\right] \tag{13.3E5d}$$

$$\sigma = 40.30\ MPa \tag{13.3E5e}$$

For a compact tension, C(T), specimen,

$$\beta = \frac{(2+x)}{(1-x)^{3/2}}\left(0.886+4.64x-13.32x^2+14.72x^3-5.60x^4\right) = 9.6591 \tag{13.3E6a}$$

$$\sigma = K_{IC}/\left(\beta\sqrt{\pi a}\right) \tag{13.3E6b}$$

$$\sigma = \left(41\ MPa\sqrt{m}\right)/\left[9.6591\sqrt{\pi\left(5*10^{-3}\ m\right)}\right] \tag{13.3E6c}$$

$$\sigma = 33.87\ MPa \tag{13.3E6d}$$

Therefore, the least allowable applied stress is the C(T), specimen with $\sigma = 33.87\ MPa$.

Example 13.4 A large plate $(2a \ll w)$ containing a through-thickness central crack 40-*mm* long is subjected to a tension stress. If the crack growth rate is 10 *mm/month* and fracture is expected at 10*months* from now, calculate the fracture stress. Data: $K_{IC} = 30\ MPa\sqrt{m}$.

Solution The crack growth rate can be defined by

$$\frac{da}{dt} = \frac{2a_c - 2a}{t} \tag{13.4E1}$$

Then,

$$2a_c = 2a + t \left(\frac{da}{dt} \right) \tag{13.4E2a}$$

$$2a_c = 40 \; mm + (10 \; month)(10 \; mm/month) \tag{13.4E2b}$$

$$2a_c = 140 \; mm \tag{13.4E2c}$$

$$a_c = 70 \; mm \tag{13.4E2d}$$

Therefore, $a_c = 3.5a$. Now, the fracture stress can be calculated using Eq. (13.15b). A large plate implies that $a/w \to 0$, and from Table 13.1, the geometric-correction factor approaches unity; $\beta = \sqrt{\sec(\pi a/w)} \to 1$. Thus, the fracture stress is

$$\sigma = \frac{K_{IC}}{\beta \sqrt{\pi a_c}} \tag{13.4E3a}$$

$$\sigma = \frac{30 M Pa \sqrt{m}}{(1) \sqrt{\pi \left(70 x 10^{-3} \; m \right)}} = 64 \; MPa \tag{13.4E3b}$$

Despite that the yield strength (σ_{ys}) is not given, one can conclude that $\sigma = 64$ $MPa << \sigma_{ys}$ due to the presence of a crack.

Example 13.5 Use the given load-displacement curve to determine the plane-strain fracture toughness K_{IC} for a quenched and tempered compact tension C(T) specimen made out of a hypothetical steel having $\sigma_{ys} = 900 \; MPa$, $v = 1/3$, and $E = 207 \; GPa$. Assume that during fatigue cracking, the machine notch produced a very sharp crack prior to testing the specimen. The total original crack length, including the notch depth and the fatigue crack, is 29 mm. The specimen dimensions are $B = 30 \; mm$, $w = 60 \; mm$, and $h/w = 0.6$. Consider the ASTM E399 requirements $P_{max}/P_Q < 1.1$ and $(a, b, B)_{ASTM} \geq 2.5 \left(K_Q/\sigma_{ys} \right)^2$. Find K_{IC}. and $r << a$ for a hypothetical material.

Solution Specimen dimensions: $0.45 \leq a/w \leq 0.55$, $b = w - a = 31 \; mm$, $B/w = 0.52$, and $h = 0.6w = 36 \; mm$. From the given figure,

$$\frac{P_{max}}{P_Q} = \frac{20}{19.2} = 1.04 \tag{13.5E1a}$$

$$\frac{P_{max}}{P_Q} < 1.1 \tag{13.5E1b}$$

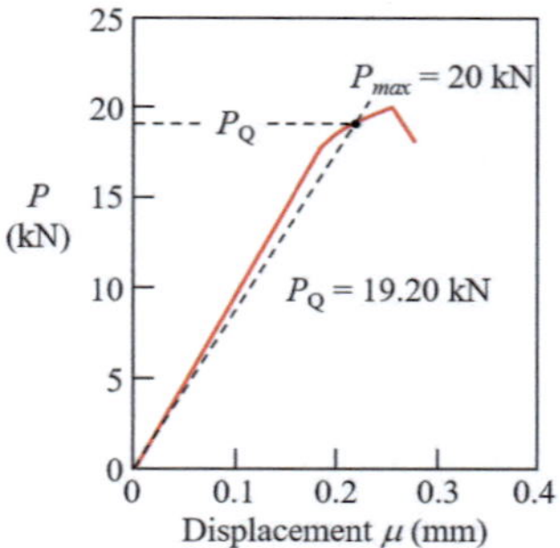

Proceed the computations since $P_{max}/P_Q < 1.1$. The applied stress is

$$\sigma = \frac{P_Q}{A_o} = \frac{19.2 x 10^{-3} \; MN}{\left(30 x 10^{-3} \; m\right)\left(60 x 10^{-3} \; m\right)} \tag{13.5E2a}$$

$$\sigma = 10.67 \; MPa \tag{13.5E2b}$$

From Table 13.1 and $x = a/w = 29/60 = 0.48333$, the finite specimen geometry-correction factor is

$$\beta = \frac{(2+x)}{(1-x)^{3/2}}\left(0.886 + 4.64x - 13.32x^2 + 14.72x^3 - 5.6x^4\right) = 9.1837 \tag{13.5E3}$$

Then, the possible fracture toughness becomes

$$K_Q = \beta\sigma\sqrt{\pi a} = (9.1837)(10.67 \; MPa)\sqrt{\pi \left(29 x 10^{-3} \; m\right)} \tag{13.5E4a}$$

$$K_Q = 29.58 \; MPa\sqrt{m} \tag{13.5E4b}$$

The size requirements, including $b = w - a$, are

$$(a, b, B)_{ASTM} \geq 2.5\left(K_Q/\sigma_{ys}\right)^2 = 2.5\left(\frac{29.58 \; MPa\sqrt{m}}{900 \; MPa}\right)^2 \tag{13.5E5a}$$

$$(a, b, B)_{ASTM} \geq 2.70 \; mm \; \text{(minimum requirement)} \tag{13.5E5b}$$

$$a = 29 \; mm, b = 31 \; mm \text{ and } B = 30 \; mm \; \text{(original)} \tag{13.5E5c}$$

Both $ASTM \; E399$ requirements, $P_{max}/P_Q < 1.1$ and $a, b, B \geq 2.70 \; mm$, are met, and therefore, the test is valid. Finally, the plane-strain fracture toughness is

$$K_{IC} = K_Q = 29.58 \; MPa\sqrt{m} \tag{13.5E6}$$

From Eq. (13.12c), the plastic zone size is

$$r = \frac{a}{2}\left(\frac{\sigma}{\sigma_{ys}}\right)^2 = \left(\frac{29\,mm}{2}\right)\left(\frac{10.67\,MPa}{900\,MPa}\right)^2 \tag{13.5E7a}$$

$$r = 0.002\,mm \tag{13.5E7b}$$

The size of the plastic zone under plane strain conditions is much smaller than that the crack length, ($r << a$) which is the constraint imposed by LEFM.

13.5.1   Elliptical Cracks

It is common to encounter embedded or surface elliptical or semielliptical cracks in certain structures. Assume that a semielliptical crack can grow into a semicircle as shown in Fig. 13.5. The goal in this type of problem is to determine the stress-intensity factor $K_I = f(\sigma)$ on the perimeter of the crack at angle λ. The crack length a is the depth of the semiellipse.

According to Irwin's analysis [6] on an infinite plate containing an embedded elliptical or semielliptical crack being remotely loaded in tension, the stress-intensity factor is defined by (Broek [7, section 3.6])

$$K_I = \frac{\sigma\sqrt{\pi a}}{\Phi}\left[\sin^2\lambda + \left(\frac{a}{c}\right)^2\cos^2\lambda\right]^{1/4} \tag{13.16}$$

where a,c are as defined in Fig. 13.5, and the geometry-correction factor Φ accounting for the crack shape effects is an elliptic integral of the second kind (Irwin [6])

$$\Phi = \int_o^{\pi/2}\left[1 - \left(\frac{c^2 - a^2}{c^2}\right)\sin^2\lambda\right]^{1/2}d\lambda \tag{13.17}$$

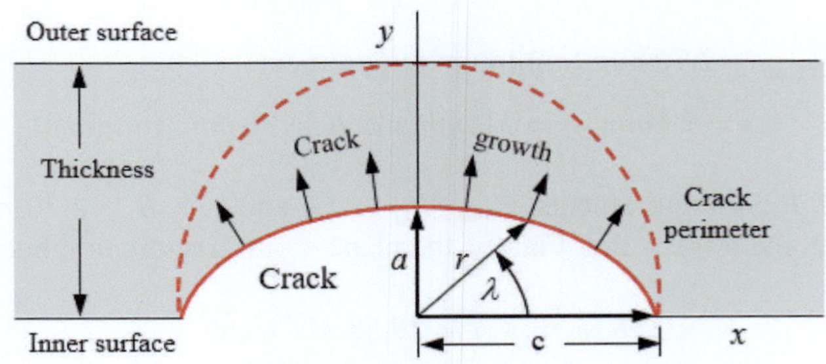

Fig. 13.5 Embedded semielliptical crack and a semicircular crack

For a circular or semicircular crack, $a = c$ and Eq. (13.17) becomes

$$\Phi = \int_{o}^{\pi/2} d\lambda = \frac{\pi}{2} \qquad (13.18)$$

Substituting $\beta = 1/\Phi = 2/\pi$ and $\lambda = 90°$ into Eq. (13.16) yields the stress-intensity factor for a circular crack of radius a, known as a penny-shaped crack (Sneddon [8])

$$K_I = \frac{2}{\pi}\sigma\sqrt{\pi a} \qquad (13.19)$$

Values of Φ can be found in mathematical tables, or it can be approximated by expanding Eq. (13.17) in a Taylor's series form. Hence, Φ becomes (Broek [7, p. 81])

$$\Phi = \frac{\pi}{2}\left[1 - \frac{1}{4}\left(\frac{c^2 - a^2}{c^2}\right) - \frac{3}{64}\left(\frac{c^2 - a^2}{c^2}\right)^2 - \cdots\cdots\right] \qquad (13.20)$$

Using the first two terms in the series given by Eq. (13.20) yields

$$\Phi = \frac{\pi}{2}\left[1 - \frac{1}{4}\left(\frac{c^2 - a^2}{c^2}\right)\right] = \frac{\pi}{2}\left[\frac{3}{4} + \left(\frac{a}{2c}\right)^2\right] \qquad (13.21)$$

Inserting Eq. (13.21) into (13.16) gives

$$K_I = \frac{2\sigma\sqrt{\pi a}}{\pi\left[3/4 + (a/2c)^2\right]}\left[\sin^2\lambda + \left(\frac{a}{c}\right)^2\cos^2\lambda\right]^{1/4} \qquad (13.22)$$

Evaluating Eq. (13.22) on the perimeter of the ellipse yields the stress-intensity factor for two extreme cases. Thus

$$K_I = \frac{2\sigma\sqrt{\pi a}}{\pi\left[3/4 + (a/2c)^2\right]} \qquad @\ \lambda = \pi/2\ \text{(Maximum)} \qquad (13.23a)$$

$$K_I = \frac{2\sigma\sqrt{\pi a}}{\pi\left[3/4 + (a/2c)^2\right]}\sqrt{\frac{a}{c}} \qquad @\ \lambda = 0\ \text{(Minimum)} \qquad (13.23b)$$

Here, K_I at $\lambda = \pi/2 = 90°$ can be predicted for crack instability when $.K_I \to K_{IC}$.

13.5.2   *Cylindrical Pressure Vessels*

In this section, the influence of internal surface cracks is analyzed in a thin cylindrical pressure vessel as schematically depicted in Fig. 13.6.

For a semielliptical or elliptical crack, the stress-intensity factor can be written in the following forms

$$K_I = F\,(a/B, a/c, a/w, \lambda, Q)\,\sigma\sqrt{\pi a} \tag{13.24a}$$

$$K_I = f\,(a/B, a/c, a/w)\,\sigma\sqrt{\frac{\pi a}{Q}}\left[\sin^2\lambda + \left(\frac{a}{c}\right)^2\cos^2\lambda\right]^{1/4} \tag{13.24b}$$

$$K_I = \sigma\sqrt{\frac{\pi a}{Q}} \quad (\text{maximum at } \lambda = \pi/2,\ f\,(a/B, a/c, a/w) = 1) \tag{13.24c}$$

Here, $F\,(a/B, a/c, a/w, \lambda, Q)$ is the boundary-correction factor that accounts for the influence of dimensionless variable known as the crack front face (a/B), back face (a/c), finite width (a/w), the angle λ, and the elastic shape factor Q. The reader is encouraged to consult Newman's review paper [9] on the chronological order of the development of the boundary-correction factor for surface cracks. Nonetheless, $f\,(a/B, a/c, a/w)$ can be taken as a generalized reduced boundary-correction factor, while the shape factor is mathematically dependent on the elliptic integral of the second kind and the stress ratio

$$Q = \Phi^2 - \frac{2}{3\pi}\left(\frac{\sigma}{\sigma_{ys}}\right)^2 \tag{13.25}$$

For a cylindrical pressure vessel (Fig. 13.6), the principle of superposition assumes that K_I arises from the hoop stress (σ_h) and the internal pressure (P_i). Thus,

$$K_I\,(\sigma) = K_I\,(\sigma_h) + K_I\,(P_i) \tag{13.26}$$

The total stress $\sigma = \sigma_h + P_i$ and $K_I\,(\sigma) = K_I$ will cause fracture if the critical crack depth is $a_c < B$. Conversely, if $a = B$, then the criterion that governs the failure mode is known as the leak-before-break.

Fig. 13.6 Semielliptical internal surface crack in a thin-wall pressure vessel

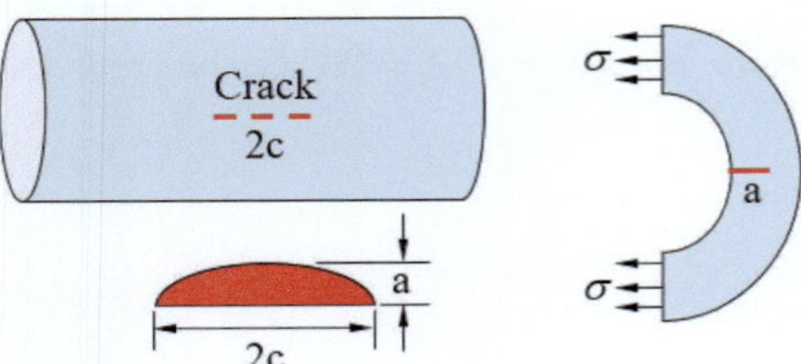

13.5.3   *Dimensional-Dependence Pressure Vessels*

The Hoop stress equations for thin-wall and thick-wall pressure vessels with d_i and d_o with internal and outer diameters, respectively, are

$$\sigma_h = \frac{P_i d_i}{2B} \quad \text{for thin-wall vessels, } d_o/d_i < 1.1 \ \& \ B \leq \frac{d_i}{20} \tag{13.27}$$

and

$$\sigma_h = \left[\frac{(d_o/d_i)^2 + 1}{(d_o/d_i)^2 - 1} \right] P_i \quad \text{for thick-wall vessels, } d_o/d_i > 1.1 \tag{13.28}$$

Combining Eqs. (13.21) and (13.25) yields a convenient mathematical expression for predicting the shape factor

$$Q = \left(\frac{\pi}{2}\right)^2 \left[\frac{3}{4} + \left(\frac{a}{2c}\right)^2\right]^2 - \frac{7}{33}\left(\frac{\sigma}{\sigma_{ys}}\right)^2 \tag{13.29}$$

The shape factor has also been reported to have the form (Murakami [10])

$$Q = 1 + 1.464 \left(\frac{a}{c}\right)^{1.65} \quad \text{for } a/c \leq 1 \tag{13.30}$$

Manipulating Eq. (13.29) and rearranging it yield

$$\frac{a}{2c} = \left[\frac{2}{\pi}\sqrt{Q + \frac{7}{33}\left(\frac{\sigma}{\sigma_{ys}}\right)^2} - \frac{3}{4} \right]^{1/2} \tag{13.31}$$

Notice that Eq. (13.31) can be used to plot a series of curves as shown in Fig. 13.7 for several σ/σ_{ys} stress ratio values.

If a back free-surface correction factor of 1.12 and plastic deformation are considered, then Eq. (13.24) with the plastic zone size (r) is further corrected as

$$K_I = 1.12\sigma\sqrt{\pi(a+r)/Q} \quad \text{For } \lambda = \pi/2 \tag{13.32}$$

Correction factors are used to improve the accuracy of a particular model. Consequently, a reliable pressure vessel design must include all possible flaw-generation sites to ensure the structural integrity of the pressure vessels.

Example 13.6 A steel thin-wall pressure vessel (Fig. 13.6) is subjected to a stress of $\sigma = \sigma_h + P = 420\,MPa$ perpendicular to the crack depth. The vessel has an internal semielliptical surface crack of dimensions $a = 3\,mm$, $2c = 10\,mm$. **(a)** Use Eq. (13.25) to calculate K_I. **(b)** Will the pressure vessel leak? Explain.

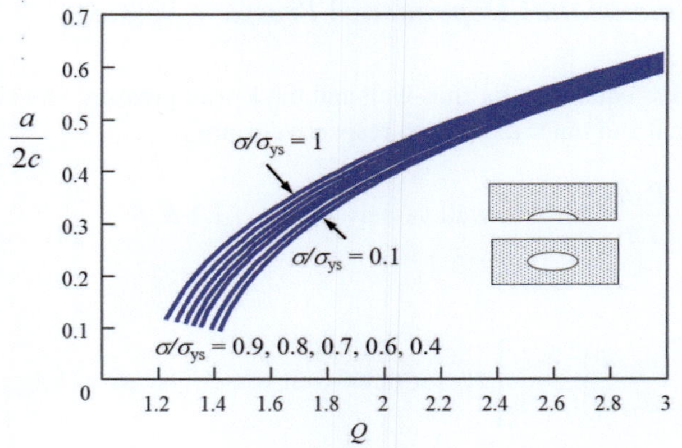

Fig. 13.7 Shape factor Q for internal and surface flaws

(c) Determined the pressure and the hoop stress. Use the following data: $\sigma_{ys} = 700\,MPa$ and $K_{IC} = 60\,MPa\sqrt{m}$, $B = 6\,mm$, $d_i = 500\,mm$.

Solution (a) Let

$$\frac{a}{2c} = 0.30, \quad \frac{a}{B} = 0.50, \quad \frac{\sigma}{\sigma_{ys}} = 0.60 \tag{13.6E1a}$$

$$Q = 1.7 \quad \text{[from Eq. (13.29) and Fig. 13.7]} \tag{13.6E1b}$$

so that

$$\beta = 1.12/\sqrt{Q} = 1.12/\sqrt{1.7} = 0.859 \tag{13.6E2a}$$

$$K_I = \beta\sigma\sqrt{\pi a} = (0.859)\,(420\,MPa)\sqrt{\pi\left(3 \times 10^{-3}\,m\right)} \tag{13.6E2b}$$

$$K_I = 35.03\,MPa\sqrt{m} \tag{13.6E2c}$$

(b) The pressure vessel will not leak because $K_I < K_{IC}$. The safety factor is

$$S_F = \frac{K_{IC}}{K_I} = \frac{60}{35.03} = 1.71 \tag{13.6E3}$$

(c) Using the total stress expression yields

$$\sigma = \sigma_h + P_i = \frac{P_i d_i}{2B} + P_i \tag{13.6E4a}$$

$$\sigma = \left(\frac{d_i}{2B} + 1\right) P_i \tag{13.6E4b}$$

Then,

$$P_i = \sigma \left(\frac{d_i}{2B} + 1\right)^{-1} = (420\ MPa)\left(\frac{500\ mm}{2 \times 6\ mm} + 1\right)^{-1} \tag{13.6E5a}$$

$$P_i = 9.84\ MPa \tag{13.6E5b}$$

$$\sigma_h = 420\ MPa - 9.84\ MPa = 410.16\ MPa \tag{13.6E5c}$$

Notice that the applied stress $\sigma = 420\ MPa$ and the hoop stress $\sigma_h = 410.16\ MPa$ are similar. This implies that $K_I\,(\sigma) \simeq K_I\,(\sigma_h)$.

Example 13.7 This example covers the analytical procedure for deriving the stress state in a pressure vessel holding a fluid under pressure. Derive the polar stress components associated with internal and external pressures for a cylindrical pressure vessel schematically shown in the given figure below.

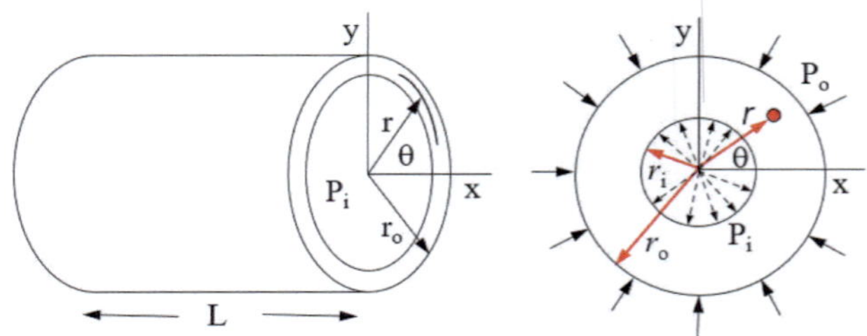

The theory of elasticity provides the mathematical definitions of the stress components in polar coordinates. Thus,

$$\sigma_r = \frac{1}{r}\frac{\partial \phi}{\partial r} + \frac{1}{r^2}\frac{\partial^2 \phi}{\partial \theta^2} \tag{13.8E1a}$$

$$\sigma_\theta = \frac{\partial^2 \phi}{\partial r^2} \tag{13.8E1b}$$

$$\tau_{r\theta} = \frac{1}{r^2}\frac{\partial \phi}{\partial \theta} - \frac{1}{r}\frac{\partial^2 \phi}{\partial r\,\partial \theta} \tag{13.8E1c}$$

Let the Airy stress function be (Dally and Riley [11, p. 74])

$$\phi = a_o + b_o \ln r + c_o r^2 \tag{13.7E2a}$$

$$\frac{\partial \phi}{\partial r} = \frac{b_o}{r} + 2c_o r \tag{13.7E2b}$$

$$\frac{\partial^2 \phi}{\partial r^2} = -\frac{b_o}{r^2} + 2c_o \tag{13.7E2c}$$

The boundary conditions for this problem are

$$\sigma_r = -P_i \text{ and } \tau_{r\theta} = 0 @ r = r_i \tag{13.7E3a}$$

$$\sigma_r = -P_o \text{ and } \tau_{r\theta} = 0 @ r = r_o \tag{13.7E3b}$$

Derive Eq. (13.27).

Solution Combining Eqs. (1.58) and (13.7E3) yields the radial stress (σ_r) and the tangential or hoop stress ($\sigma_\theta = \sigma_h$)

$$\sigma_r = \frac{1}{r}\frac{\partial \phi}{\partial r} + \frac{1}{r^2}\frac{\partial^2 \phi}{\partial \theta^2} = \frac{b_o}{r^2} + 2c_o \tag{13.7E4a}$$

$$\sigma_\theta = \frac{\partial^2 \phi}{\partial r^2} = -\frac{b_o}{r^2} + 2c_o \tag{13.7E4b}$$

$$\tau_{r\theta} = \frac{1}{r^2}\frac{\partial \phi}{\partial \theta} - \frac{1}{r}\frac{\partial^2 \phi}{\partial r \partial \theta} = 0 \tag{13.7E4c}$$

The boundary conditions on the inside and outside the wall surfaces of the circular model are

$$\sigma_r = -P_i \qquad \tau_{r\theta} = 0 \qquad @ r = r_i \tag{13.7E5}$$

$$\sigma_r = -P_o \qquad \tau_{r\theta} = 0 \qquad @ r = r_o$$

Combining Eqs. (13.7E4) and (13.7E5) yields the constants a_o, b_o and c_o, and the resultant expressions substituted back into (13.7E4) gives the radial stress (σ_r) and the tangential or hoop stress as

$$\sigma_t = \sigma_r = \frac{r_i^2 r_o^2 (P_o - P_i)}{(r_o^2 - r_i^2) r^2} + \frac{r_i^2 P_i - r_o^2 P_o}{r_o^2 - r_i^2} \tag{13.7E6a}$$

$$\sigma_h = \sigma_\theta = -\frac{r_i^2 r_o^2 (P_o - P_i)}{(r_o^2 - r_i^2) r^2} + \frac{r_i^2 P_i - r_o^2 P_o}{r_o^2 - r_i^2} \tag{13.7E6b}$$

$$\tau_{r\theta} = 0 \tag{13.7E6c}$$

If $P_o = 0$ or $P_i >> P_o$ and $r = r_i$, then Eq. (13.7E6) gives the stress distribution over the wall thickness for pipes and cylinders

$$\sigma_t = \left[\frac{1 - (r_o/r_i)^2}{(r_o/r_i)^2 - 1}\right] P_i < 0 \text{ (compression)} \tag{13.7E7a}$$

$$\sigma_h = \left[\frac{(r_o/r_i)^2 + 1}{(r_o/r_i)^2 - 1}\right] P_i > 0 \text{ (tension)} \tag{13.7E7b}$$

If the diameter of a circle is $d = 2r$, then Eq. (13.7E7) becomes

$$\sigma_h = \left[\frac{(d_o/d_i)^2 + 1}{(d_o/d_i)^2 - 1}\right] P_i \qquad \text{(for thick walls)} \qquad (13.7\text{E}8)$$

which is the hoop stress for the thick-walled cylinder subjected to internal pressure. Rearrange Eq. (13.7E7) along with $B = r_o - r_i$ so that

$$\sigma_h = \left[\frac{r_o^2 + r_i^2}{r_o^2 - r_i^2}\right] P_i \qquad\qquad (13.7\text{E}9\text{a})$$

$$\sigma_h = \left[\frac{r_o^2 + r_i^2}{(r_o + r_i)\,B}\right] P_i \qquad\qquad (13.7\text{E}9\text{b})$$

Mathematically, letting $r_o = r_i$ in Eq. (13.8E9) leads to (13.27) along with $r_i = d_i/2$. Thus, the hoop stress expression becomes

$$\sigma_h = \frac{r_i\,P_i}{B} = \frac{d_i\,P_i}{2B} \qquad \text{(thin walls)} \qquad (13.7\text{E}10)$$

Example 13.8 A cylindrical vessel made of AISI 4147 steel having an internal semielliptical surface crack 5-mm deep and 10-mm long is pressurized at 80 MPa. The vessel diameter and thickness are 50 cm and 25 cm, respectively. In order to account for the effects of both the applied pressure and the hoop stress, calculate **(a)** the stress-intensity factor. Will the pressure vessel fracture? If not, determine **(b)** the maximum pressure that would cause fracture. Data: $K_{IC} = 120\ MPa\sqrt{m}$, $\sigma_{ys} = 945\ MPa$ and $\sigma_{ts} = 1062\ MPa$.

Solution
Given data $a = 5\ mm$, $2c = 10\ mm$, $d_i = 0.5\ m$, $B = 25\ cm$, $P_i = 80\ MPa$ and $a/2c = 0.50$. Which cylindrical pressure vessel theory should be used? Let

$$B = \frac{(d_o - d_i)}{2} \qquad\qquad (13.8\text{E}1\text{a})$$

$$d_o = d_i + 2B = 50\ cm + 2 \times 25cm = 100\ cm \qquad (13.8\text{E}1\text{b})$$

$$\frac{d_o}{d_i} = \frac{100}{50} = 2 \qquad\qquad (13.8\text{E}1\text{c})$$

Therefore, the thick-wall theory must be used because $d_o/d_i > 1.1$.

(a) Using Eq. (13.28) yields the hoop stress and the total applied stress

$$\sigma_h = \left[\frac{(d_o/d_i)^2 + 1}{(d_o/d_i)^2 - 1}\right] P_i = \left[\frac{4 + 1}{4 - 1}\right] P_i \qquad (13.8\text{E}2\text{a})$$

$$\sigma_h = \frac{5}{3} P_i = \frac{5}{3} (80 \, MPa) = 133.33 \, MPa \qquad (13.8E2b)$$

$$\sigma = \sigma_h + P_i = \frac{5}{3} P_i + P_i = \frac{8}{3} P_i = \frac{8}{3} (80 \, MPa) \qquad (13.8E2c)$$

$$\sigma = 213.33 \, MPa \qquad (13.8E2d)$$

Thus, the stress ratio becomes

$$\frac{\sigma}{\sigma_{ys}} = \frac{213.33 \, MPa}{945 \, MPa} = 0.23 \qquad (13.8E3)$$

Hence, the magnitude of the total applied stress is 23% of the yield strength. Using Eq. (13.29) yields the shape factor

$$Q = \left(\frac{\pi}{2}\right)^2 \left[\frac{3}{4} + \left(\frac{a}{2c}\right)^2\right]^2 - \frac{7}{33} \left(\frac{\sigma}{\sigma_{ys}}\right)^2 = \left(\frac{\pi}{2}\right)^2 \left[\frac{3}{4} + (0.5)^2\right]^2$$

$$- \frac{7}{33} (0.23)^2 = 2.46 \qquad (13.8E4)$$

From Eq. (13.25), the total stress-intensity factor is

$$K_I = \sigma \sqrt{\frac{\pi a}{Q}} = \frac{8}{3} P_i \sqrt{\frac{\pi a}{Q}} = \frac{8}{3} (80 \, MPa) \sqrt{\frac{\pi \left(5 \times 10^{-3} \, m\right)}{2.46}} = 17.05 \, MPa\sqrt{m}$$

$$(13.8E5)$$

Therefore, fracture will not occur because $K_I < K_{IC} = 120 \, MPa\sqrt{m}$.

(b) From Eq. (13.8E5), the maximum pressure is

$$P_{max} = \frac{3K_{IC}}{8} \sqrt{\frac{Q}{\pi a}} = \frac{3}{8} \left(120 \, MPa\sqrt{m}\right) \sqrt{\frac{2.46}{\pi \left(5 \times 10^{-3} \, m\right)}} = 563.14 \, MPa$$

$$(13.8E6)$$

This implies that the critical stress becomes

$$\sigma_c = \sigma_h = \frac{5}{3} P_{max} \approx 939 MPa \approx \sigma_{ys} \qquad (13.8E7)$$

This would cause catastrophic failure if a pressure surge occurs. However, if the applied pressure is constant and crack growth is time-dependent, then leakage would occur due to lack of sufficient thickness for the crack to run through. This can be determined by calculating the critical crack length as

$$a_c = \frac{Q}{\pi} \left(\frac{3K_{IC}}{8P_i}\right)^2 = \frac{2.46}{\pi} \left(\frac{3 \times 120}{8 \times 80}\right)^2 = 24.78 \, cm \approx B = 25 \, cm$$

$$(13.8E8)$$

The remaining ligament, b, of the cylindrical vessel wall is

$$b = B - a_c = 25 \; cm - 24.78 \; cm = 0.22 \; cm = 2.2 \; mm \qquad (13.8E9)$$

Consequently, the vessel wall would act as a metallic sheet since $b << B = 250 \; mm$ and the pressure vessel would fail. Now, assume that the pressure vessel plate containing the crack is treated as a specimen. In this case, the specimen and crack dimensions required for a valid fracture mechanics test can be calculated using Eq. (13.14). Thus,

$$(a, b, B)_{ASTM} \geq 2.5 \left(K_{IC}/\sigma_{ys}\right)^2 = 2.5 \left(\frac{120 \; MPa\sqrt{m}}{945 \; MPa}\right)^2 = 40.31mm$$

$$(13.8E10)$$

Notice that the initial values of the ligament b and thickness B are in accord with Eq. (13.8E10). However, the original crack length does not meet the ASTM requirement since $a = 5 \; mm < a_{ASTM} = 40.31 \; mm$. Moreover, the theoretical plastic zone size, Eq. (13.12c), is

$$r = \frac{a}{2} \left(\frac{\sigma}{\sigma_{ys}}\right)^2 = \left(\frac{5 \; mm}{2}\right) \left(\frac{213.33 \; MPa}{945 \; MPa}\right)^2 = 0.13 \; mm \qquad (13.8E10)$$

Hence, $r = 0.13 \; mm << a = 5 \; mm$ meets the theoretical requirements imposed by linear-elastic fracture mechanics.

13.6   Mixed-Mode Fracture Mechanics

13.6.1   Crack Configurations

Regarding mixed-mode fracture mechanics, Fig. 13.8 schematically illustrates the some crack configurations. For instance, Fig. 13.8a shows a plate with a surface or through-the-thickness crack, Fig. 13.8b illustrates a hollow or solid cylindrical specimen with a circumferential, and (Fig. 13.8c) depicts a spiral surface crack.

The main objective in this section is to develop a fracture criterion based on the strain energy release rate for mixed-mode I–II or I–III (Perez [12, Chapter 8]).

13.6.2   Mixed Mode I and II

For a mixed mode I–II shown in Fig. 13.8a, the total stress in polar coordinates is the sum of the stress components for each loading mode. Thus,

Fig. 13.8 Mixed-mode fracture mechanics specimens. (**a**) Surface or through-the-thickness incline crack in a plate under tension, (**b**) circumferential surface crack in a cylinder under tension or torsion mode, and (**c**) spiral surface crack on a cylinder under mixed loading

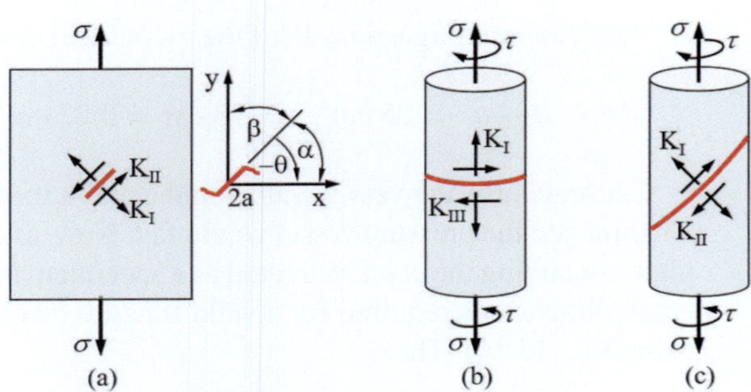

$$\sigma_r = \frac{K_I}{4\sqrt{2\pi r}}\left(5\cos\frac{\theta}{2} - 5\cos\frac{3\theta}{2}\right) + \frac{K_{II}}{4\sqrt{2\pi r}}\left(-5\sin\frac{\theta}{2} + 3\sin\frac{3\theta}{2}\right)$$

$$(13.33a)$$

$$\sigma_\theta = \frac{K_I}{4\sqrt{2\pi r}}\left(3\cos\frac{\theta}{2} + \cos\frac{3\theta}{2}\right) + \frac{K_{II}}{4\sqrt{2\pi r}}\left(-3\sin\frac{\theta}{2} - 3\sin\frac{3\theta}{2}\right)$$

$$(13.33b)$$

$$\tau_{r\theta} = \frac{K_I}{4\sqrt{2\pi r}}\left(\sin\frac{\theta}{2} + \sin\frac{3\theta}{2}\right) + \frac{K_{II}}{4\sqrt{2\pi r}}\left(\cos\frac{\theta}{2} + 3\cos\frac{3\theta}{2}\right) \qquad (13.33c)$$

For mode III,

$$\tau_{zr} = \frac{K_{III}}{2}\sqrt{\frac{2}{\pi r}}\sin\frac{\theta}{2} \qquad (13.34a)$$

$$\tau_{\theta z} = \frac{K_{III}}{2}\sqrt{\frac{2}{\pi r}}\cos\frac{\theta}{2} \qquad (13.34b)$$

Assume that the three loading modes interact on a elastic component, so that the elastic strain energy release rate and the corresponding stress-intensity factors can be combined as shown below

$$G_i = G_I + G_{II} + G_{III} \qquad (13.35a)$$

$$G_i = \frac{K_I^2}{E'} + \frac{K_{II}^2}{E'} + \frac{(1+v)\,K_{III}^2}{E} \qquad (13.35b)$$

For pure mode I tension loading at fracture, Eq. (13.35b) becomes a G-criterion for plane-strain fracture toughness

$$G_{IC} = \frac{K_{IC}^2}{E'} \qquad (13.36)$$

Substituting Eq. (13.36) into (13.35b) gives the fracture criterion equation (K-criterion)

$$K_{IC}^2 = K_I^2 + K_{II}^2 + \frac{E'\,(1+v)\,K_{III}^2}{E} \quad (G\text{-criterion}) \tag{13.37}$$

This equation indicates that any combination of the stress-intensity factors may give the value for fracture toughness K_{IC}.

Consider the mixed mode I-II configuration shown in Fig. 13.8a. The stress components are defined along with $\alpha + \beta = \pi/2$ and Mohr's circle by (Sih et al. [13])

$$\sigma_x = \sigma \sin^2 \beta = \sigma \cos^2 \alpha \tag{13.38a}$$

$$\sigma_y = \sigma \cos^2 \beta = \sigma \sin^2 \alpha \tag{13.38b}$$

$$\tau_{xy} = \sigma \sin \beta \cos \beta = \sigma \sin \alpha \cos \alpha \tag{13.38c}$$

These are the main stress equations in the preceding mixed-mode fracture mechanics analytical procedure based on an externally and remotely applied tensile stress σ (Fig. 13.8a). It should be mentioned that practical or engineering structures may be subjected to tension, shear, and torsion loading that induce mixed-mode interactions.

The corresponding stress-intensity factors become

$$K_I = \sigma_y \sqrt{\pi a} = \sigma \sqrt{\pi a} \sin^2 \beta \tag{13.39a}$$

$$K_{II} = \tau_{xy} \sqrt{\pi a} = \sigma \sqrt{\pi a} \sin \beta \cos \beta \tag{13.39b}$$

For a mixed-mode I-II interaction, Eq. (13.37) reduces to

$$K_{IC}^2 = K_I^2 + K_{II}^2 \quad \text{(Circle)} \tag{13.40}$$

This fracture criterion is named so, because Eq. (13.40) is the equation of a circle for a constant radius K_{IC}. Now, inserting Eqs. (13.39a), (13.39b) into (13.40) yields the plane-strain fracture toughness as a function of the fracture or critical stress and inclined angle

$$K_{IC} = \sigma_f \sqrt{\pi a} \sin^2 \beta \tag{13.41}$$

Notice that the fracture stress σ_f, Eq. 13.41, depends on the incline angle β at a fixed crack length a and various levels of the plane-strain fracture toughness K_{IC}. For example, letting $a = 4\,mm$, solving Eq. (13.41) for $\sigma_f = f(\beta)$ yield Fig. 13.9. According to the scale used to construct Fig. 13.9, it is clear that σ_f strongly depends on the crack inclined angle $\beta > 40°$ for specimens under plane-strain conditions.

For a constant crack length $a = 4\,mm$, the trendlines in Fig. 13.9 are displaced upward as K_{IC} increases.

Fig. 13.9 Variation of
fracture stress for a fixed
crack length and varying
fracture toughness

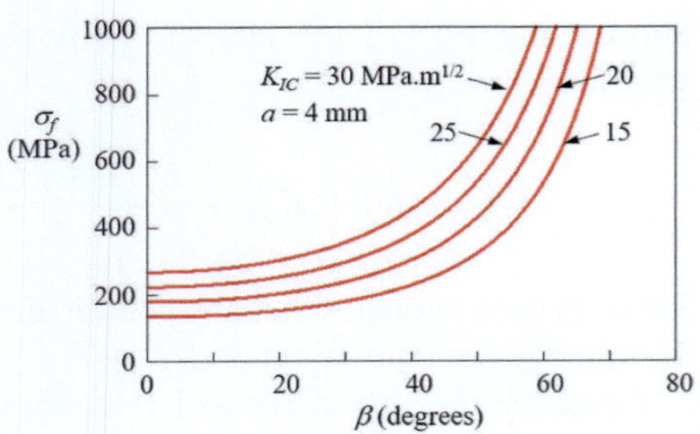

Example 13.9 A hypothetical large steel plate has one 8-*mm* long and inclined
central crack at $20°$ with respect to the direction of the applied axial load
(Fig. 13.8a). If the applied stress is 200 MPa, determine whether or not the plate
will fracture. Assume a Poisson's ratio $v = 1/3$, plane-strain fracture toughness
$K_{IC} = 30\ MPa\sqrt{m}$, yield strength $\sigma_{ys} = 1030\ MPa$, and modulus of elasticity
$E = 200\ GPa$ for the steel.

Solution From Eqs. (13.39a), (13.39b), the mode I stress-intensity factor at $\beta = 20° = 20\pi/180 = \pi/9$ is

$$K_I = \sigma\sqrt{\pi a}\sin^2\beta = (200\ MPa)\left(\sqrt{4\pi \times 10^{-3}\ m}\right)\sin^2(\pi/9) \tag{13.9E1a}$$

$$K_I = 2.62\ MPa\sqrt{m} \tag{13.9E1b}$$

From Eq. (13.39b),

$$K_{II} = \sigma\sqrt{\pi a}\sin\beta\cos\beta \tag{13.9E2a}$$

$$K_{II} = (200\ MPa)\left(\sqrt{4\pi \times 10^{-3}\ m}\right)\sin(\pi/9)\cos(\pi/9) \tag{13.9E2b}$$

$$K_{II} = 7.21\ MPa\sqrt{m} \tag{13.9E2c}$$

Thus,

$$\sqrt{K_I^2 + K_{II}^2} = 7.67\ MPa\sqrt{m} < K_{IC} \tag{13.9E3}$$

From Eq. (13.41), the fracture stress is

$$\sigma_f = \frac{K_{IC}}{\sqrt{\pi a}\sin^2\beta} = \frac{30MPa\sqrt{m}}{\left(\sqrt{4\pi \times 10^{-3}\ m}\right)\sin^2(\pi/9)} = 2287.80\ MPa$$

$$\tag{13.9E4}$$

Therefore, the plate will not fracture because $K_{IC} > 7.67\ MPa\sqrt{m}$ and $\sigma < \sigma_f$
for a fixed crack length. Despite that there is no information on mode II fracture

toughness for determining the shear fracture stress τ_f, it can be assumed that $K_{IIC} < K_{IC}$. Moreover, Eq. (13.36) gives the critical strain-energy release rate as $G_{IC} = 0.004 \; MJ/m$.

13.6.3 Fracture Angle

Assume that the incline crack shown in Fig. 13.8a changes direction under the influence of an applied tension stress and that crack propagation occurs along the x-axis (tangent plane). Thus, the fracture angle with a clockwise rotation is

$$\theta_o = \frac{\pi}{2} - \beta \tag{13.42}$$

Using the trigonometric function $\sin \beta = \sin (\pi/2 - \theta_o) = \cos \theta_o$ in Eq. (13.41) and arranging the resultant expression yield the plane-strain fracture toughness as a function of the fracture angle θ_o. Hence,

$$K_{IC} = \sigma \sqrt{\pi a} \cos \theta_o \tag{13.43}$$

This expression indicates that fracture occurs when $K_I = K_{IC}$ at a fracture angle $\theta = \theta_o$ for pure mode I loading. Conversely, $K_{II} = K_{IIC}$ for pure mode II. The fact remains that K_{IIC} vanishes when crack growth changes direction along the x-axis.

If the circular fracture criterion does not explain some data, then the elliptical fracture condition with $K_{IC} = \sqrt{2/3}K_{IIC}$ is a practical approach for the mixed I–II interaction (Hellan [14, p. 157]). Thus,

$$\left(\frac{K_I}{K_{IC}}\right)^2 + \left(\frac{K_{II}}{K_{IIC}}\right)^2 = 1 \quad \text{(Ellipse)} \tag{13.44a}$$

$$K_I^2 + \frac{2}{3}K_{II}^2 = K_{IC}^2 \quad \text{(Ellipse)} \tag{13.44b}$$

For instance, Fig. 13.10 shows the trendlines of Eqs. (13.40) and (13.44b) for a quarter of an ellipse. The elliptical fracture criterion is based on the assumption that the crack propagates on a self-similar manner (extension along the plane of the original crack).

Substituting Eqs. (13.39a), (13.39b) into (13.44b) gives

$$K_{IC} = \sigma \sqrt{\pi a} \sin \beta \sqrt{\sin^2 \beta + \frac{4}{9}\cos^2 \beta} \quad \text{(Ellipse)} \tag{13.45}$$

where σ is the fracture stress when $K_I = K_{IC}$.

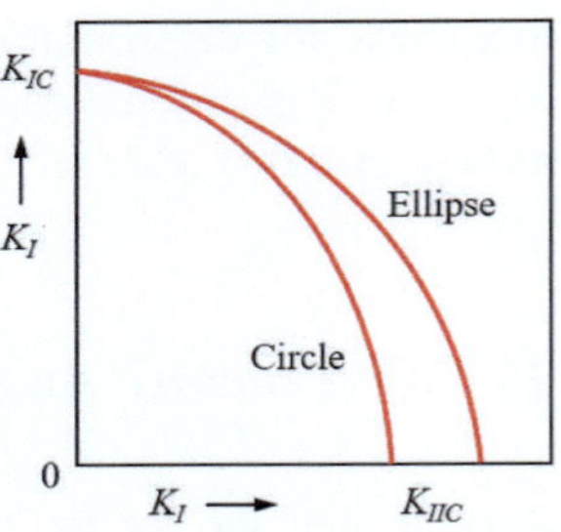

Fig. 13.10 Locus of the circle and ellipse criteria

Example 13.10 A brittle steel plate containing a 6-*mm* long and inclined central crack (Fig. 13.8a) is loaded in tension. Use both the circular and the elliptical fracture criteria to determine **(a)** the fracture stress when $\beta = \pi/4$ and $\pi/3$, and **(b)** β and θ_o when the fracture stress is $\sigma_f^{(c)}$ (circle) $= \sigma_f^{(e)}$ (Ellipse). Given data: $K_{IC} = 40 \, MPa\sqrt{m}$, $E = 206{,}850 \, MPa$ and $n = 0.30$.

Solution

(a) This part of the problem is for $\sigma_f^{(c)}$ (circle) $\neq \sigma_f^{(e)}$ (ellipse). For $\beta = \pi/4$, Eqs. (13.41) and (13.45) yield, respectively,

$$\sigma_f^{(c)} = \frac{K_{IC}}{\sqrt{\pi a}\,\sin^2\beta} = \frac{40\,MPa\sqrt{m}}{\sqrt{\pi\left(3x10^{-3}\,m\right)}\sin^2\left(\pi/4\right)} \qquad (13.10E1.a)$$

$$\sigma_f^{(c)} = 824.05 \, MPa \qquad (13.10E1.a)$$

and

$$\sigma_f^{(e)} = \frac{K_{IC}}{\sqrt{\pi a}\,\sin\beta\sqrt{\sin^2\beta + \tfrac{4}{9}\cos^2\beta}} \qquad (13.10E2a)$$

$$\sigma_f^{(e)} = \frac{\left(40\,MPa\sqrt{m}\right)}{\sqrt{\pi\left(3\times10^{-3}\,m\right)}\sin\left(\pi/4\right)\sqrt{\sin^2\left(\pi/4\right) + (4/9)\cos^2\left(\pi/4\right)}}$$
$$(13.10E2b)$$

$$\sigma_f^{(e)} = 685.65 \, MPa < \sigma_f^{(c)} \qquad (13.10E2c)$$

For $\beta = \pi/3$,

$$\sigma_f^{(e)} = \frac{\left(40\,MPa\sqrt{m}\right)}{\sqrt{\pi\left(3\times10^{-3}\,m\right)}\sin\left(\pi/4\right)\sqrt{\sin^2\left(\pi/3\right) + (4/9)\cos^2\left(\pi/3\right)}}$$
$$(13.10E3a)$$

$$\sigma_f^{(e)} = 627.93 \, MPa < \sigma_f^{(c)} \qquad (13.10E3b)$$

Therefore, the fracture stress decreases as the incline angle decreases.

(b) For $\sigma_f^{(c)}$ (circle) $= \sigma_f^{(e)}$ case,

$$\frac{K_{IC}}{\sqrt{\pi a}\,\sin^2\beta} = \frac{K_{IC}}{\sqrt{\pi a}\,\sin\beta\sqrt{\sin^2\beta + \frac{4}{9}\cos^2\beta}} \tag{13.10E4a}$$

$$\sin\beta = \sqrt{\sin^2\beta + \frac{4}{9}\cos^2\beta} \tag{13.10E4b}$$

$$\sin^2\beta = \sin^2\beta + \frac{4}{9}\cos^2\beta \tag{13.10E4c}$$

$$\cos\beta = 0 \tag{13.10E4d}$$

$$\beta = \frac{\pi}{2} = 90° \tag{13.10E4e}$$

Therefore, $\beta = \pi/2$ implies that the crack should be located along the x-axis and the fracture angle should be zero, because crack motion should occur on its tangent plane. This is known in the literature as a self-similar crack growth. This can be proven by using Eq. (13.42); $\theta_o = \pi/2 - \beta = 0$.

13.6.4   Maximum Circumferential Stress Criterion

This mixed-mode principal-stress criterion, σ_θ-criterion, postulates that crack growth takes place in a direction perpendicular to the maximum principal stress (maximum circumferential stress). It has been shown (Erdogan and Sih [15]) to be equivalent to the mixed-mode strain energy release rate criterion. Nevertheless, this fracture criterion requires that the maximum principal stress be a tension stress for opening the crack along the crack plane (Fig. 13.8a) and that

$$\left.\frac{\partial\sigma_\theta}{\partial\theta} = \tau_{r\theta}\right|_{\theta_o} = 0 \tag{13.46a}$$

$$\frac{\partial^2\sigma_\theta}{\partial\theta^2} < 0 \quad \text{for}\,\sigma_\theta > 0 \tag{13.46b}$$

This theory states that (1) the crack will grow in the direction of the maximum or critical circumferential stress $\sigma_\theta \rightarrow \sigma_c = \sigma_{\theta,\max}$ and that (2) the crack growth direction is defined by the kink angle θ_o.

Setting $\partial\sigma_\theta/\partial\theta = \tau_{r\theta} = 0$ in Eq. (13.33c) at θ_o and using Eqs. (13.39a), (13.39b) yield

$$K_I\sin\theta_o + K_{II}(3\cos\theta_o - 1) = 0 \tag{13.47a}$$

$$\sin\theta_o + (3\cos\theta_o - 1)\cot\beta = 0 \quad \text{for } \beta \neq 0 \tag{13.47b}$$

Let $x = K_I/K_{II} = \cot\beta$ in Eq. (13.47b), and solve for the clockwise-rotation crack growth direction, kink angle, or the angle of crack propagation θ_o

$$-\cos(\theta_o) = \frac{1}{3}\left[1 - \frac{x\left(x - 3\sqrt{x^2+8}\right)}{x^2+9}\right] \tag{13.48a}$$

$$\tan(\theta_o/2) = \frac{1}{4}\left(x - \sqrt{x^2+8}\right) \tag{13.48b}$$

For pure mode II loading, the stress-intensity factor for mode I vanishes ($K_I = 0$), and consequently, Eqs. (13.47b) and (13.48e) yield the fracture angle as negative variable due to its counterclockwise rotation (Fig. 13.8a). Thus, the crack growth direction is

$$\cos\theta_o = \frac{1}{3} \quad \rightarrow \quad \theta_o = 70.53° \tag{13.49a}$$

$$\tan(\theta_o/2) = \frac{\sqrt{2}}{2} \quad \rightarrow \quad \theta_o = 70.53° \tag{13.49b}$$

Manipulating the trigonometric functions in Eq. (13.33b) gives the maximum principle stress

$$\sigma_1 = \sigma_\theta\,(\theta = \theta_o) = \frac{1}{\sqrt{2\pi r}}\cos^2\frac{\theta_o}{2}\left[K_I\cos\frac{\theta_o}{2} - 3K_{II}\sin\frac{\theta_o}{2}\right] \tag{13.50}$$

For pure mode I loading at fracture along with $\theta_o = 0$, $K_I = K_{IC}$ and $K_{II} = 0$ in Eq. (13.50) yield

$$\sigma_1 = \frac{K_{IC}}{\sqrt{2\pi r}} \tag{13.51}$$

Equating Eqs. (13.50) and (13.51) yields the principal stress criterion as a trigonometric function

$$K_{IC} = K_I\cos^3\frac{\theta_o}{2} - 3K_{II}\sin\frac{\theta_o}{2}\cos^2\frac{\theta_o}{2} \quad (\sigma_\theta\text{-criterion}) \tag{13.52}$$

The fracture toughness ratio can be determined by substituting the calculated fracture angle $\theta_o = -70.53°$ into Eq. (13.52). At fracture, $K_{II} = K_{IIC}$ and $K_I = K_{IC}$

$$K_{IIC} \simeq \sqrt{\frac{5}{7}}K_{IC} = 0.85K_{IC} \quad (\sigma_\theta\text{-criterion}) \tag{13.53}$$

For convenience, squaring Eq. (13.52), manipulating and simplifying the resultant trigonometric identities yield

$$K_{IC}^2 = b_{11} K_I^2 - 2b_{12} K_I K_{II} + b_{22} K_{II}^2 \quad (\sigma_\theta\text{-criterion}) \tag{13.54}$$

with

$$b_{11} = \frac{1}{8}\,(1 + \cos\theta_o)^3 \tag{13.55a}$$

$$b_{12} = \frac{3}{8}\,\sin\theta_o\,(1 + \cos\theta_o)^2 \tag{13.55b}$$

$$b_{22} = \frac{9}{8}\,\sin^2\theta_o\,(1 + \cos\theta_o) \tag{13.55c}$$

The distribution of each constant b_{ij} is shown in Fig. 13.11.

Considering the intercepting points in Fig. 13.11, one can determine that

$$K_{IC}^2 = b_{11}\left(K_I^2 + K_{II}^2\right) - 2b_{12} K_I K_{II} \quad \text{for } b_{11} = b_{22} \tag{13.56a}$$

$$K_{IC}^2 = b_{11} K_I^2 - 2b_{22}\left(K_I K_{II} + K_{II}^2\right) \quad \text{for } b_{12} = b_{22} \tag{13.56b}$$

$$K_{IC}^2 = b_{11} K_I\,(K_I - 2b_{12} K_{II}) + b_{22} K_{II}^2 \quad \text{for } b_{11} = b_{12} \tag{13.56c}$$

which are convenient expressions for determining K_{IC}.

Example 13.11 Suppose that a thin-wall cylindrical pressure vessel made of AISI 4340 steel with $K_{IC} = 50\ MPa\sqrt{m}$, $\sigma_{ys} = 1793\ MPa$ and $v = 1/3$ contains an internal semielliptical surface crack ($2a = 4\ mm$ deep and $2c = 12\ mm$ long) inclined at $30°$ with respect to the circumferential (hoop) stress σ_θ. Assume a fracture angle θ_o (crack growth direction) as shown in the given figure below. The vessel thickness and inside diameter are $B = 6\ mm$ and $d = 500\ mm$, respectively. Calculate the fracture internal pressure P.

Fig. 13.11 Distribution of b_{ij} as functions of fracture angle θ_o

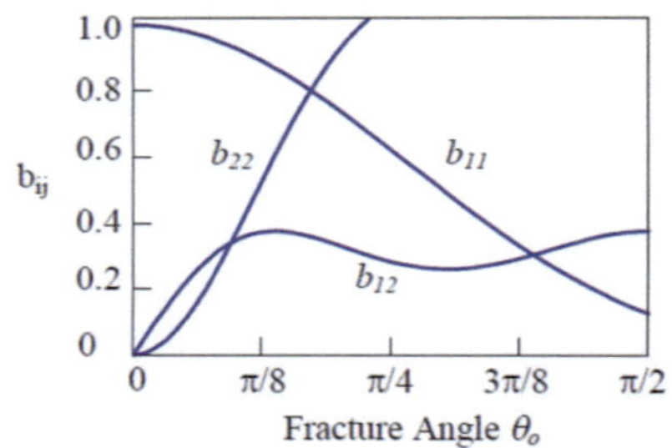

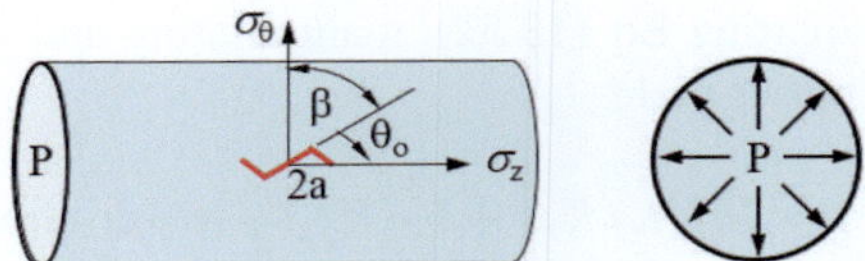

Solution The circumferential and longitudinal stresses are

$$\sigma_\theta = \frac{Pd}{2B} \quad \& \quad \sigma_z = \frac{\sigma_\theta}{2} \tag{13.11E1}$$

From Eqs. (13.39a), (13.39b) along with $\beta = 30° = \pi/6$, $\sin\beta = 1/2$, $\sin^2\beta = 1/4$, $\cos\beta = \sqrt{3}/2$, and $\sin\beta\cos\beta = \sqrt{3}/4$, the stress-intensity factor is

$$K_I = \sigma_\theta\sqrt{\pi a}\sin^2\beta = \frac{Pd}{2B}\sqrt{\pi a}\sin^2\beta = \frac{Pd}{8B}\sqrt{\pi a} \tag{13.11E2a}$$

$$K_{II} = \sigma_\theta\sqrt{\pi a}\sin\beta\cos\beta = \frac{Pd}{2B}\sqrt{\pi a}\sin\beta\cos\beta = \frac{\sqrt{3}Pd}{8B}\sqrt{\pi a} \tag{13.11E2b}$$

$$x = \frac{K_I}{K_{II}} = \tan\beta = \frac{\sqrt{3}}{3} \tag{13.11E2c}$$

From Eq. (13.53), the mode II fracture toughness is

$$K_{IIC} \simeq \sqrt{\frac{5}{7}}K_{IC} = 0.85K_{IC} = 0.85\left(50\ MPa\sqrt{m}\right) = 42.50\ MPa\sqrt{m} \tag{13.11E3}$$

From Eq. (13.48a) along with $x = \sqrt{3}/3$,

$$-\cos(\theta_o) = \frac{1}{3}\left[1 - \frac{x\left(x - 3\sqrt{x^2+8}\right)}{x^2+9}\right] \tag{13.11E4a}$$

$$-\theta_o = \arccos\left[\frac{1}{3}\left(1 - \frac{x\left(x - 3\sqrt{x^2+8}\right)}{x^2+9}\right)\right] \tag{13.11E4b}$$

$$-\theta_o = \arccos\left[\frac{1}{3}\left(1 - \frac{\left(\sqrt{3}/3\right)\left(\sqrt{3}/3 - 3\sqrt{\left(\sqrt{3}/3\right)^2+8}\right)}{\left(\sqrt{3}/3\right)^2+9}\right)\right] = -60° \tag{13.11E4c}$$

and from Eq. (13.48b),

$$\tan(\theta_o/2) = \frac{1}{4}\left(x - \sqrt{x^2 + 8}\right) \tag{13.11E5a}$$

$$\theta_o = 2\arctan\left[\frac{1}{4}\left(x - \sqrt{x^2 + 8}\right)\right] = 2\arctan\left[\frac{1}{4}\left(\sqrt{3} - \sqrt{\left(\sqrt{3}\right)^2 + 8}\right)\right] \tag{13.11E5b}$$

$$\theta_o = -60° \tag{13.11E5c}$$

Using Eq. (13.55) along with $\theta_o = -60° = -\pi/3$ yields

$$b_{11} = \frac{1}{8}(1 + \cos\theta_o)^3 = 0.42 \tag{13.11E6a}$$

$$b_{12} = \frac{3}{8}\sin\theta_o(1 + \cos\theta_o)^2 = -0.27 \tag{13.11E6b}$$

$$b_{22} = \frac{9}{8}\sin^2\theta_o(1 + \cos\theta_o) = 1.13 \tag{13.11E6c}$$

Then, Eq. (13.54) becomes

$$K_{IC}^2 = b_{11}K_I^2 - 2b_{12}K_I K_{II} + b_{22}K_{II}^2 \tag{13.11E7a}$$

$$K_{IC}^2 = 0.42K_I^2 + 0.27K_I K_{II} + 1.13K_{II}^2 \tag{13.11E7b}$$

Substituting Eqs. (13.11E1) into (13.11E7b) gives

$$K_{IC}^2 = 0.42\left(\frac{Pd}{8B}\sqrt{\pi a}\right)^2 + 0.27\left(\frac{Pd}{8B}\sqrt{\pi a}\right)\left(\frac{\sqrt{3}Pd}{8B}\sqrt{\pi a}\right)$$
$$+ 1.13\left(\frac{\sqrt{3}Pd}{8B}\sqrt{\pi a}\right)^2 \tag{13.11E8a}$$

$$K_{IC}^2 = \left(\frac{Pd}{8B}\sqrt{\pi a}\right)^2\left[0.42 + 0.27\sqrt{3} + 1.13(3)\right] \tag{13.11E8b}$$

$$K_{IC} = \frac{\sqrt{4.2777}\,Pd}{8B}\sqrt{\pi a} = \frac{0.45824\,Pd\sqrt{a}}{B} \tag{13.11E8c}$$

so that the critical pressure becomes

$$P = \frac{BK_{IC}}{0.45824d\sqrt{a}} = \frac{(6\,mm)\left(50\,MPa\sqrt{m}\right)}{0.45824\,(600\,mm)\sqrt{2 \times 10^{-3}\,m}} \tag{13.11E9a}$$

$$P = 24.40\,MPa \tag{13.11E9b}$$

Now, the applied stress-intensity factors at fracture are

$$K_I = \frac{Pd}{8B} \sqrt{\pi a} = \frac{(24.40 \; MPa)\,(600 \; mm)}{8\,(6 \; mm)} \sqrt{\pi \left(2 \times 10^{-3} \; m\right)} = 24.18 \; MPa\sqrt{m}$$

$$\text{(13.11E10a)}$$

$$K_{II} = \frac{\sqrt{3}\,Pd}{8B} \sqrt{\pi a} = \frac{\sqrt{3}\,(24.40 \; MPa)\,(600 \; mm)}{8\,(6 \; mm)} \sqrt{\pi \left(2 \times 10^{-3} \; m\right)}$$

$$\text{(13.11E10b)}$$

$$K_{II} = 41.88 \; MPa\sqrt{m} \qquad\qquad \text{(13.11E10c)}$$

and the maximum circumferential stress or hoop stress is

$$\sigma_\theta = \frac{Pd}{8B} = \frac{(24.40 \; MPa)\,(600 \; mm)}{8\,(6 \; mm)} = 305 \; MPa \qquad \text{(13.11E11)}$$

Therefore, the results $\sigma_\theta = 305 \; MPa$, $K_I = 24.18 \; MPa\sqrt{m}$, and $K_{II} = 41.88$ $MPa\sqrt{m}$ represent the upper limits for mixed-mode fracture mechanics conditions. In order to prevent fracture, use a safety factor $S_F > K_I/K_{IC} = 24.18/50 = 0.48$, say, $S_F = 2$ so that the design pressure and the design stress become $P_d = P/S_F = 12.20 \; MPa$ and $\sigma_d = \sigma_\theta/S_F = 152.50 \; MPa$. Moreover, the reader is encouraged to consult the American Society of Mechanical Engineers (ASME) code for designing pressure vessels.

13.6.5   Mixed-Mode I and III

In practice, structures are not only subjected to tension loading (mode I) but to shear (mode II) or torsion (mode III) loading. When a structural component is subjected to mixed-mode I–III loading, the crack simultaneously experiences a mixed tension and torsion interactions, and as a result, the component is exposed to a mixed-mode cracking failure mode.

In principle, the direction of crack growth and the orientation of the crack tip in a mixed-mode I–III interaction in a cylindrical specimen are expected to be orthogonal due to the nature of the specimen configuration. In such a case, the crack grows in mode I, but the principal stress field is expected to rotate, inducing crack splitting at a critical angle of rotation θ_R.

In any event, it is expected that crack growth or fracture of elastic and elastic–plastic solids under mixed-mode I/III loading starts along the radial direction. This implies that crack propagation occurs along the crack plane where the maximum tensile and shear stresses must remain perpendicular and parallel to the crack plane, respectively.

The dependency of fracture toughness on the stress-based mixed-mode I/III plastic mixity parameter (M_x) and displacement-based I/III mixity parameter (μ_x) of an elastic-plastic solid material are fundamentally inherent in mixed-mode experiments using, for example, symmetric and asymmetric bend specimens. The (μ_x parameter arises due to the out-of-plane displacement in torsional loading (mode III).

Consider the crack configuration for a solid cylinder shown in Fig. 13.8b. For a mixed mode I–III, the cylinder is subjected to a biaxial loading and Eq. (13.37) reduces to

$$K_{IC}^2 = K_I^2 + \frac{E'\,(1+v)\,K_{III}^2}{E} \tag{13.57a}$$

$$K_{IC}^2 = K_I^2 + \frac{K_{III}^2}{1-v} \qquad \text{(plane strain)} \tag{13.57b}$$

$$K_C^2 = K_I^2 + (1+v)\,K_{III}^2 \qquad \text{(plane stress)} \tag{13.57c}$$

Denote that K_{IC} is changed to K_C for plane stress. These expressions indicate that crack growth occurs in a self-similar manner or crack motion manifests itself along its tangent plane, so that the crack fracture angle becomes $\theta_o = 0$, which is the relevant theoretical condition for deriving Eq. (13.37). Moreover, the stress-intensity factor K_{III} and Poisson's ratio v relate to deformation of a specimen under an applied stress.

The loading conditions are as shown in Fig. 13.8b for a circumferentially cracked cylinder, which remotely induces a coupled mixed-mode I and III interaction or a competition of the local (internal) tension and shear stresses ahead of a crack tip during crack growth, which in turn is an inward deformation process. In this particular mixed-mode fracture mechanics case, microscopy plays an important part of the assessment of the fracture mechanics work using fractography for revealing fracture surface features.

13.7 Fatigue Crack Growth Rate

Since fatigue is a cyclic dissipation of energy process related to a cumulative damage process, the elapsed time for damage is usually expressed in terms of the number of fatigue cycles (N) at a particular load frequency.

The control parameter that is used to evaluate cracked specimens is the fatigue crack growth rate (da/dN), which depends on the applied stress-intensity factor range (ΔK) and the crack length function $a = f(N)$. Only mode I fatigue crack growth rate is considered in this section.

Figure 13.12a schematically shows a $da/dN = f(\Delta K)$ diagram based on sigmoidal curves for estimating da/dN in an actual laboratory exercise. Notice that the tangential slope of $a = f(N)$ for a constant stress ratio R is da/dN, which increases rapidly as crack growth occurs.

Notice that the effect of the stress ratio R on crack growth is remarkably noticeable since increasing R increases da/dN, and fracture is accelerated. Plotting

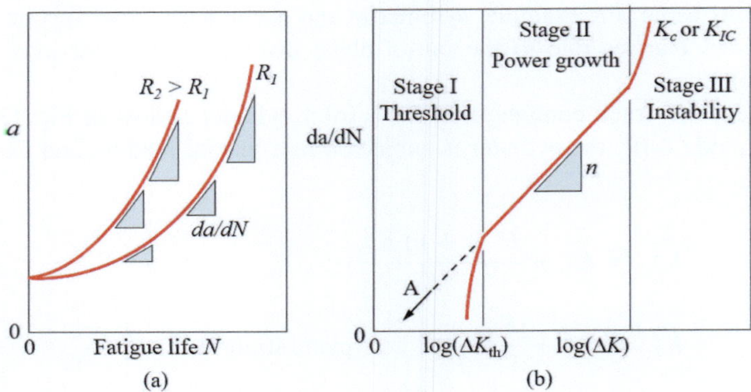

Fig. 13.12 (a) Schematic fatigue crack growth curves and (b) stages of fatigue crack growth rate from the threshold (ΔK_{th}) to fracture (K_c or K_{IC}) interval

$da/dN = f(\Delta K)$ yields a fatigue crack growth rate diagram as schematically shown in Fig. 13.12b. The ASTM E647 standard provides the guidelines on how to conduct fatigue crack growth rate tests.

Analysis of Fig. 13.12

Stage I This is a slow crack growth region in which the fatigue threshold stress-intensity factor range (ΔK_{th}) is usually a small value; $\Delta K_{th} << K_{IC}$. This implies that there is no observable crack growth at $\Delta K < \Delta K_{th}$. Thus, the fatigue process in this region strongly depends on

- Microstructural features such as grain size, precipitates, dislocation density, etc.
- Mean tress and stress ratio parameters
- Aggressiveness of the environment
- Surface damage initiation

For crack initiation in stage I, the threshold stress-intensity factor range is defined by

$$\Delta K_{th} = \beta \Delta \sigma_{th} \sqrt{\pi a} \tag{13.58}$$

Here, β is the geometry-correction factor and $\Delta \sigma_{th}$ is the threshold stress range, which is analogous to the fatigue limit S_L introduced in Fig. 12.24. This equation indicates that if $\Delta \sigma < \Delta \sigma_{th}$ crack growth does not occur.

Stage II This is a fatigue process referred to as the power growth behavior, which is usually characterized by the Paris et al. [16] power law equation of the form

$$\frac{da}{dN} = A(\Delta K)^n \tag{13.59}$$

where $\Delta K = K_{\max} - K_{\min} = (1 - R)K_{\max}$ for $R \geq 0$

$\Delta K = K_{\max}$ for $R \leq 0$

$A = \text{Constant } (MPa^{-n}.m^{1-n/2}/cycles)$

$n = \text{Exponent}$

$\Delta\sigma = \sigma_{\max} - \sigma_{\min} = \text{Stress range}$

Denote that A and n in Eq. (13.59) are material constants determined empirically from fatigue crack growth rate data and are independent of crack length. On the other hand, ΔK depends on the crack length and stress

$$\Delta K = \Delta\sigma\sqrt{\pi a} \qquad \text{(Uncorrected)} \tag{13.60a}$$

$$\Delta K = \beta\Delta\sigma\sqrt{\pi a} \qquad \text{(Corrected)} \tag{13.60b}$$

It should be clear that Eqs. (13.58) through (13.60) exclude the effects of crack closure related to residual stresses at the crack tip and to crack tip blunting.

Stage III This fatigue process strongly depends on the microstructural parameters cited in stage I and on the specimen thickness. In this stage, da/dN is high, and the fatigue damage process is represented by the instability region in Fig. 13.12b. Eventually, fracture occurs when the applied stress-intensity factor reaches a critical value; $K_{I,\max} = K_{IC}$ for plane strain or $K_{I,\max} = K_C$ for plane stress condition.

13.8 Fatigue Life Calculations

The goal here is to develop a mathematical model that predicts fatigue life (N) for a given stress range at a constant load amplitude. Integrating Eq. (13.59) yields the sought fatigue lifetime N

$$\int_0^N dN = \int_{a_o}^a \frac{da}{A\,(\Delta K)^n} = \frac{1}{A\,\left(\beta\Delta\sigma\sqrt{\pi}\right)^n} \int_{a_o}^a \frac{da}{a^{n/2}} \tag{13.61a}$$

$$N = \frac{2\left(a_o^{1-n/2} - a^{1-n/2}\right)}{A\,(n-2)\left(\alpha\Delta\sigma\sqrt{\pi}\right)^n} \qquad \text{for } n \neq 2 \tag{13.61b}$$

$$N = \frac{\ln\,(a/a_o)}{\pi A\,(\alpha\Delta\sigma)^2} \qquad \text{for } n = 2 \tag{13.61c}$$

Here, a_o denotes the original (initial) crack length and $a > a_o$ denotes the the crack length at N. Eventually, fracture occurs when the crack length reaches a critical size so that $a = a_c$ and $K_{\max} = K_{IC}$ or $K_{\max} = K_C$. For plane strain conditions, the critical crack length along with $K_{\max} = K_{IC}$ and $\Delta\sigma$ can be predicted using a modified form of Eq. (13.60b). Thus,

$$a_c = \frac{1}{\pi}\left(\frac{K_{IC}}{\beta}\right)^2 \Delta\sigma^{-2} \tag{13.62}$$

Substituting Eq. (13.62) along with $a = a_c$ into (13.61b) gives

$$C_1 \, (\Delta\sigma)^n + C_2 \, (\Delta\sigma)^{n-2} - C_3 = 0 \tag{13.63}$$

where the constants C_i with $i = 1, 2, 3$ are written as

$$C_1 = A \, (n-2) \, (N - N_o) \, (K_{IC})^n \tag{13.64a}$$

$$C_2 = \frac{2}{\pi} \left(\frac{K_{IC}}{\beta}\right)^2 \tag{13.64b}$$

$$C_3 = 2 \, (a)^{1-n/2} \left(\frac{1}{\pi}\right)^{n/2} \left(\frac{K_{IC}}{\beta}\right)^n \tag{13.64c}$$

Example 13.12 A high-strength steel string has a miniature round surface crack of 0.09 mm deep and an outer diameter of 1.08 mm. The string is subjected to a repeated fluctuating load ($\sigma_{\min} = 0$, $\sigma_{\max} > 0$) at a stress ratio $R = 0$.

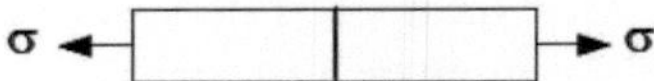

The threshold stress-intensity factor is $\Delta K_{th} = 5 \, MPa\sqrt{m}$ and the fatigue crack growth rate equation along with the boundary-correction factor are given by

$$\frac{da}{dN} = \left(5x10^{-14} \, \frac{MN^{-4}.m^{-1}}{cycles}\right) (\Delta K)^4 \tag{13.12E1a}$$

$$f \, (d/D) = \frac{1}{2}\sqrt{\frac{D}{d}} \left[\frac{D}{d} + \frac{1}{2} + \frac{3}{8}\left(\frac{d}{D}\right) - \frac{5}{14}\left(\frac{d}{D}\right)^2 + \frac{11}{15}\left(\frac{d}{D}\right)^3\right] \tag{13.12E1b}$$

Determine **(a)** the threshold stress $\Delta\sigma_{th}$ range the string can tolerate without crack growth, **(b)** the maximum applied stress range $\Delta\sigma$, **(c)** the maximum (critical) crack length for a fatigue life of $N = 10^4$ cycles, and **(d)** Will the string fracture? Use the following steel properties: $K_{IC} = 25 \, MPa\sqrt{m}$ and $\sigma_{ys} = 880 \, MPa$.

Solution It is assumed that the plastic zone with a cyclic range ΔK is smaller than that for K_I applied monotonically and that the surface crack can be treated as a single-edge crack configuration. Note that $N_o = 0$ since a_o already exists.

(a) The diameter ratios are

$$a = \frac{D - d}{2} \tag{13.12E1a}$$

$$d = D - 2a = 1.08 \, mm - 2 \, (0.09 \, mm) = 0.90 \, mm \tag{13.12E1b}$$

$$\frac{d}{D} = 0.8333 \tag{13.12E1c}$$

$$\frac{D}{d} = 1.20 \tag{13.12E1d}$$

From Eqs. (13.12E1c), (13.12E1d), the geometry-correction factor is

$$f\,(d/D) = \frac{1}{2}\sqrt{\frac{D}{d}}\left[\frac{D}{d} + \frac{1}{2} + \frac{3}{8}\left(\frac{d}{D}\right) - \frac{5}{14}\left(\frac{d}{D}\right)^2 + \frac{11}{15}\left(\frac{d}{D}\right)^3\right] \tag{13.12E2a}$$

$$\beta = f\,(d/D) = 1.1989 \tag{13.12E2b}$$

$$\Delta\sigma_{th} = \frac{\Delta K_{th}}{\beta\sqrt{\pi a_o}} = \frac{5\,MPa\sqrt{m}}{(1.1989)\sqrt{\pi\,(0.09x10^{-3}\,m)}} = 248.02\,MPa \tag{13.12E2c}$$

$$\Delta\sigma_{\min} < \Delta\sigma_{th} \tag{13.12E2d}$$

(b) The constants C_i for Eq. (13.63) are

$$C_1 = \left(5x10^{-14}\,\frac{MN^{-4}.m^7}{cycles}\right)(4-2)\left(10^4\,cycles\right)\left(25\,\frac{MN}{m^2}\sqrt{m}\right)^4 \tag{13.12E3a}$$

$$C_1 = 3.9063 \times 10^{-4}\,m \tag{13.12E3b}$$

$$C_2 = \frac{2}{\pi}\left(\frac{K_{IC}}{\beta}\right)^2 = \frac{2}{\pi}\left(\frac{25\,MPa\sqrt{m}}{1.1989}\right)^2 = 276.82\,MPa^2.m \tag{13.12E3c}$$

$$C_3 = 2\left(0.09x10^{-3}\,m\right)^{1-4/2}\left(\frac{1}{\pi}\right)^{4/2}\left(\frac{25\,MPa\sqrt{m}}{1.1989}\right)^4 \tag{13.12E3d}$$

$$C_3 = 4.2571 \times 10^8\,MPa^4.m \tag{13.12E3e}$$

Substituting these constants into Eq. (13.63) yields a fourth degree polynomial (biquadratic equation)

$$\left(3.9063 \times 10^{-4}\right)\Delta\sigma^4 + (276.82)\Delta\sigma^2 - 4.2571 \times 10^8 = 0 \tag{13.12E4a}$$

$$\Delta\sigma^4 + \left(7.0865 \times 10^5\right)\Delta\sigma^2 - 1.0898 \times 10^{12} = 0 \tag{13.12E4b}$$

Solving the above biquadratic equation yields four roots. The relevant positive root and the deduced maximum stress are, respectively,

$$\Delta\sigma = 864.93 \, MPa \tag{13.12E5a}$$

$$\sigma_{max} = \Delta\sigma = 864.93 \, MPa \quad \text{since} \quad \sigma_{min} = 0 \tag{13.12E5b}$$

$$\sigma_{max} < \sigma_{ys} \tag{13.12E5c}$$

Notice that Eq. (13.12E5a) represents a logical condition, since the specimen under a fluctuating stress state contains a crack; therefore, it is expected that the specimen fractures at σ_{max} when the crack reaches a critical length a_c. Thus,

$$K_{max} = \beta\sigma_{max}\sqrt{\pi a} = (1.1989)\,(864.93 \, MPa)\,\sqrt{\pi\left(0.09 \times 10^{-3} \, m\right)} \tag{13.12E6a}$$

$$K_{max} = 17.44 \, MPa\sqrt{m} \tag{13.12E6b}$$

(c) From Eq. (13.62), the critical crack size is

$$a_c = \frac{1}{\pi}\left[\frac{K_{IC}}{\beta}\right]^2 \Delta\sigma^{-2} = \frac{1}{\pi}\left[\frac{25 \, MPa\sqrt{m}}{(1.1989)\,(864.93 \, MPa)}\right]^2 = 0.185 \, mm \tag{13.12E7a}$$

$$\Delta a = a_c - a_o = 0.095 \, mm \tag{13.12E7b}$$

Therefore, $\Delta a = 0.095 \, mm$ represents 5.56% increment at a maximum fluctuating stress $\sigma_{max} < \sigma_{ys}$ for a fatigue life of 10^4 cycles.

(d) The stress-intensity factor range is

$$\Delta K = \beta\Delta\sigma\sqrt{\pi a} \tag{13.12E8a}$$

$$\Delta K = (1.1989)\,(864.93 \, MPa)\,\sqrt{\pi\left(0.90x10^{-3} \, m\right)} \tag{13.12E8b}$$

$$\Delta K = 55.14 \, MPa\sqrt{m} > 25 \, MPa\sqrt{m} \tag{13.12E8c}$$

Hence, the string fractures because $K_{max} = \Delta K > K_{IC}$.

13.8.1 Crack Closure

During crack closure, the crack surfaces remain in contact at the crack tip during a portion of the unloading part of the fatigue cycle, affecting, to an extent, the fatigue growth rate da/dN. Accordingly, Elber [17] crack-closure model is similar to Paris power law model

$$\frac{da}{dN} = C \left(\Delta K_{eff}\right)^m \quad \text{(Elber's model)} \tag{13.65a}$$

$$\frac{da}{dN} = A \left(\Delta K\right)^n \quad \text{(Paris's model)} \tag{13.65b}$$

with

$$\Delta K_{eff} = \delta \Delta K \tag{13.66}$$

where $\delta < 1$ denotes the crack closure factor and ΔK_{eff} defines the effective stress-intensity factor range related to crack closure due to opposite crack faces in contact during fatigue loading. Mathematically, the exponents m and n may have different values, but they have the same physical meaning. It is clear that the crack closure factor $\delta < 1$ is a condition for an effective crack closure.

Letting $R_{eff} = K_{op}/K_{max}$ the factor δ becomes

$$\delta = \frac{K_{max} - K_{op}}{K_{max} - K_{min}} = \frac{1 - K_{op}/K_{max}}{1 - K_{min}/K_{max}} \tag{13.67a}$$

$$\delta = \frac{1 - R_{eff}}{1 - R} \quad \text{for } R \neq 1 \text{ and } R_{eff} \neq 1 \tag{13.67b}$$

Denote that $da/dN = f(\Delta K)$ gives a conservative fatigue life prediction by using crack closure-free data. If crack closure is included in a suitable da/dN expression, then ΔK is replaced by an effective stress-intensity factor range $\Delta K_{eff} < \Delta K$. Here, $\Delta K_{eff} = K_{max} - K_{op}$ and $K_{op} > K_{min}$ is the stress-intensity factor that opens the crack.

Mathematically, if $K_{min} = 0$ the factor δ becomes independent of the nominal stress ratio $R = K_{min}/K_{max} = 0$ and acquires the following definition

$$\delta = 1 - \frac{K_{op}}{K_{max}} = 1 - R_{eff} \tag{13.68}$$

It is clear now that the phenomenon of crack closure vanishes when $R_{eff} = 0$ because $\delta = 1$ and $\Delta K_{eff} = \Delta K$.

Example 13.13 (a) Derive an expression for the stress-intensity factor K_{op} using Elber's model $\Delta K_{eff} = \delta \Delta K$, and (b) calculate K_{op} and da/dN for $\delta = 0.8$, $R = 0$, $K_{max} = 100 \ MPa\sqrt{m}$, $n = 4$ and $A = 5 \times 10^{-16} MN^{-4}m^{-1}/cycle$. Explain.

Solution

(a) The derivation of K_{op} equation is

$$\Delta K_{eff} = \delta \Delta K \tag{13.13E1a}$$

$$K_{\max} - K_{op} = \delta \left(K_{\max} - K_{\min}\right) \qquad (13.13\text{E1b})$$

$$1 - \frac{K_{op}}{K_{\max}} = \delta \left(1 - \frac{K_{\min}}{K_{\max}}\right) \qquad (13.13\text{E1c})$$

$$1 - \frac{K_{op}}{K_{\max}} = \delta \left(1 - R\right) \qquad (13.13\text{E1d})$$

$$K_{op} = K_{\max} \left(1 - \delta + \delta R\right) \qquad (13.13\text{E1e})$$

(b) The calculation of K_{op} along with $\delta = 0.8$, $R = 0$, $K_{\max} = 100\ MPa\sqrt{m}$, $n = 4$, and $A = 5x10^{-16}\ MN^{-4}m^{-1}/cycle$ yields

$$K_{op} = K_{\max} \left(1 - \delta + \delta R\right) = K_{\max} \left(1 - \delta\right) \qquad (13.13\text{E2a})$$

$$K_{op} = (100\ MPa)\ (1 - 0.8) \qquad (13.13\text{E2b})$$

$$K_{op} = 20\ MPa \qquad (13.13\text{E2c})$$

For comparison, the stress-intensity factor range for Paris and Elber equations are

$$\Delta K = K_{\max} - K_{\min} = 100\ MPa\sqrt{m} - 0\ MPa\sqrt{m} = 100\ MPa\sqrt{m} \qquad (13.13\text{E3a})$$

$$\Delta K_{eff} = K_{\max} - K_{op} = 100\ MPa\sqrt{m} - 20\ MPa\sqrt{m} = 80\ MPa\sqrt{m} \qquad (13.13\text{E3b})$$

The crack growth rates are

$$\frac{da}{dN} = A\left(\Delta K\right)^4 = 5.00x10^{-6}\ m/cycle\ @\Delta K = 100\ MPa\sqrt{m} \qquad (13.13\text{E4a})$$

$$\frac{da}{dN} = A\left(\Delta K_{eff}\right)^4 = 2.05x10^{-6}\ m/cycle\ @\ \Delta K_{eff} = 80\ MPa\sqrt{m} \qquad (13.13\text{E4b})$$

In conclusion, the crack truncates the unloading process by reducing the $Delta K$ to an effective mechanical driving force ΔK_{eff} and da/dN.

13.9　Impact Testing

The impact testing is a destructive method used to measure the amount of energy absorbed by a material during dynamic fracture and to determine the effect of temperature on the dynamic mechanical behavior of materials. Principally, this tool reveals the probable temperature-dependent brittle-to-ductile transition of a material, such as steel used in seagoing vessels exposed to a range of temperature during service. Basically, there are two common types of impact testing, pendulum and drop weight. The pendulum method includes the Izod and Charpy impact testing techniques under bending mode.

13.9.1 Charpy Impact Testing

The Charpy or Izod notched specimens are used for this purpose, the Charpy V-notch specimen being the most common. This technique became a conventional testing method when it was revealed in the 1940s that welded ships, large pipelines, and other steel structures fractured at notch roots. Accordingly, the impact testing technique is a dynamic fracture process being recommended by the ASTM E23 standard test method. The applied impact load (P) is through an impact blow from a falling pendulum hammer (striker).

Impact tests, in general, are used to measure the response of a material to dynamic loading. The most common laboratory test configurations are the pendulum machine and the drop tower. For instance, Fig. 13.13 shows a conventional V-notch impact specimen configuration along with dimensions (Fig. 13.13a), the Izod specimen (Fig. 13.13b), the Charpy specimen (Fig. 13.13c), and the testing machine (Fig. 13.13d) used to measure fracture toughness of a three-point bending specimen (3PB) under an impact loading condition at relatively high velocity (Callister and Rethwisch [18, p. 226]).

The results obtained from standard impact tests are usually a single value of the impact energy or energy spent on a single specimen. This is of limited value in describing the dynamic behavior of a particular sample material. Therefore,

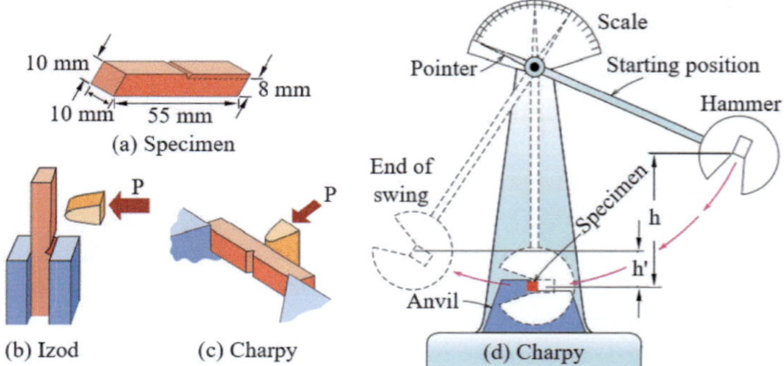

Fig. 13.13 (**a**) The three-point bending (3PB) specimen and the conventional (**b**) Izod and (**c**) Charpy specimens, and (**d**) Charpy impact testing machine (after Callister and Rethwisch [18, p. 226])

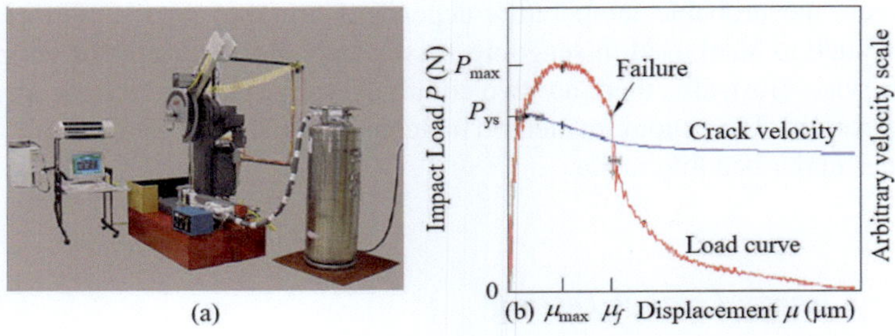

Fig. 13.14 (**a**) Tinius Olsen Model 84 Instrumented Charpy impact machine equipped with in situ heating and cooling system, optical encoder for measuring the impact velocity, and motorized hammer (striker) return and (**b**) impact loading diagram for single-edge cracked brittle material using a MPM ImpactTM v3.0 software. After Manahan and Stonesifer [19]

instrumenting an impact machine yields information on the impact forces, impact velocities, displacements, and strain energies of the striker at any time during the dynamic test.

Impact loads generate high strain rates in solid materials. For instance, the instrumented Charpy impact testing machine shown in Fig. 13.14a along with a particular impact or dynamic loading diagram (Fig. 13.14b) imparts low strain rates at low velocity when compared to ballistic impact velocity (Manahan and Stonesifer [19]). The latter technique is out of the scope of this textbook.

The instrumented Charpy impact machine remains a key means for fracture toughness testing due to its low cost, convenience, reliability based on certification standards, and simple use. Thus, the transient load history during a Charpy test is readily obtained by placing strain gages on the striker, so that it becomes the load cell.

Figure 13.14b illustrates a typical load history for a relevant case using the instrument Charpy machine setup. The interpretation of Fig. 13.14b is very vital for characterizing the toughness characteristics of a specimen or dynamic behavior with respect to the dynamic load response to dynamic displacement caused by the impact process in a Charpy specimen.

The load-displacement curve, $P = f(\mu)$, for a three-point bending (3PB) Charpy specimen can be divided into fracture initiation and fracture propagation regions expressed in terms of areas under the curve (Fig. 13.14b). These areas are measures of the elastic strain energy (U_e) and plastic strain energy (U_p). These energies are strongly dependent on the temperature, specimen size, and impact velocity imparted by the kinetic energy of the striker.

The total impact strain energy U that a specimen can absorb during the dynamic deformation is defined as a time-dependent or displacement-dependent quantity. Excluding energy losses due to specimen/anvil friction and pendulum windage and letting $P = f(t)$ or $P = f(\mu)$ the impact strain energy can be defined as the amount of energy a testing specimen absorbs during fracture. This implies that the impact test is a tool for determining the impact toughness of a material at a temperature T.

Mathematically, the total impact strain energy or impact toughness (U) is the sum of its elastic (U_e) and plastic (U_p) components

$$U = U_e + U_p \tag{13.69a}$$

$$U = U(t) = v \int P dt \tag{13.69b}$$

$$v = \sqrt{2gh} \tag{13.69c}$$

$$U = U(\mu) = \int P d\mu \tag{13.69d}$$

where v = denotes the striker velocity (m/s) on contact with the specimen, P = denotes the impact load (N), $\int P dt$ = defines the impulse ($N.s$), h = denotes the pendulum initial height (m), and g = 9.81 m/s^2 denotes the gravitational acceleration.

The resultant energy measurement is commonly referred to as Charpy impact energy (U), which is a measure of the fracture toughness of a material at testing temperatures. Thus, $U = f(T)$ is normally determined experimentally in order to reveal the effects of impact loads on the dynamic behavior of materials at relatively low and high temperatures. As a result, a material may or may not exhibit a ductile-to-brittle transition.

Using a suitable software during an impact one can record the displacements by integrating the acceleration twice with respect to time. The accuracy of these measurements may be affected by the inertial forces in the striker, variations in the contact force distribution between the striker and the specimen, striker geometry, and by strain gage location on the striker.

13.9.2 Ductile-to-Brittle Transition

Ideally, Fig. 13.15 shows complete S-shape curves, $U = f(T)$, for quasi-static (slow-bend) and dynamic testing conditions.

Note that the dynamic ductile-to-brittle transition, known as the nil-ductility-transition (NDT) temperature, which is denoted as T_o, is shifted to the left for the quasi-static testing, and the upper region is shifted downward.

The temperature shift in Fig. 13.15 is defined as $T_o^* = T_o - \Delta T$, and it represents the upper limit of the plane-strain condition for valid K_{IC} values. This clearly indicates that the loading rate affects the transition region.

With respect to the schematic dynamic curve, if $U \leq U_{NDT}$, then the specimen shows an elastic behavior (brittle appearance) denoted as the lower shelf in Fig. 13.16, and if $U > U_{NDT}$, the specimen exhibits an elastic-plastic

Fig. 13.15 Schematic quasi-static and impact (dynamic) fracture toughness

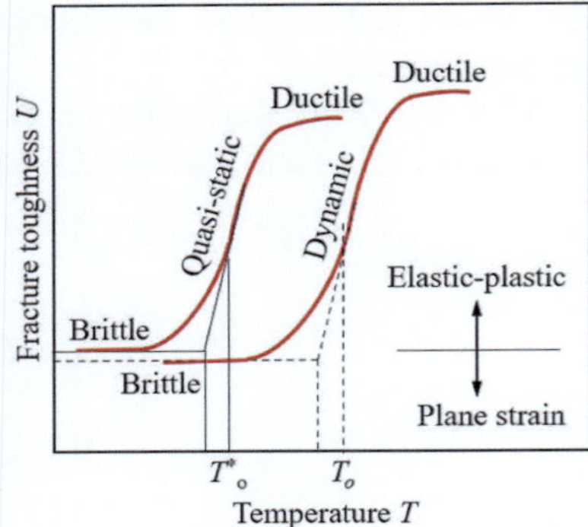

Fig. 13.16 Schematic Charpy-V notch energy, $U = f(T)$, curve showing regression parameters (Oldfield [21])

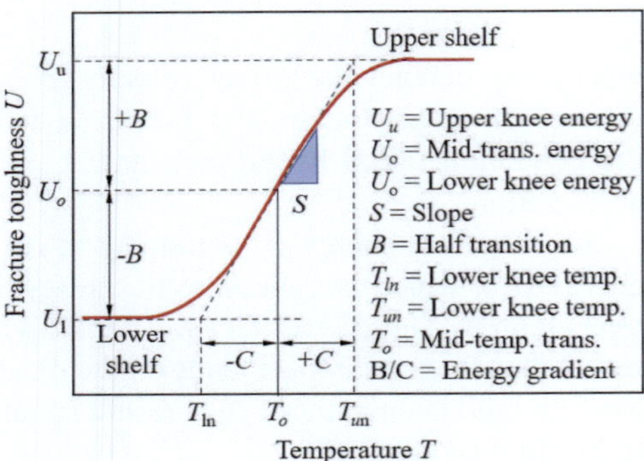

behavior (brittle-to-ductile appearance) in the transition region, which depends on the microstructure, testing temperature, strain rate, and notch acuity (Barsom and Rolfe [20, p. 100]). Therefore, this transition behavior causes a wide data scatter as encountered in most BCC low carbon steels. At $T > T_{un}$, the materials show a ductile behavior.

Thus, the lower shelf energy in Fig. 13.16 represents a brittle behavior in which plane strain condition should exist. In theory, brittle fracture requires a low consumption of energy. On the other hand, the upper shelf energy is a ductile response related to fracture by tearing, which is a fracture process that absorbs a large amount of strain energy.

Figure 13.16 shows a schematic $U = f(T)$ reference curve for a nonlinear regression procedure. Among several mathematical models that may fit U data, a model based on the hyperbolic tangent function can give reasonable results. For instance, Oldfield [21] used such a function to interpret the physical meaning of the nonlinear curve fitting parameters. Further, a common mathematical model describing the impact energy is of the form

$$U = A + B \tanh\left(\frac{T - T_o}{C}\right) \tag{13.70}$$

where A, B, C, T_o are regression constants and T is the testing temperature. In essence, T_o represents the average transition temperature in the closed temperature interval $[T_{\text{ln}} \leq T_o \leq T_{un}]$, which represents a region of data scatter.

Accordingly, the slope of a transition region can be defined as an energy gradient (energy per unit temperature). From Eq. (13.70), the slope of the impact strain energy is evaluated at the midpoint where $T = T_o$

$$\left.\frac{dU}{dT}\right|_{T=T_o} = \frac{B}{C}\left[1 - \tanh^2\left(\frac{T - T_o}{C}\right)\right]_{T=T_o} = \frac{B}{C} \tag{13.71}$$

The general dU/dT term is applicable in the transition region only. For comparison, below are given other functions that may give good fitting results

$$U = A + B \arctan\left(\frac{T - T_o}{C}\right) \tag{13.72a}$$

$$U = \frac{D}{1 + F \exp(-ET)} \tag{13.72b}$$

Accordingly, the nonlinear least-squares fitting curves (Fig. 13.17) are drawn as per Eqs. (13.70) and (13.72a) (Perez [12, pp. 398–399]).

The curve fitting equations along with T in $^\circ C$ and U in joules (J) are

$$U = 64.07 + 53.55 \tanh\left(\frac{T - 70.62}{10.67}\right) \tag{13.73a}$$

$$U = 64.34 + 36.66 \arctan\left(\frac{T - 70.57}{5.38}\right) \tag{13.73b}$$

$$U = \frac{119.46}{1 + 23{,}703.73 \exp(-0.15T)} \tag{13.73c}$$

Extensive efforts have been devoted to correlate impact fracture toughness (U) and plane-strain fracture toughness (K_{IC}) data (Barsom and Rolfe [20, Chapter 5]).

Fig. 13.17 Notch-toughness of a cold rolled (25%CR) ASTM A710 steel (Perez [12, pp. 398–399])

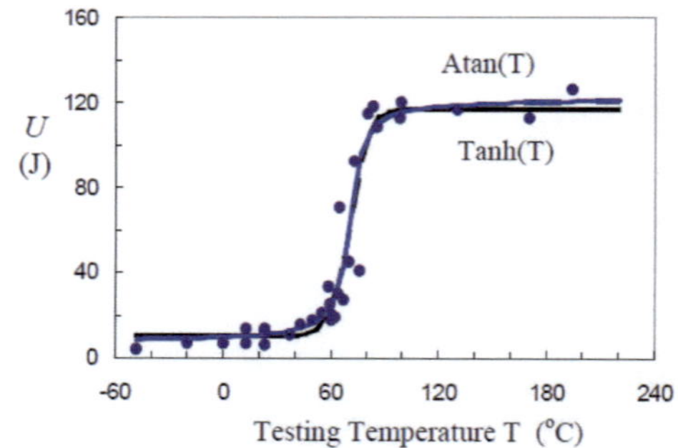

Let the plane-strain fracture toughness and the Charpy impact energy be empirically defined by

$$K_{IC} = A + B \tanh\left(\frac{T - T_o}{C}\right) \tag{13.74a}$$

$$U = D + E \tanh\left(\frac{T - T_o}{F}\right) \tag{13.74b}$$

These two expressions give a similar S-shape trendlines and the transition temperature (T_o). Letting $\tanh x \simeq x$ and solving Eq. (13.74b) for T_o and substituting the resultant expression into Eq. (13.74a) yield

$$K_{IC} = A + B \tanh\left[H\left(U - D\right)\right] \tag{13.75}$$

where $H = F/(CE)$.

13.10   Summary

Quasi-static and impact fracture toughness are two well-documented fields, where (1) strength of materials associates with crack-free specimens and area under stress-strain curves, (2) linear-elastic fracture mechanics (LEFM) of brittle materials relates to pre-cracked specimens and plane-strain fracture toughness K_{IC}, (3) fatigue crack growth rate da/dN pertains to the stress-intensity factor range ΔK, and (4) Charpy impact testing leads to aspects of brittle-to-ductile transition at a temperature range ΔT and impact energy commonly known as Charpy energy U absorbed by the specimen containing a notch or a sharp crack.

In fracture mechanics, the crystallography of slip and mechanical twinning are the mechanisms related to the formation of a plastic zone ahead of a crack tip during crack growth. However, certain materials may exhibit a ductile-to-brittle behavior, and certain datasets may be suitable for characterizing correlations between K_{IC} and U so that $K_{IC} = f(U)$ at ΔT.

Problems

13.1 Suppose that a structure made of plates has one cracked plate. If the crack reaches a critical size, will that plate fracture or the entire structure collapse? Explain.

13.2 What is crack instability based on according to Griffith criterion?

13.3 Can the Griffith Theory be applied for a quenched steel containing 1.2%C, if a penny-shaped crack is detected?

13.4 Will the Irwin Theory (or modified Griffith Theory) be valid for a changing plastic zone size as the crack advances?

13.5 What are the major roles of the surface energy and the stored elastic energies in a crack growth situation?

13.6 What does happen to the elastic strain energy when crack growth occur?

13.7 A stock of steel plates with $G_{IC} = 130\ kJ/m^2$, $\sigma_{ys} = 2200\ MPa$, $E = 207\ GPa$, $v = 1/3$ are used to fabricate a cylindrical pressure vessel ($d_i = 5\ m$ and $B = 25.4\ mm$). The vessel fractured at a pressure of 20 MPa. Subsequent failure analysis revealed an internal semielliptical surface crack of $a = 2.5\ mm$ and $2c = 10\ mm$. **(a)** Use a fracture mechanics approach to predict the critical crack length this steel would tolerate. Based on this catastrophic failure, another vessel was constructed with $d_o = 5.5\ m$ and $d_i = 5\ m$. **(b)** Calculate the applied stress-intensity factor K_I. Will this new vessel fracture at a pressure of 20 MPa if an internal semielliptical surface crack having the same dimensions as in part (a)? [Solution: (a) $a_c = 2.49\ mm$, (b) $K_I = 13.69\ MPa\sqrt{m}$. The new vessel will not fracture because $K_I < K_{IC}$].

13.8 A cylindrical pressure vessel with $B = 25.4\ mm$ and $d_i = 800\ mm$ is subjected to an internal pressure P_i. The material has $K_{IC} = 31\ MPa\sqrt{m}$ and $\sigma_{ys} = 600\ MPa$. Assume that there exists a semielliptical surface crack with $a = 5\ mm$ and $2c = 25\ mm$ and that a pressure surge occurs causing fracture of the vessel. Calculate **(a)** the actual pressure P_i using a safety factor of 2.5, **(b)** the fracture internal pressure P_f, and **(c)** the critical crack length a_c. [Solution: (a) $P_i = 15.24\ MPa$, (b) $P_f = 18.09\ MPa$, (c) $a_c = 7.04\ mm$].

13.9 This is a problem that involves strength of materials and fracture mechanics. An AISI 4340 steel is used to design a cylindrical pressure having an inside diameter and an outside diameter of 6.35 cm and 12.07 cm, respectively. The hoop stress is not to exceed 80% of the yield strength of the material. **(a)** Is the structure a thin-wall vessel or a thick-wall pipe? **(b)** What is the internal pressure? **(c)** Assume that an internal semielliptical crack exist with $a = 2\ mm$ and $2c = 6\ mm$. Will the vessel fail? **(d)** Will you recommend another steel for the pressure vessel? Why? Or why not? **(e)** What is the maximum crack length the AISI 4340 steel can tolerate? Explain. Given data: $\sigma_{ys} = 1476\ MPa$ and $K_{IC} = 81\ MPa\sqrt{m}$. [Solution: (a) Thick-wall because $B > d_i/20$, (b) $P_i = 101.51\ MPa$, (c) the pressure vessel will not fail, because $K_I < K_{IC}$, (d) safety precautions should be taken, (e) $a_c = 2.11\ mm$].

13.10 A project was carried out to measure the elastic-strain energy release rate as a function of normalized stress (σ/σ_{ys}) of large plates made out a hypothetical brittle solid. All specimens had a single-edge crack of 3-mm long. It is required to use an effective crack length a_e with a plastic zone size defined by $r = (a/2) \left(\sigma/\sigma_{ys}\right)^2$. (a) Curve fit the resulting data

σ/σ_{ys}	0	0.10	0.20	0.30	0.40	0.50	0.60	0.70	0.80
G_I $(kPa.m)$	0	0.34	1.37	1.16	6.00	9.44	14.26	20.47	28.35

Calculate **(b)** the maximum allowable σ/σ_{ys} ratio and the fracture stress for $G_{IC} = 30 \, kPa.m$ and **(c)** K_{IC} in $MPa\sqrt{m}$. Given data: $v = 0.3$, $\sigma_{ys} = 900 \, MPa$, $E = 207 \, GPa$. [Solution: (b) $\sigma/\sigma_{ys} = 0.8183$ and $\sigma_f = 736.47 \, MPa$, (c) $K_{IC} = 82.61 \, MPa\sqrt{m}$].

13.11 Determine the critical crack length of the material described in problem 13.10. [Solution: $a_c = 3 \, mm$].

13.12 A 1-mm × 15-mm × 100-mm steel strap has a 3-mm long central crack is loaded to failure. Assume that the steel is brittle and has $E = 207,000 \, MPa$, $v = 1/3$, $\sigma_{ys} = 1500 \, MPa$, and $K_{IC} = 70 \, MPa\sqrt{m}$. Calculate **(a)** the critical stress σ_c and **(b)** the critical strain energy release rate G_{IC}. [Solution: (a) $\sigma_c = 994.45 \, MPa$, (b) $G_{IC} = 21.04 \, kJ/m^2$].

13.13 A steel tension bar 8-mm thick and 50-mm wide with an initial single-edge crack of 10-mm long and plane-strain fracture toughness $K_{IC} = 60 \, MPa\sqrt{m}$ is subjected to a uniaxial stress $\sigma = 140 \, MPa$. Calculate **(a)** the stress-intensity factor K_I, **(b)** the critical crack size a_c, and **(c)** the critical load P_c. Is the crack stable? [Solution: (a) $K_I = 34.02 \, MPa\sqrt{m}$, (b) $a_c = 31.11 \, mm$, (c) $P_c = 987.71 \, kN$].

13.14 A very sharp penny-shaped crack with a diameter of 11-mm is completely embedded in a highly brittle solid. Assume that catastrophic fracture occurs when a stress of 600 MPa is applied. **(a)** What is the plane-strain fracture toughness K_{IC} for this solid? **(b)** If a sheet 5-mm thick is prepared for fracture toughness testing. Would the K_{IC} value be an acceptable number according to the ASTM E399 standard? Use $\sigma_{ys} = 1342 \, MPa$. **(c)** What thickness would be required for the K_{IC} test to be valid? [Solution: (a) $K_{IC} = 71 \, MPa\sqrt{m}$, (b) the test is not valid because $B = 5$-$mm < B_{ASTM} = 7$-mm, (c) $B \geq B_{ASTM} \geq 7 \, mm$].

13.15 An aluminum alloy plate (10-mm × 100-mm × 500-mm) having a 10-mm single-edge through-the-thickness crack is loaded in tension. Determine **(a)** the validity of the fracture mechanics test, **(b)** the maximum allowable tension stress, **(c)** is it necessary to correct K_I due to crack-tip plasticity? Why? or Why not? and **(d)** the design stress and the stress-intensity factor if the safety factor is 1.5. Data: $\sigma_{ys} = 586 \, MPa$ and $K_{IC} = 33 \, MPa\sqrt{m}$. [Solution: (a) $B = 10$-$mm > B_{ASTM} = 7.93 \, mm$, (b) $\sigma = 157.55 \, MPa$, (c) there is no need for crack-tip plastic correction since $K_I \simeq K_{IC}$, (d) $K_d = 22 \, MPa\sqrt{m}$].

13.16 Consider a steel vessel containing an internal circular crack (20-mm diameter) perpendicular to the hoop stress of $\sigma = 200\ MPa$. Calculate (a) K_I including the effect of plasticity ahead of the crack tip using Irwin's plastic zone corrected expression, **(b)** the plane-strain plastic zone size r, **(c)** the critical crack length a_c, **(d)** the critical (maximum allowable) stress, and **(e)** the strain energy release rate. When will the pressure vessel explode? Data: $K_{IC} = 60\ MPa\sqrt{m}$, $E = 200\ MPa$, $v = 1/3$ and $\sigma_{ys} = 700\ MPa$. [Solution: (a) $K_I = 32.56\ MPa\sqrt{m}$, (b) $r = 0.11$ mm, (c) $a_c = 67.91\ mm$, (d) $\sigma_c = 354\ MPa$, (e) $G_{IC} = 0.016\ MN/m$].

13.17 Consider a steel plate ($B = 10\ mm$, $w = 100\ mm$ and $L = 2w$) containing through-the-thickness double-edge cracks ($a = 3\ mm$) being subjected to an applied stress of $\sigma = 423\ MPa$. Calculate the plane-strain stress-intensity factor K_I. Will the plate fracture at this stress level? Data: $K_{IC} = 60\ MPa\sqrt{m}$, $E = 200\ MPa$, $v = 1/3$, and $\sigma_{ys} = 700\ MPa$. [Solution: $K_I = 46\ MPa\sqrt{m}$].

13.18 A large steel ship deck (30-mm thick, 12-m wide and 20-m long) having a 60-mm long through-the-thickness central crack is stressed perpendicular to the crack plane. It is operated below its ductile-to-brittle transition temperature (with $K_{IC} = 28.3\ MPa\sqrt{m}$). Calculate **(a)** the tensile stress for catastrophic failure and **(b)** the safety factor for avoiding failure. $K_{IC} = 30\ MPa\sqrt{m}$ and $\sigma_{ys} = 500\ MPa$. [Solution: (a) $\sigma_f = 184.26\ MPa$, (b) $S_f > 2.71$].

13.19 A large and thick plate containing a 4-mm long through-thickness central crack fractures when it is subjected to an external tensile stress of $7\ MPa$. Calculate the strain-energy release rate using **(a)** the Griffith's theory and **(b)** the LEFM approach. Should there be a significant difference between results? Explain. Data: $E = 62{,}000\ MPa$ and $v = 1/3$. [Solution: (a) $G_{IC} = 4.41\ J/m^2$, (b) $G_{IC} = 4.50$ J/m^2].

13.20 A long steel pipe having a diameter ratio of 1.25 and an internal diameter of 80 mm is subjected to a pressure of 60 MPa. Assume that a semielliptical surface crack, $a = 2\ mm$ and $2c = 4\ mm$, develops during service and that it is perpendicular to the hoop stress as shown in Fig. 13.6. Assume that the design lifetime of the pressurized steel pipe is 20 years and that an undesired pressure surge will burst the pipe. Calculate **(a)** the possible fracture pressure if the steel has $K_{IC} = 30\ MPa\sqrt{m}$ and $\sigma_{ys} = 550\ MPa$, and **(b)** the time it takes for a pressure surge if the crack velocity is defined as $v = c_1 K_I^n$, where v is in m/s, K_I is in $MPa\sqrt{m}$, $c_1 = 2x10^{-13.6}\ (m/sec)(MPa\sqrt{m})^{-n}$ and $n = 1.5$. [Solution: (a) $P_f = 105.23\ MPa$, (b) $t = 23.31\ years$].

13.21 Determine the elastic impact Charpy energy (U_e) from the given load vs. displacement, $P = f(\mu)$, diagram for a hypothetical steel specimen. [Solution: $U_e = 7.23\ J$].

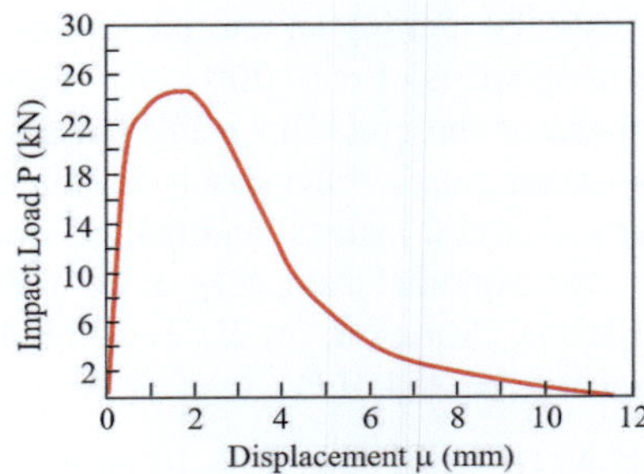

References

1. A.A. Griffith, *The Phenomena of Rupture and Flow in Solids*. Philosophical transactions of the Royal Society of London. Series A. Containing Papers of a Mathematical or Physical Character 221, No. 582–593, 1921
2. E. Orowan, *Fatigue and Fracture of Metals* (MIT Press, Cambridge, 1950)
3. G.R. Irwin, Fracture, in *Handbuch der Physik, Vol. VI*, translated as Encyclopedia of Physics, S. Flugge editor, Group 3 Mechanical and Thermal Behaviour of Matter (Springer, New York, 1958)
4. R.I. Stephens, A. Fatemi, R.R. Stephens, H.O. Fuchs, *Metal Fatigue in Engineering*, 2nd edn. (A Wiley-Interscience Publication, John Wiley & Sons, New York, 2001)
5. H. Tada, P.C. Paris, G.R. Irwin, *The Stress Analysis of Cracks Handbook*, 3rd edn. (ASME Press, New York, 2000)
6. G.R. Irwin, The crack extension force for a part-through crack in a plate. Trans. ASME, J. Appl. Mech. **29**(4), 651–654 (1962)
7. D. Broek, *Elementary Engineering Fracture Mechanics*, 4th edn. (Martinus Nijhoff Publishers, The Netherlands, 1982)
8. I.N. Sneddon, The distribution of stress in neighborhood of a crack in a elastic solid. Proc. R. Soc. Lond. A **187**, 229–260 (1946)
9. J.C. Newman, Jr., *A Review and Assessment of the Stress-Intensity Factors for Surface Cracks*. NASA Technical Memorandum 78805 (NASA-TM-78805) (ASTM International, 1978) pp. 16–42
10. Y Murakami (ed.), *Stress-Intensity Factors Handbook*, vol. 2 (Pergamon Press, New York, 1987)
11. J.W. Dally, W.F. Riley, *Experimental Stress Analysis*, 3rd edn. (McGraw-Hill, New York, 1991)
12. N. Perez, *Fracture Mechanics*, 2nd edn. (Springer International Publishing, Switzerland, 2017)
13. G.C. Sih, P.C. Paris, F. Erdogan, Crack-tip, stress-intensity factors for plane extension and plate bending problems. J. Appl. Mech. **29**(2), 306–312 (1962)
14. K. Hellan, *Introduction to Fracture Mechanics* (McGraw-Hill Book Company, New York, 1984)
15. F. Erdogan, G.C. Sih, On the crack extension in plates under plane loading and transverse shear. J. Basic Eng. **85**, 519–527 (1963)
16. P.C. Paris, M.P. Gomez, W.E. Anderson, A rational analytical theory of fatigue. Trend Eng. **13**, 9–14 (1961)
17. W. Elber, Fatigue crack closure under cyclic tension. Eng. Fract. Mech. **2**(1), 37–45 (1970)
18. W.D. Callister, Jr., D.G. Rethwisch, *Materials Science and Engineering: An Introduction*, 10th edn. (John Wiley & Sons, New York, 2018)

19. M.P. Manahan, R.B. Stonesifer, *Studies Toward Optimum Instrumented Striker Designs* (MPM Technologies, 2003). www.mpmtechnologies.com
20. J.M. Barsom, S.T. Rolfe, *Fracture and Fatigue Control in Structures: Applications of Fracture Mechanics*, 3rd edn. (American Society for Testing and Materials, West Conshohocken, 1999)
21. W. Oldfield, Curve fitting impact test data: a statistical procedure. ASTM Stand. News 3(11), 24–29 (1975)

Chapter 14
Physical Properties of Solids

14.1 Introduction

This chapter focuses on the fundamentals of electrical and magnetic behavior of solids using a combined macro- and microscales. On the basis of electrical behavior, solids can be classified as metallic, ionic, semiconducting, and superconducting materials. In conformity with magnetic behavior, solids are characterized as ferromagnetic, ferrimagnetic, paramagnetic, and diamagnetic materials.

When an active material in an electrical circuit being exposed to an electric field strength $\vec{E}$ having magnitude and direction may exhibit a conductive or nonconductive behavior, considering the extreme cases, the former is manifested by an electron (e^-) flow from the negative-to-positive terminals and, conventionally, by a current flow in the opposite direction. The latter, on the other hand, refers to an insulator.

Intrinsic (pure) and extrinsic (doped with impurity atoms) semiconductors show an intermediate electrical behavior. For these reasons, this chapter is devoted to the fundamentals of electrical conduction, magnetization, and optical physical properties of materials (or physical behavior of solids).

The effect of mechanical deformation on the electrical conduction is an interesting topic to study, since dislocations as well as impurities act as barriers to continuous and smooth conduction. As a result, the material atomic structure dictates the electrical and magnetic properties of conductive materials. The use of simple mathematical equations and graphs is the most effective manner to present the electrical conduction and magnetic susceptibility of materials.

By definition, the general description of electricity is the movement of electrons in the conduction energy band (electron energy states), while magnetism originates from the atoms acting as tiny permanent magnets with magnetic domains related to electron spin motion around discrete orbitals. This implies that the packed arrangement of atoms and the balanced interatomic forces dictate the properties of solids.

N. Perez, *Materials Science: Theory and Engineering*,
https://doi.org/10.1007/978-3-031-57152-7_14

14.2   Electrical Properties

14.2.1   Electrical Conduction

The atomic arrangement (FCC, BCC, HCP, etc.) and type of atomic bonding in solid metals and nonmetals have significant effects on electrical conduction. For metals, this is due to free valence electrons (e^-) in the electronic configuration outermost shell that form an electron cloud as schematically shown in Fig. 14.1, where the blue circles with the $+$ sign represent the positively charged ion cores.

If an electrical conductive material is exposed to an electric field, then the electrical conduction is said to be due to electrically charged particles in motion at a drift velocity v. The movement of free electrons in the outermost shell is known as the electric current I. For instance, Fig. 14.2 illustrates electrons around the nucleus of beryllium (Be).

Initially, the Be is an atom with filled $1s^2$ and $2s^2$ shells and an empty conduction shell (Fig. 14.2a). After an electrical excitation with sufficient energy, one valence electron jumps to the next highest energy level in the conduction shell or band (Fig. 14.2b). Consequently, the electron becomes a free particle under the influence of an electric field strength. As a result, Be acquires a relatively high electrical conductivity.

Fig. 14.1 Metallic bonding and free valence electron cloud

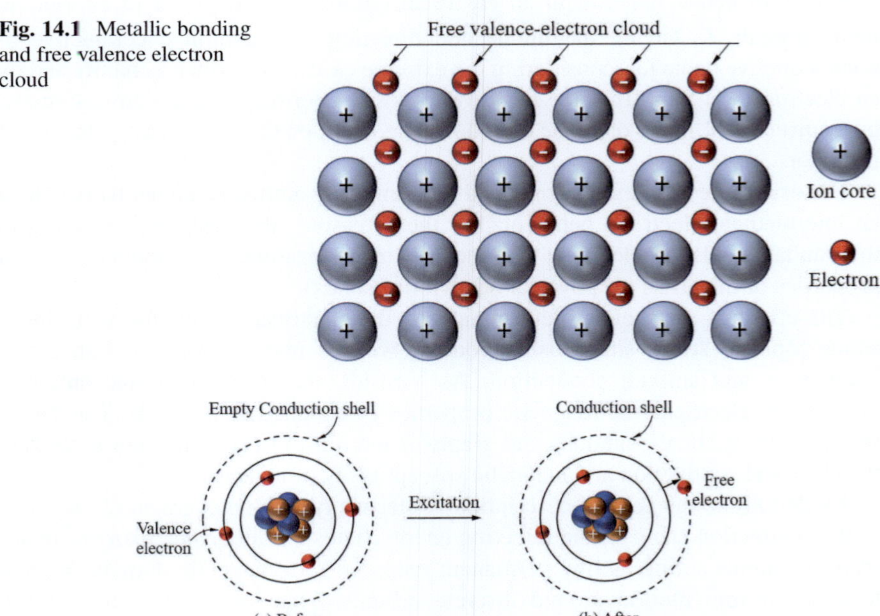

Fig. 14.2 Valence and free electrons in beryllium Be. (**a**) Before and (**b**) after an electrical excitation

Electrical conduction induced by an applied electrical potential (bias voltage) is a measure of the movement of negatively charged free electrons (e^-) at a particular drift velocity (v) within the domain of an electrical field at a temperature T in a particular direction. Reasonably, free electrons move to adjacent atoms and momentarily position themselves at different energy states during electrical conduction. Hence, the electric field accelerates the free electrons, which are those with energies greater than the Fermi level (E_F). The Fermi energy is the maximum kinetic energy an electron can attain at $T = 0K$ and the Fermi level (chemical potential of electrons) is the energy at $T > 0\,K$.

For a nonmetal such as silicon (Si), it is a covalently bonded crystal (metalloid) with shared electron pairs, and consequently, it does not have free electrons (valence electrons) available in the lattice to contribute to electrical conduction, making it a poor electrical conductor. At least one free valence electron in the outermost shell enhances the electrical conduction (Schaffer et al. [1, p. 427]). Moreover, silicon is an important metalloid in the computer-chip industry due to its semiconductor characteristic.

14.2.2   Ohm's Law

The macroscopic aspects of the electrical conduction can be characterized by using Ohm's law, which states that the current (I) through a conductor is directly proportional to the electrical potential or voltage (V) across two points on a specimen of length L as schematically illustrated in Fig. 14.3 (see Fig. 5.6). Notice that Fig. 14.3 contains the macroscopic engineering and microscopic quantum mechanics aspects of the electrical-conduction theory.

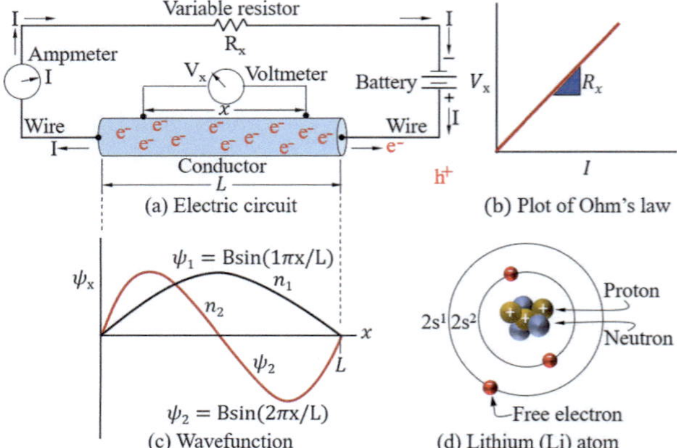

Fig. 14.3 Schematic (**a**) electrical circuit for measuring the electrical resistivity ρ, (**b**) plot of Ohm's equation $V_x = IR$, (**c**) wavefunctions ψ_x and (**d**) free-electron in a lithium (Li) atom

Electrical Engineering The electrical circuit in Fig. 14.3a is a simple macroscale setup for measuring the geometry-independent electrical resistivity ρ of a hypothetical specimen (wire or rod). Once the specimen is connected to a battery, an electrical potential difference (V_x) is imposed on the specimen. Consequently, free electron (e^-) flow occurs toward the positive terminal and current (I) flow related to an electric field strength occurs in the opposite direction. Also, free electrons flow from lower to higher potential level and current flows in the opposite direction in an electric field. The common graphical representation of the Ohm's law $V_x = f(I)$ is schematically shown in Fig. 14.3b, where the electrical resistance is defined as the slope of the straight line, $R_x = dV_x/dI$.

For one-dimensional analysis, this physical event is due to the electric field intensity (E_x) and the electric force (F) acting on free electrons. Mathematically, the geometry-dependent E_x and F are defined as

$$E_x = \frac{V_x}{L} \tag{14.1a}$$

$$F = q_e E_x = ma \tag{14.1b}$$

where V_x is the voltage across the specimen length L, $q_e = 1.602 \times 10^{-19}\ C\ (= A.s)$ is the electronic charge, m is the electron mass, a is the electron acceleration, and E_x is the potential gradient or voltage gradient. Note that Eq. (14.1b) is the classical Newton's second law.

Quantum Mechanics The microscopic scenario for free electron flow can be described by the first and second wavefunctions, $\psi_1\ \psi_2$, depicted in Fig. 14.3c. For one-dimensional analysis, the general wavefunction related to the quantum number k can be written as [see Eq. (3.1E5)]

$$\psi_n = B_n \sin\left(\frac{n\pi x}{L}\right) = B_n \sin(kx) \tag{14.2a}$$

$$k = \frac{n\pi}{L} \tag{14.2b}$$

Here, $n = 1, 2, 3, \ldots$ denotes an integer number, and Eq. (14.2) is part of the free-electron theory that describes the electrical conduction due to the motion of a free electron classified as a valence electron (Fig. 14.3d) within the quantum mechanical domain. The concept of free electron is illustrated in Fig. 14.3d for lithium (Li) due to its simple electronic configuration.

14.2.3   Classical Mathematical Models

For one-dimensional analysis, the macroscopic electric field intensity E_x acting onto an electrical conductor induces an electrical conduction due to the electron drift velocity υ_x and mobility μ_x in a medium at a specific temperature T. According to the macroscopic form of the Ohm's law (classical macroscale approach), the geometry-dependent potential difference V_x (V = Volt) along the specimen length L and the fundamental properties [the electrical resistivity ρ_x ($ohm.m = \Omega.m$) and the electrical conductivity σ] are written as a set of equations

$$V_x = IR_x = \frac{dQ}{dt} R_x \tag{14.3a}$$

$$I = \frac{V_x}{R_x} = \frac{dQ}{dt} \quad \& \quad R_x = \left(\frac{L}{A_c}\right) \rho \tag{14.3b}$$

$$\rho = \left(\frac{A_c}{L}\right) R_x = \frac{A_c}{L} \frac{V_x}{I} \tag{14.3c}$$

$$\sigma = \frac{1}{\rho} = n_e q_e \mu_x \quad \text{(metallic)} \tag{14.3d}$$

$$\sigma = \frac{1}{\rho} = n_e q_i \mu_x \quad \text{(ionic)} \tag{14.3e}$$

$$\upsilon = E_x \mu_x \quad \text{(drift velocity)} \tag{14.3f}$$

where dQ denotes the amount of charge that crosses a plane in a time interval dt, A_c denotes the cross-sectional area of the specimen, R_x denotes the specimen electrical resistance (Ω), n_e denotes the electron density (e^-/m^3), $q_e = 1.602 \times 10^{-19}\ C/e^-$ denotes the electron charge in metallic materials with C = Coulombs unit, $q_i = q_e Z$ denotes the electron charge in ionic materials, μ_x denotes the mobility of charge carriers ($m^2/V.s$), and E_x denotes the electric field gradient (V/m). Moreover, V_x in Eq. (14.13a) is analogous to pressure.

For convenience, Table 14.1 lists the electrical conductivity of some selected metals, alloys, and nonmetals in a descending order (Schaffer et al. [1, p. 430]). Moreover, the electrical conductivity is a physical property used to qualitatively compare electrical conductors, and it has units of siemens per meter (S/m) or inverse ohms-meter ($\Omega.m$)$^{-1}$, where $1\ S/m = 1\ (\Omega.m)^{-1}$.

Notice that $\sigma_{Ag} \gg \sigma_{Si}$, making Si a poor electrical conductor because it does not have free valence electrons. Consequently, silicon is classified as a semiconductor, and normally it is doped with other elements for enhancing its

Table 14.1 Average electrical conductivity of some solid materials (Schaffer et al. [1, p. 430])

Metals	$\sigma\ (\Omega.m)^{-1}$	Nonmetals	$\sigma\ (\Omega.m)^{-1}$
Ag	6.3×10^7	CrO_3	3.3×10^6
Cu	6.0×10^7	SiC	1.0×10^1
Au	4.3×10^7	Ge	2.3×10^0
Al	3.8×10^7	Si	1.0×10^{-2}
Li	1.1×10^7	Acetal	5.0×10^{-13}
1020 Steel	1.0×10^7	Polycarbonate	1.0×10^{-15}

electrical conduction characteristic and makes it useful in the electronic industry. Silver has one free valence electron in the d-orbital and possesses a high electrical conductivity, since it is less resistant to current flow than silicon.

One-dimensional analysis, the electron drift velocity (υ_x) described in Chap. 5, is related to the electron mobility μ_x and current density i_x. Thus,

$$\upsilon_x = \mu_x E_x \tag{14.4a}$$

$$\upsilon_x = \frac{i_x}{n_e Q} \tag{14.4b}$$

where υ_x is the velocity component of the random motion of an electron and Q is the electrical charge on the carriers, such as electrons. Moreover, the magnitude of the electric field vector is $\left|\overrightarrow{E}\right| = \sqrt{E_x^2 + E_y^2 + E_z^2} = E_x$ is the electric field intensity along the x-direction with volt/length (V/L) units.

From Chap. 4, Eq. (1.32b), the atom density per volume (n_a) and the number of free electrons per volume (n_e), can be defined by

$$n_a = \frac{\rho N_a}{A_w} \tag{14.5a}$$

$$n_e = \frac{\rho_e}{n_a} \tag{14.5b}$$

$$\rho_e = \frac{n_e \rho N_a}{A_w} \tag{14.5c}$$

where $N_a = 6.022 \times 10^{23}$ $particle/mol$ denotes the Avogadro's number and ρ_e denotes the electron density

Example 14.1 Consider an aluminum (Al) rod with $L = 1\text{-}m$, $A_c = 1\ mm^2$, $n_e = 1.82 \times 10^{29}\ e^-/m^3$, and a current of $I = 5\ A$. Calculate ρ, R_x, μ_x, V_x, and N_e (the number of mobile electrons per atom). Given data for aluminum: electrical conductivity $\sigma = 3.8 \times 10^7\ (\Omega.m)^{-1}$, mass density $\rho_m = 2.71\ g/cm^3$, and atomic weight $A_w = 26.98\ g/mol$.

Solution First of all, $\Omega = Ohm$, $\Omega = V/A$, $C = Coulomb = A.s$ with $V = Volt$ and $A = Ampre = Amp$, and $\Omega.C = V.s$. Using Eq. (14.3d) and Table 14.1, Eq. (14.3c) yields the electric resistivity and the electric resistance of the aluminum rod, respectively,

$$\rho = \frac{1}{\sigma} = \frac{1}{3.8 \times 10^7\ (\Omega.m)^{-1}} \doteq 2.63 \times 10^{-8}\ (\Omega.m) \tag{14.1E1a}$$

$$R_x = \left(\frac{L}{A_c}\right)\rho = \left(\frac{1\ m}{1 \times 10^{-6}\ m^2}\right)\left(2.63 \times 10^{-8}\ \Omega.m\right) \tag{14.1E1b}$$

$$R_x = 0.0263\ \Omega = 0.0263\ V/A \tag{14.1E1c}$$

From Eq. (14.3e),

$$\mu_x = \frac{\sigma}{n_e q_e} = \frac{3.8 \times 10^7 \ (\Omega.m)^{-1}}{\left(1.82 \times 10^{29} \ e^-/m^3\right)\left(1.602 \times 10^{-19} \ C/e^-\right)} \tag{14.1E2a}$$

$$\mu_x = 1.30 \times 10^{-3} \ m^2/(\Omega.C) = 1.30 \times 10^{-3} \ m^2/(V.s) \tag{14.1E2b}$$

From Eq. (14.3a), the electrical potential or voltage required to produce a current of 5 amps (5 A) is

$$V_x = I R_x = (5 \ A)(0.0263 \ V/A) = 0.13 \ V \tag{14.1E3}$$

From Eq. (1.29c), the atomic density for *FCC* aluminum with atomic radius of $R = 0.143 \ nm$ is

$$\rho_v = \frac{N_c}{V_c} = \frac{4 \ atoms}{a^3} = \frac{4 \ atoms}{\left(4R/\sqrt{2}\right)^3} \tag{14.1E4a}$$

$$\rho_v = \frac{4 \ atoms}{\left[\left(4 \times 0.143 \times 10^{-9} \ m\right)/\sqrt{2}\right]^3} = 6.05 \times 10^{28} \ atoms/m^3 \tag{14.1E4b}$$

Using the modified mass density ρ_m expression yields the atomic density for *FCC* aluminum

$$\rho_m = \left(\frac{A_w}{N_a}\right)\rho_v \tag{14.1E5a}$$

$$\rho_v = \frac{N_a \rho_m}{A_w} = \frac{\left(6.023 \times 10^{23} \ atoms/mol\right)\left(2.71 \ g/cm^3\right)\left(10^6 \ cm^3/m^3\right)}{26.98 \ g/mol} \tag{14.1E5b}$$

$$\rho_v = 6.05 \times 10^{28} \ atoms/m^3 \tag{14.1E5c}$$

Thus, the number of free electrons per aluminum atom is

$$N_e = \frac{n_e}{\rho_v} = \frac{1.82 \times 10^{29} \ e^-/m^3}{6.05 \times 10^{28} \ atoms/m^3} \simeq 3 \ e^-/atom \tag{14.1E6}$$

which is equal to the valence electrons of aluminum. For clarity, the number of free electrons in an electrical conductor depends on the types of atoms. These electrons are loosely bound in the space between atoms.

Ohm's law can be derived by combining Eqs. (14.1a), (14.3c), and (14.3d). The resultant expression defines Ohm's law in terms of electric current density J_x (in A/m^2 units), electrical conductivity σ (in $A/m.V$ units), and electric field intensity E_x (in V/m units). Thus,

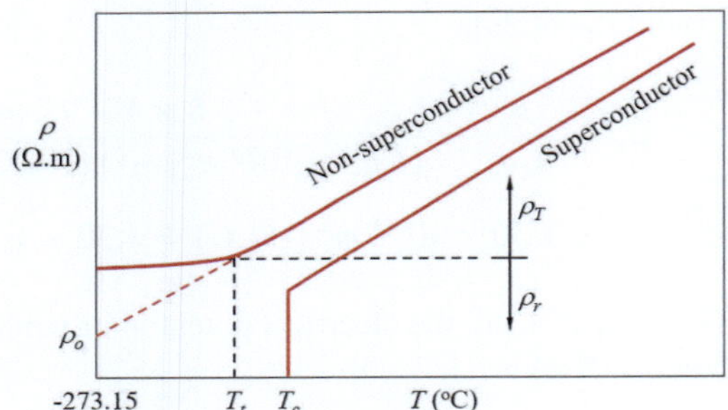

Fig. 14.4 Ideal electrical resistivity profile for (**a**) non-superconductor and (**b**) superconductor

$$J_x = \frac{I}{A_c} = \frac{V_x}{R_x}\frac{1}{A_c}\frac{L}{L} = \frac{V_x}{L}\frac{1}{\rho} \tag{14.6a}$$

$$J_x = \frac{E_x}{\rho} = \sigma E_x \tag{14.6b}$$

Here, the electric field potential E_x represents the voltage gradient across the L-spacing in Fig. 14.3a. Nonetheless, J_x is also an electron flux term which has a positive magnitude similar to the electric current; $J_x, I > 0$.

The current density can be generalized in vector notation

$$\mathbf{J} = \frac{\mathbf{E}}{\rho} = \sigma\mathbf{E} > \mathbf{0} \tag{14.7}$$

since $\mathbf{J}$ and $\mathbf{E}$ have magnitude and direction, while the electrical resistivity $\rho = 1/\sigma$ being inversely proportional to the electrical conductivity $\sigma = 1/\rho$ acts as multipliers or constants at a particular temperature.

Most non-superconductors may exhibit constant electrical resistivity ρ at $T < T_t$, where T_t is the transition temperature. For these types of materials, the electrical resistivity ρ_o at $T = 0\,K = -273.15°C$ is determined by extrapolation as indicated in Fig. 14.4. Conversely, superconductors conduct electricity without resistance, so that the electrical resistivity $\rho \rightarrow 0$ at a very low temperature $T \leq T_c \rightarrow 0\,K = -273.15°C$ with T_c being the critical temperature. For example, titanium (Ti) and beryllium (Be) becomes superconductors when $T \rightarrow 0\,K = -273.15°C$.

The electrical resistivity ρ and the electrical conductivity σ of a specimen, as used in the above equations, represent the total quantities for a pure electrical conductor. However, if a specimen contains impurities and secondary components (elements), then ρ and σ at a particular temperature T become

$$\rho = \rho_T + \rho_r \tag{14.8a}$$

$$\sigma = \sum_{k}^{m} \sigma_k \tag{14.8b}$$

where $\rho_T > 0$ is the temperature-dependent term related to positive-ion core vibration and $\rho_r \geq 0$ is the impurity-dependent term in the crystal lattice. Moreover, σ_k with $k = 1, 2, 3, ..$ represents the electrical conductivity of each component in the specimen.

Nonetheless, Fig. 14.4 schematically shows ideal temperature-dependent electrical resistivity ρ profiles for non-superconductor and superconductor solids.

14.2.4 Energy Band Theory of Solids

The energy band theory of solids is the key to understanding electrical properties of semiconductor materials, since it initially describes the electrons having discrete energy levels in individual atoms. In this case, the solution of the Schrodinger's equation provides the corresponding wavefunction ψ_n that describes the motion of an electron at a discrete energy level, where $n = 1, 2, 3, \ldots$ is the quantum number. However, as atoms are brought together to form a solid, the interatomic spacing r and lattice constants a, b, c are reduced, and the wavefunction ψ_n for each electron in the outermost orbitals overlap. As a result, the discrete energy levels split or simply curve up or down near a junction x into continuous quantum states, leading to energy bands in a crystal. In the literature, this phenomenon is referred to as energy band bending.

14.2.5 Electronic Configuration

Consider the electronic configuration of atoms in a gas. In this case, the atoms can be treated as isolated atoms, since they are separated by a large average distance r. For example, Fig. 14.5a shows an isolated sodium (Na) atom with its corresponding discrete energy levels.

In essence, an isolated Na-pair can be treated as a molecule-like (Fig. 14.5b), so that each discrete energy level splits into two levels.

Fig. 14.5 Energy levels of electrons in sodium Na. (**a**) discrete energy levels in one atom (gas), (**b**) discrete energy levels in two atoms (molecule), and (**c**) continuous energy levels within an energy band forming a crystal

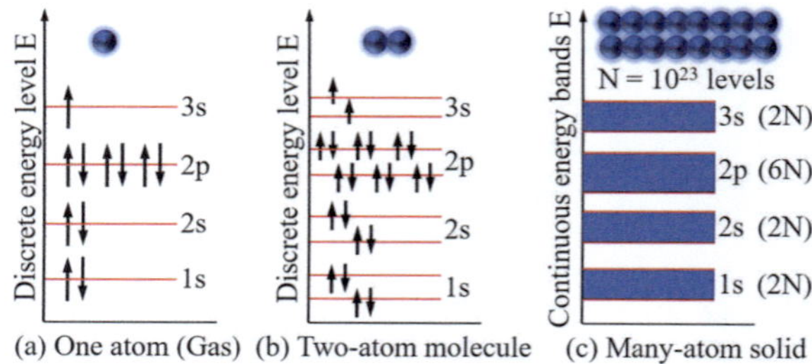

14.2.6 Energy Bands

For an N-atom system forming a solid, each initial energy level splits into $N = 10^{23}$ continuous-like energy levels, which are treated as energy bands. This is shown in Fig. 14.5c for sodium. Here, the number of electrons used to determine the number of states in an energy band is deduced from the exponent of the Avogadro's number $N_a = 6.022 \times 10^{23} \, electrons/mol$ for one mole of sodium.

According to the energy band theory of solids, many Na atoms form a crystal structure with continuous energy bands (Fig. 14.5c) colored in blue with the lower and upper energy band limits colored in red. Moreover, Fig. 14.5c also shows the number of energy states in parenthesis for each energy band. The reader is encouraged to visit the link for additional insights on this subject matter (https://www.youtube.com/watch?v=3npADYVtQOM&t=1s).

14.2.7 Energy Band Diagrams

The energy band concept is mathematically denoted as an energy $E = f(r)$ function related to many wavefunctions ψ_n and graphically represented as an energy-band diagram schematically in Fig. 14.6a for lithium (Li) and in Fig. 14.6b for silicon (Si) (Park [2]).

For the purpose of clarity, $E = f(r)$ function produces an electron energy-band diagram with r being the position in space, whereas the $E = f(k)$ function produces an electron energy-band structure plot with k being the magnitude of the wavevector $\vec{k}$ of an electron described or defined by Eq. (3.2a).

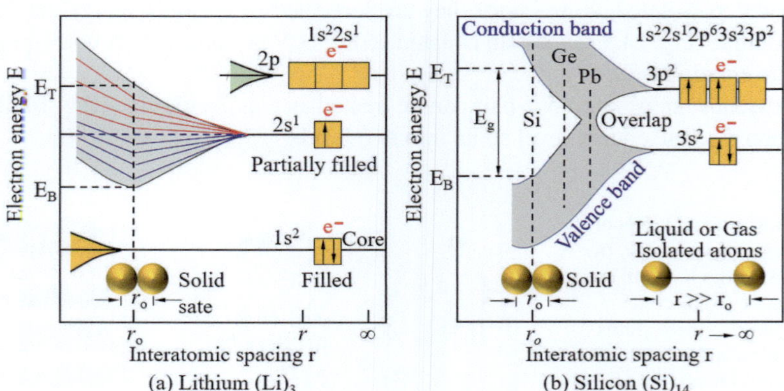

Fig. 14.6 Electron energy-band diagrams for (**a**) lithium (Li) and (**b**) silicon (Si). See video posted by Park [2]

This energy band diagram, specifically, presents the splitting energy levels in a semiconductor due to a set of wavefunctions ψ_n, which can be defined by the Schrodinger's equation, describing the electron motion and setting the energy band in a continuous form (Park [2], Shockley [3, p. 133]). Specifically, the upper and lower wavefunctions define the energy bandwidth.

Figure 14.6a schematically shows the energy bands for lithium (Li), where r_o is the equilibrium interatomic separation or interatomic spacing between two atom in the solid state. Two valence electrons reside in the valence band (filled core $1s^2$ K-shell) and one electron in the conduction band (partly filled $2s^1$ L-shell), which has certain mobility to induce electric current. Notice that the electron energy for the $2s^1$ L-shell splits as atoms get closer together to reach the equilibrium interatomic spacing r_o in the solid state. Splitting of the energy levels is consistent with the Pauli exclusion principle, which states that two electrons cannot occupy the same quantum state simultaneously. Similarly, Fig. 14.6b illustrates the $3s$ and $3p$ energy bands in silicon (Si) with a conduction-valence energy band overlap as atoms are brought together.

The primary characteristics underlying band splitting or bending near a junction x are given below as theoretical guidelines in order to interpret an energy band diagram. Thus,

- The conduction band lower edge, denoted as E_c, is used to characterize the electrons in n-type semiconductors, and the valence band upper edge, denoted as E_v, is used to analyze the electrons or holes in p-type semiconductors.
- There is a forbidden energy bandgap (E_g) between the electron-valence and the electron-conduction bands at the equilibrium interatomic spacing r_o.
- Splitting of the energy levels produces $2N$ states in the $2s$ band and $6N$ states in the $2p$ band, where N is the number of atoms in the solid.
- For comparison purposes, Fig. 14.6b shows the position for semiconductor germanium (Ge) with $E_{g,Ge} < E_{g,Si}$ and metal lead (Pb) at the cross-over point with $E_g = 0$, making lead a poor conductor and a corrosion-resistant metal. Actually, Pb is a poor conductor attributed to the $6p^2$ electrons in the electron configuration being unable to jump to the conduction energy band because Pb reacts with oxygen ($\frac{1}{2}O_2$) to form lead oxide (PbO).
- Below are some characteristics of Li, Si, Ge, and Pb with respect to the ascending order of the number of electrons and their number of valence electrons for assessing their conductance. The electron configurations of interest in this section are

$$\text{Lithium } (Li),\ 3\ e^-:\ 1s^2 2s^1 \rightarrow 1 \text{ valence } e^-$$

$$\text{Carbon}(C),\ 6\ e^-:\ [He]^2\ 2s^2 2p^2 \rightarrow 4 \text{ valence } e^-$$

$$\text{Silicon } (Si),\ 14\ e^-:\ [Ne]^{10}\ 3s^2 3p^2 \rightarrow 4 \text{ valence } e^-$$

$$\text{Germanium } (Ge),\ 32\ e^-:\ [Ar]^{18}\ 3d^{10} 4s^2 4p^2 \rightarrow 4 \text{ valence } e^-$$

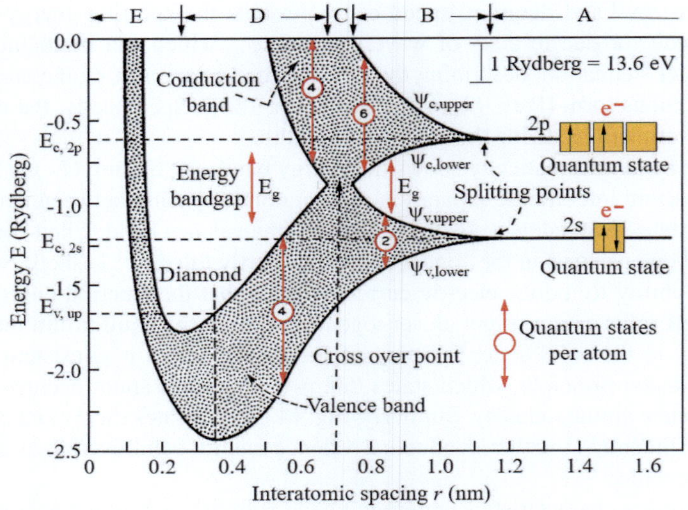

Fig. 14.7 Energy band diagram for diamond. The numbers in the circles represent the number of electrons per atom in the respective energy state or energy band. Energy unit: 1 Rydberg unit $= 13.5\,eV$ (after Shockley [3, p. 133])

$$\text{Lead } (Pb)\; 82\; e^- : [Xe]^{54}\, 4f^{14}, 5d^{10}, 6s^2, 6p^2 \to 4 \text{ valence } e^-$$

These outermost electrons are responsible for the chemical binding and for the classification of engineering solids. Hence,

$$\text{Lithium } (Li),\; ..2s^1 \to 1 \text{ valence } e^- \to \text{ Conductor}$$

$$\text{Carbon}(C),\; ..2s^2 2p^2 \to 4 \text{ valence } e^- \to \text{ Poor conductor}$$

$$\text{Silicon } (Si),\; ..3s^2 3p^2 \to 4 \text{ valence } e^- \to \text{ Semiconductor}$$

$$\text{Germanium } (Ge),\; ..4s^2 4p^2 \to 4 \text{ valence } e^- \to \text{ Semiconductor}$$

$$\text{Lead } (Pb)\; ..6s^2 6p^2 \to 4 \text{ valence } e^- \to \text{ Poor conductor}$$

- Both Si and Ge have their respective energy bandgap $E_g > 0$ and Pb has $E_g = 0$. Theoretically, $E_g = 0$ implies that there is a valence-conduction energy band overlap.

Figure 14.7 exhibits the electron energy band diagram showing the position of the interatomic spacing $r = 0.35\ nm$ for carbon (C) with the diamond structure. This diagram notice that the energy band (in Rydberg energy, $1\ Ry = 13.6\ eV$) for carbon arises from the quantum states occupied by the valence electrons (Shockley [3, p. 133]).

Analysis of Fig. 14.7

- Region A is for isolated atoms with electrons having discrete quantum states. This region ends at the splitting points.
- Region B is for the formation of solids when atoms are near to each other. As a result, electron interactions from different atoms cause the energy of some electrons in the outermost orbital increases their energy level forming the conduction band. Some decrease their energy level forming the valence band. Hence, the conduction and valence energy bands are separated by an energy gap E_g.
- Region C represents a narrow energy band overlap, which is suitable for conductors having closer atoms with a smaller lattice constant in their unit cells. Hence, there exists a mixed energy state.
- Region D represents solids with smaller lattice constant. Thus, the conduction and valence bands are separated by an energy bandgap E_g. This is the region suitable for semiconductors such a silicon and germanium. However, if the energy bandgap E_g is too large, then the solid becomes an insulator, such as diamond.
- Region E represents an increase in the electron energy states.
- For carbon atoms with $r_o = 0.35\ nm$, the electrons in the valance energy band have a lower energy level than the discrete $2s$-energy level; $E_{v,up} < E_{2s}$ and $E_{v,up} < E_{2p}$. This energy drop is attributed to the binding energy of the carbon atoms forming a stable crystal structure. According to Shockley [3, p. 154], the energy bandgap for carbon (diamond) is in the range of $6\ eV \leq E_g \leq 7\ eV$ and that for silicon is $E_g = 1.11\ eV$; therefore diamond is treated as an insulator, since four electrons in the outer shell per atom are used in the covalent bonding, since there are no delocalized electrons.

Simplified Energy Band Model For a practical purpose, the energy bands are illustrated in Fig. 14.8 in a simplified form. Each energy band is an approximation of the quantum state of a conductor, semiconductor, or insulator, and it represents a range of energy for the electronic configuration of a solid. Hence, the energy band is a group of electron energy levels in the outermost orbital of atoms.

Conventional specimens subjected to an electrical potential intensity in an electric circuit under thermal equilibrium can be classified as follows:

Fig. 14.8 Schematic bands of energy states for conductor, semiconductor, and insulator solids

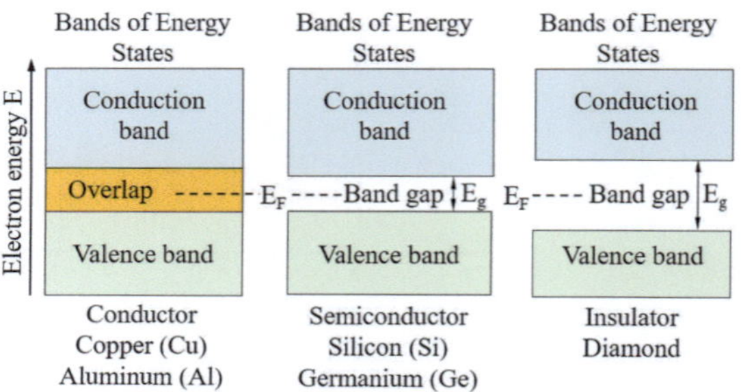

- In electrical conductors, such as copper (Cu) and silver (Ag), there is an overlap between the valence and conduction energy bands. This leads to a motion of the free electrons in the conduction band.
- In semiconductors, such as silicon (Si) and germanium (Ge), there is a small energy bandgap (E_g) between the conduction and valence bands. Some valence electrons can jump to the conduction band to become free electrons responsible for the electric current or electron current. In practice, a semiconductor energy bandgap can be reduced by a high doping density (ρ_d) of an appropriate metal, such as boron (B) into silicon.
- In insulators, there is a large energy bandgap, and consequently, electrons do not jump to the conduction band.

Another common graphical representation of the energy bandgap is illustrated in the next example using $E = f(k)$ function introduced in Chap. 3 as Eq. (3.22a).

Example 14.2 Consider plotting Eq. (3.22a) for an electron in free space and plot two wavefunctions and Eq. (3.22a) for an electron in a semiconductor crystal using superposition. **(a)** Plot $E = k^2$ for an electron in free space and show the first Brillouin zone (BZ). Derive **(b)** the theoretical electron velocity (v) and **(c)** the electron effective mass (m^*) equations.

Solution

(a) From Chap. 3, the electron energy E and the electron momentum p equations are

$$E = \frac{p^2}{2m} = \left(\frac{\hbar^2}{2m} \right) k^2 \quad \text{(in space)} \tag{14.2E1a}$$

$$E = \frac{p^2}{2m} = \left(\frac{\hbar^2}{2m^*} \right) k^2 \quad \text{(in a crystal)} \tag{14.2E1b}$$

$$p = \hbar k \tag{14.2E1c}$$

Let $\hbar^2/(2m) = 1$ and plot $E = k^2$ for an electron in free space. The resultant graph is a parabola shown in Fig. (a).

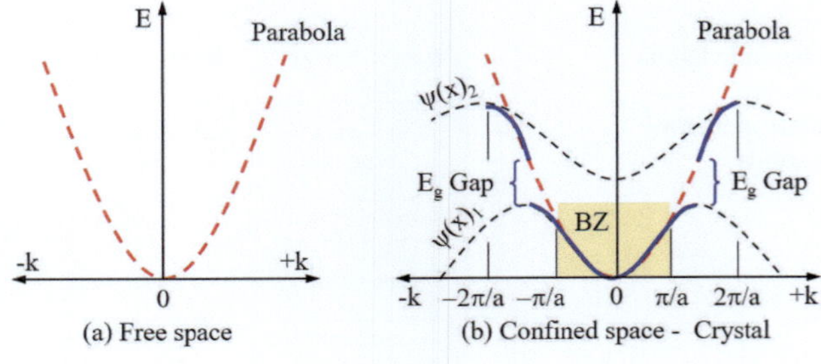

However, an electron in a semiconductor crystal with a lattice constant "a" is in a confined space, and consequently, the parabola shown in Fig. (b) in red color is discontinued (blue curves) initially at a distance $k = \pi/a$. This generates an energy bandgap E_g. The distance between $-\pi/a$ and $+\pi/a$ defines the first Brillouin zone (BZ), which corresponds to the Wigner-Seitz primitive cell in the reciprocal lattice.

(b) The electron velocity in a crystal can be predicted from Eq. (14.2E1b) as

$$v = \frac{dE}{dk} = \frac{\hbar^2 k}{m} \quad \text{(free space)} \tag{14.2E2a}$$

$$v = \frac{dE}{dk} = \frac{\hbar^2 k}{m^*} \quad \text{(in a crystal)} \tag{14.2E2b}$$

$$m^* < m \tag{14.2E2c}$$

By deduction, the effective electron mass $m^* < m$ is due to the constraint imposed by the semiconductor crystal lattice. This is attributed to interactions of electrons. Conversely, an electron free mass refers to the rest mass ($m = 9.11 \times 10^{31}\ Kg$) without the action of forces or constraints.

(c) From Eq. (14.2E2), the effective mass of the electron is defined by

$$\frac{d^2 E}{dk^2} = \frac{\hbar^2}{m^*} \tag{14.2E3a}$$

$$m^* = \hbar^2 \left(\frac{d^2 E}{dk^2} \right)^{-1} \tag{14.2E3b}$$

Hence, m^* is inversely proportional to the energy curvature $d^2 E/dk^2$.

Temperature-Dependent Bandgap The temperature dependence of the energy bandgap can quantitatively be determined using the following empirical equation (Li [4, p. 115])

$$E_g = E_o - \frac{\alpha T^2}{\beta + T} \tag{14.9}$$

where E_o is the electron energy at $T = 0\ K$, α, β are curve fitting constants and T is the absolute temperature. Table 14.2 lists these constants for most frequently used semiconductors in the electronic industry, germanium (Ge) and silicon (Si) single crystals and compound gallium arsenide ($GaAs$) semiconductors.

Plotting Eq. (14.9) using the data listed in Table 14.2 yields Fig. 14.9 as the forbidden energy bandgap diagram. This implies that the electron transfer from the valence to a conduction band produces "holes," and it is the main source of measurable electric current.

Table 14.2 Curve-fitting parameters (Li [4, p. 116])

Parameters/Material	Ge	Si	GaAs
E_o (eV)	0.744	1.170	1.519
α (m.V/K)	4.77×10^{-4}	4.73×10^{-4}	5.41×10^{-4}
β (K)	235	636	204

Fig. 14.9 Temperature dependence of the electron energy bandgap for most common semiconductors

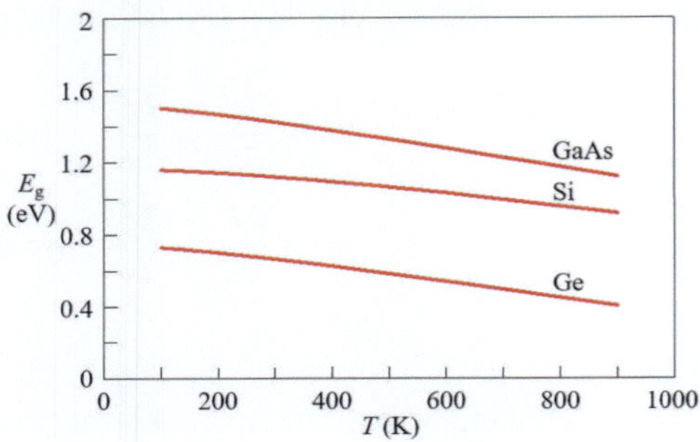

It is clearly evident in Fig. 14.9 that $E_g = f(T)$ decreases almost linearly with increasing temperature T. Mathematically, $E_g \rightarrow E_o$ as $T \rightarrow T_m$ due to crystal expansion and interatomic bond weakening. Here, T_m denotes the melting temperature of the solid.

Electron Density of States The electron and hole concentrations (N_e and N_h, respectively) in semiconductors (intrinsic or extrinsic) strongly depend on the position of the Fermi level at $T > 0\ K$. This position is located at an energy level between the valence and conduction band edges as shown in Fig. 14.8. The goal now is to derive N_e and N_h using the Fermi-Dirac function (also known as the Fermi-Dirac probability function or Fermi-Dirac distribution), which provides the distribution of elementary particles or carriers over energy states (energy levels) under thermal equilibrium.

Electrons Consider the electron density of state $\rho(E)$ function in the conduction band in the energy interval dE and the statistical Fermi-Dirac distribution $f(E)$ function defining the probability of occupation of energy levels by electrons. For an infinitesimal electron energy change dE (Kasap [5, pp. 308, 316], Pillai and Pillai [6, p. 104]), the concentration of electrons (N_e) is

$$N(E) = \rho(E) f(E) \tag{14.10a}$$

$$N(E)\, dE = \rho(E) f(E)\, dE \tag{14.10b}$$

$$\int_0^{N_c} N(E)\, dE = \int_{E_c}^{\infty} \rho(E) f(E)\, dE \tag{14.10c}$$

$$N_e = \int_{E_c}^{\infty} \rho(E) f(E)\, dE \tag{14.10d}$$

The density of states of electrons $\rho(E)\,dE$ along with $E \to E - E_F$ in Eq. (3.1f) and the Pauli's exclusion principle is written as (Pillai and Pillai [6, p. 104])

$$\rho(E)\,dE = 2\left(\frac{\pi}{2}n^2 dn\right) = \frac{\pi}{2}\left(\frac{8m_e L^2}{h^2}\right)^{3/2}(E - E_F)^{1/2}\,dE \qquad (14.11)$$

and the relevant Fermi-Dirac function which describes of carriers, such as electrons and holes, is of the form

$$f(E) = \frac{1}{1 + \exp\left[(E - E_F)/k_B T\right]} = \left[1 + \exp\left(\frac{E - E_F}{k_B T}\right)\right]^{-1} \qquad (14.12a)$$

$$f(E) \simeq \exp\left(-\frac{E - E_F}{k_B T}\right) \qquad (14.12b)$$

since $(E - E_F) >> k_B T$ and $(E - E_F)/k_B T >> 1$. In the literature, Eq. (14.12b) is known as the Maxwell-Boltzmann distribution or Boltzmann statistics. Moreover, in quantum mechanics, the probability that an energy state is occupied by an electron is determined by $f(E)$ in Eq. (14.12a) and that for an unoccupied energy state is determined by $1 - f(E)$.

The number of electrons per unit volume or the electron density n_e at the bottom (edge) of the conduction energy band is (Pillai and Pillai [6, pp. 230-231])

$$N_e = \frac{\pi}{2}\left(\frac{8m_e L^2}{h^2}\right)^{3/2}\int_E^\infty (E - E_c)^{1/2}\exp\left(-\frac{E - E_F}{k_B T}\right)dE \qquad (14.13a)$$

$$N_e = 2\left(\frac{2\pi m_e L^2 k_B T}{h^2}\right)^{3/2}\exp\left(-\frac{E_c - E_F}{k_B T}\right) \qquad (14.13b)$$

$$n_e = \frac{N_e}{L^3} = \frac{N_e}{V} = 2\left(\frac{2\pi m_e k_B T}{h^2}\right)^{3/2}\exp\left(\frac{E_F - E_c}{k_B T}\right) \qquad (14.13c)$$

Recall from Chap. 3 that an electron cell volume is defined as $V = L^3 = L_x L_y L_z$ (cube). This means that the vector $\vec{L}$ in a cube has a direction from $(0, 0, 0)$ to an arbitrary point (L_x, L_y, L_z) in the L-space. Similarly, for a spherical volume $(4/3)\pi n^3$ of radius n, the n-space has a vector $\vec{n}$ and n_x, n_y, n_z coordinates at an arbitrary point. Nonetheless, Eq. (14.13c) is a generalized expression for determining N_e using a cubic or a spherical volume element V. For convenience, Eq. (14.13c) is simplified to an Arrhenius-type equation of the form

$$n_e = N_c \exp\left(\frac{E_F - E_c}{k_B T}\right) \qquad (14.14a)$$

$$N_c = 2 \left(\frac{2\pi m_e k_B T}{h^2} \right)^{3/2} \tag{14.14b}$$

where N_c denotes the number of electrons at bottom of the conduction band, and it is a temperature-dependent constant called the effective density of states at the conduction band edge.

Holes Similarly, the number of holes (N_h) at the top of the valence energy band E_v is induced by the allowable electrons jumping to the conduction energy band upon an appropriate excitation. Thus, the hole density is

$$n_h = N_v \exp \left(\frac{E_v - E_F}{k_B T} \right) \tag{14.15a}$$

$$N_v = 2 \left(\frac{2\pi m_p k_B T}{h^2} \right)^{3/2} \tag{14.15b}$$

Position of the Fermi Level For an intrinsic semiconductor, let $n_e = n_h$ so that

$$N_c \exp \left[\frac{E_F - E_c}{k_B T} \right] = N_v \exp \left[\frac{E_v - E_F}{k_B T} \right] \tag{14.16a}$$

$$\exp \left[\frac{2E_F - (E_c + E_v)}{2k_B T} \right] = \frac{N_v}{N_c} \tag{14.16b}$$

$$\frac{2E_F - (E_c + E_v)}{k_B T} = \ln \left(\frac{N_v}{N_c} \right) = \ln \left(\frac{m_p}{m_n} \right)^{3/2} \tag{14.16c}$$

from which the Fermi level E_F is written as

$$E_F = \frac{E_c + E_v}{2} + \frac{1}{2} k_B T \ln \left(\frac{N_v}{N_c} \right) = \frac{E_c + E_v}{2} + \frac{3}{4} k_B T \ln \left(\frac{m_p}{m_n} \right) \tag{14.17a}$$

$$E_F = \frac{E_c + E_v}{2} + \frac{1}{2} k_B T \ln \left(\frac{N_v}{N_c} \right) = E_i + \frac{3}{4} k_B T \ln \left(\frac{m_p}{m_n} \right) \tag{14.17b}$$

where E_i is known as the intrinsic Fermi level (Li [4, p. 114]). Evaluating Eq. (14.17) at absolute zero temperature ($T = 0\ K$) yields the Fermi level in the middle of the bandgap $E_g = E_c - E_v$. Thus,

$$E_F = \frac{E_c + E_v}{2} \qquad \text{for } T = 0\ K \tag{14.18a}$$

$$E_v + 3k_B T \leq E_F \leq E_v - 3k_B T \quad \text{for } T > 0\ K \tag{14.18b}$$

The Fermi-Dirac function is evaluated graphically, Eq. (14.12a), as illustrated in Fig. 14.10 for undoped germanium (Ge) with $E_F = 0.72 eV$ and for undoped silicon

Fig. 14.10 Plots of
Fermi-Dirac function at
thermal equilibrium

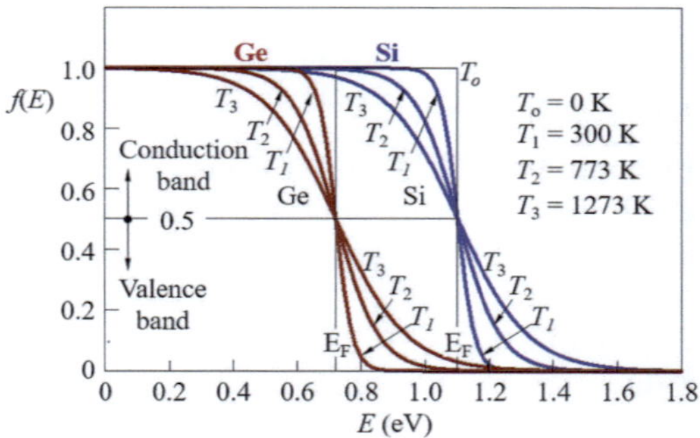

(Si) with $E_F = 1.1\ eV$ to reveal that $f(E) \to 1$ at $E >> E_F$, $f(E) = 0.5$ at
$E = E_F$, and $f(E) = 0$ at $E << E_F$ for all given temperature $T \geq 0\ K$.

These plots are given for four selected temperatures. Notice that the Fermi-Dirac
function is symmetric about the Fermi Level E_F, which represents the chemical
potential related to the probability of occupation of electrons at a particular energy
level E.

The limits of a Fermi-Dirac function are mathematically deduced as indicated
below:

- If $E << E_F$, then $(E - E_F) < 0$, $(E - E_F)/k_B T < 0$, and the upper limit
  of $f(E)$ is imposed by $\exp[(E - E_F)/k_B T] = \exp(-\infty) = 0$. Therefore,
  $f(E) = 1$.
- If $E = E_F$, then $(E - E_F) = 0$, $(E - E_F)/k_B T = 0$, and the midway limit of
  $f(E)$ is imposed by $\exp[(E - E_F)/k_B T] = \exp(0) = 1$. Therefore, $f(E) =$
  $1/2$.
- If $E >> E_F$, then $(E - E_F > 0$, and the lower limit of $f(E)$ is imposed by
  $\exp[(E - E_F)/k_B T] = \exp(\infty) = \infty$. Therefore, $f(E) = 0$.

The behavior of $f(E)$ function is that (1) $f(E) \to 1$ for $E < (E_F - 3k_B T)$ in
the valence energy band, (2) $f(E) \to 0$ for $E > (E_F - 3k_B T)$ in the conduction
energy band at $T > 0$. Moreover, Fig. 14.10 clearly reveals how the temperature
influences the behavior of the $f(E)$ function; $f(E)$ curve is displaced downward at
$E > E_F$ and upward at $E < E_F$ as T increases. From the this plot, $0 < f(E) < 1$
at $E \neq E_F$, but $f(E) = 1/2$ at $E = E_F$.

The analysis in bullet form can be taken as a simple Fermi-Dirac statistics based
on Eq. (14.12), which in turn predicts the probability to find a particle in the interval
$(E - E_F)$. However, the Fermi-Dirac function for statistical purposes indicates that
it reaches a maximum probability of one when $E << E_F$.

The Fermi energy E_F depends on the concentration of electrons and holes and
on the temperature T, since electrons can be exchanged between the valence and
conduction bands upon thermal excitation at a finite temperature T. For instance,
an electron transition related to the Fermi-Dirac function defined by Eq. (14.12)
indicates that the probability of finding the excited electron in the conduction and

valence bands can be set to $f_c(E) = 0$ and $f_v(E) = 1$, respectively. This means that the electron is energetically excited into a higher energy level or energy state at a finite temperature.

Example 14.3 Consider the energy level of silicon (Si) lying at $E - E_F = 1.10$ eV. Calculate **(a)** the probability of finding an electron at this energy level, **(b)** the number of electrons and **(c)** the electric conductivity σ_{Si} of an intrinsic silicon specimen at $T = 300\ K$. Data: The mobility of electrons is $\mu_x = 0.14 m^2/(\Omega.C) = 0.14\ m^2/(V.s)$.

Data

$m_e = 9.11 \times 10^{-31}\ Kg, \qquad k_B = 1.38 \times 10^{-23} J/K = 8.62 \times 10^{-5}\ eV/K$

$h = 6.63 \times 10^{-34}\ kg.m^2/s = 6.63 \times 10^{-34}\ J.s = 4.14 \times 10^{-15}\ eV.s$

Electron charge: $q_e = 1.602 \times 10^{-19}\ C$

Mobility of electrons: $\mu_x = 0.14\ m^2/(\Omega.C) = 0.14 m^2/(V.s)$

Given energy: $E - E_F = 1.10\ eV$

Thermal energy: $k_B T = (8.62 \times 10^{-5}\ eV/K)(300\ K) = 0.02586\ eV$

Solution

(a) From Eq. (14.12a), the probability of finding an electron at the energy range $E - E_F = 1.10\ eV$ is

$$f(E) = \left[1 + \exp\left(\frac{E - E_F}{k_B T}\right)\right]^{-1} = \left[1 + \exp\left(\frac{1.10}{0.02586}\right)\right]^{-1}$$

$$\tag{14.3E1a}$$

$$f(E) = 3.36 \times 10^{-19} \tag{14.3E1b}$$

(b) From Eq. (14.14b), the N_c constant is

$$N_c = 2\left(\frac{2\pi m_e k_B T}{h^2}\right)^{3/2} \tag{14.3E2a}$$

$$N_c = 2\left[\frac{2\pi\,(9.11 \times 10^{-31}\ Kg)\,(1.38 \times 10^{-23}\ J/K)\,(300\ K)}{(6.63 \times 10^{-34}\ J.s)^2}\right]^{3/2} \tag{14.3E2b}$$

$$N_c = 2.50 \times 10^{25}\left(\frac{Kg}{J.s^2}\right)^{3/2} = 2.50 \times 10^{25}\left[\frac{Kg}{(Kg.m^2/s^2)\,s^2}\right]^{3/2} \tag{14.3E2c}$$

$$N_c = 2.50 \times 10^{25}\ e^-/m^3 = 2.50 \times 10^{25}\ m^{-3} \tag{14.3E2d}$$

The symbol e^- is added to the units to indicate that this result is in electrons/volume. From Eq. (14.14b), the number of electrons N_e are

$$n_e = N_c \exp\left[-\frac{E_F - E}{k_B T}\right] \tag{14.3E3a}$$

$$n_e = \left(2.50 \times 10^{25} \, e^-/m^3\right) \exp\left[-\frac{1.10 \, eV}{0.02586 \, eV}\right] \tag{14.3E3b}$$

$$n_e = 8.40 \times 10^6 \, e^-/m^3 \tag{14.3E3c}$$

(c) From Eq. (14.3d),

$$\sigma = n_e q_e \mu_x \tag{14.3E4a}$$

$$\sigma = \left(8.40 \times 10^6 \, m^{-3}\right)\left(1.602 \times 10^{-19} \, C\right)\left(0.14 \, m^2/\Omega.C\right) \tag{14.3E4b}$$

$$\sigma = 1.88 \times 10^{-13} \, (\Omega.m)^{-1} \tag{14.3E4c}$$

Example 14.4 Determine the number of electrons at the Fermi level so that $E_c = 0$, $E_F = 1.10 \, eV$, and $f\,(E) = 1$.

Solution In this case, Eqs. (14.11) and (14.10d) are written as

$$\rho\,(E) = \frac{\pi}{2}\left(\frac{8m_e L^2}{h^2}\right)^{3/2}\sqrt{E_F} \tag{14.4E1a}$$

$$N_e = \int_{E_c}^{\infty} \rho\,(E)\,dE = \frac{\pi}{2}\left(\frac{8m_e L^2}{h^2}\right)^{3/2}\int \sqrt{E_F}\,dE \tag{14.4E1b}$$

$$N_e = \frac{\pi}{2}\left(\frac{8m_e L^2}{h^2}\right)^{3/2}\frac{2}{3}(E_F)^{\frac{3}{2}} = \frac{\pi}{3}\left(\frac{8m_e L^2}{h^2}E_F\right)^{3/2} \tag{14.4E1c}$$

$$n_e = \frac{N_e}{L^3} = \frac{\pi}{3}\left(\frac{8m_e E_F}{h^2}\right)^{3/2} \tag{14.4E1d}$$

For $1 \, J = 1 \, N.m = 1 \, Kg.m^2/s^2$ and $1 \, eV = 1.6 \times 10^{-19} \, J$,

$$E_F = (1.10 \, eV)\left(1.6 \times 10^{-19} \, J/eV\right) = \left(1.76 \times 10^{-19} \, J\right) \tag{14.4E2a}$$

$$E_F = 1.76 \times 10^{-19} \, Kg.m^2/s^2 \tag{14.4E2b}$$

$$h = 6.6261 \times 10^{-34} \, J.s = 6.6261 \times 10^{-34} \, Kg.m^2/s \tag{14.4E2c}$$

and

$$n_e = \frac{\pi}{3}\left[\frac{8\left(9.11 \times 10^{-31} \, Kg\right)\left(1.76 \times 10^{-19} \, Kg.m^2/s^2\right)}{\left(6.6261 \times 10^{-34} \, Kg.m^2/s\right)^2}\right]^{3/2} \tag{14.4E3a}$$

$$n_e \simeq 5.23 \times 10^{27} \; e^-/m^3 \qquad\qquad (14.4E3b)$$

which is a large number of electrons in a unit volume.

14.2.8   Types of Semiconductors

Among solid materials, there are two types of semiconductors already mentioned above. For instance, the semiconductor possessing more electrons than holes is classified as a n-type semiconductor, otherwise, a p-type semiconductor. Both electrons and holes are treated as carriers associated with the position of the Fermi level on an energy band diagram.

For the purpose of clarity, Fig. 14.11a shows the undoped silicon (Si) structure, while Fig. 14.11b illustrates the phosphorous P-doped silicon structure (wafer), which becomes a semiconductor due to the available extra electron (e^- red dot) from the "P" atoms. In effect, phosphorus is an n-type dopant.

By analogy, the concentration for electrons (C_e) and holes (C_h) is the same at the Fermi level E_F and the probability of finding these carriers at this energy level are $f(E) = 0.5$ at a temperature T (thermal equilibrium).

The density of states $\rho(E)$ defined by Eq. (14.11) is used in a modified form in order to consider the permissible density of carriers in the conduction band. Thus, the density of state equations for electrons and holes are

$$\rho(E)_e = \frac{\pi}{2}\left(\frac{8m_e}{h^2}\right)^{3/2}\left(E_c - E_g\right)^{1/2} = C_e\left(E_c - E_g\right)^{1/2} \qquad (14.19a)$$

$$\rho(E)_h = \frac{\pi}{2}\left(\frac{8m_h}{h^2}\right)^{3/2}\left(0 - E_v\right)^{1/2} = C_h\left(-E_v\right)^{1/2} \qquad (14.19b)$$

Solving for E_c and E_v yields

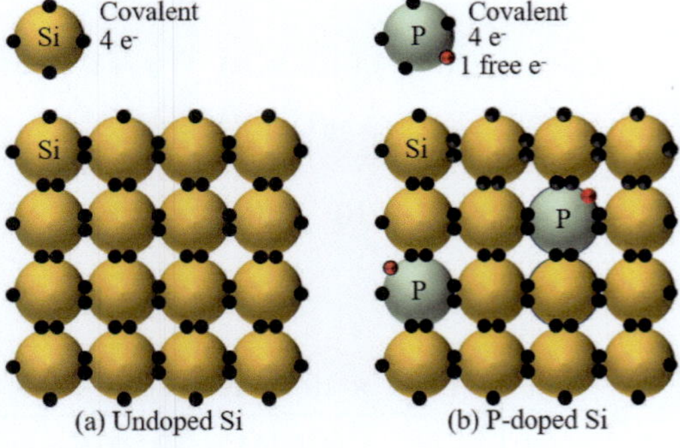

Fig. 14.11 The making of a semiconductor silicon (Si) structure. (**a**) Undoped Si and (**b**) P-doped Si structure, where each P-atom has 1 free electron (e^-)

Fig. 14.12 Energy band diagram $E = f\,[\rho\,(E)]$ and Fermi-Dirac function $E = f\,(E)$ for semiconductors. (**a**) Intrinsic semiconductor, (**b**) n-type extrinsic semiconductor and (**c**) p-type extrinsic semiconductor

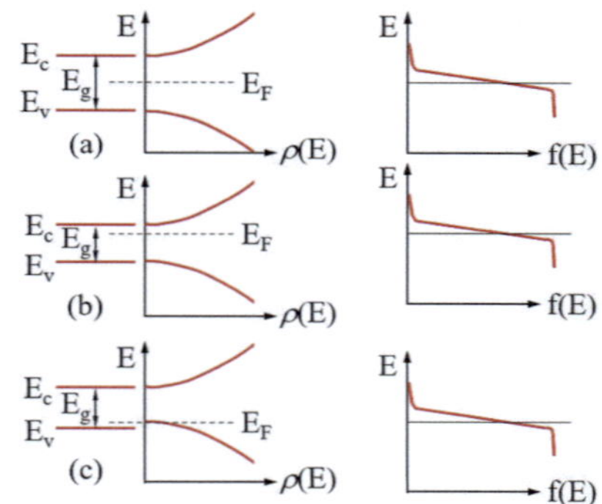

$$E_c = E_g + \left[\frac{\rho\,(E)_e}{C_e}\right]^2 \tag{14.20a}$$

$$E_v = -\left[\frac{\rho\,(E)_h}{C_h}\right]^2 \tag{14.20b}$$

which are compared to the Fermi-Dirac function

$$E = E_F + k_B T \ln\left[\frac{1}{f\,(E)} - 1\right] \tag{14.21}$$

Note that Fig. 14.12 shows the energy band diagrams as per Eqs. (14.20) and (14.21). This is a convenient way for graphically describing the types of semiconductors. In this context, for example, when excited atoms in a silicon specimen lose electrons from their outermost partially filled four-electron M-shells, the specimen becomes a p-type semiconductor leaving behind electron holes. Conversely, when the specimen gains electrons, it becomes an n-type semiconductor.

An intrinsic semiconductor has the Fermi level E_F in the center of the energy bandgap E_g so that $E_g = E_c - E_v$ and $E_F = 0.5E_g$. In this case, the number of electrons in the conduction energy band is equal to the number of holes in the valence energy band. This means that the electron density n_e represents the valence electrons that jump to the conduction band leaving behind a hole density n_h.

An extrinsic semiconductor has $n_e \neq n_h$. If $E_F \to E_c$ or $E_F > 0.5E_g$, the solid is an n-type extrinsic semiconductor with $n_e > n_h$ and if $E_F \to E_v$ or $E_F < 0.5E_g$ the solid is a p-type extrinsic semiconductor with $n_e < n_h$.

14.3 Magnetic Properties

This section includes the fundamental properties of magnetism and magnetic field (H-field) due to electron spin directions (Fig. 14.13).

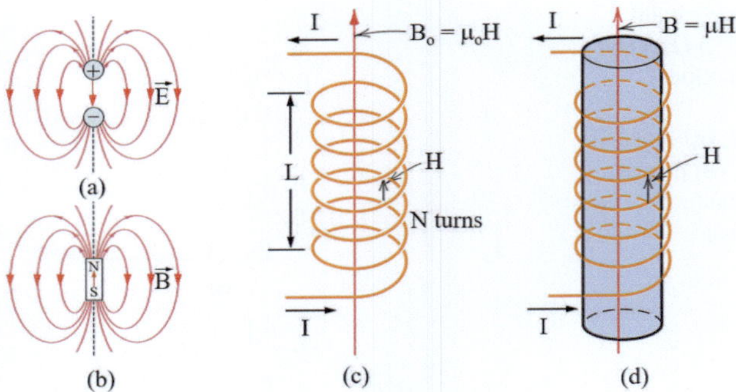

Fig. 14.13 Sketches of electric and magnetic field lines and related line directions. (**a**) Electric dipole, (**b**) magnetic dipole, (**c**) magnetic strength H and (**d**), magnetic flux B. After Callister-Rethwisch [7, p. 805]

An applied H-field gives rise to a magnetic induction B (B-field). Hence, magnetic domains in a magnetized solid, where a magnet exerts attractive and repulsive forces on magnetic materials, such as iron and steel, due to a magnetic field. The engineering application of magnetic materials can be found in electric motors, electrical power generators, loud speakers, computers, magnetic recording devices, and so forth.

In classical electromagnetism, Fig. 14.13a shows an electric dipole in the E-field and related distribution of the electric field strength $\vec{E}$ between two electrically charged particles. Moreover, Fig. 14.13b illustrates a magnetic dipole in a B-field and related distribution of the continuous magnetic field lines of force $\vec{B}$. In fact, the $\vec{B}$ line in the bar magnet is upward, where "N" stands for north and "S" stands for south poles.

The direction of the magnetic field strength H (Fig. 14.13c) and magnetic flux density B is induced by the cylindrical coil depicted in Fig. 14.13d (Callister and Rethwisch [7, p. 805]).

14.3.1   Magnetic Domain

Consider an externally imposed magnetic field strength H on an initially unmagnetized material. Mathematically, the magnetization process is described by the magnetic flux density or magnetic induction (B). The magnetization of a material described by the $B = f(H)$ function is illustrated schematically in Fig. 14.14a for ferromagnetic, ferrimagnetic, paramagnetic, and diamagnetic materials. The schematic $B = f(H)$ profiles represent the spontaneous magnetization from the alignment of the electron spins in the atomic structure of a single crystal or polycrystal Wyatt and Dew-Hughes [8, p. 598]).

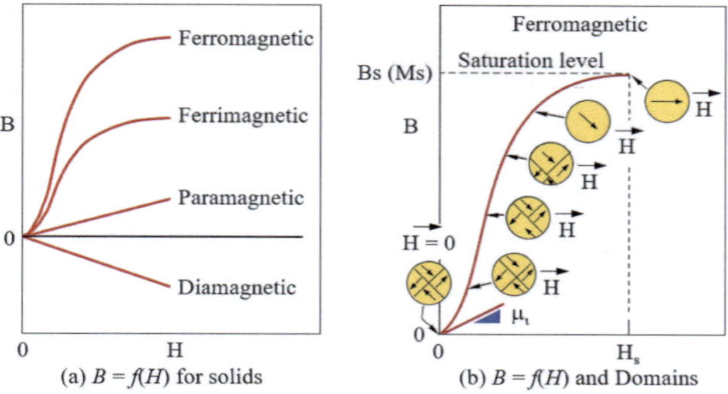

Fig. 14.14 The $B = f(H)$ relationship (**a**) for types of magnetization of materials and (**b**) ferromagnetic solid and its magnetic domain orientations (taken from Wyatt and Dew-Hughes [8, p. 598])

For a hypothetical ferromagnetic material, the magnetization curve described by the sigmoidal-shaped function $B = f(H)$ is graphically shown in Fig. 14.14b as the Wyatt and Dew-Hughes [8, p. 598] magnetization model related to the magnetic domain configuration denoted by the insets. For the sake of clarity, the sigmoidal-shaped curve in Fig. 14.14b was taken from Callister and Rethwisch book [7, p. 816].

From Fig. 14.14b, one can deduce the following:

- The electron alignment represents the magnetic domain in solids, and it is attributed to an internal magnetic field H, which in turn is induced by electron spin interactions.
- For a ferromagnetic solid at $H = 0$, the schematic magnetic domains (arrows in the insets) are randomly oriented and separated by domain walls (straight lines acting as interfaces in the insets). The phrase magnetic domain boundary is most appropriate, but the term Bloch wall is commonly used to indicate an interface between magnetic domains.
- The initial magnetic permeability μ_i is the slope of the $B = f(H)$ function at the early stages of magnetization.
- Increasing the H-field induces the creation of the B-field. Thus, the magnetic behavior is described by the general function $B = f(H)$. Curve fitting datasets may provide the best fit using a hyperbolic tangent (with constants a_i) or arctangent sigmoidal function (with constants b_i) of the form

$$B = a_1 + a_2 \tanh\left(\frac{H - a_3}{a_4}\right) \tag{14.22a}$$

$$B = b_1 + b_2 \arctan\left(\frac{H - b_3}{b_4}\right) \tag{14.22b}$$

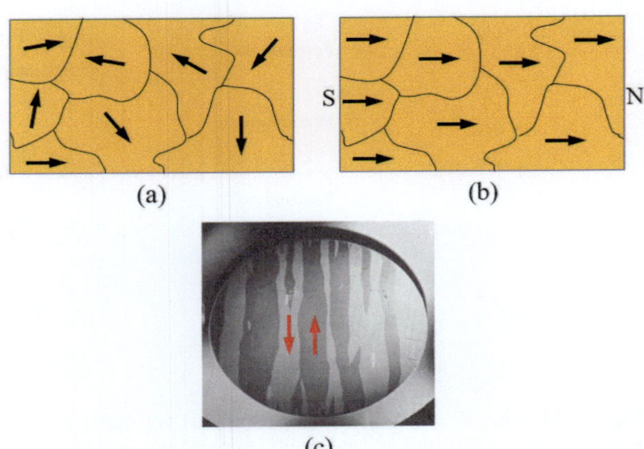

Fig. 14.15 Microscopic magnetic domain structures. (**a**) Schematic random magnetic domain configuration model, (**b**) aligned magnetic domain structure, and (**c**) actual magnetic domain alignment in 81 permalloy (Cullity and Graham [9, p. 290])

- Magnetic domains with aligned magnetic moment M with the H-field direction grow, while some domain walls vanish as $B = f(H)$ approaches a saturation magnetization level at a relatively strong H-field.
- The saturation magnetization reaches a maximum value denoted as $B_s = B_{\max}$ at H_s. This phenomenon is attributed to aligned magnetic dipole moments or electron spins. In other words, the magnet or solid is in a state of maximum magnetization when all or most electron spins are aligned in the direction of the applied external magnetic field H. Beyond this point, $B_s = f(H_s)$ is constant.

Most magnetic materials are polycrystalline, and each grain is a single crystal having its own random magnetic domain as shown in Fig. 14.15a.

Notice the randomly oriented electron domains denoted by arrows in Fig. 14.15a and the aligned electron domains in one direction from south to north magnet poles in Fig. 14.15b, which represents an ideal magnetic domain pattern due to a strong magnetic field H as in ferromagnetic polycrystalline solids.

Figure 14.15c depicts an actual magnetic domain strip structure, as seen using the Kerr effect, in a 81-permalloy specimen (Cullity and Graham [9, p. 290]). This magnetic domain structure can be referred to as magnetic ordering of the magnetic dipole alignment, specifically in ferromagnetic materials that are susceptible to a spontaneous magnetization. In other words, ferromagnetic materials magnetize readily.

14.3.2 Hysteresis Loop

Figure 14.16a shows a schematic magnetization diagram in the form of a magnetic sigmoidal-shaped hysteresis loop for a ferromagnetic solid, which is treated as a hard magnet due to a relatively large loop area. This diagram exhibits schematically the relationship between the B-field and the H-field, again, using the general

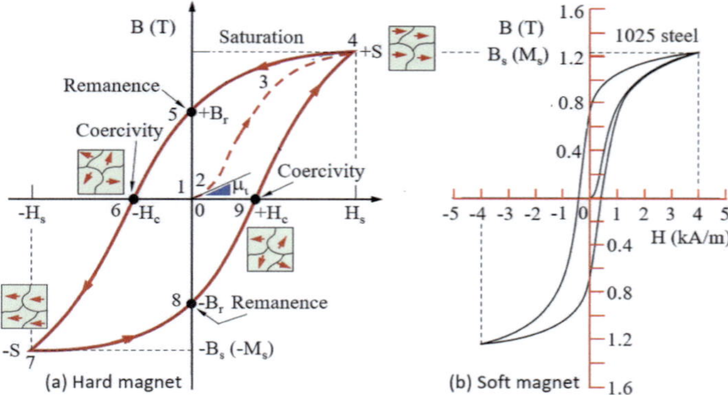

Fig. 14.16 Magnetic hysteresis loops for (**a**) a hypothetical solid as a hard magnet and (**b**) actual for an AISI 1025 carbon steel (Jiles [10, p. 90])

function $B = f(H)$ indicating the initial magnetization process from zero to the saturation point "+S" where the magnetic domains are aligned with the increasing H-field. A reversal process from "+S" down to "-S" point induces aligned domains in the opposite direction (Cullity and Graham [9, p. 326]).

Figure 14.16b exhibits an experimental sigmoidal-shaped hysteresis loop for an AISI 1025 carbon-steel specimen. In essence, this steel is treated as a soft magnet due to a small loop area (Jiles [10, p. 90]). Moreover, the sigmoidal-shaped hysteresis loop in Fig. 14.16a or b is a magnetization diagram commonly used to extract magnetic properties of a magnetized solid.

The magnetic sigmoidal-shaped hysteresis loop or magnetic diagram is unique for a certain ferromagnetic or ferrimagnetic solid under specific conditions. Notice the inset showing a magnetic domain orientations at the corresponding remanence and coercivity points, which are used to extract the most common magnetic properties of the material. For instance,

- The initial imposed external H-field generates a B-field in the early stages where the H-B data is suitable for determining the initial permeability μ_i.
- The saturation point "S" is used to define the saturation magnetization $M_s = B_s$ at $H \geq H_s$. Comparing the hysteresis areas in Fig. 14.16 yields the conclusion that a broad loop is a specific condition for a material to become a permanent magnet, since it is not easy to demagnetize it. This means that a broad hysteresis loop is for hard magnetic materials used as permanent magnets. Conversely, a narrow hysteresis loop is for soft magnetic materials used as magnetic storage media (computer memory) and as cores of electrical transformers.
- The hypothetical material reaches the saturation level denoted at "+S" point, which is called the saturation magnetization or maximum magnetic induction denoted as $B_s = B_{max}$ at H_s. Again, beyond this point, $B_s = f(H_s)$ is constant. Moreover, $B \rightarrow B_s = B_{max}$ as $T \rightarrow 0\ K$ with minimum thermal vibration and $B \rightarrow 0$ as $T \rightarrow T_c$, which is called the Curie temperature.

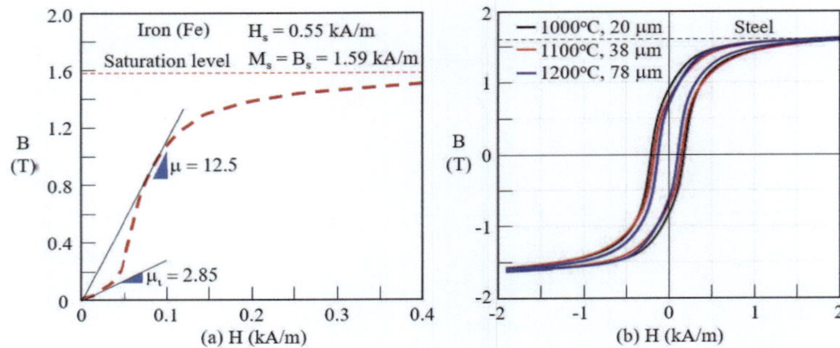

Fig. 14.17 (**a**) Magnetization of iron (Fe) and (**b**) magnetic hysteresis loops for annealed steel specimens with different average grain sizes (Zhou et al. [8])

- The corresponding physical properties from Fig. 14.16b are

$$\text{Permeability } \mu_i = 1/6 = 0.13$$
$$\text{Remanence } B_r = +0.72\,Tesla \text{ at } H = 0$$
$$B_r = -0.72\,Tesla \text{ at } H = 0$$
$$\text{Coercivity } B_r = 0 \text{ at } H_r = -0.8\,kA/m$$
$$B_r = 0 \text{ at } H_r = +0.8\,kA/m$$
$$\text{Saturation } M_s = B_s \simeq +1.22\,Tesla$$
$$M_s = B_s = -1.22\,Tesla$$

Magnetization datasets are illustrated in Fig. 14.17. For instance, Fig. 14.17a clearly shows the room-temperature magnetization curve for iron (Fe) with (1) magnetic permeability terms $\mu_i = 2.85$ and $\mu_{max} = 12.5$ and (2) saturation magnetization $M_s = B_s = 1.59\,kA/m$ at $H_s = 0.55\,kA/m$. Most experimental data is obtained at a temperature T in the range $T_o < T < T_c$, where $T_o = 0\,K$ is the absolute temperature and T_c is the Curie temperature below which a material undergoes a sharp change in their magnetic properties.

Figure 14.17b depicts the magnetic behavior of three annealed low-carbon steels at temperatures having different ferritic grain sizes (Zhou et al. [11]). The general perception from these magnetic hysteresis loops is that there is not a significant difference with respect to the hysteresis loop area (energy loss) for each specimen. However, the hysteresis loop for the annealed steel at $1000°C$ having the smallest grain size ($d = 20\,\mu m$) is slightly larger than the other loops. Therefore, the smaller the grain size, the larger the number of grain boundaries and the larger the number of obstacles to the magnetic domain boundary motion. Other crystal defects cited in Chap. 6 are assumed to have the a similar effect on this type of motion.

A magnetic field of force surrounding a magnetized material or current-carrying conductor induces a magnetic induction quantified by the internal magnetic flux density or magnetic induction (B), which is related to the magnetic permeability (μ) of a solid and the applied magnetic field strength (H). Thus, the magnetic behavior can be quantified by

$$B = \mu H \tag{14.23a}$$

$$B_o = \mu_o H \quad \text{(vacuum)} \tag{14.23b}$$

$$B = \mu_o H + \mu_o M \tag{14.23c}$$

$$B = \mu_o M \quad \text{for} \quad H << M \tag{14.23d}$$

$$H = \frac{NI}{L} \tag{14.23e}$$

where H is in A/m and the vacuum permeability μ_o is in $T.m/A$ units. The corresponding magnetic force vector $\vec{F}$ (without derivation) is

$$\vec{F} = q_e \, \vec{v} \, \vec{B} \, \sin(\theta) . \vec{n} \tag{14.24a}$$

$$B = \frac{F}{q_e v \sin(\theta)} \tag{14.24b}$$

$$\theta \angle \vec{v}, \vec{B} \tag{14.24c}$$

Here, $q_e = 1.602 \times 10^{-19}$ C ($= A.s$) is the electronic charge, $\vec{v}$ is the charged particle velocity, $\vec{B}$ is the magnetic flux vector, and $\vec{n}$ is the unit vector. Also, B represents the magnitude of the internal magnetic field strength of a specimen in a B-field, and it has units of tesla T

$$1 \, tesla = 1 \, T = 1 \, \frac{webers}{meter^2} = 1 \, \frac{Wb}{m^2} = 1 \, \frac{V.s}{m^2} \tag{14.25a}$$

$$1 \, Wb = 1 \, V.s \tag{14.25b}$$

$$1 \, tesla = \frac{N}{C.m/s} = \frac{N.s}{C.m} \tag{14.25c}$$

where N denotes Newton unit, C denotes coulomb, s denotes second, and m denotes meter. The permeability of free space (vacuum permeability) is

$$\mu_o = 4\pi \times 10^{-7} \, \frac{T.m}{A} = 4\pi \times 10^{-7} \, \frac{Wb}{m.A} = 4\pi \times 10^{-7} \, \frac{V.s}{m^3.A} \tag{14.26}$$

Now, V denotes volts and A denotes ampere, N denotes the number of turns of a coil, I denotes the electric current, and L denotes the bar length.

In electromagnetism, the magnetic susceptibility χ is a quantitative that measures the extent to which a material may be magnetized under the influence of an applied magnetic field H. Mathematically,

$$\chi = \frac{M}{H} > 0 \quad \text{(ferromagnetism, paramagnetism)} \tag{14.27a}$$

Table 14.3 Common room-temperature magnetic susceptibility χ values for diamagnetic and paramagnetic materials (Callister and Rethwisch [7, p. 809]

Diamagnetic		Paramagnetic	
Material	χ	Material	χ
Al_3O_4	-1.81×10^{-5}	Al	2.07×10^{-5}
Cu	-0.96×10^{-5}	Cr	3.13×10^{-4}
Au	-3.44×10^{-5}	$CrCl_2$	1.51×10^{-3}
Hg	-2.85×10^{-5}	$MnSO_4$	3.70×10^{-3}
Si	-0.41×10^{-5}	Mo	1.19×10^{-4}
Ag	-2.38×10^{-5}	Na	8.48×10^{-6}
$NaCl$	-1.41×10^{-5}	Ti	1.81×10^{-4}
Zn	-1.56×10^{-5}	Zr	1.09×10^{-4}

$$M = \chi H > 0 \tag{14.27b}$$

$$\chi = \frac{M}{H} < 0 \quad (diamagnetism) \tag{14.27c}$$

$$M = \chi H < 0 \tag{14.27d}$$

where χ is a dimensionless constant that dictates the degree of magnetization and M is the magnetization (magnetic moment per unit volume) of the specimen.

Despite that $\chi\,(ferromagnetism) >> \chi\,(paramagnetism)$, the induced magnetic moment M weakens the applied magnetic H-field by a factor defined as the magnetic susceptibility χ. Table 14.3 lists values of the magnetic susceptibility χ for some diamagnetic and paramagnetic solids. This dataset can be found elsewhere (Callister and Rethwisch [7, p. 809]).

A maximum magnetic moment per unit volume for a magnetic material is called the saturation magnetization (M_s), which is an intrinsic property. Thus,

$$M_s = \mu_B N_x \tag{14.28a}$$

$$N_x = \frac{\rho N_a}{A_w} \tag{14.28b}$$

and the corresponding saturation magnetic flux density is

$$B_s = \mu M_s \tag{14.29}$$

where $\mu_B = 9.27 \times 10^{-24}\ A.m^2$ is the Bohr magneton (fundamental magnetic moment), N_x is the number of atoms per volume deduced from Eq. (1.32b), $N_a = 6.022 \times 10^{23}\ atom/mol$ is the Avogadro's number, ρ is the mass density, and A_w is the atomic weight.

In magnetism, it is common to use a relative permeability defined by

$$\mu_r = \frac{\mu}{\mu_o} \tag{14.30}$$

Here, μ_r represents the magnetic permeability relative to the vacuum permeability μ_o.

14.3.3 Magnetism

The magnetic behavior of solids includes ferromagnetism, paramagnetism, and diamagnetism. For a magnetic system, the principal characteristics of the magnetic behavior are related to the magnetic moment vector $\vec{M}$, which measures the strength and direction of corresponding magnetism. Nonetheless, the following classification of solids is based on the electronic configuration (Fig. 14.18) and their response to an external magnetic field H. Bear in mind that crystals are magnetically anisotropic and the magnetic medium is assumed to be magnetically isotropic or uniform.

Ferromagnetism It is induced by an applied magnetic field H on an unpaired electrons in solids like Fe, Ni, Co, Gd (gadolinium). These materials are attracted to an applied magnetic field to become a permanent magnet. Iron (Fe) is generally associated with magnetism. For a practical purpose, paramagnetism is stronger than diamagnetism since $\chi_p > \chi_d$, but weaker than ferromagnetism being quantified by $\chi_f > \chi_p$. The strong magnetic effect on permanent magnets can be attribute to the randomly oriented ferromagnetic domains with respect to each other. In particular, Fig. 14.18 shows aligned magnetic moment p of atoms pointing in the same arbitrary direction.

Ferrimagnetism It is similar to ferromagnetism but occurs in ceramic materials. For example, magnetite Fe_3O_4 is a particular ceramic material that experiences this type of magnetism.

Antiferromagnetism It is induced by antiparallel alignments of magnetic moment coupling between adjacent atoms or ions. In this case, the opposing magnetic

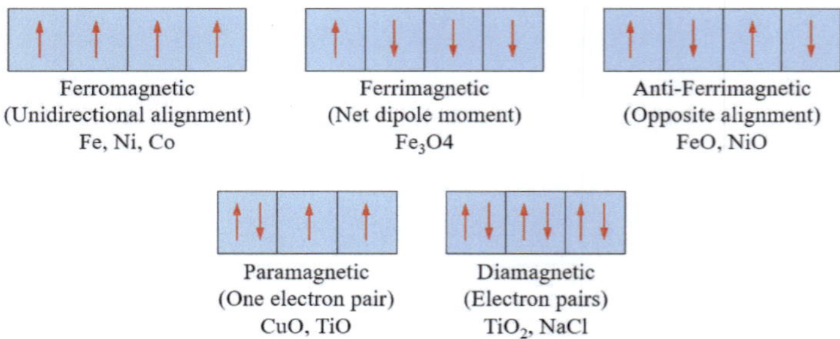

Fig. 14.18 Magnetism and related electronic configurations of the outer shells

moments cancel one another, and therefore, the net magnetic moment is zero. For instance, manganese oxide (MnO) is a ceramic material that falls in this category.

Paramagnetism   It is induced by an applied magnetic field. This magnetic mechanism produces a weak interaction between adjacent atomic magnetic dipoles, leading to small values of magnetic susceptibilities in certain solids with unpair electrons in the electronic configuration outermost shell, such as the d-shell in most transition metals ($Ti, Cr, Ni, Cu,\ldots$). This physical phenomenon disappears upon removal of the magnetic field. Consequently, the effect of the magnetic field on paramagnetic solids is considered weak.

Diamagnetism   It is induced by an applied magnetic field with a quantum mechanical effect in the opposite direction. As a result, it produces negative magnetic susceptibilities, Eq. (14.27c), in certain solids containing electron-pair orbital; that is, electrons are spin-paired in diamagnetic elements with completely filled electron shells. This implies that electron positive and negative spins cancel out, and therefore the electron magnetic moment is zero, leading to a weak magnetic interaction. Consequently, some atoms are slightly repelled from a magnetic field due to cancellation of the electron spin effect. Again, this physical phenomenon disappears upon removal of the magnetic field. For instance, copper, ionic, and covalent crystals fall in this category.

Example 14.5   Calculate **(a)** the number of atoms per volume N_x, **(b)** the saturation magnetization M_s, and **(c)** the saturation magnetic flux density B_s for iron (Fe) with a mass density of 7.87 g/cm^3. Given data: $N_a = 6.022 \times 10^{23}$ $atom/mol$ (Avogadro's number), $\mu_B = 9.27 \times 10^{-24}$ $A.m^2$ (Bohr magneton) and $\mu_o = 4\pi \times 10^{-7}$ $T.m/A$ (vacuum permeability).

Solution

(a) From Eq. (14.28b), the number of atoms per volume along with $A_{w,Fe} = 55.845$ g/mol

$$N_x = \frac{\rho N_a}{A_w} = \frac{\left(7.87 \times 10^6 \ g/m^3\right)\left(6.022 \times 10^{23} \ atom/mol\right)}{55.845 \ g/mol} \tag{14.5E1a}$$

$$N_x \simeq 8 \times 10^{28} \ atoms/m^3 \tag{14.5E1b}$$

(b) From Eq. (14.6a), the saturation magnetization is

$$M_s = \mu_B N_x = \left(9.27 \times 10^{-24} \ A.m^2\right)\left(8 \times 10^{28} \ atoms/m^3\right) \tag{14.5E2a}$$

$$M_s = 7.42 \times 10^5 \ A/m \tag{14.5E2b}$$

(c) From Eq. (14.23d), the saturation magnetic flux density is

$$B_s = \mu_o M_s = \left(4\pi \times 10^{-7} \ T.m/A\right)\left(7.42 \times 10^5 \ A/m\right) \tag{14.5E3a}$$

$$B_s = 0.93 \ T = 0.93 \ Tesla = 0.93 \ \frac{N.s}{C.m} \tag{14.5E3b}$$

Hence, the saturation magnetic flux density levels off at $B_s = 0.93 \ Tesla$.

Example 14.6 Consider the Cu-wire coil around a nonmagnetic and low permeability Al-bar shown below. Assume that the Cu-coil is connected to a battery (not shown). Calculate **(a)** the magnetic field strength H, **(b)** the flux density B_o in vacuum, **(c)** the flux density B around the coil, and **(d)** the magnitude of the magnetization M. See Fig. 14.13d (Callister and Rethwisch [7, p. 805]).

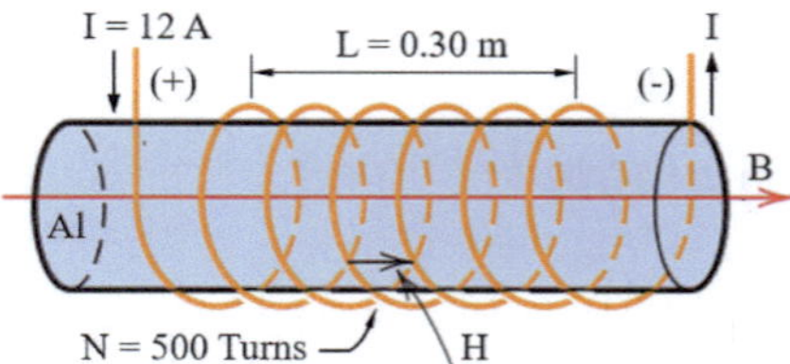

Solution

(a) From Eq. (14.23e), the magnetic field strength H is

$$H = \frac{NI}{L} = \frac{(500)\,(12\ A)}{0.30\ m} = 20,000\ A/m \tag{14.6E1}$$

(b) From Eq. (14.23b), the flux density B_o in vacuum is

$$B_o = \mu_o H = \left(4\pi \times 10^{-7} \ \frac{T.m}{A}\right)(20,000\ A/m) \tag{14.6E2a}$$

$$B_o = 2.51 \times 10^{-2} \ Tesla = 2.51 \times 10^{-2} \ \frac{N.s}{C.m} \tag{14.6E2b}$$

(c) Combining Eqs. (14.23c) and (14.27a) yields the total flux density

$$B = \mu_o H + \mu_o M = \mu_o H + \mu_o \chi H \simeq \mu_o H \tag{14.6E3a}$$

$$B = \mu_o\,(1 + \chi)\,H = \left(4\pi \times 10^{-7} \ \frac{T.m}{A}\right)\left(1 + 2.07 \times 10^{-5}\right)(20,000\ A/m) \tag{14.6E3b}$$

$$B = 2.51 \times 10^{-2} \ Tesla = 2.51 \times 10^{-2} \ \frac{N.s}{C.m} \tag{14.6E3c}$$

This calculation could have been omitted due to $\chi \ll 1$ and $\mu_o H \gg \mu_o \chi H$ and as a result, $B \simeq B_o$. This result is considered low, because the aluminum bar is nonmagnetic. However, for a ferromagnetic bar, such as iron (Fe), the flux density B around the Fe-coil would be high. The reader is encouraged to repeat this example using Fe to confirm that this is true. In this context, ferromagnetic materials, such as soft iron (Fe), steel (Fe-C-X), and some nickel (Ni) alloys, can be magnetized very easily.

(d) Using Eq. (14.27a) and Table 14.3 gives the magnetization as

$$M = \chi H = \left(2.07 \times 10^{-5}\right)(20,000\ A/m) \tag{14.6E4a}$$

$$M = 0.41\ A/m \tag{14.6E4b}$$

This result represents the density of magnetic dipole moment.

14.3.4 Magnetic Anisotropy

It is clearly evident in Fig. 14.19 (taken from Callister and Rethwisch [7, p. 819]) that the magnetization of crystalline solids, such as iron (Fe) and nickel (Ni) single crystals, exhibit magnetic anisotropy, since the magnetic flux $B_{[uvw]}$ behavior along crystallographic directions is different.

The magnetic anisotropy is less pronounced in Ni than in Fe before the magnetic flux reaches a maximum level $B_{\max} = M$ for both single crystals. Therefore, the data in Fig. 14.19 for Fe and Ni single crystals clearly represent magnetic anisotropy at the "knee" with respect to the crystallographic directions and the applied H-field at room temperature (Honda and Kaya [12], Kaya [13]).

This data clearly indicates that the magnetization along different crystallographic directions requires additional energy called magnetocrystalline anisotropy energy. Thus, the magnetizing fields (H-fields) applied in the indicated crystallographic directions induce the $B = f(H)$ curves to reach maximum at H-field values (Fig. 14.19).

The read-off values are

$$B_{\max}[100]_{Fe} = M_{Fe} = 1.7 \times 10^6\ A/m \text{ at } H = 0.5 \times 10^4\ A/m$$

$$B_{\max}[110]_{Fe} = M_{Fe} \simeq 1.7 \times 10^6\ A/m \text{ at } H = 4.0 \times 10^4\ A/m$$

$$B_{\max}[111]_{Fe} = M_{Fe} \gtrsim 1.7 \times 10^6\ A/m \text{ at } H = 4.8 \times 10^4\ A/m$$

and

$$B_{\max}[111]_{Ni} = M_{Ni} = 0.5 \times 10^6\ A/m \text{ at } H = 0.3 \times 10^4\ A/m$$

$$B_{\max}[110]_{Ni} = M_{Ni} = 0.5 \times 10^6\ A/m \text{ at } H = 1.2 \times 10^4\ A/m$$

$$B_{\max}[100]_{Ni} = M_{Ni} = 0.5 \times 10^6\ A/m \text{ at } H = 1.8 \times 10^4\ A/m$$

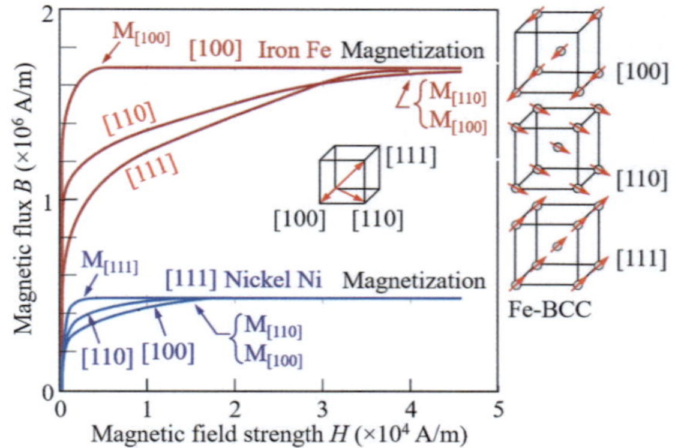

Fig. 14.19 Magnetic anisotropy of Fe and Ni single crystals. Taken from Callister-Rethwisch [7, p. 819]. Data for Fe from Honda and Kaya [12] and that for Ni from Kaya [13]

The saturation level in Fe metal is fast (easy magnetization) along the [100] direction because $B_{\max}[100]_{Fe} < B_{\max}[110]_{Fe}$, $B_{\max}[111]_{Fe}$ and in Ni metal is along the [111] direction due to $B_{\max}[111]_{Ni} < B_{\max}[110]_{Ni}$, $B_{\max}[100]_{Ni}$. Moreover, the B-H diagram in Fig. 14.19 indicates that Fe is a hard and Ni is a soft magnetic material because $M_{Fe} > M_{Ni}$ at $B = B_{max}$.

14.3.5 Temperature-Dependence Magnetic Behavior

The effect of temperature on the saturation magnetization, $M_s = f(T)$, is depicted in Fig. 14.20 for natural iron-oxide mineral called magnetite Fe_3O_4, which is classified as a ferrimagnetic compound. The saturation magnetization, $M_s = f(T)$, for pure iron Fe is also included in Fig. 14.20 for comparison purposes (taken from Callister-Rethwisch [7, p. 815]).

The insets in Fig. 14.20 are purposely added to indicate the possible alignment of the magnetic dipole moments at different temperatures. Notice that the alignment of the magnetic dipole moments or electron spins changes with increasing temperature. This phenomenon is attributed to the increasing frequency of the lattice vibrations induced at high temperatures.

Regarding the magnetic phase diagram shown in Fig. 14.21, the temperature-dependent magnetic behavior of magnetite Fe_3O_4 and pure iron (Fe) is clearly denoted by a decreasing saturation magnetization M_s with increasing temperature T. This is attributed to the transition of the magnetic domain from a well-organized to a randomly oriented magnetic domain field. For instance, the drastic decrease of M_s for Fe at $T > 400^oC$ may be induced by the instability of Fe-ferrite BCC-phase, which in turn changes to austenite FCC phase at 912^oC (see Figs. 10.30 and 10.32). Nonetheless, both magnetite and pure iron are sensitive to changes in magnetic properties, and consequently, they are not magnetic when $M_s = 0$.

Fig. 14.20
Temperature-dependent
magnetic behavior of natural
magnetite (Fe_3O_4), mineral,
and pure iron (Fe). Taken
from Callister and Rethwisch
book [7, p. 815]

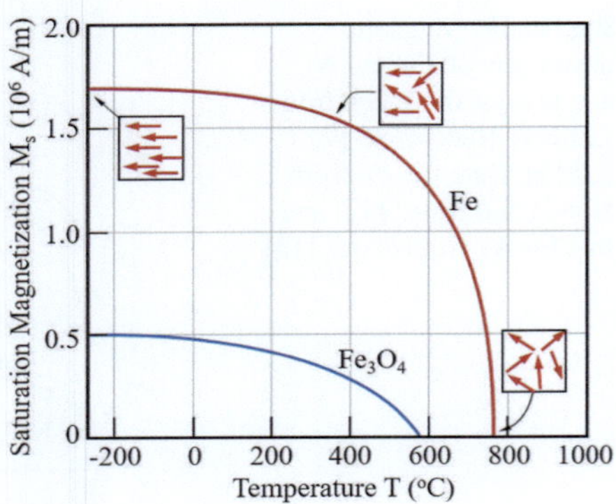

Temperature Limits There are two common temperature limits of significant value. The Curie temperature (T_c) at which a material undergoes a sharp change in their magnetic properties attributed to the alignment of atoms that behave as tiny magnets. Accordingly,

- If $T < T_c$, the material remains as a ferromagnetic or ferrimagnetic.
- If $T = T_c$, the magnetism is destroyed.
- If $T > T_c$, the ferromagnetic and ferrimagnetic behavior transforms to paramagnetic behavior. From Fig. 14.20, $Tc \simeq 770°C$ for pure iron and $Tc \simeq 580°C$ for natural magnetite Fe_3O_4 at $M_s = 0$.
- If $T = T_n$ (Neel temperature), the antiferromagnetism vanishes.
- If $T > T_{,n}$, the antiferromagnetism behavior transforms to paramagnetic behavior.

Regarding the Curie temperature T_c, it is very sensitive to microstructural features and chemical composition of a magnetic materials. This may be attribute to grains and particles with different magnetic domains (Fabian et al. [14]).

14.4 Optical Properties

Optics is the study of the behavior and properties of light, and it is classified as atmospheric or natural optics, atomic optics, ion optics, laser optics, lens optics, and so forth. For instance, Fig. 14.21a depicts an image showing the natural optics of the sunset, Fig. 14.21b shows an image showing the artificial optics of light waves from a flashlight reflections on a black leather-like chair, Fig. 14.21c depicts the natural rainbow, and Fig. 14.21d illustrates the artificial rainbow coming out of a prism (https://www.sciencephoto.com/media/f0106002/view). These images exhibit the light reflection in different colors due to different photon wavelengths. This means that light is electromagnetic radiation with a visible electromagnetic spectrum of different wavelengths in the range 380 $nm \leq \lambda \leq 760\ nm$.

Fig. 14.21 Light optics. (**a**) Natural optics of the sunset, (**b**) artificial optics induced by a flashlight, (**c**) natural optics of the rainbow after a rainy day, and (**d**) artificial optics induced by a prism (from a website, see text)

The theoretically interactions of a material with light govern the optical properties, and by definition, optics is the science of light induced by electromagnetic wavelike or light radiation due to magnetic field oscillations. Hence, it is appropriate to introduce a relevant theoretical background on the spectrum of light related to photon motion and photon-electron interactions.

This will help understand the importance of optical properties of materials in applications like lenses, pulse laser, and fiber optics.

14.4.1 Electromagnetic Spectrum

James Clerk Maxwell in 1860s developed the theory electromagnetic (EM) waves traveling along the x-axis in the orthogonal $\vec{E}$-field and $\vec{B}$-fields in a oscillatory manner at constant velocity υ. Subsequently, Heinrich Hertz in 1980s confirmed that an EM wave travels at a finite speed $\upsilon < c$ in a particular medium as shown in Fig. 14.22. The wave motion is modeled as a sinusoidal wave-like component on the $\vec{E}$-plane or on the $\vec{B}$-plane. For clarity, the image in Fig. 14.22 can be found at Chemistry LibreTexts, electromagnetic radiation (April 16, 2022), or at https://en.wikipedia.org/wiki/Electromagnetic_radiation.

The source of electromagnetic waves may be the sun, radiation, dispersion of white light, and so forth. Nonetheless, electromagnetic radiation is manifested as

Fig. 14.22 Electromagnetic
wave in a combined
electric-magnetic field
components

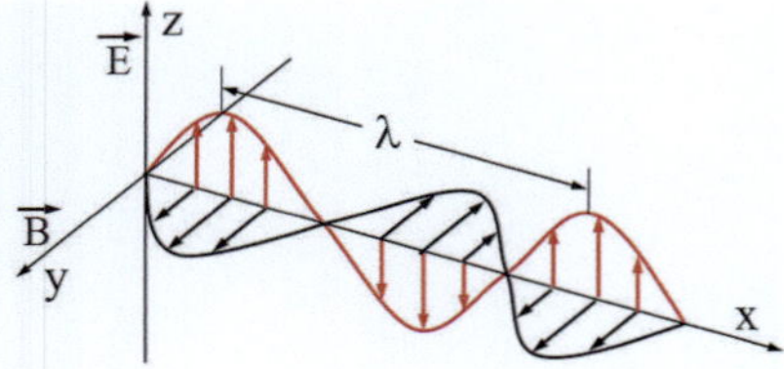

synchronized sinusoidally oscillations of electric and magnetic fields, producing electromagnetic waves (EM waves) that behave like standing waves (SW) due to charge motion.

Light particles are massless photons that carry energy and momentum; therefore, light is a type of EM. From Eq. (3.1e), the photon energy ΔE is proportional to its frequency v and inversely proportional to its wavelength λ.

Mathematically,

$$\Delta E = hv = h\frac{c}{\lambda} \tag{14.31a}$$

$$c = v\lambda \tag{14.31b}$$

where $h = 6.6261 \times 10^{-34}\ m^2.kg/s = 6.6261 \times 10^{-34}\ J.s$ is the Planck's constant, v is the photon frequency, $c = 3 \times 10^8\ m/s$ is the speed of light, and λ is the photon wavelength.

The momentum (p) of a photon is also proportional to its frequency and inversely proportional to its wavelength

$$p = \frac{\Delta E}{c} = \frac{hv}{c} = \frac{h}{\lambda} \tag{14.32}$$

Notice that $p \to 0$ as $\lambda \to \infty$ and $p = 0$ when radiation is absorbed by a surface.

Optical properties of solids are characterized by color, transparency, index of refraction (n), and reflectivity. Specifically, each color has its own wavelength as shown by the optical spectrum of visible light (Fig. 14.23) to the human eye.

Example 14.7 Calculate (a) the photon energy change ΔE from the conduction to the valence energy bands in a semiconductor and (b) the valence energy band E_v if the conduction energy band is $E_c = 1.40\ eV$. Use the wavelength $\lambda = 580\ nm$ for visible yellow light.

Solution

(a) Using Eq. (14.31a) with the conversion factor $1\ eV = 1.60 \times 10^{-19}\ J$ yields

$$\Delta E = h\frac{c}{\lambda} = \frac{\left(6.6261 \times 10^{-34}\ J.s\right)\left(3 \times 10^8\ m/s\right)}{580 \times 10^{-9}\ m} \tag{14.7E1a}$$

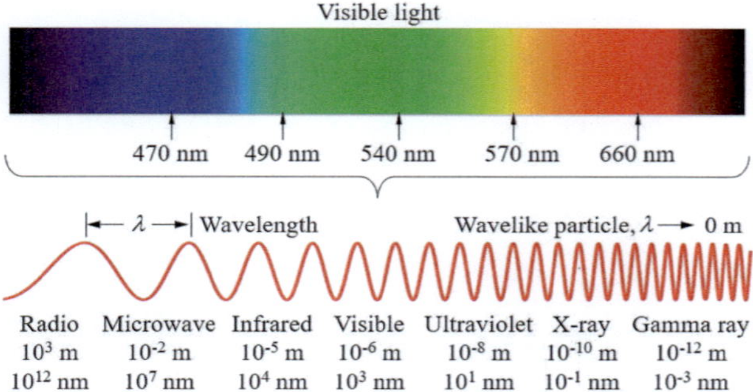

Fig. 14.23 Electromagnetic radiation spectrum for visible light colors and electromagnetic wavelength range

$$\Delta E = 3.43 \times 10^{-19} \, J = 2.14 \, eV \tag{14.7E1b}$$

(b) The valence energy state is

$$\Delta E = E_v - E_c \tag{14.7E2a}$$

$$E_v = \Delta E + E_c = 2.14 \, eV + 1.40 \, eV \tag{14.7E2b}$$

$$E_v = 3.54 \, eV \tag{14.7E2c}$$

The sequence of visible colors to the human eye is a rainbow of white light undergoing optical refraction within a wavelength range. This spectrum of colors can be seen when shining white light through a prism. This is an optical refraction phenomenon attributed to wavelength bending at slightly different angles.

Figure 14.24 depicts the emission line spectrum of hydrogen H (Fig. 14.24a) and of iron Fe (Fig. 14.24b). The emission spectra for pure elements are attributed to electrons in their initial energy states falling back down to lower energy states at different wavelengths.

The optical properties of a material or matter as considered in optical physics are broad, and only the most basic engineering properties are included henceforth.

14.4.2   *Refraction and Refractive Index*

Refraction of light is a phenomenon related to the change in direction of a photon wave passing through a medium at a certain speed as schematically illustrated in Fig. 14.25. The practical application of this optical phenomenon is found in optical prisms and lenses.

Fig. 14.24 Emission line spectrum of (**a**) hydrogen (H) and (**b**) iron (Fe)

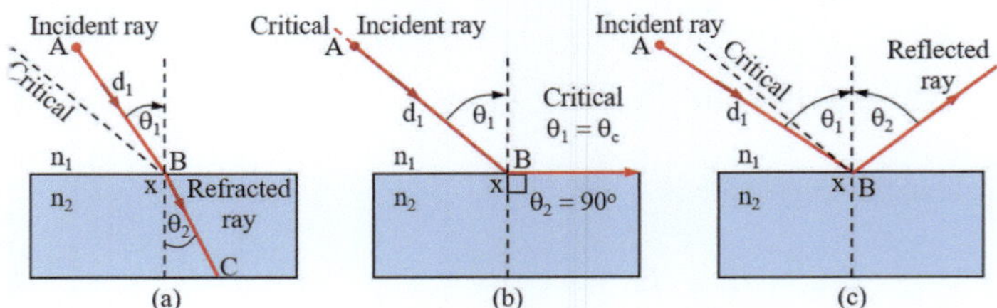

Fig. 14.25 Light ray (**a**) refraction at $\theta_2 < 90°$ when $\theta_1 < \theta_c$ and $n_2 > n_1$ (**b**) refraction at $\theta_2 = 90°$ when $\theta_1 = \theta_c$ and $n_2 > n_1$ and (**c**) internal reflection at a phase boundary where $\theta_1 > \theta_c$ and $n_1 < n_2$

Consider an incident light ray (Fig. 14.25) traveling at an incident angle θ_1 with respect to a point x located on the surface boundary between a medium with a refractive index n_1 and a medium with a refractive index n_2.

Light ray refraction occurs at an incident angle within the range $0 < \theta \leq \theta_c$. However, if $\theta > \theta_c$ (critical angle), the light ray is reflected within the denser material. For simplicity,

- If $\theta_1 < \theta_c$ in Fig. 14.25a, then the incoming light ray is refracted (transmitted) at $\theta_2 < 90°$ because medium 2 with $n_2 > n_1$ is transparent.
- If $\theta_1 = \theta_c$ in Fig. 14.25b, then the incident light ray is reflected at $\theta_2 = 90°$ along the n_1-n_2 interface, because the medium 2 with $n_2 < n_1$ may be transparent or not.
- If $\theta_1 > \theta_c$ in Fig. 14.25c, then the incoming light ray is reflected at $\theta_2 < 90°$, because medium 2 is not transparent. This is an internal reflection phenomenon in the medium 1 with $n_1 < n_2$.

Certain materials have the capacity to reflect and refract the incident light ray. This case is elaborated in the next section. The general index of refraction n (refractive index) of a material or medium is defined by the speed ratio or wavelength ratio

Table 14.4 Refractive indices (Smith [15, p. 835])

Medium	n	Medium	n	Material	n
Air	1.000	Silica glass	1.458	Polystyrene	1.595
Water	1.333	Quartz, SiO_2	1.555	Corundum, Al_2O_3	1.760
Acetone	1.360	Epoxy	1.580	Diamond	2.410

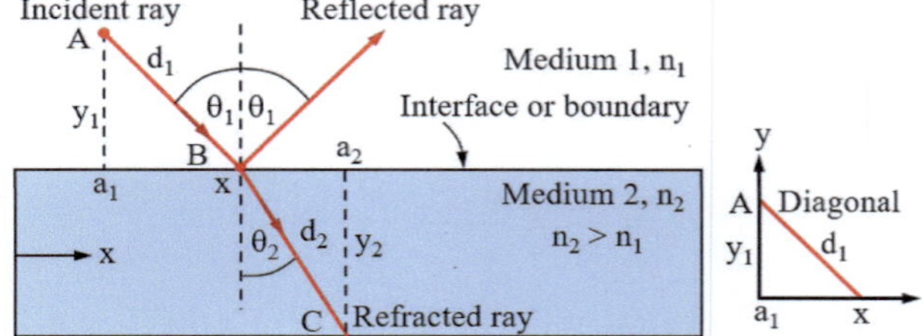

Fig. 14.26 Light reflection and refraction at a phase boundary

$$n = \frac{c}{\upsilon} = \frac{\lambda_{vac}}{\lambda_{mat}} \tag{14.33}$$

Here, $\upsilon < c = 3 \times 10^8 \ m/s$ is the speed of light ray or light beam in a dense material, where the light ray is commonly refracted at a certain angle $\theta < 90°$ with respect to a smooth surface normal. In effect, this expression, Eq. (14.33), can be applied to any medium as illustrated in Fig. 14.25. Moreover, Table 14.4 lists average refractive indices for some media (Schaffer et al. [1, p. 495], Smith [15, p. 835]).

14.4.3   Coupled Reflection-Refraction of Light

For a material being exposed to electromagnetic radiation, the light reflection and refraction are schematically shown in Fig. 14.26. This sketch represents the Pierre de Fermat's Principle of Least Time model for the path of light (diagonals d_1 and d_2) or optical path between two isotropic or homogeneous media.

The optical path denoted as diagonal d_1 in the medium 1 is the distance traveled by the light ray from point A to point B (interface or phase boundary) at an angle θ_1 with respect to the normal to the surface at point B. If the light ray is refracted into medium 2, then the light ray is bent and travels in the medium 2 along the optical path denoted by d_2 from point B to point C at angle θ_2.

In this ideal case, the light ray travels in straight lines in medium 1 and medium 2 having index of refraction n_1 and $n_2 > n_1$, respectively. Moreover, assume that the light ray speed in medium 1 is $\upsilon_1 > \upsilon_2$, where υ_2 is the speed in medium 2.

14.4.4　The Refraction Phenomenon

The analysis of a refraction phenomenon is a simple analytical procedure for deriving the so-called the **Snell's Law of Refraction**. From Fig. 14.26, the speed of light in medium 1 and medium 2 can be defined as

$$v_1 = \frac{d_1}{t_1} = \frac{\sqrt{y_1^2 + (x - a_1)^2}}{t_1} \tag{14.34a}$$

$$v_2 = \frac{d_2}{t_2} = \frac{\sqrt{y_2^2 + (x - a_2)^2}}{t_2} \tag{14.34b}$$

Rearranging and manipulating Eqs. (14.34a) and (14.34b) yield the total time for the light ray that travels from point A (source) to point C

$$t = t_1 + t_2 = \frac{\sqrt{y_1^2 + (x - a_1)^2}}{v_1} + \frac{\sqrt{y_2^2 + (a_2 - x)^2}}{v_2} \tag{14.35}$$

The time for this refraction phenomenon is simply an addition from A to B and from B to C points.

Letting $dt/dx = 0$ yields

$$\frac{x - a_1}{v_1 \sqrt{y_1^2 + (x - a_1)^2}} - \frac{a_2 - x}{v_2 \sqrt{y_2^2 + (a_2 - x)^2}} = 0 \tag{14.36}$$

Evaluating d^2t/dx^2 dictates either a minimum time $t_{\min}$ or maximum time $t_{\max}$ for the light ray to reach point x. Assume the former case as predicted by the Fermat's principle, which states that light takes a minimum path to the surface of a medium. Moreover, the prove of this statement is not necessary at this moment, since the main goal in this section is to derive the Snell's law using Eq. (14.36) and the geometry given in Fig. 14.26.

The corresponding trigonometric function are

$$\sin \theta_1 = \frac{x - a_1}{d_1} = \frac{x - a_1}{\sqrt{y_1^2 + (x - a_1)^2}} \tag{14.37a}$$

$$\sin \theta_2 = \frac{a_2 - x}{d_2} = \frac{a_2 - x}{\sqrt{y_2^2 + (x - a_2)^2}} \tag{14.37b}$$

Combining Eqs. (14.36), (14.37a), and (14.37b) yields the Snell's law of refraction

$$\frac{\sin \theta_1}{\sin \theta_2} = \frac{\upsilon_1}{\upsilon_2} = \frac{n_2}{n_1} \tag{14.38}$$

Rearranging Eq. (14.38) yields the Snell's law used in optics for designing curved mirrors and lenses

$$n_1 \sin \theta_1 = n_2 \sin \theta_2 \tag{14.39}$$

The physical meaning of the Snell's law, Eq. (14.39), can be interpreted as a constant sine-ratio equals the refractive index-ratio when the incident light ray crosses a media-boundary or travels from the less dense medium to a denser medium to get totally or partially refracted at an incident angle $\theta_1 \leq \theta_c$ (Fig. 14.25a), where θ_c is the critical incident angle of refraction.

The incident **critical angle**, $\theta_1 \to \theta_c$, is attained if the transmitted or reflected angle becomes $\theta_2 = 90°$ with respect to the boundary surface normal or interface normal. Mathematically, Eq. (14.39) along with $\sin \theta_2 = \sin(90°) = 1$ becomes

$$\sin \theta_c = \frac{n_2}{n_1} \quad \text{for } n_1 > n_2 \tag{14.40a}$$

$$\theta_c = \sin^{-1}\left(\frac{n_2}{n_1}\right) \quad \text{for } n_1 > n_2 \tag{14.40b}$$

Notice that $(n_1/n_2) \sin \theta_c \leq 1$ in Eq. (14.40a) since *sine* is bounded by ± 1 and its maximum value is used to calculate θ_c. In fact, θ_c represents the maximum angle for light ray refraction. If $\theta_1 > \theta_c$, then the incident light ray is reflected as shown in Fig. 14.25c.

If the light ray is reflected to an extent, the amount of reflected light can be predicted by the Fresnel's formula (Schaffer et al. [1, p. 497]). For a normal incidence light ray, the reflection coefficient or reflectivity is

$$R_f = \left(\frac{n_2 - n_1}{n_2 + n_1}\right)^2 \tag{14.41}$$

which is applicable to transparent materials. However, if $n_2 >> n_1$, then the light ray is $R_f = 100\%$ reflected and if $n_2 = n_1$, the light ray is $R_f = 0\%$ reflected and 100% refracted.

Example 14.8 Consider a light beam traveling **(a)** from air ($n_1 = 1$) into water ($n_2 = 1.333$) at an incident angle $\theta_1 = 30°$ and **(b)** from water into air at $\theta_1 = 30°$.

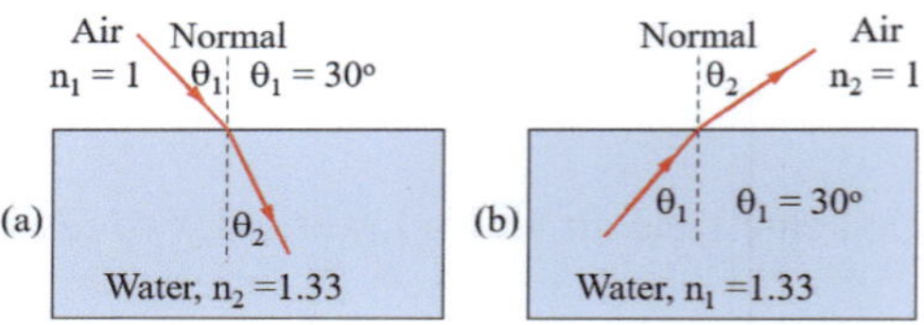

Use the Snell's law for both cases to calculate the critical light refraction angle θ_c, the percentage of reflected light R_f, and the speed of light in water. Assume that the speed of light in air is approximately equals to the speed of light in vacuum; $c = 3 \times 10^8\ m/s$.

Solution

(a) Using Snell's law for an incident light beam traveling from air into water yields

$$n_1 \sin \theta_1 = n_2 \sin \theta_2 \tag{14.8E1a}$$

$$\theta_2 = \sin^{-1}\left[\left(\frac{n_1}{n_2}\right) \sin \theta_1\right] = \sin^{-1}\left[\left(\frac{1}{1.33}\right) \sin (30°)\right] \tag{14.8E1b}$$

$$\theta_2 = \sin^{-1}\left[\left(\frac{1}{1.33}\right)(0.5)\right] = 0.38541 = 22.08° \tag{14.8E1c}$$

Similarly, the critical angle becomes

$$\theta_c = \sin^{-1}\left(\frac{n_1}{n_2}\right) = \sin^{-1}(1/1.33) = 0.85091\ rad = 48.75° \tag{14.8E2}$$

From Eq. (14.41), the percentage of reflected light is

$$R_f = \left(\frac{n_2 - n_1}{n_2 + n_1}\right)^2 = 100 \times \left(\frac{1.33 - 1}{1.33 + 1}\right)^2 \simeq 2\% \tag{14.8E3}$$

Therefore, approximately 2% of the incident light ray is reflected at the water-air boundary and 98% is refracted into the air medium. From Eq. (14.38), the speed of light in water is

$$v_2 = \frac{n_1}{n_2} v_1 = \left(\frac{1}{1.33}\right)\left(3 \times 10^8\ m/s\right) = 2.26 \times 10^8\ m/s \tag{14.8E4}$$

(b) Using Snell's law for incident light beam traveling from water into air yields

$$n_1 \sin \theta_1 = n_2 \sin \theta_2 \tag{14.8E5a}$$

$$(1) \sin \theta_1 = (1.33) \sin (30°) \tag{14.8E5b}$$

$$\theta_1 = \sin^{-1}\left[\left(\frac{1.33}{1}\right) \sin (30°)\right] = \sin^{-1}\left[\left(\frac{1.33}{1}\right)(0.5)\right] \tag{14.8E6a}$$

$$\theta_1 = 0.72749 = 41.68° \tag{14.8E6a}$$

The light beam enters the air with a lower index of refraction in air at $\theta_1 = 41.68°$.

From Eqs. (14.38) and (14.41), the critical angle and the percentage of reflected light are

$$\theta_c = \sin^{-1}\left(\frac{n_1}{n_2}\right) = \sin^{-1}(1/1.33) = 0.85091 \; rad = 48.75° \qquad (14.8E7a)$$

$$R_f = \left(\frac{n_2 - n_1}{n_2 + n_1}\right)^2 = 100 \times \left(\frac{1.33 - 1}{1.33 + 1}\right)^2 \simeq 2\% \qquad (14.8E7b)$$

From Eq. (14.38), the speed of light in water is

$$\frac{\upsilon_1}{\upsilon_2} = \frac{n_2}{n_1} \qquad (14.8E8a)$$

$$\upsilon_1 = \frac{n_2}{n_1}\upsilon_2 = \left(\frac{1.33}{1}\right)\left(2.26 \times 10^8 \; m/s\right) \qquad (14.8E8b)$$

$$\upsilon_1 = 3 \times 10^8 \; m/s \qquad (14.8E8c)$$

Example 14.9 Use the Snell's law to calculate the angle β for the light path shown below.

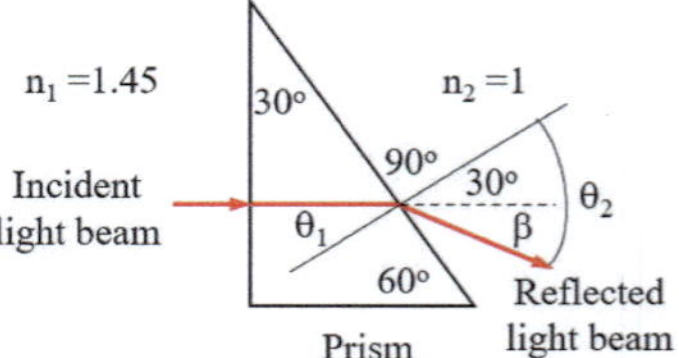

Solution The Snell's law with $\theta_1 = 30°$ gives

$$n_1 \sin\theta_1 = n_2 \sin\theta_2 \qquad (14.9E1a)$$

$$\sin\theta_2 = \frac{n_1}{n_2}\sin\theta_1 = \left(\frac{1.45}{1}\right)\sin(30°) = 0.725 \qquad (14.9E1b)$$

$$\theta_2 = \sin^{-1}(0.725) = 46.47° \qquad (14.9E1c)$$

Thus,

$$\beta = \theta_2 - 30° = 46.47° - 30 = 16.47° \qquad (14.9E2)$$

This result is shown in the given sketch.

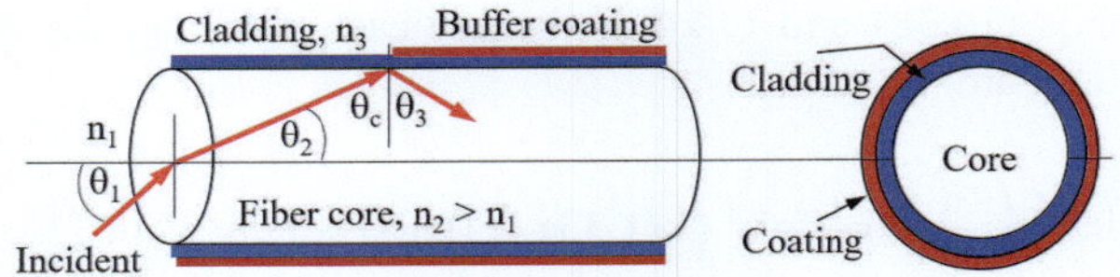

Fig. 14.27 Fiber optics and light path

14.4.5 Laser and Fiber Optics

The basic light sources for fiber optic or optical fiber are lasers and light-emitting diodes. The most common application of a fiber-optic network begins with glass fiber optic cables running from the network hub to an internet connection, which is the foundation of communications and computer networks. Hence, optical fiber technology is a transmission medium.

For clarity, the word "Laser" is very common in technological applications as a light source, and it is an acronym for Light Amplification by Stimulated Emission of Radiation. The most relevant characteristics of a laser are directionality, brightness, and monochromaticity (one wavelength). The most common device is the pulse ruby laser.

A light-emitting diode (LED) is a semiconductor that emits light of a particular color due to the photon wavelength. In order for a LED to work, current must flow through it. Technologically, LED applications include aviation lighting, traffic signals, automotive headlamps, and so forth.

The characterization of a light path within an optical fiber is schematically illustrated in Fig. 14.27. This is a simple model for representing the optical fiber technology, in which the refractive index of the optical fiber must be higher than the one for the cladding material; $n_2 > n_3$. This forces the light to bounce in a zigzag motion within the optical fiber.

This technology uses a single-mode fiber of approximately 10 μm in diameter or multiple-mode fiber with an overall diameter of 100 μm or more. Essentially, the optical fiber acts as the light-carrying core surrounded by cladding along a suitable length.

14.5 Summary

The physical properties of solid materials are normally based on electrical behavior or magnetic behavior. Regarding electrical behavior, solids can be classified as metallic, ionic, semiconducting, and superconducting materials. On the other hand, magnetic behavior can be classified into ferromagnetic, ferrimagnetic, paramagnetic, and diamagnetic materials.

A material in an electrical circuit being exposed to an electric field strength $\vec{E}$ may exhibit a conductive behavior, provided that there exists an electron (e^-) flow from the negative-to-positive terminals. If a material shows a nonconductive behavior, then it is considered as an insulator. Moreover, semiconductors fall between these extreme cases and are classified as intrinsic semiconductors (pure material) or as extrinsic semiconductors containing impurities deliberately added (doped).

The material atomic structure dictates the electrical and magnetic properties of conductive materials. Thus, electricity is attributed to the movement of electrons in the conduction energy band (electron energy states), while magnetism is attributed to atoms acting as tiny permanent magnets with magnetic domains related to electron spin-motion around discrete orbitals.

Optical properties of solid materials are related to the interaction of light. In general, optical properties are determined based on the absorption, dispersion, and emission of light. Specifically, the refraction and reflection of an incident light beam can be assessed by using Snell's law.

Problems

14.1 Consider an aluminum (Al) rod with $L = 1.2\text{-}m$, $A_c = 1.2\ mm^2$, and $n_e = 1.92 \times 10^{29}\ e^-/m^3$. Assume a current of $I = 5\ A$ carried by electrons. Calculate ρ_v and N_e (the number of mobile electrons per atom). Given data: for aluminum, electrical conductivity $\sigma = 4 \times 10^7\ (\Omega.m)^{-1}$, mass density $\rho_m = 2.71\ g/cm^3$, and atomic weight $A_w = 26.98\ g/mol$. [Solution: $\rho_v = 6.05 \times 10^{28}\ atoms/m^3$, $N_e \simeq 3\ e^-/atom$].

14.2 Consider an aluminum (Al) rod with $L = 1.2\text{-}m$, $A_c = 1.2\ mm^2$, and $n_e = 2 \times 10^{29}\ e^-/m^3$. Assume a current of $I = 2\ A$ carried by electrons and an electric resistance of $R_x = 5\ \Omega = 5\ V/A$. Calculate ρ, μ_x, E_x, and N_e (the number of mobile electrons per atom). Given data: mass density $\rho_m = 2.71\ g/cm^3$ and atomic weight $A_w = 26.98\ g/mol$. [Solution: $\rho = 0.005\ \Omega.m$, $\mu_x = 6.24 \times 10^{-9}\ m^2/(V.s)$, $E_x = 8.33\ V/m$, $N_e \simeq 3\ e^-/atom$].

14.3 Consider the energy level of silicon (Si) lying at $E - E_F = 1.10\ eV$. Calculate **(a)** the probability of finding an electron at this energy level, **(b)** the number of electrons per unit volume, and **(c)** the electric conductivity σ of an intrinsic silicon specimen at $T = 303\ K$. Data: The mobility of electrons is $\mu_x = 0.14\ m^2/(\Omega.C) = 0.14\ m^2/(V.s)$. [Solution: (a) $f(E) = 5.13 \times 10^{-19}$, (b) $n_e = 1.30 \times 10^7\ e^-/m^3$, (c) $\sigma = 2.92 \times 10^{-13}\ (\Omega.m)^{-1}$].

14.4 This problem requires the use of Eqs. (14.2a) and (14.2b) along with $n = 1$ and $B_n = \sqrt{2/L}$ to calculate the probability of finding a particle at $0 \le x \le L/4$ interval within an energy well. [Solution: $P(x) = 0.0909$].

14.5 Assume that the atom density of iron (Fe) at room temperature is $N \simeq 8 \times 10^{28}\ atoms/m^3$. Calculate the magnitude of the saturation magnetization M_s. Given data: $\rho = 7.87\ g/cm^3$ (mass density), $N_a = 6.022 \times 10^{23}\ atom/mol$ (Avogadro's number), $\mu_B = 9.27 \times 10^{-24}\ A.m^2$ (Bohr magneton) and $\mu_o = 4\pi \times 10^{-7}\ T.m/A$ (vacuum permeability). [Solution: $M_s = 7.42 \times 10^5\ A/m$].

14.6 Assume the classical electromagnetism conditions for a coil of Cu-wire shown in Example 14.6. In this case, double the current for calculating **(a)** the magnetic field strength H, **(b)** the flux density B_o in vacuum, **(c)** the flux density B inside a bar of paramagnetic aluminum (Al), and **(d)** the magnitude of the magnetization M. [Solution: (a) $H = 40,000\ A/m$, (b) $B_o = 5.03 \times 10^{-2}\ Tesla$, (c) $B = 5.03 \times 10^{-2}\ Tesla$, (d) $M = 0.83\ A/m$].

14.7 Consider a hypothetical semiconductor undergoing energy change from the conducting energy band to the valence energy band. Calculate **(a)** the photon energy ΔE and **(b)** the valence energy band E_v if the conduction energy band is $E_c = 1.40$ eV. The wavelength for a visible red light is $660\ nm$. [Solution: (a) $\Delta E = 1.88\ eV$, (b) $E_v = 3.28\ eV$].

14.8 Assume that a light ray travels from liquid parafin ($n_1 = 1.48$) to air ($n_2 = 1$). Determine **(a)** the critical light refraction angle θ_c and **(b)** the percentage of reflected light R_f. [Solution: (a) $\theta_c = 42.51°$, (b) $R_f = 3.75\%$].

14.9 Assume that the classical electromagnetism conditions for a coil of Cu-wire shown in Example 14.6. In this case, let the current and the magnetization be $I = 10$ A and $M = 0.78\ A/m$, respectively. Calculate **(a)** the magnetic field strength H, **(b)** the flux density B inside a bar of paramagnetic aluminum (Al) and **(c)** the number of turns of the Cu-wire for a coil length of $L = 0.30\ m$. [Solution: (a) $H = 37,681$ A/m, (b) $B = 4.74 \times 10^{-2}\ Tesla$, (c) $N = 1,130\ turns$].

14.10 Assume that the classical electromagnetism conditions for a coil of Cu-wire shown in Example 14.6. In this case, let the current and the magnetization be $I = 10$ A and $M = 0.78\ A/m$, respectively. Calculate **(a)** the magnetic field strength H, **(b)** the flux density B inside a bar of paramagnetic chromium (Cr), and **(c)** the number of turns of the Cu-wire for a coil length of $L = 0.30\ m$. [Solution: (a) $H = 2,492$ A/m, (b) $B = 3.13 \times 10^{-3}\ Tesla$, (c) $N = 75\ turns$].

14.11 Assume that a light ray travels from yellowish vegetable oil ($n_1 = 1.47$) to air ($n_2 = 1$). Use the given refractive indices to determine **(a)** the critical light refraction angle θ_c and **(b)** the percentage of reflected light R_f. [Solution: (a) $\theta_c = 42.87°$, (b) $R_f = 3.62\%$].

14.12 Suppose light travels from air to a window glass and then to air as shown below along the index of refraction for air and glass.

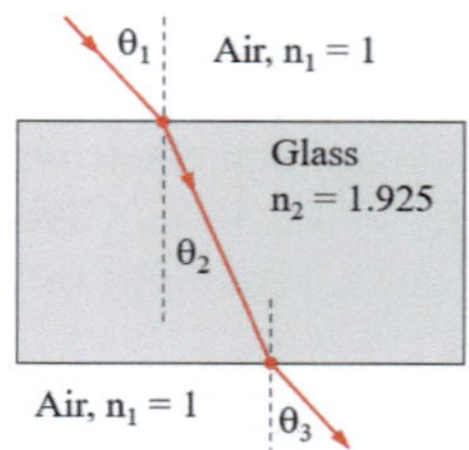

If the percentage of reflected light at the air-glass surface is $R_f = 10\%$, then calculate **(a)** the refractive index of the window glass, **(b)** the critical light refraction angle θ_c and the reflected angle θ_2 using the Snell's law, and **(c)** the reflected light angle θ_3 at the glass-air surface. [Solution: (a) $n_2 = 1.925$, (b) $\theta_c = 31.30°$ and $\theta_2 = 89.82°$, (c) $\theta_3 = 31.30°$].

14.13 Assume that light travels from air (with an index of refraction of $n_1 = 1$) into a yellowish vegetable oil with an index of refraction of $n_2 = 1.47$, and consequently, light will bend toward the normal. **(a)** If the light angle of incidence is $20°$, then calculate the angle of reflection inside the vegetable oil. **(b)** Sketch the path of the light. [Solution: (a) $\theta_2 = 0.23482\ rad = 13.08°$].

14.14

(a) Show that the critical angle for light reflection between two different media is

$$\theta_c = \sin^{-1}\left[\frac{\sin(\theta_2)}{\sin(\theta_1)}\right] \tag{1}$$

(b) If $\theta_1 = 30°$ and $\theta_2 = 50°$, then find θ_c. [Solution: (b) $\theta_c = 35.26°$].

14.15 Consider the energy level of silicon (Si) lying at $E - E_F = 1.12\ eV$. Calculate **(a)** the probability of finding an electron at this energy level, **(b)** the number of electrons per unit volume and **(c)** the electric conductivity σ of an intrinsic silicon specimen at $T = 300\ K$. Data: The mobility of electrons is $\mu_x = 0.14\ m^2/(\Omega.C) = 0.14\ m^2/(V.s)$. [Solution: (a) $f(E) = 1.55 \times 10^{-19}$, (b) $n_e = 3.88 \times 10^6\ e^-/m^3$, (c) $\sigma = 8.70 \times 10^{-14}\ (\Omega.m)^{-1}$].

14.16 Calculate the number of electrons per unit volume at the Fermi level, so that $E = 0$, $E_F = 1.12\ eV$ and $f(E) = 1$. [Solution: $n_e = 0.33\ e^-/m^3$].

14.17 Calculate the probability that a state located at the bottom of the silicon conduction band with a energy gap $E_g = 1.12\ eV$ and Fermi level $E_F = (E + E_v)/2$ is filled at $T = 300\ K$. Let $E = E_c$ and $k_B = 8.62 \times 10^{-5}\ eV/K$. [Solution: $f(E) = 3.94 \times 10^{-10}$].

14.18 Use Eqs. (14.2a) and (14.2b) along with $n = 1$ and $B_n = \sqrt{2/L}$ to calculate the probability of finding the particle at $0 \leq x \leq L/3$ within an energy well. [Solution: $P(x) = 0.1955$].

14.19 Assume that the electric conductivity of silver and the number of conduction electrons are $\sigma = 7 \times 10^7 \, ohm/m \, (= A/m.V)$ and $n_e = 7 \times 10^{28} \, e^-/m^3$. Calculate (a) the electron mobility of the conduction electrons μ_x and (b) the drift velocity υ in the electric field gradient of $E_x = 1.5 \, V/m$. Given data: $m_e = 9.11 \times 10^{-31} \, Kg$ and $q_e = 1.602 \times 10^{-19} \, C \, (= A.s)$. [Solution: (a) $\mu_x = 6.24 \times 10^{-3} \, m^2/(V.s)$, (b) $\upsilon = 9.36 \, mm/s$].

14.20 This problem requires the use of Eqs. (14.2a) and (14.2b) along with $n = 1$ and $B_n = \sqrt{2/L}$ to calculate the probability of finding the particle at $0 \le x \le L/2$ within an energy well. [Solution: $P(x) = 0.50$].

References

1. J. Schaffer, A. Saxena, S.D. Antolovich, T.H. Sanders, Jr., S.B. Warner, *The Science and Design of Engineering Materials*, 2nd edn. (McGraw-Hill Companies, Inc., New York, 1999)
2. W. Park, *Module 1.4 video*. Department of Electrical, Computer & Energy Engineering, University of Colorado, Boulder, Colorado State. The video is at shorturl.at/FIUW3
3. W. Shockley, *Electrons and Holes in Semiconductors* (D. Van Nostrand Company, Inc., New York, 1950)
4. S.S. Li, *Semiconductor Physical Electronics*, 2nd edn. (Springer Science+Business Media, LLC, Berlin, 2006)
5. S.O. Kasap, *Principles of Electronic Materials and Devices*, 4th edn. (McGraw-Hill, New York, 2018)
6. S.O. Pillai, S. Pillai, *Rudiments of Material Science* (New Age International Publisher, New Delhi, 2012)
7. W.D. Callister, D.G. Rethwisch, *Materials Science and Engineering: An Introduction*, 9th edn. (Wiley, New York, 2014)
8. O.H. Wyatt, D. Dew-Hughes, *Metals, Ceramics and Polymers* (Cambridge University, Cambridge, 1974)
9. B.D. Cullity, C.D. Graham, *Introduction to Magnetic Material* (Wiley, Hoboken, 2009)
10. D. Jiles, *Introduction to Magnetism and Magnetic Materials* (Springer-Science+Business Media, B.V., Switzerland, 1991)
11. L. Zhou, C. Davis, P. Kok, F. Van Den Berg, S. Labbe, A. Martinez-De-Guerenu, D. Jorge-Badicla, I. Gutierrez, *Magnetic NDT for steel microstructure characterisation—modeling the effect of ferrite grain size on magnetic properties*, in *19th World Conference of Non-Destructive Testing, WCNDT 2016*, vol. 21 (2016)
12. K. Honda, S. Kaya, On the magnetization of single crystals of iron. Sci. Rep. Tohoku Univ. **15**, 721 (1926)
13. S. Kaya, On the magnetization of single crystals of nickel. Sci. Rep. Tohoku Univ. **17**, 639 (1928)
14. K. Fabian, V.P. Shcherbakov, S.A. McEnroe, Measuring the Curie temperature. Geochem. Geophys. Geosyst. **14**, 947–961 (2013)
15. W.F. Smith, *Principles of Materials Science and Engineering*, 3rd edn. (McGraw-Hill Inc., New York, 1996)

Chapter 15
Nondestructive Methods

15.1   Introduction

This chapter contains fundamentals of physics related to nondestructive evaluation (NDE) of structural components using ultrasonic waves, ultrasonic thermography (UT), radiography, magnetic particle, eddy currents, and dye penetrant or liquid penetrant. Actually, NDE is also known as nondestructive testing (NDT), nondestructive inspection (NDI), and nondestructive examination (NDE).

In particular, sound waves are vibrations carrying vibrational energy that is transmitted through a medium (gases, liquids or solids) and are classified as pleasant sound (speech, music, sound of birds, etc.) or an unpleasant (noise) audible waves. On the other hand, ultrasonic waves are high-frequency vibrations and are inaudible sound waves to humans.

Commonly, audible sound waves propagate in space having a frequency (f) range related to the speed of sound (v) in decibels (dB) units. Specifically, the frequency range of human hearing (audible range of sound) is $20\ Hz \leq f \leq 20 \times 10^3\ Hz$, which is converted to $0\ dB \leq L_p \leq 140\ dB$ at ambient temperature. The extreme values represent the threshold of hearing and the threshold of pain, respectively. Conversely, non-audible sound waves are treated as ultrasonic waves that are suitable for quality control inspection, say, for crack detection in metals. Fundamentally, ultrasonic waves are the principal source for nondestructive testing evaluation (NDE) of objects.

Practical soundness testing (NDT) methods are in great demand for quality control and corrective actions (if any) to assure the integrity of structural components. A quality control protocol may require testing samples or products from a manufacturing line, inspect, or evaluate the physical condition (soundness) of old or in-place structural components subjected to develop cracks or corrode in aggressive environments.

Nondestructive methods are used in the engineering field to assure the quality, reliability, and soundness of structural parts that must meet specifications and

N. Perez, *Materials Science: Theory and Engineering*,
https://doi.org/10.1007/978-3-031-57152-7_15

tolerances. Actually, some resultant findings from a nondestructive procedure are related to design codes, theoretical requirements, or customer needs.

15.2   Waves

In general, waves are rhythmic disturbances of matter that carry energy in the form of oscillations without carrying matter through a medium (gas, liquid, or solid). Whereas, matter is referred to as particles, such as atoms or molecules. For clarity,

Mechanical Waves These are matter-dependent oscillations for propagating at a certain speed, frequency, and amplitude. This means that mechanical waves can propagate through matter, not through space (vacuum). The most common mechanical waves are water waves induced by water ripples moving up and down, audible sound waves produced by clapping hands, rumbling noise caused by engines, roaring thunders induced by rapid expansion of atmospheric gases, and inaudible sound waves produced at very high frequencies for nondestructive evaluation of solid discontinuities (flaws or cracks).

Electromagnetic Waves Electromagnetic radiation is a synonym for electromagnetic waves, which are classified as radio waves, infrared waves, visible light waves, and so forth In essence, electromagnetic waves can travel through space and matter.

The radiant energy from the sun is the most important type of electromagnetic waves suitable for life on earth and subsequently, the frequency of electromagnetic waves, in general, represents the number of repetitive occurrences per unit of time and its unit is normally in $Hertz = Hz = 1/sec$.

15.2.1   Classification of Waves

The speed of any wave depends upon the elastic properties (modulus of elasticity E) and inertial properties (mass density ρ) of the phase of a medium (solid, liquid, and gas). It is commonly characterized by its amplitude (A), wavelength (λ), and frequency (f). For the sake of clarity, a homogeneous elastic medium has a constant density, and an isotropic system has elastic constants along the principal axes in Cartesian coordinates.

Below is a list of mechanical waves related to the concept of longitudinal waves, transverse waves, or surface waves having particular characteristics, (such as frequency, wavelength, wave velocity) in certain media:

- **Plane waves** are waves with imaginary wavefronts perpendicular to the direction of propagation.
- **Sound waves** are a manifestation of a disturbance caused by the motion of energy traveling through a medium.

- **Acoustic waves** are mechanical waves having the same direction of vibration and propagation.
- **Ultrasonic waves** are inaudible sound waves with high frequency (f) for humans to hear beyond the threshold of pain; $f \geq 20\ kHz$. These type of waves are suitable for nondestructive evaluation (NDE) of solids.
- **Standing waves** are regular and evenly spaced waves moving back and forth causing an interference.
- **Rayleigh waves** are traveling surface waves.
- **Lamb waves** are waves that propagate in a plate.
- **Love waves** are propagating waves in a layer of a substrate.
- **Stoneley waves** are those that propagate along a solid-solid boundary or interface.
- **Guided waves** can propagate long distances from a transducer.

Despite that these waves have different characteristics, they are treated as either longitudinal or transverse waves carrying energy.

15.2.2 Traveling Waves

Consider one harmonic wave shown in Fig. 15.1a and assume that the harmonic wave travels along the $+x$-axis with constant wavelength λ, velocity υ, and pressure P.

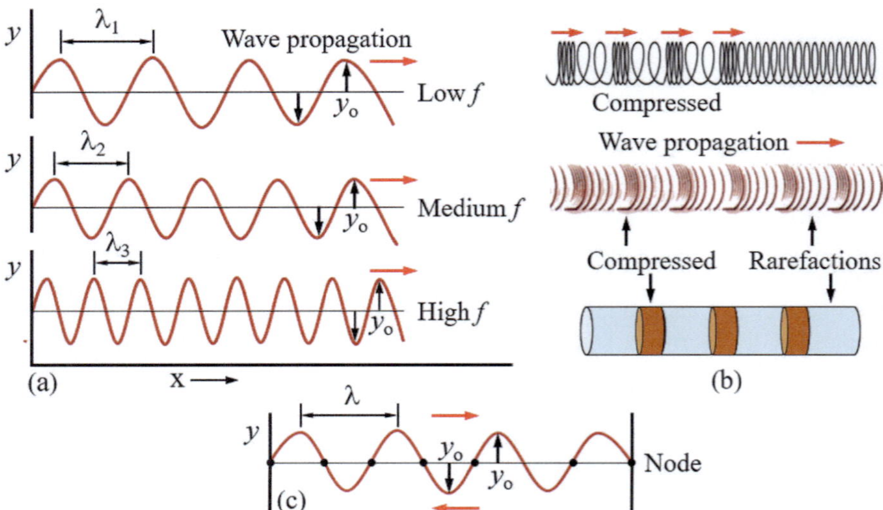

Fig. 15.1 (**a**) Transverse sound waves vibrating perpendicular to the direction of energy transfer, (**b**) longitudinal sound waves vibrating along the direction of energy transfer, and (**c**) stationary wave with fixed ends

Sound, fundamentally, is a manifestation of a mechanical disturbance that travels through an elastic medium at a certain speed due to sound pressure waves. The wave sound speed is characteristic of the medium (gas, liquid, or solid), and the sound wave behavior is commonly characterized using intensity-time diagrams for a particular sound frequency (f) and wavelength λ as shown in Fig. 15.1a for transverse waves (sound, water, radio, light, heat waves) based on up and down motion perpendicular to the direction of the propagating waves with an amplitude y_o.

Further, Fig. 15.1b for a longitudinal wave (speech, loud speaker waves) is based on a compressed element motion parallel to the direction of the propagating wave. Lastly, Fig. 15.1c illustrates the stationary wave as standing wave with fixed ends and common nodes along the horizontal line, where the $y = y_o = 0$.

The wave equation (Appendix 15A) for a traveling wave of the form $y = y(x, t)$ and its general solution without attenuation (loss or reduction of signal strength) are

$$\frac{\partial^2 y}{\partial t^2} = v^2 \frac{\partial^2 y}{\partial x^2} \tag{15.1a}$$

$$y = y_o \sin(kx - \omega t) \quad \text{(traveling)} \tag{15.1b}$$

$$y = y_o \sin(kx) \qquad \text{(stationary)} \tag{15.1c}$$

with the wavenumber k, angular frequency ω, frequency f, velocity v, and wavelength λ being defined by

$$k = \frac{2\pi}{\lambda} \tag{15.2a}$$

$$\omega = \frac{2\pi}{f} \tag{15.2b}$$

$$v = \frac{\omega}{k} = \frac{\lambda}{f} \tag{15.2c}$$

Notice that the stationary $\sin(kx)$ wave (standing wave with fixed ends) in Eq. (15.1c) becomes a traveling wave if kx is replaced by $k(x - vt)$ so that $\sin(kx) \rightarrow \sin(kx - \omega t)$ with the aid of Eq. (15.2c). Here, $y = y(x, t)$ denotes the one-dimensional spatial (space) and temporal (time) dependent variable, such as particle displacement, sound pressure or particle velocity in an elastic medium. Here, y_o is the amplitude of the wave, and the velocity of the traveling wave is at $v < c = 3 \times 10^8 \ m/s$ (speed of light in vacuum).

The solution to Eq. (15.1a) can be derived using the separation of variables method (see Appendices in Chaps. 7 and 11), which requires that $y(x, t) = H(x) K(t)$, where $H(x)$ and $K(t)$ are unknown functions related to the boundary conditions (BCs)

- $y = y(0, t) = 0$ for all $t > 0$
- $y = y(L, t) = 0$ for all $t > 0$

and to the initial conditions (ICs)

- $y = y(x, 0) = 0$ for all $0 < x < L$
- $y = y(x, 0) = f(x)$ for all $0 < x < L$

The analytical procedure for solving Eq. (15.1a) using the above BCs and ICs can be found at https://shorturl.at/9VO0c.

Example 15.1 Show that the given exponential function in complex form is also a solution of Eq. (15.1a).

$$y = y_o \exp\left[i\left(kx - \omega t\right)\right] \qquad (15.1\text{E}1)$$

with $i = \sqrt{-1}$ and $t > 0$. Assume that this complex exponential function depends on any real number t.

Solution The first- and second-order partial derivatives are

$$y = y_o \exp\left[i\left(kx - \omega t\right)\right] \qquad (15.1\text{E}2a)$$

$$\frac{\partial y}{\partial x} = iky_o \exp\left[i\left(kx - \omega t\right)\right] \qquad (15.1\text{E}2b)$$

$$\frac{\partial^2 y}{\partial x^2} = -k^2 y_o \exp\left[i\left(kx - \omega t\right)\right] = -k^2 y \qquad (15.1\text{E}2c)$$

$$\frac{\partial y}{\partial t} = -i\omega y_o \exp\left[i\left(kx - \omega t\right)\right] \qquad (15.1\text{E}2d)$$

$$\frac{\partial^2 y}{\partial t^2} = -\omega^2 y_o \exp\left[i\left(kx - \omega t\right)\right] = -\omega^2 y \qquad (15.1\text{E}2e)$$

Manipulating Eqs. (15.1E2c) and (15.1E2e) gives

$$y = -\frac{1}{k^2}\frac{\partial^2 y}{\partial x^2} \qquad (15.1\text{E}3a)$$

$$y = -\frac{1}{\omega^2}\frac{\partial^2 y}{\partial t^2} \qquad (15.1\text{E}3b)$$

Equating Eq. (15.1E3a) and (15.1E3b), rearranging the resultant expression, and using Eq. (15.1E1a) yield

$$\frac{\partial^2 y}{\partial t^2} = \frac{\omega^2}{k^2}\frac{\partial^2 y}{\partial x^2} \qquad (15.1\text{E}4a)$$

$$\frac{\partial^2 y}{\partial t^2} = v^2\frac{\partial^2 y}{\partial x^2} \qquad (15.1\text{E}4b)$$

Hence, Eq. (15.1E1) is also the solution of Eq. (15.1a).

15.3   Sound Waves

In the ultrasonic field, ultrasonic waves are generated by a probe (source), and the wave (motion) propagation into a material is studied as an oscillation of waves called vibration. For instance, the speed of any mechanical wave (longitudinal or transverse) depends on an elastic property of the medium, such as the modulus of elasticity or shear modulus, to store potential energy and on an inertial property of the medium, such as density, to store kinetic energy. Specifically, the speed of sound is a highly medium-dependent scalar, and it is mathematically derived in Appendix 15A for solids and in Appendix 15B for air or gases. Additional details on the speed of sound, known as acoustic velocity, are also given below. Nonetheless, speed is a scalar quantity described by its magnitude, while velocity or force is a vector with magnitude and direction. Sound waves are normally characterized by their particular frequency range as illustrated below:

- Infrasound at $f < 20\ Hz$
- Acoustic sound at $20\ Hz \leq f \leq 20 \times 10^3\ Hz$
- Ultrasound at $f > 20 \times 10^3\ Hz = 20\ kHz$

Assume that a high-frequency incident longitudinal wave strikes a metal surface (medium) at an incident angle $\theta_i = \theta_1$ (Fig. 15.2). According to Snell's law introduced in Chap. 14, the incident wave is split at the point of impact between media interface. A portion of it may be refracted at an angle θ_2 within the sample, and a portion of it may be reflected at an angle $\theta_r = \theta_3$ (Bray and Stanley book [1, pp. 66–58]).

The wave-split model shown in Fig. 15.2 is an ultrasonic mode conversion due to the presence of an interface with different characteristics, such as impedance. Consequently, the ultrasonic wave may reflect many times at the point of impact due to at least one defect such as a crack or pore in a structural component.

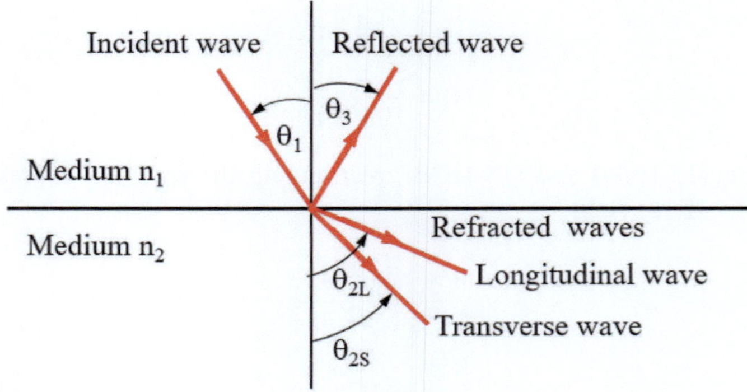

Fig. 15.2   Snell's law for a reflected or refracted ultrasound wave at an angle of incidence relative to a medium interface

Insights on harmonic and ultrasound waves can be found elsewhere (Shull and Tittmann [2, Chapter 3], Raichel [3, Sec. 5.2], Kinsler et al. [4, Chapter 5]).

15.3.1 *Speed of Sound in a Medium*

The speed of sound or wave speed depends upon the density, temperature, pressure, and elasticity of the medium, and it is produced by vibrational energy in a medium.

Mathematically, the speed of sound in various media is defined below along with specific values related to solid, liquid, and vapor (gas) phases

$$\upsilon = \sqrt{\frac{C_{ij}}{\rho}} \tag{15.3a}$$

$$\upsilon = \sqrt{\frac{E}{\rho}}; \quad \upsilon \simeq 5048 \ m/s \text{ in ASTM A36 steel} \tag{15.3b}$$

$$\upsilon = \sqrt{\frac{B}{\rho}}; \quad \upsilon \simeq 1449 \ m/s \text{ in liquid water} \tag{15.3c}$$

$$\upsilon = \sqrt{\frac{3k_B T}{m}}; \quad \upsilon \simeq 350 \ m/s \text{ in } N_2 \text{ gas at } 20\,^{\circ}C \tag{15.3d}$$

$$\upsilon = \sqrt{\frac{\gamma P}{\rho}}; \quad \upsilon \simeq 343 \ m/s \text{ in air at } 20\,^{\circ}C \tag{15.3e}$$

where C_{ij} = Elastic stiffness constant (MPa) (See Appendix 15A) $E = \sigma/\varepsilon =$ Modulus of elasticity (GPa)

$\sigma =$ Elastic stress (MPa)
$\varepsilon =$ Elastic strain
$B = -\Delta P/(\Delta V/V_o) =$ Adiabatic bulk modulus
$k_B = 8.62 \times 10^{-5} eV/K = 1.38 \times 10^{-23} \ J/K =$ Boltzmann constant
$P =$ Pressure $(Pa \text{ or } kPa)$
$\Delta P =$ Change in pressure $(Pa \text{ or } kPa)$
$V_o =$ Initial volume (m^3)
$\Delta V =$ Change in volume (m^3)
$\rho = m/V = 1/v =$ Mass density of medium (Kg/m^3)
$m =$ Mass (Kg)
$v =$ Specific volume (m^3/Kg)
$\gamma =$ Specific heat capacity ratio or adiabatic index for an ideal gas
$c_p, c_v =$ Specific heat capacities at constant P and constant V, respectively

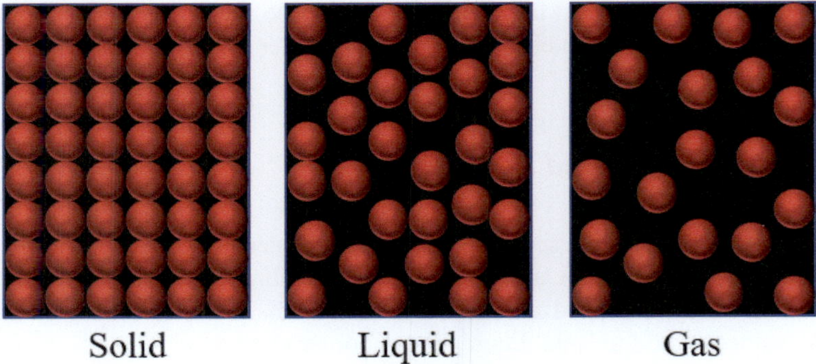

Solid   Liquid   Gas

Fig. 15.3 Atomic models for sound traveling as a wave through a medium such as solid, liquid, or gas. The spheres represent an arrangement of atoms in the medium

The elastic compliance constant C_{ij} in Eq. (15.3a) is the solid resistance to deformation along a certain crystallographic direction, while the modulus of elasticity E in Eq. (15.3b) is the average resistance to elastic deformation.

Moreover, Eq. (15.3e) is known as Laplace's adiabatic equation, which is derived in Appendix 15B as Eq. (15B.7). The reader is encouraged to consult suitable references for additional insights on sound (Ling et al. [5, Chapter 17], Krout and Sohrab [6]).

15.3.2   Medium Atomic Arrangement

The speed of sound is strongly dependent on the atomic arrangement of matter or medium as schematically shown in Fig. 15.3.

In the solid phase, the atoms are more compacted than in the liquid and gas phases, and therefore, the speed of sound is high in solids because the energy of the sound wave is transferred to adjacent atoms more readily. In summary, the atomic rearrangement of atoms in the lattice is a major contribution to the speed of sound.

Sound travels faster through a solid than through liquids and gases due to the massive vibrations of atoms in a compact arrangement, where the interatomic distance between adjacent atoms is very short. Although sound travels fastest through solids, the speed of sound in all solids is not the same because of the atomic structure. For instance, sound travels faster through copper than through lead despite that they have the same FCC structure.

For an adiabatic process without heat transfer, the equation of state for an ideal gas along with the mass density (ρ) and specific volume (v) relationship ($v = 1/\rho$) is written as

$$Pv = RT \tag{15.4a}$$

$$Pv^\gamma = P\left(\frac{1}{\rho}\right)^\gamma = C \tag{15.4b}$$

$$\gamma = \frac{c_p}{c_v} \tag{15.4c}$$

Combining Eqs. (15.3e) and (15.4b) for air at T yields the temperature-dependent speed of sound of an adiabatic wave without the influence of heat transfers by conduction or radiation

$$v = \sqrt{\frac{\gamma P}{\rho}} = \sqrt{\gamma RT} = \sqrt{\frac{\gamma RT}{M_{gas}}} \tag{15.5}$$

where $R = 8.3145 \; J/(mol.K)$ is the universal gas constant and M_{gas} is its molecular mass.

Example 15.2 Calculate and plot the speed of sound (v) for air at the temperatures given in the table below using Laplace's adiabatic theory. The given thermodynamics dataset is taken from Lemmon et al. [7, Table A2, p. 366] at 101.325 kPa. Explain.

T	c_p	c_v	v	v_{cal}
(K)	$(J/mol.K)$	$(J/mol.K)$	(m/s)	(m/s)
280	29.13	20.77	335.6	335.47
300	29.15	20.80	347.4	347.25
320	29.18	20.83	358.7	358.64
340	29.21	20.87	369.7	369.67
360	29.26	20.92	380.3	380.39

Solution If air contains $f_{m,N_2} = 78\%$, $f_{m,O_2} = 21\%$, and $f_{m,Ar} = 1\%$, then use the molecular weight of these elements (28 g/mol, 32 g/mol, and 40 g/mol, respectively) to calculate the molecular of air. Thus,

$$M_{air} = \sum f_m M_i = 28.02 \times 0.78 + 32 \times 0.21 + 40 \times 0.01 \tag{15.2E1a}$$

$$M_{air} = 28.96 \; g/mol = 0.02896 \; Kg/mol \tag{15.2E1b}$$

If $\gamma = c_p/c_v = 1.40$ at all selected temperatures, then Eq. (15.5) yields

$$v = v_{cal} = \sqrt{\frac{\gamma RT}{M_{gas}}} = \sqrt{\frac{(1.40)\,(8.3145 \; J/mol.K)\,T}{0.02896 \; Kg/mol}} \tag{15.2E2a}$$

$$v_{cal} = \sqrt{\left(401.94 \; \frac{N.m}{Kg.K}\right) T} = \sqrt{\left(401.94 \; \frac{Kg.m^2/s^2}{Kg.K}\right) T} \tag{15.2E2b}$$

$$v_{cal} = \sqrt{\left(401.94\,\frac{m^2/s^2}{K}\right) T} \qquad\qquad (15.2E2c)$$

The calculated speed of sound (v_{cal}) values are tabulated above. Plotting the given and calculated datasets using Eq. (15.2E2c) gives a linear relationship mathematically defined the by a general function $v = f(T)$. Thus,

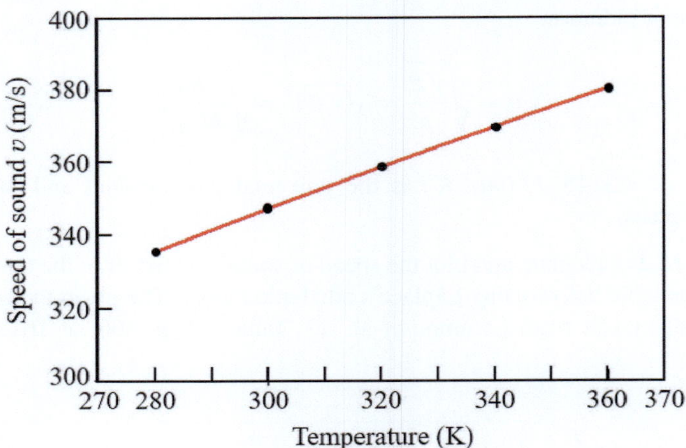

Notice that the given v and calculated v_{cal} data differ very slightly within the selected temperature range.

15.4   Ultrasonic Testing

Consider the ultrasonic wave technique for detection of an internal defect in a solid specimen as schematically shown in Fig. 15.4.

The basic principles of ultrasonic testing (UT) are the propagation and reflection of sound waves in materials. Commonly, the ultrasonic wave approach is a pulse-echo technique for detecting defects in solids. The wave propagation is induced by a transmitter, and the wave reflection is induced by an echo at a defect (crack or pore in solids) or at back wall of a sample, such as a bar.

The common components of a beam transducer are the transmitter, the receiver, and a computer monitor or oscilloscope for capturing the signal spectrum. For instance, Fig. 15.4a schematically shows a right-angle transducer emitting an ultrasonic wave as a pulse. The position of the probe does not detect the internal defect (crack or inclusion), but a back wall signal is reflected from the back wall as an echo, shown in the screen diagram as a time-dependent peak intensity.

If the probe is moved to the right (Fig. 15.4b), then the defect is detected, and the signal is a defect echo with a small signal intensity. Actually, ultrasonic probes can

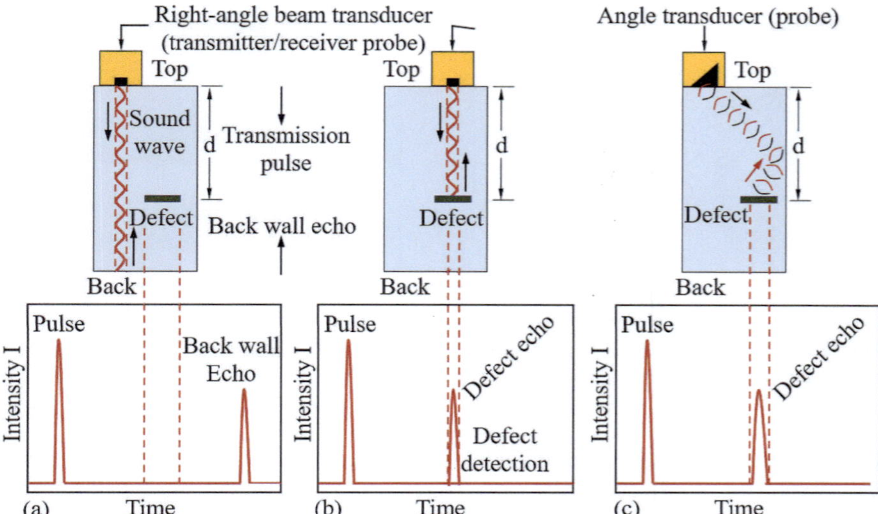

Fig. 15.4 Ultrasonic wave technique for detection of internal defects using a right-angle beam transducer (transmitter/receiver probe). (**a**) Ultrasonic pulse wave and back wall echo (reflection wave) and (**b**) ultrasonic pulse wave and defect reflection echo from a depth d and (**c**) ultrasonic wave pulse and reflection echo at an angle

be designed to send wave pulses at an angle as shown in Fig. 15.4c, which contains arcs representing the transmitting (black) and receiving (red) wave signals (Bray and Stanley book [1, pp. 113–115]).

Some ultrasonic testing equipment may contain a very useful feature for plotting intensity and detecting defect depth. Thus, an signal intensity and defect depth are used to construct an ultrasonic diagram for defect location. For instance, Fig. 15.5a illustrates the use of a coupled transmitter-receiver ($T_x - R_x$), and Fig. 15.5b depicts the signal diagram for crack detection (Lozev et al. [8], Her and Lin [9], Mak [10]). Here, the symbol S_i with $i = 1, 2, 3$ stands for shear waves.

The plot in Fig. 15.5b is a frequency diagram showing the arrival times (t_i with $i = 1, 2, \ldots$) as mean values. The arrival times are properly identified, and it can be assumed that $\Delta t = t_3 - t_2$ is small in magnitude, but essential for crack detection.

Other crack orientations with respect to the location of the probe are summarized by Mak [10], who has defined Δt as the difference in the wave travel time from the tip and root at the ith scan position.

For one-dimensional analysis of the propagation of sound in a homogeneous isotropic medium, the sound wave equation or acoustic wave equation representing the scalar time-dependent quantity $y = y(x, t)$, such as the velocity of sound, displacement or pressure P, is defined by the second-order partial differential equation (PDE) defined by Eq. (15.1a), where the speed of sound can be converted to frequency through Eq. (15.2c). Additional insights on nondestructive evaluation (NDE) using the ultrasonic technique are given in the next example.

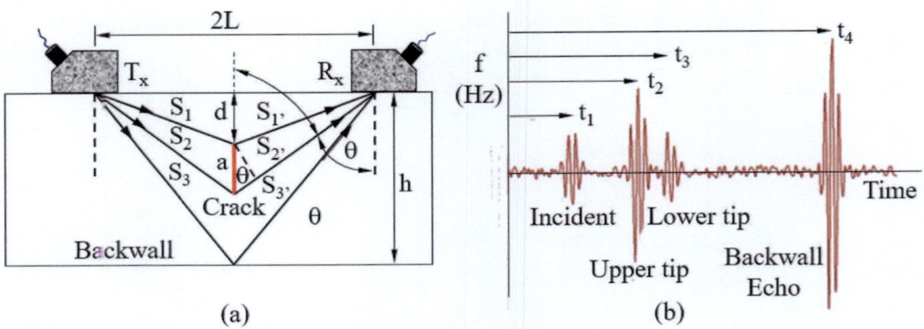

Fig. 15.5 Schematic ultrasonic crack detection. **(a)** Specimen thickness with an embedded crack and **(b)** the corresponding signal diagram (after Lozev et al. [8])

Example 15.3 Consider combining NDE and fracture mechanics fields for evaluating a hypothetical polycrystalline plate containing an embedded crack "a." Assume that the ultrasonic pulse-echo method with a straight-beam transducer produces the information shown in the sketches below. **(a)** Use the right triangle trigonometry for deriving expressions for the crack length "a."

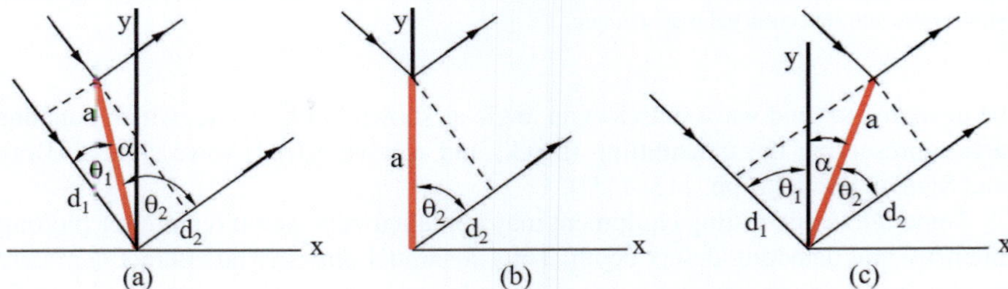

(b) For the hypothetical case shown in Fig. (a), the speed of sound in the plate is $\upsilon = 5500 \ m/s$, the incident sound path is $d_1 = 10 \ mm$ and the refracted angle is $\theta_2 = 20°$. Calculate the echo time (Δt) of the pulse, and the crack length (a). **(c)** Assume that the cracked plate is subjected to a remote stress $\sigma = 200 \ MPa$, determine if the plate will fracture. Evaluate the applied stress intensity factor K_I and the allowable crack growth Δa. Will fracture occur? Assume a plane-strain fracture toughness of $K_{IC} = 40 \ MPa\sqrt{m}$ and a large enough plane so that the geometric-correction factor is $\beta = f(a/w) = 1$.

Solution

(a) The sound path is given by

$$d_i = \frac{\upsilon \Delta t_i}{2} \quad \text{where } i = 1, 2 \qquad (15.3\text{E}1)$$

For Fig. (a),

$$\cos(\theta_2) = \frac{d_2}{a} = \frac{\upsilon \Delta t}{2a} \tag{15.3E2a}$$

$$a = \frac{\upsilon \Delta t}{2 \cos(\theta_2)} \tag{15.3E2b}$$

or

$$\cos(\theta_1) = \frac{d_1}{a} = \frac{\upsilon \Delta t}{2a} \tag{15.3E3a}$$

$$a = \frac{\upsilon \Delta t}{2 \cos(\theta_1)} \tag{15.3E3b}$$

For Fig. (b),

$$\cos\theta_2 = \frac{d_2}{a} = \frac{\upsilon \Delta t}{2a} \tag{15.3E4a}$$

$$a = \frac{\upsilon \Delta t}{2 \cos(\theta_2)} \tag{15.3E4b}$$

Thus far there are more than one equation for predicting the crack length and subsequently the stress intensity factor.

For Fig. (c),

$$\cos(\theta_1 + \alpha) = \frac{d_1}{a} = \frac{\upsilon \Delta t}{2a} \tag{15.3E5a}$$

$$a = \frac{\upsilon \Delta t}{2 \cos(\theta_1 + \alpha)} \tag{15.3E5b}$$

or

$$\cos(\theta_2 - \alpha) = \frac{d_2}{a} = \frac{\upsilon \Delta t}{2a} \tag{15.3E6a}$$

$$a = \frac{\upsilon \Delta t}{2 \cos(\theta_2 - \alpha)} \tag{15.3E6b}$$

(b) For $\upsilon = 5500\ m/s$ and the sound path is $d_1 = 10\ mm$, the echo time of the pulse is

$$d_1 = \frac{\upsilon \Delta t_1}{2} \tag{15.3E7a}$$

$$\Delta t = \frac{2 d_1}{\upsilon} = \frac{2\left(10 \times 10^{-3}\ m\right)}{5500\ m/s} = 3.64 \times 10^{-6}\ s = 3.64\ \mu s \tag{15.3E7b}$$

From Eq. (15.3E2a) at $\theta_2 = 20°$, the crack length is

$$a = \frac{\upsilon \Delta t}{2 \cos (\theta_2)} = \frac{(5500)\left(3.64 \times 10^{-6}\, s\right)}{2 \cos (20°)} \tag{15.3E8a}$$

$$a = 1.07 \times 10^{-2}\, m = 10.70\, mm \tag{15.3E8b}$$

(c) If $\sigma = 200\, MPa$, then the applied stress intensity factor (Chap. 13) with $\beta = 1$ is

$$K_I = \beta \sigma \sqrt{\pi a} \tag{15.3E9a}$$

$$K_I = (1)\,(200\, MPa)\sqrt{\pi \left(1.07 \times 10^{-2}\, m\right)} \tag{12A.1}$$

$$K_I = 36.67\, MPa\sqrt{m} \tag{15.3E9b}$$

Therefore, fracture will not occur since $K_I \lessdot K_{IC}$. Assume that the critical crack length a_c in the plate can be approximated by Eq. (13.15b). Thus,

$$a_c = \frac{1}{\pi}\left(\frac{K_{IC}}{\beta \sigma}\right)^2 = \frac{1}{\pi}\left[\frac{40\, MPa\sqrt{m}}{(1)\,(200\, MPa)}\right]^2 \tag{15.3E10a}$$

$$a_c = 1.27 \times 10^{-2}\, m = 12.70\, mm \tag{15.3E10b}$$

The crack extension is

$$\Delta a = a_c - a = 12.70 - 10.70 = 2\, mm \tag{15.3E11}$$

This is a very small amount of crack growth, and in conclusion, the cracked plate must be repaired or replaced. In fact, a critical crack length is an unstable physical defect at a certain external mechanical load. In most cases, the critical crack length a_c induces crack propagation, which is a fast and an undesirable scenario that may lead to catastrophic failure. Moreover, the critical stress intensity factor, say, for mode I loading, is defined as the plane-strain fracture toughness K_{IC}, which is a material property analogous to the yield strength σ_{ys}. This implies that a cracked specimen becomes unstable when $a \to a_c$ and $K_I \to K_{IC}$.

One-Dimensional Longitudinal Ultrasonic Wave Consider an ultrasonic wave traveling along the x-axis in a cylindrical rod (Fig. 15.6a) at a high frequency not audible by humans (Shull and Tittmann [2, p. 77]). In this case, Fig. 15.6a shows a selected segment dx for the one-dimensional analysis, and Fig. 15.6b illustrates the segment as a free-body diagram (FBD) in order to show the force F or stress σ acting on the rod volume $V = A dx$. The goal is to derive the wave equation in terms of displacement u as a PDE.

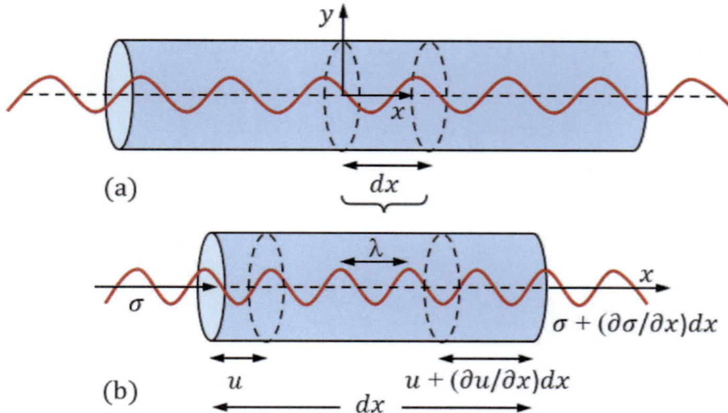

Fig. 15.6 (**a**) Ultrasonic longitudinal wave propagating in a long rod showing a segment dx and (**b**) expanded segment dx indicating the corresponding positions of related displacements along the x-direction. After Shull and Tittmann [2, p. 77]

The relevant procedure for deriving the PDE is borrowed from Shull and Tittmann [2, p. 77]. According to Newton's second law of motion, an imbalance net force acting on the dx element with a volume $V = Adx$ causes motion due to a force defined by

$$F = ma = m\frac{\partial \upsilon}{dt} \tag{15.6}$$

where m denotes the mass, a denotes the acceleration, υ denotes the velocity of a particle, and t is usual time variable.

From the FBD in Fig. 15.6b, the net force acting on the dx element and the imbalance force F are, respectively,

$$F + \frac{\partial F}{\partial x}dx - F = \frac{\partial F}{\partial x}dx \tag{15.7a}$$

$$\sigma A + \frac{\partial (\sigma A)}{\partial x}dx - \sigma A = \frac{\partial (\sigma A)}{\partial x}dx \tag{15.7b}$$

$$\sigma + \frac{\partial \sigma}{\partial x}dx - \sigma = \frac{\partial \sigma}{\partial x}dx \tag{15.7c}$$

$$F = \frac{\partial (\sigma A)}{\partial x}dx \tag{15.7d}$$

$$F = A\left(\frac{\partial \sigma}{\partial x}dx\right) \tag{15.7e}$$

Here, A is the cross-sectional area of the rod which remains constant though the passage of the ultrasonic wave known as the stress wave.

Combining Eqs. (15.6) and (15.7d) along with density equation $\rho = m/V$ or mass equation $m = \rho V = \rho (Adx)$ yields the stress gradient

$$\left(\frac{\partial \sigma}{\partial x} dx\right) A = m \frac{\partial v}{\partial t} = \rho (Adx) \frac{\partial v}{\partial t} \tag{15.8a}$$

$$\left(\frac{\partial \sigma}{\partial x} dx\right) A = \rho (Adx) \frac{\partial^2 u_x}{dt^2} \tag{15.8b}$$

$$\frac{\partial \sigma}{\partial x} = \rho \frac{\partial^2 u_x}{dt^2} \tag{15.8c}$$

This expression, Eq. (15.8c), defines the stress gradient or load density, which is not a common term in crack analytical analysis.

Using Hooke's law $\sigma = E\varepsilon$, where E is the modulus of elasticity, the elastic uniaxial strain ε along the x-axis is defined by

$$\varepsilon = \frac{u + (\partial u_x/\partial x)\, dx - u_x}{\partial x} = \frac{\partial u_x}{\partial x} \tag{15.9a}$$

$$\frac{\partial \varepsilon}{\partial x} = \frac{\partial^2 u_x}{\partial x^2} \tag{15.9b}$$

Differentiating Hooke's law gives

$$\sigma = E\varepsilon \tag{15.10a}$$

$$\frac{\partial \sigma}{\partial x} = E \frac{\partial \varepsilon}{\partial x} = E \frac{\partial^2 u_x}{\partial x^2} \tag{15.10b}$$

Equating Eqs. (15.8c) and (15.10b) yields the sought PDE (Appendix 15A) for an ultrasonic wave in terms of displacement μ

$$\frac{\partial^2 u}{\partial t^2} = \frac{E}{\rho} \frac{\partial^2 u}{\partial x^2} \tag{15.11a}$$

$$\frac{\partial^2 u}{\partial t^2} = v^2 \frac{\partial^2 u}{\partial x^2} \tag{15.11b}$$

$$v = \sqrt{\frac{E}{\rho}} \tag{15.11c}$$

where ρ is the density, E is the modulus of elasticity of the solid rod, and v is the wave velocity. Notice that Eqs. (15.1a) and (15.11b) are similar.

One-Dimensional Transverse Ultrasonic Wave This is a traveling shear wave generated by applying time-varying forces to produce shear stresses and a shear displacement (u_y) along the y-axis. Thus, the shear wave equation is of the form

(Shull and Tittmann [2, p. 79])

$$\frac{\partial^2 u_y}{\partial t^2} = \frac{G}{\rho} \frac{\partial^2 u_y}{\partial x^2} \tag{15.12a}$$

$$\upsilon = \sqrt{\frac{G}{\rho}} \tag{15.12b}$$

where G is the shear modulus of elasticity and ρ is the bulk mass density of an isotropic and crystalline or amorphous material exposed to ultrasonic waves at finite temperature. For clarity, a crystalline solid has a regular or periodic, and an amorphous solid has an irregular arrangement of atoms or molecules. On the other hand, an ultrasonic wave is a highly energetic sound wave at a high frequency greater than 20 kHz.

The shear modulus measures stiffness, and the G/ρ radio is known as specific stiffness. In fact, stiffness is considered as a material property commonly used in the structural design field, which is out of the scope of this textbook.

Example 15.4 Assume that a transverse wave with a constant amplitude is described by the function

$$y\,(x, t) = (4\ mm) \cos\left[2\pi \left(\frac{t}{0.025\ s} - \frac{x}{100\ mm} \right) \right] \tag{15.4E1}$$

Determine **(a)** the wave amplitude (y_o), angular velocity (ω), and the wavenumber (k), **(b)** the wavelength of the wave $y\,(x, t)$, **(c)** the wave frequency f, and **(d)** the speed of wave propagation.

Solution

(a) From the given function,

$$y = y_o \cos\,(\omega t - kx) \tag{15.4E2a}$$

$$y_o = 4\ mm = 0.40\ cm \tag{15.4E2b}$$

$$\omega = \frac{1}{0.025\ sec} = 40\ sec^{-1} = 40\ Hz \tag{15.4E2c}$$

$$k = \frac{1}{100\ mm} = 0.01\ mm^{-1} = 0.10\ cm^{-1} \tag{15.4E2d}$$

(b) From Eqs. (15.2a,b), the wavelength λ and the wavenumber k are

$$\lambda = \frac{\upsilon}{f} = \frac{2\pi\,\upsilon}{\omega} \tag{15.4E3a}$$

$$k = \frac{\omega}{\upsilon} = \frac{2\pi}{\lambda} \tag{15.4E3b}$$

Then,

$$\lambda = \frac{2\pi}{k} = \frac{2\pi}{0.01\ mm^{-1}} = 628.32\ mm = 0.62832\ m \qquad (15.4\text{E}4)$$

(c) From Eq. (15.2b), the frequency is

$$\frac{\upsilon}{f} = \frac{2\pi\upsilon}{\omega} \qquad (15.4\text{E}5\text{a})$$

$$f = \frac{\omega}{2\pi} = \frac{40\ \sec^{-1}}{2\pi} = 6.37\ \sec^{-1} = 6.37\ Hz \qquad (15.4\text{E}5\text{b})$$

(d) The speed of wave propagation is

$$\lambda = \frac{\upsilon}{f} \qquad (15.4\text{E}6\text{a})$$

$$\upsilon = \lambda f = (0.62832\ m)\left(6.37\ \sec^{-1}\right) = 4\ m/s \qquad (15.4\text{E}6\text{b})$$

Therefore, the wave propagates very slowly at $\upsilon = 4\ m/s$.

15.5   Acoustical Sound Pressure

Consider the acoustic field in an arbitrary-shaped domain of a homogeneous isotropic medium. For one-dimensional spatial (space) and temporal (time) analysis, the governing linear wave equation and its corresponding solution for the evolved acoustic pressure $P = P(x, t)$ are (Kirkup [11, p. 12] and Raichel [3, pp. 25–27])

$$\frac{\partial^2 P}{\partial t^2} = \upsilon^2 \frac{\partial^2 P}{\partial x^2} \qquad (15.13\text{a})$$

$$P = P_o \exp\left[i\left(\omega t - kx\right)\right] \qquad (15.13\text{b})$$

$$P = P_o \sin\left(\omega t - kx\right) \qquad (15.13\text{c})$$

Notice that the sound pressure (P) depends on the distance x between the sound source and the place of measurement. This implies that the loudness of sound is essentially a measure of sound intensity related to the energy of the pressure wave defined by Eq. (15.13c). Moreover, Eqs. (15.13b), (15.13c) can be taken as a mathematical guideline to measure sound or noise source and to identify the sound pressure for effective corrections.

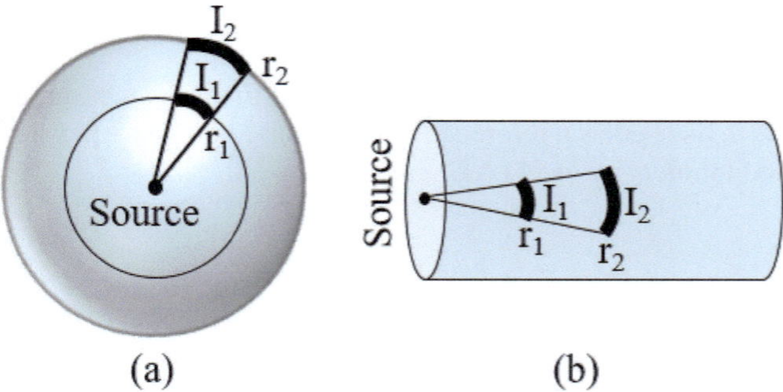

Fig. 15.7 Sound as a function of distance and sound levels (**a**) in a sphere and (**b**) in a cylinder

15.5.1 *Levels of Sound*

Firstly, sound spreads from the source in a spherical, cylindrical, or any other shape in an open area; $A = 4\pi r^2$ for a spherical surface area, $A = \pi r^2$ for a cylindrical cross-sectional area, and so forth. This implies that the sound intensity (I) is proportional to $I \propto 1/r^2$ in these shapes. Hence, the larger the radius, the less intensive the sound becomes from the source. These scenarios are illustrated in Fig. 15.7, representing the spreading of sound energy as a result of the expansion of the sound wavefront being independent of frequency. For instance, Fig. 15.7a schematically illustrates the sound intensity level related to certain areas at a distance r. On the other hand, Fig. 15.7b shows a similar pattern in a cylindrical environment.

Secondly, decibel (dB) is a logarithmic base-10 scale used to compare acoustic sound intensity ratio based on two measurements, such as intensity ratio I_2/I_1, pressure ratio P_2/P_1, power ratio E_2/E_1, or voltage ratio V_2/V_1. In general, sound is acoustic energy transmitted through matter (gases, liquids and solids) at a finite temperature. However, if sound is produced by an oscillatory concussive pressure being mathematically defined by Eq. (15.13c), it is a manifestation of mechanical disturbance of energy.

Understanding sound emission, perception or disturbance in all life forms (humans, plants, animals) is a complicated effort at a specific range of sound frequencies. For humans, the audible range of sound frequency is $20\ Hz \leq f \leq 20 \times 10^3\ Hz$, and sound level is $0\ dB \leq L_p \leq 140\ dB$ at ambient temperature, and a frequency above $20 \times 10^3\ Hz$ is ultrasonic.

In the light of recent developments, sound plays a significant role in the behavioral response of life forms. In this context, studies of acoustic sound effects on humans are quite abundant in the literature. However, in general, the biological relevance of sound waves on plants and living organisms is an interesting topic for determining the response of plants to sound and for expanding the application of audible sound and acoustic sound. Unfortunately, this topic is out of the scope of this textbook.

In the sound wave field, the ratio of two sound intensity measurements, I_1 and I_2 where $I_1 = I_o$ is used as a reference quantity based on some physical limitation, defines the sound intensity level (ΔL_I) as

$$\Delta L_I = (10\,dB)\log\left(\frac{I_2}{I_1}\right) \tag{15.14a}$$

$$I = \frac{E}{A} = \frac{Power}{Area} \quad (\text{in } W/m^2) \tag{15.14b}$$

$$E = \frac{AP^2}{\rho\upsilon} \quad (\text{sphere}) \tag{15.14c}$$

$$I = \frac{P^2}{\rho\upsilon} \tag{15.14d}$$

$$E = AI = (2\pi rh)\,I \quad (\text{cylinder}) \tag{15.14e}$$

$$E = AI = \left(2\pi r^2\right)I \quad (\text{sphere}) \tag{15.14f}$$

where the factor of 10 is used for conversion from bel (B) to decibel (dB), E denotes the power crossing a sphere or cylinder surrounding the source, A denotes the area at a particular distance r, P denotes the sound pressure amplitude for a spherical sound wave, ρ denotes the density of the medium, and υ denotes the speed of sound.

The decibel scale is also used in electrical engineering when comparing power measurements, $E = V^2/R$, where V is the voltage and R is the electrical resistance. By definition, the sound loudness can be described in accordance with the type of level. Thus,

- **Sound Intensity Level** (ΔL_{SIL}) measures the amount of acoustical energy per unit area and it decreases in proportion to the inverse distance ($1/r$ for a cylinder and $1/r^2$ for a sphere). This implies that as sound radiates away from its source, it creates a sphere of acoustical energy.
- **Sound Pressure Level** (ΔL_{SPL}) measures the force per unit area, and it decreases in proportion to the inverse of the listening distance. Actually, an acoustical sound wave has pressure and velocity, where pressure is easier to measure than velocity. Commonly, a SPL meter is used for this purpose.
- **Sound Energy (Sound Power) level** (ΔL_{SEL}) measures the amount of acoustical energy created by the sound source.

Substituting Eqs. (15.14b)–(15.14f) into (15.14a) and letting the first measurement become the reference variable ($I_1 = I_o$, $P_1 = P_o$ and $E_1 = E_o$,) in order to start from the threshold of hearing yield useful acoustic sound-level equations for single pair and multiple measurements

$$L_I = \Delta L_{SIL} = (10 \, dB) \log \left(\frac{I}{I_o} \right) \quad \text{(pair)} \tag{15.15a}$$

$$L_p = \Delta L_{SPL} = (20 \, dB) \log \left(\frac{P}{P_o} \right) \quad \text{(pair)} \tag{15.15b}$$

$$L_e = \Delta L_{SEL} = (20 \, dB) \log \left(\frac{E}{E_o} \right) \quad \text{(pair)} \tag{15.15c}$$

$$L_{r2} = L_{r1} - (20 \, dB) \log \left(\frac{r_2}{r_1} \right) \quad \text{(pair)} \tag{15.15d}$$

$$L_p = (10 \, dB) \log \left[\sum_{i=1}^{N} \left(\frac{P}{P_o} \right)^2 \right] \quad \text{(multiple)} \tag{15.15e}$$

$$L_p = (10 \, dB) \log \left[\sum_{i=1}^{N} 10^{L_{pi}/10} \right] \quad \text{(multiple)} \tag{15.15f}$$

The analogy of Eq. (15.15) reveals that:

- If $\log(I) = 0$ in Eq. (15.15a), then there is still sound at a very low sound intensity $I_o = 10^{-12} \, W/m^2$ as the threshold intensity a human can hear.
- If $\log(P) = 0$ in Eq. (15.15b), then there is still sound at a very low sound pressure $P_o = 2 \times 10^{-5} \, Pa$.
- L_p, Eq. (15.15c), is defined in terms of power E (Watts) along with a reference power E_o. Nonetheless, the sound power (E) is proportional to the square of sound pressure (P); $E/E_0 \propto (P/Po)^2$
- Lastly, Eq. (15.15d) is used to predict the sound level at a distance $r_2 > r_1$, provided that L_{p1} is known at r_1.
- Equations (15.15e), (15.15f) defined the sound intensity level for multiple measurements.

The acoustic domain for L_p is subjectively based on the normal human hearing pressure range $20 \, \mu Pa \leq L_p \leq 20 \, Pa$ or $0 \, dB \leq L_p \leq 120 \, dB$, where:

- $L_p = 0 \, dB_{SPL}$ is the threshold of hearing
- $L_p \simeq 30 \, dB_{SPL}$ or $P \simeq 0.0006 \, Pa$ is the ideal background noise in a study room
- $L_p \simeq 60 \, dB_{SPL}$ or $P \simeq 0.02 \, Pa$ at $1 \, m$ distance is the normal conversation level
- $L_p \simeq 90 \, dB_{SPL}$ or $P \simeq 0.63 \, Pa$ is the musical concert level
- $L_p \simeq 100 \, dB_{SPL}$ or $P \simeq 2 \, Pa$ at $10 \, m$ distance is the traffic noise level
- $L_p = 120 \, dB_{SPL}$ or $P \simeq 20 \, Pa$ is the threshold of pain
- $L_p > 447 \, dB_{SPL}$ or $P > 101 \, kPa$ is the shock wave level

Example 15.5 Calculate **(a)** the total sound pressure level (L_{SPL}) due to $L_{SPL1} = 70$ dB, $L_{SPL2} = 50$ dB, $L_{SPL3} = 80$ dB, and $L_{SPL4} = 90$ dB dataset, **(b)** the average sound pressure level ($\overline{L}_{SPL}$), and **(c)** the noise level L_{PS} if all the above L_{SPL} sources are shut down and a new sound-level measurement reads 15 dB.

Solution

(a) First of all, $L_p \neq \sum_{i=1}^{N} L_{pi}$ because of the $(P/P_o)^2$ term defined by

$$\left(\frac{P_i}{P_o}\right)^2 = \log^{-1}\left(\frac{L_{pi}}{10}\right) \tag{15.5E1}$$

Instead, take the antilog of the SPL, Eq. (15.15f), so that the total sound pressure level (L_{SPL}) becomes

$$L_{SPL} = (10\,dB) \log\left[\sum_{i=1}^{4} 10^{L_{pi}/10}\right] \tag{15.5E2a}$$

$$L_{SPL} = (10\,dB) \log\left[10^{70/10} + 10^{50/10} + 10^{80/10} + 10^{90/10}\right] \tag{15.5E2b}$$

$$L_{SPL} \simeq 90\,dB \tag{15.5E2c}$$

(b) The average $\overline{L}_{SPL}$ is based on four measurements

$$\overline{L}_{SPL} = (10\,dB) \log\left[\frac{1}{4}\sum_{i=1}^{4} 10^{L_{pi}/10}\right] \tag{15.5E3a}$$

$$\overline{L}_{SPL} = (10\,dB) \log\left[\frac{1}{4}\left(10^{70/10} + 10^{50/10} + 10^{80/10} + 10^{90/10}\right)\right] \tag{15.5E3b}$$

$$\overline{L}_{SPL} \simeq 84\,dB \tag{15.5E3c}$$

(c) The noise level is

$$L_{PS} = (10\,dB) \log\left[10^{90/10} - 10^{15/10}\right] \tag{15.5E4a}$$

$$L_{PS} = 90\,dB \tag{15.5E4b}$$

Calculations indicate that $\overline{L}_{SPL} < L_{PS}$. Therefore, the sound pressure level, also known as sound power level in industry, is assumed to measure the acoustic energy emitted a sound or noise source.

15.5.2   *Reflection and Transmission at Boundaries*

The intensity of an ultrasound beam during the passage of an ultrasound pulse is a measure of the rate of energy flow or power through an area. However, if a sound wave strikes a boundary or interface between two different solid phases, then the intensity (I) of the signal reflected strongly depends on the acoustic pressure (P), the wave velocity (v), and the specific acoustic impedance (Z) of the two materials.

Mathematically, the incident, reflected, and transmitted particle velocities are defined in complex variable form ($j = \sqrt{-1}$) representing harmonic traveling waves in phase 1 and phase 2 (Bray and Stanley [1, p. 64])

$$v_i = A_i \exp\left[j\left(\omega_1 t - k_1 x\right)\right] \tag{15.16a}$$

$$v_r = A_r \exp\left[j\left(\omega_1 t + k_1 x\right)\right] \tag{15.16b}$$

$$v_t = A_t \exp\left[j\left(\omega_2 t - k_2 x\right)\right] \tag{15.16c}$$

Accordingly, the sound pressure (stress) expressions are defined in a similar manner

$$P_i = B_i \exp\left[j\left(\omega_1 t - k_1 x\right)\right] \tag{15.17a}$$

$$P_r = B_r \exp\left[j\left(\omega_1 t + k_1 x\right)\right] \tag{15.17b}$$

$$P_t = B_t \exp\left[j\left(\omega_2 t - k_2 x\right)\right] \tag{15.17c}$$

The continuity terms of the particle velocity and pressure at the boundary are defined by

$$v_i + v_r = v_t \tag{15.18a}$$

$$P_i + P_r = P_t \tag{15.18b}$$

so that

$$A_i \exp\left[j\left(\omega_1 t - k_1 x\right)\right] + A_r \exp\left[j\left(\omega_1 t + k_1 x\right)\right] = A_t \exp\left[j\left(\omega_2 t - k_2 x\right)\right] \tag{15.19a}$$

$$B_i \exp\left[j\left(\omega_1 t - k_1 x\right)\right] + B_r \exp\left[j\left(\omega_1 t + k_1 x\right)\right] = B_t \exp\left[j\left(\omega_2 t - k_2 x\right)\right] \tag{15.19b}$$

Assuming that the angular frequencies $\omega_1 = \omega_2$ and placing the boundary at $x = 0$ yield

$$A_i + A_r = A_t \tag{15.20a}$$

$$B_i + B_r = B_t \tag{15.20b}$$

Manipulating Eqs. (15.20a), (15.20b) gives the acoustic impedance expressions as

$$Z_1 = Z_i = \frac{B_t}{A_i} = \frac{B_t - B_r}{A_t - A_r} \tag{15.21a}$$

$$Z_1 = Z_r = \frac{Br}{A_r} = \frac{B_t - B_i}{A_t - A_i} \tag{15.21b}$$

$$Z_2 = Z_t = \frac{B_t}{A_t} = \frac{B_i + B_r}{A_i + A_r} \tag{15.21c}$$

These impedance equations are generalized statements, and they are conveniently redefined in the most practical form. Substituting Eqs. (15.16) and (15.17) into (15.21) along with $\omega_1 = \omega_2$ and $x = 0$ yields the specific acoustic impedance for phase 1 and phase 2. Thus,

$$Z_1 = \frac{P_i / \exp\left[j\left(\omega_2 t - k_2 x\right)\right]}{v_i / \exp\left[j\left(\omega_2 t - k_2 x\right)\right]} = \frac{P_i}{v_i} \tag{15.22a}$$

$$Z_1 = \frac{P_r / \exp\left[j\left(\omega_2 t - k_2 x\right)\right]}{v_r / \exp\left[j\left(\omega_2 t - k_2 x\right)\right]} = \frac{P_r}{v_r} \tag{15.22b}$$

$$Z_2 = \frac{P_t / \exp\left[j\left(\omega_2 t - k_2 x\right)\right]}{v_t / \exp\left[j\left(\omega_2 t - k_2 x\right)\right]} = \frac{P_t}{v_t} \tag{15.22c}$$

Now, the signal intensity (I) and the acoustic impedance (Z) are mathematically defined in a general form

$$I = \frac{P^2}{\rho v} = \frac{P^2}{Z} \tag{15.23a}$$

$$Z = \rho v \tag{15.23b}$$

Thus,

$$Z_1 = Z_i = (\rho v)_i \tag{15.24a}$$

$$Z_1 = Z_r = (\rho v)_r \tag{15.24b}$$

$$Z_2 = Z_t = (\rho v)_t \tag{15.24c}$$

Equating Eqs. (15.22) and (15.24) yields the wave velocity or the speed of sound in phase-medium 1 and phase-medium 2

$$v_i = \sqrt{\left(\frac{P}{\rho}\right)_i} \tag{15.25a}$$

$$v_r = \sqrt{\left(\frac{P}{\rho}\right)_r} \tag{15.25b}$$

$$v_t = \sqrt{\left(\frac{P}{\rho}\right)_t} \tag{15.25c}$$

The reflection coefficient (R_c) and the transmission coefficient (T_c) are defined as acoustic pressure or impedance ratios

$$R_c = \frac{P_r}{P_i} = \frac{Z_2 - Z_1}{Z_2 + Z_1} \tag{15.26a}$$

$$T_c = \frac{P_t}{P_i} = \frac{2Z_2}{Z_2 + Z_1} \tag{15.26b}$$

with

$$T_c^2 + R_c^2 = 1 \tag{15.27}$$

Substituting Eq. (15.23b) into (15.26) gives

$$R_c = \frac{(\rho v)_2 - (\rho v)_1}{(\rho v)_2 + (\rho v)_1} \tag{15.28a}$$

$$T_c = \frac{2\,(\rho v)_2}{(\rho v)_2 + (\rho v)_1} \tag{15.28b}$$

$$R_c = \left[\frac{1}{2} - \frac{(\rho v)_1}{(\rho v)_2}\right] T_c \tag{15.28c}$$

It is worth noting that R_c and T_c depend on the density of phases and speed of sound and have units of $Kg/(m^2.s)$. Conceptually, the R_c measures quantitatively the reflected wave, and T_c describes the wave behavior on a barrier, such as a rigid boundary. Moreover, If $(\rho v)_2 \gg (\rho v)_1$, then Eq. (15.28c) yields $R_c = 0.5 T_c$. Also, Eq. (15.28c) holds for $(\rho v)_1 / (\rho v)_2 \neq 1/2$, where $R_c > 0$ or $R_c < 0$ and $T_c > 0$.

For convenience of the reader, Table 15.1 lists some interesting acoustic properties of different media taken from online sources.

Notice that the ascending order of the acoustic impedance values is based on the type of medium and that the engineering prefix $\times 10^6$ prevails in liquid an solids. The reason for this observation is indirectly illustrated in Fig. 15.3 with respect to the atomic arrangement of atoms in a medium. Apparently, the sound velocity in solid media,(FCC, BCC, hexagonal, and so forth) is a temperature-dependent variable.

Table 15.1 Speed of sound, density, and acoustic impedance

Medium	ρ (Kg/m^3)	υ (m/s)	Z, Eq. (15.23b) $[Kg/(m^2.s)]$
Air	1.210	343	415.03
Water	1000	1449	1.45×10^6
Brain	1030	1560	1.61×10^6
Muscles	1060	1580	1.67×10^6
Eye	1140	1670	1.90×10^6
Bone	1600	4080	6.53×10^6
steel	7800	5048	39.37×10^6

15.6   Acoustic Attenuation

By definition, acoustic attenuation is the degradation or loss of the intensity strength of an incident sound wave. Thus, the signal strength ratio $I/I_o < 1$ introduced in Chap. 2 is also used in the science of sound. This ratio measures the attenuation of the sound wave in a medium by the scattering and absorption of the sound wave energy. Physically, the degree of attenuation is defined by the linear attenuation coefficient μ, which, in turn, is a measure of the penetrability of the incident energy sound waves into a material.

For comparison purposes, attenuation is the loss, and amplification is the gain (boost) of the strength of a signal, respectively. For instance, liquids and gases attenuate both light and sound waves, lead (Pb) attenuates X-rays, dark glasses attenuate sunlight, and metals attenuate sound waves. Hence, attenuation affects the propagation of waves in a certain medium.

Attenuation must be prevented or reduced in the field of telecommunications and ultrasound applications in order to maintain the signal strength at its optimum level as a function of distance. For example, radio frequencies, electrical currents, connectors, and wire leakage may interfere with the signal and, consequently, cause attenuation.

15.6.1   Attenuation Coefficient

Consider now the scattering and absorption of traveling sound waves in a medium other than vacuum. By comparison, the speed of light is $\upsilon_{light} = 3 \times 10^8 \ m/s$ and that of the traveling sound wave is $\upsilon_{sound} = 0$ in vacuum.

Assume that $\upsilon_{sound} > 0$ in a medium, where scattering and absorption of traveling sound waves cause damping of sound or fluctuation of sound due to a reduction or loss of sound intensity or sound energy. This phenomenon is called attenuation, which is related to the linear attenuation coefficient μ.

In the field of sound, acoustic attenuation refers specifically to the loss of intensity of sound waves traveling in a mediums, such as air, liquid, or solid having polycrystalline or amorphous structure. Mathematically, the reduction in acoustic sound intensity (energy) or sound pressure over a distance $\Delta r = r_2 - r_1$ is represented as a decaying exponential function, which is derived from a homogeneous ordinary differential equation (ODE) (Raichel [3, p. 418])

$$dI + (k\,dr)\,I = 0 \tag{15.29a}$$

$$\int_{I_1}^{I_2} \frac{dI}{I} = -k \int_{r_1}^{r_2} dr \tag{15.29b}$$

$$\ln\left(\frac{I_2}{I_1}\right) = -k\,(r_2 - r_1) = -k\Delta r \tag{15.29c}$$

$$I_2 = I_1 \exp\left(-k\Delta r\right) \tag{15.29d}$$

where k is the proportionality constant, Δr is the path length of the sound waves in a given medium, the $k\Delta r$ may be treated as the intensity level of the sound waves, and I_2 (*emergent*) $< I_1$ (*incident*) is the condition for a decay behavior being defined by Eq. (15.29d), which is used to derive the linear attenuation coefficient (μ) equation as a log base-10 expression.

The sound intensity level (L_{SIL}) is linearly related to the linear attenuation coefficient (μ); it depends on an average distance. Thus,

$$\log\left(I_2\right) = \log\left(I_1\right) - k\Delta r \log\left(e\right) \tag{15.30a}$$

$$10\log\left(I_2\right) = 10\log\left(I_1\right) - 10k\Delta r \log\left(e\right) \tag{15.30b}$$

$$\mu = 10k \log\left(e\right) \tag{15.30c}$$

$$\mu = \frac{(10\,dB)}{\Delta r}\,\log\left(\frac{I_2}{I_1}\right) \tag{15.30d}$$

$$L_{SIL} = \mu\Delta r \tag{15.30e}$$

Here, $e = \sum_{n=0}^{\infty}(1/n!) \simeq 2.7183\ldots$ is the exponential constant, and L_{SIL} is the sound intensity level. Manipulating Eq. (15.30d) yields the linear attenuation coefficient μ due to dispersion or scattering and absorption of ultrasonic waves

$$\mu = \frac{L_{SIL}}{\Delta r} \quad \text{(in } dB/m \text{ units)} \tag{15.31a}$$

$$\mu = \mu_s + \mu_x \tag{15.31b}$$

where μ_s and μ_x are the scattering and absorption components, respectively. Usually, when $\mu_x >> \mu_s$ the linear attenuation coefficient is called the linear absorption coefficient.

Regarding Eq. (15.30e), L_{SIL} can be treated as the loss of signal amplitude with increasing wave propagation distance Δr. On the other hand, Eq. (15.31a) indicates that $\mu \to \infty$ as $\Delta r \to 0$ land consequently; the sound wave is significantly attenuated while passing through a medium.

Additional theoretical background on sound attenuation and related attenuation coefficient μ can be found in the vast literature on the science of sound.

Example 15.6 Calculate **(a)** the linear attenuation coefficient (μ) in a certain medium when the sound intensity is $I_2 = 10^{12}\ W/m^2$, **(b)** the sound intensity level (L_{SIL}) if the incident sound intensity, and the source distance are $I_1 = 10^2\ W/m^2$ and $\Delta r = 20\ m$, respectively. **(c)** The distance for the upper hearing limit using the calculated μ.

Solution

(a) From Eq. (15.30d),

$$\mu = \frac{10}{\Delta r} \log\left(\frac{I_2}{I_1}\right) = \frac{10\,dB}{20\,m} \log\left(\frac{10^{12}}{10^2}\right) \tag{15.6E1a}$$

$$\mu = 11.51\ dB/m \tag{15.6E1b}$$

This is the quantity or extent to which the incident sound intensity is reduced as it passes through a specific material at 20-m distance. In this context, $\mu = 11.51\ dB/m$ characterizes how easily the unknown material or medium can be penetrated by the incident sound waves.

(b) From Eq. (15.30e),

$$L_{SIL} = \mu \Delta r = (11.51\ dB/m)\,(20\ m) \simeq 230\ dB \tag{15.6E2}$$

which is above the threshold of pain in humans.

(c) The distance for the upper hearing limit, $L_{SIL} = 140\ dB$, is

$$\Delta r = L_{SIL}/\mu = (140\ dB)\,/\,(11.51\ dB/m) = 12.16\ m$$

15.7 Physiological Acoustics

The auditory system in humans and animals is the ear, which is the receptor organ of mechano-acoustic energy evolved in the form of sound pressure waves. The sound pressure wave equation defined by Eq. (15.13a) can be used as a simple or ideal mathematical model to characterize the sound behavior related to hearing.

Technologically, the sound behavior from loudspeakers and earphones are different since both systems have different calibration procedures. For instance, the latter may give rise to unwanted physiological noise that may interfere with low-frequency sound waves associated with sound pressure variants. Consequently, physiological discomfort may be an issue for the receptor.

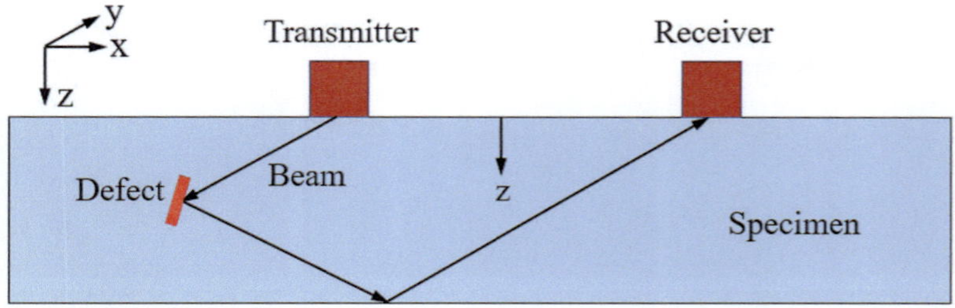

Fig. 15.8 Ultrasound setup showing the traveling wave and defect detection

15.7.1   Ultrasonic Imaging

Consider the sketch shown in Fig. 15.8 for a block diagram for an ultrasonic defect inspection as a non-destructive testing (NDT) used to detect the location of a defect, such as a crack, weld discontinuity and so forth (Raichel [3, 482]).

This is an increasingly important imaging technique in the medical and industrial fields. This technique uses sound waves to produce images of human or animal internal organs and metal parts that may contain defects. Essentially, ultrasound is safe and noninvasive when used to examine a baby in pregnant women and to diagnose the causes of pain, swelling, and infection in the body's internal organs. The basic equipment includes ultrasound scanners connected to a computer console, video display screen, and an attached transducer (a hand-held device that sends and receives ultrasound signals).

The basic ultrasound mechanism is based on a traveling beam being intercepted and reflected by a discontinuity. Subsequently, the return signal is captured on a oscilloscope screen, followed by analysis, processing, and interpretation.

15.8   Infrared Thermography

Consider the infrared thermography (IT) images shown in Fig. 15.9 used as a tool for quality assurance, forensic investigations, and research & development.

Nowadays, high accuracy and precision crack detection is achieved by ultrasonic thermography (UT) based on infrared technology (IT), which provides thermal emission from the specimen surface. The thermal emission strongly depends on the solid heat conduction, and it is classified as ultrasound-excited thermography (UET) or vibro-thermography (VT).

Infrared technology provides an infrared imaging system to detect, display, and record thermal patterns and corresponding temperature variations across a tested solid surface. For instance, Fig. 15.9 shows a numerical sequence of IT images of a crack growth process. The IT technique provides colorful patterns related to local temperature patterns and stress distribution ahead of the crack tip. These images

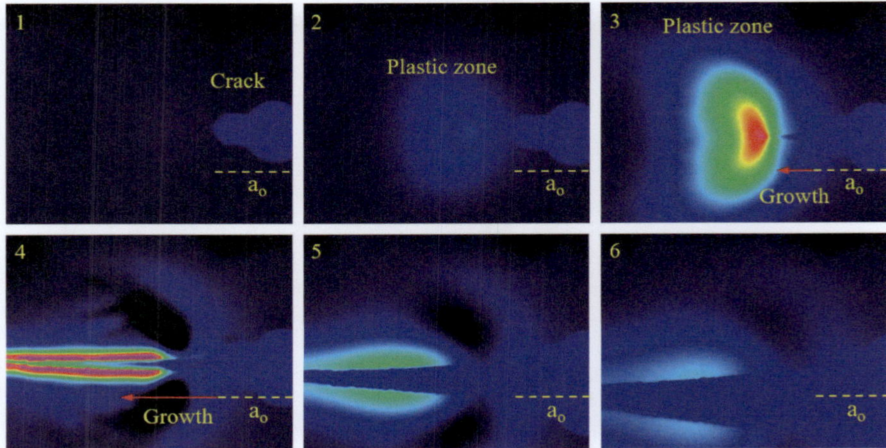

Fig. 15.9 Infrared thermography images captured from a video posted online on crack growth in superelastic Ni-Ti smart alloy at https://www.youtube.com/watch?v=L80PMrctK4cInfrared thermography images captured from a video posted online on crack growth in superelastic Ni-Ti smart alloy at https://www.youtube.com/watch?v=L80PMrctK4c

were captured from a video posted online on crack growth in superelastic Ni-Ti smart alloy at https://www.youtube.com/watch?v=L80PMrctK4c.

Image 1 A Ni-Ti specimen containing a crack (light blue, right-hand side surface) being loaded in tension. The dashed line is added to show the original crack length a_o in light blue color. Actually, the crack is an empty space, but its slightly blunt-crack tip grows from right to left upon loading.

Image 2 A blue circle ahead of the crack tip represents the plastic zone, which, in turn, is the mechanical process region where crystal defects evolve upon loading. Once the plastic zone reaches a critical size or critical state, crack growth is continuous and relatively fast.

Image 3 There is a small crack growth denoted by the red arrow, and the plastic zone is mechanically more active. The red color within the plastic zone arises due to a localized increase in temperature. Notice that the slightly blunt crack tip is now a sharp crack tip, and the colorful plastic zone is larger than before.

Image 4 Crack growth is evident and the specimen crack mouth is larger than before. Notice how the plastic zone has changed its shape during crack propagation. The sharp crack tip between the red crack edges (bands in red) is at high temperature.

Image 5 Additional crack growth and larger crack length. Notice that the crack tip is out of the image area because the crack is propagating very fast.

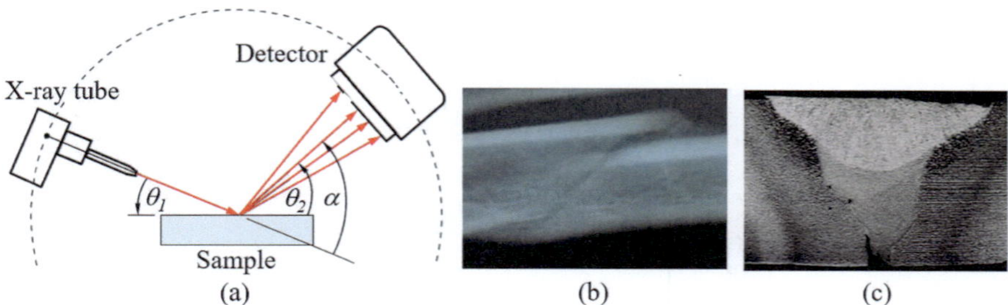

Fig. 15.10 X-ray optics in X-ray diffraction. (**a**) Schematic parts of an X-ray instrument (He [12, p. 63]), (**b**) human bone crack [13], and (**c**) lack of fusion in a weldment [14]

Image 6 Additional crack propagation and separation of the crack edges. The mechanical test is done. The light blue bands above and below the crack edges represent the wake, where evolved internal stresses remain in the $Ni\text{-}Ti$ smart alloy after testing.

15.9 Radiography

Radiography is another nondestructive technique (NDT) which uses radiation to provide images of defects in metals and images of human body parts or organs. In essence, radiography includes X-rays or gamma rays with short wavelengths as shown schematically in Fig. 2.1.

Note that Fig. 15.10a depicts schematically the most relevant parts of the X-ray optics machine for revealing defects, such as cracks (He [12, p. 63], [13], [14]), Fig. 15.10b depicts a bone crack, and Fig. 15.10c shows a lack of fusion in a weldment.

Digital X-ray (DXR) systems are more faster, efficient, and convenient than traditional or conventional X-ray film methodology. The principal function of the DXR system for nondestructive evaluation (NDE) is the high-quality digitization and the convenient storage of radiographic information in image media.

Qualitative and quantitative evaluation of cracks and related crack growth in solids provide significant information for assessing the integrity of engineering structures. For instance, the quantitative characterization of crack growth using radiography is a useful technique in the NDT field. For crack growth analysis, corresponding details on radiography can be found elsewhere in cited references [12, 13, 14].

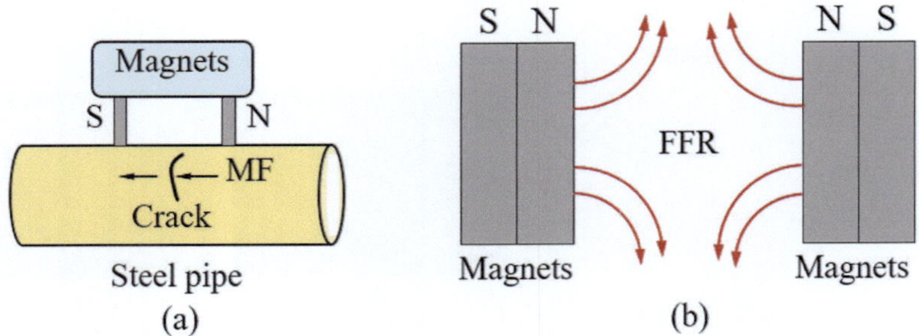

Fig. 15.11 MPI techniques. (**a**) Magnetic particle inspection and (**b**) magnetic particle imaging

15.10 Magnetic Particle Inspection

The magnetic particle inspection (MPI) is a nondestructive testing method suitable for revealing fine surface discontinuities such as cracks, seams, and the like. The most common application of magnetic particles is in the form of dry powder, which is dusted over the surface of a ferromagnetic workpiece. However, wet particles (suspension) may be used in some applications. Nonetheless, Fig. 15.11a schematically shows the MPI technique used to detect surface or slightly subsurface flaws.

Despite that this chapter in dedicated to nondestructive evaluation of defects in metals, it is worth mentioning that the field-free region (FFR) method is also used and designed for human scales in the biomedical field. This field-free region is located at the center point between magnets (Fig. 15.11b). For instance, magnetic hyperthermia using magnetic nanoparticles is a method for application in cancer therapy. This implies that the strength of the magnetic field for generating the FFR must be controlled in order to avoid severe secondary effects on human cells.

15.11 Fluorescent-Penetrant Inspection

Fluorescent-penetrant inspection (FPI) is a highly sensitive method used to detect discontinuities or flaws as schematically illustrated in Fig. 15.12.

This technique includes three conventional spray dye cans (Fig. 15.12a) for cleaning, penetrating, and developing the site of a crack on a welded steel plate (Fig. 15.12b). The sketch in Fig. 15.12c depicts the three required steps for applying the spray cans in the shown sequence.

A fluorescent dye is simply a colored chemical substance that bonds to a substrate surface as shown in Fig. 15.12b by the red color along the steel weld. There are two important aspects of fluorescent-penetrant inspection (FPI) method for detecting

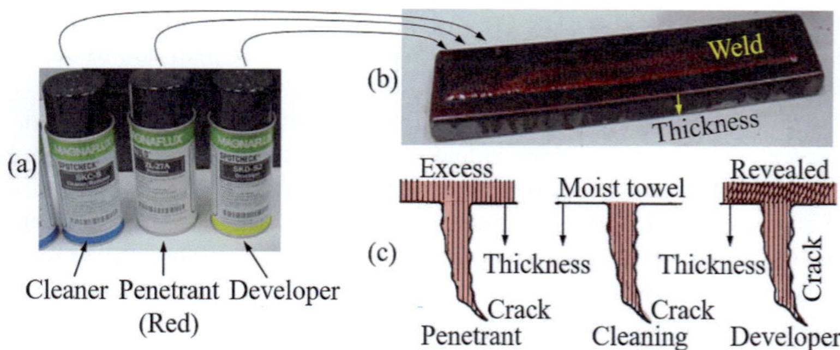

Fig. 15.12 Fluorescent-penetrant inspection (FPI) for detecting surface defects. (**a**) Commercial products in spraying cans, (**b**) specimen, and (**c**) schematic cracks

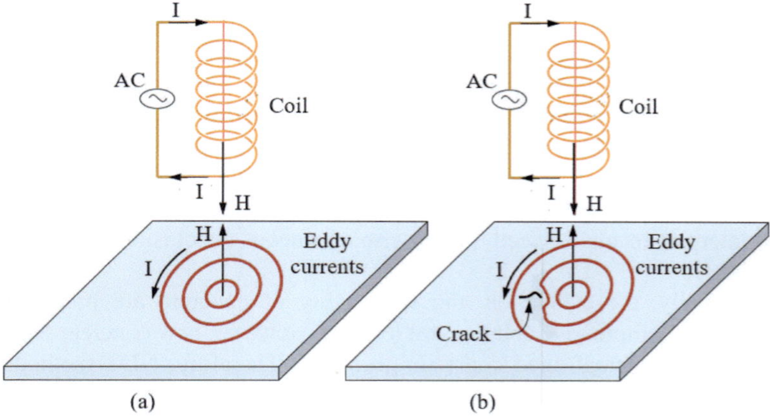

Fig. 15.13 Sketch of a plate-magnet forming a magnetic field (H-field) that induces a magnetic flux B perpendicular to the eddy-current plane (blue and black vertical arrows). (**a**) Crack-free plate and (**b**) cracked plate

defects. (1) The material to be inspected has to be impermeable in nature, and (2) it is limited to external surface defects.

15.12   Eddy Current Inspection

According to Faraday's law of induction, eddy currents are electrical current loops induced by an H-field from a coil as shown in Fig. 15.13.

Eddy current inspection (ECI) is another nondestructive testing (NDT) technique used to inspect structural components for a variety of purposes, including cracks, microstructural features due to a certain heat treatment procedure, coating thickness measurements, and the like. The principal characteristic of eddy currents is manifested as a magnetic field (H-field) disturbance during its path due to defects in a conductive solid material.

For a crack-free plate (Fig. 15.13a), the eddy currents are continuous loops emitting their own H-field, and for a plate-containing a crack (Fig. 15.13b), the crack induces a disruption in the circular current-flow patterns.

The nature of crack formation may be static or dynamic, but its existence in a component may be critical regarding the assurance of a reliable structure, say, an aircraft component. Thus, ECI is widely used because it provides a relatively fast electrical changes in the exciting coil.

15.13   Nondestructive Testing of Concrete

Nondestructive testing (NDT) methods are used to evaluate the condition of concrete structures, such as bridges, buildings, and roads. Any suitable NDT method mandates the evaluation of reinforced-concrete mechanical properties and the corrosion of reinforcement steel bars, permeability, cracking, void structure, and spalling.

It is significant to consider the use of NDT methods that are based on their potential, limitations, and inspection techniques, and most importantly, one must be able to interpret results correctly; otherwise, erroneous conclusions are susceptible to be reported.

Traditionally, the evaluation and the quality of concrete are based on cast specimens for compressive and flexural loading. However, new concrete specimens may differ from actual aged concrete structures. Therefore, NDT methods have been developed for assessing the physical and mechanical conditions of reinforced concrete structures. Among NDT methods,

- The Windsor probe system (WPS) is a penetration resistance of hardened concrete. It can also be described as a surface hardness method used to determine the compressive strength of concrete.
- The Schmidt rebound hammer test (RHT) is a surface hardness method also used to determine the compressive strength of concrete and compression wave propagation in concrete. This type of NDT method is a relatively simple and cost-effective approach, where curve fitting or regression analysis between average rebound number and compressive strength is performed on representative datasets.
- The ultrasonic pulse velocity (USPV) method is a coupled pulse-echo process used to detect flaws. In effect, a transducer emitting vibration at its fundamental frequency is the main component required in this method.
- Ultrasonic monitoring is also a technique that uses high-frequency acoustic waves (sound waves) to measure variations in thickness of pipes, vessels, and tanks. Moreover, this method requires two transducers for transmitting and receiving the ultrasonic pulse.

Table 15.2 Some attributes of sound

Physical	Amplitude	Frequency	Spectrum	Duration
Perceptual	Loudness	Pitch	Timbre	Length

Significant background on procedures for nondestructive testing of concrete can be found in American Society for testing materials (ASTM) standard test methods for concrete and elsewhere (Malhotra and Carino [15], Helal and Mendis [16]).

15.14 Music Waves

A music wave is sound consisting of vibrational energy caused by the rapid motion of atoms and molecules in a medium, such as air. Thus, sound can be described as a cyclic or sinusoidal behavior related to fast oscillations in air pressure, and musical sound can be defined as time domain- and pressure range-dependent sine wave. This implies that sound waves are longitudinal waves with certain amplitudes.

The simplest mathematical model for the sound pressure is written as a sine wave that represents the perception of pitch as an approximation to the equation of motion at a particular distance. Thus,

$$P = A \sin(\omega t) = A \sin(2\pi f t) \tag{15.32a}$$

$$\omega = 2\pi f \tag{15.32b}$$

where A denotes the wave amplitude, ω denotes the angular frequency, and f denotes the frequency or pitch.

The main characteristics of sound waves or attributes of sound are related to the physical and perception of the sound as indicated in Table 15.2 (Benson [17, p. 2]).

A musical wave moves as a succession of compression through the air with a frequency-dependent wavelength. Hence, the higher the pitch, the shorter the wavelength. Moreover, music and noise are mixtures of sound waves of different frequencies. The former is based on discrete frequencies representing polyphonic music (simultaneous combination of two or more tones), and the latter represents random frequencies related to incoherent or disturbing effect to the human hearing system.

15.14.1 Music Staff

Audio vibrations produced by singers or musical instruments are classified as musical notes having specific frequencies. For example, notice the music staff shown in Fig. 15.14 for a piano and related frequencies.

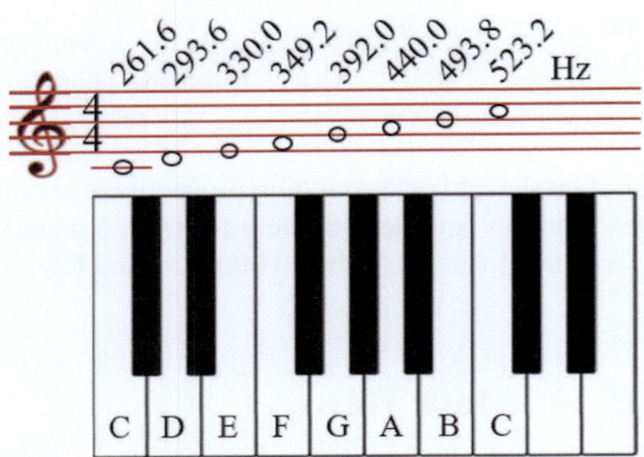

Fig. 15.14 Musical staff for piano and corresponding note frequencies

The musical notes related to an organized melodic music are placed on paper using a set of five lines and four spaces representing the musical pitch. The simultaneous scattering and absorption of sound waves can be defined as a sound attenuation, which is a distance-dependent physical property related to a reduction in sound level (volume) over a distance r with respect to a source within a spherical, cylindrical (capsule), cone, or rectangular shape. Despite that sound or music wave spreads in one of these geometric shapes, a plucked string, as an example, vibrates emitting sine waves with different amplitudes.

Consonance of an Interval of an Octave An interval of one octave corresponds to doubling the frequency of the vibration. For example, the left C key corresponds to a frequency of 261.6 Hz, while the right C corresponds to a frequency of 523.2 Hz.

Dissonance of an Interval of Two Notes The presence of components at 440 Hz and 445 Hz causes a sensation of roughness which is interpreted by the ear as dissonance.

15.14.2   Musical Instruments

For one-dimensional analysis, assume that a vibrating string produces a transverse displacement $u_y = u(x, t)$ in the y-direction related to the string force $F(x, t)$ acting on the string and the density $\rho(x, t)$ per unit length L of the string. Thus, the corresponding string sound wave is

$$\frac{\partial^2 u}{\partial t^2} = v^2 \frac{\partial^2 u}{\partial x^2} + \frac{F}{\rho} \tag{15.33a}$$

$$\frac{\partial^2 u}{\partial t^2} \simeq v^2 \frac{\partial^2 u}{\partial x^2} \quad \text{if } \frac{F}{\rho} \to 0 \tag{15.33b}$$

$$v = \sqrt{\frac{\tau}{\mu}} \qquad (15.33c)$$

where τ is the shear stress due to the transverse mechanical action of the string.

Example 15.7 Consider the frequency of sound in air at $f = 20\ Hz$ and $f = 20,000\ Hz$ from two loud speakers. Calculate the wavelength (λ) of sound for each case at $25\,^{\circ}C$. Explain.

Solution From Eqs. (15.2E1a), (15.2E1b) and (15.2b) at $f = 20\ Hz = 20\ 1/s$ and $T = 25\,^{\circ}C = 298\ K$,

$$v = \sqrt{\left(401.94\ \frac{m^2/s^2}{K}\right)T} = \sqrt{\left(401.94\ \frac{m^2/s^2}{K}\right)(298\ K)} = 346.09\ m/s$$

$$\hspace{12cm} (15.6E1a)$$

$$\lambda = \frac{v}{f} = \frac{346.09\ m/s}{20\ 1/s} = 17.31\ m \qquad (15.6E1b)$$

At $f = 20,000\ Hz = 20,000\ 1/s$ and $v = 346.09\ m/s$,

$$\lambda = \frac{v}{f} = \frac{346.09\ m/s}{20,000\ 1/s} = 0.02\ m \qquad (15.eqn6E2)$$

The sound wavelength is inversely proportional to its frequency. Therefore, the larger the frequency f, the smaller the wavelength λ and vice versa.

15.15 Summary

Sound is a traveling acoustical wave bounded by frequency in oscillating systems (gas, liquid, or solid), while its properties are related to frequency, wave propagation velocity, wavelength, and pressure. In essence, the sound wave is based on audible or inaudible oscillations propagating in space with a specific intensity.

Mathematically, sound is characterized by an linear wave equation with a wide range of use in certain engineering and physiological applications. The former is relevant in fluid acoustics, recording music, digital games, television, radio, and film industry. The latter is important in hearing and speech medical fields.

Nondestructive evaluation (NDE) methods, such as ultrasonic testing (UT), eddy current, and the like, are widely used in engineering for detecting discontinuities (cracks, inclusions, voids, etc.) and in the medical fields for health-related conditions.

Appendix 15A Elastic Wave Propagation

Consider a bar or rod (Fig. 15.15) being subjected to an impulse disturbance (impact by a hammer or laser pulse) in the form of a stress wave propagation along the x-axis.

The disturbed element of thickness dx is positioned between planes A_1 and A_2. The initial impact induces a small displacement μ due to an impulse force F_x acting perpendicularly to the plane A_1. Similarly, plane A_2 moves a $\mu + d\mu$ distance due to a reactive force $F_x + (\partial F_x/\partial x)\,dx$.

One-Dimensional Wave Equation

The goal now is to use a disturbed element (Fig. 15.15) to derive a one-dimensional equation for a plane longitudinal stress wave parallel to the impulse direction.

Using the displacement condition $\mu = \mu(x,t)$ of plane A_2 in Fig. 15.15 and approximating it by Taylor's series expansion, the small displacements and small changes in the bar length L are written as (Bray and Stanley book [1, pp. 49–51] and Collins' book [18, pp. 533–535])

$$\mu + d\mu \simeq \mu + \frac{\partial \varepsilon}{\partial x}dx + \ldots \tag{15.34a}$$

$$dL \simeq (\mu + d\mu) - \mu = d\mu \tag{15.34b}$$

$$dL = \frac{\partial \mu}{\partial x}dx \tag{15.34c}$$

The resulting strain is

$$d\varepsilon = \frac{dL}{dx} = \frac{(\partial \mu/\partial x)\,dx}{dx} = \frac{\partial \mu}{\partial x} \tag{15.35}$$

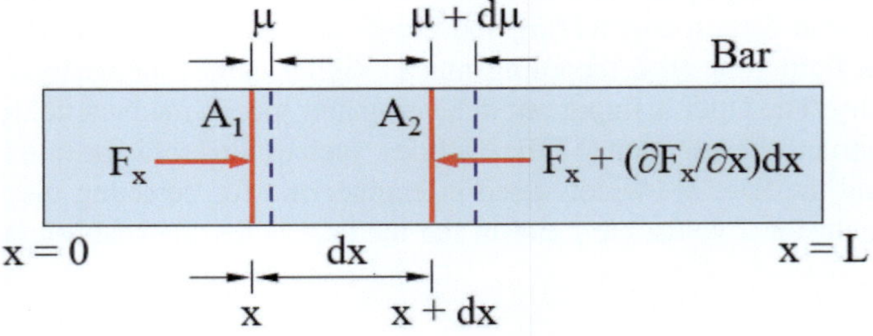

Fig. 15.15 Disturbed element showing the force and displacement balance

and the corresponding force balance is defined by

$$dF_x = \left(F_x + \frac{\partial F_x}{\partial x}dx\right) - F_x = \frac{\partial F_x}{\partial x}dx \tag{15A.3}$$

Using Hooke's law due to the elastic deformation caused by the impulse force yields

$$E = \frac{d\sigma}{d\varepsilon} = \frac{F_x/A}{\partial \mu/\partial x} \tag{15A.4a}$$

$$F_x = AE\frac{\partial \mu}{\partial x} \tag{15A.4b}$$

$$\frac{\partial F_x}{\partial x} = AE\frac{\partial^2 \mu}{\partial x^2} \tag{15A.4c}$$

Inserting Eq. (15A.4c) into (15A.3) gives

$$dF_x = AE\frac{\partial^2 \mu}{\partial x^2}dx \tag{15A.5}$$

According to Newton's law and the mass balance,

$$dF_x = ma = (\rho A dx)\left(\frac{\partial^2 \mu}{\partial t^2}\right) \tag{13A.6}$$

Combining Eqs. (15A.5) and (13A.6) gives the sought one-dimensional wave equation for a plane longitudinal wave

$$\frac{\partial^2 \mu}{\partial t^2} = \frac{E}{\rho}\frac{\partial^2 \mu}{\partial x^2} \tag{15A.7a}$$

$$\frac{\partial^2 \mu}{\partial t^2} = v^2\frac{\partial^2 \mu}{\partial x^2} \tag{15A.7b}$$

which is identical to Eq. (15.1a) with $y = \mu$ and $y(x,t) = \mu(x,t)$.

The Speed of Sound in Solids

According to Collins [18, pp. 538–539], the impulse (I) is just the change in momentum (ΔM)

$$I = Pt = A\sigma t \tag{15A.8a}$$

$$\Delta M = m\upsilon_x = (\rho A L_e)\,\upsilon_x = (\rho A\upsilon t)\,\upsilon_x \tag{15A.8b}$$

Notice that the change in momentum of the element is defined by the product of mass and velocity of the impulse (υ_x), while the impulse is the load or force applied to the element for a short interval of time t. Thus, the impulse or impact velocity υ_x is

$$I = \Delta M \tag{15A.9a}$$

$$A\sigma t = (\rho A\upsilon t)\,\upsilon_x \tag{15A.9b}$$

$$\upsilon_x = \frac{\sigma}{\rho\upsilon} \tag{15A.9c}$$

where P denotes the impulse load, A denotes the cross-sectional area of plane A_1, m denotes mass, t denotes time, σ denotes the impulse stress, and $L_e = dx$ denotes the length of the compressed element (Fig. 15.15). From Hooke's law, the axial strain due to the elastic deformation of the element is

$$\varepsilon = \frac{\Delta L_e}{L_e} = \frac{\sigma}{E} \tag{15A.10a}$$

$$L_e = \upsilon t \tag{15A.10b}$$

$$\Delta L_e = \frac{\sigma}{E} L_e = \frac{\sigma}{E}\upsilon t \tag{15A.10c}$$

Subsequently, the impulse velocity can be defined as

$$\upsilon_x = \frac{\Delta L_e}{t} = \frac{1}{t}\frac{\sigma}{E}\upsilon t = \frac{\sigma}{E}\upsilon \tag{15A.11}$$

Equating Eqs. (15A.9c) and (15A.11) yields the speed of sound equation, which is compared with the corresponding expression for transverse waves

$$\upsilon = \sqrt{\frac{E}{\rho}} \quad \text{(longitudinal)} \tag{15A.12a}$$

$$\upsilon = \sqrt{\frac{G}{\rho}} \quad \text{(transverse)} \tag{15A.12b}$$

Hence, υ refers to the fast motion of a sound wave from atom to atom in the medium (crystal lattice).

Stiffness Constants

In general, the speed of sound dependency on elastic constants and density suggests that pure longitudinal waves can travel along the high symmetry crystallographic $[uvw]$ directions (major directions), such as [100], [110], and [111]. On the other hand, transverse waves can propagate in any $[uvw]$ crystallographic directions (Adachi [19], Daoud et al. [20], Kittel [21, p. 80]).

The speed of sound can be defined in terms of the general elastic compliance constant C_{ij}

$$v = \sqrt{\frac{C_{ij}}{\rho}} \tag{15A.13}$$

For elastic waves in cubic crystals and related adiabatic elastic stiffness constants, Eq. (15A.13) is an excellent source for an analytical procedure on the current topic (Kittel's book [21, p. 80]). Further analysis on this matter is out of the scope of this chapter.

Appendix 15B Speed of Sound in Air

Consider a mass flow rate (dm/dt) of a fluid entering and exiting a particular plane (A) in a certain medium as shown in Fig. 15.16.

Mathematically, the general continuity equation is

$$m = \rho V = \rho A x \tag{15B.1a}$$

$$\frac{dm}{dt} = \frac{d\,(\rho V)}{dt} = \frac{d\,(\rho A x)}{dt} = \rho A \frac{dx}{dt} = \rho A v \tag{15B.1b}$$

$$\left(\frac{dm}{dt}\right)_{in} = \left(\frac{dm}{dt}\right)_{out} \tag{15B.1c}$$

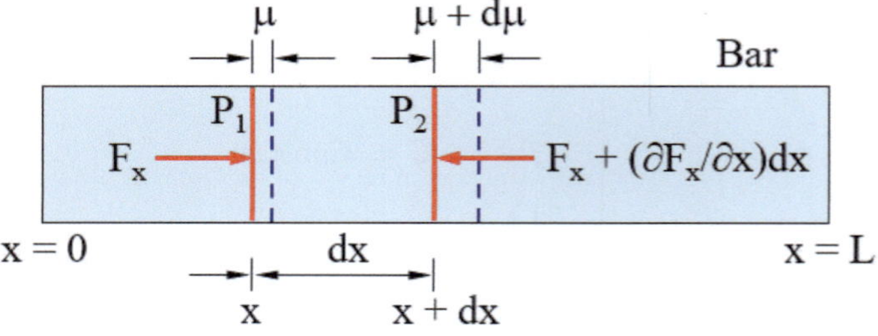

Fig. 15.16 Sound wave through a plane A in a three-dimensional element

$$(\rho A v)_{in} = (\rho A v)_{out} \tag{15B.1d}$$

For the speed of sound v,

$$(\rho A v)_{in} = [(\rho + d\rho)(A)(v + dv)]_{out} \tag{15B.2a}$$

$$\rho dv = -v d\rho \tag{15B.2b}$$

since $d\rho dv \simeq 0$. Using the net force on the volume element yields

$$F = PA \tag{15B.3a}$$

$$F_{net} = PA - (P + dP)A \tag{15B.3b}$$

$$F_{net} = PA - PA - AdP = -AdP \tag{15B.3c}$$

From Newton's second law,

$$F_{net} = ma = m\frac{dv}{dt} \tag{15B.4a}$$

$$ma = -AdP \tag{15B.4b}$$

$$a = -\frac{AdP}{m} = -\frac{AdP}{\rho V} = -\frac{AdP}{\rho A dx} \tag{15B.4c}$$

$$\frac{dv}{dt} = -\frac{dP}{\rho dx} \tag{15B.4d}$$

$$dv = -\frac{dP}{\rho dx}dt = -\frac{dP}{\rho v} \tag{15B.4e}$$

$$\rho v dv = -dP \tag{15B.4f}$$

Combining Eqs. (15B.2b) and (15B.4f) gives

$$v = \sqrt{\frac{dP}{d\rho}} \tag{15B.5}$$

For an adiabatic process (no heat flow) or isentropic process (adiabatic with constant entropy), the governing equation is

$$Pv^\gamma = C = \text{Constant} \tag{15B.6a}$$

$$P\left(\frac{1}{\rho}\right)^\gamma = C \tag{15B.6b}$$

$$\frac{1}{P}\frac{dP}{d\rho} - \frac{\gamma}{\rho} = 0 \tag{15B.6c}$$

$$\frac{dP}{d\rho} = \frac{\gamma P}{\rho} \tag{15B.6d}$$

Substituting Eq. (15B.6d) into (15B.5) yields the sought speed of sound equation defined above as Eq. (15.3e) or (15.5)

$$\upsilon = \sqrt{\frac{\gamma P}{\rho}} \tag{15B.7}$$

with the specific heat capacity ratio

$$\gamma = \frac{c_p}{c_v} \tag{15B.8}$$

Note that the relation defined by Eq. (15B.6a) is valid for isentropic processes only due to reversible and adiabatic thermodynamic conditions.

Problems

15.1 Assume that a transverse wave with a constant amplitude is described by the function

$$y(x,t) = (6\ mm)\cos\left[2\pi\left(\frac{t}{0.025\ s} - \frac{x}{100\ mm}\right)\right]$$

Find **(a)** the wave amplitude (y_o), angular velocity (ω) and the wavenumber (k), **(b)** the wavelength of the wave $y(x,t)$, **(c)** the wave frequency f, and **(d)** the speed of wave propagation. [Solution: (a) $k = 0.10\ cm^{-1}$, (b) $\lambda = 0.62832\ m$, (c) $f = 6.37\ Hz$, (d) $\upsilon = 4\ m/s$].

15.2 Calculate the speed of sound of a longitudinal wave in the [100] direction in a gold (Au) cubic crystal at $300\ K$. The elastic constant C_{11} and density of gold are $195\ GPa$ and $19{,}450\ Kg/m^3$, respectively. [Solution: $\upsilon = 3166.3\ m/s$].

15.3 Calculate the speed of sound **(a)** in air and **(b)** in seawater at $27\,^\circ C$ given the bulk modulus and the density of the latter medium are $2.30\ GPa$ and $1030\ Kg/m^3$. [Solution: (a) $\upsilon = 347.25\ m/s$, (b) $\upsilon = 1494.30\ m/s$].

15.4 Calculate the speed of sound for copper single crystals using the elasticity moduli **(a)** $E_{[100]} = 67\ MPa$ and **(b)** $E_{[111]} = 192\ MPa$. Assume a constant mass density of $\rho = 8.96\ Kg/m^3$. Is $\upsilon_{[111]} > \upsilon_{[100]}$? Why? [Solution: (a) $\upsilon_{[100]} = 86{,}474\ m/s$, (b) $\upsilon_{[111]} = 146{,}390\ m/s$].

15.5 Why is it necessary to use a lubricant in an ultrasonic testing?

15.6 Assume that a gas tank resting on a concrete floor explodes and consequently, the bang radiates $200\ dB$ at 20 meters away from it. Compute **(a)** the sound intensity of the source, **(b)** the sound power in a hemispherical field, and **(c)** the sound pressure, and **(d)** is the sound level due to disturbing explosion beyond the threshold of pain? [Solution: (a) $I = 10^8\ W/m^2$, (b) $W_I = 2.51 \times 10^{11}$ watts, (c) $P = 2 \times 10^{15}\ Pa$].

15.7 Suppose that a measured sound pressure is exactly equals to that of the reference pressure level. Is there any sound at all?

15.8 Suppose that a sound of $120\ dB$ is measured at a distance of $x_1 = 10\ m$ from the source (machinery). Predict the sound pressure level at a distance $x_2 = 100\ m$. [Solution: $L_{p,2} = 100\ dB$].

15.9 Assume that the acoustic wave radiated by a particular source is measured some distance away from the source. If the source generates a sound pressure of $516\ \mu Pa$, then calculate the sound level in dB unit. [Solution: $L_p = 65\ dB$].

15.10 Assume that two speakers located nearby radiate a total disturbing sound power level. Speaker 1 and speaker 2 operate $130\ dB$ and $120\ dB$, respectively. In this case, the reference source W_o for each speaker is unknown. Based on this information, calculate **(a)** the combined and **(b)** the average sound power levels. [Solution: (a) $L_p \simeq 300\ dB$, (b) $L_p \simeq 293\ dB$].

15.11 Suppose two sound waves from the same source radiate a sound intensity level of $50\ dB$ each at a distance r. Calculate the total sound intensity level. [Solution: $L_I \simeq 122\ dB$].

15.12 Consider a sound wave with frequency $f = 20\ Hz$ and wavelength $\lambda = 18\ m$ conditions. Calculate the temperature of the measurements. [Solution: $T = 49.44\,^\circ C$].

15.13 Ultrasound can be used to image internal organs in humans. For ultrasonic waves with a constant $4\text{-}MHz$ frequency and a constant $0.4\text{-}mm$ wavelength, calculate the speed of sound a human cannot hear. The given data values are common in the field of sonography or ultrasonic imaging. Would a normal and young human be able to hear this type of wave sound? [Solution: $\upsilon = 1600\ m/s$].

15.14 Given a plane sound wave in air with a frequency of $120\ Hz$ and a peak sound pressure of $3\ Pa$, let the density and speed of sound of air be $1.20\ Kg/m^3$ and $343\ m/s$ at $20\,^\circ C$. Calculate **(a)** the intensity of the sound pressure and **(b)** the sound intensity level. [Solution: (a) $I = 2.17 \times 10^{-2}\ W/m^2$, (b) $L_I = 103.36\ dB$].

15.15 Consider two waves reaching a particular point x in space. If $L_{I1} - L_{I2} = 3\ dB$, then determine **(a)** the sound pressure ratio P_1/P_2 for a constant air density and **(b)** the speed of sound. [Solution: (a) $I_1/I_2 = 2$, (b) $P_1/P == 2$].

15.16 Compute the sound intensity level at some location where the sound intensity is $2 \times 10^{-3}\ W/m^2$. [Solution: $L_I = 93\ dB$].

15.17 Assume that a $0.15\text{-}m^2$ solid radiates a power of 2 W. Calculate **(a)** the intensity and **(b)** the sound intensity level at the object surface. [Solution: (a) $I = 10\ W/m^2$, (b) $L_I = 299.34\ dB$].

15.18 Assume that a noisy lawn mower motor resting on a concrete floor radiates 85 dB at 2 meters away from it. Compute **(a)** the sound intensity of the source, **(b)** the sound power in a hemispherical field, and **(c)** the sound pressure, and **(d)** is the noisy sound radiated by the lawn mower too loud for you? Can you tolerate this level of noise? [Solution: (a) $I = 3.16 \times 10^{-4}\ W/m^2$, (b) $W_I = 7.94 \times 10^{-3}$ watts, (c) $W_I = 7.94 \times 10^{-3}$ watts, (d) $P = 6324.6\ Pa$].

15.19 Assume that the sound pressure 134 dB at a 2-m distance. Calculate the reduction of the sound pressure level at a 50-m distance. [Solution: $\Delta L_p = 60\ dB$].

15.20 The noise of a machine at 0.80-m location is 120 dB. Calculate the distance for reducing the sound level to 80 dB. [Solution: $r_2 = 80\ m$].

References

1. D.E. Bray, R.K. Stanley, *Nondestructive Evaluation* (McGraw-Hill, New York, 1989)
2. P.J. Shull, B.R. Tittmann, *Ultrasound*, in *Nondestructive Evaluation: Theory, Techniques, and Applications*, ed. by P.J. Shull (Marcel Dekker, New York, 2001)
3. D.R. Raichel, *The Science and Applications of Acoustics* (Springer, New York, 2006)
4. L.E. Kinsler, A.R. Frey, A.B. Coppens, J.V. Sanders, *Fundamentals of Acoustics*, 4th edn. (John Wiley & Sons, New York, 2000)
5. S.J. Ling, J. Sanny, W. Moebs, *University Physics Volume 1* (an OpenStax resource, OpenStax.org, Houston, Texas, Rice University, 2021)
6. K.A. Krout, S.H. Sohrab, On the speed of sound. Int. J. Thermodyn. **19**(1), 29–34 (2016)
7. E.W. Lemmon, R.T Jacobsen, S.G. Penoncello, D.G. Friend, Thermodynamic properties of air and mixtures of nitrogen, argon, and oxygen from 60 to 2000 K at pressures to 2000 MPa. J. Phys. Chem. Ref. Data **29**(3), 331–385 (2000)
8. M. Lozev, B. Grimmett, E. Shell, R. Spencer, *Final Report Evaluation of Methods for Detecting and Monitoring of Corrosion and Fatigue Damage in Risers*, EWI Project No. 45891GTH, Edison Welding Institute, Columbus, OH, USA (2003)
9. S.C. Her, S.T. Lin, Non-destructive evaluation of depth of surface cracks using ultrasonic frequency analysis. Sensors (Basel) **14**(9), 17146–17158 (2014)
10. D.K. Mak, Ultrasonic measurement of crack-like defects in metals. J. Mech. Behav. Mater. **2**(1–2), 1–18 (1989)
11. S.M. Kirkup, *The Boundary Element Method in Acoustics*. Integrated Sound Software, Stephen Kirkup 1998–2007 (Hebden Bridge, England, 2007). ISBN 0 953 4031 06
12. B.B. He, *Two-Dimensional X-Ray Diffraction* (John Wiley & Sons, New York, 2009)
13. https://www.youtube.com/watch?v=1gD_0OeDUWI
14. https://www.youtube.com/watch?v=tlE3eK0g6vU
15. V.M. Malhotra, N.J. Carino (eds.), *Handbook on Nondestructive Testing of Concrete*, 2nd edn. (CRC Press, New York, 2004)
16. M.S. Helal, P. Mendis, Non-destructive testing of concrete: a review of methods. Electron. J. Struct. Eng. **14**(1), 97–105 (2015)
17. D. Benson, *Music: A Mathematical Offering* (Cambridge University Press, Cambridge, 1995–2008). http://www.maths.abdn.ac.uk/~bensondj/html/maths-music.html

18. J.A. Collins, *Failure of Materials in Mechanical Design*, 2nd edn. (John Wiley & Sons, New York, 1993)
19. S. Adachi, *Handbook on Physical Properties of Semiconductors*, vol. 2 (Kluwer Academic Publishers, Boston, 2004)
20. S. Daoud, K. Loucif, N. Bioud, N. Lebgaa, First-principles study of structural, elastic and mechanical properties of zinc-blende boron nitride (B3 BN). Acta Phys. Polonica A **122**(1), 109–115 (2012)
21. C. Kittel, *Introduction to Solid State Physics*, 8th edn. (John Wiley & Sons, New York, 2005)

Chapter 16
Electrochemical Corrosion

16.1 Introduction

Electrochemistry is a science that studies chemical reactions that involve electron transfer (e^-) at the interface between an electric conductor (electrode) and an ionic conductor (electrolyte), which is a solution that conducts electricity due to ion motion. Moreover, if an electrochemical reaction occurs on a metal surface, then this leads to metal deterioration or degradation, and the electrochemical process is called corrosion, which is represented by and oxidation reaction, where a metal M losses electrons (ze^-) to the immediate interface. As a result, M becomes a metallic ion M^{z+} known as a cation having a positive electric charge.

Corrosion is a natural process due to the instability of refined (man made) metals or alloys, specifically in the presence of water and oxygen. In essence, corrosion is a natural degradation or deterioration of the lattice structure of a solid material due to electrochemical or chemical interactions with its environment. This implies that corrosion is a form of alteration or surface destruction of the crystalline or amorphous atomic structure of a refined metal M, which reverts to its natural ore state as a chemically stable oxide (cuprite Cu_2O), hydroxide [rust $FeO(OH) \cdot nH_2O$)], carbonate (siderite $FeCO_3$), or sulfide (pyrite FeS_2). Materials like aluminum and iron alloys can undergo a mechanism of self-protection by metal-oxide passive films with a few nanometers (nm) in thickness.

Corrosion prevention can be achieved by adding a corrosion inhibitor (silicate or phosphate compound) to an electrolyte for decreasing or avoiding corrosion from taking place. Basically, electrochemical corrosion through an anodic reaction $M \rightarrow M^{z+} + ze^-$ and its counterpart electroplating through a cathodic reaction $M^{z+} + ze^- \rightarrow M$ are governed by the rate of electron transfer process at the metal-electrolyte or dissimilar metal-metal interface (galvanic corrosion).

For comparison, ceramic and polymers do degrade thermally, physically, or mechanically, and they may be considered as corrosion-resistant materials. Further, corrosion studies may include chemical corrosion of nonmetals and quantum corrosion based on corrosion inhibitors, but this chapter only considers aspects of electrochemical corrosion of metals due to lack of space.

16.2   Aspects of Electrochemistry

The word **corrosion** is a surface damage in an aggressive environment (liquid, gas, or hybrid soil-liquid) known as an electrolyte.

The rate of corrosion is governed by the rate of electron transfer process at the metal-electrolyte interface denoted by the generalized electrochemical anodic reaction $M \rightarrow M^{z+} + ze^-$, where M is the metal, M^{z+} is the positively charged metal ion (cation) that has been oxidized anodically, z is the valence electron occupying the outermost shell in the electronic configuration of M, and ze^- is the number of released electrons.

A simultaneous cathodic reaction, such as $zH^+ + ze^- \rightarrow zH$ for atomic hydrogen formation, takes place by accepting the ze^- electrons being released by the oxidizing metal through a forward electrochemical reaction $M \rightarrow M^{z+} + ze^-$ for metal oxidation.

Metallic corrosion is an electrochemical oxidation process that occurs at relatively low and high temperatures, whereas metallic reduction is an electroplating or electrodeposition process that occurs at relatively low temperatures. For instance,

- In effect, ambient (low) temperature corrosion is the most common form of metal dissolution being studied for decades.
- Anodic oxidation, known as anodizing, at relatively near ambient temperature produces a hard metal oxide film, which acts a protective film, usually termed passive film.
- High-temperature corrosion is a high-temperature oxidation process that forms a porous and thick enough metal oxide layer used to evaluate the protectiveness of a metal oxide (CuO, FeO, etc.) in high-temperature applications, such as boilers, nuclear reactors, and so forth.
- An electrochemical reduction reaction, $M^{z+} + ze^- \rightarrow M$, represents metal production by electrodeposition from a solution containing the M^{z+} ions.
- During electrodeposition, an external power supply must be used to provide the electrons so that M^{z+} ions deposited on cathode surfaces.
- Cathodic protection (CP) is an electrochemical technique used to control or avoid corrosion by supplying electrons (e^-) to a metallic structure.
- During cathodic protection, the structure becomes a cathodic half-cell in an electrochemical cell.
- Designing against corrosion is an industrial practice, where organic and metallic coatings are applied to a structure for protecting metal structures from corrosion.

In conclusion, electrochemical reactions describing corrosion, metal reduction, or metal protection involve electron transfer related to the passage of electric current in an active electrochemical cell at a certain temperature and pressure. Actually, an active electrochemical cell refers to a closed electric circuit composed of electrodes and electrolytes containing relevant positively and negatively charged ions.

Most of teaching material in this chapter is based on galvanic corrosion and related electrochemical reactions. In essence, a galvanic cell is based on dissimilar metals and their alloys or a single metal part with different microstructural features induced by cold working and heat treatment. Simply stated, metal corrosion depends on aggressive environments, temperatures, and microstructure features induced by a certain metal processing technique.

16.3 Corrosion Reactions

An electrolyte is analogous to an electrically conductive solution containing positively and negatively charged ions called cations $(+)$ and anions $(-)$. An ion is an atom that has lost or gained one or more electrons and carries an electric charge. Thus, corrosion is related to a coupled anodic-cathodic process represented by a redox reaction, which in turn depends on simultaneous anodic and cathodic partial reactions.

Consider a hypothetical metal M immersed in a water-sulfuric acid (H_2SO_4) solution containing H^+ and SO_4^- ions. The possible electrochemical reactions are

$$M \rightarrow M^{z+} + ze^- \qquad \text{(anodic)} \qquad (16.1a)$$

$$zH^+ + zSO_4^- + ze^- \rightarrow \frac{z}{2}H_2SO_4 \qquad \text{(cathodic)} \qquad (16.1b)$$

$$M + zH^+ + zSO_4^- \rightarrow M^{z+} + \frac{z}{2}H_2SO_4 \qquad \text{(redox, overall)} \qquad (16.1c)$$

The electrochemical reactions defined by Eq. (16.1) occur on the surface of solid electrodes (known as substrates). Thus,

- Equation (16.1a) denotes the anodic partial reaction (anodic half-cell). For example, if $M = Cu$ (copper), then the electrochemical reaction $Cu \rightarrow Cu^{2+} + 2e^-$ occurs by releasing $ze^- = 2e^-$ electrons from the $[Ar]3d^{10}4s^1$ electronic configuration, which becomes $[Ar]3d^9$ upon oxidation related to an anodic Gibbs energy ΔG_a and an anodic electric potential or voltage potential E_a.
- Equation (16.1b) denotes the cathodic partial reaction (cathodic half-cell), which represents a reduction process, where H^+ and SO_4^- gain the ze^- electrons lost by M and react to form H_2SO_4. This cathodic reaction is also related to a cathodic Gibbs energy ΔG_c and the cathodic electric potential E_c.
- Equation (16.1c) denotes the overall or simply the **redox** (red = reduction and ox = oxidation) electrochemical reaction defined by the sum of anodic and cathodic reactions. Thus, the redox electric potential becomes $E_{redox} = E_a + E_c$ and $\Delta G_{redox} < 0$; otherwise, the redox reaction will not occur as written. The

concepts of Gibbs energy ΔG and electric potential E in electrochemical cells are dealt with in a later section.

Chemical corrosion is due to chemical reactions, and electrochemical corrosion is due to electrochemical reactions related to electron transfer which refers to surface damage.

Furthermore, it is convenient now to distinguish the symbols used to classify the types of reactions. For instance, a reversible reaction is symbolized with an equal sign ($=$) or double arrows ($\rightleftarrows$), and an irreversible reaction is denoted with a forward arrow ($\rightarrow$) or backward arrow ($\leftarrow$). Accordingly,

$$M = M^{z+} + ze^- \quad \text{(reversible)} \tag{16.2a}$$

$$M \rightleftarrows M^{z+} + ze^- \quad \text{(reversible)} \tag{16.2b}$$

$$M \rightarrow M^{z+} + ze^- \quad \text{(irreversible)} \tag{16.2c}$$

Mitigating corrosion may be accomplished by using sacrificial anodes, which is a common industrial practice.

16.4   Classification of Corrosion

There is not a unique classification of the types of corrosion, but the following classification is adapted hereafter (Fontana [1, Chapter 3]).

16.4.1   Localized Corrosion

It is clearly evident that images of corrosion are not appealing to the naked eye, but they are common structural deterioration that may lead to failure. Moreover, the images in Fig. 16.1 are purposely included side by side, so that the reader can visualize and compare the characteristics of corroded solid surfaces and understand the implications of corrosion related to mechanical properties relevant materials [2–7].

Corrosion, in general, occurs due to different degradation mechanisms and appears in many forms. Among many images of corrosion available in the literature, Fig. 16.1 shows some corrosion images ranging (1) from localized corrosion, such as pitting corrosion in Fig. 16.1a (Alexander [2]); crevice corrosion in Fig. 16.1b [3]; galvanic corrosion in Fig. 16.1c [4]; biological corrosion in Fig. 16.1d [5]; leaching of zinc from brass in Fig. 16.1e [6] (2) to general corrosion in Fig. 16.1f for steel gears [7].

Localized corrosion is the more difficult surface damage to control than the general corrosion. Localized corrosion can be classified as

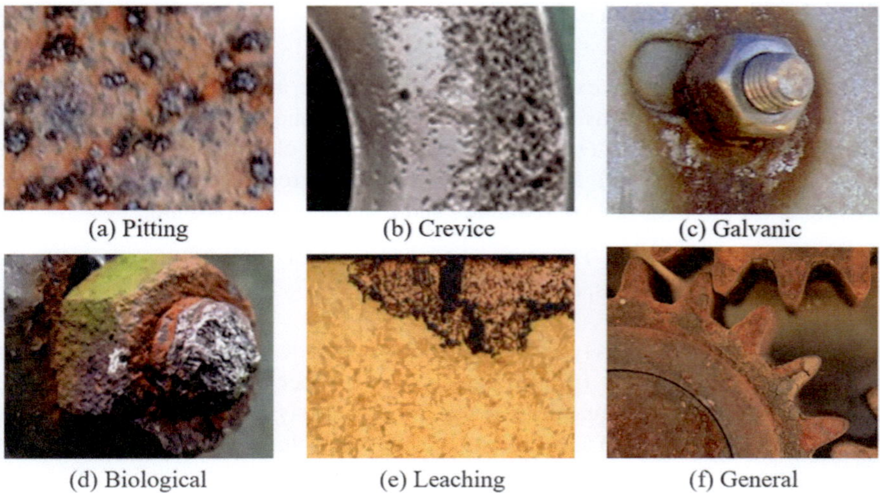

Fig. 16.1 Types of corrosion. (**a**) Pitting (Alexander [2]), (**b**) crevice [3], (**c**) galvanic [4], (**d**) biological [5], (**e**) leaching of zinc from brass (dezincification) [6], and (**f**) general [7]

Pitting Corrosion This is an extremely localized corrosion mechanism that causes destructive pits on a solid surface.

Crevice Corrosion This is associated with a stagnant electrolyte such as dirt, corrosion product, sand, etc. It occurs on a metal/alloy surface holes, underneath a gasket, lap joints under bolts, under rivet heads.

Filiform Corrosion This is basically a special type of crevice corrosion, which occurs under a protective film. It is common on food and beverage cans being exposed to the atmosphere.

Oral Corrosion This type of corrosion occurs on dental alloys exposed to saliva.

Microbiological Corrosion This is due to fouling organisms (microbes) adhered on steel in marine environments. Some common microbes are fungi (yeast and molds), algae, and bacteria.

Selective Leaching Corrosion This is a metal removal process from the base alloy matrix, such as dezincification where Zn is removed from Cu-Zn alloys and graphitization where Fe is removed from cast irons (ferrous alloys). This type of corrosion is also known as dealloying of solid solution alloys.

16.4.2   General Corrosion

This is the case when an exposed metal/alloy surface area is uniformly corroded in an environment, such as a liquid electrolyte (chemical solution, liquid metal), gaseous electrolyte (air, CO_2, SO_2^-, etc.), or a hybrid electrolyte (solid and water, biological organisms, etc.). Some types of general corrosion and their description, excluding the effect of rates of oxidation and reduction reactions, are:

Atmospheric Corrosion This general corrosion occurs on steel tanks, steel containers, Zn parts, Al plates, etc.

Galvanic Corrosion This type of corrosion occurs between dissimilar metal/alloys or microstructural phases (pearlitic steels, α-β copper alloys, α-β lead alloys). Moreover, galvanic cell reactions occur spontaneously at the anodic electrodes immersed in or exposed to aqueous and nonaqueous solutions.

High-Temperature Corrosion This occurs on carburized steels that forms a porous scale of several iron oxide phases.

Liquid-Metal Corrosion This occurs on stainless steel exposed to a sodium chloride ($NaCl$) environment.

Molten-Salt Corrosion This occurs on stainless steels due to molten fluorides (LiF, BeF_2, and so forth).

Biological Corrosion This is a biologically driven corrosion that occurs on steel, Cu-alloys, Zn-alloys in seawater.

Stray-Current Corrosion This occurs on pipelines near a railroad. Normally, a function generator is directly connected to a reference electrode for minimizing the negative effect of the stray currents.

Stress-Corrosion Cracking (SCC) This occurs on metals being stressed in an aggressive environment.

Erosion Corrosion This occurs on metals due to aqueous or gaseous flow.

   Each form of corrosion cited above requires specific conditions responsible for its initiation and subsequent propagation. Only specific types of corrosion mechanisms are described next.

(a) Steel bridge   (b) Steel pipeline

Fig. 16.2 Uniform corrosion considered as a form of atmospheric attack caused by airborne salinity and moisture in sea coastal areas. (**a**) A steel bridge and (**b**) a steel pipeline located on a concrete pier above the ocean water (Perez [8, pp. 6–7])

16.4.3 Atmospheric Corrosion

This is a uniform and general attack of an entire metal structure being gradually degraded by oxidation to form a metal oxide (MO or MxO_y) film. For example, Fig. 16.2 shows uniform atmospheric corrosion on common steel structures due to airborne salinity and moisture. The steel bridge (Fig. 16.2a) is located approximately $2\ miles$ ($322\ Km$) from the seashore, and the pipeline (Fig. 16.2b) is located on an ocean pier in the Caribbean (Perez [8, pp. 6–7]).

Corrosion of the Steel Bridge Atmospheric corrosion is manifested as a brown ferric hydroxide, $Fe(OH)_3$ or $Fe_2O_3 \cdot 3H_2O$, surface layer known as brown, which is due to a series of reactions (Broomfield [9, p. 8]). Thus, the most common reactions are

$$(Fe \rightarrow Fe^{2+} + 2e^-)(x2) \tag{16.3a}$$

$$O_2 + 2H_2O + 4e^- \rightarrow 4OH^- \tag{16.3b}$$

$$2Fe^{2+} + 4OH^- \rightarrow 2Fe(OH)_2 \tag{16.3c}$$

$$2Fe(OH)_2 + \frac{1}{2}O_2 + H_2O \rightarrow 2Fe(OH)_3 \equiv Fe_2O_3 \cdot 3H_2O \tag{16.3d}$$

$$\rightarrow Fe_2O_3 \cdot H_2O + 2H_2O \tag{16.3e}$$

where ($x2$) in Eq. (16.3a) denotes a multiplying factor for balancing the number of electrons, $Fe(OH)_2$ denotes the initial ferrous hydroxide film (unstable compound with ferrous Fe^{2+} cations), $Fe(OH)_3$ denotes the trihydroxide layer (with Fe^{3+} cations), and $Fe_2O_3 \cdot 3H_2O$ denotes the ferric hydroxide layer called rust, which is a form of porous hydrated ferric oxide.

The iron oxide Fe_2O_3 compound, in general, can be present in a complex corrosion layer as hematite ($\alpha\text{-}Fe_2O_3$) or maghemite ($\gamma\text{-}Fe_2O_3$).

Corrosion of Steel Pipelines For steel pipelines in a marine environment, it is assumed that the steel rust gives rise to the formation of ferrous chloride ($FeCl_2$) due to the airborne salinity and moisture. This is a common occurrence of corrosion on other steel structures exposed to this particular environment. Actually, pipelines are common structures in the petroleum and oil refinery industry.

Cathodic protection is normally used to protect or eliminate the corrosion process of these structures. Specifically, steels have iron as the main chemical element, which in turn has a natural tendency to combine with oxygen or other chemical elements to reach its natural lowest energy state.

The possible reactions related to rust formation, $Fe(OH)_3$, in the presence of chloride ions, oxygen, and water are

$$NaCl \rightarrow Na^+ + Cl^- \tag{16.4a}$$

$$Fe \rightarrow Fe^{2+} + 2e^- \tag{16.4b}$$

$$Fe^{2+} + 2Cl^- \rightarrow FeCl_2 \tag{16.4c}$$

$$FeCl_2 + 2OH^- \rightarrow Fe(OH)_2 + 2Cl^- \tag{16.4d}$$

$$2Fe(OH)_2 + \frac{1}{2}O_2 + H_2O \rightarrow 2Fe(OH)_3 (\text{Rust}) \tag{16.4e}$$

It is clearly evident in Fig. 16.2 that corrosion results in a reddish-brown and flaky rust (solid phase). Consequently, the steel bridge and the steel pipeline are no longer in service, since $Fe(OH)_3$ is not a self-healing oxide layer.

Corrosion of Zinc Basically, zinc (Zn) on galvanized steel can uniformly corrode forming a zinc carbonate, $Zn_4CO_3 \cdot (OH)_6$, as white rust. In this case, the relevant reactions are

$$(Zn \rightarrow Zn^{2+} + 2e^-)x2 \tag{16.5a}$$

$$O_2 + 2H_2O + 4e^- \rightarrow 4OH^- \tag{16.5b}$$

$$2Zn + O_2 + 2H_2O \rightarrow 2Zn^{2+} + 4OH^- \tag{16.5c}$$

$$2Zn^{2+} + 4OH^- \rightarrow 2Zn(OH)_2 \tag{16.5d}$$

$$2Zn(OH)_2 + CO_2 + O_2 + H_2O \rightarrow Zn_4CO_3 \cdot (OH)_6 \tag{16.5e}$$

Corrosion of Aluminum Alloys Atmospheric corrosion of aluminum alloys is due to the formation of a protective passive oxide film. In effect, the natural gray/black-color film may form according to surface reactions on an aluminum alloy substrate. Consider that aluminum is the principal chemical element in Al alloys exposed to oxygen and moisture. Thus, the Al-oxide film forms due to the following reactions

$$(Al \rightarrow Al^{3+} + 3e^-)x2 \tag{16.6a}$$

$$\frac{3}{2}O_2 + 3H_2O + 6e^- \rightarrow 6OH^- \tag{16.6b}$$

$$2Al + \frac{3}{2}O_2 + 3H_2O \rightarrow 2Al^{3+} + 6OH^- \rightarrow Al_2O_3 \cdot 3H_2O \tag{16.6c}$$

Regarding degradation of aluminum and its alloys, once the passive film breaks down in aggressive environments, corrosion takes place mainly along grain boundaries containing segregated precipitates or intermetallic compounds. This leads to intergranular corrosion referred to as grain boundary corrosion.

Welding Al-alloy parts are susceptible to undergo a localized melting and solidification processes at the joining parts. Eventually, the molten weld pool solidifies having a weldment structure dependent on the heating/cooling cycle and dimensional geometry. Further, welded zones may become susceptible to chemical or electrochemical corrosion due to variations of the welded microstructural features interacting with the environment. Moreover, the reader is encouraged to search the corrosion field for detailed insights on the mechanisms of corrosion in relevant environments.

Natural Galvanic Corrosion This is due to an electric potential difference between two different metals or electrodes. Evidently, atmospheric galvanic corrosion is clearly shown in Fig. 16.3a for a steel chain holding an aluminum gate and in Fig. 16.3b for a coupled steel nut bolt holding a painted steel plate (organic coating). Both chain and nut bolt exhibit the common brown rust layer as evidence of corrosion, since they act as anodes. The anodes have very small surface areas, while the gate and the coated steel plate have very large cathodic surface areas. Therefore, there is an area ratio effect on the galvanic corrosion process.

Despite that both Figs. 16.1c and 16.3b represent galvanic corrosion between the coupled nut bolt plates, the former exhibits a cathodic nut bolt and latter shows an anodic nut bolt. Therefore, the organic coating (paint) in Fig. 16.3b reverts the corrosion process making the nut bolt a particular fastener corrosion case.

(a) Steel-aluminum coupling   (b) Steel-coating coupling

Fig. 16.3 Natural galvanic corrosion. (**a**) a steel alloy chain and (**b**) carbon steel nut bolt-plate coupling

Prevention of Uniform Corrosion This can be accomplished by selecting an adequate (1) material having a uniform microstructure, (2) coating or paint, (3) inhibitor used to retard or suppress corrosion, and (4) cathodic protection to suppress corrosion in large steel structures.

An inhibitor is the adsorption-type hydrogen-evolution substance that acts as an ionic layer between the metal surface and the electrolyte. As a result, an inhibitor slows down or prevents a particular chemical or electrochemical reaction from taking place.

16.5   Concrete Corrosion

In this section, concrete corrosion is referred to as corrosion of steel bars embedded in concrete. Embedding carbon steel or stainless steel bars in concrete containing chemical admixtures (chemicals or additives) is a common technique for reinforcing the mechanical strength of the resultant cementitious composite known as **reinforced concrete**.

Application of reinforced concrete in severe environments, such as sea water and deicing salts, requires a maintenance plan to control or avoid corrosion of the steel bars due to the presence of chloride ions (Cl^-), carbon dioxide (CO_2) gas, oxygen, and water. Specifically, Cl^- ions and CO_2 molecules can diffuse into the concrete pores or defects, and eventually, they reach the steel bars to initiate the corrosion process.

From a macroscale, Fig. 16.4 illustrates the physical conditions of steel bars used to reinforce concrete. Notice the steel bar goes from corrosion-free in Fig. 16.4a to heavily corroded condition in Fig. 16.4d. The intermediate stages show slightly passivated (Fig. 16.4b) and completely passivated (Fig. 16.4c) steel bar surfaces by a reddish-brown protective film, which is initially an Fe-oxide protective coating.

The formation of rust on steel reinforced bars in concrete (Fig. 16.4d) is described by the series of reactions defined by Eq. (16.3). According to Mansfield [10] and Roberge [11, p. 159], the possible relative volume (dimensionless quantity) of iron and its oxides and hydroxides on reinforced-concrete steel bars (also known as rebars) are as depicted in Fig. 16.5.

Oxides and hydroxides are the buildup of corrosion products that induce internal pressure in the immediate reinforced concrete interface and cause subsequent cracking and spalling.

Notice that Fig. 16.6a clearly shows a rust staining on a house ceiling (concrete layer). Rust staining is referred to as a localized spotty corrosion, which must be eliminated because it may lead to failure of the concrete by cracking and spalling. The initial spotty corrosion is attributed to diffusion of chloride ions (Cl^-) along with oxygen and moisture in the concrete slab (roof). Moreover, Fig. 16.6b depicts spotty corrosion adjacent to a major branched crack, Fig. 16.6c illustrates corroded steel bars, and Fig. 16.2d elucidates an excavated away spot revealing heavily corroded steel bars responsible for concrete spalling.

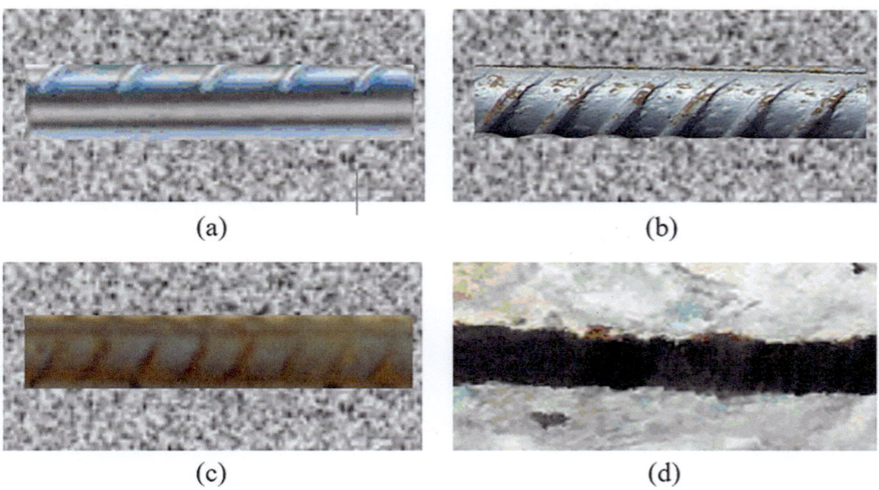

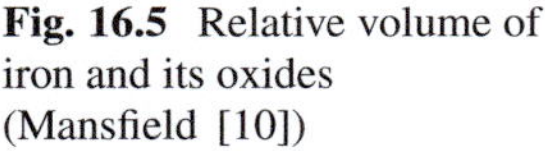

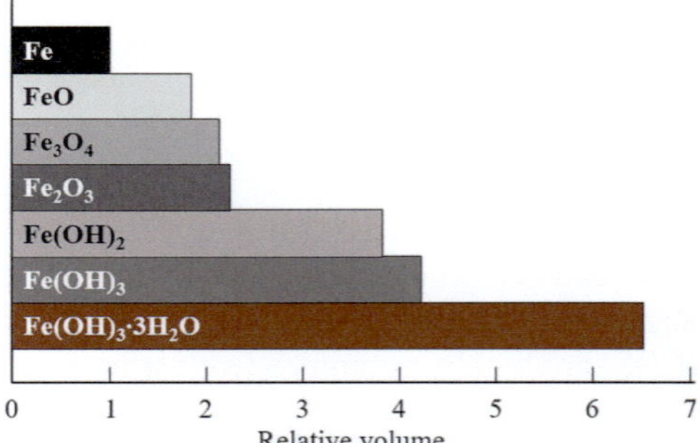

Fig. 16.4 Four stages of a structural carbon steel bar embedded in concrete. (a) Corrosion-free steel, (b) slightly passivated steel, (c) passivated steel, and (d) heavily corrode steel bars

Fig. 16.5 Relative volume of iron and its oxides (Mansfield [10])

Actually, the extent of structural damage of reinforced concrete depends on the carbonation of concrete (which reduces pH), content of chloride ions $[Cl^-]$, properties of the concrete, microstructural features of the steel, sustained mechanical stresses, and environment.

Corrosion Studies Thus far corrosion has been presented as an overview at macro- and microscales, including some general electrochemical reactions for assumed mechanisms of oxidation and reduction processes. In essence, characterizing the initiation of the corrosion phenomena as part of surface science is most appropriate at an atomic level, specifically at metal-substrate interfaces using in situ or ex situ methods.

One particular aspect of corrosion studies is the characterization of an electrode surface being polarized in aqueous solutions containing known ions at temperatures. The goal is to determine the corrosion damage through microscopy accompanied

Fig. 16.6 Spotty corrosion in four different residences in the Caribbean. (**a**) Spotty corrosion, (**b**) spotty corrosion and branched crack, (**c**) corroded steel bars, and (**d**) severe reinforced steel bars in the concrete slab

with energy-dispersive X-ray spectroscopy (EDX) for elemental analysis of a specimen is highly recommended for corrosion studies. Besides a visual inspection, a scanning electron microscope (SEM) and an energy-dispersive spectrometer (EDS) are highly recommended instruments for characterizing corrosion.

16.6 Galvanic Cells

Figure 16.7 schematically shows two different galvanic cells and their macro-components. For instance, Fig. 16.7a contains one electrolyte and two iron (Fe) electrodes having different microstructures.

Figure 16.7b, on the other hand, shows a two-component artificial galvanic macro-cell having a diaphragm separating the α-M_1 and β-M_2 half-cells. This galvanic cell is a suitable electrochemical system for corrosion studies.

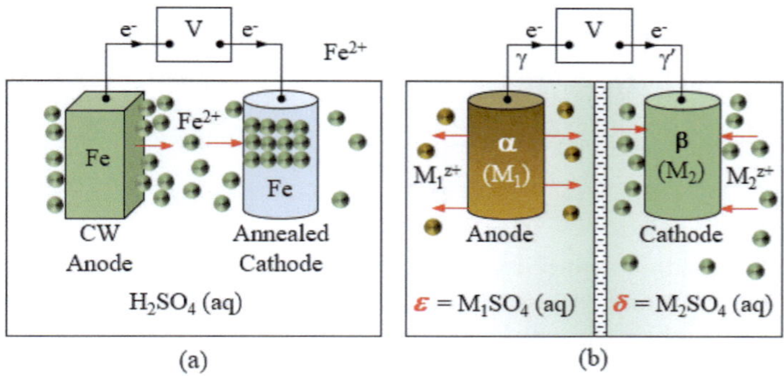

Fig. 16.7 Schematic galvanic cells. (**a**) One-component cell having an anodic cold worked and a cathodic annealed electrodes and (**b**) two-component cell having an anodic α-M_1 and a cathodic β-M_2 metallic electrodes

16.6.1 Galvanic Cell Components

Figure 16.7b can be used as the point of departure for developing the thermodynamics of electrochemical cells containing at least two half-cells being identified as α-phase/ε-phase and β-phase/δ-phase. Moreover, in a conventional galvanic cell setup, the cathode is the positive electrode kept on the right-hand side, while the anode is the negative electrode on the left-hand side of the cell. Thus, electron flow occurs from left to right through the terminals. Nonetheless, the macro-components of this galvanic cell are the following:

Anode Half-Cell It is an electric conductor containing a metallic α-phase called the anode electrode (plate or cylinder) immersed into an electrolyte ε-phase (liquid, paste, gel, or moist soil). The α-phase is a metal M_1, such as Zn, and ε-phase is the electrolyte, such as a metal sulfate M_1SO_4.

Cathode Half-Cell It is also an electric conductor containing metallic β-phase called the cathode electrode (plate or cylinder) immersed into an electrolyte δ-phase (liquid, paste, gel or moist soil). The β-phase is a metal M_2, such as Cu, and δ-phase is the electrolyte, such as a metal sulfate M_2SO_4.

Wiring System Both half-cells are connected using a metallic wire, usually Cu in order to complete an electric circuit. Denote the direction (arrow) of the electron flow, which produces a direct current (DC).

Diaphragm or Salt Bridge This separates the half-cells and allows the flow of ions for charge balance between the oxidation and reduction processes.

Assume that the electrodes in Fig. 16.7b are immersed in their own sulfate-base electrolytes, M_1-$M_1SO_4(aq)$ and M_2-$M_2SO_4(aq)$. Once the electric circuit is complete, the cell electric potential (E) is measured at a temperature T and pressure

P. It is assumed now that the reactions proceed as written in an irreversible manner

$$M_1 \rightarrow M_1^{z+} + ze^- \qquad \text{(Anode)} \qquad\qquad (16.7a)$$

$$M_2^{z+} + ze^- \rightarrow M_2 \qquad \text{(Cathode)} \qquad\qquad (16.7b)$$

$$M_1 + M_2^{z+} \rightarrow M_1^{z+} + M_2 \qquad \text{(Redox)} \qquad\qquad (16.7c)$$

The redox reaction defined by Eq. (16.7c) is the sum of Eqs. (16.7a) and (16.7b) in which the electrons ze^- cancel out. Thus, the metallic cation M_1^{z+} produced at the anodic α-phase (negative terminal) goes into the ε-phase solution, while the number of electrons ze^- move along the wires and arrive at the cathodic electrode-electrolyte interface (positive terminal). This means that M_1^{z+} ions detach themselves from the metal lattice and become part of the electrolyte ionic particles. Note that the cation M_2^{z+} gains ze^- electrons released by the metal M_1 atoms and reduces on the the cathode β-phase to become part of the metal M_2 lattice or react there with oxygen to form a metal oxide film, such as M_2O.

16.6.2 Copper-Zinc Galvanic Cell

The concept of electrode phases in a galvanic cell is also illustrated in Fig. 16.8, where the half-cells are connected through a salt bridge. This cell is also known as a galvanic cell or Daniell Cell (invented in 1836). Assume that the oxidation on the anode and reduction on the cathode processes are spontaneous and that they are the representation of ionic mass transport being dependent upon the strength of the electric field among other factors.

The reactions for the electrochemical cell in Fig. 16.8 are

$$Zn = Zn^{2+} + 2e^- \qquad \text{(Anode)} \qquad\qquad (16.8a)$$

$$Cu^{2+} + 2e^- = Cu \qquad \text{(Cathode)} \qquad\qquad (16.8b)$$

$$Zn + Cu^{2+} = Zn^{2+} + Cu \qquad \text{(Redox)} \qquad\qquad (16.8c)$$

Notice that $\phi_{Zn/Cu}$ and $\phi_{Cu/Cu}$ are contact electric potentials and ϕ_j is the liquid junction electric potential. The standard cell electric potential is defined as the sum of interfacial electric potentials. Thus,

$$E_{cell}^o = \left(E_{Zn}^o + E_{Cu}^o\right) + \left[\phi_{Zn/Cu} + \phi_{Cu/Cu} + \phi_j\right] \qquad\qquad (16.9)$$

In real cases, nonstandard conditions exist, and the cell electric potential can be predicted by

$$E_{cell} = E_{cell}^o + [E_{Nernst} - E_{Ohmic} - \eta] \qquad\qquad (16.10)$$

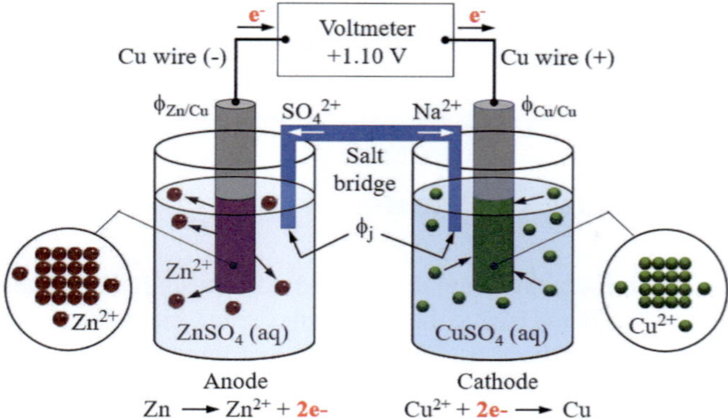

Fig. 16.8 The Zn/Cu galvanic (voltaic) cell, also known as Daniell cell. The insets show atomic arrangements on the electrode surfaces

where E_{Nernst}, E_{Ohmic}, $\eta = \eta_{overpotential}$ are electric potential terms for correcting the standard electric potential expression defined by Eq. (16.10).

Moreover, the electrochemical coupling in Fig. 16.8 undergoes several events which can be described as follows:

- When zinc (Zn) is immersed into aqueous zinc sulfate, $ZnSO_4(aq)$, the subsystem Zn-$ZnSO_4(aq)$ becomes the anode half-cell. Similarly, copper (Cu) immersed in aqueous copper sulfate, Cu-$CuSO_{4(aq)}$, becomes the cathode half-cell.
- Both $ZnSO_{4(aq)}$ and $CuSO_{4(aq)}$ are electrolyte solutions that act as ionic conductors.
- Normally, Cu wire is used to connect the Zn and Cu electrodes, so that no electric potential difference develops between the wires, which are called terminals.
- The salt bridge is filled with an agar gel (polysaccharide agar or polymer gel) and concentrated aqueous salt ($NaCl$ or KCl) or sodium sulfate ($NaSO_4$) solution in order to maintain electric neutrality between the half cells. The gel allows diffusion of ions but eliminates convective currents due to turbulent transport of electric charges (Markson et al. [12]).
- If a salt bridge is used, then a voltmeter measures the electric potential difference (E_{cell}) between the Zn and Cu electrodes.

Thus far it has been assumed that all electrochemical reactions proceed as written. In this case, the Gibbs energy change must be $\Delta G < 0$ in order for a reaction to proceed as written; otherwise, the reaction would be reversed.

The goal in this section is to elucidate the fundamental characteristics and technological significance of electrochemical cells. A comprehensive review of the subject is excluded, since the intention hereafter is to describe the principles of electrochemistry, which should provide the reader with relevant key definitions and concepts on metal reduction in electrolytic cells and metal oxidation in galvanic couplings. In general, electrochemistry deals with the chemical response of an electrode-electrolyte system, known as a half-cell, to an electrical stimulation.

16.7   Standard Electric Potential

The reference electrode, in general, selected to measure the standard electric potential (E^o) of a metal has to be reversible, since classical thermodynamics applies to all reversible processes. The standard hydrogen electrode (SHE) is used for this purpose. Actually, E^o is also known as the electromotive force (emf) under equilibrium conditions, which are unit activity, $25\,^{\circ}C$, and $1\,atm$ $(101\,kPa)$ pressure.

The standard electric potential E^o measurements are for reducing metallic cations (M^{z+}) as indicated by the SHE diagram in Fig. 16.9a and illustrated by the SHE electrochemical cell in Fig. 16.9b, which has means for hydrogen gas flow onto the inert platinum (Pt) anodic electrode, which provides the surface for oxidation of hydrogen gas (H_2).

The spontaneous reactions for the SHE cell in Fig. 16.9b are

$$H_2 = 2H^+ + 2e^- \qquad \text{(Anode)} \qquad\qquad (16.11a)$$

$$M^{z+} + 2e^- = M \qquad \text{(Reduction)} \qquad\qquad (16.11b)$$

$$M^{z+} + H_2 = M + 2H^+ \qquad \text{(Redox)} \qquad\qquad (16.11c)$$

The SHE is a gas electrode that consists of a platinum foil suspended in sulfuric acid solution $(H^+SO_4^-)$ having H^+ unit activity $(a_{H+} = 1\,mol/l)$ at $1\,atm$ and $25\,^{\circ}C$. In order to maintain $a_{H+} = 1\,mol/l$, purified hydrogen (H_2) gas is injected into the anode half-cell for removing any dissolved oxygen in solution

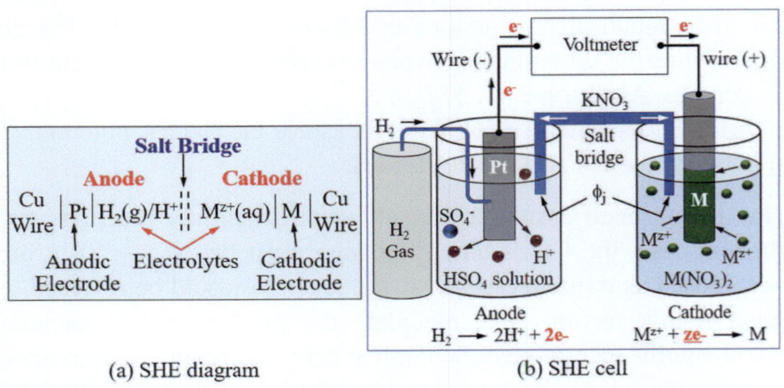

(a) SHE diagram   (b) SHE cell

Fig. 16.9 Metal/SHE standard hydrogen electrode. (**a**) SHE diagram and (**b**) SHE cell

Table 16.1 Standard electric potential series for metal reduction (Jones [13, p. 44])

	Reduction reaction	E^o (V_{SHE})
Noble	$Au^{3+} + 3e^- = Au$	+1.498
$\uparrow$	$O_2 + 4H^+ + 4e^- = 2H_2O$	+1.229
	$Pt^{2+} + 2e^- = Pt$	+1.200
	$Pd^{2+} + 2e^- = Pd$	+0.987
	$Ag^+ + e^- = Ag$	+0.799
	$Fe^{3+} + e^- = Fe^{2+}$	+0.770
	$Cu^{2+} + 2e^- = Cu$	+0.337
$\updownarrow$	$2H^+ + 2e^- = H_2$	0.000
	$Fe^{3+} + 3e^- = Fe$	−0.036
	$Pb^{2+} + 2e^- = Pb$	−0.126
	$Ni^{2+} + 2e^- = Ni$	−0.250
	$Co^{2+} + 2e^- = Co$	−0.277
	$Cd^{2+} + 2e^- = Cd$	−0.403
	$Fe^{2+} + 2e^- = Fe$	−0.440
	$Cr^{3+} + 3e^- = Cr$	−0.744
	$Zn^{2+} + 2e^- = Zn$	−0.763
	$Ti^{2+} + 2e^- = Ti$	−1.630
	$Al^{3+} + 3e^- = Al$	−1.662
	$Mg^{2+} + 2e^- = Mg$	−2.363
	$Na^+ + e^- = Na$	−2.714
$\downarrow$	$K^+ + e^- = K$	−2.925
Active	$Li^+ + +e^- = Li$	−3.045

(Jones [13, p. 65]). The platinum foil is an inert material in this solution, and it allows the hydrogen molecules to oxidize, providing the electrons needed by the metal M^{z+} ions to be reduced on the cathode surface. The concentration of M^{z+} ions is also kept at unit activity.

As can be seen in Fig. 16.9b, the electric potential (E) of the metal M at the surface is measured against the SHE, which is a reference electrode having an arbitrary standard electric potential equals to zero; $E_H^o = 0$. Actually, E is the result of several interfacial electric potentials described by Eq. (16.9), and it is also known as the **interfacial cell electric potential**, which depends on the **chemical potential** (μ) of the ions in solution at equilibrium. On the other hand, the electrons flow toward the cathodic electrode M, where the metal cations M^{z+} gain these electrons and enter the electrode lattice.

Galvanic corrosion can be predicted by using the electromotive force (emf) or standard electric potential E^o series for metal reduction listed in Table 16.1, which conforms to the convention adopted by the International Union of Pure and Applied Chemistry (IUPAC).

The usefulness of the galvanic series in Table 16.1 is illustrated in Fig. 16.10, which illustrates the assessment of iron Fe-based couplings to form galvanic cells. Basically, iron (Fe) is the base metal for steel; therefore, Fe is to be protected against corrosion.

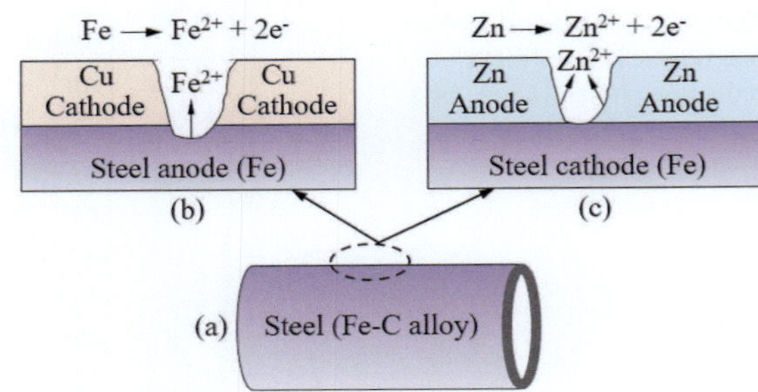

Fig. 16.10 Galvanic cells. (a) Steel pipe, (b) anodic steel leading to pitting corrosion, and (c) cathodic steel inducing Zn-anode dissolution for steel protection

For instance,

- If the steel pipe (Fig. 16.10a) is coated with a copper (Cu) film, an Fe-Cu galvanic cell (Fig. 16.10b) develops naturally. However, if the Cu coating breaks down, then the steel pipe is exposed to an electrolyte, Fe becomes the anode, and Cu the cathode. Therefore, Fe oxidizes, and the steel pipe corrodes.
- If the steel pipe is coated with Zn, an Fe-Zn galvanic cell (Fig. 16.10c) forms with Fe being the cathode and zinc Zn the sacrificial anode. This is the principle of galvanic coupling for galvanized steel sheets and pipes to be protected against corrosion.

When coupling two different metals, the metal with the lowest standard electric potential acts as the anode, and its standard electric potential sign is changed. For example,

- The sign of E^o_{Fe} in the Fe-Cu coupling is reversed ($E^o_{Fe} = +0.44\ V_{SHE}$), because Fe is below the Cu position in the galvanic series (Table 16.1) with $E^o_{Fe} < E^o_{Cu}$.
- The sign of E^o_{Zn} in the Fe-Zu coupling is reversed ($E^o_{Zn} = +0.763\ V$), since it acts as the anode, because Zn is below Fe in the galvanic series with $E^o_{Zn} < E^o_{Fe}$.

Mathematically, the cell electric potentials for the cited cases (Fig. 16.10) are

$$Fe\text{-}Cu:\ E^o_{cell} = E^o_{Fe} + E^o_{Cu} = +0.44\ V + 0.337\ V = 0.777\ V \qquad (16.12a)$$

$$Fe\text{-}Zn:\ E^o_{cell} = E^o_{Fe} + E^o_{Zn} = -0.44\ V + 0.763\ V = 0.323\ V \qquad (16.12b)$$

For comparison, Table 16.2 lists some reference electrodes (RE) used to measure the corrosion electric potentials of metals and alloys in specific environments. The purpose of RE is to provide a stable and reproducible electric potential (voltage).

Most reference electrodes are suitable over a limited range of conditions, such as pH or temperature; otherwise, an electrode behavior becomes unpredictable. Apparently, the mercury/mercurous chloride (Hg_2Cl_2-calomel), the copper/copper sulfate ($Cu/CuSO_4$), and silver/silver chloride ($Ag/AgCl$) reference electrodes are common in industrial applications, such as corrosion and cathodic protection (cathodic polarization) on buried and submerged structures.

Table 16.2 Reference electrode electric potentials (Jones [13, p. 66])

Name	Half-cell reaction	$E(V)\ vs.SHE$
Mercury sulfate	$HgSO_4 + 2e^- = 2Hg + SO_4^{2-}$	0.615
Copper sulfate	$CuSO_4 + 2e^- = 2Cu + SO_4^{2-}$	0.318
Saturated calonel	$Hg_2Cl_2 + 2e^- = 2Hg + 2Cl^-$	0.241
Silver chloride	$AgCl + e^- = Ag + Cl^-$	0.222
Standard hydrogen	$2H^+ + 2e^- = H_2$	0.000

Use of Table 16.2 If the measured electric potential is $E = -0.541\ V_{SCE}$, then convert V_{SCE} to V_{SHE}. Thus, $E = E_{measure} + E_{Table}$ and

$$E = (-0.541 + 0.241)\ V_{SHE} = -0.30\ V_{SHE}$$

In selecting two metals for a galvanic coupling, both metals should have similar electric potentials or be close to each other in the standard electric potential (E^o) series in order to suppress galvanic corrosion. For example, Fe-Cr or Cu-Sn (bronze) couplings develop a very small electric potential differences, since they are close to each other in their respective standard electric potential series. The given data in Table 16.1 is very appealing in designing against galvanic corrosion of pure metals. The closer the standard electric potentials of two metals, the weaker the galvanic effect; otherwise, the galvanic effect is enhanced.

Other types of galvanic coupling are batteries and fuel cells. Both are electrochemical power sources in which chemical energy is converted into electrochemical energy through controlled redox electrochemical reactions. Subsequently, these electrochemical devices represent the beneficial application of galvanic corrosion.

16.8   Thermodynamics of Electrochemical Cells

The subsequent analytical procedure leads to the derivation of the Gibbs energy change (ΔG) and the Nernst equation , which is suitable for determining the cell electric potential (E) when ion activities are less than unity under nonstandard conditions.

16.8.1   Phases of an Electrochemical Cell

Consider the electrochemical cell shown in Fig. 16.7b, and treat α-phase/ε-phase half-cell and β-phase/δ-phase half-cell as isolated subsystems.

For the α-phase/ε-phase half-cell subsystem, ϕ_α is the electric potential of the metallic α-phase, and ϕ_ε is the electric potential of the electrolyte ε-phase. Once the

α-phase is in contact with the ε-phase, an exchange of charged particles causes a change in the total energy of the subsystem, where a strong electric field develops and an electric double layer of charged particles disturbs the charge on the α-phase/ε-phase interface.

The electric double layer is a normal balance (equal number) of positive and negative ions, which diffuse to the interface causing such as disturbance in energy. Eventually, the system reaches the state of equilibrium within the electrolyte (Evans and De Jonghe [14, p. 146]).

Accordingly, the general electrochemical reactions for the electrochemical cell in Fig. 16.7b are defined by

$$(M_1)_\alpha = \left(M_1^{z+}\right)_\varepsilon + \left(ze^-\right)_\gamma \qquad \text{(anode)} \qquad (16.13a)$$

$$\left(M_2^{z+}\right)_\delta + \left(ze^-\right)_{\gamma'} = (M_2)_\beta \qquad \text{(Cathode)} \qquad (16.13b)$$

$$(M_1)_\alpha + \left(M_2^{z+}\right)_\delta = \left(M_1^{z+}\right)_\varepsilon + (M_2)_\beta \qquad \text{(Redox)} \qquad (16.13c)$$

where $\left(ze^-\right)_\gamma = \left(ze^-\right)_{\gamma'}$ for electron balance and γ, γ' are the wire phases, usually Pt or Cu metals. If one type of wire material is used, then $\gamma = \gamma'$. Recall that the redox reaction is the sum of anodic and cathodic reactions, where the redox electric potential is simply defined by the sum of anodic and cathodic electric potentials; $E_{redox} = E_a + E_c$.

The corresponding chemical potential equations for each half-cell are

$$\left(\mu_{M_1}\right)_\alpha = \left(\mu_{M_1^{z+}}\right)_\varepsilon + (z\mu_{e^-})_\gamma \qquad (16.14a)$$

$$\left(\mu_{M_2}\right)_\beta = \left(\mu_{M_2^{z+}}\right)_\delta + (z\mu_{e^-})_{\gamma'} \qquad (16.14b)$$

The mathematical procedure for deriving Eq. (16.14) can be found elsewhere (Dehoff [15, Sec. 15.2]). Nonetheless, the electric work for the redox reaction is

$$W_e = FE = F\left(\phi_\gamma - \phi_{\gamma'}\right) \qquad (16.15a)$$

$$FE = (\mu_{e^-})_\gamma - (\mu_{e^-})_{\gamma'} \qquad (16.15b)$$

where $E = \left(\phi_\gamma - \phi_{\gamma'}\right)$ is the electric potential difference between α and β phases being connected through the voltmeter V and $F = 96{,}500 \ C/mol$ is the Faraday's constant.

Multiplying Eq. (16.15b) by z and combining the resultant expression with Eq. (16.14) yields the Gibbs energy change (ΔG) for the redox reaction, Eq. (16.13c),

$$zFE = \left(\mu_{M_1}\right)_\alpha - \left(\mu_{M_1^{z+}}\right)_\varepsilon - \left(\mu_{M_2}\right)_\beta + \left(\mu_{M_2^{z+}}\right)_\delta \qquad (16.16a)$$

$$\Delta G = -\left[\left(\mu_{M_1} \right)_\alpha - \left(\mu_{M_1^{z+}} \right)_\varepsilon - \left(\mu_{M_2} \right)_\beta + \left(\mu_{M_2^{z+}} \right)_\delta \right] \qquad (16.16b)$$

For the α-phase or β-phase, the electrochemical potential ($\overline{\mu}_j$) of a species j, measured in J/mol, is defined as the sum of the chemical potential and Gibbs energy change

$$\overline{\mu}_j = \mu_j + z_j F E \qquad (16.17a)$$

$$\mu_j = \left(\frac{\partial G}{\partial n_j} \right)_{T,P} \qquad (16.17b)$$

$$\overline{\mu}_j = \left(\frac{\partial \overline{G}}{\partial n_j} \right)_{T,P} \qquad (16.17c)$$

where n_j is the number of moles of j in the α-phase or β-phase and $\overline{G}$ is the electrochemical Gibbs energy. Combining Eqs. (16.16a) and (16.16b) at standard and nonstandard conditions gives the Gibbs energy equations defined by

$$\Delta G^o = -zFE^o \qquad \text{(standard)} \qquad (16.18a)$$

$$\Delta G = -zFE \qquad \text{(nonstandard)} \qquad (16.18b)$$

Accordingly, Eq. (16.18b) is used to define three Gibbs energy criteria. That is,

- $\Delta G > 0$ for a non-spontaneous process
- $\Delta G < 0$ for a spontaneous process
- $\Delta G = 0$ for equilibrium condition

Similarly, ΔG^o in Eq. (16.18a) undergoes the same conditions. Moreover, ΔG^o and E^o are extensive (dependent on the amount of a species) and intensive (independent of the amount of a species) properties, respectively.

The standard Gibbs energy change (ΔG) can be defined using the chemical potentials (μ_j) and combined first and second laws of thermodynamics for a species j. Thus,

$$\Delta G = \sum v_j \mu_j = \left(\sum v_j \mu_j \right)_{product} - \left(\sum v_j \mu_j \right)_{reactant} \qquad (16.19a)$$

$$\Delta G = \Delta H - T\Delta S \qquad (16.19b)$$

$$\Delta H = zFE + zF \frac{\partial E}{\partial T} \qquad (16.19c)$$

$$\Delta S = zF \frac{\partial E}{\partial T} \qquad (16.19d)$$

$$Q = T\Delta S = zFT \frac{\partial E}{\partial T} \qquad (16.19e)$$

Table 16.3 Standard chemical potentials at $25\,^{\circ}C$ (Pourbaix [16]). Abbreviations: aqueous (aq), gaseous (g), liquid (liq), solid (s)

Name	μ^o (kJ/mol)	Name	μ^o (kJ/mol)
$Al(s)$	0	$Cu\,(s)$	0
$Al_2O_3(s)$	-1576.41	$CuO\,(s)$	-127.19
$Al(OH)_3(s)$	-1137.63	$Cu(OH)_2\,(s)$	-356.90
$Al^{3+}(aq)$	-481.16	$CuSO_4\,(s)$	-661.91
$AlO_2^-(aq)$	-839.77	$Cu^{2+}\,(aq)$	$+64.98$
$Cr(s)$	0	$Fe\,(s)$	0
$Cr(OH)_2\,(s)$	-587.85	$Fe(OH)_2\,(s)$	-483.54
$Cr(OH)_3\,(s)$	-900.82	$Fe(OH)_3\,(s)$	-694.54
$Cr^{3+}\,(aq)$	-215.48	$FeSO_4\,(s)$	-829.69
$CrO_2^-\,(aq)$	-835.93	$Fe^{2+}\,(aq)$	-84.94
$H_2(g)$	0	$Fe^{3+}\,(aq)$	-10.59
$H^+\,(aq)$	0	$Zn\,(s)$	0
$OH^-\,(aq)$	-157.30	$Zn(OH)_2\,(s)$	-559.09
$H_2O\,(liq)$	-237.19	$Zn^{2+}\,(aq)$	-147.21

where $j = 1, 2, 3..$ species and ν_j denotes the number of ions or molecules, ΔH denotes the change in enthalpy, ΔS denotes the change in entropy, and Q denotes the heat transfer.

The standard chemical potentials at $25\,^{\circ}C$ reported by Pourbaix [16] in 1960 can be found elsewhere (Shreir et al. [17, pp. 21:12-21:24]). For clarity, some standard chemical potentials at $25\,^{\circ}C$ are listed in Table 16.3.

Combining Eqs. (16.18b) and (16.19a) yields the standard electric potential in terms of standard chemical potential

$$E^o = -\frac{\sum \nu_j \mu_j}{zF} \tag{16.20}$$

Applying Eq. (16.20) to the redox reaction given by Eq. (16.13c) yields

$$E^o = -\frac{1}{zF}\left\{\left[(\mu_{M^{z+}})^\varepsilon + (\mu_{M_2})^\beta\right] - \left[(\mu_{M_1})^\alpha + (\mu_{M^{z+}})^\delta\right]\right\} \tag{16.21}$$

For $M + 2H^+ \rightarrow M^{z+} + H_2$, Eq. (16.20) gives

$$E^o_{M^{z+}/M} = \frac{1}{zF}\left\{\left[(1)\mu^o_M + (2)\mu^o_{H^+}\right] - \left[(1)\mu^o_{M^{z+}} + (1)\mu^o_{H_2}\right]\right\} \tag{16.22a}$$

$$E^o_{M^{z+}/M} = \frac{\mu^o_{M^{z+}}}{zF} \quad \text{since } \mu^o_M = \mu^o_{H^+} = \mu^o_{H_2} = 0 \tag{16.22b}$$

Hence, this is a simplified form of the electromotive force measured in volts (V).

Example 16.1 Assume that the cell in Fig. 16.7b is in equilibrium and that the cell phases are $M_1^\alpha = Zn^\alpha$ with $z_1 = +2$, $\varepsilon = ZnSO_4$, $M_2^\beta = Cu^\beta$ with $z_2 = +2$, and $\delta = CuSO_4$. Calculate (a) $E^o_{Cu^{2+}/Cu}$, and (b) determine if the redox reaction will occur.

Solution

(a) If $M^{z+} = Cu^{2+}$, then from Eq. (16.22b) and Table 16.3, the standard electric potential $E^o_{M^{z+}/M} = E^o_{Cu^{2+}/Cu}$ is

$$E^o_{Cu^{2+}/Cu} = \frac{\mu^o_{Cu^{2+}}}{zF} = \frac{+64.98\ kJ/mol}{(2)(96.50\ kJ/mol.V_{SHE})} \tag{16.1E1a}$$

$$E^o_{Cu^{2+}/Cu} = +0.337\ V_{SHE} \tag{16.1E1b}$$

which is exactly the value given in Table 16.1 for the reduction of copper ions; $Cu^{+2} + 2e^- \rightarrow Cu$.

(b) From Table 16.1, $E^o = E^o_{Zn} + E^o_{Cu} = 0.763\ V_{SHE} + 0.337\ V_{SHE} = 1.10\ V_{SHE}$. Using Eq. (16.18a) along with and $z_1 = z_2 = z = 2$ yields

$$\Delta G^o = -zFE^o = (2)(96.50\ kJ/mol.V_{SHE})(1.10\ V_{SHE}) \tag{16.1E2a}$$

$$\Delta G^o = -212.30\ kJ/mol \tag{16.1E2b}$$

Therefore, the redox reaction, $Zn^\alpha + (Cu^{2+})^\delta = (Zn^{2+})^\varepsilon + Cu^\beta$, will occur spontaneously because $\Delta G^o < 0$.

Example 16.2 Let the anodic phases in Fig. 16.7b be $(M_1)_\alpha = Cr$ with $z = +3$ in $\varepsilon = CrSO_4$, whereas the cathodic phases are $(M_2)_\beta = Cu$ with $z = +2$ in $\delta = CuSO_4$. Determine (a) the standard electric potential E^o_{cell} for the redox reaction, and (b) calculate the Gibbs energy change. Will the redox reaction occur?

Solution

(a) The half-cell reactions from Table 16.1 must be balanced, but the standard cell electric potentials do not change. Thus,

$$2Cr \rightarrow 2Cr^{3+} + 6e^- \qquad E^o_{Cr} = 0.744\ V \tag{16.2E1a}$$

$$3Cu^{2+} + 6e^- \rightarrow 3Cu \qquad E^o_{Cu} = 0.337\ V \tag{16.2E1b}$$

$$2Cr + 3Cu^{2+} \rightarrow 2Cr^{3+} + 3Cu \qquad E^o_{cell} = E^o_{Cr} + E^o_{Cu} \tag{16.2E1c}$$

$$\rightarrow \qquad E^o_{cell} = 1.081\ V$$

(b) The Gibbs energy change for the Cr/Cr^{3+} half-cell is

$$\Delta G^o_{Cr/Cr^{3+}} = -zFE^o_{Cr/Cr^{3+}} = -(6)\,[96.50\,kJ/(mol.V)]\,(0.744\,V)$$
$$(16.2E2a)$$

$$\Delta G^o_{Cr/Cr^{3+}} = -430.78\,kJ/mol \tag{16.2E2b}$$

that for Cu^{2+}/Cu is

$$\Delta G^o_{Cu^{2+}/Cu} = -zE^o_{Cu^{2+}/Cu} = (6)\,[96.50\,kJ/(mol.V)]\,(0.337\,V)$$
$$(16.2E3a)$$

$$\Delta G^o_{Cu^{2+}/Cu} = -195.12\,kJ/mol \tag{16.2E3b}$$

For the entire cell,

$$\Delta G^o_{Cr/Cu} = \Delta G^o_{Cr/Cr^{3+}} + \Delta G^o_{Cu^{2+}/Cu} \tag{16.2E4a}$$

$$\Delta G^o_{Cr/Cu} = -430.78\,kJ/mol - 195.12\,kJ/mol \tag{16.2E4b}$$

$$\Delta G^o_{Cr/Cu} = -625.90\,kJ/mol \tag{16.2E4c}$$

In addition, $\Delta G^o_{Cr/Cu}$ can also be determined in the following way

$$\Delta G^o_{Cr/Cu} = \Delta G^o_{Cr/Cr^{3+}} + \Delta G^o_{Cu^{2+}/Cu} = -zFE^o_{Cr/Cr^{3+}} - zE^o_{Cu^{2+}/Cu}$$
$$(16.2E5a)$$

$$\Delta G^o_{Cr/Cu} = -zF\left(E^o_{Cr/Cr^{3+}} + E^o_{Cu^{2+}/Cu}\right) = -zFE_{cell} \tag{16.2E5b}$$

$$\Delta G^o_{Cr/Cu} = -(6)\,[96.50\,kJ/(mol.V)]\,(1.081\,V) \tag{16.2E5c}$$

$$\Delta G^o_{Cr/Cu} = -625.90\,kJ/mol \tag{16.2E5d}$$

Therefore, the redox reaction occurs spontaneously because $\Delta G^o < 0$.

16.8.2 The Nernst Equation

The Nernst equation is frequently used to relate electrode potential and activities of reactants and products of half-cell or redox reaction. The activity $a_j = [j]$ of a dissolved species j in a electrolyte solution is fundamentally the concentration in such a solution. Its mathematical definition is based on the chemical potential (μ_j) of the species j

$$a_j = \exp\left[\frac{\mu_j - \mu^o}{RT}\right] \tag{16.23}$$

Table 16.4 Some useful definitions

a_j = Activity $(mol/liter)$	$a_j = [\,j\,] = \gamma_j X_j$ (Henry's law)
C_j = Concentration (g/l)	$C_j = a_j A_{w,j}$
V_j = Molar volume (l)	$a_j = P_j/P_o = \gamma \left[C_j/C^o \right]$
P_j = Pressure (MPa)	$H_2O = H^+ + OH^-$
$P_o = 1\,atm = 101\,kPa$	$K_w = \left[H^+ \right]\left[OH^- \right] = 10^{-14}$
γ_j = Activity coefficient	$pH = -\log\left[H^+ \right]$
A_w = Atomic weight (g/mol)	$pH = 14 + \log\left[OH^- \right]$
X_j = Mole fraction	μ^o = Standard chemical potential

Table 16.4 illustrates some important definitions, which are needed to solve electrochemical problems at standard conditions, $T = 25\,°C$, $P_o = 1\,atm$ and $a_j = 1\,mol/liter = 1\,mol/l$.

From the second law of thermodynamics, the change in Gibbs energy can also be defined as a temperature- and pressure-dependent entity. Thus,

$$dG = -SdT + VdP \tag{16.24}$$

which can be reduced for a isothermal process with $\Delta T = 0$ or isobaric with $\Delta P = 0$.

For an isothermal $(T = \text{Constant})$ and isometric $(V = \text{Constant})$ system, Eq. (16.24) yields

$$\int_{\Delta G_j^o}^{\Delta G_j} dG = V \int_{P_o}^{P} dP \tag{16.25a}$$

$$\Delta G_j - \Delta G^o = V\,(P - P_o) \tag{16.25b}$$

Dividing this equation by RT and using the natural exponential function give

$$\exp\left[\frac{\Delta G - \Delta G^o}{RT} \right] = \exp\left[\frac{V\,(P - P_o)}{RT} \right] \tag{16.26}$$

where P = Pressure
$\quad P_o$ = Standard pressure = $1\,atm$ $(101\,kPa)$
$\quad R = 8.3145\ J.mol^{-1}K^{-1}$ (Universal gas constant)
$\quad T$ = Absolute temperature (K)

Now, the equilibrium constant K_e (reaction quotient) for partial reactions, Eqs. (16.13a), (16.13b), can be defined in terms of activities. Thus,

$$K_e = \frac{\sum a_{\text{Product}}}{\sum a_{\text{Reactants}}} = \frac{\sum [\text{Product}]}{\sum [\text{Reactants}]} \tag{16.27a}$$

$$K_{e,\alpha-\varepsilon} = \frac{\left[\left(M_1^{z+}\right)^\varepsilon\right]}{\left[M_1^\alpha\right]} \qquad \text{(Anodic half-cell)} \qquad (16.27b)$$

$$K_{e,\beta-\delta} = \frac{\left[M_2^\beta\right]}{\left[\left(M_2^{z+}\right)^\delta\right]} \qquad \text{(Cathodic half-cell)} \qquad (16.27c)$$

and for the redox reaction, Eq. (16.13c),

$$K_e = K_{e,\alpha-\varepsilon}K_{e,\beta-\delta} = \frac{\left[\left(M_1^{z+}\right)^\varepsilon\right] \cdot \left[M_2^\beta\right]}{\left[M_1^\alpha\right] \cdot \left[\left(M_2^{z+}\right)^\delta\right]} \qquad (16.28)$$

For the dissolution of a compound such as $AgCl$ salt, the reaction constant is called solubility product (K_{sp}) instead of equilibrium constant. From Eq. (16.26), K_{sp} for a simple redox reaction is defined by

$$K_{sp} = \exp\left(\frac{\Delta G - \Delta G^o}{RT}\right) \qquad (16.29)$$

Solving Eq. (16.29) for the nonequilibrium Gibbs energy change ΔG gives

$$\Delta G = \Delta G^o + RT \ln\left(K_{sp}\right) \qquad (16.30)$$

At standard conditions, $\Delta G = 0$ and Eq. (16.30) becomes

$$\Delta G^o = -RT \ln\left(K_{sp}\right) \qquad (16.31)$$

Thus, the equilibrium constant K_{sp} can be determined from measurements of the electric potential (E^o) or chemical potential (μ^o).

Substituting Eq. (16.18) into (16.30) yields the **Nernst Equation** for calculating the nonequilibrium electric potential (E) of an electrochemical cell. Hence,

$$E = E^o - \frac{RT}{zF} \ln\left(K_{sp}\right) \qquad (16.32)$$

From a macroscale, the Nernst equation, Eq. (16.32), can be linearized as $E = f\left[\ln\left(K_{sp}\right)\right]$ function, where E^o is the intercept and $b = -RT/zF$ is the slope.

The interpretation of the Nernst equation suggests that the current resulting from the change of oxidation state of the metals M_1 and M_2 is known as the **Faradaic current**, which is a measure of the rate of redox reaction.

The term zF in Eq. (16.32) stands for the number of coulombs needed for reacting 1 *mol* of an electroactive species. Therefore, Eq. (16.32) embodies the Faraday's law of electrolysis (defined in a later section), since the redox reaction involves electron transfer between phases having different electric potentials. For instance, electrochemical equilibrium is achieved if current flow ceases in a reversible galvanic cell (Fig. 16.7b, (16.8) or (16.9b).

Despite that the Nernst equation is derived assuming thermodynamic equilibrium of a reversible cell, K_{sp} in Eq. (16.32) is not exactly the same as the chemical equilibrium constant K_e due to a current flow. The Nernst equation gives the open-circuit potential difference for a reversible galvanic cell and excludes liquid junction potentials, which are always present, but small in magnitude.

Example 16.3 Apply the preceding procedure to the galvanic cell shown in Fig. 16.7b at $25\,^\circ C$ for determining the solubility product K_{sp} and the Nernst cell potential E. Let $\alpha = Zn$ with $\left[Zn^{2+}\right] = 0.0002\ mol/l$ and $\beta = Cu$ with $\left[Cu^{2+}\right] = 0.04\ mol/l$ at $298K$.

Solution

The reactions and the corresponding standard electric potentials are

$$Zn \rightarrow Zn^{2+} + 2e^- \quad E^o_{Zn} = 0.763\ V_{SHE} \tag{16.3E1a}$$

$$Cu^{2+} + 2e^- \rightarrow Cu \quad E^o_{Cu} = 0.337\ V_{SHE} \tag{16.3E1b}$$

$$Zn + Cu^{2+} \rightarrow Zn^{2+} + Cu \quad E^o = E^o_{Zn} + E^o_{Cu} \tag{16.3E1c}$$

$$\rightarrow \quad E^o = 1.10\ V_{SHE} \tag{16.3E1d}$$

From Eq. (16.3E1c) or (16.27a), the equilibrium constant is

$$K_{sp} = \frac{[Cu]\left[Zn^{2+}\right]}{\left[Cu^{2+}\right][Zn]} = \frac{\left[Zn^{2+}\right]}{\left[Cu^{2+}\right]} = \frac{0.0002}{0.04} = 0.005 \tag{16.3E2}$$

Now, the cell potential defined by Eq. (16.32) along with $z = 2$ and $E^o = 1.10\ V$ gives

$$E = E^o - \frac{RT}{zF} \ln K_{sp} \tag{16.3E3a}$$

$$E = 1.10\ V - (0.0296\ V)\ln(0.005) = 1.26\ V \tag{16.3E3b}$$

Hence, $K_{sp} = 0.005$ and $E = 1.26\ V$ represent the characteristics of the cell when it deviates from equilibrium.

16.9 Pourbaix Diagrams

Historically, Pourbaix in 1938 developed the potential $E = f(pH)$ function or E-pH plot for thermodynamically stable phases. Nowadays, the pH-E plot is known as a Pourbaix diagram.

A compilation of Pourbaix diagrams is available in the *Atlas of Electrochemical Equilibria in Aqueous Solutions* (Pourbaix [18]) and is suitable for corrosion studies, electrowinning, electroplating, hydrometallurgy, electrolysis, electric cells, and water treatment, since they are electrochemical maps indicating the domain of stability of ions, oxides, and hydroxides. These electrochemical maps provide the oxidizing power in an electrochemical field measured as potential (voltage), and the acidity and alkalinity of species measured as a function of pH. Specifically, a reaction involving hydroxyl (OH^-) ions is written in terms of H^+ ion concentration, which in turn is converted into $pH = -\log[H^+]$ (Table 16.4).

Despite that a Pourbaix diagram does not include any information about kinetic studies for determining the corrosion rate, it is a useful tool for predicting the spontaneous direction of electrochemical reactions.

16.9.1   The Pourbaix Diagram for Iron-Water System

Consider Fig. 16.11 which illustrates the pH-E equilibrium diagram or Pourbaix diagram for the Fe-H_2O system containing 10^{-6} mol/l of Fe at room temperature $T = 25\,^\circ C$ and atmospheric pressure $P = 1$ atm (Hoar [19, p. 88]).

A Pourbaix diagram provides important regions for designing and analyzing electrochemical systems. These regions are electrochemical domains known as corrosion, passivation, and immunity.

This particular diagram shows graphically the $E = f(pH)$ relationship and the electrochemical field boundaries separating the phase fields or regions.

The pH-E diagram shows the corrosion fields due to the production of ferrous-iron ion Fe^{2+} and ferric-iron ion Fe^{2+} when $E > -0.6\ V_{SHE}$ at a pH range. Notice that Fe is immune (stable) to H_2O at $E < -0.6\ V_{SHE}$ and iron compounds [$Fe(OH)_2$ and $HFeO_2^-$] are formed in alkaline water at relatively high pH values.

For the purpose of clarity, the field boundaries separating the electrochemical domains can be altered by different [Fe^{2+}] concentrations.

Fig. 16.11 Simplified E-pH diagram for iron-water system [(Hoar [19, p. 88])

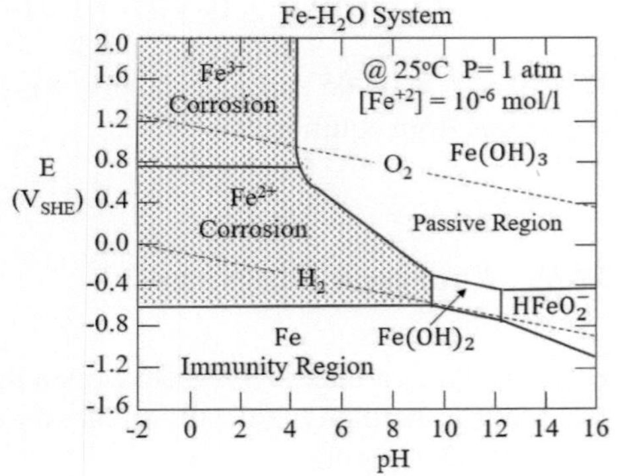

Fig. 16.12 Simplified E-pH diagram for the copper-water system showing domains of corrosion, immunity, and passivation of copper (Cu) at $T = 25\,^{\circ}C$ and $P = 1\ atm = 101\ kPa$ (De Zoubov et al. [20, p. 387])

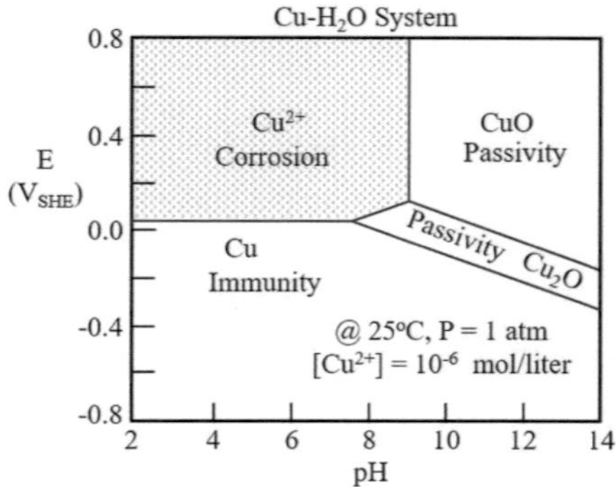

16.9.2   The Pourbaix Diagram for Copper-Water System

Furthermore, Fig. 16.12 depicts the Pourbaix diagram for the copper-water (Cu-H_2O) system with a concentration of $10^{-6}\ mol/liter$ of Cu^{2+} ions in solution (De Zoubov et al. [20, p. 387]).

This diagram shows that copper (Cu) corrodes at $E > 0\ V_{SHE}$ and $pH < 10$. In particular, the CuO and Cu_2O compounds form in alkaline water environments. However, CuO in the form of a film is the principal passivation metal oxide at high pH values. The field boundaries separating the electrochemical domains can also be altered by different $[Cu^{2+}]$ concentrations.

16.10   Kinetics of Electrochemical Polarization

In the corrosion field, polarization is an electrochemical process induced by a deviation of the electrochemical equilibrium potential or corrosion potential (E_{corr}) due to an electric current (I) passing through the electrochemical cell. This deviation is known as the electric overpotential (n), which is the essential driving force for polarization.

Fontana's classical models [1, pp. 20, 459] for concentration polarization and activation polarization are combined in Fig. 16.13 for a metal-hydrogen system at a temperature T and pressure P.

A polarization diagram is a practical electrochemical map for assessing the kinetic characteristics of an electrochemical cell for metal corrosion or metal deposition studies and for designing against corrosion. Actually, electrochemical kinetic studies lead to the determination of the corrosion rate and the electrodeposition rate.

Electrochemical kinetics is an important field for determining the rate of corrosion of a metal M exposed to a corrosive medium (electrolyte) . On the other hand, thermodynamics predicts the possibility of corrosion, but it does not provide information on how slow or fast corrosion occurs. The kinetics of a reaction on a

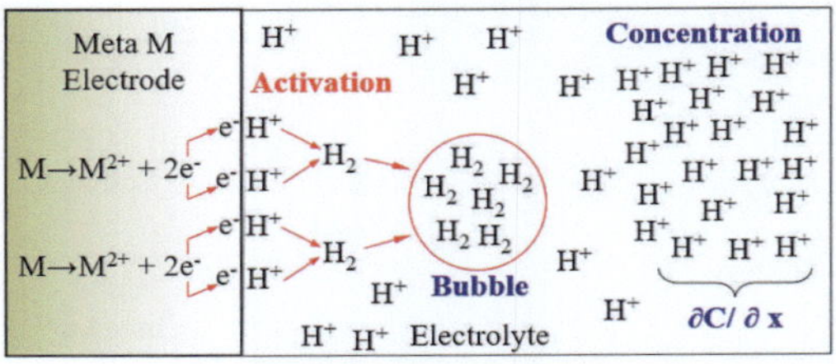

Fig. 16.13 Polarization model related to (**a**) concentration of H^+ ions at the right side of the electrolyte and (**b**) activation due to metal anodic oxidation on the electrode surface. Notice the formation of hydrogen molecules (H_2) and a hydrogen bubble. After Fontana [1, pp. 20, 459]

electrode surface depends on the electrode potential, while the reaction rate strongly depends on the rate of electron flow to or from a metal-electrolyte interface. If the electrochemical system (electrode and electrolyte) is at equilibrium, then the net rate of reaction is zero ($r_{net} = 0$), which leads to zero corrosion rate ($C_R = 0$). In comparison, reaction rates are governed by chemical kinetics, while corrosion rates are primarily governed by electrochemical kinetics.

16.10.1 Concentration Polarization

For concentration polarization, the ionic concentration of H^+ ions in the bulk of solution (C_b) is higher than that on the electrode surface (C_o). For one-dimensional analysis, $C_b > C_o$ imparts a concentration gradient $\partial C / \partial x$ as the driving force for H^+ ionic diffusion toward the electrode surface, where cathodic reactions take place, and the diffusion molar flux (J_x) is determined by Fick's first law equation under steady-state conditions. Despite that ionic diffusion is a natural ionic motion mechanism, it can be coupled with migration of ions due to a current flow within an electric field in an electrochemical cell.

In addition, microcircuit failures due to metal degradation or corrosion are a concern in the electronic industry. This type of metal degradation is also referred to as "metal migration" within the microcircuit environment, such as a moisture acting as an electrolyte.

16.10.2 Activation Polarization

For activation polarization, an electrochemical polarization diagram is a suitable representation of the deviation process from equilibrium into the active electrochemical domain, where a series of reaction steps may take place. Thus, activation polarization is an anodic process during oxidation, which is controlled by the slowest reaction step. Consequently, the rate of oxidation defines the corrosion rate C_R either in terms of electric current density $i = i_{corr}$ (in units of amp/cm^2) or weight loss per unit time (in units of $mm/year$).

16.10.3   Polarization Diagram

Polarization diagrams are characterized as concentration polarization or activation polarization. Commonly, the electric current is converted into electric current density $i = I/A$, where A is the specimen exposed area and i represents the corrosion rate, which in turn can be converted into common units of $mm/year$ or $\mu m/year$ using faraday's law of electrolysis. Thus, the electrochemical behavior of a working electrode (specimen) exposed to an electrolyte can be disclosed by a polarization curve (E $vs.$ $\log i$) or by the $E = f(i)$ function.

For example, Fig. 16.14 schematically shows superimposed Evans and Stern partial polarization diagrams which provide the common corrosion point denoted as the (i_{corr}, E_{corr}) closed interval defining the electrochemical equilibrium condition of an electrode in a hydrogen-containing electrolyte.. A deviation from equilibrium imparts a polarization condition due to an anodic overpotential n_a or cathodic overpotential n_c (Fontana [1, pp. 458–462]).

Electrochemical kinetics of a corroding metal can be characterized by determining, at least, three polarization parameters, such as corrosion electric current density (i_{corr}), corrosion electric potential (E_{corr}), and Tafel slopes (β_a and/or β_c). Evaluation of these parameters leads to the determination of the polarization resistance (R_p) and the corrosion rate as i_{corr}, which is often converted into **Faradaic corrosion rate** C_R having units of mm/y. Moreover, C_R is mathematically defined in a later section.

The electric overpotential for polarization and equilibrium of galvanic cells is written as

Fig. 16.14 Schematic Evans and Stern polarization diagrams and related kinetic parameters

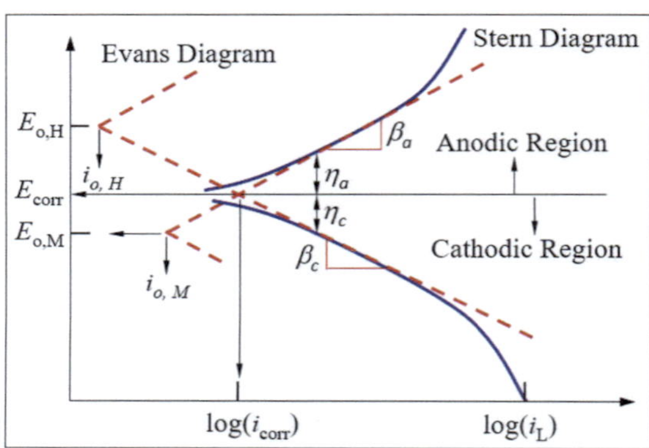

$$\eta_a = (E - E_{corr}) > 0 \quad \text{(anodic)} \tag{16.33a}$$

$$\eta_c = (E - E_{corr}) < 0 \quad \text{(cathodic)} \tag{16.33b}$$

$$\eta = 0 \qquad\qquad\qquad \text{(equilibrium)} \tag{16.33c}$$

Hence, $n_a > 0$ causes electrochemical corrosion, and $n_c < 0$ induces electrochemical reduction of at least one ionic species. Moreover, these overpotentials are defined by (Perez [8, p. 208])

$$\eta_a = \beta_a \log \left(\frac{i_a}{i_{corr}} \right) > 0 \tag{16.34a}$$

$$\eta_c = - |\beta_c| \log \left(\frac{i_c}{i_{corr}} \right) < 0 \tag{16.34b}$$

From Fig. 16.14, $E_{o,H}$ and $E_{o,M}$ are the open-circuit potentials for hydrogen and metal M, respectively, $i_{o,H}$ $i_{o,M}$ are the exchange current densities, and i_L is the limiting electric current density.

For a reversible electrode, Evans diagram allows the determination of a corrosion point where both the hydrogen cathodic and the metal anodic lines intercept. Other parameters illustrated in Fig. 16.14 will become relevant as the electrochemical analysis advances in this chapter. With respect to the Stern diagram, this represents a polarization behavior that can be determined experimentally using potentiostatic and potentiodynamic methods.

Analysis of Fig. 16.14 yields the following summary:

- The solid curve can be obtained statically or dynamically.
- This nonlinear curve is divided into two parts. If $E > E_{corr}$, the upper curve represents an anodic polarization behavior for oxidation of the metal M. On the contrary, if $E < E_{corr}$, the lower curve is a cathodic polarization for hydrogen reduction as molecular gas (hydrogen evolution). Both polarization cases deviate from the electrochemical equilibrium potential (E_{corr}) due to the generation of anodic and cathodic overpotentials, which are arbitrarily shown in Fig. 16.14 as η_a and η_c, respectively.
- Both anodic and cathodic polarization curves exhibit small linear parts known as Tafel lines, which are used for determining the Tafel slopes β_a and β_c. These slopes can be determined using either the Evans or Stern diagram.
- Extrapolating the Tafel or Evans straight lines until they intersect for defining the (E_{corr}, i_{corr}) point.
- The disadvantage in using the Evans diagram is that the exchange electric current density (i_o), the open-circuit potential E_o (no external circuit is applied), and the Tafel slopes for the metal and hydrogen have to be known quantities prior to determining the (E_{corr}, i_{corr}) point, where $i_{corr} = i_a = -i_c$ at $E = E_{corr}$.
- The advantage of the Stern diagram over the Evans diagram is that it can easily be obtained using the potentiodynamic polarization technique at a constant potential sweep (scan rate) and no prior knowledge of the above kinetics parameter is

necessary for determining the (E_{corr}, i_{corr}) point. The resultant curve is known as a potentiodynamic polarization curve.

At electrochemical equilibrium, the corrosion potential E_{corr} is a reversible potential also known as a mixed potential at equilibrium, where the electric potentials and the current densities are $E_{corr} = E_a = E_c$ and $i_{corr} = i_a = -i_c$, respectively. For electrochemical processes, the correct use of the concept of corrosion potential E_{corr} is graphically illustrated in the above polarization diagrams.

16.11   Energy Distribution for Anodic Polarization

Thermodynamically, the corresponding Gibbs energy change terms for anodically polarizing an electrochemical cell are schematically depicted in Fig. 16.15 after Hoar [21] polarization energy-distance diagram being presented at the 59th AES Annual Convention in 1972. This energy-distance, $\Delta G = f(x)$, diagram can also be found in Jones' book [13, p. 83]. From thermodynamics, the energy diagram in Fig. 16.15 resembles the Boltzmann or Maxwell-Boltzmann distribution law for polarization energy distribution of the reacting ions. Moreover, $\Delta G_{max} = f(\delta)$ at $\delta = x/2$ defines the **activation state** as the transition point (saddle point or hump), where the rates of dissolution (oxidation) and reduction are equal; $R_f = R_r$.

For an anodic overpotential with $\eta_a > 0$, the polarization energy (ΔG) decreases by an amount $\Delta G_p = \Delta G_a - \alpha z F \eta_c$, where α is a fraction in the range $0 < \alpha < 1$ and ΔG_a is the anodic energy responsible the anodic reaction $M \rightarrow M^{z+} + ze^-$.

The electrochemical and chemical rates of reactions due to either anodic or cathodic overpotentials can be predicted using both Faraday's and Arrhenius' equations, respectively

Fig. 16.15 Schematic polarization energy distribution diagram for activation overpotential (η) between two electrodes. After Jones [13, p. 83] and Hoar [21]

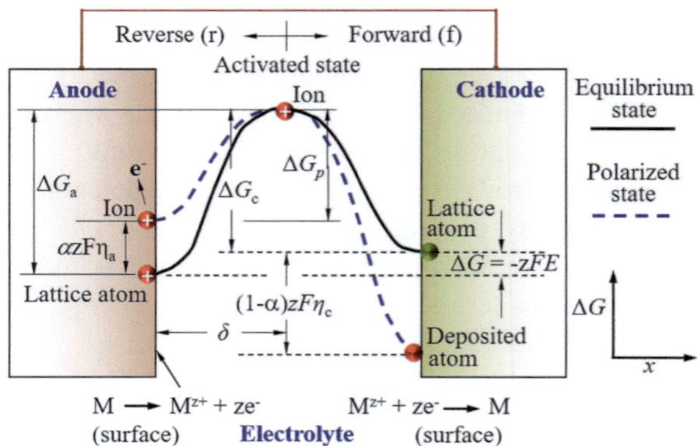

$$R_F = \frac{i\,A_{w,j}}{zF} \tag{16.35a}$$

$$R_A = \gamma_a \exp\left(-\frac{\Delta G_p}{RT}\right) \tag{16.35b}$$

where i = Applied electric current density $\left(A/cm^2\right)$
$A_{w,j}$ = Atomic weight of species j (g/mol)
$A_{w,alloy} = \sum f_j A_{w,j}/z_j$ in units of g/mol
z = Oxidation state or valence number
z_j = Valence number of element j
f_j = Weight fraction of element j
γ_a = Chemical reaction constant
$\Delta G^* $ = Activation energy or free energy change (V/mol)

At equilibrium, Faraday's and Arrhenius' rate equations are equal $(R_F = R_A)$, and consequently, the electric current density is defined by

$$i = \gamma_o \exp\left(-\frac{\Delta G}{RT}\right) \tag{16.36}$$

Here, the term $\gamma_o = zF\gamma_a/A_w$ may be defined as the electrochemical rate constant having unit of electric current density $\left(A/cm^2\right)$.

16.12   Exchange Current Density

For a reversible electrode at equilibrium, the electric current density in Eq. (16.36) becomes the exchange electric current density $i = i_o$. Table 16.5 gives typical experimental values for i_o.

Denote that values of Pb and Pt in 1 N HCl solution. Thus, $i_{o,Pt}/i_{o,Pb} = 5x10^9$ means that hydrogen evolution,$2H^+ + 2e^- = H_2$, is 5 trillion times faster on the Pt surface electrode than on the Pb surface. If Pt-Pb are galvanically coupled in $1N$ HCl solution, Pt becomes the cathode where hydrogen evolution (H_2) occurs and Pb is the anode for providing the electrons.

If an electrode is polarized by an overpotential under steady-state conditions, then the rates of reactions are not equal $(R_F \neq R_A)$, and consequently, the forward (cathodic) and reverse (anodic) electric current density components must be defined in terms of the polarization energy ΔG_p deduced from Fig. 16.15 (Jones [13, Sec. 3.2]).

Hence, the anodic and cathodic current densities are, respectively,

$$i_a = k'_f \exp\left[-\frac{\Delta G_p}{RT}\right] \quad \text{(anodic)} \tag{16.37a}$$

Table 16.5 Exchange electric current density at $25\,^{\circ}C$ (Fontana [1, 457])

Electrode	Electrolyte	Reaction	$i_o\ (A/cm^2)$
Al	$2N\ H_2SO_4$	$2H^+ + 2e^- = H_2$	$1x10^{-10}$
Au	$1N\ HCl$	$2H^+ + 2e^- = H_2$	$1x10^{-6}$
Cu	$0.1N\ HCl$	$2H^+ + 2e^- = H_2$	$2x10^{-7}$
Fe	$2N\ H_2SO_4$	$2H^+ + 2e^- = H_2$	$1x10^{-6}$
Ni	$1N\ HCl$	$2H^+ + 2e^- = H_2$	$4x10^{-6}$
Ni	$05N\ NiSO_4$	$Ni = Ni^{2+} + 2e^-$	$1x10^{-6}$
Pb	$1N\ HCl$	$2H^+ + 2e^- = H_2$	$2x10^{-13}$
Pt	$1N\ HCl$	$2H^+ + 2e^- = H_2$	$1x10^{-3}$
Pt	$0.1N\ NaOH$	$O_2 + 4H^+ + 4e^- = 2H_2O$	$4x10^{-13}$
Pd	$06N\ HCl$	$2H^+ + 2e^- = H_2$	$2x10^{-4}$
Sn	$1N\ HCl$	$2H^+ + 2e^- = H_2$	$1x10^{-8}$

$$i_c = k_r' \exp\left[-\frac{\Delta G_p^*}{RT}\right] \quad \text{(cathodic)} \tag{16.37b}$$

where $\Delta G_p = \Delta G_a - \alpha z F \eta_a$ for anodic polarization, $\Delta G_p^* = \Delta G_c + (1 - \alpha)\, z F \eta_c$ for cathodic polarization, and α denotes the symmetry coefficient.

For a cathodic case, the net current and the overpotential are $i = i_a - i_c$ and $\eta_c < 0$, respectively. Thus,

$$i = k_a' \exp\left(-\frac{\Delta G_p}{RT}\right) \exp\left(\frac{\alpha z F \eta}{RT}\right)$$
$$- k_c' \exp\left(-\frac{\Delta G_p^*}{RT}\right) \exp\left[-\frac{(1 - \alpha)\, z F \eta}{RT}\right] \tag{16.38}$$

Evaluating Eq. (16.38) at $\eta = 0$ yields the exchange current density $i = i_o$ defined as

$$i_o = k_a' \exp\left(-\frac{\Delta G_p}{RT}\right) = k_c' \exp\left(-\frac{\Delta G_p^*}{RT}\right) \tag{16.39}$$

Here, the exchange current density i_o is an electrochemical kinetic parameter that balances out the cathodic and anodic current densities at $\eta = 0$. This implies that i_o flows equally in both directions during equilibrium and that it controls the reaction rate. At equilibrium, the cell electrodes do not experience any gain or loss of electrons; therefore, there is no electron transfer.

16.13   Polarization from the Open-Circuit Potential

Substituting Eq. (16.39) into (16.38) for one-step reaction yields the well-known **Butler-Volmer equation** for polarizing an electrode from the open-circuit potential E_o under steady-state conditions

$$i = i_o \left\{ \exp\left[\frac{\alpha z F \eta_a}{RT} \right]_a - \exp\left[-\frac{(1-\alpha) z F \eta_a}{RT} \right]_c \right\} \tag{16.40a}$$

$$i = z F J (x, t) \tag{16.40b}$$

where $\eta_a = (E - E_o) > 0 =$ Overpotential $(Volts)$

$\quad E =$ Applied potential $(Volts)$

$\quad i_o =$ Exchange electric current density (A/cm^2)

$\quad J (x, t) =$ Molar flux along the x-axis at time $t > 0$ $[mol/\left(cm^2 . s\right)]$

At equilibrium conditions , the potential is $E = E_o$ and $\eta = 0$, and Eq. (16.40a) yields $i_a = i_c = i_o$. This implies that the rates of metal dissolution (oxidation) and deposition (reduction) are equal. However, a deviation from **equilibrium of the half-cells** is the normal case in real situations.

The Butler-Volmer equation, Eq. (16.40a), defines two electrochemical cases: high activation polarization for high electric current density and low activation polarization for low electric current density studies. This means that i defined by Eq. (16.40a) measures the electric current density response due to an applied overpotential potential on an electrode surface.

For convenience, Eq. (16.40a) can be rearranged as a hyperbolic sine function along with $\alpha = 0.5$. Thus,

$$i = 2 i_o \sinh\left(\frac{z F \eta}{2 RT} \right) \tag{16.41}$$

and from which the overpotential becomes

$$\eta = \left(\frac{2 RT}{z F} \right) \sinh^{-1}\left(\frac{i}{2 i_o} \right) \tag{16.42}$$

The Arrhenius-type equation, Eq. (16.39), can be redefined using a reference temperature T_o. Thus,

$$i_o (T) = k_a' \exp\left(-\frac{\Delta G}{RT} \right) = k_c' \exp\left(-\frac{\Delta G}{RT} \right) \tag{16.43a}$$

$$i_o (T_o) = k_a' \exp\left(-\frac{\Delta G}{RT_o} \right) = k_c' \exp\left(-\frac{\Delta G}{RT_o} \right) \tag{16.43b}$$

$$\frac{i_o(T)}{i_o(T_o)} = \frac{\exp\left(-\frac{\Delta G}{RT}\right)}{\exp\left(-\frac{\Delta G}{RT_o}\right)} \tag{16.43c}$$

$$i_{o,T} = i_{o,T_o} \exp\left[-\frac{\Delta G}{R}\left(\frac{1}{T} - \frac{1}{T_o}\right)\right] \tag{16.43d}$$

where $i_{o,T}$ is the exchange electric current density at temperature T and i_{o,T_o} is the reference exchange electric current density at temperature T_o. In general, the exchange current density i_o is analogous to the rate of chemical reaction.

If one is interested in determining the total ionic molar flux J due to the mass transfer of a species j, then combine Eqs. (16.40a), (16.40b) to get

$$J = \frac{i}{zF}\left\{\exp\left[\frac{\alpha z F\eta}{RT}\right]_f - \exp\left[-\frac{(1-\alpha)z F\eta}{RT}\right]_r\right\} \tag{16.44}$$

For any reaction, Eq. (16.40a) can be approximated by letting $x = \alpha z F\eta/RT$ and $y = (1-\alpha)z F\eta/RT$. Expanding the exponential functions according to Taylor's series yields

$$\exp(x) = \sum_{k=0}^{\infty} \frac{x^k}{k!} \simeq 1 + x \tag{16.45a}$$

$$\exp(-y) = \sum_{k=0}^{\infty} \frac{(-y)^k}{k!} \simeq 1 - y \tag{16.45b}$$

Mathematically, Eq. (16.40a) can be expressed as a linear approximation when the overpotential is small in magnitude. Inserting Eqs. (16.45a) and (16.45b) into (16.40a) yields

$$i = i_o(x + y) = [\alpha z F\eta/RT + (1-\alpha)z F\eta/RT] \tag{16.46a}$$

$$i \simeq i_o\left(\frac{z F\eta}{RT}\right) \tag{16.46b}$$

If

$$\exp\left[\frac{\alpha z F\eta}{RT}\right]_f >> \exp\left[-\frac{(1-\alpha)z F\eta}{RT}\right]_r \tag{16.47}$$

then, Eq. (16.46a) yields

$$i = i_o \exp\left(\frac{\alpha z F\eta}{RT}\right) \tag{16.48}$$

Solving Eq. (16.48) for the overpotential yields the **Tafel equation**

$$\eta = \frac{RT}{\alpha z F} \ln\left(\frac{i}{i_o}\right) = \frac{RT}{\alpha z F} \ln(i) - \frac{RT}{\alpha z F} \ln(i_o) \tag{16.49a}$$

$$\eta = \frac{2.303 RT}{\alpha z F} \log(i) - \frac{2.303 RT}{\alpha z F} \log(i_o) \tag{16.49b}$$

$$\eta = a + b \log(i) \tag{16.49c}$$

with

$$a = -\frac{2.303 RT}{\alpha z F} \log(i_o) \tag{16.50a}$$

$$b = \frac{2.303 RT}{\alpha z F} \tag{16.50b}$$

According to Faraday's law of electrolysis, the quantity of charge transferred (Q) at time t and known current (I) or current density (i) is

$$Q = It = i A_s t \tag{16.51}$$

The quantity of moles for zF **charge transfer** at t and i is

$$N = \frac{Q}{zF} \tag{16.52}$$

According to Faraday's law of electrolysis, the mass loss or gain during the electrochemical process can be defined by

$$m = N A_w = \frac{It A_w}{zF} = \frac{It A_w N}{Q} \tag{16.53a}$$

$$A_w = \sum \frac{f_j A_{w,j}}{z_j} \tag{16.53b}$$

Here, f_j is the mass or weight fraction and $A_{w,j}$ is the atomic weight of an element j in the alloy.

Example 16.4 Two identical zinc rods are individually exposed to HCl-base electrolyte. One rod is immersed at $25\,^\circ C$ for 2 *hours*, while the other rod is at $50\,^\circ C$ for 1 *hour*. Calculate the time of exposure of a third identical rod immersed into the electrolyte at $30\,^\circ C$, and plot the reaction rate R_x at $200\ K \leq T \leq 400\ K$.

Solution
The zinc oxidation reaction is

$$Zn \longrightarrow Zn^{2+} + 2e^- \tag{16.4E1}$$

which requires a chemical activation energy to proceed in the written direction. The zinc rods are not polarized through an external circuit. Thus, the Arrhenius equation, Eq. (16.35b), yields

$$R_1 = 1/t_1 = \gamma_a \exp\left(-\frac{\Delta G^*}{RT_1}\right) \tag{16.4E2a}$$

$$R_2 = 1/t_2 = \gamma_a \exp\left(-\frac{\Delta G^*}{RT_2}\right) \tag{16.4E2b}$$

Dividing these equations gives

$$\frac{t_2}{t_1} = \exp\left[-\frac{\Delta G^*}{R}\left(\frac{1}{T_1} - \frac{1}{T_2}\right)\right] \tag{16.4E3}$$

Then, the reaction rates at $T_1 = 25\,°C = 298\ K$, $t_1 = 2\ h$ and at $T_2 = 50\,°C = 323°K$, $t_2 = 1\ h$ along with $R = 8.3145\ J.mol^{-1}K^{-1}$ are

$$R_1 = 1/t_1 = 0.50\ h^{-1} \tag{16.4E4a}$$

$$R_2 = 1/t_2 = 1.00\ h^{-1} \tag{16.4E4b}$$

Hence, the activation energy becomes

$$\Delta G^* = -R\left(\frac{1}{T_1} - \frac{1}{T_2}\right)^{-1}\ln\left(\frac{t_2}{t_1}\right) \tag{16.4E5a}$$

$$\Delta G^* = -\left(8.3145\ J.mol^{-1}K^{-1}\right)\left(\frac{1}{298\ K} - \frac{1}{323\ K}\right)^{-1}\ln\left(\frac{1\ h}{2\ h}\right) \tag{16.4E5b}$$

$$\Delta G^* = 22,189\ J/mol \tag{16.4E5c}$$

This result can be obtained by linearizing the Arrhenius equation and subsequently using a graphical method, where ΔG^* is simply the slope of a line. Substituting this energy into Eq. (16.4E2a) and solving for γ_a yield

$$\gamma_a = \frac{1}{t_1}\exp\left(\frac{\Delta G^*}{RT_1}\right) \tag{16.4E6a}$$

$$\gamma_a = \frac{1}{2\ h}\exp\left[\frac{22,189\ J/mol}{\left(8.3145\ J.mol^{-1}K^{-1}\right)(298\ K)}\right] \tag{16.4E6b}$$

$$\gamma_a = 3874.80\ h^{-1} \tag{16.4E6c}$$

For $T_3 = 30\,°C = 303\ K$, the rate of reaction can be described by the empirical Arrhenius equation, which is used to assess the temperature dependence of reaction rates. In this context, the general form of the Arrhenius equation is defined by

$$R_3 = 1/t_3 = \gamma_a \exp\left(-\frac{\Delta G^*}{RT_3}\right) \tag{16.4E7}$$

This expression, Eq. (16.4E7), contains a the frequency factor γ_a that accounts for atom or molecule collisions and defines the temperature-dependent reaction rates. Nevertheless, the magnitudes of time and rate of reaction for the current problem are

$$t_3 = \frac{1}{\gamma_a} \exp\left(\frac{\Delta G^*}{RT_3}\right) \tag{16.4E8a}$$

$$t_3 = \frac{1}{3874.80\ h^{-1}} \exp\left[\frac{22{,}189\ J/mol}{\left(8.3145\ J.mol^{-1}K^{-1}\right)(303\ K)}\right] \tag{16.4E8b}$$

$$t_3 = 1.73\ h \tag{16.4E8c}$$

$$R_3 = 1/t_3 = 0.58\ h^{-1} \tag{16.4E8d}$$

Therefore, the reaction rates are $R_1 < R_3 < R_2$. Now that ΔG^* and γ_a are known, plot the reaction rate equation as

$$R_x = \gamma_a \exp\left(-\frac{\Delta G^*}{RT_3}\right) \tag{16.4E9a}$$

$$R_x = \left(3874.80\ h^{-1}\right)\exp\left(-\frac{22{,}189}{(8.3145)\,(T)}\right) \tag{16.4E9b}$$

$$R_x = 3874.80\exp\left(-\frac{2668.70}{T}\right) \tag{16.4E9c}$$

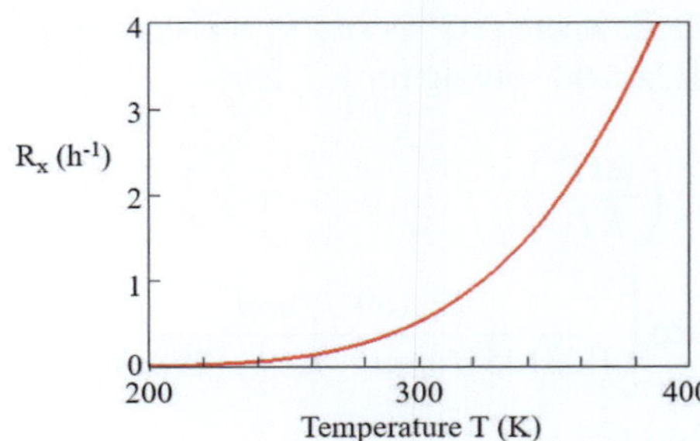

Example 16.5 Use the given dataset (table below) for a hypothetical metal undergoing chemical oxidation in a aqueous solution **(a)** to plot the data (Arrhenius plot)

and determine the activation energy from the plot and **(b)** to calculate the chemical
rate of reaction R_A described by the Arrhenius rate equation at $308\,^\circ C$ temperature.

$T\ (K)$	$R_A\ (x10^{-2}\ h^{-1})$
298	50.01
300	50.26
305	50.88
310	51.48
315	52.08

For your information, a temperature-dependent reaction rate measures the speed at
which a chemical reaction proceeds.

Solution

(a) Performing regression analysis on the given data yields the relevant curve
fitting equation for the reaction rate R_A defined by the Arrhenius expression,
Eq. (16.35b). Thus,

$$R_A = \gamma_a \exp\left(-\frac{\Delta G^*}{RT}\right) \tag{16.5E1a}$$

$$\ln(R_A) = \ln(\gamma_a) - \frac{\Delta G^*}{RT} \tag{16.5E1b}$$

$$\ln(R_A) = 4.3046 \times 10^{-2}\ h^{-1} - \frac{219.20\ K}{T} \tag{16.5E1c}$$

where the intercept and the slope are $\ln(\gamma_a) = 4.3046 \times 10^{-2}\ h^{-1}$ and $S = 219.20\ K$, respectively. Thus, the activation energy ΔG^* and the constant γ_a
are

$$\Delta G^* = -RS = -(8.3145\ J/mol.K)\,(219.20\ K) \tag{16.5E2a}$$

$$\Delta G^* = -1822.50\ J/mol \tag{16.5E2b}$$

$$\ln(\gamma_a) = 4.3046 \times 10^{-2}\ h^{-1} \tag{16.5E2c}$$

$$\gamma_a = \exp\left(4.3046 \times 10^{-2}\right) = 1.0440\ h^{-1} \tag{16.5E2d}$$

The corresponding Arrhenius plot is

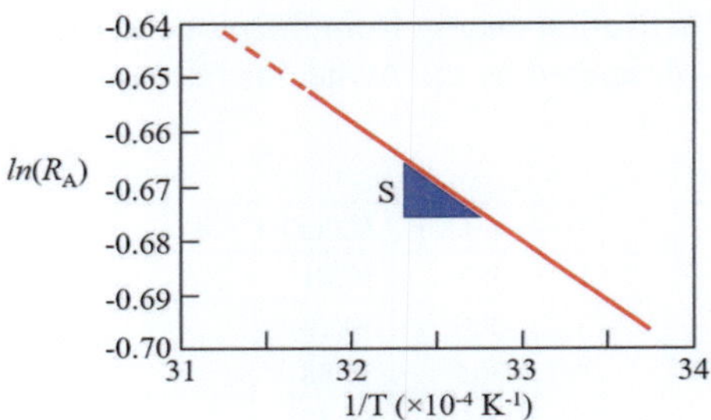

(b) If $T = 308 \ K$, then Eq. (16.5E2) gives

$$\ln(R_A) = 4.3046 \times 10^{-2} - \frac{219.20}{308} = -0.66864 \tag{16.5E3a}$$

$$R_A = \exp(-0.66864) = \simeq 0.51 \ h^{-1} \tag{16.5E3b}$$

which is a reasonable result since $\Delta G^* < 0$ for the unknown chemical oxidation.

16.14   Polarization from the Corrosion Potential

Evaluation of the corrosion behavior is normally done through a function that depends on kinetic parameters depicted in Fig. 16.14. For one-step reaction under steady-state condition, the electric current density function for polarizing from the corrosion point (i_{corr}, E_{corr}) is defined by

$$i = i_{corr} \left\{ \exp\left[\frac{\alpha z F \eta}{RT} \right]_f - \exp\left[-\frac{(1-\alpha) z F \eta}{RT} \right]_r \right\} \tag{16.54}$$

where $\eta = E - E_{corr}$
$\qquad E = $ Applied potential (V)
$\qquad \beta_a = $ Tafel anodic slope (V)
$\qquad \beta_c = $ Tafel cathodic slope (V)

This expression is suitable for corrosion studies at a temperature T.

The anodic polarization caused by an overpotential $\eta = \eta_a = (E - E_{corr}) > 0$ and $E > E_{corr}$ is referred to as an electrochemical process in which a metal surface oxidizes (corrodes) by losing electrons. Consequently, the metal surface is positively charged due to the loss of electrons. This electrochemical polarization is quantified by η_a.

Cathodic polarization, on the other hand, requires that electrons be supplied to the metal surface at an negative overpotential, $\eta = \eta_c = (E - E_{corr}) < 0$, which implies that $E < E_{corr}$. Moreover, Eq. (16.54) indicates that $\eta = \eta_a = 0$ for an unpolarized and $\eta = \eta_a \neq 0$ for a polarized electrode surface.

16.15 Potentiostatic Polarization Technique

Consider the potentiostatic polarization diagram shown in Fig. 16.16 for a 1080 eutectoid steel in deaerated 1N sulfuric acid solution at room temperature (Bandy and Jones [22]).

Corrosion of metals and alloys can be evaluated using potentiostatic and potentiodynamic anodic polarization measurements in a suitable electrolyte. For instance, Fig. 16.16 shows a potentiostatic polarization diagram for a eutectoid AISI 1080 steel in deaerated sulfuric acid solution (Bandy and Jones [22]). This polarization diagram is clearly divided into anodic and cathodic regions along with the Tafel β_a and β_c slopes, and it exhibits the corrosion point (i_{corr}, E_{corr}).

The pertinent experimental data extracted from Fig. 16.16 are included as legend. Notice that the polarization diagram covers data near the corrosion point defined at $(1180\ \mu A/cm^2, -0.51\ V_{SCE})$. In essence, the corrosion current density $i_{corr} = 1180\ \mu A/cm^2$ at $E_{corr} = -0.51\ V_{SCE}$ is a measure of the corrosion rate (C_R) of the eutectoid 1080 steel in deaerated $1N\ H_2SO_4$ solution at $25\,^\circ C = 298\ K$. Moreover, $1N$ stands for normality, which describes the mass (grams or moles) equivalent of solute dissolved in one liter of solution.

Converting the kinetic parameter i_{corr} to C_R is a commonly practice in the field of corrosion science. This conversion is done in a later section. At any rate, i_{corr} and $E_{corr} = -0.51\ V_{SCE}$ are the most important parameters to be determined, because they provide the transition point for anodic and cathodic polarization studies.

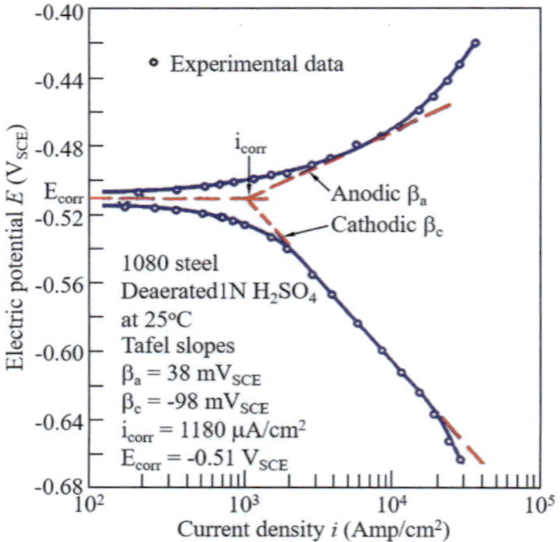

Fig. 16.16 Experimental stepwise polarization curve for 1080 eutectoid steel in deaerated 1N sulfuric acid (Bandy and Jones [22])

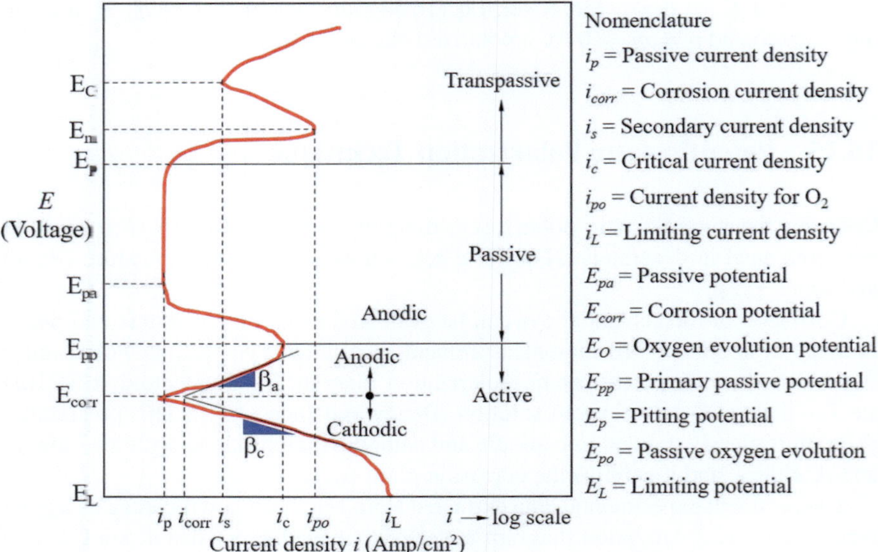

Fig. 16.17 Hypothetical potentiodynamic polarization curve, electrochemical regions, and nomenclature for kinetic studies of an active-passive electrochemical system

16.16   Potentiodynamic Polarization Technique

Consider a complete polarization diagram in Fig. 16.17 along with relevant polarization regions as schematically shown in Fig. 16.17 shows. This type of $E = f(i)$ curve can be obtained in a relatively short time using the potentiodynamic polarization technique.

Regarding electrochemical cells, a polarization curve simply displays the voltage output referred to as the electric potential (E) of the cell for a given current (I) or current density (i) loading. In order to obtain a polarization curve, a potentiostat/galvanostat equipment is a common and practical voltage or current source.

The polarization curve illustrated in Fig. 16.17 is the basic kinetic approach for characterizing or identifying electrochemical reactions and for quantifying the behavior of metals under relevant electrochemical conditions, including the effect of certain ions in solution, temperature, acidity, and alkalinity of solutions. Notice that potentiostatic and potentiodynamic procedures are useful in determining the electrochemical corrosion behavior of an active-passive material. Thus, the **passivity** of a metal and the **corrosivity** of an electrolyte can be characterized as functions of temperature, concentration of an ionic species, scan rate, convection, and even pressure. For example, Fig. 16.18 shows the effect of scan rate on a RSA 304 specimen immersed in sulfuric acid (H_2SO_4) solutions (Perez [8, pp. 212–213]).

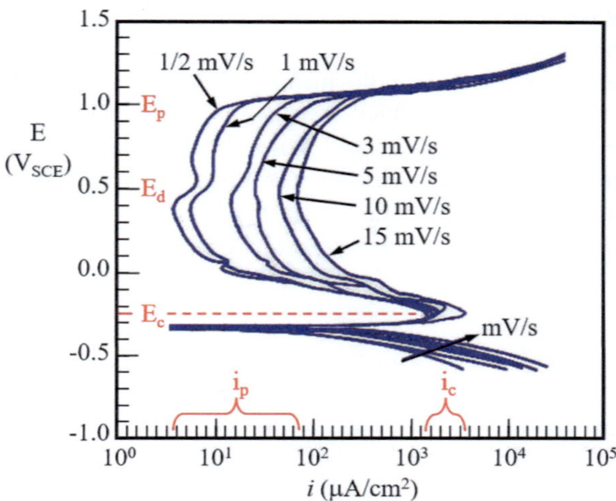

Fig. 16.18 Room temperature experimental potentiodynamic polarization curves for (50%CR) RSA 304 as a function of scan rate in 10N sulfuric acid solution at pH = -1.2 (Perez [8, p. 213])

Despite that the RSA 304 alloy exhibits an active-passive behavior, the potentiodynamic polarization curves show nearly constant passive potential range (ΔE). However, the corresponding passive current density i_p is affected by the scan rate $\Delta E/dt$.

Note that the electric potentials E_{corr}, E_c and E_p approximately remain constant, and the passive range is considered sufficiently wide enough for applications in the field of anodic protection, where a practical design electric potential at a certain electric potential scan rate can be defined as $E_d = (E_p - E_{pp})/2 \simeq 0.5\ V_{SCE}$.

Among the electrochemical kinetic parameters that can be extracted from the potentiodynamic polarization curves in Fig. 16.18, it is clearly evident that the scan rate dE/dt affects the passive current density i_p very significantly. However, the corrosion potential and the corrosion current density remain nearly constant in the order of $E_{corr} = -0.4\ V_{SCE}$ and $i_{corr} = 3 \times 10^2\ \mu A/cm^2$, respectively. Similarly, the critical potential and the corresponding critical current density for passivation are $E_c \simeq -0.25\ V_{SCE}$ and $i_{corr} = 2 \times 10^3\ \mu A/cm^2$. Remarkably, the pitting electric potential is $E_p \simeq 1\ V_{SCE}$ at all scan rates, but the pitting current density increases with increasing scan rate at a fixed temperature.

A polarization curve can provide evidence on whether or not a material is active, passive, or active-passive. Hence, the corrosion behavior of an electrode can be characterized in a particular environment. For instance, potentiostatic and potentiodynamic testing methods can give fairly similar results; however, the latter method requires a slow scan rate (mV/s) in order to maintain a stable corrosion potential and a steady-state behavior of a metal or alloy since service conditions are susceptible to changes in corrosion potential. Nevertheless, both potentiostatic and potentiodynamic procedures are useful in determining the electrochemical corrosion behavior of an active-passive material.

Passivation A distinction between surface layers and passive oxide films should be made. If a metal is insoluble in an environment, it may form an insoluble oxide corrosion product on the metal surface as a crystalline and poorly adherent surface layer (nonpassive thick layer), such as the blue/green layer on corroded copper plumbing and well-known brown "Rust" layer on iron and iron alloys. On the other hand, if a metal is soluble in a solution, it oxidizes and dissolves in the solution as cations. These cations react with dissolved oxygen and are deposited back on the metal as oxides, forming a passive oxide film under the influence of an electric field.

16.16.1 Thermodynamics and Kinetics of Polarization

Thermodynamics (non-kinetic approach) describes the property behavior and the equilibrium state of a system, and kinetics describes the rate at which a particular process will occur. Combining the $E = f(pH)$ and $E = f(i)$ functions for iron in water, as an example, provides a kinetic and non-kinetic information on the electrochemical states of an electrode. This is shown in Fig. 16.19a for $[Fe^{+2}] = 10^{-6}\ mol/l$, $pH = 6$ and $25\,^{\circ}C$ along with reference A, B, C points, the hydrogen (H_2) and oxygen (O_2) evolution lines for comparison (Pourbaix [18, p. 88]).

Figure 16.19b, on the other hand, illustrates a schematic potentiodynamic diagram containing the relevant reference A, B, C points. Observe the correspondence of potential between both diagrams using the reference A, B, C points. For instance, point A corresponds to the corrosion potential (E_{corr}) and corrosion electric current density (i_{corr}).

Corrosion occurs along AB line, and passivation takes place along the BC line for anodically protecting iron (Fe) though an oxide film. Point B is a transition potential at the critical electric current density, and point C corresponds to oxygen evolution and iron pitting potential (E_p).

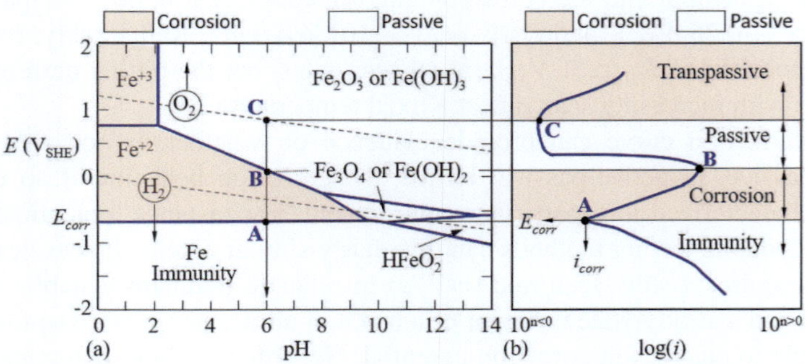

Fig. 16.19 Pourbaix diagram, $E = f(pH)$, in conjunction with a polarization curve, $E = f(i)$. (a) Iron (Fe) or carbon steel $(Fe\text{-}C$ alloy) in water containing $[Fe^{+2}] = 10^{-6}\ mol/l$ at $T = 25^{\circ}C$, $pH = 6$ and $P = 101\ kPa$ (Pourbaix [18, p. 88]) and (b) schematic potentiodynamic curve showing the reference A, B, C points

The formation of the passive film is through the hematite (Fe_2O_3) chemical reaction

$$2Fe + 3H_2O \rightarrow Fe_2O_3 + 6H^+ + 6e^- \tag{16.55a}$$

$$2Fe^{2+} + 3O^{-2} \rightarrow 2Fe_2O_3 \tag{16.55b}$$

The $pH\text{-}E$ diagram (Fig. 16.19a) and the polarization curve (Fig. 16.19b) show the corrosion region for iron or carbon steel at a $pH = 6$, but only the latter exhibits the corrosion point A, from which the corrosion rate (C_R) is determined using Faraday's law of electrolysis.

Figure 16.19b shows the regions for cathodic protection (CP) at $E < E_{corr}$ and anodic protection (AP) at $E_B < E < E_C$. However, the anodic potential for corrosion, $E_A \leq E < E_B$, can lie anywhere between A and B points.

The electric potential E and current density i range between A and B points may be small or large in magnitude. This may be a concern in certain applications related to potentiostatic or potentiodynamic anodic polarization schemes.

16.17 Polarization Energy

For **activation polarization**, the generalized half-cell redox reaction for metal oxidation or corrosion is written as

$$O + ze^- \rightleftarrows R \tag{16.56}$$

where O is the oxidized species, R is the reduced species, and ze^- is the number of electron transfer. The activation energy for this redox reaction under an applied anodic overpotential η_a is crudely approximated as

$$U_a = q_e\eta_a \ \& \ U_c = q_e\eta_c \tag{16.57}$$

Assume that the oxidation of iron, $Fe \rightarrow Fe^{2+} + 2e^-$, occurs at $\eta_a = 0.2\ V$. Then, Eq. (16.57) gives

$$U_a = q_e\eta_a = \left(1.6x10^{-19}\ C\right)(0.2V) = 3.2\ J = 0.2\ eV \tag{16.58}$$

Notice that $0.2\ eV$ is needed for oxidation of iron (Fe). This calculation serves the purposes of illustrating the possible magnitude of a redox reaction. Conversely, **concentration polarization** is the electrochemical process that requires a negative deviation from equilibrium so that $\eta_c = (E - E_{corr}) < 0$. If $\eta_c = -0.2\ V$ for the reduction of iron, $Fe^{2+} + 2e^- \rightarrow Fe$, then $U_c = -0.2\ eV$.

16.18   Corrosion Rate

During corrosion (oxidation) process, both anodic and cathodic reaction rates are coupled together on the electrode surface at a specific electric current density known as i_{corr}. This is an electrochemical phenomenon which dictates that both reactions must occur on different sites on the metal-electrolyte interface. At equilibrium, the current densities are related as $i_a = -i_c = i_{corr}$ @ E_{corr}.

When polarizing from the corrosion potential with respect to anodic or cathodic electric current density, the overpotential expressions given by Eqs. (16.34a), (16.34b) can be solved for the current densities. Thus,

$$i_a = i_{corr} \exp\left[\frac{2.303\,(E - E_{corr})}{\beta_a}\right] \tag{16.59a}$$

$$i_c = i_{corr} \exp\left[-\frac{2.303\,(E - E_{corr})}{|\beta_c|}\right] \tag{16.59b}$$

Assuming that the applied electric current density is $i = i_a - i_c$. Substituting Eqs. (16.59a),(16.59b) into $i = i_a - i_c$ yields the **Butler-Volmer equation** that quantifies the kinetics of the electrochemical corrosion

$$i = i_{corr}\left\{\exp\left[\frac{2.303\,(E - E_{corr})}{\beta_a}\right] - \exp\left[-\frac{2.303\,(E - E_{corr})}{|\beta_c|}\right]\right\} \tag{16.60}$$

This expression resembles Eq. (16.54), but it is a convenient expression, since the inverse polarization resistance is easily obtainable by deriving Eq. (16.60) with respect to the applied potential E. Thus,

$$\left(\frac{di}{dE}\right) = 2.303 i_{corr}\left\{\begin{array}{l} \beta_a^{-1} \exp\left[2.303\,(E - E_{corr})/\beta_a\right] \\ -\left|\beta_c^{-1}\right| \exp\left[-2.303\,(E - E_{corr})/\beta_c\right] \end{array}\right\} \tag{16.61}$$

The second derivative of Eq. (16.60) is

$$\left(\frac{d^2 i}{dE^2}\right) = 5.3038 i_{corr}\left\{\begin{array}{l} \beta_a^{-2} \exp\left[2.303\,(E - E_{corr})/\beta_a\right] \\ -\left|\beta_c^{-2}\right| \exp\left[-2.303\,(E - E_{corr})/\beta_c\right] \end{array}\right\} \tag{16.62}$$

This expression provides useful information on the cell potential

$$\left(\frac{d^2 i}{dE^2}\right) \Rightarrow \left\{\begin{array}{ll} < 0 & \text{for } E = E\max > E_{corr} \\ = 0 & \text{for an inflation point: } E = E_{corr} \\ > 0 & \text{for } E = E_{\min} < E_{corr} \end{array}\right. \tag{16.63}$$

Evaluating Eq. (16.62) at the inflation point yields

$$\left(\frac{d^2 i}{dE^2}\right)_{E=E_{corr}} = 5.3038 i_{corr}\left(\beta_a^{-2} - \left|\beta_c^{-2}\right|\right) \tag{16.64}$$

Now, evaluating Eq. (16.61) at $E = E_{corr}$ yields the polarization resistance R_p

$$\left(\frac{di}{dE}\right)_{E=E_{corr}} = 2.303 i_{corr}\left(\frac{\beta_a + |\beta_c|}{\beta_a |\beta_c|}\right) \tag{16.65a}$$

$$R_p = \left(\frac{di}{dE}\right)_{E=E_{corr}}^{-1} \tag{16.65b}$$

then,

$$R_p = \frac{\beta_a |\beta_c|}{2.303 i_{corr}\left(\beta_a + |\beta_c|\right)} = \frac{\beta}{i_{corr}} \tag{16.66a}$$

$$i_{corr} = \frac{\beta}{R_p} \tag{16.66b}$$

$$\beta = \frac{\beta_a |\beta_c|}{2.303\left(\beta_a + |\beta_c|\right)} \tag{16.66c}$$

where β is the proportionality constant and R_p is inversely proportional to i_{corr} as indicated by Eq. (16.66a). For convenience, Eq. (16.66a) can be linearized as

$$\log\left(R_p\right) = \log\left(\beta\right) - \log\left(i_{corr}\right) \tag{16.67}$$

The expressions defined by Eqs. (16.66a) and (16.67) are plotted in Fig. 16.20a and b, respectively.

Now, dividing Faraday's rate of reaction , Eq. (16.35a), by the metal density ρ defines the rate of metal dissolution as the electrochemical corrosion rate. Thus,

$$C_R = \frac{R_F}{\rho} \tag{16.68a}$$

$$C_R = \frac{i A_w}{z F \rho} = \frac{i_{corr} A_w}{z F \rho} \tag{16.68b}$$

Fig. 16.20 (a) Nonlinear and (b) linear R_p plots

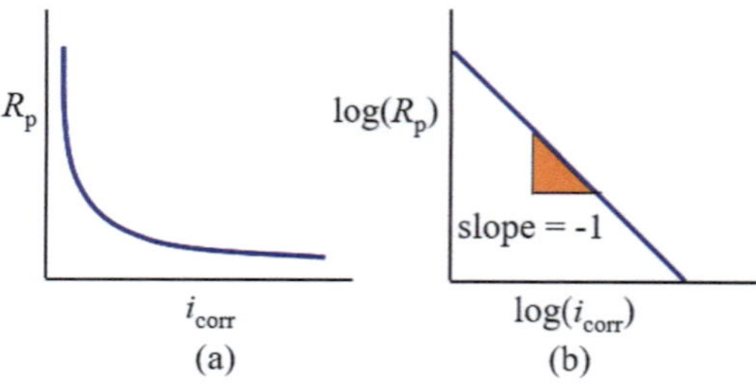

This expression, Eq. (16.68b), is a convenient mathematical model for determining the rate of metal dissolution in terms of penetration per year in units of mm/y, which are common units to the engineers or designers. Additional C_R equations can be found elsewhere (Shreir et al. [23, pp. 2:5, 2:104, 2:111]). Thus,

$$C_R = f(C_i) = A_1 C_j^n \quad \text{at } T, P \tag{16.69a}$$

$$C_R = f(pH) = A_2 (pH)^n \quad \text{at } T, P \tag{16.69b}$$

$$C_R = f(T) = A_3 \exp\left(-\frac{A_4}{T}\right) \quad \text{at } P \tag{16.69c}$$

where A_1-A_4 and n denote constants, C_j denotes the concentration of species j and $pH = -\log(H^+)$.

Example 16.6 Calculate **(a)** the corrosion rate C_R in units of mm/y and **(b)** the electrochemical rate constant in $\mu g/cm^2.s$ and $mol/cm^2 s$ for a low carbon steel plate ($1cm \times 1cm \times 4cm$) immersed in sea water. Given data: $I = 110\ \mu A$, $\rho = 7.87 g/cm^3$ and $A_w = 55.85\ g/mol$.

Solution

(a) Since the carbon steel is basically iron (Fe), the oxidation reaction involved in this problem is $Fe \longrightarrow Fe^{2+} + 2e^-$ with $z = 2$ and $F = 96,500\ A \cdot s/mol$, and the exposed area is $A = 1cm \times 1cm = 1\ cm^2$. Then, the electric current density is $i = I/A = 110 \times 10^{-6}\ amps/cm^2$. Using Eq. (16.68b), the corrosion rate is

$$C_R = \frac{i A_w}{z F \rho} = \frac{\left(110 \times 10^{-6}\ A/cm^2\right)(55.85\ g/mol)}{(2)\,(96,500\ A \cdot s/mol)\left(7.87\ g/cm^3\right)} \tag{16.6E1a}$$

$$C_R = 4.04 \times 10^{-9}\ cm/s = 1.28\ mm/y \tag{16.6E1b}$$

(b) From Eq. (16.35a), the electrochemical rate constant is

$$R_F = \frac{i A_w}{z F} = \frac{\left(110\ \mu A/cm^2\right)(55.85\ g/mol)}{(2)\,(96,500\ A \cdot s/mol)} \tag{16.6E2a}$$

$$R_F = 0.0318\ \mu g/cm^2 s \tag{16.6E2b}$$

Eliminating the atomic weight in Eq. (16.35a) yields

$$R_F = \frac{i}{z F} = \frac{\left(110\ \mu A/cm^2\right)}{(2)\,(96,500\ A \cdot s/mol)} \tag{16.6E3a}$$

$$R_F = 5.70 \times 10^{-10}\ mol/\left(cm^2 \cdot s\right) \tag{16.6E3b}$$

Example 16.7 Assume that the rates of oxidation of Zn and reduction of H^+ are controlled by activation polarization. Use the data given below to **(a)** plot the appropriate polarization curves and determine E_{corr} and i_{corr} from the plot. Calculate **(b)** both E_{corr} and i_{corr}, **(c)** the corrosion rate C_R in mm/y, and the polarization resistance R_p in $ohm \cdot cm^2$. Given data:

$Zn = Zn + 2e^- \Longrightarrow$	$E_{Zn} = -0.80\ V_{SHE}$	$i_{o,Zn} = 10^{-7}\ A/cm^2$
	$\beta_a = 0.10\ V_{SHE}$	$\rho = 7.14\ g/cm^3$
	$A_w = 65.37\ g/mol$	
$2H^+ + 2e^- = H_2 \Longrightarrow$	$E_H = 0.10\ V_{SHE}$	$i_{o,H} = 10^{-10}\ A/cm^2$
	$\beta_c = -0.10\ V_{SHE}$	

Solution

(a) E vs. $\log(i)$ plot:

(1) Plot $(i_{o,H}, E_H) = (\log 10^{-10}\ A/cm^2, 0.10\ V_{SHE})$

$\qquad (i_{o,Zn}; E_{o,Zn}) = (\log 10^{-7}\ A/cm^2, -0.80\ V_{SHE})$

(2) Draw the hydrogen reduction and zinc oxidation lines with β_c and β_a slopes, respectively

(3) Locate the corrosion point $(i_{corr}, E_{corr}) = (10^{-4}\ A/cm^2, -0.50\ V_{SHE})$ for the unpolarized Zn metal with $i_{corr} = i_a = i_c$ at $E = E_{corr}$. Moreover, $E_{o,Zn}$, $E_{o,H}$ are the open-circuit potentials, which can be determined using the Nernst equation, Eq. (16.32). The plot is

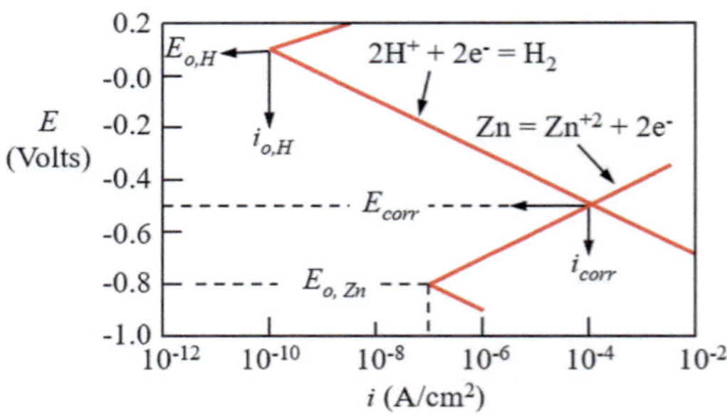

(b) Using Eqs. (16,34) with $i_a = i_c = i_{corr}$, $\eta_a = E - E_{o,Zn}$ and $\eta_c = E - E_{o,H}$ yields

$$E = E_{o,Zn} + \beta_a \log \left(i_{corr}/i_{o,Zn}\right) \qquad (16.7E1a)$$

$$E = E_{o,H} - |\beta_c| \log \left(i_{corr}/i_{o,H}\right) \qquad (16.7E1b)$$

Equating Eqs. (16.7E1a) and (16.7E1b) and solving for i_{corr} yield the corrosion electric current density

$$\log i_{corr} = \frac{1}{\beta_a + |\beta_c|} \left[E_{o,H} - E_{o,Zn} + \beta_a \log i_{o,Zn} + |\beta_c| \log i_{o,H} \right]$$

$$(16.7E2a)$$

$$\log i_{corr} = \frac{0.1 + 0.8 + 0.1 \log 10^{-7} + 0.1 \log 10^{-10}}{0.2}$$

$$(16.7E2b)$$

$$\log i_{corr} = -4 \tag{16.7E2c}$$

$$i_{corr} = 10^{-4} \; A/cm^2 \tag{16.7E2d}$$

Letting $E = E_{corr}$ and using Eq. (16.7E1a) or (16.7E1b) gives

$$E_{corr} = E_{o,Zn} + \beta_a \log \left(i_{corr}/i_{o,Zn} \right) \tag{16.7E3a}$$

$$E_{corr} = -0.8 \; V_{SHE} + (0.1 \; V_{SHE}) \log \left(\frac{10^{-4} \; A/cm^2}{10^{-7} \; A/cm^2} \right) \tag{16.7E3b}$$

$$E_{corr} = -0.50 \; V_{SHE} \tag{16.7E3c}$$

(c) From Eq. (16.68b)

$$C_R = \frac{i_{corr} A_w}{zF\rho} = \frac{\left(10 x 10^{-4} \; A/cm^2 \right) (65.37 \; g/mol)}{(2) (96{,}500 \; A \cdot s/mol) \left(7.14 \; g/cm^3 \right)} \tag{16.7E4a}$$

$$C_R = 1.50 \; mm/y \tag{16.7E4b}$$

From Eq. (16.66c),

$$\beta = \frac{\beta_a |\beta_c|}{2.303 \left(\beta_a + |\beta_c| \right)} \tag{16.7E5a}$$

$$\beta = \frac{(0.10 \; V_{SHE}) (0.10 \; V_{SHE})}{2.303 (0.10 \; V_{SHE} + 0.10 \; V_{SHE})} \tag{16.7E5b}$$

$$\beta = 0.0217 \; V_{SHE} \tag{16.7E5c}$$

and from Eq. (16.66a),

$$R_p = \frac{\beta}{i_{corr}} = \frac{0.0217 \; V_{SHE}}{10^{-4} \; A/cm^2} \tag{16.7E6a}$$

$$R_p = 217 \; ohm \cdot cm^2 \tag{16.7E6b}$$

Example 16.8 A copper surface area, $A = 100 \; cm^2$, is exposed to an acid solution. After 24 hours, the loss of copper due to corrosion (oxidation) is $15 x 10^{-3}$ grams. Calculate **(a)** the electric current density i in A/cm^2 and **(b)** the corrosion rate C_R in

mm/y. Data for copper: $A_w = 63.55\ g/mol$, $\rho = 8.94\ g/cm^3$, $q_e = 1.602x10^{-19}$ $C/reaction$.

Solution
First of all, the corrosion process of copper can be represented by the following anodic reaction:

$$Cu \longrightarrow Cu^{2+} + 2e^- \qquad (16.8E1)$$

where the valence is $Z = +2$.

(a) According to Faraday's law of electrolysis, the electric current density is

$$i = \frac{I}{A_s} = \frac{zFm}{A_s t A_w} \qquad (16.8E2a)$$

$$i = \frac{(2)\,(96,500\ A.s/mol)\,(15x10^{-3}\ g)}{(100\ cm^2)\,(24x3600\ s)\,(63.54\ g/mol)} \qquad (16.8E2b)$$

$$i = 5.27\ A/cm^2 \qquad (16.8E2c)$$

$$I = iA = \left(5.27\ A/cm^2\right)\left(100\ cm^2\right) = 527.34\ A \qquad (16.8E2d)$$

(b) From Eq. (16.68b),

$$C_R = \frac{iA_w}{zF\rho} \qquad (16.8E3a)$$

$$C_R = \frac{\left(5.27\ A/cm^2\right)(63.54\ g/mol)}{(2)\,(96,500\ A.s/mol)\,\left(8.94\ g/cm^3\right)} \qquad (16.8E3b)$$

$$C_R = 1.94 \times 10^{-4}\ cm/s = 1.94 \times 10^{-3}\ mm/s \qquad (16.8E3c)$$

$$C_R = 6118\ cm/y = 61,180\ mm/y \qquad (16.8E3d)$$

Hence, this result represents a high corrosion rate.

16.19 Passive Oxide Films

Passive oxide films act as barriers against corrosion, but they can break down leading to pitting corrosion of the substrate due to localized anodic reactions driven by an anodic overvoltage $\eta_a > 0$. For instance, the following redox reactions indicate the types of copper oxide films susceptible to pitting corrosion

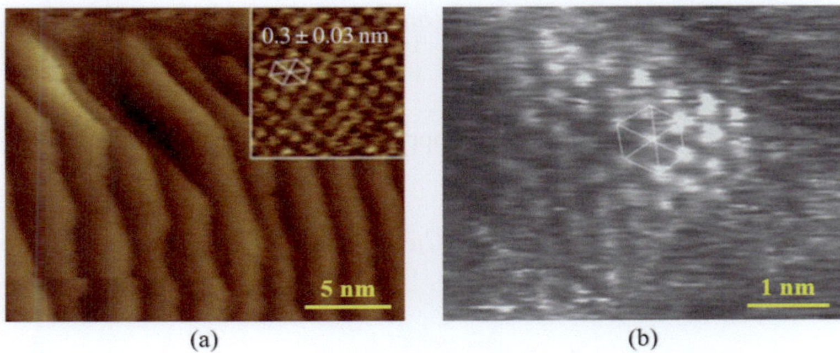

(a)   (b)

Fig. 16.21 STM images of metallic films showing the hexagonal sublattice of oxygen. **(a)** Cuprite Cu_2O grown on Cu (111) passive film in 0.1 M $NaOH(aq)$ at -0.20 V_{SHE}. Inset corresponds to the Cu sublattice in $Cu_2O(111)$ plane (Kunze et al. [24]). **(b)** Oxide passive film grown on Fe-18Cr-13Ni stainless steel in 0.5 M $H_2SO_4(aq)$ at 0.5 V_{SHE} for2 h (Maurice et al. [25]). All images are taken from Marcus and Maurice article [26]

$$2Cu + \frac{1}{2}O_2 \rightarrow Cu_2O \quad [\text{red } Cu(I)] \tag{16.70a}$$

$$Cu_2O + \frac{1}{2}O_2 \rightarrow CuO \quad [\text{black } Cu(II)] \tag{16.70b}$$

A particular copper oxide film is shown in Fig. 16.21a for cuprite Cu_2O as reported by Kunze et al. [24] in their in situ electrochemical scanning tunneling microscopy (STM) measurements on anodic oxidation of $Cu(111)$ in 0.1 M $NaOH$ electrolyte at -0.20 V_{SHE}. The notation Cu (111) implies that Cu is the substrate surface and $(hkl) = (111)$ are the Miller indices of the lattice surface plane.

The inset in Fig. 16.21a depicts the Cu_2O hexagonal crystal structure with the basal plane lattice parameter of approximately 0.3 nm. The Cu_2O passive film is formed with a faceted surface, and it is clearly evident that this film is a pit-free structure. Additional insights on metal oxide films can be found elsewhere (Kunze et al. [24] as cited in (Maurice et al. [25] and Marcus and Maurice article [26]).

A similar STM image of an hexagonal crystallographic pattern is depicted in Fig. 16.21b on Fe-18Cr-13Ni stainless steel (Maurice et al. [25] as cited in Marcus and Maurice article [26]). This image was obtained after passivation in 0.5 M $H_2SO_4(aq)$ at 0.5 V_{SHE} for 2 hours.

It can be concluded that the metal oxide films shown in Fig. 16.21 are relevant examples of passivation for growing protective metallic coatings against corrosion.

16.20 Thermodynamics of Metal Oxides

Thermodynamically, the driving force for the oxidation of a metal in gaseous environments is known as the Gibbs free energy of formation (ΔG), and consequently, the occurrence of an oxidation chemical reaction depends on the magnitude of ΔG. The Ellingham diagram along with nomographic scales for partial oxygen pressure P_{O_2}, P_{H_2}/P_{H_2O} and P_{CO}/P_{CO_2} pressure ratios shown in Fig. 16.22 is very valuable for determining the standard Gibbs energy change ΔG^o for metal oxide reductions under stable conditions so that $\Delta G^o < 0$. If $\Delta G^o = 0$ metal oxides are at their equilibrium state and if $\Delta G^o > 0$ the metal oxides are unstable to be reduced to their metal form (Gaskell [27, p. 429]).

The Ellingham diagram includes thermodynamics information on chemical reactions with respect to their occurrence, but it excludes any kinetics information on the rate of reactions at temperatures. Nonetheless, the Ellingham diagram is commonly used in the fields of high-temperature corrosion and extractive metallurgy.

Mathematically, the standard ΔG^o at a constant pressure is defined by

$$\Delta G^o = -RT \ln\left(K_{sp}\right) \tag{10.66a}$$

$$\Delta G^o = \Delta H^o - T\Delta S^o \tag{10.66b}$$

The Gibbs energy change (ΔG^o) is taken as the driving force for the chemical reactions cited in Fig. 16.22. Actually, it depends on the enthalpy change (ΔH^o) as the heat energy between the reaction and its surroundings and the entropy change (Δs^o) of a reaction.

Thermodynamically, endothermic reactions absorb heat from the surroundings with $\Delta S^o > 0$ and $\Delta H^o > 0$, whereas exothermic reactions release heat to the surrounding with $\Delta S^o < 0$ and $\Delta H^o < 0$. According to Eq. (10.66b), $T ==\Delta H^o/\Delta S^o$ at equilibrium since $\Delta G^o = 0$.

The variations of both ΔH and ΔS with temperature are commonly written as

$$\Delta H(T) = \Delta H^o\left(T_o\right) + \int_{T_o}^{T} C_p(T)dT \tag{10.67a}$$

$$\Delta S(T) = \Delta S^o\left(T_o\right) + \int_{T_o}^{T} \frac{C_p(T)}{T}dT \tag{10.67b}$$

where $\Delta G^o\left(T_o\right)$ denotes the standard Gibbs energy change, $\Delta H^o\left(T_o\right)$ denotes the standard enthalpy change which is a measure of the heat absorbed or released at constant gas pressure, $\Delta S^o\left(T_o\right)$ denotes the standard entropy change which is a measure of monoatomic disorder, $T_o = 298\ K = 25\,^\circ C$ denotes the standard temperature, and C_p denotes the heat capacity at constant pressure P.

Notice that the Ellingham diagram illustrates (Fig. 16.22) linear relationships, $\Delta G = f(T)$, for the oxidation of metals over a range of temperature for a fixed phase. See the legend in Fig. 16.22 for $\Delta G = RT \ln(P)$. The most detailed

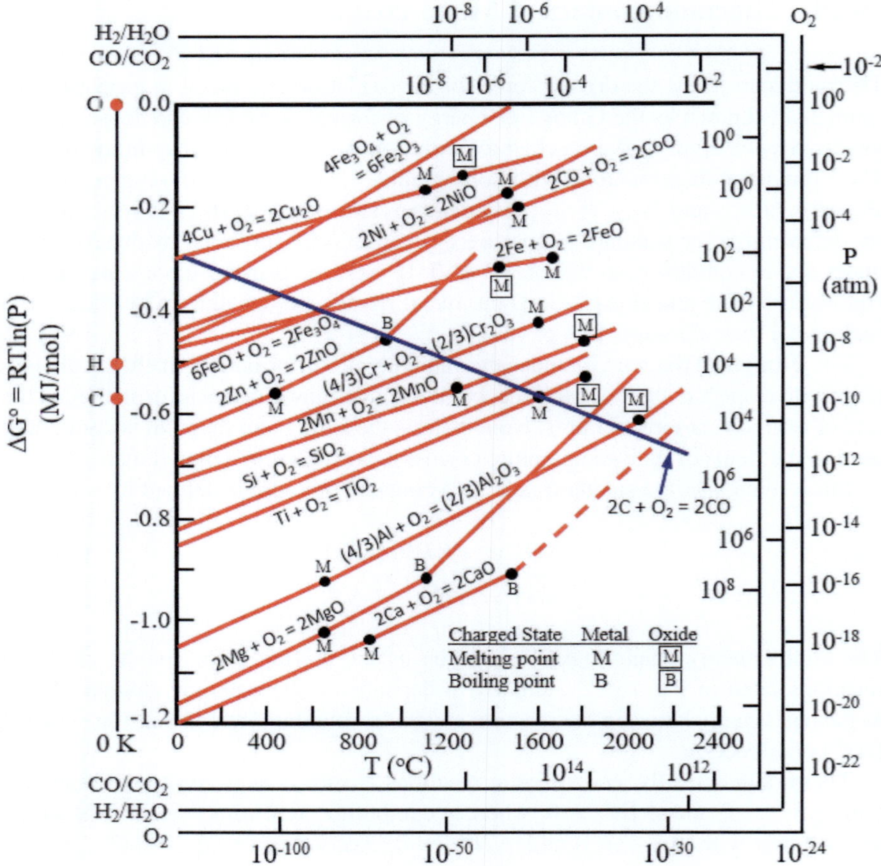

Fig. 16.22 Ellingham standard energy change diagram (Gaskell [27, p. 429])

Ellingham diagram for metal oxides in oxygen O_2, carbon monoxide CO, and carbon dioxide CO_2 gases can be found in Gaskell's book [27, p. 412]. Moreover, the Ellingham diagram predicts the high-temperature conditions for metal ores undergoing reduction to their metal forms or for pure metals undergoing oxidation to form metal oxides. Nevertheless, Eq. (10.66b) predicts the slope of $\Delta G = f(T)$ as $\Delta G^o/dT = -\Delta S^o$, where $\Delta S^o > 0$ for all metal chemical reactions depicted in Fig. 16.22.

Consider some generalized high-temperature reactions for the formation of metal oxides in an oxygen-rich O_2 environment having different chemical formulae. These reactions represent the metal-gas interactions being influenced by the gas pressure (oxygen O_2, carbon dioxide CO_2, and water H_2O) in the melt. Thus,

Reaction 1:

$$\frac{2x}{y} M + O_2 = \frac{2}{y} M_x O_y \tag{10.68a}$$

$$\Delta G^o = -RT \ln \left(\frac{\left[M_x O_y \right]^{2/y}}{[M]^{2x/y} [O_2]} \right) \tag{10.68b}$$

$$\Delta G^o = RT \ln \left(P_{O_2} \right) \tag{10.68c}$$

Reaction 2:

$$x M + y C O_2 = M_x O_y + y C O \tag{10.69a}$$

$$\Delta G^o = -RT \ln \left(\frac{\left[M_x O_y \right] [CO]^y}{[M]^x [CO_2]^y} \right) \tag{10.69b}$$

$$\Delta G^o = -yRT \ln \left(\frac{P_{CO}}{P_{CO_2}} \right) \tag{10.69c}$$

Reaction 3:

$$x M + y H_2 O = M_x O_y + y H_2 \tag{10.70a}$$

$$\Delta G^o = -RT \ln \left(\frac{\left[M_x O_y \right] [H_2]^y}{[M]^x [H_2 O]^y} \right) \tag{10.70b}$$

$$\Delta G^o = -yRT \ln \left(\frac{P_{H_2}}{P_{H_2 O}} \right) \tag{10.70c}$$

where $[j] = a_j$ denotes the activity of species j, $[M] = 1$, $[M_x O_y] = 1$ and P_{O_2}, P_{H_2}, $P_{H_2 O}$ denote the pressures kPa). Here, $[j] = P_j / P_o$.

Gibbs Energy Criterion If $\Delta G^o < 0$,, then it is a measure of negative deviation from equilibrium condition and the reaction proceeds from left to right as written in the above reactions. On the contrary, if $\Delta G^o > 0$, then a positive deviation from equilibrium implies that a reaction occurs in the reverse direction from right to left. However, if $\Delta G^o = 0$, the system is at equilibrium. These particular conditions or criteria are very important in assessing chemical reactions at high temperatures as indicated in Fig. 16.22. The example below considers some metal reactions in an oxygen-rich environment.

Example 16.9 Firstly, assume that an amount of copper ore (cuprous oxide, $Cu_2 O$), is to be reduced to its metal form in an oxygen-rich environment. Secondly, assume that copper is oxidized as a $Cu_2 O$ product in a similar environment. Both cases are characterized by the given reversible reaction (equal sign, =) at $T = 887\,^\circ C$.

$$4Cu + O_2 = 2Cu_2O$$

Determine **(a)** ΔG^o and the oxygen pressure P_{O_2} for the oxidation of copper to form cuprous oxide Cu_2O, **(b)** Will iron Fe or magnetite Fe_3O_4 reduce Cu_2O to its metal state? **(c)** How useful is the data in Fig. 16.22?

Solution

(a) Using a simplified version of Fig. 16.22 given below, draw a straight line (blue) from point "O" through the $Cu - 2O\text{-}O$ reaction line at $T = 887\,^\circ C$ until the line intercepts the scale line for O_2 pressure. This is shown below.

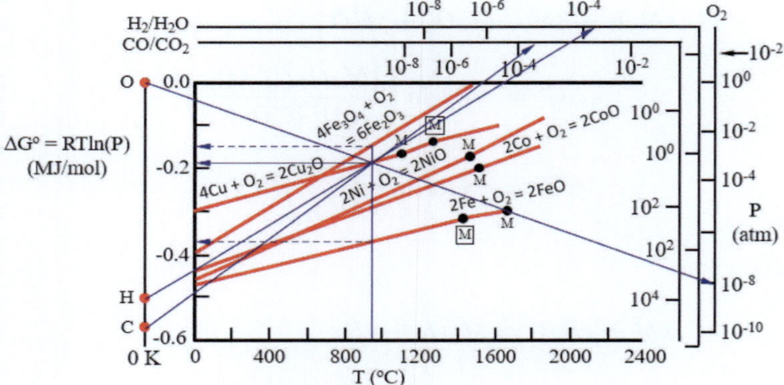

Then, read-off the corresponding pressure and Gibbs energy change values

$$\Delta G^o = -0.19 \; MJ/mol \; of \; O_2$$

$$P_{O_2} = 2.78 \times 10^{-9} \; atm$$

(b) For reducing cuprous oxide Cu_2O to its metal form Cu in an oxygen-rich environment,

$$2Cu_2O \rightarrow 4Cu + O_2 \qquad \Delta G^o_{Cu_2O} = +0.19 \; MJ/mol$$

$$2Fe + O_2 \rightarrow 2FeO \qquad \Delta G^o_{Fe} = -3.80 \; MJ/mol$$

$$2Cu_2O + 2Fe \rightarrow 4Cu + 2FeO \qquad \Delta G^o = \Delta G^o_{Cu_2O} + \Delta G^o_{Fe}$$

$$\Delta G^o = 0.19 \; MJ/mol - 3.80 \; MJ/mol = -3.61 \; MJ/mol$$

$$\Delta G^o < 0 \qquad \text{(spontaneous)}$$

This result means that the chemical reactions proceed in the written direction, and fortunately, iron (Fe) reduces copper (Cu_2O). Moreover, mixing cuprous oxide Cu_2O and magnetite Fe_3O_4 yields

$$2Cu_2O \rightarrow 4Cu + O_2 \qquad \Delta G^o_{Cu_2O} = +0.19 \ MJ/mol$$

$$4Fe_3O_4 + O_2 \rightarrow 6Fe_2O_3 \qquad \Delta G^o_{Fe} = -0.15 \ MJ/mol$$

$$2Cu_2O + 4Fe_3O_4 \rightarrow 4Cu + 6Fe_2O_3 \quad \Delta G^o = \Delta G^o_{Cu_2O} + \Delta G^o_{Fe_3O_4}$$

$$\Delta G^o = 0.19 \ MJ/mol - 0.15 \ MJ/mol = 0.04 \ MJ/mol$$

$$\Delta G^o > 0 \qquad \text{(non-spontaneous)}$$

Therefore, only Fe reduces Cu_2O because $\Delta G^o < 0$. This means that the chemical reactions proceed in the opposite direction.

(c) Despite that Fig. 16.22 is normally used for reducing mineral compounds, such as oxides, it provides a basis for evaluating the possibility of chemical separation by oxidation. For instance,

$$4Cu + O_2 \rightarrow 2Cu_2O \quad \text{at } \Delta G^o < 0 \quad \text{(reduction)}$$

$$4Cu + O_2 \leftarrow 2Cu_2O \quad \text{at } \Delta G^o > 0 \quad \text{(oxidation)}$$

These are reversible reactions which can be unified using an equal sign instead; that is,

$$4Cu + O_2 = 2Cu_2O \quad \text{at } \Delta G^o < 0 \quad \text{(reduction)}$$

$$4Cu + O_2 = 2Cu_2O \quad \text{at } \Delta G^o > 0 \quad \text{(oxidation)}$$

Notice that the reactions in Fig. 16.22 represent the sum of reactants equals the product.

16.21 Summary

Corrosion is a form of degradation or deterioration of the surface lattice structure, and it occurs due to electrochemical or chemical interactions with the environment. In this chapter, only electrochemical corrosion of metals is considered as the most relevant approach to assess metal deterioration. However, corrosion studies may include chemical corrosion of nonmetals and corrosion based on corrosion inhibitors.

Corrosion is a form of alteration of the lattice atomic structure of a refined metal M. This implies that M reverts to its natural ore state as a chemically stable oxide (cuprite Cu_2O), hydroxide [rust $FeO(OH) \cdot nH_2O$)], carbonate (siderite $FeCO_3$), or sulfide (pyrite FeS_2).

Corrosion of aluminum and iron alloys can be beneficial when it undergoes a mechanism of self-protection by the formation of metal-oxide passive films with a few nanometers (nm) in thickness.

Electrochemical corrosion of a metal M can be represented by an anodic reaction $M \rightarrow M^{z+} + ze^-$, which indicates the emission of electrons to its intermediate surrounding.

Problems

16.1 What's the action of hydrogen on a steel blade exposed to an aqueous solution of hydrogen sulfide if the following reaction takes place $Fe + H_2S = FeS + 2H$? Assume that the steel blade gets damaged during service. [Solution: Hydrogen embrittlement].

16.2 An aerated acid solution containing 10^{-2} mol/l of dissolved oxygen (O_2) moves at 2×10^{-4} cm/s in a stainless steel pipe when a critical current density of 10^3 $\mu A/cm^2$ passivates the pipe. The pertinent reaction and data for this problem are

$$O_2 + 2H_2O + 4e^- = 4OH^- \quad \text{(Cathode)}$$

$D_{O_2} = 10^{-5}$ cm^2/s	$T = 25\,°C$	$F = 96,500$ $A.s/mol$

Calculate **(a)** the thickness of the ionic structural layer (δ), **(b)** the limiting current density (i_L) if the three modes of flux are equal in magnitude, and **(c)** Will the pipe corrode under the current conditions? Why? or Why not? [Solution: (a) $\delta = 0.05$ $cm = 0.5$ mm, (b) $i_L = 2.32 \times 10^3$ $\mu A/cm^2$, (c) No corrosion].

16.3 What are the two mechanisms for oxidation of iron? Write down the suitable electrochemical reactions.

16.4 Investigate the differences between dry and wet corrosion. Write down at least three different corrosion cases.

16.5 Calculate the standard potential for the formation of ferric hydroxide $Fe(OH)_3$ (brown rust). [Solution: $E^o = 1.172$ V].

16.6 Calculate the E^o values for each of the half-cells responsible the redox reaction given below, and determine which half-cell reaction is the cathode and anode. [Solution: $E^o_{Cd/Cd^{2+}} = +0.403 V_{SHE}$].

$$2Fe^{3+} + Cd \rightarrow Cd^{2+} + 2Fe^{2+}$$

16.7 Consider the following electrochemical cell $\left|Cu^{+2}, Cu\right|\left|H_2, H^+\right| Pt$ to calculate the maximum activity of copper ions Cu^{+2} in solution due to oxidation of a copper strip immersed in sulfuric acid H_2SO_4 at $25\,^\circ C$, $101\ kPa$, and $pH = 2$. [Solution: $a_{Cu^{+2}} = \left[Cu^{+2}\right] = 4.06x10^{-16}\ mol/l$].

16.8 Consider a galvanic cell $\mid Zn\left|0.10MZnSO_4\right|\left|0.10\ M\ NiSO_4\right| Ni\mid$. Calculate the cell potential using the Nernst equation at $T = 25^\circ C$. [Solution: $E = 0.513\ V$].

16.9 The dissociation constant of silver hydroxide, $AgOH$, is $1.10x10^{-4}$ at $25\,^\circ C$ in an aqueous solution. **(a)** Write down the chemical reaction, and **(b)** determine the Gibbs energy change ΔG^o. [Solution: (b) $\Delta G^o \approx -22,585\ J/mol$].

16.10 Calculate **(a)** the concentration of $AgOH$ and Ag^+ in g/l at $25\,^\circ C$ and **(b)** the pH if $\left[OH^-\right] = 4.03x10^{-3}\ mol/l$ and $K_e = 1.10x10^{-4}$. [Solution: (a) $C_{OH^-} = 6.85x10^{-2}\ g/l$, (b) $pH = 11.61$].

16.11 For an electrochemical copper reduction reaction at $25\,^\circ C$ and $101\ kPa$, calculate **(a)** the Gibbs energy change ΔG^o as the driving force, **(b)** the equilibrium constant K_e and **(c)** the copper ion concentration $\left[Cu^{+2}\right]$ in g/l. [Solution: (a) $\Delta G^o = -65,041\ J/mol$, (b) $K_e = 2.52x10^{11}$, (c) $C_{Cu^{+2}} = 2.52x10^{-10}\ g/l$].

16.12 Use the electrochemical cell $\mid Cd\left|0.05\ M\ Cd^{+2}\right|\left|0.25\ M\ Cu^{+2}\right| Cu\mid$ to calculate **(a)** the temperature for an electric potential of $0.762\ V_{SHE}$, and **(b)** ΔS, ΔG, ΔH, and Q. [Solution: (a) $T = 44.30\,^\circ C$, (b) $Q = -4,245.50\ J/mol$].

16.13 Pure copper and pure cobalt electrodes are separately immersed in solutions of their respective divalent ions, making up a galvanic cell that yields a cell potential of $0.65\ V_{SHE}$. Calculate **(a)** the cobalt (Co) activity if the activity of copper ions is $0.85\ mol/l$ at $25\,^\circ C$ and **(b)** the change in entropy ΔS^o and Gibbs energy ΔG^o. [Solution: (a) $a_{Co^{+2}} = 0.051\ mol/l$, (b) $\Delta G = -125.45\ kJ/mol$].

16.14 Show that the activities are $\left[M_1^{+2}\right] = \left[M_2^{+2}\right]$ in a galvanic cell, provided that the potentials is $E = E^o$.

16.15 Derive an expression for $\ln\left[Cu^{+2}\right] = f(T)$ for the given cells below. Assume that the current ceases to flow and that the cells are not replenished with fresh solutions. Plot the expressions for a temperature range of $0\,^\circ C \leq T \leq 60\,^\circ C$, and draw some suitable conclusions to explain the electrochemical behavior based on the resultant trend.

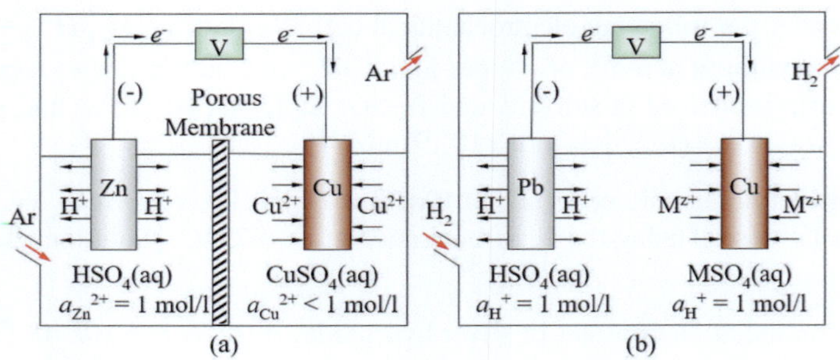

16.16 Use the data given in Table 16.3 to determine **(a)** the half-cells and **(b)** the redox standard potentials for the galvanic cell shown in Fig. 16.8 with $\alpha = Zn$ and $\beta = Cu$. Compare the results with the data given in Table 16.1. [Solution: (a) $E^o_{Zn/Zn^{2+}} = +0.763\ V$, (b) $E^o = +1.10\ V$].

16.17 If the state of equilibrium of an electrochemical cell is disturbed by an applied current density i_x, then $i_a = -i_c = i_{corr}$ no longer holds. Let i_x be a cathodic current density and the slope of the corresponding polarization curve be $dE/d\log(i_x)$, which increases approaching the Tafel constant β_c. Determine **(a)** the value of i_x when $i_c = 10^{-3}\ A/cm^2$ and $i_a = 10^{-9}\ A/cm^2$ and **(b)** the value of β_c when $i_{corr} = 10^{-5}\ A/cm^2$ and $\eta = -0.20\ V$. [Solution: (a) $i_x = 10^{-3}\ A/cm^2$, (b) $\beta_c = 0.10\ V$].

16.18 According to the Stockholm Convention Cell $Fe\,|\,Fe^{+2}\,|\,|\,H^+, H_2\,|\,Pt$, the corrosion potential of iron (Fe) is $-0.70\ V_{SCE}$ at $pH = 4.4$ and $25\,°C$ in a deaerated (no oxygen is involved) acid solution. Calculate **(a)** the corrosion rate in mm/y when

$$i_{o,H} = 10^{-6}\ A/cm^2 \qquad \beta_c = -0.10\ V_{SHE}$$
$$i_{o,Fe} = 10^{-8}\ A/cm^2 \qquad E_{Fe} = -0.50\ V_{SHE}$$

and **(b)** the Tafel anodic slope β_a and **(c)** draw the kinetic diagram E vs. $\log i$. [Solution: (a) $C_R = 1.16\ mm/y$, (b) $\beta_a = 0.01\ V$].

16.19 Let the following anodic and cathodic reactions be, respectively,

$M = M^{+2} + 2e$	$E_M = -0.941\ V_{SCE}$	$i_{o,M} = 10^{-2}\ \mu A/cm^2$
$2H^+ + 2e^- = H_2$	$E_H = -0.200\ V_{SHE}$	$i_{o,H} = 0.10\ \mu A/cm^2$

where $i_{corr} = 10^2\ \mu A/cm^2$ and $E_{corr} = -0.741\ V_{SCE}$. **(a)** Construct the corresponding kinetic diagram and **(b)** determine both β_a and β_c from the diagram and from the definition of overpotential equations. [Solution: (b) $\beta_a = 0.05\ V$ and $\beta_c = 0.10\ V$].

16.20 A steel tank is hot dipped in a deaerated acid solution of $5x10^{-4}\ mol/cm^3$ molality zinc chloride $(ZnCl_2)$ so that a 0.15-*mm* zinc coating is deposited on the steel surface. This process produces a galvanized steel tank. Calculate the time it takes for the zinc coating to corrode completely at a $pH = 4$. Data: $E_{Zn} = -0.8$ V, $i_{o,Zn} = 10\ \mu A/cm^2$, $i_{o,H} = 10^{-3}\ \mu A/cm^2$, $\beta_a = 0.08$ V, $\beta_c = 0.12$ V, $T = 25\,°C$. [Solution: $t = 146\ days$].

16.21 Calculate **(a)** the activity and **(b)** the corrosion rate of iron (Fe) immersed in an aerated aqueous solution of $pH = 9$. The dissociation constant for ferrous hydroxide, $Fe\,(OH)_2$, is $1.64x10^{-14}$. Given data:

$i_{o,Fe} = 10^{-2}\ \mu A/cm^2$	$E_{corr} = -0.30$ V	$\beta_a = 0.10$ V
$A_{w,Fe} = 55.85\ g/mol$	$\rho_{Fe} = 7.86\ g/cm^3$	

[Solution: (a) $\left[Fe^{+2}\right] = 1.64x10^{-4}\ mol/l$, (b) $C_R = 0.0473\ \mu m/y$].

16.22 Plot the anodic data given below, and determine the polarization resistance $\left(R_p\right)$ and the anodic Tafel slope (β_a) for a metal M. Use $\beta_c = 0.07$ V and $i_{corr} = 0.019\ A/cm^2$. [Solution: $\beta_a = 0.12\ V$].

$$M = M^{+2} + 2e$$

$$2H^+ + 2e^- = H_2$$

$i\ \left(\mu A/cm^2\right)$	+0.08	+0.10	+0.15	+0.18	+0.20
$E\ (V_{SHE})$	-0.32	-0.30	-0.25	-0.22	-0.20

16.23 An electrolyte contains a very low activity $\left(8x10^{-9}\ mol/l\right)$ of silver cations $\left(Ag^+\right)$ and an unknown concentration of copper cations. If the cell potential difference between the copper anode and the silver cathode is -0.04 V, determine **(a)** the Gibbs energy change ΔG^o and **(b)** the concentration of $\left[Cu^{+2}\right]$ in g/l at $40\,°C$. Neglect the effects of other ions that might react with silver. [Solution: (a) $\Delta G^o = -89,166\ J/mol$, (b) $C_{Cu^{+2}} = 59.74\ g/l$].

References

1. M.G. Fontana, *Corrosion Engineering* (McGraw-Hill Book Company, New York, 1986)
2. J. Alexander, *Understand the Danger: Pitting Corrosion.* Efficient Plant (2017). https://www.efficientplantmag.com/2017/06/understand-danger-pitting-corrosion/
3. https://steemit.com/steemstem/@akeelsingh/the-curious-case-of-corrosion-part-2
4. https://civilengi.com/types-of-corrosion/

5. https://www.nace.org/resources/what-is-corrosion/forms-of-corrosion/microbiologically-influenced-corrosion-mic
6. https://nanopdf.com/download/dezincification-selective-corrosion_pdf
7. https://civilengi.com/types-of-corrosion/
8. N. Perez, *Electrochemistry and Corrosion Science*, 2nd edn. (Springer, Switzerland, 2016)
9. J.P. Broomfield, *Corrosion of Steel in Concrete*, 2nd edn. (Taylor & Francis, Abingdon, 2007)
10. F. Mansfield, Recording and analysis of ac impedance data for corrosion studies. Corrosion **37**(5), 301–307 (1981)
11. P.R. Roberge, *Handbook of Corrosion Engineering* (McGraw-Hill, New York, 2000)
12. R. Markson, J. Sedlacek, C.W. Fairall, Turbulent transport of electric charge in the marine atmospheric boundary layer. J. Geophys. Res. **86**(C12), 12115–12121 (1981)
13. D.A. Jones, *Principles and Prevention of Corrosion* (Prentice-Hall, Hoboken, 1996)
14. J.W. Evans, L.C. De Jonghe, *The Production and Processing of Inorganic Materials* (Springer, Switzerland, 2016)
15. R, Dehoff, *Thermodynamics in Materials Science*, 2nd edn. (CRC Press, Taylor & Francis Group, Boca Raton, 2006)
16. M. Pourbaix, *Enthalpies Libres de Formations Standards*. Cebelcor Rapport Technique, No. 87 (1960)
17. L.L. Shreir, R.A. Jarman, G.T. Burstein (eds.), *Corrosion: Corrosion Control*, vol. 2, 3rd edn. (Butterworth Heinemann (BH), New York, 1998)
18. M. Pourbaix, *Atlas of Electrochemical Equilibria in Aqueous Solution*. Translated from the French by J.A. Franklin (NACE International, Houston, Texas, 1974)
19. T.P. Hoar, Electrodeposition, in *Atlas of Electrochemical Equilibria in Aqueous Solution by Pourbaix*, Translated from the French by J.A. Franklin (NACE International, Houston, 1974)
20. N. De Zoubov, C. Vanleugenhaghe, M. Pourbaix, Copper, in *Atlas of Electrochemical Equilibria in Aqueous Solution by Pourbaix*. Translated from the French by J.A. Franklin (NACE International, Houston, Texas, 1974)
21. T.P. Hoar, Polarization, in *The 13th William Blum Lecture, Presented at the 59th AES Annual Convention in Cleveland* (Ohio, 1972). https://www.pfonline.com/cdn/cms/1605_Printable-Version.pdf
22. R. Bandy, D.A. Jones, *Analysis of Errors in Measuring Corrosion Rates by Linear Polarization*, vol. 32, No. 4 (1976)
23. L.L. Shreir, R.A. Jarman, G.T. Burstein (eds.), *Corrosion: Metal & Environment Reactions*, vol. 1, 3rd edn. (Butterworth Heinemann (BH), New York, 1998)
24. J. Kunze, V. Maurice, L.H. Klein, H-H Strehblow, P. Marcus, In-situ scanning tunneling microscopy study of the anodic oxidation of Cu (111) in 0.1 M NaOH. J. Phys. Chem. B **105**(19), 4263–4269 (2001)
25. V. Maurice, W. Yang, P. Marcus, XPS and STM study of passive films formed on Fe–22Cr(110) single-crystal surface. J. Electrochem. Soc. **143**, 1182–1200 (1996)
26. P. Marcus, V. Maurice, *Atomic Level Characterization in Corrosion Studies*. Philosophical Transactions of the Royal Society A: Mathematical, Physical and Engineering Sciences 375, No. 2098 (2017)
27. D.R. Gaskell, *Introduction to Metallurgical Thermodynamics*, 4th edn. (Taylor & Francis Books, London, 2003)

Appendix A
Conversion Tables

This appendix contains some useful unit conversion tables.

Prefixes

Prefix	Symbol	Factor	Prefix	Symbol	Factor
exa	E	10^{18}	milli	m	10^{-3}
peta	P	10^{15}	micro	μ	10^{-6}
tera	T	10^{12}	nano	n	10^{-9}
giga	G	10^{9}	pico	p	10^{-12}
mega	M	10^{6}	femto	f	10^{-15}
kilo	k	10^{3}	atto	a	10^{-18}

Abbreviations

atm = atmosphere	gal = gallon	MPa = megapascal
A = ampere	hp = horsepower	N = Newton
$\bar{A}$ = angstrom	h = hour	nm = nanometer
Btu = British energy	$in.$ = inch	Pa = pascal
C = coulomb	J = joule	psi = pounds/square inch
$^{\circ}C$ = degrees Celsius	$^{\circ}K$ = degrees Kelvin	s = second
cal = calorie	Kg = kilogram	T = temperature
cm = centimeter	L = liter	W = watt or weight
eV = electron volt	min = minute	μ = micron
$^{\circ}F$ = degrees Fahrenheit	mm = millimeter	μm = micrometer
g = gram	mol = mole	Ω = ohm = V/A

© The Author(s), under exclusive license to Springer Nature Switzerland AG 2024
N. Perez, *Materials Science: Theory and Engineering*,
https://doi.org/10.1007/978-3-031-57152-7

Greek Alphabet

Alpha	α	A	Iota	ι	I	Rho	ρ	P
Beta	β	B	Kappa	κ	K	Sigma	σ	Σ
Gamma	γ	Γ	Lambda	λ	Λ	Tau	τ	Υ
Delta	δ	Δ	Mu	μ	M	Upsilon	υ	Y
Epsilon	ϵ	E	Nu	ν	N	Phi	ϕ	Φ
Zeta	ζ	Z	Xi	ξ	Ξ	Chi	$\varkappa$	χ
Eta	η	H	Omicron	o	O	Psi	ψ	Ψ
Theta	θ	Θ	Pi	π	Π	Omega	ω	Ω

Area

$1\ m^2 = 10^4\ cm^2$	$1\ cm^2 = 10^{-4}\ m$
$1\ m^2 = 10.76\ ft^2$	$1\ ft^2 = 0.0929\ m^2$
$1\ cm^2 = 0.155\ in.^2$	$1\ in^2 = 0.4516\ nm^2$
$1\ cm^2 = 10^2\ mm.^2$	$1\ mm^2 = 10^{-2}\ cm^2$

Corrosion Rate

$1\ mm/y = 39.37\ mils/y\ (mpy)$	$1\ mil/y = 2.54 \times 10^{-2}\ mm/y$
$1\ \mu m/y = 3.937 \times 10^{-2}\ mils/y$	$1\ \mu m/y = 25.40\ mils/y$

Current Rate

$1\ A/m^2 = 9.29 \times 10^{-2}\ A/ft^2$	$1\ A/ft^2 = 10.764\ A/m^2$
$1\ A/cm^2 = 6.452\ A/in^2$	$1\ A/in^2 = 0.155\ A/cm^2$
$1\ A/mm^2 = 6.452 \times 10^2\ A/in^2$	$1\ A/in^2 = 1.55 \times 10^{-3}\ A/mm^2$
$1\ \mu A/cm^2 = 6.452\ \mu A/in^2$	$1\ \mu A/in^2 = 0.155\ \mu A/cm^2$

Density

$1\ Kg/m^3 = 10^{-3}\ g/cm^3$	$1\ g/cm^3 = 10^3\ Kg/m^3$
$1\ Kg/m^3 = 0.0624\ lb_m/ft^3$	$1\ ft^3 = 16.02\ Kg/m^3$
$1\ g/cm^3 = 62.40\ lb_m/ft^3$	$1\ lb_m/ft^3 = 0.016\ g/cm^3$
$1\ g/cm^3 = 0.0361\ lb_m/in^3$	$1\ lb_m/in^3 = 27.70\ g/cm^3$

Electricity and Magnetism

$d\phi/dx$ = V/cm	V = volts
1 mho = 1 S	1 S = 1 mho
1 S = ohm^{-1} = Ω^{-1}	1 ohm = 1 S^{-1}
1 Ω.cm = 1.00 $\times$ 10^{-2} Ω.m	1 Ω.m = 100 Ω.cm
1 Ω.cm = 1 ohm.cm	1 ohm = V/A
1 maxwell = 10^{-2} μweber	1 μweber = 100 maxwell

Energy

1 J = 2.778 $\times$ 10^{-7} kW.h	1 kW.h = 3.60 $\times$ 10^6 J
1 J = 0.239 cal	1 cal = 4.184 J
1 J = 9.48 $\times$ 10^{-4} Btu	1 Btu = 1.05 $\times$ 10^3 J
1 J = 0.7376 ft.lb$_f$	1 ft.lb$_f$ = 1.3558 J

Flow Rate

1 L/min = 2.1189 ft^3/h	1 ft^3/h = 0.4719 L/min
1 L/min = 3.53$\times$10^{-2} ft^3/min	1 ft^3/min = 28.31
1 L/min = 15.85 gal/h	1 gal/h = 6.31$\times$10^{-2} L/min
1 L/min = 0.2642 gal/min	1 gal/min = 3.7854 L/min
1 m^3/s = 2.1189$\times$10^3 ft^3/min	1 ft^3/min = 4.7195$\times$10^{-4} m^3/s
1 m^3/s = 35.315 ft^3/s	1 ft^3/s = 2.8317$\times$10^{-2} m^3/s
1 m^3/s = 3.66$\times$10^6 in^3/min	1 in^3/min = 2.73$\times$10^{-7} m^3/s

Force

1 N = 10^5 dynes	1 dyne = 10^{-5} N
1 N = 0.2248 lb$_f$	1 lb$_f$ = 4.448 N
1 N = 2.248$\times$10^{-4} kips	1 kip = 4.448$\times$10^3 N
1 N = 0.1019 Kg$_f$	1 Kg$_f$ = 9.81 N

Length

1 m = 10 $\bar{A}$	1 $\bar{A}$ = 10^{-10} m	1 mil = 25.40 μm
1 m = 10^9 nm	1 nm = 10^{-9} m	1 yd = 0.9144 m
1 m = 10^6 μ	1 μ = 10^{-6} m	1 mile = 1.61 Kg
1 m = 10^3 mm	1 mm = 10^{-3} m	
1 m = 10^2 cm	1 cm = 10 mm	
1 m = 39.36 in.	1 in. = 25.4 mm	
1 m = 3.28 ft	1 ft. =12 in.	

Mass

$1\ Kg = 10^3\ g$	$1\ g = 10^{-3}\ Kg$
$1\ Kg = 2.205\ lb_m$	$1\ lb_m = 0.4536\ Kg$
$1\ g = 2.205 \times 10^{-3}\ lb_m$	$1\ lb_m = 45.36\ g$
$1\ Kg = 10^{-3}$ metric ton	1 metric ton $= 10^3\ Kg$
$1\ Kg = 1.1023 \times 10^{-3}$ short ton	1 short ton $= 2 \times 10^3\ lb_m$
$1\ Kg = 9.8421 \times 10^{-4}$ long ton	1 long ton $= 1.016 \times 10^3\ Kg$
1 metric ton $= 1.1023$ short ton	$1\ lb_m = 5 \times 10^{-4}$ short ton
1 metric ton $= 0.98421$ long ton	1 long ton $= 1.106$ metric ton
$1\ Kg = 6.85 \times 10^{-2}$ slug	1 slug $= 14.59$ kg

Physical Constants

Quantity	Symbol	Value
Acceleration of Gravity	g	$= 9.81\ m^2/s = 32.2\ ft/s^2$
Avogadro's number	N_A	$= 6.022 \times 10^{23}\ particle/mol$
Boltzmann's constant	$k = R/N_A$	$= 1.38 \times 10^{-23}\ J/K = 8.62 \times 10^{-5}\ eV/K$
Electronic charge	q_e	$= 1.602 \times 10^{-19}\ C$
Faraday's constant	$F = q_e N_A$	$= 96{,}500\ C/mol = 96{,}500\ A.s/mol$
Gas constant	$R = k N_A$	$= 8.314\ J/(mol.K) = 1.987\ cal/(mol.K)$
Planck's constant	h	$= 6.626 \times 10^{-34}\ J.s = 4.136 \times 10^{-15}\ eV.s$
Mass of electron	m_e	$= 9.11 \times 10^{-31}\ Kg$
Speed of light	c	$= 3 \times 10^8\ m/s$

Power

$1\ W = 1\ J/s$	$1\ Btu/s = 1\ ft.lb_f/s$
$1\ kW = 0.9478\ Btu/s$	$1\ Btu/s = 1.0551\ kW$
$1\ kW = 56.869\ Btu/min$	$1\ Btu/min = 1.758 \times 10^{-2}\ kW$
$1\ kW = 3.4121 \times 10^3\ Btu/h$	$1\ Btu/h = 2.9307 \times 10^{-4}\ kW$
$1\ kW = 1.3405 \times 10^{-3}\ hp$	$1\ hp = 746\ kW$

Pressure (Fluid)

$1\ Pa = 1\ N/m^2$	$1\ atm = 760\ mm.Hg,\ ^\circ C$
$1\ Pa = 9.87 \times 10^{-6}\ atm$	$1\ atm = 1.0133 \times 10^5\ Pa$
$1\ MPa = 9.87 \times 10^{-3}\ atm$	$1\ atm = 101.33\ MPa$
$1\ Pa = 1.00 \times 10^{-5}\ bar$	$1\ bar = 1.00 \times 10^5\ Pa$
$1\ Pa = 1.45 \times 10^{-4}\ psi$	$1\ psi = 6.895 \times 10^3\ Pa$
$1\ MPa = 0.145\ ksi$	$1\ ksi = 6.895\ MPa$
$1\ Pa = 7.501 \times 10^{-3}\ torr\ (mm\ Hg,\ ^\circ C)$	$1\ torr = 1.333 \times 10^2\ Pa$
$1\ Pa = 2.953 \times 10^{-4}\ in.\ Hg,\ 32^\circ F$	$1\ in.\ Hg,\ 32^\circ F = 3.386 \times 10^3\ Pa$
$1\ Pa = 2.0886 \times 10^{-2}\ lb_g/ft^2$	$1\ atm = 14.7\ psi$

Temperature

$T(K) = T(^\circ C) + 273.15$	$T(R) = T(^\circ F) + 459.67$
$\Delta T(K) = \Delta T(^\circ C)$	$T(R) = (9/5)T(^\circ K) + 459.67$
	$T(^\circ F) = (9/5)T(^\circ C) + 32$
	$T(^\circ C) = (5/9)[T(^\circ F) - 32]$

Velocity

$1\ m/s = 1.1811 \times 10^4\ ft/h$	$1\ ft/h = 8.4667 \times 10^{-5}\ m/s$
$1\ m/s = 1.9685 \times 10^2\ ft/min$	$1\ ft/min = 5.08 \times 10^{-3}\ m/s$
$1\ m/s = 3.281\ ft/s$	$1\ ft/s = 0.305\ m/s$
$1\ m/s = 39.37\ in/s$	$1\ in/s = 2.54 \times 10^{-2}\ in/s$

Volume

$1\ m^3 = 10^6\ cm^3$	$1\ cm^3 = 10^{-6}\ m^3$
$1\ cm^3 = 10^3\ mm^3$	$1\ mm^3 = 10^{-3}\ cm^3$
$1\ m^3 = 35.32\ ft^3$	$1\ ft^2 = 0.0283\ m^3$
$1\ cm^3 = 0.061\ in^3$	$1\ in^3 = 16.39\ in^3$
$1\ L = 10^3\ cm^3\ (cc)$	$1\ cm^3 = 10^{-3}\ L$
$1\ US\ gal = 3.785\ L$	$1\ L = 0.264\ gal$

Index

A
Acoustic attenuation, 814
Acoustic energy, 807
Acoustic sound frequency, 794
Activation energy, 210, 276, 355, 356, 365, 868
Activation polarization, 863, 881
Activation state, 867
Active-passive behavior, 878, 879
Activity, 310, 323, 850, 851, 858, 859
AFM assembly, 155
AFM deflection, 155
AFM load, 155
    graphene, 155
Aggressive environment, 654, 720
Airy stresses, 252
Airy stress function, 224, 252
    polar coordinates, 703
AISI 304 microstructure, 156
Allotropy, 523
Alloy steel, 528
Alpha function, 333
Ambipolar diffusion, 295
Amount of diffusing solute, 280, 286
Amount of heat transfer, 306
Amount of removed energy, 388
Amount of thermal energy, 406
Angular frequency, 103
Anion, 209
Anisotropic material, 1
Annealing twinning, 624
Anodic electrode, 847
Anodic overpotential, 866, 876
Anodic polarization, 866, 876
Anodic reaction, 837

Antiferromagnetism, 769
Aqua regia, 156
Arrhenius equation, 210, 292, 867, 870
Arrhenius-type equation, 275, 354, 356
ASTM standard test method
    ASTM A247, 530
    ASTM A576, 529
    ASTM A710, 731
    ASTM E8, 636
    ASTM E23, 727
    ASTM E139, 638
    ASTM E255, 578
    ASTM E292, 638
    ASTM E399, 685
    ASTM E647, 719
Atmospheric corrosion, 840, 841
Atom, 2
Atomic mismatch, 156
Atomic orbital, 3
Atomic packing factor, 61, 187
Atomic scale, 2
Atomic weight, 210
Atom model, 2
Atom pair, 314, 315, 322
Attenuation, 56
Attenuation coefficient, 814, 815
Austenite cooling rate, 561
Avogadro's number, 210, 269
Avrami equation, 575

B
Bain distorsion model, 566
Ballistic impact velocity, 728